MINERAL PHYSICS

Treatise on Geophysics

MINERAL PHYSICS

Editor-in-Chief

Professor Gerald Schubert

Department of Earth and Space Sciences and Institute of Geophysics and Planetary Physics,
University of California Los Angeles, Los Angeles, CA, USA

Volume Editor

Dr. G. David Price

University College London, London, UK

ELSEVIER

Amsterdam • Boston • Heidelberg • London • New York • Oxford
Paris • San Diego • San Francisco • Singapore • Sydney • Tokyo

Elsevier B.V.
Radarweg 29, 1043 NX Amsterdam, the Netherlands

First edition 2009

Notice
No responsibility is assumed by the publisher for any injury and/or damage to persons
or property as a matter of products liability, negligence or otherwise, or from any use
or operation of any methods, products, instructions or ideas contained in the material
herein. Because of rapid advances in the medical sciences, in particular, independent
verification of diagnoses and drug dosages should be made

British Library Cataloguing in Publication Data
A catalogue record for this book is available from the British Library

Library of Congress Control Number: 2009929977

ISBN: 978-0-444-53458-3

For information on all Elsevier publications
visit our website at elsevierdirect.com

Printed and bound by CPI Group (UK) Ltd, Croydon, CR0 4YY

Contents

Preface   vii

Contributors   xi

Editorial Advisory Board   xiii

1   Overview – Mineral Physics: Past, Present, and Future   1
G. D. Price, *University College London, London, UK*

2   Properties of Rocks and Minerals – Seismic Properties of Rocks and Minerals, and
Structure of the Earth   7
L. Stixrude, *University of Michigan, Ann Arbor, MI, USA*

3   Mineralogy of the Earth – Phase Transitions and Mineralogy of the Lower Mantle   33
T. Irifune and T. Tsuchiya, *Ehime University, Matsuyama, Japan*

4   Mineralogy of the Earth – Trace Elements and Hydrogen in the Earth's Transition Zone
and Lower Mantle   63
B. J. Wood and A. Corgne, *Macquarie University, Sydney, NSW, Australia*

5   Mineralogy of the Earth – The Earth's Core: Iron and Iron Alloys   91
L. Vočadlo, *University College London, London, UK*

6   Theory and Practice – Thermodynamics, Equations of State, Elasticity, and Phase
Transitions of Minerals at High Pressures and Temperatures   121
A. R. Oganov, *ETH Zurich, Zurich, Switzerland*

7   Theory and Practice – Lattice Vibrations and Spectroscopy of Mantle Phases   153
P. F. McMillan, *University College London, London, UK*

8   Theory and Practice – Multianvil Cells and High-Pressure Experimental Methods   197
E. Ito, *Okayama University, Misasa, Japan*

9   Theory and Practice – Diamond-Anvil Cells and Probes for High P–T Mineral Physics Studies   231
H. -K. Mao, *Carnegie Institution of Washington, Washington, DC, USA*
W. L. Mao, *Los Alamos National Laboratory, Los Alamos, NM, USA*

10   Theory and Practice – Techniques for Measuring High P/T Elasticity   269
J. D. Bass, *University of Illinois at Urbana-Champaign, Urbana, IL, USA*

11   Theory and Practice – Measuring High-Pressure Electronic and Magnetic Properties   293
R. J. Hemley, V. V. Struzhkin and R. E. Cohen, *Carnegie Institution of Washington,
Washington, DC, USA*

12   Theory and Practice – Methods for the Study of High P/T Deformation and Rheology   339
D. J. Weidner and L. Li, *State University of New York (SUNY), Stony Brook, NY, USA*

13 Theory and Practice – The *Ab Initio* Treatment of High-Pressure and -Temperature
   Mineral Properties and Behavior   359
   D. Alfè, *University College London, London, UK*

14 Properties of Rocks and Minerals – Constitutive Equations, Rheological Behavior, and
   Viscosity of Rocks   389
   D. L. Kohlstedt, *University of Minnesota, Minneapolis, MN, USA*

15 Properties of Rocks and Minerals – Diffusion, Viscosity, and Flow of Melts   419
   D. B. Dingwell, *University of Munich, Munich, Germany*

16 Seismic Anisotropy of the Deep Earth from a Mineral and Rock Physics Perspective   437
   D. Mainprice, *Université Montpellier II, Montpellier, France*

17 Properties of Rocks and Minerals – Physical Origins of Anelasticity and Attenuation in Rock   493
   I. Jackson, *Australian National University, Canberra, ACT, Australia*

18 Properties of Rocks and Minerals – High-Pressure Melting   527
   R. Boehler, *Max Planck Institute für Chemie, Mainz, Germany*
   M. Ross, *University of California, Livermore, CA, USA*

19 Properties of Rocks and Minerals – Thermal Conductivity of the Earth   543
   A. M. Hofmeister and J. M. Branlund, *Washington University, St. Louis, MO, USA*
   M. Pertermann, *Rice University, Houston, TX, USA*

20 Properties of Rocks and Minerals – Magnetic Properties of Rocks and Minerals   579
   R. J. Harrison, R. E. Dunin-Borkowski, T. Kasama, E. T. Simpson and J. M. Feinberg,
   University of Cambridge, Cambridge, UK

21 Properties of Rocks and Minerals – The Electrical Conductivity of Rocks, Minerals, and the Earth 631
   J. A. Tyburczy, *Arizona State University, Tempe, AZ, USA*

Preface

Geophysics is the physics of the Earth, the science that studies the Earth by measuring the physical consequences of its presence and activity. It is a science of extraordinary breadth, requiring 10 volumes of this treatise for its description. Only a treatise can present a science with the breadth of geophysics if, in addition to completeness of the subject matter, it is intended to discuss the material in great depth. Thus, while there are many books on geophysics dealing with its many subdivisions, a single book cannot give more than an introductory flavor of each topic. At the other extreme, a single book can cover one aspect of geophysics in great detail, as is done in each of the volumes of this treatise, but the treatise has the unique advantage of having been designed as an integrated series, an important feature of an interdisciplinary science such as geophysics. From the outset, the treatise was planned to cover each area of geophysics from the basics to the cutting edge so that the beginning student could learn the subject and the advanced researcher could have an up-to-date and thorough exposition of the state of the field. The planning of the contents of each volume was carried out with the active participation of the editors of all the volumes to insure that each subject area of the treatise benefited from the multitude of connections to other areas.

Geophysics includes the study of the Earth's fluid envelope and its near-space environment. However, in this treatise, the subject has been narrowed to the solid Earth. The *Treatise on Geophysics* discusses the atmosphere, ocean, and plasmasphere of the Earth only in connection with how these parts of the Earth affect the solid planet. While the realm of geophysics has here been narrowed to the solid Earth, it is broadened to include other planets of our solar system and the planets of other stars. Accordingly, the treatise includes a volume on the planets, although that volume deals mostly with the terrestrial planets of our own solar system. The gas and ice giant planets of the outer solar system and similar extra-solar planets are discussed in only one chapter of the treatise. Even the *Treatise on Geophysics* must be circumscribed to some extent. One could envision a future treatise on Planetary and Space Physics or a treatise on Atmospheric and Oceanic Physics.

Geophysics is fundamentally an interdisciplinary endeavor, built on the foundations of physics, mathematics, geology, astronomy, and other disciplines. Its roots therefore go far back in history, but the science has blossomed only in the last century with the explosive increase in our ability to measure the properties of the Earth and the processes going on inside the Earth and on and above its surface. The technological advances of the last century in laboratory and field instrumentation, computing, and satellite-based remote sensing are largely responsible for the explosive growth of geophysics. In addition to the enhanced ability to make crucial measurements and collect and analyze enormous amounts of data, progress in geophysics was facilitated by the acceptance of the paradigm of plate tectonics and mantle convection in the 1960s. This new view of how the Earth works enabled an understanding of earthquakes, volcanoes, mountain building, indeed all of geology, at a fundamental level. The exploration of the planets and moons of our solar system, beginning with the Apollo missions to the Moon, has invigorated geophysics and further extended its purview beyond the Earth. Today geophysics is a vital and thriving enterprise involving many thousands of scientists throughout the world. The interdisciplinarity and global nature of geophysics identifies it as one of the great unifying endeavors of humanity.

The keys to the success of an enterprise such as the *Treatise on Geophysics* are the editors of the individual volumes and the authors who have contributed chapters. The editors are leaders in their fields of expertise, as distinguished a group of geophysicists as could be assembled on the planet. They know well the topics that had to be covered to achieve the breadth and depth required by the treatise, and they know who were the best of their colleagues to write on each subject. The list of chapter authors is an impressive one, consisting of geophysicists who have made major contributions to their fields of study. The quality and coverage achieved by this group of editors and authors has insured that the treatise will be the definitive major reference work and textbook in geophysics.

Each volume of the treatise begins with an 'Overview' chapter by the volume editor. The Overviews provide the editors' perspectives of their fields, views of the past, present, and future. They also summarize the contents of their volumes and discuss important topics not addressed elsewhere in the chapters. The Overview chapters are excellent introductions to their volumes and should not be missed in the rush to read a particular chapter. The title and editors of the 10 volumes of the treatise are:

Volume 1: Seismology and Structure of the Earth

> Barbara Romanowicz
> University of California, Berkeley, CA, USA
> Adam Dziewonski
> Harvard University, Cambridge, MA, USA

Volume 2: Mineral Physics

> G. David Price
> University College London, UK

Volume 3: Geodesy

> Thomas Herring
> Massachusetts Institute of Technology, Cambridge, MA, USA

Volume 4: Earthquake Seismology

> Hiroo Kanamori
> California Institute of Technology, Pasadena, CA, USA

Volume 5: Geomagnetism

> Masaru Kono
> Okayama University, Misasa, Japan

Volume 6: Crust and Lithosphere Dynamics

> Anthony B. Watts
> University of Oxford, Oxford, UK

Volume 7: Mantle Dynamics

> David Bercovici
> Yale University, New Haven, CT, USA

Volume 8: Core Dynamics

> Peter Olson
> Johns Hopkins University, Baltimore, MD, USA

Volume 9: Evolution of the Earth

> David Stevenson
> California Institute of Technology, Pasadena, CA, USA

Volume 10: Planets and Moons

> Tilman Spohn
> Deutsches Zentrum für Luft-und Raumfahrt, GER

In addition, an eleventh volume of the treatise provides a comprehensive index.

The *Treatise on Geophysics* has the advantage of a role model to emulate, the highly successful *Treatise on Geochemistry*. Indeed, the name *Treatise on Geophysics* was decided on by the editors in analogy with the geochemistry compendium. The *Concise Oxford English Dictionary* defines treatise as "a written work dealing formally and systematically with a subject." Treatise aptly describes both the geochemistry and geophysics collections.

The *Treatise on Geophysics* was initially promoted by Casper van Dijk (Publisher at Elsevier) who persuaded the Editor-in-Chief to take on the project. Initial meetings between the two defined the scope of the treatise and led to invitations to the editors of the individual volumes to participate. Once the editors were on board, the details of the volume contents were decided and the invitations to individual chapter authors were issued. There followed a period of hard work by the editors and authors to bring the treatise to completion. Thanks are due to a number of members of the Elsevier team, Brian Ronan (Developmental Editor), Tirza Van Daalen (Books Publisher), Zoe Kruze (Senior Development Editor), Gareth Steed (Production Project Manager), and Kate Newell (Editorial Assistant).

G. Schubert
Editor-in-Chief

Contributors

D. Alfè
University College London, London, UK

J. D. Bass
University of Illinois at Urbana-Champaign, Urbana, IL, USA

R. Boehler
Max Planck Institute für Chemie, Mainz, Germany

J. M. Branlund
Washington University, St. Louis, MO, USA

R. E. Cohen
Carnegie Institution of Washington, Washington, DC, USA

A. Corgne
Macquarie University, Sydney, NSW, Australia

D. B. Dingwell
University of Munich, Munich, Germany

R. E. Dunin-Borkowski
University of Cambridge, Cambridge, UK

J. M. Feinberg
University of Cambridge, Cambridge, UK

R. J. Harrison
University of Cambridge, Cambridge, UK

R. J. Hemley
Carnegie Institution of Washington, Washington, DC, USA

A. M. Hofmeister
Washington University, St. Louis, MO, USA

T. Irifune
Ehime University, Matsuyama, Japan

E. Ito
Okayama University, Misasa, Japan

I. Jackson
Australian National University, Canberra, ACT, Australia

T. Kasama
University of Cambridge, Cambridge, UK

D. L. Kohlstedt
University of Minnesota, Minneapolis, MN, USA

L. Li
State University of New York (SUNY), Stony Brook, NY, USA

D. Mainprice
Université Montpellier II, Montpellier, France

H. -K. Mao
Carnegie Institution of Washington, Washington, DC, USA

W. L. Mao
Los Alamos National Laboratory, Los Alamos, NM, USA

P. F. McMillan
University College London, London, UK

A. R. Oganov
ETH Zurich, Zurich, Switzerland

M. Pertermann
Rice University, Houston, TX, USA

G. D. Price
University College London, London, UK

M. Ross
University of California, Livermore, CA, USA

E. T. Simpson
University of Cambridge, Cambridge, UK

V. V. Struzhkin
Carnegie Institution of Washington, Washington, DC, USA

L. Stixrude
University of Michigan, Ann Arbor, MI, USA

T. Tsuchiya
Ehime University, Matsuyama, Japan

J. A. Tyburczy
Arizona State University, Tempe, AZ, USA

L. Vočadlo
University College London, London, UK

D. J. Weidner
State University of New York (SUNY), Stony Brook, NY, USA

B. J. Wood
Macquarie University, Sydney, NSW, Australia

EDITORIAL ADVISORY BOARD

1 Overview – Mineral Physics: Past, Present, and Future

G. D. Price, University College London, London, UK

References	5

Mineral physics involves the application of physics and chemistry techniques in order to understand and predict the fundamental behavior of Earth materials (e.g., Kieffer and Navrotsky, 1985), and hence provide solutions to large-scale problems in Earth and planetary sciences. Mineral physics, therefore, is relevant to all aspects of solid Earth sciences, from surface processes and environmental geochemistry to the deep Earth and the nature of the core. In this volume, however, we focus only on the geophysical applications of mineral physics (see also Ahrens (1995), Hemley (1998), and Poirier (2000)). These applications, however, are not just be constrained to understanding structure the Earth and its evolution, but also will play a vital role in our understanding of the dynamics and evolution of other planets in our solar system (see Oganov *et al.* (2005)).

As a discipline, mineral physics as such has only been recognized for some 30 years or so, but in fact it can trace its origins back to the very foundations of solid Earth geophysics itself. Thus, for example, the work of Oldham (1906) and Gutenberg (1913), that defined the seismological characteristics of the core, led to the inference on the basis of materials physics that the outer core is liquid because of its inability to support the promulgation of shear waves.

A landmark paper in the history of the application of mineral physics to the understanding of the solid Earth is the *Density of the Earth* by Williamson and Adams (1923). Here the elastic constants of various rock types were used to interpret the density profile as a function of depth within the Earth that had been inferred from seismic and gravitational data. Their work was marked by taking into account the gravitationally induced compression of material at depth within the Earth, which is described by the Williamson–Adams relation that explicitly links geophysical observables ($g(r)$, the acceleration due to gravity as a function of radius, r, and the longitudinal and shear seismic wave velocities V_p and V_s) with mineral properties (K_s, the adiabatic bulk modulus and density, ρ), via

$$d\rho(r)/\rho(r) = -g(r)dr/\varphi(r) \qquad [1]$$

where $\varphi(r)$ is the seismic parameter as a function of radius, and is given by

$$\varphi(r) = V^2_p(r) - (4/3)V^2_s(r) = K_s(r)/\rho(r) \qquad [2]$$

Further progress in inferring the nature of Earth's deep interior rested upon the experimental determination of the elastic properties of rocks and minerals as a function of pressure and temperature. Notably, this work was pioneered over several decades by Bridgman (1958). In parallel with experimental studies, however, a greater understanding of the theory behind the effect of pressure on compressibility was being made by Murnaghan (1937) and Birch (1938). These insights into the equations of state of materials enabled Birch (1952) (see **Figure 1**) to write his classic paper entitled *Elasticity and the Constitution of the Earth's Interior*, which laid the foundations of our current understanding of the composition and structure of our planet.

One notable outcome from the investigation of the effect of pressure and temperature on material properties was the discovery of new high-density polymorphs of crustal minerals. Thus, Coes (1953) synthesized a new high-density polymorph of SiO_2 (subsequently named coesite), and Ringwood (1959) reported the synthesis of the spinel structured Fe_2SiO_4 (that had previously been predicted by Bernal (1936)). Ringwood and colleagues went on to make a variety of other high-density silicate polymorphs, including the phases, which are now thought to make up the transition zone of the mantle, namely the spinelloids wadsleyite (β-Mg_2SiO_4) and ringwoodite (γ-Mg_2SiO_4), and the garnet-structured polymorph of $MgSiO_3$ (majorite). Further insights into the probable nature of deep Earth minerals came from Stishov and Popova (1961) who synthesized the rutile-structured polymorph of SiO_2 (stishovite) that is characterized by having Si in

Scanned at the American
Institute of Physics

Figure 1 Francis Birch (1903–92), Royal Astronomical
Society Medalist 1960; Bowie Medalist 1960.

octahedral coordination and from Takahashi and
Bassett (1964) who first made the hexagonal close-
packed polymorph of Fe, which is today thought to
be the form of Fe to be found in the Earth's core (but
see Chapter 5). As high-pressure and -temperature
experimental techniques evolved, still further phases

were discovered, the most important of which was
the 'postspinel, perovskite-structured polymorph of
$MgSiO_3$ (Liu, 1975). It was thought for sometime that
this discovery and the subsequent work on the details
of the high-pressure phase diagrams of silicate minerals had enabled a robust mineralogical model for the
mantle to be established. This view, however, has had
to be revised in the past few years, after the recent
discovery of a 'postperovskite' phase (Murakami *et al.*,
2004; Oganov and Ono, 2000), which may be stable in
the deepest part of the lower mantle.

Notwithstanding, however, the possibility of
further new discoveries, the mineralogy, and composition of the mantle and the core are now relatively
well defined. The current view of the mineralogy of
the mantle is summarized in **Figure 2**, while as
suggested by Birch (1952), the core is considered to
be composed of iron (with minor amounts of nickel)
alloyed with light elements (probably O, S, and or Si).
The solid inner core is crystallizing from the outer
core, and so contains less light elements. The current
status of our understanding of the nature of the deep
Earth is reviewed in detail in Chapters 2, 3, 4, and 5.
Chap Stixrude provides a general overview of the
structure of the mantle. The nature of the lower mantle
is still relatively controversial, since generating lower
mantle pressures ($\sim$25–130 GPa) and temperatures
($\sim$2000–3000 K) is still experimentally challenging,
and mineral physics data and phase relations for the

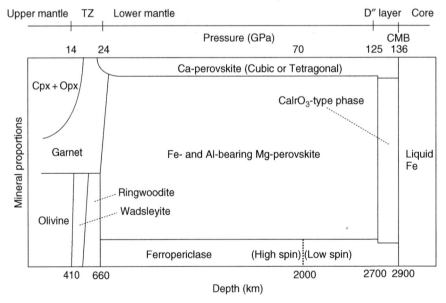

Figure 2 Phase relations of pyrolitic mantle composition as a function of depth. From Ono S and Oganov AR (2005) *In situ*
observations of phase transition between perovskite and CaIrO₃-type phase in MgSiO₃ and pyrolitic mantle composition.
Earth and Planetary Science Letters 236: 914–932.

minerals thought to be found here are less robust. Furthermore, the recent discovery of the postperovskite phase has added even greater uncertainty to the nature of the D″ zone and the core–mantle boundary. The problems of the lower mantle and the core–mantle boundary are therefore reviewed in Chapter 3. Although the major element chemistry of the mantle is quite well studied, it is probably fair to say that the understanding of the trace elements chemistry and the role of volatiles in the deep mantle is still in its infancy. This aspect of mantle chemistry, however, is vital if we are to fully understand the processes involved in planetary formation, core segregation, and subsequent evolution of the Earth. The Chapter 4 provides a review of our understanding of this aspect of the mantle, while the considerable progress in our understanding of the nature and evolution of the core is provided in Chapter 5.

As indicated above, our understanding of the lower mantle and core are limited to some extent by our inability easily to reproduce the high-pressure and -temperature conditions to be found in planetary interiors. To obtain greater insight, theory and experiment must be used together, and Chapters 6 and 13 present reviews of the theory underlying high-pressure, high-temperature physics, and the major experimental methods that are being developed to probe this parameter space. The Chapter 6 outlines the thermodynamic basis behind high-pressure–high-temperature behavior, and expands in greater detail on equations of state and the way in which the density and elastic properties of materials respond to changes in pressure and temperature. The macroscopic behavior of minerals depends upon the microscopic or atomistic interactions within the mineral structure. Thus, for example, free energy (and eventually phase stability) depends in part upon entropy, which in turn is dominated (for silicates at least) by lattice vibrations. Hence, in Chapter 7, a detailed analysis of lattice vibrations and spectroscopy of mantle minerals is presented. For the past 20 years, advances in computing power have enabled computational mineral physics to make a contribution to our understanding of the thermodynamic, thermo-elastic, and dynamical properties of high-pressure minerals. Initially, the results of simulations based on inter-atomic potentials provided semiquantitative insights into, for example, the lattice vibrations and the thermodynamics of mantle phases (e.g., Price *et al.*, 1987; Wall and Price, 1988). But more recently, quantum mechanical simulations of mantle and core phases have been able to achieve a precision

and accuracy that is comparable with that achievable experimentally, and as such *ab initio* modeling must now be seen as a legitimate complement to experimental study. Therefore in this volume, the theory and results from *ab initio* studies of some deep Earth phases are reviewed in Chapter 13.

Despite the power and insight provided by theory, mineral physics is dependent upon quantitative high-pressure and -temperature experimental results. Such work was pioneered by Williamson, Adams (**Figure 3**), and Bridgman, and then taken up by others such as Ringwood, Bassett, Liu, and many groups in Japan (see, e.g., Akimoto (1987) and **Figure 4**), but in the past 30 years huge advances have been made and today, for example, laser-heated diamond anvil cells can be used to access temperatures and pressures up to ~6000 K and 250 GPa. Such experiments, however, can only be carried out on very small sample volumes and for very short periods of time. Multianvil experiments can be used to study much larger volumes of material and are stable over a longer time interval. These techniques are, therefore, complementary and are both developing rapidly. Chap Ito provides a review of multianvil cell

Figure 3 Leason Heberling Adams (1887–1969), American Geophysical Union (AGU) President 1944–47, Bowie Medalist 1950.

Figure 4   Syun-iti Akimoto (1925–2004), Royal Astronomical Society Medalist 1983, Bowie Medalist 1983.

methods, while Chap Mao gives an analysis of diamond-anvil cell techniques. A third technique for producing high pressures and temperatures is via shock compression. In this approach a high-velocity impact is produced by firing a projectile at a mineral target. Very high pressure and shock-produced heating can be obtained, but only for a few 100 ns. Analysis of the experimental results is not always straightforward, as a knowledge of some high P–T physical properties is usually needed to infer peak shock temperatures, and shock results are not always in agreement with static experiments (see, e.g., the debate on the high-pressure melting of Fe in Chapters 18 and 13). For more details on this techniques see Ahrens (1995), Poirier (2000), and Nellis (2002).

Achieving the high pressures and temperatures that occur in the deep Earth is, however, just the beginning of the challenge that faces experimental mineral physicists, since it is also necessary to measure a variety of physical properties under these conditions. Therefore in Chapters 10, 11, and 12 the specific techniques for measuring elastic and acoustic properties, electronic and magnetic properties, and rheological properties are described in greater detail.

The long-term goal of mineral physics research is to enable a full interpretation of seismic tomographic data, and to provide a three-dimensional description of the mineralogy, composition, and thermal structure of the Earth's deep interior. To this end, we need not only full descriptions of the seismic properties of rocks and minerals, as a function of composition, pressure, and temperature (*see* Chapter 2), but also an understanding of the nature and origin of anisotropy in the Earth (*see* Chapter 16) and the significance and reasons for seismic wave attenuation.

In addition to being able to invert seismic data to provide an instantaneous picture of the Earth's interior, we desire to be able to describe the dynamic evolution of our planet. As such we need to be able to constrain, for example, geodynamical models of mantle convection, with the appropriate rheological descriptions of mantle phases (*see* Chapter 14), and self-consistent descriptions of heat flux and planetary thermal structure. Key to the latter is our understanding of thermal conductivity at high pressures and temperatures, which is an area where theory and experiment are still evolving and much progress is still required (*see* Chapter 19). Similarly, our ability to describe the dynamics of core convection, and hence the geodynamo, is critically dependent on our understanding of the physical properties of liquid and solid iron under core conditions (*see* Chapters 5, 13, 18, and Price *et al.*, 2004).

One particular problem in our understanding of the Earth is the nature of the core–mantle boundary. It has been variously suggested, for example, that it is a 'grave yard' for subducted slabs; a reaction zone with the core; associated with a postperovskite phase change; has pods of partially molten material; is enriched in iron; and is the source of deep mantle plumes. To establish the importance of some of these hypotheses requires high-quality seismic data, but also tight constraints on the thermal structure of the core and the high-pressure melting behavior of mantle and core material (*see* Chapter 18). The melting of mantle material also obviously plays a vital role in volcanism and planetary evolution, and the migration of melts and their eruption or injection into the crust is central to our understanding of the dynamics of the upper mantle. The properties of melts are discussed in Chapter 15.

Another long-standing problem in geophysics is the energetics of the core and the driving force behind the geodynamo. To be able to model this process, accurate descriptions of the melting of iron and the chemical compositions of the inner and outer core are essential (e.g., Nimmo *et al.*, 2004). The possible presence of radioactive elements in the core (such as K^{40}) would be highly significant as such internal sources of heat would greatly influence the cooling rate of the core and hence the age and rate of growth of the inner core; but it is still an open question. The stability of the geodynamo may also be affected by thermal and electrical coupling with the mantle; hence, we need to fully understand the electrical properties of mantle phases (*see* Chapter 21). Finally, for example, insights into the past nature of the geodynamo, as its palaeo-intensity, can only be obtained from paleomagnetic data, which requires a detailed understanding of rock minerals and magnetism (*see* Chapter 20).

In the following chapters in this volume, the great progress that has been made in our understanding of the physics and chemistry of minerals is clearly laid out. However, each author also highlights a number of issues which are still outstanding or that need further work to resolve current contradictions. The resolution of some of the problems outlined above, and in the subsequent chapters, will depend on more precise or higher-resolution geophysical measurements (e.g., the exact density contrast between the inner and outer core), while others will be solved as a result of new experimental or computational techniques. The combination, for example, of intense synchrotron radiation and neutron sources with both diamond anvil and multianvil cell devices promises to yield much more information on high pressure seismic (or equivalently elastic) properties of deep Earth phases. In the foreseeable future, developments in anvil design will see large-volume devices that can be used to determine phase equilibria to 60 GPa, or to study high-pressure rheology to 30 GPa and beyond. But more predictably, the continued development of increasingly powerful computers will see the role of computational mineral physics grow still further, as the study of more complex problems (like rheology and thermal transport) becomes routine. In addition, the use of more sophisticated *ab initio* methods (like quantum Monte Carlo techniques) will become possible, and accurate studies of the band structure and electrical properties of iron bearing silicates will be performed. Finally, the pressures of the Earth's interior (up to 360 GPa) will not represent the limit of interest for mineral physicists. Already extra-solar system, giant, Earth-like planets are being considered, and understanding their internal structure will open up even greater challenges (e.g., Umemoto and Wentzcovitch, 2006 Umemoto *et al.*, 2006).

In conclusion, therefore, this volume contains a comprehensive review of our current state of understanding of mineral physics, but without doubt there is still much

that is unknown, but the prospect for further progress is excellent, and it is certain that in the next decades many issues, which are still controversial today, will have been resolved.

References

Ahrens TJ (1995) *AGU Reference Shelf 2: Mineral Physics and Crystallography: A Handbook of Physical Constants*. Washington, DC: American Geophysical Union.

Akimoto S (1987) High-pressure research in geophysics: Past, present, and future. In: Manghnani MH and Syono Y (eds.) *High-Pressure Research in Mineral Physics*, pp. 1–13. Washington, DC: American Geophysical Union.

Bernal JD (1936) Discussion. *Observatory* 59: 268.

Birch F (1938) The effect of pressure upon the elastic properties of isotropic solids according to Murnaghan's theory of finite strain. *Journal of Applied Physics* 9: 279–288.

Birch F (1952) Elasticity and constitution of the Earth's interior. *Journal of Geophysical Research* 57: 227–286.

Bridgman PW (1958) *Physics of High Pressure*. London: G Bell and Sons.

Coes L (1953) A new dense crystalline silica. *Science* 118: 131–132.

Gutenberg B (1913) Uber die Konstitution der Erdinnern, erschlossen aus Erdbebenbeobachtungen. *Physika Zeitschrift* 14: 1217–1218.

Hemley RJ (1998) *Reviews in Mineralogy*, Vol. 37, *Ultrahigh-Pressure Mineralogy: Physics and Chemistry of the Earth's Deep Interior,* Washington, DC: Mineralogical Society of America.

Liu LG (1975) Post-oxide phases of olivine and pyroxene and the mineralogy of the mantle. *Nature* 258: 510–512.

Murakami M, Hirose K, Kawamura K, Sata N, and Ohishi Y (2004) Post-perovskite phase transition in $MgSiO_3$. *Science* 304: 855–858.

Murnaghan FD (1937) Finite deformations of an elastic solid. *American Journal of Mathematics* 59: 235–260.

Nellis WJ (2002) Dynamic experiments: An overview. In: Hemely RJ and Chiarotti GL (eds.) *High Pressure Phenomena*, pp. 215–238. Amsterdam: IOS Press.

Nimmo F, Price GD, Brodholt J, and Gubbins D (2004) The influence of potassium on core and geodynamo evolution. *Geophysical Journal International* 156: 363–376.

Oldham RD (1906) The constitution of the Earth as revealed by earthquakes. *Quarternary Journal of Geological Society* 62: 456–475.

Oganov AR and Ono S (2000) Theoretical and experimental evidence for a post-perovskite phase of $MgSiO_3$ in Earth's D″ layer. *Nature* 430: 445–448.

Oganov AR, Price GD, and Scandolo S (2005) *Ab initio* theory of planetary materials. *Zietschrift fur Kristallographie* 220: 531–548.

Ono S and Oganov AR (2005) *In situ* observations of phase transition between perovskite and $CaIrO_3$-type phase in $MgSiO_3$ and pyrolitic mantle composition. *Earth and Planetary Science Letters* 236: 914–932.

Poirier JP (2000) *Introduction to the Physics of the Earth's Interior*. Cambridge: Cambridge University Press.

Price GD, Alfè D, Vočadlo D, and Gillan MJ (2004) The Earth's core: An approach from first principles. In: Sparks RSJ and Hawksworth CJ (eds.) *AGU Geophysical Monograph Series*, Vol. 150. *The State of the Planet: Frontiers and Challenges in Geophysics*, pp. 1–12. Washington, DC: American Geophysical Union.

Price GD, Parker SC, and Leslie M (1987) The lattice dynamics and thermodynamics of the Mg_2SiO_4 polymorphs. *Physics and Chemistry of Minerals* 15: 181–190.

Ringwood AE (1959) The olivine-spinel inversion in fayalite. *American Mineralogist* 44: 659–661.

Stishov SM and Popova SV (1961) A new dense modification of silica. *Geochemistry* 10: 923–926.

Takahashi T and Bassett WA (1964) A high pressure polymorph of iron. *Science* 145: 483–486.

Umemoto K and Wentzcovitch RM (2006) Potential ultrahigh pressure polymorphs of ABX(3)-type compounds. *Physical Review B* 74: 224105.

Umemoto K, Wentzcovitch RM, and Allen PB (2006) Dissociation of $MgSiO_3$ in the cores of gas giants and terrestrial exoplanets. *Science* 311: 983–986.

Wall A and Price GD (1988) A computer simulation of the structural, lattice dynamical and thermodynamic properties of $MgSiO_3$ ilmenite. *American Mineralogist* 73: 224–231.

Williamson ED and Adams LH (1923) Density distribution in the Earth. *Journal of Washington Academy of Sciences* 13: 413–432.

2 Properties of Rocks and Minerals – Seismic Properties of Rocks and Minerals, and Structure of the Earth

L. Stixrude, University of Michigan, Ann Arbor, MI, USA

2.1	Introduction	7
2.2	Radial Structure	8
2.2.1	Overview	8
2.2.2	Upper Mantle	10
2.2.3	Transition Zone	12
2.2.4	Lower Mantle	15
2.2.5	Core	17
2.3	Lateral Heterogeneity	19
2.3.1	Overview	19
2.3.2	Temperature	19
2.3.3	Composition	20
2.3.4	Phase	21
2.4	Anisotropy	22
2.4.1	Overview	22
2.4.2	Upper Mantle	22
2.4.3	D″ Layer	23
2.5	Attenuation and Dispersion	24
2.5.1	Overview	24
2.5.2	Influence of Temperature	24
2.5.3	Speculations on the Influence of Pressure	25
2.6	Conclusions	26
References		26

2.1 Introduction

The theoretical problem of computing the radial structure of a hydrostatic, gravitationally self-compressed body like Earth is well developed. In its simplest form, one solves the Poisson equation together with a constitutive relation for the variation of pressure and density with depth. While such solutions are important for our understanding of the structure of stars and giant planets, they have not played a major role in our understanding of Earth structure for at least two important reasons. First is the still overwhelming complexity of the constitutive relation in the case of the Earth and the other terrestrial planets. There are at least six essential elements and they are distributed inhomogeneously with depth, most evidently in the separation between the crust, mantle, and core. Moreover, there are more than 10 essential phases in the mantle alone, as compared with the single fluid phase that suffices in the case of the gas giants. Second is the power of seismological observations to reveal Earth structure. Adams and Williamson were able to estimate the radial density profile of Earth's mantle from a very different starting point, the observed variation of seismic wave velocities with depth, as embodied in their famous equation. The number and quality of seismic observations continues to grow as does our ability to interpret them, so that over the past few decades, seismologists have been able to constrain not only radial, but also lateral variations in seismic wave velocities and to produce three-dimensional (3-D) models of Earth structure.

One of the goals of geophysics is to relate the structure of Earth's interior as revealed primarily by seismology, to its thermal and chemical state, its dynamics and evolution. It is not generally possible to construct this more complete picture from knowledge of the seismologically revealed structure alone. The limitations are not so much practical – there is every indication that our knowledge of Earth structure will continue to improve for some time with further observation – but fundamental. The relationship of seismic wave velocity to composition and temperature cannot be inverted uniquely without further information.

Our primary interest in the seismic properties of Earth materials derives from this overarching goal, to place the current structural snapshot provided by seismology in the larger context of Earth evolution. Knowledge of material properties and behavior is the essential link between seismological observations and the temperature and composition of the interior. Indeed, the combination of geophysical observation and knowledge of materials properties derived from experimental and theoretical studies provide our most important constraints on the composition of the mantle and its thermal structure.

It is the aim of this contribution to illustrate the role of mineral physics as the interpretive link between seismic structure and dynamics, and to review what studies of material properties have taught us about the dynamics and composition of Earth. Our focus will be on Earth's mantle because the problem of determining the chemical and thermal state is particularly rich and because substantial progress has been made in the last decade.

The organization of this contribution takes its cue from Earth's seismic structure. We begin with a review of the 1-D elastic structure that accounts for most of the observed seismic signal. Special topics here will include the origin of high gradient zones and the origin of discontinuities. We continue with a discussion of lateral variations in seismic structure and their interpretation in terms of lateral variations in temperature, composition, and phase. A discussion of anisotropy will highlight the many important advances that have been made, particularly in our knowledge of the full elastic constant tensor of minerals, and also the formidable difficulties remaining, including a need for a better understanding of deformation mechanisms. We end with that part of Earth's seismic structure that is currently least amenable to interpretation, the anelastic part, indicating important developments in our understanding of the relevant material properties, and the gaps in our knowledge that remain.

2.2 Radial Structure

2.2.1 Overview

Perhaps the most remarkable and informative feature of Earth's radial structure is that it is not smooth

(**Table 1**). The variation of seismic wave velocities with depth is broken up at several depths by rapid changes in physical properties. These are generally referred to as discontinuities, although in all probability, they represent regions where physical properties change very rapidly over a finite depth interval. Each discontinuity is associated with a mean depth, a range of depth due to topography on the discontinuity, and a contrast in physical properties, most directly the impedance contrast

$$\frac{\Delta I}{I} = \frac{\Delta \rho}{\rho} + \frac{\Delta V}{V} \qquad [1]$$

where ρ is the density and V is either the shear or longitudinal-wave velocity and Δ represents the difference across the discontinuity. For isochemical changes in physical properties that follow Birch's law (Anderson *et al.*, 1968), the velocity contrast is approximately two-thirds of the impedance contrast for S-waves and approximately three-fourths for P-waves. Of the 14 discontinuities reported, some (e.g., 410, 660) are much more certain than others (e.g., 1200, 1700), both in terms of their existence and their properties.

Discontinuities are important because they tie seismological observations to the Earth's thermal and chemical state in an unusually precise and rich way. Many discontinuities in the mantle occur at depths that correspond closely to the pressure of phase transformations that are known from experiments to occur in plausible mantle bulk compositions (**Figure 1**). The pressure at which the phase transformation occurs generally depends on temperature via the Claussius–Clapeyron equation. The mean depth of a discontinuity then anchors the geotherm. Lateral variations in the depth of the discontinuity constrain lateral variations in temperature.

Phase transformations also influence mantle dynamics. For example, in a subducting slab, the perovskite forming reaction will be delayed as compared with the surroundings, tending to impede the slab's descent. The extent to which a phase transformation alters dynamics scales with the phase buoyancy parameter (Christensen and Yuen, 1985):

$$\Phi = \frac{\gamma \Delta \rho}{g \alpha \rho^2 h} \qquad [2]$$

where γ is the Clapeyron slope, g is the acceleration due to gravity, α is the thermal expansivity, and h is the depth of the mantle. Phase transformations with negative Clapeyron slopes, such as the perovskite forming reaction, tend to impede radial mass transfer,

Table 1   Mantle discontinuities

Discontinuity	Depth (km)	Contrast	Affinity	References	Origin	References
Hales	59(12)	+7.0%	Global?	Revenaugh and Jordan (1991a, 1991b)	Anisotropy	Bostock (1998)
Gutenberg	77(14)	−9.0%	Oceans	Revenaugh and Jordan (1991a, 1991b)	Anisotropy	Gung, *et al.* (2003)
Lehmann	230(30)	+3.8%	Continents	Revenaugh and Jordan (1991a, 1991b)	Anisotropy	Gung, *et al.* (2003)
X	313(21)	+3.3%	Subduction	Revenaugh and Jordan (1991a, 1991b)	opx = hpcpx	Woodland (1998)
–	410(22)	+8.5%	Global	Flanagan and Shearer (1998), Dziewonski and Anderson (1981)	ol = wa	Ringwood and Major (1970)
–	520(27)	+3.0%	Global	Flanagan and Shearer (1998), Shearer (1990)	wa = ri	Rigden, *et al.* (1991)
–	660-	+8%?	Subduction	Simmons and Gurrola (2000)	ak = pv	Hirose (2002)
–	660(38)	+15.8	Global	Flanagan and Shearer (1998), Dziewonski and Anderson (1981)	ri = pv + mw	Ito and Takahashi (1989)
–	720(24)	+2.0	Global?	Revenaugh and Jordan (1991)	gt = pv	Stixrude (1997)
–	900(150)	?	Subduction?	Kawakatsu and Niu (1994)	Comp.?	Kawakatsu and Niu (1994)
–	1200	?	Global?	Vinnik, *et al.* (2001)	st = hy?	Karki, *et al.* (1997a, 1997b)
–	1700	?	?	Vinnik, *et al.* (2001)	capvc = capvt?	Stixrude, *et al.* (1996)
D″	2640(100)	+2.0%	Fast	Lay, *et al.* (1998b)	pv = ppv	Oganov and Ono (2004)
ULVZ	2870(20)	−10%	Slow	Garnero and Helmberger (1996)	melt	Williams and Garnero (1996)

Contrast is the S-wave impedance contrast, except for the D″ and ULVZ discontinuities (2640, 2870 km) which are reported as the S-wave velocity contrast. Value in parentheses following depth is the peak-to-peak (410, 520, 660), or rms (others) variation in the depth. The estimated impedance contrast at 660- is half of that of the 660 itself based on subequal reflectivity from 660- and 660 in the study of Simmons and Gurrola (2000). Affinity designations fast and slow refer to regions of anomalously high and low wave speeds in tomographic models, respectively. Origin designation "comp." refers to a contrast in bulk composition possibly associated with subducted crust. Abbreviations capvc and capvt refer to the cubic and tetragonal phases of $CaSiO_3$ perovskite, respectively. Entries with a question mark are either unknown or uncertain.

while those with positive Clapeyron slopes, such as the olivine to wadsleyite transformation, tend to encourage it. In application to Earth's mantle, the phase buoyancy parameter must be generalized to account for the fact that only a fraction of the mantle undergoes the phase transformation (i.e., 50–60% in the case of the olivine to wadsleyite transition), and that nearly all phase transformations are at least divariant and occur over a finite range of depth.

Phase transformations also depend sensitively on bulk composition, which means that the Earth's discontinuity structure also constrains mantle chemistry. For example, the sequence of phase transformations with increasing pressure in olivine would be quite different if the mantle had twice as much FeO, and would no longer resemble the Earth's radial structure (Akaogi *et al.*, 1989).

Not all discontinuities can be explained by phase transformations. In some cases, no phase transformations occur near the appropriate depth. In others, phase transformations do occur, but the change in physical properties caused by the transition is far too subtle to explain the seismic signal. In the case of the D″ discontinuity, a phase transformation was only recently found that finally appears to explain most of the properties of this previously enigmatic

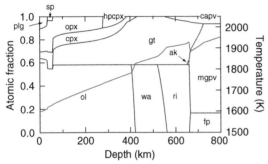

Figure 1 Phase proportions (blue, left axis) in a typical mantle bulk composition: Depleted MORB Mantle (DMM) of Workman and Hart (2005), along an isentrope (red, right axis) consistent with MORB genesis: potential temperature of 1600 K (Klein and Langmuir, 1987; McKenzie and Bickle, 1988). The isentrope and phase proportions are computed self-consistently from a unified thermodynamic formalism (Stixrude and Lithgow-Bertelloni, 2005a, 2005b). Phases are: plagiolcase (plg), spinel (sp), olivine (ol), orthopyroxene (opx), clinopyroxene (cpx), high-pressure Mg-rich clinopyroxene (hpcpx), garnet (gt), wadsleyite (wa), ringwoodite (ri), akimotoite (ak), Calcium silicate perovskite (capv), Magnesium silicate perovsite (pv), and ferropericlase (fp). From Stixrude L and Lithgow-Bertelloni C (2007) Influence of phase transformations on lateral heterogeneity and dynamics in Earth's mantle. *Earth and Planetary Science Letters*.

structure. Rapid variation with depth in chemical composition (e.g., the Moho), the pattern or strength of anisotropy, or in the magnitude of attenuation and dispersion may also cause discontinuities.

2.2.2   Upper Mantle

We take the upper mantle to be that region of the Earth between the Mohorovicic discontinuity and the 410 km discontinuity. This part of the Earth is challenging from a mineral physics point of view because heterogeneity, anisotropy, and attenuation are all large. Discussions of globally average radial structure are prone to mislead, which is why regional 1-D profiles, as well as 3-D tomographic models are important to keep in mind. This part of the Earth is remarkably rich in structure with features that have resisted explanations in terms of mineralogical models.

One of the most remarkable features of upper mantle structure is the presence of a low velocity zone (Gutenberg, 1959). A pattern of decreasing velocity with increasing depth appears unique to the boundary layers of the mantle (see also discussion

of D″ below). The low velocity zone is more prominent under oceans than under continents; and is more prominent under young oceanic lithosphere than old (Anderson, 1989).

The origin of the low velocity zone is readily understood in terms of Earth's thermal state and the properties of minerals. Negative velocity gradients will appear whenever the thermal gradient exceeds a critical value such that the increase of velocity with compression is overcome by the decrease of velocity on heating (Anderson *et al.*, 1968; Birch, 1969). The critical thermal gradient

$$\beta_S = -\left(\frac{\partial V_S}{\partial P}\right)_T \Bigg/ \left(\frac{\partial V_S}{\partial T}\right)_P \qquad [3]$$

for shear waves is modest: $2\,\mathrm{K\,km^{-1}}$ for olivine. One then expects negative velocity gradients to be widespread in the shallow upper mantle on the basis of heat flow observations. There is a critical thermal gradient for the density as well

$$\beta_\rho = \frac{1}{\alpha K_T} \qquad [4]$$

where α is the thermal expansivity and K_T is the isothermal bulk modulus, which is just the inverse of the thermal pressure gradient that has been discussed by Anderson (Anderson, 1995). The value for olivine $\sim 10\,\mathrm{K\,km^{-1}}$, is less than that in most of the Earth's upper thermal boundary layer. The low velocity zone should be associated with locally unstable stratification in which more dense material overlies less dense material.

Comparison to the critical thermal gradient also explains much of the lateral variations in the structure of the low velocity zone. In the half-space cooling model of mantle, the geothermal gradient decreases with lithospheric age, producing a weaker negative velocity gradient as seen seismologically (Graves and Helmberger, 1988; Nataf *et al.*, 1986; Nishimura and Forsyth, 1989). Under continents, the geothermal gradient is much less steep, but even under cratons exceeds the critical value; for example, the gradient is $4\,\mathrm{K\,km^{-1}}$ under the Abitibi province (Jaupart and Mareschal, 1999). Indeed, the low velocity zone appears to be much less prominent, but not absent under some continental regions (Grand and Helmberger, 1984).

Are other factors required to quantitatively explain the values of the lowest velocities seen in the low velocity zone? It has long been suggested that partial melt may be present in this region of the mantle (Anderson and Sammis, 1969; Birch,

1969; Green and Liebermann, 1976; Lambert and Wyllie, 1968; Ringwood, 1969; Sato *et al.*, 1989). While partial melt is not required to explain the existence of a low velocity zone *per se* as the argument regarding the critical thermal gradient above demonstrates, it is difficult to rule out its presence on the basis of quantitative comparisons of seismological observations to the elastic properties of upper mantle assemblages. Recent such comparisons have concluded that partial melt is not required to explain the seismological observations, except in the immediate vicinity of the ridge (Faul and Jackson, 2005; Stixrude and Lithgow-Bertelloni, 2005a) (**Figure 2**). These studies also find it essential to account for the effects of attenuation and dispersion, which is unusually large in this region (see below).

The low velocity zone has sometimes been seen as being bounded above by the Gutenberg (G) discontinuity near 65 km. This discontinuity has also been associated with the base of the lithosphere. The impedance contrast is variable, large, and negative, that is, velocities are less below the transition (Gaherty *et al.*,

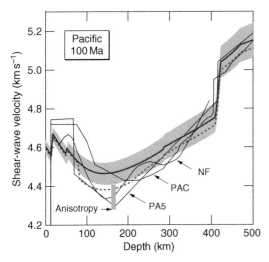

Figure 2 The shear-wave velocity of pyrolite along a 100 Ma conductive cooling geotherm in the elastic limit (bold line), and including the effects of dispersion according to the seismological attenuation model QR19 (Romanowicz, 1998) and $\alpha = 0.25$ (bold dashed). The shading represents the uncertainty in the calculated velocity. The mineralogical model is compared with seismological models (light lines) PAC (Graves and Helmberger, 1988), NF110+ (Nishimura and Forsyth, 1989), and PA5 (Gaherty, *et al.*, 1999a). The approximate magnitude of SH–SV anisotropy is indicated by the vertical bar. From Stixrude L and Lithgow-Bertelloni C, (2005a) Mineralogy and elasticity of the oceanic upper mantle: Origin of the low-velocity zone *Journal of Geophysical Research, Solid Earth* 110: B03204 (doi:10.1029/2004JB002965).

1999a; Revenaugh and Jordan, 1991b). Phase transformations occurring near this depth do not produce velocity discontinuities of nearly sufficient magnitude, or even the right sign. A rapid change in attenuation produces a velocity change of the right sign (Karato and Jung, 1998), but far too small in magnitude (Stixrude and Lithgow-Bertelloni, 2005a). A rapid change in partial melt fraction, from a melt-free lithosphere to a partially molten low velocity zone, also produces a velocity change that is too small as modeling indicates that the discontinuity is made up equally of anomalously high velocity above the discontinuity and anomalously low velocity below (Stixrude and Lithgow-Bertelloni, 2005a) (**Figure 2**).

The Lehmann discontinuity, near 200 km depth, also does not have a counterpart in mantle phase transformations. In some global models (Dziewonski and Anderson, 1981) this discontinuity is enormous, exceeding in magnitude the much more widely studied 410 km discontinuity. The Lehmann discontinuity is observed preferentially under continents (Gu *et al.*, 2001), while the G discontinuity is found preferentially under oceans (Revenaugh and Jordan, 1991b). It has been proposed that the Lehmann discontinuity represents a change in anisotropy caused by a change in the dominant deformation mechanism from dislocation dominated to diffusion dominated creep (Karato, 1992). The Lehmann and Gutenberg discontinuities may actually be the same feature, occurring at shallower depths under oceans and greater depths under continents. Both may be caused by a change in anisotropy via a change in the pattern of preferred orientation associated with the base of the lithosphere and a transition to more deformable asthenosphere where preferred orientation is more likely to develop (Gung *et al.*, 2003) (**Figure 3**). This model is in apparent disagreement with those based on ScS reverberation, which require an increase in SV with increasing depth across the Lehmann (Gaherty and Jordan, 1995; Revenaugh and Jordan, 1991b).

A smaller discontinuity is seen locally called the X discontinuity near 300 km depth (Revenaugh and Jordan, 1991b). A phase transition appears to be the most likely explanation (Woodland, 1998), involving a subtle modification of the orthopyroxene structure to a slightly denser monoclinic form with the same space group as diopside (clinopyroxene) (Angel *et al.*, 1992; Pacalo and Gasparik, 1990) (**Figure 1**). The elastic properties of this unquenchable high pressure phase have only recently been measured and appear to confirm the connection to the seismic discontinuity (Kung *et al.*, 2005). The velocity contrast

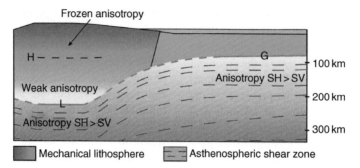

Figure 3 Sketch illustrating the pattern of anisotropy found in a recent tomographic model in relation to lithospheric thickness, and to the Lehmann (L) and Gutenberg (G) discontinuities. The Hales discontinuity (H) is also indicated. From Gung YC, Panning M, and Bomanowicz B (2003) Global anisotropy and the thickness of continents. *Nature* 422: 707–711.

in the pure composition is 8%, reduced to less than 0.5% in a typical pyrolite composition along a normal mantle geotherm (Stixrude and Lithgow-Bertelloni, 2005a). The proportion of orthopyroxene, and thus the magnitude of the velocity change, should increase with decreasing temperature and with depletion; indeed, the X discontinuity is preferentially observed near subduction zones.

The velocity gradient between the low velocity zone and the 410 km discontinuity is very large (**Figure 2**). In some regional models, the gradient exceeds that of the transition zone (Gaherty *et al.*, 1999a; Graves and Helmberger, 1988; Nishimura and Forsyth, 1989). According to the usual Bullen analysis, this means that this region is either nonadiabatic, or inhomogeneous in bulk composition or phase (Stixrude and Lithgow-Bertelloni, 2005a). In this context, it is important to remind ourselves that when we discuss radial gradients in velocities, we are not discussing seismological observations, but rather nonunique models that are consistent with those observations. An approach that has seen only little application to date is the direct comparison of mineralogical models of mantle structure with the seismological observations themselves. This could be accomplished, for example, in the manner of a hypothesis test: computing seismological observables (e.g., body wave travel times and normal mode frequencies) from a mineralogical model and comparing with the observations. It is probably only through approaches like these that many of the more difficult issues in upper mantle structure will be resolved.

2.2.3 Transition Zone

This region is so named because of the phase transformations that were anticipated and subsequently

found to take place within it. While phase transformations are not restricted to the transition zone as our discussion of the upper and lower mantles make clear, the two largest ones in terms of changes in physical properties do occur here. This region encompasses the transformation of the mantle from uniformly fourfold to uniformly sixfold coordinated silicon.

The two largest discontinuities in the mantle occur at 410 and 660 km depth. These have been explained by phase transformations from olivine to its high pressure polymorph wadsleyite for the 410 (Ringwood and Major, 1970) and from ringwoodite, the next highest pressure olivine polymorph, to the assemblage perovskite plus periclase for the 660 according to the reaction (Liu, 1976)(**Figure 1**).

$$\text{Mg}_2\text{SiO}_4 \text{ (ringwoodite)} = \text{MgSiO}_3 \text{ (perovskite)} \\ + \text{MgO (periclase)} \quad [5]$$

These explanations have received considerably scrutiny over the past several years and have held up well. In addition to the issues discussed in more detail below has been vigorous debate concerning the pressure at which the two transformations occur and whether these in fact match the depth of the discontinuities (Chudinovskikh and Boehler, 2001; Fei *et al.*, 2004; Irifune *et al.*, 1998; Shim *et al.*, 2001), and whether the Clapeyron slope of the transitions can account for the seismologically observed topography on the discontinuities (Helffrich, 2000).

In the case of the 410 km discontinuity, the magnitude and the sharpness of the velocity jump have also been the subject of considerable debate. Several studies have argued that the magnitude of the velocity change at the olivine to wadsleyite transition in a typical pyrolitic mantle composition is too large and that the mantle at this depth must have a relatively olivine poor bulk composition (Duffy *et al.*, 1995; Gwanmesia *et al.*, 1990; Li *et al.*, 1998; Liu *et al.*, 2005; Zha *et al.*, 1997). Phase equilibrium studies (Akaogi *et al.*, 1989; Frost 2003; Katsura and Ito, 1989; Katsura *et al.*, 2004) have shown that the olivine to wadsleyite transition in the binary Mg_2SiO_4–Fe_2SiO_4 system occurs over a depth range too wide to account for the observed reflectivity of this transition, at frequencies as high as 1 Hz (Benz and Vidale, 1993).

There have been several proposals to explain the apparent discrepancy in sharpness (Fujisawa, 1998; Solomatov and Stevenson, 1994). Perhaps the simplest recognizes that the mantle is not a binary system: chemical exchange of olivine and wadsleyite with nontransforming phases sharpens the transition considerably (Irifune and Isshiki, 1998; Stixrude, 1997) (**Figure 4**). Nonlinearities inherent in the form of coexistence loops also tend to sharpen the transition (Helffrich and Bina, 1994; Helffrich and Wood, 1996; Stixrude, 1997). Modeling of finite frequency wave propagation shows that the form of the transition predicted by equilibrium thermodynamics and a pyrolitic bulk composition can account quantitatively for the observations (Gaherty *et al.*, 1999b). This study also shows that when the shape of the transition is properly accounted for, the magnitude of the velocity jump in pyrolite is consistent with those observations most sensitive to it, that is, the reflectivity.

The detailed form of the 410 km discontinuity may become a very sensitive probe of mantle conditions. The width of the discontinuity is found to increase with decreasing temperature (Katsura and Ito, 1989), which would affect the visibility of transition (Helffrich and Bina, 1994). Phase equilibria depend strongly on iron content such that the transition is broadened with iron enrichment. Iron enrichment beyond $x_{Fe} = 0.15$ at normal mantle temperatures causes the transition to become univariant (infinitely sharp in a pure olivine composition), with ringwoodite coexisting with olivine and wadsleyite (Akaogi *et al.*, 1989; Fei *et al.*, 1991). The transition is also sensitive to water content as wadsleyite appears to have a much higher solubility than olivine (Wood, 1995) although more recent results suggest that water partitioning between these two phases, and the influence of water on the form of the 410, is not as large as previously assumed (Hirschmann *et al.*, 2005). Portions of the mantle are found to have broad and nonlinear 410 discontinuities, consistent with the anticipated effects of water enrichment (van der Meijde *et al.*, 2003).

In some locations, the 410 is overlain by low velocity patches that have been interpreted as regions of partial melt, perhaps associated with water enrichment. These patches appear to be associated with subduction zones, suggesting the slab as a possible source of water (Obayashi *et al.*, 2006; Revenaugh and Sipkin 1994; Song *et al.*, 2004). Other interpretations involving much more pervasive fluxing of water through the transition zone have also been advanced (Bercovici and Karato, 2003), although it has been argued that this scenario is inconsistent with the thermodynamics of water-enhanced mantle melting (Hirschmann *et al.*, 2005). Measurements of the density of hydrous melt at the pressure of the 410 indicate that it may be gravitationally stable (Matsukage *et al.*, 2005; Sakamaki *et al.*, 2006).

Initial studies of the influence of iron on the ringwoodite to perovskite plus periclase transition also seemed to indicate that this transition was too broad to explain reflectivity observations (Jeanloz and Thompson, 1983; Lees *et al.*, 1983). But in a seminal

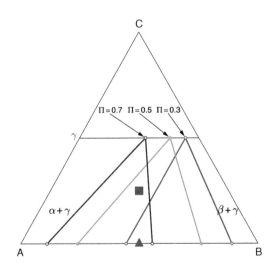

Figure 4 The influence of nontransforming phase(s) (γ) on the sharpness of the transformation from phase α to phase β. Colored lines indicate the compositions of the three coexisting phases with increasing scaled pressure (Π) within the coexistence interval. When the bulk composition lies on the A–B join, the transition is broad: the bulk composition represented by the solid triangle lies within the coexistence region over the entire pressure range shown ($\Pi = 0.3$–0.7). When component C is added, the transition occurs over a narrower pressure interval: the bulk composition represented by the solid square lies within the coexistence region over a fraction of the pressure range shown ($\Pi = 0.5$–0.7). After Stixrude L (1997) Structure and sharpness of phase transitions and mantle discontinuities. *Journal of Geophysical Research, Solid Earth* 102: 14835–14852.

study Ito and Takahashi (1989) showed that this transition was remarkably sharp, occurring over a pressure range less than experimental resolution (0.1 GPa). The unusual sharpness was subsequently explained in terms of Mg–Fe partitioning and nonideal Mg–Fe mixing in the participating phases (Fei *et al.*, 1991; Wood, 1990). While the central result of Ito and Takahashi's study remains unchallenged, it will be important to replicate these results in realistic mantle compositions (Litasov *et al.*, 2005), and to further explore the thermochemistry of the phases involved.

Recent studies have indicated complexity in the structure of the 660 (**Figure 5**). In relatively cold mantle, the transition (eqn [5]) is preceded by (Weidner and Wang, 1998)

$$MgSiO_3 \text{ (akimotoite)} = MgSiO_3 \text{ (perovskite)} \quad [6]$$

while in hot mantle the amount of ringwoodite is diminished prior to the transition (eqn [5]) via (Hirose, 2002)

$$Mg_2SiO_4 \text{ (ringwoodite)} = MgSiO_3 \text{ (majorite)} \\ + MgO \text{ (periclase)} \quad [7]$$

In cold and hot mantle, the reactions (eqns [5]–[7]) will be followed by a further reaction

$$MgSiO_3 \text{ (majorite)} = MgSiO_3 \text{ (perovskite)} \quad [8]$$

The sequence of reactions [5] and [6] should produce a 'doubled' 660 in which a single velocity jump is replaced by two that are closely spaced in depth and of similar magnitude. There is some seismological evidence for this doubling in some locations (Simmons and Gurrola, 2000). The relative importance of reactions [5]–[8] will also depend on the bulk composition, particularly the Al content (Weidner and Wang, 1998).

Below 660 is a steep velocity gradient that may be considered a continuation of the transition zone. This feature is seen in mineralogical models as the signature of the garnet to perovskite transition (eqn [8]), which is gradual and spread out over more than 100 km (Weidner and Wang, 1998). The gradual nature of this transition, as opposed to the much sharper ones discussed so far, is readily understood. The difference in composition between Al-rich garnet, and relatively Al-poor perovskite is large, and this creates a broad pressure–composition coexistence region (phase loop)

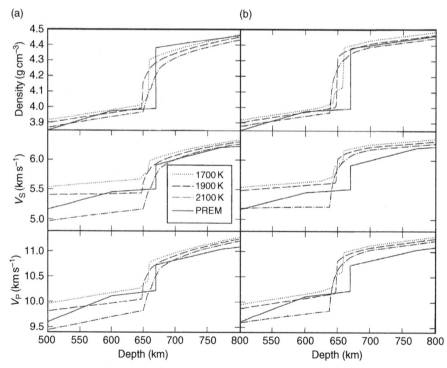

Figure 5 Calculated density and elastic wave velocities for different geotherms for pyrolite-like compositions with (a) 3% Al_2O_3 and (b) 5% Al_2O_3. Calculations are compared with the PREM seismological model (Dziewonski and Anderson, 1981). From Weidner DJ and Wang YB (1998) Chemical- and Clapeyron-induced buoyancy at the 660 km discontinuity. *Journal of Geophysical Research, Solid Earth* 103: 7431–7441.

(Akaogi *et al.*, 2002; Hirose *et al.*, 2001). Although it is broad, this transition may be responsible for a seismic discontinuity seen in some studies near 720 km depth (Deuss *et al.*, 2006; Revenaugh and Jordan, 1991b; Stixrude, 1997). Apparent bifurcation of the 660 may then have two distinct sources: eqns [5] and [6] in cold mantle, and eqns [5] and [8] in hotter mantle. It is conceivable that three reflectors may be observable in cold mantle (eqns [5], [6], and [8]), although this has not yet been seen.

The 520 km discontinuity is seen in global stacks of SS precursors (Shearer, 1990); there is evidence that it is only regionally observable (Gossler and Kind, 1996; Gu *et al.*, 1998; Revenaugh and Jordan, 1991a). Phase transformations that may contribute to this discontinuity include wadsleyite to ringwoodite (Rigden *et al.*, 1991) and the exsolution of calcium perovskite from garnet (Ita and Stixrude, 1992; Koito *et al.*, 2000). Both of these transitions are broad, which may account for the intermittent visibility of the 520. Seismological evidence for doubling of this discontinuity in some regions has renewed interest in the structure and sharpness of these transitions (Deuss and Woodhouse, 2001). The 520 km discontinuity has also been attributed to chemical heterogeneity, possibly associated with subduction (Sinogeikin *et al.*, 2003).

In addition to velocity discontinuities, the transition zone is distinguished by an unusually large velocity gradient, which exceeds that of any individual transition zone phase (**Figure 6**). Most explanations have focused on the series of phase transformations that occur in the transition zone as the cause of this gradient. These include not only those already mentioned in connection with discontinuities, but also broader features such as the dissolution of pyroxenes into garnet. The standard model, that the transition zone is adiabatic and chemically homogeneous with a pyrolitic composition, has been tested successfully against seismological constraints on the bulk sound velocity (Ita and Stixrude, 1992). Quantitative tests against the longitudinal- and shear-wave velocity are still hampered by uncertainties in the physical properties of key phases, particularly garnet-majorite (Liu *et al.*, 2000; Sinogeikin and Bass, 2002) and $CaSiO_3$ perovskite.

2.2.4 Lower Mantle

The most prominent seismic anomaly in the lower mantle is the division between the D'' layer and

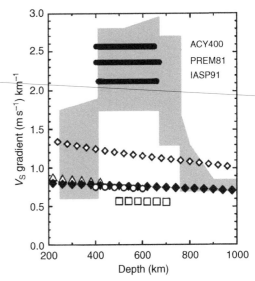

Figure 6 The velocity gradient from seismic studies and mineral physics. The shaded area is from Kennett (1993). The average gradients from global seismic models PREM81, ACY400, IASP91 (Dziewonski and Anderson, 1981; Kennett and Engdahl, 1991; Montagner and Anderson, 1989) are shown by horizontal solid lines. Solid diamonds are extrapolation of Brillouin spectroscopy data on a garnet of 1:1 majorite:pyrope composition, and open diamonds are an extrapolation of data of (Liu, *et al.*, 2000) on the same composition. Triangles – olivine, circles – wadsleyite, squares – ringwoodite. All velocity extrapolations are done along a 1673 K adiabat. From Sinogeikin SV and Bass JD (2002) Elasticity of majorite and a majorite–pyrope solid solution to high pressure: Implications for the transition zone. *Geophysical Research Letters* 29: 1017 (doi:10.1029/2001GL013937).

the rest. The D'' layer is distinguished by a change in radial velocity gradient, and a velocity discontinuity at its top (Lay and Helmberger, 1983). The velocity discontinuity is prominent where it is seen, although it is not present everywhere and is larger in S than in P (Lay *et al.*, 1998a; Wysession *et al.*, 1998; Young and Lay, 1987). Below this discontinuity most models show velocity decreasing with increasing depth: the D'' represents the mantle's second low velocity zone. In analogy with that in the shallow mantle, the D'' velocity gradient may also be caused by superadiabatic temperature gradients associated with the lower dynamical boundary layer of mantle convection.

Mineralogical explanations of the D'' discontinuity now focus on the transformation from $MgSiO_3$ perovskite to the postperovskite phase (**Figure 7**) (Murakami *et al.*, 2004; Oganov and Ono, 2004; Tsuchiya *et al.*, 2004a). The transition appears to

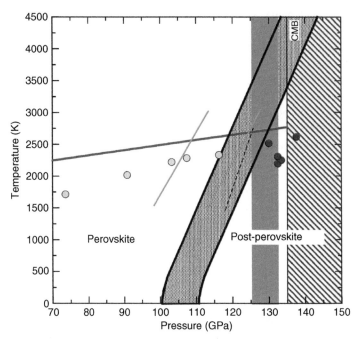

Figure 7   Post-perovskite transition from Tsuchiya *et al.* (2004a) (blue band), and Oganov and Ono (2004) (green lines) with lower (LDA) and upper (GGA) bounds, compared with experiment (Murakami, *et al.*, 2004) (blue circles: pv, red circles: ppv), and the Clapeyron slope inferred from D″ topography (Sidorin, *et al.*, 1999) (dashed), a mantle geotherm (orange) (Brown and Shankland, 1981) and the pressure range of D″ topography (gray shading) (Sidorin, *et al.*, 1999). After Tsuchiya T, Tsuchiya J, Umemoto K, and Wentzcovitch RM (2004a) Phase transition in MgSiO$_3$ perovskite in the earth's lower mantle *Earth and Planetary Science Letters* 224: 241–248.

occur at the right pressure and to have the right Clapeyron slope to explain the intermittent visibility. Predictions of the elastic constants show a larger contrast in V_S than in V_P, consistent with most seismological observations of the discontinuity (Iitaka *et al.*, 2004; Tsuchiya *et al.*, 2004b; Wentzcovitch *et al.*, 2006). Indeed, a phase transformation with very similar properties was anticipated on the basis of a combined seismological and geodynamical study (Sidorin *et al.*, 1999).

The ultralow velocity zone is a narrow intermittent layer less than 40 km thick with longitudinal-wave velocities approximately 10% lower than that of the rest of D″ (Garnero and Helmberger, 1996). It is still unclear what could cause such a large anomaly. A search for explanations is hampered by uncertainties concerning the relative magnitude of density and longitudinal and shear velocity anomalies. Many studies have focused on the possibility of partial melt in this region (Stixrude and Karki, 2005; Williams and Garnero, 1996). Others have emphasized the role of iron enrichment, possibly related to chemical reaction between mantle and core (Garnero and Jeanloz, 2000), or subduction (Dobson and Brodholt, 2005). In

this view the ultralow velocity zone may be more properly regarded as the outermost layer of the core, rather than the bottom-most layer of the mantle (Buffett *et al.*, 2000).

There have been persistent reports of reflection or scattering from within the lower mantle (Castle and van der Hilst, 2003; Deuss and Woodhouse, 2002; Johnson, 1969; Kaneshima and Helffrich, 1998; Kawakatsu and Niu, 1994; Lestunff, *et al.*, 1995; Paulssen, 1988; van der Meijde *et al.*, 2005; Vinnik *et al.*, 2001). These are only locally observed, and so may either be associated with chemical heterogeneity, possibly related to subduction (Kaneshima and Helffrich, 1998; Kawakatsu and Niu, 1994), or a phase transition that has either a weak velocity signal or a large Clapeyron slope. Phase transformations that are expected to occur in the lower mantle and that are associated with significant velocity anomalies include (**Figure 8**): (1) Phase transformations in silica, from stishovite, to CaCl$_2$ to α-PbO$_2$ structured phases (Carpenter *et al.*, 2000; Karki *et al.*, 1997b). Free silica would be expected only in enriched compositions, such as deeply subducted oceanic crust. The first of these transitions has

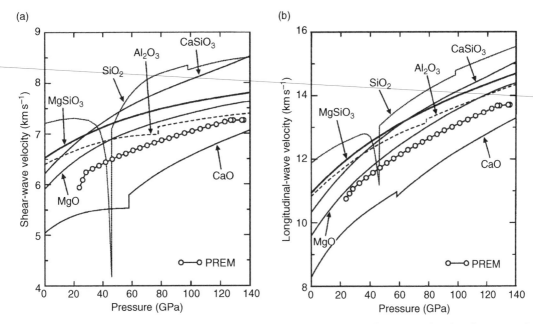

Figure 8 Comparisons of the static shear (a) and longitudinal (b) wave velocities of lower mantle minerals computed via density functional theory compared with the PREM model (Dziewonski and Anderson, 1981). The discontinuous changes in SiO_2 are caused by phase transformations from stishovite to the $CaCl_2$ structure (45 GPa) and then to the α-PbO_2 structure (100 GPa). After Karki BB and Stixrude L (1999) Seismic velocities of major silicate and oxide phases of the lower mantle. *Journal of Geophysical Research* 104: 13025–13033.

a particularly large elastic anomaly. (2) A phase transformation in calcium perovskite from tetragonal to cubic phases (Stixrude *et al.*, 1996).

2.2.5 Core

It is known that the properties of the outer core do not match those of pure iron, which is significantly denser (Birch, 1964; Brown and McQueen, 1986). The search for the light element is still in its infancy, with the number of candidate light elements seeming to grow, rather than narrow with time (Poirier, 1994). It has been pointed out that there is no *a priori* reason to believe that the composition of the core is simple or that it contains only one or a few elements other than iron (Stevenson, 1981). In any case, further experimental and theoretical constraints on the phase diagrams and physical properties of iron light element systems will be essential to the solution.

Two relatively new approaches to constraining the nature and amount of light element in the core show considerable promise. The first relates the partitioning of candidate light elements between solid and liquid phases to the density contrast at the inner-core boundary (Alfe *et al.*, 2002a). While sulfur and

silicon are found to partition only slightly, producing a density jump that is too small, oxygen strongly favors the liquid phase, producing a density jump that is too large. Within the accuracy of the theoretical calculations used to predict the partitioning, this study then appears to rule out either of these three as the sole light element in the inner core, and suggests a combination of oxygen with either silicon or sulfur as providing a match to the seismologically observed properties. Another approach is to test core formation hypotheses via experimental constraints on element partitioning between core and mantle material. In the limit that proto-core liquid approached equilibrium with the mantle as it descended, experiments point to sulfur as the only one sufficiently siderophile to remain in significant amounts in iron at pressures thought to be representative of core formation (~30 GPa) (Li and Agee, 1996). On the other hand, experimental simulations of the core–mantle boundary (136 GPa) show substantial reaction between core and mantle material and incorporation of large amounts of Si and O in the metal (Knittle and Jeanloz, 1989; Takafuji *et al.*, 2005) (**Figure 9**).

The temperature in the core may be constrained by the melting temperature of iron and iron alloys.

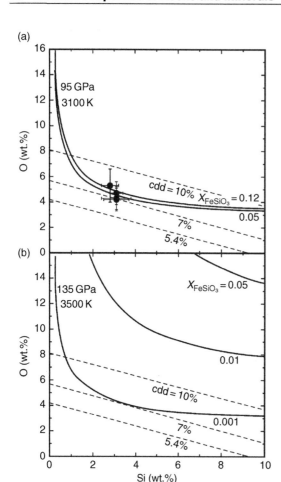

Figure 9 The simultaneous solubilities of O and Si in liquid iron in equilibrium with (Mg, Fe)SiO$_3$ perovskite. Solid curves represent values calculated from a thermodynamic model at (a) 95 GPa and 3100 K and (b) 135 GPa and 3500 K as a function of FeSiO$_3$ component in coexisting perovskite (X$_{FeSiO3}$). The model is constrained by experimental measurements, a subset of which are shown in (a) at 93–97 GPa and 3050–3150 K in contact with (Mg, Fe)SiO$_3$ perovskite (X$_{FeSiO3}$ = 0.04–0.12) (circles). The broken lines show the amounts of O and Si that are required to account for a core density deficit (cdd) relative to pure iron of 10%, 7%, and 5.4% (Anderson and Isaak, 2002). From Takafuji N, Hirose K, Mitome M, and Bando Y (2005) Solubilities of O and Si in liquid iron in equilibrium with (Mg, Fe)SiO$_3$ perovskite and the light elements in the core. *Geophysical Research Letters* 32: L06313 (doi:10.1029/2005GL022773).

Consensus has been building in recent years on the melting temperature of pure iron near the inner-core boundary. First-principles theory (Alfe *et al.*, 2002b) and temperature measurements in dynamic compression (Nguyen and Holmes, 2004), both converge on the melting temperature proposed by Brown and McQueen (1986) on the basis of dynamic compression and modeled temperatures. The largest

remaining uncertainty appears to be the influence of the light element on the melting temperature. Density functional theory predicts freezing point depression of 700 K for light element concentrations required to match the density deficit (Alfe *et al.*, 2002a), similar to estimates based on the van Laar equation (Brown and McQueen, 1982). An alternative approach to estimating the temperature at Earth's center is to compare the elasticity of iron to that of the inner core. This leads to an estimate of 5400 K at the inner-core boundary, consistent with estimates based on melting temperatures (Steinle-Neumann *et al.*, 2001).

The outer core is homogeneous and adiabatic to within seismic resolution (Masters, 1979) and represents possibly the clearest opportunity for estimating the temperature increment across any layer in the Earth. The adiabatic temperature gradient

$$\left(\frac{\partial \ln T}{\partial P}\right)_S = \frac{\gamma}{K_S} \qquad [9]$$

depends on K_S, which is known as a function of depth from seismology, and the Grüneisen parameter γ, which has been measured for pure iron (Brown and McQueen, 1986). Integrating this equation across the outer core produces a temperature difference of 1400 K (Steinle-Neumann *et al.*, 2002), yielding a temperature of 4000 K at the core–mantle boundary. The influence of the light element on the Grüneisen parameter is not necessarily negligible and will need to be measured to test this prediction.

One of the most remarkable discoveries of core structure in recent years has been that of inner-core anisotropy. Longitudinal waves travel a few percent faster along a near polar axis than in the equatorial plane (Morelli *et al.*, 1986; Woodhouse *et al.*, 1986). Lattice-preferred orientation seems a natural explanation because all plausible crystalline phases of iron have single-crystal elastic anisotropy at the relevant conditions that exceeds that of the inner core (Stixrude and Cohen, 1995; Söderlind *et al.*, 1996), notwithstanding disagreements about the values of the high-temperature elastic constants as predicted by different approximate theories (Gannarelli *et al.*, 2005; Steinle-Neumann *et al.*, 2001). The identity of the stable phase is critical since the pattern of anisotropy in body-centered cubic and hexagonal candidates are very different. Although the body-centered cubic phase appears to be elastically unstable for pure iron, favoring a hexagonal close-packed inner core, there is some evidence that light elements and nickel tend to stabilize this structure (Lin *et al.*, 2002a; Lin *et al.*, 2002b; Vocadlo *et al.*, 2003). Still unknown is the source of stress that might produce preferred orientation, or knowledge of the dominant slip planes and critical resolved shear stresses that would permit prediction of texture given an applied stress. Further seismological investigations have shown that the structure of the inner core is also laterally heterogeneous (Creager, 1997; Su and Dziewonski, 1995; Tanaka and Hamaguchi, 1997), and that there may be a distinct innermost layer (Ishii and Dziewonski, 2002), which may be due to a phase transformation.

2.3   Lateral Heterogeneity

2.3.1   Overview

Lateral variations in seismic wave velocities have three sources: lateral variations in temperature, chemical composition, and phase assemblage (**Figure 10**). The first of these has received most attention. Indeed, comparison of tomographic models with surface tectonic features indicates that temperature may be the largest source of lateral variations: arcs are slow, cratons and subducting slabs are fast, and there is a good correlation between the history of past subduction and the location of fast velocity

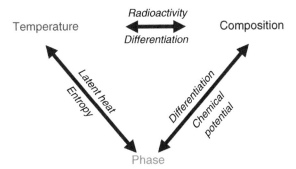

Figure 10   The three sources of lateral heterogeneity in Earth's mantle: lateral variations in temperature, in chemical composition, and phase. Arrows indicate that these three sources may influence each other. Increasing temperature alters the relative stability of phases via contrasts in entropy, and may generate changes in composition via differentiation caused by partial melting. Changes in composition may include changes in the concentration of radioactive heat producing elements and therefore temperature, and influence phase stability through the chemical potential. Changes in phase release or absorb latent heat, thus altering the temperature and may produce density contrasts that lead to differentiation, potentially altering the composition.

anomalies throughout the mantle. Quantifying these interpretations has proved challenging. One reason is that temperature is almost certainly not the only important factor. For example, it is recognized that continental mantle lithosphere must be depleted in order to remain dynamically stable, and that this depletion contributes along with temperature, to the velocity anomaly (Jordan, 1975). Throughout the upper 1000 km of the mantle, lateral variations in phase assemblage may contribute as much as the influence of temperature alone. In the lower mantle, it can be demonstrated that sources other than temperature are required to explain the lateral heterogeneity.

2.3.2   Temperature

Several studies have sought to estimate lateral variations in temperature in the upper mantle on the basis of tomographic models and measured properties of minerals (Godey *et al.*, 2004; Goes *et al.*, 2000; Sobolev *et al.*, 1996). These estimates are consistent with the results of geothermobarometry, surface tectonics, and heat flow observations. In one sense, the upper mantle is an ideal testing ground for the physical interpretation of mantle tomography because the elastic constants of the constituent minerals are relatively well known. But the upper mantle is also remarkably complex. A potentially important factor that is not accounted for in these studies is the lateral variations in phase proportions. Another source of uncertainty is anelasticity, which magnifies the temperature dependence of the velocity by an amount that is poorly constrained experimentally.

The temperature dependence of the seismic wave velocities of mantle minerals decreases rapidly with increasing pressure (**Figure 11**). If we ascribe a purely thermal origin to lateral structure, we would then expect the amplitude of tomographic models to decrease rapidly with depth. But tomographic models show a different pattern. The amplitude tends to decrease rapidly in the upper boundary layer, reaches a minimum at mid-mantle depths, and then increases with depth in the bottom half of the mantle. Part of the increase in amplitude with depth may be associated with the lower thermal boundary layer, where we would expect lateral temperature variations to exceed that in the mid-mantle. While concerns regarding radial resolution of tomographic models cannot be dismissed, it seems likely that this aspect of lower mantle structure cannot be explained by temperature alone.

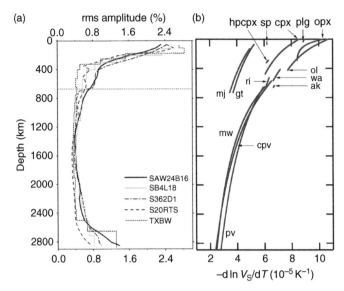

Figure 11   The amplitude of tomographic models as a function of depth (a) from (Romanowicz, 2003) as compared with the temperature derivative of the shear-wave velocity for mantle minerals along a typical isentrope (b) from (Stixrude and Lithgow-Bertelloni, 2005b).

Another indication that temperature alone cannot explain lower mantle structure is the comparison of lateral variations in S- and P-wave velocities. The ratio

$$R = \left(\frac{\delta \ln V_S}{\delta \ln V_P}\right)_z \qquad [10]$$

where V_S is the shear-wave speed and V_P is the longitudinal-wave speed and the subscript z indicates that variations are at constant depth, reaches values as large as 3.5 in the lower mantle and appears to increase systematically with increasing depth (Masters et al., 2000; Ritsema and van Heijst, 2002; Robertson and Woodhouse, 1996). This is to be compared with

$$R_{thermal} = \left[\frac{(1-A)(\delta_S-1)}{\Gamma-1} + A\right]^{-1} \qquad [11]$$

where $A = 4/3 V_S^2/V_P^2$, $\delta_S = (\partial \ln K_S/\partial \ln \rho)_P$, $\Gamma = (\partial \ln G/\partial \ln \rho)_P$. The ratio takes on its limiting value $R_{thermal} \rightarrow A^{-1} \approx 2.5$ as $\delta_S \rightarrow 1$. In some portions of the mantle, lateral variations in shear and bulk sound velocities appear to be anticorrelated: regions that are slow in V_S are fast in V_B (Masters et al., 2000; Su and Dziewonski, 1997; Vasco and Johnson, 1998). Large values of R and anticorrelated V_S and V_B both require $\delta_S < 1$ if they are to be explained by temperature alone, a condition not met by lower mantle minerals. Lower mantle phases all have $\delta_S > 1$ so

that longitudinal, shear, and bulk sound velocities all decrease with increasing temperature (Agnon and Bukowinski, 1990; Karki et al., 1999; Oganov et al., 2001; Wentzcovitch et al., 2004). Anelasticity increases the value of $R_{thermal}$ (Karato, 1993), but apparently not sufficiently to account for the tomographic value (Masters et al., 2000), and in any case cannot produce anticorrelation in V_S and V_B.

2.3.3   Composition

Of the five major cations in the mantle, iron has the largest influence on the density and the elastic wave velocities. The influence of iron content on the shear velocity differs substantially among the major mantle minerals. This means that different bulk compositions (e.g., MORB, pyrolite) will have different sensitivities to lateral variations in iron content (**Figure 12**).

What sort of lateral variations in composition might explain the unusual features of lower mantle structure? A definitive answer is prevented by our ignorance of the elastic properties of lower mantle phases. Lateral variations in iron content do not appear capable of producing anticorrelation in bulk- and shear-wave velocity, as both velocities are decreased by addition of iron to $MgSiO_3$ perovskite (Kiefer et al., 2002). It has been argued that lateral variations in Ca content can produce anticorrelation since $CaSiO_3$ perovskite has a greater shear-wave

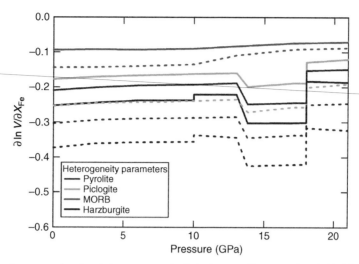

Figure 12 Sensitivity of compressional and shear-wave velocity heterogeneity parameters to Fe-Mg substitution calculated for pyrolite, piclogite, harzburgite, and MORB, along an isentropic pressure/temperature path with a potential temperature of 1673 K. Solid lines: $\partial \ln v_P / \partial X_{Fe}$, dashed lines: $\partial \ln v_S / \partial X_{Fe}$. From Speziale S, Fuming J, and Duffy TS (2005a) Compositional dependence of the elastic wave velocities of mantle minerals: Implications for seismic properties of mantle rocks. In: Matas J, van der Hilst RD, Bass JD, and Trampert J *Earth's Deep Mantle: Structure, Composition, and Evolution*, pp. 301–320. Washington, DC: American Geophysical Union.

velocity but lesser bulk sound velocity that $MgSiO_3$ perovskite (Karato and Karki, 2001). However, this argument is based on first-principles calculations of $CaSiO_3$ perovskite that assumed a cubic structure (Karki and Crain, 1998). One of the remarkable features of this phase is that it undergoes a slight noncubic distortion with decreasing temperature (Adams and Oganov, 2006; Ono *et al.*, 2004; Shim *et al.*, 2002; Stixrude *et al.*, 1996) that nevertheless produces a very large 15% softening of V_S, while leaving the density and V_B virtually unaffected (Stixrude *et al.*, 2007).

Dynamical models suggest that lateral variations in composition in the lower mantle may be associated with segregation of basalt (Christensen and Hofmann, 1994; Davies, 2006; Nakagawa and Buffett, 2005; Xie and Tackley, 2004). *In situ* equation of state studies indicate that basalt is denser than average mantle and would tend to pile up at the core–mantle boundary (Hirose *et al.*, 2005). Basalt-rich piles may be swept up by mantle convection, and may explain lower mantle anomalies with boundaries that are sharp, that is, of apparently nonthermal origin (Ni *et al.*, 2002), and primitive signatures in mantle geochemistry (Kellogg *et al.*, 1999). A definitive test of such scenarios awaits determination of the elastic properties of minerals in basalt that are currently unmeasured, such as a calcium-ferrite structured oxide.

2.3.4 Phase

The mantle is made of several different phases, with distinct elastic properties. As temperature varies, the relative proportions and compositions of these phases change, producing an additional contribution to laterally varying structure (**Figure 13**). This effect has been discussed in terms of thermally induced phase transformations (Anderson, 1987). The magnitude of the effect is

$$\left(\frac{\partial V}{\partial T}\right)_{P,\text{phase}} \approx f\left(\frac{\partial P}{\partial T}\right)_{\text{eq}} \frac{\Delta V}{\Delta P} \qquad [12]$$

where f is the volume fraction of the mantle composed of the transforming phases, ΔV is the velocity contrast between them, ΔP is the pressure range over which the transition occurs, and the derivative on the right-hand side is the effective Clapeyron slope. The effect is largest for sharp transitions, such as the olivine to wadsleyite transition, and in this case may also be described in terms of the topography of the transition. It is sensible to describe lateral variations in velocity due to phase equilibria in terms of topography when the topography exceeds the width of the phase transformation

$$\left(\frac{\partial P}{\partial T}\right)_{\text{eq}} \Delta T > \Delta P \qquad [13]$$

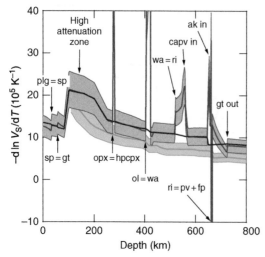

Figure 13 The variation of shear-wave velocity with temperature in pyrolite (red), including the effects of pressure, temperature, phase transformations, and anelasticity according to the model QR19 (Romanowicz, 1995) and $\alpha = 0.25$, and variation of Q with temperature given by $H^* = 430\,kJ\,mol^{-1}$. The blue curve shows the derivative neglecting the influence of phase transformations. The green curve shows a similar calculation (Cammarano, et al., 2003) except that the effect of phase transformations is neglected. From Stixrude L and Lithgow-Bertelloni C (2007) Influence of phase transformations on lateral heterogeneity and dynamics in Earth's mantle. *Earth and Planetary Science Letters* (submitted).

where ΔT is the anticipated magnitude of lateral temperature variations. But the influence of phase transformations is also important for broad transitions such as pyroxene to garnet. In the upper mantle, most phase transitions have positive Clapeyron slopes. This means that, except in the vicinity of the 660 km discontinuity, phase transitions systematically increase the temperature dependence of seismic wave velocities throughout the upper 800 km of the mantle.

Lateral variations in phase have been invoked to explain some of the unusual features of the deep lower mantle, including the anticorrelation in bulk- and shear-wave velocities mentioned earlier (Iitaka et al., 2004; Oganov and Ono, 2004; Tsuchiya et al., 2004b; Wentzcovitch et al., 2006). As perovskite transforms to postperovskite, the shear modulus increases while the bulk modulus decreases. Lateral variations in the relative abundance of these two phases caused, for example, by lateral variations in temperature, will then produce anticorrelated variations in V_S and V_B. Since the stability of the postperovskite phase is apparently restricted to the D″ layer, it is not clear

whether this mechanism can explain all of the anomalous features of lower mantle tomography which appear to extend over a much greater depth interval. The high-spin to low-spin transition in iron may have a significant effect on the physical properties of ferropericlase and perovskite (Badro et al., 2003, 2004; Cohen et al., 1997; Li et al., 2005; Pasternak et al., 1997). Recent theoretical results indicate that the transition is spread out over hundreds of kilometers of depth and should affect laterally varying structure over much of the lower mantle (Tsuchiya et al., 2006). The high-spin to low-spin transition increases the bulk modulus in ferropericlase (Lin et al., 2005; Speziale et al., 2005b). If it also increases the shear modulus, as Birch's law would predict, the transition would not account for anticorrelation of V_B and V_S.

2.4 Anisotropy

2.4.1 Overview

Anisotropy is a potentially powerful probe of mantle dynamics because it is produced by deformation. In dislocation creep, the easiest crystallographic slip plane tends to align with the plane of flow, and the easiest slip direction with the direction of flow. Since all mantle minerals are elastically anisotropic (**Figure 14**), this preferred orientation or texturing produces bulk anisotropy that is measurable by seismic waves. This mechanism for producing anisotropy is often referred to as lattice preferred orientation. Another mechanism is shape-preferred orientation in which the rheological contrast between the materials making up a composite cause them to be arranged inhomogeneously on deformation. Heterogeneous arrangements familiar from structural geology, including foliation and lineation, lead to anisotropy for seismic waves with wavelengths much longer than the scale length of heterogeneity.

Anisotropy is pervasive in the continental crust, and significant in the upper mantle, the D″ layer, and the inner core, the latter of which was discussed above in the section on the core. Anisotropy may also exist in the transition zone as indicated by global (Montagner and Kennett, 1996; Trampert and van Heijst, 2002) and regional (Fouch and Fischer, 1996) seismic models.

2.4.2 Upper Mantle

Anisotropy is well developed in the uppermost mantle. Studies of the seismic structure of this region, the

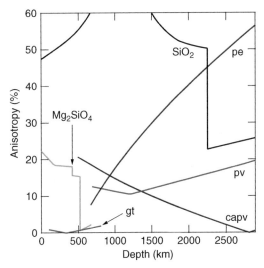

Figure 14 Single crystal azimuthal anisotropy of S-wave velocity in mantle minerals plotted, except for silica, over their pressure range of stability in pyrolite. The discontinuous changes in silica and Mg_2SiO_4 are due to phase transformations. The breaks in slope in garnet (pyrope), periclase, and calcium perovskite are due to interchanges in the fastest and slowest directions where the anisotropy passes through zero (Karki, *et al.*, 1997a). After Mainprice D, Barruol G, and Ismail WB (2000) The seismic anisotropy of the Earth's mantle: From single crystal to polycrystal. In: Karato SI, Forte A, Liebermann R, Masters G, and Stixrude L (eds.) *Earth's Deep Interior: Mineral Physics and Tomography from the Atomic to the Global Scale.* Washington, DC: American Geophysical Union.

texture of mantle samples, and the elastic properties of the constituent minerals have led to a simple first order picture. Convective flow in the upper mantle preferentially aligns olivine crystals such that the seismically fast direction is parallel to the flow direction (Christensen and Salisbury, 1979; Mainprice *et al.*, 2000). Two seismological signals are often distinguished, associated with different data sets and modeling strategies. Horizontally propagating shear waves travel at two different velocities depending on the direction of polarization: the faster is horizontal with V_{SH} and the slower is vertical with V_{SV}. The difference in velocity of the two polarizations is also referred to as shear-wave splitting. The shear-wave velocity also depends on the direction of propagation, and this is usually referred to as azimuthal anisotropy. Azimuthal anisotropy is also detected along vertically propagating paths via shear-wave splitting (Silver, 1996).

In detail, the picture of upper-mantle anisotropy is much richer than this first-order pattern. There are unexplained anomalies even in regions where the pattern of flow is presumably simple and large scale such as the Pacific basin. For example, the fast direction does not align everywhere with the direction of present-day plate motions (Montagner and Guillot, 2002; Tanimoto and Anderson, 1984). The magnitude of V_{SH}–V_{SV} varies spatially in a way that does not correlate simply with lithosphere age: the difference is maximum near the center of the north Pacific (Ekstrom and Dziewonski, 1998). At subduction zones, the fast direction is sometimes parallel, rather than normal to the trench (Fouch and Fischer, 1996; Russo and Silver, 1994). Explanations of these features will come from a better understanding of the development of texture in large-scale mantle flow (Chastel *et al.*, 1993; Ribe 1989), the preferred slip planes and critical resolved shear stresses in olivine and how these depend on stress, temperature, pressure, and impurities (Jung and Karato, 2001; Mainprice *et al.*, 2005), how stress and strain are partitioned between lithosphere and asthenosphere, and the relative importance of dislocation versus diffusion creep (Karato, 1992; van Hunen *et al.*, 2005).

2.4.3 D″ Layer

The D″ layer is also substantially anisotropic (Lay *et al.*, 1998b; Panning and Romanowicz, 2004). The minerals that make up this region all have large single-crystal anisotropies, with ferropericlase having the largest (Karki *et al.*, 2001; Stackhouse *et al.*, 2005; Wentzcovitch *et al.*, 2006) (**Figure 14**). The anisotropy of periclase depends strongly on pressure and actually reverses sense near 20 GPa: the directions of fastest and slowest elastic wave propagation interchange (Karki *et al.*, 1997a). This behavior emphasizes the importance of experimental measurements and first-principles predictions of elasticity at the relevant pressures. The easy glide planes of the minerals in D″ are unconstrained by experiment or first-principles theory. If one assumes that the easy glide planes are the same as those seen in ambient pressure experiments or analog materials, anisotropy produced by lattice preferred orientation in neither ferropericlase nor perovskite can explain the dominant pattern $V_{SH} > V_{SV}$ seen in seismology (Karato, 1998; Stixrude, 1998), while postperovskite does (Merkel *et al.*, 2006). One must be cautious, because the sense and magnitude of anisotropy depends strongly on the easy glide plane.

2.5   Attenuation and Dispersion

2.5.1   Overview

The Earth is not a perfectly elastic medium, even at seismic frequencies. There are two important consequences. Attenuation refers to the loss of amplitude of the elastic wave with propagation. Dispersion refers to the dependence of the elastic-wave velocity on the frequency. It can be shown that attenuation and dispersion are intimately related. Indeed, the importance of attenuation in the Earth first alerted seismologists to the dispersion that should yield a difference in velocities between low-frequency normal mode oscillations and higher-frequency body waves. The magnitude of attenuation is measured by the inverse quality factor Q^{-1}, the fractional energy loss per cycle.

From the mineralogical point of view, attenuation is potentially important as a very sensitive probe of temperature and minor element composition (Karato, 1993; Karato and Jung, 1998). Because attenuation is caused at the atomic scale by thermally activated processes, its magnitude is expected to depend exponentially on temperature. By analogy with the viscosity, the attenuation is expected also to depend strongly on the concentration of water, although there are as yet no experimental constraints on this effect. But with this promise come formidable challenges. Dissipative processes depend on defect concentrations and dynamics and/or the presence of multiple scales, such as grains and grain boundaries. These are much more challenging to study in the laboratory than equilibrium thermodynamic properties such as the elastic response.

2.5.2   Influence of Temperature

The most thorough study of attenuation is of olivine aggregates, specially fabricated to have approximately uniform grain size and minimal dislocation density (Gribb and Cooper, 1998; Jackson *et al.*, 2002). Whether such samples are representative of the mantle is an important question, although at this stage, when experimental data are so few, the need for precise and reproducible results is paramount. The key results of Jackson *et al.* (2002) within the seismic frequency band ($\sim$1–100 s) and in the limit of small attenuation, are consistent with the following relations:

$$Q^{-1}(P, T, \omega) = A d^{-m} \omega^{-\alpha} \exp\left(-\alpha \frac{E^* + PV^*}{RT}\right) \quad [14]$$

$$V(P, T, \omega) = V(P, T, \infty)$$
$$\times \left[1 - \frac{1}{2}\cot\left(\frac{\alpha\pi}{2}\right)Q^{-1}(P, T, \omega)\right] \quad [15]$$

where ω is the frequency, d is the grain size, $m = 0.28$ is the grain size exponent, $\alpha = 0.26$, $E^* = 430\,\text{kJ}\,\text{mol}^{-1}$ is the activation energy, V^* is the activation volume, and R is the gas constant. For the experimental value of α, the factor multiplying Q^{-1} in eqn [15] has the value 1.16. The activation volume is currently unconstrained, which means the influence of pressure on attenuation is highly uncertain, although this quantity has been measured for other rheological properties such as viscosity and climb-controlled dislocation creep in olivine. Partial melt in the amount of 1% increases Q^{-1} by a factor that depends weakly on frequency and grain size and is approximately 2 in the seismic band and for plausible grain sizes (1–10 mm) (Faul *et al.*, 2004). The influence of crystallographically bound hydrogen on the attenuation is currently unconstrained, although it is known that the viscosity decreases with increasing hydrogen concentration (Mei and Kohlstedt, 2000a, 2000b). The second relation makes the relationship explicit between attenuation and dispersion. Equations [14] and [15] are consistent with earlier theories developed primarily on the basis of seismological observations and the idea of a distribution of dissipitave relaxation times in solids (Anderson, 1989).

An important consequence of eqn [15] is that, if the value of the frequency exponent is known, the magnitude of the dispersion depends only on the value of Q^{-1}. This means that it is possible directly to relate seismological velocity models to experimental measurements in the elastic limit. Knowing the seismological value of Q^{-1} and V, one may compute $V(\infty)$ and compare this directly with high-frequency experiments or first-principles calculations. Alternatively, one may correct elastic wave velocities according to the seismological value of Q^{-1} in order to compare with the seismological velocity model (Stixrude and Lithgow-Bertelloni, 2005a). These considerations are most important in the low velocity zone where $1000/Q$ reaches 20 and the difference in velocity between the elastic limit and the seismic band may be as great as 2%.

Attenuation enhances the temperature dependence of seismic wave velocities and is important to consider in the interpretation of seismic tomography (Karato, 1993). The temperature derivative of eqn [15] is

$$\frac{\partial \ln V(\omega)}{\partial T} = \frac{\partial \ln V(\infty)}{\partial T} - \frac{1}{2}\cot\left(\frac{\alpha\pi}{2}\right)Q^{-1}\frac{H^*}{RT^2} \quad [16]$$

where $H^* = E^* + PV^*$ is the activation enthalpy. For $\alpha = 0.26$, $H^* = 430\,\mathrm{kJ\,mol^{-1}}$, $1000/Q = 10$, and $T = 1600\,\mathrm{K}$, the second term is $-6 \times 10^{-5}\,\mathrm{K^{-1}}$, nearly as large as the elastic temperature derivative neglecting the influence of phase transformations (**Figure 11**).

Laboratory studies predict large lateral variations in attenuation in the mantle. Over a range of temperature 1500–1600 K, the attenuation varies by a factor of 2. This extreme sensitivity promises powerful constraints on the temperature structure of the mantle that are complimentary to those derived from the elastic wave speeds. Attenuation tomography reveals a decrease in Q^{-1} with increasing lithospheric age from 0–100 Ma in the Pacific low velocity zone of approximately one order of magnitude (Romanowicz, 1998). Arcs appear to have exceptionally high attenuation (Barazangi and Isacks, 1971; Roth et al., 2000). Unraveling the relative contributions of partial melt, water concentration, and other factors (grain size?) to this observation will provide unique constraints on the process of arc magma generation. This program is still limited by the lack of experimental data on the influence of water content on Q.

2.5.3 Speculations on the Influence of Pressure

Comparisons of laboratory determinations to radial seismological models of Q^{-1} reveal some surprises. The comparison is hampered by the fact that there are currently no experimental data at elevated pressure. Two models have been explored in the literature, eqn [14] with the activation volume assumed to be independent of pressure and temperature, and the homologous temperature formulation in which activated properties are assumed to scale with the ratio of the absolute temperature to melting temperature (Weertman, 1970)

$$Q^{-1}(P, T, \omega) = Ad^{-m}\omega^{-\alpha}\exp\left[-\alpha\frac{gT_m(P)}{T}\right] \quad [17]$$

where $T_m(P)$ is the melting temperature as a function of pressure, and the parameter g is assumed to be constant. Neither of these models resembles the variation of Q^{-1} in the vicinity of the low velocity zone (**Figure 15**). The high attenuation zone in the mantle is much more sharply defined at its lower boundary than either eqn [14] or [17]. While the limited vertical resolution of the seismological Q models cannot be overlooked, the comparison seems to imply either (1) eqns [14] and [17] do not represent the pressure

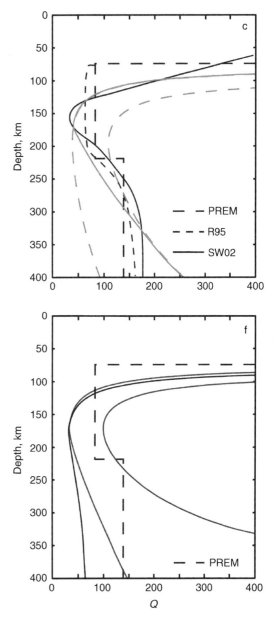

Figure 15 Quality profiles calculated assuming a range of temperatures and activation volumes and grain size either independent of depth (dashed curves in upper figure, rightmost pink curve in lower figure), or increasing by a factor of 5 (solid curve in upper figure) or 2 (red and blue curves in lower figure) as a function of depth, compared with seismological attenuation models (Dziewonski and Anderson, 1981; Romanowicz, 1995; Selby and Woodhouse, 2002). From Faul UH and Jackson I (2005) The seismological signature of temperature and grain size variations in the upper mantle. *Earth and Planetary Science Letters* 234: 119–134.

dependence of activated processes, at least not those responsible for dissipation in the seismic band in the mantle; (2) there is a change in the composition or phase of the mantle at this depth, for example, a change in water content or partial melt fraction; (3) there is a change in the physical mechanism of attenuation at the base of the low velocity zone. Faul and Jackson (2005) have proposed that the base of the low velocity zone represents a rapid change in grain size with increasing depth, although the mechanism by which such a change could occur is uncertain.

2.6   Conclusions

Over the past decade, studies of the seismic properties of major earth forming materials have dramatically expanded in scope. The field has moved from measurements of the elastic properties of individual minerals at elevated pressure and temperature to now include studies of deformation mechanisms and their relationship to the development of anisotropy, attenuation, and dispersion, and the interplay between phase equilibria and elasticity in mantle assemblages approaching realistic bulk compositions. Major challenges remain in each of the mantle's regions. Upper-mantle structure is remarkably rich and surprisingly full of features that resist explanation, such as the discontinuity structure and the high velocity gradient. Phase transformations are ubiquitous in the transition zone: there is no depth at which a transformation is not in progress, so that understanding this region will require the continued development of thermodynamic models of the mantle that are simultaneously able to capture phase equilibria and physical properties. The lower mantle still lies beyond the reach of many experimental techniques for measuring the elastic constants and most of our knowledge of elasticity in this region comes from first-principles theory. New physics in the lower mantle, including high-spin to low-spin transitions, the postperovskite transition, and valence state changes, promise new solutions to problems such as the anticorrelation between bulk- and shear-wave velocity variations, and fertile ground for new ideas. Many classic problems of the core, including its composition and temperature, seem to be yielding to new approaches and promise new constraints on the origin and evolution of Earth's deepest layer. At the same time, the seismic structure of the inner core, including its anisotropy, heterogeneity, and layering, remains enigmatic.

References

Adams DJ and Oganov AR (2006) *Ab initio* molecular dynamics study of $CaSiO_3$ perosvkite at P–T conditions of Earth's lower mantle. *Physical Review B* 73: 184106.

Agnon A and Bukowinski MST (1990) Delta s at high-pressure and dlnvs/dlnvp in the lower mantle. *Geophysical Research Letters* 17: 1149–1152.

Akaogi M, Ito E, and Navrotsky A (1989) Olivine-modified spinel–spinel transitions in the system Mg_2SiO_4–Fe_2SiO_4 – Calorimetric measurements, thermochemical calculation, and geophysical application. *Journal of Geophysical Research, Solid Earth and Planets* 94: 15671–15685.

Akaogi M, Tanaka A, and Ito E (2002) Garnet–ilmenite–perovskite transitions in the system $Mg_4Si_4O_{12}$–$Mg_3Al_2Si_3O_{12}$ at high pressures and high temperatures: Phase equilibria, calorimetry and implications for mantle structure. *Physics of the Earth and Planetary Interiors* 132: 303–324.

Alfe D, Gillan MJ, and Price GD (2002a) Composition and temperature of the Earth's core constrained by combining *ab initio* calculations and seismic data. *Earth and Planetary Science Letters* 195: 91–98.

Alfe D, Price GD, and Gillan MJ (2002b) Iron under Earth's core conditions: Liquid-state thermodynamics and high-pressure melting curve from *ab initio* calculations. *Physical Review B* 65: 165118.

Anderson DL (1987) Thermally induced phase-changes, lateral heterogeneity of the mantle, continental roots, and deep slab anomalies. *Journal of Geophysical Research, Solid Earth and Planets* 92: 13968–13980.

Anderson DL (1989) *Theory of the Earth*, 366pp. Boston: Blackwell Scientific.

Anderson DL and Sammis C (1969) The low velocity zone. *Geofisica Internacional* 9: 3–19.

Anderson OL (1995) *Equations of State of Solids for Geophysics and Ceramic Science*. Oxford: Oxford University Press.

Anderson OL and Isaak DG (2002) Another look at the core density deficit of Earth's outer core. *Physics of the Earth and Planetary Interiors* 131: 19–27.

Anderson OL, Schreiber E, Liebermann RC, and Soge N (1968) Some elastic constant data on minerals relevant to geophysics. *Reviews of Geophysics* 6: 491–524.

Angel RJ, Chopelas A, and Ross NL (1992) Stability of high-density clinoenstatite at upper-mantle pressures. *Nature* 358: 322–324.

Badro J, Fiquet G, Guyot F, *et al.* (2003) Iron partitioning in Earth's mantle: Toward a deep lower mantle discontinuity. *Science* 300: 789–791.

Badro J, Rueff J-P, Vankó G, Monaco G, Fiquet G, and Guyot F (2004) Electronic transitions in perovskite: Possible nonconvecting layers in the lower mantle. *Science* 305: 383–386.

Barazangi M and Isacks B (1971) Lateral variations of seismic-wave attenuation in upper mantle above inclined earthquake zone of Tonga Island Arc – Deep anomaly in upper mantle. *Journal of Geophysical Research* 76: 8493–8515.

Benz HM and Vidale JE (1993) Sharpness of upper-mantle discontinuities determined from high-frequency reflections. *Nature* 365: 147–150.

Bercovici D and Karato S (2003) Whole-mantle convection and the transition-zone water filter. *Nature* 425: 39–44.

Birch F (1964) Density and composition of the mantle and core. *Journal of Geophysical Research* 69: 4377–4388.

Birch F (1969) Density and compostion of the upper mantle: First approximation as an olivine layer. In: Hart PJ (ed.) *The Earth's Crust and Upper Mantle*, pp. 18–36. Washington, DC: American Geophysical Union.

Bostock MG (1998) Mantle stratigraphy and evolution of the slave province. *Journal of Geophysical Research, Solid Earth* 103: 21183–21200.

Brown JM and McQueen RG (1982) The equation of state for iron and the Earth's core. In: Akimoto S and Manghnani MH (eds.) *High Pressure Research in Geophysics*, pp. 611–623. Dordrecht: Reidel.

Brown JM and McQueen RG (1986) Phase transitions, Grüneisen parameter, and elasticity for shocked iron between 77 GPa and 400 GPa. *Journal of Geophysical Research* 91: 7485–7494.

Brown JM and Shankland TJ (1981) Thermodynamic parameters in the Earth as determined from seismic profiles. *Geophysical Journal of the Royal Astronomical Society* 66: 579–596.

Buffett BA, Garnero EJ, and Jeanloz R (2000) Sediments at the top of Earth's core. *Science* 290: 1338–1342.

Cammarano F, Saskia G, Pierre V, and Domenico G (2003) Inferring upper-mantle temperatures from seismic velocities. *Physics of the Earth and Planetary Interiors* 138: 197–222.

Carpenter MA, Hemley RJ, and Ho-kwang M (2000) High-pressure elasticity of stishovite and the P4(2)/mnm reversible arrow Pnnm phase transition. *Journal of Geophysical Research, Solid Earth* 105: 10807–10816.

Castle JC and van der Hilst RD (2003) Searching for seismic scattering off mantle interfaces between 800 km and 2000 km depth. *Journal of Geophysical Research, Solid Earth* 108: 2095.

Chastel YB, Dawson PR, Wenk H-R, and Bennett K (1993) Anisotropic convection with implications for the upper-mantle. *Journal of Geophysical Research, Solid Earth* 98: 17757–17771.

Christensen UR and Hofmann AW (1994) Segregation of subducted oceanic-crust in the convecting mantle. *Journal of Geophysical Research, Solid Earth* 99: 19867–19884.

Christensen NI and Salisbury MH (1979) Seismic anisotropy in the upper mantle: Evidence from the Bay of Islands ophiolite complex. *Journal of Geophysical Research* 84: 4601–4610.

Christensen UR and Yuen DA (1985) Layered convection induced by phase transitions. *Journal of Geophysical Research* 90: 10291–10300.

Chudinovskikh L and Boehler R (2001) High-pressure polymorphs of olivine and the 660-km seismic discontinuity. *Nature* 411: 574–577.

Cohen RE, Mazin II, and Isaak DG (1997) Magnetic collapse in transition metal oxides at high pressure: Implications for the Earth. *Science* 275: 654–657.

Creager KC (1997) Inner core rotation rate from small-scale heterogeneity and time-varying travel times. *Science* 278: 1284–1288.

Davies GF (2006) Gravitational depletion of the early Earth's upper mantle and the viability of early plate tectonics. *Earth and Planetary Science Letters* 243: 376–382.

Deuss A and Woodhouse JH (2001) Seismic observations of splitting of the mid-transition zone discontinuity in Earth's mantle. *Science* 294: 354–357.

Deuss A and Woodhouse JH (2002) A systematic search for mantle discontinuities using SS-precursors. *Geophysical Research Letters* 29: doi:10.1029/2002GL014768.

Deuss A, Redfern SAT, Chambers K, and Woodhouse JH (2006) The nature of the 660-kilometer discontinuity in Earth's mantle from global seismic observations of PP precursors. *Science* 311: 198–201.

Dobson DP and Brodholt JP (2005) Subducted banded iron formations as a source of ultralow-velocity zones at the core–mantle boundary. *Nature* 434: 371–374.

Duffy TS, Zha C-S, Downs RT, Ho-kwang M, and Russell JH (1995) Elasticity of forsterite to 16 GPa and the composition of the upper-mantle. *Nature* 378: 170–173.

Dziewonski AM and Anderson DL (1981) Preliminary reference earth model. *Physics of the Earth and Planetary Interiors* 25: 297–356.

Ekstrom G and Dziewonski AM (1998) The unique anisotropy of the Pacific upper mantle. *Nature* 394: 168–172.

Faul UH, Fitz Gerald JD, and Jackson I (2004) Shear wave attenuation and dispersion in melt-bearing olivine polycrystals: 2. Microstructural interpretation and seismological implications. *Journal of Geophysical Research, Solid Earth* 109: B06202 (doi:10.1029/2003JB002407).

Faul UH and Jackson I (2005) The seismological signature of temperature and grain size variations in the upper mantle. *Earth and Planetary Science Letters* 234: 119–134.

Fei Y, Ho-Kwang M, and Mysen BO (1991) Experimental-determination of element partitioning and calculation of phase-relations in the MgO–FeO–SiO$_2$ system at high-pressure and high-temperature. *Journal of Geophysical Research, Solid Earth and Planets* 96: 2157–2169.

Fei Y, Van Orman J, Li J, et al. (2004) Experimentally determined postspinel transformation boundary in Mg$_2$SiO$_4$ using MgO as an internal pressure standard and its geophysical implications. *Journal of Geophysical Research, Solid Earth* 109: B02305 (doi:10.1029/2003JB002562).

Flanagan MP and Shearer PM (1998) Global mapping of topography on transition zone velocity discontinuities by stacking SS precursors. *Journal of Geophysical Research, Solid Earth* 103: 2673–2692.

Fouch MJ and Fischer KM (1996) Mantle anisotropy beneath northwest Pacific subduction zones. *Journal of Geophysical Research, Solid Earth* 101: 15987–16002.

Frost DJ (2003) The structure and sharpness of (Mg, Fe)$_2$SiO$_4$ phase transformations in the transition zone. *Earth and Planetary Science Letters* 216: 313–328.

Fujisawa H (1998) Elastic wave velocities of forsterite and its beta-spinel form and chemical boundary hypothesis for the 410-km discontinuity. *Journal of Geophysical Research, Solid Earth* 103: 9591–9608.

Gaherty JB and Jordan TH (1995) Lehmann discontinuity as the base of an anisotropic layer beneath continents. *Science* 268: 1468–1471.

Gaherty JB, Kato M, and Jordan TH (1999a) Seismological structure of the upper mantle: A regional comparison of seismic layering. *Physics of the Earth and Planetary Interiors* 110: 21–41.

Gaherty JB, Yanbin W, Jordan TH, and Weidner DJ (1999b) Testing plausible upper-mantle compositions using fine-scale models of the 410-km discontinuity. *Geophysical Research Letters* 26: 1641–1644.

Gannarelli CMS, Alfe D, and Gillan MJ (2005) The axial ratio of hcp iron at the conditions of the Earth's inner core. *Physics of the Earth and Planetary Interiors* 152: 67–77.

Garnero EJ and Helmberger DV (1996) Seismic detection of a thin laterally varying boundary layer at the base of the mantle beneath the central-Pacific. *Geophysical Research Letters* 23: 977–980.

Garnero EJ and Jeanloz R (2000) Fuzzy patches on the Earth's core–mantle boundary? *Geophysical Research Letters* 27: 2777–2780.

Godey S, Deschamps F, Trampert J, and Snieder R (2004) Thermal and compositional anomalies beneath the North American continent. *Journal of Geophysical Research, Solid Earth* 109: B01308 (doi:10.1029/2002JB002263).

Goes S, Govers R, and Vacher P (2000) Shallow mantle temperatures under Europe from P and S wave tomography. *Journal of Geophysical Research, Solid Earth* 105: 11153–11169.

Gossler J and Kind R (1996) Seismic evidence for very deep roots of continents. *Earth and Planetary Science Letters* 138: 1–13.

Grand SP and Helmberger DV (1984) Upper mantle shear structure of North-America. *Geophysical Journal of the Royal Astronomical Society* 76: 399–438.

Graves RW and Helmberger DV (1988) Upper mantle cross-section from Tonga to Newfoundland. *Journal of Geophysical Research, Solid Earth and Planets* 93: 4701–4711.

Green DH and Liebermann RC (1976) Phase-equilibria and elastic properties of a pyrolite model for oceanic upper mantle. *Tectonophysics* 32: 61–92.

Gribb TT and Cooper RF (1998) Low-frequency shear attenuation in polycrystalline olivine: Grain boundary diffusion and the physical significance of the Andrade model for viscoelastic rheology. *Journal of Geophysical Research, Solid Earth* 103: 27267–27279.

Gu Y, Dziewonski AM, and Agee CB (1998) Global de-correlation of the topography of transition zone discontinuities. *Earth and Planetary Science Letters* 157: 57–67.

Gu YJ, Dziewonski AM, and Ekström G (2001) Preferential detection of the Lehmann discontinuity beneath continents. *Geophysical Research Letters* 28: 4655–4658.

Gung YC, Panning M, and Bomanowicz B (2003) Global anisotropy and the thickness of continents. *Nature* 422: 707–711.

Gutenberg B (1959) Wave velocities below the mohorovicic discontinuity. *Geophysical Journal of the Royal Astronomical Society* 2: 348–352.

Gwanmesia GD, Rigden S, Jackson I, and Liebermann RC (1990) Pressure-dependence of elastic wave velocity for beta-Mg_2SiO_4 and the composition of the Earths mantle. *Science* 250: 794–797.

Helffrich G (2000) Topography of the transition zone seismic discontinuities. *Reviews of Geophysics* 38: 141–158.

Helffrich G and Bina CR (1994) Frequency-dependence of the visibility and depths of mantle seismic discontinuities. *Geophysical Research Letters* 21: 2613–2616.

Helffrich GR and Wood BJ (1996) 410 km discontinuity sharpness and the form of the olivine alpha–beta phase diagram: Resolution of apparent seismic contradictions. *Geophysical Journal International* 126: F7–F12.

Hirose K (2002) Phase transitions in pyrolitic mantle around 670-km depth: Implications for upwelling of plumes from the lower mantle. *Journal of Geophysical Research, Solid Earth* 107: B42078 (doi:10.1029/2001JB000597).

Hirose K, Komabayashi T, Murakami M, and Funakoshi K-i (2001) *In situ* measurements of the majorite–akimotoite–perovskite phase transition boundaries in $MgSiO_3$. *Geophysical Research Letters* 28: 4351–4354.

Hirose K, Takafuji N, Sata N, and Ohishi Y (2005) Phase transition and density of subducted MORB crust in the lower mantle. *Earth and Planetary Science Letters* 237: 239–251.

Hirschmann MM, Abaud C, and Withers AC (2005) Storage capacity of H_2O in nominally anhydrous minerals in the upper mantle. *Earth and Planetary Science Letters* 236: 167–181.

Iitaka T, Hirose K, Kawamura K, and Murakami M (2004) The elasticity of the $MgSiO_3$ post-perovskite phase in the Earth's lowermost mantle. *Nature* 430: 442–445.

Irifune T and Isshiki M (1998) Iron partitioning in a pyrolite mantle and the nature of the 410-km seismic discontinuity. *Nature* 392: 702–705.

Irifune T, Nishiyama N, Kuroda K, *et al.* (1998) The postspinel phase boundary in Mg_2SiO_4 determined by *in situ* X-ray diffraction. *Science* 279: 1698–1700.

Ishii M and Dziewonski AM (2002) The innermost inner core of the Earth: Evidence for a change in anisotropic behavior at the radius of about 300 km. *Proceedings of the National Academy of Sciences of the United States of America* 99: 14026–14030.

Ita J and Stixrude L (1992) Petrology, elasticity, and composition of the mantle transition zone. *Journal of Geophysical Research, Solid Earth* 97: 6849–6866.

Ito E and Takahashi E (1989) Postspinel transformations in the system Mg_2SiO_4–Fe_2SiO_4 and some geophysical implications. *Journal of Geophysical Research, Solid Earth and Planets* 94: 10637–10646.

Jackson I, *et al.* (2002) Grain-size-sensitive seismic wave attenuation in polycrystalline olivine. *Journal of Geophysical Research, Solid Earth* 107: B122360 (doi:10.1029/2001JB001225).

Jaupart C and Mareschal JC (1999) The thermal structure and thickness of continental roots. *Lithos* 48: 93–114.

Jeanloz R and Thompson AB (1983) Phase transitions and mantle discontinuities. *Reviews of Geophysics* 21: 51–74.

Johnson LR (1969) Array measurements of P velocities in lower mantle. *Bulletin of the Seismological Society of America* 59: 973–1008.

Jordan TH (1975) Continental tectosphere. *Reviews of Geophysics* 13: 1–12.

Jung H and Karato S (2001) Water-induced fabric transitions in olivine. *Science* 293: 1460–1463.

Kaneshima S and Helffrich G (1998) Detection of lower mantle scatterers northeast of the Marianna subduction zone using short-period array data. *Journal of Geophysical Research, Solid Earth* 103: 4825–4838.

Karato S (1992) On the Lehmann discontinuity. *Geophysical Research Letters* 19: 2255–2258.

Karato S (1993) Importance of anelasticity in the interpretation of seismic tomography. *Geophysical Research Letters* 20: 1623–1626.

Karato S and Jung H (1998) Water, partial melting and the origin of the seismic low velocity and high attenuation zone in the upper mantle. *Earth and Planetary Science Letters* 157: 193–207.

Karato S and Karki BB (2001) Origin of lateral variation of seismic wave velocities and density in the deep mantle. *Journal of Geophysical Research, Solid Earth* 106: 21771–21783.

Karato S-I (1998) Seismic anisotropy in the deep mantle, boundary layers and the geometry of mantle convection. *Pure and Applied Geophysics* 151: 565–587.

Karki BB and Crain J (1998) First-principles determination of elastic properties of $CaSiO_3$ perovskite at lower mantle pressures. *Geophysical Research Letters* 25: 2741–2744.

Karki BB and Stixrude L (1999) Seismic velocities of major silicate and oxide phases of the lower mantle. *Journal of Geophysical Research* 104: 13025–13033.

Karki BB, Stixrude L, Clark SJ, Warren MC, Ackland GJ, and Crain J (1997a) Structure and elasticity of MgO at high pressure. *American Mineralogist* 82: 51–60.

Karki BB, Stixrude L, and Crain J (1997b) *Ab initio* elasticity of three high-pressure polymorphs of silica. *Geophysical Research Letters* 24: 3269–3272.

Karki BB, Stixrude L, and Wentzcovitch RM (2001) High-pressure elastic properties of major materials of Earth's mantle from first principles. *Reviews of Geophysics* 39: 507–534.

Karki BB, Wentzcovitch RM, de Gironcoli S, and Baroni S (1999) First-principles determination of elastic anisotropy and wave velocities of MgO at lower mantle conditions. *Science* 286: 1705–1707.

Katsura T and Ito E (1989) The system Mg_2SiO_4–Fe_2SiO_4 at high-pressures and temperatures – Precise determination of stabilities of olivine, modified spinel, and spinel. *Journal of Geophysical Research, Solid Earth and Planets* 94: 15663–15670.

Katsura T, Yamada H, Nishikawa O, *et al.* (2004) Olivine–wadsleyite transition in the system (Mg, Fe)$_2$SiO$_4$. *Journal of*

Geophysical Research, Solid Earth 109: B02209 (doi:10.1029/2003JB002438).

Kawakatsu H and Niu FL (1994) Seismic evidence for a 920-km discontinuity in the mantle. *Nature* 371: 301–305.

Kellogg LH, Hager BH, and van der Hilst RD (1999) Compositional stratification in the deep mantle. *Science* 283: 1881–1884.

Kennett BLN (1993) Seismic structure and heterogeneity in the upper mantle. In: Aki K and Dmowska R (eds.) *Relating Geophysics Structures and Processes: The Jeffreys Volume; Geophysical Monograph 76*. Washington, DC: American Geophysical Union and International Union of Geodosy and Geophysics.

Kennett BLN and Engdahl ER (1991) Traveltimes for global earthquake location and phase identification. *Geophysical Journal International* 105: 429–465.

Kiefer B, Stixrude L, and Wentzcovitch RM (2002) Elasticity of (Mg, Fe)SiO$_3$–Perovskite at high pressures. *Geophysical Research Letters* 29: 1539 (doi:10.1029/2002GL014683).

Klein EM and Langmuir CH (1987) Global correlations of ocean ridge basalt chemistry with axial depth and crustal thickness. *Journal of Geophysical Research, Solid Earth and Planets* 92: 8089–8115.

Knittle E and Jeanloz R (1989) Simulating the core-mantle boundary; an experimental study of high-pressure reactions between silicates and liquid iron. *Geophysical Research Letters* 16: 609–612.

Koito S, Akaogi M, Kubota O, and Suzuki T (2000) Calorimetric measurements of perovskites in the system CaTiO$_3$–CaSiO$_3$ and experimental and calculated phase equilibria for high-pressure dissociation of diopside. *Physics of the Earth and Planetary Interiors* 120: 1–10.

Kung J, Li B, Uchida T, and Wang Y (2005) *In situ* elasticity measurement for the unquenchable high-pressure clinopyroxene phase: Implication for the upper mantle. *Geophysical Research Letters* 32: L01307 (doi:10.1029/2004GL021661).

Lambert IB and Wyllie PJ (1968) Stability of hornblende and a model for low velocity zone. *Nature* 219: 1240–1241.

Lay T and Helmberger DV (1983) A lower mantle s-wave triplication and the shear velocity structure of D″. *Geophysical Journal of the Royal Astronomical Society* 75: 799–837.

Lay T, Williams Q, and Garnero EJ (1998a) The core–mantle boundary layer and deep Earth dynamics. *Nature* 392: 461–468.

Lay T, Garnero EJ, William Q, *et al.* (1998b) Seismic wave anisotropy in the D″ region and its implications. In: Gurnis M, Wysession ME, Knittle E, and Buffett BA (eds.) *The Core–Mantle Boundary Region*, pp. 299–318. Washington, DC: American Geophysical Union.

Lees AC, Bukowinski MST, and Jeanloz R (1983) Reflection properties of phase-transition and compositional change models of the 670-km discontinuity. *Journal of Geophysical Research* 88: 8145–8159.

Lestunff Y, Wicks CW, Jr., and Romanowicz B (1995) P′P′ precursors under Africa – Evidence for mid-mantle reflectors. *Science* 270: 74–77.

Li B, Liebermann RC, and Weidner DJ (1998) Elastic moduli of wadsleyite (beta-Mg$_2$SiO$_4$) to 7 gigapascals and 873 Kelvin. *Science* 281: 675–677.

Li J and Agee CB (1996) Geochemistry of mantle–core differentiation at high pressure. *Nature* 381: 686–689.

Li L, Brodholt JP, Stackhouse S, Weidner DJ, Alfredsson M, and Price GD (2005) Electronic spin state of ferric iron in Al-bearing perovskite in the lower mantle. *Geophysical Research Letters* 32: L17307 (doi:10.1029/2005GL023045).

Lin JF, Heinz DL, Campbell AJ, Devine JM, Mao WL, and Shen G (2002a) Iron–nickel alloy in the Earth's core.

Geophysical Research Letters 29: 1471 (doi:10.1029/2002GL015089).

Lin JF, Heinz DL, Campbell AJ, Devine JM, and Shen G (2002b) Iron–silicon alloy in Earth's core? *Science* 295: 313–315.

Lin JF, Struzhkin VV, Jacobsen SD, *et al.* (2005) Spin transition of iron in magnesiowustite in the Earth's lower mantle. *Nature* 436: 377–380.

Litasov K, Ohtani E, Suzuki A, and Funakoshi K (2005) *In situ* X-ray diffraction study of post-spinel transformation in a peridotite mantle: Implication for the 660-km discontinuity. *Earth and Planetary Science Letters* 238: 311–328.

Liu J, Chen G, Gwanmesia GD, and Liebermann RC (2000) Elastic wave velocities of pyrope-majorite garnets (Py(62)Mj(38) and Py(50)Mj(50)) to 9 GPa. *Physics of the Earth and Planetary Interiors* 120: 153–163.

Liu LG (1976) Orthorhombic perovskite phases observed in olivine, pyroxene and garnet at high-pressures and temperatures. *Physics of the Earth and Planetary Interiors* 11: 289–298.

Liu W, Kung J, and Li B (2005) Elasticity of San Carlos olivine to 8 GPa and 1073 K. *Geophysical Research Letters* 32: L16301 (doi:10.1029/2005GL023453).

Mainprice D, Barruol G, and Ismail WB (2000) The seismic anisotropy of the Earth's mantle: From single crystal to polycrystal. In: Karato SI, Forte A, Liebermann R, Masters G, and Stixrude L (eds.) *Earth's Deep Interior: Mineral Physics and Tomography from the Atomic to the Global Scale*. Washington, DC: American Geophysical Union.

Mainprice D, Tommasi A, Couvy H, Cordier P, and Frost DJ (2005) Pressure sensitivity of olivine slip systems and seismic anisotropy of Earth's upper mantle. *Nature* 433: 731–733.

Masters G, Laske G, Bolton H, and Dziewonski A (2000) The relative behavior of shear velocity, bulk sound speed, and compressional velocity in the mantle: Implications for chemical and thermal structure. In: Karato SI, Forte A, Liebermann R, Masters G, and Stixrude L (eds.) *Earth's Deep Interior: Mineral Physics and Tomography from the Atomic to the Global Scale*, pp. 63–88. Washington, DC: American Geophysical Union.

Masters TG (1979) Observational constraints on the chemical and thermal structure of the Earth's deep interior. *Geophysical Journal of the Royal Astronomical Society* 57: 507–534.

Matsukage KN, Jing Z, and Karato S (2005) Density of hydrous silicate melt at the conditions of Earth's deep upper mantle. *Nature* 438: 488–491.

McKenzie D and Bickle MJ (1988) The volume and composition of melt generated by extension of the lithosphere. *Journal of Petrology* 29: 625–679.

Mei S and Kohlstedt DL (2000a) Influence of water on plastic deformation of olivine aggregates 1. Diffusion creep regime. *Journal of Geophysical Research, Solid Earth* 105: 21457–21469.

Mei S and Kohlstedt DL (2000b) Influence of water on plastic deformation of olivine aggregates 2. Dislocation creep regime. *Journal of Geophysical Research, Solid Earth* 105: 21471–21481.

Merkel S, Kubo A, Miyagi L, *et al.* (2006) Plastic deformation of MgGeO$_3$ post-perovskite at lower mantle pressures. *Science* 311: 644–646.

Montagner JP and Anderson DL (1989) Constrained reference mantle model. *Physics of the Earth and Planetary Interiors* 58: 205–227.

Montagner JP and Guillot L (2002) Seismic anisotropy and global geodynamics. In: Karato S-i and Wenk H-R (eds.) *Plastic Deformation of Minerals and Rocks*, pp. 353–385.

Montagner JP and Kennett BLN (1996) How to reconcile body-wave and normal-mode reference Earth models. *Geophysical Journal International* 125: 229–248.

Morelli A, Dziewonski AM, and Woodhouse JH (1986) Anisotropy of the inner core inferred from PKIKP travel times. *Geophysical Research Letters* 13: 1545–1548.

Murakami M, Hirose K, Kawamura K, Sata N, and Ohishi Y (2004) Post-perovskite phase transition in MgSiO$_3$. *Science* 304: 855–858.

Nakagawa T and Buffett BA (2005) Mass transport mechanism between the upper and lower mantle in numerical simulations of thermochemical mantle convection with multicomponent phase changes. *Earth and Planetary Science Letters* 230: 11–27.

Nataf HC, Nakanishi I, and Anderson DL (1986) Measurements of mantle wave velocities and inversion for lateral heterogeneities and anisotropy.3. Inversion. *Journal of Geophysical Research, Solid Earth and Planets* 91: 7261–7307.

Nguyen JH and Holmes NC (2004) Melting of iron at the physical conditions of the Earth's core. *Nature* 427: 339–342.

Ni SD, Tan E, Gurnis M, and Helnberger D (2002) Sharp sides to the African superplume. *Science* 296: 1850–1852.

Nishimura CE and Forsyth DW (1989) The anisotropic structure of the upper mantle in the Pacific. *Geophysical Journal, Oxford* 96: 203–229.

Obayashi M, Sugioka H, Yoshimitsu J, and Fukao Y (2006) High temperature anomalies oceanward of subducting slabs at the 410-km discontinuity. *Earth and Planetary Science Letters* 243: 149–158.

Oganov AR, Brodholt JP, and Price GD (2001) The elastic constants of MgSiO$_3$ perovskite at pressures and temperatures of the Earth's mantle. *Nature* 411: 934–937.

Oganov AR and Ono S (2004) Theoretical and experimental evidence for a post-perovskite phase of MgSiO$_3$ in Earth's D″ layer. *Nature* 430: 445–448.

Ono S, Ohishi Y, and Mibe K (2004) Phase transition of Ca-perovskite and stability of Al-bearing Mg-perovskite in the lower mantle. *American Mineralogist* 89: 1480–1485.

Pacalo REG and Gasparik T (1990) Reversals of the orthoenstatite–clinoenstatite transition at high-pressures and high-temperatures. *Journal of Geophysical Research, Solid Earth and Planets* 95: 15853–15858.

Panning M and Romanowicz B (2004) Inferences on flow at the base of Earth's mantle based on seismic anisotropy. *Science* 303: 351–353.

Pasternak MP, Taylor RD, Jeanloz R, Li X, Nguyen JH, and McCammon CA (1997) High pressure collapse of magnetism in Fe$^{0.94}$O: Mossbauer spectroscopy beyond 100 GPa. *Physical Review Letters* 79: 5046–5049.

Paulssen H (1988) Evidence for a sharp 670-km discontinuity as inferred from P- to S-converted waves. *Journal of Geophysical Research, Solid Earth and Planets* 93: 10489–10500.

Poirier J-P (1994) Light elements in the Earth's outer core: A critical review. *Physics of the Earth and Planetary Interiors* 85: 319–337.

Revenaugh J and Jordan TH (1991a) Mantle layering from SCS reverberations.2. The transition zone. *Journal of Geophysical Research, Solid Earth* 96: 19763–19780.

Revenaugh J and Jordan TH (1991b) Mantle layering from SCS reverberations.3. The upper mantle. *Journal of Geophysical Research, Solid Earth* 96: 19781–19810.

Revenaugh J and Sipkin SA (1994) Seismic evidence for silicate melt atop the 410 km mantle discontinuity. *Nature* 369: 474–476.

Ribe NM (1989) Seismic anisotropy and mantle flow. *Journal of Geophysical Research, Solid Earth and Planets* 94: 4213–4223.

Rigden SM, Gwanmesia GD, Fitzgerald JD, Jackson L, and Liebermann RC (1991) Spinel elasticity and seismic structure of the transition zone of the mantle. *Nature* 354: 143–145.

Ringwood AE (1969) Composition and evolution of the upper mantle. In: Hart PJ (ed.) *The Earth's Crust and Upper Mantle*, pp. 1–17. Washington, DC: American Geophysical Union.

Ringwood AE and Major A (1970) The system Mg$_2$SiO$_4$–Fe$_2$SiO$_4$ at high pressures and temperatures. *Physics of the Earth and Planetary Interiors* 3: 89–108.

Ritsema J and van Heijst HJ (2002) Constraints on the correlation of P- and S-wave velocity heterogeneity in the mantle from P, PP, PPP and PKPab traveltimes. *Geophysical Journal International* 149: 482–489.

Robertson GS and Woodhouse JH (1996) Ratio of relative S to P velocity heterogeneity in the lower mantle. *Journal of Geophysical Research* 101: 20041–020052.

Romanowicz B (1995) A global tomographic model of shear attenuation in the upper-mantle. *Journal of Geophysical Research, Solid Earth* 100: 12375–12394.

Romanowicz B (1998) Attenuation tomography of the Earth's mantle: A review of current status. *Pure and Applied Geophysics* 153: 257–272.

Romanowicz B (2003) Global mantle tomography: Progress status in the past 10 years. *Annual Review of Earth and Planetary Sciences* 31: 303–328.

Roth EG, Wiens DA, and Zhao D (2000) An empirical relationship between seismic attenuation and velocity anomalies in the upper mantle. *Geophysical Research Letters* 27: 601–604.

Russo RM and Silver PG (1994) Trench-parallel flow beneath the nazca plate from seismic anisotropy. *Science* 263: 1105–1111.

Sakamaki T, Suzuki A, and Ohtani E (2006) Stability of hydrous melt at the base of the Earth's upper mantle. *Nature* 439: 192–194.

Sato H, Sacks IS, and Murase T (1989) The use of laboratory velocity data for estimating temperature and partial melt fraction in the low-velocity zone – Comparison with heat-flow and electrical-conductivity studies. *Journal of Geophysical Research, Solid Earth and Planets* 94: 5689–5704.

Selby ND and Woodhouse JH (2002) The Q structure of the upper mantle: Constraints from Rayleigh wave amplitudes. *Journal of Geophysical Research, Solid Earth* 107: 2097 (doi:10.1029/2001JB000257).

Shearer PM (1990) Seismic imaging of upper-mantle structure with new evidence for a 520-km discontinuity. *Nature* 344: 121–126.

Shim SH, Duffy TS, and Shen G (2001) The post-spinel transformation in Mg$_2$SiO$_4$ and its relation to the 660-km seismic discontinuity. *Nature* 411: 571–574.

Shim SH, Jeanloz R, and Duffy TS (2002) Tetragonal structure of CaSiO$_3$ perovskite above 20 GPa. *Geophysical Research Letters* 29: 2166 (doi:10.1029/2002GL016148).

Sidorin I, Gurnis M, and Helmberger DV (1999) Evidence for a ubiquitous seismic discontinuity at the base of the mantle. *Science* 286: 1326–1331.

Silver PG (1996) Seismic anisotropy beneath the continents: Probing the depths of geology. *Annual Review of Earth and Planetary Sciences* 24: 385–432.

Simmons NA and Gurrola H (2000) Multiple seismic discontinuities near the base of the transition zone in the Earth's mantle. *Nature* 405: 559–562.

Sinogeikin SV and Bass JD (2002) Elasticity of majorite and a majorite-pyrope solid solution to high pressure: Implications for the transition zone. *Geophysical Research Letters* 29: 1017 (doi:10.1029/2001GL013937).

Sinogeikin SV, Bass JD, and Katsura T (2003) Single-crystal elasticity of ringwoodite to high pressures and high temperatures: Implications for 520 km seismic discontinuity. *Physics of the Earth and Planetary Interiors* 136: 41–66.

Sobolev SV, Zeyen H, Stoll G, and Friederike WAF (1996) Upper mantle temperatures from teleseismic tomography of French Massif Central including effects of composition, mineral reactions, anharmonicity, anelasticity and partial melt. *Earth and Planetary Science Letters* 139: 147–163.

Solomatov VS and Stevenson DJ (1994) Can sharp seismic discontinuities be caused by nonequilibrium phase-transformations. *Earth and Planetary Science Letters* 125: 267–279.

Song TRA, Helmberger DV, and Grand SP (2004) Low-velocity zone atop the 410-km seismic discontinuity in the northwestern United States. *Nature* 427: 530–533.

Speziale S, Fuming J, and Duffy TS (2005a) Compositional dependence of the elastic wave velocities of mantle minerals: Implications for seismic properties of mantle rocks. In: Matas J, van der Hilst RD, Bass JD, and Trampert J (eds.) *Earth's Deep Mantle: Structure, Composition, and Evolution*, pp. 301–320. Washington, DC: American Geophysical Union.

Speziale S, Milner A, Lee VE, Clark SM, Pasternak MP, and Jeanloz R (2005b) Iron spin transition in Earth's mantle. *Proceedings of the National Academy of Sciences of the United States of America* 102: 17918–17922.

Stackhouse S, Brodholt JP, Wookey J, Kendall J-M, and Price GD (2005) The effect of temperature on the seismic anisotropy of the perovskite and post-perovskite polymorphs of MgSiO$_3$. *Earth and Planetary Science Letters* 230: 1–10.

Steinle-Neumann G, Stixrude L, and Cohen RE (2002) Physical properties of iron in the inner core. In: Dehant V, Creager KC, Karato S-i, and Zatman S (eds.) *Core Structure, Dynamics, and Rotation*, pp. 137–161. Washington, DC: American Geophysical Union.

Steinle-Neumann G, Stixrude L, Cohen RE, and Glseren O (2001) Elasticity of iron at the temperature of the Earth's inner core. *Nature* 413: 57–60.

Stevenson DJ (1981) Models of the Earths core. *Science* 214: 611–619.

Stixrude L (1997) Structure and sharpness of phase transitions and mantle discontinuities. *Journal of Geophysical Research, Solid Earth* 102: 14835–14852.

Stixrude L (1998) Elastic constants and anisotropy of MgSiO$_3$ perovskite, periclase, and SiO$_2$ at high pressure. In: Gurnis M, Wysession ME, Knittle E, and Buffett BA (eds.) *The Core–Mantle Boundary Region*, pp. 83–96. Washington, DC: American Geophysical Union.

Stixrude L and Cohen RE (1995) High pressure elasticity of iron and anisotropy of Earth's inner core. *Science* 267: 1972–1975.

Stixrude L, Cohen RE, Yu R, and Krakauer H (1996) Prediction of phase transition in CaSiO$_3$ perovskite and implications for lower mantle structure. *American Mineralogist* 81: 1293–1296.

Stixrude L and Karki B (2005) Structure and freezing of MgSiO$_3$ liquid in Earth's lower mantle. *Science* 310: 297–299.

Stixrude L and Lithgow-Bertelloni C (2005a) Mineralogy and elasticity of the oceanic upper mantle: Origin of the low-velocity zone. *Journal of Geophysical Research, Solid Earth* 110: B03204 (doi:10.1029/2004JB002965).

Stixrude L and Lithgow-Bertelloni C (2005b) Thermodynamics of mantle minerals. I: Physical properties. *Geophysical Journal International* 162: 610–632.

Stixrude L and Lithgow-Bertelloni C (2007) Influence of phase transformations on lateral heterogeneity and dynamics in Earth's mantle. *Earth and Planetary Science Letters*, (submitted).

Stixrude L, Lithgow-Bertelloni C, Kiefer B, and Fumagalli P (2007) Phase stability and shear softening in CaSiO$_3$ perovskite at high pressure. *Physical Review B,* 75: 024108.

Su WJ and Dziewonski AM (1995) Inner-core anisotropy in 3 Dimensions. *Journal of Geophysical Research, Solid Earth* 100: 9831–9852.

Su Wj and Dziewonski AM (1997) Simultaneous inversion for 3-D variations in shear and bulk velocity in the mantle. *Physics of the Earth and Planetary Interiors* 100: 135–156.

Söderlind P, Moriarty JA, and Wills JM (1996) First-principles theory of iron up to earth-core pressures: Structural, vibrational, and elastic properties. *Physical Review B* 53: 14063–14072.

Takafuji N, Hirose K, Mitome M, and Bando Y (2005) Solubilities of O and Si in liquid iron in equilibrium with (Mg,Fe)SiO$_3$ perovskite and the light elements in the core. *Geophysical Research Letters* 32: L06313 (doi:10.1029/2005GL022773).

Tanaka S and Hamaguchi H (1997) Degree one heterogeneity and hemispherical variation of anisotropy in the inner core from PKP(BC)-PKP(DF) times. *Journal of Geophysical Research, Solid Earth* 102: 2925–2938.

Tanimoto T and Anderson DL (1984) Mapping convection in the mantle. *Geophysical Research Letters* 11: 287–290.

Trampert J and van Heijst HJ (2002) Global azimuthal anisotropy in the transition zone. *Science* 296: 1297–1299.

Tsuchiya T, Tsuchiya J, Umemoto K, and Wentzcovitch RM (2004a) Phase transition in MgSiO$_3$ perovskite in the Earth's lower mantle. *Earth and Planetary Science Letters* 224: 241–248.

Tsuchiya T, Tsuchiya J, Umemoto K, Wentzcovitch RM, *et al.* (2004b) Elasticity of post-perovskite MgSiO$_3$. *Geophysical Research Letters* 31: L14603 (doi:10.1029/2004GL020278).

Tsuchiya T, Wentzcovitch RM, da Silva CRS, and de Gironcoli S (2006) Spin transition in magnesiowüstite in Earth's lower mantle. *Physical Review Letters* 96: 198501.

van der Meijde M, Marone F, Giardini D, and van der Lee S (2003) Seismic evidence for water deep in Earth's upper mantle. *Science* 300: 1556–1558.

van der Meijde M, van der Lee S, and Giardini D (2005) Seismic discontinuities in the Mediterranean mantle. *Physics of the Earth and Planetary Interiors* 148: 233–250.

van Hunen J, Zhong S, Shapiro NM, and Ritzwoller MH (2005) New evidence for dislocation creep from 3-D geodynamic modeling of the Pacific upper mantle structure. *Earth and Planetary Science Letters* 238: 146–155.

Vasco D and Johnson LR (1998) Whole Earth structure estimated from seismic arrival times. *Journal of Geophysical Research* 103: 2633–2672.

Vinnik L, Kato M, and Kawakatsu H (2001) Search for seismic discontinuities in the lower mantle. *Geophysical Journal International* 147: 41–56.

Vocadlo L, Alfe D, Gillan MJ, Wood IG, Brodholt JP, and Price GD (2003) Possible thermal and chemical stabilization of body-centred-cubic iron in the Earth's core. *Nature* 424: 536–539.

Weertman J (1970) Creep strength of Earth's mantle. *Reviews of Geophysics and Space Physics* 8: 145.

Weidner DJ and Wang YB (1998) Chemical- and Clapeyron-induced buoyancy at the 660 km discontinuity. *Journal of Geophysical Research, Solid Earth* 103: 7431–7441.

Wentzcovitch RM, Karki BB, Cococcioni M, and de Gironcoli S (2004) Thermoelastic properties of MgSiO$_3$-perovskite: Insights on the nature of the Earth's lower mantle. *Physical Review Letters* 92: 018501.

Wentzcovitch RM, Tsuchiya T, and Tsuchiya J (2006) MgSiO$_3$ postperovskite at D″ conditions. *Proceedings of the National Academy of Sciences of the United States of America* 103: 543–546.

Williams Q and Garnero EJ (1996) Seismic evidence for partial melt at the base of Earth's mantle. *Science* 273: 1528.

Wood BJ (1990) Postspinel transformations and the width of the 670-km discontinuity – A comment on postspinel transformations in the system Mg_2SiO_4–Fe_2SiO_4 and some geophysical implications by Ito,E. And Takahashi,E. *Journal of Geophysical Research, Solid Earth and Planets* 95: 12681–12685.

Wood BJ (1995) The effect of H_2O on the 410-kilometer seismic discontinuity. *Science* 268: 74–76.

Woodhouse JH, Giardini D, and Li X-D (1986) Evidence for inner core anisotropy from free oscillations. *Geophysical Research Letters* 13: 1549–1552.

Woodland AB (1998) The orthorhombic to high-P monoclinic phase transition in Mg–Fe pyroxenes: Can it produce a seismic discontinuity? *Geophysical Research Letters* 25: 1241–1244.

Workman RK and Hart SR (2005) Major and trace element composition of the depleted MORB mantle (DMM). *Earth and Planetary Science Letters* 231: 53–72.

Wysession ME, Lay T, Revenaugh J, *et al.* (1998) The D″ discontinuity and its implicatinos. In: Gurnis M, Wysession ME, Knittle E, and Buffett BA (eds.) *The Core–Mantle Boundary Region*. Washington, DC: American Geophysical Union.

Xie SX and Tackley PJ (2004) Evolution of helium and argon isotopes in a convecting mantle. *Physics of the Earth and Planetary Interiors* 146: 417–439.

Young CJ and Lay T (1987) The core–mantle boundary. *Annual Review of Earth and Planetary Sciences* 15: 25–46.

Zha C-s, Duffy TS, Ho-kwang M, Downs RT, Hemley RJ, and Weidner DJ (1997) Single-crystal elasticity of beta-Mg_2SiO_4 to the pressure of the 410 km seismic discontinuity in the Earth's mantle. *Earth and Planetary Science Letters* 147: E9–E15.

3 Mineralogy of the Earth – Phase Transitions and Mineralogy of the Lower Mantle

T. Irifune and T. Tsuchiya, Ehime University, Matsuyama, Japan

3.1	Introduction	33
3.2	Experimental and Theoretical Backgrounds	34
3.2.1	High-Pressure Technology	34
3.2.2	*Ab Initio* Calculation	36
3.2.3	Pressure Scale	37
3.3	Mineral-Phase Transitions in the Lower Mantle	38
3.3.1	Major Minerals in the Mantle and Subducted Slab	38
3.3.1.1	$MgSiO_3$	38
3.3.1.2	$MgSiO_3-FeSiO_3$	39
3.3.1.3	$MgSiO_3-Al_2O_3$	40
3.3.1.4	$MgO-FeO$	41
3.3.1.5	$CaSiO_3$	42
3.3.1.6	SiO_2	43
3.3.1.7	Al-rich phase	44
3.3.2	Minor Minerals	44
3.3.2.1	$MgAl_2O_4$, $NaAlSiO_4$	44
3.3.2.2	$KAlSi_3O_8$, $NaAlSi_3O_8$	45
3.3.2.3	CAS phase	45
3.3.2.4	Phase D, δ-AlOOH	45
3.3.2.5	$MgCO_3$, $CaCO_3$	46
3.4	Phase Transitions and Density Changes in Mantle and Slab Materials	46
3.4.1	Chemical Compositions and Density Calculations	46
3.4.2	Phase and Density Relations	48
3.4.2.1	Pyrolite	48
3.4.2.2	Harzburgite	49
3.4.2.3	MORB	50
3.5	Mineralogy of the Lower Mantle	50
3.5.1	The 660 km Discontinuity	50
3.5.2	Middle Parts of the Lower Mantle	51
3.5.3	Postperovskite Transition and the D″ Layer	53
3.6	Summary	55
References		56

3.1 Introduction

Until recently, experimental high-pressure mineral physics studies mainly focused on materials in the upper part of the mantle, because of technical restrictions in pressure and temperature generation and also in precise measurements of crystal structures and physical properties under the lower-mantle conditions. However, developments in technologies of both laser-heated diamond-anvil cell (LHDAC) and large-volume Kawai-type multianvil apparatus (KMA), combined with synchrotron radiation, have enabled us to quantitatively study phase transitions and some key physical properties of mantle minerals under the P,T conditions encompassing those of the whole mantle.

Progress in computational mineral physics based on *ab initio* calculations has also been dramatic in the last decade in conjunction with the rapid advancement of computer technologies. Classical molecular dynamics calculations required *a priori* assumptions on the parameters for interatomic model potentials, which largely rely on available experimental data. Quantum mechanical Hamiltonians of many-body electron systems can be efficiently and quantitatively evaluated on the basis of the density functional theory (DFT) (Hohenberg and Kohn, 1964; Kohn and Sham, 1965). Practical calculations of minerals having complex crystal structures can be achieved using various methods and techniques developed following the DFT. As a result of such advancement in *ab initio* calculations, it is now possible to predict stability and some physical properties of high-pressure forms quantitatively with uncertainties that are even comparable to those attached in experimentally derived data, as was dramatically shown in a series of recent articles relevant to the postperovskite phase transition (Tsuchiya *et al.*, 2004a, 2004b; Oganov and Ono, 2004; Iitaka *et al.*, 2004).

Another reason for the relatively scarce mineral physics studies for the lower part of the mantle may originate from the fact that there has been no indication of the occurrence of major phase transitions under these conditions, except for those corresponding to uppermost ($\sim$660–800 km) and lowermost ($\sim$2700–2900 km; the D$''$ layer) parts of the lower mantle. This is in marked contrast to the nature of the mantle transition region, best described as "the key to a number of geophysical problems" by Birch (1952). Recent seismological studies, however, demonstrated that there are regions that significantly scatter seismic waves, which may be related to some unknown phase transitions and/or chemical boundaries at certain depths in the upper to middle regions of the lower mantle (e.g., Niu and Kawakatsu, 1996). In addition, both geochemical considerations and mantle dynamics simulations suggest that the mantle may be divided into chemically distinct regions by a boundary at 1500–2000 km depths in the lower mantle (e.g., Kellogg *et al.*, 1999; Tackley, 2000; Trampert *et al.*, 2004). Moreover, seismological studies with various methods of analysis have shown detailed structures in the D$''$ layer, demonstrating marked heterogeneity in both horizontal and vertical directions (Lay *et al.*, 2004). Thus, there is growing evidence that the lower mantle is not featureless any more, and many mineral physicists have started to develop experimental and computational techniques for higher pressure and temperature conditions (with improved accuracy) in order to elucidate the mineralogy of the lower mantle.

In spite of the great efforts in both experimental and theoretical studies (mostly conducted in the last decade), our knowledge on the stability and mineral physics properties of high-pressure phases relevant to the lower mantle is still very limited compared with that on the shallower parts of the mantle. Although LHDACs are now capable of generating pressures and temperatures corresponding to, or even beyond, the mantle–core boundary, there remain a number of disagreements in phase equilibrium studies using this method with various techniques in heating, temperature/pressure measurements, phase identification, etc. Mineral physics parameters that constrain densities of high-pressure phases in the lower mantle have been accumulated thanks to the intense synchrotron sources, particularly those available at third-generation synchrotron facilities, such as ESRF, APS, and SPring-8, combined with both LHDAC and KMA, but those related to elastic wave velocities are scarce to date.

In the present chapter, we briefly review recent progress and limitations in experimental and computational techniques in studying phase transitions and some key physical properties under the lower-mantle conditions. Then we summarize current knowledge on these properties obtained experimentally and/or theoretically using these techniques. We rather focus on the phase transitions and phase relations in simple silicates, oxides, carbonates, and hydrous systems closely related to the mantle and slab materials, because experimental measurements or theoretical predictions of other properties such as shear moduli and their pressure/temperature dependency are still limited for lower-mantle conditions. In contrast, equation of state (EoS) parameters, that is, zero-pressure densities, bulk moduli, and their pressure/temperature dependencies of some of the high-pressure phases are summarized, as far as reliable data are available. We also review the phase transitions and density changes in lithologies associated with the subduction of slabs and also those in the surrounding model mantle materials on the basis of experimental results on multicomponent systems. Some implications for the mineralogy of the lower mantle are discussed based on these data and those obtained for the simpler systems.

3.2　Experimental and Theoretical Backgrounds

3.2.1　High-Pressure Technology

Two kinds of high-pressure devices, LHDAC and KMA, have been used to realize static high-pressure and high-temperature conditions of the lower mantle in the laboratory. The upper limit of high-pressure generation in LHDAC has been dramatically expanded using a smaller anvil top (culet) with various shapes for utilizing the potential hardness of single-crystal diamond. As a result, generation of pressures of multimegabars can now be comfortably produced in some laboratories. Moreover, quasihydrostatic pressures are also available by introducing gas pressure media, such as Ar or He, which also serve as thermal insulators in heating samples with laser beams.

The quality of heating of samples in LHDAC has also been substantially improved using various laser sources, such as YAG, CO_2, and YLF, with higher powers and more sophisticated computer-controlled feedback systems, as compared with the laser heating in early stages of its development in the 1970s and 1980s. These can now yield stable high temperature in a sample as small as $\sim$100 to several tens of micrometers in diameter and even smaller thickness.

Pressures produced in KMA studies (Kawai and Endo, 1970) using conventional tungsten carbide anvils have long been limited to 25–30 GPa, although the relatively large sample volume in this apparatus made it possible to precisely determine the phase transitions, some physical properties, melting temperatures, element partitioning, etc., of high-pressure phases. Temperatures up to 3000 K are also produced stably for hours or even a few days using various forms of heaters, such as C, $LaCrO_3$, TiC, WC, and some refractory metals. In addition, both temperature and pressure gradients within the sample in KMA are believed to be far smaller than those in LHDAC.

Introduction of harder materials, that is, sintered bodies of polycrystalline diamonds (SD) with some binders, such as Co and Si, as the second-stage anvils of KMA has dramatically changed this situation. Using relatively large SD anvil cubes of over 10 mm in edge length, some laboratories are now able to produce pressures approaching 60 GPa at high temperatures using KMA (e.g., Ito *et al.*, 2005), without sacrificing the advantage of the relatively large sample volumes in this apparatus. It is expected that pressures as high as $\sim$100 GPa may be produced in KMA, if SD anvils with larger dimensions are supplied on a commercial basis.

Applications of synchrotron radiation to both LHDAC and KMA started in the mid-1980s, when the second-generation synchrotron sources became available worldwide (e.g., Shimomura *et al.*, 1984). A combination of white X-ray and an energy-dispersive system has been used for KMA experiments at synchrotron facilities, because geometrical constraints

imposed by the tungsten carbide anvils and surrounding guide-block systems make it difficult to conduct angle-dispersive diffraction measurements. Use of the energy-dispersive method combined with a multichannel analyzer has the merit of rapid acquisition and analysis of X-ray diffraction data, so that realtime observations of phase transitions are possible under high pressure and high temperature. Although this method is not very suitable for the precise determination of crystal structures (due to relatively low spatial resolutions in the diffraction peak position and also to significant variations in background X-ray with the energy range), some attempts have been successfully made to make crystal structure refinements using a combined step scanning and energy-dispersive measurements (Wang *et al.*, 2004).

Identification of the phases present and precise determinations of the lattice parameters, and hence unit-cell volumes, of the high-pressure phases can be made by *in situ* X-ray diffraction measurements. The *in situ* pressure can also be monitored by the unit-cell volume changes in some reference materials, such as NaCl, Au, Pt, and MgO, using an appropriate EoS. Thus the phase boundaries and the P,V,T relations of a number of high-pressure phases relevant to lower-mantle mineralogy have been determined by *in situ* X-ray measurements using KMA, although there remain some uncertainties in the estimated pressures due to the lack of reliable pressure scales, as reviewed later, particularly at pressures of the deeper parts of the lower mantle. The effect of pressure on the electromotive force of a thermocouple is another unresolved issue, which may yield an additional uncertainty in the pressure estimation based on these EoS's.

Corresponding *in situ* X-ray observations have also been made using LHDAC. As the geometrical restrictions on the X-ray paths are not so severe in this device, X-ray diffraction is measured with an angle-dispersion method using monochromatized X-ray. The X-ray beam is focused to generally $\sim$10–20 μm with collimating mirrors, and directed to the disk-shape sample with diameters of $\sim$20–200 μm, depending on pressure ranges. YAG or ILF lasers with beam sizes of $\sim$10 to several tens of micrometers are used in most of the synchrotron facilities. By adopting imaging plates (IPs) for X-ray exposure combined with data processing systems, rapid data acquisition and reductions are also possible. Thus, the phase identification and measurements of unit-cell parameters of high-pressure phases can be made by the combination of LHDAC and synchrotron radiation at pressure and temperature conditions corresponding to the entire mantle (e.g., Murakami *et al.*, 2004a).

The major uncertainty in the *in situ* X-ray observations with LHDAC arises from the possible large temperature gradients in the small thin sample. Some studies demonstrated that variations of temperature are not very large in the radial direction of the disk-shape sample, suggesting that the temperature uncertainty may be of the order of ∼10% or less of the nominal values, if the diameter of the laser beam is significantly larger than that of the X-ray beam (Shen *et al.*, 2001). However, the temperature gradient in the axial direction of the sample can be substantially larger than this estimation (Irifune *et al.*, 2005), depending on the nature and thickness of the thermal insulator or pressure medium, as diamond has very high thermal conductivity. It should also be noted that the high thermal conductivity of diamond makes it difficult to maintain high temperatures greater than ∼2000 K at pressures of the Earth's core ($P > 300$ GPa) in LHDAC, as the thickness of the thermal insulator becomes so thin that the sample cannot be efficiently heated by laser.

3.2.2 *Ab Initio* Calculation

Ab initio approaches are those that solve the fundamental equations of quantum mechanics with a bare minimum of approximations. DFT is, in principle, an exact theory for the ground state and allows us to reduce the interacting many-electron problem to a single-electron problem (the nuclei being treated as an adiabatic background). A key to the application of DFT in handling the interacting electron gas was given by Kohn and Sham (1965) by splitting the kinetic energy of a system of interacting electrons into the kinetic energy of noninteracting electrons plus some remainder, which can be conveniently incorporated into the exchange-correlation energy.

The local density approximation (LDA) replaces the exchange-correlation potential at each point by that of a homogeneous electron gas with a density equal to the local density at the point. The LDA works remarkably well for a wide variety of materials, especially in the calculations of EoS's, elastic constants, and other properties of silicates. Cell parameters and bulk moduli obtained from well-converged calculations often agree with the experimental data within a few percent and ∼10%,

respectively. Agreement with the laboratory data is not perfect, however, and some systematic discrepancies are noted for some materials.

Attempts to improve LDA via introducing non-local corrections have yielded some success. The generalized gradient approximation (GGA; Perdew *et al.*, 1992, 1996) is a significantly improved method over LDA for certain transition metals (Bagno *et al.*, 1989) and hydrogen-bonded systems (Hamann, 1997; Tsuchiya *et al.*, 2002, 2005a). There is some evidence, however, that GGA improves the energetics of silicates and oxides but the structures can be underbound. The volume and bulk modulus calculated with GGA tend to be larger and smaller, respectively, than those measured experimentally (Hamann, 1996; Demuth *et al.*, 1999; Tsuchiya and Kawamura, 2001). Considering the thermal effect with zero-point motion, LDA provides the structural and elastic quantities much closer (typically within a few percent) to experimental values than those obtained with GGA. In addition, a discrepancy of about $10 \sim 15$ GPa is usually seen in transition pressures calculated with LDA and GGA (Hamann, 1996; Tsuchiya *et al.*, 2004a, 2004c), which provide lower and upper bounds, respectively. Experimental transition pressures are usually found between the values obtained with LDA and GGA, although GGA tends to provide the pressure with better fit to the experimental value than LDA (Hamann, 1996; Tsuchiya *et al.*, 2004a, 2004c). The main source of computational error can be attributed to how to treat the exchange-correlation potential.

The standard DFT has limitations in applying to Fe-bearing oxides and silicates in the following case. One-electron approximation with the standard DFT approaches fails to describe the electronic structure of Fe–O bonding correctly due to its strongly correlated behavior. Both LDA and GGA usually produce metallic bands for Fe–O bonding in silicates. They also do not provide the correct crystal field effects that break the d-orbital degeneracy. More sophisticated classes of technique, such as LDA + U, LDA + DMFT (dynamical mean-field theory), multireference configurational interaction, etc., are needed to treat the many-body effect of electrons more accurately and to investigate geophysically important iron-bearing systems.

Among these schemes, LDA + U (Anisimov *et al.*, 1991) is the most practical method for minerals under the current state of computer technology. The main problem of applying LDA + U to materials under pressure is the determination of the effective

Hubbard U parameter, meaning screened on-site Coulomb interaction. Tsuchiya *et al.* (2006) computed the effective U in magnesiowüstite $(Mg_{1-x}Fe_x)O$ in a nonempirical and internally consistent way based on a linear response approach for the occupancy matrix (Cococcioni and de Gironcoli, 2005). Thus, the *ab initio* LDA + U technique appears to open a new way to explore the mineral physics properties of iron-bearing systems relevant to the Earth's deep interior (*see* Chapter 13).

3.2.3   Pressure Scale

The construction of an accurate pressure standard is a critical issue in the quantitative measurements of mineral physics properties under high pressure. The pressure (*P*)–volume (*V*)–temperature (*T*) equation of state (PVT-EoS) of materials is most useful in evaluating the experimental pressures under the *P*,*T* conditions of the Earth's deep mantle. The EoS of a pressure standard is usually derived on the basis of a conversion of dynamical shock Hugoniot data to isothermal compression data. In principle, some parameters specifying the thermal properties of a solid are necessary for this conversion process of the Hugoniot. However, measurements of these parameters without any pressure and also temperature standards are virtually impossible. In all EoS's presently used as primary pressure standards, the conversion of the Hugoniot was therefore performed with some simple assumptions about the unknown high-pressure behavior of the conversion parameters. The most critical issue on the validity of such assumptions is the volume dependence of the thermodynamic Grüneisen parameter, which is a fundamental quantity characterizing the thermal effect on the material.

The characteristic properties of gold (Au), namely its low rigidity, simple crystal structure, chemical inertness, and structural stability, make it particularly suitable as a pressure standard under high *P*,*T* conditions, and it has therefore been used as a primary standard in many *in situ* X-ray diffraction studies (e.g., Mao *et al.*, 1991; Funamori *et al.*, 1996). However, some recent *in situ* experiments noted that the pressure values estimated by different thermal EoS's of gold show significant discrepancies. Using the EoS of gold proposed by Anderson *et al.* (1989), which is frequently used as the pressure scale in experiments using multianvil apparatus, Irifune *et al.* (1998a) first reported that the postspinel phase boundary of Mg_2SiO_4 shifted to about 2.5 GPa lower

than the pressure corresponding to the depth of the 660 km seismic discontinuity ($\sim$23.5 GPa and $\sim$2000 K). Similar results were obtained for other various minerals as summarized in Irifune (2002).

Tsuchiya (2003) predicted the thermal properties and the PVT-EoS of gold based on the *ab initio* theory including thermal effect of electrons. The state-of-the-art *ab initio* calculation showed that the relationship $\gamma/\gamma_0 = (V/V_0)^q$, assumed in some studies, is adequate for gold, at least up to $V/V_0 = 0.7$ and the predicted value of *q* was 2.15, which is intermediate between the values used in Heinz and Jeanloz (1984) and Anderson *et al.* (1989). According to this study, the *ab initio* EoS model reduced the discrepancies between the observed phase boundaries of spinel, ilmenite, and garnet, and the seismic discontinuity. However, a gap of about 0.7 GPa still remains between the postspinel transition pressure and that of the 660 km discontinuity. The similar conclusions with a slightly larger (1.0–1.4 GPa) discrepancy have also been obtained by an experimental study using an empirical PVT-EoS model of MgO to determine pressure (Matsui and Nishiyama, 2002).

Even for gold, there are several PVT models derived by different groups, which are still not mutually consistent. Such a problem is also found in EoS's of platinum (Jamieson *et al.*, 1982; Holmes *et al.*, 1989). Moreover, platinum appears to be unsuitable for the pressure scale in some cases, because of its reactivity with the sample or materials of the experimental cell at high *P*,*T* conditions (Ono *et al.*, 2005a). A similar problem is also encountered with the pressures obtained using MgO. The pressures obtained using EoS's of different materials therefore show a significant scatter. Akahama *et al.* (2002) determined the pressures based on room-temperature EoS's of several materials compressed simultaneously in a diamond-anvil cell and reported that the EoS of platinum proposed by Holmes *et al.* (1989) provided pressures more than 10 GPa higher than those calculated using EoS of gold as proposed by Anderson *et al.* (1989) at pressures of a megabar even at 300 K. A similar inconsistency has also been reported in pressure determination based on gold and silver EoS's (Akahama *et al.*, 2004).

Examples of the difference in pressures based on different EoS's of gold and platinum are shown in **Figure 1**, along a temperature (2300 K) close to the typical lower-mantle geotherm. The pressures evaluated based on various EoS's of gold differ by $\sim$2–3 GPa at the pressures of the uppermost parts of the lower mantle (25–30 GPa), and are within

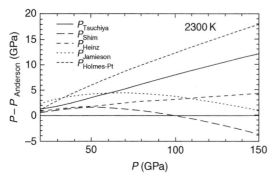

Figure 1 Differences between the pressures calculated using various EoS's of gold (Tsuchiya, 2003; Shim et al., 2002a; Heinz and Jeanloz, 1984; Jamieson et al., 1982) and those based on Anderson et al. (1989, horizontal line) as a function of pressure at 2300 K. The difference between the pressures using the Anderson scale and those using EoS of platinum by Holmes et al. (1989) is also shown for comparison.

5 GPa at pressures up to ∼60 GPa. However, the differences become substantially larger at higher pressures, reaching ∼15 GPa at the base of the lower mantle (136 GPa). It is also seen that the pressures based on an EoS of Pt (Holmes et al., 1989) are even higher than the highest pressure estimation by Tsuchiya (2003) by 5 GPa at 136 GPa. Thus the establishment of the mutually consistent pressure scales is urgently needed for more accurate experimental evaluation of phase transition and mineral physics under the lower-mantle P, T conditions.

3.3   Mineral-Phase Transitions in the Lower Mantle

3.3.1   Major Minerals in the Mantle and Subducted Slab

Phase transitions in Earth-forming materials dominate the structure and dynamics of the Earth. Major changes in seismic velocities traveling through the Earth's mantle can be generally attributed to the phase transitions of the constituent minerals, although some of them may be closely related to some chemical changes. Exploration and investigation of high-pressure phase transitions in mantle minerals have therefore been one of the major issues in studying the Earth's deep interior.

Here, we summarize the phase transitions in major minerals under lower-mantle conditions. Phase transitions in some relatively minor minerals relevant to the subducted slab lithologies are also reviewed in

the following section. As the experimental data on the phase transitions are still limited and controversial at P, T conditions of the lower mantle, we have tried to construct the most likely phase diagrams for the minerals with simple chemical compositions based on available laboratory data and *ab initio* calculations. Thermoelastic properties of some key high-pressure phases are also reviewed and summarized here, as far as experimental data or theoretical predictions are available.

3.3.1.1   MgSiO₃

The high-pressure orthorhombic perovskite polymorph of $MgSiO_3$ (Mg-Pv) is believed to be the most abundant mineral in the Earth's lower mantle. The possibility of a further phase transition of this phase under the lower-mantle P, T conditions has been controversial. Some studies suggested that Mg-Pv dissociates into an assemblage of SiO_2 and MgO at 70–80 GPa and 3000 K (Meade et al., 1995; Saxena et al., 1996) or that it undergoes a subtle phase change above 83 GPa and 1700 K (Shim et al., 2001b), while others claimed that Mg-Pv is stable almost throughout the lower mantle (e.g., Fiquet et al., 2000). However, more recent studies suggested that the result of Shim et al. (2001b) were due to misidentification of the diffraction peaks of a newly formed platinum carbide (Ono et al., 2005a). The dissociation of Mg-Pv into the oxides is also unlikely to occur in the Earth's mantle according to the subsequent experimental (Murakami et al., 2004a, 2005; Oganov and Ono, 2004) studies. Theoretical investigations also suggest that the dissociation should occur at extremely high pressure above 1 TPa (Umemoto et al., 2006).

Recently, the Pv to postperovskite (PPv) transition in $MgSiO_3$ was found to occur by *in situ* X-ray diffraction experiment using LHDAC and *ab initio* calculations at ∼2500 K and ∼125 GPa (Murakami et al., 2004a; Tsuchiya et al., 2004a; Oganov and Ono, 2004) (**Figure 2**), close to the $P–T$ conditions of the D″ layer near the core–mantle boundary (CMB). The Mg-PPv phase has a crystal structure identical to that of $CaIrO_3$ with a space group *Cmcm*. This structure consists of silica layers stacking along the *b*-direction and intercalated Mg ions. In the silica layers, SiO_6 octahedra connect sharing edges along the *a*-direction and sharing corners along the *c*-direction. Thus, this structure is more favorable at high pressure than the Pv structure, although the cation coordinations are basically same in both structures.

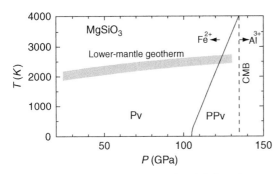

Figure 2 High P,T phase diagram for $MgSiO_3$ at lower-mantle pressures summarized based on the LHDAC experiments by Murakami *et al.* (2004a) and *ab initio* calculations by Tsuchiya *et al.* (2004a). Shaded area represents a typical lower-mantle geotherm based on Brown and Shankland (1981). Pv, perovskite; PPv, postperovskite; CMB, core–mantle boundary.

The PPv structure is expected to be highly anisotropic. It is more compressible along the *b*-direction perpendicular to the silica layers because only ionic Mg–O bonding exists in the interlayer spacing. The Pv and PPv structures therefore look very different in this respect. However, the structural relationship between Pv and PPv is quite simple and by applying shear strain ε_6, Pv can change to PPv directly (Tsuchiya *et al.*, 2004a). According to this relation, the *c*-direction remains unchanged via the structural transition. This suggests that nonhydrostaticity could significantly affect the transition kinetics of the Pv-to-PPv transition.

PPv's thermodynamic properties and the position and slope of the phase boundary were investigated by means of *ab initio* quasiharmonic free energy calculations (Tsuchiya *et al.*, 2004a, 2005b). The predicted Clapeyron slope of the Pv–PPv transition was $\sim 7.5 \, MPa \, K^{-1}$, which is remarkably close to that required for a solid–solid transition to account for the D″ discontinuity (Sidorin *et al.*, 1999). Thus the results of both experimental and theoretical studies suggest that the PPv should be the most abundant high-pressure phase in the D″ region.

Over the past decades, there have been a number of experiments to determine elastic property of Mg-Pv (*e.g.*, Yagi *et al.*, 1982; Mao *et al.*, 1991; Yeganeh-Haeri *et al.*, 1989; Ross and Hazen, 1990; Funamori *et al.*, 1996; Fiquet *et al.*, 2000). Among these experiments, early studies yielded relatively large variations for zero-pressure bulk modulus K_0 of Mg-Pv ranging from 254 GPa (Ross and Hazen, 1990) to 273 GPa (Mao *et al.*, 1991). However, more recent experiments of static compression and

Brillouin spectroscopy yielded mutually consistent K_0 values of $\sim 253–264$ GPa. Density functional calculations have reported similar but slightly smaller room-temperature bulk modulus for Mg-Pv of about 250 GPa (Wentzcovitch *et al.*, 2004; Tsuchiya *et al.*, 2004b, 2005b). This underbinding tendency for Mg-Pv is seen in standard density functional calculations but is much more prominent in GGA than in LDA (Wentzcovitch *et al.*, 2004).

In contrast to the extensive investigations of the EoS of Mg-Pv, the EoS of Mg-PPv has not been well constrained experimentally. According to a density functional prediction, the zero-pressure volume of Mg-PPv is very close to that of Mg-Pv (Tsuchiya *et al.*, 2004a, 2005b). However, K_0 and its pressure derivative of Mg-PPv are significantly smaller and larger than those of Mg-Pv, respectively, which implies the volume of Mg-PPv should be smaller than that of Mg-Pv at the relevant pressure range. The volume decrease associated with the Pv–PPv transition is estimated to be $\sim 1.5\%$ on the basis of *ab initio* calculations (Tsuchiya *et al.*, 2004a), which is consistent with those estimated based on LHDAC experiments.

3.3.1.2 $MgSiO_3$–$FeSiO_3$

Mg-Pv is supposed to incorporate 5–10 mol.% of $FeSiO_3$ in peridotitic compositions in the lower mantle (Irifune, 1994; Wood and Rubie, 1996; Katsura and Ito, 1996). However, experimental studies for the system $MgSiO_3$–$FeSiO_3$ have been limited to the pressures of the uppermost part of the lower mantle, except for an LHDAC study (Mao *et al.*, 1991). Recent developments in KMA with sintered diamond anvils substantially extend the pressure and temperature ranges for phase equilibrium studies (as reviewed earlier), and Tange (2006) extensively studied the phase relations and the Mg–Fe partitioning between coexisting Mg-Pv and magnesiowüstite (Mw) at pressures up to ~ 50 GPa and temperatures to 2300 K.

Immediately after the discovery of the Pv–PPv transition in $MgSiO_3$, the effects of iron on this transition were studied using LHDAC (Mao *et al.*, 2004, 2005). It was noted that the presence of iron significantly reduces the stability pressure of PPv, and PPv with up to ~ 80 mol.% of the $FeSiO_3$ component was synthesized at a pressure of ~ 140 GPa and at temperatures of ~ 2000 K (Mao *et al.*, 2005). Thus the phase relations in the system $MgSiO_3$–$FeSiO_3$ at pressure up to ~ 130 GPa and at a typical lower-mantle temperature can be drawn as illustrated in **Figure 3**, according to the results of these and

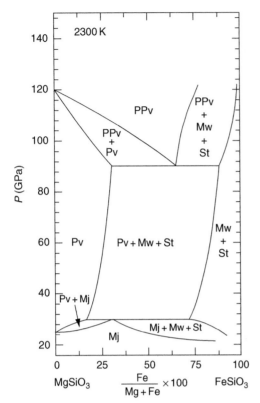

Figure 3 A predicted phase diagram for the system MgSiO$_3$–FeSiO$_3$ at lower-mantle pressures and at a representative lower-mantle temperature (2300 K). The relations below ∼30 GPa are estimated on the basis of Sawamoto (1987) and Ohtani et al. (1991) using KMA, while those above 80 GPa are based on LHDAC experiments by Mao et al. (2004, 2005). The boundaries between these pressures are depicted based on recent experimental data using KMA with SD anvils (Tange, 2006). Mj, majorite garnet; Mw, magnesiowüstite; St, stishovite.

Tange's studies. Although a significant effect of ferrous iron on the stability of Mg-PPv has been predicted by some theoretical works using GGA (Caracas and Cohen, 2005b; Stackhouse et al., 2006), the latest experiment showed much smaller effects (Sinmyo et al., 2006). Nevertheless, this issue has not been fully resolved yet and there remains large room for further theoretical and experimental investigations.

Effect of iron on the PPv transition has also been studied in the light of the high-pressure phase changes of hematite, corundum-type Fe$_2$O$_3$. Hematite transforms first to the high-pressure phase with the *Pbnm* Pv structure or Rh$_2$O$_3$(II) structures at 30 GPa, whose X-ray diffraction patterns are very similar to each other (Ono et al., 2005b). Further transitions in Fe$_2$O$_3$ to the structure assigned as the CaIrO$_3$-type structure has been reported to occur at a

transition pressure of ∼50 GPa (Ono et al., 2005b). This transition pressure is significantly lower than the Pv–PPv transition pressure in MgSiO$_3$. Thus the presence of iron in the trivalent state is also suggested to lower the pressure of the Pv–PPv transition, as is found for divalent iron.

3.3.1.3 MgSiO$_3$–Al$_2$O$_3$

Irifune (1994) demonstrated that Mg-Pv is the major host of aluminum in a pyrolite composition at pressures and temperatures of the uppermost parts of the lower mantle, possessing ∼4 mol.% of Al$_2$O$_3$. Phase relations in the system MgSiO$_3$–Al$_2$O$_3$ under the lower-mantle conditions have since been studied (Irifune et al., 1996a; Ito et al., 1998) with an emphasis on the MgSiO$_3$–Mg$_3$Al$_2$Si$_3$O$_{12}$ system. Irifune et al. (1996a) showed that majorite garnet with less than ∼15 mol.% Al$_2$O$_3$ transforms to the Pv structure via a mixture of these two phases, while this assemblage further changes to an assemblage of majorite plus corundum at pressures about ∼28 GPa, at 1800 K. This assemblage with the Mg$_3$Al$_2$Si$_3$O$_{12}$ composition was later shown to form almost pure Pv at ∼38 GPa using KMA with SD anvils (Ito et al., 1998).

Phase transitions in Al$_2$O$_3$ corundum under the lower-mantle condition were first studied by Funamori and Jeanloz (1997) using LHDAC based on earlier *ab initio* predictions (Marton and Cohen, 1994; Thomson et al., 1996), which demonstrated that corundum transforms to a new phase with the Rh$_2$O$_3$(II) structure at ∼100 GPa and at ∼1000 K. More recently, it has been predicted by *ab initio* studies that Al$_2$O$_3$ has a similar high-pressure phase relation to Fe$_2$O$_3$ (Caracas and Cohen, 2005a; Tsuchiya et al., 2005c; Stackhouse et al., 2005b). The Pv-to-PPv transition was thus suggested to occur in Al$_2$O$_3$ at about 110 GPa at 0 K, although the Pv phase is eclipsed by the stability field of the Rh$_2$O$_3$(II) phase. This Pv-to-PPv transition pressure in Al$_2$O$_3$ is about 10 GPa higher but fairly close to that in MgSiO$_3$, though Akber-Knutson et al. (2005) predicted a much larger effect of Al by estimating the solid solution energy. *In situ* X-ray diffraction experiments on pyrope compositions by LHDAC (Tateno et al., 2005) demonstrated a consistent result about the effect of aluminum incorporation on the PPv transition pressure. Therefore, we expect that though aluminum tends to increase the PPv transition pressure in MgSiO$_3$, the effect is not very significant. The plausible phase diagram of the system MgSiO$_3$–Al$_2$O$_3$ in the lower-mantle *P,T* condition is illustrated in **Figure 4**.

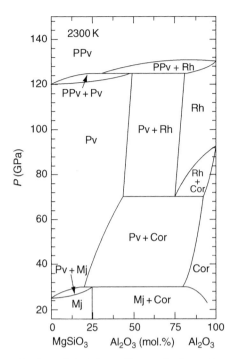

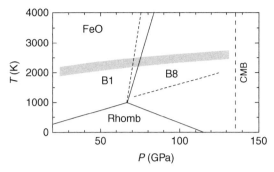

Figure 5 High P,T phase diagram for FeO at lower-mantle pressures summarized based on the *in situ* measurement by Fei and Mao (1994), Murakami *et al.* (2004b), and Kondo *et al.* (2004). B1, NaCl structure; B8, NiAs structure; Rhomb, rhombohedral B1 phase.

Figure 4 A predicted phase diagram for the system MgSiO$_3$-Al$_2$O$_3$ at lower mantle pressures and at 2300K. The phase relations below ∼40 GPa are based on KMA experiments (Irifune *et al.*, 1996a; Ito *et al.*, 1998), while those on Al$_2$O$_3$ are from Funamori *et al.* (1998) using LHDAC and *ab initio* calculations by Tsuchiya *et al.* (2005c). The results of Tateno *et al.* (2005) and Tsuchiya *et al.* (2005c) were also taken into account to illustrate the phase relations at 120-130 GPa. Cor, corundum; Rh, Rh$_2$O$_3$(II).

3.3.1.4 MgO–FeO

Mw, $(Mg_{1-x}Fe_x)O$, is believed to be the next major mineral phase in Earth's lower mantle after ferrosilicate Pv, $(Mg_{1-x}Fe_x)SiO_3$ (e.g., Helffrich and Wood, 2001). The magnesium end member of Mw, periclase, possessing the B1 (NaCl) structure is known to be an extraordinarily stable phase. No phase transitions in this material have been observed or predicted under the P,T conditions of the entire mantle (Duffy *et al.*, 1995; Alfé *et al.*, 2005), primarily due to substantially smaller ionic radius of magnesium relative to that of oxygen. FeO wüstite, on the other hand, transforms to an antiferromagnetic phase accompanied by a small rhombohedral distortion. This transition is a typical magnetic order–disorder transition and therefore the transition temperature corresponds to the Neel point. Although the rhombohedral B1 (rB1) phase is stable at low temperatures up to about 110 GPa, at higher pressure over 65 GPa and high temperatures, the rB1 phase transforms to the normal or inverse B8 (NiAs) structure, as shown in **Figure 5** (Fei and Mao, 1994;

Murakami *et al.*, 2004b; Kondo *et al.*, 2004). However, several important properties of this high-pressure phase such as high-temperature stability, structural details, and electronic property are still in debate (Fei and Mao, 1994; Mazin *et al.*, 1998; Murakami *et al.*, 2004b).

For the compositions between these two end members, some controversial experimental results have emerged: Mw with $X_{Fe} = 50\%$ was reported to dissociate into two components, Fe-rich and Mg-rich Mw's at 86 GPa and 1000 K (Dubrovinsky *et al.*, 2000). In contrast, another study using LHDAC found no dissociation of Mw with even higher X_{Fe} of 0.61 and 0.75 up to 102 GPa and 2550 K, though the sample with $X_{Fe} = 0.75$ showed a displacive transition to the rB1 structure at low temperature similar to FeO wüstite (Lin *et al.*, 2003). Thus, further experimental and theoretical studies are required to address the possible dissociation of Mw.

Another issue relevant to Mw, as well as the ferrosilicate Pv, which should affect thermoelastic properties of these phases, is the occurrence of electron spin transitions under the lower-mantle conditions. High spin (HS) to low spin (LS) transitions in iron have been observed by *in situ* X-ray emission spectroscopy (XES) and Mössbauer spectroscopy from 40 to 70 GPa in Mw containing about 18% of iron (Badro *et al.*, 2003; Lin *et al.*, 2005) and from 70 and 120 GPa in (Mg,Fe)SiO$_3$ Pv (Badro *et al.*, 2004; Li *et al.*, 2004; Jackson *et al.*, 2005) at room temperature. Several significant effects on thermochemical state of the lower mantle can be inferred by the spin transition of iron. The spin transition in Mw is accompanied by significant volume reductions (Lin *et al.*, 2005; Tsuchiya *et al.*, 2006) and changes in these minerals' optical absorption spectrum (Badro *et al.*, 2004). These can produce (1) seismic velocity anomalies, (2) variations in $Mg-Fe^{2+}$ partitioning between Mw and Pv, (3) changes in radiative heat conductivity, and (4) compositional layering (Gaffney and Anderson, 1973; Badro *et al.*, 2003, 2004). The elastic signature of this transition in Mw has been partially explored (Lin *et al.*, 2005), but there is still much uncertainty. In contrast, anomalous compression behavior has not yet been observed in $(Mg,Fe)SiO_3$ Pv.

The strongly correlated behaviour of iron oxide has deterred the quantification of these changes by density functional calculations based on the local spin density (LSDA) and spin polarized generalized gradient approximations (σ-GGA). These approaches incorrectly predict a metallic HS ground state and then successive spin collapses across the transition as reported for FeO (Sherman and Jansen, 1995; Cohen *et al.*, 1997). More recently, a new model explaining the mechanism of HS-to-LS transition of iron in Mw has been proposed based on calculations using more sophisticated $LDA + U$ technique that describe the electronic structure of strongly correlated system more correctly (Tsuchiya *et al.*, 2006). In this study, the effective Hubbard U parameter has been optimized at each volume and at each iron concentration up to X_{Fe} of 18.75% in an internally consistent way. As a result, it has been demonstrated that the large stability field of HS/LS mixed state appears at high temperatures instead of the intermediate spin state, due to the contributions of coexisting HS/LS mixing entropy and magnetic entropy. According to this transition mechanism, Mw is expected to be in this HS/LS mixed state for almost the entire range of lower-mantle P,T conditions, with the proportion of HS iron decreasing continuously with increasing pressure (**Figure 6**). No discontinuous change in any physical properties would appear associated with the spin transition in Mw within this range.

It has been reported that the spin transition pressure significantly increases with increasing X_{Fe}. A volume decrease associated with the spin transition was observed in $(Mg_{0.4}Fe_{0.6})O$ at 95 GPa (Lin *et al.*, 2005), while no spin transition has so far been reported in pure FeO up to 143 GPa (Badro *et al.*, 1999). Iron–iron interactions, which are no longer negligible at X_{Fe} higher than 20%, might also reinforce the magnetic moment significantly, although the mechanism of this tendency has not been fully understood to date.

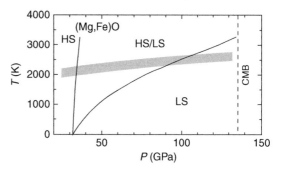

Figure 6 High P,T spin transition diagram for Mw with iron concentration of 18.75% at lower-mantle pressures predicted on the basis of $LDA + U$ calculations by Tsuchiya *et al.* (2006). HS, high spin; LS, low spin.

3.3.1.5   CaSiO₃

The lower mantle is believed to consist mainly of $(Mg,Fe)SiO_3$ Pv and Mw, with some $CaSiO_3$ perovskite (Ca-Pv) up to 7–8 vol.% (e.g., Irifune, 1994). Despite its importance, there are many unanswered questions about the structure, stability, the EoS, and other physical properties of Ca-Pv under pressure and temperature, which complicate some attempts to model the mineralogy of the lower mantle (Stacey and Isaak, 2001).

$CaSiO_3$ crystallize to the Pv structure over 10–13 GPa, depending on temperature, and is known to be unquenchable at ambient conditions. At lower-mantle conditions, $CaSiO_3$ has an ideal cubic Pv structure, while at lower temperatures it is suggested to be slightly distorted. The small degree of the possible distortion is hardly observed by current high-temperature and high-pressure X-ray techniques, and several orthorhombic and tetragonal structures have been proposed, based on *in situ* X-ray diffraction measurements (Shim *et al.*, 2002b; Kurashina *et al.*, 2004; Ono *et al.* 2004) or theoretical calculations (Stixrude *et al.*, 1996; Chizmeshya *et al.*, 1996; Magyari-Köpe *et al.*, 2002; Caracas *et al.*, 2005).

EoS of Ca-Pv has been determined up to CMB pressures by different groups (e.g., Mao *et al.*, 1989; Tamai and Yagi, 1989; Wang *et al.*, 1996; Shim *et al.*, 2000a, 2000b, 2002b; Kurashina *et al.*, 2004; Ono *et al.*, 2004; Shieh *et al.*, 2004). The fitting of the experimental results by third-order Birch–Murnaghan EoS, yielded a unit-cell volume, $V_0 = 45.54$ Å^3, bulk modulus, K_0, ranging from 232 to 288 GPa, and its pressure derivative, K_0', within 3.9–4.5. Most of the recent results with careful removing of the effect of deviatoric stress on the produced pressure, however, yielded the lower end values (232–236 GPa; Wang *et al.*, 1996; Shim *et al.*, 2000b) in this range. The results of corresponding *ab initio* studies (e.g., Wentzcovitch *et al.*, 1995; Chizmeshya *et al.*, 1996; Stixrude *et al.*, 1996; Karki and Crain, 1998) are similar to those obtained for the experimental data. Most of these studies, both experimental and theoretical, have focused on the behavior of the cubic modification of CaSiO$_3$ under pressure.

Caracas *et al.* (2005) performed a detailed investigation of the major symmetry-allowed modifications of CaSiO$_3$ obtained as distortions from the parent cubic phase by means of *ab initio* pseudopotential theory. They examined nine modifications having different symmetries and reported that the I4/mcm phase is the most likely stable static atomic configuration up to about 165 GPa. Enthalpy difference between this I4/mcm and the cubic Pv phase increased with increasing pressure, indicating that the I4/mcm structure becomes more stable relative to the cubic structure at higher pressure. The bulk modulus was estimated to be about 250 GPa for all modifications with the exception of R-3c structure. This theoretical K_0 is fairly similar to recent experimental values (Wang *et al.*, 1996; Shim *et al.*, 2002b), but much smaller than those reported in earlier nonhydrostatic experiments (Mao *et al.*, 1989; Tamai and Yagi, 1989).

Some studies focused on the high-temperature phase change from low symmetry phase to cubic Ca-Pv. Ono *et al.* (2004) reported that this transition occurs at about 600–1200 K at 25–120 GPa, where the transition temperature increased with increasing pressure, consistent with the theoretical prediction (Caracas *et al.*, 2005). These temperatures are much lower than the typical lower-mantle geotherm of about 2000–2500 K, suggesting that CaSiO$_3$ may have the cubic form throughout the actual lower mantle (**Figure 7**). However, most recent *ab initio*

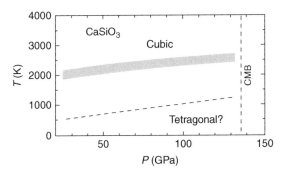

Figure 7 High *P,T* phase diagram for CaSiO$_3$ at lower-mantle pressures summarized based on the LHDAC experiments by Kurashina *et al.* (2004) and Ono *et al.* (2004).

molecular dynamics studies (Li *et al.*, 2006a, 2006b) reported very different results. They found the tetragonal phase stable even at the mantle temperatures in addition to the low-temperature stability of the orthorhombic phase. Their calculated elasticity of Ca-Pv is also very different from earlier results (Karki and Crain, 1998), particularly with respect to the shear modulus. The cause of the discrepancies is unclear.

3.3.1.6 SiO$_2$

Recent theoretical studies suggested that a second-order displacive phase transition from stishovite to the CaCl$_2$-type structure occurs at 50–60 GPa at room temperature (Kingma *et al.*, 1995). It has also been predicted that the CaCl$_2$-type silica undergoes a further structural transition to the α-PbO$_2$ phase (Dubrovinsky *et al.*, 1997; Karki *et al.*, 1997a). The results of the experimental studies on these issues, however, have been controversial. LHDAC studies reported that the CaCl$_2$-type phase persists at least up to 120 GPa (Andrault *et al.*, 1998), and the transition to the α-PbO$_2$ phase occurs at 121 GPa and 2400 K (Murakami *et al.*, 2003). In contrast, another similar experiment showed that the α-PbO$_2$-like phase was formed from cristobalite above 37 GPa at room temperature, and that stishovite directly transformed to the α-PbO$_2$-type structure above 64 GPa at 2500 K with a negative Clapeyron slope (Dubrovinsky *et al.*, 2001). In addition, Sharp *et al.* (1999) found the α-PbO$_2$-type phase in a natural meteorite sample, which would have experienced very low shock pressure below 30 GPa. These discrepancies may come from various difficulties in LHDAC experiments, such as associated with kinetic problems, temperature or pressure uncertainties, effect of different starting materials, etc.

During the last decade, a series of theoretical studies (Kingma *et al.*, 1995; Dubrovinsky *et al.*, 1997; Karki *et al.*, 1997a) also addressed this issue of the post-stishovite phase transitions in the framework of the *ab initio* calculations. These early theoretical studies were limited to static conditions, the calculations being performed at $T = 0$ K, without considering the zero-point energy. To investigate the contradictory experimental results on the high-temperature phase stability of SiO$_2$ under high pressure, finite temperature thermal effect on the transitions obtained by these static calculations should be taken into account.

Tsuchiya *et al.* (2004c) predicted the high-pressure and high-temperature phase equilibrium of three ordered modifications of SiO_2 using *ab initio* density functional perturbation theory and the quasiharmonic approximation. The predicted stishovite-$CaCl_2$ phase transition boundary is $P = 56 + 0.0059\,T$ (K) GPa, which is consistent with the results of LHDAC experiment by Ono *et al.* (2002). This predicted slope of the stishovite–$CaCl_2$ boundary of about $5.9\,MPa\,K^{-1}$ is close to but slightly larger than the earlier rough estimate of $4\,MPa\,K^{-1}$ (Kingma *et al.*, 1995). The LHDAC experiment resulting in a stishovite–α-PbO_2 boundary with a negative Clapeyron slope (Dubrovinsky *et al.*, 2001), which disagrees with the phase diagram based on Tsuchiya *et al.* (2004c), as shown in **Figure 8**. This disagreement may be due to experimental uncertainties and/or to kinetic problems in the former LHDAC experiment. These are supported by later calculations by Oganov *et al.* (2005), though they proposed substantially lower transition pressures primarily due to the application of LDA (see Section 3.2.2).

On the other hand, the phase transition boundary between $CaCl_2$ and α-PbO_2 in SiO_2 is predicted to be $P = 106.3 + 0.005\,79\,T$ (K) GPa based on *ab initio* calculations (Tsuchiya *et al.*, 2004c), which locates near the lowermost-mantle P,T conditions. This calculation also indicates that the α-PbO_2-type phase is the stable form of silica at depths down to the CMB, consistent with the result of an LHDAC experiment (Murakami *et al.*, 2003). These theoretical and experimental results also suggest that the α-PbO_2-type phase silica recently discovered in the meteorite sample might have been formed by a metastable reaction.

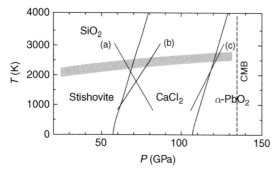

Figure 8 High P,T phase diagram for SiO_2 at lower-mantle pressures predicted by the *ab initio* calculations by Tsuchiya *et al.* (2004c). The results of *in situ* X-ray experiments using LHDAC by (a) Dubrovinsky *et al.* (2001), (b) Ono *et al.* (2002), and (c) Murakami *et al.* (2003) are also shown.

3.3.1.7 *Al-rich phase*

Majorite garnet is the main host of aluminum in the mantle transition region in both pyrolite and basaltic compositions. It transforms to an assemblage of Mg-Pv + Ca-Pv at pressures corresponding to the uppermost lower mantle. Aluminum is incorporated mostly in Mg-Pv in pyrolite composition (Irifune, 1994), whereas a separate aluminous phase is formed in basaltic compositions under the P,T conditions of the lower mantle, as demonstrated by Irifune and Ringwood (1993).

The aluminous phase, named as 'Al-rich phase' by these authors, was suggested to have a crystal structure similar to but not completely identical to the calcium ferrite structure. This Al-rich phase was later proposed to have a hexagonal structure or NAL-phase (Miyajima *et al.*, 2001; Akaogi *et al.*, 1999; Sanehira *et al.*, 2005), whereas others proposed that this phase possesses the calcium ferrite structure (Kesson *et al.*, 1998; Hirose *et al.*, 1999; Ono *et al.*, 2005d). Although the stability relations of these two phases in MORB compositions are not very clear to date, only very minor effects on the mineralogy and dynamics in the lower mantle is expected by the possible misidentification of these two phases under the lower-mantle conditions (Sanehira *et al.*, 2005; Shinmei *et al.*, 2005), as the crystal structures of these two phases are quite similar and yield only slight difference in densities.

3.3.2 Minor Minerals

3.3.2.1 *$MgAl_2O_4$, $NaAlSiO_4$*

$MgAl_2O_4$ spinel is known to decompose to simple oxides of MgO and Al_2O_3 at pressure and temperatures of mantle transition region, which recombine to form a calcium ferrite ($CaFe_2O_4$, CF) type phase at about 25 GPa (Irifune *et al.*, 1991; Funamori *et al.*, 1998; Akaogi *et al.*, 1999). Although the formation of an unknown phase named ε-phase (Liu, 1978) was reported at a similar pressure at 1300 K in LHDAC experiments, none of the subsequent LHDAC and KMA experiments using both quench and *in situ* X-ray measurements have confirmed this phase. Instead, it has been shown that the CF-phase further transforms to a calcium titanate ($CaTiO_4$, CT) phase at pressures of $\sim$40–45 GPa (Funamori *et al.*, 1998). An *ab initio* periodic LCAO (linear combination of atomic orbitals) calculation by Catti (2001) demonstrated that the calcium titanate structure is indeed stable relative to the calcium ferrite structure at pressures greater than ~39–57 GPa.

NaAlSiO$_4$ also adopts the calcium ferrite structure at pressures above 25 GPa (Liu, 1977; Akaogi *et al.*, 1999), but the formation of the complete solid solutions between this and MgAl$_2$O$_4$ has not been reported, as there is a region of the hexagonal phase in the intermediate region of these end-member compositions (Akaogi *et al.*, 1999; Shinmei *et al.*, 2005). It was demonstrated that the CF-type NaAlSiO$_4$ is stable at least at pressures up to 75 GPa and temperatures to 2500 K on the basis of LHDAC experiments (Tutti *et al.*, 2000).

3.3.2.2 KAlSi$_3$O$_8$, NaAlSi$_3$O$_8$

KAlSi$_3$O$_8$-rich feldspar, an important mineral in K-rich basalt (e.g., Wang and Takahashi, 1999) and continental crust and marine sediment lithologies (Irifune *et al.*, 1994), transforms to the hollandite structure via a mixture of K$_2$Si$_2$O$_5$ wadite + Al$_2$O$_3$ kyantite + SiO$_2$ coesite at about 9 GPa. KAlSi$_3$O$_8$ hollandite plays an important role in fractionation of some trace elements because of its peculiar tunnel structure that accommodates large ion lithophile elements (Irifune *et al.*, 1994). Although an LHDAC study suggested that this structure is stable almost throughout the lower-mantle *P,T* conditions (Tutti *et al.*, 2001), recent *in situ* X-ray diffraction studies using DAC with a helium pressure medium at room temperature (Ferroir *et al.*, 2006) and KMA at high pressure and high temperature demonstrated that the hollandite transforms to an unquenchable phase, named as hollandite II (Sueda *et al.*, 2004), at about 22 GPa at room temperature with a positive Clapeyron slope. Although only a slight modification in crystal structures between these phases was noted (Ferroir *et al.*, 2006), this transition may significantly affect partitioning of some trace elements between the K-hollandite and coexisting melts in the lower mantle.

NaAlSi$_3$O$_8$-rich hollandite was found in shock veins of some meteorites (e.g., Tomioka *et al.*, 2000; Gillet *et al.*, 2000). However, attempts to reproduce the hollandite with such compositions by high-pressure experiments have failed (Yagi *et al.*, 1994; Liu, 2006), as the solubility of this component is limited to about 50 mol.% at pressures of ~22 GPa and at temperatures up to 2500 K. Thus this phase could have been formed metastably in a very short period of time under shock compression in the parental bodies of these meteorites.

3.3.2.3 CAS phase

The 'CAS phase' was first described by Irifune *et al.* (1994) as a new Ca- and Al-rich high-pressure phase in a continental crust composition at pressures above ~15 GPa, which was suggested to be a major host for aluminum and calcium in the subducted marine sediments in the mantle transition region. Subsequent experimental studies demonstrated that this phase has the ideal composition of CaAl$_4$Si$_2$O$_{11}$, possessing a hexagonal barium ferrite-type structure with space group P6$_3$/mmc (Gautron *et al.*, 1999). Moreover, this high-pressure phase with a composition of (Ca$_x$Na$_{1-x}$)Al$_{3+x}$Si$_{3-x}$O$_{11}$ was recently discovered in a shergottite shocked Martian meteorite in association with stishovite and/or K, Na-rich hollandite (Beck *et al.*, 2004), both of which are known to be stable only at pressures above ~9 GPa.

The CAS phase is suggested to have silicon in fivefold coordination in a trigonal bipyramid site at high pressure and high temperature, which is supposed to decompose into fourfold and sixfold coordinations upon quenching and subsequent release of pressure (Gautron *et al.*, 1999). *In situ* Raman spectroscopy and X-ray diffraction measurements actually indicated the formation of fivefold coordinated silicon under pressure, which should play an important role in the transport properties of minerals through the formation of oxygen vacancies (Gautron *et al.*, 2005).

3.3.2.4 Phase D, δ-AlOOH

Phase D was first noted by Liu (1986) as a new dense hydrous magnesium (DHMS) phase in serpentine, which was later confirmed by *in situ* X-ray diffraction (Irifune *et al.*, 1996b). Both X-ray power diffraction profile and the chemical composition of this phase were also refined on the quenched sample (Irifune *et al.*, 1996b; Kuroda and Irifune, 1998). Two groups subsequently succeeded in refining its crystal structure independently (Yang *et al.*, 1997; Kudoh *et al.*, 1997). The stability of phase D has since been studied experimentally using both KMA (Ohtani *et al.*, 1997; Irifune *et al.*, 1998b; Frost and Fei, 1998) and LHDAC (Shieh *et al.*, 1998), which demonstrated that this phase has a wide stability field up to 40–50 GPa, at temperatures to ~1800 K, whereas it dehydrates to form an assembly containing Mg-Pv and Mw at higher temperatures (Shieh *et al.*, 1998).

Serpentine is the major hydrous mineral in the subducted slab, and phase D should be the only possible DHMS present in the upper part of the lower mantle transported via the subduction of slabs

(Irifune *et al.*, 1998b; Shieh *et al.*, 1998; Ohtani *et al.*, 2004), although the newly found δ-AlOOH (Suzuki *et al.*, 2000) could be an alternative water reservoir in the lower mantle under very limited Al-rich circumstances. For both phase D and δ-AlOOH, structural changes associated with hydrogen bond symmetrization are expected to occur on the basis of *ab initio* calculations (Tsuchiya *et al.*, 2002, 2005a), which should affect the compressional behavior and hence density changes of these hydrous phases in the lower mantle.

3.3.2.5 *MgCO$_3$, CaCO$_3$*

Carbonates are important constituents of pelagic sediments, parts of which are supposed to subduct into the mantle. It has been shown that $MgCO_3$ magnesite is the major carbonate in the mantle (e.g., Biellmann *et al.*, 1993), whose stability under high pressure has been studied using LHDAC. Magnesite was reported to be stable at pressures up to 80 GPa, almost throughout the lower mantle (Gillet, 1993, Fiquet *et al.*, 2002), but a recent *in situ* X-ray diffraction study demonstrated it transforms to an unknown phase (magnesite II) at pressures above $\sim$115 GPa, at 2000–3000 K (Isshiki *et al.*, 2004). Although the dissociation of magnesite into assemblages of $MgO + CO_2$ (Fiquet *et al.*, 2002) or $MgO + C + O_2$ (Liu, 1999) was suggested on the basis of thermodynamic considerations, such reactions are unlikely to occur along appropriate geotherms in the lower mantle (Isshiki *et al.*, 2004).

In contrast, certain amounts of $CaCO_3$ could survive in the subducted slabs without decarbonation or reaction with surrounding minerals, due to low temperatures of the slabs, and thus are delivered into the lower mantle. $CaCO_3$ adopts the aragonite structure under *P,T* conditions of the uppermost mantle, which was recently found to transform to a high-pressure form with an orthorhombic symmetry at pressures greater than $\sim$40 GPa and at temperatures 1500–2500 K using LHDAC (Ono *et al.*, 2005c). Although the possibility of the transformation of $CaCO_3$ aragonite to a trigonal phase is suggested on the basis of DAC experiments at room temperature (Santillán and Williams, 2004), this phase could have been metastably formed, and the orthorhombic post-aragonite phase should be stable at least to depths of $\sim$2000 km in the lower mantle (Ono *et al.*, 2005c). Thus, $MgCO_3$ and possibly $CaCO_3$ are the potential hosts of CO_2 throughout most parts of the lower mantle, except for the bottom parts of the D$''$ layer (Isshiki *et al.*, 2004).

3.4 Phase Transitions and Density Changes in Mantle and Slab Materials

3.4.1 Chemical Compositions and Density Calculations

Subducting oceanic lithosphere is modeled by layers of basaltic oceanic crust of $\sim$6 km thickness, underlain by thicker layers ($\sim$50–100 km) of residual harzburgite and fertile lherzolite, which are covered with thin ($\sim$1 km) terrigeneous and/or pelagic sediments. Typical chemical compositions of these lithologies are listed in **Table 1**. Most parts of the sedimentary materials are believed to be trapped to form accretion terrains underneath island arcs upon subduction of slabs at ocean trenches, although geochemical evidence suggests that certain parts of such materials may be subducted deeper into the mantle (e.g., Loubet *et al.*, 1988). At least part of the bottom warmer lherzolite layer of a slab may also be assimilated to the surrounding mantle during subduction in the upper mantle and mantle transition region, and thus the slab approaching the 660 km seismic discontinuity can reasonably be modeled by a layered structure of basaltic and harzburgitic rocks (Ringwood and Irifune, 1988).

The chemical composition of the lower mantle has been a major controversial issue in the mineralogy of the Earth's interior. Some (e.g., Ringwood, 1962) believe peridotitic or pyrolitic materials are dominant in the whole mantle, while others (e.g., Liu, 1982; Anderson, 1989; Hart and Zindler, 1986) claim that more Si-rich chondritic materials should be representative for the composition of the lower mantle (**Table 1**). The difference is based on rather philosophical arguments on the origin and subsequent differentiation processes of the Earth, which are critically dependent on the models of condensation/evaporation processes of elements and compounds in the primordial solar system and the possible formation of deep magma ocean in the early stage of the formation of the Earth. As the elastic properties, particularly those related to shear moduli, of high-pressure phases have not been well documented under the pressure and temperature conditions of the lower mantle, it is hard to unambiguously evaluate the feasibility of these two alternative composition models in the light of mineral physics and seismological observations (e.g., Bina, 2003; Mattern *et al.*, 2005). Moreover, the knowledge of variation of temperature with depth is vital to address this issue, which also has not been well constrained in the lower mantle.

Table 1 Representative chemical compositions of lower mantle and those related to subducting slabs

| | Lower mantle | | | | |
	Chondrite	Pyrolite	Harzburgite	MORB	Continental crust
SiO_2	53.8	44.5	43.6	50.4	66.0
TiO_2	0.2	0.2		0.6	0.5
Al_2O_3	3.8	4.3	0.7	16.1	15.2
Cr_2O_3	0.4	0.4	0.5		
FeO	3.5	8.6	7.8	7.7	4.5
MgO	35.1	38.0	46.4	10.5	2.2
CaO	2.8	3.5	0.5	13.1	4.2
Na_2O	0.3	0.4		1.9	3.9
K_2O		0.1		0.1	3.4

MORB, mid-ocean ridge basalt.
Chondrite, Liu (1982); pyrolite, Sun (1982); harzburgite, Michael and Bonatti (1985); MORB, Green *et al.* (1979); continental crust, Taylor and McLennan (1985).

Table 2 PVT-EoS parameters of lower-mantle phases determined from various experimental and theoretical data and their systematics

	Mg-Pv	Fe-Pv	Al_2O_3-Pv	Mg-PPv	Ca-Pv	MgO	FeO	SiO_2 (St)	SiO_2 (α-PbO_2)
V_0 ($cm^3 mol^{-1}$)	24.45	25.48	24.77	24.6	27.45	11.36	12.06	14.02	13.81
B_0 (GPa)	257	281	232	226	236	158	152	314	325
B'	4.02	4.02	4.3	4.41	3.9	4.4	4.9	4.4	4.2
Θ_D (K)	1054	854	1020	1040	984	725	455	1044	1044
γ	1.48	1.48	1.48	1.55	1.53	1.52	1.28	1.34	1.34
q	1.2	1.2	1.2	1.2	1.5	1.5	1.5	2.4	2.4

Mg-Pv (Shim and Duffy, 2000; Fiquet *et al.*, 2000; Sinogeikin *et al.*, 2004; Tsuchiya *et al.*, 2004a, 2005b), Fe-Pv (Jeanloz and Thompson, 1983; Parise *et al.*, 1990; Mao *et al.*, 1991; Kiefer *et al.*, 2002), Al_2O_3-Pv (Thomson *et al.*, 1996; Tsuchiya *et al.*, 2005c), Mg-PPv (Tsuchiya *et al.*, 2004a, 2005b), Ca-Pv (Wang *et al.*, 1996; Shim *et al.*, 2000b; Karki and Crain, 1998), MgO (Fiquet *et al.*, 1999; Sinogeikin and Bass, 2000), FeO (Jackson *et al.*, 1990; Jacobsen *et al.*, 2002; systematics), SiO_2 (Ross *et al.*, 1990; Andrault *et al.*, 2003; Karki *et al.*, 1997a; Tsuchiya *et al.*, 2004c).
High-*T* Birch–Murnaghan equation was applied only for the hexagonal aluminous phase with parameters $V_0 = 110.07 \, cm^3 mol^{-1}$, $B_0 = 185.5$ GPa, B' = 4 (fix), $dB/dT = -0.016$ GPa K^{-1} and $\alpha_0 = 3.44 \times 10^{-5} K^{-1}$ (Sanehira *et al.*, 2005).

Here, we assume the whole mantle is of a pyrolitic composition to address the phase transitions and associated density changes in the lower mantle, as there are significant variations in the chondritic/cosmochemical models and also because high-pressure experimental data on the latter compositions have been scarce to date. We also assume the major lithologies transported into the lower mantle via subduction of slabs are of mid-ocean ridge basalt (MORB) and harzburgite compositions. Phase transitions in MORB have been extensively studied down to the depths near the mantle–core boundary. In contrast, although virtually no experimental data exist on harzburgite compositions in this depth region, they can be reasonably estimated from those available on the high-pressure phases with simple chemical compositions, as harzburgitic

compositions have a minor amount of Fe and only very small amounts of Ca, and Al, and are well approximated by the MgO (FeO)–SiO_2 system.

We calculated the density changes in pyrolite, harzburgite, and MORB compositions using available experimental data as follows. The densities of the individual high-pressure phases that appeared in these lithologies were calculated at given pressures using the thermal EoS combining third-order Birch–Murnaghan EoS and Debye theory along an appropriate geotherm, using the PVT-EoS parameters given in **Table 2**. The resultant density changes of individual phases along the geotherm are depicted in **Figure 9**. The density changes in the bulk rocks were then calculated using the proportions of the individual phases with pressure along the geotherm.

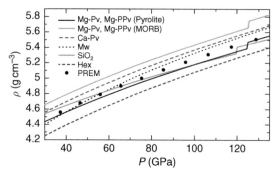

Figure 9 Density changes in major minerals constituting pyrolite and MORB compositions as a function of pressure along the adiabatic geotherm, calculated with EoS using the mineral physics parameters listed in **Table 2**. The density changes of Mg-Pv and Mw in the harzburgite composition are very close to but slightly lower than those of the corresponding phases in pyrolite, which are not shown in this figure. Dots represent the densities in PREM. Hex, hexagonal aluminous phase.

Although some recent studies suggested that sub-adiabatic temperature gradients are required to match the observed and calculated density and bulk sound velocity for pyrolitic compositions (Bina, 2003; Mattern *et al.*, 2005), we simply assumed adiabatic temperature changes throughout the lower mantle (i.e., 1900 K at 660 km and 2450 K at 2890 km with an averaged gradient of $dT/dz = {\sim}0.3\,K\,km^{-1}$; e.g., Brown and Shankland, 1981) as such conclusions are not robust, given the uncertainties in both mineral physics measurements and seismological observations and also due to inaccessibility of shear properties of high-pressure phases. In addition, significantly sharp temperature increases are expected to occur near the mantle–core boundary and presumably near the 660 km discontinuity, as these regions are accompanied by chemical changes and form thermal boundary layers (which are also not taken into account in the present calculations).

The mineral proportion changes in pyrolite, harzburgite, and MORB compositions are shown in **Figure 10**, while the calculated density changes of these lithologies are depicted in **Figure 11**. The density changes at pressures lower than 30 GPa are based on an earlier estimate of Irifune (1993), using the similar method and mineral physics parameters. Although the density change in pyrolite seems to agree well with that of PREM (Dziewonski and Anderson, 1981), the latter estimate inevitably has significant uncertainties, as this density profile is rather indirectly determined from seismic velocities with some assumptions. The calculated density

values may also have significant errors mainly due to the uncertainty in the geotherm stated in the above. Nevertheless, mineral physics parameters to constrain the density have been reasonably well determined, mostly on the basis of *in situ* X-ray diffraction measurements, and the differences among these calculated density profiles are regarded as robust results.

3.4.2 Phase and Density Relations

3.4.2.1 Pyrolite

Figure 10(a) illustrates the phase transitions in pyrolite as a function of depth. Pyrolite transforms from an assemblage of ringwoodite (Rw) + majorite garnet (Mj) + Ca-Pv under the P,T conditions of the mantle transition region to that of Mg-Pv + Ca-Pv + Mj + Mw at depths near the 660 km seismic discontinuity. The spinel to post-spinel transition in this composition has actually been believed to occur at pressures near 23.5 GPa, at a temperature of ${\sim}2000\,K$, on the basis of quench experiments (e.g., Irifune, 1994). The slope of this phase transition boundary has also been determined by both quench experiments (Ito and Takahashi, 1989) and calorimetric measurements (Akaogi and Ito, 1993), yielding a value of $c.\ -3\,MPa\,K^{-1}$. Although some recent studies suggested somewhat lower transition pressures and larger Clapeyron slope, the spinel to postspinel transition is generally believed to be the main cause of the 660 km discontinuity, which is followed by the smeared-out transition of majorite to Pv over a pressure interval of ${\sim}2\,GPa$ (Irifune, 1994; Nishiyama *et al.*, 2004).

Mg-Pv is stable throughout most regions of the lower mantle, but it is found to transform to the Mg-PPv with the $CaIrO_3$ structure in a peridotite composition at pressures close to those near the top of the D″ layer. A small density jump of about 1–1.5% is expected to occur associated with this transition in the lower mantle. Ca-Pv, on the other hand, is likely to remain in the Pv structure throughout the lower mantle. On the other hand, Mw remains in the rock salt (B1) structure at pressures to ${\sim}80\,GPa$, where it may transform to the NiAs (B8) structure if iron concentration is very high (Fei and Mao, 1994). An HS-to-LS transition has also been suggested to occur in this phase over the pressure range of the entire lower mantle, as shown in **Figure 6**, which may yield an additional gradual density increase over this interval and also affect partitioning of iron between this phase and the ferrosilicate Pv.

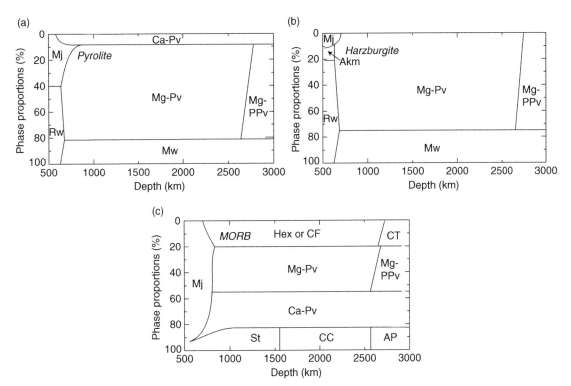

Figure 10 Mineral proportion changes in (a) pyrolite, (b) harzburgite, and (c) MORB as a function of depth along the adiabatic geotherm. Data source: Irifune and Ringwood (1987, 1993); Irifune (1994); Hirose *et al.* (1999, 2005); Murakami *et al.* (2004a, 2005); Ono *et al.* (2001, 2005d). Akm, akimotoite; Rw, ringwoodite; CF, calcium-ferrite phase; CT, calcium-titanite phase; CC, $CaCl_2$ phase; AP, α-PbO_2 phase.

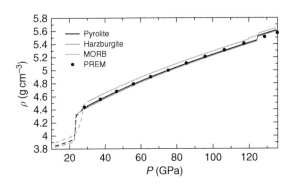

Figure 11 Bulk density variations of pyrolite, hartzburgite, and MORB calculated, based on the PVT-EoS of constituent mineral phases (**Table 1** and **Figure 9**) and their proportions (**Figure 10**). Broken lines at pressures lower than 30 GPa are results in Irifune (1993).

Although element partition data among the co-existing phases under the lower-mantle conditions have been limited to those below $\sim$30 GPa in KMA for the pyrolite composition (e.g., Irifune, 1994; Wood, 2000), some results have also been obtained using LHDAC combined with ATEM at higher pressures

(e.g., Murakami *et al.*, 2005; Ono *et al.*, 2005d). These studies demonstrated that the presence of a minor amount ($\sim$5 wt.%) of Al_2O_3 in Mg-Pv has dramatic effects on the partitioning of iron between Mg-Pv and Mw, in addition to those on the compressibility of Mg-Pv. Nevertheless, the variations in iron partitioning between the two phases would not cause any significant changes in the bulk density of the pyrolitic mantle (e.g., Bina, 2003). Some changes in the iron partitioning between Mg-Pv and Mw have also been suggested upon the Pv–PPv transition in a recent LHDAC study (Murakami *et al.*, 2005), but it may also yield invisible effects on the density change in the pyrolite bulk composition.

3.4.2.2 Harzburgite

Harzburgite, with Mg# ($= Mg/(Mg + Fe) \times 100$) $= \sim 92$ as listed in **Table 1**, crystallizes to form an assemblage of $\sim$80% olivine and $\sim$20% orthopyroxene at depths of the uppermost mantle, which transforms to Mg-Pv and Mw near the 660 km discontinuity via an assemblage of Rw + Mj + akimotoite (Akm, ilmenite form of $MgSiO_3$) as shown by Irifune and Ringwood (1987). Although no experimental data have been available at pressures higher than 26 GPa, the nature of the phase transition in this composition can be evaluated based on the changes in these two phases under the lower-mantle conditions. The result estimated along the adiabatic geotherm is shown in **Figure 10(b)**, which shows that this lithology is less dense than pyrolite throughout the lower mantle except for a very limited region immediately below the 660 km discontinuity (Irifune and Ringwood, 1987), where the density relation reverses due to the completion of the spinel to post-spinel transition at lower pressures in the harzburgite composition.

It is likely that Fe significantly lowers the Pv–PPv transition pressure (Mao *et al.*, 2004, 2005), while the presence of Al seems to increase this transition pressure (Tsuchiya *et al.*, 2005c; Tateno *et al.*, 2005), as shown in **Figures 2–4**. In harzburgite compositions, Mg-Pv should have less alumina and iron contents as compared with those of Mg-Pv in pyrolite under the lower-mantle conditions. Thus the pressures of the Pv–PPv transitions in these two compositions may be close to each other due to the opposite effects of Fe and Al, although further detailed experimental studies on these compositions are needed to resolve this issue.

3.4.2.3 MORB

Phase transitions in basaltic compositions, such as illustrated in **Figure 10(c)** for a MORB composition, are quite different from those expected in pyrolite and harzburgite compositions. Basaltic compositions are shown to crystallize to Mj + small amounts of stishovite (St) in the mantle transition region (Irifune and Ringwood, 1987, 1993; Hirose *et al.*, 1999; Ono *et al.*, 2001), which progressively transform to an assemblage of Ca-Pv + Mg-Pv + St + Al-rich phase (hexagonal or CF/CT structures) over a wide pressure range of ~3 GPa (from ~24 to ~27 GPa). Although the garnetite facies of MORB, composed mainly of Mj, is substantially denser than pyrolite, a density crossover is expected to occur in a limited depth range (660 to ~720 km) of the uppermost lower mantle due to this smeared-out nature of the garnetite-to-Pv transition in MORB.

Once Ca-Pv and Mg-Pv are formed in basaltic compositions, it is shown that they become denser than pyrolite or peridotite throughout almost the entire region of the lower mantle (Irifune and Ringwood, 1993; Hirose *et al.*, 1999, 2005; Ono *et al.*, 2001, 2005d). As St is highly incompressible

(**Figure 9**), the density of the perovskite facies of basaltic compositions may approach that of the pyrolitic composition with increasing pressure. However, the transition of St to $CaCl_2$ (CC) and α-PbO_2 (AP) structures should keep this lithology denser than pyrolite throughout the lower mantle, as shown in **Figure 11**. In fact, most recent experimental studies using LHDAC (Ono *et al.*, 2005d; Hirose *et al.*, 2005) conclude that densities of basaltic compositions are higher than those in the representative model mantle compositions throughout the lower mantle by about 0.02–0.08 g cm^{-3}, depending on the adopted pressure scale for gold.

3.5 Mineralogy of the Lower Mantle

3.5.1 The 660 km Discontinuity

The 660 km seismic discontinuity is a globally recognized feature, and is the sharpest among the proposed discontinuities throughout the whole mantle. The cause of this discontinuity, chemical or phase transition boundary, has been a major controversial issue in Earth sciences as stated earlier, but the detailed experimental study based on quench experiments using KMA strongly suggested that this is caused by the spinel to post-spinel transition in a pyrolite or peridotitic mantle (Ito and Takahashi, 1989; Irifune, 1994). The experimental data on the sharpness, pressure, and Clapeyron slope of this phase transition all seemed to be consistent with those estimated from seismological observations.

However, some recent experimental, seismological, and geodynamics studies cast some doubt on the simple idea of the phase transition. Precise *in situ* X-ray diffraction measurements using KMA demonstrated that this transition occurs at pressures somewhat (~2.5–1 GPa) lower than that correspond to the 660 km discontinuity (Irifune *et al.*, 1998a; Matsui and Nishiyama, 2002; Nishiyama *et al.*, 2004; Fei *et al.*, 2004; Katsura *et al.*, 2003), although some LHDAC experiments reported that the phase transition pressure is consistent with that of the discontinuity (Shim *et al.*, 2001a; Chudinovskikh and Boehler, 2001). Seismological observations also suggest this discontinuity is divided into several discontinuities in some areas (Simmons and Gurrola, 2000), which cannot be explained by the spinel to post-spinel transition alone.

Moreover, recent *in situ* X-ray diffraction measurements have suggested that the Clapeyron slope of the spinel to post-spinel transition is significantly larger (-2 to $-0.4\,\mathrm{MPa\,K^{-1}}$; Katsura *et al.*, 2003; Fei *et al.*, 2004; Litasov *et al.*, 2005) than previously thought (*c.* $-3\,\mathrm{MPa\,K^{-1}}$; Ito and Takahashi, 1989; Akaogi and Ito, 1993) and *ab initio* prediction (*c.* $-2\,\mathrm{MPa\,K^{-1}}$; Yu *et al.*, 2006). If this is the case, it should be difficult to trap the subducted slabs underneath many subduction zones near this boundary, known as 'stagnant slabs', (e.g., Fukao *et al.*, 1992, 2001; Zhao, 2004), in the light of geodynamics calculations (e.g., Davis, 1998). Moreover, the observed depth variations of the 660 km discontinuity sometimes reach $\sim$30–40 km (e.g., Flanagan and Shearer, 1998), which corresponds to a temperature difference of $\sim$1000 K if we adopt $\mathrm{d}P/\mathrm{d}T = -1\,\mathrm{MPa\,K^{-1}}$ for the spinel–post-spinel transition boundary. Such a temperature difference between subducting slabs and the surrounding mantle near the 660 km discontinuity should be too large to be accounted for by seismological tomographic observations and also by any reasonable models of thermal structures of the subducting slabs.

Nevertheless, considering the uncertainties in pressure and temperature measurements in the *in situ* X-ray observations in KMA due to unresolved problems on the pressure scales and temperature measurements at high pressure using thermocouple emf, it is generally accepted that the 660 km discontinuity can be explained by the spinel–post-spinel transition in a pyrolitic mantle. Recent studies also suggested possibilities of elucidating the multiple nature and the depth variation of the 660 km discontinuity by reactions involving akimotite at relatively low temperatures (Weidner and Wang, 2000; Hirose, 2002). In this case, the high velocity/density gradients shown in some representative seismological models at depths of 660–750 km may be largely explained by the smeared-out transition of majorite to perovskite in pyrolite over this depth interval (Irifune, 1994).

Models with chemical composition changes may reconcile the contradictory experimental and seismological observations regarding the 660 km discontinuity. If the spinel–post-spinel transition does occur at pressures lower than that of 660 km ($\sim$23.5 GPa) by, for instance, 1 GPa, the phase discontinuity should locate at about 630 km in a pyrolite mantle. As the oceanic crust component of the subducted slab, modeled by the MORB composition, becomes less dense than the surrounding pyrolite mantle immediately below this depth as shown in **Figure 11**, it should have been buoyantly trapped

on this primordial '630 km discontinuity' at the initial stage of the onset of operation of plate tectonics.

The delamination and accumulation of the former oceanic crust, transformed to garnetite in the mantle transition region, can be enhanced by its plausible different viscosity relative to the surrounding mantle (Karato, 1997) and extremely slow reaction kinetics upon transition to the denser phase assemblage including Mg-Pv (Kubo *et al.*, 2002). Continuation of this process of trapping basaltic crust over a couple of billion years should yield a thick (on average $\sim$50–100 km, depending on the production rates of oceanic crust and efficiency of the trapping) layer of garnetite near the 630 km discontinuity. The garnetite layer develops both upward and downward from this primordial discontinuity if it isostatically floats on this boundary. The bottom of this garnetite layer eventually reaches 660 km, under which the harzburgite portion of the subducted slab should be present. In fact, a recent mineral physics study suggests the uppermost lower mantle can be Mg rich as compared with the deeper regions (Bina, 2003).

Accumulation of oceanic crust above the 660 km seismic discontinuity was originally proposed by Anderson (1979) to form an eclogite layer in the lower half of the upper mantle and throughout the mantle transition region, who later confined it to the latter region and favored more mafic lithology named 'piclogite' (e.g., Anderson and Bass, 1986; Duffy and Anderson, 1989). Ringwood (1994) also proposed the presence of a thin (*c.* <50 km) layer of former basaltic crust immediately above the lower mantle, as a result of accumulation of relatively young and warm slab materials at the 660 km discontinuity. The above model of thicker garnetite layer is thus in between those proposed by these authors. Precise measurements of elastic-wave velocities on these lithologies under the P,T conditions of the mantle transition region are needed to further test these hypotheses.

3.5.2 Middle Parts of the Lower Mantle

The uppermost part of the lower mantle is believed to be structurally and chemically heterogeneous due to the presence of stagnant slabs (Fukao *et al.*, 1992, 2001; Zhao, 2004) which were originally proposed as 'megaliths' on the basis of high-pressure experimental studies (Ringwood and Irifune, 1988; Ringwood, 1994). Moreover, there are some areas where slabs seem to penetrate deep into the lower mantle by tomographic images (e.g., van der Hilst *et al.*, 1997; Fukao *et al.*, 2001), which should contribute to the observed reflection, refraction, or conversion of seismic waves at these depths. Thus, the presence of subducted slabs in the uppermost lower mantle may at least partly contribute to the transitional nature of the seismic velocities for a depth interval between 660 and ~750 km in some global models of the velocity profiles.

In the deeper parts of the lower mantle, discontinuous changes in seismic velocities have also been suggested to occur, particularly at depths 900–1200 km beneath some subduction areas (Kawakatsu and Niu, 1994; Niu and Kawakatsu, 1996; Kruger et al., 2001; Kaneshima, 2003). As there are no major phase transitions in the mantle material modeled by pyrolite as shown in **Figure 10(a)**, such discontinuous changes are most likely to be related to the subducted slab materials. Both basaltic and underlain harzburgite layers of such slabs should yield locally high acoustic impedance to produce seismic discontinuities, because these basaltic and harzburgite materials seem to have seismic velocities higher than those of the mantle material (e.g., Bina, 2003), in addition to their lower temperatures compared to the surrounding mantle.

In further deeper parts of the lower mantle (~1200–1850 km in depth), the presence of seismic scattering bodies with a low-velocity signature was recognized (Kaneshima and Helffrich, 1999, 2003; Vinnik et al., 2001). As shown earlier, the most notable phase transition in major minerals in subducted slab and the surrounding mantle materials is the rutile to $CaCl_2$ transition of SiO_2 in subducted former oceanic crust under the P,T conditions of the middle part (~1600 km, corresponding to pressures of ~70 GPa; Tsuchiya et al., 2004c) of the lower mantle. A significant shear softening is expected to occur in stishovite associated with this phase transition (Karki et al., 1997b; Andrault et al., 1998; Shieh et al., 2002). Thus, this transition in subducted basaltic material may explain the signatures of the scatterers observed in the middle part of the lower mantle.

In contrast, tomographic imaging and other seismological studies demonstrate that only little anomalous changes in seismic velocities exist in the middle to lower part of the lower mantle except for regions related to subducted slabs and rising plumes (e.g., Grand et al., 1997; Zhao, 2004; Mattern et al., 2005). As shown in the previous section, no major phase changes have been reported in Mg-Pv, Ca-Pv, and Mw at pressures up to ~120 GPa, except for the slight distortion of cubic Ca-Pv to a tetragonal structure and the possible HS-to-LS transition and the dissociation of Mw. It is expected that the transition in Ca-Pv may not occur at temperatures of the lower mantle, while the spin transition in Mw should occur continuously over a wide pressure range as shown in **Figure 7** and **Figure 6**, respectively. Accordingly, both of these transitions would not cause any notable seismic velocity changes in this region, although further study is required to address the effects of the possible dissociation of Mw. Thus, the relatively homogeneous nature in seismic velocity distributions in the middle to lower part of the lower mantle is consistent with the absence of major phase transitions in the mantle material.

On the other hand, seismological, geochemical, and geodynamical studies suggest that there should be a chemically distinct region at depths below 1500–2000 km, presumably due to iron enrichment related to the primitive mantle materials or interaction with the rising hot plumes (e.g., Kellogg et al., 1999; Ishii and Tromp, 1999; Trampert et al., 2004). Attempts have been made to estimate the chemical composition of the lower mantle by mineral physics tests (e.g., Jackson and Rigden, 1998; Bina, 2003; Mattern et al., 2005), comparing calculated densities and bulk sound velocities of various compositions with those obtained seismologically. However, it is difficult to constrain the chemical composition of the lower mantle without reasonable estimations of thermal structures in this region and of shear properties of the relevant high-pressure phases. Actually, a wide range of chemical compositions from peridotite to chondrites can be accommodated for the acceptable geotherms, although anomalously high X_{Mg} ($=Mg/(Mg+Fe)$) and low X_{Pv} ($=Si/(Mg+Fe)$) and low X_{Mg} and high X_{Pv} are suggested in the uppermost and lowermost ~200 km of the lower mantle, respectively (Bina, 2003).

The possible iron and silica rich nature in the bottom part of the lower mantle may reflect the heterogeneous chemical composition and mineralogy of this region. These silicon-rich signatures can be explained by the presence of subducted oceanic crust materials in this layer, which is denser than the surrounding pyrolitic or harzburgitic lower-mantle materials, as shown in **Figure 11**. Actually, the $CaCl_2-\alpha-PbO_2$ transition in SiO_2 is calculated to take place at 120–125 GPa and 2000–2500 K (Tsuchiya et al., 2004c), which should stabilize the SiO_2-bearing basaltic material in the D'' layer. Moreover, SiO_2 is also expected to exist in this region as a product of the reaction between the solid silicate Pv (or PPv) of the mantle and the molten Fe of the outer core (Knittle and Jeanloz, 1991).

In contrast, the Mg-rich and Si-poor refractory nature of the uppermost part of the lower mantle suggested by Bina (2003) could be due to the presence of accumulated harzburgite-rich bodies of the stagnant slabs. This is mainly because harzburgite is less dense than pyrolite throughout the lower mantle, as is seen in **Figure 11**. Nevertheless, the density difference is not so large, and the thermal anomalies due to subduction of slabs or rising plumes may be more effective in circulation of the harzburgitic materials within the lower mantle.

3.5.3   Postperovskite Transition and the D″ Layer

The bottom ∼200 km of the lower mantle is known as a highly heterogeneous region on the basis of seismological observations, named as the D″ layer to distinguish it from grossly homogeneous part of the lower mantle above (region D; Bullen, 1949). The horizontally and vertically heterogeneous nature of the D″ layer is primarily because this region is the chemical boundary layer between convective rocky mantle and molten iron core, which inevitably produces a large thermal boundary layer ($\Delta T > \sim 1000$ K; e.g., Williams, 1998) and accordingly yields active reactions of mantle and core materials (e.g., Knittle and Jeanloz, 1991), partial melting of mantle and slab materials (Lay et al., 2004), generation of hot mantle plumes (Garnero et al., 1998), etc. Various models have been proposed to account for the observed complex seismological signatures of D″ as reviewed by Garnero et al. (2004).

The recent finding of the Pv–PPv transition in MgSiO₃ has dramatically affected the conventional interpretations of the complex and heterogeneous features of the D″ layer. Many researchers in different research fields of Earth sciences, including mineral physics, seismology, geochemistry, mantle dynamics, etc., have started studies relevant to these topics, and a number of papers have been published immediately after the appearance of the first report on this issue (Murakami et al., 2004a). As the rates of accumulating data are so fast and the relevant research fields so vast, we are unable to thoroughly evaluate all of these studies at this moment. So, instead of discussing the origin of the D″ layer based on rather immature knowledge in these broad research fields, we herein summarize and examine currently available phase relations and some physical properties of the PPv in terms of high-pressure experimental and theoretical points of view. We will make some discussion and speculations on the nature of the D″ layer within these mineral physics frameworks.

Now, what is most robust is that the Pv–PPv transition in MgSiO₃ does occur at pressures near the D″ layer on the basis of both high-pressure experiments and ab initio calculations. Also true is that the latter phase adopts the crystal structure of CaIrO₃ type, as evident from all of these studies. The density increase in Mg-Pv associated with this transition is most likely to be only 1.2–1.5% as constrained experimentally and predicted by ab initio calculations. The remarkable agreements among these independent studies on both experimental and theoretical bases strongly suggest that these should be regarded as facts beyond any doubts.

On the other hand, there are some uncertainties in the phase transition pressure and its temperature dependency. The pressures calculated with different scales in DAC experiments, even for those based on the EoS of gold, may differ by as much as 10–15 GPa at pressures of D″ layer (∼120–136 GPa) due to the inaccuracy of the EoS's of pressure reference materials as discussed earlier (see **Figure 1**), and accordingly this transition could be realized only at pressures of the outer core (Shim et al., 2004). Moreover, uncertainties in temperature measurements in LHDAC at these very high pressure regions are fairly large, and are generally ±10–20% of the nominal values. These yield additional uncertainties of greater than 2–5 GPa in pressure measurements in LHDAC. Thus the errors of the transition pressure can be c. ±20 GPa for the Pv–PPv transition in MgSiO₃ within the current state-of-the-art LHDAC technology. Moreover, it is fair to say the Clapeyron slope of this transition has been virtually unconstrained by LHDAC experiments.

In contrast, remarkably good agreement in phase transition pressures (∼110 GPa, at 0 K) and the Clapeyron slopes (7–10 MPa K⁻¹) have been obtained in some independent ab initio calculations (Tsuchiya et al., 2004a; Oganov and Ono, 2004), which are consistent with the experimental results and also with what are expected by seismological observations (Sidorin et al., 1999). The mutual agreement among the results of the ab initio calculations, however, does not warrant the validity of these values, as these authors used basically the same technique with only some relatively minor difference in computational methods and techniques. Thus these results on the Pv–PPv phase boundary based on ab initio calculations should be further tested on the basis of experimental studies, before they are regarded as robust ones.

Nevertheless, there is no strong evidence against the occurrence of the Pv–PPv transition at pressures about 120 GPa, with a Clapeyron slope of $7-10\,\mathrm{MPa\,K^{-1}}$, and it is reasonable to tentatively take these values as realistic ones. It has been reported that such a large Clapeyron slope is expected to affect the thermal structure of the lower mantle and thus enhances the mantle convection and plume dynamics (Nakagawa and Tackley, 2004). For actual mantle compositions, effects of relatively minor elements such as Fe and Al should be taken into account, but, as we mentioned earlier, these may have opposite effects in modifying the transition pressure, which should be cancelling out each other. Actually, recent experimental results demonstrated that the Pv–PPv transition in peridotitic compositions occurs at pressures close to 120 GPa (Murakami *et al.*, 2005; Ono *et al.*, 2005d). Although a smeared-out transition is expected for this transition when some iron is incorporated in $MgSiO_3$ (Mao *et al.*, 2004, 2005), these studies on the rock samples indicate that the pressure interval of the mixed phase region of Pv and PPv is not so large, suggesting the occurrence of a sharp boundary (Wysession *et al.*, 1998). It is also demonstrated that the PPv phase in the peridotite composition is highly magnesium rich relative to Mw (Murakami *et al.*, 2005), whereas PPv is reported to favor iron compared with Pv in other experimental studies (Mao *et al.*, 2004, 2005; cf. Kobayashi *et al.*, 2005).

Another striking feature of PPv is its elasticity, as predicted by *ab initio* calculations (Tsuchiya *et al.*, 2004b; Oganov and Ono, 2004; Iitaka *et al.*, 2004). The predicted seismic wave speeds of PPv are slightly faster in V_P and V_S and slower in V_Φ than those of Pv at the transition P,T condition as typically observed at the D″ discontinuities (Lay *et al.*, 2004). These velocity changes across the transition boundary seem to explain the enigmatic anticorrelated anomaly between V_S and V_Φ observed at the bottom of the mantle (Wookey *et al.*, 2005; Wentzcovitch *et al.*, 2006). Also, they can vary with changing the iron and aluminum contents in Pv and PPv (Tsuchiya and Tsuchiya, 2006).

Because of the peculiar crystal structure of PPv, which is highly compressible along the *b*-direction, this phase is suggested to be elastically quite anisotropic. The *ab initio* studies also reported that PPv polycrystalline aggregates with lattice-preferred orientation around some crystallographic directions produced by macroscopic stress appear to explain the observed anisotropic propagation of seismic waves in some regions of the D″ layer. In particular, some *ab initio* studies on high-temperature elasticity (Stackhouse *et al.*, 2005a; Wentzcovitch *et al.*, 2006)

showed that the transversely isotropic aggregates of PPv can yield horizontally polarized S wave (SH) which travels faster than vertically poralized S wave (SV) at the lower-mantle P,T conditions, as observed underneath Alaska and Central America (Lay *et al.*, 2004). However, the peculiar nature of its shear elasticity suggests that the simple shear motion along the layered structure may not be applicable to the PPv phase at relevant pressures (Tsuchiya *et al.*, 2004b). Moreover, it was predicted that Mw would have substantially large elastic anisotropy under the lower-mantle conditions (Yamazaki and Karato, 2002), suggesting that the PPv phase may not be important in producing anisotropic nature in the D″ layer.

The most likely region, where the Pv–PPv transition plays important roles in the D″ layer, is probably underneath some of the subduction zones, where the temperatures are expected to be relatively lower than that of the surrounding mantle. In these relatively cold regions in the D″ layer, it is known that a sharp discontinuity exists on the top of this layer, under which the highly anisotropic nature of propagation of SV and SH has also been recognized. If this discontinuity corresponds to the Pv–PPv transition, then one of the most likely features of this transition, that is, its large Clapeyron slope of $\mathrm{d}P/\mathrm{d}T = 7-10\,\mathrm{MPa\,K^{-1}}$, suggests that such discontinuity may not exist in warmer regions. For instance, if the temperature in the warmer region is higher than the cold areas by several hundred degrees (such a lateral temperature variation is very likely to exist in the CMB region), this transition would occur in the middle of the D″ layer. However, as the temperature is expected to increase sharply toward the CMB within the D″ layer, the geotherm in the warmer region may not cross the Pv–PPv phase boundary as illustrated in **Figure 12**. In contrast, it is interesting to see that there is a possibility that the geotherm in the colder region might cross the boundary twice, as suggested by recent computational modeling (Hernlund *et al.*, 2005; Wookey *et al.*, 2005).

Thus, if the transition pressure and the Clapeyron slope of the Pv–PPv transition predicted by *ab initio* calculations are correct, many of the observed features in the CMB region can be explained. In warmer regions such as underneath Hawaii or South Africa, this transition would not have any significant roles in producing the peculiar features, such as the presence of ultra-low-velocity region or a large low-velocity structure in these regions, which should be attributed to other phenomena, such as partial melting of the D″ layer material or chemical changes (e.g., Lay *et al.*,

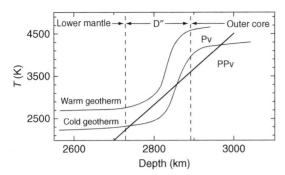

Figure 12 A schematic illustration of the plausible thermal structure at the bottom of the lower mantle, where a steep temperature gradient should be realized due to the formation of the thermal boundary layer. Only the subadiabatic cold geotherm may cross the Pv–PPv phase boundary, and the PPv phase may be sandwiched by Pv because of the shallow slope of this transition and the sharp temperature increase in this region (Hernlund *et al.*, 2005).

2004). Reactions between Mg-Pv in the warm regions with liquid iron may produce an iron-rich PPv phase in these regions, which could also contribute to the observed low velocities in these regions, as suggested by Mao *et al.* (2005). Such reactions between Pv (or PPv) and molten iron should be further explored experimentally to address the nature of the (ultra-)low-velocity zones within the D″ layer.

3.6 Summary

Recent advances in both experimental and computational techniques have enabled the quantitative study of phase transitions and mineral physics properties under the lower-mantle P, T conditions. Although uncertainties in pressure and temperature are of the order of ~10% of the nominal values in typical experiments of LHDAC, this apparatus can now produce pressures and temperatures equivalent to those of the entire mantle of the Earth. In contrast, pressures available in KMA have long been limited to ~30 GPa in spite of its superiority in the accuracy of $P-T$ measurements over LHDAC. However, recent developments in KMA using SD anvils doubled this pressure limit, allowing detailed mineral physics studies down to the middle part of the lower mantle.

The accuracy of the mineral physics studies based on *ab initio* calculations has also been dramatically improved in the last decade. A variety of methods and techniques have been developed for the practical applications of DFT to mineral physics studies at very high pressure and temperature. As a result,

remarkable agreements among the results from different research groups have been obtained for the phase transition pressures, elastic properties, etc., of some high-pressure phases under the lower-mantle pressures. Thus, such mineral physics properties of high-pressure phases in the lower mantle can now be evaluated and cross-checked on the basis of *ab initio* calculations and the independent experimental studies using LHDAC and KMA.

Phase transitions in major and minor minerals relevant to the mantle and subducted slabs have been studied by using the above independent methods, which clarified structural phase transitions in most of these minerals at pressures and temperatures characteristic of the lower mantle, although there remain some controversial results on the phase transition pressures, their temperature dependencies, element partitioning among the coexisting phases, etc. Although the elastic properties of high-pressure phases, particularly shear properties, have not well documented under the lower-mantle conditions, EoS parameters of some of these phases have successfully been determined by a combination of synchrotron source and KMA-LHDAC or by *ab initio* methods. Thus the density changes in representative mantle and slab compositions can be reasonably evaluated on the basis of the thermoelastic data on individual high-pressure phases and mineral proportion changes in these compositions.

The 660 km discontinuity has conventionally been interpreted in terms of the spinel–post-spinel phase transition in Mg_2SiO_4. However, results of the recent experimental, seismological, and geodynamics studies do not seem to be totally consistent with this interpretation. The presence of a basaltic garnetite layer, as a result of accumulation of subducted oceanic crust materials, may explain such inconsistency and the complex structure near the 660 km discontinuity. The validity of this and other classes of mineralogical models relevant to the origin of the 660 km discontinuity can be evaluated on the basis of mineral physics tests, when accurate elastic-wave velocity data become available for the pyrolite and basaltic compositions under the P, T conditions corresponding to these depths.

The middle part of the lower mantle is believed to be generally homogeneous in both mineralogy and chemistry, as compared with those in the uppermost and lowermost parts of the lower mantle. No major phase transitions are expected to occur in the subducted slab lithologies and the surrounding mantle under the P, T conditions of this part of the lower mantle, except for the rutile-to-$CaCl_2$ transition in SiO_2 and HS-to-LS transition in Mw (and perhaps Pv). The seismic scatterers of presumably flat bodies found in the middle part of the lower mantle should be related to this phase transition in the former subducted oceanic crust, which may subduct into the deep lower mantle once it passes the density crossover near the 660 km discontinuity. Thus, some parts of the basaltic component of the slab may ultimately reach the bottom of the lower mantle because of its higher density relative to the mantle materials throughout the lower mantle. In contrast, harzburgite-rich layer may be present in the upper part of the lower mantle as a result of accumulation of main bodies of stagnant slabs in this region. However, as the density difference between this layer and the surrounding mantle is not large, thermal effects should be predominant over the chemical effects in gravitational stability of such a layer in the upper part of the lower mantle.

The recently discovered Pv–PPv transition should have significant implications for the structure, properties, and dynamics of the D″ layer, although its actual presence in this region has not been fully proved. Nevertheless, some properties, such as Clapeyron slope, transition pressure, elastically anisotropic nature, and density increase, relevant to the PPv transition seem to be consistent with the seismologically observed signatures in some places of the D″ layer. Further detailed and more accurate experimental studies on this transition are required based on independent techniques using KMA, or LHDAC with improved accuracy, in addition to the theoretical studies and establishment of reliable pressure scales, to understand the nature of this region of the lower mantle.

References

Akahama Y, Kawamura H, and Singh AK (2002) The equation of state of Bi and cross-checking of Au and Pt scales to megabar pressure. *Journal of Physics, Condensed Matter* 14: 11495–11500.

Akahama Y, Kawamura H, and Singh AK (2004) A comparison of volume compressions of silver and gold to 150 GPa. *Journal of Applied Physiology* 95: 4767–4771.

Akaogi M and Ito E (1993) Refinement of enthalpy measurement of $MgSiO_3$ perovskite and negative pressure-temperature slopes for perovskite-forming reactions. *Geophysical Research Letters* 20: 1839–1842.

Akaogi M, Hamada Y, Suzuki T, Kobayashi M, and Okada M (1999) High pressure transitions in the system $MgAl_2O_4$–$CaAl_2O_4$: A new hexagonal aluminous phase with implication for the lower mantle. *Physics of Earth and Planetary Interiors* 115: 67–77.

Akber-Knutson S, Steinle-Neumann G, and Asimow PD (2005) Effect of Al on the sharpness of the $MgSiO_3$ perovskite to post-perovskite phase transition. *Geophysical Research Letters* 32: L14303 (doi:10.1029/2005GL023192).

Alfé D, Alfredsson M, Brodholt J, Gillan MJ, Towler MD, and Needs RJ (2005) Quantum Monte Carlo calculations of the structural properties and the B1–B2 phase transition of MgO. *Physical Review B* 72: 014114.

Anisimov VI, Zaanen J, and Andersen OK (1991) Band theory and Mott insulators: Hubbard U instead of Stoner I. *Physical Review B* 44: 943–954.

Anderson DL (1979) The upper mantle: Eclogite? *Geophysical Research Letters* 6: 433–436.

Anderson DL and Bass JD (1986) Transition region of the Earth's upper mantle. *Nature* 320: 321–328.

Anderson DL (1989) Composition of the Earth. *Science* 243: 367–370.

Anderson OL, Isaak DG, and Yamamoto S (1989) Anharmonicity and the equation of state for gold. *Journal of Applied Physiology* 65: 1534–1543.

Andrault D, Fiquet G, Guyot F, and Hanfland M (1998) Pressure-induced Landau-type transition in stishovite. *Science* 282: 720–724.

Andrault D, Angel RJ, Mosenfelder JL, and Bihan TLe (2003) Equation of state of stishovite to lower mantle pressures. *American Mineralogist* 88: 1261–1265.

Badro J, Struzhkin VV, Shu J, *et al.* (1999) Magnetism in FeO at megabar pressures from X-ray emission spectroscopy. *Physical Review Letters* 83: 4101–4104.

Badro J, Fiquet G, Guyot F, *et al.* (2003) Iron partitioning in Earth's mantle: Toward a deep lower mantle discontinuity. *Science* 300: 789–791.

Badro J, Rueff J-P, Vankó G, Monaco G, Fiquet G, and Guyot F (2004) Electronic transitions in perovskite: Possible nonconvecting layers in the lower mantle. *Science* 305: 383–386.

Bagno P, Jepsen O, and Gunnarsson O (1989) Ground-state properties of third-row elements with nonlocal density functionals. *Physical Review B* 40: 1997–2000.

Beck P, Gillet P, Gautron L, Daniel I, and El Goresy A (2004) A new natural high-pressure (Na,Ca)-hexaluminosilicate [$(Ca_xNa_{1-x})Al_{3+x}Si_{3-x}O_{11}$] in shocked Martian meteorites. *Earth and Planetary Science Letters* 219: 1–12.

Biellmann C, Gillet P, Guyot F, Peyronneau J, and Reyard B (1993) Experimental evidence for carbonate stability in Earth's lower mantle. *Earth and Planetary Science Letters* 118: 31–41.

Bina CR (2003) Seismological constraints upon mantle composition. In: Carlson R (ed.) *Treatise on Geochemistry*, vol. 2, pp. 39–59. Amsterdam: Elsevier.

Birch F (1952) Elasticity and constitution of the Earth's interior. *Journal of Geophysical Research* 57: 227–286.

Brown JM and Shankland TJ (1981) Thermodynamic parameters in the Earth as determined from seismic profiles. *Geophysical Journal of the Royal Astronomical Society* 66: 579–596.

Bullen KE (1949) Compressibility-pressure hypothesis and the Earth's interior. *MNRAS. Geophysical Supplement* 5: 355–368.

Caracas R and Cohen RE (2005a) Prediction of a new phase transition in Al_2O_3 at high pressures. *Geophysical Research Letters* 32: L06303 (doi:10.1029/2004GL022204).

Caracas R and Cohen RE (2005b) Effect of chemistry on the stability and elasticity of the perovskite and post-perovskite phases in the $MgSiO_3$–$FeSiO_3$–Al_2O_3 system and implications for the lowermost mantle. *Geophysical Research Letters* 32: L16310 (doi:10.1029/2005GL023164).

Caracas R, Wentzcovitch R, Price GD, and Brodholt J (2005) $CaSiO_3$ perovskite at lower mantle pressures. *Geophysical Research Letters* 32: L06306 (doi:10.1029/2004GL022144).

Catti M (2001) High-pressure stability, structure and compressibility of $Cmcm$-MgAl$_2$O$_4$: An ab $initio$ study. $Physics$ and $Chemistry$ of $Minerals$ 28: 729–736.

Chizmeshya AVG, Wolf GH, and McMillan PF (1996) First principles calculation of the equation-of-state, stability, and polar optic modes of CaSiO$_3$ perovskite. $Geophysical$ $Research$ $Letters$ 23: 2725–2728.

Chudinovskikh L and Boehler R (2001) High-pressure polymorphs of olivine and the 660-km seismic discontinuity. $Nature$ 411: 574–577.

Cococcioni M and de Gironcoli S (2005) Linear response approach to the calculation of the effective interaction parameters in the LDA+U method. $Physical$ $Review$ B 71: 035105.

Cohen RE, Mazin, II, and Isaak DG (1997) Magnetic collapse in transition metal oxides at high pressure: implications for the Earth. $Science$ 275: 654–657.

Davis GF (1998) Plates, plumes, mantle convection, and mantle evolution. In: Jackson I (ed.) The $Earth's$ $Mantle$, pp. 228–258. New York: Cambridge University Press.

Demuth Th, Jeanvoine Y, Hafner J, and Ángyán JG (1999) Polymorphism in silica studied in the local density and generalized-gradient approximations. $Journal$ of $Physics,$ $Condensed$ $Matter$ 11: 3833–3874.

Dubrovinsky LS, Saxena SK, Lazor P, et $al.$ (1997) Experimental and theoretical identification of a new high-pressure phase of silica. $Nature$ 388: 362–365.

Dubrovinsky LS, Dubrovinskaia NA, Saxena SK, et $al.$ (2000) Stability of Ferropericlase in the lower mantle. $Science$ 289: 430–432.

Dubrovinsky LS, Dubrovinskaia NA, Saxena SK, et $al.$ (2001) Pressure-induced transformations of cristobalite. $Chemical$ $Physics$ $Letters$ 333: 264–270.

Duffy TS and Anderson DL (1989) Seismic velocities in mantle minerals and the mineralogy of the upper mantle. $Journal$ of $Geophysical$ $Research$ 94: 1895–1912.

Duffy TS, Hemley RJ, and Mao H-K (1995) Equation of state and shear strength at multimegabar pressures: Magnesium oxide to 227 GPa. $Physical$ $Review$ $Letters$ 74: 1371–1374.

Dziewonski AM and Anderson DL (1981) Preliminary reference Earth model. $Physics$ of $Earth$ and $Planetary$ $Interiors$ 25: 297–356.

Fei Y and Mao H-K (1994) In $situ$ determination of the NiAs phase of FeO at high pressure and high temperature. $Science$ 266: 1678–1680.

Fei Y, Van Orman J, Li J, et $al.$ (2004) Experimentally determined postspinel transformation boundary in Mg$_2$SiO$_4$ using MgO as an internal pressure standard and its geophysical implications. $Journal$ of $Geophysical$ $Research$ 109: B02305 1–8.

Ferroir T, Yagi T, Onozawa T, et $al.$ (2006) Equation of state and phase transition in KAlSi$_3$O$_8$ hollandite at high pressure. $American$ $Mineralogist$ 91(2–3): 327–332.

Fiquet G, Richet P, and Montagnac G (1999) High-temperature thermal expansion of lime, periclase, corundum and spinel. $Physics$ and $Chemistry$ of $Minerals$ 27: 103–111.

Fiquet G, Dewaele A, Andrault D, Kunz M, and Bihan TL (2000) Thermoelastic properties and crystal structure of MgSiO$_3$ perovskite at lower mantle pressure and temperature conditions. $Geophysical$ $Research$ $Letters$ 27: 21–24.

Fiquet A, Guyot F, Kunz M, Matas J, Andrault D, and Hanfland M (2002) Structural refinements of magnesite at very high pressure. $American$ $Mineralogist$ 87: 1261–1265.

Flanagan MP and Shearer PM (1998) Global mapping of topography on transition zone velocity discontinuities by stacking SS precursors. $Journal$ of $Geophysical$ $Research$ 103: 2673–2692.

Frost DJ and Fei Y (1998) Stability of phase D at high pressure and high temperature. $Journal$ of $Geophysical$ $Research$ 103: 7463–7474.

Fukao Y, Obayashi M, Inoue H, and Nenbai M (1992) Subducting slabs stagnant in the mantle transition zone. $Journal$ of $Geophysical$ $Research$ 97: 4809–4822.

Fukao Y, Widiyantoro S, and Obayashi M (2001) Stagnant slabs in the upper and lower mantle transition region. $Reviews$ of $Geophysics$ 39: 291–323.

Funamori N, Yagi T, Utsumi W, Kondo T, and Uchida T (1996) Thermoelastic properties of MgSiO$_3$ perovskite determined by in situ X-ray observations up to 30 GPa and 2000 K. $Journal$ of $Geophysical$ $Research$ 101: 8257–8269.

Funamori N and Jeanloz R (1997) High-pressure transformation of Al$_2$O$_3$. $Science$ 278: 1109–1111.

Funamori N, Jeanloz R, Nguyen JH, et $al.$ (1998) High-pressure transformations in MgAl$_2$O$_4$. $Journal$ of $Geophysical$ $Research$ 103: 20813–20818.

Gaffney ES and Anderson DL (1973) Effect of low-spin Fe^{2+} on the composition of the lower mantle. $Journal$ of $Geophysical$ $Research$ 78: 7005–7014.

Garnero EJ, Revenaugh JS, Williams Q, Lay T, and Kellogg LH (1998) Ultralow velocity zone at the core-mantle boundary. In: Gurnis M, Wysession M, Knittle E, and Buffett B (eds.) The $Core-Mantle$ $Boundary$ $Region$, pp. 319–334. Washington, DC: AGU.

Garnero EJ, Maupin V, Lay T, and Fouch MJ (2004) Variable azimuthal anisotropy in Earth's lowermost mantle. $Science$ 306: 259–261.

Gautron L, Angel RJ, and Miletich R (1999) Structural characterization of the high-pressure phase CaAl$_4$Si$_2$O$_{11}$. $Physics$ and $Chemistry$ of $Minerals$ 27: 47–51.

Gautron L, Daniel I, Beck P, Gulgnot N, Andrault D, and Greaux S (2005) On the track of 5-fold silicon signature in the high pressure CAS phase CaAl$_4$Si$_2$O$_{11}$. Eos $Transactions$ of the $American$ $Geophysical$ $Unioin$ 86(52): Fall Meeting Supplement, Abstract MR23C-0087.

Gillet P (1993) Stability of magnesite (MgCO$_3$) at mantle pressure and temperature conditions: A Raman spectroscopic study. $American$ $Mineralogist$ 78: 1328–1331.

Gillet P, Chen M, Dubrovinsky LS, and El Goresy A (2000) Natural NaAlSi$_3$O$_8$-hollandite in the shocked Sixiangkou meteorite. $Science$ 287: 1633–1636.

Grand S, van der Hilst R, and Widiyantoro S (1997) Global seismic tomography: a snap shot of convection in the Earth. GSA $Today$ 7: 1–7.

Green DH, Hibberson WO, and Jaques AL (1979) Petrogenesis of mid-ocean ridge basalts. In: McElhinny MW (ed.) The $Earth:$ Its $Origin,$ $Structure$ and $Evolution$, pp. 269–299. London: Academic Press.

Hamann DR (1996) Generalized gradient theory for silica phase transitions. $Physical$ $Review$ $Letters$ 76: 660–663.

Hamann DR (1997) H$_2$O hydrogen bonding in density-functional theory. $Physical$ $Review$ B 55: R10157–R10160.

Hart SR and Zindler A (1986) In search for a bulk Earth composition. $Chemical$ $Geology$ 57: 247–267.

Heinz DL and Jeanloz R (1984) The equation of state of the gold calibration standard. $Journal$ of $Applied$ $Physiology$ 55: 885–893.

Helffrich G and Wood B (2001) The Earth's mantle. $Nature$ 412: 501–507.

Hernlund JW, Thomas C, and Tackley PJ (2005) A doubling of the post-perovskite phase boundary and structure of the Earth's lowermost mantle. $Nature$ 434: 882–886.

Hirose K, Fei Y, Ma Y, and Mao H-K (1999) The fate of subducted basaltic crust in the Earth's lower mantle. $Nature$ 397: 53–56.

Hirose K (2002) Phase transition in pyrolitic mantle around 670-km depth: implications for upwelling of plumes from the lower mantle. $Journal$ of $Geophysical$ $Research$ 107: 2078 (doi:10.1029/2001JB000597).

Hirose K, Takafuji N, Sata N, and Ohishi Y (2005) Phase transition and density of subducted MORB crust in the lower mantle. *Earth and Planetary Science Letters* 237: 239–251.

Hohenberg P and Kohn W (1964) Inhomogeneous electron gas. *Physical Review* 136: B364–B871.

Holmes NC, Moriarty JA, Gathers GR, and Nellis WJ (1989) The equation of state of platinum to 660 GPa (6.6 Mbar). *Journal of Applied Physiology* 66: 2962–2967.

Iitaka T, Hirose K, Kawamura K, and Murakami M (2004) The elasticity of the MgSiO$_3$ post-perovskite phase in the Earth's lowermost mantle. *Nature* 430: 442–444.

Irifune T and Ringwood AE (1987) Phase transformations in a harzburgite composition to 26 GPa: Implications for dynamical behaviour of the subducting slab. *Earth and Planetary Science Letters* 86: 365–376.

Irifune T, Fujino K, and Ohtani E (1991) A new high-pressure form of MgAl$_2$O$_4$. *Nature* 349: 409–411.

Irifune T (1993) Phase transformations in the earth's mantle and subducting slabs: Implications for their compositions, seismic velocity and density structures and dynamics. *The Island Arc* 2: 55–71.

Irifune T and Ringwood AE (1993) Phase transformations in subducted oceanic crust and buoyancy relationships at depths of 600–800 km in the mantle. *Earth and Planetary Science Letters* 117: 101–110.

Irifune T (1994) Absence of an aluminous phase in the upper part of the Earth's lower mantle. *Nature* 370: 131–133.

Irifune T, Ringwood AE, and Hibberson WO (1994) Subduction of continental crust and terrigenous and pelagic sediments: An experimental study. *Earth and Planetary Science Letters* 126: 351–368.

Irifune T, Koizumi T, and Ando J (1996a) An experimental study of the garnet-perovskite transformation in the system MgSiO$_3$–Mg$_3$Al$_2$Si$_3$O$_{12}$. *Geophysical Research Letters* 96: 147–157.

Irifune T, Kuroda K, Funamori N, et al. (1996b) Amorphization of serpentine at high pressure and high temperature. *Science* 272: 1468–1470.

Irifune T, Nishiyama N, Kuroda K, et al. (1998a) The postspinel phase boundary in Mg$_2$SiO$_4$ determined by in situ X-ray diffraction. *Science* 279: 1698–1700.

Irifune T, Kubo N, Isshiki M, and Yamasaki Y (1998b) Phase transformations in serpentine and transportation of water into the lower mantle. *Geophysical Research Letters* 25: 203–206.

Irifune T (2002) Application of synchrotron radiation and Kawai-type apparatus to various studies in high-pressure mineral physics. *Mineralogical Magazine* 66: 769–790.

Irifune T, Naka H, Sanehira T, Inoue T, and Funakoshi K (2002) In situ X-ray observations of phase transitions in MgAl$_2$O$_4$ spinel to 40 GPa using multianvil apparatus with sintered diamond anvils. *Physics and Chemistry of Minerals* 29: 645–654.

Irifune T, Isshiki M, and Sakamoto S (2005) Transmission electron microscope observation of the high-pressure form of magnesite retrieved from laser heated diamond anvil cell. *Earth and Planetary Science Letters* 239: 98–105.

Ishii M and Tromp J (1999) Normal-mode and free-air gravity constraints on lateral variations in velocity and density of Earth's mantle. *Science* 285: 1231–1236.

Isshiki M, Irifune T, Hirose K, et al. (2004) Stability of magnesite and its high-pressure form in the lowermost mantle. *Nature* 427: 60–63.

Ito E and Takahashi E (1989) Post-spinel transformations in the system. Mg$_2$SiO$_4$–Fe$_2$SiO$_4$ and some geophysical implications. *Journal of Geophysical Research* 94: 10637–10646.

Ito E, Kubo A, Katsura T, Akaogi M, and Fujita T (1998) High-pressure phase transition of pyrope (Mg$_3$Al$_2$Si$_3$O$_{12}$) in a sintered diamond cubic anvil assembly. *Geophysical Research Letters* 25: 821–824.

Ito E, Katsura T, Aizawa Y, Kawabe K, Yokoshi S, and Kubo A (2005) High-pressure generation in the Kawai-type apparatus equipped with sintered diamond anvils: Application to the wurtzite–rocksalt transformation in GaN. In: Chen J, Wang Y, Duffy TS, Shen G, and Dobrzhinetskaya LF (eds.) *Advances in High-Pressure Technology for Geophysical Applications*, pp. 451–460. Amsterdam: Elsevier.

Jackson I, Khanna SK, Revcolevschi A, and Berthon J (1990) Elasticity, shear-mode softening and high-pressure polymorphism of Wüstite (Fe$_{1-x}$O). *Journal of Geophysical Research* 95: 21671–21685.

Jackson I and Rigden SM (1998) Composition and temperature of the Earth's mantle: Seismological models interpreted through experimental studies of the Earth materials. In: Jackson I (ed.) *The Earth's Mantle*, pp. 405–460. New York: Cambridge University Press.

Jackson JM, Sturhahn W, Shen G, et al. (2005) A synchrotron Mössbauer spectroscopy study of (Mg,Fe)SiO$_3$ perovskite up to 120 GPa. *American Mineralogist* 90: 199–205 .

Jacobsen SD, Reichmann HJ, Spetzler HA, et al. (2002) Structure and elasticity of single crystal (Fe,Mg)O and a new method of generating shear waves for gigahertz ultrasonic interferometry. *Journal of Geophysical Research* 107: 1–14 (doi:10.1029 2001JB000490).

Jamieson JC, Fritz JN, and Manghnani MH (1982) Pressure measurement at high temperature in X-ray diffraction studies: Gold as a primary standard. In: Akimoto S, and Manghnani MH (eds.) *High-Pressure Research in Geophysics*, pp. 27–48. Tokyo: Center for Academic Publications.

Jeanloz R and Thompson AB (1983) Phase transitions and mantle discontinuities. *Reviews of Geophysics Space Phys* 21: 51–74.

Kaneshima S and Helffrich G (1999) Dipping low-velocity layer in the mid-lower mantle: Evidence for geochemical heterogeneity. *Science* 283: 1888–1891.

Kaneshima S (2003) Small scale heterogeneity at the top of the lower mantle around the Mariana slab. *Earth and Planetary Science Letters* 209: 85–101.

Kaneshima S and Helffrich G (2003) Subparallel dipping heterogeneities in the mid-lower mantle. *Journal of Geophysical Research* 108: 2272 (doi:10.1029/2001JB001596).

Karato S (1997) On the separation of crustal component from subducted oceanic lithosphere near the 660 km discontinuity. *Physics of Earth and Planetary Interiors* 99: 103–111.

Karki BB, Stixrude L, and Crain J (1997a) Ab initio elasticity of three high-pressure polymorphs of silica. *Geophysical Research Letters* 24: 3269–3272.

Karki BB, Warren MC, Stixrude L, Clark SJ, Ackland GJ, and Crain J (1997b) Ab initio studies of high-pressure structural transformations in silica. *Physical Review B* 55: 3465–3471.

Karki BB and Crain J (1998) First-principles determination of elastic properties of CaSiO$_3$ perovskite at lower mantle pressures. *Geophysical Research Letters* 25: 2741–2744.

Karaki BB, Wentzcovitch RM, de Gironcoli S, and Baroni S (2000) High-pressure lattice dynamics and thermoelasticity of MgO. *Physical Review B* 61: 8793–8800.

Katsura T and Ito E (1996) Determination of Fe–Mg partitioning between perovskite and magnesiowüstite. *Geophysical Research Letters* 23: 2005–2008.

Katsura T, Yamada H, Shinmei T, et al. (2003) Post-spinel transition in Mg$_2$SiO$_4$ determined by high P-T in situ X-ray diffractometry. *Physics of Earth and Planetary Interiors* 136: 11–24.

Kawai N and Endo S (1970) The generation of ultrahigh hydrostatic pressures by a split sphere apparatus. *Review of Scientific Instruments* 41: 1178–1181.

Kawakatsu H and Niu F (1994) Seismic Evidence for a 920 km discontinuity in the mantle. *Nature* 371: 301–305.

Kellogg LH, Hager BH, and van der Hilst RD (1999) Compositional stratification in the deep mantle. *Science* 283: 1881–1884.

Kesson SE, Fitz Gerald JD, and Shelley JM (1998) Mineralogy and dynamics of a pyrolite lower mantle. *Nature* 393: 252–255.

Kiefer B, Stixrude L, and Wentzcovitch RM (2002) Elasticity of $(Mg,Fe)SiO_3$-perovskite at high pressure. *Geophysical Research Letters* 29: 1539 (doi:10.1029/2002GL014683).

Kingma KJ, Cohen RE, Hemley RJ, and Mao HK (1995) Transformation of stishovite to a denser phase at lower-mantle pressures. *Nature* 374: 243–246.

Knittle E and Jeanloz R (1991) Earth's core-mantle boundary: results of experiments at high pressure and temperatures. *Science* 251: 1438–1443.

Kobayashi Y, Kondo T, Ohtani T, *et al.* (2005) Fe-Mg partitioning between $(Mg,Fe)SiO_3$ post-perovskite, perovskite, and magnesiowüstite in the Earth's lower mantle. *Geophysical Research Letters* 32: L19301 (doi:10.1029/2005GL023257).

Kohn W and Sham LJ (1965) Self-consistent equation including exchange and correlation effects. *Physical Review* 140: A1133–A1138.

Kondo T, Ohtani E, Hirao N, Yagi T, and Kikegawa T (2004) Phase transitions of $(Mg,Fe)O$ at megabar pressures. *Physics of Earth and Planetary Interiors* 143-144: 201–213.

Kruger F, Banumann M, Scherbaum F, and Weber M (2001) Mid mantle scatterers near the Mariana slab detected with a double array method. *Geophysical Research Letters* 28: 667–670.

Kubo T, Ohtani E, Kondo T, *et al.* (2002) Metastable garnet in oceanic crust at the top of the lower mantle. *Nature* 420: 803–806.

Kudoh Y, Nagase T, Mizohata H, and Ohtani E (1997) Structure and crystal chemistry of phase G, a new hydrous magnesium silicate synthesized at 22 GPa and 1050°C. *Geophysical Research Letters* 24: 1051–1054.

Kurashina T, Hirose K, Ono S, Sata N, and Ohishi Y (2004) Phase transition in Al-bearing $CaSiO_3$ perovskite: Implications for seismic discontinuities in the lower mantle. *Physics of Earth and Planetary Interiors* 145: 67–74.

Kuroda K and Irifune T (1998) Observation of phase transformations in serpentine at high pressure and high temperature by *in situ* X ray diffraction measurements. In: Manghnani MH and Yagi T (eds.) *High Pressure–Temperature Research: Properties of Earth and Planetary Materials*, pp. 545–554. Washington DC: AGU.

Lay T, Garnero EJ, and Williams Q (2004) Partial melting in a thermo-chemical boundary layer at the base of the mantle. *Physics of Earth and Planetary Interiors* 146: 441–467.

Li J, Struzhkin VV, Mao H-K, *et al.* (2004) Electronic spin state of iron in lower mantle perovskite. *Proceedings of the National Academy of Sciences* 101: 14027–14030.

Li L, Weidner DJ, Brodholt J, *et al.* (2006a) Elasticity of $CaSiO_3$ perovskite at high pressure and high temperature. *Physics of Earth and Planetary Interiors* 155: 249–259.

Li L, Weidner DJ, Brodholt J, *et al.* (2006b) Phase stability of $CaSiO_3$ perovskite at high pressure and temperature: Insights from *ab initio* molecular dynamics. *Physics of Earth and Planetary Interiors* 155: 260–268.

Lin J-F, Heinz DL, Mao H-K, *et al.* (2003) Stability of magnesiowüstite in earth's lower mantle. *Proceedings of the National Academy of Sciences* 100: 4405–4408.

Lin J-F, Struzhkin VV, Jacobsen SD, *et al.* (2005) Spin transition of iron in magnesiowüstite in the Earth's lower mantle. *Nature* 436: 377–380.

Litasov K, Ohtani E, Sano A, Suzuki A, and Funakoshi K (2005) *In situ* X-ray diffraction study of post-spinel transformation in a peridotite mantle: Implication for the 660-km discontinuity. *Earth and Planetary Science Letters* 238: 311–328.

Liu L-G (1977) High pressure $NaAlSiO_4$: The first silicate calcium ferrite isotope. *Geophysical Research Letters* 4: 183–186.

Liu L-G (1978) A new high-pressure phase of spinel. *Earth and Planetary Science Letters* 41: 398–404.

Liu L-G (1982) Speculations on the composition and origin of the Earth. *Geochemical Journal* 16: 287–310.

Liu L-G (1986) Phase transformations in serpentine at high pressures and temperatures and implications for subducting lithosphere. *Physics of Earth and Planetary Interiors* 42: 255–262.

Liu L-G (1999) Genesis of diamonds in the lower mantle. *Contributions to Mineralogy and Petrology* 134: 170–173.

Liu X (2006) Phase relations in the system $KAlSi_3O_8$–$NaAlSi_3O_8$ at high pressure-high temperature conditions and their implication for the petrogenesis of lingunite. *Earth and Planetary Science Letters* 246: 317–325.

Loubet M, Sassi R, and Di Donato G (1988) Mantle heterogeneities: A combined isotope and trace element approach and evidence for recycled continental crust materials in some OIB sources. *Earth and Planetary Science Letters* 89: 299–315.

Magyari-Köpe B, Vitos L, Grimvall G, Johansson B, and Kollár J (2002) Low-temperature crystal structure of $CaSiO_3$ perovskite: An *ab initio* total energy study. *Physical Review B* 65: 193107.

Mao HK, Chen LC, Hemley RJ, Jephcoat AP, and Wu Y (1989) Stability and equation of state of $CaSiO_3$ up to 134 GPa. *Journal of Geophysical Research* 94: 17889–17894.

Mao HK, Hemley RJ, Fei Y, *et al.* (1991) Effect of pressure, temperature and composition on lattice parameters and density of $(Fe,Mg)SiO_3$-perovskites to 30 GPa. *Journal of Geophysical Research* 96: 8069–8080.

Mao WL, Shen G, Prakapenka VB, *et al.* (2004) Ferromagnesian postperovskite silicates in the D″ layer of the Earth. *Proceedings of the National Academy of Sciences* 101: 15867–15869.

Mao WL, Meng Y, Shen G, *et al.* (2005) Iron-rich silicates in the Earth's D″ layer. *Proceedings of the National Academy of Sciences* 102: 9751–9753.

Marton F and Cohen R (1994) Prediction of a high pressure phase transition in Al_2O_3. *American Mineralogist* 79: 789–792.

Matsui M and Nishiyama N (2002) Comparison between the Au and MgO pressure calibration standards at high temperature. *Geophysical Research Letters* 29: 1368 (doi:10.1029/2001GL014161).

Mattern E, Matas J, Ricard Y, and Bass J (2005) Lower mantle composition and temperature from mineral physics and thermodynamic modeling. *Geophysical Journal International* 160: 973–990.

Mazin II, Fei Y, Downs R, and Cohen R (1998) Possible polytypism in FeO at high pressures. *American Mineralogist* 83: 451–457.

Meade C, Mao HK, and Hu J (1995) High-temperature phase transition and dissociation of $(Mg,Fe)SiO_3$ perovskite at lower mantle pressures. *Science* 268: 1743–1745.

Michael PJ and Bonatti E (1985) Peridotite composition from the North Atlantic: Regional and tectonic variations and implications for partial melting. *Earth and Planetary Science Letters* 73: 91–104.

Miyajima N, Yagi T, Hirose K, Kondo T, Fujino K, and Miura H (2001) Potential host phase of aluminum and potassium in the Earth's lower mantle. *American Mineralogist* 86: 740–746.

Murakami M, Hirose K, Ono S, and Ohishi Y (2003) Stability of $CaCl_2$-type and α-PbO_2-type SiO_2 at high pressure and temperature determined by in-situ X-ray measurements. *Geophysical Research Letters* 30: doi:10.1029/2002GL016722.

Murakami M, Hirose K, Kawamura K, Sata N, and Ohishi Y (2004a) Post-perovskite phase transition in MgSiO$_3$. *Science* 304: 855–858.

Murakami M, Hirose K, Ono S, Isshiki M, and Watanuki T (2004b) High pressure and high temperature phase transitions of FeO. *Physics of Earth and Planetary Interiors* 146: 273–282.

Murakami M, Hirose K, Sata N, and Ohishi Y (2005) Post-perovskite phase transition and mineral chemistry in the pyrolitic lowermost mantle. *Geophysical Research Letters* 32: L03304 (doi:10.1029/2004GL021956).

Nakagawa T and Tackley PJ (2004) Effects of a perovskite–post perovskite phase change near core–mantle boundary in compressible mantle convection. *Geophysical Research Letters* 31: L16611 (doi:10.1029/2004GL020648).

Niu F and Kawakatsu H (1996) Complex structure of the mantle discontinuities at the tip of the subducting slab beneath the northeast China: A preliminary investigation of broadband receiver functions. *Journal of the Physics of the Earth* 44: 701–711.

Nishiyama N, Irifune T, Inoue T, Ando J, and Funakoshi K (2004) Precise determination of phase relations in pyrolite across the 660 km seismic discontinuity by in situ X-ray diffraction and quench experiments. *Physics of Earth and Planetary Interiors* 143-144: 185–199.

Nishihara Y (2003) *Density and Elasticity of Subducted Oceanic Crust in the Earth's mantle*, pp.169. Ph.D Thesis, Tokyo Institute of Technology.

Oganov AR, Gillan MJ, and Price GD (2005) Structural stability of silica at high pressures and temperatures. *Physical Review B* 71: 064104.

Oganov AR and Ono S (2004) Theoretical and experimental evidence for a post-perovskite phase of MgSiO$_3$ in Earth's D" layer. *Nature* 430: 445–448.

Ohtani E, Kagawa N, and Fujino K (1991) Stability of majorite (Mg,Fe)SiO$_3$ at high pressures and 1800°C. *Earth and Planetary Science Letters* 102: 158–166.

Ohtani E, Mizobata H, Kudoh Y, et al. (1997) A new hydrous silicate, a water reservoir, in the upper part of the lower mantle. *Geophysical Research Letters* 24: 1047–1050.

Ohtani E, Litasov K, Hosoya T, Kubo T, and Kondo T (2004) Water transport into the deep mantle and formation of a hydrous transition zone. *Physics of Earth and Planetary Interiors* 143–144: 255–269.

Ono S, Ito E, and Katsura T (2001) Mineralogy of subducted basaltic crust (MORB) from 25 to 37 GPa, and chemical heterogeneity of the lower mantle. *Earth and Planetary Science Letters* 190: 57–63.

Ono S, Hirose K, Murakami M, and Isshiki M (2002) Post-stishovite phase boundary in SiO$_2$ determined by in situ X-ray observations. *Earth and Planetary Science Letters* 197: 187–192.

Ono S, Ohishi Y, and Mibe K (2004) Phase transition of Ca-perovskite and stability of Al-bearing Mg-perovskite in the lower mantle. *American Mineralogist* 89: 1480–1485.

Ono S, Kikegawa T, and Ohishi Y (2005a) A high-pressure and high-temperature synthesis of platinum carbide. *Solid State Communications* 133: 55–59.

Ono S, Funakoshi K, Ohishi Y, and Takahashi E (2005b) In situ X-ray observation of the phase transition of Fe$_2$O$_3$. *Journal of Physics – Condensed Matter* 17: 269–276.

Ono S, Kikegawa T, Ohishi Y, and Tsuchiya J (2005c) Post-aragonite phase transformation in CaCO$_3$ at 40 GPa. *American Mineralogist* 90: 667–671.

Ono S, Ohishi Y, Isshiki M, and Watanuki T (2005d) In situ X-ray observations of phase assemblages in peridotite and basalt compositions at lower mantle conditions: Implications for density of subducted oceanic plate. *Journal of Geophysical Research* 110: B02208 (doi:10.1029/2004JB003196).

Parise JB, Wang Y, Yeganeh-Haeri A, Cox DE, and Fei Y (1990) Crystal structure and thermal expansion of (Mg,Fe)SiO$_3$ perovskite. *Geophysical Research Letters* 17: 2089–2092.

Perdew JP, Chevary JA, Vosko SH, et al. (1992) Atoms, molecules, solids, and surfaces: Applications of the generalized gradient approximation for exchange and correlation. *Physical Review B* 46: 6671–6687.

Perdew JP, Burke K, and Ernzerhof M (1996) Generalized gradient approximation made simple. *Physical Review Letters* 77: 3865–3868.

Ringwood AE (1962) A model for the upper mantle. *Journal of Geophysical Research* 67: 857–867.

Ringwood AE and Irifune T (1988) Nature of the 650-km seismic discontinuity: Implications for mantle dynamics and differentiation. *Nature* 331: 131–136.

Ringwood AE (1994) Role of the transition zone and 660 km discontinuity in mantle dynamics. *Physics of Earth and Planetary Interiors* 86: 5–24.

Ross NL and Hazen RM (1990) High-pressure crystal chemistry of MgSiO$_3$ perovskite. *Physics and Chemistry of Minerals* 17: 228–237.

Ross NL, Shu JF, Hazen M, and Gasparik T (1990) High-pressure crystal chemistry of stishovite. *American Mineralogist* 75: 739–747.

Sanehira T, Irifune T, Shinmei T, Brunet F, Funakoshi K, and Nozawa A (2005) In-situ X-ray diffraction study of an aluminous phase in MORB under lower mantle conditions. *Physics and Chemistry of Minerals* 33: 28–34 (doi:10.1007/s00269-005-0043-0).

Santillán J and Williams Q (2004) A high pressure X-ray diffraction study of aragonite and the post-aragonite phase transition in CaCO$_3$. *American Mineralogist* 89: 1348–1352.

Sawamoto H (1987) Phase diagram of MgSiO$_3$ at pressures up to 24 GPa and temperatures up to 2200C: Phase stability and properties of tetragonal garnet. In: Manghnani MH and Syono Y (eds.) *High-Pressure Research in Mineral Physics*, pp. 209–219. Tokyo: Terrapub/AGU.

Saxena SK, Dubrovinsky LS, Lazor P, et al. (1996) Stability of perovskite (MgSiO$_3$) in the earth's mantle. *Science* 274: 1357–1359.

Sharp TG, El Goresy A, Wopenka B, and Chen M (1999) A post-stishovite SiO$_2$ polymorph in the meteorite Shergotty: Implications for impact events. *Science* 284: 1511–1513.

Shen G, Rivers ML, Wang Y, and Sutton R (2001) Laser heated diamond anvil cell system at the Advanced Photon Source for in situ X-ray measurements at high pressure and temperature. *Review of Scientific Instruments* 72: 1273–1282.

Sherman DM and Jansen H (1995) First-principles prediction of the high-pressure phase transition and electronic structure of FeO; implications for the chemistry of the lower mantle and core. *Geophysical Research Letters* 22: 1001–1004.

Shieh SR, Mao H-K, Hemley RJ, and Ming LC (1998) Decomposition of phase D in the lower mantle and the fate of dense hydrous silicates in subducting slab. *Earth and Planetary Science Letters* 159: 13–23.

Shieh SR, Duffy TS, and Li B (2002) Strength and elasticity of SiO$_2$ across the stishovite–CaCl$_2$-type structural phase boundary. *Physical Review Letters* 89: 255507.

Shieh SR, Duffy TS, and Shen G (2004) Elasticity and strength of calcium silicate perovskite at lower mantle pressures. *Physics of Earth and Planetary Interiors* 143-144: 93–105.

Shim S-H and Duffy TS (2000) Constraints on the P–V–T equation of state of MgSiO$_3$ perovskite. *American Mineralogist* 85: 354–363.

Shim S-H, Duffy TS, and Shen G (2000a) The stability and P–V–T equation of state of CaSiO$_3$ perovskite in the earth's lower mantle. *Journal of Geophysical Research* 105: 25955–25968.

Shim S-H, Duffy TS, and Shen G (2000b) The equation of state of CaSiO$_3$ perovskite to 108 GPa at 300 K. *Physics of Earth and Planetary Interiors* 120: 327–338.

Shim S-H, Duffy TS, and Shen G (2001a) The post-spinel transformation in Mg$_2$SiO$_4$ and its relation to the 660-km seismic discontinuity. *Nature* 411: 571–573.

Shim S-H, Duffy TS, and Shen G (2001b) Stability and structure of MgSiO$_3$ perovskite to 2300-kilometer depth in earth's mantle. *Science* 293: 2437–2440.

Shim S-H, Duffy TS, and Takemura K (2002a) Equation of state of gold and its application to the phase boundaries near 660 km depth in the Earth's mantle. *Earth and Planetary Science Letters* 203: 729–739.

Shim S-H, Jeanloz R, and Duffy TS (2002b) Tetragonal structure of CaSiO$_3$ perovskite above 20 GPa. *Geophysical Research Letters* 29: 2166.

Shim S-H, Duffy TS, Jeanloz R, and Shen G (2004) Stability and crystal structure of MgSiO$_3$ perovskite to the core–mantle boundary. *Geophysical Research Letters* 31: 10.1029/2004GL019639.

Shimomura O, Yamaoka S, Yagi T, et al. (1984) *Materials Research Society Symposium Proceedings, vol. 22: Multi-Anvil Type X-ray Apparatus for Synchrotron Radiation*, 17–20 pp. Amsterdam: Elsevier.

Shinmei T, Sanehira T, Yamazaki D, et al. (2005) High-temperature and high-pressure equation of state for the hexagonal phase in the system NaAlSiO$_4$–MgAl$_2$O$_4$. *Physics and Chemistry of Minerals* 32: 594–602 (doi:10.1007/s00269-005-0029-y).

Sidorin I, Gurnis M, and Helmberger DV (1999) Evidence for a ubiquitous seismic discontinuity at the base of the mantle. *Science* 286: 1326–1331.

Simmons NA and Gurrola H (2000) Seismic evidence for multiple discontinuities near the base of the transition zone. *Nature* 405: 559–562.

Sinmyo R, Hirose K, Hamane D, Seto Y, and Fujino K (2006) Partitioning of iron between perovskite/post-perovskite and magnesiowüstite in the Lower Mantle. *Eos Transactions of the American Geophysical Unioin* 87: Fall Meeting Supplement, Abstract MR11A-0105.

Sinogeikin SV and Bass JD (2000) Single-crystal elasticity of pyrope and MgO to 20 GPa by Brillouin scattering in the diamond cell. *Physics of Earth and Planetary Interiors* 120: 43–62.

Sinogeikin SV, Zhang J, and Bass JD (2004) Elasticity of single crystal and polycrystalline MgSiO$_3$ perovskite by Brillouin spectroscopy. *Geophysical Research Letters* 31: L06620 (doi:10.1029/2004GL019559).

Stacey FD and Isaak DG (2001) Compositional constraints on the equation of state and thermal properties of the lower mantle. *Geophysical Journal International* 146: 143–154.

Stackhouse S, Brodholt JP, Wookey J, Kendall J-M, and Price GD (2005a) The effect of temperature on the seismic anisotropy of the perovskite and post-perovskite polymorphs of MgSiO$_3$. *Earth and Planetary Science Letters* 230: 1–10.

Stackhouse S, Brodholt JP, and Price GD (2005b) High temperature elastic anisotropy of the perovskite and post-perovskite polymorphs of Al$_2$O$_3$. *Geophysical Research Letters* 32: L13305 (doi:10.1029/2005GL023163).

Stackhouse S, Brodholt JP, and Price GD (2006) Elastic anisotropy of FeSiO$_3$ end-members of the perovskite and post-perovskite phases. *Geophysical Research Letters* 33: L01304 (doi:10.1029/2005GL023887).

Stixrude L, Cohen RE, Yu R, and Krakauer H (1996) Prediction of phase transition in CaSiO$_3$ perovskite and implications for lower mantle structure. *American Mineralogist* 81: 1293–1296.

Sueda Y, Irifune T, Nishiyama N, et al. (2004) A new high-pressure form of KAlSi$_3$O$_8$ under lower mantle conditions. *Geophysical Research Letters* 31: L23612 (doi:10.1029/2004GL021156).

Sun S (1982) Chemical composition and the origin of the Earth's primitive mantle. *Geochimica et Cosmochimica Acta* 46: 179–192.

Suzuki A, Ohtani E, and Kamada T (2000) A new hydrous phase δ-AlOOH synthesized at 21 GPa and 1000°C. *Physics and Chemistry of Minerals* 27: 689–693.

Tackley PJ (2000) Mantle convection and plate tectonics: toward an integrated physical and chemical theory. *Science* 288: 2002–2007.

Tamai H and Yagi T (1989) High-pressure. and high-temperature phase relations in CaSiO$_3$ and CaMgSi$_2$O$_6$ and elasticity. of perovskite-type CaSiO$_3$. *Physics of Earth and Planetary Interiors* 54: 370–377.

Tange Y (2006) The system MgO–FeO–SiO$_2$ up to 50 GPa and 2000°C: *An Application of Newly Developed Techniques Using Sintered Diamond Anvils in Kawai-Type High-Pressure Apparatus*, Doctoral Thesis, Tokyo Institute of Technology.

Tateno S, Hirose K, Sata N, and Ohishi Y (2005) Phase relations in Mg$_3$Al$_2$Si$_3$O$_{12}$ to 180 GPa: Effect of Al on post-perovskite phase transition. *Geophysical Research Letters* 32: L15306 (doi:10.1029/2005GL023309).

Taylor SR and McLennan SM (1985) *The Continental Crust: Its Composition and Evolution*, 312 pp. Oxford: Blackwell.

Thomson KT, Wentzcovitch RM, and Bukowinski M (1996) Polymorphs of Alumina predicted by first principles. *Science* 274: 1880–1882.

Tomioka N, Mori H, and Fujino K (2000) Shock-induced transition of NaAlSi$_3$O$_8$ feldspar into a hollandite structure in a L6 chondrite. *Geophysical Research Letters* 27: 3997–4000.

Trampert T, Deschamps F, Resovsky J, and Yuen D (2004) Probabilistic tomography maps chemical heterogeneities throughout the lower mantle. *Science* 306: 853–856.

Tsuchiya J, Tsuchiya T, and Tsuneyuki S (2002) First principles calculation of a high-pressure hydrous phase, δ-AlOOH. *Geophysical Research Letters* 29: 1909.

Tsuchiya J, Tsuchiya T, and Tsuneyuki S (2005a) First-principles study of hydrogen bond symmetrization of phase D under pressure. *American Mineralogist* 90: 44–49.

Tsuchiya J, Tsuchiya T, and Wentzcovitch RM (2005b) Vibrational and thermodynamic properties of MgSiO$_3$ post-perovskite. *Journal of Geophysical Research* 110: B02204 (doi:10.1029/2004JB003409).

Tsuchiya J, Tsuchiya T, and Wentzcovitch RM (2005c) Transition from the Rh$_2$O$_3$(II)-to-CaIrO$_3$ structure and the high-pressure–temperature phase diagram of alumina. *Physical Review–B* 72: 020103(R) (doi:10.1103/PhysRevB.72.020103).

Tsuchiya T and Kawamura K (2001) Systematics of elasticity: *Ab initio* study in B1-type alkaline earth oxides. *Journal of Chemical Physics* 114: 10086–10093.

Tsuchiya T (2003) First-principles prediction of the *P–V–T* equation of state of gold and the 660-km discontinuity in Earth's mantle. *Journal of Geophysical Research* 108: 2462 (doi:10.1029/2003JB002446).

Tsuchiya T, Tsuchiya J, Umemoto K, and Wentzcovitch RM (2004a) Phase transition in MgSiO$_3$ perovskite in the Earth's lower mantle. *Earth and Planetary Science Letters* 224: 241–248.

Tsuchiya T, Tsuchiya J, Umemoto K, and Wentzcovitch RM (2004b) Elasticity of post-perovskite MgSiO$_3$. *Geophysical Research Letters* 31: L14603 (doi:10.1029/2004GL020278).

Tsuchiya T, Caracas R, and Tsuchiya J (2004c) First principles determination of the phase boundaries of high-pressure polymorphs of silica. *Geophysical Research Letters* 31: L11610 (doi:10.1029/2004GL019649).

Tsuchiya T and Tsuchiya J (2006) Effect of impurity on the elasticity of perovskite and postperovskite: Velocity contrast across the postperovskite transition in (Mg,Fe,Al) (Si,Al)O$_3$.

Geophysical Research Letters 33: L12S04 (doi:10.1029/2006GL025706).

Tsuchiya T, Wentzcovitch RM, and de Gironcoli S (2006) Spin transition in magnesiowüstite in Earth's lower mantle. *Physical Review Letters* 96: 198501.

Tutti F, Dubrovinsky LS, and Saxena SK (2000) High pressure phase transformation of jadeite and stability of NaAlSiO$_4$ with calcium-ferrite type structure in the lower mantle conditions. *Geophysical Research Letters* 27: 2025–2028.

Tutti F, Dubrovinsky LS, Saxena SK, and Carlson S (2001) Stability of KAlSi$_3$O$_8$ hollandite-type structure in the Earth's lower mantle conditions. *Geophysical Research Letters* 28: 2735–2738.

Umemoto K, Wentzcovitch RM, and Allen PB (2006) Dissociation of MgSiO$_3$ in the cores of gas giants and terrestrial exoplanets. *Science* 311: 983–986.

van der Hilst R, Widiyantoro S, and Engdahl E (1997) Evidence for deep mantle circulation from global tomography. *Nature* 386: 578–584.

Vinnik L, Kato M, and Kawakatsu H (2001) Search for seismic discontinuities in the lower mantle. *Geophysical Journal International* 147: 41–56.

Wang W and Takahashi E (1999) Subsolidus and melting experiments of a K-rich basaltic composition to 27 GPa: Implication for the behavior of potassium in the mantle. *American Mineralogist* 84: 357–361.

Wang Y, Weidner DJ, and Guyot F (1996) Thermal equation of state of CaSiO$_3$ perovskite. *Journal of Geophysical Research* 101: 661–672.

Wang Y, Uchida T, Von Dreele R, *et al.* (2004) A new technique for angle-dispersive powder diffraction using an energy-dispersive setup and synchrotron radiation. *Journal of Applied Crystallography* 37: 947–956.

Weidner DJ and Wang Y (2000) Phase transformations: implications for mantle structure. In: Karato S, Forte AM, Liebermann RC, Masters G, and Stixrude L (eds.) *Geophysical Monograph*, vol. 117, pp. 215–235. Washington, DC: American Geophysical Union.

Wentzcovitch RM, Ross NL, and Price GD (1995) *Ab initio* study of MgSiO$_3$ and CaSiO$_3$ perovskites at lower-mantle pressures. *Physics of Earth and Planetary Interiors* 90: 101–112.

Wentzcovitch RM, Stixrude L, Karki BB, and Kiefer B (2004) Akimotoite to perovskite phase transition in MgSiO$_3$. *Geophysical Research Letters* 31: L10611 (doi:10.1029/2004GL019704).

Wentzcovitch RM, Tsuchiya T, and Tsuchiya J (2006) MgSiO$_3$ postperovskite at D″ conditions. *Proceedings of the National Academy of Sciences USA* 103: 543–546.

Williams Q (1998) The temperature contrast across D″. In: Gurnis M, Wysession ME, Knittle E, and Buffett BA (eds.) *The Core-Mantle Boundary Region, pp.73–81*. Washington, DC: AGU.

Wood BJ and Rubie DC (1996) The effect of alumina on phase transformations at the 660-kilometer discontinuity from Fe-Mg partitioning experiments. *Science* 273: 1522–1524.

Wood BJ (2000) Phase transformations and partitioning relations in peridotite under lower mantle conditions. *Earth and Planetary Science Letters* 174: 341–354.

Wookey J, Stackhouse S, Kendall J-M, Brodholt JP, and Price GD (2005) Efficacy of the post-perovskite phase as an explanation of lowermost mantle seismic properties. *Nature* 438: 1004–1008.

Wysession ME, Lay T, Revenaugh J, *et al.* (1998) The D″ discontinuity and its implications. In: Gurnis M, Wysession ME, Knittle E, and Buffett BA (eds.) *The Core–Mantle Boundary Region*, pp. 273–298. Washington: DC AGU.

Yagi T, Mao HK, and Bell PM (1982) Hydrostatic compression of perovskite-type MgSiO$_3$. In: Saxena K (ed.) *Advances on Physical Geochemistry*, vol.2, pp. 317–325. New York: Springer-Verlag.

Yagi T, Suzuki T, and Akaogi M (1994) High pressure transitions in the system KAlSi$_3$O$_8$-NaAlSi$_3$O$_8$. *Physics and Chemistry of Minerals* 21: 387–391.

Yang H, Prewitt CT, and Frost DJ (1997) Crystal structure of the dense hydrous magnesium silicate, phase D. *American Mineralogist* 82: 651–654.

Yamazaki D and Karato S (2002) Fabric development in (Mg,Fe)O during large strain, shear deformation: Implications for seismic anisotropy in Earth's lower mantle. *Physics of Earth and Planetary Interiors* 131: 251–267.

Yeganeh-Haeri A, Weidner DJ, and Ito E (1989) Elasticity of MgSiO$_3$ in the perovskite structure. *Science* 248: 787–789.

Yu YG, Wentzcovitch RM, Tsuchiya T, Umemoto K, and Tsuchiya J (2006) First Principles Investigation of Ringwoodite's Dissociation. *Geophysical Research Letters* (submitted).

Zhao D (2004) Global tomographic images of mantle plumes and subducting slabs: insight into deep Earth dynamics. *Physics of Earth and Planetary Interiors* 146: 3–34.

4 Mineralogy of the Earth – Trace Elements and Hydrogen in the Earth's Transition Zone and Lower Mantle

B. J. Wood and A. Corgne, Macquarie University, Sydney, NSW, Australia

4.1	Introduction	63
4.1.1	Concentrations in the Primitive Upper Mantle and Bulk Silicate Earth	64
4.1.2	Differentiation of the Silicate Earth and Formation of Deep-Mantle Reservoirs	65
4.1.3	Objectives	66
4.2	Crystal–Melt Partition Coefficients and Controlling Factors	66
4.2.1	Majoritic Garnet	67
4.2.1.1	Crystallochemical control	67
4.2.1.2	Effects of pressure, temperature, and composition	68
4.2.2	Magnesium Silicate Perovskite	69
4.2.2.1	Crystal chemistry	69
4.2.2.2	Effects of Al content on large trivalent and tetravalent cations	69
4.2.2.3	Other potential factors	70
4.2.3	Calcium Silicate Perovskite	70
4.2.3.1	Crystal chemistry	70
4.2.3.2	Substitution mechanisms in calcium silicate perovskite	71
4.2.3.3	Effects of melt composition	72
4.2.3.4	Comparison between $CaTiO_3$ and $CaSiO_3$ perovskites	73
4.2.4	Trace Elements in Other Mantle Minerals	74
4.2.4.1	Wadsleyite and ringwoodite	74
4.2.4.2	Ferropericlase	74
4.2.4.3	Post-perovskite	75
4.3	Implications for Planetary Differentiation and Trace Element Distribution in the Deep Mantle	75
4.3.1	A Primitive Majoritic or Perovskitic Reservoir?	75
4.3.2	Downward Migration and Segregation of Dense Melts?	77
4.3.3	Recycled Lithosphere in the Deep Mantle?	77
4.4	Water in the Deep Mantle	78
4.4.1	Water Solubility in Minerals from the Transition Zone	78
4.4.2	Seismic Constraints on the Water Content of the Transition Zone	79
4.4.2.1	Seismic velocity and density jump across the 410 km discontinuity	79
4.4.2.2	Water and sharpness of the 410 km discontinuity	79
4.4.3	Transition Zone 'Water Filter'	82
4.4.4	Water Solubility in Lower-Mantle Minerals	82
4.4.4.1	Magnesium silicate perovskite	83
4.4.4.2	Calcium silicate perovskite	83
4.4.4.3	Ferropericlase	84
4.5	Conclusions	84
References		85

4.1 Introduction

The aim of this chapter is to review the available experimental and theoretical evidence bearing on the distribution of trace elements and H in the transition zone and lower mantle. The first step, before considering the processes involved in the chemical differentiation of the Earth, is to consider the available constraints on the bulk composition of the silicate Earth.

4.1.1 Concentrations in the Primitive Upper Mantle and Bulk Silicate Earth

The silicate Earth is conveniently divided into continental crust (0.55% by mass), oceanic crust (0.15%), upper mantle (15.3%), transition zone (11.1%), and lower mantle (72.9%). The compositions of continental and oceanic crust are consistent with ultimate derivation from the mantle (Hofmann, 1988) by processes of near-anhydrous partial melting at mid-ocean ridges and more H_2O-rich melting above subduction zones (Tatsumi and Eggins, 1995). In order to estimate the composition of the bulk silicate Earth, therefore, we need to reconstruct the composition of the mantle prior to extraction of the crust. This is normally done using the compositions of fertile or least fractionated upper-mantle peridotites (e.g., Jagoutz *et al.*, 1989; Allègre *et al.*, 1995; McDonough and Sun, 1995; Palme and O'Neill, 2003; Walter, 2003), which occur as xenoliths in volcanic rocks and in orogenic massifs. Garnet peridotite xenoliths in kimberlites and lamproites sample the deepest levels, up to about 200 km below the continental crust (e.g., Rudnick and Nyblade, 1999). This procedure yields the concentrations of a wide range of elements in the primitive (i.e., undifferentiated) upper mantle. It is most accurate for those elements which are either compatible in mantle silicates (partition strongly into the solid during partial melting) or are only slightly incompatible. The concentrations of highly incompatible elements, which are concentrated strongly into the crust, are more accurately determined from the crustal rocks themselves. It should be noted that this primitive upper mantle does not correspond to the source region of mid-ocean ridge basalts (MORBs). The latter is more depleted in incompatible elements such as K, U, Th, and the light rare earth elements (REEs).

Having estimated the composition of the primitive upper mantle, it is important to establish any clear compositional differences between this part of the mantle and the inaccessible deeper parts. To do this, it is necessary to have a model for the bulk composition of the Earth. Chemical affinities between Earth, Moon, Mars, Mercury, Vesta, and chondritic meteorites imply that these meteorites are a suitable reference point for the compositions of terrestrial planets (e.g., Ringwood, 1979; Morgan and Anders, 1980; Wänke and Dreibus, 1988; Wasson, 1988). CI carbonaceous chondrites in particular have strong compositional similarities to the solar photosphere (e.g., Palme and Jones (2003), and references therein) and appear to represent primitive

planetary material, with the exceptions that they are volatile-enriched relative to the terrestrial planets and volatile-depleted relative to the sun.

Figure 1 shows the estimated composition of the primitive upper mantle plotted as a function of the temperature by which 50% of the element of interest would have condensed during cooling of a gas of solar composition (Wasson, 1985; Lodders, 2003). The most important point is that refractory lithophile elements such as Ca, Sc, Ti, Zr, U, and the REEs are present in the primitive upper mantle in approximately the same relative proportions as in carbonaceous chondrites (e.g., McDonough and Sun, 1995). The implication is that the bulk Earth contains the same relative proportions of refractory elements as the carbonaceous chondrites. This is the basis of the (carbonaceous) chondrite reference model. Furthermore, since refractory lithophile elements are not fractionated from one another in Earth's primitive upper mantle, one important consequence of choosing a chondritic reference is that the core, lower mantle, and transition zone do not have fractionated refractory lithophile element patterns either. The composition of the primitive upper mantle has therefore been considered for many elements to be representative of the composition of the bulk silicate Earth.

As can be seen on the left-hand side of **Figure 1**, there is a decreasing relative abundance of some lithophile elements in the silicate Earth with decreasing condensation temperature or increasing

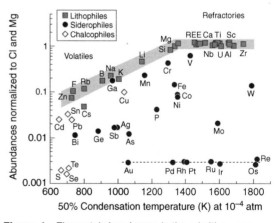

Figure 1 Elemental abundances in the primitive upper mantle normalized to CI carbonaceous chondrite and Mg are plotted relative to 50% condensation temperature at 10^{-4} atm pressure, which is used as a proxy for a nebular condensation sequence. Data for condensation temperatures are from Lodders (2003); chemical data for the chondrites are from Wasson and Kellemeyn (1988) and for the primitive upper mantle composition is from McDonough (2001).

volatility. This demonstrates that the silicate Earth is depleted in volatile elements (e.g., Li, B, Na, K, Rb) with respect to the chondritic reference. Depletions of the silicate Earth in refractory 'siderophile' elements such as W, Mo, Au, Re, and platinum-group elements are likely due to partial extraction of these elements into the core. Some of the more volatile elements such as S, Se, Pb, Ge, and P fall well below their expected positions on the 'volatility trend' of **Figure 1** and are also be assumed to be partially present in the core.

The concentrations of hydrogen species (not plotted in **Figure 1** due to their extremely high volatility) in the depleted upper mantle can be approximated from corresponding abundances in MORB glasses, while those in the whole mantle may be estimated from geochemical arguments such as those above (Wänke and Dreibus, 1988) and the K/H ratio of the exosphere (Jambon and Zimmermann, 1990). If we consider N-MORBs, which come from upper-mantle sources depleted in incompatible elements, then, assuming 10% melting, the source regions contain 80–180 ppm by weight of H_2O. For the more enriched source of E-MORBs, mantle water contents of 200–950 ppm are inferred, while basalts from back-arc basins are products of source regions containing around 1000–3000 ppm of H_2O (Michael, 1988, 1995; Sobolev and Chaussidon, 1996). Jambon and Zimmermann (1990) argue that the bulk silicate Earth water content (including crust and exosphere) is in the 550–1900 ppm range with the lower bound corresponding to the oceans plus the water concentration in the MORB source and the upper bound coming from the assumption that the K_2O/H_2O ratio of the exosphere (0.14) applies to the bulk silicate Earth. In the former case, current water concentrations in the whole mantle would be similar to those for the MORB source region (~200 ppm) and would correspond reasonably well to those derived from cosmochemical arguments (Wänke and Dreibus, 1988). In the latter case, the mantle would currently contain about 1500 ppm water and the implication is that there is a water-rich reservoir in either the transition zone (e.g., Bercovici and Karato, 2003) or the lower mantle (e.g., Murakami *et al.*, 2002). Geochemical arguments suggest that the source regions of ocean island basalts (OIBs) are water rich, with water contents between 300 and 1000 ppm (e.g., Dixon and Clague, 2001). Assuming that the sources of plumes reside in the deep mantle, this provides further evidence for a water-rich deep mantle.

Although there are considerable uncertainties in the concentrations of H species in the mantle, particularly below 410 km depth, geochemical arguments indicate that these concentrations in both upper and lower mantles are well below 1 wt.%, so that they are unlikely to affect the elastic properties or density of the mantle except where conditions are close to the solidus. However, they have impacts (1) on the transport properties of minerals (diffusion, deformation, electrical conduction), with implications for mantle viscosity and convection, and seismic anisotropy; (2) on the differentiation processes (melting behavior of mantle rocks, metasomatism); and (3) on the properties of the seismic discontinuities.

4.1.2 Differentiation of the Silicate Earth and Formation of Deep-Mantle Reservoirs

To study the distribution of trace elements and H in the deep mantle, it is helpful to know which mineralogical structures/reservoirs inherited from planetary differentiation are potentially present in the mantle. During its early history, it is probable that the Earth underwent extensive melting due to the impact energy of planetesimal bombardment, the heat produced by decay of short-lived radiogenic isotopes and the gravitational energy of core segregation (e.g., Wetherill, 1990; Chambers, 2004). Melting may have been long-lived and represented by a deep vigorously convecting magma ocean (e.g., Ohtani, 1985; Abe, 1997). Upon cooling, the magma ocean would have crystallized from the bottom up, leading, in principle, to a fractionally crystallized mantle. Although mantle convection may have significantly reduced such stratification (Tonks and Melosh, 1993; Solomatov, 2000), segregation of large amounts of deep-mantle liquidus minerals such as majoritic garnet, magnesium silicate perovskite, and calcium silicate perovskite would likely have driven element concentrations away from the initial primitive mantle values. Mantle degassing of volatile species (H, C, N, O and noble gases) and closure of the atmosphere took place soon after accretion ceased (e.g., Allègre *et al.*, 1986).

Until recently, the lack of accurate crystal–liquid partition coefficients under deep-mantle conditions has limited the ability to model properly the chemical differentiation of a molten silicate Earth. Estimated compositions of the primitive upper mantle are consistent with a whole mantle of homogeneous composition at least for refractory lithophile elements. Qualitatively, this implies either that the mantle did not differentiate early in Earth history or that subsequent solid-state convection in the mantle erased any initial stratification (e.g., van Keken *et al.*, 2002). We will investigate the possibility of stratification further in Section 4.3. Another possible mechanism of fractionation is downward migration of melts denser than the solid mantle to produce a compositionally distinct 'layer' in the deep mantle. This will also be addressed in Section 4.3. The existence of large-scale heterogeneities in the transition zone and lower mantle is supported by geochemical arguments, which suggest the presence in the deep mantle of undegassed reservoirs enriched in radiogenic heat-producing elements and with subchondritic $^{142}Nd/^{144}Nd$ ratio (e.g., Allègre *et al.*, 1996; Becker *et al.*, 1999; Kellogg *et al.*, 1999; Turcotte *et al.*, 2001; Samuel and Farnetani, 2003; Boyet and Carlson, 2005).

4.1.3   Objectives

The aim of this work is to provide a summary of experimental constraints on the distribution of trace elements in the transition zone and lower mantle, with a particular emphasis on refractory lithophile elements and the volatile species of H. We will start by briefly describing theoretical methods of systematizing the incorporation of trace elements in deep-mantle minerals as a function of ionic radius, and cation charge. Together with a review of recent experimental results on deep-mantle minerals, we will discuss the effects of crystal structure, composition, pressure, and temperature on trace element partitioning. This will allow us to describe the implications of deep-mantle melting for the distribution of trace elements in the transition zone and lower mantle. The second part of this chapter will focus on the distribution of water in the transition zone and lower mantle.

As discussed above, available samples of mantle rocks are limited to the uppermost mantle (less than ~200 km in depth). Studies of mineral inclusions in 'deep' diamonds have, however, provided additional insights into the geochemistry and mineralogy of the deep mantle. The results tend to confirm conclusions obtained from high-pressure experimental studies and will not be discussed in detail here. Readers seeking further information on inclusions in diamonds are advised to begin with the reviews by Harte *et al.* (1999) and Stachel *et al.* (2004).

4.2   Crystal–Melt Partition Coefficients and Controlling Factors

In order to understand the distribution of trace elements and volatiles in the silicate Earth, it is first necessary to consider how they dissolve in the minerals of the mantle. As shown in **Figure 2**, if we assume that the whole mantle is approximately the same major element composition as upper-mantle peridotite, then the principal mineral phases are magnesium silicates and oxides – olivine, orthopyroxene, and garnet in the upper mantle; majoritic garnet, wadsleyite $[\beta\text{-}(Mg, Fe)_2SiO_4]$, and ringwoodite $[\gamma\text{-}(Mg, Fe)_2SiO_4]$ in the transition zone; and magnesium silicate perovskite and ferropericlase in the lower mantle. It must be noted that, in addition to garnet and orthopyroxene, Ca dissolves in two other mineral phases: clinopyroxene in the upper mantle and calcium silicate perovskite in the lower mantle. The distribution of elements between these phases depends, of course, on the chemical properties of the elements of concern and the crystallochemical properties of the phases themselves. Since fractionation of elements between one part of the Earth and another (e.g., mantle–crust or upper mantle–lower mantle) almost invariably takes place through processes of

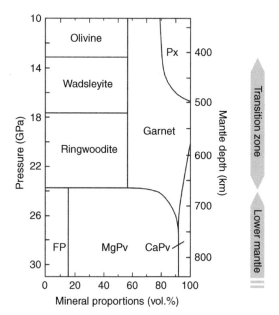

Figure 2   Mineral proportions in a peridotite composition as a function of pressure or depth (modified from Ringwood (1991) and Irifune (1994)). Symbols used are: Px, pyroxene; FP, ferropericlase; MgPv, magnesium silicate perovskite; CaPv, calcium silicate perovskite.

melting and crystallization, it is convenient to consider element distributions in terms of partitioning between solids and melts.

Recent work on solid–melt partitioning of cations (Blundy and Wood, 1994; Wood and Blundy, 2001) has quantified the effects of the ionic radius and charge on partitioning behavior. Blundy and Wood (1994) showed, following elastic strain theory (Brice, 1975), that the solid–liquid partition coefficient of ion i of radius r_i (D_i) can be calculated from that (D_o) of an ion of the same charge which enters the site without strain as follows:

$$D_i = D_o \exp\left(\frac{-4\pi E_s N_A}{RT} \left(\frac{r_o}{2}(r_o - r_i)^2 - \frac{1}{3}(r_o - r_i)^3 \right) \right) \quad [1]$$

In eqn [1], r_o is the effective radius of the site, N_A is Avogadro's number, and E_S is the effective Young's modulus of the site. Equation [1] describes a near-parabolic relationship between D_i and ionic radius for ions of fixed charge with the peak of the parabola corresponding to r_o and D_o and the tightness of the parabola depending on the 'stiffness' of the site represented by E_s. Partition coefficient D_i is defined as the ratio of weight concentration of element i in crystal phase to the weight concentration of element i in the liquid phase.

4.2.1 Majoritic Garnet

4.2.1.1 Crystallochemical control
Garnets have cubic symmetry with the general structural formula $X_3Y_2Z_3O_{12}$, where the X-, Y-, and Z-positions are dodecahedral, octahedral, and

tetragonal cation sites, respectively (e.g., Heinemann *et al.*, 1997). Majoritic garnets are rich in $MgSiO_3$ component which means that Mg is present in both X- and Y-sites, while Si occupies both Y- and Z-sites. Al and Fe^{3+} almost certainly occupy the Y-site, while Ca and Fe^{2+} occupy the X-site. Corgne and Wood (2004) have shown that crystal–melt partition coefficients for a suite of isovalent cations entering either the X- or Y-site of majoritic garnet exhibit the near-parabolic dependence on the radius of the incorporated cations predicted from eqn [1] (see **Figure 3**; Corgne and Wood, 2004). This underlines the important contribution made by the crystal structure to the control of trace element partitioning. Furthermore, as previously found for clinopyroxene and low-pressure garnet (Hill *et al.*, 2000; Van Westrenen *et al.*, 2000a), Young's modulus, E_s, increases and optimum radius r_o decreases with increasing charge of the cation incorporated at the X-site of majoritic garnet. In constructing **Figure 3**, Corgne and Wood (2004) considered that each ion enters only one site, with the exception of Mg and Si for which individual partition coefficients for the X- and Y-sites are reported. All ions with radius greater than that of Mg were allocated to the large X-site and all ions close to Al in size to the Y-site. This leaves a few ions, notably Ni, Sc, Ga, Cr, Ti, Sn, Hf, and Zr, which may, like Mg, enter both X- and Y-sites. Comparison of measured partition coefficients for these elements with the parabolae based on other elements suggests that Cr, Ga, and Ti predominantly enter the Y-site, while Sn, Hf, and Zr enter the X-site (**Figure 3**). Similarly, P and B may, like Si, enter both Y- and Z-sites. *D* values for divalent

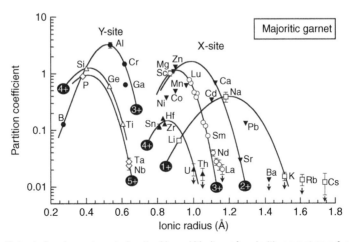

Figure 3 Partition coefficients for elements entering the X- and Y-sites of majoritic garnet as a function of ionic radius. Parabolae are fits of eqn [1] to experimental data (from Corgne and Wood, 2004). Partition coefficients for U, Th, La, Ba, K, Rb, and Cs are upper bounds.

transition metals do not fit the trend predicted from the elastic strain model applied to Ca, Mg, and Sr. These anomalies probably reflect both crystal field effects (for Ni and Co) and variable valence (for Fe, Cr, and perhaps Mn). The anomalous behaviour of Pb^{2+} is also generally observed (e.g., Blundy and Wood, 1994) and probably due to the presence of a lone pair of electrons on this ion.

4.2.1.2 Effects of pressure, temperature, and composition

Trace element partitioning must, in general, depend on pressure and temperature because partitioning from melt to crystal involves changes in fundamental thermodynamic properties of the species of concern. In addition to these fundamental effects, there are 'pressure effects' arising from changes in crystal composition as pressure and temperature are increased. For example, in the shallow mantle, garnets have high concentrations of the pyrope ($Mg_3Al_2Si_3O_{12}$) component. At greater depths, as pressure and temperature increase, the Al content of the garnet decreases and the phase becomes richer in majorite ($Mg_4Si_4O_{12}$) component (e.g., Irifune, 1987). As a result, trace element partitioning should change with pressure and temperature for both fundamental thermodynamic reasons and due to changes in crystal chemistry. **Figure 4** presents a series of partitioning parabolae obtained over a wide range of pressure conditions (3–25 GPa). Data at 3 GPa from van Westrenen *et al.* (2000a) in the CaO–MgO–Al_2O_3–SiO_2–FeO system ($Py_{78}Gr_{19}Alm_9$ garnet with Mg# of 89) are compared with data from Draper *et al.* (2003) obtained at 9 GPa on an ordinary chondrite composition ($Py_{80}Gr_{17}Alm_3$ garnet with Mg# of 83), with data from Walter *et al.* (2004) obtained at 23.5 GPa on a fertile peridotite composition (garnet Mg# of 95), with data from Corgne and Wood (2004) at 25 GPa on an Al-, Fe-rich peridotite composition (garnet Mg# of 89), and with data from Armstrong *et al.* (in preparation) obtained at 10, 15, and 17 GPa on a primitive mantle model composition with chondritic Mg/Si (garnet Mg# 94–96). As observed by Draper *et al.* (2003) between 5 and 9 GPa at approximately constant temperature (~1800°C), D_o^{3+} and E_X^{3+} for the X-site decrease with increasing pressure, that is, with increasing majorite component. Considering a wider range of pressure (**Figure 4**), D_o^{3+} decreases by about an order of magnitude from 8 at 3 GPa to 0.8 at 25 GPa. Furthermore, although there are relatively large uncertainties in D's for light REE, an increase of pressure results in more open

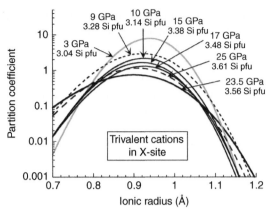

Figure 4 Lattice strain parabolae for trivalent cations (REE, Y, Sc) entering the X-site of majoritic garnet. As observed at low pressures (Draper *et al.*, 2003), D_o^{3+} decreases with increasing pressure, that is, with increasing majorite component (quantified using number of Si atoms per formula unit, Si pfu where $3.0 = 0\%$ majorite and $4.0 = 100\%$). References are van Westrenen *et al.* (2000a), 3 GPa and garnet Mg# of 89; Draper *et al.* (2003), 9 GPa and garnet Mg# of 83; Armstrong *et al.* (2006), 10–17 GPa and garnet Mg# of 94–96; Walter *et al.* (2004), 23.5 GPa garnet Mg# of 95; Corgne and Wood (2004), 25 GPa and garnet Mg# of 89. See text for complete discussion on controlling factors.

parabolae, corresponding to lower values of Young's modulus (E_X^{3+}). In other words, the X-site becomes 'softer', that is, more elastic, with increasing pressure and increasing majorite component.

Although the changes in D values shown in **Figure 4** are nominally due to pressure, it must be emphasized that temperature and compositional changes cannot be separated from those of pressure. Experiments at progressively higher pressure involve progressively higher temperatures and higher majorite contents of the resultant garnets. Thus, for example, one could envisage that the decrease of D_o^{3+} with increasing pressure (**Figure 4**) is not directly due to pressure but rather due to the increasing incorporation of Si^{4+} at the Y-site in the garnet. Increasing replacement of Al^{3+} by Si^{4+} could be expected to make it more difficult to charge-compensate the substitution of Mg^{2+} by 3+ ions at the X-site and hence lead to decreasing D_o^{3+} for this site. The interrelationships between pressure, temperature, and composition mean, therefore, that selection of partition coefficients for majoritic garnets must be performed cautiously and preferably by interpolation of values measured for bulk compositions closely corresponding to the ones of interest.

4.2.2 Magnesium Silicate Perovskite

As shown in **Figure 2**, in a peridotitic mantle composition, magnesium silicate perovskite forms from the breakdown of ringwoodite at 24 GPa and from the gradual decomposition of majoritic garnet between 24 and 27 GPa. Assuming that the lower mantle is peridotitic, it is made up, by weight, of approximately 79% $(Mg, Fe)SiO_3$ perovskite (MgPv), 16% $(Mg, Fe)O$ ferropericlase (FP), and 5% $CaSiO_3$ perovskite (CaPv) (e.g., Wood, 2000). The perovskite phases are crystallochemically unusual because of the presence of Si^{4+} in octahedral coordination (B-site), while the larger cation site (A-site) is regular 12-coordinate in CaPv and distorted 12-coordinate (with 8 close oxygens) in MgPv. When crystallized from peridotitic compositions at high temperatures, MgPv contains ~4 wt.% Al_2O_3 (Liebske *et al.*, 2005), while the alumina content of coexisting calcium perovskite is substantially less (Wood, 2000). Dissolution of alumina into MgPv has a dramatic effect on the properties of this phase, particularly on the oxidation state of the iron it contains. Al^{3+} prefers to replace Si^{4+} on the octahedral site and this is coupled, in Fe-bearing perovskites, to replacement of Mg^{2+} by Fe^{3+} (Wood and Rubie, 1996). The component $Fe^{3+}AlO_3$ is so stable in solution (McCammon, 1997; Frost and Langenhorst, 2002) that it can force the disproportionation of ferrous iron into ferric iron plus metal (Frost *et al.*, 2004), and its stability means that ~60% of the iron present in lower-mantle Al-bearing $(Mg, Fe)SiO_3$ perovskite should be Fe^{3+}. Clearly, coupled substitutions of this type may also affect the partitioning of other highly charged elements into perovskite.

Recent experimental and analytical improvements have made it possible to generate accurate MgPv–melt partition coefficients for a large number of trace elements. Since MgPv is believed to be the most abundant mineral in the mantle, these technical advancements are critical to the understanding of trace element distributions in the Earth. Hereafter, we summarize results from recent experimental studies at 23.5–26 GPa (Walter *et al.*, 2004; Corgne *et al.*, 2005; Liebske *et al.*, 2005) that constrain factors controlling MgPv–melt partitioning.

4.2.2.1 Crystal chemistry

As illustrated above for majoritic garnet, MgPv–melt partition coefficients can be parametrized using the 'lattice strain' model (eqn [1]). An example of lattice strain fits for MgPv–melt partition coefficients of elements entering the A-site is given in **Figure 5**. Note

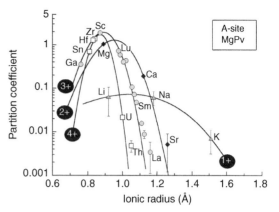

Figure 5 Partition coefficients for elements entering the A-site of magnesium silicate perovskite (MgPv) plotted as a function of ionic radius. Parabolae are fits of eqn [1] to experimental data from Corgne *et al.* (2005). In this case, MgPv contains 1.9 wt.% Al_2O_3 and 3.0 wt.% 'FeO'. Partition coefficients for divalent transition metals, Pb^{2+}, Ba^{2+}, and Ce^{3+}, are not shown because of effects in addition to lattice strain (see text for discussion). Ionic radii are values for an eightfold coordinated site taken from Shannon (1976), as no consistent set of 12-fold coordinated radii exists.

that because ionic radii for 12-fold coordination are not available for most elements, values corresponding to an eightfold coordinated site were used. These are ~0.2 Å smaller (Shannon, 1976). The successful application of the lattice strain theory confirms that crystal structure makes an important contribution to MgPv–melt partitioning. However, the partitioning behavior of some elements, notably transition elements, Pb, Ce, and Ba, is poorly modeled by the lattice strain theory. This indicates that additional factors such as crystal field effects or multiple valency complicate MgPv–melt partitioning of transition metals. For example, as shown in **Figure 6**, D_{Ce} is greater than predicted from the '3+ parabola'. Liebske *et al.* (2005) suggested that this is the result of some Ce being present as Ce^{4+}. Ba also has a partition coefficient higher than predicted from the lattice strain model. Based on the observation that Ba^{2+} has an ionic radius similar to O^{2-}, Walter *et al.* (2004) suggested that Ba^{2+} partly substitutes for O^{2-} and that charge compensation occurs by formation of an Si^{4+} vacancy.

4.2.2.2 Effects of Al content on large trivalent and tetravalent cations

As mentioned above, Al and Fe are two elements that are likely to influence the MgPv–melt partitioning of trace elements. Liebske *et al.* (2005) investigated the effects of Al and Fe contents of MgPv on trace element partitioning at similar conditions of pressure and

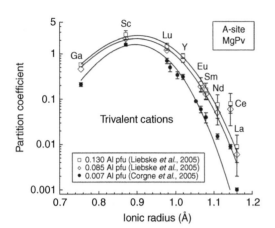

Figure 6 MgPv–melt partition coefficients for trivalent elements are plotted together with fits to eqn [1] as a function of cation radius. Note that D_{Ce} plots higher than predicted probably because of some Ce being present as Ce^{4+} (Liebske et al., 2005).

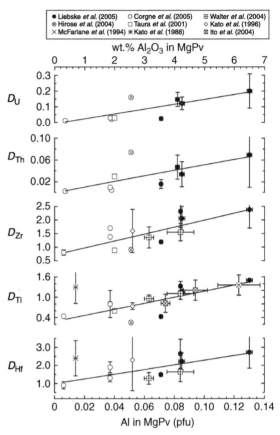

Figure 7 MgPv–melt partition coefficients for tetravalent elements plotted as a function of Al per formula unit in MgPv (normalized to two cations pfu).

temperature. They found, however, no evidence of a significant effect of bulk Fe content on trace element partitioning, even with large amounts of FeO (up to 35 wt.%). Increasing Al content of Mg-perovskite was also found not to influence partitioning of divalent cations. Partitioning of large trivalent and tetravalent cations is, however, sensitive to the Al content of perovskite, as shown in **Figures 6** and **7**. Both figures show that partition coefficients for trivalent and tetravalent cations entering the A-site increase with increasing MgPv Al content. The effect is the largest for light REE, D_{La} increasing by an order of magnitude as Al_2O_3 increases from ~0 to 6.5 wt.%. Ti, Zr, and Hf are incompatible in Al-free MgPv, but become compatible as Al is added in the crystal structure.

Substitution of trivalent and tetravalent cations into the large A-site requires charge compensation. Two main mechanisms have been proposed: (1) Al-coupled substitution into the B-site and (2) creation of Mg-vacancies (e.g., Corgne et al., 2003). The correlation between trivalent and tetravalent cation partition coefficients and MgPv Al content seems to support the first mechanism although incorporation of Al may also influence the energetics of Mg-vacancy formation. Heterovalent substitution in the large site of silicate perovskites is discussed further in Section 4.2.3.

4.2.2.3   Other potential factors

Apart from crystal composition, additional influences on partitioning behavior include pressure, temperature, and melt composition. Due to experimental

limitations, MgPv–melt partition coefficients have been generated for a very restricted range of pressure and temperature conditions. This means that it is currently impossible to separate the effects of the two intensive variables on partitioning. As for majoritic garnet, the effect of melt composition on partitioning is unknown. Although computational simulations (Corgne et al., 2003) indicate that melt composition is a significant factor, variation of measured partition coefficients can, to date, be explained without calling for a 'melt contribution' (e.g., Liebske et al., 2005).

4.2.3   Calcium Silicate Perovskite

4.2.3.1   Crystal chemistry

Figure 8 shows crystal–melt partition coefficients for trace elements entering the large site of CaPv at 25 GPa and 2300°C (Corgne et al., 2005). In this case, CaPv was crystallized from a non-peridotitic calcium-enriched composition and contained 1.6 wt.% Al_2O_3. As can be seen, the partitioning patterns of elements of different charge broadly fit the lattice strain model of Blundy and Wood (1994). As found for upper-mantle silicates (e.g., van Westrenen et al., 2000a), majoritic garnet (Corgne and Wood, 2004; see **Figure 3**) and in agreement with recent atomistic simulations (Corgne et al., 2003), r_o decreases and E_s generally increases with increasing charge of the cation incorporated at the large site of both $(Mg, Fe)SiO_3$ and $CaSiO_3$ perovskites. Comparison between the partition coefficients measured for MgPv (**Figure 5**) and those measured for CaPv (**Figure 8**) shows that the calcic phase much more readily accepts elements normally regarded as strongly incompatible. D_o^{3+} and D_o^{4+} for the large A-site are roughly 1 order of magnitude lower for MgPv than for CaPv. Furthermore, because of the differences in r_o^{3+} and r_o^{4+} between the A-site of MgPv and the A-site of CaPv, crystallization of CaPv fractionates the ratios of many refractory lithophile elements (e.g., Sc/Sm, Lu/Hf, U/Th) in the opposite sense to crystallization of MgPv. When coupled with the effect of ionic size, the differences between the two perovskites mean that, despite being volumetrically subordinate to MgPv, CaPv contains more than 90% of the heat-producing elements K, U, and Th in the lower mantle, together with most of the light rare

earths, Sr and Ba. The relative ease of entry of highly charged ions into both phases is unusual, however, and prompted experimental and theoretical investigations of the mechanisms by which charge balancing occurs in these phases (Corgne and Wood, 2005).

4.2.3.2 Substitution mechanisms in calcium silicate perovskite

Clear resolution of substitution mechanisms for $(Mg, Fe)SiO_3$ and $CaSiO_3$ perovskites are difficult tasks, requiring, in principle, large numbers of experiments at high pressures and temperatures. There are, however, potential shortcuts provided by calculations of the energetics of the different potential mechanisms (Corgne et al., 2003) and, experimentally, by use of the low-pressure analog of $CaSiO_3$, $CaTiO_3$. The patterns of partitioning of trace elements into $CaTiO_3$ and $CaSiO_3$ perovskites are virtually identical, as shown in **Figure 9** (Corgne and Wood, 2002). Furthermore, like $CaSiO_3$ perovskite, $CaTiO_3$ adopts the *Pm3m* cubic structure above ~1550 K and mechanisms of charge compensation might reasonably be expected to be the same in the two perovskites. Because large crystals of $CaTiO_3$ perovskite can be easily obtained at 1 atm, work on $CaTiO_3$ circumvents the experimental and analytical challenges related to the generation of large numbers of $CaSiO_3$ perovskite–melt partition coefficients.

Determining the concentration limits of Henry's law behavior in $CaTiO_3$ perovskite enables identification of the most likely substitution mechanisms for Ca by 3+ and 4+ cations such as Pr^{3+} and Th^{4+}. In the absence of Al^{3+} to substitute for Ti^{4+}, the energetically favorable mechanism for incorporation of Pr^{3+} is through the generation of calcium vacancies V''_{Ca} (Corgne et al., 2003). Corgne and Wood (2005)

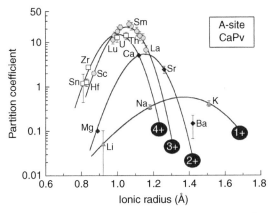

Figure 8 Partition coefficients for elements entering the A-site of calcium silicate perovskite (CaPv) plotted as a function of ionic radius. Parabolae are fits of eqn [1] to experimental data from Corgne et al. (2005). In comparison to MgPv (**Figure 5**) at identical conditions of pressure and temperature (~25 GPa, ~2300°C), CaPv more readily accepts elements normally regarded as incompatible. Ionic radii are values for an eightfold coordinated site taken from Shannon (1976), as no consistent set of 12-fold coordinated radii exists.

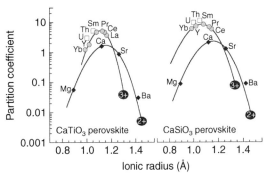

Figure 9 Partitioning behavior of 2+, 3+, and 4+ cations entering the A-site of $CaSiO_3$ and $CaTiO_3$ perovskites. Data are from Corgne and Wood (2002).

showed that if this is the only substitution mechanism, D_{Pr} should be a linear function of the reciprocal of the square root of vacancy concentration $[V''_{Ca}]$, while D_{Th} should be a linear function of the reciprocal of $[V''_{Ca}]$. Calcium vacancies can be formed in two ways. There are those vacancies generated as part of the heterovalent substitution (V_s), which depend linearly on the amount of trace ion added to the crystal and those intrinsically present in the pure crystal (V_i). If V_i dominate, then $[V''_{Ca}]$ is constant, and D_{Pr} and D_{Th} should be independent of the concentrations of Pr and Th in calcium perovskite, that is, Henry's law behavior. If, on the other hand, V_s dominate over V_i, then vacancy concentration and hence D_{Pr} and D_{Th} must depend on the concentrations of Pr and Th in the perovskite (noted $[Pr^{.}_{Ca}]$ and $[Th^{..}_{Ca}]$ below). Corgne and Wood (2005) found that the latter case leads to

$$D_{Pr}^{CaPv/Liq} \approx \text{constant} \cdot [Pr^{.}_{Ca}]^{1/2} \qquad [2]$$

$$D_{Th}^{CaPv/Liq} \approx \text{constant} \cdot [Th^{..}_{Ca}]^{-1} \qquad [3]$$

Figure 10(a) shows a plot of D_{Pr} versus the inverse of the square root of $[Pr^{.}_{ca}]$ for one specific composition (CAST) studied by Corgne and Wood (2005). At high Pr content in perovskite, D_{Pr} is approximately a linear function of the inverse of the square root of $[Pr^{.}_{Ca}]$. This agrees with a mechanism of substitution via Ca-vacancy formation with a predominance of V_s over V_i. At $[Pr^{.}_{Ca}]$ lower than ~1000 ppm, D_{Pr} is constant, independent of $[Pr^{.}_{Ca}]$ and Henry's law is obeyed. At

concentrations below ~1000 ppm Pr in perovskite, therefore, it appears that V_i dominate over V_s. Similar conclusions can be drawn for Th from **Figure 10(b)**. At high Th contents, D_{Th} is approximately a linear function of the inverse of $[Th^{..}_{Ca}]$ ($V_s \gg V_i$), while at Th contents lower than ~1000 ppm, D_{Th} is constant ($V_i \gg V_s$).

If Al is added to the system, it is possible, in addition to vacancy formation, to charge-compensate the insertion of Pr^{3+} or Th^{4+} for Ca^{2+} by coupled replacement of Al^{3+} for Ti^{4+}. Corgne and Wood (2005) showed, however, that the dependencies of partition coefficients on Pr and Th concentrations are the same in both Al-bearing and Al-free compositions. Furthermore, using similar arguments to those outlined above, they were able to show that the compositional dependence of partitioning is consistent with calcium vacancies being the dominant mechanism of charge compensation in Al-bearing as well as Al-free bulk compositions.

4.2.3.3   Effects of melt composition

Melt composition is known to be an important factor controlling the partitioning of trace elements between silicates and melts. In this case, it has been observed that partition coefficients for 3+ and 4+ ions into perovskite depend strongly on the CaO content of the coexisting melt (Hirose et al., 2004; Walter et al., 2004; Corgne et al., 2005; Corgne and Wood, 2005). Considering formation of cation vacancies as the dominant substitution mechanism for REE and Th^{4+} into the Ca-site of

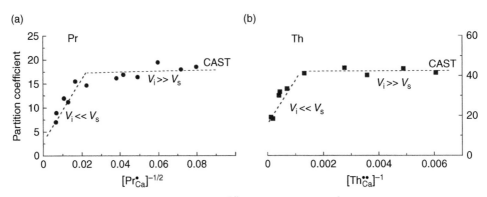

Figure 10 Figure showing plots of (a) D_{Pr} vs D_{Pr} vs $[Pr^{.}_{ca}]^{-1/2}$ and (b) D_{Th} vs $[Th^{..}_{ca}]^{-1}$ for $CaTiO_3$ perovskite. The data indicate that Ca-vacancy formation is the likely charge balance mechanism for the substitution of Pr and Th at the Ca-site. At low perovskite contents of Pr and Th or high $[Pr^{.}_{Ca}]^{-1/2}$ and $[Th^{..}_{Ca}]^{-1}$, intrinsic Ca-vacancies (V_i) dominate and partition coefficient is constant (Henry's law). Ca-vacancies formed as part of the substitution (V_s) start to become important when the V_i content is not large enough to account for high concentrations of Pr and Th. When V_s dominates, D_{Pr} and D_{Th} are linear functions of $[Pr^{.}_{ca}]^{-1/2}$ and $[Th^{..}_{ca}]^{-1}$ respectively. Data are from Corgne and Wood (2005) for a bulk composition chosen in the CaO–Al_2O_3–SiO_2–TiO_2 system (CAST).

perovskite and assuming that the activity of CaO in the melt varies as an approximately linear function of the CaO concentration, [CaO], Corgne and Wood (2005) predicted the following dependencies in the Henry's law region:

$$D_{Pr}^{CaPv/Liq} \approx \alpha \cdot [CaO_{Liq}]^{-3/2} \qquad [4]$$

$$D_{Th}^{CaPv/Liq} \approx \beta \cdot [CaO_{Liq}]^{-2} \qquad [5]$$

where α and β are constants.

As can be seen from **Figure 11**, both D_{Pr} and D_{Th} decrease with increasing CaO content of the melt and fit well to the models represented by eqns [4] and [5]. Since these equations are based on the cation vacancy mechanism of charge compensation, the fits further support this being the predominant substitution mechanism. They also indicate that charge compensation cannot be via oxygen interstitial incorporation. In the latter case, a dependence of partition coefficient on $[CaO]^{-1}$ should be observed. Such a compositional dependence is much weaker than that observed in the data of **Figure 11**, indicating that this mechanism of charge compensation is not important.

4.2.3.4 Comparison between CaTiO₃ and CaSiO₃ perovskites

Figure 9 shows that the forms and heights of the partitioning parabolae for CaTiO₃ and CaSiO₃

perovskites are very similar, which is the reason why the former was used as an analog to investigate the behavior of the latter, lower-mantle phase. When data for the two phases are inspected more closely, however, the similarities between them become even more striking. In **Figure 12**, partition coefficients for 2+, 3+ (both as D_o, the peak of the partitioning parabola), and 4+ (Th) ions entering the Ca-sites of the two perovskites are plotted as functions of the CaO content of the coexisting liquids. As can be seen, the data for each charge form one continuous line as a function of melt composition, 'independent of whether the solid is CaSiO₃ or CaTiO₃'. Thus, on the right-hand side of the diagram are data collected at 1 atm and about 1400°C on CaTiO₃ perovskite, while on the left are results obtained for CaSiO₃ at around 25 GPa and 2200–2300°C. Despite the differences in composition and pressure, temperature conditions, the trace element partitioning behavior is identical. All that seems to be important is that these elements are entering the Ca-sites in perovskites with the same crystal structure. The result demonstrates that low-pressure experiments on CaTiO₃ perovskite are near-perfect analogs of experiments on CaSiO₃ perovskite at high pressure. Note that the dependence of D_o^{2+} on the inverse of $[CaO]_{Liq}$ reflects the fixed CaO content of perovskite and does not require defect substitution (Corgne and Wood, 2005).

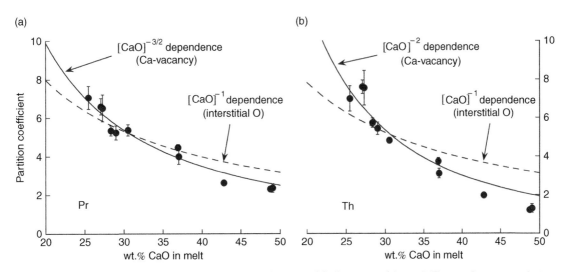

(a)   [CaO]$^{-3/2}$ dependence (Ca-vacancy)   [CaO]$^{-1}$ dependence (interstitial O)   Pr

(b)   [CaO]$^{-2}$ dependence (Ca-vacancy)   [CaO]$^{-1}$ dependence (interstitial O)   Th

Partition coefficient — wt.% CaO in melt

Figure 11   Figure showing (a) D_{Pr} and (b) D_{Th} plotted as a function of CaO content of the melt. The good agreement between data and models represented by eqns [4] and [5] indicates that Ca-vacancy formation rather than oxygen interstitial incorporation is the likely charge balance mechanism for the substitution of 3+ and 4+ cations in the A-site of CaPv. Data from Corgne A, and Wood BJ (2005) Trace element partitioning and substitution mechanisms in calcium perovskites. *Contributions to Mineralogy and Petrology* 149: 85–97.

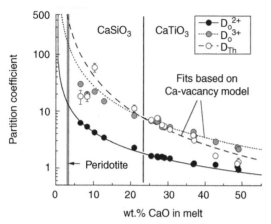

Figure 12 Figure showing D_o^{2+}, D_o^{3+}, and D_{Th} plotted as a function of CaO content in the melt for both $CaSiO_3$ and $CaTiO_3$ perovskites. Data for $CaTiO_3$ perovskite are from Corgne and Wood (2005). Data for $CaSiO_3$ perovskite are from the experiments S2738 (Corgne and Wood, 2002), LO#50, LO#111, and LO#113 (Hirose *et al.*, 2004), H2020b (Corgne *et al.*, 2005), and 266 (Kato *et al.*, 1988). Interestingly, D_o^{3+} and D_{Th} for $CaSiO_3$ perovskite closely follow the trends derived from the study of the low pressure analog.

One important implication of the partitioning data is that, despite being a volumetrically minor phase, $CaSiO_3$ perovskite is the major host for the heat-producing elements Th and U in the deep Earth, as well as being the principal reservoir for the REEs (Corgne *et al.*, 2005). Observed solid–liquid partition coefficients for Th are in the range 30–60 in $CaSiO_3$ crystallizing from Ca-rich compositions, but extrapolation of the defect model to peridotitic compositions (3–3.5 wt.% CaO) suggests that D_{Th} and D_U should be closer to 400 in the mantle (**Figure 12**). Thus, crystallization of as little as 0.25 vol.% of $CaSiO_3$ perovskite from an early Earth magma ocean could, if it had remained isolated in the lower mantle, contribute substantially to current mantle heat production.

4.2.4   Trace Elements in Other Mantle Minerals

Trace element partitioning data for other minerals of the transition zone and lower mantle are limited, even lacking in the case of ringwoodite and $MgSiO_3$-rich post-perovskite. As shown briefly below, the effects of these minerals on trace element distribution during mantle differentiation are likely to be minor, given that most of geochemically important trace elements are expected to be highly incompatible in these phases, except perhaps for post-perovskite.

4.2.4.1   Wadsleyite and ringwoodite

Crystal–melt partition coefficients for wadsleyite (Wd) have been determined by McFarlane (1994) for Cr, Sc, Ni, and Mn. $D^{Wd/Liq}$ for these elements are respectively about 0.8, 0.1, 2, and 1. Considering the value obtained for D_{Sc}, one should expect D values lower than 0.1 for larger trivalent cations such as REE and Y. Wadsleyite–olivine partition coefficients for minor elements were also determined by Gudfinnsson and Wood (1998). The results indicate that crystal–melt partition coefficients for elements entering the large Mg-site are likely to be similar in magnitude for both minerals. Therefore, given the results obtained for olivine (e.g., McDade *et al.*, 2003), 4+ and 5+ cations such as U, Th, Ta, and Nb are expected to be highly incompatible in wadsleyite. These predictions are confirmed by new partitioning data obtained for hydrous compositions at 14–16 GPa and 1400–1600°C (Mibe *et al.*, 2006; Wood, unpublished data). In these systems, $D^{Wd/Liq}$ values for REE are less than 0.002 (obtained for Lu), and D_{Sc} is ~0.04, a value similar to the one derived for the optimum partition coefficient D_o^{3+} for the Mg-site. Moreover, U, Th, Hf, Zr, Ta, Nb, Rb, Sr, and Pb are found to be highly incompatible too.

Because of structural similarities between ringwoodite and the lower-pressure polymorphs olivine and wadsleyite, it is probable that REE, U, and Th are highly incompatible in this phase. Partitioning of minor elements between wadsleyite and ringwoodite (Gudfinnsson and Wood, unpublished data) indicates that ringwoodite does not exhibit strong trace element preferences and is relatively 'sterile' with respect to incompatible elements. Note, however, that despite their apparent geochemical 'inaction', ringwoodite and wadsleyite may play an important role during melting of peridotite given that these minerals have significant water-storage capacity (see Section 4.4).

4.2.4.2   Ferropericlase

(Mg, Fe)O, ferropericlase, is an important constituent of the lower mantle (**Figure 2**), comprising about 16% by mass of peridotitic compositions under lower-mantle conditions (Wood, 2000). It crystallizes in the cubic *Fm3m* structure with one octahedral site (A) occupied nominally by divalent cations Mg and Fe. Walter *et al.* (2004) determined ferropericlase–melt partition coefficients for a series of trace elements. Despite large measurement uncertainties, it can be shown that most trace elements, with the exceptions of Cr, Fe, Mg, and Ni, are incompatible in ferropericlase. In particular, REE have D values below 0.1. U and Th were not measured but are expected to be highly incompatible based on the results obtained for other 4+ cations (Si, Ti, Zr, and Hf). Furthermore, Al, Ba, Na, and Mn are moderately incompatible ($D \sim 0.3$–0.8) and Nb is highly incompatible ($D \sim 0.002$).

4.2.4.3 Post-perovskite

A new polymorph of MgPv with a structure more stable than the *Pbnm* perovskite phase has recently been identified at pressures and temperatures corresponding to the lowermost part of the Earth's mantle by both experimental and theoretical techniques (Murakami *et al.*, 2004; Oganov and Ono, 2004). The MgSiO$_3$-rich post-perovskite (MgPP) structure belongs to space group *Cmcm* and consists of layers of SiO$_6$ octahedra alternating with layers of Mg atoms in approximate eightfold coordination. Current experimental data indicate that MgPP is unlikely to be stable in the mantle at liquidus temperatures (~ 4400 K), and hence this phase would not have been involved in crystal differentiation of an early magma ocean. Nevertheless, MgPP–melt partitioning data would help in modeling chemical interactions between the solid mantle and melts such as those suggested to explain ultra-low-velocity zones (e.g., Williams and Garnero, 1996) present at the core–mantle boundary.

Performing MgPP–melt partitioning experiments is currently impossible due to the extreme conditions of pressure and temperature that are required. In absence of experimental results, it is difficult to know how the differences in structure between MgPv and MgPP may influence trace element partitioning. One approach to this problem would be to perform computer simulations similar to those of Corgne *et al.* (2003) and compare the energies associated with the incorporation of trace elements in both phases. Iitaka *et al.* (2004) have calculated that the average Si–O and Mg–O distances in MgPP are respectively $\sim 1\%$ longer and $\sim 1\%$ shorter than in MgPv of pure MgSiO$_3$ composition. This indicates that optimum radii r_0 for the Si- and Mg-sites are respectively larger and smaller in MgPP than in MgPv. Therefore, elements larger than Si and smaller than Mg would be incorporated more easily in MgPP than in MgPv. On the other hand, elements larger than Mg (e.g., REE, K, U, and Th) should have

partition coefficients higher in MgPv than in MgPP, that is, these elements should be highly incompatible in MgPP. Addition of Al, Fe^{2+}, and Fe^{3+} should also influence trace element partitioning. Oganov and Ono (2004) argued that more Fe^{3+} can be incorporated in MgPP than in MgPv since Fe$_2$O$_3$ has the same structure as MgPP at high pressure. Recent experimental results by Murakami *et al.* (2005) show that in peridotite composition MgPP contains about half the Fe present in MgPv but about the same amount of Al. Assuming that Al is charge-balanced by Fe^{3+} in the Mg-site, this would mean that MgPP contains less Fe^{2+} than MgPv but about the same amount of Fe^{3+}. Depending on its oxidation state, the presence of Fe in the large site of MgPP could either decrease r_0 (Fe mostly Fe^{3+}) or increase r_0 (Fe mostly Fe^{2+}). Future experimental and theoretical work will have to test further these first-order considerations.

4.3 Implications for Planetary Differentiation and Trace Element Distribution in the Deep Mantle

4.3.1 A Primitive Majoritic or Perovskitic Reservoir?

Melting experiments on peridotitic compositions have shown that majoritic garnet is the liquidus phase at depths of the transition zone (e.g., Trønnes and Frost, 2002). The liquidus phase is ferropericlase between about 24 and 30 GPa and magnesium silicate perovskite above 30 GPa, followed closely down temperature by ferropericlase and calcium silicate perovskite (Trønnes and Frost, 2002; Ito *et al.*, 2004). Global melting and subsequent fractionation and gravitational settling of majoritic garnet or silicate perovskites in a magma ocean has been proposed as a mechanism for forming a primitive upper mantle with high Mg/Si and Ca/Al ratios (e.g., Ringwood, 1979; Ohtani *et al.*, 1986; Herzberg *et al.*, 1988). The difficulty with this hypothesis is that it is required to have occurred without significantly changing the ratios of refractory lithophile elements (RLEs) to one another in the residual melt, that is, in the primitive upper mantle (**Figure 1**). Crystal–melt partition coefficients for majoritic garnet and silicate perovskites obtained in high-pressure experiments enable us to calculate the maximum volume of any majoritic and/or surviving perovskitic reservoir in the lower mantle using the constraint that RLE ratios in the primitive upper mantle are chondritic.

Attempts were made to perform this calculation some time ago with limited data and dramatically different conclusions (e.g., Agee and Walker, 1988; Kato *et al.*, 1988). Recent results however (Corgne and Wood, 2004; Walter *et al.*, 2004; Corgne *et al.*, 2005; Liebske *et al.*, 2005) point to the allowable extent of a deep majoritic or perovskitic reservoir being small.

Figure 13(a) shows the variations of RLE ratios of the residual melt (primitive upper mantle, PUM) as a function of majoritic garnet fractionation as calculated by Corgne and Wood (2004). Due to low D values, Sm and Nd are not fractionated one from another significantly during crystallization of majorite, which means that the Sm/Nd ratio remains in the

PUM range. In contrast, the Pr/Sc, Gd/Yb, Lu/Hf, and Lu/La ratios are all driven outside the PUM range after less than 25% of majoritic garnet fractionation. The strongest constraint on majorite fractionation is given by the Pr/Sc ratio, for which the maximum allowable extent of majoritic garnet fractionation is 14%. In other words, less than 14% fractionation of majoritic garnet would not be detected in the composition of PUM, assuming that the PUM composition of McDonough and Sun (1995) is correct. Given chondritic ratios of refractory lithophile elements in PUM, no more than 14% majoritic garnet fractionation may have survived from an early magma ocean. This value compares with an even smaller maximum majorite fractionation of 4% estimated by Walter *et al.* (2004) using similar D values but different elemental ratios in the PUM (Ca/Al, Al/Mg, Al/Sc, and Al/Yb being the most constraining). Furthermore, isotopic ratios in modern volcanic rocks effectively preclude any significant fractionation of majoritic garnet, since such fractionation would lower considerably both ^{143}Nd/^{144}Nd and ^{176}Hf/^{177}Hf ratios of the residual peridotitic liquid (PUM) and drive its geochemical signature away from that of the observed mantle array (e.g., Blichert-Toft and Albarède, 1997; Corgne and Wood, 2004).

Figure 13(b) shows the variations of RLE ratios of the residual melt (PUM) as a function of perovskite fractionation as calculated by Corgne *et al.* (2005). The ratios Sr/Sc and Sm/Ba, which give the strongest constraints, are used to place bounds on the volume of the hypothetical perovskitic reservoir. For MgPv crystallization alone, Sr/Sc is driven outside the chondritic range after only 6.5% of crystallization. The addition of CaPv to the crystallizing assemblage increases the allowed amount of fractionation to, for instance, 9% when the ratio of MgPv:CaPv is 80:20. Taking account of both Sr/Sc and Sm/Ba ratios, it appears that the allowable extent of perovskite fractionation is a maximum for a mixture containing 5–10% CaPv. Under these circumstances, approximately 8% fractionation would not be detected in the composition of the primitive upper mantle. Similar conclusions were arrived at by Liebske *et al.* (2005) and Walter *et al.* (2004). Following a similar approach to that outlined above, and incorporating Nd and Hf isotopic constraints, Walter *et al.* (2004) concluded that 10–15% (but no more) fractionation of a lower-mantle assemblage consisting of 93:3:4 MgPv:CaPv:FP could yield the current upper mantle as residuum.

(a)

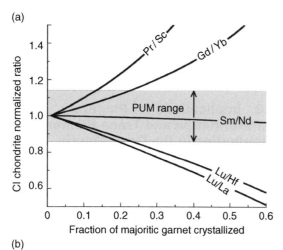

(b)

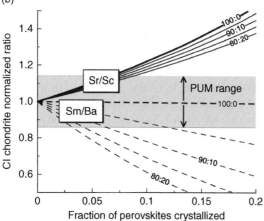

Figure 13 Variation of normalized refractory lithophile element ratios in the residual melt as a function of fractionation of (a) majoritic garnet and (b) perovskites during a magma ocean stage. Numbers on lines in (b) refer to proportions of MgPv and CaPv perovskites. Gray boxes show range of ratios in the PUM, assuming that ratios of refractory lithophile elements are known within 15% (e.g., McDonough and Sun, 1995). Partition coefficients are taken from Corgne and Wood (2004) and Corgne *et al.* (2005).

The results above mean that the PUM depletion relative to CI chondrite in Si (**Figure 1**) cannot be due to majorite or perovskite fractionation since ~60% crystallization would be required. The silicate Earth is either depleted in Si due to its volatility during accretion (e.g., Ringwood, 1979), or due to its incorporation in the core (Allègre *et al.*, 1995; Wade and Wood, 2005). The putative perovskitic reservoir would be enriched in U and Th relative to primitive upper mantle, if the CaPv proportion were greater than ~7%, assuming $D^{CaPv/Liq}$ values for U and Th of ~15 as determined experimentally by Corgne *et al.* (2005). Note that if $D^{CaPv/Liq}$ values for U and Th for a peridotite composition are ~400 as suggested by extrapolation of the Corgne and Wood (2005) data, a small fraction of CaPv (>0.25%) isolated from the residual melt would be sufficient to produce a heat-producing reservoir in the deep mantle. If the perovskitic reservoir contains enough CaPv, it could also have higher Lu/Hf, U/Pb, Sr/Rb, and Sm/Nd ratios than the primitive upper mantle, and hence have the major geochemical characteristics of the HIMU component of OIBs. Since, however, the latter can be interpreted as recycled MORBs (Hofmann and White, 1982), it would be extremely difficult to distinguish a small perovskitic reservoir in the lower mantle on the basis of its isotopic properties. Furthermore, since majoritic garnet, MgPv, and CaPv all prefer Sm to Nd, the putative majoritic or perovksitic layer can clearly not be the low Sm/Nd reservoir required to balance terrestrial samples with superchondritic $^{142}Nd/^{144}Nd$ with a chondritic silicate Earth (Boyet and Carlson, 2005).

4.3.2 Downward Migration and Segregation of Dense Melts?

Since crystallization of a magma ocean in early Earth history would have proceeded from the bottom up (e.g., Solomatov, 2000), residual melts are normally regarded as migrating upward. Caro *et al.* (2005), for example, proposed that the Nd and Hf isotopic signatures of the Hadean depleted mantle may have been acquired after formation of a proto-crust produced by melt extracted from a crystallizing magma ocean at upper-mantle pressures. An alternative suggestion is that melts at high pressures are denser than the surrounding mantle minerals and could remain isolated at great depths (e.g., Suzuki and Ohtani, 2003, and references therein). Olivine flotation could also have occurred within a limited depth interval of 400–450 km in a terrestrial magma ocean (Suzuki and Ohtani, 2003). The impact of olivine fractionation on trace element patterns would be small, however, given the low crystal–melt partition coefficients obtained for olivine and its high-pressure polymorphs. Of more interest is the possibility of dense melts segregating from majorite and silicate perovskites. Recent molecular-dynamics simulations in the $MgSiO_3$ system (Stixrude and Kirko, 2005) suggest that the density contrast between melt and crystals decreases with increasing pressure, such that a density crossover near the base of the mantle may take place. In the present-day Earth, the existence of dense melts provides a possible explanation for the ultra-low-velocity zone at the base of the mantle (e.g., Williams and Garnero, 1996). Composition is, however, a key variable and other experimental results suggest that, in a peridotite composition (as opposed pure to $MgSiO_3$), melts would be less dense than mantle minerals over the entire pressure range of the lower mantle (Agee, 1998).

Recently, Hofmann *et al.* (2005) have proposed that downward migration of dense melts formed in the deep mantle in equilibrium with CaPv could have created a deep 'missing' reservoir unradiogenic in Pb (with a U/Pb ratio lower than that of PUM) as required to balance the Pb composition of the bulk MORB and OIB-source reservoirs. This reservoir would have a complementary signature to the putative majoritic or perovskite reservoir discussed in Section 4.3.1. Thus, it could also be the missing subchondritic $^{142}Nd/^{144}Nd$ reservoir (Boyet and Carlson, 2005).

4.3.3 Recycled Lithosphere in the Deep Mantle?

Rather than being formed early in Earth's history, a deep reservoir may have formed through geologic time by accumulation of recycled oceanic crust–lithosphere subducted into the lower mantle (e.g., Hofmann and White, 1982; Albarède and van der Hilst, 2002). Experimental support for this idea comes from recent data indicating that subducted MORB crust is denser than the surrounding mantle over its entire pressure range (Hirose *et al.*, 2005). Transformation of the original low-pressure minerals to an assemblage containing lower-mantle minerals, including MgPv, CaPv, stishovite, $CaFe_2O_4$-type Al-phase, and possibly MgPP, without loss of trace elements could lead to formation of a geochemical reservoir enriched in elements normally regarded as incompatible (K, U, Th, REE, etc.).

4.4   Water in the Deep Mantle

In this section, we review the present state of our understanding of water distribution in the transition zone and lower mantle. We present a summary of water-solubility measurements in deep-mantle minerals together with theoretical considerations regarding the mechanisms of water incorporation in the crystal lattice of these minerals. The presence of minor components and volatile species is expected to broaden seismic discontinuities (e.g., Wood, 1995; Gudfinnsson and Wood, 1998). Using recent high-pressure experimental data, we present an updated model that constrains water contents in the transition zone and lower mantle based on observed widths of seismic discontinuities. We discuss the implications that such water contents have for mantle dynamics and differentiation processes.

4.4.1   Water Solubility in Minerals from the Transition Zone

Following the predictions by Smyth (1987, 1994) that wadsleyite could store a large of amount of water, Kohlstedt et al. (1996) determined the water-storage capacity of olivine, wadsleyite, and ringwoodite as a function of pressure at 1100°C using Fourier transform infrared (FTIR) spectroscopy and the spectroscopic calibration of Paterson (1982). The latter has been shown recently to underestimate the water concentration in olivine by a factor of $\sim$3–3.5, however (Bell et al., 2003; Koga et al., 2003). After applying the new correction, the experiments of Kohlstedt et al. (1996) indicate that olivine can store up to $\sim$5000 ppm water (by weight) at pressures close to the 410 km discontinuity, in agreement with the recent experimental results of Chen et al. (2002). Wadsleyite–olivine partition coefficients ($D^{Wd/Ol}$) suggest that the high-pressure polymorph has larger water-storage capacity. Applying the new correction, the FTIR results of Young et al. (1993) imply that $D^{Wad/Ol}$ is between 10 and 30, while those of Kohlstedt et al. (1996) imply that $D^{Wd/Ol}$ is $\sim$6. This compares with $D^{Wd/Ol}$ of $\sim$5 reported by Chen et al. (2002) from SIMS measurements on $(Mg_{0.91}, Fe_{0.09})_2SiO_4$ synthesized at $\sim$13 GPa and 1200°C. The range of $D^{Wd/Ol}$ values therefore suggests that wadsleyite could contain between 2 and 3 wt.% H_2O

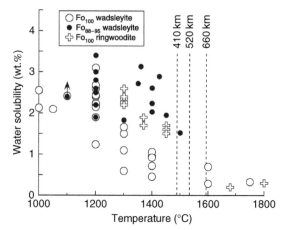

Figure 14   Water contents in wadsleyite and ringwoodite as a function of temperature. Open circles are data for Fe-free hydrous wadsleyite from Inoue et al. (1995), Kohn et al. (2002), Litasov and Ohtani (2003), Demouchy et al. (2005), and Jacobsen et al. (2005). Filled circles are data for Fe-bearing wadsleyite from Kawamoto et al. (1996), Kohlstedt et al. (1996), Smyth et al. (1997), Smyth and Kawamoto (1997), and Chen et al. (2002). Crosses are data for Fe-free hydrous ringwoodite from Inoue et al. (1998) and Ohtani et al. (2000). Estimated temperatures at the 410, 520, and 660 km discontinuities are from Katsura et al. (2004). Based on these data, at conditions of the transition zone, the water-storage capacity of wadsleyite would range between 1.5 and 3.0 wt.%. The corresponding value for ringwoodite is expected to be similar (see text for discussion).

in the transition zone. As shown in **Figure 14**, hydrous wadsleyite synthesized at 13–18.5 GPa over a range of temperature conditions contains up to $\sim$3 wt.% water, very close to the limit originally suggested on crystal structure grounds by Smyth (1987). Although insensitive to pressure, water solubility in wadsleyite decreases with increasing temperature (e.g., Demouchy et al., 2005) (**Figure 14**).

Inoue et al. (1998) and Ohtani et al. (2000) reported water solubility in Fe-free ringwoodite of >2.2 wt.% at 19–23 GPa and 1300–1450°C, in agreement with previous measurements of Kohlstedt et al. (1996) on Fe-bearing ringwoodite at 19.5 GPa and 1100°C. As shown in **Figure 14**, the water content of Fe-free ringwoodite tends to decrease with increasing temperature in a similar manner to Fe-free wadsleyite, reaching contents of 0.3 wt.% at 1800°C (Ohtani et al., 2000). The temperature dependence for Fe-bearing ringwoodite is currently unknown. Given the similarity of results for Fe-free wadsleyite and Fe-free ringwoodite, however, one may reasonably expect similar water-storage capacity for the corresponding Fe-bearing phases, that is, slightly higher than for the

Fe-free phases. Consequently, in a mantle of peridotite composition, the water-storage capacity of ringwoodite should be similar to that of wadsleyite.

4.4.2 Seismic Constraints on the Water Content of the Transition Zone

4.4.2.1 Seismic velocity and density jump across the 410 km discontinuity

The nature of the discontinuity in seismic velocities at 410 km depth has been widely used as an indicator of the content of olivine and its high-pressure polymorphs in the mantle. As shown in **Figure 2**, a peridotitic mantle should contain ~60 vol.% olivine. Using measured and extrapolated elastic parameters for mantle minerals, however, Duffy and Anderson (1989) argued that a composition containing ~40 vol.% olivine would better match seismic velocities across the 410 km discontinuity. Because of uncertainties in the temperature and pressure derivatives of the bulk modulus of wadsleyite, subsequent studies have provided a wide range for the appropriate olivine contents, from ~30 to 65 vol.% (e.g., Weidner, 1985; Bina and Wood, 1987; Gwanmesia *et al.*, 1990; Ita and Stixrude, 1992; Rigden *et al.*, 1992; Duffy *et al.*, 1995; Sinogeikin *et al.*, 1998). The presence of water in wadsleyite and ringwoodite lowers seismic velocities and could make a match with ~60 vol.% olivine easier to achieve (e.g., Inoue *et al.*, 2004; Jacobsen *et al.*, 2004). Inoue *et al.* (2004) suggested, for example, that a peridotitic mantle with ~1.5 wt.% H_2O in wadsleyite would satisfy a bulk seismic velocity contrast of ~4.6% at the 410 km discontinuity (e.g., Nolet *et al.*, 1994). Given the uncertainties in the elastic properties, a dry peridotitic mantle may still be consistent with high-resolution seismic velocity profiles of the transition zone, however (Li *et al.* 2001; Mayama *et al.* 2004). In view of the uncertainties in elastic properties, a better way of constraining the water content of the mantle would be to use the sharpness of the 410 km discontinuity (Wood, 1995).

4.4.2.2 Water and sharpness of the 410 km discontinuity

Potentially, one of the most important constraints on the water contents of the transition zone and lower mantle are the observed depth intervals over which phase transformations occur. Seismological studies using high-frequency reflected waves indicate that both 410 and 660 km discontinuities are very sharp and that the changes in physical properties associated with them occur over very small depth intervals. For the 660 km discontinuity, for example, a width of 4–5 km is consistent with the seismic data (Paulssen, 1988; Benz and Vidale, 1993; Yamazaki and Hirahara, 1994). Most authors favor a slightly broader 410 km discontinuity, on the order of 10 km wide (Paulssen, 1988), but a study of precursors to P′P′ (Benz and Vidale, 1993) implies that this too is locally extremely sharp. A maximum interval of about 8 km (0.3 GPa) is required to explain the data of Benz and Vidale (1993) (Helffrich and Wood, 1996).

As discussed above, water partitions strongly into wadsleyite relative to olivine and no known hydrous phases are stable under the *P–T* conditions of the 410 km discontinuity (~14 GPa and ~1500°C). As illustrated schematically in **Figure 15** (vertical arrow labeled 'low H_2O divariant'), the strong partitioning of H_2O into wadsleyite must broaden the olivine–wadsleyite transition interval, provided the mantle is not water saturated. At water saturation (either in a fluid or melt phase), the system Mg_2SiO_4–H_2O becomes univariant and the transition sharpens again (vertical arrow in **Figure 15**, corresponding to 'high H_2O univariant'). In the mantle, these effects of H_2O are

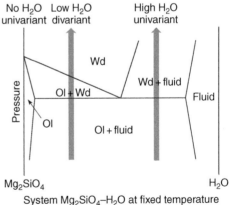

System Mg_2SiO_4–H_2O at fixed temperature

Figure 15 Schematic illustrating the effect of H_2O on the transition from olivine (O1) to wadsleyite (Wd) at about 410 km depth. In the absence of water, the transition is sharp (univariant). Addition of small amounts of water to pure Mg_2SiO_4 generates a broad two-phase loop because of the strong partitioning of H_2O into wadsleyite. This means that the phase transition would take place over an extended depth interval as can be seen by following the vertical arrow labeled 'low H_2O divariant'. This behavior continues as more H_2O is added until fluid (or melt) saturation is reached at which point the reaction becomes univariant again (horizontal line). In the latter region, olivine + fluid transforms to wadsleyite at a fixed pressure as can be seen by following the vertical arrow labelled 'high H_2O univariant'.

added on to those caused by partial substitution of Fe^{2+} for Mg which also act to broaden the transition interval (e.g., Bina and Wood, 1987).

Wood (1995) attempted to quantify the effect illustrated in **Figure 15** by developing models for the free energy of solution of H_2O in the two phases. He adopted a model based on Smyth's (1987) calculations, which suggest that the nonsilicate oxygen atoms (O1) in Mg_2SiO_4 wadsleyite are readily protonated. Given charge balance by vacancies on Mg sites, this yields a hydrated wadsleyite end member containing 3.3 wt.% H_2O of formula $Mg_7Si_4O_{14}(OH)_2$. Smyth's (1987) predictions are confirmed by crystal structure analysis of end-member $Mg_7Si_4O_{14}(OH)_2$ (Kudoh et al., 1996) and by Jacobsen et al. (2005) who used polarized FTIR spectroscopy and X-ray diffraction on oriented single crystals to find that the O1-site is the most important oxygen site for hydrogen storage in wadsleyite.

In a mantle saturated with $(Mg, Fe)_2SiO_4$ and the $(Mg, Fe)SiO_3$ component of pyroxene, the formation of the end-member $Mg_7Si_4O_{14}(OH)_2$ may be represented by the equilibrium:

$$\underset{\text{Wadsleyite}}{3Mg_2SiO_4} + \underset{\text{Pyroxene}}{MgSiO_3} + H_2O = \underset{\text{Wadsleyite}}{Mg_7Si_4O_{14}(OH)_2} \quad [6]$$

In order to calculate the effect of water on the transformation, we require the partial molar free energies of Mg_2SiO_4 and Fe_2SiO_4 components in the hydrated wadsleyite solid solution. At low concentrations of OH, the solution of this component will be in the Henry's law region, where its only effect on the major components is entropic. In that case, the chemical potentials of Mg_2SiO_4 and Fe_2SiO_4 components are given by

$$\mu_{Mg_2SiO_4}^{Wd} = \mu_{Mg_2SiO_4}^{o} + RT\ln(X_{Mg}^2 \cdot \gamma_{Mg} \cdot (1 - X_{OH})^{0.5}) \quad [7a]$$

$$\mu_{Fe_2SiO_4}^{Wd} = \mu_{Fe_2SiO_4}^{o} + RT\ln(X_{Fe}^2 \cdot \gamma_{Fe} \cdot (1 - X_{OH})^{0.5}) \quad [7b]$$

In eqns [7a] and [7b], X_{Mg} and X_{Fe} refer to atomic fractions of magnesium and iron, respectively, on the large cation positions, while X_{OH} is the fraction of protonated O1 atoms. The standard state chemical potentials $\mu_{Mg_2SiO_4}^{o}$ and $\mu_{Fe_2SiO_4}^{o}$ refer to the free energies of the pure end-member wadsleyite at the pressure and temperature of interest. The activity coefficients for the Mg–Fe sites, γ_{Mg} and γ_{Fe}, refer to nonideal Mg–Fe mixing, combined with (nominally) ideal mixing of vacancies, which have a concentration,

$$V''_{Mg} = \frac{1}{8} X_{OH} \quad [8]$$

Equations [7] and [8] enable calculation of the effect of water on the two major components, given the assumptions of random mixing of OH groups with O1 and of vacancies with Mg and Fe. The nonideal part of Mg–Fe mixing was obtained by assuming a symmetric interaction parameter W_{MgFe} of $4.0 \, kJ \, g^{-1}$ atom^{-1} (Akaogi et al., 1989).

For olivine, Bai and Kohlstedt (1993) found H-solubility data to be consistent with formation of complexes between H and interstitial O atoms. In Kröger–Vink notation (Kröger and Vink, 1956), this may be represented in simplified form as:

$$O''_i + H_2O = 2OH'_i \quad [9]$$

Calculation of the chemical potentials of Mg_2SiO_4 and Fe_2SiO_4 components in the partially hydrated olivine, relative to water-free olivine under the same conditions of intensive variables, are obtained from

$$\mu_{Mg}^{Ol} = \mu_{Mg_2SiO_4}^{o} + RT\ln\left(\frac{X_{Mg}^2 \cdot \gamma_{Mg} \cdot (1 - X_{OH})^4}{(1 - 0.5 X_{OH})^4}\right) \quad [10a]$$

$$\mu_{Fe}^{Ol} = \mu_{Fe_2SiO_4}^{o} + RT\ln\left(\frac{X_{Fe}^2 \cdot \gamma_{Fe} \cdot (1 - X_{OH})^4}{(1 - 0.5 X_{OH})^4}\right) \quad [10b]$$

Activity coefficients γ_{Mg} and γ_{Fe} were calculated from the symmetric solution model using the experimentally measured interaction parameter of Wiser and Wood (1991) adjusted to 14 GPa from the excess volumes on the forsterite–fayalite join, giving a value of $5.0 \, kJ \, g^{-1}$ atom^{-1}.

The combined effect of water and iron on the olivine–wadsleyite transition is calculated by fixing the bulk $Fe/(Fe + Mg)$ and water content of mantle olivine at a point just above the depth where wadsleyite appears and then calculating the shift in equilibrium pressure relative to the value for pure anhydrous Mg_2SiO_4. Proportions of phases through the two-phase region were calculated as follows. The $Fe/(Fe + Mg)$ ratio of the olivine was set to a value less than or equal to the value just above the transition zone. The $Fe/(Fe + Mg)$ ratio of the coexisting wadsleyite was obtained from the partitioning data of Katsura and Ito (1989). Then, from the $Fe/(Fe + Mg)$ of the initial mantle olivine, phase proportions were simply solved from the lever rule. Given proportions of olivine and wadsleyite, the concentrations of water in the two phases were calculated using a fixed water content of olivine above the transition zone and fixing the H_2O partition coefficient at either 10:1 or 5:1 in favor of wadsleyite (see above). The equilibrium pressure was then calculated from (see Wood, 1995)

$$(P - P^\circ)\Delta V^\circ = -RT\ln\left(X^2_{Mg} \cdot \gamma_{Mg} \cdot (1 - X_{OH})^{0.5}\right)_{Wd}$$
$$+ RT\ln\left(\frac{X^2_{Mg} \cdot \gamma_{Mg} \cdot (1 - X_{OH})^4}{(1 - 0.5\,X_{OH})^4}\right)_{Ol} \quad [11]$$

where P° refers to the equilibrium pressure for the end-member reaction,

$$\underset{\text{Olivine}}{Mg_2SiO_4} = \underset{\text{Wadsleyite}}{Mg_2SiO_4} \quad [12]$$

ΔV° refers to the volume change of this reaction at the P,T conditions of interest. We performed the calculations at 1773 K and took P° to be 14.5 GPa (**Figure 16**).

In calculating the results shown in **Figure 16**, we initially assumed that the ratio of water contents (by weight) in wadsleyite to olivine is 10:1 (following Wood, 1995) and repeated the calculation using a value of 5:1, consistent with that obtained recently by Chen *et al.* (2002) and the correction to Kohlstedt *et al.*'s (1996) results (see Section 4.4.1). The results indicate that small amounts of water act to lower the pressure of the transformation and broaden it. If mantle olivine contains 1000 ppm H_2O by weight, the transformation would shallow by up to 10 km and broaden substantially. The results of Wood (1995), shown in **Figure 16(a)**, indicate that, given a transition interval in the anhydrous system of about 7 km (Helffrich and Wood, 1996), 500 ppm H_2O in mantle olivine would broaden the transformation to over 20 km. He argued that, given seismological observations at high frequency which imply that the 410 km discontinuity is less than 10 km wide (Paulssen, 1988; Benz and Vidale, 1993; Yamazaki and Hirahara, 1994), the H_2O content of mantle olivine is constrained to be less than 200 ppm. This constraint is relaxed by a factor of about 2 if the partition coefficient for H_2O between wadsleyite and olivine is 5, rather than 10 (**Figure 16(b)**). Thus, using recent water partitioning data (Chen *et al.* 2002), about 400 ppm of H_2O is the calculated upper limit for mantle olivine under conditions of the 410 km discontinuity (**Figure 16(b)**) if the transformation interval is 10 km wide. Smyth and Frost (2002) have more recently confirmed experimentally that the olivine–wadsleyite transformation is broadened and shallowed by the addition of H_2O, in line with the predictions made by Wood (1995).

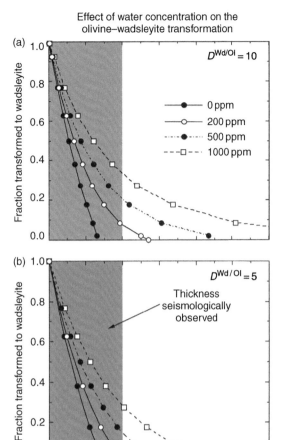

Figure 16 Figure showing the calculated depth interval of the olivine–wadsleyite phase transformation for initial H_2O contents of olivine of 0, 200, 500, and 1000 ppm by weight. Water is assumed to be concentrated in wadsleyite relative to olivine by a factor of 10 in (a) and a factor of 5 in (b). Given a seismologically determined transformation interval of less than about 10 km (gray area), maximum H_2O contents are ∼200 ppm (a) and ∼400 ppm (b).

Seismological evidence for the presence of H_2O at the 410 km discontinuity has to some extent focused on regions where broadening of the discontinuity may be detected. Van der Meije *et al.* (2003) have for example investigated P-to-S conversions from the 410 km discontinuity in the Mediterranean region which has a long history of subduction and possible introduction of water into the mantle. They find broadening of the 410 discontinuity to between 20 and 35 km, consistent with about 700 ppm ($D^{Wd/Ol} = 10$) to 1500 ppm ($D^{Wd/Ol} = 5$) H_2O in olivine. The 660 km discontinuity is, in contrast, sharp and apparently unaffected by the long period of active subduction.

If we accept that the width of the 410 km discontinuity constrains the water content of the mantle olivine then we obtain maximum values of around 400 ppm with, given the seismological study discussed above, regional variations up to a factor of ~4 greater. It is interesting to note that these values are in very good agreement with upper-mantle H_2O contents inferred from the water contents of MORBs. As discussed earlier, the latter lead to 80–180 ppm H_2O in the depleted source regions of N-MORB, 200–950 ppm in the source regions of E-MORB, and 1000–3000 ppm in the mantle sources of back-arc basin basalts.

4.4.3 Transition Zone 'Water Filter'

Bercovici and Karato (2003) used the observation that the major minerals of the transition zone, wadsleyite and ringwoodite, dissolve substantially more H_2O than those of the upper mantle (above 410 km depth) and lower mantle (below 660 km) to suggest that the transition zone is very rich in H_2O and that the 410 km acts as a 'water filter' for trace elements. They argue that, provided the transition zone is richer in H_2O than the solubility limit of the mantle at 410 km depth, upwelling mantle in the transition zone must release H_2O as it crosses from the wadsleyite to olivine stability field at 410 km. At an approximate temperature of 1500 °C, this water would not exist as a fluid, as shown schematically in **Figure 15**, but as a hydrous silicate melt. Partial melting would strip most of the incompatible trace elements out of the solid mantle traversing the 410 km discontinuity and, provided the melt was left behind 'perched' on the discontinuity, the residual solids could, in the upper mantle, provide the trace element-depleted source region of N-MORB. Furthermore, melting could generate a much sharper discontinuity than that calculated for subsolidus conditions above, even for quite high water concentrations. The melt, which must be denser than the solid mantle to be left behind at the discontinuity, would be recycled back into the lower mantle at subduction zones where it would freeze and attach itself to descending plates.

Bercovici and Karato's mechanism for generating an upper mantle depleted in incompatible elements is attractive in that it provides the possibility of producing depleted (upper) and enriched (lower) reservoirs without invoking strictly layered mantle convection. It can also be disproven, because it is based on a number of assumptions which are experimentally testable. Firstly, the partitioning of elements between the solid phases at 410 km depth and hydrous silicate melts must be consistent with

the pattern of trace element depletion of the MORB source region. The available experimental data (Mibe *et al.*, 2006) suggest that the model passes this test. Secondly, the hydrous melts must be denser than the mantle above 410 km depth, and thirdly the storage capacity of the upper mantle must be very low in order to generate the water-depleted (80–180 pm) source regions of N-MORB. Based on more recent solubility measurements for H_2O in olivine, Hirschmann *et al.* (2005) argued that the storage capacity of the mantle just above the 410 km discontinuity is >4000 ppm, so hydrous melting at 410 km would generate a 'depleted' mantle which is much richer in H_2O than the MORB source. This observation is difficult to reconcile with the 'water filter' hypothesis. Hirschmann (2006) also argued that it is very unlikely that hydrous melts would be denser than the mantle and remain behind at 410 km as the solid mantle ascends. Recent measurements at high pressure by Sakamaki *et al.* (2006) indicate, however, that the melt would be gravitationally stable provided it contains less than 6.7% H_2O. Furthermore Song *et al.* (2004) have found seismological evidence for the existence of a low-velocity layer, plausibly a dense partial melt, on top of the 410 km discontinuity in the northwestern United States.

The current status of the 'water filter' hypothesis is that it remains consistent with the available melt density and trace element partitioning data but that it cannot explain the very low H_2O concentrations in the MORB source region. We anticipate further testing through both experiment and seismological observation in the near future.

4.4.4 Water Solubility in Lower-Mantle Minerals

The water-storage capacity of the lower mantle has remained controversial due to the difficulties of synthesizing suitable samples (free of impurities and water saturated) and analyzing them. Some studies (Murakami *et al.*, 2002; Litasov *et al.*, 2003) reported large water-storage capacities for MgPv (~1400–2000 ppm), FP (~2000 ppm) and CaPv (~4000 ppm) in peridotite. On the other hand, Bolfan-Casanova *et al.* (2000, 2002, 2003) reported water contents in MgPv and FP to be <10 and 20–75 ppm, respectively, implying that, if the mantle contains several hundred ppm H_2O, this cannot be stored in the nominally anhydrous minerals of the lower mantle.

4.4.4.1 Magnesium silicate perovskite

Navrotsky (1999) suggested that the incorporation of trivalent cations (hereafter M^{3+}) such as Al^{3+} and Fe^{3+} enhances the solubility of hydrogen in magnesium silicate perovskite. Assuming that M^{3+} substitutes for Si^{4+} (B-site) in magnesium perovskite and is charge-balanced via formation of an oxygen vacancy, we can write the following reaction for dissolution of M_2O_3 (in Kröger–Vink notation):

$$M_2O_3 + 2Si_{Si} + O_o = 2M'_{Si} + V_o^{\bullet\bullet} + 2SiO_2 \quad [13]$$

Alternatively, if M^{3+} enters MgPv via coupled substitution in both A- and B-sites, we can write the reaction as

$$M_2O_3 + Si_{Si} + Mg_{Mg} = M'_{Si} + M^{\bullet}_{Mg} + MgSiO_3 \quad [14]$$

As inferred from MgPv–melt partitioning data in peridotitic compositions (Liebske *et al.*, 2005), ~90% of Al^{3+} (and ~40% Cr^{3+}) enter the B-site of MgPv. In contrast, Fe^{3+}, which is larger, would be predominantly present in the A-site, thus charge-balancing some of the Al^{3+} in the B-site according to reaction [14]. Assuming 60% of Fe in MgPv is Fe^{3+} (e.g., McCammon, 1997; Frost *et al.*, 2004) and Fe^{3+} enters only the A-site (e.g., Vanpeteghem *et al.* 2006), one can estimate that in MgPv containing 4–5 wt.% Al_2O_3, and 7 wt.% FeO^* as expected in the mantle (e.g., Wood, 2000; Nishiyama and Yagi, 2003) about 80–95% of Al^{3+} would be charge-balanced by Fe^{3+} in the A-site. Therefore, an order of 5–20% of the Al^{3+} present could be charge-balanced by formation of oxygen vacancies. This estimate leads to ~0.06–0.3% of the oxygen sites in MgPv being vacant. These vacancies would plausibly interact with H_2O as follows:

$$H_2O + V_o^{\bullet\bullet} + O_o = 2(OH)_o^{\bullet} \quad [15]$$

Then, as shown in **Figure 17**, one can derive a maximum water solubility of between 0 and 2200 ppm in MgPv taking account of uncertainty in the $Fe^{3+}/\sum Fe$ ratio ($60 \pm 5\%$). Increasing the proportion of Fe^{3+} in the B-site increases the number of oxygen vacancies and the anticipated water solubility. For example, if 20% of Fe^{3+} is actually present in the B-site, the derived water content would be ~2200–4200 ppm (**Figure 17**). The water contents of MgPv estimated by Murakami *et al.* (2002) and Litasov *et al.* (2003) are broadly consistent with this model of hydroxylation of oxygen vacancies. The results of Bolfan-Casanova *et al.* (2003) on similar bulk compositions indicate about 2 orders of magnitude lower

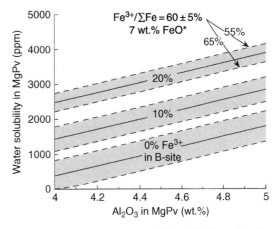

Figure 17 Maximum water solubility in MgPv as function of Al_2O_3 content. The calculation assumes that MgPv contains 7 wt.% FeO^* in the lower mantle and that hydroxyl groups occupy oxygen vacancies created to charge-balance Al^{3+} in the B-site (reaction [15]). At a given Al_2O_3 content, water solubility increases with increasing the amount of Fe^{3+} in the B-site and with increasing $Fe^{3+}/\sum Fe$ ratio. See text for discussion.

solubility of H_2O in Mg-perovskite, however. Currently, the large differences between these studies of (nominally) the same problem are unresolved, and there is clearly need for improvement of both synthesis and analytical methods.

4.4.4.2 Calcium silicate perovskite

In peridotite, CaPv contains minor amounts of Mg, Al, and Fe (e.g., Kesson *et al.*, 1998; Wood, 2000). Based on the partitioning parabolae obtained by Corgne *et al.* (2005), it is reasonable to assume that Al enters only the B-site. Therefore, the fraction of oxygen vacancies required for charge compensation depends on the proportion and site occupancy of ferric iron. Considering Al_2O_3 and FeO^* contents of 1 wt.% and 1–2 wt.%, respectively, as representative of the CaPv composition in the lower mantle (e.g., Kesson *et al.*, 1998; Wood, 2000), ~0.6–1% of oxygen positions are required to be vacant for charge compensation if Fe is assumed to be only Fe^{3+} and to enter only the B-site. In this case and if all the oxygen vacancies created are filled by hydroxyl groups according to reaction [15], we derive a maximum water content of ~3000–4300 ppm consistent with the measurements of Murakami *et al.* (2002). This value is halved (~1750 ppm) if Fe is only Fe^{2+} and located in the A-site. However, Fe is likely to be present in CaPv as both Fe^{2+} and Fe^{3+} with Fe^{2+} and probably most of Fe^{3+} in the A-site. As shown in

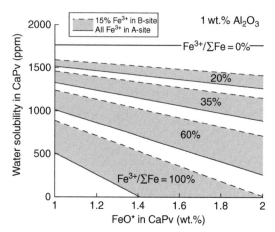

Figure 18 Estimated water solubility in CaPv as function of FeO* content. The calculation assumes that CaPv contains 1 wt.% Al_2O_3 in the lower mantle and that hydroxyl groups occupy oxygen vacancies created to charge-balance Al^{3+} in the B-site (reaction 20). At a given FeO* content, water solubility increases with increasing the amount of Fe^{3+} in the B-site and with decreasing $Fe^{3+}/\sum Fe$ ratio. See text for discussion.

Figure 18, water solubility decreases with increasing FeO* content and decreasing proportion of Fe^{3+} in the B-site, the decrease being more pronounced at high $Fe^{3+}/\sum Fe$ ratio. For example, water solubility would be zero for CaPv contents above 1.8 wt.% FeO* and $Fe^{3+}/\sum Fe$ ratio above 80%. Assuming that $\sim 60\%$ Fe is Fe^{3+}, as in MgPv, water contents would be $\sim 250–1000$ ppm if Fe^{3+} is only in the A-site and $\sim 700–1250$ ppm if 15% of Fe^{3+} is in the B-site (**Figure 18**).

4.4.4.3 Ferropericlase

Potential mechanisms of H_2O dissolution in (Mg, Fe)O ferropericlase are similar to those outlined for perovskite above. Currently, however, large discrepancies in the data for H_2O solubility in this phase (75 ppm, Bolfan-Casanova *et al.*, 2003; 2000 ppm, Murakami *et al.*, 2002) make further discussion and comparison with perovskite meaningless.

4.5 Conclusions

Partitioning of trace elements between the high-pressure phases of the transition zone and lower mantle and coexisting silicate melts has been found to obey the simple elastic strain model which has been successfully applied to a wide range of low-pressure minerals (Blundy and Wood, 1994). For elements of fixed charge, solid–liquid partition coefficients D_i are an approximately parabolic function of ionic radius, where the peak of the parabola D_o corresponds to the partition coefficient for the ion which enters the site without strain. One marked difference from low pressure minerals such as clinopyroxene and plagioclase is that many elements which are very incompatible in shallow mantle and crustal minerals are very compatible in the perovskites of the lower mantle (**Figures 5** and **8**). In **Figure 5**, for example, it can be seen that Zr and Hf have partition coefficients into MgPv which are greater than 1, while U and Th have partition coefficients into $CaSiO_3$ perovskite of larger than 10 (**Figure 8**). As can be seen from **Figure 8**, Th^{4+} and U^{4+} are markedly and very unusually more compatible in the Ca site of perovskite than Ca^{2+} itself.

Investigation of the limits of Henry's law for Ca-perovskite (in the form of the low-pressure-phase $CaTiO_3$) demonstrates that the high partition coefficients for 3+ and 4+ ions into the Ca^{2+} site arise from the ease with which excess charge is compensated in the perovskite structure. This takes place by the creation of cation vacancies. Thus, the compositional dependence of Th partitioning indicates that Th^{4+} replaces $2Ca^{2+}$ with the creation of a vacancy in the Ca sublattice.

The measured liquid–solid partition coefficients have been used to investigate the possibility that the mantle underwent gross stratification early in Earth history, with establishment of a lower layer rich in majorite or perovskite. Applying the constraint that the primitive 'upper' mantle has chondritic ratios of the refractory lithophile elements, we find that the permissible extents of high-pressure fractionation are very small. Thus, for example, fractionation of perovskite into the lower mantle is limited to about 8% of mantle volume (**Figure 13**). The limit of possible majorite fractionation (in the absence of perovskite fractionation) is 4%. This does not mean that extensive melting and fractional crystallization of the mantle did not occur in early Earth history. It means only that later rehomogenization of the mantle has erased virtually all record of any such event and that only a few percent of the lower mantle volume can, at most, remain isolated from the upper mantle. This result is robust provided that there is not a very strong negative pressure effect on partitioning of 'incompatible' trace elements such as U and Th into perovskites.

Experimentally and analytically, the behavior of H in high-pressure phases is much more difficult to constrain than that of other trace elements. Much of the available data show appreciable scatter. Nevertheless, it is apparent that wadsleyite and ringwoodite of the transition zone can dissolve H_2O at levels larger than 2 wt.% (**Figure 14**), while the maximum solubility in the low-pressure olivine polymorph is about 0.5 wt.% (Hirschmann *et al.*, 2005). Given that H_2O partitions between wadsleyite and olivine with a ratio between 5:1 and 10:1, it can be shown that the width of the 410 km discontinuity places constraints on the water content of the mantle (**Figures 15** and **16**). In the absence of melting, the seismologically determined sharpness of the 410 km discontinuity, which corresponds to the olivine–wadsleyite phase transformation, fixes the maximum water content of the mantle olivine at about 400 ppm. This is in broad agreement with mantle water contents estimated from H_2O contents in MORBs.

Water solubility in the principal minerals of the lower mantle, ferropericlase, magnesium, and calcium silicate perovskites, is very poorly known. Currently available data vary by almost 2 orders of magnitude from a few tens of ppm to a few thousand ppm. It is clear, however, that whichever data are accepted, H_2O is much more soluble in the transition zone than in the lower mantle. These observations have led to the suggestion of Bercovici and Karato (2003) that the transition zone is water rich while the upper and lower mantles are relatively dry. According to this idea, the depleted upper mantle is generated by hydrous partial melting of upwelling fertile mantle as it crosses the 410 km discontinuity. Measured densities of hydrous melts at appropriate conditions (Matsukage *et al.*, 2005; Sakamaki *et al.*, 2006) indicate that the melts could, as predicted, remain perched on the 410 km discontinuity as the depleted solid mantle rises. The depleted mantle should, however, be much richer in H_2O (4000 ppm) than appears plausible from the H_2O concentrations in MORB source region (200 ppm).

Acknowledgments

BJW and AC acknowledge support from ARC grants FF0456999 and DP066453.

References

Abe Y (1997) Thermal and chemical evolution of the terrestrial magma ocean. *Physics of the Earth and Planetary Interiors* 100: 27–39.

Agee CB (1998) Crystal-liquid density inversions in terrestrial and lunar magmas. *Physics of the Earth and Planetary Interiors* 107: 63–74.

Agee CB and Walker D (1988) Mass balance and phase density constraints on early differentiation of chondritic mantle. *Earth and Planetary Science Letters* 90: 144–156.

Akaogi M, Ito E, and Navrotsky A (1989) Olivine-modified spinel–spinel transitions in the system Mg_2SiO_4–Fe_2SiO_4: Calorimetric measurements, thermochemical calculation, and geophysical application. *Journal of Geophysical Research* 94: 15671–15685.

Albarède F and van der Hilst RD (2002) Zoned mantle convection. *Philosophical Transactions of the Royal Society of London, Series A, Mathematical Physical and Engineering Sciences* 360: 2569–2592.

Allègre CJ, Poirier J-P, Humler E, and Hofmann AW (1995) The chemical composition of the Earth. *Earth and Planetary Science Letters* 134: 515–526.

Allègre CJ, Staudacher T, and Sarda P (1986) Rare gas systematics: Formation of the atmosphere, evolution and structure of the Earth's mantle. *Earth and Planetary Science Letters* 81: 127–150.

Allègre CJ, Hofmann A, and O'Nions K (1996) The Argon constraints on mantle structure. *Geophysical Research Letters* 23: 3555–3557.

Armstrong L, Corgne A, Keshav S, Fei Y, McDonough WF, and Minarik W (2006) The effect of pressure on majoritic garnet-melt partitioning of trace elements. Manuscript in preparation.

Bai Q and Kohlstedt DL (1993) Effects of chemical environment on the solubility and incorporation mechanism for hydrogen in olivine. *Physics and Chemistry of Minerals* 19: 460–471.

Becker TW, Kellogg JB, and O'Connell RJ (1999) Thermal constraints on the survival of primitive blobs in the lower mantle. *Earth and Planetary Science Letters* 171: 351–365.

Bell DR, Rossman GR, Maldener J, Endisch D, and Rauch F (2003) Hydroxide in olivine: A quantitative determination of the absolute amount and calibration of the IR spectrum. *Journal of Geophysical Research* 108: 2105, doi:10.1029/2001JB000679.

Benz HM and Vidale JE (1993) Sharpness of upper mantle discontinuities determined from high-frequency reflections. *Nature* 365: 147–150.

Bercovici D and Karato S-I (2003) Whole-mantle convection and the transition-zone water filter. *Nature* 425: 39–44.

Bina CR and Wood BJ (1987) The olivine–spinel transitions: Experimental and thermodynamic constraints and implications for the nature of the 400 km seismic discontinuity. *Journal of Geophysical Research* 92: 4853–4866.

Blichert-Toft J and Albarède F (1997) The Lu–Hf isotope geochemistry of chondrites and the evolution of the mantle–crust system. *Earth and Planetary Science Letters* 148: 243–258.

Blundy JD and Wood BJ (1994) Prediction of crystal–melt partition coefficients from elastic moduli. *Nature* 372: 452–454.

Bolfan-Casanova N, Keppler H, and Rubie DC (2000) Partitioning of water between mantle phases in the system MgO–SiO_2–H_2O up to 24 GPa: Implications for the distribution of water in the Earth's mantle. *Earth and Planetary Science Letters* 182: 209–221.

Bolfan-Casanova N, Mackwell S, Keppler H, McCammon C, and Rubie DC (2002) Pressure dependence on H solubility in magnesiowüstite up to 25 GPa: Implications for the storage of water in the Earth's lower mantle. *Geophysical Research Letters* 29, doi:10.1029/2001GL014457.

Bolfan-Casanova N, Keppler H, and Rubie DC (2003) Water partitioning at 660 km depth and evidence for very low water solubility in magnesium silicate perovskite. *Geophysical Research Letters* 30, doi:10.1029/2003GL017182.

Boyet M and Carlson RW (2005) ^{142}Nd evidence for early (>4.53 Ga) global differentiation of the silicate Earth. *Science* 309: 576–581.

Brice JC (1975) Some thermodynamics aspects of the growth of strained crystals. *Journal of Crystal Growth* 28: 249–253.

Brodholt JP (2000) Pressure-induced changes in the compression mechanism of aluminous perovskite in the Earth's mantle. *Nature* 407: 620–622.

Cameron AGW and Benz W (1991) Origin of the Moon and the single impact hypothesis. *Icarus* 92: 204–216.

Caro G, Bourdon B, Wood BJ, and Corgne A (2005) Trace-element fractionation in Hadean mantle generated by melt segregation from a magma ocean. *Nature* 436: 246–249.

Chambers JE (2004) Planetary accretion in the inner Solar system. *Earth and Planetary Science Letters* 223: 241–252.

Chen J, Inoue T, Yurimoto H, and Weidner DJ (2002) Effect of water on olivine–wadsleyite phase boundary in the $(Mg,Fe)_2SiO_4$ system. *Geophysical Research Letters* 29, doi:10.1029/2001GL014429.

Corgne A and Wood BJ (2002) $CaSiO_3$ and $CaTiO_3$ perovskite–melt partitioning of trace elements: Implications for gross mantle differentiation. *Geophysical Research Letters* 29, doi:10.1029/2001GL014398.

Corgne A and Wood BJ (2004) Trace element partitioning between majoritic garnet and silicate melt at 25 GPa. *Physics of the Earth and Planetary Interiors* 143–144: 407–419.

Corgne A and Wood BJ (2005) Trace element partitioning and substitution mechanisms in calcium perovskites. *Contributions to Mineralogy and Petrology* 149: 85–97.

Corgne A, Allan NL, and Wood BJ (2003) Atomistic simulations of trace element incorporation into the large site of $MgSiO_3$ and $CaSiO_3$ perovskites. *Physics of the Earth and Planetary Interiors* 139: 113–127.

Corgne A, Liebske C, Wood BJ, Rubie DC, and Frost DJ (2005) Silicate perovskite–melt partitioning of trace elements and geochemical signature of a deep perovskitic reservoir. *Geochimica et Cosmochimica Acta* 146: 249–260.

Demouchy S, Deloule E, Frost DJ, and Keppler H (2005) Pressure and temperature-dependence of water solubility in iron-free wadsleyite. *American Mineralogist* 90: 1084–1091.

Dixon JE and Clague DA (2001) Volatiles in basaltic glasses from Loihi Seamount, Hawaii: Evidence for a relatively dry plume component. *Journal of Petrology* 42: 627–654.

Draper DS, Xirouchakis D, and Agee CB (2003) Trace element partitioning between garnet and chondritic melt from 5 to 9 GPa: Implications for the onset of the majorite transition in the martian mantle. *Physics of the Earth and Planetary Interiors* 139: 149–169.

Duffy TS and Anderson DL (1989) Seismic velocities in mantle minerals and the mineralogy of the upper mantle. *Journal of Geophysical Research* 94: 1895–1912.

Duffy TS, Zha CS, Downs RT, Mao HK, and Hemley RJ (1995) Elasticity of forsterite to 16 GPa and the composition of the upper mantle. *Nature* 378: 170–173.

Frost DJ and Langenhorst F (2002) The effect of Al_2O_3 on Fe–Mg partitioning between magnesiowüstite and magnesium silicate perovskite. *Earth and Planetary Science Letters* 199: 227–241.

Frost DJ, Liebske C, Langenhorst F, McCammon CA, Trønnes RG, and Rubie DC (2004) Experimental evidence for the existence of iron-rich metal in the Earth's lower mantle. *Nature* 428: 409–412.

Gudfinnsson GH and Wood BJ (1998) The effect of trace elements on the olivine–wadsleyite transformation. *American Mineralogist* 83: 1037–1044.

Gwanmesia GD, Rigden SM, Jackson I, and Liebermann RC (1990) Pressure dependence of elastic wave velocity for β-Mg_2SiO_4 and the composition of the Earth's mantle. *Science* 250: 794–797.

Harte B, Harris JW, Hutchison MT, Watt GR, and Wilding MC (1999) Lower mantle mineral associations in diamonds from Sao Luiz, Brazil. In: Fei Y, Bertka C, and Mysen BO (eds.) *Mantle Petrology; Field Observations and High-Pressure Experimentation: A Tribute to Francis R. (Joe) Boyd*, vol. 6, pp. 125–153. Houston: Geochemical Society.

Heinemann S, Sharp TG, Seifert F, and Rubie DC (1997) The cubic–teragonal phase transition in the system majorite $(Mg_4Si_4O_{12})$, pyrope $(Mg_3Al_2Si_3O_{12})$, and garnet symmetry in the earth's transition zone. *Physics and Chemistry of Minerals* 24: 206–221.

Helffrich GR and Wood BJ (1996) 410 km discontinuity sharpness and the form of the olivine alpha–beta phase diagram: Resolution of apparent seismic contradictions. *Geophysical Journal International* 126: F7–F12.

Herzberg C, Feigenson M, Skuba C, and Ohtani E (1988) Majorite fractionation recorded in the geochemistry of peridotites from South Africa. *Nature* 332: 823–826.

Hill E, Wood BJ, and Blundy JD (2000) The effect of Ca–Tschermaks component on trace element partitioning between clinopyroxene and silicate melt. *Lithos* 53: 203–215.

Hirose K, Shimizu N, van Westrenen W, and Fei Y (2004) Trace element partitioning in Earth's lower mantle and implications for geochemical consequences of partial melting at the core–mantle boundary. *Physics of the Earth and Planetary Interiors* 146: 249–260.

Hirose K, Takafuji N, Sata N, and Ohishi Y (2005) Phase transition and density of subducted MORB crust in the lower mantle. *Earth and Planetary Science Letters* 237: 239–251.

Hirschmann MM, Aubaud C, and Withers AC (2005) Storage capacity of H_2O in nominally anhydrous minerals in the upper mantle. *Earth and Planetary Science Letters* 236: 167–181.

Hirschmann MM (2006) Water, melting and the deep earth H_2O cycle. *Annual Review of Earth and Planetary Sciences.* 34: 629–653.

Hofmann AW (1988) Chemical differentiation of the Earth: The relationship between mantle, continental crust, and oceanic crust. *Earth and Planetary Science Letters* 90: 297–314.

Hofmann AW and White WM (1982) Mantle plumes from ancient oceanic crust. *Earth and Planetary Science Letters* 57: 421–436.

Hofmann AW, Hemond C, Sarbas B, and Jochum KP (2005) Yes, there really is a lead paradox. *EOS Transactions of the American Geophysical Union Fall Meeting Supplement* 86, Abstract V23D–05.

Iitaka T, Hirose K, Kawamura K, and Murakami M (2004) The elasticity of the $MgSiO_3$ post-perovskite phase in the Earth's lowermost mantle. *Nature* 430: 442–445.

Inoue T, Yurimoto H, and Kudoh Y (1995) Hydrous modified spinel, $Mg_{1.75}SiH_{0.5}O_4$: A new water reservoir in the mantle transition region. *Geophysical Research Letters* 22: 117–120.

Inoue T, Weidner DJ, Northrup PA, and Parise JB (1998) Elastic properties of hydrous ringwoodite (γ-phase) in Mg_2SiO_4. *Earth and Planetary Science Letters* 160: 107–113.

Inoue T, Tanimoto Y, Irifune T, Suzuki T, Fukui H, and Ohtaka O (2004) Thermal expansion of wadsleyite, ringwoodite, hydrous wadsleyite and hydrous ringwoodite. *Physics of the Earth and Planetary Interiors* 143–144: 279–290.

Irifune T (1987) An experimental investigation of the pyroxene–garnet transformation in a pyrolite composition and its bearing on the constitution of the mantle. *Earth and Planetary Science Letters* 45: 324–336.

Irifune T (1994) Absence of an aluminous phase in the upper part of the Earth's lower mantle. *Nature* 370: 131–133.

Ita J and Stixrude L (1992) Petrology, elasticity, and composition of the mantle transition zone. *Journal of Geophysical Research* 97: 6849–6866.

Ito E, Kubo A, Katsura T, and Walter MJ (2004) Melting experiments of mantle materials under lower mantle conditions with implications for magma ocean differentiation. *Physics of the Earth and Planetary Interiors* 143–144: 397–406.

Jacobsen SD, Smyth JR, Spetzler H, Holl CM, and Frost DJ (2004) Sound velocities and elastic constants of iron-bearing hydrous ringwoodite. *Physics of the Earth and Planetary Interiors* 143–144: 47–56.

Jacobsen SD, Demouchy S, Frost DJ, Ballaran TB, and Kung J (2005) A systematic study of OH in hydrous wadsleyite from polarized FTIR spectroscopy and single-crystal X-ray diffraction: Oxygen sites for hydrogen storage in Earth's interior. *American Mineralogist* 90: 61–70.

Jagoutz E, Palme H, Baddenhausen H, *et al.* (1989) The abundances of major, minor and trace elements in the earth's mantle as derived from primitive ultramafic nodules. *Proceedings 10th Lunar Planetary Science Conference* 2: 2031–2050.

Jambon A and Zimmermann JL (1990) Water in oceanic basalts: Evidence for dehydration of recycled crust. *Earth and Planetary Science Letters* 101: 323–331.

Jochum KP, Hofmann AW, Ito E, Seufert HM, and White WM (1986) K, U and Th in mid-ocean ridge basalt glasses and heat production, K/U and K/Rb in the mantle. *Nature* 306: 431–436.

Kato T, Ohtani E, Ito Y, and Onuma K (1996) Element partitioning between silicate perovskites and calcic ultrabasic melt. *Physics of the Earth and Planetary Interiors* 96: 201–207.

Kato T, Ringwood AE, and Irifune T (1988) Experimental determination of element partitioning between silicate perovskites, garnets and liquids constraints on early differentiation of the mantle. *Earth and Planetary Science Letters* 89: 123–145.

Katsura T and Ito E (1989) The system Mg_2SiO_4–Fe_2SiO_4 at high pressures and temperatures: Precise determination of stabilities of olivine, modified spinel, and spinel. *Journal of Geophysical Research* 94: 15663–15670.

Katsura T, Yamada H, Nishikawa O, *et al.* (2004) Olivine–wadsleyite transition in the system $(Mg, Fe)_2SiO_4$. *Journal of Geophysical Research* 109, doi:10.1029/2003JB002438.

Kawamoto T, Hervig RL, and Holloway JR (1996) Experimental evidence for a hydrous transition zone in the early Earth's mantle. *Earth and Planetary Science Letters* 142: 587–592.

Kellogg LH, Hager BH, and van der Hilst RD (1999) Compositional stratification in the deep mantle. *Science* 283: 1881–1884.

Kesson SE, Fitz Gerald JD, and Shelley JM (1998) Mineralogy and dynamics of a pyrolite lower mantle. *Nature* 393: 252–255.

Koga K, Hauri E, Hirschmann MM, and Bell D (2003) Hydrogen concentration analyses using SIMS and FTIR: Comparison and calibration for nominally anhydrous minerals. *Geochemistry Geophysics Geosystems* 4: 1019 (doi:10.1029/2002GC000378).

Kohlstedt DL, Keppler H, and Rubie DC (1996) Solubility of water in the α, β and γ phases of $(Mg, Fe)_2SiO_4$. *Contributions to Mineralogy and Petrology* 123: 345–357.

Kohn SC, Brooker RA, Frost DJ, Slesinger AE, and Wood BJ (2002) Ordering of hydroxyl defects in hydrous wadsleyite (β-Mg_2SiO_4). *American Mineralogist* 87: 293–301.

Kröger FA and Vink HJ (1956) Relations between the concentrations of imperfections in crystalline solids. *Solid State Physics* 3: 307–435.

Kudoh Y, Inoue T, and Arashi H (1996) Structure and crystal chemistry of hydrous wadsleyite, $Mg_{1.75}SiH_{0.5}O_4$: Possible hydrous magnesium silicate in the mantle transition zone. *Physics and Chemistry of Minerals* 23: 461–469.

Li B, Liebermann RC, and Weidner DJ (2001) P–V–V_P–V_S–T measurements on wadsleyite to 7 GPa and 873 K: Implications for the 410-km seismic discontinuity. *Journal of Geophysical Research* 106: 30575–30591.

Liebske C, Corgne A, Frost DJ, Rubie DC, and Wood BJ (2005) Compositional effects on element partitioning between Mg–silicate perovskite and silicate melts. *Contributions to Mineralogy and Petrology* 149: 113–128.

Litasov K and Ohtani E (2003) Stability of various hydrous phases in CMAS pyrolite–H_2O system up to 25 GPa. *Physics and Chemistry of Minerals* 30: 147–156.

Litasov K, Ohtani E, Langenhorst F, Yurimoto H, Kubo T, and Kondo T (2003) Water solubility in Mg-perovskites and water storage capacity in the lower mantle. *Earth and Planetary Science Letters* 211: 189–203.

Lodders K (2003) Solar system abundances and condensation temperatures of the elements. *The Astrophysical Journal* 591: 1220–1247.

McCammon C (1997) Perovskite as a possible sink for ferric iron in the lower mantle. *Nature* 387: 694–696.

Matsukage KN, Jing Z, and Karato S-I (2005) Density of hydrous silicate melt at the conditions of Earth's deep upper mantle. *Nature* 438: 488–491.

Mayama N, Suzuki I, Saito T, Ohno I, Katsura T, and Yoneda A (2004) Temperature dependence of elastic moduli of β–$(Mg, Fe)2SiO_4$. *Geophysical Research Letters* 31, doi:10.1029/2003GL019247.

McDade P, Blundy JD, and Wood BJ (2003) Trace element partitioning on the Tinaquillo Lherzolite solidus at 1.5 GPa. *Physics of the Earth and Planetary Interiors* 139: 129–147.

McDonough WF (2001) The composition of the Earth. In: Teissseyre R and Majewski E (eds.) *Earthquake Thermodynamics and Phase Transformations in* the Earth's Interior, vol. 76, pp. 3–23. San Diego: Academic Press.

McDonough WF and Sun S-S (1995) The composition of the Earth. *Chemical Geology* 120: 223–253.

McFarlane EA (1994) *Differentiation in the Early Earth: An Experimental Investigation.* PhD Thesis, University of Arizona, 153pp.

McFarlane EA, Drake MJ, and Rubie DC (1994) Element partitioning between Mg-perovskite, magnesiowüstite, and silicate melt at conditions of the Earth's mantle. *Geochimica et Cosmochimica Acta* 58: 5161–5172.

Mibe K, Orihashi Y, Nakai S, and Fujii T (2006) Element partitioning between transition-zone minerals and ultramafic melt under hydrous conditions. *Geophysical Research Letters* 33, doi:10.1029/2006GL026999.

Michael P (1988) The concentration, behavior and storage of H_2O in the suboceanic upper mantle: Implications for mantle metasomatism. *Geochimica et Cosmochimica Acta* 52: 555–566.

Michael P (1995) Regionally distinctive sources of depleted MORB: Evidence from trace elements and H_2O. *Earth and Planetary Science Letters* 131: 301–320.

Morgan JW and Anders E (1980) Chemical composition of Earth, Venus, and Mercury. *Proceedings of the National Academy of Science of the United States of America* 77: 6973–6977.

Murakami M, Hirose K, Yurimoto H, Nakashima S, and Takafuji N (2002) Water in Earth's lower mantle. *Science* 295: 1885–1887.

Murakami M, Hirose K, Kawamura K, Sata N, and Ohishi Y (2004) Post-perovskite phase transition in $MgSiO_3$. *Science* 304: 855–858.

Murakami M, Hirose K, Sata N, and Ohushi Y (2005) Post-perovskite phase transition and mineral chemistry in the pyrolitic lowermost mantle. *Geophysical Research Letters* 32, doi:10.1029/2004GL021956.

Navrotsky A (1999) Mantle geochemistry: A lesson from ceramics. *Science* 284: 1788–1789.

Nishiyama N and Yagi T (2003) Phase relation and mineral chemistry in pyrolite to 2000 °C under the lower mantle pressures and implications for dynamics of mantle plumes.

Journal of Geophysical Research 108, doi:10.1029/2002JB002216.

Nolet G and Zielhuis A (1994) Low S velocities under the Tornquist–Teisseyre zone: Evidence for water injection into the transition zone by subduction. *Journal of Geophysical Research* 99: 15813–15820.

Nolet G, Grand SP, and Kennett BLN (1994) Seismic heterogeneity in the upper mantle. *Journal of Geophysical Research* 99: 23753–23766.

Oganov AR and Ono S (2004) Theoretical and experimental evidence for a post-perovskite phase of MgSiO₃ in Earth's D″ layer. *Nature* 430: 445–448.

Ohtani E (1985) The primordial terrestrial magma ocean and its implications for stratification of the mantle. *Physics of the Earth and Planetary Interiors* 78: 70–80.

Ohtani E, Kato T, and Sawamoto H (1986) Melting of a model chondritic mantle to 20 GPa. *Nature* 322: 352–353.

Ohtani E, Mizobata H, and Yurimoto H (2000) Stability of dense hydrous magnesium silicate phases in the systems Mg₂SiO₄–H₂O and MgSiO₃–H₂O at pressures up to 27 GPa. *Physics and Chemistry of Minerals* 27: 533–544.

Palme H and Jones A (2003) Solar system abundances of the elements. In: Holland H and Turekian KK (eds.) *Treatise on Geochemistry*, vol. 1, pp. 41–61. Amsterdam: Elsevier.

Palme H and O'Neill HSC (2003) Cosmochemical estimates of mantle composition. In: Holland H and Turekian KK (eds.) *Treatise on Geochemistry*, vol. 2, pp. 1–38. Amsterdam: Elsevier.

Paterson MS (1982) The determination of hydroxyl by infrared absorption in quartz, silicate glasses and similar materials. *Bulletin de Mineralogie* 105: 20–29.

Paulssen H (1988) Evidence for a sharp 670 km discontinuity as inferred from P-to-S converted waves. *Journal of Geophysical Research* 93: 10489–10500.

Rigden SM, Gwanmesia GD, Jackson I, and Liebermann RC (1992) Progress in high-pressure ultrasonic interferometry, the pressure dependence of elasticity of Mg₂SiO₄ polymorphs and constraints on the composition of the transition zone of the Earth's mantle. In: Syono Y and Manghnani M (eds.) *High Pressure Research: Application to Earth and Planetary Sciences*, pp. 167–182. Washington, DC: American Geophysical Union.

Ringwood AE (1979) *Origin of the Earth and Moon*, 295 p. New York: Springer.

Ringwood AE (1991) Phase transformation and their bearing on the constitution and dynamics of the mantle. *Geochimica et Cosmochimica Acta* 55: 2083–2110.

Rudnick RL and Nyblade AA (1999) The thickness and heat production of Archean lithosphere: Constraints from xenolith thermobarometry and surface heat flow. In: Fei Y, Bertka C, and Mysen BO (eds.) *Mantle Petrology: Field Observations and High Pressure Experimentation*, vol. 6, pp. 3–12. Houston: The Geochemical Society.

Sakamaki T, Suzuki A, and Ohtani E (2006) Stability of hydrous melt at the base of the Earth's upper mantle. *Nature* 439: 192–194.

Samuel H and Farnetani CG (2003) Thermochemical convection and helium concentrations in mantle plumes. *Earth and Planetary Science Letters* 207: 39–56.

Shannon RD (1976) Revised effective ionic radii and systematic studies of interatomic distances in halides and chalcogenides. *Acta Crytallographica Section A* 32: 751–767.

Sinogeikin SV, Katsura T, and Bass JD (1998) Sound velocities and elastic properties of Fe-bearing wadsleyite and ringwoodite. *Journal of Geophysical Research* 103: 20819–20825.

Smyth JR (1987) β-Mg₂SiO₄: A potential host for water in the mantle? *American Mineralogist* 72: 1051–1055.

Smyth JR (1994) A crystallographic model for hydrous wadsleyite (β-Mg₂SiO₄): An ocean in the Earths interior? *American Mineralogist* 79: 1021–1024.

Smyth JR and Bish DL (1988) *Crystal Structures and Cation Sites of the Rock-forming Minerals*, 332 pp. Boston: Allen & Unwin.

Smyth JR and Kawamoto T (1997) Wadsleyite II: A new high pressure hydrous phase in the peridotite–H₂O system. *Earth and Planetary Science Letters* 146: E9–E16.

Smyth JR and Frost DJ (2002) The effect of water on the 410-km discontinuity: An experimental study. *Geophysical Research Letters* 29, doi:10.1029/2001GL014418.

Smyth JR, Kawamoto T, Jacobsen SD, Swope RJ, Hervig RL, and Holloway JR (1997) Crystal structure of monoclinic hydrous wadsleyite β-(Mg, Fe)₂SiO₄. *American Mineralogist* 82: 270–275.

Sobolev AV and Chaussidon M (1996) H₂O concentrations in primary melts from supra–subduction zones and mid-ocean ridges: Implications for H₂O storage and recycling in the mantle. *Earth and Planetary Science Letters* 137: 45–55.

Solomatov VS (2000) Fluid Dynamics of a Terrestrial Magma Ocean. In: Canup RM and Righter K (eds.) *Origin of the Earth and Moon*, pp. 323–360. Tucson: University of Arizona Press.

Song T-RA, Helmberger DV, and Grand SP (2004) Low-velocity zone atop the 410 km seismic discontinuity in the northwestern United States. *Nature* 427: 530–533.

Stachel T, Aulbach S, Brey GP, *et al.* (2004) The trace element composition of silicate inclusions in diamonds: A review. *Lithos* 77: 1–19.

Stixrude L and Kirko B (2005) Structure and freezing of MgSiO₃ liquid in Earth's lower mantle. *Science* 310: 297–299.

Suzuki A and Ohtani E (2003) Density of peridotite melts at high pressure. *Physics and Chemistry of Minerals* 30: 449–456.

Tatsumi Y and Eggins S (1995) *Subduction zone magmatism*, 211 p. Oxford: Blackwell Science.

Taura H, Yurimoto H, Kato T, and Sueno S (2001) Trace element partitioning between silicate perovskites and ultracalcic melt. *Physics of the Earth and Planetary Interiors* 124: 25–32.

Tonks WB and Melosh HJ (1993) Magma ocean formation due to giant impacts. *Journal of Geophysical Research* 98: 5319–5333.

Trønnes RG and Frost DJ (2002) Peridotite melting and mineral–melt partitioning of major and minor elements at 22–24.5 GPa. *Earth and Planetory Science Letters* 197: 117–131.

Turcotte DL, Paul D, and White WM (2001) Thorium–uranium systematics require layered mantle convection. *Journal of Geophysical Research* 106: 4265–4276.

van der Meijde M, Marone F, Giardini D, and van der Lee S (2003) Seismic evidence for water deep in Earth's upper mantle. *Science* 300: 1556–1558.

van Keken PE, Hauri EH, and Ballentine CJ (2002) Mantle mixing: The generation, preservation, and destruction of chemical heterogeneity. *Annual Review of Earth and Planetary Sciences* 30: 493–525.

van Westrenen W, Blundy JD, and Wood BJ (2000a) Effect of Fe²⁺ on garnet–melt trace element partitioning: Experiments in FCMAS and quantification of crystal-chemical controls in natural systems. *Lithos* 53: 191–203.

van Westrenen W, Allan NL, Blundy JD, Purton JA, and Wood BJ (2000b) Atomistic simulation of trace element incorporation into garnets - comparison with experimental garnet–melt partitioning data. *Geochimica et Cosmochimica Acta* 64: 1629–1639.

Vanpeteghem CB, Angel RJ, Ross NL, *et al.* (2006) Al, Fe substitution in the MgSiO₃ perovskite structure: A single-crystal X-ray diffraction study. *Physics of the Earth and Planetary Interiors* 155: 96–103.

Wade J and Wood BJ (2005) Core formation and the oxidation state of the Earth. *Earth and Planetary Science Letters* 236: 78–95.

Walter MJ (2003) Melt extraction and compositional variability in mantle lithosphere. In: Holland H and Turekian KK (eds.) *Treatise on Geochemistry*, vol. 2, pp. 363–394. Amsterdam: Elsevier.

Walter MJ, Nakamura E, Tronnes RG, and Frost DJ (2004) Experimental constraints on crystallization differentiation in a deep magma ocean. *Geochimica et Cosmochimica Acta* 68: 4267–4284.

Wänke H and Dreibus G (1988) Chemical-composition and accretion history of terrestrial planets. *Philosophical Transactions of the Royal Society of London, Series A, Mathematical Physical and Engineering Sciences* 325: 545–557.

Wasson JT (1985) *Meteorites: Their Record of Early Solar-System History*, 267 pp. New York: W.H. Freeman.

Wasson JT (1988) The building stones of the planets. In: Vilas F, Chapman CR, and Matthews MS (eds.) *Mercury*, pp. 622–650. Tucson: University of Arizona Press.

Wasson JT and Kallemeyn GT (1988) Compositions of Chondrites. *Philosophical Transactions of the Royal Society of London, series A, Mathematical Physical and Engineering Sciences* 325: 535–544.

Weidner DJ (1985) A mineral physics test of a pyrolite mantle. *Geophysical Research Letters* 12: 417–420.

Wetherill GW (1990) Formation of the earth. *Annual Review of Earth and Planetary Sciences* 18: 205–256.

Williams Q and Garnero EJ (1996) Seismic evidence for partial melt at the base of Earth's mantle. *Science* 273: 1528–1530.

Wiser N and Wood BJ (1991) Experimental determination of activities in Fe–Mg olivine at 1400K. *Contributions to Mineralogy and Petrology* 108: 146–153.

Wood BJ (1995) The effects of H_2O on the 410 km seismic discontinuity. *Science* 268: 74–76.

Wood BJ (2000) Phase transformations and partitioning relations in peridotite under lower mantle conditions. *Earth and Planetary Science Letters* 174: 341–354.

Wood BJ and Blundy JD (2001) The effect of cation charge on crystal–melt partitioning of trace elements. *Earth and Planetary Science Letters* 188: 59–71.

Wood BJ and Rubie DC (1996) The effect of alumina on phase transformations at the 660-kilometer discontinuity from Fe–Mg partitioning experiments. *Science* 273: 1522–1524.

Yamazaki A and Hirahara K (1994) The thickness of upper mantle discontinuities, as inferred from short-period J-array data. *Geophysical Research Letters* 21: 1811–1814.

Young TE, Green HW, Hofmeister AM, and Walker D (1993) Infrared spectroscopic investigation of hydroxyl in β-(Mg, Fe)$_2$SiO$_4$ and coexisting olivine: Implications for mantle evolution and dynamics. *Physics and Chemistry of Minerals* 19: 409–422.

5 Mineralogy of the Earth – The Earth's Core: Iron and Iron Alloys

L. Vočadlo, University College London, London, UK

5.1	Introduction	91
5.2	Seismological Observations of the Earth's Core	92
5.2.1	PREM and Beyond	92
5.2.2	Anisotropy and Layering in the Inner Core	92
5.2.3	Super-Rotation of the Inner Core	94
5.2.4	Seismological Observations of the Outer Core	95
5.3	The Structure of Iron in the Inner Core	95
5.4	Thermoelastic Properties of Solid Iron	98
5.4.1	Thermodynamic Properties from Free Energies	98
5.4.1.1	Equation of state	99
5.4.1.2	Incompressibility	99
5.4.1.3	Thermal expansion	99
5.4.1.4	Heat capacity	99
5.4.1.5	Grüneisen parameter	100
5.4.2	Elasticity of Solid Iron	101
5.5	Rheology of Solid Iron	102
5.5.1	Slip Systems in Iron	102
5.5.2	Viscosity and the Inner Core	104
5.6	The Temperature in the Earth's Core	106
5.7	Thermodynamic Properties of Liquid Iron	108
5.8	Rheology of Liquid Iron	108
5.8.1	Viscosity and Diffusion	108
5.8.2	The Structure of Liquid Iron	110
5.9	Evolution of the Core	110
5.10	The Composition of the Core	111
5.10.1	Bulk Composition	111
5.10.2	The Effect of Nickel	112
5.10.3	Light Elements	112
5.10.3.1	Chemical potential calculations of FeX binary systems	112
5.10.3.2	High-temperature elasticity of FeSi and FeS	114
5.10.3.3	Rheology of liquid iron alloys	116
5.11	Summary	116
References		117

5.1 Introduction

The Earth's core plays a fundamental role in the evolution of our planet. As the Earth cools, the inner core grows, crystallizing from the outer core of liquid iron (alloyed with ~5–10% light elements) to form an inner core of solid iron (alloyed with ~2–3% light elements (Jephcoat and Olson, 1987; Poirier, 1994; Stixrude *et al.*, 1997; Alfè *et al.*, 2002a)). The release of latent heat of fusion, together with chemical buoyancy arising from the enrichment of the outer core with light elements, provide driving forces for the fluid flow responsible for the geodynamo, and hence for the Earth's magnetic field. The heat released from the core helps drive mantle convection which leads in turn to surface features such as volcanism and plate tectonics. Since the inner core is sitting within the liquid outer core, it is isolated from the rest of the Earth; however, coupling with the outer core and mantle (e.g., geomagnetic and gravitational) prevent its motion from being entirely independent, allowing it to super-rotate and wobble.

The only direct observations of the Earth's core come from seismology; therefore, any credible mineralogical model has to match exactly the seismic observations. Increasingly accurate seismic observations have shown the Earth's inner core, in particular, to be far more complex than had previously been thought. The current seismological models reveal an inner core which is anisotropic, layered, and possibly laterally heterogeneous, but the origins of this anisotropy and layering are not yet understood. It is generally considered that the anisotropy reflects the preferred orientation of the crystals present, which could have arisen either during inner-core crystallization or developed over time through solid state flow; these two mechanisms have vastly different implications for core evolution. The observed layering may be due to changes in chemical composition, crystal structure, preferred orientation or some combination of all three. Again, these imply very different core processes and evolution. The observed layering also implies that the upper and lower inner core could be compositionally or structurally different. The origin of anisotropy and layering is fundamental to understanding and constraining evolutionary models for the Earth's inner core, and the mineralogical model must reflect this complex structure.

Mineral physics also has an important role to play in understanding the processes going on in the outer core. Knowledge of the thermoelastic properties of candidate liquid iron alloys can be compared with seismological observations and therefore lead to models for the composition of the outer core. Estimates for the dynamic properties of liquids such as diffusivity and viscosity can be incorporated into the magnetohydrodynamics equations that quantify the magnetic field. Furthermore, mineral physics can provide values for thermodynamic quantities that can be put into thermal evolution models of core formation leading to timescales for inner core growth and quantification of the heat budget of the Earth.

In order to place fundamental constraints on the properties of the Earth's core, it is essential to know the behavior of iron and iron alloys at core conditions. In particular, the key questions to resolve are: what is the most stable phase(s) of iron present in the inner core, what are the elastic properties of the stable phase(s) (at core pressures and temperatures), what are the rheological properties of these candidate solid phases, do the combined thermoelastic and rheological properties of iron alloys lead to a comprehensive model for inner

core composition, evolution, anisotropy, and layering; in addition, what is the temperature of the Earth's core, what are the thermodynamic properties of candidate liquid iron alloy phases, what are the rheological properties of these phases, what is the composition of the outer core. Notwithstanding the difficulties in answering any single one of these questions, the combined answers of them all should exactly match the models inferred from seismology.

5.2 Seismological Observations of the Earth's Core

5.2.1 PREM and Beyond

One-dimensional global seismological models for the Earth's core, such as the 'preliminary reference Earth model' (PREM; Dziewonski and Anderson, 1981), give bulk elastic properties as a function of depth via compressional and shear wave velocity profiles and free oscillation data. Although these models tell us nothing about exact composition or temperature, they do give fairly robust measurements for pressure, density, and also elastic moduli, such as incompressibility and shear modulus, from which compositional models can be derived. Since PREM there have been a number of refinements of the seismological model such as IASP91 (Kennett and Engdahl 1991) and AK135 (Kennett et al., 1995) which show subtle differences in the detailed velocity profiles, but are generally consistent with each other. More complex structures of the Earth have been observed with three-dimensional (3-D) imaging generated using seismic tomography; this shows how the wave velocities vary laterally as well as radially generating a 3-D representation of wave velocities throughout the Earth (see, e.g., Romanowicz (2003) for a review). It is the job of mineral physics to provide estimates for core properties which best match these seismological models and which thereby provide an explanation for the more detailed seismic structure of the inner core.

5.2.2 Anisotropy and Layering in the Inner Core

It is well established that the inner core exhibits significant anisotropy, with P-wave velocities ~3% faster along the polar axis than in the equatorial plane (Creager, 1992; Song and Helmberger, 1993; Song, 1997; Song and Xu, 2002; Sun and Song, 2002; Beghein and Trampert, 2003; Oreshin and Vinnik, 2004). Early models for the inner core suggested that this seismic anisotropy had cylindrical symmetry, with the symmetry axis parallel to the Earth's rotation axis. More recent seismic observations suggest that, as well as being cylindrically anisotropic, the inner core may in fact be layered (**Figure 1(a)**). The evidence is for a seismically isotropic or weakly anisotropic upper layer, with lateral variations in thickness of ~100–400 km, overlaying an irregular, nonspherical transition region to an anisotropic lower layer (Song and Helmberger, 1998; Ouzounis and Creager, 2001; Song and Xu, 2002; Ishii and Dziewonski, 2003), although variations on this model have also been suggested (e.g., **Figure 1(b)**). The existence of an isotropic upper layer implies that the magnitude of the seismic anisotropy in the lower inner core should be significantly greater than previously thought, possibly as much as 5–10% (Ouzounis and Creager, 2001; Song and Xu, 2002). The observed layering also implies that the upper and lower inner core could be compositionally or structurally different.

With this recently observed complex structure of the inner core, the question arises as to the mechanisms by which such anisotropy and layering can occur, and the nature of the seismic boundary. The isotropic layer could be due either to randomly oriented crystals (containing, e.g., b.c.c. and/or h.c.p. iron) or to a material with a low intrinsic anisotropy. The anisotropic lower layer is thought to be due to the preferred orientation of aligned crystals (Anderson, 1989), possibly, although not

necessarily, of a different iron phase to the upper layer. The inner core could therefore, for example, be comprised entirely of b.c.c. iron, entirely of h.c.p. iron, or some combination of the two.

There are three hypotheses of how the anisotropy and layering could have occurred:

1. Yoshida *et al.* (1996) have suggested a driving mechanism whereby the inner core grows more rapidly at the equator than the poles, giving rise to a flattened sphere. In order to maintain hydrostatic equilibrium, this ellipsoidal inner core would have to continuously deform as it grows, resulting in stress-induced recrystallization. **Figure 2** shows the flow lines resulting from their calculations. In this model, the older material will have undergone the most deformation, and hence show the greatest preferential orientation, consistent with the observation that the center of the inner core has a greater degree of anisotropy.

2. Buffett and Wenk (2001) suggested instead that the depth dependence of anisotropy can be explained by plastic deformation under the influence of electromagnetic shear stresses (**Figure 3**). This would produce a toroidal type flow, quite different from that of the previous model. Again, the high anisotropy in the lower inner core results from the fact that it is older and has had more time to develop.

3. The third possibility is that the preferred orientation was locked into the crystal on cooling from the

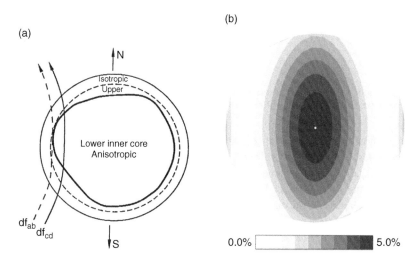

Figure 1 (a) Schematic of isotropic upper-inner-core and anisotropic lower-inner-core structure. The boundary is speculated to be irregular, which may explain recent reports of large scatter in inner core travel times. (b) Cross section through the axisymmetric anisotropic inner-core model. The contour levels show the compressional-velocity perturbations relative to PREM for waves traveling parallel to the rotation axis. Note that this inner-core model is highly anisotropic near its center. The ability of the model to fit the inner-core spectral data degrades appreciably if an isotropic layer thicker than 100–200 km is imposed at the top of the inner core. (a): Song XD and Helmberger DV (1998) Seismic evidence for an inner core transition zone. *Science* 282: 924–927; (b): Durek JJ and Romanowicz B (1999) Inner core anisotropy by direct inversion of normal mode spectra. *Geophysical Journal International* 139: 599–622.

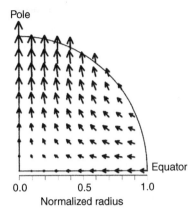

Figure 2 Flow field showing predominantly radial flow. From Yoshida S, Sumita I, and Kumazawa M (1996) Growth model of the inner core coupled with outer core dynamics and the resulting anisotropy. *Journal of Geophysical Research* 101: 28085–28103.

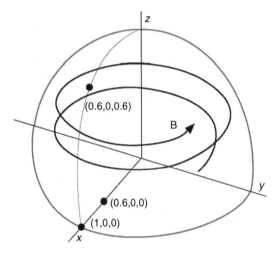

Figure 3 Direction of magnetic field causing preferred crystal alignment in model of Buffett and Wenk (2001).

liquid, perhaps due to flow in the outer core (Karato, 1993). If the alignment was frozen in on crystallization, there would have to be an, as yet unknown, mechanism for the formation of the outer isotropic layer. One possibility is simply that the iron phase in the outer layer of the inner core is different to that in the deeper inner core and does not have slip systems which cause anisotropy.

The first model predicts the *c*-axis of h.c.p.-iron to be aligned parallel to the pole directions, while the second mode predicts it to be in the equatorial plane. Despite the completely different underlying dynamics and resultant flow fields, both studies were able to show at the time of publication (1996 and 2001, respectively) that the predicted lattice

preferred orientation gave an anisotropy consistent with that observed for the inner core. The reason they were able to do this was because they used different sets of elastic constants to calculate the anisotropy (see Section 5.4.2), and they used different mechanisms for producing the preferred orientation in the first place. An additional, but fundamental uncertainty in these models is that they only considered the h.c.p. phase of iron. As shown in Section 5.3, there is growing evidence that b.c.c. may in fact be stable in all or part of the inner core.

5.2.3 Super-Rotation of the Inner Core

In addition to anisotropy and layering, there is another inference from seismology which provides a constraint on the dynamics of the inner core, in particular, inner-core viscosity. Seismic observations suggest that the Earth's inner core may be rotating with respect to the bulk of the Earth. Although there have been several seismic studies, until recently, there appeared to be no consensus as to the amount of differential rotation of the Earth's core with respect to the mantle. Observations suggest rotation rates which range from marginally detectable to 3° per year (Song and Richards, 1996; Su *et al.*, 1996; Creager, 1997; Song, 2000; Song and Li, 2000; Souriau, 1998; Poupinet *et al.*, 2000; Collier and Hellfrich, 2001). More recently, however, the debate seems to have settled on the low end of the spectrum at 0.3–0.5° per year (Zhang *et al.*, 2005). If the rotation rate is high, and if seismic anisotropy is due to preferred alignment, the inner core would have to adjust on a relatively short timescale in order to maintain its texture as it rotates. Collier and Helffrich (2001) suggested that the relative motion of the inner core and the mantle might take the form of an inner-core oscillation on a timescale of ~280 days rather than a simple relative rotation of the two; this type of oscillation could be caused by the heterogeneous distribution of mass in the Earth's mantle exerting a significant gravitational pull on the inner core, which will tend to keep the inner core aligned. The extent to which the inner core adjusts during differential rotation or oscillation will therefore depend critically on the inner-core viscosity. Buffett (1997) modeled the viscous relaxation of the inner core by calculating numerically the relaxation time for the inner core to adjust, as it rotates, back to its equilibrium shape after small distortions due to perturbations in gravitational potential imposed by the overlying mantle. He suggested that the viscosity has to be constrained to be either less than 10^{16} Pa s or greater than 10^{20} Pa s depending on the dynamical regime. The latter value is supported by Collier and Hellfrich (2001), who concluded that the attenuation of the observed inner-core torsional oscillations with time are consistent with a viscosity of 3.9×10^{19} Pa s. Clearly, an independent measure of inner-core viscosity would place constraints on these values which will depend not only on the crystal structure adopted by the iron alloy of the inner core but also on the mechanism(s) by which deformation occurs.

5.2.4 Seismological Observations of the Outer Core

In general, the outer core is considered to be homogeneous due to the continual mixing of the liquid via convection. Despite this, there is some seismological evidence to suggest that there is some structure and heterogeneity in the outer core. For example, lateral velocity heterogeneities, possibly associated with the topography at the core–mantle boundary (CMB), have been observed with seismic tomography (Soldati et al., 2003). However, at the present time, it is impossible to be definitive as to whether or not these lateral anomalies are an artifact of the methodology (see, e.g., Lei and Zhao, 2006). There is also some evidence for some structure at the top of the outer core. A very small (a few kilometers across) and thin (150 m) outer core rigidity zone may exist just below the core–mantle boundary (Garnero and Jeanloz, 2000); it is possible that this anomaly is responsible for the wobbling of the Earth's rotation axis (Rost and Revenaugh, 2001). Finally, there may be a mushy layer (crystals suspended in liquid) or slurry zone (solidification via dendrites) at the base of the inner core, the latter being more likely (Shimizu et al., 2005).

Accurate mineral physics data can be used to resolve many of the uncertainties about the structure and composition of the Earth's core; however, as mentioned in Section 5.1, the mineralogical model that comes out of combining theoretical and experimental studies must match the seismological observations exactly. Furthermore, the seismological observations themselves must be sufficiently detailed and robust to enable accurate and meaningful comparisons to be made.

5.3 The Structure of Iron in the Inner Core

The properties of the inner core are determined by the materials present; although we know that the inner core is made or iron alloyed with nickel and other lighter elements (see Section 5.10), we must first understand the behavior of pure iron before we are able to understand the behavior of iron alloys. However, the properties of pure iron relevant to the Earth's core depend upon the structure iron adopts at core conditions, and it is this we need to address first.

Experimentalists have put an enormous effort over the last 15–20 years into obtaining a phase diagram of pure iron, but above relatively modest pressures and temperatures there is still much uncertainty (Brown, 2001). Under ambient conditions, Fe adopts a b.c.c. structure, that transforms with temperature to an f.c.c. form, and with pressure transforms to an h.c.p. phase, ε-Fe. The high P/T phase diagram of pure iron itself however is still controversial (see **Figure 4** and also the discussion in Stixrude and Brown (1998)). Various diamond anvil cell (DAC)-based studies have been interpreted as showing that h.c.p. Fe transforms at high temperatures to a phase which has variously been described as having a double hexagonal close packed structure (dh.c.p.) (Saxena et al., 1996) or an orthorhombically distorted h.c.p. structure (Andrault et al., 1997). Furthermore, high-pressure shock experiments have also been interpreted as showing a high-pressure solid–solid phase transformation (Brown and McQueen, 1986; Brown, 2001), which has been suggested could be due to the development of a b.c.c. phase (Matsui and Anderson, 1997). Other experimentalists, however, have failed to detect such a post-h.c.p. phase (e.g., Shen et al., 1998; Nguyen and Holmes, 2004), and have suggested that the previous observations were due to either minor impurities or metastable strain-induced behavior. Nevertheless, the experiments, together with theoretical calculations of the static, zero-Kelvin solid (Stixrude et al., 1997; Vočadlo et al., 2000), suggested that the h.c.p. phase of iron is the most likely stable phase in the inner core. Experiments at moderate pressure and ambient temperature (Antonangeli et al., 2004), and calculations (Stixrude and Cohen, 1995a) at high pressures but at 0 K, showed that the compressional wave velocity along the c-axis of h.c.p.-Fe is significantly faster than that in the basal plane; this led to the conclusion that the inner core is made up of oriented h.c.p.-Fe crystals with the c-axis parallel to the rotation axis. However, more recent calculations (using a particle-in-cell (PIC) method) found that the elastic properties of h.c.p.-Fe change dramatically as a function of temperature (Steinle-Neumann et al., 2001), and that compressional waves travel faster in the basal plane of h.c.p.-Fe at high temperatures and pressures, in

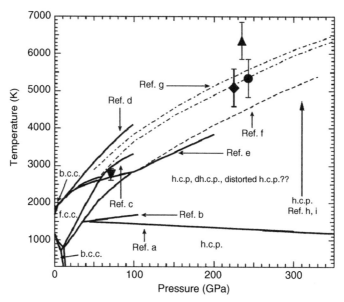

Figure 4 Phase diagram of pure iron. Solid lines represent phase boundaries and melt lines from DAC experiments; symbols with error bars are points on the melting curve from shock experiments; broken lines are melting curves from first-principles calculations. DAC data: Ref. a: Saxena and Dubrovinsky (2000); b: Andrault *et al.* (2000); c: Shen *et al.* (1998); d: Williams *et al.* (1987); e: Boehler (1993). Shock data: triangle: Yoo *et al.* (1993); circles: Brown and McQueen (1986); reverse triangle: Ahrens *et al.* (2002); diamond: Nguyen and Holmes (2004). First principles calculations; f: Laio *et al.* (2000); g: Alfè *et al.* (2002a, 2002b); h: Vočadlo *et al.* (2000); i: Vočadlo *et al.* (2003a, 2003b). Adapted from Nguyen JH and Holmes NC (2004) Melting of iron at the physical conditions of the Earth core. *Nature* 427: 339–342.

complete contrast to the low-temperature results. These results can account for the observed seismic anisotropy if the preferred orientation of h.c.p.-Fe in the inner core is with the *c*-axis aligned parallel to the equatoral plane instead of parallel to the poles. Obviously, if the seismic anisotropy is caused by deformation, the two studies imply very different stresses and flow fields in the inner core (as described in Section 5.2.2). There are two main issues with these conclusions that need to be considered.

First, the assumption that iron must have the h.c.p. structure at core conditions has been recently challenged, especially in the presence of lighter elements (Beghein and Trampert, 2003 and **Figure 5**; Lin *et al.*, 2002) – it now seems possible or even probable that a b.c.c. phase might be formed (Vočadlo *et al.*, 2003a).

Previously, the b.c.c. phase of iron was considered an unlikely candidate for a core-forming phase because it is elastically unstable at high pressures, with an enthalpy considerably higher than that of h.c.p. (Söderlind *et al.*, 1996; Stixrude and Cohen, 1995b; Vočadlo *et al.*, 2000). However, *ab initio* molecular dynamics calculations to obtain free energies at core pressures and temperatures have found that the b.c.c. phase of iron does, in fact, become entropically

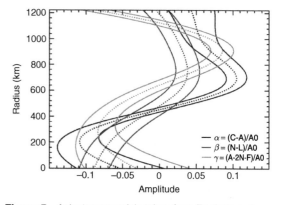

Figure 5 Anisotropy models taken from Beghein and Trampert (2003). The thin dotted line represents the mean model, and the thick surrounding lines correspond to two standard deviations. α, β, and γ are the three anisotropic parameters describing P-wave anisotropy, S-wave anisotropy, and the anisotropy of waves that do not travel along the vertical or horizontal directions, respectively. A comparison with the elasticity of h.c.p. iron at inner-core conditions (Steinle-Neumann *et al.*, 2001) shows that some of their models can be explained by progressively tilted h.c.p. iron in the upper half of the inner core, with their symmetry axes oriented at 45° from Earth's rotation axis at $r = 900$ km and at 90° in the middle of the inner core. In the deepest inner core ($r = 0$–400 km), 'none' of their models is compatible with published data of h.c.p. iron. This result might suggest the presence of another phase from these depths. Such a phase of iron could indeed be stable in the presence of impurities.

stabilized at core temperatures (Vočadlo *et al.*, 2003a). In an earlier paper (Vočadlo *et al.*, 2000) spin-polarized simulations were initially performed on candidate phases (including a variety of distorted b.c.c. and h.c.p. structures and the dh.c.p. phase) at pressures ranging from 325 to 360 GPa. These revealed, in agreement with Söderlind *et al.* (1996), that under these conditions only b.c.c. Fe has a residual magnetic moment and all other phases have zero magnetic moments. It should be noted, however, that the magnetic moment of b.c.c. Fe disappears when simulations are performed at core pressures and an electronic temperature of >1000 K, indicating that even b.c.c. Fe will have no magnetic stabilization energy under core conditions. At these pressures, both the b.c.c. and the suggested orthorhombic polymorph of iron (Andrault *et al.*, 1997) are mechanically unstable (**Figure 6**). The b.c.c. phase can be continuously transformed to the f.c.c. phase (confirming the findings of Stixrude and Cohen (1995a)), while the orthorhombic phase spontaneously transforms to the h.c.p. phase, when allowed to relax to a state of isotropic stress.

In contrast, h.c.p., dh.c.p., and f.c.c. Fe remain mechanically stable at core pressures, and we were therefore able to calculate their phonon frequencies and free energies. These showed that the h.c.p. phase was the more stable phase. However, it must be remembered that the free energies were obtained from phonon frequencies calculated at 0 K. More recently, Vočadlo *et al.* (2003a) used the method of thermodynamics integration combined with *ab initio* molecular dynamics to calculate the free energy at core pressures and temperatures of b.c.c. and h.c.p. iron. The conclusion was that, although the thermodynamically most stable phase of pure iron is still the h.c.p. phase, the free energy difference is so very small (**Table 1**) that a small amount of light element impurity could stabilize the b.c.c. phase at the expense of the h.c.p. phase (Vočadlo *et al.*, 2003a).

Second, considerable doubts have now been cast over the PIC method used to calculate the high-temperature elastic constants (Gannarelli *et al.*, 2003). Of particular importance is use of the correct

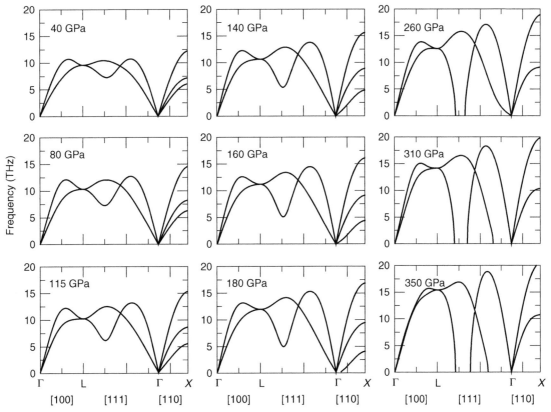

Figure 6 The calculated phonon dispersion curves for b.c.c.-Fe at nine different volumes (corresponding to ~40–350 GPa) (Vočadlo *et al.*, 2003a). The softening in the [110] direction at ~180 GPa indicates the onset of the b.c.c. → f.c.c. transition, and that in the [111] direction above ~260 GPa corresponds to the b.c.c. → ω transition.

Table 1   The *ab initio* Helmholtz free energy per atom of the b.c.c. and h.c.p. phases of Fe at state points along (* and below) the calculated melting curve

V (Å³)	T (K)	F_bcc (eV)	F_hcp (eV)	ΔF (meV)
9.0	3500	−10.063	−10.109	46
8.5	3500	−9.738	−9.796	58
7.8	5000	−10.512	−10.562	50
7.2	6000	−10.633	−10.668	35
6.9	6500	−10.545	−10.582	37
6.7	6700	−10.288	−10.321	33
7.2*	3000	−7.757	−7.932	175

Vočadlo L, Alfe D, Gillan MJ, Wood IG, Brodholt JP, and Price GD (2003a) Possible thermal and chemical stabilisation of body-centred-cubic iron in the Earth's core. *Nature* 424: 536–539. Note that at core conditions, ΔF is only ~35 meV.

equilibrium c/a ratio in h.c.p.-Fe; indeed, there has been an evolving story in the literature just on this property of iron alone, as it has a significant effect on the nature of the elastic anisotropy. In their work on the high-pressure, high-temperature elastic properties of h.c.p.-Fe, Steinle-Neumann *et al.* (2001) reported an unexpectedly large c/a ratio of almost 1.7. However, the work of Gannarelli *et al.* (2005) casts doubt on the robustness of these calculations and they found that the c/a ratio ranges from 1.585 at zero pressure and temperature to 1.62 at 5500 K and 360 GPa (see **Figure 7**), a result confirmed by further calculations (Vočadlo, 2007).

It should be clear from this section that neither the stable phase(s) nor the elasticity of iron in the Earth's inner core are known with any certainty; the studies outlined above serve to highlight the need to perform more detailed calculations under the appropriate conditions of pressure and temperature. Although the free energy of h.c.p.-Fe is lower than that of b.c.c.-Fe, even for a hypothetical pure iron core, the free energy difference is so small that, at core temperatures, both phases are likely to exist. In addition, we need to consider the effect of light elements, not only on the stability of the phases, but also on seismic anisotropy and slip systems.

5.4   Thermoelastic Properties of Solid Iron

The only experimental data on solid iron under true core pressures and temperatures comes from either simultaneously high-P/high-T static experiments or shock experiments; however, although the

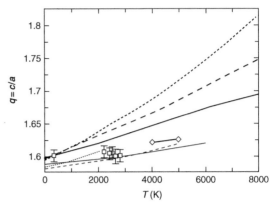

Figure 7   Calculated equilibrium axial ratio as a function of temperature for different volumes (light curves). Atomic volumes are 6.97 Å³ (solid curve), 7.50 Å³ (dashed curve), and 8.67 Å³ (dotted curve). For Steinle-Neumann *et al.* (2001) (heavy curves) volumes are 6.81 Å³ (solid curve), 7.11 Å³ (dashed curve), and 7.41 Å³ (dotted curve). Also shown are diffraction measurements due to Ma *et al.* (2004) at 7.73 Å³/atom (open squares with error bars) and *ab initio* MD calculations (open diamonds) at 6.97 Å³ (Gannarelli *et al.*, 2005). After Gannarelli CMS, Alfé D, and Gillan MJ (2005) The axial ration of hcp iron at the conditions of the Earth's inner core. *Physics of the Earth and Planetary Interiors* 139: 243–253.

measurement of temperature in shock experiments has been attempted, it is problematic (e.g. Yoo *et al.*, 1993). There are a number of static experimental studies on pure iron at ambient pressures and high temperatures, and high pressures and ambient temperatures (see below). Simultaneously high-pressure/high-temperature experiments are very challenging, although they have been attempted (e.g., the iron phase diagram above). In this section, we look at the thermoelastic properties of iron as obtained from *ab initio* calculations, and, where possible, compare them to experimental data.

5.4.1   Thermodynamic Properties from Free Energies

Computational mineral physics can play an extremely important role in quantifying some of the key thermodynamics properties that determine the state and evolution of the inner core. These calculations both compliment experimental data, and extend our knowledge where no experimental data exist. In Section 5.3 we have already shown that 'pure' iron is likely to take the h.c.p. structure at core conditions, so for the properties we present here, we focus only on h.c.p.-Fe. The results are presented as a function of pressure along isotherms. At each temperature, the results are only shown for the pressure range where, according to the calculations, the h.c.p. phase is thermodynamically stable. The calculated thermodynamic properties at a given temperature, T, and pressure, P, can be determined from the Gibbs free energy, $G(P, T)$. In practice, we calculate the Helmholtz free energy, $F(V, T)$, as a function of volume, V, and hence obtain the pressure through the relation $P = -(\partial F / \partial V)T$ and G through its definition $G = F + PV$. To calculate the free energies we use the method of thermodynamic integration which allows us to calculate the difference in free energy, $F - F_0$, between our *ab initio* system and a reference system whose potential energies are U and U_0 respectively. Technical details of the methodology can be found in Alfè *et al.* (2001).

5.4.1.1 Equation of state

The equation of state for h.c.p.-Fe has been studied both experimentally (e.g., Mao *et al.*, 1990; Brown and McQueen, 1986) and theoretically (e.g., Söderlind *et al.*, 1996). In **Figure 8** we show the density as a function of pressure from *ab initio* calculations together with that from static compression measurements at 300 K (Mao *et al.*, 1990), shock experiments (Brown and McQueen, 1986), theoretical calculations (Stixrude

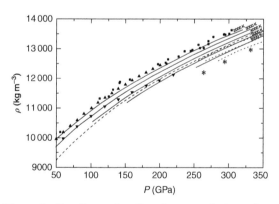

Figure 8 Density as a function of pressure for h.c.p. iron from *ab initio* calculations along isotherms between 2000 and 6000 K. Filled circles are the static compression measurements at 300 K (Mao *et al.*, 1990); dashed line is the calculations for the 4000 K isotherm of Stixrude *et al.* (1997); dotted lines are the 5000 K, 6000 K, and 7000 K isotherms of Steinle-Neumann *et al.*, (2002); stars are from an analysis of a thermal equation of state at 6000 K (Isaak and Anderson, 2003); triangles-up and -down are the experiments of Brown and McQueen (1986) along the 300 K isotherm and Hugoniot respectively. Modified from Vočadlo L, Alfe D, Gillan MJ, and Price GD (2003b) The properties of iron under core conditions from first principles calculations. *Physics of the Earth and Planetary Interiors* 140: 101–125.

et al., 1997; Steinle-Neumann *et al.*, 2002) and an analysis at 6000 K based on a thermal equation of state (Isaak and Anderson, 2003). All the results from both theory and experiment are in comforting agreement and provide a successful basis from which to make further comparisons of other properties.

5.4.1.2 Incompressibility

The isothermal and adiabatic incompressibility (K_T and K_S, respectively) are shown in **Figure 9**. The incompressibility increases significantly (and almost linearly) with pressure, and decreases with increasing temperature (more so in the case of K_T than K_S).

5.4.1.3 Thermal expansion

The thermal expansivity, α, is shown in **Figure 10** together with the value determined from shock data at 5200 K (Duffy and Ahrens, 1993), the theoretical calculations of Stixrude *et al.* (1997), and an analysis using a thermal equation of state (Anderson and Isaak, 2002; Isaak and Anderson, 2003). It is clear that α decreases strongly with increasing pressure and increases significantly with temperature.

5.4.1.4 Heat capacity

The total constant-volume specific heat per atom C_v (**Figure 11**) emphasizes the importance of electronic excitations in our calculations. In a purely harmonic system, C_v would be equal to $3k_B$, and it is striking that C_v is considerably greater than that even at

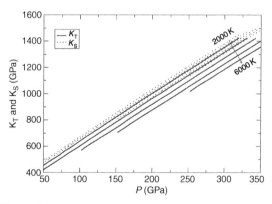

Figure 9 Isothermal incompressibility, K_T (solid lines), and adiabatic incompressibility, K_S (dotted lines), as a function of pressure for h.c.p. iron from *ab initio* calculations along isotherms between 2000 and 6000 *K*. From Vočadlo L, Alfe D, Gillan MJ, and Price GD (2003b) The properties of iron under core conditions from first principles calculations. *Physics of the Earth and Planetary Interiors* 140: 101–125.

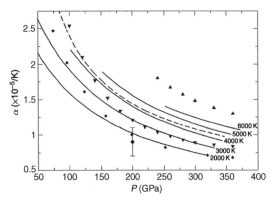

Figure 10 Thermal expansion as a function of pressure for h.c.p. iron from *ab inito* calculations along isotherms between 2000 and 6000 K (solid lines). The circle is the value of Duffy and Ahrens (1993) determined from shock experiments at 5200 ± 500 K; the dashed line is from theoretical calculations for the 4000 K isotherm (Stixrude et al., 1997); the diamonds are from an analysis of room temperature compression data at 300 K (Anderson and Isaak, 2002); the triangles (up: 6000 K, down: 2000 K) are from an analysis using a thermal equation of state applied to room temperature compression data (Isaak and Anderson, 2003). Modified from Vočadlo L, Alfe D, Gillan MJ, and Price GD (2003b) The properties of iron under core conditions from first principles calculations. *Physics of the Earth and Planetary Interiors* 140: 101–125.

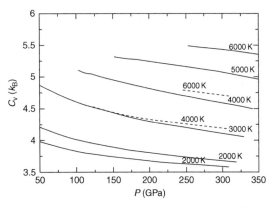

Figure 11 Constant volume heat capcity as a function of pressure for h.c.p. iron from *ab inito* calculations along isotherms between 2000 and 6000 K (solid lines). The much lower dashed lines are from Stixrude et al. (1997). Modified from Vočadlo L, Alfe D, Gillan MJ, and Price GD (2003b) The properties of iron under core conditions from first principles calculations. *Physics of the Earth and Planetary Interiors* 140: 101–125.

the modest temperature of 2000 K, while at 6000 K it is nearly doubled. The decrease of C_v with increasing pressure evident in **Figure 9** comes from the suppression of electronic excitations by high compression, and to a smaller extent from the

suppression of anharmonicity. Our C_v values are significantly higher than those of Stixrude *et al.* (1997); this is likely to be due to the inclusion of anharmonic corrections via *ab initio* molecular dynamics and the temperature dependence of harmonic frequencies.

5.4.1.5 *Grüneisen parameter*

The Grüneisen parameter, γ, is an important quantity in geophysics as it often occurs in equations which describe the thermoelastic behavior of materials at high pressures and temperatures. The value for γ is used to place constraints on geophysically important parameters such as the pressure and temperature dependence of the thermal properties of core, the adiabatic temperature gradient, and the geophysical interpretation of Hugoniot data. The Grüneisen parameter has considerable appeal to geophysicists because it is an approximately constant, dimensionless parameter that varies slowly as a function of pressure and temperature. It has both a microscopic and macroscopic definition, the former relating it to the vibrational frequencies of atoms in a material, and the latter relating it to familiar thermodynamic properties such as heat capacity and thermal expansion. Unfortunately, the experimental determination of γ, defined in either way, is extremely difficult; the microscopic definition requires a detailed knowledge of the phonon dispersion spectrum of a material, whereas the macroscopic definition requires experimental measurements of thermodynamic properties at simulatneously high pressures and temperatures.

The microscopic definition of the Grüneisen parameter (Grüneisen, 1912) is written in terms of the volume dependence of the *i*th mode of vibration of the lattice (ω_i) and is given by

$$\gamma_i = -\frac{\partial \ln \omega_i}{\partial \ln V} \quad [1]$$

However, evaluation of all γ_i throughout the Brillouin zone is impossible without some lattice dynamical model or high pressure inelastic neutron scattering data. It can be shown (e.g., Barron, 1957) that the sum of all γ_i throughout the first Brillouin zone leads to a macroscopic or thermodynamic definition of g which may be written as

$$\gamma_{tb} = \frac{\alpha V K_T}{C_v} \quad [2]$$

where α is the thermal expansion, V is the volume, K_T is the isothermal bulk modulus, and C_v is the heat capacity at constant volume. Evaluation of γ_i is also very difficult, however, because it requires experimental measurements of α, K_T, etc., at extreme conditions of pressure and temperature which are not readily attainable. Integrating the above equation with respect to temperature at constant volume leads to the Mie-Grüneisen expression for γ (see, e.g., Poirier 2000):

$$\gamma_{th} = \frac{P_{th}V}{E_{th}} \qquad [3]$$

where P_{th} is the thermal pressure and E_{th} is the thermal energy. This too is difficult to determine because the thermal energy is not readily obtained experimentally. At low temperatures, where only harmonic phonons contribute to E_{th} and P_{th}, γ should indeed be temperature independent above the Debye temperature, because $E_{th}=3k_B T$ per atom, and $P_{th}V=-3k_B T(\mathrm{d}\ln\omega/\mathrm{d}\ln V)=3k_B T\gamma_{ph}$, so that $\gamma_{th}=\gamma_i$ (the phonon Grüneisen parameter above). But in iron at high temperatures, the temperature independence of γ will clearly fail, because of electronic excitations and anharmonicity. Our results for γ (**Figure 12**) indicate that it varies rather little with either pressure or temperature in the region of interest. At temperatures below $\sim$5000 K, it decreases with increasing pressure, as expected from the behavior of γ_i. This is also expected from the often-used empirical rule of thumb $\gamma = (V/V_0)^q$, where V_0 is a reference volume and q is a constant exponent usually taken to be roughly unity. Since V decreases by a factor of about 0.82 as P goes

from 100 to 300 GPa, this empirical relation would make γ decrease by the same factor over this range at lower temperatures, which is roughly what we see. However, the pressure dependence of γ is very much weakened as T increases, until at 6000 K, γ is almost constant.

5.4.2   Elasticity of Solid Iron

A fundamental step toward resolving the structure and composition of the Earth's inner core is to obtain the elastic properties of the candidate phases that could be present.

The elastic constants of h.c.p.-Fe at 39 and 211 GPa have been measured in an experiment reported by Mao *et al.* (1999). Calculations of athermal elastic constants for h.c.p.-Fe have been reported by Stixrude and Cohen (1995b), Söderlind *et al.* (1996), Steinle-Neumann *et al.* (1999), and Vočadlo *et al.* (2003b). These values are presented in **Table 2**, and plotted as a function of density in **Figures 13(a)** and **13(b)**. Although there is some scatter on the reported values of c_{12}, overall the agreement between the experimental and various *ab initio* studies is excellent.

The resulting bulk and shear moduli and the seismic velocities of h.c.p.-Fe as a function of pressure are shown in **Figures 14** and **15**, along with experimental data. The calculated values compare well with experimental data at higher pressures, but discrepancies at lower pressures are probably due to the neglect of magnetic effects in the simulations (see Steinle-Neumann *et al.*, 1999).

The effect of temperature on the elastic constants of Fe was reported by Steinle-Neumann *et al.* (2001) based on calculations using the approximate 'particle in a cell' method, as discussed in Section 5.3. With increasing temperature, they found a significant change in the c/a axial ratio of the h.c.p. structure, which in turn caused a marked reduction in the elastic constants c_{33}, c_{44}, and c_{66} (**Figure 16(a)**). This led them to conclude that increasing temperature reverses the sense of the single-crystal longitudinal anisotropy of h.c.p.-Fe, and that the anisotropy of the core should now be viewed as being due to h.c.p.-Fe crystals having their c-axis preferably aligned equatorially, rather than axially as originally suggested by Stixrude and Cohen (1995a) (**Figures 16(b)** and **17**).

However, the work of Gannarelli *et al.* (2005) casts doubt on the robustness of the calculations of Steinle-Neumann *et al.* (2001); they found that the c/a ratio ranges from 1.585 at zero pressure and temperature

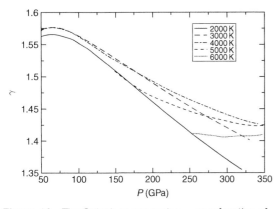

Figure 12   The Grüneisen parameter, γ, as a function of pressure for h.c.p. iron from *ab inito* calculations along isotherms between 2000 and 6000 K (solid lines). From Vočadlo L, Alfe D, Gillan MJ, and Price GD (2003b) The properties of iron under core conditions from first principles calculations. *Physics of the Earth and Planetary Interiors* 140: 101–125.

Table 2 A compilation of elastic constants (c_{ij}, in GPa), bulk (K), and shear (G) moduli (in GPa), and longitudinal (V_P) and transverse (V_S) sound velocity (in km s^{-1}) as a function of density (ρ, in g cm^{-3}) and atomic volume (in Å^3 per atom)

	V	ρ	c_{11}	c_{12}	c_{13}	c_{33}	c_{44}	c_{66}	K	G	V_P	V_S
Stixrude & Cohen	9.19	10.09	747	301	297	802	215	223	454	224	8.64	4.71
	7.25	12.79	1697	809	757	1799	421	444	1093	449	11.50	5.92
Steinle-Neumann *et al.*	8.88	10.45	930	320	295	1010	260	305	521	296	9.36	5.32
	7.40	12.54	1675	735	645	1835	415	470	1026	471	11.49	6.13
	6.66	13.93	2320	1140	975	2545	500	590	1485	591	12.77	6.51
Mao *et al.*	9.59	9.67	500	275	284	491	235	113	353	160	7.65	4.06
	7.36	12.60	1533	846	835	1544	583	344	1071	442	11.48	5.92
Söderlind *et al.*	9.70	9.56	638	190	218	606	178	224	348	200	8.02	4.57
	7.55	12.29	1510	460	673	1450	414	525	898	448	11.03	6.04
	6.17	15.03	2750	893	1470	2780	767	929	1772	789	13.70	7.24
Vočadlo *et al.*	9.17	10.12	672	189	264	796	210	242	397	227	8.32	4.74
	8.67	10.70	815	252	341	926	247	282	492	263	8.87	4.96
	8.07	11.49	1082	382	473	1253	309	350	675	333	9.86	5.38
	7.50	12.37	1406	558	647	1588	381	424	900	407	10.80	5.74
	6.97	13.31	1810	767	857	2007	466	522	1177	500	11.77	6.13
	6.40	14.49	2402	1078	1185	2628	580	662	1592	630	12.95	6.59

$K = (<c_{11}> + 2<c_{12}>)/3$ and $G = (<c_{11}> - <c_{12}> + 3<c_{44}>)/5$, where $<c_{11}> = (c_{11} + c_{22} + c_{33})/3$, etc. Previous calculated values are from Stixrude and Cohen (1995b), Steinle-Neumann *et al.* (1999), Söderlind *et al.* (1996), and Vočadlo *et al.* (2003b). The experimental data of Mao *et al.* (1999) are also presented.

to 1.62 at 5500 K and 360 GPa. These lower values for the c/a ratio are confirmed by the work of Vočadlo (2007), who reported elastic constants for both b.c.c.- and h.c.p.-iron using *ab initio* finite temperature molecular dynamics calculations (**Table 3**).

A further analysis of the high-temperature behavior of iron can be made by considering Birch's law. In the past, Birch's law has been used to make inferences about the elastic properties of the inner core. Birch's law suggests a linear relationship between V_Φ and ρ, and in the absence of reliable experimental data at very high pressures and temperatures, it has been assumed that this linearity may be extrapolated to the conditions of the inner core. In their very recent work, Badro *et al.* (2006) used inelastic X-ray scattering in a diamond-anvil-cell at high pressures, but ambient temperatures, to demonstrate a linear relationship between V_p and ρ for a number of systems including h.c.p.-Fe. **Figure 18** shows the fit to their results extrapolated to core conditions, together with the results from Vočadlo (2007) (where the uncertainties lie within the symbols) and also from shock experiments (Brown and McQueen, 1986). The agreement is generally outstanding; it is noteworthy that the calculated b.c.c.-Fe and h.c.p.-Fe velocity–density systematics are indistinguishable.

Figure 19 shows how V_Φ varies with density for both athermal and hot *ab initio* calculations (Vočadlo,

2007), together with values from PREM (Dziewonski and Anderson, 1981), experiments of Brown and McQueen (1986) and from a previous computational study (Steinle-Neumann *et al.*, 2001). It is clear that Birch's law holds for the systems studied in the present work, and that the velocities at constant density are almost temperature independent.

The calculated P-wave anisotropy for the h.c.p. and b.c.c. phases of Fe at core conditions is $\sim$6% and $\sim$4%, respectively, the former being close to the experimentally determined value of 4–5% (Antonangeli *et al.*, 2004). The seismically observed anisotropy (3–5%; Song and Helmberger, 1998) and layering in the inner core could, therefore, be accounted for both phases if the crystals were randomly oriented in the isotropic upper layer and partially aligned in the anisotropic lower layer. In order to make further conclusions about the elasticity of the inner core, the effect of light elements must first be taken into account (see Section 5.10).

5.5 Rheology of Solid Iron

5.5.1 Slip Systems in Iron

In order to understand anisotropy and layering, we need to understand the deformation mechanism and processes which could be responsible for textural

(a)

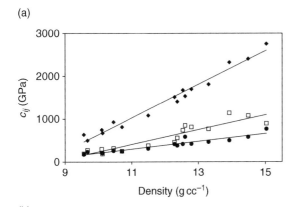

(b)

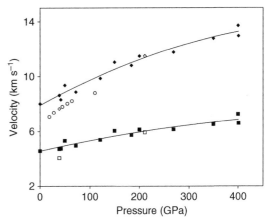

Figure 15 Plot of aggregate V_P (diamonds) and V_S (squares) wave velocity for h.c.p.-Fe as a function of pressure, with values taken from Stixrude and Cohen (1995a), Steinle-Neumann *et al.* (1999), Söderlind *et al.* (1996), Mao *et al.* (1999), and Vočadlo *et al.* (2003). Black diamonds and squares represent *ab initio* values, while white diamonds and squares represent values obtained from experimentally determined elastic constants. White circles are the experimental data of Fiquet *et al.* (2001).

Figure 13 Plot of the elastic constants of h.c.p.-Fe given in **Table 2** as a function of density from Stixrude and Cohen (1995b), Söderlind *et al.* (1996), Steinle-Neumann *et al.* (1999), and Vočadlo *et al.* (2003b): (a) c_{11} black diamonds, c_{12} white squares, c_{44} black circles; (b) c_{33} white diamonds, c_{13} black squares and c_{66} white circles.

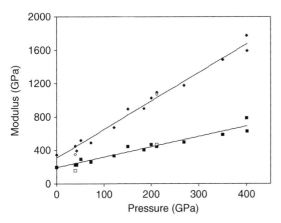

Figure 14 Plot of bulk modulus (diamonds) and shear modulus (squares) for h.c.p.-Fe as a function of pressure, with values taken from Stixrude and Cohen (1995a), Steinle-Neumann *et al.* (1999), Söderlind *et al.* (1996), Mao *et al.* (1999), and Vočadlo *et al.* (2003b). Black diamonds and squares represent *ab initio* values, while white diamonds and squares represent values obtained from experimentally determined elastic constants.

development in iron at core conditions. Seismic anisotropy will only be developed during deformation if this deformation leads to lattice-preferred orientation. A dominant slip system may lead to the development of the necessary texture and fabric. It is commonly assumed that glide in crystals occurs along the densest plane of atoms. However, there are exceptions to this rule which prevent this criterion from being predictive, and therefore, in order to determine the favored deformation mechanism in iron, we need to know the primary slip systems in candidate structures under core conditions. Poirier and Price (1999) calculated the elastic constants of h.c.p.-Fe at 0 K together with stacking fault energies for partial dislocations separated by a ribbon of stacking faults lying in the chosen slip plane. They concluded that slip should occur on the basal, rather than prismatic, plane, although it is, however, far from certain that this result will be valid at high temperatures.

If the primary slip system in h.c.p.-Fe becomes prismatic at high temperatures, this will have significant implications for the direction and extent of anisotropy; possible slip systems in h.c.p.-Fe include the basal, prismatic, pyramidal-*a* and pyramidal-*c*+*a* (Merkel *et al.*, 2004). In b.c.c. crystals, primary slip is likely to occur along the plane of the body diagonal because the shortest atomic distance is along <111>. Although the slip direction is always <111–>, the slip

(a)

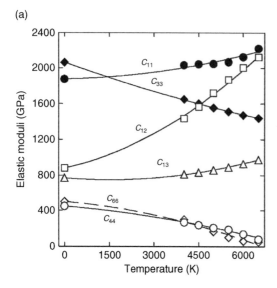

(b)

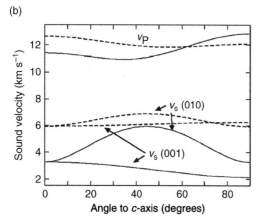

Figure 16 (a) Single-crystal elastic constants as a function of temperature. Note the marked reductions in c_{33}, c_{44}, and c_{66}. (b) Single-crystal velocities in h.c.p.-iron as a function of propagation direction with respect to the c-axis. Results at 6000 K (solid lines) are compared to static results (dashed lines). Note that at 6000 K, V_P (90°) > V_P (0°). After Steinle-Neumann G, Stixrude L, Cohen RE, and Gulseren O (2001) Elasticity of iron at the temperature of the Earth's inner core. *Nature* 413: 57–60.

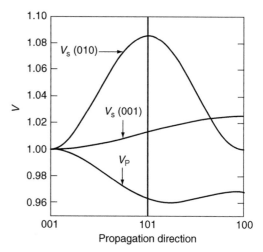

Figure 17 Single-crystal velocities in h.c.p.-iron as a function of propagation direction with respect to the c-axis. Note that V_P (90°) < V_P (0°). After Stixrude L and Cohen RE (1995a) High pressue elasticity of iron and anisotropy of Earth's inner core. *Science* 267: 1972–1975.

anisotropy. If the observed seismic anisotropy in the lower inner core is consistent with that calculated from the elastic constants of high-P/T h.c.p.-Fe phase, and if the h.c.p. phase retains a primary slip system at high P/T leading to preferred orientation, then the anisotropy could have developed either through deformation or on crystallization. If, however, the observed seismic anisotropy in the lower inner core is consistent with that calculated from the elastic constants of the high-P/T b.c.c.-Fe phase, then the anisotropy is most unlikely to be as a result of texture development, but is far more likely to be established on crystallization. However, as we shall see in the next section, recent studies suggest that the inner-core viscosity is low and therefore any textural development on crystallization will have been lost due to the subsequent deformation.

5.5.2 Viscosity and the Inner Core

The inner core is not perfectly elastic and has a finite viscosity with deformation occurring over long timescales. Placing numerical constraints on the viscosity of the inner core is fundamental to understanding important core processes such as differential inner-core rotation, inner-core oscillation, and inner-core anisotropy (see Bloxham, 1988).

High-temperature experiments on solid iron at ambient pressure lead to estimates for viscosities of ~10^{13} Pa s (Frost and Ashby, 1982); however, this

plane is normally {110}; however, since the b.c.c. structure is not close packed, other slip planes are possible, namely {112} and {123}. The variety of possible slip systems in b.c.c. metals and the expected modest strain rates in the inner core suggest that b.c.c.-Fe is highly unlikely to develop deformation-driven crystalline alignment. However, very little is known about the dominant slip systems and creep mechanism at core conditions that it is impossible to be definitive.

The existence of a dominant active slip system has consequences for the establishment and magnitude of

Table 3   Isothermal (adiabatic) elastic constants and sound velocities of h.c.p.-Fe and b.c.c.-Fe at different densities and temperatures, together with values taken from PREM

	ρ (kgm^{-3})	T (K)	c_{11} (GPa)	c_{12} (GPa)	c_{44} (GPa)	c_{23} (GPa)	c_{33} (GPa)	V_P (km s^{-1})	V_S (km s^{-1})
h.c.p.	11628.1	4000	1129 (1162)	736 (769)	155	625 (658)	1208 (1240)	9.91	4.15
	13155	5500	1631 (1730)	1232 (1311)	159	983 (1074)	1559 (1642)	11.14	4.01
b.c.c.	11592.91	750	1100 (1106)	712 (718)	287			10.11	4.64
	11592.91	1500	1066 (1078)	715 (727)	264			9.98	4.44
	11592.91	2250	1011 (1029)	740 (758)	250			9.88	4.20
	13155	5500	1505 (1603)	1160 (1258)	256			11.29	4.11
	13842	2000	1920 (1967)	1350 (1397)	411			12.22	5.1
	13842	4000	1871 (1966)	1337 (1431)	167			11.66	3.87
	13842	6000	1657 (1795)	1381 (1519)	323			11.83	4.24
PREM	12760							11.02	3.5
	13090							11.26	3.67

Taken from Vočadlo L (2007) *Ab initio* calculations of the elasticity of iron and iron alloys at inner core conditions: evidence for a partially molten inner core? *Earth and Planetary Science Letters*. 254: 227–232.

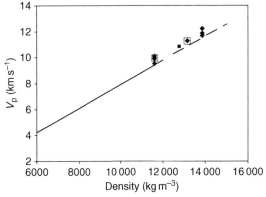

Figure 18 P-wave velocity as a function of density for pure iron compared with the high-*P* ambient-T DAC experiments of Badro *et al.* (2006); solid line is the fit to data, dashed line is an extrapolation of the fit. Diamonds: calculated b.c.c.-Fe *V*$_P$ at different temperatures; open squares: calculated h.c.p.-Fe at different temperatures (Vočadlo, 2006); filled square: shock experiments of Brown and McQueen (1986).

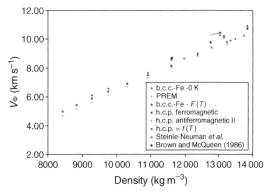

Figure 19 Calculated bulk sound velocity as a function of density for pure iron (Vočadlo, 2006) compared with PREM (Dziewonski and Anderson, 1981), the calculations of Steinle-Neumann *et al.* (2001), and the shock experiments of Brown and McQueen (1986).

is likely to be a lower limit as the value may increase at higher pressures. Seismological and geodetic observations have led to a number of estimates for inner-core viscosity ranging from 10^{11} to 10^{20} Pa s (see Dumberry and Bloxham, 2002). In particular, as we have already seen in Section 5.2.3, Buffett (1997) modeled the viscous relaxation of the inner core by calculating the relaxation time for the inner core to adjust, as it rotates, back to its equilibrium shape after small distortions due to perturbations in gravitational potential imposed by the overlying mantle. He suggested that the viscosity has to be constrained to be either less than 10^{16} Pa s (if the whole inner core is involved in

the relaxation) or greater than 10^{20} Pa s (if there is no relaxation of the inner core), although the latter case may lead to gravitational locking and hence no differential rotation. In a more recent study, Van Orman (2004) used microphysical models of the flow properties of iron and showed that the dominant deformation process was via Harper-Dorn creep leading to a viscosity of 10^{11} Pa s, at the lowest end of previous estimates. Such a low viscosity would allow the core to adjust its shape and maintain alignment with the mantle on a minute timescale; furthermore, the strain required to produce significant lattice-preferred orientation (LPO) could develop very quickly (years to hundreds of years) suggesting that such deformation could produce the observed anisotropy and all memory if primary crystallization on solidification is lost.

Quantifying the viscosity of the phase(s) present in the inner core at a microscopic level is a very difficult problem. At temperatures close to the melting point (as expected in the inner core), viscous flow is likely to be determined either by dislocation creep (Harper-Dorn creep) or by diffusion creep (Nabarro-Herring creep).

The overall viscosity of inner-core material has diffusion-driven and dislocation-driven contributions:

$$\eta = \left(\left(\frac{1}{\eta_{\text{diff}}} \right) + \left(\frac{1}{\eta_{\text{disl}}} \right) \right)^{-1} \quad [4]$$

Diffusion-controlled viscosity, whereby the material strain is caused by the motion of lattice defects (e.g., vacancies) under applied stress, is given by (Frost and Ashby, 1982)

$$\eta_{\text{diff}} = \frac{d^2 RT}{\alpha D_{\text{sd}} V} \quad [5]$$

where d is the grain size, R is the gas constant, T is the temperature, α is a geometric constant and V is the volume. The self-diffusion coefficient, D_{sd}, is given by (Frost and Ashby, 1982)

$$D_{\text{sd}} = D_0 \exp\left(-\frac{\Delta H}{RT} \right) \quad [6]$$

where D_0 is a pre-exponential factor and ΔH is the activation enthalpy for self-diffusion.

For simple materials, dislocation-controlled viscosity, whereby material strain is caused by the movement of linear defects along crystallographic planes, is given by:

$$\eta_{\text{disl}} = \frac{RT}{\rho D_{\text{sd}} V} \quad [7]$$

where ρ is the dislocation density.

Both dislocation- and diffusion-controlled creep mechanisms are thermally activated and the thermally controlled parameter in both cases is the self-diffusion coefficient, D_{sd}. A commonly used empirical relation for metals assumes that ΔH is linearly proportional to the melting temperature, T_{m}, and hence that

$$D_{\text{sd}} = D_0 \exp\left(-\frac{g T_{\text{m}}}{T} \right) \quad [8]$$

where g is a constant taking a value of ~ 18 for metals (Poirier, 2002).

Considering iron close to its melting point at core pressures (~ 5500 K), and using reasonable estimates for other quantities ($\alpha \sim 42$, $D_0 \sim 10^{-5}\,\text{m}^2\,\text{s}^{-1}$, $V \sim 5 \times 10^{-6}\,\text{m}^3\,\text{mol}^{-1}$), we obtain values for η_{diff} and η_{disl} of $\sim 10^{21}\,d^2\,\text{Pa s}$ and $\sim 6 \times 10^{22}/\rho\,\text{Pa s}$,

respectively. Unfortunately, the strong dependence of the viscosity expressions on the completely unknown quantities of grain size and dislocation density means that it is extremely difficult to produce reliable final numerical values. Grain sizes in the inner core could be anything from 10^{-3} to 10^3 m, resulting in diffusion viscosities in the range 10^{15}–10^{27} Pa s; dislocation densities could be as low as $10^6\,\text{m}^{-2}$ or nearer to the dislocation melting limit of $10^{13}\,\text{m}^{-2}$, resulting in dislocation-driven viscosities of 10^9–10^{16} Pa s. Thus, even the relative contributions from dislocation-controlled and diffusion-controlled viscosity are as yet unknown.

Clearly, inner-core viscosity is not a well-constrained property, with estimates varying over many orders of magnitude. Future microscopic simulations, combined with high-resolution seismic and geodetic data, should constrain this quantity further and thereby improve our understanding of inner core dynamics.

5.6   The Temperature in the Earth's Core

Having shown how mineral physics can be used to understand the properties of solid iron, we turn now to its melting behavior. An accurate knowledge of the melting properties of Fe is particularly important, as the temperature distribution in the core is relatively uncertain and a reliable estimate of the melting temperature of Fe at the pressure of the inner-core boundary (ICB) would put a much-needed constraint on core temperatures. As with the subsolidus behavior of Fe, there is much controversy over its high-P melting behavior (e.g., see Shen and Heinz, 1998). Static compression measurements of the melting temperature, T_{m}, with the DAC have been made up to ~ 200 GPa (e.g., Boehler, 1993), but even at lower pressures results for T_{m} disagree by several hundred Kelvin. Shock experiments are at present the only available method to determine melting at higher pressures, but their interpretation is not simple, and there is a scatter of at least 2000 K in the reported T_{m} of Fe at ICB pressures (see Nguyen and Holmes, 2004).

An alternative to experiment is theoretical calculations which can, in principle, determine accurate melting curves to any desired pressure. Indeed, *ab initio* methods have successfully been used to calculate the melting behavior of transition metals such as aluminum (Vočadlo and Alfè, 2002; **Figure 20**) and copper (Vočadlo *et al.*, 2004). Using the technique of thermodynamic integration, Alfe *et al.* (1999, 2004)

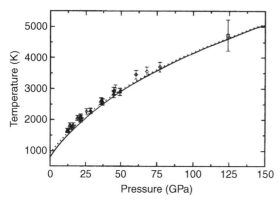

Figure 20 *Ab initio* calculations for the high-pressure melting curve of aluminum (Vočadlo and Alfè, 2002). Solid curve: without pressure correction. Dotted curve: with pressure correction to account for ~2% difference (as a result of using DFT) between calculated and experimental zero pressure equilibrium volume. Diamonds and triangles: DAC measurements of Boehler and Ross (1997) and Hanstrom and Lazor (2000), respectively. Square: shock experiments of Shaner *et al.* (1984).

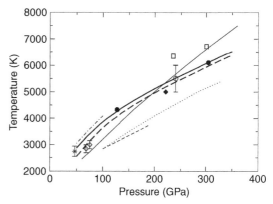

Figure 21 Comparison of the melting curves of Fe from experimental and *ab initio* results from Alfe *et al.* (2004) – heavy solid and long dashed curves: results from Alfe *et al.* (2004) without and with free-energy correction; filled circles: results of Belonoshko *et al.* (2000) corrected for errors in potential fitting; dotted curve: *ab initio* results of Laio *et al.* (2000); light curve: uncorrected *ab initio* results of Belonoshko *et al.* (2000); chained curve and short dashed curve: DAC measurements of Williams *et al.* (1987) and Boehler (1993); open diamonds: DAC measurements of Shen *et al.* (1998); star: DAC measurements of Jephcoat and Besedin (1996); open squares, open circle, and full diamond: shock experiments of Yoo *et al.* (1993), Brown and McQueen (1986), and Nguyen and Holmes (2004).

calculated the melting curve of iron. The condition for two phases to be in thermal equilibrium at a given temperature, T, and pressure, P, is that their Gibbs free energies, $G(P, T)$, are equal. To determine T_m at any pressure, therefore, Alfè *et al.* (2004) calculated G for the solid and liquid phases as a function of T and determined where they are equal. They first calculated the Helmholtz free energy, $F(V, T)$, as a function of volume, V, and hence obtained the pressure through the relation $P = -(\partial F/\partial V)_T$ and G through its definition $G = F + PV$. The free energy of the harmonic solid was calculated using lattice dynamics, while that of the anharmonic solid and the liquid was calculated with molecular dynamics using thermodynamic integration.

Since their first *ab initio* melting curve for Fe was published (Alfè *et al.*, 1999), the authors have improved their description of the *ab initio* free energy of the solid, and have revised their estimate of T_m of Fe at ICB pressures to be ~6250 K (see **Figure 21** and Alfè *et al.* (2004)), with an error of ±300 K. For pressures $P < 200$ GPa (the range covered by DAC experiments) their curve lies ~900 K above the values of Boehler (1993) and ~200 K above the more recent values of Shen *et al.* (1998) (who stress that their values are only a lower bound to T_m). The *ab inito* curve falls significantly below the shock-based estimates for T_m of Yoo *et al.* (1993); the latter deduced the temperature by measuring optical emission (however, the difficulties of obtaining temperature by this method in shock experiments are well known), but accords almost

exactly with the shock data value of Brown and McQueen (1986) and the new data of Nguyen and Holmes (2004). The *ab initio* melting curve of Alfe *et al.* (2004) differs somewhat from the calculations of both Belonoshko *et al.* (2000) and Laio *et al.* (2000). However, the latter two used model potentials fitted to *ab inito* simulations, and so their melting curves are those of the model potential and not the true *ab initio* one (it is important to note here that Laio *et al.* (2000) used a more sophisticated approach in which the model potential had an explicit dependence on thermodynamic state that exactly matched the *ab initio* result). In their paper, Alfe *et al.* (2004) illustrated this ambiguity by correcting for the errors associated with the potential fitting of Belonoshko *et al.* (2000); the result is in almost exact agreement with the true *ab initio* curve (**Figure 21**). The addition of light elements reduces the melting temperature of pure iron by ~700 K making the likely temperature at the inner-core boundary to be ~5500 K (Alfe *et al.*, 2002a). An independent measure of the likely temperature in the inner core was made by Steinle-Neumann *et al.* (2001) who performed first-principles calculation of the structure and elasticity of iron at high temperatures; they found that the temperature for which the elastic moduli best matched those of the inner core was ∼5700 K.

5.7 Thermodynamic Properties of Liquid Iron

The outer core of the Earth is liquid iron alloyed with 5–10% light elements. Experiments on liquid iron at core conditions are prohibitively challenging. However, once more, from the *ab initio* simulation of the free energy of the pure iron liquid, we can obtain first-order estimates for a range of thermodynamic properties at the conditions of the Earth's outer core. **Figures 22(a)–22(f)** show the results of *ab initio* calculations for values of density, adiabatic and isothermal bulk moduli, thermal expansion coefficient, heat capacity (C_v), Grüneisen parameter and bulk sound velocity, respectively, over a range of pressures and temperatures (see Vočadlo *et al.*, 2003b). Results from the experimental analysis of Anderson and Ahrens (1994) for density and adiabatic incompressibility are also shown (as gray lines; upper: 5000 K, lower: 8000 K). The calculations reproduce the experimentally derived density and incompressibility values to within a few percent. It is worth noting that the bulk sound velocity is almost independent of temperature, confirming the conclusion of Anderson and Ahrens (1994).

5.8 Rheology of Liquid Iron

5.8.1 Viscosity and Diffusion

Viscosity is a very important parameter in geophysics since the viscosity of materials in the Earth's core are a contributing factor in determining overall properties of the core itself, such as convection and heat transfer; indeed, the fundamental equations governing the dynamics of the outer core and the generation and sustention of the magnetic field are dependent, in part, on the viscosity of the outer core fluid. Quantifying viscosity at core conditions is far from straightforward, especially as the exact composition of the outer core is not known. Furthermore, although there have been many estimates made for outer-core viscosity derived from geodetic, seismological, geomagnetic, experimental, and theoretical studies, the values so obtained span 14 orders of magnitude (see Secco, 1995).

Geodetic observations (e.g., free oscillations, the Chandler wobble, length of day variations, nutation

of the Earth, tidal measurements, and gravimetry) lead to viscosity estimates ranging from 10 mPa s (observations of the Chandler wobble; Verhoogen, 1974) to 10^{13} mPa s (analysis of free oscillation data; Sato and Espinosa, 1967). Theoretical geodetic studies (e.g., viscous coupling of the core and mantle, theory of rotating fluids, inner-core oscillations, and core nutation) lead to viscosity estimates ranging from 10 mPa s (evaluation of decay time of inner-core oscillations; Won and Kuo, 1973) to 10^{14} mPa s (secular deceleration of the core by viscous coupling; Bondi and Lyttleton, 1948). Generally, much higher values for viscosity (10^{10}–10^{14} mPa s) are obtained from seismological observations of the attenuation of P- and S-waves through the core (e.g., Sato and Espinosa, 1967; Jeffreys, 1959), and from geomagnetic data (e.g., 10^{10} mPa s; Officer, 1986).

The viscosities of core-forming materials may also be determined experimentally in the laboratory and theoretically through computer simulation. Empirically, viscosity follows an Ahrrenius relation of the form (see Poirier, 2002)

$$\eta \propto \exp\left(\frac{Q_V}{k_B T}\right) \quad [9]$$

where Q_V is the activation energy. Poirier (1988) analyzed data for a number of liquid metals and found that there is also an empirical relation between Q_V and the melting temperature, consistent with the generalized relationship of Weertman (1970):

$$Q_V \cong 2.6RT_m \quad [10]$$

This very important result implies that the viscosity of liquid metals remains constant (i.e., independent of pressure) along the melting curve and therefore equal to that at the melting point at ambient pressure, which is generally of the order of a few mPa s. Furthermore, Poirier went on to state that the viscosity of liquid iron in the outer core would, therefore, be equal to that at ambient pressure (∼6 mPa s; Assael *et al.*, 2006).

On a microscopic level, an approximation for the viscosity of liquid metals is given by the Stokes–Einstein equation, which provides a relationship between diffusion and viscosity of the form

$$D\eta = \frac{k_B T}{2\pi a} \quad [11]$$

where a is an atomic diameter, T is the temperature, k_B is the Boltzmann constant, and D is the diffusion coefficient.

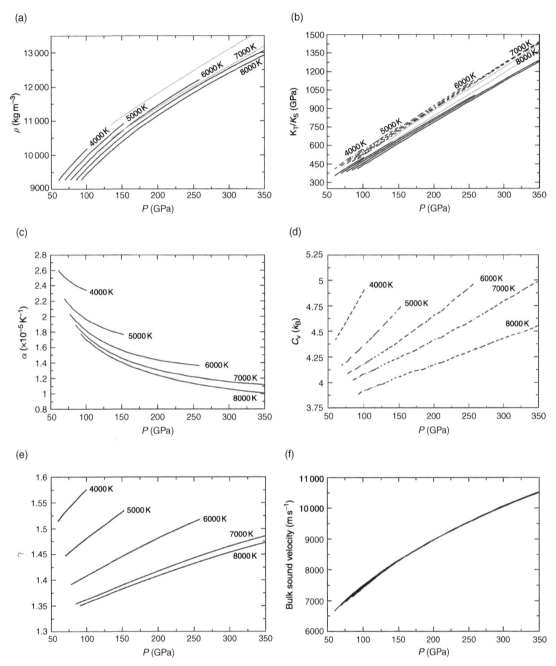

Figure 22 (a–f) Density, incompressibility, thermal expansion, heat capacity, Grüneisen parameter, and bulk sound velocity as a function of pressure for liquid iron. Black lines: *ab initio* calculations (Vočadlo *et al.*, 2003b); gray lines: isentropic model determined from experimental data (upper: 5000 K; lower 8000 K. Anderson and Ahrens (1994)).

Theoretical values for diffusion coefficients have been obtained from *ab initio* molecular dynamics simulations on liquid iron at core conditions, leading to a predicted viscosity of ~12–15 mPa s using the Stokes–Einstein relation above. However, although the Stokes–Einstein equation has proved successful in establishing a link between viscosity and diffusion for a number of monatomic liquids, it is not necessarily the case that it should be effective for alloys or at high pressures and temperatures. More recently, Alfe *et al.* (2000a) used the more rigorous Green–Kubo functions to determine viscosities for a range of thermodynamic states relevant to the Earth's core (**Table 4**). Throughout this range, the results show

Table 4   The diffusion coefficient D and the viscosity from *ab initio* simulations of 'liquid iron' at a range of temperatures and densities (Alfe *et al.*, 2000a)

	T (K)	ρ (kg m^{-3})				
		9540	10 700	11 010	12 130	13 300
D (10^{-9} m^3 s^{-1})	3000	4.0 ± 0.4				
	4300		5.2 ± 0.2			
	5000		7.0 ± 0.7			
	6000	14 ± 1.4	10 ± 1	9 ± 0.9	6 ± 0.6	5 ± 0.5
	7000		13 ± 1.3	11 ± 1.1	9 ± 0.9	6 ± 0.6
η (mPa s)	3000	6 ± 3				
	4300		8.5 ± 1			
	5000		6 ± 3			
	6000	2.5 ± 2	5 ± 2	7 ± 3	8 ± 3	15 ± 5
	7000		4.5 ± 2	4 ± 2	8 ± 3	10 ± 3

that liquid iron has a diffusion coefficient and viscosity similar to that under ambient conditions, a result already suggested much earlier by Poirier (1988). Both *ab initio* calculations and experiments consistently give viscosities of the order of a few mPa s. This suggests that viscosity changes little with homologous temperature, and it is now generally accepted that the viscosity of the outer core is likely to be a few mPa s (comparable to that of water on the Earth's surface).

5.8.2   The Structure of Liquid Iron

It has long been established that liquid structure is approximately constant along the melting curve (e.g., Ross, 1969) and here, in the case of iron, there appears to be a truly remarkable simplicity in the variation of the liquid properties with thermodynamic state; indeed, not only are viscosities and diffusivities almost invariant (as described above), but structural properties show similarly consistent behavior. Alfe *et al.* (2000a) calculated the radial distribution function of liquid iron as a function of temperature at a density of 10 700 kg m^{-3}, representative of that in the upper outer core (**Figure 23**); between 4300 and 8000 K, the effect of varying temperature is clearly not dramatic, and consists only of the expected weakening and broadening of the structure with increasing T.

Furthermore, Shen *et al.* (2004) determined the structure factors of liquid iron as a function of pressure along the melting curve using X-ray scattering in a laser-heated diamond anvil cell up to 58 GPa (**Figure 24**). Once more, the structure factor preserves

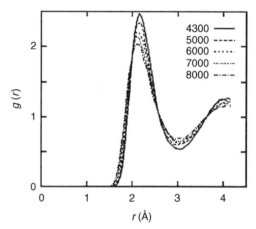

Figure 23   Variation of radial distribution function with temperature from *ab initio* simulations of liquid iron at the fixed density 10 700 kg m^{-3}. Adapted from Alfè D, Kresse G, and Gillan MJ (2000a) Structure and dynamics of liquid iron under Earth's core conditions. *Physics Review B* 61: 132–142.

essentially the same shape; the behavior is consistent with that of close-packed liquid metals. These results provide structural verification of the theoretical predictions given in **Figure 23**, and also confirm that it is justifiable to extrapolate viscosities measured under ambient conditions to high pressures.

5.9   Evolution of the Core

The Earth's paleomagnetic record suggests that the magnetic field has been operating for over three billion years; the fact that this magnetic field still exists today and has not decayed through Ohmic dissipation (possible on a timescale of $\sim10^4$ years)

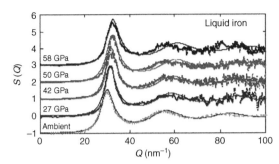

Figure 24 Structure factors of liquid iron along the melting curve determined by X-ray scattering at high pressures (Shen *et al.*, 2004).

suggests that it is being regenerated and sustained by a process such as a geodynamo. For the geodynamo to work, it requires convection of the core fluid (thermal and/or compositional); whether the core is stable to convection depends on the rate of cooling of the core and thus on the evolution of the Earth. As the core cools and solidifies, it releases energy, much of which is transported across the core via conduction. The thermal conductivity of the core material limits the amount of heat which can be transported in this way, and any excess energy may be used to power the geodynamo. The conductivity (both thermal and electrical) of the core material is therefore crucial: if the core material has too high a conductivity, there is no excess heat and therefore no dynamo. A further consideration is the heat flow into the base of the mantle. If this is too large, the core would cool too quickly, and the inner core would grow too rapidly; if the heat flow into the mantle is too low, the core would cool too slowly, convection would be difficult to initiate and sustain, and magnetic field generation would be difficult.

As a result of inner-core growth, the relative contributions of thermal convection to compositional convection have changed over geological time. Thermal convection would have been dominant in the early stages of Earth's evolution, and would have been the sole power source before the formation of the inner core (possibly ~2 billion years ago). However, as the inner core grew, compositionally driven convection (from the preferential partitioning of light elements into the outer core and their subsequent buoyancy) became more important (now responsible for ~80% power to geodynamo). Two other important factors which affect the thermal evolution of the core are the possible contribution to

the heat budget from radioactive decay of, for example, ^{40}K, and the ability of the mantle to remove heat away from the core. A reliable thermal evolution model has to reproduce: (1) the correct present-day inner-core size; (2) the present-day heat flux (~42 TW); (3) the heat flux through the core–mantle boundary (estimates range from 2 to 10 TW); (4) enough entropy to drive a dynamo; and (5) reasonable mantle temperatures.

There have been a number of studies that have developed both analytical and numerical models for the thermal evolution of the core based on calculating heat flux across the core–mantle boundary coupled with heat balance relations associated with core convection and inner-core growth (e.g., Buffett *et al.*, 1992, 1996; Labrosse *et al.*, 2001; Labrosse, 2003; Nakagawa and Tackley, 2004, Gubbins *et al.*, 2003, 2004; Nimmo *et al.*, 2004). The requirement for there to be a radiogenic heat source in order to maintain sufficient power to the geodynamo remains a controversial subject. While some models require a radiogenic heat source, such as potassium, to power the geodynamo (Labrosse, 2003; Nimmo *et al.*, 2004; Costin and Butler, 2006), others suggest that convective processes alone are sufficient to maintain the geodynamo (e.g., Buffett *et al.*, 1996; Christensen and Tilgner, 2004).

However, key to all these thermal evolution models is the need for reliable data for material properties at inner-core conditions. While mineral physics has made some progress (e.g., the quantification of the viscosity of the outer core described in Section 5.8.1 and the thermodynamic properties for pure iron described in Sections 5.4 and 5.7), key properties in the heat transfer relations, such as the electrical and thermal conductivity of iron and iron alloys at core conditions, remain unknown, and are presently only estimated from extrapolations to experimental data.

5.10 The Composition of the Core

5.10.1 Bulk Composition

The exact composition of the Earth's inner core is not very well known. On the basis of cosmochemical and geochemical arguments, it has been suggested that the core is an iron alloy with possibly as much as ~5 wt.% Ni and very small amounts (only fractions of a wt.% to trace) of other siderophile elements such as Cr, Mn, P, and Co (McDonough and Sun, 1995). On the basis of materials-density/sound-wave velocity systematics, Birch (1964) further concluded that the core is composed of iron that is alloyed with a small fraction of lighter elements. The light alloying elements most commonly suggested include S, O, Si, H, and C, although minor amounts of other elements, such K, could also be present (e.g., Poirier, 1994; Gessmann and Wood, 2002). From seismology it is known that the density jump across the inner-core boundary is between ~4.5% and 6.7% (Shearer and Masters, 1990; Masters and Gubbins, 2003), indicating that there is more light element alloying in the outer core. The evidence thus suggests that the outer core contains ~5–10% light elements, while the inner core has ~2–3% light elements. Our present understanding is that the Earth's solid inner core is crystallizing from the outer core as the Earth slowly cools, and the partitioning of the light elements between the solid and liquid is therefore crucial to understanding the evolution and dynamics of the core.

5.10.2   The Effect of Nickel

It is generally assumed that the small amount of nickel alloyed to iron in the inner core is unlikely to have any significant effect on core properties as nickel and iron have sufficiently similar densities to be seismically indistinguishable, and addition of small amounts of nickel is unlikely to appreciably change the physical properties of iron. However, very recent *ab initio* calculations at 0 K show that this may not be the case (Vočadlo *et al.*, 2006). The addition of small amounts of nickel (a few atomic percent) by atomic substitution stabilizes the h.c.p. structure with respect to the b.c.c. structure by up to ~20 GPa. Experiments at modestly high pressures and temperatures (72 GPa and 3000 K) show that the presence of nickel stabilizes the f.c.c. phase over the h.c.p. phase (Mao *et al.*, 2006). Clearly, full free energy calculations at core temperatures and pressures or further high-P/T experiments are required to resolve this matter. Nevertheless, the previously held assumption that nickel has little or no effect on the first-order elastic properties of iron may not necessarily be valid.

5.10.3   Light Elements

In contrast, it has long been known that the presence of light elements in the core does have an effect on core properties. Cosmochemical abundances of the elements, combined with models of the Earth's history,

limit the possible impurities to a few candidates. The light element impurities most often suggested are sulfur, oxygen, and silicon. These alloying systems have been experimentally studied up to pressures of around 100 GPa (e.g., Li and Agee, 2001; Lin *et al.*, 2003; Rubie *et al.*, 2004), and with rapid developments in *in situ* techniques we eagerly anticipate experimental data for iron alloys at the highly elevated pressures and temperatures of the Earth's inner core in the near future. In a study combining thermodynamic modeling with seismology, Helffrich and Kaneshima (2004) modeled the ternary Fe–O–S liquid system at core conditions. It is well known that iron alloy systems exhibit liquid immiscibility, and they wanted to see if such immiscibility could occur in the outer core. If this was the case, layering would occur in the outer core which could be seismologically observable. However, they failed to find any such layering and concluded that, if the outer core was an Fe–S–O alloy, it must exist outside of the two liquid field. This would therefore constrain the composition of the outer core to have < 6 wt.% oxygen and 2–25 wt.% sulfur.

5.10.3.1   Chemical potential calculations of FeX binary systems

An alternative approach to understanding the composition of the inner core is to simulate the behavior of these iron alloys with *ab initio* calculations which are readily able to access the pressures and temperatures of the inner core. Alfè *et al.* (2000b, 2002a, 2002b) calculated the chemical potentials of iron in binary systems alloyed with sulfur, oxygen, and silicon. They developed a strategy for constraining both the impurity fractions and the temperature at the ICB based on the supposition that the solid inner core and liquid outer core are in thermodynamic equilibrium at the ICB. For thermodynamnic equilibrium the chemical potentials of each species must be equal to both sides of the ICB, which fixes the ratio of the concentrations of the elements in the liquid and in the solid, which in turn fixes the densities. The mole fractions required to reproduce the liquid core density are 16%, 14%, and 18%, respectively, for S, Si, and O (**Figure 25(a)**). If the core consisted of pure iron, equality of the chemical potential (the Gibbs free energy in this case) would tell us only that the temperature at the ICB is equal to the melting temperature of iron at the ICB pressure of 330 GPa. With impurities present, the *ab initio* results reveal a major qualitative difference between oxygen and the other two impurities: oxygen partitions strongly into the liquid, but sulfur and silicon both partition equally in

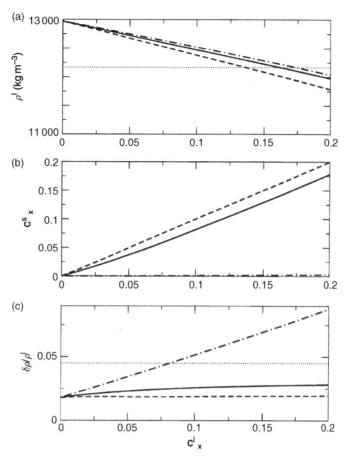

Figure 25 Liquid and solid impurity mole fractions c_x^l and c_x^s of impurities x = S, Si, and O, and resulting densities of the inner and outer core predicted by *ab initio* simulations. Solid, dashed, and chain curves represent S, Si, and O, respectively. (a) Liquid density ρ^l (kg m^{-3}); horizontal dotted line shows density from seismic data; (b) mole fractions in solid resulting from equality of chemical potentials in solid and liquid; (c) relative density discontinuity ($\delta\rho/\rho^l$) at the ICB; horizontal dotted line is the value from free oscillation data. Taken from Alfé D, Gillan MJ, Vočadlo L, Brodholt JP, and Price GD (2002b) The *ab initio* simulation of the Earth's Core. *Philosophical Transactions of the Royal Society Series A* 360: 1227–1244.

the solid and liquid. This is shown in **Figure 25(b)**: the calculated chemical potentials in the binary liquid and solid alloys give the mole fractions in the solid that would be in equilibrium with these liquids (i.e., at liquid impurity concentrations corresponding to the seismically observed density in **Figure 25(a)**) of 14%, 14%, and 0.2%, respectively, for S, Si, and O. Having established the partitioning coefficients, Alfé *et al.* (2000b, 2002a, 2002b) then investigated whether the known densities of the outer and inner core, estimated from seismology, could be matched by one of their calculated binary systems. For sulfur and silicon, their ICB density discontinuities were considerably smaller than the known seismological value at that time of $4.5 \pm 0.5\%$ (Shearer and Masters, 1990); for oxygen, the discontinuity was markedly greater than that from seismology. The

partial volumes in the binary solids give ICB density discontinuities of $2.7 \pm 0.5\%$, $1.8 \pm 0.5\%$, and $7.8 \pm 0.2\%$, respectively (**Figure 25(c)**). Therefore, none of these binary systems are plausible, that is, the core cannot be made solely of Fe/S, Fe/Si, or Fe/O. However, the seismic data can clearly be matched by a ternary/quaternary system of iron and oxygen together with sulfur and/or silicon. A system consistent with seismic data could contain 8 mol.% oxygen and 10 mol.% sulfur and/or silicon in the outer core, and 0.2 mol.% oxygen and 8.5 mol.% sulfur and/or silicon in the inner core (Alfé *et al.*, 2002a). However, it should be remembered that it is likely that several other light elements could exist in the inner core and would therefore have to be considered before a true description of inner-core composition could be claimed.

5.10.3.2 High-temperature elasticity of FeSi and FeS

A fundamental step toward resolving the structure and composition of the Earth's inner core is to obtain the elastic properties of the candidate phases that could be present. Previous work has already suggested that oxygen (see Section 5.10.3.1) and carbon (Vočadlo *et al.*, 2002) are unlikely to be the light element in the inner core, while the presence of hydrogen seems questionable as the quantities needed to produce the required density deficit are improbably high (Poirier, 1994), although this certainly cannot be ruled out. Furthermore, experiments (Lin *et al.*, 2002) and theory (Côté *et al.*, 2007) show that the presence of silicon has a significant effect on the phase diagram of iron (**Figure 26**), significantly stabilizing the b.c.c. phase with respect to the h.c.p. phase.

From experimental and theoretical work, we already know that the shear wave velocities of h.c.p. iron at high pressures are significantly higher than those of the inner core as inferred from seismology (Antonangeli *et al.*, 2004; Mao *et al.*, 2001). We have already seen in Section 5.4.2 (**Table 3**) the elastic properties of pure iron as obtained from *ab initio*

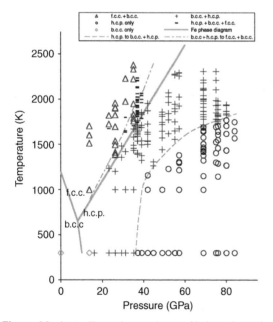

Figure 26 Iron–silicon phases observed in laser-heated diamond anvil cell experiments with the starting materials of Fe (7.9 wt.% Si). The presence of silicon stabilizes the b.c.c. phase with respect to the h.c.p. phase by up tp ~40 GPa at high temperatures. After Lin JF, Heinz DL, Campbell AJ, Devine JM, and Shen GY (2002) Iron – Silicon alloy in the Earth's core. *Science* 295: 313–315.

calculations at high temperatures and pressures, and in **Table 5** are shown those for FeS and FeSi from Vočadlo (2007). **Figure 27** shows the P-wave velocity of FeSi as a function of density compared to the experiments of Badro *et al.* (2006). The agreement is excellent, confirming that this phase too exhibits Birch's law type behavior (see Section 5.4.2).

Figure 28 shows how the bulk sound velocity varies for both FeS and FeSi as a function of density for both athermal and hot *ab initio* calculations together with values from PREM (Dziewonski and Anderson, 1981). Once again there seems to be little dependence on temperature of $V_\Phi(\rho)$.

The calculated P-wave anisotropy for both FeS and FeSi is ~6%. The seismically observed anisotropy (3–5%; Song and Helmberger, 1998) and layering in the inner core could, therefore, be accounted for by any of the phases studied if the crystals were randomly oriented in the isotropic upper layer and partially aligned in the anisotropic lower layer. However, the fundamental conclusion of these calculations is that, for all candidate core phases, V_S at viable core temperatures (i.e., >5000 K) is more than 10% higher than that inferred from seismology (PREM values between 3.5 and 3.67 km s^{-1}; Dziewonski and Anderson, 1981). **Table 3** shows that the calculated values of V_S for pure iron phases are >4.0 km s^{-1} (in agreement with inferences drawn from the extrapolation of lower pressure experimental data; Antonangeli *et al.* (2004) and also with the value of 4.04 inferred from shock experiments at a density of 12 770 kg m^{-3} (Brown and McQueen, 1986)), while the effect of light elements is to increase the shear wave velocities to over 5 km s^{-1}. If the uncertainties in the seismological values are well constrained, the difference between these observations and the results from both theory and experiment suggests that a simple model for the inner core based on the commonly assumed phases is wrong.

An important consideration is the effect of anelasticity. The reduction in shear wave velocity due to shear wave attenuation is given by

$$V(\omega, T) = V_0(T)\left(1 - \frac{1}{2}\cot\left(\frac{\pi\alpha}{2}\right)Q^{-1}(\omega, T)\right) \quad [12]$$

where $V(\omega, T)$ and $V_0(T)$ are the attenuated and unattenuated shear wave velocities respectively, Q is the quality factor, and α is the frequency dependence of Q. For the inner core $Q = 100$ (Resovsky *et al.*, 2005) and $\alpha = 0.2$–0.4 (Jackson *et al.*, 2000), which result in

Table 5   Isothermal (adiabatic) elastic constants and sound velocities of FeSi and FeS at different densities and temperatures, together with values taken from PREM

ρ (kg m^{-3})	T (K)	c_{11} (GPa)	c_{12} (GPa)	c_{44} (GPa)	c_{23} (GPa)	c_{33} (GPa)	V_P (km s^{-1})	V_S (km s^{-1})
FeSi								
6969.44	1000	488 (489)	213 (214)	125			8.32	4.32
6969.44	2000	425 (428)	238 (241)	150			8.29	4.28
8199.34	1000	938 (942)	413 (417)	263			10.74	5.66
8199.34	2000	863 (871)	431 (439)	263			10.58	5.45
8199.34	3500	788 (803)	469 (484)	250			10.42	5.11
10211.74	5500	1643 (1732)	1030 (1119)	462			13.53	6.26
10402.15	1000	2025 (2043)	1007 (1025)	625			14.34	7.46
10402.15	2000	1909 (1944)	1029 (1064)	583			14.08	7.11
10402.15	3500	1904 (1972)	1117 (1185)	603			14.36	7.06
10402.15	5000	1780 (1874)	1132 (1226)	563			14.12	6.71
FeS								
8587.14	1000	788 (793)	531 (536)	213			10.03	4.56
8587.14	2000	763 (772)	519 (528)	175			9.79	4.23
10353	5500	1294 (1371)	1050 (1127)	257			12.02	4.43
10894.13	1000	1513 (1533)	1400 (1420)	575			13.38	5.81
10894.13	2000	1571 (1613)	1386 (1428)	532			13.43	5.72
10894.13	3500	1545 (1617)	1360 (1432)	492			13.34	5.52
10894.13	5000	1558 (1666)	1379 (1487)	458			13.41	5.34
PREM								
12760							11.02	3.5
13090							11.26	3.67

Taken from Vočadlo L (2007) *Ab initio* calculations of the elasticity of iron and iron alloys at inner core conditions: evidence for a partially molten inner core? *Earth and Planetary Science Letters*. 254: 227–232.

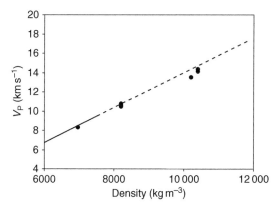

Figure 27   P-wave velocity as a function of density for FeSi compared with the high-*P* ambient-*T* DAC experiments of Badro *et al.* (2006); solid line is the fit to data, dashed line is an extrapolation of the fit. Circles: calculated FeSi–V_P at different temperatures (Vočadlo, 2006).

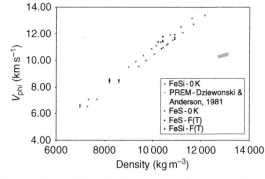

Figure 28   Calculated bulk sound velocity as a function of density for FeS and FeSi (Vočadlo, 2006) compared with PREM (Dziewonski and Anderson, 1981).

a decrease in the shear velocity of only 0.5–1.5%, nowhere near the >10% difference between the seismological observations and the calculated materials properties. It is important to note that the above analysis is necessarily approximate; anelasticity is a very complex issue that requires material data at the conditions of the Earth's inner core in order to draw irrefutable conclusions – clearly such data are unavailable at present.

Another possible explanation for the difference between the observations and the results from both theory and experiment is that parts of the inner core may be partially molten, with solute rich liquid pockets trapped between solid grains. The amount of melt can be estimated by taking the Hashin–Shtrikman bound for the effective shear modulus of two-phase media leading to a minimum amount of melt in the inner core is estimated to be ~8%. These liquid pockets are not necessarily concentrated in the upper part of the inner core; the observed PKJKP waves go right through the center of the Earth (Cao *et al.*, 2005) so the difference in V_S between seismology (e.g., Dziewonski and Anderson, 1981) and theory suggest that melt may exist throughout the inner core. However, more detailed models involving liquid inclusions can only be tested when more exact, spatially resolved seismological data become available. Whatever the reason for the discrepancy, having shown that attenuation is likely to be small (~1%), the current seismological and standard mineralogical models cannot, at present, be reconciled.

5.10.3.3 Rheology of liquid iron alloys

A number of experimental and theoretical studies have been performed on the Fe–FeS system in order to obtain diffusivities and viscosities. High-pressure tracer diffusion experiments (Dobson, 2000) have been carried out on liquid Fe–FeS alloys at 5 GPa resulting in high diffusivities (10^{-5} cm^2 s^{-1}) in excellent agreement with *ab initio* molecular dynamics calculations performed at the same conditions (Vočadlo *et al.*, 2000). When incorporated into the Stokes–Einstein relation (eqn [11]), these diffusivities lead to values for viscosity of a few mPa s, that is, of the same order as that of pure iron.

Direct viscosity measurements (Dobson *et al.*, 2000) of Fe–FeS alloys by means of the falling-sphere technique have been made at similar pressures and temperatures to those used in the diffusion experiments above; these resulted in values for viscosities in excellent agreement with those derived experimentally using the Stokes–Einstein relation. Furthermore, these results are in excellent agreement with *ab initio* molecular dynamics calculations of viscosity based on rigorous Green–Kubo functions of the stresses obtained directly from the simulations (Vočadlo *et al.*, 2000). All of these results thus provide both experimental and theoretical verification of the Stokes–Einstein relation (eqn [11]) and also show that the introduction of light elements into Fe liquid does not appear to significantly affect the values.

Atomistic classical molecular dynamics simulations on the Fe–Ni system (Zhang and Guo, 2000) also show that at the conditions of the Earth's outer core, the viscosities are, again, of the order of that of pure iron indicating that nickel has little or no effect.

5.11 Summary

It is clear that mineral physics has a very important role to play in the understanding of the structure, composition, and evolution of the Earth's core. With increasing computer power and the advancing sophistication and precision of experimental techniques, reaching ever high pressures and temperatures, more complex systems will be able to be explored at the conditions of the Earth's core. Results from this research will enable many of the questions concerning the Earth's core to be resolved. The key issues are as follows:

1. Seismological evidence for anisotropy and layering in the inner core is strong, but the mechanism by which this is occurring is unclear at present. With the presence of light elements, it is distinctly possible that the inner-core phase is not just h.c.p.-Fe, but has two or more phases of iron present contributing to the anisotropy and layering. Conversely, both these phenomena could be entirely down to growth mechanisms and therefore only one phase need be present, which, in the presence of light elements, is likely to be b.c.c.-Fe. Full quantification of the thermoelastic properties of multicomponent iron alloy systems at the conditions of the Earth's inner core would enable a compositional model to be developed that is consistent with increasingly accurate seismological data.

2. More precise seismological data, particularly shear wave velocities, along with a better understanding of anelasticity in the inner core, would answer the question as to whether or not there are melt pockets distributed throughout the inner core.

3. Super-rotation of the inner core now seems to be marginal, yet a better understanding of the rheology of the multicomponent systems present would both confirm this and allow an evolutionary model to be developed that is consistent with the present-day Earth.

4. The temperature profile of the core is still unknown; in particular, while the question of the melting temperature of pure iron at inner-core boundary pressures seems to be resolved, it is far from clear exactly what the effect of light elements will be on this quantity.

5. Mineral physics constraints on the thermoelastic properties and processes of multicomponent liquids are essential to both the magnetohydrodynamics relations governing the geodynamo, and also to the core evolution models that determine

the age and heat budget of the Earth; of particular import are better estimates for key parameters for thermal evolution models such as electrical and thermal conductivity.

6. Composition models for the outer core can also help clarify the possible structure of the outer core both at the crystallization surface at the inner-core boundary and also in the iron-silicate reaction zone at the core–mantle boundary;

Mineral physics may soon have many of the answers, but results from such advanced theoretical and experimental techniques are nothing without well-constrained seismological data. Models of core composition, structure, and evolution can only be believed when the mineral physics data exactly matches the primary seismological observations. This multidisciplinary approach is the only way forward to a full and thorough understanding of the Earth's core.

References

Ahrens TJ, Holland KG, and Chen GQ (2002) Phase diagram of iron, revised core temperatures. *Geophysical Research Letters* 29: 1150.

Alfè D (2005) Melting curve of MgO from first principles simulations. *Physical Review Letters* 94: 235701.

Alfè D, Gillan MJ, and Price GD (1999) The melting curve of iron at the pressure of the Earth's core from *ab initio* calculations. *Nature* 401: 462–464.

Alfè D, Gillan MJ, and Price GD (2000b) Constraints on the composition of the Earth's core from *ab initio* calculations. *Nature* 405: 172–175.

Alfè D, Gillan MJ, and Price GD (2002a) *Ab initio* chemical potentials of solid and liquid solutions and the chemistry of the Earth's core. *Journal Chemical Physics* 116: 7127–7136.

Alfè D, Gillan MJ, Vočadlo L, Brodholt JP, and Price GD (2002b) The *ab initio* simulation of the Earth's core. *Philosophical Translation of the Royal Society of London Series A* 360: 1227–1244.

Alfè D, Kresse G, and Gillan MJ (2000a) Structure and dynamics of liquid iron under Earth's core conditions. *Physical Review B* 61: 132–142.

Alfè D, Price GD, and Gillan MJ (2001) Thermodynamics of hexagonal close-packed iron under Earth's core conditions. *Physical Review B* 64: 045123.

Alfè D, Vočadlo L, Price GD, and Gillan MJ (2004) Melting curve of materials: theory versus experiment. *Journal of Physics-Condensed Matter* 16: S973–S982.

Anderson DL (1989) *Theory of the Earth*. Blackwell.

Anderson OL and Isaak DG (2002) Another look at the core density deficit of Earth's outer core. *Physics of the Earth and Planetary Interiors* 131: 19–27.

Anderson WW and Ahrens TJ (1994) An equation of state for liquid iron and implications for the Earth's core. *Journal Geophysical Research* 99: 4273–4284.

Andrault D, Fiquet G, Charpin T, and le Bihan T (2000) Structure analysis and stability field of beta-iron at high P and T. *American Mineralogist* 85: 364–371.

Andrault D, Fiquet G, Kunz M, Visocekas F, and Hausermann D (1997) The orthorhombic structure of iron: and in situ study at high temperature and high pressure. *Science* 278: 831–834.

Antonangeli D, Occelli F, and Requardt H (2004) Elastic anisotropy in textured hcp-iron to 112 GPa from sound wave propagation measurements. *Earth and Planetary Science Letter* 225: 243–251.

Assael MJ, Kakosimos K, Banish RM, *et al.* (2006) Reference data for the density and viscosity of liquid aluminium and liquid iron. *Journal of Physical and Chemical Reference Data* 35: 285–300.

Badro J, Fiquet G, Guyot F, *et al.* (2006) Effect of light elements on the sound velocities in solid iron: Implications for the composition of the Earth's core – Short communication. *Earth and Planetary Science Letters* 254(1–2): 233–238.

Barron THK (1957) Grüneisen parameters for the equation of state of solids. *Annals of Physics* 1: 77–90.

Beghein C and Trampert J (2003) Robust normal mode constraints on inner core anisotropy from model space search. *Science* 299: 552–555.

Belonoshko AB, Ahuja R, and Johansson B (2000) Quasi-*ab initio* molecular dynamics study of Fe melting. *Physical Review Letters* 84: 3638–3641.

Birch F (1964) Density and composition of the mantle and core. *Journal of Geophysical Research* 69: 4377–4388.

Bloxham J (1988) Dynamics of angular momentum in the Earth's core. *Annual Reviews in Earth and Planetary Science* 26: 501–517.

Boehler R (1993) Temperatures in the Earth's core from melting point measurements of iron at high static pressures. *Nature* 363: 534–536.

Boehler R and Ross M (1997) Melting curve of aluminium in a diamond anvil cell to 0.8 Mbar: implications for iron. *Earth and Planetary Science Letter* 153: 223–227.

Bondi H and Lyttleton RA (1948) On the dynamical theory of the rotation of the Earth. *Proceedings of the Cambridge Philosophical Society* 44: 345–359.

Brown JM (2001) The equation of state of iron to 450 GPa: another high pressure phase? *Geophysical Research Letters* 28: 4339–4342.

Brown JM and McQueen RG (1986) Phase transitions, Grüneisen parameter and elasticity of shocked iron between 77 GPa abd 400 GPa. *J. Geophys. Res* 91: 7485–7494.

Buffett BA (1997) Geodynamic estimates of the viscosity of the inner core. *Nature* 388: 571–573.

Buffett BA (2002) Estimates of heat flow in the deep mantle based on the power requirements for the geodynamo. *Geophysical Research Letter* 29: 1566.

Buffett BA, Huppert HE, Lister JR, and Woods AW (1992) Analytical model for the solidification fo the Earth's core. *Nature* 356: 329–331.

Buffett BA, Huppert HE, Lister JR, and Woods AW (1996) On the thermal evolution of the Earth's core. *Journal of Geophysical Research* 101: 7989–8006.

Buffett BA, Huppert HE, Lister JR, *et al.* (1992) Analytical model for solidification of the Earth's core. *Nature* 356(6367): 329–331.

Buffett BA and Wenk HR (2001) Texturing of the Earth's inner core by Maxwell stresses. *Nature* 413: 60–63.

Cao AM, Romanowicz B, and Takeuchi N (2005) An observation of PKJKP: inferences on inner core shear properties. *Science* 308: 1453–1455.

Christensen UR and Tilgner A (2004) Power requirement for the geodynamo from ohmic losses in numerical and laboratory dynamics. *Nature* 429: 169–171.

Collier JD and Helffrich G (2001) Estimate of inner core rotation rate from United Kingdom regional seismic netweork data and consequences for inner core dynamical behaviour. *Earth and Planetary Science Letter* 193: 523–537.

Costin SO and Butler SL (2006) Modelling the effects of internal heating in the core and lowermost mantle on the Earth's magnetic history. *Physics of the Earth and Planetary Interiors* 157: 55–71.

Côté AS, Vočadlo L, and Brodholt J (2007) The effect of silicon impurities on the phase diagram of iron and possible implications for the Earth's core structure (submited).

Creager KC (1992) Anisotropy of the inner core from differential travel times of the phases PKP and PKIKP. *Nature* 356: 309–314.

Creager KC (1997) Inner core rotation rate from small-scale heterogeneity and time-varying travel times. *Science* 278: 1284–1288.

Dobson DP (2000) Fe-57 and Co tracer diffusion in liquid Fe-FeS at 2 and 5 GPa. *Physics of the Earth and Planetary Interiors* 120: 137–144.

Dobson DP, Brodholt JP, Vočadlo L, and Chrichton W (2001) Experimental verification of the Stokes-Einstein relation in liquid Fe-FeS at 5 GPa. *Molecular Physics* 99: 773–777.

Dobson DP, Chrichtom WA, Vočadlo L, *et al.* (2000) In situ measurements of viscosity of liquids in the Fe-FeS system at high pressures and temperatures. *American Mineralogist* 85: 1838–1842.

Duffy TS and Ahrens TJ (1993) Thermal expansion of mantle and core materials at very high pressures. *Geophysical Research Letter* 20: 1103–1106.

Dumberry M and Bloxham J (2002) Inner core tilt and polar motion. *Geophysical Journal International* 151: 377–392.

Durek JJ and Romanowicz B (1999) Inner core anisotropy by direct inversion of normal mode spectra. *Geophysical Journal International* 139: 599–622.

Dziewonski AM and Anderson DL (1981) Preliminary reference Earth model. *Physics of the Earth and Planetary Interiors* 25: 297–356.

Fiquet G, Badro J, Guyot F, Requardt H, and Kirsch M (2001) Sound velocities in iron to 110 gigapascals. *Science* 291: 468–471.

Frost HJ and Ashby MF (1982) *Deformation Mechanism Maps.* Oxford, UK: Pergamon.

Gannarelli CMS, Alfè D, and Gillan MJ (2003) The particle-in-a-cell model for *ab initio* thermodynamics: Implications for the elastic anisotropy of the Earth's inner core. *Physics of the Earth and Planetary Interiors* 139: 243–253.

Gannarelli CMS, Alfè D, and Gillan MJ (2005) The axial ration of hcp iron at the conditions of the Earth's inner core. *Physics of the Earth and Planetary Interior* 152: 67–77.

Garnero EJ and Jeanloz R (2000) Fuzzy patches on the Earth's core-mantle boundary? *Geophys. Res. Lett* 27: 2777–2780.

Gessmann CK and Wood BJ (2002) Potassium in the Earth's core? *Earth and Planetery Science Letter* 200: 63–78.

Grüneisen E (1912) Theorie des festen zuxtandes einatomiger element. *Annals Physik* 12: 257–306.

Gubbins D, Alfè D, Masters G, Price GD, and Gillan MJ (2003) Can the Earth's dynamo run on heat alone? *Geophysical Journal International* 155: 609–622.

Gubbins D, Alfè D, Masters G, Price GD, and Gillan MJ (2004) Gross thermodynamics of two-component core convection. *Geophysical Journal International* 157: 1407–1414.

Gubbins D, Masters TG, and Jacobs JA (1979) Thermal evolution of the Earth's core. *Geophysical Journal of the Royal Astronomical Society* 59: 57–99.

Hanstrom A and Lazor P (2000) High pressure melting and equation of state of aluminium. *Journal of Alloys and Compounds* 305: 209–215.

Helffrich G and Kaneshima S (2004) Seismological constraints on core composition from Fe-O-S liquid immiscibility. *Science* 306: 2239–2242.

Isaak DG and Anderson OL (2003) Thermal expansivity of hcp iron at very high pressure and temperature. *Physica B* 328: 345–354.

Ishii M and Dziewonski AM (2003) Distinct seismic anisotropy at the centre of the Earth. *Physics of the Earth and Planetary Interiors* 140: 203–217.

Jackson I, Fitz Gerald JD, and Kokkonen H (2000) High temperature viscoelastic relaxation in iron and its implications for the shear modulus and attenuation of the Earth's inner core. *Journal of Geophysical Research* 105: 23605–23634.

Jeffreys H (1959) *The Earth, its Origin, History and Physical Constitution.* Cambridge, UK: Cambridge University Press.

Jephcoat AP and Besedin SP (1996) Temperature measurement and melting determination in the laser heated diamond anvil cell. *Philosophical Transactions of the Royal Society of London* 354: 1333–1390.

Jephcoat A and Olson P (1987) Is the inner core of the Earth pure iron. *Nature* 325: 332–335.

Karato S (1993) Inner core anisotropy due to the magnetic field induced preferred orientation of iron. *Science* 262: 1708–1711.

Kennett BLN, Engdahl ER, and Buland R (1995) Constraints on seismic velocities in the Earth from travel-times. *Geophysical Journal International* 122: 108–124.

Kennett BLN and Engdahl ER (1991) Travel times for global earthquake location and phase identification. *Geophysical Journal International* 105: 429–465.

Labrosse S (2003) Thermal and magnetic evolution of the Earth's core. *Physics of the Earth and Planetary Interiors* 140: 127–143.

Labrosse S, Poirier JP, and Le Mouel JL (2001) The age of the inner core. *Earth and Planetary Science Letters* 190: 111–123.

Laio A, Bernard S, Chiarotti GL, Scandolo S, and Tosatti E (2000) Physics of iron at Earth's core conditions. *Science* 287: 1027–1030.

Lei J and Zhao D (2006) Global P-wave tomography: on the effect of various mantle and core phases. *Physics of the Earth and Planetary Interiors* 154: 44–69.

Li J and Agee CB (2001) Element partitioning constraints on the light element composition of the Earth's core. *Geophysical Research Letters* 28: 81–84.

Lin JF, Heinz DL, Campbell AJ, Devine JM, and Shen GY (2002) Iron–silicon alloy in the Earth's core? *Science* 295: 313–315.

Lin JF, Struzhkin VV, Sturhahn W, *et al.* (2003) Sound velocities of iron-nickel and iron-silicon alloys at high pressures. *Geophysical Research Letters* 30: 2112.

Ma YZ, Somayazulu M, Shen GY, Mao HK, Shu JF, and Hemley RJ (2004) In situ X-ray diffraction studies of iron to Earth core conditions. *Physics of the Earth and Planetary Interiors* 143: 455–467.

Mao HK, Shu JF, Shen GY, Hemley RJ, Li BS, and Singh AK (1999) Elasticity and rheology of iron above 220 GPa and the nature of the Earth's inner core. *Nature* 396: 741–743 (and correction in *Nature* 399: 280–280.).

Mao HK, Wu Y, Chen LC, Shu JF, and Jephcoat AP (1990) Static compression of iron to 300 GPa and Fe0.8Ni0.2 alloy to 260 GPa – implications for composition of the core. *Journal of Geophysical Research* 95: 21737–21742.

Mao WL, Campbell AJ, Heinz DL, and Shen G (2006) Phase relations of Fe-Ni alloys at high pressure and temperature. *Physics of the Earth and Planetary Interiors* 155: 146–151.

Mao HK, Xu J, Struzhkin VV, *et al.* (2001) Phonon density of states of iron up to 153 GPa. *Science* 292: 914–916.

Masters G and Gubbins D (2003) On the resolution of density within the Earth. *Physics of the Earth and Planetary Interiors* 140: 159–167.

Matsui M and Anderson OL (1997) The case for a body-centred cubic phase (alpha ') for iron at inner core conditions. *Physics of the Earth and Planetary Interiors* 103: 55–62.

McDonough WF and Sun S-s (1995) The composition of the Earth. *Chemical Geology* 120: 223–253.

Merkel S, Wenk HR, Gillet P, Mao HK, and Hemley RJ (2004) Deformation of polycrystalline iron up to 30 GPa and 1000 K. *Physics of the Earth and Planetary Interiors* 145: 239–251.

Nakagawa T and Tackley PJ (2004) Effects of thermo-chemical mantle convection on the thermal evolution of the Earth's core. *Earth and Planetary Science Letter* 220: 107–109.

Nguyen JH and Holmes NC (2004) Melting of iron at the physical conditions of the Earth's core. *Nature* 427: 339–342.

Nimmo F, Price GD, Brodholt J, and Gubbins D (2004) The influence of potassium on core and geodynamo evolution. *Geophysical Journal International* 156: 363–376.

Officer CB (1986) A conceptual model of core dynamics and the Earth's magnetic field. *Journal of Geophysics* 59: 89–97.

Oreshin SI and Vinnik LP (2004) Heterogeneity and anisotropy of seismic attenuation in the inner core. *Geophysical Research Letters* 31: L02613.

Ouzounis A and Creager KC (2001) Isotropy overlying anisotropy at the top of the inner core. *Geophysical Research Letter* 28: 4221–4334.

Poirier JP (1988) Transport properties of liquid metals and viscosity of the Earth's core. *Geophysical Journal International* 92: 99–105.

Poirier JP (1994) Light elements in the Earth's outer core: a critical review. *Physics of the Earth and Planetary Interiors* 85: 319–337.

Poirier JP (2000) *Introduction to the Physics of the Earth's Interior*. Cambridge, UK: Cambridge University Press.

Poirier JP (2002) Rheology: Elasticity and viscosity at high pressure. In: Hemley RJ, Chiarotti GL, Bernasconi M, and Ulivi L (eds.) *Proceedings of the International School of Physics "Enrico Fermi"*, Course CXLVII, Amsterdam: IOS Press.

Poirier JP and Price GD (1999) Primary slip system of epsilon iron and anisotropy of the Earth's inner core. *Physics of the Earth and Planetary Interiors* 110: 147–156.

Poupinet G, Souriau A, and Coutant O (2000) The existence of an inner core super-rotation questioned by teleseismic doublets. *Physics of the Earth and Planetary Interiors* 118: 77–88.

Resovsky J, Trampert J, and Van der Hilst RD (2005) Error bars in the global seismic Q profile. *Earth and Planetary Science Letter* 230: 413–423.

Romanowicz B (2003) Global mantle tomography: progress status in the past 10 years. *Annual Review Earth and Planetary Science* 31: 303–328.

Ross M (1969) Generalised Lindemann melting law. *Physical Review* 184: 233–242.

Rost S and Revenaugh J (2001) Seismic detection of rigid zones at the top of the core. *Science* 294: 1911–1914.

Rubie DC, Gessmann CK, and Frost DJ (2004) Partitioning of oxygen during core formation of the Earth and Mars. *Nature* 429: 58–61.

Sato R and Espinosa AF (1967) Dissipation factor of the torsional mode $_0T_2$ for a homogeneous mantle Earth with a soft-solid or viscous-liquid core. *Journal of Geophysical Research* 72: 1761–1767.

Saxena SK and Dubrovinsky LS (2000) Iron phases at high pressures and temperatures: Phase transitions and melting. *American Mineralogist* 85: 372–375.

Saxena SK, Dubrovinsky LS, and Haggkvist P (1996) X-ray evidence for the new phase of beta-iron at high temperature and high pressure. *Geophysical Research Letter* 23: 2441–2444.

Secco RA (1995) Viscosity of the outer core. In: Ahrens TJ (ed.) *Mineral Physics and Crystallography: A Handbook of Physical Constants, AGU Reference Shelf 2*, pp. 218–226. Washington: AGU.

Shaner JW, Brown JM, and McQueen RG (1984) In: *High Pressure in Science and Technology*, Homan C, Mac-Crone RK, and Whalley E (eds.) p. 137. Amsterdam: North Holland.

Shearer P and Masters G (1990) The density and shear velocity contrast at the inner core boundary. *Geophysical Journal International* 102: 491–498.

Shen GY and Heinz DL (1998) High pressure melting of deep mantle and core materials. *Reviews in Mineralogy* 37: 369–396.

Shen GY, Mao HK, Hemley RJ, Duffy TS, and Rivers ML (1998) Melting and crystal structure of iron at high pressures and temperatures. *Geophysical Research Letters* 25: 373–376.

Shen GY, Prakapenka VB, Rivers ML, and Sutton SL (2004) Structure of liquid iron at pressures up to 58 GPa. *Physical Review Letters* 92: 185701.

Shimizu H, Poirier JP, and Le Mouel JL (2005) On the crystallisation at the inner core boundary. *Physics of the Earth and Planetary Interiours* 151: 37–51.

Söderlind P, Moriarty JA, and Wills JM (1996) First principles theory of iron up to earth's core pressures: structural, vibrational and elastic properties. *Physical Review B* 53: 14063–14072.

Soldati G, Boschi L, and Piersanti A (2003) Outer core density heterogeneity and the discrepancy between PKP and PcP travel time observations. *Geophysical Research Letter* 30: 1190.

Song X (1997) Anisotropy of the Earth's inner core. *Reviews of Geophysics* 35: 297–313.

Song XD (2000) Joint inversion for inner core rotation, inner core anisotropy and mantle heterogeneity. *Journal of Geophysical Research* 105: 7931–7943.

Song XD and Helmberger DV (1993) Anisotropy of Earth's inner core. *Geophysical Research Letter* 20: 2591–2594.

Song XD and Helmberger DV (1998) Seismic evidence for an inner core transition zone. *Science* 282: 924–927.

Song XD and Li AY (2000) Support for differential inner core superrotation from earthquakes in Alaska recorded at South Pole station. *Journal of Geophysical Research* 105: 623–630.

Song XD and Richards PG (1996) Seismological evidence for differential rotation of the Earth's inner core. *Nature* 382: 221–224.

Song XD and Xu XX (2002) Inner core transition zone and anomalous PKP(DF) waveforms from polar paths. *Geophysical Research Letters* 29: 1042.

Souriau A (1998) New seismological constraints on differential rotation of the inner core from Novaya Zemlya events recorded at DRV, Antartica. *Geophysical Journal International* 134: F1–F5.

Steinle-Neumann G, Stixrude L, and Cohen RE (1999) First principles elastic constants for the hcp transition metals Fe, Co and Re at high pressure. *Physical Review B* 60: 791–799.

Steinle-Neumann G, Stixrude L, Cohen RE, and Gulseren O (2001) Elasticity of iron at the temperature of the Earth's inner core. *Nature* 413: 57–60.

Steinle-Neumann G, Stixrude L, and Cohen RE (2002) Physical properties of iton in the inner core. In: Dehant V, Creager K, Zatman S, and Karato S-I (eds.) *Core Structure, Dynamics and Rotation*, pp. 137–161. Washington, DC: AGU.

Stixrude L and Brown JM (1998) The Earth's core. *Review in Mineralogy* 37: 261–282.

Stixrude L and Cohen RE (1995a) High pressure elasticity of iron and anisotropy of Earth's inner core. *Science* 267: 1972–1975.

Stixrude L and Cohen RE (1995b) Constraints on the crystalline structure of the inner core – mechanical stability of bcc iron at high pressure. *Geophysical Research Letters* 22: 125–128.

Stixrude L, Wasserman E, and Cohen RE (1997) Composition and temperature of the Earth's inner core. *Journal of Geophysical Research* 102: 24729–24739.

Su WJ, Dziewonski AM, and Jeanloz R (1996) Planet within a planet: rotation of the inner core of the Earth. *Science* 274: 1883–1887.

Sun XL and Song XD (2002) PKP travel times at near antipodal distances: implications for inner core anisotropy and lowermost mantle structure. *Earth and Planetary Science Letters* 199: 429–445.

Van Orman JA (2004) On the viscosity and creep mechanism of Earth's inner core. *Geophysical Research Letter* 31: L20606.

Verhoogen J (1974) Chandler wobble and viscosity in the Earth's core. *Nature* 249: 334–335.

Vočadlo L (2007) *Ab initio* calculations of the elasticity of iron and iron alloys at inner core conditions: evidence for a partially molten inner core? *Earth and Planetary Science Letters* 254: 227–232.

Vočadlo L and Alfe D (2002) *Ab initio* melting curve of the fcc phase of aluminium. *Physical Review B* 65: 214105: 1–12.

Vočadlo L, Alfe D, Gillan MJ, Wood IG, Brodholt JP, and Price GD (2003a) Possible thermal and chemical stabilisation of body-centred-cubic iron in the Earth's core. *Nature* 424: 536–539.

Vočadlo L, Alfe D, Gillan MJ, and Price GD (2003b) The properties of iron under core conditions from first principles calculations. *Physics of the Earth and Planetary Interiors* 140: 101–125.

Vočadlo L, Alfe D, Price GD, and Gillan MJ (2004) *Ab initio* melting curve of copper by the phase coexistence approach. *Journal of Chemical Physics* 120: 2872–2878.

Vočadlo L, Brodholt JP, Alfe D, Gillan MJ, and Price GD (2000) *Ab initio* free energy calculations on the polymorphs of iron at core conditions. *Physics of the Earth and Planetary Interiors* 117: 123–137.

Vočadlo L, Brodholt J, Dobson D, Knight KS, Marshal WG, Price GD, and Wood IG (2002) The effect of ferromagnetism on the equation of state of Fe3C studied by first principles calculations. *Earth and Planetary Science Letters* 203: 567–575.

Vočadlo L, Dobson DP, and Wood IG (2006) An *ab initio* study of nickel substitution into iron. *Earth and Planetary Science Letters* 248: 132–137.

Weertman J (1970) The creep strength of the Earth's mantle. *Reviews of the Geophysics and Space Physics* 8: 145–168.

Williams Q, Jeanloz R, Bass J, Svendsen B, and Ahrens TJ (1987) The melting curve of iron to 250 Gigapascals: a constraint on the temperature at the Earth's centre. *Science* 236: 181–182.

Won IJ and Kuo JT (1973) Oscillation of the Earth's inner core and its relation to the generation of geomagnetic field. *Journal of Geophysical Research* 78: 905–911.

Yoo CS, Holmes NC, Ross M, Webb DJ, and Pike C (1993) Shock temperature and melting of iron at Earth's core conditions. *Phys. Rev. Lett* 70: 3931–3934.

Yoshida S, Sumita I, and Kumazawa M (1996) Growth model of the inner core coupled with the outer core dynamics and the resulting anisotropy. *Journal of Geophysical Research* 101: 28085–28103.

Zhang YG and Guo Gj (2000) Molecular dynamics calculation of the bulk viscosity of liquid iron-nickel alloy and the mechanism for the bulk attenuation of seismic waves in the Earth's outer core. *Physics of the Earth and Planetary Interiors* 122: 289–298.

Zhang J, Song XD, Li YC, *et al.* (2005) Inner core differential motion confirmed by earthquake waveform doublets. *Science* 309(5739): 1357–1360.

6 Theory and Practice – Thermodynamics, Equations of State, Elasticity, and Phase Transitions of Minerals at High Pressures and Temperatures

A. R. Oganov, ETH Zurich, Zurich, Switzerland

6.1	Thermodynamics of Crystals	122
6.1.1	Thermodynamic Potentials	122
6.1.2	Differential Relations	122
6.1.3	Partition Function	123
6.1.4	Harmonic Approximation	124
6.1.4.1	Debye model	125
6.1.4.2	General harmonic potential	126
6.1.5	Quantum Effects in Thermodynamics	127
6.1.6	Thermodynamic Perturbation Theory	128
6.1.7	Quasiharmonic Approximation	129
6.1.8	Beyond the QHA	129
6.2	Equations of State and Elasticity	131
6.2.1	Equations of State	131
6.2.1.1	Mie–Grüneisen EOS	131
6.2.1.2	Analytical static EOS	132
6.2.1.3	Anharmonicity in static EOS	134
6.2.1.4	EOS, internal strain, and phase transitions	134
6.2.2	Elastic Constants	135
6.2.2.1	Cauchy relations	138
6.2.2.2	Mechanical stability	139
6.2.2.3	Birch's law and effects of temperature on the elastic constants	139
6.2.2.4	Elastic anisotropy in the Earth's interior	140
6.3	Phase Transitions of Crystals	140
6.3.1	Classifications of Phase Transitions	140
6.3.2	First-Order Phase Transitions	141
6.3.3	Landau Theory of First- and Second-Order Transitions	141
6.3.4	Shortcomings of Landau Theory	142
6.3.5	Ginzburg–Landau Theory	143
6.3.6	Ising Spin Model	144
6.3.7	Mean-Field Treatment of Order–Disorder Phenomena	145
6.3.8	Isosymmetric Transitions	145
6.3.9	Transitions with Group–Subgroup Relations	146
6.3.10	Pressure-Induced Amorphization	146
6.4	A Few Examples of the Discussed Concepts	147
6.4.1	Temperature Profile of the Earth's Lower Mantle and Core	147
6.4.2	Polytypism of $MgSiO_3$ Post-Perovskite and Anisotropy of the Earth's D″ layer	147
6.4.3	Spin Transition in (Mg, Fe)O Magnesiowüstite	148
References		149

6.1   Thermodynamics of Crystals

Thermodynamics provides the general basis for the theory of structure and properties of matter. This chapter presents only as much thermodynamics as needed for good comprehension of geophysics, and at a relatively advanced level. For further reading the reader is referred to Landau and Lifshitz (1980), Chandler (1987), Wallace (1998), and Bowley and Sánchez (1999).

6.1.1   Thermodynamic Potentials

If one considers some system (e.g., a crystal structure) at temperature $T = 0\,\mathrm{K}$ and pressure $P = 0$, the equilibrium state of that system corresponds to the minimum of the internal energy E:

$$E \rightarrow \min \qquad [1]$$

that is, any changes (e.g., atomic displacements) would result in an increase of energy. The internal energy itself is a sum of the potential and kinetic energies of all the particles (nuclei, electrons) in the system.

The principle [1] is valid in only two situations: (1) at $T = 0\,\mathrm{K}$, $P = 0$, and (2) at constant V (volume) and S (entropy), that is, if we impose constraints of constant S,V, the system will adopt the lowest-energy state. Principle [1] is a special case of a more general principle that the thermodynamic potential W describing the system be minimum at equilibrium:

$$W \rightarrow \min \qquad [2]$$

As already mentioned, at constant V,S: $W_{V,S} = E \rightarrow \min$.

At constant P,S, the appropriate thermodynamic potential is the enthalpy H:

$$W_{P,S} = H = E + PV \rightarrow \min \qquad [3]$$

At constant V,T, the Helmholtz free energy F is the thermodynamic potential:

$$W_{V,T} = F = E - TS \rightarrow \min \qquad [4]$$

At constant P,T (the most frequent practical situation), the relevant thermodynamic potential is the Gibbs free energy G:

$$W_{P,T} = G = E + PV - TS \rightarrow \min \qquad [5]$$

The minimum condition implies that

$$\frac{\partial W}{\partial x_i} = 0 \qquad [6]$$

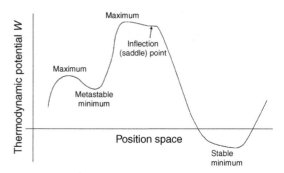

Figure 1 Extrema and saddle points in a one-dimensional representation of the (free) energy surface.

However, this condition is also satisfied for maxima of the thermodynamic potential, and for saddle points (**Figure 1**). To exclude saddle points and maxima, one has to make sure that the matrix of second derivatives of W with respect to all the degrees of freedom (in case of a crystal structure, with respect to atomic coordinates and lattice parameters):

$$H_{ij} = \frac{\partial^2 W}{\partial x_i \partial x_j} \qquad [7]$$

be positive definite:

$$\det \mathbf{H}_{ij} > 0 \qquad [8]$$

Still, there may be a large (or infinite) number of minima. The equilibrium state corresponds to the lowest minimum of W (the global minimum), whereas all the other minima are called local and correspond to metastable states. Local minima have the property of stability to an infinitesimal displacement (after any such displacement the system returns to the initial state), but one can always find a sufficiently large energy fluctuation that will irreversibly destroy the metastable state.

6.1.2   Differential Relations

From the first law of thermodynamics one has

$$dE = -P\,dV + T\,dS \qquad [9]$$

Applying Legendre transformations, the following relations can be obtained:

$$dH = V\,dP + T\,dS \qquad [10]$$

$$dF = -P\,dV - S\,dT \qquad [11]$$

$$dG = V\,dP - S\,dT \qquad [12]$$

When there is thermodynamic equilibrium between two phases (denoted 1 and 2) at given P and T,

$G_1 = G_2$. Moving along the two-phase equilibrium line in P–T space requires $dG_1 = dG_2$, that is,

$$\Delta V \, dP - \Delta S \, dT = 0 \qquad [13]$$

or, in a different form,

$$\frac{dP}{dT} = \frac{\Delta S}{\Delta V} \qquad [14]$$

This is the famous Clausius–Clapeyron equation.

Using eqns [9]–[12], one can express various thermodynamic parameters:

$$P = -\left(\frac{\partial E}{\partial V}\right)_S = -\left(\frac{\partial F}{\partial V}\right)_T \qquad [15]$$

$$V = \left(\frac{\partial H}{\partial P}\right)_S = \left(\frac{\partial G}{\partial P}\right)_T \qquad [16]$$

$$T = \left(\frac{\partial E}{\partial S}\right)_V = \left(\frac{\partial H}{\partial S}\right)_P \qquad [17]$$

$$S = -\left(\frac{\partial F}{\partial V}\right)_S = -\left(\frac{\partial G}{\partial T}\right)_P \qquad [18]$$

Taking second derivatives, Maxwell relations are obtained (see, e.g., Poirier (2000)):

$$\left(\frac{\partial S}{\partial P}\right)_T = -\left(\frac{\partial V}{\partial T}\right)_P \qquad [19]$$

$$\left(\frac{\partial S}{\partial V}\right)_T = \left(\frac{\partial P}{\partial T}\right)_V \qquad [20]$$

$$\left(\frac{\partial T}{\partial P}\right)_S = \left(\frac{\partial V}{\partial S}\right)_P \qquad [21]$$

$$\left(\frac{\partial T}{\partial V}\right)_S = -\left(\frac{\partial P}{\partial S}\right)_V \qquad [22]$$

Using the Maxwell relations, a number of important thermodynamic relations are derived, for example,

$$\left(\frac{\partial S}{\partial V}\right)_T = \alpha K_T \qquad [23]$$

$$\left(\frac{\partial S}{\partial V}\right)_P = \frac{C_P}{\alpha V T} \qquad [24]$$

$$\left(\frac{\partial S}{\partial P}\right)_T = -\alpha V \qquad [25]$$

$$\left(\frac{\partial T}{\partial P}\right)_S = \frac{\alpha V T}{C_P} \qquad [26]$$

$$\left(\frac{\partial V}{\partial T}\right)_S = -\frac{C_P}{\alpha K_S T} \qquad [27]$$

$$\left(\frac{\partial P}{\partial T}\right)_V = \alpha K_T \qquad [28]$$

In eqns [23]–[28] we used thermal expansion

$$\alpha = \frac{1}{V}\left(\frac{\partial V}{\partial T}\right)_P \qquad [29]$$

isothermal bulk modulus

$$K_T = -\frac{1}{V}\left(\frac{\partial V}{\partial P}\right)_S \qquad [30]$$

and isobaric heat capacity

$$C_P = \left(\frac{\partial E}{\partial T}\right)_P \qquad [31]$$

We note, on passing, that the bulk modulus and the heat capacity depend on the conditions of measurement. There are general thermodynamic equations relating the heat capacity at constant pressure (isobaric) and constant volume (isochoric):

$$C_P = C_V\left(1 + \frac{\alpha^2 K_T V}{C_V}\right) \qquad [32]$$

and bulk modulus at constant temperature (isothermal) and at constant entropy (adiabatic):

$$K_S = K_T\left(1 + \frac{\alpha^2 K_T V}{C_P}\right) \qquad [33]$$

The most interesting of eqns [23]–[28] are eqn [26], describing the increase of the temperature of a body on adiabatic compression (e.g., in shock waves, and also inside rapidly convecting parts of planets), and eqn [28], describing thermal pressure. These equations are important for thermal equations of state and for calculating the temperature distributions inside planets.

6.1.3 Partition Function

Let us consider a system with energy levels E_i corresponding to the ground state and all the excited states. The probability to find the system in the ith state is proportional to $e^{-\beta E_i}$, where $\beta = 1/(k_B T)$ (k_B is the Boltzmann constant).

More rigorously, this probability p_i is given as

$$p_i = \frac{e^{-\beta E_i}}{\sum_i e^{-\beta E_i}} \qquad [34]$$

The denominator of this equation is called the partition function Z:

$$Z = \sum_i e^{-\beta E_i} \qquad [35]$$

where the summation is carried out over all discrete energy levels of the system. The partition function is much more than a mere normalization factor; it plays a fundamental role in statistical physics, providing a link between the microscopic energetics and the macroscopic thermodynamics. Once Z is known, all thermodynamic properties can be obtained straightforwardly (e.g., Landau and Lifshitz, 1980). For instance, the internal energy

$$E = \sum_i p_i E_i = \frac{\sum_i E_i \mathrm{e}^{-\beta E_i}}{\mathbf{z}} = -\frac{1}{Z}\left(\frac{\partial Z}{\partial \beta}\right)_V$$
$$= -\left(\frac{\partial \ln Z}{\partial \beta}\right)_V \qquad [36]$$

From this one can derive a very important expression for the Helmholtz free energy

$$F = -\frac{1}{\beta}\ln Z = -k_B T \ln Z \qquad [37]$$

entropy

$$S = K_B \ln Z - \frac{k_B \beta}{Z}\left(\frac{\partial Z}{\partial \beta}\right)_V \qquad [38]$$

and the heat capacity at constant volume (From eqn [39] one can derive (see Dove, 2003) the following important formula: $C_V = K_B \beta^2 (\langle E^2 \rangle - \langle E \rangle^2)$.):

$$C_V = -\frac{k_B \beta^2}{Z^2}\left(\frac{\partial Z}{\partial \beta}\right)_V^2 + \frac{k_B \beta^2}{Z}\left(\frac{\partial^2 Z}{\partial \beta^2}\right)_V \qquad [39]$$

Unfortunately, in many real-life cases it is practically impossible to obtain all the energy levels – neither experimentally nor theoretically, and therefore the partition function cannot be calculated exactly. However, for some simplified models it is possible to find the energy levels and estimate the partition function, which can then be used to calculate thermodynamic properties.

Below we consider the harmonic approximation, which plays a key role in the theory of thermodynamic properties of crystals. It gives a first approximation to the distribution of the energy levels E_i, which is usually accurate for the most-populated lowest excited vibrational levels. The effects not accounted for by this simplified picture can often be included as additive corrections to the harmonic results.

6.1.4 Harmonic Approximation

The harmonic oscillator is a simple model system where the potential energy (U) is a quadratic

function of the displacement x from equilibrium, for example, for a simple diatomic molecule

$$U(x) = U_0 + \frac{1}{2}kx^2 \qquad [40]$$

where U_0 is the reference energy and k is the force constant.

The energy levels of the harmonic oscillator can be found by solving the Schrödinger equation with the harmonic potential [40]; the result is an infinite set of equi-spaced energy levels:

$$E_n = \left(\frac{1}{2} + i\right)\hbar\omega \qquad [41]$$

where $\hbar$ is Planck's constant, ω is the vibrational frequency of the oscillator, and integer i is the quantum number: $i = 0$ for the ground state, and $i \geq 1$ for excited states. Energy levels in a true vibrational system are well described by [41] only for the lowest quantum numbers n, but these represent the most populated, and thus the most important vibrational excitations.

A very interesting feature of [41] is that even when $i = 0$, that is, when there are no vibrational excitations (at 0 K), there is still a vibrational energy equal to $\hbar\omega/2$. This energy is called zero-point energy and arises from quantum fluctuations related to the Heisenberg uncertainty principle.

With [41] the partition function for the harmonic oscillator is rather simple:

$$Z = \frac{1}{1 - \mathrm{e}^{-\hbar\omega/k_B T}} \qquad [42]$$

This allows one to calculate thermodynamic functions of a single harmonic oscillator (as was first done by Einstein):

$$E_{vib}(\omega, T) = \frac{1}{2}\hbar\omega + \frac{\hbar\omega}{\exp(\hbar\omega/k_B T) - 1} \qquad [43]$$

$$C_{V,vib}(\omega, T) = k_B \left(\frac{\hbar\omega}{k_B T}\right)^2 \frac{\exp(\hbar\omega/k_B T)}{(\exp(\hbar\omega/k_B T) - 1)^2} \qquad [44]$$

$$S_{vib}(\omega, T) = -k_B \ln[1 - \exp(-\hbar\omega/k_B T)] + \frac{1}{T}\frac{\hbar\omega}{\exp(\hbar\omega/k_B T) - 1} \qquad [45]$$

$$F_{vib}(\omega, T) = \frac{1}{2}\hbar\omega + k_B T \ln\left[1 - \exp\left(-\frac{\hbar\omega}{k_B T}\right)\right] \qquad [46]$$

The first term in [43] is the zero-point energy originating from quantum motion of atoms discussed above. The second, temperature-dependent term gives the thermal energy according to the Bose–Einstein distribution. The thermal energy (or heat content) gives the energy absorbed by the crystal upon heating from 0 K to the temperature T. In the harmonic approximation, the isochoric C_V and isobaric C_P heat capacities are equal: $C_V = C_P$.

The number of phonons in a crystal containing N atoms in the unit cell is $3N$ (per unit cell). In the harmonic approximation, lattice vibrations do not interact with each other (in other words, propagation of one vibration does not change the energy or momentum of other vibrations), and their contributions to thermodynamic properties are additive. If all of the phonons had the same frequency (the assumption of the Einstein model), then, multiplying the right-hand sides of [43]–[46] by the total number of vibrations $3N$, all thermodynamic properties would be obtained immediately. However, normal mode frequencies form a spectrum (called the phonon spectrum, or phonon density of states $g(\omega)$); and an appropriate generalization [43]–[46] involves integration over all frequencies:

$$
\begin{aligned}
E_{\text{vib}}(T) &= \int_0^{\omega_{\max}} E_{\text{vib}}(\omega, T) g(\omega) \mathrm{d}\omega \\
&= \int_0^{\omega_{\max}} \left(\frac{1}{2}\hbar\omega + \frac{\hbar\omega}{\exp(\hbar\omega/k_{\text{B}}T) - 1} \right) \\
&\quad \times g(\omega)\mathrm{d}\omega
\end{aligned}
\tag{47}
$$

$$
\begin{aligned}
C_{V,\text{vib}}(T) &= \int_0^{\omega_{\max}} C_{V,\text{vib}}(\omega, T) g(\omega) \mathrm{d}\omega \\
&= \int_0^{\omega_{\max}} \left(k_{\text{B}} \left(\frac{\hbar\omega}{k_{\text{B}}T} \right)^2 \frac{\exp(\hbar\omega/k_{\text{B}}T)}{(\exp(\hbar\omega/k_{\text{B}}T) - 1)^2} \right) \\
&\quad \times g(\omega)\mathrm{d}\omega
\end{aligned}
\tag{48}
$$

$$
\begin{aligned}
S_{\text{vib}}(T) &= \int_0^{\omega_{\max}} S_{\text{vib}}(\omega, T) g(\omega) \mathrm{d}\omega \\
&= \int_0^{\omega_{\max}} \left(-k_{\text{B}} \ln\left[1 - \exp\left(-\frac{\hbar\omega}{k_{\text{B}}T} \right) \right] \right. \\
&\quad \left. + \frac{1}{T}\frac{\hbar\omega}{\exp(\hbar\omega/k_{\text{B}}T) - 1} \right) g(\omega)\mathrm{d}\omega
\end{aligned}
\tag{49}
$$

$$
\begin{aligned}
F_{\text{vib}}(T) &= \int_0^{\omega_{\max}} F_{\text{vib}}(\omega, T) \\
&= \int_0^{\omega_{\max}} \left(\frac{1}{2}\hbar\omega + k_{\text{B}}T \ln\left\{ 1 - \exp\left(-\frac{\hbar\omega}{k_{\text{B}}T} \right) \right\} \right) \\
&\quad \times g(\omega)\mathrm{d}\omega
\end{aligned}
\tag{50}
$$

6.1.4.1 Debye model

In early works, the phonon density of states $g(\omega)$ had often been simplified using the Debye model. For the

acoustic modes the phonon spectrum can be described, to a first approximation, by a parabolic function:

$$
g(\omega) = 9N \left(\frac{\hbar}{k_{\text{B}}\theta_{\text{D}}} \right)^3 \omega^2
\tag{51}
$$

truncated at the maximum frequency $\omega_{\text{D}} = (k_{\text{B}}\theta_{\text{D}})/\hbar$, where θ_{D} is the Debye temperature.

With this $g(\omega)$ thermodynamic functions take the following form:

$$
E_{\text{vib}} = \frac{9}{8}k_{\text{B}}N\theta_{\text{D}} + 3k_{\text{B}}NTD\left(\frac{\theta_{\text{D}}}{T} \right)
\tag{52}
$$

$$
C_V(T) = \left(\frac{\mathrm{d}E_{\text{vib}}}{\mathrm{d}T} \right)_V = 3k_{\text{B}}N\left[4D\left(\frac{\theta_{\text{D}}}{T} \right) - \frac{3(\theta_{\text{D}}/T)}{\mathrm{e}^{\theta_{\text{D}}/T} - 1} \right]
\tag{53}
$$

$$
S(T) = \int_0^T \frac{C_p}{T}\mathrm{d}T = k_{\text{B}}N[4D(\theta_{\text{D}}/T) - 3\ln(1 - \mathrm{e}^{\theta_{\text{D}}/T})]
\tag{54}
$$

where

$$
D(x) = \frac{3}{x^3} \int_0^x \frac{x^3\mathrm{d}x}{\mathrm{e}^x - 1}, \quad x = \frac{\theta_{\text{D}}}{T}
$$

The first term in [52] is the zero-point energy in the Debye model, the second term is the heat content. The Debye temperature is determined by the elastic properties of the solid or, more precisely, its average sound velocity $\langle v \rangle$:

$$
\theta_{\text{D}} = \frac{\hbar}{k_{\text{B}}} \left(\frac{6\pi^2 N}{V} \right)^{1/3} \langle v \rangle
\tag{55}
$$

The mean sound velocity can be accurately calculated from the elastic constants tensor (Robie and Edwards, 1966). Usually, however, an approximate formula is used:

$$
\langle v \rangle = \left(\frac{1}{v_P^3} + \frac{2}{v_S^3} \right)^{-1/3}
\tag{56}
$$

where v_P and v_S are the longitudinal and transverse sound velocities, respectively. Later in this chapter we shall see as to how to calculate these velocities. The advantages of the Debye model are its relative simplicity and correct low- and high-temperature limits for all thermodynamic properties. The crucial disadvantage is that it is hardly capable of giving accurate entropies for anything other than monatomic lattices. Deep theoretical analyses of this model and its critique can be found in Seitz (1949) and Kieffer (1979). In **Figure 2** we compare the phonon spectra and C_V obtained in the Debye model and in full-phonon harmonic calculations done with the same model

(a) (b)

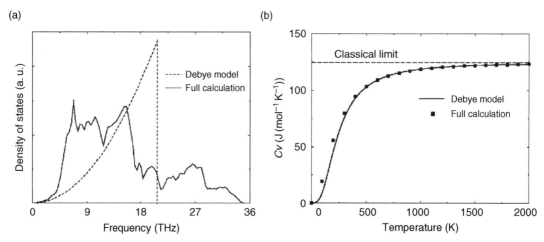

Figure 2 Phonon density of state (a) and heat capacity C_V of MgSiO$_3$ perovskite. Reproduced from Oganov AR, Brodholt JP, and Price GD (2000) Comparative study of quasiharmonic lattice dynamics, molecular dynamics and Debye model in application to MgSiO$_3$ perovskite. *Physics of the Earth and Planetary Interiors* 122: 277–288, with permission from Elsevier.

interatomic potential (see Oganov *et al.* (2000)). The phonon spectra are very different, but heat capacities are reasonably close (only below $\sim$500 K, the disagreement is appreciable).

6.1.4.2 General harmonic potential

Let us now come back to the general harmonic case. First of all, the potential [40] describing a simple elastic spring or a diatomic molecule can be generalized to the case of three-dimensional structures. One can expand the crystal potential energy U around the equilibrium configuration in terms of displacements $u_\alpha^i(l)$ of ith atoms in the lth unit cell along each αth coordinate (Cartesian) axis:

$$U = U_0 + \sum_{l,i,\alpha} \Phi_\alpha^i(l) u_\alpha^i(l)$$

$$+ \frac{1}{2!} \sum_{l<l',i<j,a,b} \Phi_{\alpha\beta}^{ij}(ll') u_\alpha^i(l) u_\beta^j(l')$$

$$+ \frac{1}{3!} \sum_{l<l'<l'',j<j<k,\alpha,\beta,\gamma} \Phi_{\alpha\beta\gamma}^{ijk}(ll'l'') u_\alpha^i(l) u_\beta^j(l') u_\gamma^k(l'')$$

$$+ \cdots \qquad [57]$$

where

$$\Phi_\alpha^i(l) = \frac{\partial U}{\partial u_\alpha^i(l)}$$

$$\Phi_{\alpha\beta}^{ij}(ll') = \frac{\partial^2 U}{\partial u_\alpha^i(l)\partial u_\beta^j(l')} \qquad [58]$$

$$\Phi_{\alpha\beta\gamma}^{ijk}(ll'l'') = \frac{\partial^3 U}{\partial u_\alpha^i(l)\partial u_\beta^j(l')\partial u_\gamma^k(l'')}$$

At equilibrium $\Phi_\alpha^i(l) = 0$, so neglecting third- and higher-order terms (called anharmonic terms), we obtain the harmonic expansion of the potential energy:

$$U = U_0 + \frac{1}{2} \sum_{l<l',i<j,a,b} \Phi_{\alpha\beta}^{ij}(ll') u_\alpha^i(l) u_\beta^j(l') \qquad [59]$$

The generalized harmonic potential [59] includes noncentral forces, due to which directions of the displacement and force may differ. In spite of the complicated mathematical form of [59], it is really analogous to [40]. It also corresponds to a set of phonons, which are again noninteracting and have the same quantization as given by [41]. For each vibrational mode, the partition function is expressed as [42], and thermodynamic properties are described by [47]–[50].

The use of the harmonic approximation, neglecting third- and higher-order terms in the interatomic potential, leads to a number of fundamental errors. The phonon frequencies in this approximation do not depend on temperature or volume, and are noninteracting. This leads to a simple interpretation of experimentally observed vibrational spectra and greatly simplifies the calculation of thermodynamic properties [47]–[50], but noninteracting phonons can freely travel within the crystal, leading to an infinite thermal conductivity of the harmonic crystal. In a real crystal, thermal conductivity is, of course, finite due to phonon–phonon collisions, scattering on defects, and finite crystal size. In the harmonic approximation, the energy needed to remove an atom from the crystal is infinite – therefore, diffusion and melting cannot be explained within this approximation. The same can be said about displacive phase transitions – even though the harmonic approximation can indicate such a transition by showing imaginary phonon frequencies, calculation of properties of the high-temperature dynamically disordered phase is out of reach of the harmonic approximation. In the harmonic approximation there is no thermal expansion, which obviously contradicts experiment. Related to this is the equality $C_V = C_P$, whereas experiment indicates $C_V < C_P$ (see [32]). In a harmonic crystal, at high temperatures C_V tends exactly to the Dulong–Petit limit of $3Nk_B$, whereas for anharmonic crystals this is not the case (see Section 6.1.8 for more details on anharmonicity).

The first approximation correcting many of these drawbacks, the quasiharmonic approximation, as well as methods to account for higher-order anharmonicity will be discussed later in this chapter, but now let us explore some more fundamental aspects of thermodynamics.

6.1.5 Quantum Effects in Thermodynamics

Quantum effects are of fundamental importance for thermodynamic properties. Insufficiency of classical mechanics is apparent in any experimental determinations of the heat capacity at low temperatures.

According to classical mechanics, every structural degree of freedom has $(k_B T)/2$ worth of kinetic energy. In a harmonic solid, there is an equal amount of potential energy, so the total vibrational energy equals $3NRT$, and the heat capacity C_V is then $3NR$. In a stark contrast, experiment shows C_V going to zero as T^3 at low temperatures. Similarly, thermal expansion goes to zero at low temperatures – in contrast to classical theory, predicting a finite value. A very important consequence is for the entropy: if, as the classical approximation claims, $C_V = 3NR$ at all temperatures, then the entropy $(S = \int_0^T (C_V/T) \mathrm{d}T)$ is infinite.

The partition function [35] includes the relevant quantum effects, and so do harmonic expression [43]–[50] for thermodynamic functions. In the classical approximation, the partition function is

$$Z_{\text{class}} = \frac{1}{N! \lambda^{3N}} \int \int e^{-\beta[U(r) + E_{\text{kin}}(p)]} \mathrm{d}r\, \mathrm{d}p \qquad [60]$$

The denominator in this definition already accounts for some quantum effects. There, one has $N!$ to account for indistinguishability of same-type particles, and λ^{3N} that takes into account the fact that quantum states are discrete and very small differences in coordinates/momenta of particles may correspond to the same quantum state. Nevertheless, this definition is classical – since it involves integration in the phase space, rather than summation over discrete quantum states and since some essentially quantum effects (such as exchange) are not present in [60].

According to the uncertainty principle, quantum particles are never at rest and there is quantum motion of atoms even at 0 K (zero-point motion). The corresponding energy, arising from quantum motion in a potential field, is called the zero-point energy, which we already encountered in harmonic expressions [43], [46], [47], and [50]. The magnitude of zero-point motion is significant – it can contribute more than 50% of the total experimentally observed atomic mean-square displacements at room temperature.

It is important that at temperatures significantly exceeding the characteristic temperatures θ of all the vibrational modes $(\theta = (\hbar\omega/k_B))$, classical expressions will be correct. This circumstance justifies the application of methods based on classical mechanics (molecular dynamics, Monte Carlo, etc.) in simulations of materials at high temperatures. At low temperatures, where quantum effects dominate, one could use the harmonic approximation (or, better, the quasiharmonic approximation – see below) or include quantum corrections to classical results.

The classical free energy can be calculated as

$$F_{\text{class}} = E_0 - k_B T \ln Z_{\text{class}} \qquad [61]$$

where E_0 is the internal energy of a static crystal structure. The quantum correction to [61] per atom in the lowest order is (Landau and Lifshitz, 1980)

$$\begin{aligned} \Delta F = F - F_{\text{class}} &= \frac{\hbar^2}{24 k_B^2 T^2} \left\langle \sum_i \frac{(\nabla_i U)^2}{m_i} \right\rangle \\ &= \frac{\hbar^2}{24 k_B T} \left\langle \sum_i \frac{\nabla_i^2 U}{m_i} \right\rangle \end{aligned} \qquad [62]$$

where is ∇_i^2 is the Laplacian with respect to the coordinates of the ith atom. Higher-order ($\hbar^3$ and higher) corrections are needed only at temperatures below $\sim (\theta_D/2)$. Quantum corrections to other properties can be worked out by differentiating [62] (see Matsui (1989) and **Figure 3**).

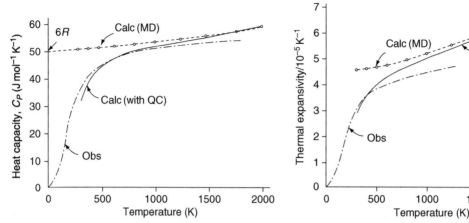

Figure 3 Heat capacity C_P and thermal expansion of MgO from classical molecular dynamics and with quantum corrections [5.3]. After Matsui M (1989) Molecular dynamics study of the structural and thermodynamic properties of MgO crystal with quantum correction. *Journal of Chemical Physics* 91: 489–494.

Other possibilities to incorporate quantum corrections into classical results can be done using (1) path integral formalism (see Allen and Tildesley, 1987), (2) phonon density of states $g(\omega)$, which can be calculated classically, and quasiharmonic formulas. Montroll (1942, 1943) has formulated a method of calculating thermodynamic properties of a solid without the knowledge of $g(\omega)$ but using moments of the frequency distribution instead.

Defining the moments as

$$\mu_{2k} = \frac{1}{3N} \int_0^\infty \omega^{2k} g(\omega) \, d\omega \qquad [63]$$

when $T > \hbar\omega_{max}/k_B$ one can write

$$E(T) = 3Nk_B T - 3Nk_B T$$
$$\times \sum_{n=1}^\infty \frac{(-1)^n B_n}{(2n)!} \left(\frac{\hbar}{2k_B T}\right)^{2n} \mu_{2n} \qquad [64]$$

$$C_V(T) = 3Nk_B - 3Nk_B$$
$$\times \sum_{n=1}^\infty \frac{(-1)^n B_n (1-2n)}{(2n)!} \left(\frac{\hbar}{2k_B T}\right)^{2n} \mu_{2n} \qquad [65]$$

where B_n are Bernoulli numbers. First terms in [64] and [65] are classical contributions, the second terms (sums) can be considered as quantum corrections. Taking only the first few terms, [65] takes the following form:

$$C_V(T) \approx 3Nk_B \left[1 - \left(\frac{\hbar}{k_B T}\right)^2 \frac{\mu_2}{12} + \left(\frac{\hbar}{k_B T}\right)^4 \frac{\mu_4}{240} \right.$$
$$\left. - \left(\frac{\hbar}{k_B T}\right)^6 \frac{\mu_6}{6048} + \cdots \right] \qquad [66]$$

The lowest-order quantum term is, as expected, of order $\hbar^2$.

6.1.6 Thermodynamic Perturbation Theory

It can be demonstrated (Landau and Lifshitz, 1980) that by modifying the potential energy of the system from U_0 to U_1 so that $V = U_1 - U_0$ is a small perturbation, to first order the free energy becomes

$$F_1 = F_0 + \langle V \rangle_0 \qquad [67]$$

where subscript '0' means that averaging is performed over the configurations of the unperturbed system. This means that the free energy of a system with the potential U_1 can be found by thermodynamic integration from (any) system U_0, the free energy of which is known:

$$F_1 = F_0 + \int_{\lambda=0}^1 U_\lambda \, d\lambda \qquad [68]$$

where $U_\lambda = (1-\lambda)U_0 + \lambda U_1$. The same ideas can be used to calculate the free energy profile along the chemical reaction coordinate, or generally the free energy surface – as done in metadynamics simulations (Laio and Parrinello, 2002; Iannuzzi et al., 2003).

To second order, we have

$$F_1 = F_0 + \langle V \rangle_0 - \frac{1}{2k_B T} \langle (V - \overline{V})^2 \rangle_0 \qquad [69]$$

where $\overline{V}$ is the averaged perturbing potential. Note that the expressions [67] and [69] are classical, but quantum extensions are available (Landau and

Lifshitz, 1980). Thermodynamic perturbation theory plays an important role in methods to calculating free energies.

6.1.7 Quasiharmonic Approximation

In this approximation, it is assumed that the solid behaves like a harmonic solid at any volume, but the phonon frequencies depend on volume. It is assumed that they depend only on volume – that is, heating at constant volume does not change them.

In the quasiharmonic approximation (QHA) phonons are still independent and noninteracting. Thermodynamic functions at constant volume, as before, are given by [47]–[50], C_V still cannot exceed $3NR$. Melting, diffusion, and dynamically disordered phases are beyond the scope of this approximation, which breaks down at high temperatures. Thermal conductivity is still infinite.

However crude, this approximation heals the biggest errors of the harmonic approximation. Introducing a volume dependence of the frequencies is enough to create nonzero thermal expansion and account for $C_V < C_P$ [32]. Thermal pressure contributes to all constant-pressure thermodynamic functions (enthalpy H, Gibbs free energy G, isobaric heat capacity C_P, etc.). This is the first approximation to the thermal equation of state of solids, which can be effectively used in conjunction with realistic interatomic potentials (Parker and Price, 1989; Kantorovich, 1995; Gale, 1998) or quantum-mechanical approaches such as density-functional perturbation theory (Baroni *et al.*, 1987, 2001). For instance, using the QHA and calculating phonon frequencies using density-functional theory, Karki and co-authors calculated high-pressure thermal expansion and elastic constants of MgO (Karki *et al.*, 1999) and thermal expansion of MgSiO$_3$ perovskite (Karki *et al.*, 2000). Using similar methodology, Oganov and colleagues calculated a number of mineral phase diagrams – MgO, SiO$_2$, MgSiO$_3$, Al$_2$O$_3$. They found that MgO retains the NaCl-type structure at all conditions of the Earth's mantle (Oganov *et al.*, 2003) and that phase transitions of SiO$_2$ do not correspond to any observed seismic discontinuities in the mantle (Oganov *et al.*, 2005a). For MgSiO$_3$ (Oganov and Ono, 2004) and Al$_2$O$_3$ (Oganov and Ono, 2005), new high-pressure 'post-perovskite' phases with the CaIrO$_3$-type structure were found to be stable, and their P–T stability fields were predicted and, in the same papers, experimentally verified. Also using the QHA and density-functional perturbation theory, Tsuchiya *et al.* (2004) studied stability of MgSiO$_3$ post-perovskite and confirmed previous experimental (Murakami *et al.*, 2004; Oganov and Ono, 2004) and theoretical (Oganov and Ono, 2004) findings. Oganov and Price (2005) confirmed that MgSiO$_3$ perovskite and post-perovskite remain stable against decomposition at all conditions of the Earth's mantle, but their decomposition into MgO and SiO$_2$ was predicted to occur at conditions of cores of extraterrestrial giant planets (Umemoto *et al.*, 2006).

6.1.8 Beyond the QHA

At temperatures roughly below one-half to two-thirds of the melting temperature, QHA is quite accurate. Only at higher temperatures do its errors become significant. All the effects beyond the QHA are known as 'intrinsic anharmonicity'. For instance, phonon–phonon interactions, displacive phase transitions, and explicit temperature dependence of the vibrational frequencies (which is experimentally measurable) are intrinsic anharmonic phenomena. Here we focus on the role of intrinsic anharmonicity in thermodynamics and equations of state of solids, rather than on aspects related to thermal conductivity and phonon–phonon interactions.

This simplest way of treating intrinsic anharmonicity takes advantage of the fact that in the high-temperature expansion of the anharmonic free energy, the lowest-order term is quadratic (Landau and Lifshitz, 1980; Zharkov and Kalinin, 1971; Gillet *et al.*, 1999). Explicit molecular dynamics simulations for MgO (**Figure 4**) show that third- and fourth-order terms still play some role, but overall the T^2-term dominates. Limiting ourselves to this term, we write

$$\frac{F_{\text{anh}}(V, T)}{3Nk_B} = \frac{1}{2}aT^2 \qquad [70]$$

where a is intrinsic anharmonicity parameter, usually of order $10^{-5}\,\text{K}^{-1}$. Equation [70] assumes that intrinsic anharmonic contributions from different modes are additive. This is clearly a simplification, but it finds some justification in the arguments of Wallace (1998). Intrinsic anharmonicity normally decreases with pressure, which can be accounted for by a simple volume dependence (Zharkov and Kalinin, 1971):

$$a = a_0 \left(\frac{V}{V_0}\right)^m \qquad [71]$$

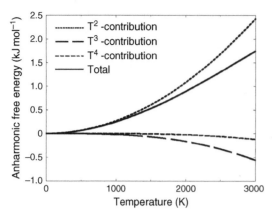

Figure 4 Intrinsic anharmonic free energy of MgO. From Oganov AR and Dorogokupets PI (2004) Intrinsic anharmonicity in thermodynamics and equations of state of solids. *Journal of Physics – Condensed Matter* 16: 1351–1360.

where a_0 is the intrinsic anharmonicity parameter at standard conditions, and $m = (\mathrm{d}\ln a/\mathrm{d}\ln V)$ is a constant.

One can easily find other anharmonic thermodynamic properties, such as the entropy, energy, isochoric heat capacity, thermal pressure, and bulk modulus:

$$\frac{S_{\text{anh}}}{3Nk_{\text{B}}} = -aT, \quad \frac{E_{\text{anh}}}{3Nk_{\text{B}}} = -\frac{1}{2}aT^2$$

$$\frac{C_{V\text{anh}}}{3Nk_{\text{B}}} = -aT, \quad \frac{P_{\text{anh}}}{3Nk_{\text{B}}} = -\frac{1}{2}a\frac{m}{V}T^2 \qquad [72]$$

$$K_{Ta} = P_a(1-m)$$

This model works well at high temperatures. However, at low temperatures there are problems: linear anharmonic heat capacity [72] overwhelms the harmonic term, leading to large errors in the thermal expansion coefficient below ~100 K. The problem is that [70] and [72] are classical equations and completely ignore quantum vibrational effects, which determine low-temperature thermodynamics.

Wallace (1998) has shown that in the first approximation intrinsic anharmonic effects can be incorporated by using the true (i.e., temperature-dependent) vibrational frequencies ω (or characteristic temperatures $\theta = \hbar\omega/k_{\text{B}}$) and substituting them into the (quasi)harmonic expression for the entropy for a harmonic oscillator [45]. The result will contain both quasiharmonic and intrinsic anharmonic contributions. We follow Gillet *et al.* (1999) and define the temperature-dependent characteristic temperature as

$$\Theta_{VT} = \theta \exp(aT) \qquad [73]$$

where θ is the quasiharmonic (only volume-dependent) characteristic temperature. Equation [73] thus defines the physical meaning of this parameter as the logarithmic derivative of the vibrational frequency (or characteristic temperature) with respect to volume:

$$a = \left(\frac{\partial \ln \omega_{VT}}{\partial T}\right)_V = \left(\frac{\partial \ln \Theta_{VT}}{\partial T}\right)_V \qquad [74]$$

In the classical limit $(\Theta_{VT}/T \to 0)$ eqns [70] and [72] are easily derived from [74].

Another approach to include quantum corrections in anharmonic properties is offered by thermodynamic perturbation theory of an anharmonic oscillator (see Oganov and Dorogokupets (2004)). Consider a general anharmonic potential

$$U_1 = \frac{1}{2}kx^2 + a_3x^3 + a_4x^4 + \cdots \qquad [75]$$

with $k > 0$.

As a reference system we take a harmonic oscillator

$$U_0 = \frac{1}{2}kx^2 \qquad [76]$$

Using first-order thermodynamic perturbation theory [69] anharmonic free energy can be calculated as follows:

$$\begin{aligned} F_{\text{anh}} &= \langle U - U_0 \rangle_0 = \langle a_3x^3 + a_4x^4 + \cdots \rangle_0 \\ &= a_4\langle x^4 \rangle_0 + a_6\langle x^6 \rangle_0 + a_8\langle x^8 \rangle_0 \\ &\quad + \cdots \end{aligned} \qquad [77]$$

This expression is remarkable in that the moments of atomic displacements used are those of a harmonic oscillator, and can be easily calculated. Since the harmonic reference potential is symmetric, only even-order terms are retained in [77]. Truncating at the $\langle x^4 \rangle_0$ term, Oganov and Dorogokupets (2004) found

$$\frac{F_{\text{anh}}}{3n} = \frac{a}{6k_{\text{B}}}[\langle E \rangle^2 + 2k_{\text{B}}C_VT^2] \qquad [78]$$

Other thermodynamic functions are easy to derive from F_{anh} by differentiation. From [78], one trivially obtains anharmonic zero-point energy:

$$\frac{E_{\text{anh}}^{z.p.}}{3n} = \frac{a}{24}k_{\text{B}}\theta^2 \qquad [79]$$

For typical values of parameters ($a = 2 \times 10^{-5}$ K^{-1}, $\theta = 1000$ K), this value amounts to only 0.17% of the harmonic zero-point energy. For more details on this formalism, see Oganov and Dorogokupets (2004).

Computationally, intrinsic anharmonic effects can be fully accounted for by the use of Monte Carlo or molecular dynamics simulations (Allen and Tildesley, 1987): these methods involve a full sampling of the potential hypersurface without any assumptions regarding its shape or the magnitude of atomic vibrations; these methods are also applicable to liquids and gases. Free energies of significantly anharmonic systems can be calculated using thermodynamic integration technique (e.g., Allen and Tildesley, 1987). For example, using this technique Alfè *et al.* (1999) calculated the melting curve of Fe at conditions of the Earth's core and provided first-order estimates of core temperatures (more accurate estimates were later obtained taking into account the effects of alloying elements, see Alfè *et al.* (2002)).

6.2 Equations of State and Elasticity

Equations of state (EOSs) (i.e., the *P–V–T* relationships) of Earth-forming minerals are of special interest – indeed, accurate EOSs of minerals are necessary for the interpretation of seismological observations. The importance of the elastic constants for Earth sciences springs from the fact that most of the information about the deep Earth is obtained seismologically, by measuring the velocities of seismic waves passing through the Earth. Seismic wave velocities, in turn, are related to the elastic constants of Earth-forming rocks and minerals. Acoustic anisotropy of the Earth, measurable seismologically, is related to the elastic anisotropy of Earth-forming minerals and the degree of their alignment.

6.2.1 Equations of State

Generally, thermodynamics gives

$$P = -\left(\frac{\partial F}{\partial V}\right)_T \quad \text{and} \quad V = \left(\frac{\partial G}{\partial P}\right)_T \quad \text{(Isothermal EOS)}$$

$$T = \left(\frac{\partial H}{\partial S}\right)_P \quad \text{and} \quad S = -\left(\frac{\partial G}{\partial T}\right)_P \quad \text{(Isobaric EOS)}$$

$$P = -\left(\frac{\partial E}{\partial V}\right)_S \quad \text{and} \quad V = -\left(\frac{\partial H}{\partial P}\right)_S \quad \text{(Adiabatic EOS)}$$

An explicit analytical EOS can only be written for an ideal gas (where interatomic interactions are absent; in the case, there are no problems in the analytical representation of the interatomic potential, and entropy can be easily and exactly calculated using the Sackur–Tetrode relation). For solids and liquids

interatomic interactions are essential, and all existing analytical EOSs are by necessity approximate. Even worse, interactions between atoms make phase transitions possible, and EOS becomes discontinuous (i.e., nonanalytical) at phase transitions. All the approximate EOS formulations are valid only for one phase (though for a phase transition involving only small structural changes it is possible to formulate a single EOS describing two or more phases – see, e.g., Tröster *et al.* (2002)), and generally the accuracy of the EOS is best at conditions far from phase transitions.

6.2.1.1 Mie–Grüneisen EOS

To advance further, consider the isothermal EOS $P = -(\partial F/\partial V)_T$, taking the QHA as the starting point. Using indices i and $\mathbf{k}$ to denote the number of the phonon branch and the wave vector $\mathbf{k}$, we can write a formula analogous to [50]:

$$F(T) = E_0 + \frac{1}{2}\sum_{i,\mathbf{k}} \hbar\omega_{i\mathbf{k}} + k_B T \sum_{i,\mathbf{k}} \ln\left[1 - \exp\left(-\frac{\hbar\omega_{i\mathbf{k}}}{k_B T}\right)\right] \quad [80]$$

From this, we have

$$P(V, T) = P_{st}(V) + \frac{1}{2}\sum_{i,\mathbf{k}} \hbar\frac{\gamma_{i\mathbf{k}}\omega_{i\mathbf{k}}}{V} + \sum_{i,\mathbf{k}} \frac{\gamma_{i\mathbf{k}}}{V}\frac{\hbar\omega_{i\mathbf{k}}}{\exp(\hbar\omega_{i\mathbf{k}}/k_B T) - 1} \quad [81]$$

where $P_{st}(V)$ is the static pressure, and the mode Grüneisen parameter $\gamma_{i\mathbf{k}}$ is defined as

$$\gamma_{i\mathbf{k}} = -\left(\frac{\partial \ln \omega}{\partial \ln V}\right)_T \quad [82]$$

In the QHA, the Grüneisen parameter is temperature independent.
At high temperatures or when all $\gamma_{i\mathbf{k}}$ are equal, [81] can be simplified:

$$P(V, T) = P_{st}(V) + \gamma\frac{E_{vib}(V, T)}{V} \quad [83]$$

where

$$\gamma = \langle\gamma_{i\mathbf{k}}\rangle \quad [84]$$

Equation [83] is the famous Mie–Grüneisen thermal EOS. It should be noted that in the classical approximation, which is put in the basis of the standard molecular dynamics and Monte Carlo simulations, the thermodynamic Grüneisen parameter will always be close to $<\gamma_{ik}>$ (Welch *et al.*, 1978), but it will also include a temperature-dependent correction due to intrinsic anharmonic effects.

As shown by Holzapfel (2001), the three common definitions of the Grüneisen parameter (via the thermal pressure, thermal expansion, and volume derivatives of the phonon frequencies)

$$\gamma_P(V, T) = \frac{P_{\text{vib}}}{E_{\text{vib}}} V \qquad [85a]$$

$$\gamma_\alpha(V, T) = \alpha \frac{K_T V}{C_V} \qquad [85b]$$

$$\gamma_{\text{qh}}(V) = \left\langle \frac{-\text{d} \ln \omega_i}{\text{d} \ln V} \right\rangle \qquad [85c]$$

are all identical for a classical quasiharmonic solid, and all different for a system with intrinsic anharmonicity. Very roughly, $\gamma_P(V, T)$ is halfway between $\gamma_{\text{qh}}(V)$ and $\gamma_\alpha(V, T)$, that is, anharmonic effects are much pronounced in thermal expansion than in thermal pressure. We stress that care must be taken as to which definition of the Grüneisen parameter is used when analyzing experimental and theoretical results. **Figure 5** shows the different definitions of the Grüneisen parameter and that the differences are small at low temperatures, but significantly increase with temperature; also shown is the volume dependence of the parameter q:

$$q = \left(\frac{\partial \ln \gamma}{\partial \ln V} \right)_T \qquad [86]$$

Often, the volume dependence of γ is described by a power law:

$$\gamma(V) = \gamma_0 \left(\frac{V}{V_0} \right)^q \qquad [87]$$

where parameter q is usually assumed to be constant. However, this form becomes poor at high compression. A much better function was proposed by Al'tshuler *et al.* (1987) (see also Vorobev (1996)):

$$\gamma = \gamma_\infty + (\gamma_0 - \gamma_\infty) \left(\frac{V}{V_0} \right)^\beta \qquad [88]$$

where γ_0 and γ_∞ are Grüneisen parameters at $V = V_0$ and at infinite compression ($V = 0$), respectively.

6.2.1.2 Analytical static EOS

Good discussions of this issue can be found in many sources, including Holzapfel (1996, 2001), Sutton (1993), Hama and Suito (1996), Cohen *et al.* (2000), Poirier (2000), and Vinet *et al.* (1986, 1989). Over the decades, many different EOS forms have been generated, but here we discuss only the ones that are most interesting from the theoretical and practical points of view.

The simplest approach is based on elasticity theory. Assuming that the bulk modulus K varies linearly with pressure and denoting $K'_0 = (\partial K / \partial P)_{P=0}$, we obtain the Murnaghan EOS:

$$P = \frac{K_0}{K'_0} \left[\left(\frac{V}{V_0} \right)^{-K'_0} - 1 \right] \qquad [89]$$

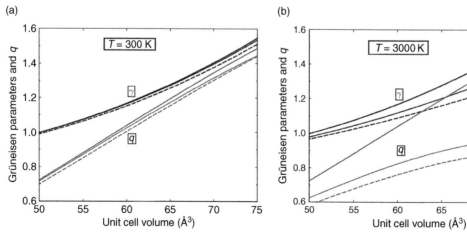

Figure 5 Grüneisen parameters and q parameters of MgO as a function of volume. (a) At 300 K, (b) at 3000 K. Soild lines indicate quasiharmonic results, dotted lines (middle) indicate γ_P, dashed lines indicate γ_α. From Oganov AR and Dorogokupets PI (2003) All-electron and pseudopotential study of MgO: Equation of state, anharmonicity, and stability. *Physical Review* B67 (uppermost for γ and q) (art. 224110).

This simple EOS works reasonably well only in a very limited compression range. A better approach (in terms of the accuracy relative to the number of parameters of the mathematical formulation) is provided by the effective potential methods, where an approximate model for the energy as a function of $x = V/V_0$, or some other measure of strain, is used.

For example, starting from a polynomial

$$E = E_0 + af^2 + bf^3 + cf^4 + \cdots \qquad [90]$$

in terms of the Eulerian strain f_E: $f_E = (1/2)[x^{-2/3} - 1]$, one arrives at the family of Birch–Murnaghan EOSs. (It is advantageous to use the Eulerian finite strain rather than the Lagrangian strain $f_L = (1/2)[1 - x^{2/3}]$, because the Eulerian strain leads to a better description of the correct $E(V)$ dependence with fewer terms in the expansion [90]. At infinite pressure, Eulerian strain is infinite, whereas Lagrangian strain remains finite and will require an infinite-order expansion. However, for infinitesimal strains both definitions become equivalent.) The often used third-order Birch–Murnaghan EOS is

$$P = \frac{3}{2}K_0\left[x^{-7/3} - x^{-5/3}\right]\left\{1 + \xi\left[x^{-2/3} - 1\right]\right\} \qquad [91]$$

$$E = E_0 + \frac{3}{2}K_0V_0\left[\frac{3}{2}(\xi - 1)x^{-2/3} + \frac{3}{4}(1 - 2\xi)x^{-4/3} \right.$$
$$\left. + \frac{1}{2}\xi x^{-6/3} - \frac{2\xi - 3}{4}\right] \qquad [92]$$

where $\xi = (3/4)(K_0' - 4)$.

It is possible to derive systematically higher-order BM EOSs, but this appears to be of little use since the number of parameters involved becomes too large; only the fourth-order BM EOS

$$P = 3K_0 f_E (1 + 2f_E)^{5/2}\left\{1 + \frac{3}{2}(K_0' - 4)f_E \right.$$
$$\left. + \frac{3}{2}\left[K_0 K_0'' + (K_0' - 4)(K_0' - 3) + \frac{35}{9}\right]f_E^2\right\} \qquad [93]$$

is sometimes used when ultrahigh pressures are studied.

The Vinet EOS (Vinet et al., 1986, 1989) is sometimes considered as one of the most impressive recent achievements in solid-state physics (Sutton, 1993). In fact, this is a whole family of EOSs of different orders. The most remarkable feature is its very fast convergence with respect to the order of EOS – one seldom needs to use beyond the third-order Vinet EOS.

This EOS is based on a universal scaled binding energy curve

$$E = E_0(1 + a)\exp(-a) \qquad [94]$$

where E_0 is the bond energy at equilibrium, $a = (R - R_0)/l$, $l = \sqrt{E_0/(\partial^2 E/\partial R^2)}$ being a scaling length roughly measuring the width of the potential well, and R the Wigner–Seitz radius (the average radius of a sphere in the solid containing one atom). The potential [94] was invented and first used by Rydberg (1932) for fitting potential curves of molecules and obtaining their anharmonic coefficients; it turned out (Vinet et al., 1986) that it describes very accurately systems with different types of chemical bonding in solids, molecules, adsorbates, etc.

The third-order Vinet EOS is (Vinet et al., 1989; Hama and Suito, 1996)

$$P = 3K_0\frac{1 - x^{1/3}}{x^{2/3}}\exp[\eta(1 - x^{1/3})] \qquad [95]$$

$$E(V) = E(V_0) + \frac{9K_0V_0}{\eta^2}\left\{1 - [1 - \eta(1 - x^{1/3})]\right.$$
$$\left. \times \exp[\eta(1 - x^{1/3})]\right\} \qquad [96]$$

where $\eta = (3/2)(K_0' - 1)$. The resulting expression for the bulk modulus is

$$K = \frac{K_0}{x^{2/3}}[1 + (1 + \eta x^{1/3})(1 - x^{1/3})]\exp[\eta(1 - x^{1/3})] \qquad [97]$$

From this one has (Vinet et al., 1989)

$$K_0'' = -\frac{1}{K_0}\left[\left(\frac{K_0'}{2}\right)^2 + \left(\frac{K_0'}{2}\right) - \frac{19}{36}\right] \qquad [98]$$

The Vinet EOS proved to be very accurate for fitting EOS of solid hydrogen (Loubeyre et al., 1996; Cohen et al., 2000) throughout the whole experimentally studied pressure range 0–120 GPa, roughly to the eightfold compression.

In very rare cases a higher-order Vinet EOS may be needed; such higher-order versions of the Vinet EOS already exist (Vinet et al., 1989):

$$P = \frac{3K_0}{x^{2/3}}(1 - x^{1/3})\exp[\eta(1 - x^{1/3}) + \beta(1 - x^{1/3})^2 $$
$$+ \gamma(1 - x^{1/3})^3 + \cdots] \qquad [99]$$

and the fourth-order Vinet EOS, where $\beta = (1/24)(36K_0K'' + 9K_0'^2 + 18K_0' - 19)$ and $\gamma = 0$, has been successfully applied to solid H_2 at extreme compressions (Cohen et al., 2000) and has led to significant improvements of the description of experimental PV-data.

In the limit of extreme compressions ($x \rightarrow 0$) the Vinet EOS fails to reproduce the correct free-electron limit and gives a finite (rather than positive infinite) energy equal to $(9K_0V_0/\eta^2)[1-(1-\eta)\exp(\eta)]$ (we do not consider here nuclear forces, which become important at densities $\sim 10^{15}\,\mathrm{g\,cm^{-3}}$ ($P \sim 10^{20}\,\mathrm{GPa}$ corresponding to $x < 10^{-12}$ (Holzapfel, 2001)). EOSs, manifesting the correct Thomas–Fermi behavior at extreme compressions, have been developed and discussed in detail by Holzapfel (1996, 2001) and Hama and Suito (1996).

Holzapfel (1996, 2001) has modified the Vinet EOS so as to make it satisfy the electron-gas limit at extreme compressions. His APL EOS (also a family of Lth-order EOSs) is as follows (Holzapfel, 2001):

$$P = \frac{3K_0}{x^{-5/3}}\left(1-x^{1/3}\right)\exp[c_0(1-x^{1/3})]$$
$$\times \left\{1 + x^{1/3}\sum_{k=2}^{L}c_k(1-x^{1/3})^{k-1}\right\} \qquad [100]$$

where $c_0 = -\ln(3K_0/P_{FG0})$, $P_{FG0} = a_{FG}(Z/V_0)^{5/3}$, $a_{FG}=0.02337\,\mathrm{GPa\,\AA^5}$, and Z the total number of electrons per volume V_0.

This EOS correctly predicts that at infinite compression $K'_\infty = 5/3$ (while at $x=1$ $K'_0 = 3 + (3/2)(c_0 + c_2)$), but becomes very similar to the Vinet EOS at moderate compressions. The mathematical similarity between [99] and [100] is obvious, and it is easy to generalize these EOSs into one family. For a third-order generalized Vinet–Holzapfel EOS, one has (Kunc *et al.*, 2003)

$$P = \frac{3K_0}{x^{n/3}}\left(1-x^{1/3}\right)\exp[\eta(1-x^{1/3})] \qquad [101]$$

where $\eta = (3K'_0/2) + (1/2) - n$. The Vinet EOS is recovered when $n=2$, and the Holzapfel EOS is obtained when $n=5$. Kunc *et al.* (2003) found that theoretical EOS of diamond is best represented by the EOS [101] with an intermediate value $n = 7/2$. In this case, the energy can be expressed analytically:

$$E(V) = E(V_0) + 9K_0V_0[f(V)-f(V_0)]\frac{\exp(\eta)}{\sqrt{\eta}} \qquad [102]$$

where

$$f(V) = \sqrt{\pi}(2\eta+1)\mathrm{erf}(\sqrt{\eta}x^{2/3}) + \frac{\left(2\sqrt{\eta}\exp(-\eta x^{1/3})\right)}{x^{2/3}}$$

However, it remains to be seen how accurate [102] is for other materials.

6.2.1.3 Anharmonicity in static EOS

Since both K' and γ come from anharmonic interactions, an intriguing possibility arises to establish a general relation between these parameters. This possibility has been widely discussed since 1939, when J. Slater suggested the first solution of the problem:

$$\gamma_s = \frac{1}{2}K' - \frac{1}{6} \qquad [103]$$

Later approaches resulted in very similar equations, the difference being in the value of the constant subtracted from $(1/2)K':1/2, 5/6$, or 0.95. If any of the relations of the type [103] were accurate, it would greatly simplify the construction of thermal EOS. Although some linear correlation between γ and K' does exist, the correlation is too poor to be useful (Wallace, 1998; Vočadlo *et al.*, 2000).

6.2.1.4 EOS, internal strain, and phase transitions

All the EOSs discussed in the previous section implicitly assumed that crystal structures compress uniformly, and there is no relaxation of the unit cell shape or of the atomic positions. For some solids (e.g., MgO) this is definitely true. For most crystals and all glasses, however, this is an approximation, sometimes crude. Classical EOSs are less successful for crystals with internal degrees of freedom and perform particularly poorly in the vicinity of phase transitions. In the simplest harmonic model, Oganov (2002) obtained the following formula:

$$P(V) = P_{\mathrm{unrelaxed}}(V) + \sum_i m_i^2(V - V_0) \qquad [104]$$

with parameters $\mathbf{m}_i$. $P_{\mathrm{unrelaxed}}$ is well described by the conventional EOSs, for example, Vinet EOS, whereas the total EOS is not necessarily so (see line 2 in **Figure 6**). The bulk modulus is always lowered by the relaxation effects, in the simplest approximation [104]:

$$K(V) = K_{\mathrm{unrelaxed}}(V) - \sum_i m_i^2 V \qquad [105]$$

which implies the tendency of K' to be higher than the corresponding unrelaxed value:

$$K'(V) = K'_{\mathrm{unrelaxed}}(V) + \sum_i m_i^2\frac{V}{K} \qquad [106]$$

This simple model explains qualitatively correctly the real effects of internal strain. Complex structures are usually relatively 'soft' and usually have large K'_0

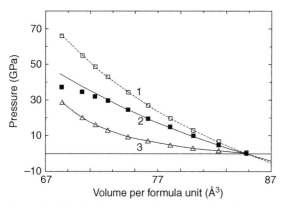

Figure 6 Effects of internal strains on equation of state. At the highest pressures shown, the structure is on the verge of an isosymmetric phase transition. 1, unrelaxed EOS; 2, correct EOS including relaxation; 3, the difference caused by relaxation. Note that in the pretransition region the full EOS is poorly fit, while the unrelaxed EOS is very well represented by analytical EOSs (in this case BM3).

(often significantly exceeding 'normal' $K_0' = 4$), in agreement with the prediction [106]. For example, quartz SiO_2, although consisting of extremely rigid SiO_2 tetrahedra, has a very low bulk modulus $K_0 = 37.12$ GPa and high $K_0' = 5.99$ (Angel *et al.*, 1997): its structure is very flexible due to relaxation of the internal degrees of freedom. Perhaps, the highest known $K_0' = 13$ was found in amphibole grunerite (Zhang *et al.*, 1992) with a very complicated structure having many degrees of freedom.

As an illustration, consider two series of *ab initio* calculations on sillimanite, Al_2SiO_5 (based on results from Oganov *et al.* (2001b)). In one series all structural parameters were optimized, while in the other series the zero-pressure structure was compressed homogeneously (i.e., without any relaxation). Results are shown in **Figure 6**, where a very large relaxation effect can be seen.

It is well known that internal strains always soften the elastic constants (e.g., Catti, 1989). In extreme cases, the softening can be complete, leading to a phase transition. In such cases, the simplified model [104] is not sufficient. To study EOS in the vicinity of the phase transition, one needs to go beyond the harmonic approximation built in this model. This can be done using the Landau expansion of the internal energy in powers of Q including the full elastic constants tensor and allowed couplings of the order parameter and lattice strains (see, e.g., Tröster *et al.* (2002)).

6.2.2 Elastic Constants

A number of excellent books and reviews exist, especially, Nye (1998), Sirotin and Shaskolskaya (1975), Wallace (1998), Alexandrov and Prodaivoda (1993), Born and Huang (1954), Belikov *et al.* (1970), Barron and Klein (1965), and Fedorov (1968). Elastic constants characterize the ability of a material to deform under small stresses. They can be described by a fourth-rank tensor C_{ijkl}, relating the second-rank stress tensor σ_{ij} to the (also second-rank) strain tensor e_{kl} via the generalized Hooke's law:

$$\sigma_{ij} = C_{ijkl} e_{kl} \qquad [107]$$

where multiplication follows the rules of tensor multiplication (see Nye, 1998). Equation [107] can be simplified using the Voigt notation (Nye, 1998), which represents the fourth-rank tensor C_{ijkl} by a symmetric 6×6 matrix C_{ij}. In these notations, indices '11', '22', '33', '12', '13', '23' are represented by only one symbol – 1, 2, 3, 6, 5, and 4, respectively. So we write instead of [107]

$$\sigma_i = C_{ij} e_j \qquad [108]$$

Note that infinitesimal strains are being used; in this limit, all definitions of strain (e.g., Eulerian, Lagrangian, Hencky, etc.) become equivalent. Under a small strain, each lattice vector $a_{ij}{'}$ of the strained crystal is obtained from the old lattice vector a_{ij}^0 and the strain tensor e_{ij} using the relation

$$a_{ij}{'} = (\delta_{ij} + e_{ij}) a_{ij}^0 \qquad [109]$$

In the original tensor notation and in the Voigt notation (Nye, 1998), the $(\delta_{ij} + e_{ij})$ matrix is represented as follows:

$$\begin{bmatrix} 1 + e_{11} & e_{12} & e_{13} \\ e_{12} & 1 + e_{22} & e_{23} \\ e_{13} & e_{23} & 1 + e_{33} \end{bmatrix} = \begin{bmatrix} 1 + e_1 & e_6/2 & e_5/2 \\ e_6/2 & 1 + e_2 & e_4/2 \\ e_5/2 & e_4/2 & 1 + e_3 \end{bmatrix}$$

$$[110]$$

Voigt notation is sufficient in most situations; only in rare situations such as a general transformation of the coordinate system, the full fourth-rank tensor representation must be used to derive the transformed elastic constants.

The number of components of a fourth-rank tensor is 81; the Voigt notation reduces this to 36. The thermodynamic equality $C_{ij} = C_{ji}$ makes the 6×6

matrix of elastic constants symmetric, reducing the number of independent constants to the well-known maximum number of 21, possessed by triclinic crystals. Crystal symmetry results in further reductions of this number: 13 for monoclinic, 9 for orthorhombic, 6 or 7 (depending on the point group symmetry) for trigonal and tetragonal, 5 for hexagonal, and 3 for cubic crystals; for isotropic (amorphous) solids there are only two independent elastic constants.

One can define the inverse tensor S_{ijkl} (or, in the Voigt notation, S_{ij}), often called the elastic compliance tensor:

$$\{S_{ijkl}\} = \{C_{ijkl}\}^{-1} \quad \text{or} \quad \{S_{ij}\} = \{C_{ij}\}^{-1} \quad [111]$$

(Note that in Voigt notation $C_{ijkl} = C_{mn}$, but $S_{ijkl} = S_{mn}$ only when m and $n = 1,2$, or 3; when either m or $n = 4,5$, or $6: 2S_{ijkl} = S_{mn}$; when both m and $n = 4,5$, or $6: 4S_{ijkl} = S_{mn}$ (Nye, 1998).) The S_{ij} tensor can be defined via the generalized Hooke's law in its equivalent formulation:

$$e_i = S_{ij}\sigma_j \quad [112]$$

Linear compressibilities can be easily derived from the S_{ij} tensor. Full expressions for an arbitrary direction can be found in Nye (1998); along the coordinate axes, the linear compressibilities are

$$\beta_x = -\frac{1}{l_x}\left(\frac{\partial l_x}{\partial P}\right)_T = \sum_{j=1}^{3} S_{1j} = S_{11} + S_{12} + S_{13}$$

$$\beta_y = -\frac{1}{l_y}\left(\frac{\partial l_y}{\partial P}\right)_T = \sum_{j=1}^{3} S_{2j} = S_{12} + S_{22} + S_{23} \quad [113]$$

$$\beta_z = -\frac{1}{l_z}\left(\frac{\partial l_z}{\partial P}\right)_T = \sum_{j=1}^{3} S_{3j} = S_{13} + S_{23} + S_{33}$$

where l_x, l_y, l_z are linear dimensions along the axes of the coordinate system. (These axes may not coincide with the lattice vectors for nonorthogonal crystal systems. Coordinate systems used in crystal physics are always orthogonal.) For the bulk compressibility, we have

$$\beta = -\frac{1}{V}\left(\frac{\partial V}{\partial P}\right)_T = \beta_x + \beta_y + \beta_z = \sum_{i=1}^{3}\sum_{j=1}^{3} S_{ij}$$
$$= S_{11} + S_{22} + S_{33} + 2(S_{12} + S_{13} + S_{23}) \quad [114]$$

The values of the elastic constants depend on the orientation of the coordinate system. There are two particularly important invariants of the elastic constants tensor – bulk modulus K and shear modulus G, obtained by special averaging of the individual elastic constants. There are several different schemes of such averaging. Reuss averaging is based on the assumption of a homogeneous stress throughout the crystal, leading to the Reuss bulk modulus:

$$K_R = \frac{1}{S_{11} + S_{22} + S_{33} + 2(S_{12} + S_{13} + S_{23})} = \frac{1}{\beta} \quad [115]$$

and shear modulus:

$$G_R = \frac{15}{4(S_{11} + S_{22} + S_{33}) - 4(S_{12} + S_{13} + S_{23}) + 3(S_{44} + S_{55} + S_{66})} \quad [116]$$

It is important to realize that it is the Reuss bulk modulus, explicitly related to compressibility, that is used in constructing EOSs and appears in all thermodynamic equations involving the bulk modulus.

Another popular scheme of averaging is due to Voigt. It is based on the assumption of a spatially homogeneous strain, and leads to the following expressions for the Voigt bulk and shear moduli:

$$K_V = \frac{C_{11} + C_{22} + C_{33} + 2(C_{12} + C_{13} + C_{23})}{9} \quad [117]$$

$$G_V = \frac{C_{11} + C_{22} + C_{33} - (C_{12} + C_{13} + C_{23}) + 3(C_{44} + C_{55} + C_{66})}{15} \quad [118]$$

For an isotropic polycrystalline aggregate the Voigt moduli give upper and the Reuss moduli lower bounds for the corresponding moduli. More accurate estimates can be obtained from Voigt–Reuss–Hill averages:

$$K_{VRH} = \frac{K_V + K_R}{2}, \qquad G_{VRH} = \frac{G_V + G_R}{2} \quad [119]$$

The most accurate results (and tighter bounds) are given by the Hashin–Shtrikman variational scheme, which is much more complicated, but leads to results similar to the Voigt–Reuss–Hill scheme (see Watt *et al.* (1976) for more details).

There are two groups of experimental methods for measuring the elastic constants: (1) static and low-frequency methods (based on determination of stress–strain relations for static stresses) and (2) high-frequency or dynamic methods (e.g., ultrasonic methods and Brillouin spectroscopy). High-frequency methods generally enable much higher accuracy. Static measurements yield isothermal elastic constants (the timescale of the experiment allows thermal equilibrium to be attained within the sample); high-frequency measurements give adiabatic constants (Belikov *et al.*, 1970). The difference, which is entirely due to anharmonic effects (see below), vanishes at 0 K. Adiabatic C_{ij} are larger, usually by a few percent. The following thermodynamic equation gives the difference in terms of thermal pressure tensor b_{ij} (Wallace, 1998):

$$C_{ijkl}^S = C_{ijkl}^T + \frac{TV}{C_V} b_{ij} b_{kl} \qquad [120]$$

where $b_{ij} = \left(\partial \sigma_{ij} / \partial T\right)_V$ is related to the thermal expansion tensor. Equation [120] implies, for the bulk moduli, the already-mentioned formula [33]:

$$K_S = K_T(1 + \alpha \gamma T) = K_T\left(1 + \frac{\alpha^2 K_T V}{C_V}\right)$$

where α and γ are the thermal expansion and Grüneisen parameter, respectively. Adiabatic and isothermal shear moduli are strictly equal for cubic crystals and usually practically indistinguishable for crystals of other symmetries.

Acoustic wave velocities measured in seismological experiments and ultrasonic determinations of elastic constants are related to the adiabatic elastic constants. Isothermal constants, on the other hand, are related to the compressibility and EOS.

The general equation for the calculation of velocities of acoustic waves with an arbitrary propagation direction, the Christoffel equation (Sirotin and Shaskolskaya, 1975), is

$$C_{ijkl}^S \mathbf{m}_j \mathbf{m}_k p_l = \rho v^2 p_i \qquad [121]$$

where ρ is the polarization vector of the wave (of unit length), $\mathbf{m}$ the unit vector parallel to the wave vector, and ρ the density of the crystal. It can also be represented in the form of a secular equation:

$$\det\|C_{ijkl}^S \mathbf{m}_j \mathbf{m}_k - \rho v^2 \delta_{il}\| = 0 \qquad [122]$$

This equation has three solutions, one of which corresponds to a longitudinal, and the other two to transverse waves (see, e.g., **Figure 7**). For example, one can obtain the following velocities for a cubic crystal along high-symmetry directions:

$$(a)\ \mathbf{m} = [100]:\ v_1 = \sqrt{\frac{C_{11}}{\rho}}\ (\mathbf{p} = [100])$$

$$v_2 = \sqrt{\frac{C_{44}}{\rho}}\ (\mathbf{p} = [010])$$

$$v_3 = \sqrt{\frac{C_{44}}{\rho}}\ (\mathbf{p} = [001])$$

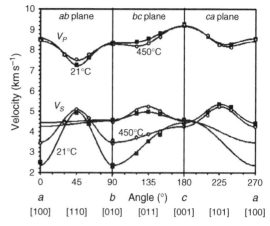

Figure 7 Acoustic velocities as a function of the propagation direction in lawsonite CaAl$_2$(Si$_2$O$_7$)(OH)$_2$*H$_2$O. Solid squares, at 21°C; open circles, 450°C. Reproduced from Schilling FR, Sinogeikin SV, and Bass JD (2003) Single-crystal elastic properties of lawsonite and their variation with temperature. *Physics of the Earth and Planetary Interiors* 136: 107–118, with permission from Elsevier.

$$(b)\ \mathbf{m} = [110]:\ v_1 = \sqrt{\frac{C_{11} + C_{12} + 2C_{44}}{2\rho}}\ (\mathbf{p} = [110])$$

$$v_2 = \sqrt{\frac{C_{44}}{\rho}}\ (\mathbf{p} = [001])$$

$$v_3 = \sqrt{\frac{C_{11} - C_{12}}{\rho}}\ (\mathbf{p} = [1\bar{1}0])$$

The average velocities are given by famous equations (Belikov *et al.*, 1970)

$$v_P = \sqrt{\frac{3K + 4G}{3\rho}} \qquad [123]$$

and

$$v_S = \sqrt{\frac{G}{\rho}} \qquad [124]$$

where the adiabatic Voigt–Reuss–Hill (or Hashin–Shtrikman) values are used for the bulk and shear moduli.

At constant P, T, the elastic constants describing stress–strain relations [107] are given by

$$C_{ijkl}^T = \frac{1}{V}\left(\frac{\partial^2 G}{\partial e_{ij} \partial e_{kl}}\right)_T \qquad [125]$$

while at constant P, S, they are

$$C_{ijkl}^S = \frac{1}{V}\left(\frac{\partial^2 H}{\partial e_{ij} \partial e_{kl}}\right)_S \qquad [126]$$

Now let us derive from [125] an expression for the elastic constants in terms of the second derivatives of the internal energy; in this derivation, we follow Ackland and Reed (2003). The unit cell of a crystal can be represented by a matrix $\overset{\leftrightarrow}{\mathbf{V}} = (\mathbf{a}_1, \mathbf{a}_2, \mathbf{a}_3)$, and the volume of the equilibrium unit cell is then $V_0 = \det \overset{\leftrightarrow}{\mathbf{V}}$. Using [109], for the volume V of a strained cell we obtain

$$\frac{V}{V_0} = \frac{\det \overset{\leftrightarrow}{\mathbf{V}}}{\det \overset{\leftrightarrow}{\mathbf{V}}} = 1 + e_1 + e_2 + e_3 + e_1 e_2 + e_2 e_3$$
$$+ e_1 e_3 - \frac{e_4^2}{4} - \frac{e_5^2}{4} - \frac{e_6^2}{4}$$
$$+ e_1 e_2 e_3 - \frac{e_1 e_4^2}{4} - \frac{e_2 e_5^2}{4}$$
$$- \frac{e_3 e_6^2}{4} + \frac{e_4 e_5 e_6}{4} \qquad [127]$$

Then one has in the standard tensor notation

$$\frac{\Delta V}{V_0} = e_{ii} + \frac{1}{4}\left(2\delta_{ij}\delta_{kl} - \delta_{ik}\delta_{jl} - \delta_{il}\delta_{jk}\right)e_{ij}e_{kl} + O(e^3) \qquad [128]$$

Then, the change of the Gibbs free energy associated with strain is, to the second order,

$$\Delta G = \Delta F + P e_{ii} + \frac{PV}{4}\left(2\delta_{ij}\delta_{kl} - \delta_{ik}\delta_{jl} - \delta_{il}\delta_{jk}\right)e_{ij}e_{kl} \qquad [129]$$

From this one has

$$C_{ijkl}^T = \frac{1}{V}\left(\frac{\partial^2 F}{\partial e_{ij}\partial e_{kl}}\right)_T + \frac{P}{2}\left(2\delta_{ij}\delta_{kl} - \delta_{il}\delta_{jk} - \delta_{jl}\delta_{ik}\right) \qquad [130a]$$

and, by analogy,

$$C_{ijkl}^S = \frac{1}{V}\left(\frac{\partial^2 F}{\partial e_{ij}\partial e_{kl}}\right)_S + \frac{P}{2}\left(2\delta_{ij}\delta_{kl} - \delta_{il}\delta_{jk} - \delta_{jl}\delta_{ik}\right) \qquad [130b]$$

It is well known (Barron and Klein, 1965; Wallace, 1998) that under nonzero stresses there can be several different definitions of elastic constants. The constants C_{ijkl}^T and C_{ijkl}^S defined by eqns [130a] and [130b] are those appearing in stress–strain relations and in the conditions of mechanical stability of crystals (see below), whereas the long-wavelength limit of lattice dynamics is controlled by

$$\frac{1}{V}\left(\frac{\partial^2 E}{\partial e_{ij}\partial e_{kl}}\right)_S$$

These two definitions (via stress–strain relations and from long-wavelength lattice dynamics) become identical at zero pressure.

Calculating the second derivatives with respect to the finite Lagrangian strains η_{ij}, different equations are obtained (Wallace, 1998) for the case of hydrostatic pressure:

$$C_{ijkl}^S = \frac{1}{V}\left(\frac{\partial^2 E}{\partial \eta_{ij}\partial \eta_{kl}}\right)_S + P\left(\delta_{ij}\delta_{kl} - \delta_{il}\delta_{jk} - \delta_{jl}\delta_{ik}\right) \qquad [131a]$$

$$C_{ijkl}^T = \frac{1}{V}\left(\frac{\partial^2 F}{\partial \eta_{ij}\partial \eta_{kl}}\right)_T + P\left(\delta_{ij}\delta_{kl} - \delta_{il}\delta_{jk} - \delta_{jl}\delta_{ik}\right) \qquad [131b]$$

For a general stress the analogous equations are

$$C_{ijkl}^S = \frac{1}{V}\left(\frac{\partial^2 E}{\partial \eta_{ij}\partial \eta_{kl}}\right)_S$$
$$- \frac{1}{2}\left(2\sigma_{ij}\delta_{kl} - \sigma_{ik}\delta_{jl} - \sigma_{il}\delta_{jk} - \sigma_{jl}\delta_{ik} - \sigma_{jk}\delta_{il}\right) \qquad [132a]$$

$$C_{ijkl}^T = \frac{1}{V}\left(\frac{\partial^2 F}{\partial \eta_{ij}\partial \eta_{kl}}\right)_T$$
$$- \frac{1}{2}\left(2\sigma_{ij}\delta_{kl} - \sigma_{ik}\delta_{jl} - \sigma_{il}\delta_{jk} - \sigma_{jl}\delta_{ik} - \sigma_{jk}\delta_{il}\right) \qquad [132b]$$

Cauchy relations, originally derived with the definition via the energy density, can be elegantly formulated in this definition as well (see below). Note, however, that the elastic constants C_{ijkl}, defined from stress–strain relations, have the full Voigt symmetry only at hydrostatic pressure. It is essential to distinguish between different definitions of elastic constants under pressure.

6.2.2.1   Cauchy relations

For crystals where all atoms occupy centrosymmetric positions, and where all interatomic interactions are central and pairwise (i.e., depend only on the distances between atoms, and not on angles), in the static limit Cauchy relations (Born and Huang, 1954; but take into account eqns [130a] and [130b]) hold:

$$C_{23} - C_{44} = 2P; \ C_{31} - C_{55} = 2P; \ C_{12} - C_{66} = 2P$$
$$C_{14} - C_{56} = 0; \ C_{25} - C_{64} = 0; \ C_{36} - C_{45} = 0 \qquad [133]$$

These relations would reduce the maximum number of independent elastic constants to 15; however, they never hold exactly because there are always noncentral and many-body contributions to crystal energy. Violations of the Cauchy relations can serve as a useful indicator of the importance of such interactions. While for many alkali halides Cauchy relations hold reasonably well, for alkali earth oxides (e.g., MgO) they are grossly violated. This is because the free O^{2-} ion is unstable and can exist only in the crystalline environment due to the stabilizing Madelung potential created by all atoms in the crystal; the charge density around O^{2-} is thus very susceptible to the changes of structure, including strains. Consequently, interactions of the O^{2-} ion with any other ion depend on the volume of the crystal and location of all other ions; this is a major source of many-body effects in ionic solids. This point of view is strongly supported by the success of potential induced breathing (PIB; see Bukowinski (1994) and references therein) and similar models in reproducing the observed Cauchy violations. In these models, the size of an O^{2-} ion (more precisely, the radius of the Watson sphere stabilizing the O^{2-}) depends on the classical electrostatic potential induced by other ions.

6.2.2.2 Mechanical stability

One of the most common types of instabilities occurring in crystals is the so-called mechanical instability, when some of the elastic constants (or their special combinations) become zero or negative. The condition of mechanical stability is the positive definiteness of the elastic constants matrix:

$$
\begin{array}{cccccc}
C_{11} & C_{12} & C_{13} & C_{14} & C_{15} & C_{16} \\
C_{21} & C_{22} & C_{23} & C_{24} & C_{25} & C_{26} \\
C_{31} & C_{32} & C_{33} & C_{34} & C_{35} & C_{36} \\
C_{41} & C_{42} & C_{43} & C_{44} & C_{45} & C_{46} \\
C_{51} & C_{52} & C_{53} & C_{54} & C_{55} & C_{56} \\
C_{61} & C_{62} & C_{63} & C_{64} & C_{65} & C_{66}
\end{array}
$$

This is equivalent to positiveness of all the principal minors of this matrix (principal minors are square submatrices symmetrical with respect to the main diagonal – they are indicated by dashed lines in the scheme above). All diagonal elastic constants C_{ii} are principal minors, and, therefore, must be positive for all stable crystals. Mechanical stability criteria were first suggested by Max Born (Born and Huang, 1954) and are sometimes called Born conditions. In general form, they are analyzed in detail in Sirotin and Shaskolskaya (1982) and Fedorov (1968), and cases of different symmetries have been thoroughly analyzed by Cowley (1976) and by Terhune *et al.* (1985). Mechanical stability criteria for crystals under stress must employ the C_{ij} derived from the stress–strain relations (Wang *et al.*, 1993, 1995; Karki, 1997). Violation of any of the mechanical stability conditions leads to softening of an acoustic mode in the vicinity of the Γ-point, inducing a phase transition.

6.2.2.3 Birch's law and effects of temperature on the elastic constants

The famous Birch's law (Birch, 1952, 1961; Poirier, 2000) states that compressional sound velocities depend only on the composition and density of the material:

$$v_P = a(\bar{M}) + b(\bar{M})\rho \qquad [134]$$

where $\bar{M}$ is the average atomic mass, a and b constants, ρ the density. Thus, for the mantle materials (average atomic mass between 20 and 22),

$$v_P = -1.87 + 3.05\rho \qquad [135]$$

Similar relations hold for the bulk sound velocity $v_\Phi = \sqrt{K/\rho}$; for mantle compositions

$$v_\Phi = -1.75 + 2.36\rho \qquad [136]$$

Birch's law implies that for a given material at constant volume, the elastic constants are temperature independent. This can be accepted only as a first (strictly harmonic) approximation. Thermal contributions to the bulk modulus can be represented as additive corrections to the zero-temperature result:

$$
\begin{aligned}
K^T(V, T) = K_{0\,\mathrm{K}}(V) &+ \Delta K^T_{\mathrm{qha}}(V, T) \\
&+ \Delta K^T_{\mathrm{a}}(V, T)
\end{aligned} \qquad [137]
$$

$$\Delta K^T_{\mathrm{qha}}(V, T) = p_{\mathrm{th,qha}}(1 + \gamma - q) - \gamma^2 T C_V / V \qquad [138]$$

$$\Delta K^T_{\mathrm{a}}(V, T) = p_{\mathrm{a}}(1 + \gamma_{\mathrm{a}} - q_{\mathrm{a}}) \qquad [139]$$

where

$$p_{\mathrm{a}} = \gamma_{\mathrm{a}} E_{\mathrm{a}}/V, \qquad \gamma_{\mathrm{a}} = -\left(\frac{\partial \ln a}{\partial \ln V}\right)_T, \qquad q_{\mathrm{a}} = \left(\frac{\partial \ln \gamma_{\mathrm{a}}}{\partial \ln V}\right)_T$$

For the adiabatic bulk modulus

$$
\begin{aligned}
K^S(V, T) = K_{0\,\mathrm{K}}(V) &+ p_{\mathrm{th,qha}}(1 + \gamma - q) \\
&+ p_{\mathrm{a}}(1 + \gamma_{\mathrm{a}} - q_{\mathrm{a}})
\end{aligned} \qquad [140]
$$

These results can be generalized for the individual elastic constants. Garber and Granato (1975), differentiating the free energy, expressed in the QHA as a sum of mode contributions over the whole Brillouin zone:

$$F = E_{\mathrm{st}} + \frac{1}{2}\sum_{i,\mathbf{k}} \hbar\omega_{i\mathbf{k}} + \sum_{i,\mathbf{k}} k_{\mathrm{B}} T \ln\left[1 - \exp\left(-\frac{\hbar\omega_{i\mathbf{k}}}{k_{\mathrm{B}} T}\right)\right]$$

and obtained the following result, which can be used in calculations of the elastic constants at finite temperatures:

$$\frac{1}{V}\left(\frac{\partial^2 F}{\partial \eta_{ij}\partial \eta_{kl}}\right)_V = \frac{1}{V}\left(\frac{\partial^2 E_{st}}{\partial \eta_{ij}\partial \eta_{kl}}\right)_v + \frac{1}{V}\sum_{i,k}$$
$$\times \left[\left(\gamma_{ij}^{ik}\gamma_{kl}^{ik} - \frac{\partial \gamma_{ij}}{\partial \eta_{kl}}\right)E_{vib,ik} - \gamma_{ij}^{ik}\gamma_{kl}^{ik}C_{V,ik}\,T\right]$$

$$[141]$$

6.2.2.4 Elastic anisotropy in the Earth's interior

While most of the lower mantle and the entire outer core are elastically isotropic, seismological studies have indicated seismic anisotropy amounting to a few percent in the upper mantle, lowermost mantle (D″ layer), and in the inner core. This anisotropy can be due to lattice-preferred orientation (e.g., appearing due to plastic flow orienting crystallites in a rock), or due to their reasons such as shape-preferred orientation or macroscopic-scale ordered arrangements of crystals of different minerals and/or molten rock. The most directly testable case is lattice-preferred orientation. Elastic anisotropy causes splitting of seismic waves — much akin to birefringence of light waves in anisotropic crystals. For an overview, see Anderson (1989).

One would expect that crystals will orient their easiest plastic slip planes parallel to the direction of the plastic flow (e.g., in convective streams). The selection of a single dominant slip plane is, of course, a simplification — which, however, leads to a most useful model of a transversely isotropic aggregate (where crystallites have parallel slip planes, but within the slip plane their orientations are random). For the case of a transversely isotropic aggregate with a small degree of anisotropy, Montagner and Nataf (1986) considered the following parameters (the unique axis of the transversely isotropic aggregates is set to be c-axis):

$$A = \frac{3}{8}(C_{11} + C_{22}) + \frac{1}{4}C_{12} + \frac{1}{2}C_{66}$$
$$C = C_{33}$$
$$F = \frac{1}{2}(C_{13} + C_{23})$$
$$L = \frac{1}{2}(C_{44} + C_{55})$$
$$N = \frac{1}{8}(C_{11} + C_{22}) - \frac{1}{4}C_{12} + \frac{1}{2}C_{66}$$

$$[142]$$

From these, they derived the velocities of the shear vertically (v_{SV}) and horizontally (v_{SH}), and compressional vertically (v_{PV}) and horizontally (v_{PH}) polarized waves:

$$v_{PH} = \sqrt{\frac{A}{\rho}}, \quad v_{PV} = \sqrt{\frac{C}{\rho}}$$
$$v_{SH} = \sqrt{\frac{N}{\rho}}, \quad v_{SV} = \sqrt{\frac{L}{\rho}}$$

$$[143]$$

What determines the dominant slip system? Strictly speaking, the dislocations — their number and the activation energy for their migration — should be the smallest for the best slip system. However, on the example of h.c.p.-metals, Legrand (1984) has demonstrated that a simplified criterion works very well. The product of the stacking fault enthalpy γ calculated per area S_{sf} ($\gamma = \Delta H_{sf}/S_{sf}$) and the shear elastic constant relevant for the motion of this stacking fault is smallest for the preferred slip plane. For example, comparing basal {0001} and prismatic {10$\bar{1}$0} slip for h.c.p.-metals, the ratio

$$R = \frac{\gamma_{0001}\,C_{44}}{\gamma_{10\bar{1}0}\,C_{66}}$$

$$[144]$$

is greater than 1 in cases of prismatic slip and smaller than 1 for materials with basal slip. This criterion was used by Poirier and Price (1999) in their study of the anisotropy of the inner core and, in an extended form, by Oganov et al. (2005b) in their revision of the nature of seismic anisotropy of the Earth's D″ layer (see also Section 6.4.2).

6.3 Phase Transitions of Crystals

The study of phase transitions is of central importance to modern crystallography, condensed matter physics, and chemistry. Phase transitions are a major factor determining the seismic structure of the Earth and thus play a special role in geophysics (e.g., Ringwood, 1991).

6.3.1 Classifications of Phase Transitions

A popular classification of phase transitions was proposed by Ehrenfest in 1933 (for a historical and scientific discussion, see Jaeger (1998)), distinguishing between first-, second-, and higher-order phase transitions. For the 'first-order' transitions the 'first' derivatives of the free energy with respect to P and T (i.e., volume and entropy) are discontinuous at the transition point; for 'second-order' transitions the 'second' derivatives (compressibility, heat capacity, and thermal expansion) are discontinuous, and so forth. In some cases, the order of the same phase transition is different at different P–T conditions: isosymmetric transitions must be first order, but become completely continuous (infinite-order) transitions at and above the critical temperature. Some transitions change under pressure/temperature from first to second order; the crossover point is called the tricritical point. Among the examples of systems with tricritical crossover are NH_4Cl (Garland and Weiner, 1971), zone-center cubic-tetragonal transition in $BaTiO_3$ perovskite, possibly the transition from calcite to metastable calcite (II) in $CaCO_3$ (see Hatch and Merrill (1981)) and, possibly the α–β transition in quartz (SiO_2). For example, the order–disorder transition in NH_4Cl from a phase with a complete orientational disordering of the NH_4-group $(Pm\bar{3}m)$ to an ordered phase $(P4\bar{3}m)$ is first order at 1 atm and 242 K, but becomes second order at the tricritical point, 0.15 GPa and 256 K. Therefore, the order of the transition is not something fundamentally inherent to the transition.

The first structural classification was due to Buerger (1961), who distinguished two main types of phase transitions – those with and without changes of the first coordination number, respectively. Each of these types was further classified into reconstructive (i.e., requiring formation/breaking of bonds), displacive, order–disorder, electronic, etc., transitions.

Even though Buerger's classification is purely structural, it naturally gives some insight into thermodynamics and kinetics of phase transitions. For instance, reconstructive transitions are first order and require activation (and, hence, are kinetically controlled) (Polymorphs of carbon (graphite, diamond) and Al_2SiO_5 (minerals kyanite, andalusite, and sillimanite, see Kerrick (1990)) are classical examples. All the transitions between these minerals are first-order reconstructive and require substantial activation energies to proceed; therefore, all the three minerals can coexist at not very high temperatures for millions of years in nature.) Also, as recognized by L. D. Landau in 1937 (see Landau and Lifshitz, 1980), for a second-order transition the two phases must be structurally related, and their symmetry groups must conform to certain group–subgroup relations.

6.3.2 First-Order Phase Transitions

Thermodynamics of first-order transitions are based on the Clausius–Clapeyron relation:

$$\frac{dP}{dT} = \frac{\Delta S}{\Delta V} \qquad [145]$$

where ΔS and ΔV are the entropy and volume differences, respectively, between the phases. Using [145] one can calculate the slopes of the equilibrium lines of phase coexistence. This relation is valid only for first-order transitions, because for second-order transitions both ΔV and ΔS are equal to zero. The transition temperatures and pressures can be found from accurate atomistic or quantum-mechanical total energy calculations (e.g., Alfè et al., 1999; Oganov et al., 2003, 2005a; Oganov and Ono, 2004, 2005; Umemoto et al., 2006). Only when the two phases are structurally similar can one apply approximate analytical theories, such as Landau theory (which was initially devised to study second-order phase transitions).

A relation, analogous to [145], for second-order transitions was derived by Ehrenfest:

$$\frac{dP}{dT} = \frac{\Delta C_P}{TV\Delta\alpha} \qquad [146]$$

where ΔC_P and $\Delta\alpha$ are the jumps of the heat capacity and thermal expansion at the transition. However, precise experiments, computer simulations, and accurate theories indicate a qualitatively different behavior of the heat capacity – instead of having a finite jump, it logarithmically diverges to infinity on both sides of the transition. This 'λ-behavior' invalidates the Ehrenfest relation.

6.3.3 Landau Theory of First- and Second-Order Transitions

When the structural changes occurring upon transition are small, it is usually possible to define an order parameter (or several order parameters), whose continuous change describes all the intermediate structures on the transition pathway. The simplest expression for the free energy is the Landau potential

$$G(Q) = G_0 + \frac{1}{2}A(T - T_C)Q^2$$
$$+ \frac{1}{3}BQ^3 + \frac{1}{4}CQ^4 + \cdots \qquad [147]$$

where T_C is the critical temperature, and G_0 the free energy of the phase with $Q=0$ (e.g., high-temperature high-symmetry disordered phase). Landau's assumption that the second term of [147] is simply proportional to $(T-T_C)$ was analyzed and justified mathematically by Sposito (1974). The entropy as a function of the order parameter is simply

$S(Q) = -\partial G(Q)/\partial T = S_0 - (1/2)AQ^2$. This dependence of the entropy on the order parameter is most appropriate for displacive phase transitions. (For order–disorder transitions, the entropy is more accurately expressed as $S(Q) = S_0 - R[(1+Q)\ln(1+Q) + (1-Q)$ $(Q) = S_0 - R[(1+Q)\ln(1+Q) + (1-Q)\ \ln(1-Q)].)$ The internal energy is then $E(Q) = E_0 - (1/2)AT_CQ^2 + (1/3)BQ^3 + (1/4)CQ^4 + \cdots$. In the case $A > 0, B > 0, C > 0$, this corresponds to a double-well potential $E(Q)$. (More than two minima can exist for higher-order polynomials [147].) For second-order transitions the odd-order terms in [147] must be zero, making the double well symmetric. (This is only one of the necessary conditions. Other necessary conditions were formulated by Birman (1966) using group theory.)

Consider a second-order transition

$$G(Q) = G_0 + \frac{1}{2}A(T - T_c)Q^2 + \frac{1}{4}CQ^4 + \cdots \quad [148]$$

One can observe that at the transition point ($T = T_C$, $Q = 0$) the second derivative of F with respect to Q changes sign, corresponding to freezing in of a soft mode below T_C and a corresponding structural distortion. For first-order transitions, complete mode softening does not occur at $T = T_C$.

Second-order phase transitions are always characterized by group–subgroup relations: the symmetry group of one ('ordered', usually low-temperature) phase is a subgroup of the symmetry group of the other ('disordered', usually high-temperature) phase. The two symmetrically equivalent minima then correspond to the same ordered phase, and can be considered as 'twin domains', related by a symmetry element present in the disordered phase, but absent in the ordered one (**Figure 8**).

The potential [147] is often complicated by the coupling of the order parameter to lattice strains. In such cases, the potential will be

$$G(Q) = \left(G_0 + \frac{1}{2}A(T - T_C)Q^2 + \frac{1}{3}BQ^3 + \frac{1}{4}CQ^4 + \cdots\right)$$
$$+ a_1Q\varepsilon + a_2Q\varepsilon^2 + a_3Q^2\varepsilon + \frac{1}{2}C\varepsilon^2 + \cdots \quad [149]$$

where a_1, a_2, a_3 are coupling coefficients, and C is an elastic constant. Coupling of the order parameter to strains can cause a first-order behavior even for a symmetric $E(Q)$.

In some cases, more than one order parameter is required to describe a phase transition. Then, for the

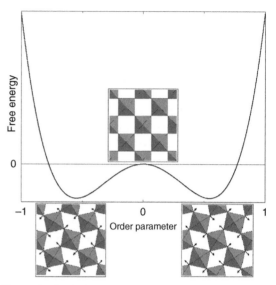

Figure 8 Symmetric Landau potential at $T < T_C$. The two distorted perovskite-type structures shown on the bottom are equivalent (they are mirror reflections of each other); arrows show the directions of octahedral rotations away from the cubic structure. The undistorted structure is shown in the center. At temperatures higher than T_C, the stable structure will be locally distorted, but on average will have the symmetry of the undistorted phase. From Oganov AR, Brodholt JP, and Price GD (2002) *Ab initio* theory of thermoelasticity and phase transitions in minerals. In: Gramaccioli CM (ed.) *EMU Notes in Mineralogy, Vol. 4: Energy Modeling in Minerals*, pp. 83–170. Bolin: Springer.

case of two order parameters, the Landau potential looks like

$$G(Q) = G(Q_1) + G(Q_2) + \xi_1Q_1Q_2$$
$$+ \xi_2Q_1{}^2Q_2 + \xi_3Q_1Q_2{}^2 + \cdots \quad [150]$$

where ξ_1, ξ_2, and ξ_3 are coupling coefficients for the Q_1–Q_2 coupling. In cases where odd-order terms of the kind $\xi Q_1Q_2Q_3$ are present, the transition must be first order. For a detailed general account of Landau theory, see Landau and Lifshitz (1980), Dove (1993, 1997), Carpenter *et al.* (1998), and Carpenter and Salje (1998, 2000).

6.3.4 Shortcomings of Landau Theory

Landau theory belongs to a class of approximate theories known as mean-field theories. Mean-field treatment is a common way of approximately solving complex physical problems in many areas of science. The main drawback of these methods is the neglect of short-range fluctuations (in Landau theory, the local structure and fluctuations of the order parameter are neglected). In other words, Landau theory assumes that all the neighboring unit cells have the same configuration; therefore, domain structures and fluctuations of the order parameter in space and time are not treated. This problem becomes severe in the vicinity of T_C (in the so-called Ginzburg interval). For second-order transitions Landau theory predicts $Q \sim (T_C - T)^{1/2}$, while experiments indicate $Q \sim (T_C - T)^{1/3}$. The critical exponent of $1/3$ has been confirmed many times by numerical computer simulations and could be explained only with the advent of renormalization group theory. (In fact, experiments give mean-field critical exponents far from T_C, but nearer T_C there is crossover from the mean-field to critical behavior, where the critical exponents depart significantly from mean-field predictions.) Landau theory cannot explain the logarithmic divergence of the heat capacity near the critical point – instead, it yields a finite jump. Finally, Landau theory does not consider quantum effects at low temperatures. As a consequence, it does not reproduce experimentally observed order parameter saturation at low temperatures; instead, it predicts a steady increase of the order parameter with decreasing temperature.

6.3.5 Ginzburg–Landau Theory

In 1950, V. L. Ginzburg and L. D. Landau (see Landau and Lifshitz (1980) and Bowley and Sánchez (1999)) considered the case of an order parameter slowly varying in space. This leads to the simplest theory beyond the mean field. The free energy becomes a 'functional' of the order parameter, and an additional term proportional to the square of the gradient of the order parameter appears:

$$F[Q(\mathbf{r})] = \int \left\{ f(Q(\mathbf{r})) + \frac{1}{2}\lambda[\nabla Q(\mathbf{r})]^2 \right\} \mathrm{d}r \quad [151]$$

with the stiffness parameter $\lambda > 0$. For example, for a second-order transition

$$F[Q(\mathbf{r})] = \int \left\{ \frac{1}{2}a(T - T_C)Q^2(\mathbf{r}) \right.$$
$$\left. + \frac{1}{4}bQ^4(\mathbf{r}) + \frac{1}{2}\lambda[\nabla Q(\mathbf{r})]^2 \right\} \mathrm{d}r \quad [152]$$

The order parameter is then expressed as a sum of a constant term (the average order parameter) and fluctuations, given by a Fourier series:

$$Q(\mathbf{r}) = \bar{Q} + \sum_{\mathbf{k}} Q_{\mathbf{k}} e^{i\mathbf{k}r} \quad [153]$$

Equation [152] can be rewritten as

$$F[Q(r)] = \int \left\{ f(\bar{Q}) + Q_1(\mathbf{r})f' \right.$$
$$\left. + \frac{1}{2}Q_1^2(\mathbf{r})f'' + \cdots + \frac{1}{2}\lambda[\nabla Q(\mathbf{r})]^2 \right\} \mathrm{d}r$$
$$= V\left\{ f(\bar{Q}) + \frac{1}{2}\sum_{\mathbf{k}}|Q_{\mathbf{k}}|^2(f'' + \lambda k^2) + \cdots \right\}$$
$$\quad [154]$$

Let us consider the case $f'' < 0$. In this case, the system is unstable against all fluctuations whose wave vectors satisfy $f'' + \lambda k^2 > 0$. Hence, the maximum unstable wave vector is $k_c = \sqrt{|f''|/\lambda}$. The correlation length ξ is

$$\xi = k_c^{-1} = \sqrt{\frac{\lambda}{|f''|}} \quad [155]$$

Ginzburg and Landau have proposed a criterion of the validity of Landau theory, defining the following value:

$$r(T) = \frac{f_m \xi^3}{k_B T} \quad [156]$$

where f_m is the difference of energies at the energy maximum and minimum. If $r(T) > 1$, fluctuations are not important, and Landau theory is valid. When $r(T) < 1$, fluctuations are essential and Landau theory is invalid; this occurs in the vicinity of T_C (in the temperature region called Ginzburg interval). Ginzburg intervals are usually quite narrow (of the order of ~ 10 K).

For second-order transitions, $f_m = a^2(T - T_C)/4b$ and

$$\xi = \sqrt{\frac{\lambda}{2a(T_C - T)}} \quad [157]$$

Ginzburg–Landau theory is still approximate and does not reproduce experimental critical exponents. Renormalization group theory overcomes all these difficulties and serves as the modern basis of theory of critical phenomena; it goes beyond the mean-field approximation and fully treats all possible fluctuations of the order parameter. Introductory texts on this theory can be found in Chandler (1987), Rao and Rao (1978), and Wilson (1983); the latter reference is the Nobel lecture of Kenneth Wilson, one of its main inventors. This theory has led to the prediction of new physical phenomena, for example, continuous lattice melting, experimentally found in Na_2CO_3 (Harris and Dove, 1995).

6.3.6 Ising Spin Model

This model is widely used to describe magnetic and atomic ordering processes in materials. In this model, a spin +1 or −1 is associated with each lattice site, depending on whether the magnetic moment on the site is 'up' or 'down', or whether the atom occupying the site is of the type 'A' or 'B'.

The total energy of the system is

$$U = U_0 - \mathcal{J} \sum_{i,j} S_i S_j - H \sum_i S_i \qquad [158]$$

where U_0 is the reference energy, and $\mathcal{J}$ the interaction parameter between the sites: if $\mathcal{J} < 0$, unlike spins prefer to group together, and there is a tendency to ordering at low temperatures; if $\mathcal{J} > 0$, unmixing will occur at low temperatures. Complete disorder, although unfavorable energetically, will be stabilized by the entropy at high temperatures. An external field H leads to a preferred orientation of the spins. The Ising model can be analytically solved only in one and two dimensions; for three dimensions it is solved numerically, usually by the Monte Carlo method. One-dimensional Ising model exhibits no phase transitions, and at all temperatures above 0 K yields the disordered state.

Ising-like models provide an interesting route for theoretical studies of polytypism and polysomatism (see, e.g., Price (1983), Price and Yeomans (1984) and references therein). The crucial observation is the mathematical similarity between polytypic sequences (e.g., **Figure 9**) and one-dimensional Ising models.

The Ising model is also very attractive for studies of ordering processes; for a review the reader is referred to Warren *et al.* (2001), and can be generalized for the case of more than two spins (see Yeomans, 1992) – such variants will be applicable to ordering in multicomponent solid solutions and polytypic (polysomatic) systems with more than two types of layers.

The conventional Ising models assume that spins can be only 'up' or 'down', and therefore these models cannot be applied to noncollinear magnetic materials. For these cases, various Heisenberg models are appropriate, which take into account the orientations of the spins. The simplest of these models is based on the following Hamiltonian:

$$U = U_0 - \mathcal{J} \sum_{i,j} \mathbf{S}_i \mathbf{S}_j - H \sum_i S_i^z \qquad [159]$$

involving spin vectors $\mathbf{S}_i$ and $\mathbf{S}_j$. By analogy with the one-dimensional Ising model, the Heisenberg model

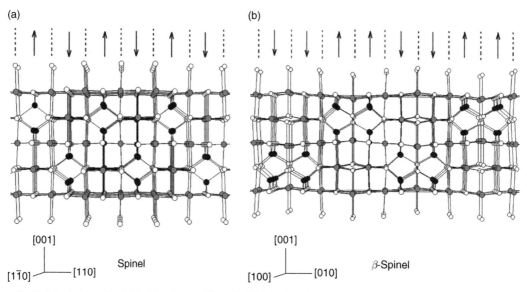

Figure 9 Polytypism in spinelloids. Structures of (a) spinel, (b) wadsleyite (β-spinel). Black circles are Si atoms (tetrahedrally coordinated), gray circles Mg atoms (octahedrally coordinated), and empty circles are O atoms. Layers of different orientations are shown by 'up' and 'down' arrows highlighting the similarity with the one-dimensional Ising spin lattice. Spinelloids are interesting for Earth sciences, because of the phases of Mg_2SiO_4 – ringwoodite (spinel-like phase) and wadsleyite (β-spinel phase), which are the major constituents of the transition zone of the Earth's mantle.

has no phase transitions for one- and two-dimensional systems.

6.3.7 Mean-Field Treatment of Order–Disorder Phenomena

The Bragg–Williams model is the simplest mean-field approach applicable to ordering phenomena. The free energy of the alloy as a function of temperature and order parameter is

$$G = G_0 - \frac{Nz}{4}\mathcal{J}Q^2 + Nk_B T[(1+Q)\ln(1+Q)$$
$$+ (1-Q)\ln(1-Q)] \qquad [160]$$

where G_0 is the free energy of the fully disordered state, N is the number of sites where disordering occurs, the order parameter $Q = X_{A,\alpha} - X_{A,\beta} = X_{B,\beta} - X_{B,\alpha} = 2X_{A,\alpha} - 1$, and the exchange energy $\mathcal{J} = E_{AA} + E_{BB} - 2E_{AB}$.

The expression [160] is analogous to the Landau potential [148] and yields the same critical exponents. In three dimensions, this model gives qualitatively reasonable results; however, even with accurate exchange energies $\mathcal{J}$, the predicted transition temperatures are usually a few times higher than the experimental ones (Redfern, 2000).

Drawbacks of the Bragg–Williams model can be corrected by explicitly considering short-range order. In the Bethe model (see Rao and Rao (1978)), apart from the long-range order parameter Q, one or more short-range order parameters are considered. These additional parameters describe the distribution of neighbors of both kinds in the nearest proximity of each atom. The resulting critical exponents and transition temperatures are much more realistic than mean-field predictions.

In the following, we discuss features of different types of phase transitions, classified by their symmetry. This gives a new viewpoint on the variety of phenomena associated with phase transitions in solids.

6.3.8 Isosymmetric Transitions

Using Landau theory, it is easy to show that isosymmetric transitions must be first order, but can disappear (i.e., become fully continuous, infinite-order transitions) above the critical temperature (Bruce and Cowley, 1981; Christy, 1995). There is a complete analogy here with the liquid–gas and liquid–liquid transitions (which are also isosymmetric). All liquid, gaseous, and conventional amorphous phases are isosymmetric, having spherical point-group symmetry. At supercritical temperatures there are generally rapid, but continuous changes in all properties along any P–T path going above the critical point (Angel, 1996).

Increasingly, many crystals are now known to exhibit isosymmetric phase transitions (i.e., those for which both phases have the same space group with the same number of atoms in the unit cell, with atoms occupying the same Wyckoff positions). Such transitions can be electronic (where the electronic structure changes, e.g., Ce and SmS), structural (where the coordination or ordering of the atomic species change discontinuously, e.g., $KTiOPO_4$), or intermediate (both electronic and structural changes are involved, e.g., Na_3MnF_6). Metallic Ce undergoes an isosymmetric phase transition Ce(I)–Ce(IV) (see Liu and Bassett (1986) and references therein), presumably due to 6s–4f (or 5d) electronic transition. Both Ce(I) and Ce(IV) have the f.c.c. structure (space group $Fm\bar{3}m$). The volume change at the transition is very large (13%) at room temperature, but it rapidly decreases along the Ce(I)–Ce(IV) equilibrium line until it disappears at the critical point (2.15 GPa and 613 K). Another famous example of an electronic transition is SmS, which transforms from the low-pressure insulating phase to the high-pressure metallic phase; both phases have an NaCl-type structure. **Figure 10** explains this transition.

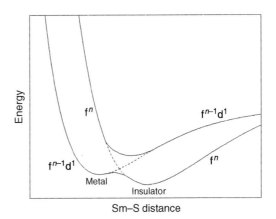

Figure 10 Illustration of the isosymmetric metal–insulator transition in SmS. Mixing of two configurations (metallic $f^{n-1}d$ and insulating f^n) produces a double-well energy curve for the ground state, where the minimum with a smaller interatomic distance corresponds to a metal. Compression triggers the insulator–metal transition. Adapted from Burdett JK (1995) *Chemical Bonding in Solids*, 319 pp. New York: Oxford University Press.

Structural isosymmetric transitions do not involve any drastic changes in the electronic structure, but are purely atomistic. $KTiOPO_4$ (KTP) is known to undergo a phase transition at 5.8 GPa with a volume decrease of 2.7% with preservation of space group $Pna2_1$ (Allan and Nelmes, 1996). Large cages, occupied by K, lose 12% of their volume upon transition. $KNO_3(II)$–$KNO_3(IV)$ phase transition, which occurs at 0.3 GPa and is accompanied by a volume decrease of 11.5%, does not alter the space group ($Pnma$) (Adams *et al.*, 1988), while for potassium atoms the coordination number changes from 9 to 11. An isosymmetric (space group $Pnam$) phase change has been observed at 9.8 GPa for PbF_2 (Haines *et al.*, 1998) and involves a change of the coordination number of Pb atoms from 9 to 10. Na_3MnF_6 (space group $P2_1/n$) is an example of a phase transition with a simultaneous change in the atomic and electronic structure. At 2.2 GPa this compound undergoes a first-order isosymmetric phase transformation, which is associated with a change of orientation of the Jahn–Teller elongation of MnF_6 octahedra (Carlson *et al.*, 1998).

6.3.9 Transitions with Group–Subgroup Relations

There are several possibilities here, stemming from different types of subgroups/supergroups of crystal symmetry. Examples are $\alpha \rightarrow \beta$ quartz, $P2_1/c$-$C2/c$ pyroxenes (see very interesting papers by Arlt and Angel (2000) and Arlt *et al.* (1998)), and $Pbnm \rightarrow Pm\bar{3}m$ transitions in perovskites.

The second type of transitions with group–subgroup transitions involve indirect symmetry relations between two phases via an intermediate archetypal phase of a higher symmetry, which is a supergroup for symmetries of both phases. An example is $BaTiO_3$, where the transition between the rhombohedral and tetragonal phases can be described with reference to the higher-symmetry cubic phase. Such transitions are usually weakly first order.

The third possibility involved a transition state of lower symmetry, which is a common subgroup of the symmetries of both phases. These transitions are usually strongly first order; often they can be described as reconstructive (see Christy (1993)). The f.c.c.→b.c.c. transition in Fe can be described with reference to lower-symmetry tetragonal or rhombohedral configurations, whose symmetries are common subgroups of both symmetry groups of the b.c.c. and f.c.c. phases.

6.3.10 Pressure-Induced Amorphization

This phenomenon, discovered in 1984 in experiments on compression of ice (Mishima *et al.*, 1984) to 1 GPa at 77 K, is still poorly understood. For detailed reviews, see excellent papers (Sharma and Sikka, 1996; Richet and Gillet, 1997).

A great number of crystals undergoing pressure-induced amorphization are known (e.g., Quartz SiO_2, coesite SiO_2, berlinite $AlPO_4$, GeO_2, zeolites scolecite $Ca_8Al_{16}Si_{24}O_{80}*24H_2O$ and mesolite $Na_{16}Ca_{16}Al_{48}Si_{72}O_{240}*64H_2O$, anorthite $CaAl_2Si_2O_8$, wollastonite $CaSiO_3$, enstatite $MgSiO_3$, muscovite $KAl_3Si_3O_{10}(OH)_2$, serpentine $Mg_3Si_2O_5(OH)_4$, portlandite $Ca(OH)_2$), as well as a few substances undergoing pressure-release amorphization, whereby high-pressure phases, when decompressed to pressures well below their stability fields, become dynamically unstable and amorphize. (This happens to the perovskite-type modification of $CaSiO_3$, one of the main minerals of the Earth's lower mantle, which at ambient conditions turns to a glass within a few hours.)

Pressure-induced amorphization is always a metastable first-order transition. It occurs in the limit of dynamical stability of the crystal. Behavior of pressure-induced amorphous phases on decompression can be very different: some compounds (e.g., $Ca(OH)_2$) recrystallize, others (e.g., SiO_2, ice) remain amorphous. Elastic anisotropy was found in pressure-amorphized quartz by Brillouin spectroscopy (McNeil and Grimsditch, 1991) and molecular dynamics simulations (Tse and Klug, 1993). The latter study found no structural relationships between pressure-amorphized quartz and silica glass.

The mechanisms driving pressure-induced amorphization are still not quite clear. The necessary conditions are (1) higher density of the amorphous phase relative to the crystal and (2) presence of soft modes in the crystalline phase. Softening of a vibrational mode at a single point of the Brillouin zone should drive a transition to a crystalline (if the soft wave vector is rational) or incommensurate (if the wave vector is irrational) phase. Simultaneous or nearly simultaneous softening of a phonon branch at a range of k-vectors could produce an amorphous phase (Keskar *et al.*, 1994; Binggeli *et al.*, 1994; Hemmati *et al.*, 1995). Any atomic displacement, expressible as a combination of soft modes, lowers the energy; the multitude of possible combinations gives rise to the disorder. However, a large degree of order should remain because the displacements are expected to be small and because only displacements related to the softening phonon branch are allowed to freeze in. Simultaneous softening of a phonon branch along a direction in the Brillouin zone implies weak dispersion of this branch, which is most naturally achieved when the unit cell is large. Indeed, crystals with complicated open structures and large unit cells are more prone to pressure-induced amorphization.

6.4 A Few Examples of the Discussed Concepts

Very briefly, we will discuss some recent results illustrating the use of the notions and theories discussed above. These include the calculation of the temperature profile of the Earth's lower mantle and core, polytypism of $MgSiO_3$ post-perovskite and seismic anisotropy of the Earth's D″ layer, and spin transition in (Mg, Fe)O magnesiowüstite.

6.4.1 Temperature Profile of the Earth's Lower Mantle and Core

Equation [26] can be rewritten, taking into account eqn [85b], as follows:

$$\left(\frac{\partial \ln T}{\partial \rho}\right)_S = \gamma_\alpha \qquad [161]$$

This formula describes adiabatic change of temperature upon compression and is relevant for first-order estimates of the average temperature distribution in convecting mantle (where superadiabatic effects might be non-negligible) and outer core (which is very closely adiabatic). *Ab initio* calculations of Alfè *et al.* (2002) produced an estimate of the temperature at the inner–outer core boundary (5150 km depth) of 5600 K. This was calculated from the melting curve of iron, taking into account the effect of impurities (Si, S, O). While within the solid inner core the temperature is likely to be constant, the temperature distribution in the liquid and rapidly convecting outer core is adiabatic [161]. With their estimates of the Grüneisen parameter of liquid iron at relevant pressures and temperatures, Alfè *et al.* (2002) calculated the temperature distribution in the outer core. In particular, the core temperature at the boundary with the mantle was estimated to be in the range 4000–4300 K.

Phase-equilibrium experiments of Ito and Katsura (1989) produced another 'anchor' point for the calculation of the geotherm – 1873 K at the depth of 670 km (top of the lower mantle). Taking into account the Grüneisen parameters of $MgSiO_3$ perovskite and MgO periclase obtained in their *ab initio* simulations. Oganov *et al.* (2002) have calculated the adiabatic geotherm of the lower mantle. The resulting mantle temperature at the boundary with the core (2891 km depth) is 2700 K, indicating a strong thermal boundary layer with large temperature variations at the bottom of the mantle. Lateral temperature variations in the lower mantle have been estimated (Oganov *et al.*, 2001a) by combining seismic tomography images and computed elastic constants of $MgSiO_3$ perovskite as a function of pressure and temperature (Oganov *et al.*, 2001a). These variations were found to increase from 800 K at the depth of 1000 km to ∼2000 K close to the bottom of the lower mantle.

6.4.2 Polytypism of MgSiO₃ Post-Perovskite and Anisotropy of the Earth's D″ layer

The original findings of the post-perovskite phase of $MgSiO_3$ (Murakami *et al.*, 2004; Oganov and Ono, 2004) came as a big surprise. The unusual crystal structure of post-perovskite and its elastic properties naturally explained most of the anomalies of the D″ layer – the D″ discontinuity and its variable depth, the anticorrelation of shear and bulk sound velocities and seismic anisotropy of the D″ layer (see Oganov and Ono (2004), Murakami *et al.* (2004), and Oganov *et al.* (2005b)).

Recently, Oganov *et al.* (2005b) found that $MgSiO_3$ perovskite and post-perovskite can be considered as end members of an infinite polytypic series (**Figure 11**) – this is a case of nontraditional polytypism (for another illustration, see **Figure 9**), since the 'layers', whose shifting produces all the structures in the polytypic series, are not weakly bound and are not even immediately obvious in the structure. All these structures are energetically very similar, and since intermediate structures have only marginally higher enthalpies than perovskite or post-perovskite, these phases could be stabilized by temperature and/or impurities in the Earth's lowermost mantle. This polytypism with low-energy stacking faults has interesting implications for plasticity of $MgSiO_3$ post-perovskite and for seismic anisotropy of the D″ layer.

Initially, {010} slip planes parallel to the silicate sheets of the post-perovskite structure were expected

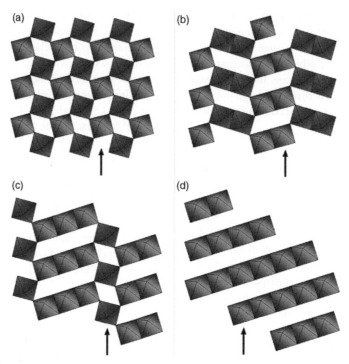

Figure 11 MgSiO$_3$ polytypes: (a) Perovskite (space group *Pbnm*); (b) and (c) intermediate structures 2 × 2 (*Pbnm*) and 3 × 1 (*P2$_1$/m*), respectively and (d) post-perovskite (*Cmcm*). Only silicate octahedra are shown; Mg atoms are omitted for clarity. Arrows indicate the predicated slip planes in these structures. From Oganov AR, Martoňák R, Laio A, Raiteri P, and Parrinello M (2005b) Anisotropy of Earth's D″ layer and stacking faults in the MgSiO$_3$ post-perovskite phase. *Nature* 438: 1142–1144.

to be dominant in post-perovskite. However, *ab initio* simulations (Oganov *et al.*, 2005b) found that the {110} slip planes are much more favorable. In particular, this conclusion was supported by applying Legrand's criterion – generalization of eqn [144]. With these slip planes and using the method of Montagner and Nataf (1986), eqns [142] and [143], one obtains a more consistent interpretation of seismic anisotropy of the D″ layer than with the {010} slip planes. In particular, much smaller degrees of lattice-preferred orientation are needed to explain the observed seismic anisotropy and there is now a possibility to explain the observed (Garnero *et al.*, 2004; Wookey *et al.*, 2005) inclined character of anisotropy. Subsequent radial diffraction experiments on analog MgGeO$_3$ post-perovskite (Merkel, personal communication) have confirmed the prediction of Oganov *et al.* (2005b) on the dominant role of the {110} slip planes in MgSiO$_3$ post-perovskite. Furthermore, recent seismological studies (Wookey, personal communication) found that only {110} slip is consistent with observations.

6.4.3 Spin Transition in (Mg, Fe)O Magnesiowüstite

Iron impurities play a large role in determining the properties of Earth-forming minerals. One particular complication arising from the presence of these impurities is the possible pressure-induced transition of Fe^{2+} (or Fe^{3+}) impurities from the high-spin into the low-spin state. Typically, crystal fields induced by the O^{2-} ions at low pressures are weak, and transition metal ions prefer to adopt the high-spin configurations (like in free ions). However, under pressure the increasing crystal field and the additional *PV* term in the free energy prompt these ions to adopt much more compact low-spin forms (it is well documented that ionic radii are much larger for high-spin ions than for low-spin ones – for example, Shannon and Prewitt (1969). Recent studies of such a transition in (Mg, Fe)O magnesiowüstite (Badro *et al.*, 2003; Lin *et al.*, 2005) demonstrated that this transition might have large effects on physical properties of minerals.

This spin transition is isosymmetric, and as such it must (see Section 6.3) be first -order at low temperatures and fully continuous above some critical temperature T_{cr}. While at 0 K the low-pressure phase will contain only high-spin Fe^{2+} ions, and only low-spin ions will be present in the high-pressure phase, on increasing temperature there will be an increased degree of coexistence of the two spin states in the same phase – as a consequence, the first-order character of the transition decreases with temperature. At T_{cr} the miscibility of high- and low-spin ions become complete and the transition becomes fully continuous (infinite order, rather than first or second order); T_{cr} is proportional to the enthalpy that arises from the deformation of the structure due to insertion of a 'wrong'-spin ion. While quantitative aspects of this transition are actively studied by several groups, the most important qualitative features (in addition to those mentioned above) are immediately clear:

1. Large positive Clapeyron slope (since at high temperatures magnetic entropy is large for high-spin Fe^{2+} and zero for low-spin Fe^{2+}).
2. Low T_{cr}, perhaps several hundred kelvin, since for relevant compositions (e.g. $Mg_{0.8}Fe_{0.2}O$) the energetic effects of Fe incorporation and the enthalpy of 'spin mixing' will be rather small.

At lower-mantle temperatures the transition is likely to be continuous. A schematic phase diagram is shown in **Figure 12**. Simplified theory of the spin transition in (Mg, Fe)O was developed by Sturhahn *et al.* (2005).

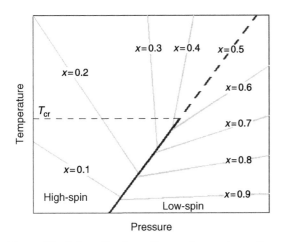

Figure 12 Schematic phase diagram for a system with a spin transition, for example, (Mg,Fe)O. Thick solid line indicates first-order transition, and thick dashed line indicates continuous transition. Gray lines show the degree of spin miscibility (numbers: concentration of low-spin species relative to the total number of iron atoms). Above T_{cr} the transition is fully continuous.

References

Ackland GJ and Reed SK (2003) Two-band second moment model and an interatomic potential for caesium. *Physical Review* B67: 174108.

Adams DM, Hatton PD, Heath AE, and Russell DR (1988) X-ray diffraction measurements on potassium nitrate under high pressure using synchrotron radiation. *Journal of Physics* C21: 505–515.

Al'tshuler LV, Brusnikin SE, and Kuzmenkov EA (1987) Isotherms and Grüneisen functions of 25 metals. *Journal of Applied Mechanics and Technical Physics* 161: 134–146.

Alexandrov KS and Prodaivoda GT (1993) Elastic properties of minerals. *Crystallografiya* 38: 214–234 (in Russian).

Alfè D, Gillan MJ, and Price GD (1999) The melting curve of iron at the pressures of the Earth's core from *ab initio* calculations. *Nature* 401: 462–464.

Alfè D, Gillan MJ, and Price GD (2002) Composition and temperature of the Earth's core constrained by combining ab initio calculations and seismic data. *Earth and Planetary Science Letters* 195: 91–98.

Allan DR and Nelmes RJ (1996) The structural pressure dependence of potassium titanyl phosphate (KTP) to 8 GPa. *Journal of Physics – Condensed Matter* 8: 2337–2363.

Allen MP and Tildesley DJ (1987) *Computer Simulation of Liquids*, 385 pp. Oxford: Clarendon Press.

Anderson DL (1989) *Theory of the Earth*, 366 pp. Boston: Blackwell.

Angel RJ (1996) New phenomena in minerals at high pressures. *Phase Transitions* 59: 105–119.

Angel RJ, Allan DR, Milletich R, and Finger LW (1997) The use of quartz as an internal pressure standard in high-pressure crystallography. *Journal of Applied Crystallography* 30: 461–466.

Arlt T and Angel RJ (2000) Displacive phase transitions in C-centred clinopyroxenes: Spodumene, $LiScSi_2O_6$ and $ZnSiO_3$. *Physics and Chemistry of Minerals* 27: 719–731.

Arlt T, Angel RJ, Miletch R, Armbruster T, and Peters T (1998) High-pressure $P2_1/c$-$C2/c$ phase transitions in clinopyroxenes: Influence of cation size and electronic structure. *American Mineralogist* 83: 1176–1181.

Badro J, Fiquet G, Guyot F, *et al.* (2003) Iron partitioning in Earth's mantle: Toward a deep lower mantle discontinuity. *Science* 300: 789–791.

Baroni S, de Gironcoli S, Dal Corso A, and Gianozzi P (2001) Phonons and related crystal properties from density-functional perturbation theory. *Reviews of Modern Physics* 73: 515–562.

Baroni S, Gianozzi P, and Testa A (1987) Green-function approach to linear response in solids. *Physical Review Letters* 58: 1861–1864.

Barron THK and Klein ML (1965) Second-order elastic constants of a solid under stress. *Proceedings of the Physical Society* 85: 523–532.

Belikov BP, Aleksandrov KS, and Ryzhova TV (1970) *Elastic Constants of Rock-Forming Minerals*, 276 pp. Moscow: Nauka (in Russian).

Binggeli N, Keskar NR, and Chelikowsky JR (1994) Pressure-induced amorphisation, elastic instability, and soft modes in α-quartz. *Physical Review B* 49: 3075–3081.

Birch F (1952) Elasticity and constitution of the Earth's interior. *Journal of Geographical Research* 57: 227–286.

Birch F (1961) Composition of the Earth's mantle. *Geophysical Journal of the Royal Astronomical Society* 4: 295–311.

Birman JL (1966) Simplified theory of symmetry change in second-order phase transitions: Application to V_3Si. *Physical Review Letters* 17: 1216–1219.

Born M and Huang K (1954) *Dynamical Theory of Crystal Lattices*, 420 pp. Oxford: Oxford University Press.

Bowley R and Sánchez M (1999) *Introductory Statistical Mechanics*, 2nd edn., 352 pp. Oxford: Oxford University Press.

Bruce AD and Cowley RA (1981) *Structural Phase Transitions*, 326 pp. London: Taylor and Francis.

Buerger MJ (1961) Polymorphism and phase transformations. *Fortschritte Der Mineralogie* 39: 9–24.

Bukowinski MST (1994) Quantum geophysics. *Annual Reviews of Earth and Planetary Science* 22: 167–205.

Burdett JK (1995) *Chemical Bonding in Solids*, 319 pp. New York: Oxford University Press.

Carlson S, Xu YQ, Halenius U, and Norrestam R (1998) A reversible, isosymmetric, high-pressure phase transition in Na_3MnF_6. *Inorganic Chemistry* 37: 1486–1492.

Carpenter MA and Salje EKH (1998) Elastic anomalies in minerals due to structural phase transitions. *Eurpoean Journal of Mineralogy* 10: 693–812.

Carpenter MA and Salje EKH (2000) Strain and Elasticity at Structural Phase Transitions in Minerals. *Reviews in Mineralogy and Geochemistry* 39: 35–64.

Carpenter MA, Salje EKH, and Graeme-Barber A (1998) Spontaneous strain as a determinant of thermodynamic properties for phase transitions in minerals. *European Journal of Mineralogy* 10: 621–691.

Catti M (1989) Crystal elasticity and inner strain – A computational model. *Acta Crystallographica* A45: 20–25.

Chandler D (1987) *Introduction to Modern Statistical Mechanics*, 274 pp. NY: Oxford University Press.

Christy AG (1993) Multistage diffusionless pathways for reconstructive phase transitions: application to binary compounds and calcium carbonate. *Acta Crystallographica B* 49: 987–996.

Christy AG (1995) Isosymmetric structural phase transitions: Phenomenology and examples. *Acta Crystallographica B* 51: 753–757.

Cohen RE, Gulseren O, and Hemley RJ (2000) Accuracy of equation-of-state formulations. *American Mineralogist* 85: 338–344.

Cowley RA (1976) Acoustic phonon instabilities and structural phase transitions. *Physical Review* B13: 4877–4885.

Dove M (2003) *Structure and Dynamics*. Oxford: Oxford University Press.

Dove MT (1993) *Introduction to Lattice Dynamics*, 258 pp. Cambridge: Cambridge University Press.

Dove MT (1997) Theory of displacive phase transitions in minerals. *American Mineralogist* 82: 213–244.

Fedorov FI (1968) *Theory of Elastic Waves In Crystals*, 375 pp. New York: Plenum.

Gale JD (1998) Analytical free energy minimization of silica polymorphs. *Journal of Physical Chemistry B* 102: 5423–5431.

Garber JA and Granato AV (1975) Theory of the temperature dependence of second-order elastic constants in cubic materials. *Physical Review B* 11: 3990–3997.

Garland CW and Weiner BB (1971) Changes in the thermodynamic character of the NH_4Cl order–disorder transition at high pressures. *Physical Review B* 3: 1634–1637.

Garnero EJ, Maupin V, Lay T, and Fouch MJ (2004) Variable azimuthal anisotropy in Earth's lowermost mantle. *Science* 306: 259–261.

Gillet P, Matas J, Guyot F, and Ricard Y (1999) Thermodynamic properties of minerals at high pressures and temperatures from vibrational spectroscopic data. In: Wright K and Catlow R

(eds.) *Miscroscopic Properties and Processes in Minerals*, NATO Science Series, v. C543, pp. 71–92. Dordrecht: Kluwer.

Haines J, Leger JM, and Schulte O (1998) High-pressure isosymmetric phase transition in orthorhombic lead fluoride. *Physical Review B* 57: 7551–7555.

Hama J and Suito K (1996) The search for a universal equation of state correct up to very high pressures. *Journal of Physics – Condensed Matter* 8: 67–81.

Harris MJ and Dove MT (1995) Lattice melting at structural phase transitions. *Modern Physics Letters* 9: 67–85.

Hatch DM and Merrill L (1981) Landau description of the calcite-$CaCO_3(II)$ phase transition. *Physical Review B* 23: 368–374.

Hemmati M, Czizmeshya A, Wolf GH, Poole PH, Shao J, and Angell CA (1995) Crystalline-amorphous transition in silicate perovskites. *Physical Review B* 51: 14841–14848.

Holzapfel WB (1996) Physics of solids under strong compression. *Reports on Progress in Physics* 59: 29–90.

Holzapfel WB (2001) Equations of state for solids under strong compression. *Zeitschrift Fur Kristallographie* 216: 473–488.

Iannuzzi M, Laio A, and Parrinello M (2003) Efficient exploration of reactive potential energy surfaces using Car-Parrinello molecular dynamics. *Physical Review Letters* 90: (art. no. 238302).

Hahn Th (*ed.*) (1994) *International Tables for Crystallography, Vol. A., Space-Group Symmetry*. Dordrecht: Kluwer.

Ito E and Katsura T (1989) A temperature profile of the mantle transition zone. *Geophysical Research Letters* 16: 425–428.

Jaeger G (1998) The Ehrenfest classification of phase transitions: Introduction and evolution. *Archives of Histology and Exact Science* 53: 51–81.

Kantorovich LN (1995) Thermoelastic properties of perfect crystals with nonprimitive lattices: 1: General theory. *Physical Review B* 51: 3520–3534.

Karki BB (1997) *High-Pressure Structure and Elasticity of the Major Silicate and Oxide Minerals of the Earth's Lower Mantle*, 170 pp. PhD Thesis, University of Edinburgh.

Karki BB, Wentzcovitch RM, de Gironcoli S, and Baroni S (1999) First-principles determination of elastic anisotropy and wave velocities of MgO at lower mantle conditions. *Science* 286: 1705–1707.

Karki BB, Wentzcovitch RM, de Gironcoli S, and Baroni S (2000) Ab initio lattice dynamics of $MgSiO_3$ perovskite at high pressure. *Physical Review B* 62: 14750–14756.

Kerrick DM (1990) The Al_2SiO_5 Polymorphs. *Reviews in Mineralogy* 22: 406.

Keskar NR, Chelikowsky JR, and Wentzcovitch RM (1994) Mechanical instabilities in $A1PO_4$. *Physical Review B* 50: 9072–9078.

Kieffer SW (1979) Thermodynamics and lattice vibrations of minerals. 1: Mineral heat capacities and their relationship to simple lattice vibrational models. *Reviews of Geophysics and Space Physics* 17: 1–19.

Kunc K, Loa I, and Syassen K (2003) Equation of state and phonon frequency calculations of diamond at high pressures. *Physical Review B* 68: (art. 094107).

Laio A and Parrinello M (2002) Escaping free-energy minima. *Proceedings of the National Academy of Sciences* 99: 12562–12566.

Landau LD and Lifshitz EM (1980) *Statistical Physics. Part I*. (Course of Theoretical Physics, v. 5), 3rd edn., 544 pp. Oxford: Butterworth and Heinemann.

Legrand B (1984) Relations entre la structure èlectronique et la facilitè de glissement dans les métaux hexagonaux compacts. *Philosophical Magazine* 49: 171–184.

Lin JF, Struzhkin VV, Jacobsen SD, *et al.* (2005) Spin transition of iron in magnesiowusite in Earth's lower mantle. *Nature* 436: 377–380.

Liu L-G and Bassett WA (1986) *Elements, Oxides, and Silicates. High-Pressure Phases with Implications for the Earth's Interior*, 250 pp. New York: Oxford University Press.

Loubeyre P, LeToullec R, Hausermann D, *et al.* (1996) X-ray diffraction and equation of state of hydrogen at megabar pressures. *Nature* 383: 702–704.

Matsui M (1989) Molecular dynamics study of the structural and thermodynamic properties of MgO crystal with quantum correction. *Journal of Chemical Physics* 91: 489–494.

McNeil LE and Grimsditch M (1991) Pressure-amorphized SiO_2 α-Quartz : An anisotropic amorphous solid. *Physical Review Letters* 68: 83–85.

Mishima O, Calvert LD, and Whalley E (1984) Melting of ice I at 77 K and 10 kbar: A new method for making amorphous solids. *Nature* 310: 393–394.

Montagner J-P and Nataf H-C (1986) A simple method for inverting the azimuthal anisotropy of surface waves. *Journal of Geophysical Research* 91: 511–520.

Montroll EW (1942) Frequency spectrum of crystalline solids. *Journal of Chemical Physics* 10: 218–229.

Montroll EW (1943) Frequency spectrum of crystalline solids. II: General theory and applications to simple cubic lattices. *Journal of Chemical Physics* 10: 481–495.

Murakami M, Hirose K, Kawamura K, Sata N, and Ohishi Y (2004) Post-perovskite phase transition in $MgSiO_3$. *Science* 304: 855–858.

Nye JF (1998) *Physical Properties of Crystals. Their Representation by Tensors and Matrices*, 329 pp. Oxford: Oxford University Press.

Oganov AR and Dorogokupets PI (2003) All-electron and pseudopotential study of MgO: Equation of state, anharmonicity, and stability. *Physical Review B* 67: (art. 224110).

Oganov AR and Dorogokupets PI (2004) Intrinsic anharmonicity in thermodynamics and equations of state of solids. *Journal of Physics – Condensed Matter* 16: 1351–1360.

Oganov AR (2002) *Computer Simulation Studies of Minerals*, 290 pp. PhD Thesis, University of London.

Oganov AR, Brodholt JP, and Price GD (2000) Comparative study of quasiharmonic lattice dynamics, molecular dynamics and Debye model in application to $MgSiO_3$ perovskite. *Physics of the Earth and Planetary Interiors* 122: 277–288.

Oganov AR, Brodholt JP, and Price GD (2001a) The elastic constants of $MgSiO_3$ perovskite at pressures and temperatures of the Earth's mantle. *Nature* 411: 934–937.

Oganov AR, Brodholt JP, and Price GD (2002) Ab initio theory of thermoelasticity and phase transitions in minerals. In: Gramaccioli CM (ed.) *EMU Notes in Mineralogy, Vol. 4: Energy Modeling in Minerals*, pp. 83–170. Bolin: Springer.

Oganov AR, Gillan MJ, and Price GD (2003) Ab initio lattice dynamics and structural stability of MgO. *Journal of Chemical Physics* 118: 10174–10182.

Oganov AR, Gillan MJ, and Price GD (2005a) Structural stability of silica at high pressures and temperatures. *Physical Review B*71: (art. 064104).

Oganov AR, Martoňák R, Laio A, Raiteri P, and Parrinello M (2005b) Anisotropy of Earth's D″ layer and stacking faults in the $MgSiO_3$ post-perovskite phase. *Nature* 438: 1142–1144.

Oganov AR and Ono S (2004) Theoretical and experimental evidence for a post-perovskite phase of $MgSiO_3$ in Earth's D″ layer. *Nature* 430: 445–448.

Oganov AR and Ono S (2005) The high-pressure phase of alumina and implications for Earth's D″ layer. *Proceedings of the National Academy of Sciences* 102: 10828–10831.

Oganov AR and Price GD (2005) Ab initio thermodynamics of $MgSiO_3$ perovskite at high pressures and temperatures. *Journals of Chemical Physics* 122: (art. 124501).

Oganov AR, Price GD, and Brodholt JP (2001b) Theoretical investigation of metastable Al_2SiO_5 polymorphs. *Acta Crystallographica* A57: 548–557.

Parker SC and Price GD (1989) Computer modelling of phase transitions in minerals. *Advances in Solid State Chemistry* 1: 295–327.

Poirier JP (1999) Equations of state. In: Wright K and Catlow R (eds.) *Miscroscopic Properties and Processes in Minerals*, NATO Science Series, v. C543, pp. 19–42. Dordrecht: Kluwer.

Poirier J-P (2000) *Introduction to the Physics of the Earth's Interior* 2nd edn. 326 pp. Cambridge: Cambridge University Press.

Poirier J-P and Price GD (1999) Primary slip system of ε-iron and anisotropy of the Earth's inner core. *Physics of the Earth and Planetary Interiors* 110: 147–156.

Price GD and Yeomans J (1984) The application of the ANNNI model to polytypic behavior. *Acta Crystallographica B* 40: 448–454.

Price GD (1983) Polytypism and the factors determining the stability of spinelloid structures. *Physics and Chemistry of Minerals* 10: 77–83.

Rao CNR and Rao KJ (1978) *Phase Transitions in Solids: An Approach to the Study of the Chemistry and Physics of Solids*, 330 pp. NY: McGraw-Hill.

Redfern SAT (2000) Order-disorder phase transitions. *Reviews in Mineralogy and Geochemistry* 39: 105–133.

Richet P and Gillet P (1997) Pressure-induced amorphisation of minerals: A review. *European Journal of Mineralogy* 9: 907–933.

Ringwood AE (1991) Phase transformations and their bearing on the constitution and dynamics of the mantle. *Geochimica et Cosmochimica Acta* 55: 2083–2110.

Robie RA and Edwards JL (1966) Some Debye temperatures from single crystal elastic constant data. *Journal of Applied Physics* 37: 2659–2663.

Rydberg VR (1932) Graphische Darstellungeiniger banden-spektroskopischer Ergebnisse. *Journal of Applied Physics* 73: 376–385.

Schilling FR, Sinogeikin SV, and Bass JD (2003) Single-crystal elastic properties of lawsonite and their variation with temperature. *Physics of the Earth and Planetary Interiors* 136: 107–118.

Seitz F (1949) *Modern Theory of Solids*, 736 pp. Moscow: GITTI (Russian translation).

Shannon RD and Prewitt CT (1969) Effective ionic radii in oxides and fluorides. *Acta Crystallographia B* 25: 925–946.

Sharma SM and Sikka SK (1996) Pressure-induced amorphization of materials. *Progress in Materials Science* 40: 1–77.

Sirotin Yu I and Shaskolskaya MP (1975) *Fundamentals of Crystal Physics*, 680 pp. Moscow: Nauka (in Russian).

Sposito G (1974) Landau's choice of the critical-point exponent β. *American Journal of Physics* 42: 1119–1121.

Sturhahn W, Jackson JM, and Lin J-F (2005) The spin state of iron in minerals of Earth's lower mantle. *Geophysical Research Letters* 32: L12307.

Sutton AP (1993) *Electronic Structure of Materials*, 260 pp. Oxford: Oxford University Press.

Terhune RW, Kushida T, and Ford GW (1985) Soft acoustic modes in trigonal crystals. *Physical Review* B32: 8416–8419.

Tröster A, Schranz W, and Miletich R (2002) How to couple Landau theory to an equation of state. *Physical Review Letters* 88: art. 055503.

Tse JS and Klug DD (1993) Anisotropy in the structure of pressure-induced disordered solids. *Physical Review Letters* 70: 174–177.

Tsuchiya T, Tsuchiya J, Umemoto K, and Wentzcovitch RM (2004) Phase transition in $MgSiO_3$ perovskite in the earth's lower mantle. *Earth and Planetary Science Letters* 224: 241–248.

Umemoto K, Wentzcovitch RM, and Allen PB (2006) Dissociation of $MgSiO_3$ in the cores of gas giants and terrestrial exoplanets. *Science* 311: 983–986.

Vinet P, Ferrante J, Smith JR, and Rose JH (1986) A universal equation of state for solids. *Journal of Physics* C19: L467–L473.

Vinet P, Rose JH, Ferrante J, and Smith JR (1989) Universal features of the equation of state of solids. *Journal of Physics – Condensed Matter* 1: 1941–1963.

Vočadlo L, Price GD, and Poirier J-P (2000) Grüneisen parameters and isothermal equations of state. *American Mineralogist* 85: 390–395.

Vorob'ev VS (1996) On model description of the crystalline and liquid states. *Teplofizika Vysokih Temperatur (High-Temperature Thermophysics)* 34: 397–406 (in Russian).

Wallace DC (1998) *Thermodynamics of Crystals*, 484 pp. New York: Dover Publications.

Wang Y, Li J, Yip S, Phillpot S, and Wolf D (1995) Mechanical instabilities of homogeneous crystals. *Physical Review B* 52: 12627–12635.

Wang Y, Yip S, Phillpot S, and Wolf D (1993) Crystal instabilities at finite strain. *Physical Review Letters* 71: 4182–4185.

Warren MC, Dove MT, Myers ER, *et al.* (2001) Monte Carlo methods for the study of cation ordering in minerals. *Mineralogical Magazine* 65: 221–248.

Watt JP, Davies GF, and O'Connell RJ (1976) The elastic properties of composite materials. *Reviews of Geophysics and Space Physics* 14: 541–563.

Welch DO, Dienes GJ, and Paskin A (1978) A molecular dynamical study of the equation of state of solids at high temperature and pressure. *Journal of Physics and Chemistry of Solids* 39: 589–603.

Wilson KG (1983) The renormalization group and related phenomena. *Reviews of Modern Physics* 55: 583–600.

Wookey J, Kendall J-M, and Rümpker G (2005) Lowermost mantle anisotropy beneath the north Pacific from differential *S-ScS* splitting. *Geophysical Journal International* 161: 829–838.

Yeomans JM (1992) *Statistical Mechanics of Phase Transitions*, 168 pp. Oxford: Oxford University Press.

Zhang L, Ahsbahs H, Kutoglu A, and Hafner SS (1992) Compressibility of grunerite. *American Mineralogist* 77: 480–483.

Zharkov VN and Kalinin VA (1971) *Equations of state of solids at high pressures and temperatures*, 257 pp. New York: Consultants Bureau.

7 Theory and Practice – Lattice Vibrations and Spectroscopy of Mantle Phases

P. F. McMillan, University College London, London, UK

7.1	Introduction	153
7.1.1	Deep Earth Mineralogy and Mineral Physics	153
7.1.2	Experimental Vibrational Spectroscopy of Mantle Minerals	155
7.1.3	Mineral Lattice Dynamics Calculations	156
7.2	Mineral Lattice Dynamics and Vibrational Spectroscopy	157
7.2.1	Molecular Vibrations	157
7.2.2	Vibrational Dispersion Relations	159
7.2.3	Quantized Vibrations in Molecules and Crystals (Phonons)	161
7.2.4	Vibrational Frequency Shifts at High P and T; Quasi-Harmonic Model and Phonon Anharmonicity	162
7.2.5	Mineral Thermodynamic Quantities from Vibrational Spectra	165
7.2.6	Theoretical Calculations of Mantle Mineral Lattice Vibrations	166
7.3	Lattice Dynamics of Mantle Minerals	168
7.3.1	Diamond and $(Mg, Fe)O$ Magnesiowüstite	168
7.3.2	$(Mg, Fe)SiO_3$ Perovskite, Post-Perovskite, and $CaSiO_3$ Perovskite	171
7.3.3	Stishovite and Post-Stishovite SiO_2 Polymorphs	177
7.3.4	$(Mg, Fe)_2SiO_4$ Olivine, β- and γ-$(Mg, Fe)_2SiO_4$, and $(Mg, Fe)SiO_3$ Pyroxenes	180
7.3.5	$MgSiO_3$ Ilmenite and Majoritic Garnets	182
7.3.6	OH in Mantle Minerals	184
References		185

7.1 Introduction

7.1.1 Deep Earth Mineralogy and Mineral Physics

Mantle mineralogy is sampled directly by xenoliths brought to the surface, and by phases contained inside inclusions within high-P minerals, such as diamonds (Boyd and Meyer, 1979; Dawson, 1980; Moore and Gurney, 1985; Nixon, 1987; Sautter *et al.*, 1991; Harris, 1992; Harte and Harris, 1993, 1994; Sautter and Gillet, 1994; McCammon *et al.*, 1997; Liou *et al.*, 1998; McCammon, 2001). However, our main understanding of the mineralogy and the thermophysical properties of the deep Earth is derived from seismology studies and constraints provided by cosmology, combined with laboratory experiments carried out under high pressure–high temperature conditions and theoretical modeling of mantle minerals (Birch, 1952; Ringwood, 1975; Anderson, 1989; Hemley, 1998).

The structure of the mantle is defined in terms of major seismic discontinuities occurring near 410 and 660 km. These delineate regions known as the 'upper mantle' (20–410 km), 'transition zone' (410–660 km), and 'lower mantle' (660–2900 km) (Ringwood, 1962a; Jeanloz and Thompson, 1983; Anderson, 1989; Gillet, 1995; Agee, 1998; Bina, 1998). Based on laboratory studies and theoretical modeling, it is likely that the primary mineral phase in the lower mantle is a silicate perovskite with composition near $(Mg_{0.9} Fe_{0.1}) SiO_3$, that also likely contains a few percent Al_2O_3 component (Hemley and Cohen, 1992). That phase coexists with $(Mg, Fe)O$ (magnesiowüstite), that maintains its rock salt (B1) structure throughout the lower mantle (Lin *et al.*, 2002), and also a $CaSiO_3$ perovskite phase (Hemley and Cohen, 1992). It has been thought for some time now that $(Mg, Fe)SiO_3$ perovskite would remain stable throughout the entire range of high-P,T conditions within the lower mantle (Knittle and Jeanloz, 1987, 1991; Hemley and Cohen, 1992; Serghiou *et al.*, 1998). However, it now appears that a further transformation is likely to occur to a $CaIrO_3$-structured phase of $MgSiO_3$ at the base of the lower mantle (Murakami *et al.*, 2004a; Oganov and Ono, 2004; Shim *et al.*, 2004; Tsuchiya *et al.*, 2004b). A free silica phase with the stishovite structure (rutile-structured SiO_2) could be present in basaltic regions trapped near the top of the lower mantle derived from subducted oceanic crust. The possible existence of an SiO_2 phase near the base of the lower mantle could also occur due to reactions among mineral phases under high-P,T conditions (Knittle and Jeanloz, 1991; Jeanloz, 1993; Kesson et al., 1994). SiO_2 might then transform to $CaCl_2$- or α-PbO_2 structured phases below 1250 km ($P > 50$ GPa), via displacive phase transitions driven by the vibrational lattice dynamics (Kingma et al., 1995; Dubrovinsky et al., 1997; Andrault et al., 1998; Shieh et al., 2005).

The upper mantle is dominated by ultramafic minerals, principally olivine (α-$(Mg, Fe)_2SiO_4$), $(Mg, Fe)SiO_3$, and (Ca, Al)-containing pyroxenes and garnets. Phases including plagioclase feldspar, kaersutite amphibole, and phlogopite mica are also likely to be present (McDonough and Rudnick, 1998). Some of these phases are volatile-rich, providing hosts and sinks for OH/H_2O species and CO_2, as well as F, Cl, and S, either being returned to the mantle via subduction processes or degassing from primary materials deep within the Earth (Gasparik, 1990; Bell, 1992; Bell and Rossman, 1992; Thompson, 1992; Gasparik, 1993; Carroll and Holloway, 1994; Bose and Navrotsky, 1998). Partial melting events can also occur within the upper mantle, especially when considered throughout the past history of the planet that might involved a deep magma ocean. Understanding the structures and dynamical properties of molten silicates at high pressure are important for geophysics studies (Wolf and McMillan, 1995; Dingwell, 1998). Melting could also be occurring at the base of the present-day lower mantle (Hemley and Kubicki, 1991; Wolf and McMillan, 1995; Shen and Heinz, 1998; Akins et al., 2004).

The mantle transition zone is dominated by the mineralogy of refractory ultramafic minerals including β-$(Mg, Fe)_2SiO_4$ wadsleyite (also known in earlier works as 'beta-phase', or 'modified spinel') and spinel-structured γ-$(Mg, Fe)_2SiO_4$ (ringwoodite), along with $MgSiO_3$ garnet (majorite) or $MgSiO_3$ ilmenite (akimotoite) (Gasparik, 1990; Agee, 1998). Both $MgSiO_3$ ilmenite and garnet contain [6]-coordinated silicon species. Majoritic garnets exist as solid solutions with Mg^{2+}, Fe^{2+}, and Ca^{2+} in the large (8–12 coordinated) site, Al^{3+}, Si^{4+}, and Fe^{3+} in octahedral sites, and Si^{4+}/Al^{3+} cations on tetrahedral positions.

All of these mantle minerals are nominally anhydrous, but they might contain significant quantities of dissolved H as 'defect' OH species (Bell, 1992; Bell and Rossman, 1992). In addition, various high-pressure hydrous phases, often with compositions expressed in the system $MgO–SiO_2–H_2O$, could be present especially within the transition zone (Liu, 1987; Kanzaki, 1991b; Prewitt and Finger, 1992; Bose and Navrotsky, 1998).

The development of mineral physics has largely paralleled that of the fields of geophysics, high-pressure mineralogy, and geochemistry. It involves the application of physics and chemistry techniques to understand and predict the fundamental behavior of Earth materials (Kieffer and Navrotsky, 1985; Navrotsky, 1994; Anderson, 1995; Hemley, 1998; Gramaccioli, 2002a), and extend these to large-scale problems in Earth and planetary sciences. Studies of mineral lattice vibrations are central to mineral physics research. The specific heat among solids is primarily stored and transported via vibrational excitations, so that studying mineral lattice dynamics helps us understand the thermal budget within the mantle (Jeanloz and Richter, 1979; Kieffer, 1979a, 1979b, 1979c, 1980, 1985; Jeanloz and Morris, 1986; Chai et al., 1996; Hofmeister, 1999). Lattice vibrations contribute fundamentally to the thermal free energy of minerals. Also, mineral phase transitions can be determined by vibrational excitations, in addition to chemical driving forces. Mineral physics vibrational studies thus play a key role in understanding phase transitions among mantle mineral phases. The vibrational modes in solids and liquids are controlled by the same interatomic forces that determine their elastic properties and seismic propagation velocities within the Earth, and they contribute to the fracture kinetics and rheology of mineral assemblages (Birch, 1952; Weidner, 1975; Brown et al., 1989; Hofmeister, 1991a, 1991b; Bass, 1995; Bina, 1998; Liebermann and Li, 1998; Weidner, 1998; Oganov et al., 2005b; Stixrude and Lithgow-Bertelloni, 2005), The occurence of lattice dynamical instabilities as a function of P and T can lead to unexpectedly large changes in mineral elastic properties at depth within the Earth (Scott, 1974; Salje, 1990; Kingma et al., 1995; Murakami et al., 2003; Iitaka et al., 2004; Nakagawa and Tackley, 2004; Oganov et al., 2005b). Consideration of vibrational lattice instabilities of minerals can also allow us to understand and predict melting under high-P,T conditions (Wolf and Jeanloz, 1984; Shen and Heinz, 1998; Alfé et al., 1999, 2004; Alfé 2005; Akins et al., 2004).

Vibrational spectroscopy is a complementary technique to X-ray diffraction that is used to identify and carry out in situ structural studies of high-pressure mineral phases, including samples contained inside xenoliths, either as separate particles or within inclusions, or within products of high-P,T

experimental runs. Microbeam Raman and IR spectroscopic techniques have proved useful for such studies (McMillan, 1985, 1989; McMillan and Hofmeister, 1988; McMillan et al., 1996a, 1996b; Gillet et al., 1998). Vibrational spectroscopy is particularly useful when applied to highly disordered or amorphous materials, including melts, and minerals containing light-element (C, O, H) 'volatile' species (McMillan, 1985, 1989, 1994; Bell, 1992; Bell and Rossman, 1992; Ihinger et al., 1994; McMillan and Wolf, 1995; Rossman, 1996). The optical spectroscopy experiments are ideally matched with the transparent window-anvils in diamond anvil cell (DAC) studies, used to carry out *in situ* studies of mineral behavior under the entire range of high-P, T conditions existing throughout the Earth's mantle, and also into the deep interiors of other planets, including the gas giants Neptune, Saturn and Jupiter, and their moons (Ferrarro, 1984; Challener and Thompson, 1986; Hemley et al., 1987b; Chervin et al., 1992; Gillet et al., 1998; Hemley, 1998; Hemley and Mao, 2002).

7.1.2 Experimental Vibrational Spectroscopy of Mantle Minerals

Many experimental studies of mineral lattice vibrations have been carried out at ambient conditions primarily using infrared (IR) spectroscopy, especially via Fourier transform (FTIR) techniques, usually employing powder transmission methods (Farmer, 1974; McMillan, 1985; McMillan and Hofmeister, 1988). Reflectance IR studies, that can be carried out on single crystals or pressed powder samples provide valuable additional information on the lattice vibrations (Spitzer et al., 1962; Piriou and Cabannes, 1968; Gervais, et al., 1973a; Gervais and Piriou, 1975; Gervais, 1983; McMillan, 1985; Hofmeister, 1987; McMillan and Hofmeister, 1988; Hofmeister, 1996, 1997). *In situ* measurements of mantle minerals at high-P conditions in the diamond anvil cell have generally been carried out via IR powder transmission methods (Williams et al., 1986, 1987; Hofmeister et al., 1989; Hofmeister and Ito, 1992; Hofmeister, 1996, 1997; McMillan et al., 1996b; Gillet et al., 1998). IR emission spectroscopy has been applied to *in situ* studies of minerals and melts at high-T; this technique is also useful for remote sensing studies of planetary surfaces and their atmospheres (Gervais, 1983; Christensen et al., 1992; Efimov, 1995; McMillan and Wolf, 1995). Access to synchrotron IR sources now allows greatly increased flux for mineral

physics studies, especially in the far-IR region (Hemley et al., 1998).

The other primary laboratory technique for measuring mineral vibrational spectra is Raman spectroscopy, that relies on the inelastic scattering of light by vibrational excitations in molecules and solids (Long, 1977; McMillan, 1985, 1989; Hemley et al., 1987b; McMillan and Hofmeister, 1988). Brillouin scattering is a related method to Raman spectroscopy, that is used to determine sound waves propagating within solids at very low energies via high-resolution interferometric techniques, and thus provides information on the elastic properties of mantle minerals, including under high-P, T conditions (Weidner, 1975; Bass, 1995).

The IR and Raman spectroscopies provide information on vibrational waves propagating inside solids with very long wavelengths (e.g., $\lambda_{vib} \sim 500$ nm in the case of Raman scattering, and $\lambda_{vib} \sim 1-50$ μm for IR studies) (Farmer, 1974; Decius and Hexter, 1977; McMillan, 1985; McMillan and Hofmeister, 1988). In order to conduct experimental studies of the full lattice dynamics, including the shortest wavelength lattice modes that exist down to $\lambda_{vib} \sim a_o$ (i.e., on the order of a lattice constant), it is necessary to use a technique like inelastic neutron scattering (INS) (Ghose, 1985, 1988; Dove, 1993, 2002; Chaplot et al., 2002a; Choudhury et al., 2002).

INS studies are implemented at national or international facilities, that can provide appropriate neutron beams derived from reactors or spallation sources (Ghose, 1985, 1988; Choudhury et al., 2002; Dove, 2002; Winkler, 2002). The neutron flux is generally very low compared with that of the incident photons provided in laboratory optical spectroscopy studies, so that INS experiments demand much larger sample sizes and long data collection times. Also, access to the instruments is limited by competition for beam time among various international scientific communities. For these reasons, systematic INS studies of representative mineral structures, including olivine, pyroxene, and garnet phases that are so important in the Earth's upper mantle, are only just beginning to be carried out (Rao et al., 1988, 1999, 2002; Price et al., 1991; Choudhury et al., 1998, 2002; Mittal et al., 2000; Chaplot et al., 2002b).

Most INS experiments have been carried out on powdered minerals, to yield the vibrational density of states (the VDOS, or $g(\nu)$ function) that represents the mineral lattice vibrational spectrum averaged over all crystallographic directions. In selected cases where sufficiently large samples are available for study, single-crystal INS measurements have revealed the dependence of vibrational energies on the wave vector for phonon propagation along various crystallographic directions in reciprocal space. Such studies have so far been restricted to lower crustal and upper mantle minerals, and simple dense materials with the diamond and rock-salt structures, including diamond itself and MgO (Bilz and Kress, 1979; Dove, 1993; Chaplot et al., 2002b; Choudhury et al., 2002). Those investigations have mainly been carried out ambient P and T. To date, there have been no INS studies of the principal lower mantle or transition zone mineral phases, including $MgSiO_3$ perovskite or ilmenite, majorite garnet, or $(Mg, Fe)_2SiO_4$ spinel, mainly due to lack of availability of powdered samples in sufficient quantity (or sufficiently large single crystals) from the high-P,T syntheses. In particular, there have only been limited INS measurements of any materials in high-P,T environments, that might correspond to mantle conditions.

In situ neutron studies of materials under high-P,T conditions have been enabled by coupling neutron diffraction and inelastic scattering techniques with specially designed 'large volume' high-pressure devices, such as the 'Paris–Edinburgh' cell based on a toroidal press design (Besson et al., 1992; Dove, 2002; Redfern, 2002; Khvostantsev et al., 2004; Loveday, 2004). The lattice dynamics of simple crystals including Ge and b.c.c.-Fe have been studied up to $P \sim 10\,GPa$ using such a system (Klotz et al., 1997; Klotz and Braden, 2000). This pressure range currently constitutes an operational limit for *in situ* high-P,T INS experiments. Neutron diffraction studies can now be carried out to $P \sim 30\,GPa$ at low T, and up to $P \sim 10\,GPa$ under simultaneous high-P,T conditions (Redfern, 2002; Loveday, 2004). Those limits are likely to be exceeded in the near future, through the use of new materials including sintered diamond for the pressurisation anvils and other parts in the high-P,T assembly, combined with advances in the high-P,T cell designs (Irifune, 2002; Loveday, 2004). These technical advances will be coupled with improved access to high-intensity neutron sources, including new facilities that are currently being developed worldwide (e.g., the Spallation Neutron Source (SNS) in the USA and Target Station 2 (TS2) at ISIS, UK). Neutron focusing optics are also being developed, that are expected to greatly reduce the sample size requirements and data collection times that are required for high-P,T

experiments and studies of mantle minerals. The predicted availability of new large diamond windows for high-P,T cells grown by chemical vapor deposition (CVD) techniques is also expected to greatly expand the pressure range for neutron diffraction and INS experiments on minerals, perhaps even into the multi-megabar range (Hemley and Mao, 2002; Vohra and Weir, 2002; Yan et al., 2002).

New experimental approaches to determining the lattice dynamics of mantle minerals and other materials under high-P,T conditions are also being developed at synchrotron facilities, using techniques such as high-resolution inelastic X-ray scattering or nuclear resonant spectroscopies (Kao et al., 1996; Schwoerer-Böhning et al., 1998; Fiquet et al., 2001; Mao et al., 2001; Hemley and Mao, 2002; Zhao et al., 2004; Duffy, 2005). These techniques are enabled by the availability of tightly focused and highly collimated high-intensity X-ray beams, and they are well adapted to DAC investigations, so that they can be readily applied into the multi-megabar pressure regime, including measurements at high T. The nuclear resonant techniques are specific to Mössbauer-absorbing atoms with appropriate γ-ray resonances such as Fe, Sn, Eu, I, etc. They have now been mainly applied to study the phases of iron present within the Earth's core (Mao et al., 2001, 2005; Zhao et al. 2004), However, they are expected to become generally useful for vibrational studies for various Fe-bearing mantle phases, and for studies of various phases including Mössbauer-absorbing nuclei, under high-P,T conditions. High-resolution inelastic X-ray scattering (IXS) experiments provide density of states and phonon dispersion data that are similar to the information provided by inelastic neutron scattering studies (Fiquet et al., 2001; Mao et al., 2001; Hemley and Mao, 2002; Zhao et al., 2004; Duffy, 2005). These techniques are expected to revolutionize lattice dynamic studies of mantle minerals, especially under the extreme high-P,T conditions of the lower mantle and transition zone. (Hemley and Mao, 2002; Duffy, 2005).

7.1.3 Mineral Lattice Dynamics Calculations

Vibrational spectroscopy experiments are complemented and extended by various theoretical calculations of mineral lattice dynamics under high-P,T conditions. A wide range of empirical and semi-empirical methods largely based on the ionic model of crystal structures and interatomic bonding have been developed to study inorganic solids. These methods have been applied to predict and understand the energetics, structures, lattice dynamics and thermoelastic behavior of mantle minerals under high-P,T conditions (Matsui and Busing, 1984; Burnham, 1985; Matsui et al., 1987; Price et al., 1987; Matsui, 1988; Burnham, 1990; Catlow and Price, 1990; Dove, 1993; Tse, 2004; Winkler, 2004). Calculations using empirical potential models are currently being used to simulate defect formation and migration in mantle minerals (Watson et al., 2000).

Mineral lattice vibrations have also long been studied using empirical force constant models (Wilson et al., 1955; Shimanouchi et al., 1961; Nakamoto, 1978; Dowty, 1987a, 1987b, 1987c; Gramaccioli, 2002b). By the mid-1980s, ab initio calculations in molecular chemistry were established using Hartree–Fock methods (Hehre et al., 1996). Those methods were applied to various silicate clusters that are important for mineralogy (Gibbs, 1982; Sauer, 1989; Tossell and Vaughan, 1992). Calculations of first and second derivatives of the energy with respect to atomic displacements led to ab initio force constants that were then incorporated into lattice dynamics calculations for minerals, or used to construct potential functions for use in molecular dynamics simulations (Hess et al., 1986; Lasaga and Gibbs, 1987; Tsuneyuki et al., 1988; McMillan and Hess, 1990; van Beest et al., 1994).

The lattice dynamics of dense oxide phases, including several important mantle minerals, were also modeled using a first-principles modified electron gas (MEG) approach, based upon the ionic model (Hemley and Gordon, 1985; Wolf and Jeanloz, 1985; Mehl et al., 1986; Cohen, 1987; Cohen et al., 1987a, 1987b; Hemley et al., 1987a; Wolf and Bukowinski, 1987; Hemley and Cohen, 1992; Chizmeshya et al., 1994; Ita and Cohen, 1998). Such calculations immediately provided new insights into the lattice dynamics and the thermodynamic behavior of rocksalt-structured MgO, and $MgSiO_3$ and $CaSiO_3$ perovskites, that led to a new understanding of lower mantle mineralogy and thermophysical properties (Hemley and Cohen, 1992).

Recent advances in theoretical methods and computational methodology are now making it possible to carry out accurate ab initio or first-principles calculations of mineral lattice dynamics, as well as their structures and thermoelastic properties, into high-P,T ranges where vibrational spectroscopic data are not currently available, and may even never be possible to collect experimentally (Bukowinski, 1994;

Karki et al., 2001; Oganov et al., 2002; Oganov, 2004; Wentzcovich et al., 2004b; Jung and Oganov, 2005; Oganov et al., 2005b). Such calculations are revolutionizing our understanding of the properties of mantle minerals and their phase relationships (Stixrude et al., 1996, 1998; Oganov et al., 2005b; Stixrude and Lithgow-Bertelloni, 2005).

New experimental results continue to be required to complement the theoretical calculations, however. The anharmonic vibrational properties of minerals are not yet fully accounted for theoretically, so that the predicted behavior at high T and high P must be tested by experiments whenever possible (Guyot et al., 1996; Oganov et al., 2000; Karki and Wentzcovich, 2002; Wentzcovich et al., 2004a). The behavior of mineral solid solutions and disordered systems, including defect formation, atomic/ionic diffusion processes, and melting phenomena, are difficult to model accurately using nonempirical theoretical methods. Empirical simulation approaches to these problems require experimental data as input and to calibrate the findings. Further advances in mineral physics studies thus require a continued close interaction between theoretical and experimental studies, to investigate and determine the behavior of materials that occur deep within the Earth, as well as in other planetary systems.

7.2 Mineral Lattice Dynamics and Vibrational Spectroscopy

7.2.1 Molecular Vibrations

The vibrational dynamics of molecules and crystalline solids are determined by the relative positions of atoms and their interatomic bonding (Born and Huang, 1954; Wilson et al., 1955; McMillan, 1985; Dove, 1993). Within isolated diatomic molecules in the gas phase, the characteristic vibrational frequency (ν_o) is defined within the harmonic oscillator model by the atomic masses m_1 and m_2 along with the bond force constant (f) by (Herzberg, 1950; Nakamoto, 1978; McMillan, 1985):

$$\nu_o = \frac{1}{2\pi}\sqrt{\frac{f}{m^*}} \qquad [1]$$

Here, m^* is the 'reduced mass' ($m^* = m_1 m_2/(m_1 + m_2)$), that is introduced to eliminate rotations or translations among gas-phase molecules from the problem: its use leads to transformation of the diatomic vibration into that of a particle with mass m^* held between fixed walls. That model is readily extended to a chain of atoms, that provides the simplest model to understand vibrations within solids (**Figure 1**). ν_o in eqn [1] has units of Hz (s^{-1}); typical values lie within the THz range ($\sim10^{12}\,s^{-1}$). The vibrational 'frequencies' determined by IR or Raman spectroscopy are usually reported in wave number units (cm^{-1}), obtained from multiplying ν_o by $2\pi\nu/c$ ($c = 2.998 \times 10^{10}\,cm^{-1}s$). The force constant f corresponds to the second derivative of the potential energy ($E(r)$), evaluated at the bonded minimum (r_o). Within polyatomic molecules and solids, the potential energy function forms a multidimensional surface ($E(\zeta)$), where ζ represents the various interatomic displacement coordinates, that are commonly expressed as bond-stretching, angle-bending motions, etc. (Wilson *et al.*, 1955; Decius and Hexter, 1977; Nakamoto, 1978; McMillan, 1985; Schrader, 1994). The derivatives $\partial^2 E/\partial\zeta^2$ represent the curvature of the $E(\zeta)$ function in the vicinity of the global energy minimum with respect to these various motions, so that they constitute a set of local force constants expressed within the chosen coordinates of the force field. For example, a 'valence' force field model usually has only covalent bond-stretching and angle-bending force constants; these might be supplemented by repulsive interactions in other models (Wilson *et al.*, 1955; Shimanouchi, 1970; Nakamoto, 1978; Dowty, 1987a, 1987b, 1987c). In polar solids that have their

structures determined by ionic bonding, the short-range interactions are supplemented by long-range Coulombic forces, that also contribute to the lattice vibrations (Born and Huang, 1954; Bilz and Kress, 1979; McMillan, 1985; Burnham, 1990; Catlow and Price, 1990; Dove, 1993; Gramaccioli, 2002b).

Among the vibrational modes of polyatomic molecules, that include the molecular groupings found within minerals such as SiO_4 tetrahedra, SiO_6 octahedra, and SiOSi linkages, the various vibrational 'modes' are usefully classified using point group symmetry (Cotton, 1971; Harris and Bertollucci, 1978; McMillan and Hess, 1988; Pavese, 2002) (**Figure 2**). The molecular vibrations are given symmetry labels according to the Schönflies scheme (e.g., A_1, B_{2g}, E_g, F_{1u} (or T_{1u})) depending upon their transformation characteristics under the symmetry operations of the point group, and the degeneracy of the vibration. For example, E modes are doubly degenerate (i.e., they have two components that are equal in energy, and that form part of the same representation; an example is the bending vibration of tetrahedral $SiO_4{}^{4-}$ units), and T (or F) representations are triply degenerate. The group theoretical analysis allows prediction of the IR and Raman activities of the vibrational modes (Hamermesh, 1962; Cotton, 1971;

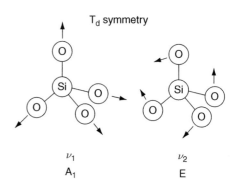

T_d symmetry

ν_1
A_1

ν_2
E

(a)

r

m_1 m_2

(b)

$f/2$ $f/2$

m^*

(c)

a m f

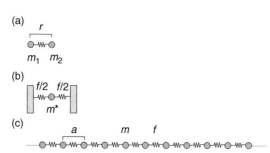

Figure 1 Simple vibrational oscillator models for molecules and crystals. (a) In a diatomic molecule, the two atoms can be approximated by point masses (m_1, m_2) and the bonding between them as a classical spring with force constant f. The separation between the atoms is r. (b) Molecular vibrations are usefully treated in center-of-mass coordinates, using the reduced mass ($m^* = m_1m_2/(m_1 + m_2)$). The diatomic vibration then corresponds to that of a single particle with reduced mass m^*, held between fixed walls by equal springs, each with force constant $f/2$. (c) That model of a diatomic vibration is readily extended to that of a monatomic chain, to provide a useful approach for understanding the vibrations of crystals. The lattice repeat distance is a.

Figure 2 Two vibrational modes of the $SiO_4{}^{4-}$ tetrahedron with point group symmetry T_d, that is found as a molecular species within mantle minerals such as silicate spinels (e.g., γMg_2SiO_4). The symmetric stretching vibration (ν_1) belongs to symmetry species A_1: it is predicted to appear in the Raman spectrum, but not in the IR spectrum (Herzberg 1945; Basile *et al.* 1973; Nakamoto 1978). A ν_1 asymmetric stretching vibration also occurs (T_2 symmetry), that is IR but not Raman active, The bending vibration ν_2 is doubly degenerate (E symmetry): it is both IR and Raman active. The final bending vibration (ν_4) has T_2 symmetry. In olivine-structured minerals, the $SiO_4{}^{4-}$ units are distorted from tetrahedral symmetry (C_s point group), resulting in modifications to the IR and Raman symmetry selection rules (Piriou and McMillan, 1983; Hofmeister, 1987).

McMillan, 1985; McMillan and Hofmeister, 1988; Schrader, 1994). The symmetry analysis is readily extended to crystalline minerals, where the point group of the unit cell is used instead of that of the finite molecule (Fateley *et al.*, 1972; Farmer, 1974; Decius and Hexter, 1977; McMillan, 1985).

7.2.2 Vibrational Dispersion Relations

In crystalline solids, the atoms are all inter-connected so that the vibrational modes take the form of atomic displacement waves traveling throughout the solid. Transverse (T) and longitudinal (L) modes are distinguished by the orientation of atomic displacement vectors relative to the propagation direction (**Figure 3**). Within each unit cell, vibrational modes with different frequencies (ν_i) occur as atoms oscillate about their equilibrium positions within the local force field. Because of the propagation of the vibrational displacements throughout the periodic solid, the lattice vibrational frequencies are also determined by the relative phase of vibrations between adjacent unit cells, so that a wavelength parameter (λ_{vib}) must also be defined for each vibrational mode. It is convenient to define a vibrational 'wave vector' appearing in reciprocal space (**k** or **q**, with magnitude $1/\lambda_i$, or $2\pi/\lambda_i$, that is usually expressed as a fraction of the primitive lattice constants. The result is the first Brillouin zone (BZ), that extends from $q = 0$ (i.e., $\lambda_{vib} \to \infty$) to $q = 1/2a_o$ ($\lambda_{vib} = \pi/a_o$, where a_o is a lattice constant).

Each vibrational mode within the crystal has a specific frequency associated with a particular wave vector value, $\nu_i(\mathbf{q})$, for propagation along a given direction in reciprocal space. Because macroscopic crystals contain a very large number of atoms, and hence vibrational frequencies, the $\nu_i(\mathbf{q})$ relations are usually represented as continuous curves (Bilz and Kress, 1979; McMillan, 1985; Dove, 1993; Choudhury *et al.*, 2002). A crystalline solid containing N atoms within its primitive unit cell contains $3N$ 'branches' in its vibrational dispersion relations (**Figure 4**). Three of these rise from $\nu = 0$ at $q = 0$: these constitute the acoustic branches that determine the sound speeds (Weidner, 1975, 1987; Bass, 1995; Liebermann and Li, 1998). The slopes of $\nu_i(\mathbf{q})$ curves as $q \to 0$ determine the elastic constants in various crystallographic directions: these are often directionally averaged to produce the longitudinal, shear and bulk elastic moduli (Weidner, 1975; Kieffer, 1979a, 1979b, 1979c; Bass, 1995; Liebermann and Li, 1998). The remaining $3n$-3 branches are associated with

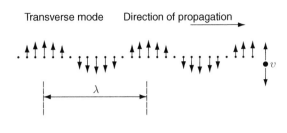

Transverse mode   Direction of propagation

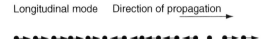

Longitudinal mode   Direction of propagation

Figure 3 Within crystals, the atomic vibrations are all coupled together to constitute lattice vibrational waves with wavelength λ, that is described relative to the structural periodicity imposed by the crystallographic (primitive) unit cell. Different types of lattice modes can be distinguished. For 'acoustic' vibrations, all of the atoms within each unit cell are displaced approximately in phase with each other, especially at long wavelength, to constitute the excitations that result in propagation of sound waves, and that determine the mineral elasticity. The 'optic' modes have atoms moving in opposition to each other, to generate dielectric perturbations that can interact with light, and thus give rise to IR or Raman activity. Transverse lattice modes have a vibrational oscillation of atoms about their equilibrium position that is generally normal to the direction of propagation. Because light is a transverse electromagnetic wave, it is the TO vibrations that are observed in IR absorption experiments. Longitudinal optic or acoustic modes (LO, LA) have their atomic displacements aligned with the direction of propagation. The LO mode frequencies for IR-active vibrations are determined by the positions of poles in the reflectivity function in IR reflectance experiments (Gervais et al., 1973a, 1973b; Gervais and Piriou, 1975; Decius and Hexter, 1977; Gervais, 1983; Hofmeister, 1987; McMillan and Hofmeister, 1988; Efimov, 1995). LO modes can appear be observed directly in Raman scattering spectroscopy. LA and TA mode frequencies are determined very close to the Brillouin zone centre via high-resolution Brillouin scattering spectroscopy. All modes throughout the Brillouin zone are observed by techniques like inelastic neutron scattering (INS) spectroscopy.

atomic displacements that can interact with light and thus lead to IR and Raman activity: these are the 'optic' branches. Only vibrational excitations with $q \to 0$ can usually be observed in IR or Raman spectra, due to the long wavelength nature of the incident light waves (McMillan, 1985; McMillan and Hofmeister, 1988). However, all of the modes can interact with 'slow' neutrons (i.e., neutron beams thermally equilibrated near $T = 300\,\mathrm{K}$) in INS experiments, to provide a more complete picture of the lattice dynamics (Bilz and Kress, 1979;

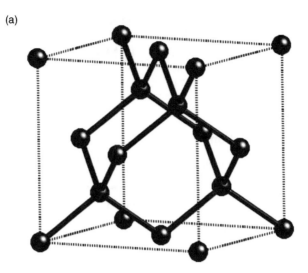

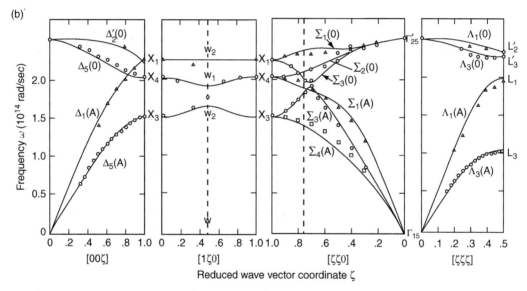

Figure 4 Structure and lattice vibrations of diamond. The various points and lines correspond to experimental INS determinations and results of lattice dynamics calculations. (a) The cubic diamond structure contains two atoms in its primitive unit cell. For that reason, its lattice dynamics correspond to those of a diatomic, rather than a monatomic chain of atoms, extended to three dimensions. (b) The lattice dynamics of a single crystal of diamond measured along various directions in reciprocal space, using INS (Warren *et al.*, 1967). Three acoustic modes rise from zero frequency (two TA + one LA modes); these determine the sound speed and the elastic properties, and also the thermal conductivity. Three additional modes constitute the optic branches (two TO + one LO) that converge to provide the single Raman active frequency (T_{2g} symmetry) at the Brillouin zone center. The diagrams in part (b) are reprinted from Warren JL, Yarnell JL, Dolling G, and Cowley RA (1967) Lattice dynamics of diamond. *Physical Review* 158: 158.

Ghose, 1988; Dove, 1993; Chaplot *et al.*, 2002b; Choudhury *et al.*, 2002; Winkler, 2002).

The BZ for a given crystal structure contains various special symmetry points and directions (Decius and Hexter, 1977; Birman, 1984; Choudhury *et al.*,

2002). The BZ center lies at the 'Γ' point in reciprocal space, that possesses the full point group symmetry of the unit cell (i.e., all the unit cell vibrations are in phase). For crystal structures including B1-structured MgO and the ideal cubic perovskite, the special symmetry '*K*' point occurs at the BZ boundary along the (111) direction, and the '*M*' point occurs along (110): at these points, the vibrational displacements within adjacent cells are exactly out of phase, for vibrational excitations along these crystallographic directions (Wolf and Jeanloz, 1985; Hemley and Cohen, 1992) (**Figure 3**). Within solid-state physics, the symmetry labels for crystal lattice vibrations are usually designated following a numbering scheme at each special symmetry point. For example, Γ_1, Γ_3 or Γ_{25}' refer to different vibrational modes occurring at the BZ center; these are readily correlated with the Schönflies symmetry designations, by considering the point group symmetry of the unit cell (Decius and Hexter, 1977). Vibrational modes at the BZ boundary (i.e., when vibrations in adjacent cells are exactly out of phase) have designations like R_{25}, L_3, M_{15}', etc. The dispersion relations connect vibrations at special symmetry points within the BZ to constitute 'branches'. For example, the triply degenerate Raman active mode Γ_{25}' (i.e., T_{2g} symmetry) for the diamond structure (space group *Fd3m*, point group $O_h{}^7$) occurs at the BZ center. For propagation along the Γ–L or Λ ($\zeta\zeta\zeta$) direction in reciprocal space to the L point at (0.5, 0.5, 0.5), the degeneracy is partly lifted to yield the singly degenerate L_2' and doubly degenerate L_3' modes. Likewise, the transverse and acoustic branches (2TA + LA) rise from zero frequency at Γ to give rise to the L_3 and L_1 modes at the BZ boundary along [111] (**Figure 4**).

The $\nu_i(\mathbf{q})$ relations averaged throughout the BZ result in a vibrational density-of-states function (VDOS, or $g(\nu)$). That information is provided directly by INS measurements on powders, taking into account the different contributions of atoms involved in each vibration to the scattering function (Ghose, 1988; Chaplot *et al.*, 2002b; Choudhury *et al.*, 2002). Similar information on mineral lattice dynamics is provided by high-resolution inelastic X-ray scattering, where the incident light wavelength is on the order of the unit cell dimensions (Fiquet *et al.*, 2001; Mao *et al.*, 2001; Duffy, 2005). Theoretical studies calculate the $\nu_i(\mathbf{q})$ and $g(\nu)$ relations directly, using lattice dynamics or MD simulation techniques with empirical potentials adapted to the mineral system of interest, or increasingly using highly accurate first principles methods (Hemley *et al.*, 1987a; Price *et al.*, 1987; Cohen, 1991, 1992b; Chizmeshya *et al.*, 1996; Mao *et al.*, 2001; Choudhury *et al.*, 2002; Oganov *et al.*, 2002, 2005b; Oganov, 2004; Tse, 2004; Wentzcovich *et al.*, 2004a, 2004b).

7.2.3 Quantized Vibrations in Molecules and Crystals (Phonons)

Vibrations of molecules and solids are properly treated using quantum mechanics, with excitations occurring between vibrational energy levels (E_{vib}) and states described by the corresponding vibrational wave functions (ϕ) (Herzberg, 1950; Pauling and Wilson, 1963; McMillan, 1985). The relationship between E_{vib} and ϕ is given by the Schrödinger equation ($\hat{H}\phi = E\phi$), where $\hat{H}$ is the Hamiltonian operator appropriate to the vibrational problem, that gives rise to a differential equation in ϕ_{vib} as a function of the vibrational displacement coordinate. This is solved to return the mathematical forms of the vibrational wave functions and their corresponding energies. The solutions are found to contain an integer n ($n = 0, 1, 2, 3, \ldots$) known as the vibrational quantum number. For a diatomic molecule within the harmonic approximation (i.e., the potential energy is a parabolic function of the vibrational displacement), the vibrational energies are $E_{vib} = (n + \frac{1}{2})h\nu_0$, where ν_0 is the classical oscillator frequency. The $1/2$ term arises from the wave equation solution. This means that even in the lowest vibrational state ($n = 0$), there is vibrational energy present ($E_{vib} = (n + \frac{1}{2})h\nu_0$), known as the 'zero point' energy (ZPE). For large molecules with many vibrational modes including crystalline minerals, the ZPE can provide a substantial contribution to the total energy. Transitions between solid-state phases including mantle minerals could be affected by differences in the ZPE between different phases. However, such contributions are found to change phase boundaries by only up to a few GPa, that is on the order of errors associated with experimental pressure determination in high-*P,T* studies (Li and Jeanloz, 1987; Wentzcovich *et al.*, 2004a). For large molecules containing several vibrational modes (ν_i), series of vibrational levels exist associated with each mode (Herzberg, 1945; Wilson *et al.*, 1955; McMillan, 1985). As the temperature is raised, higher energy levels become populated according to Maxwell–Boltzmann statistics, and the vibrational energy is a weighted sum over populated levels. The result is a vibrational contribution to the thermal free energy (F) and its derivatives, including the heat capacity (C_v) and the entropy (S), that can be evaluated using methods of statistical mechanics (see below).

The units of vibrational excitation within crystals are termed 'phonons'; each represents a vibrational energy ($E_i = h\nu_i(\mathbf{q})$ that corresponds to the energy difference between adjacent vibrational energy states

(Wallace, 1972; Reissland, 1973; Decius and Hexter, 1977; Ghose, 1985; McMillan, 1985, 1989; Ghose, 1988; Dove, 1993; Chaplot *et al.*, 2002b; Choudhury *et al.*, 2002). Because of the dependence of vibrational excitations upon the wave vector, the quantity *hq* associated with each phonon has units of momentum (i.e., kg m s^{-1}), so that phonons possess an 'effective' mass and thus behave as 'quasi-particles'. This property gives rise to the inelastic scattering phenomena observed to occur with neutrons in INS experiments, and also with photons in Raman, Brillouin, and inelastic X-ray scattering spectroscopies. The phonon 'quasi-particles' can also scatter among themselves (this is a consequence of vibrational anharmonicity: see below), and these processes are *T*-dependent, so that phonon–phonon interactions can play an important role in determining the high-*T* properties of mantle minerals (Wallace, 1972; Reissland, 1973; Wolf and Jeanloz, 1984; Gillet *et al.*, 1991, 1996a, 1997, 1998, 2000; Oganov *et al.*, 2000; Wentzcovich *et al.*, 2004a, 2004b).

The parabolic relationship between the potential energy and vibrational displacement assumed within the harmonic oscillator model for molecules holds approximately for low degrees of vibrational excitation, close to the minimum energy. Real diatomic $E(r)$ curves deviate significantly from this relation and are thus anharmonic, because of the nature of interatomic bonding interactions (Herzberg, 1945, 1950; Nakamoto, 1978; McMillan, 1985; McMillan and Hofmeister, 1988). Vibrations within polyatomic molecules give rise to more complex modes of vibration, including angle bending and torsional motions that are often intrinsically anharmonic. In crystals, significant effects of anharmonicity can be encountered at both low and high *T* for even the simplest solids (including diamond and diamond-structured Si and Ge), because of the complex interactions that determine the vibrational potential function (Brüesch, 1982; Gillet *et al.*, 1998). The anharmonicity has various consequences for molecules and solids, at different levels of complexity. In the case of diatomic molecules, the spacing between adjacent vibrational levels ($\Delta E = h\nu_0$ for a harmonic oscillator) decreases with increasing quantum number (*n*). In addition, transitions between levels spaced by $\Delta n = \pm 2, \pm 3$, etc. become allowed in IR and Raman spectra (Herzberg, 1950; McMillan, 1985). Among polyatomic molecules, transitions between the energy level progressions associated with different vibrational modes become allowed. These effects give rise to 'overtone' and 'combination' modes that are observed in the vibrational

spectra. They are usually much weaker than the 'fundamental' transitions that appear within the harmonic model, that occur between the vibrational ground state (i.e., all $n_i = 0$) and the first excited state for each vibration. The identification and characterization of overtone and combination bands can be extremely useful in determining the potential energy function or surface for molecules and crystals, however (Herzberg, 1945; Decius and Hexter, 1977).

7.2.4 Vibrational Frequency Shifts at High *P* and *T*; Quasi-Harmonic Model and Phonon Anharmonicity

Within crystals, the anharmonic nature of the interatomic potential generally causes the interatomic distances (*r*) to shift to larger values as *T* is increased: this results in the phenomenon of thermal expansion that is usually observed among minerals (however, some exceptions are well known, especially among framework mineral structures such as SiO_2 cristobalite and ZrW_2O_6 (Mary *et al.*, 1996; David *et al.*, 1999; Heine *et al.*, 1999; Mittal and Chaplot, 1999; Dove *et al.*, 2002)). Vibrational frequencies generally shift to lower wave number with increasing *T*, and they increase with *P*, due to the asymmetric form of the interatomic potential energy function (Herzberg, 1950; McMillan, 1985). Within the 'quasi-harmonic' (QH) model of mineral lattice dynamics, the lattice expands or contracts according to the high-*P*, *T* conditions, and a new set of harmonic phonons is considered to exist at each new set of distances and interatomic configurations. With the QH model, no phonon–phonon scattering events are considered to occur: that has implications for the high-*T* heat capacity and other thermodynamic parameters (see below). The vibrational linewidths measured in IR, Raman or INS spectroscopy are predicted to be independent of temperature.

Vibrational frequency shifts measured as a function of *P* or *T* (i.e., $(\partial \nu_i / \partial P)_T$ and $(\partial \nu_i / \partial T)_P$) are often expressed in terms of the volume (*V*) as mode Grüneisen parameters: i.e.,

$$\gamma_i = \frac{d \ln \nu_i}{dV} \qquad [2]$$

For experimental data measured at constant *T* but variable *P* (e.g., in a diamond anvil cell experiment at ambient *T*), these parameters are obtained by

$$\gamma_{iT} = \frac{K_T}{\nu_i} \left(\frac{\partial \nu_i}{\partial P} \right)_T = K_T \left(\frac{\partial \ln \nu_i}{\partial P} \right)_T \qquad [3]$$

K_T is the isothermal bulk modulus. For data determined at variable T and constant P, the corresponding parameters are obtained using the volume thermal expansion coefficient (α):

$$\gamma_{iP} = -\frac{1}{\alpha \nu_i}\left(\frac{\partial \nu_i}{\partial T}\right)_P = -\frac{1}{\alpha}\left(\frac{\partial \ln \nu_i}{\partial T}\right)_P \qquad [4]$$

Measured mode Grüneisen parameters can be averaged to give an approximate mean value ($<\gamma>$), to be compared with the 'thermal' Grüneisen parameter (γ_{th}) that appears in mineral equation-of-state formulations (Jeanloz, 1985; Anderson, 1995; Duffy and Wang, 1998; Gillet *et al.*, 1998; Holzapfel, 2004):

$$\gamma_{th} = \frac{\alpha K_T V}{C_v} \qquad [5]$$

This can be a useful exercise, because vibrational spectra can be measured into high-P,T ranges where thermodynamic data are often inaccessible. Comparisons between γ_{th} and $<\gamma>$ have now been described for important mantle minerals, including $MgSiO_3$ perovskite, garnet and ilmenite, SiO_2 stishovite, and $(Mg, Fe)_2SiO_4$ polymorphs (Williams *et al.*, 1986, 1987, 1993; Hemley, 1987; Hemley *et al.*, 1989; Hofmeister *et al.* 1989; Chopelas 1990, 1991; Gillet *et al.* 1990, 1991, 1997, 1998, 2000; Hofmeister and Ito 1992; Reynard *et al.* 1992; Chopelas *et al.* 1994; Liu *et al.* 1994; Hofmeister 1996; McMillan *et al.* 1996b; Reynard and Rubie 1996).

Carrying out independent studies for the same mineral under (1) variable-T, constant P, and (2) high-P, constant T conditions provides an important experimental test of the quasi-harmonic model, for the behavior of individual vibrational mode frequencies as a function of the volume or strain parameters induced by changes in the T and P. Observed $\nu_{i,T}(P)$ and $\nu_{i,P}(T)$ data are converted to $\nu_i(V)$ relations using measured values for the bulk modulus and the thermal expansion coefficient, respectively, and the results for $(d\nu_i/dV)$ are compared on a single plot. If the QH approximation held true, then both data sets should be collinear. This is not found to be the case for the vibrational modes of mantle minerals examined to date, including simple structures like diamond (Gillet *et al.*, 1990, 1991, 1993, 1996a, 1997, 1998, 2000; Reynard *et al.* 1992; Reynard and Rubie 1996) (**Figure 5**). The general conclusion is that most vibrational modes for most minerals exhibit some degree of anharmonicity. The main question for mantle behavior is how this anharmonicity affects the phase relations and thermophysical properties under high-P,T conditions.

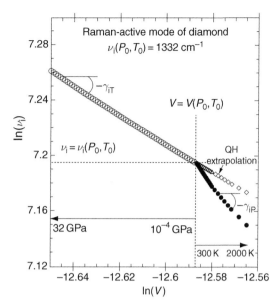

Figure 5 Anharmonicity of the Raman active mode of diamond, from high-T, ambient-P and high-P, ambient-T measurements. Here, the results from high-P, ambient-T and variable-T, P = 1 atm experiments are combined. The P, T dependencies of the Raman-active mode at the Brillouin zone center ($\Gamma_{vib} = T_{2g}$), that is, $(\partial \nu_i/\partial P)_T$ and $(\partial \nu_i/\partial T)_P$), are transformed into volume dependencies using the bulk modulus and thermal expansion coefficients. The two $\partial \nu_i/\partial V$ relationships should be collinear, within a QH model. The observed mismatch between the high-/low-T and high-P relationships demonstrates that intrinsic phonon anharmonicity is present, and this must be considered in evaluation of the lattice dynamics and thermophysical properties. The diagram is reprinted from Gillet P, Hemley RJ, and McMillan PF (1998) Vibrational properties at high pressures and temperatures. In: Hemley RJ (ed.) *Ultra-High Pressure Mineralogy*, vol. 37, p. 559. Mineralogical Society of America.

The intrinsic mode anharmonicity represents a dependence of the vibrational frequencies on T at constant volume, that arises due to T-dependent phonon–phonon scattering events. The vibrational energy associated with anharmonic phonons is usually expressed as a series expansion, with terms derived from perturbation theory carried out to various orders, that involve scattering among various numbers of phonons (Cowley, 1963; Wallace, 1972; Reissland, 1973; Brüesch, 1982). The phonon scattering events and the resulting series terms are usefully evaluated and depicted using graphical techniques, that are related to those developed by R. Feynman for describing subatomic processes (Reissland, 1973; Wolf and Jeanloz, 1984). The intrinsic mode anharmonicity parameters obtained from the comparison of experimental data for $\nu_{i,T}(P)$ and $\nu_{i,P}(T)$ are usefully expressed as (Gillet *et al.*, 1991, 1997, 1998):

$$a_i = \left(\frac{\partial \ln \nu_i}{\partial T}\right)_V = \alpha(\gamma_{iT} - \gamma_{iP}) \qquad [6]$$

The a_i anharmonicity parameters defined in this way have units K^{-1}. Information on intrinsic vibrational mode anharmonicities are also obtained directly from analysis of IR reflectance data via damped oscillator models, in which a related anharmonicity parameter (Γ_i) for each IR active mode is empirically adjusted to fit the observed reflectivity line shape (Piriou and Cabannes, 1968; Gervais *et al.*, 1973a, 1973b; Gervais and Piriou, 1975; Gervais, 1983; McMillan, 1985; Hofmeister, 1987, 1997; McMillan and Hofmeister, 1988). Here, the Γ_i parameter represents the width at half-maximum of the peak in the dielectric loss function (i.e., $\varepsilon_2(v)$) associated with infrared absorption, and it is a measure of the phonon lifetime (i.e., it has units of s^{-1}). The a_i and Γ_i parameters can be directly compared between IR and Raman studies, for minerals that have modes that are both IR- and Raman active (e.g., quartz), that do not possess inversion symmetry. This is not often the case for mantle minerals, however, and anharmonicity comparisons between IR and Raman active vibrations must be made among different members of families of modes that have similar origins.

The presence of intrinsic phonon anharmonicity has several potential implications for the thermoelastic behavior of mantle minerals, under deep Earth conditions. First, the anharmonic contribution causes the heat capacity (C_v, C_p) and the vibrational entropy (S) to exceed the classical limit at high T, the definition of the thermal expansion coefficient is affected, and the thermal pressure is augmented during heating under constant volume conditions: the analysis of phase transition boundaries within the mantle can be affected, as can the thermoelastic properties of mantle minerals. These various effects have all been recorded experimentally during high-T, P studies of mantle minerals (Gillet *et al.*, 1990, 1991, 1996a, 1997, 1998, 2000; Fiquet *et al.*, 1998; Wentzcovich *et al.*, 2004a).

Most current first-principles (or *ab initio*) calculations of mantle minerals do not evaluate the effects of intrinsic phonon anharmonicity explicitly. Instead, the QH model is used. The static lattice energy (E) phonon frequencies (ν_i) and vibrational density of states ($g(\nu)$) are computed as a function of the volume at $T = 0$ K. The $E(V)$ data are then fitted to an appropriate equation of state (e.g., a Birch–Murnaghan function) to obtain the isothermal compressibility ($K_o = (\partial \ln V/\partial P)_T$), and hence the phonon frequency shifts with pressure (i.e., $\partial \ln \nu_i/\partial P$). The vibrational contribution to the thermal free energy (F_{th}) is computed at each volume (or pressure) as an integral over the phonon density of states using the methods of statistical thermodynamics (Wallace, 1972; Gillet *et al.*, 1998; Oganov *et al.*, 2000, 2002), resulting in a QH approximation to the thermal expansion coefficient ($\alpha = \partial \ln V/\partial T$). The high-temperature phonon behavior and thermoelastic properties can then be derived (Guyot *et al.*, 1996; Duffy and Wang, 1998; Oganov *et al.*, 2000, 2002; Oganov, 2004; Tse, 2004).

The quasi-harmonic model is now known to reproduce many of the important thermoelastic properties of mantle minerals under the range of high-P, T conditions encountered within the lower mantle and transition zone (Guyot *et al.*, 1996; Gillet *et al.*, 2000; Oganov *et al.*, 2000, 2002; Karki and Wentzcovich, 2002; Oganov, 2004; Wentzcovich *et al.*, 2004a), It has been used to predict phase transitions among important lower mantle candidate phases (e.g., stishovite, Al_2O_3 and $MgSiO_3$ and $CaSiO_3$ perovskites) that have been verified by experiment (Cohen, 1992b; Marton and Cohen, 1994; Kingma *et al.*, 1995; Chizmeshya *et al.*, 1996; Stixrude *et al.*, 1996; Funamori and Jeanloz, 1997; Murakami *et al.*, 2003, 2004a; Oganov and Ono, 2004; Shim *et al.*, 2004; Tsuchiya *et al.*, 2004b). However, phase equilibrium calculations for phase transitions involving major structural transformations, such as that between $MgSiO_3$ ilmenite (akimotoite) and perovskite, have shown significant differences from the experimental P–T transition line: such discrepancies have been assigned to intrinsic anharmonic phonon effects (Wentzcovich *et al.*, 2004a).

A full theoretical treatment of intrinsic anharmonicity in minerals is achieved using molecular dynamics (MD) or Monte Carlo (MC) simulation techniques (Allen and Tildesley, 1987; Dove, 1993; Tse, 2004; Winkler, 2004). These methods are usually implemented using empirical potential energy functions, because they involve such large system sizes (i.e., number of atoms) (Matsui and Busing, 1984; Matsui *et al.*, 1987; Price *et al.*, 1987, 1989, 1991; Matsui, 1988; Dove, 1993; Ghose *et al.*, 1994; Choudhury *et al.*, 1998, 2002; Chaplot *et al.*, 2002b). They are thus subject to possible errors when the calculations are applied outside the range of validity of the data used to construct the potential function.

However, simulations of anharmonic lattice dynamics of mantle minerals carried out using empirical MD simulations have given many valuable insights into the thermodynamic properties and phase relations (Matsui and Busing, 1984; Matsui et al., 1987; Price et al., 1987, 1989; Matsui, 1988; Matsui and Tsuneyiki, 1992; Winkler and Dove, 1992; Dove, 1993). Several MD simulations and lattice dynamics calculations of mineral properties, including mineral phase transitions as well as melting under high-P conditions, have been carried out using potential energy models derived using *ab initio* calculations for molecular units that provide models for the structural units present, or using first-principles calculations for solids (Gibbs, 1982; Lasaga and Gibbs, 1987; Tsuneyuki et al., 1988, 1989; McMillan and Hess, 1990; Kobayashi et al., 1993; van Beest et al., 1994; Ita and Cohen, 1997, 1998; Kolesov and Geiger, 2000). Such calculations provide a more fundamental approach to MD studies of mineral anharmonic properties and phase transitions. They can suffer from problems associated with transferability of the potential functions between molecules and solids, or among condensed phases of very different structure type.

Anharmonic first-principles calculations of solids, including lattice dynamic properties of mantle minerals under high-P conditions, can now be computed using a quantum mechanical perturbation theory approach, using linear response theory and Green's function methods (Baroni et al., 1987; Stixrude et al., 1996; Karki et al., 2000a, 2000c; Baroni et al., 2001). A fully *ab initio*/first-principles treatment including intrinsic anharmonic effects can also be achieved using quantum molecular dynamics (QMD) or Monte Carlo methods (Car and Parrinello, 1985; Remler and Madden, 1990; Oganov et al., 2002). Such calculations were previously limited to quite small system sizes (e.g., 20–30 atoms). However, following advances in both computational techniques and computer speed, memory capacity and architecture, they are being applied to increasingly large and complex problems. In addition, the original theoretical methods are being improved, especially as a result of mineral physics studies of the deep Earth and other planetary systems (Ancilotto et al., 1997; Alfé et al., 1999; Cavazzoni et al., 1999; Oganov et al., 2002). Such calculations have now been applied to several important mantle mineral systems, and they hold great promise for obtaining accurate thermoelastic properties of planetary materials under high-P,T conditions (Kobayashi et al., 1993; Wentzcovich et al.,

1993, 2004b; Oganov et al., 2000, 2001, 2002, 2005b; Marton et al., 2001; Brodholt et al., 2002; Marton and Cohen, 2002; Trave et al., 2002; Alfé et al., 2004; Alfé, 2005).

Anharmonicity among mineral vibrations also results in the occurrence of displacive structural phase transitions, as the vibrational excitations cause atomic displacements to occur along pathways that connect the low- and high-P,T structures (Raman and Nedungadi, 1940; Scott, 1974; Ghose, 1985; McMillan, 1985; Kingma et al., 1995). If a single vibrational mode is involved, its restoring force vanishes at the phase transition so that the vibrational frequency goes asymptotically to zero: this is known as 'soft mode' behavior (Scott, 1974; McMillan, 1985; Ghose, 1985; Salje, 1990). The phenomenon was first observed to occur for quartz, in which the α–β transition at 858 K is associated with a vibrational mode occurring at ~200 cm^{-1} at ambient T and P (Raman and Nedungadi, 1940; Scott, 1974; Ghose, 1985). Soft mode behavior has also been predicted and observed for important mantle minerals including $CaSiO_3$ perovskite (in which a cubic-tetragonal transition occurs) and SiO_2 stishovite, where it leads to a transformation into a $CaCl_2$-structured phase (Cohen, 1992b; Hemley and Cohen, 1992; Kingma et al., 1995; Chizmeshya et al., 1996; Stixrude et al., 1996; Andrault et al., 1998). Soft-mode-driven phase transitions among orthorhombic-tetragonal-cubic phases were predicted for $MgSiO_3$ perovskite (Wolf and Jeanloz, 1985; Hemley et al., 1987a, 1989; Bukowinski and Wolf, 1988; Hemley and Cohen, 1992); however, it is now thought unlikely that these will occur under the P, T conditions achieved within the lower mantle.

7.2.5 Mineral Thermodynamic Quantities from Vibrational Spectra

Within the harmonic model, the vibrational contribution to the Helmholtz free energy (F) at constant V and variable T is

$$F_{th} = \int \frac{h\nu_i}{2} + kT \ln\left(1 - e^{-h\nu_i/kT}\right)g(\nu_i)d\nu$$

where the integration runs over the vibrational density of states ($g(\nu)$). Experimental or theoretical determinations of $g(\nu)$ thus lead to prediction and rationalization of important thermodynamic quantities derived from F_{th}, including the heat capacity (C_v or C_p), the 'third law' or vibrational entropy (S_{vib}), and the thermal pressure (P_{th}), using the methods of statistical thermodynamics (Wallace, 1972; Kieffer, 1979a, 1979b, 1979c, 1985; Hofmeister *et al.*, 1989; Chopelas, 1991; Gillet *et al.*, 1991, 1997, 1998, 2000; Fiquet *et al.*, 1992; Hofmeister and Ito, 1992; Chopelas *et al.*, 1994; Guyot *et al.*, 1996; Hofmeister, 1996).

The contribution of a given vibrational mode ($\nu_i(q)$) to the specific heat at constant volume (C_{vi}) is

$$C_{Vi} = T \frac{\partial S_i}{\partial T}\bigg|_V \qquad [7]$$

where the vibrational entropy is obtained by integration over $g(\nu)$:

$$S_i(V, T) = k_B \left[-\ln\left(1 - \exp\left(-\frac{h\nu_i}{k_B T}\right)\right) \right.$$
$$\left. + \frac{h\nu_i}{k_B T}\frac{1}{(\exp(h\nu_i/k_B T)-1)} \right] \qquad [8]$$

The contribution to the thermal pressure (P_{th}^i) is determined by

$$P_{th}^i(V, T) - P_{th}^i(V, T_0) = \int_{T_0}^{T} \frac{\partial S_i}{\partial V}\bigg|_{T'} dT' \qquad [9]$$

with

$$\frac{\partial S_i}{\partial V}\bigg|_T = k_B \frac{\gamma_{iT}}{V}\left(\frac{h\nu}{k_B T}\right)^2 \frac{\exp(h\nu/k_B T)}{(\exp(h\nu/k_B T)-1)^2} \qquad [10]$$

Within the QH model of mineral lattice vibrations, the specific heat at constant pressure (C_p) is related to that at constant volume by

$$C_p = C_v + \alpha^2 V K_T T \qquad [11]$$

where α is the coefficient of volume thermal expansion.

Following the approach described by Kieffer (1979a, 1979b, 1979c, 1980, 1985), VDOS models for mantle minerals have been constructed from the experimental vibrational spectra combined with elastic data, and these have been used to predict the thermoelastic properties of important mantle minerals and their phase transitions (Akaogi *et al.*, 1984, 1989; Chopelas, 1990, 1991; Gillet *et al.*, 1991, 1996b, 1997, 2000; Hofmeister and Chopelas, 1991a; Chopelas *et al.*, 1994; Lu *et al.*, 1994; Hofmeister 1996). Empirical $g(\nu)$ functions have also been used to calculate mantle mineral thermal conductivity (Hofmeister, 1999). VDOS functions determined experimentally by INS techniques have also been used to calculate C_v and S directly (Ghose, 1985, 1988; Price *et al.*, 1991; Choudhury *et al.*, 1998, 2002; Mittal *et al.*, 2000; Chaplot *et al.*, 2002b). Similar

studies have been carried out for Fe under deep mantle and core conditions, using $g(\nu)$ functions determined from inelastic X-ray scattering techniques (Mao *et al.*, 2001, 2005; Zhao *et al.*, 2004). There are still no experimental data on the VDOS functions of important mantle mineral phases such as $MgSiO_3$ perovskite, ilmenite and garnet, or β- and γ-Mg_2SiO_4, at ambient or under mantle high-P,T conditions. That information is currently obtained from *ab initio* or first-principles theoretical calculations of the lattice dynamics, usually within the context of the quasi harmonic model (Wentzcovich *et al.*, 1993; Chizmeshya *et al.*, 1996; Stixrude *et al.*, 1996; Karki *et al.*, 2000b, 2000c; Karki and Wentzcovich, 2002; Oganov *et al.*, 2002, 2005b; Oganov, 2004; Wentzcovich *et al.*, 2004b).

7.2.6 Theoretical Calculations of Mantle Mineral Lattice Vibrations

Since the earliest days of molecular vibrational spectroscopy and studies of covalent crystals, empirical force field models were used to supplement the experimental data and to deduce the form of the vibrational normal modes (Mills, 1963; Nakamoto, 1978). Such empirical force field calculations were readily extended to crystalline minerals, including models that incorporated long-range Coulombic forces into the vibrational calculations (Zulumyan *et al.*, 1976; Dowty, 1987b, 1987a, 1987c). Most of that work has been focused on the tetrahedrally coordinated silicates occurring within the crust and upper mantle; however, calculations of perovskite- and spinel-structured phases have also been carried out in this way (Shimanouchi *et al.*, 1961; Dowty, 1987a, 1987b, 1987c).

The lattice dynamics of ionic crystals such as NaCl and MgO were first modeled using hard-sphere ions and potential energy functions such as the Born–Mayer formulation; those calculations have been extended to study the structures and energetics of various mineral mineral structures (Burnham, 1990; Catlow and Price, 1990). With the advent of INS techniques, it became necessary to interpolate between the experimental data points to establish the phonon dispersion relations. The rigid ion model did not take proper account of the ionic polarisation effects, so that 'shell' models were developed in which the ionic charges (usually centered on the anions) were separated into inner and outer spheres that could be relaxed independently (Bilz and Kress,

1979; Dove, 1993, 2002; Chaplot *et al.*, 2002b; Choudhury *et al.*, 2002).

By the mid- to late 1970s, Hartree–Fock *ab initio* methods were being routinely applied to gas-phase molecules, including various species and molecular fragments that are relevant to mineralogy and geochemistry such as $Si(OH)_4$, H_6Si_2O, and $H_6Si_2O_7$ (Gibbs, 1982; Sauer, 1989; Tossell and Vaughan, 1992). Development of methods to yield the first and second derivatives of the Hartee–Fock energy with respect to the atomic displacements led to *ab initio* predictions of the vibrational mode frequencies and eigenvectors (i.e., the displacement patterns associated with each vibrational mode) (Hess *et al.*, 1986; Yamaguchi *et al.*, 1994; Hehre *et al.*, 1996). These methods were used to obtain force fields that could be transferred into mineral vibrational calculations, or to construct potentials for molecular dynamics (MD) simulations ((Lasaga and Gibbs, 1987; Tsuneyuki *et al.*, 1988; McMillan and Hess, 1990; van Beest *et al.*, 1994). Periodic Hartree–Fock methods were later extended to study crystalline solids, and these have now been applied to obtain the structures, energetics, phase ransitions, and bonding within mantle minerals (D'Arco *et al.*, 1993; D'Arco *et al.*, 1994; Pisani, 1996).

The Hartee–Fock *ab initio* method leads to approximate solutions to Schrödinger's equation by modeling the electronic wave functions of participating atoms, for example, usually via H-like orbitals, and then forming linear combinations to provide the molecular wave function. The total energy and its derivatives are obtained using wave mechanics methods. For crystalline structures, periodic electronic wave functions are also constructed using linear combinations of electronic plane waves, that can be 'augmented' by rapidly varying functions to model the atomic cores (i.e., APW methods). Pseudo-potentials were introduced to avoid unnecessarily detailed calculations of the electronic structure in the atomic core regions. Within the Hartree–Fock approach, dynamical electron correlations that result from electron–electron repulsions are not included. These can be related to the classical 'many-body' problem that involves determination of the relative motions of three or more objects, such as the nucleus and >1 electrons, that is not yet solved analytically. The mathematical problem is carried over into quantum mechanics. Various methods devised to correct for 'dynamic electron correlation' in molecular calculations include perturbation theory, or methods based on 'configuration interaction' (CI) or 'coupled-

cluster' approaches. Such techniques have been successfully implemented for molecules, but they are not readily applied to the solid state (Hehre *et al.*, 1996; Pisani, 1996; Oganov *et al.*, 2002; Oganov, 2004; Tse, 2004).

Density functional theory (DFT) was developed as an alternative methodology for electronic structure calculations, as it was recognized that physical observables such as the energy and its derivatives could be expressed as functionals (i.e., functions of a function) of the electron density, ρ (Kohn and Sham, 1965; Parr and Yang, 1989; Kohn, 1999). If the proper functionals were known, derivatives of the energy functionals with respect to the atomic positions would immediately allow exact first-principles calculations of mineral lattice dynamics and their thermoelastic properties to be carried out (Stixrude *et al.*, 1998; Oganov *et al.*, 2002, 2005b; Oganov, 2004; Tse, 2004; Winkler, 2004). The key problem to be solved is that of establishing the functionals associated with the electron kinetic energy (T), and with electronic exchange and correlation, $E_{xc}[\rho]$ (Perdew and Zunger, 1981; Perdew and Wang, 1992). The kinetic energy is usually obtained by constructing a set of Kohn–Sham 'orbitals', that have been interpreted in chemical studies as effective molecular orbitals and associated energy levels as in Hartree–Fock studies (Baerends and Gritsenko, 1997; Stowasser and Hoffman, 1999). The formalism for $E_{xc}[\rho]$ is known exactly for a homogeneous electron gas (Perdew and Wang, 1992). However, difficulties arise in determining the proper exchange-correlation functionals for 'real' molecules and solids, including minerals, in which large fluctuations in the electron density (ρ) occur in the vicinity of atomic cores.

The electron gas exchange-correlation functionals are usually applied to molecules and solids within the local density approximation (LDA), in which a spatially averaged electron density is considered to exist throughout the system. Although it obviously does not account for the electron density fluctuations that must occur at the atom positions, the LDA approach has been remarkably successful in describing most of the important structural and thermodynamic properties of minerals and molecules, and their chemical bonding (Stixrude *et al.*, 1998; Oganov *et al.*, 2002, 2005b). This is presumably because most of these properties are largely independent of the electron density fluctuations occurring within the region of the atomic cores. The vibrational frequencies of minerals are usually obtained using

DFT–LDA methods to within a few per cent (i.e., 5–15%) of experimental values (Stixrude *et al.*, 1998; Oganov *et al.*, 2000, 2002; Winkler, 2004; Oganov *et al.*, 2005b). Various 'generalized gradient approximations' (GGA) have been developed that provide functionals that take better account of local fluctuations in the electron density, and that have been applied to minerals (Wang and Perdew, 1991; Perdew and Burke, 1996; Perdew *et al.*, 1996; Stixrude *et al.*, 1998; Oganov *et al.*, 2002). Although use of these advanced functionals leads to a general improvement in the quality of the calculation, and in certain of the predicted structural and thermoelastic properties of minerals under high-P, T conditions, they do not always yield better predictions of the vibrational dynamics, the thermoelastic properties, or electronic properties including the bandgap (Oganov *et al.*, 2002).

For most solid-state calculations, the lattice dynamics are obtained by considering finite x-, y-, z-displacements (ξ) of atoms occurring about their equilibrium positions and analyzing the resulting energy variations ($E(\xi)$) to construct a dynamical matrix **D**, that has the second-derivative elements ($\partial^2 E/\partial \xi^2$). **D** is then combined with an 'inverse mass' matrix constituted from the elements $1/m_i$, and the result is then diagonalised to obtain vibrational mode frequencies and their corresponding atomic displacement patterns (i.e., the eigenvalues and eigenvectors of the dynamical matrix). The results thus yield the calculated vibrational dispersion relations ($v_i(q)$ values), obtained at a given level of the electronic structure theory. This approach constitutes the 'frozen phonon' model, that is usually evaluated within the quasi-harmonic approximation. Many of the first-principles or *ab initio* calculations of mineral lattice dynamics have been calculated within the DFT–LDA model using this prescription. Modern electronic structure calculations usually result in good models for the electronic energies that provide a realistic potential energy surface for the lattice vibrations, and the thermoelastic properties obtained from the calculations can be extrapolated to high-P, T conditions using thermodynamics and statistical mechanics methods (Stixrude *et al.*, 1998; Oganov *et al.*, 2002, 2005a, 2005b; Wentzcovich *et al.*, 2004b; Stixrude and Lithgow-Bertelloni, 2005). However, such extrapolations are subject to the limitations of the quasi-harmonic model, and an accurate representation of high-T mineral properties should include a proper anharmonic treatment of phonon behaviour, that is

achieved using a perturbation approach based on linear response theory aplied to the lattice dynamics (Baroni *et al.*, 1987; Stixrude *et al.*, 1996; Karki *et al.*, 2000b; Oganov *et al.*, 2000; Baroni *et al.*, 2001; Karki and Wentzcovich, 2002), or via quantum molecular dynamics methods (Oganov *et al.*, 2000, 2001; Marton *et al.*, 2001; Marton and Cohen 2002).

7.3 Lattice Dynamics of Mantle Minerals

7.3.1 Diamond and (Mg, Fe)O Magnesiowüstite

Diamond is only a minor mineral by abundance within the Earth; however, it is critically important to mantle research. Because of its gteat mechanical and chemical resistance, diamonds that grow under high-P, T conditions deep within the mantle encapsulate mineral and fluid species formed at depth that are subsequently brought to the surface as inclusions during explosive eruptions: they thus provide a unique window into mantle mineralogy and geochemistry (Sautter *et al.*, 1991; Harte and Harris, 1993, 1994; Sautter and Gillet, 1994). Diamonds also provide unique high-strength and ultra-transparent anvil-windows for diamond anvil cell (DAC) experiments, to study mineral lattice dynamics and other properties *in situ* under mantle high-P, T conditions (Hemley *et al.*, 1987b; Hemley and Mao, 2002). Diamond also provides the simplest example among covalently bonded solids as a model for understanding their lattice dynamics.

Diamond possesses a cubic structure with space group $Fd3m(O_h^7)$, with two atoms in its primitive cell (**Figure 4**). The bonding is entirely determined by covalent interactions. Two transverse and one longitudinal branches rise from $\nu = 0\,\text{cm}^{-1}$ at $q = 0$ to provide the acoustic branches ($2\text{TA} + \text{LA}$). The slopes of the TA/LA branches determine the elastic properties and the sound speed; diamond has the highest measured sound speeds of any material, and also the lowest compressibility and highest bulk modulus value (K_o); it is the hardest known material among simple elements and compounds, that is correlated with its high atomic density and cohesive energy (Brazhkin *et al.*, 2002). It also has the highest thermal conductivity (κ) of all solids, that is correlated with the high-frequency values achieved by the acoustic modes at the BZ boundary (Brüesch, 1982).

Diamond contains two crystallographically distinct atoms within its primitive cell, so that its dispersion relations are best understood from those for a diatomic chain, rather than a monatomic chain model. Compared with the monatomic case, the extent of the BZ is halved (i.e., $q'_{max} = 2q_{max}$), so that lattice modes that were previously present at q'_{max} now appear for the diatomic crystal at $q = 0$. A break occurs within the transverse and longitudinal branches at $q_{max} = \pi/a_o$, and the upper (optic mode) dispersion relations return along the 2TO + LO optical branches to give a triply degenerate point at $q = 0$ (T_{2g} symmetry, or labelled in solid state physics texts as Γ_{25}' (Warren *et al.*, 1967; Decius and Hexter, 1977) that is observed as a single peak in the Raman spectrum of diamond at $1332 \, cm^{-1}$. This unit cell doubling process relative to vibrational dispersion relations is termed 'BZ folding', and it provides a useful concept for understanding relationships between the IR and Raman spectra and unit cell size among mineral polytypes, as well as order–disorder relations (Decius and Hexter, 1977; McMillan, 1985; Choudhury *et al.*, 1998, 2002).

Diamond is a nonpolar crystal, and the vibrational mode that gives rise to its single Raman peak at $q = 0$ involves a symmetric relative displacement of the C atoms in the unit cell about the inversion center. That vibration does not result in any dipole moment change, and so the vibration is not IR active. Additional broad features also occur in both the Raman and IR spectra of diamond at approximately twice the value of the zone center Raman mode, in the $2200–2600 \, cm^{-1}$ range (Gillet *et al.*, 1998). These occur due to second-order processes involving mainly overtone vibrations, that are enabled by the anharmonicity of the interatomic potential function. These features can be used to deduce the form of $g(\nu)$ (Decius and Hexter, 1977). Other weak features observed in the IR spectra are usually due to the presence of various impurities (principally N and H) (Clark *et al.*, 1992). Diamonds can also contain inclusions of mantle minerals or C–O–H fluids that are studied by microbeam FTIR or Raman spectroscopy: these provide valuable information on the nature of fluids at depth and the redox state of the mantle, and they also yield the deepest examples of mantle mineralogy brought to the surface (Navon and Hutcheon, 1988; Liu *et al.*, 1990; Navon, 1991; Harte and Harris, 1993, 1994; Kagi *et al.*, 2000). Some inclusions contain magnesiowüstite ((Mg, Fe)O), and also (Mg, Fe)SiO_3 phases that may be derived via back-transformation from silicate perovskite, that are only stable below 670 km (Harris, 1992; Harte and Harris, 1993, 1994; McCammon, 2001).

The dispersion relations of diamond have been measured experimentally along various directions in reciprocal space using INS techniques on a large (254 ct) single crystal, at ambient P and T, and the data were fit using a shell model calculation (Warren *et al.*, 1967) (**Figure 4**). The dispersion relations have also been calculated using various empirical local force fields or sets of force constants obtained from Hartree–Fock calculations on clusters, and also directly using first-principles or *ab initio* techniques for the solid-state structure (Guth *et al.*, 1990; Stoneham, 1992). The P and T dependence of the zone center Raman-active phonon mode has been determined to $P = 40 \, GPa$ and $T = 1900 \, K$ in several studies (Boppart *et al.*, 1985; Hanfland *et al.*, 1985; Zouboulis and Grimsditch, 1991; Muinov *et al.*, 1994). Gillet *et al.* (1998) analyzed the $(\partial v/\partial P)_T$ and $(\partial v/\partial T)_P$ data to obtain the volume dependence of the diamond Raman mode frequency, and thus demonstrated its large intrinsic anharmonicity, that likely extends to other vibrational modes of diamond.

(Mg, Fe)O magnesiowüstite is the second most abundant phase in the Earth's mantle. It is formed by decomposition reactions involving (Mg, Fe)$_2SiO_4$ and (Mg, Fe)SiO_3 phases below 660–670 km. Both MgO and FeO possess the cubic B1 (NaCl) structure (space group $Fm3m$) at ambient P and T. Pure MgO (periclase) retains the B1 structure to at least 227 GPa, whereas FeO (wüstite) transforms first into a rhombohedral structure at $P > 18 \, GPa$, and then into the inverse NiAs structure (Mazin *et al.*, 1998; Murakami *et al.*, 2004b). The wüstite phase is usually slightly oxygen deficient (i.e., FeO$_{1-x}$). Experiments indicate that Mg-rich (Mg, Fe)O is stable within the B1 structure throughout the P, T conditions of the lower mantle, however (Lin *et al.*, 2002).

As for diamond, there are two atoms in the primitive unit cell, so that the dispersion curves form both (TA, LA) and (TO, LO) branches (**Figure 6**). However, the bonding within these ionic crystals is now highly 'polar', so that long-range Coulombic interactions also contribute substantially to the lattice dynamics. In the absence of any such electrostatic effects, the IR-active vibrational mode for MgO or FeO at the BZ center would be triply degenerate, that is, a single mode with F_{1u} symmetry. In fact, a single peak is observed in IR absorption studies, at $\sim$380 cm^{-1} (Gillet *et al.*, 1998). However, the IR absoprtion technique only probes the TO lattice vibrations close to the BZ center. IR reflectance studies at ambient P indicate that a corresponding

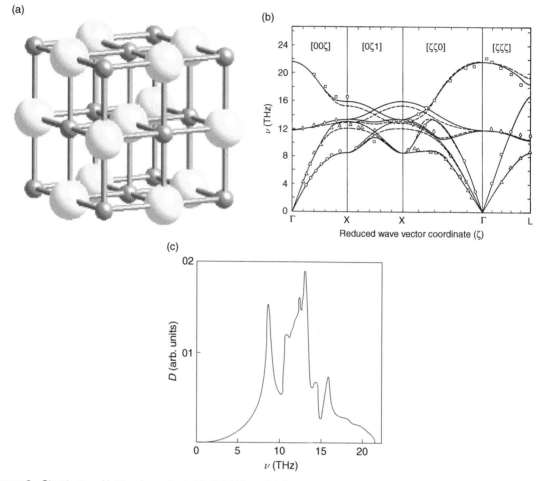

Figure 6 Structure and lattice dynamics in MgO. (a) The cubic B1-structured mineral MgO is dominated by ionic bonding interactions. (b) The ionic nature of the crystal results in highly polar optic modes, and there is a large contribution from long-range Coulombic forces to the lattice dynamics. This results in a large TO–LO splitting among the IR-active modes at the Brillouin zone center. The TO–LO splitting is not observed directly by IR transmission spectroscopy, because only the TO modes cause IR absorption. The LO mode frequencies appear as poles in the IR reflectance spectra, however, and they also cause band broadening in powder transmission experiments (Piriou and Cabannes, 1968; McMillan, 1985; McMillan and Hofmeister, 1988; Hofmeister et al., 1990; Hofmeister, 1997). (c) The vibrational dispersion relations are averaged throughout the Brillouin zone to provide the vibrational density of states (VDOS) or $g(\nu)$ function. This constitutes the complete vibrational spectrum, which is deduced from INS experiments, or obtained directly from *ab initio*, first-principles, or empirical potential model calculations. Note that the vibrational frequencies in MgO are given here in THz units. The diagrams in parts (b) and (c) are reprinted from Bilz H and Kress W (1979) *Phonon Dispersion Relations in Insulators. Springer Series on Solid-State Science* 10, p. 50. Berlin: Springer.

LO mode exists at higher frequency ($\sim$720 cm^{-1}; Piriou and Cabannes, 1968). That observation results from the contribution of long-range Coulombic forces to the lattice dynamics, that cause the LO mode to experience an additional restoring force. The phenomenon is termed 'TO-LO splitting', and it occurs for all 'ionic' crystals, in response to the macroscopic electric field developed within polar solids (Born and Huang, 1954; Ashcroft and Mermin, 1976; Ferrarro, 1984; Burns, 1985;

McMillan, 1985; Hofmeister, 1987; McMillan and Hofmeister, 1988; Hofmeister, 1997). The magnitude of TO–LO splitting is correlated with the measured 'static' (i.e., ε_o, in the limit that $\varepsilon \to 0$ frequency, or $\varepsilon(0)$), and high-frequency (ε_∞) dielectric constants (with a limit in the high-frequency or optical range), via the Lyddane–Sachs–Teller (LST) relation:

$$\frac{\omega_{\text{LO}}}{\omega_{\text{TO}}} = \sqrt{\frac{\varepsilon(0)}{\varepsilon(\infty)}} \qquad [12]$$

For complex minerals with n atoms in the primitive cell, the generalized LST relation becomes:

$$\prod_{i=1}^{3n-3} \frac{\omega_{\text{LO},\,i}}{\omega_{\text{TO},\,i}} = \sqrt{\frac{\varepsilon(0)}{\varepsilon(\infty)}} \qquad [13]$$

Here the product runs over all optic modes (Gervais *et al.*, 1973a, 1973b; Gervais and Piriou, 1975; Hofmeister, 1987; McMillan and Hofmeister, 1988; Hofmeister *et al.*, 1989). TO–LO splitting does not only occur for typically 'ionic' solids, but for any crystals that contain heteropolar bonding. For example, cubic moissanite (SiC) is isostructural with diamond and is usually considered to be covalently bonded; however, this phase exhibits a large TO–LO splitting of $\sim$180 cm^{-1} (Bilz and Kress, 1979).

The phonon dispersion relations of MgO and FeO are well known from INS measurements (Bilz and Kress, 1979), and also from various *ab initio* and first-principles theoretical calculations that have been carried out to model the thermoelastic properties under the high-P,T conditions of the deep mantle (Hemley *et al.*, 1985; Cohen *et al.*, 1987b; Wolf and Bukowinski, 1988; Agnon and Bukowinski, 1990; Isaak *et al.*, 1990; Chizmeshya *et al.*, 1994; Bukowinski *et al.*, 1996; Stixrude *et al.*, 1998; Karki *et al.*, 2000c; Oganov *et al.*, 2005b; Stixrude and Lithgow-Bertelloni, 2005).

The TO and LO mode anharmonic parameters of pure MgO have been measured experimentally to $T \sim 1000\,^\circ$C, using IR reflectance techniques at ambient P (Piriou and Cabannes, 1968). Ambient T pressure shifts have been determined directly by IR transmission studies up to 22 GPa (Gillet *et al.*, 1998) (**Figure 7**). Information on the VDOS in MgO has also been obtained at high-P by examining the phonon sidebands on the Cr^{3+} luminescence in Cr^{3+}: MgO doped crystals that were studied up to $P = 37$ GPa (Chopelas and Nicol, 1982; Chopelas, 1992, 1996). The TO mode Grüneisen parameter obtained from the measurements was $\gamma \sim 1.5$–1.6, that lies close to the thermal value obtained from analysis of the thermodynamic properties ($\gamma_{\text{th}} = 1.52$) (Gillet *et al.*, 1998). The TO and LO mode frequencies for FeO$_{1-x}$ crystals occur at lower frequencies than for MgO; that is, at $\nu_{\text{TO}} \sim 300$ cm^{-1}; $\nu_{\text{LO}} \sim 530$ cm^{-1} at ambient P and T (Bilz and Kress, 1979). It will be important to determine the effects of Mg/Fe substitution, and also the presence of O^{2-} vacancies, on the phonon spectrum and thermoelastic properties in future studies of the magnesiowüstite phase.

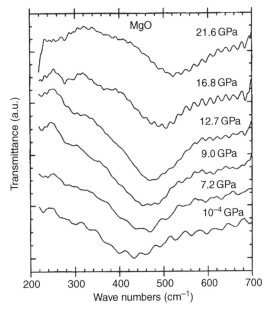

Figure 7 High-pressure IR transmission spectra of MgO to $P \sim 22$ GPa (Gillet *et al.*, 1998). Here, the vibrational energy scale is shown in cm^{-1} (1 cm^{-1} = 2.998 THz, for comparison with the data in **Figure 6**). The main band is broad, and corresponds to the IR absorption associated with the TO mode. The weak 'fringes' in the data set are due to IR interference effects, associated with the thickness of the sample. The data were obtained by A. Grzechnik and P. Simon (Grzechnik *et al.*, 1999), and the diagram shown here is reprinted from Gillet P, Hemley RJ, and McMillan PF (1998) Vibrational properties at high pressures and temperatures. In: Hemley RJ (ed.) *Ultra-High Pressure Mineralogy*, vol. 37, p. 560. Mineralogical Society of America.

7.3.2 (Mg, Fe)SiO₃ Perovskite, Post-Perovskite, and CaSiO₃ Perovskite

MgSiO₃ perovskite is formed by pressure-induced structural transformations and reactions among lower pressure silicate minerals at above $P = 22$–24 GPa; its appearance along with that of (Mg, Fe)O magnesiowüstite is thought to mark the 660–670 km seismic discontinuity that defines the onset of the lower mantle (Ringwood, 1962b; Liu, 1974, 1975; Ringwood, 1975; Jeanloz and Thompson, 1983; Liu and Bassett, 1986; Knittle and Jeanloz, 1987; Anderson, 1989; Bina, 1998). It then likely remains stable throughout nearly the entire P–T range of the lower mantle, to result in the predominant mineral phase within the Earth (Ringwood, 1962b). MgSiO₃ perovskite has not yet been observed in any natural samples brought to the surface, although (Mg, Fe)SiO₃ phases that might constitute its decompression products have been identified in deep diamond inclusions (Harris, 1992; Harte and Harris, 1993, 1994; Sautter and Gillet, 1994; McCammon *et al.*, 1997; Liou *et al.*, 1998; McCammon, 2001). The silicate perovskite phase has been prepared in laboratory experiments and recovered to ambient conditions in various studies designed to establish its high-P,T stability range, and its vibrational, elastic, and thermodynamic properties (Williams *et al.*, 1987; Hemley *et al.*, 1989; Durben and Wolf, 1992; Hemley and Cohen, 1992). The lattice dynamics and thermophysical behavior of $MgSiO_3$ perovskite has been studied extensively using various theoretical approaches, beginning with models based on the modified electron gas (MEG) theory, and most recently using highly accurate first-principles methods to determine the phonons and thermoelastic properties under a wide range of mantle P, T conditions (Wolf and Jeanloz, 1985; Hemley *et al.*, 1987a; Wolf and Bukowinski, 1987; Hemley *et al.*, 1989; Karki *et al.*, 2000b; Oganov *et al.*, 2000; Parlinski and Kawazoe, 2000; Karki *et al.*, 2001; Wentzcovich *et al.*, 2004b).

The type mineral phase that gives rise to the name perovskite is $CaTiO_3$. This was originally thought to possess a cubic structure (*Fm3m*), known as the ideal perovskite aristotype. However, $CaTiO_3$ is now determined to possess an orthorhombic structure with space group *Pbnm*. This is the typical 'GdFeO$_3$' structure that is found for $MgSiO_3$ perovskite at ambient conditions, and also obtained under high-P,T conditions relevant to the lower mantle (Hemley and Cohen, 1992; Serghiou *et al.*, 1998; Shim *et al.*, 2001). Perovskite-structured solids have a general composition ABX_3; they are constructed from corner-linked BX_6 octahedral units, with the A cations occupying 12-coordinated sites within the ideal cubic structure (**Figure 8**). The A site coordination becomes lowered for distorted perovskite structures; for example, the Mg^{2+} cations in $MgSiO_3$ perovskite are eightfold coordinated.

Perovskite-structured materials exhibit a wide range of structure types and symmetries derived from the structural distortional parameters. The largest family of these occur due to concerted rotations of the linked BX_6 octahedra about their highly flexible B–X–B linkage (i.e., the inter-octahedral Mg–O–Mg linkage, in the case of $MgSiO_3$ perovskite) (**Figure 8**). Various coupled rotations in '+' or '−' directions result in a large variety of tetragonal, orthorhombic and trigonal structures (Glazer, 1972,

1975). Other structural distortions involve displacements of the A or B cations away from their central positions within the octahedral or larger polyhedral sites. These off-center cation displacements give rise to ferroelectric behavior that result in large variations in the dielectric constant (ε) as a function of T and P, and they result in important ferroelectric ceramic materials including perovskites based on $BaTiO_3$ and $PbZrO_3$ that are used in capacitors and mobile telecommunications applications (Burfoot and Taylor, 1979; Navrotsky and Weidner, 1989; Cohen, 1992a; Ghosez *et al.*, 1999; Choudhury *et al.*, 2005). These structural distortions are driven by soft phonon modes that mainly occur at the BZ boundaries, for example, at the R or M points of the cubic perovskite unit cell (Wolf and Jeanloz, 1985; Wolf and Bukowinski, 1987; Hemley and Cohen, 1992) (**Figure 8**).

Early theoretical studies of $MgSiO_3$ perovskite using techniques derived from the modified electron gas (MEG) theory, suggested that dynamically driven transitions might occur at high T between orthorhombic–tetragonal–cubic phases of silicate perovskite, under the high-P conditions of the Earth's lower mantle (Wolf and Jeanloz, 1985; Hemley *et al.*, 1987a, 1989; Wolf and Bukowinski, 1987; Bukowinski and Wolf, 1988; Hemley and Cohen, 1992). These predictions had important potential implications for seismic wave propagation velocities and their attenuation characteristics within the deep lower mantle (Bukowinski and Wolf, 1988; Hemley and Cohen, 1992). More recent theoretical calculations coupled with experimental X-ray determinations of $MgSiO_3$ perovskite under deep mantle high-P,T conditions now indicate that such phase transitions associated with lattice dynamical instabilities are unlikely to be present throughout the P, T range of the lower mantle (Karki *et al.*, 2000b; Oganov *et al.*, 2000, 2005b; Shim *et al.*, 2001, 2004). The *in situ* X-ray experiments carried out to date might still not have sufficient resolution to absolutely rule out such possible high-P,T phase transitions among orthorhombic–tetragonal–trigonal–cubic polymorphs of $MgSiO_3$ perovskite, and theoretical studies have not always taken full account of intrinsic phonon anharmonicity (Gillet *et al.*, 1996a, 2000; Oganov *et al.*, 2000, 2005b).

It now seems that any displacive transformations occurring within orthorhombic $MgSiO_3$ perovskite would occur at such high-P,T conditions that they would be first intersected by first-order phase transitions, including melting (Hemley and Kubicki, 1991;

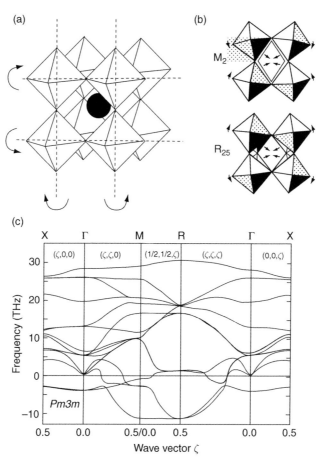

Figure 8 Structure and lattice dynamics of the cubic perovskite aristotype (*Pm3m* symmetry). This structure is taken by $SrTiO_3$ at ambient *P* and *T*, and by $CaSiO_3$ perovskite throughout the *P,T* range of the mantle. (a) The interoctahedral linkages are flexible, and low-frequency vibrations are associated with coupled rotational modes around various crystallographic axes that result in soft mode behavior and structural distortions away from cubic symmetry. (b) The octahedral rotation modes propagate along various different directions in reciprocal space, to yield different symmetry points at the Brillouin zone boundary (Wolf and Jeanloz, 1985; Hemley and Cohen, 1992). (c) Lattice dynamics calculated for the cubic perovskite structure of $MgSiO_3$. Several phonon branches are unstable (i.e., occur with negative frequency values) along various directions in reciprocal space, especially along the Γ-M $(\xi,\xi, 0)$ and Γ-R (ξ,ξ, ξ) directions. The M_2 and R_{25} modes shown above correspond to unstable phonons at the Brillouin zone boundary: the resulting lattice distortions result in MgSiO3 perovskite with orthorhombic (*Pbnm*) symmetry. The diagram in part (a) is reprinted from Hemley RJ and Cohen RE (1992) Annual Review of Earth and Planetary Sciences 20: 555. Part (b) is reprinted from Parlinski K and Kawazoe Y (2000) *European Physics Journal B* 16: 53. (c) From Parlinski K and Kawazoe Y (2000) *Ab initio* study of phonons and structural stabilities of the dynamics of the perovskite-type $MgSiO_3$. *European Physics Journal B* 16: 53.

Zerr and Boehler, 1993; Akins *et al.*, 2004), or transformation into a newly recognized $CaIrO_3$-structured postperovskite phase (Oganov and Ono, 2004; Shim *et al.*, 2004; Tsuchiya *et al.*, 2004b, 2005).

Raman and IR spectra at ambient *P* and *T* have been determined experimentally for pure $MgSiO_3$ perovskite for various samples recovered from high-*P, T* synthesis experiments (Weng *et al.*, 1983; Wolf and Jeanloz, 1985; Williams *et al.*, 1987; Hemley *et al.*, 1989; Hemley and Cohen, 1992; Lu *et al.*, 1994) (**Figure 9**). Recent first-principles calculations

reproduce the experimentally determined frequencies to within a few percent, that is typical for such experiment-theory comparisons (Oganov *et al.*, 2000, 2002; Parlinski and Kawazoe, 2000; Karki *et al.*, 2001; Wentzcovich *et al.*, 2004b). The experiments are generally carried out at $T = 300$ K, whereas the calculations are enabled at $T = 0$ K; also, both the theoretical results and the experimental measurements are subject to perhaps $\sim$3–5% uncertainty. However, significant progress has been made since our previous reviews of the lattice dynamics of mantle minerals

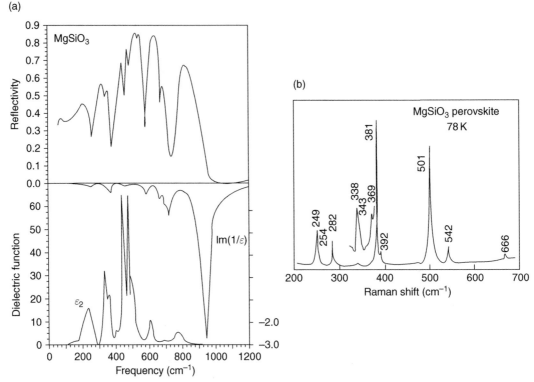

Figure 9 Infrared and Raman spectra of $MgSiO_3$ perovskite recorded at ambient pressure. (a) The infrared reflectivity data are from Lu et al. (1994). Peaks and minima in the dielectric functions $\varepsilon_2(v)$ and Im $(1/\varepsilon(v))$ indicate the positions of TO and LO mode frequencies, respectively. (b) The Raman spectrum at top was obtained by Durben and Wolf (1992), for a sample cooled to 78 K. The appearance of a first-order Raman spectrum indicates the presence of an orthorhombic distortion of the perovskite structure. Part (a) is reprinted from Lu R, Hofmeister AM, and Wang YB (1994). *Journal of Geophysical Research* 99: 11798. Part (b) is reprinted with permission from Durben D and Wolf GH (1992) *American Mineralogist* 77: 891.

including $MgSiO_3$ perovskite (McMillan et al., 1996b; Gillet et al., 1998), in that all of the Raman and IR modes at the BZ center at ambient P, T conditions have now been definitely identified (Parlinski and Kawazoe, 2000; Karki et al., 2001; Oganov et al., 2002; Wentzcovich et al., 2004a, 2004b). Raman and IR spectra of the $MgSiO_3$ perovskite phase have been measured experimentally to $P = 65$ GPa at ambient T, at high T (metastably) at ambient P to $T = 600$ K, and now under simultaneous high-P,T conditions (Williams et al., 1987; Hemley et al., 1989; Durben and Wolf, 1992; Liu et al., 1994; Wang et al., 1994; Chopelas, 1996; Gillet et al., 2000). These data have been used to determine mode Gruneisen parameters. Gillet et al. (1996a, 1998, 2000) have combined results from high- and low-T Raman measurements with high-P data to estimate the intrinsic mode anharmonicities.

The cubic perovskite aristotype with space group $Fm3m$ has no Raman active modes; only three triply degenerate IR-active vibrations exist at the BZ center. Symmetry analysis yields:

$$\Gamma_{vib} = 3F_{1u}(\text{IR}) + F_{2u} \quad (\text{inactive}) \qquad [14]$$

Among cubic perovskites such as $BaTiO_3$ and $NaMgF_3$, the highest-frequency IR modes at $750–1000$ cm^{-1} involve asymmetric stretching of the BX_6 octahedra, the mid-frequency modes correspond to octahedral deformation vibrations $(500–700$ cm$^{-1})$, and the lowest-frequency modes involve translations of M^{2+} cations within the large (12-coordinate) cages. The low-frequency vibrations are often coupled to various 'soft' modes that occur due to inter-octahedral rotations, within the $50–400$ cm^{-1} range (Shimanouchi et al., 1961; Cowley, 1963; Perry et al., 1964b, 1964a; Perry and Young, 1967; Scott, 1974; Williams et al., 1987). These vibrational mode assignments generally agree with the results obtained from first-principles calculations of the lattice dynamics for $MgSiO_3$ perovskite (Hemley and Cohen, 1992; Karki et al., 2000b; Oganov et al., 2000; Parlinski and Kawazoe, 2000).

Reduction of the perovskite symmetry to the orthorhombic space group *Pbnm* with $Z = 4MgSiO_3$ units in the primitive cell causes a splitting of the IR-active modes, the occurrence of additional peaks in the IR spectrum, and the appearance of a first order Raman spectrum (**Figure 9**):

$$\Gamma_{vib} = 7A_g(R) + 7B_{1g}(R) + 5B_{2g}(R)$$
$$+ 5B_{3g}(R) + 8A_u(\text{inactive}) + 7B_{1u}(IR)$$
$$+ 9B_{2u}(IR) + 9B_{3u}(IR) \qquad [15]$$

The first observation of a Raman spectrum for $MgSiO_3$ perovskite thus immediately confirmed its reduction in symmetry to an orthorhombic structure (space group *Pbnm*) that had been suggested by X-ray diffraction (Williams *et al.*, 1987). Previously, some discrepancies between theory and experiment or missing IR and Raman modes appeared to be present,

especially among the low-frequency vibrations (McMillan *et al.*, 1996b); however, these anomalies have now been resolved by accurate first-principles calculations (Karki *et al.*, 2000a; Oganov *et al.*, 2000; Parlinski and Kawazoe, 2000).

The phonon dispersion relations or $g(\nu)$ function for $MgSiO_3$ perovskite have not yet been measured experimentally at ambient or under high-P,T conditions; however, the lattice dynamics have been calculated theoretically by first-principles methods over a wide P, T range (**Figure 10**), and the results have been used to predict the thermophysical properties of the lower mantle (Hemley and Cohen, 1992; Karki *et al.*, 2000b; Oganov *et al.*, 2000, 2002, 2005; Parlinski and Kawazoe, 2000; Wentzcovich *et al.*, 2004b). These calculations are known to be accurate at high P, however, they do not fully take into account the thermodynamic properties

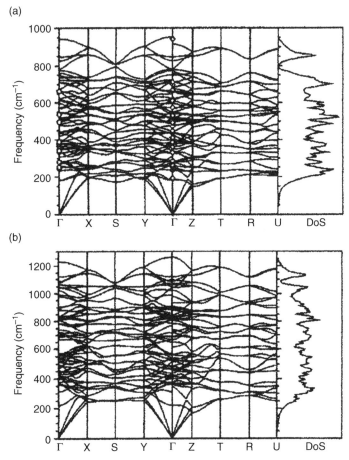

Figure 10 Lattice dynamics calculated using accurate first-principles methods for $MgSiO_3$ perovskite (a) at ambient pressure and (b) at $P = 100\,GPa$ (Karki *et al.*, 2000b). Experimentally observed frequency values taken from IR and Raman data at ambient P are shown as circles on the axis at the Γ point. Reprinted from Karki B, Wentzcovitch RM, de Gironcoli S, and Baroni S (2000) *Ab initio* lattice dynamics of $MgSiO_3$ perovskite at high pressure. *Physical Review B* 62: 14752, with permission from American Physical Society.

at high T, because the intrinsic phonon anharmonicity is ignored within the quasi-harmonic approximation (Gillet *et al.*, 2000; Oganov *et al.*, 2000; Karki and Wentzcovich, 2003). Gillet *et al.* (2000) have analyzed the Raman spectra obtained at high P and low T to evaluate the intrinsic anharmonicity contribution to the lattice dynamics and thermophysical properties of $MgSiO_3$ perovskite. Also, Lu *et al.* (1994) obtained data on the intrinsic anharmonicity of the IR-active modes, from analysis of the IR reflectivity. The spectroscopy data and first-principles calculations have been used to construct $g(\nu)$ functions in order to calculate the heat capacity (C_v, C_p) and the vibrational entropy (S) as a function of T; the thermodynamic data were used to demonstrate the general nature of negative $P–T$ Clapeyron slopes involving perovskite-forming reactions in the mantle (Navrotsky, 1980, 1989; Akaogi and Ito, 1993; Lu *et al.*, 1994).

Most experimental and theoretical studies of the lattice dynamics of $MgSiO_3$ perovskite have focused on the end-member phase. However, the mineral present within the lower mantle is likely to exist as a solid solution within the $(Mg, Fe, Al)SiO_3$ system, that might contain vacancies and chemical substitutions on cation and anion sites. These substitutions could have large effects on the stability and theormophysical properties of silicate perovskites, as well as the properties of mantle materials in the vicinity of the 660–670 km seismic anomaly and their associated phase transitions (Wood and Rubie, 1996). Lu *et al.* (1994) and Wang *et al.* (1994) determined IR and Raman spectra for $(Mg, Fe)SiO_3$ perovskites recovered from high-P,T experiments; they found only small changes in the spectra upon Fe^{2+}/Mg^{2+} substitution, but the Raman peaks were broadened, presumably due to Mg/Fe structural disorder on the large cation sites. The frequencies were all slightly shifted to lower wavenumbers, consistent with the small lattice expansion that occurs within the Fe/Mg perovskite series (Kudoh *et al.*, 1990; Mao *et al.*, 1991; Wang *et al.*, 1994).

It was thought until recently that orthorhombic $MgSiO_3$ perovskite would remain stable throughout the entire $P–T$ range of the lower mantle (Knittle and Jeanloz, 1987; Hemley and Cohen, 1992). However, analysis of various anomalies in the seismic features associated with the deepest mantle and the D'' layer, combined with results of high-P,T experiments and theoretical calculations, now indicate that $MgSiO_3$ transforms into a new post-perovskite phase with the $CaIrO_3$ structure, that contains edge-shared SiO_6 units arranged into layers (Shim *et al.*, 2001; Murakami *et al.*, 2004a; Oganov and Ono, 2004; Shim *et al.*, 2004;

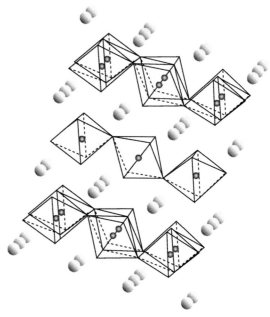

Figure 11 The $CaIrO_3$ structure determined recently to be adopted by $MgSiO_3$ above $P = 125$ GPa. That pressure corresponds to a depth ~ 200 km above the base of the mantle.

Tsuchiya *et al.*, 2004b; Hernlund *et al.*, 2005) (**Figure 11**). The formation of this layered structure is expected to have a large effect on the elastic anisotropy and the thermophysical properties developed at the base of the mantle (Iitaka *et al.*, 2004; Akber-Knutson *et al.*, 2005), and also within the D' layer (Iitaka *et al.*, 2004; Nakagawa and Tackley, 2004). The post-perovskite transition is also observed to occur for perovskite-structured materials with $(Mg, Fe)SiO_3$ compositions: Al^{3+} included in the silicate perovskite also affects the P and T regime for the transition (Mao *et al.*, 2004, 2005).

The lattice dynamics of the $CaIrO_3$-structured phase of $MgSiO_3$ have not yet been determined experimentally; however, its vibrational and thermoelastic properties have been predicted theoretically using first-principles methods, extending up to the high-P conditions existing at the base of the mantle (Tsuchiya *et al.*, 2005).

A $CaSiO_3$ perovskite phase is also expected to occur within the lower mantle (Liu and Ringwood, 1975; Hemley and Cohen, 1992). This material was originally expected to remain cubic throughout all mantle $P–T$ conditions; however, it is now recognized that a slight tetragonal distortion should occur at pressures below $P \sim 14$ GPa, driven by phonon mode instabilities (Hemley and Cohen, 1992;

Chizmeshya *et al.*, 1996; Stixrude *et al.*, 1996; Jung and Oganov, 2005). The $CaSiO_3$ phase cannot normally be recovered to ambient *P*, *T* conditions; instead, it reverts to lower-*P* phases or amorphises during decompression. MD simulations combined with first-principles calculations indicate that the phase undergoes amorphisation during decompression due to lattice dynamic instabilities occurring among low-lying vibrational modes and phonon branches, involving Ca^{2+} vibrations coupled with SiO_6 octahedral distortions (Hemmati *et al.*, 1995; Bukowinski *et al.*, 1996). The ferroelectric properties associated with these mode instabilities have now been harnessed to provide new dielectric materials, based on related $SrGeO_3$ perovskites decompressed following high-*P,T* syntheses (Grzechnik *et al.*, 1997, 1998). Molecular dynamics simulations using empirical potentials have proved useful in gaining a general understanding of the softening of phonon branches in $CaSiO_3$ perovskite as a function of crystal volume, and developing a model for the amorphisation event (Hemmati *et al.*, 1995). The empirical potential approach has also begun to be used the energetics of point defect formation in $MgSiO_3$ and $CaSiO_3$ perovskites (Watson *et al.*, 2000).

7.3.3 Stishovite and Post-Stishovite SiO_2 Polymorphs

The α-quartz phase of SiO_2 is abundant at the Earth's surface. Its lattice dynamics have been thoroughly studied by theory and experimental studies, including IR, Raman and INS techniques, carried out at high *T* and *P* (Shapiro *et al.*, 1967; Gervais and Piriou, 1975; Dolino and Bachheimer, 1977; Dolino *et al.*, 1983; Ghose, 1985, 1988). It has been studied extensively as a type phase to develop the principles of soft mode lattice dynamical behaviour (Scott, 1974; Ghose, 1985). A low-lying Raman-active mode at 206 cm^{-1} at ambient *P* and *T* shifts rapidly and asymptotically to approach zero frequency at *T* = 858 K, at the temperature of the displacive α–β quartz transition (Raman and Nedungadi, 1940). The A_1 symmetry mode has atomic displacement vectors that involve Si and O motions around the $\bar{3}$ axis, that relates the α- and β-quartz structures. As *T* is increased and the anharmonic vibrational mode is excited, the atoms move closer to the new positions that define the high-*T* structure, and the restoring force is diminished. At the *T* of the phase transition, the atoms now occupy a new set of potential energy minima that correspond to the positions within the β-quartz

phase. That process was first described by Raman and Nedungadi (1940); the work has now become a classic example of what constitutes a second-order displacive phase transformation driven by a vibrational soft mode (Raman and Nedungadi, 1940; Scott, 1974; Decius and Hexter, 1977; Ghose, 1985; McMillan, 1985, 1989). The interaction between the lattice dynamics and the structural and elastic properties of quartz at high *T* are now known to be considerably more complex than this simple picture (Dolino and Bachheimer, 1977; Dolino *et al.*, 1983). As with all such 'soft mode' transitions, the phase transformation begins to take on some order–disorder character as the vibrational frequency is lowered, and the effects of finite barriers in the potential surface are encountered, along with zero-point energy effects (Ghose, 1985, 1988; Salje, 1990). Also, the low-frequency A_1 soft mode of α-quartz begins to interact anharmonically with second-order phonons as the frequency is lowered on approach to the transition, thus, further complicating the lattice dynamical interpretation of the transition (Scott, 1968). Finally, the A_1 mode softening causes the appearance of lattice instabilities in acoustic branches slightly away from the BZ center, to result in incommensurate structures within the quartz phase (Dolino and Bachheimer, 1977; Dolino *et al.*, 1983). The high-*T* SiO_2 phase cristobalite also exhibits a dynamically driven α–β phase transition, that has not yet been as thoroughly studied as that for quartz.

At above *P* ~ 9 GPa, SiO_2 transforms into the mineral phase stishovite, that contains octahedrally coordinated Si^{4+} ions, and that is isostructural with rutile (TiO_2) (Stishov and Popova, 1961). Its high-*P,T* stability means that stishovite could be present as a free SiO_2 phase throughout the mantle, formed by disproportionation reactions occurring among silicate minerals at high pressure (Kesson *et al.*, 1994; Kingma *et al.*, 1995; Funamori *et al.*, 2000). The natural occurrence of stishovite was first described as a product of impact metamorphism (Chao *et al.*, 1962); it has now also been recorded in highly metamorphised deep crustal rocks (Liou *et al.*, 1998).

The zone-center vibrational modes for rutile-structured SiO_2 stishovite are

$$\begin{aligned}
\Gamma_{\text{vib}} = {} & A_{1g}(R) + A_{2g}(\text{inactive}) + B_{1g}(R) + B_{2g}(R) \\
& + E_g(R) + 2B_{1u}(\text{IR}) + 3E_u(\text{IR})
\end{aligned} \qquad [16]$$

Raman spectra obtained by Hemley *et al.* (1986) for both natural and synthetic stishovite samples showed all four expected modes, and their pressure derivatives have been studied (Hemley, 1987) (**Figure 12**). Early IR transmission data (Lyon, 1962) were distorted due to optical effects within the highly refractive mineral, as shown by subsequent reflectivity studies (McMillan and Hofmeister, 1988; Hofmeister *et al.*, 1990; Hofmeister, 1996). (**Figure 12**). Pressure shifts of the IR-active modes were recorded by Williams *et al.* (1993) and Hofmeister (1996), and the *T* dependence of the Raman spectrum was investigated by Gillet *et al.* (1990, 1998).

The Raman spectrum of stishovite at ambient *P* and *T* contains a low-frequency mode with B_{1g} symmetry (231 cm^{-1}), that shifts to lower wavenumber with increasing *P* (Hemley, 1987; Kingma *et al.*, 1995), analogous to the behavior observed for rutile-structure TiO_2 and SnO_2 (Samara and Peercy, 1973; Peercy and Morosin, 1973). The atomic displacements associated with this mode correspond to those required to transform the structure into a $CaCl_2$-structured polymorph, via a second-order displacive phase transition (Nagel and O'Keeffe, 1971; Peercy and Morosin, 1973; Samara and Peercy, 1973; Hemley, 1987; Cohen, 1992b; Kingma *et al.*, 1995). The phonon-driven rutile-$CaCl_2$ phase transition for SiO_2 was predicted by theory (Cohen, 1991, 1992b; Matsui and Tsuneyiki, 1992; Lacks and Gordon, 1993; Lee and Gonze, 1995) and it has now been

observed to occur, using a combination of *in situ* Raman spectroscopy and X-ray diffraction studies (Kingma *et al.*, 1995; Andrault *et al.*, 1998; Murakami *et al.*, 2003; Shieh *et al.*, 2005) (**Figure 12**). The transition occurs at $P \sim 50\,GPa$, corresponding to the high-*P* conditions attained at the base of the mantle.

A further phonon-assisted transition to a fluorite-structured phase could also occur at $P > 150\,GPa$; however, that transition is predicted to result in a phase that is always dynamically unstable (Bukowinski and Wolf, 1986). Instead, a transformation to a cubic phase related to the pyrite structure (FeS_2) with $Pa\overline{3}$ symmetry might be achieved at very high pressure: that transition is observed to occur among dense transition metal dioxides such as RuO_2, and the group 14 oxides PbO_2, SnO_2 including the close silica analog GeO_2 (Haines and Léger, 1993; Haines *et al.*, 1996a, 1996b; Ono *et al.*, 2002). This phase has now been observed to occur for SiO_2 under high-*P* conditions (Kuwayama *et al.*, 2005). In this case, however, the relative positions of the O^{2-} anions indicate that no O···O bonding interactions occur, unlike the observed structure of FeS_2 pyrite.

An SiO_2 polymorph with the orthorhombic α-PbO_2 structure has also been identified in run products from high-*P,T* experiments and also in natural meteorite

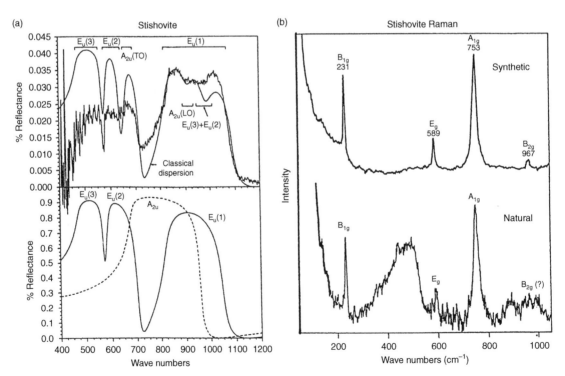

Figure 12   *(continued)*

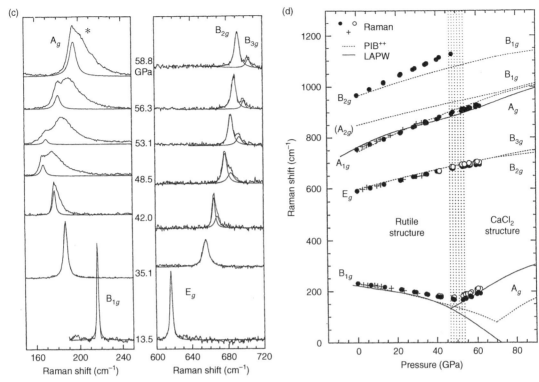

Figure 12 Infrared and Raman spectra of SiO_2 stishovite, and Raman data for the $CaCl_2$-structured phase (Hemley et al., 1986; Hemley, 1987; Hofmeister et al., 1990; Kingma et al., 1995; Hofmeister, 1996). (a) The unpolarized IR reflectance spectrum of a polycrystalline sample of SiO_2 stishovite is shown at top, along with the TO and LO mode frequencies obtained from analysis of the reflectivity data. The synthetic spectrum at bottom was then reconstructed using these parameters, for both predicted mode symmetries (A_{2u} and E_u) (Hofmeister et al., 1990). (b) The Raman spectra of synthetic and natural samples were first obtained by Hemley et al. (1986). The four expected modes are clearly visible in the spectrum of the sample at top. The natural sample also shows broad bands corresponding to amorphous SiO_2, that is also present within the sample. (c) Raman spectra obtained in situ at high pressure by Kingma et al. (1995) showed evidence for softening of the lowest frequency B_{1g} mode, resulting in a transition to $CaCl_2$-structured SiO_2 at $P \sim 50\,GPa$ pressure. (d) The mode softening in SiO_2 stishovite and its transformation to the $CaCl_2$ structure was predicted by first-principles theoretical calculations, and was confirmed by experimental measurements (Cohen, 1992b; Kingma et al., 1995). Part (a) is reprinted from Hofmeister AM, Xu J, and Akimoto S (1990) *American Mineralogist* 75: 952. Part (b) is reprinted from Hemley RJ (1987) *High Pressure Research in Mineral Physics* Manghnani MH and Syono Y (eds.), p. xxx. Part (c) is reprinted from Gillet P, Hemley RJ, and McMillan PF (1998) Vibrational properties at high pressures and temperatures. In: Hemley RJ (ed.) *Ultra-High Pressure Mineralogy*, vol. 37, p. 575. Mineralogical Society of America. The original data for this figure were presented by Kingma et al. (1995).

samples: it appears to become stable above $P \sim 90\,GPa$ (Tsuchida and Yagi, 1989, 1990; Dubrovinsky et al., 1997, 2003; El Goresy et al., 1998; Sharp et al., 1999; Dubrovinskaia et al., 2001; Dera et al., 2002; Murakami et al., 2003; Shieh et al., 2005). Micro-Raman spectroscopy has been usefully combined with X-ray diffraction and transmission electron microscopy to experimentally characterize the various new polymorphs of SiO_2. Another monoclinic phase with the baddeleyite (ZrO_2) structure has now also been identified in meteorites (El Goresy et al., 2000). The likely stability ranges of $CaCl_2$- and α-PbO_2-structured SiO_2 under deep mantle conditions have recently been discussed by Shieh et al. (2005). Other high-density forms

of SiO_2 have also been predicted by theoretical calculations, often formed via lattice dynamical instabilities occurring at low T within low-pressure structures (Badro et al., 1997; Teter et al., 1998; Wentzcovich et al., 1998; Tsuchiya et al., 2004a; Oganov et al., 2005b). However, it is unlikely that these metastable phases would be achieved, or if formed, that they could persist over geological timescales, under the high-P,T conditions attained within the mantle (Shieh et al., 2005).

S. W. Kieffer first developed her models of mineral lattice dynamics related to their thermodynamic properties in order to understand the formation of stishovite found within Meteor Crater.

Reliable IR and Raman data were not available at that time, However. Hofmeister (1996) has now measured and interpreted the IR spectra of stishovite at ambient and at high P, and has used her data to construct a $g(\nu)$ model to estimate the heat capacity and entropy under mantle P, T conditions. The lattice dynamics and thermoelastic properties of stishovite have also been predicted at high P using first-principles calculations (Cohen, 1992b; Karki *et al.*, 1997; Oganov *et al.*, 2005a). The intrinsic mode anharmonicity is not yet taken into account in the theoretical calculations. That could affect the C_p and vibrational entropy by 3–5% (Gillet *et al.*, 1990).

Stishovite recovered to ambient P possesses important material properties. The phase exhibits extremely high hardness, related to its dense atomic packing that results in a high cohesive energy (Léger *et al.*, 1996; Brazhkin *et al.*, 2002). It also possesses a high dielectric constant associated with the highly ionic charge distribution (Cohen, 1991; Lee, 1996; Lee and Gonze, 1997; Oganov *et al.*, 2005a, 2005b).

7.3.4 (Mg, Fe)$_2$SiO$_4$ Olivine, β- and γ-(Mg, Fe)$_2$SiO$_4$, and (Mg, Fe)SiO$_3$ Pyroxenes

An olivine-structured phase with composition near $(Mg_{0.9}Fe_{0.1})_2SiO_4$ dominates the mineralogy of the upper mantle (Jeanloz and Thompson, 1983; Ito and Takahashi, 1987; Agee, 1998; McDonough and Rudnick, 1998). The vibrational modes of end member Mg_2SiO_4 (forsterite) and Fe_2SiO_4 (fayalite), along with other natural and synthetic olivine compositions, have been studied extensively using IR and Raman spectroscopy. Vibrational dynamics in forsterite and fayalite samples have also been investigated by INS techniques, and theoretical calculations of the lattice dynamics have been carried out (Servoin and Piriou, 1973; Iishi, 1978; Piriou and McMillan, 1983; McMillan, 1985; Hofmeister, 1987, 1997; Price *et al.*, 1987, 1991; McMillan and Hofmeister, 1988; Rao *et al.*, 1988; Hofmeister *et al.*, 1989; Durben *et al.*, 1993; McMillan *et al.*, 1996b; Choudhury *et al.*, 2002). Several studies have been carried out *in situ* at high P, at high T, and under simultaneous high-P, T conditions; the results were used to predict and interpret the high-P, T thermoelastic properties and vibrational mode anharmonicity (Hofmeister, 1987; Hofmeister *et al.*, 1989; Chopelas, 1990; Gillet *et al.*, 1991, 1997, 1998; Reynard *et al.*, 1992; Wang *et al.*, 1993; Hofmeister, 1997).

The zone centre vibrational modes for the orthorhombic olivine (*Pbnm*) structure are determined to be

$$\begin{aligned} \Gamma_{\text{vib}} = {} & 11A_g(R) + 11B_{1g}(R) + 7B_{2g}(R) + 7B_{3g}(R) \\ & + 10A_u(\text{inactive}) + 9B_{1u}(\text{IR}) + 13B_{2u}(\text{IR}) \\ & + 13B_{3u}(\text{IR}) \end{aligned} \qquad [17]$$

The full set of IR- and Raman-active modes for Mg_2SiO_4 forsterite and Fe_2SiO_4 fayalite have been assigned experimentally via single-crystal measurements (Servoin and Piriou, 1973; Iishi, 1978; Piriou and McMillan, 1983; McMillan, 1985; Hofmeister, 1987; McMillan and Hofmeister, 1988). The $g(\nu)$ functions are determined from INS measurements (Rao *et al.*, 1988; Price *et al.*, 1991; Chaplot *et al.*, 2002b; Choudhury *et al.*, 2002). Both experimentally modeled and theoretically predicted VDOS functions have been used to carry out calculations of the high-P, T specific heat and entropy, leading to various predictions and a new understanding of phase transformations among the Mg_2SiO_4 polymorphs, and also the thermal conductivity of mantle phases, throughout the conditions of the upper mantle (Akaogi *et al.*, 1984; Rao *et al.*, 1988; Hofmeister *et al.*, 1989; Gillet *et al.*, 1997, 1998; Choudhury *et al.*, 1998, 2002; Hofmeister, 1999).

At P above $\sim$18–20 GPa, (Mg, Fe)$_2$SiO$_4$ transforms into spinel-structured polymorphs (Bina and Wood, 1987; Akaogi *et al.*, 1989; Gasparik, 1990; Rigden *et al.*, 1991; Agee, 1998). These transitions mark the onset of the mantle 'transition zone' at $\sim$450 km depth, associated with an abrupt increase in the seismic velocities. Pure Fe_2SiO_4 transforms directly between olivine and spinel structures at $P \sim$ 6–7 GPa. However, the end-member Mg_2SiO_4 does not transform directly into the spinel-structured γ-Mg_2SiO_4 phase (ringwoodite), but it first exhibits a transformation to a β-Mg_2SiO_4 (wadsleyite) phase (Akaogi *et al.*, 1984, 1989; Bina and Wood, 1987; Katsura and Ito, 1989; Wu *et al.*, 1993). The Fe content of the (Mg, Fe)$_2$SiO$_4$ component is an important parameter in determining the depth and width of the α–(β)–γ transitions occurring within the mantle (Agee, 1998).

In the first application of Kieffer's model to mantle mineralogy, Akaogi *et al.* (1984) obtained IR and Raman data for the α-(olivine), β- and γ-polymorphs of Mg_2SiO_4 and constructed a model $g(\nu)$ function to calculate the heat capacity and entropy. The results were combined with calorimetric data to predict P, T boundaries between the Mg_2SiO_4 phases. The early results have now been improved upon in later studies (Akaogi *et al.*, 1989; Hofmeister *et al.*, 1989; Chopelas, 1990, 1991; Gillet *et al.*, 1997).

The zone-center vibrational modes for cubic ($Fd3m$) spinel-structured γ-Mg$_2$SiO$_4$ or γ-Fe$_2$SiO$_4$ are

$$\Gamma_{vib} = A_{1g}(R) + E_g(R) + T_{1g}(\text{inactive}) + 3T_{2g}(R)$$
$$+ 2A_{2u}(\text{inactive}) + 2E_u(\text{inactive}) + 4T_{1u}(\text{IR})$$
$$+ 2T_{2u}(\text{inactive}) \qquad [18]$$

Jeanloz (1980) first obtained the powder IR spectrum for Fe$_2$SiO$_4$ spinel; he observed three peaks at 848, 503 and 344 cm^{-1}. IR reflectance data were recorded by A. Hofmeister over a similar frequency range (McMillan and Hofmeister, 1988; McMillan et al., 1996b). The highest frequency mode is due to the v_3 asymmetric stretching vibration of tetrahedral SiO$_4$ units: this highly polar vibrational mode is associated with a large TO–LO splitting, and it gives rise to broad bands in the IR reflectance and powder transmission spectra. Akaogi et al. (1984) observed a similar broad peak with maximum at 830 cm^{-1} occurring for γ-Mg$_2$SiO$_4$. The mid-range mode corresponds to the IR-active SiO$_4$ deformation (v_4), with mainly octahedral MgO$_6$/FeO$_6$ stretching vibrations in the lower-frequency range; some coupling is expected to occur among these various vibrational modes (Jeanloz, 1980; Akaogi et al., 1984; McMillan and Hofmeister, 1988; McMillan et al., 1996b). v_4 occurs at 503 cm^{-1} for γ-Fe$_2$SiO$_4$, and at 445 cm^{-1} for γ-Mg$_2$SiO$_4$. A further peak is just observed at 240 cm^{-1} in the IR transmission spectrum for γ-M$_2$SiO$_4$ (Akaogi et al., 1984).

Raman spectra for Mg$_2$SiO$_4$ spinel were first obtained by Akaogi et al. (1984) and McMillan and Akaogi (1987). The spectra showed strong peaks at 794 and 836 cm^{-1}, determined to correspond to the symmetric (v_1) and asymmetric (v_3) stretching vibrations of the SiO$_4$ tetrahedra (Chopelas et al., 1994). Additional weak peaks at 600, 370 and 302 cm^{-1} were assigned to Raman active modes associated with v_2 and v_4 SiO$_4{}^{4-}$ deformation modes and MgO$_6$ vibrations (McMillan and Akaogi, 1987; Chopelas et al., 1994). The pressure shifts were measured in the study to 20 GPa in the study by Chopelas et al. (1994). The lattice dynamics of γ-Mg$_2$SiO$_4$ have been studied theoretically using empirical potential methods (Matsui and Busing, 1984; Price et al., 1987).

The β-(Mg, Fe)$_2$SiO$_4$ polymorph (i.e., 'wadsleyite'; also referred to as 'modified spinel', or the 'β-phase', in previous studies) is expected to be a dominant mineral within the mantle transition zone. The transition between olivine and β-(Mg, Fe)$_2$SiO$_4$ occurs at $P \sim 13$ GPa and $T \sim 1700$ K; it marks the first onset of the 400 km seismic discontinuity (Agee, 1998). The IR spectrum of β-Co$_2$SiO$_4$ was first obtained by Jeanloz (1980), and that of β-Mg$_2$SiO$_4$ by Akaogi et al. (1984). The IR and Raman spectra of β-Mg$_2$SiO$_4$ are now characterized both at ambient P, T and under high-T and high-P conditions, and its lattice dynamics have been investigated by MD simulations using empirical potentials (McMillan and Akaogi, 1987; Price et al., 1987; Chopelas, 1991; Cynn and Hofmeister, 1994; Reynard and Rubie, 1996) (**Figure 13**). The IR and Raman spectra of wadsleyite-structured phases show unexpected peaks in the 600–700 cm^{-1} range, that occur due to the presence of SiOSi linkages contained within Si$_2$O$_7$ units in the structure (McMillan and Akaogi, 1987; Chopelas, 1991; Cynn and Hofmeister, 1994; McMillan et al., 1996b; Reynard et al., 1996). The resulting crystallographic sites could provide important hosts for OH species in the mantle (Smyth, 1987; McMillan and Hofmeister, 1988; Cynn and Hofmeister, 1994; Kohlstedt et al., 1996).

Compared with (Mg, Fe)$_2$SiO$_4$ structures, there have been fewer vibrational data obtained, or theoretical lattice dynamics calculations carried out, for (Mg, Fe)SiO$_3$ pyroxenes, that provide the next most important group of upper mantle minerals. The chain silicate phase MgSiO$_3$ (enstatite) exhibits complex polymorphism among various crystal structures as a function of P and T (Gasparik, 1990; Pacalo and Gasparik, 1990; Kanzaki, 1991a; Angel et al., 1992; Angel and Hugh-Jones, 1994; Yang and Ghose, 1994, 1995; McMillan et al., 1996b; Prewitt and Downs, 1998; Choudhury et al., 2002). The assignment of the various phase boundaries and recognition of stable vs. metastable phases has been quite controversial. Studies of the lattice dynamics using INS methods combined with MD simulations has now advanced our understanding of the system (Choudhury and Chaplot, 2000; Chaplot and Choudhury, 2001; Chaplot et al., 2002b; Choudhury et al., 2002). It now appears likely that the stable phase at ambient P and T is orthoenstatite (space group $Pbca$). Polarized Raman spectra and INS data for this phase have been obtained by Choudhury et al. (1998). At high temperature ($T > 800$ K), a phonon-assisted displacive transformation into the protoenstatite phase ($Pbcn$ symmetry) occurs at ambient P (Choudhury and Chaplot, 2000). There still remain discrepancies between experimental and theoretically-determined phase relations (Choudhury et al., 2002). At high pressures ($P > 4$–6 GPa), the structure transforms into a monoclinic ($C2/c$) clinoenstatite

(a)

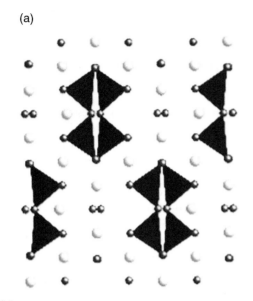

(b)

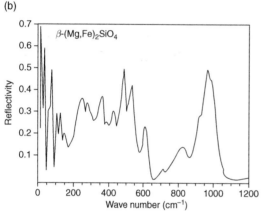

Figure 13 (Mg,Fe)$_2$SiO$_4$ wadsleyite, formed from mantle olivines above $P > 18$–20 GPa, marking the onset of the mantle transition zone. (a) The β-(Mg, Fe)$_2$SiO$_4$ structure contains Si$_2$O$_7^{6-}$ units, and one O^{2-} ion per asymmetric unit is bonded only to (Mg, Fe)$^{2+}$ cations. This results in voids occuring within the structure that are electron-rich, that might accomodate H$^+$ cations (Smyth, 1987; McMillan and Hofmeister, 1988; Cynn et al., 1996; Kohlstedt et al., 1996). (b) The IR and Raman spectra show sharp peaks in the 600–700 cm^{-1} range that are unexpected for orthosilicate structures, and that are due to SiOSi bending vibrations within the Si$_2$O$_7$ groups (Jeanloz, 1980; Akaogi et al., 1984; McMillan and Akaogi, 1987; Chopelas, 1991; Cynn and Hofmeister, 1994; Reynard and Rubie, 1996). Here we show the IR reflectance spectrum obtained for a powdered sample by Cynn and Hofmeister (1994). Part (b) is reprinted from Cynn H and Hofmeister AM (1994) *Journal of Geophysical Research B* 99: 17178.

phase. (Chopelas and Boehler, 1992) prepared a sample of that phase during laser heating experiments on MgSiO$_3$ in the diamond anvil cell, and obtained its Raman spectrum at high pressure. The high-density

clinoenstatite reverted to the orthorhombic polymorph during decompression to below $P \sim 5$ GPa. Fe-rich (Mg, Fe)SiO$_3$ phases also transform into the $C2/c$ structure type at high pressure (Yang and Ghose, 1995). The Raman and IR spectra of the monoclinic pyroxene diopside (CaMgSi$_2$O$_6$), present within the upper mantle, are well known from single-crystal measurements, combined with empirical force field calculations (Etchepare, 1970; Zulumyan et al., 1976; Dowty, 1987b): Raman spectra for this phase have been measured to $P = 16$ GPa (Chopelas and Boehler, 1992).

7.3.5 MgSiO$_3$ Ilmenite and Majoritic Garnets

Ilmenite-structured MgSiO$_3$ is stable within a narrow high-P range that could occur within cold subducted regions at 600–700 km depth toward the bottom of the transition zone, perhaps extending into the top of the lower mantle (Ito and Yamada, 1982; Gasparik, 1990; Chen and Brudzinski, 2001; Karki and Wentzcovich, 2002; Wentzcovich et al., 2004a). An ilmenite-structured (Mg, Fe)SiO$_3$ mineral has now been identified in natural meteorite samples, and it has been named 'akimotoite' (Tomioka and Fujino, 1999). Laboratory experiments also indicate that (Mg, Fe)SiO$_3$ ilmenite can enter into solid solution with Al$_2$O$_3$ component under mantle high-P,T conditions (Liu, 1977).

The vibrational modes at the BZ center are

$$\Gamma_{vib} = 5A_g(\mathrm{R}) + 5E_g(\mathrm{R}) + 4A_u(\mathrm{IR}) + 4E_u(\mathrm{IR}) \quad [19]$$

The Raman spectrum is dominated by an intense peak at 799 cm^{-1}, that was assigned by Ross and McMillan (1984) and McMillan and Ross (1987) as due to the A_g symmetric Si–O stretching mode of the octahedral SiO$_6$ groups in the structure (**Figure 14**). In their first-principles phonon calculations for MgSiO$_3$ ilmenite, Karki and Wentzcovitch (2002) apparently predicted an A_g mode at 783 cm^{-1} that was not observed experimentally, and assigned the strong Raman peak to an E_g SiO$_6$ asymmetric stretching vibration. However, that is most likely due to a typographic error in the published table, and the experimental assignment is likely correct. The vibrational spectra and elastic properties of MgSiO$_3$ ilmenite have now been determined by Raman, IR and Brillouin spectroscopy, including spectra obtained at high T and high P, and the lattice dynamics have been studied using both empirical

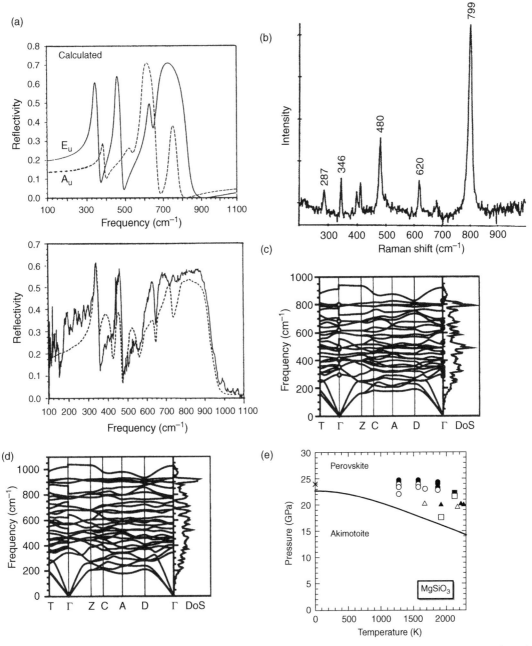

Figure 14 Lattice dynamics of MgSiO₃ ilmenite (akimotoite): experiment and theory. (a) The IR reflectance spectrum shown here was obtained for a polycrystalline sample by Hofmeister and Ito (1992). The synthetic spectrum below was reconstructed using both mode symmetries (A_u and E_u). All expected zone center vibrational modes have been assigned (Hofmeister and Ito, 1992). (b) The Raman spectrum at right is from McMillan and Ross (1987). More recent measurements by Reynard and Rubie (1995) indicate an additional weak peak occuring at 499 cm⁻¹. (c) and (d) First-principles calculations of the lattice dynamics (phonon dispersion curves) for MgSiO₃ ilmenite at (c) low pressure ($P = 1$ atm) and (d) high pressure ($P = 33$ GPa) (Karki and Wentzcovich, 2002; Wentzcovich *et al.*, 2004a). The experimentally determined Raman-active modes are marked at the Brillouin zone center (Γ point) by circles. (e) Theoretically predicted MgSiO₃ akimotoite–perovskite high-P,T phase boundary calculated using a first-principles phonon density of states ($g(\nu)$ function) within the QH approximation to determine the thermal free energy and related thermodynamic parameters (Karki and Wentzcovich, 2002; Wentzcovich *et al.*, 2004a). The calculated P, T phase boundary is shown as a continuous line. Available experimental data are shown as points. The experimental phase boundary passes between the filled symbols (MgSiO₃ ilmenite) and open symbols (perovskite). Part (a) is reprinted from Hofmeister AM and Ito E (1992) *Physics and Chemistry of Minerals* 18: 426. Part (b) is reprinted from McMillan PF and Ross NL (1987) *Physics and Chemistry of Minerals* 16. Parts (c) and (d) are reprinted from Karki B and Wentzcovitch RM (2002) First-principles lattice dynamics and thermoelasticity of MgSiO₃ ilmenite at high pressure. *Journal of Geophysical Research B* 107: 11, 2267. Part (e) is reprinted from Wentzcovitch RM, Stixrude L, Karki B, and Kiefer B (2004) Akimotoite to perovskite phase transition in MgSiO₃. *Geophysical Research Letters* 31: L10611, doi:10,101029/2004GL019704.

simulations and first principles calculations (Ross and McMillan, 1984; Weidner and Ito, 1985; McMillan and Ross, 1987; Wall and Price, 1988; Madon and Price, 1989; Hofmeister and Ito, 1992; Liu *et al.*, 1994; McMillan *et al.*, 1996b; Reynard and Rubie, 1996; Chopelas, 1999; Karki and Wentzcovich, 2002; Wentzcovich *et al.*, 2004a). The full lattice dynamics of the related corundum-structured α-Al_2O_3 have been studied experimentally by INS techniques, and they have been calculated using *ab initio* methods (Heid *et al.*, 2000).

In their high-pressure Raman spectroscopic study, Reynard and Rubie (1996) determined Grüneisen shifts for the Raman-active modes ranging between 1.7–3.7 cm^{-1}/GPa. These authors also observed the T-dependent shifts of the Raman peaks and determined the intrinsic mode anharmonicity parameters. The experimental and theoretical results have been used to estimate the P-T Clapeyron slopes for the garnet-ilmenite and garnet-perovskite phase boundaries within the mantle (McMillan and Ross, 1987; Hofmeister and Ito, 1992; Reynard and Guyot, 1994; Reynard *et al.*, 1996; Chopelas, 1999; Wentzcovich *et al.*, 2004a). The entropy calculated from vibrational models has also been combined with calorimetric data to determine phase boundaries within the $MgSiO_3$–Mg_2SiO_4–SiO_2 system over the P-T range of the transition zone (Ashida *et al.*, 1988).

A majoritic garnet based within the $(Mg,Fe)SiO_3$ system containing octahedrally coordinated Si^{4+} could also be present within the transition zone under high-T conditions, and like $MgSiO_3$ ilmenite, it could extend into the lower mantle (Ringwood, 1967; Akaogi and Akimoto, 1977; Kato and Kumazawa, 1985; Kato, 1986; Akaogi *et al.*, 1987; Yusa *et al.*, 1993). As with other high-density silicate mineral phases, majorititic garnet was identified and characterised in samples from meteorites (Coorara and Tenham), using a combination of optical microscopy, X-ray diffraction, and IR and Raman spectroscopy (Ringwood, 1967; Smith and Mason, 1970; Jeanloz, 1981; McMillan *et al.*, 1989). X-ray diffraction and vibrational spectroscopy of the end-member garnet $MgSiO_3$ indicate that the phase is tetragonally-distorted (Kato and Kumazawa, 1985; Akaogi *et al.*, 1987; McMillan *et al.*, 1989). At high-T,P solid solutions exist between majorite and pyrope $(Mg_3Al_2Si_3O_{12})$ end members. Members of the series have been studied by vibrational spectroscopy at high pressure (Liu *et al.*, 1995).

The predicted vibrational modes for cubic garnets, including pyrope, almandine and grossular

species that exist throughout the upper mantle, are given by symmetry analysis as

$$
\begin{aligned}
\Gamma_{vib} = {} & 3A_{1g}(\text{R}) + 5A_{2g}(\text{inactive}) + 8E_g(\text{R}) \\
& + 14T_{1g}(\text{inactive}) + 14T_{2g}(\text{IR}) + 5A_{1u}(\text{inactive}) \\
& + 5A_{2u}(\text{inactive}) + 10E_u(\text{inactive}) + 17T_{1u}(\text{IR}) \\
& + 16T_{2u}(\text{inactive})
\end{aligned}
\qquad [20]
$$

The IR and Raman modes have now been assigned from IR and Raman studies (Hofmeister and Chopelas, 1991a, 1991b; McMillan *et al.*, 1996b; Kolesov and Geiger, 1998, 2000; Geiger and Kolesov, 2002) (**Figure 15**). The vibrational data have been used to construct model VDOS functions to predict the specific heat and entropy relations of mantle garnets under high-P,T conditions (Hofmeister and Chopelas, 1991a; Geiger and Kolesov, 2002). The lattice dynamics and VDOS of upper mantle garnets have now been determined using ionic model calculations combined with INS measurements (Artioli *et al.*, 1996; Pavese *et al.*, 1998; Mittal *et al.*, 2000, 2001).

7.3.6 OH in Mantle Minerals

A long-standing question within mantle mineralogy concerns the amount and potential repositories for volatile components especially within the H–O–C–N system, within the deep mantle. It has long been recognized that nominally anhydrous minerals such as olivine, pyroxene and garnet can incorporate up to several tens to hundreds of ppm OH into their structure, and that this could contribute substantially to the water budget of the mantle (Aines and Rossman, 1984c, 1984b, 1984a; Bell, 1992; Bell and Rossman, 1992; Rossman, 1996). These experimental determinations of OH incorporation into the mineral structure were made by IR spectroscopy. A further advance in understanding the potential sources and sinks for OH species within the mantle was derived from the determination of the crystal structure of β-Mg_2SiO_4, that is formed at the boundary between the upper mantle and the transition zone (Agee, 1998). This phase has an orthosilicate composition, but it contains Si_2O_7 groups (Horiuchi and Sawamoto, 1981). That observation requires the presence of O^{2-} ions bound only to Mg^{2+} within the structure, that could then explain the unusual ability of this nominally anhydrous phase to accept large quantities of H into its structure, perhaps even determining the overall water budget of the mantle (Smyth, 1987; McMillan and Hofmeister, 1988; Downs, 1989; Young *et al.*, 1993; Kohlstedt *et al.*, 1996). Now several

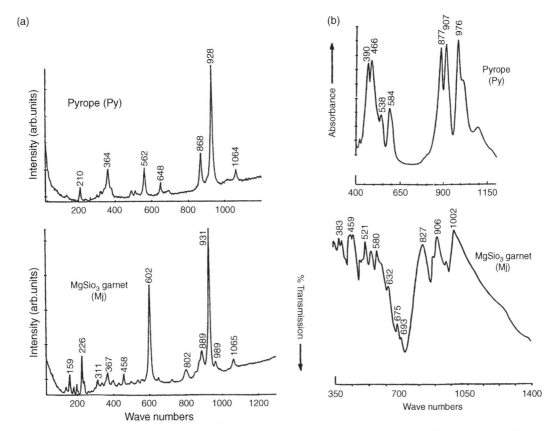

Figure 15 (a) Raman spectra and (b) powder transmission infrared spectra of pyrope and MgSiO$_3$ garnets (McMillan *et al.*, 1989). The Raman and IR data of MgSiO$_3$ garnet show the additional weak peaks that appear due to the tetragonal distortion of the structure compared with cubic pyrope garnet. Reprinted from McMillan PF, Akaogi M, Ohtani E, Williams Q, Nieman R, and Sato R (1989) *Physics and Chemistry of Minerals* 16.

compounds within the MgO–SiO$_2$–H$_2$O system have been discovered to be stable under the high-P,T conditions of the mantle transition zone, especially 'phase A' (Akaogi and Akimoto, 1986; Liu, 1987; Kanzaki, 1991b; Bose and Navrotsky, 1998). These could also provide important hosts or repositories for H$_2$O in the mantle. Deeper within the Earth, lower mantle minerals including SiO$_2$ stishovite and MgSiO$_3$ perovskite have been found to contain dissolved OH species in experimental runs carried out under high-P,T conditions (Pawley *et al.*, 1993; Meade *et al.*, 1994).

Acknowledgments

The work of PFM in mineral spectroscopy and research under high-P,T conditions has been supported by the NSF, EPSRC and by a Royal Society-Wolfson Foundation Research Merit Award Fellowship.

References

Agee CB (1998) Phase transformations and seismic structure in the upper mantle and transition zone. In: Hemley RJ (ed.) *Ultra-High Pressure Mineralogy: Physics and Chemistry of the Earth's Deep Interior*, vol. 37, pp. 165–203. Washington DC: Mineralogical Society of America.

Agnon A and Bukowinski MST (1990) Thermodynamic and elastic properties of a many-body model for simple oxides. *Physical Review B* 41: 7755–7766.

Aines RD and Rossman GR (1984a) Water in minerals? A peak in the infrared. *Journal of Geophysical Research* 89: 4059–4071.

Aines RD and Rossman GR (1984b) Water content of mantle garnets. *Geology* 12: 720–723.

Aines RD and Rossman GR (1984c) The hydrous component in garnets: Pyralspites. *American Mineralogist* 69: 1116–1126.

Akaogi M and Akimoto S (1977) Pyroxene–garnet solid solution equilibria in the systems Mg$_4$Si$_4$O$_{12}$–Mg$_3$Al$_2$Si$_3$O$_{12}$ and Fe$_4$Si$_4$O$_{12}$–Fe$_3$Al$_2$Si$_3$O$_{12}$ at high pressures and temperatures. *Physics of the Earth and Planetary Interiors* 15: 90–106.

Akaogi M and Akimoto S (1986) Infrared spectra of high-pressure hydrous silicates in the system MgO–SiO$_2$–H$_2$O. *Physics and Chemistry of Minerals* 13: 161–164.

Akaogi M and Ito E (1993) Refinement of enthalpy measurement of MgSiO$_3$ perovskite and negative pressure–temperature slopes for perovskite-forming reactions. *Geophysical Research Letters* 20: 1839–1842.

Akaogi M, Ito E, and Navrotsky A (1989) Olivine-modified spinel–spinel transitions in the system Mg_2SiO_4–Fe_2SiO_4: Calorimetric measurement, thermochemical calculation, and geophysical application. *Journal of Geophysical Research B* 94: 15671–15685.

Akaogi M, Navrotsky A, Yagi T, and Akimoto S-I (1987) Pyroxene–garnet transformations: Thermochemistry and elasticity of garnet solid solutions, and application to a pyrolite mantle. In: Manghnani MH and Syono Y (ed.) *High-Pressure Research in Mineral Physics*, pp. 251–260. Washington DC: Terra Scientific Publishing Company, Tokyo/American Geophysical Union.

Akaogi M, Ross NL, McMillan PF, and Navrotsky A (1984) The Mg_2SiO_4 polymorphs (olivine, modified spinel, spinel) – Thermodynamic properties from oxide melt solution calorimetry, phase relations, and models of lattice vibrations. *American Mineralogist* 69: 499–512.

Akber-Knutson S, Steinle-Neumann G, and Asimow PD (2005) Effects of Al on the sharpness of the $MgSiO_3$ perovskite to post-perovskite phase transition. *Geophysical Research Letters* 32: L14303–7.

Akins JA, Luo S-N, Asimow PD, and Ahrens TJ (2004) Shock-induced melting of $MgSiO_3$ perovskite and implications for melts in Earth's lowermost mantle. *Geophysical Research Letters* 31: L14612–5.

Alfé D (2005) Melting curve of MgO from first-principles simulations. *Physical Review Letters* 94: 235701.

Alfé D, Gillan MJ, and Price GD (1999) The melting curve of iron at the pressures of the Earth's core from *ab initio* calculations. *Nature* 401: 462–464.

Alfé D, Vocadlo L, Price GD, and Gillan MJ (2004) Melting curve of materials: Theory versus experiments. *Journal of Physics: Condensed Matter* 16: S973–S982.

Allen MP and Tildesley DJ (1987) *Computer Simulation of Liquids*. Oxford: Clarendon Press.

Ancilotto F, Chiarotti GL, Scandolo S, and Tosatti E (1997) Dissociation of methane into hydrocarbons at extreme (planetary) pressure and temperature. *Science* 275: 1288–1290.

Anderson DL (1989) *Theory of the Earth*. Boston: Blackwell.

Anderson OL (1995) *Equations of State of Solids for Geophysics and Ceramic Science*. Oxford: Oxford University Press.

Andrault D, Fiquet G, Guyot F, and Hanfland M (1998) Pressure-induced Landau-type transition in stishovite. *Science* 282: 720–724.

Angel RJ, Chopelas A, and Ross NL (1992) Stability of high-density clinoenstatite at upper-mantle pressures. *Nature* 358: 322–324.

Angel RJ and Hugh-Jones DA (1994) A compressional study of $MgSiO_3$ orthoenstatite up to 8.5 GPa. *American Mineralogist* 79: 19777–19783.

Artioli G, Pavese A, and Moze O (1996) Dispersion relation of acoustic phonons in pyrope garnet: Relationship between the vibrational properties and elastic constants. *American Mineralogist* 81: 19–25.

Ashcroft NW and Mermin ND (1976) *Solid State Physics*. Philadelphia: Holt-Saunders.

Ashida T, Kume S, Ito E, and Navrotsky A (1988) $MgSiO_3$ ilmenite: Heat capacity, thermal expansivity, and enthalpy of transformation. *Physics and Chemistry of Minerals* 16: 239–245.

Badro J, Teter DM, Downs RT, Gillet P, Hemley RJ, and Barrat JL (1997) Theoretical study of a novel five-coordinated silica polymorph. *Physical Review B* 56: 5797–5806.

Baerends EJ and Gritsenko OV (1997) A quantum chemical view of density functional theory. *Journal of Physical Chemistry A* 101: 5383–5403.

Baroni S, de Gironcoli S, Dal Corso A, and Gianozzi P (2001) Phonons and related crystal properties from density-functional perturbation theory. *Review of Modern Physics* 73: 515–562.

Baroni S, Gianozzi P, and Testa A (1987) Green-function approach to linear response in solids. *Journal of Physical Chemistry A* 101: 5383–5403.

Basile LJ, Ferraro JR, LaBonville P, and Wall MC (1973) A study of force fields for tetrahedral molecules and ions. *Coordination Chemistry Reviews* 11: 21–69.

Bass JD (1995) Elasticity of minerals, glasses, and melts. In: Ahrens TJ (ed.) *Mineral Physics and Crystallography. A Handbook of Physical Constants*, pp. 45–63. Washington DC: American Geophysical Union.

Bell DR (1992) Water in mantle minerals. *Nature* 357: 646–647.

Bell DR and Rossman GR (1992) Water in the Earth's mantle: The role of anhydrous minerals. *Science* 255: 1391–1397.

Besson JM, Nelmes RJ, Hamel G, Loveday JS, Weill G, and Hull S (1992) Neutron powder diffraction above 10 GPa. *Physica B* 180: 907–910.

Bilz H and Kress W (1979) *Phonon Dispersion Relations in Insulators*. Berlin: Springer.

Bina CR (1998) Lower mantle mineralogy and the geophysical perspective. In: Hemley RJ (ed.) *Ultra-High Pressure Mineralogy: Physics and Chemistry of the Earth's Deep Interior*, Vol. 37, pp. 205–239. Washington DC: Mineralogical Society of America.

Bina CR and Wood BJ (1987) Olivine–spinel transitions: Experimental and thermodynamic constraints and implications for the nature of the 400-km seismic discontinuity. *Journal of Geophysical Research B* 92: 4835–4866.

Birch F (1952) Elasticity and constitution of the Earth's interior. *Journal of Geophysical Research* 57: 527–552.

Birman JL (1984) *Theory of Crystal Space Groups and Lattice Dynamics*. New York: Springer.

Boppart H, Straaten JV, and Silvera IF (1985) Raman spectra of diamond at high pressure. *Physical Review B* 32: 1423–1425.

Born M and Huang K (1954) *Dynamical Theory of Crystal Lattices*. Oxford: Clarendon Press.

Bose K and Navrotsky A (1998) Thermochemistry and phase equilibria of hydrous phases in the system MgO–SiO_2–H_2O: Implications for volatile transport to the mantle. *Journal of Geophysical Research B* 103: 9713–9719.

Boyd FR and Meyer HOA (ed.) (1979) *The Mantle Sample: Inclusions in Kimberlites and Other Volcanics*, Washington, DC: American Geophysical Union.

Brazhkin V V, Lyapin AG, and Hemley RJ (2002) Harder than diamond: Dreams and reality. *Philosophical Magazine A* 82: 231–253.

Brodholt JP, Oganov AR, and Price GD (2002) Computational mineral physics and the physical properties of perovskite. *Philosophical Transaction of the Royal Society A* 360: 2507–2520.

Brown JM, Slutsky LJ, Nelson KA, and Cheng L-T (1989) Single-crystal elastic constants for San Carlos peridot: An application of impulsive stimulated scattering. *Journal of Geophysical Research* 94: 9485–9492.

Brüesch P (1982) *Phonons: Theory and Experiments*. Berlin: Springer.

Bukowinski MST (1994) Quantum geophysics. *Annual Review of Earth and Planetary Science* 22: 167–205.

Bukowinski MST, Chizmeshya A, Wolf GH, and Zhang H (1996) Advances in electron-gas potential models: Applications to some candidate lower mantle minerals. In: Silvi B and D'Arco P (ed.) *Modelling of Minerals and Silicate Structures*, pp. 81–112. Dordrecht: Kluwer Academic.

Bukowinski MST and Wolf GH (1986) Equation of state and stability of fluorite-structured SiO_2. *Journal of Geophysical Research B* 91: 4704–4710.

Bukowinski MST and Wolf GH (1988) Equation of state and possible critical phase transitions in $MgSiO_3$ perovskite at lower-mantle conditions. In: Ghose S, Coey JM, and Salje E (ed.) (eds.) *Structural and Magnetic Phase Transitions*, vol. 7, pp. 91–112. New York: Springer.

Burfoot JC and Taylor GW (1979) *Polar Dielectrics and their Applications*. New York: Macmillan.

Burnham CW (1985) Mineral structures energetics and modelling using the ionic approach. In: Kieffer SW and Navrotsky A (ed.) *Microscopic to Macroscopic. Atomic Environments to Mineral Thermodynamics*, Vol. 14, pp. 347–388. Washington DC: Mineralogical Society of America.

Burnham CW (1990) The ionic model: Perceptions and realities in mineralogy. *American Mineralogist* 75: 443–463.

Burns G (1985) *Solid State Physics*. San Diego: Academic Press.

Car R and Parrinello (1985) Unified approach for molecular dynamics and density-functional theory. *Physical Review Letters* 55: 2471–2474.

Carroll M and Holloway JR (ed.) (1994) *Volatiles in Magmas. Reviews in Mineralogy,* vol. 30: Mineralogical Society of America, Washington DC.

Catlow CRA and Price GD (1990) Computer modelling of solid-state inorganic materials. *Nature* 347: 243–248.

Cavazzoni C, Chiarotti GL, Scandolo S, Tosatti E, Bernasconi M, and Parrinello M (1999) Superionic and metallic states of water and ammonia at giant planet conditions. *Science* 283: 44–46.

Chai M, Brown JM, and Slutsky LJ (1996) Thermal diffusivity of mantle minerals. *Physics and Chemistry of Minerals* 23: 470–475.

Challener WA and Thompson JD (1986) Far-infrared spectroscopy in diamond anvil cells. *Applied Spectroscopy* 40: 298–303.

Chao ECT, Fahey JJ, Littler J, and Milton DJ (1962) Stishovite, a very high pressure new mineral from Meteor Crater, Arizona. *Journal of Geophysical Research* 67: 419–421.

Chaplot SL and Choudhury N (2001) Molecular dynamics simulations of seismic discontinuities and phase transitions of $MgSiO_3$ from 4 to 6-coordinated silicon via a novel 5-coordinated phase. *American Mineralogist* 86: 752–761.

Chaplot SL, Choudhury N, Ghose S, Rao MN, Mittal R, and Goel P (2002a) Inelastic neutron scattering and lattice dynamics of minerals. *European Journal of Mineralogy* 142: 291–330.

Chaplot SL, Choudhury N, Ghose S, Rao MN, Mittal R, and Goel P (2002b) Inelastic neutron scattering and lattice dynamics of minerals. *European Journal of Mineralogy* 14: 291–329.

Chen WR and Brudzinski MR (2001) Evidence for a large-scale remnant of subducted lithosphere beneath Fiji. *Science* 292: 2475–2479.

Chervin JC, Canny B, Gauthier M, and Pruzan P (1992) Micro-Raman at variable low-temperature and very high pressure. *Review of Scientific Instruments* 64: 203–206.

Chizmeshya A, Zimmerman FM, LaViolette R, and Wolf GH (1994) Variational charge relaxation in ionic crystals: An efficient treatment of statics and dynamics. *Physical Review B* 50: 15559–15574.

Chizmeshya AVG, Wolf GH, and McMillan PF (1996) First principles calculation of the equation of state, stability, and polar optic modes of $CaSiO_3$ perovskite. *Geophysical Research Letters* 23: 2725–2728.

Chopelas A (1990) Thermochemical properties of forsterite at mantle pressures derived from vibrational spectroscopy. *Physics and Chemistry of Minerals* 17: 149–156.

Chopelas A (1991) Thermal properties of β-Mg_2SiO_4 at mantle pressures derived from vibrational spectroscopy: Implications for the mantle at 400 km depth. *Journal of Geophysical Research B* 96: 11817–11829.

Chopelas A (1992) Sound velocities of MgO to very high compression. *Earth and Planetary Science Letters* 114: 195–202.

Chopelas A (1996) Thermal expansivity of lower mantle phases MgO and $MgSiO_3$ perovskite at high pressure derived from vibrational spectroscopy. *Physics of the Earth and Planetary Interiors* 98: 3–15.

Chopelas A (1999) Estimates of mantle relevant Clapeyron slopes in the $MgSiO_3$ system from high-pressure spectroscopic data. *American Mineralogist* 84: 233–244.

Chopelas A and Boehler R (1992) Raman spectroscopy of high pressure $MgSiO_3$ phases synthesized in a CO_2 laser heated diamond anvil cell: Perovskite and pyroxene. In: Syono Y and Manghnani MH (eds.) *High-Pressure Research: Application to Earth and Planetary Sciences*, pp. 101–108. Washington DC: Terra Scientific Publishing Co., Tokyo/American Geophysical Union.

Chopelas A, Boehler R, and Ko T (1994) Thermodynamics and behaviour of γ-Mg_2SiO_4 at high pressure: Implications for Mg_2SiO_4 phase equilibrium. *Physics and Chemistry of Minerals* 21: 351–359.

Chopelas A and Nicol MF (1982) Pressure dependence to 100 kbar of the phonons of MgO at 90 and 295 K. *Journal of Geophysical Research B* 87: 8591–8597.

Choudhury N and Chaplot SL (2000) Free energy and relative stability of the enstatite $Mg_2Si_2O_6$ polymorphs. *Solid State Communications* 114: 127–132.

Choudhury N, Chaplot SL, Ghose S, Rao MN, and Mittal R (2002) Lattice dynamics, inelastic neutron scattering and thermodynamic properties of minerals. In: Gramaccioli CM (ed.) *Energy Modelling in Minerals*, Vol. 4, pp. 211–243. Budapest: Eötvös University Press (European Mineralogical Union).

Choudhury N, Ghose S, Chowdhury CP, Loong CK, and Chaplot SL (1998) Lattice dynamics, Raman scattering and inelastic neutron scattering of orthoenstatite, $MgSiO_3$. *Physical Review B* 58: 756–765.

Choudhury N, Wu Z, Walter EJ, and Cohen RE (2005) *Ab initio* linear response and frozen phonons for the relaxor $PbMg_{1/3}Nb_{2/3}O_3$. *Physical Review B* 71: 125134.

Christensen PR, Anderson DL, Chase SC, Clark RN, Kieffer HH, Malin MC, et al. (1992) Thermal emission spectrometer experiment: Mars observer mission. *Journal of Geophysical Research E* 97: 7719–7734.

Clark CD, Collins AT, and Woods GS (1992) Absorption and luminescence spectroscopy. In: Field JE (ed.) *The Properties of Natural and Synthetic Diamond*, pp. 35–79. London: Academic Press.

Cohen RE (1987) Calculation of elasticity and high pressure instabilities in corundum and stishovite with the potential induced breathing model. *Geophysical Research Letters* 14: 37–40.

Cohen RE (1991) Bonding and elasticity of stishovite at high pressure: Linearized augmented plane wave calculations. *American Mineralogist* 76: 733–742.

Cohen RE (1992a) Origin of ferroelectricity in oxide ferroelectrics and the difference in ferroelectric behavior of $BaTiO_3$ and $PbTiO_3$. *Nature* 358: 136–138.

Cohen RE (1992b) First-principles predictions of elasticity and phase transitions in high pressure SiO_2 and geophysical applications. In: Syono Y and Manghnani MH (eds.) *High-Pressure Research: Application to Earth and Planetary Sciences*, pp. 425–431. Tokyo/Washington DC: Terra Scientific Publishing Co.,/American Geophysical Union.

Cohen RE, Boyer LL, and Mehl MJ (1987a) Lattice dynamics of the potential induced breathing model: Phonon dispersion in the alkaline Earth oxides. *Physical Review B* 35: 5749–5760.

Cohen RE, Boyer LL, and Mehl MJ (1987b) Theoretical studies of charge relaxation effects on the static and lattice dynamics of oxides. *Physics and Chemistry Minerals* 14: 294–302.

Cotton FA (1971) *Chemical Applications of Group Theory.* New York: Wiley.

Cowley RA (1963) The lattice dynamics of the anharmonic crystal. *Advances in Physics* 12: 421–480.

Cynn H and Hofmeister AM (1994) High-pressure IR spectra of lattice modes and OH vibrations in Fe-bearing wadsleyite. *Journal of Geophysical Research B* 99: 17717–17727.

Cynn H, Hofmeister AM, and Burnley PC (1996) Thermodynamic properties and hydrogen speciation from vibrational spectra of dense hydrous magnesium silicates. *Physics and Chemistry of Minerals* 23: 361–376.

D'Arco P, Sandrone G, Dovesi R, Orlando R, and Saunders VR (1993) A quantum mechanical study of the perovskite structure type of $MgSiO_3$. *Physics and Chemistry of Minerals* 20: 407–414.

D'Arco P, Sandrone G, Dovesi R, Apra E, and Saunders CR (1994) A quantum-mechanical study of the relative stability under pressure of $MgSiO_3$–ilmenite, $MgSiO_3$–perovskite, and MgO–periclase+SiO_2–stishovite assemblage. *Physics and Chemistry of Minerals* 21: 285–293.

David WL, Evans JSO, and Sleight AW (1999) Direct evidence for a low-frequency phonon mode mechanism in the negative thermal expansion compound ZrW_2O_6. *Europhysics Letters* 46: 661–666.

Dawson JB (1980) *Kimberlites and Their Xenoliths.* Berlin: Springer.

Decius JC and Hexter RM (1977) *Molecular Vibrations in Crystals.* New York: McGraw-Hill.

Dera P, Prewitt CT, Boctor NZ, and Hemley RJ (2002) Characterization of a high-pressure phase of silica from the Martian meteorite Shergotty. *American Mineralogist* 87: 1018–1023.

Dingwell DB (1998) Melt viscosity and diffusion under elevated pressures. In: Hemley RJ (ed.) *Ultra-High Pressure Mineralogy: Physics and Chemistry of the Earth's Deep Interior,* Vol. 37, pp. 397–424. Washington DC: Mineralogical Society of America.

Dolino G and Bachheimer JP (1977) Fundamental and second-harmonic light scattering by the α-β coexistence state of quartz'. *Physica Status Solidi (A)* 41: 674–677.

Dolino G, Bachheimer JP, Gervais F, and Wright AF (1983) La transition α-β du quartz: le point sur quelques problemes actuales: Transition order-desordre ou displacive: Comptement thermodynamique. *Bulletin de Mineralogie* 106: 267–285.

Dove MT (1993) *Introduction to Lattice Dynamics.* Cambridge: Cambridge University Press.

Dove MT (2002) An introduction to the use of neutron-scattering methods in mineral sciences. *European Journal of Mineralogy* 142: 203–224.

Dove MT, Tucker MG, Wells SA, and Keen DA (2002) Reverse Monte Carlo methods. In: Gramaccioli CM (ed.) *Energy Modelling in Minerals,* Vol. 4, pp. 59–82. Budapest: Eötvös University Press.

Downs JW (1989) Possible sites for protonation in β-Mg_2SiO_4 from an experimentally-derived electrostatic potential. *American Mineralogist* 74: 1124–1129.

Dowty E (1987a) Vibrational interactions of tetrahedra in silicate glasses and crystals. I: Calculations on ideal silicate–aluminate–germanate structural units. *Physics and Chemistry of Minerals* 14: 80–93.

Dowty E (1987b) Vibrational interactions of tetrahedra in silicate glasses and crystals. II: Calculations on melilites, pyroxenes, silica polymorphs and feldspars. *Physics and Chemistry of Minerals* 14: 122–138.

Dowty E (1987c) Vibrational interactions of tetrahedra in silicate glasses and crystals. III: Calculations on simple alkali silicates, thortveitite and rankinite. *Physics and Chemistry of Minerals* 14: 542–552.

Dubrovinskaia NA, Dubrovinsky LS, Saxena SK, Tutti F, Rekhi S, and Le Bihan T (2001) Direct transition from cristobalite to post-stishovite α-PbO_2-like silica phase. *European Journal of Mineralogy* 13: 479–483.

Dubrovinsky L, Saxena SK, Lazor P, Ahuja R, Eriksson O, Wills JM, and Johansson B (1997) Experimental and theoretical identification of a new high-pressure phase of silica. *Nature* 388: 362–365.

Dubrovinsky LS, Dubrovinskaia NA, Prakapenka V, *et al.* (2003) High-pressure and high-temperature polymorphism in silica. *High Pressure Research* 23: 35–39.

Duffy TS (2005) Synchrotron facilities and the study of the Earth's deep interior. *Reports on Progress in Physics* 68: 1811–1859.

Duffy TS and Wang Y (1998) Pressure–volume–temperature equations of state. In: Hemley RJ (ed.) *Ultra-High Pressure Mineralogy: Physics and Chemistry of the Earth's Deep Interior,* vol. 37, pp. 425–457. Washington DC: Mineralogical Society of America.

Durben DJ, McMillan PF, and Wolf GH (1993) Raman study of the high pressure behavior of forsterite Mg_2SiO_4 crystal and glass. *American Mineralogist* 78: 1143–1148.

Durben DJ and Wolf GH (1992) High-temperature behavior of metastable $MgSiO_3$ perovskite: A Raman spectroscopic study. *American Mineralogist* 77: 890–893.

Efimov AM (1995) *Optical Constants of Inorganic Glasses.* Boca Raton, FL: CRC Press.

El Goresy A, Dubrovinsky L, Saxena SK, and Sharp TS (1998) A new poststishovite silicon dioxide polymorph with the baddelyite structure (zircon oxide) in the SNC meteorite Shergotty: Evidence for extreme shock pressure. *Meteoritics and Planetary Science* 33: A45.

El Goresy A, Dubrovinsky L, Sharp TS, Saxena SK, and Chen M (2000) A monoclinic post-stishovite polymorph of silica in the Shergotty meteorite. *Science* 288: 1632–1634.

Etchepare J (1970) Spectres Raman du diopside cristallisé et vitreux. *Comptes Rendus de l'Academie des Sciences de Paris, Série II* 270: 1339–1342.

Farmer VC (1974) *The Infrared Spectra of Minerals.* London: The Mineralogical Society.

Fateley WG, Dollish FR, McDevitt NT, and Bentley FF (1972) *Infrared and Raman Selection Rules for Molecular and Lattice Vibrations: The Correlation Method.* New York: Wiley-Interscience.

Ferrarro JR (1984) *Vibrational Spectroscopy at High External Pressures. The Diamond Anvil Cell.* Orlando, FL: Academic Press.

Fiquet G, Andrault D, Dewaele A, Charpin T, Kunz M, and Hausermann D (1998) P–V–T equation of state of $MgSiO_3$ perovskite. *Physics of the Earth and Planetary Interiors* 105: 21–32.

Fiquet G, Badro J, Guyot F, Requardt H, and Krisch M (2001) Sound velocities in iron to 110 gigapascals. *Science* 291: 468–471.

Fiquet G, Gillet P, and Richet P (1992) Anharmonicity and high-temperature heat capacity of crystals: The examples of Ca_2GeO_4, Mg_2GeO_4 and $CaMgGeO_4$ olivines. *Physics and Chemistry of Minerals* 18: 469–479.

Funamori N and Jeanloz R (1997) High pressure transformation of Al_2O_3. *Science* 278: 1109–1111.

Funamori N, Jeanloz R, Miyajima N, and Fujino K (2000) Mineral assemblages of basalt in the lower mantle. *Journal of Geophysics Research B* 105: 26037 26034.

Gasparik T (1990) Phase relations in the transition zone. *Journal of Geophysical Research B* 95: 15751–15769.

Gasparik T (1993) The role of volatiles in the transition zone. *Journal of Geophysical Research B* 98: 4287–4299.

Geiger CA and Kolesov BA (2002) Microscopic–macroscopic relationships in silicates: Examples from IR and Raman spectroscopy and heat capacity measurements. In: Gramaccioli CM (ed.) *Energy Modelling in Minerals*, vol. 4, pp. 347–387. Budapest: Eötvös University Press.

Gervais F (1983) High-temperature infrared reflectivity spectroscopy by scanning interferometry. In: Burton KJ (ed.) *Infrared and Millimeter Waves. Part 1: Electromagnetic Waves in Matter*, vol. 8, pp. 279–339. Orlando: Academic Press.

Gervais F and Piriou B (1975) Temperature dependence of transverse and longitudinal optic modes in the α and β phases of quartz. *Physical Review B* 11: 3944–3950.

Gervais F, Piriou B, and Cabannes F (1973a) Anharmonicity in silicate crystals: Temperature dependence of A_u type vibrational modes in $ZrSiO_4$ and $LiAlSi_2O_6$. *Journal of Physics and Chemistry of Solids* 34: 1785–1796.

Gervais F, Piriou B, and Cabannes F (1973b) Anharmonicity of infrared vibrational modes in the nesosilicate Be_2SiO_4. *Physical status solidi (B)* 55: 143–154.

Ghose S (1985) Lattice dynamics, phase transitions and soft modes. In: Kieffer SW and Navrotsky S (eds.) *Microscopic to Macroscopic. Atomic Environments to Mineral Thermodynamics*, vol. 14, pp. 127–163. Washington DC: Mineralogical Society of America.

Ghose S (1988) Inelastic neutron scattering. In: Hawthorne F (ed.) *Spectroscopic Methods in Mineralogy and Geology*, vol. 18, pp. 161–192. Washington DC: Mineralogical Society of America.

Ghose S, Choudhury N, Pal Chowdhury C, and Sharma SK (1994) Lattice dynamics and Raman spectroscopy of protoenstatite $Mg_2Si_2O_6$. *Physics and Chemistry of Minerals* 20: 469–471.

Ghosez P, Cockayne E, Waghmare UV, and Rabe KM (1999) Lattice dynamics of $BaTiO_3$, $PbTiO_3$, and $PbZrO_3$: A comparative first-principles study. *Physical Review B* 60: 836–843.

Gibbs GV (1982) Molecules as models for bonding in silicates. *American Mineralogist* 67: 421–450.

Gillet P (1995) Mineral physics, mantle mineralogy and mantle dynamics. *Comptes Rendus de l'Académie des Sciences, Série II* 220: 341–356.

Gillet P, Daniel I, and Guyot F (1997) Anharmonic properties of Mg_2SiO_4–forsterite measured from the volume dependence of the Raman spectrum. *European Journal of Mineralogy* 9: 255–262.

Gillet P, Daniel I, Guyot F, Matas J, and Chervin JC (2000) A thermodynamic model for $MgSiO_3$ perovskite derived from the pressure, temperature and volume dependence of the Raman mode frequencies. *Physics of the Earth and Planetary Interiors* 117: 361–384.

Gillet P, Fiquet G, Daniel I, and Reynard B (1993) Raman spectroscopy at mantle pressure and temperature conditions. Experimental set-up and the example of $CaTiO_3$ perovskite. *Geophysical Research Letters* 20: 1931–1934.

Gillet P, Guyot F, and Wang Y (1996a) Microscopic anharmonicity and the equation of state of $MgSiO_3$–perovskite. *Geophysical Research Letters* 23: 3043–3046.

Gillet P, Hemley RJ, and McMillan PF (1998) Vibrational properties at high pressures and temperatures. In: Hemley RJ (ed.) *Ultra-High Pressure Mineralogy*, vol. 37: pp. 525–590. Mineralogical Society of America. Washington DC: Mineralogical Society of America.

Gillet P, Le Cléach A, and Madon M (1990) High-temperature Raman spectroscopy of SiO_2 and GeO_2 polymorphs: Anharmonicity and thermodynamic properties at high-temperatures. *Journal of Geophysical Research B* 95: 21635–21655.

Gillet P, McMillan P, Schott J, Badro J, and Grzechnik A (1996b) Thermodynamic properties and isotopic fractionation of calcite from vibrational spectroscopy of ^{18}O-substituted calcite. *Geochimica et Cosmochimica Acta* 60: 3471–3485.

Gillet P, Richet P, Guyot F, and Fiquet G (1991) High-temperature thermodynamic properties of forsterite. *Journal of Geophysical Research B* 96: 11805–11816.

Glazer AM (1972) The classification of tilted octahedra in perovskites. *Acta Crystallographica Series B* 28: 3384–3392.

Glazer AM (1975) Simple ways of determining perovskite structures. *Acta Crystallographica Series A* 31: 756–762.

Gramaccioli CM (2002a) Lattice dynamics: Theory and application to minerals. In: Gramaccioli CM (ed.) *Energy Modelling in Minerals*, vol. 4, pp. 245–270. Budapest: Eötvös University Press.

Gramaccioli CM (2002b) *Energy Modelling in Minerals*, EMU Notes in Mineralogy, vol. 4. Budapest: Eotvos University Press, EMU Notes in Mineralogy.

Grzechnik A, Chizmeshya AVG, Wolf GH, and McMillan PF (1998) An experimental and theoretical investigation of phonons and lattice instabilities in metastable decompressed $SrGeO_3$ perovskite. *Journal of Physics: Condensed Matter* 10: 221–233.

Grzechnik A, McMillan P, and Simon P (1999) An infrared study of $MgCO_3$ at high pressure. *Physica B* 262: 67–73.

Grzechnik A, McMillan PF, Chamberlain R, Hubert H, and Chizmeshya AVG (1997) $SrTiO_3$–$SrGeO_3$ perovskites obtained at high pressure and high temperature. *European Journal of Solid State and Inorganic Chemistry* 34: 269–281.

Guth JR, Hess AC, McMillan PF, and Petuskey WT (1990) An *ab initio* valence force field for diamond. *Journal of Physics C, Solid State Physics* 2: 8007–8014.

Guyot F, Wang Y, Gillet P, and Ricard Y (1996) Quasi-harmonic computations of thermodynamic parameters of olivines at high-pressure and high-temperature. A comparison with experiment data. *Physics of the Earth and Planetary Interiors* 98: 17–29.

Haines J and Léger JM (1993) Phase transitions in ruthenium dioxide up to 40 GPa: Mechanism for the rutile-to-fluorite transformation and a model for the high-pressure behavior of stishovite SiO_2. *Physical Review B* 48: 13344–13350.

Haines J, Léger JM, and Schulte O (1996a) Pa-3 modified fluorite-type structures in metal dioxides at high pressure. *Science* 271: 629–631.

Haines J, Léger JM, and Schulte O (1996b) Modified fluorite-type structures in metal dioxides at high pressure. *Science* 271: 629–631.

Hamermesh M (1962) *Group Theory and Its Application to Physical Problems*. London: Constable and Co, (Reprinted by Dover Publications).

Hanfland M, Syassen K, Fahy K, Louie SG, and Cohen ML (1985) Pressure dependence of the first-order Raman mode of diamond. *Physical Review B* 31: 6896–6899.

Harris DC and Bertolucci MD (1978) *Symmetry and Spectroscopy. An Introduction to Vibrational and Electronic Spectroscopy*. Oxford: Oxford University Press, (Re-printed in 1989 by Dover Press).

Harris JW (1992) Diamond geology. In: Field J E (ed.) *The Properties of Natural and Synthetic Diamond*, pp. 345–393. London: Academic Press.

Harte B and Harris J (1993) Lower mantle inclusions from diamonds. *Terra Nova (supplement 1)* 5: 101–107.

Harte B and Harris JW (1994) Lower mantle associations preserved in diamonds. *Mineralogical Magazine* 58A: 386–387.

Hehre WJ, Radom L, Schleyer VR, and Pople JA (1996) *Ab Initio Molecular Orbital Theory*. New York: Wiley.

Heid R, Strauch D, and Bohnen K-P (2000) *Ab initio* lattice dynamics of sapphire. *Physical Review B* 61: 8625–8627.

Heine V, Welche PRL, and Dove MT (1999) Geometric origin and theory of negative thermal expansion in framework structures. *Journal of the American Ceramic Society* 82: 1793–1802.

Hemley RJ and Kubicki JD (1991) Deep mantle melting. *Nature* 349: 283–284.

Hemley RJ (1987) Pressure dependence of Raman spectra of SiO_2 polymorphs: α-quartz, coesite and stishovite. In: Manghnani MH and Syono Y (ed.) High Pressure Research in Mineral Physics, pp. 347–360. Tokyo/ Washington DC: Terra Scientific/American Geophysical Union.

Hemley RJ (ed.) (1998) *Ultra-High Pressure Mineralogy: Physics and Chemistry of the Earth's Deep Interior. Reviews in Mineralogy.* Vol. 37. Washington: DC: Mineralogical Society of America.

Hemley RJ, Bell PM, and Mao H-K (1987b) Laser techniques in high-pressure geophysics. *Science* 237: 605–611.

Hemley RJ and Cohen RE (1992) Silicate perovskite. *Annual Review of Earth and Planetary Science* 20: 553–600.

Hemley RJ, Cohen RE, Yeganeh-Haeri A, Mao H-K, and Weidner DJ (1989) Raman spectroscopy and lattice dynamics of $MgSiO_3$ perovskite. In: Navrotsky A and Weidner DJ (ed.) Perovskite: A Structure of Great Interest to Geophysics and Materials Science, pp. 35–44. Washington DC: American Geophysical Union.

Hemley RJ, Goncharov AF, Lu R, Li M, Struzhkin V V, and Mao H-K (1998) High-pressure synchrotron infrared spectroscopy at the National Synchrotron Light Source. *Il Nuevo Cim. D* 20: 539–551.

Hemley RJ and Gordon RG (1985) Theoretical study of solid NaF and NaCl at high pressures and temperatures. *Journal of Geophysical Research* B90: 7803–7813.

Hemley RJ, Jackson MD, and Gordon RG (1985) First-principles theory for the equations of state of minerals to high pressures and temperatures: Application to MgO. *Geophysical Research Letters* 12: 247–250.

Hemley RJ, Jackson MD, and Gordon RG (1987a) Theoretical study of the structure, lattice dynamics and equations of state of perovskite-type $MgSiO_3$ and $CaSiO_3$. *Physics and Chemistry of Minerals* 14: 2–12.

Hemley RJ and Mao H-K (2002) New windows on Earth and planetary interiors. *Mineralogical Magazine* 66: 791–811.

Hemley RJ, Mao H-K, and Chao ECT (1986) Raman spectrum of natural and synthetic stishovite. *Physics and Chemistry of Minerals* 13: 285–290.

Hemmati M, Chizmeshya AVG, Wolf GH, Poole PH, Shao JS, and Angell CA (1995) Crystalline-amorphous transition in silicate perovskites. *Physical Review B* 51: 14841–14848.

Hernlund JW, Thornas C, and Tackley PJ (2005) A doubling of the post-perovskite phase boundary and structure of the Earth's lowermost mantle. *Nature* 434: 882–886.

Herzberg G (1945) *Molecular Spectra and Molecular Structure. II: Infrared and Raman Spectra of Polyatomic Molecules.* New York: Van Nostrand Reinhold.

Herzberg G (1950) *Molecular Spectra and Molecular Structure. I: Spectra of Diatomic Molecules.* New York: Van Nostrand Reinhold.

Hess AC, McMillan PF, and O'Keeffe M (1986) Force fields for SiF_4 and H_4SiO_4: *Ab initio* molecular orbital calculations. *Journal of Physical Chemistry* 90: 5661–5665.

Hofmeister A (1999) Mantle values of thermal conductivity and the geotherm from phonon lifetimes. *Science* 283: 1699–1706.

Hofmeister A, Xu J, and Akimoto S (1990) Infrared spectroscopy of synthetic and natural stishovite. *American Mineralogist* 75: 951–955.

Hofmeister AM (1987) Single-crystal absorption and reflection infrared spectroscopy of forsterite and fayalite. *Physics and Chemistry of Minerals* 14: 499–513.

Hofmeister AM (1991a) Calculation of bulk moduli and their pressure derivatives from vibrational frequencies and mode Grüneisen parameters: Solids with high symmetry or one nearest-neighbor distance. *Journal of Geophysical Research B* 96: 16181–16203.

Hofmeister AM (1991b) Pressure derivatives of the bulk modulus. *Journal of Geophysical Research* 96: 21893–21907.

Hofmeister AM and Ito E (1992) Thermodynamic properties of $MgSiO_3$ ilmenite from vibrational spectra. *Physics and Chemistry of Minerals* 18: 423–432.

Hofmeister AM (1996) Thermodynamic properties of stishovite at mantle conditions determined from pressure variations of vibrational modes. In: Dyar D, McCammon CA, and Schaefer MW (ed.) Mineral Spectroscopy: A Tribute to Roger G. Burns, pp. 215–227. Houston: Geochemical Society.

Hofmeister AM (1997) Infrared reflectance spectra of fayalite, and absorption spectra from assorted olivines, including pressure and isotope effects. *Physics and Chemistry of Minerals* 24: 535–546.

Hofmeister AM and Chopelas A (1991a) Thermodynamic properties of pyrope and grossular from vibrational spectra. *American Mineralogist* 76: 880–891.

Hofmeister AM and Chopelas A (1991b) Vibrational spectroscopy of end-member silicate garnets. *Physics and Chemistry of Minerals* 17: 503–526.

Hofmeister AM, Xu J, Mao H-K, Bell PM, and Hoering TC (1989) Thermodynamics of Fe–Mg olivines at mantle pressures: Mid-and far-infrared spectroscopy at high pressures. *American Mineralogist* 74: 281–306.

Holzapfel WB (2004) Equations of state and thermophysical properties of solids under pressure. In: Katrusiak A and McMillan PF (ed.) High-Pressure Crystallography, pp. 217–236. Dordrecht: Kluwer Academic Publishers.

Horiuchi H and Sawamoto H (1981) ®-Mg_2SiO_4: Single crystal X-ray diffraction study. *American Mineralogist* 66: 568–575.

Ihinger P, Hervig RL, and McMillan PF (1994) Analytical methods for volatiles in glasses. In: Carroll M and Holloway JR (ed.) Volatiles in Magmas, vol. 30, pp. 67–121. Washington DC: Mineralogical Society of America.

Iishi K (1978) Lattice dynamics of forsterite. *American Mineralogist* 63: 1198–1208.

Iitaka T, Hirose K, Kawamura K, and Muramaki M (2004) The elasticity of the $MgSiO_3$ post-perovskite phase in the Earth's lowermost mantle. *Nature* 430: 442–445.

Irifune T (2002) Application of synchrotron radiation and Kawai-type apparatus to various studies in high-pressure mineral physics. *Mineralogical Magazine* 66: 769–790.

Isaak DG, Cohen RE, and Mehl MJ (1990) Calculated elastic and thermal properties of MgO at high pressure and temperatures. *Journal of Geophysical Research B* 95: 7055–7067.

Ita J and Cohen RE (1997) Effects of pressure on diffusion and vacancy formation in non-empirical free-energy integrations. *Physical Review Letters* 79: 3198–3201.

Ita J and Cohen RE (1998) Diffusion in MgO at high pressure: Implications for lower mantle rheology. *Geophysical Research Letters* 25: 1095–1098.

Ito E and Takahashi E (1987) Ultrahigh-pressure phase transformations and the constitution of the deep mantle. In: Manghani MH and Syono Y (ed.) High-Pressure Research in Mineral Physics, pp. 221–229. Washington DC: Terra Scientific Publishing Company, Tokyo/American Geophysical Union.

Ito E and Yamada H (1982) Stability relations of silicate spinels, ilmenites, and perovskites. In: Akimoto S and Manghnani MH (ed.) High-Pressure Research in Geophysics, pp. 405–419. Massachussetts: D Reidel, Norwell.

Jeanloz R (1980) Infrared spectra of olivine polymorphs: α-, β-phase and spinel. *Physics and Chemistry of Minerals* 5: 327–341.

Jeanloz R (1981) Majorite: Vibrational and compressional properties of a high-pressure phase. *Journal of Geophysical Research B* 86: 6171–6179.

Jeanloz R (1985) Thermodynamics of phase transitions. In: Kieffer SW and Navrotsky A (ed.) *Microscopic to Macroscopic. Atomic Environments to Mineral Thermodynamics*, vol. 14, pp. 389–428. Washington DC: Mineralogical Society of America.

Jeanloz R and Morris S (1986) Temperature distribution in the crust and mantle. *Annual Review of Earth and Planetary Science* 14: 377–415.

Jeanloz R (1993) Chemical reactions at the Earth's core–mantle boundary: Summary of evidence and geomagnetic implications. In: Aki K and Dmowska R (eds.) *Relating Geophysical Structures and Processes: The Jeffreys*, vol. 76, pp. 121–127. Washington DC: American Geophysical Union.

Jeanloz R and Richter FM (1979) Convection, composition and the thermal state of the lower mantle. *Journal of Geophysical Research* 84: 5497–5504.

Jeanloz R and Thompson AB (1983) Phase transitions and mantle discontinuities. *Review of Geophysics and Space Physics* 21: 51–74.

Jung DY and Oganov AR (2005) *Ab initio* study of the high-pressure behavior of $CaSiO_3$ perovskite. *Physics and Chemistry of Minerals* 32: 146–153.

Kagi H, Lu R, Hemley RJ, and al E (2000) Evidence for ice VI as an inclusion in cuboid diamonds from high P–T near infrared spectroscopy. *Mineralogical Magazine* 64: 1089–1097.

Kanzaki M (1991a) Ortho to clinoenstatite transition. *Physics and Chemistry of Minerals* 17: 726–730.

Kanzaki M (1991b) Stability of hydrous magnesium silicates in the mantle transition zone. *Physics of the Earth and Planetary Interiors* 66: 307–312.

Kao CC, Kaliebe WA, Hastings JB, Hämäläinen K, and Krisch MH (1996) Inelastic X-ray scattering at the National Synchrotron Light Source. *Reviews of Scientific Instruments* 67: 1–5.

Karki BB, Warren JL, Stixrude L, Ackland GJ, and Crain J (1997) *Ab initio* studies of high-pressure structural transformations in silica. *Physical Review B* 55: 3465–3471.

Karki BB, Wentzcovich RM, Gironcoli D, and Baroni S (2000a) First-principles determination of elastic anisotropy and wave velocities of MgO at lower mantle conditions. *Science* 286: 1705–1707.

Karki BB, Wentzcovich RM, Gironcoil D, and Baroni S (2000b) *Ab initio* lattice dynamics of $MgSiO_3$–perovskite at high pressure. *Physical Review B* 62: 14750–14756.

Karki BB, Wentzcovich RM, Gironcoli D, and Baroni S (2000c) High-pressure lattice dynamics and thermoelasticity of MgO. *Physical Review B* 61: 8793–8800.

Karki BB, Wentzcovich RM, Gironcoli D, and Baroni S (2001) First principles thermoelasticity of $MgSiO_3$–perovskite: Consequences for the inferred properties of the lower mantle. *Geophysical Research Letters* 28: 2699–2702.

Karki BB and Wentzcovich RM (2002) First-principles lattice dynamics and thermoelasticity of $MgSiO_3$ ilmenite at high pressure. *Journal of Geophysical Research B* 107: 2267.

Karki BB and Wentzcovich RM (2003) Vibrational and quasi-harmonic thermal properties of CaO under pressure. *Physical Review B* 68: 224304.

Kato T (1986) Stability relation of $(Mg, Fe)SiO_3$ garnets, major constituents in the Earth's interior. *Earth and Planetary Science Letters* 77: 399–408.

Kato T and Kumazawa M (1985) Garnet phase of $MgSiO_3$ filling the pyroxene–ilmenite gap at very high temperature. *Nature* 316: 803–804.

Katsura T and Ito E (1989) The system Mg_2SiO_4–Fe_2SiO_4 at high pressures and temperatures: Precise determination of stabilities of olivine, modified spinel, and spinel. *Journal of Geophysical Research B* 94: 15663–15670.

Kesson SE, Fitzgerald JD, and Shelley JMG (1994) Mineral chemistry and density of subducted basaltic crust at lower-mantle pressures. *Nature* 372: 767–769.

Khvostantsev LG, Slesarev VN, and Brazhkin V V (2004) Toroid type high-pressure device: History and prospects. *High Pressure Research* 24: 371–383.

Kieffer SW (1979a) Thermodynamics and lattice vibrations of minerals. 2: Vibrational characteristics of silicates. *Review of Geophysics and Space Physics* 17: 20–34.

Kieffer SW (1979b) Thermodynamics and lattice vibrations of minerals. 1: Mineral heat capacities and their relationships to simple lattice vibrational models. *Review of Geophysics and Space Physics* 17: 1–19.

Kieffer SW (1979c) Thermodynamics and lattice vibrations of minerals. 3: Lattice dynamics and an approximation for minerals with application to simple substances and framework silicates. *Review of Geophysics and Space Physics* 17: 35–39.

Kieffer SW (1980) Thermodynamics and lattice vibrations of minerals. 4: Application to chain and sheet silicates and orthosilicates. *Review of Geophysics and Space Physics* 18: 862–886.

Kieffer SW (1985) Heat capacity and entropy: Systematic relations to lattice vibrations. In: Kieffer SW and Navrotsky A (eds.) *Microscopic to Macroscopic. Atomic Environments to Mineral Thermodynamics*, vol. 14, pp. 65–126. Washington DC: Mineralogical Society of America.

Kieffer SW and Navrotsky A (ed.) (1985) *Microscopic to macroscopic. Atomic environments to mineral thermodynamics. Reviews in Mineralogy*, vol. 4. Washington DC. Mineralogical Society of America.

Kingma KJ, Cohen RE, Hemley RJ, and Mao H-K (1995) Transformation of stishovite to a denser phase at lower mantle pressures. *Nature* 374: 243–245.

Klotz S, Besson JM, Braden M, *et al.* (1997) Pressure induced frequency shifts of transverse acoustic phonons in germanium to 9.7 GPa. *Physical Review Letters* 79: 1313–1317.

Klotz S and Braden M (2000) Phonon dispersion of bcc iron to 10 GPa. *Physical Review Letters* 85: 3209–3212.

Knittle E and Jeanloz R (1987) Synthesis and equation of state of $(Mg, Fe)SiO_3$ perovskite to over 100 gigapascals. *Science* 235: 668–670.

Knittle E and Jeanloz R (1991) Earth's core–mantle boundary: Results of experiments at high pressures and temperatures. *Science* 251: 1438–1443.

Kobayashi K, Kokko K, Terakura K, and Matsui M (1993) First-principles molecular dynamics study of pressure induced structural phase transition of silica. In: Doyama M, Kihara J, Tanaka M, and Yamamoto R (eds.) *Computer Aided Innovation of New Materials II*, pp. 121–124. New York: Elsevier.

Kohlstedt DL, Keppler H, and Rubie DC (1996) Solubility of water in the α, β and γ phases of $(Mg, Fe)_2SiO_4$. *Contributions to Mineralogy and Petrology* 123: 345–357.

Kohn W (1999) Nobel lecture: Electronic structure of matter - wave functions and density functionals. *Reviews of Modern Physics* 71: 1253–1266.

Kohn W and Sham LJ (1965) Self-consistent equations including exchange and correlation effects. *Physical Review A* 140: 1133–1138.

Kolesov BA and Geiger CA (1998) Raman spectra of silicate garnets. *Physics and Chemistry of Minerals* 25: 142–151.

Kolesov BA and Geiger CA (2000) Low-temperature single-crystal Raman spectrum of pyrop. *Physics and Chemistry of Minerals* 27: 645–649.

Kudoh Y, Prewitt CT, Finger LW, Darovskikh A, and Ito E (1990) Effect of iron on the crystal structure of $(Mg, Fe)SiO_3$ perovskite. *Geophysical Research Letters* 17: 1481–1484.

Kuwayama Y, Hirose K, Sata N, and Ohishi Y (2005) The pyrite-type high-pressure form of silica. *Science* 309: 923–925.

Lacks DJ and Gordon RG (1993) Calculations of pressure-induced phase transitions in silica. *Journal of Geophysical Research B* 98: 22147–22155.

Lasaga AC and Gibbs GV (1987) Applications of quantum mechanical potential surfaces to mineral physics calculations. *Physics and Chemistry of Minerals* 15: 588–596.

Lee C (1996) Analyses of the *ab initio* harmonic interatomic force constants of stishovite. *Physical Review B* 54: 8973–8976.

Lee C and Gonze X (1995) The pressure-induced ferroelastic phase transition of SiO_2 stishovite. *Journal of Physics: Condensed Matter* 7: 3693–3698.

Lee C and Gonze X (1997) SiO_2 stishovite under high pressure: Dielectric and dynamical properties and the ferroelastic phase transition. *Physical Review B* 56: 7321–7330.

Léger JM, Haines J, and Schmidt M (1996) Discovery of hardest known oxide. *Nature* 383: 401.

Li XY and Jeanloz R (1987) Measurement of the B1–B2 transition pressure in NaCl at high temperatures. *Physical Review B* 36: 474–479.

Liebermann RC and Li B (1998) Elasticity at high pressures and temperatures. In: Hemley RJ (ed.) *Ultra-High Pressure Mineralogy: Physics and Chemistry of the Earth's Deep Interior*, vol. 37, pp. 459–492. Washington DC: Mineralogical Society of America.

Lin J-F, Heinz DL, Mao H-K, et al. (2002) Stability of magnesiowüstite in Earth's lower mantle. *Proceedings of the National Academy of Sciences of the United States of America* 10,1073/pnas.252782399: 1–4 1073/pnas.252782399.

Liou JG, Zhang RY, Ernst WG, and Maruyama S (1998) High-pressure minerals from deeply-subducted metamorphic rocks. In: Hemley RJ (ed.) *Ultra-High Pressure Mineralogy: Physics and Chemistry of the Earth's Deep Interior*, vol. 37, pp. 33–96. Washington DC: Mineralogical Society of America.

Liu L-G (1974) Silicate perovskite from phase transformations of pyrope–garnet at high pressure and temperature. *Geophysical Research Letters* 1: 277–280.

Liu L-G (1975) Post-oxide phases of olivine and pyroxene and mineralogy of the mantle. *Nature* 258: 510–512.

Liu L-G (1977) Ilmenite-type solid solutions between $MgSiO_3$ and Al_2O_3 and some structural systematics among ilmenite compounds. *Geochimica et Cosmochimica Acta* 41: 1355–1361.

Liu L-G (1987) Effects of H_2O on the phase behaviour of the forsterite–enstatite system at high pressures and temperatures and implications for the Earth. *Physics of the Earth and Planetary Interiors* 49: 142–167.

Liu L-G and Bassett WA (1986) *Elements, Oxides, Silicates*. New York: Oxford University Press.

Liu L-G, Mernagh TP, and Irifune T (1994) High pressure Raman spectra of β-Mg_2SiO_4, γ-Mg_2SiO_4, $MgSiO_3$–ilmenite and $MgSiO_3$–perovskite. *Journal of the Physics and Chemistry of Solids* 55: 185–193.

Liu L-G, Mernagh TP, and Irifune T (1995) Raman spectra of pyrope and $MgSiO_3$–10 Al_2O_3 garnet at various pressures and temperatures. *High Temperature-High Pressure* 26: 363–374.

Liu L-G, Mernagh TP, and Jaques AL (1990) A mineralogical Raman spectroscopy study on eclogitic garnet inclusions in diamonds from Argyle. *Contributions to Mineralogy and Petrology* 105: 156–161.

Liu L-G and Ringwood AE (1975) Synthesis of a perovskite-type polymorph of $CaSiO_3$. *Earth and Planetary Science Letters* 28: 209–211.

Long DA (1977) *Raman Spectroscopy*. New York: McGraw-Hill.

Loveday JS (2004) Neutron diffraction studies of ices and ice mixtures. In: Katrusiak A and McMillan P F (eds.) *High-Pressure Crystallography*, pp. 69–80. Dordrecht: Kluwer Academic Publishers.

Lu R, Hofmeister AM, and Wang YB (1994) Thermodynamic properties of ferromagnesium silicate perovskites from vibrational spectroscopy. *Journal of Geophysical Research B* 99: 11795–11804.

Lyon R JP (1962) Infra-red confirmation of 6-fold co-ordination of silicon in stishovite. *Nature* 196: 266–267.

Madon M and Price GD (1989) Infrared spectroscopy of the polymorphic series (enstatite, ilmenite and perovskite) of $MgSiO_3$, $MgGeO_3$ and $MnGeO_3$. *Journal of Geophysical Research B* 94: 15701–15867.

Mao H-K, Hemley RJ, Fei Y, et al. (1991) Effect of pressure, temperature, and composition on lattice parameters and density of (Fe,Mg)SiO_3-perovskites to 30 GPa. *Journal of Geophysical Research B* 96: 8069–8079.

Mao H-K, Xu J, Struzhkin V V, et al. (2001) Phonon density of states of iron up to 153 Gigapascals. *Science* 292: 914–916.

Mao W, Meng Y, Shen G, et al. (2005) Iron-rich silicates in the Earth's D″ layer. *Proceedings of the National Academy of Sciences of the United States of America* 102: 9751–9753.

Mao W, Shen G, Prakapenka VB, et al. (2004) Ferromagnesian postperovskite silicates in the D″ layer of the Earth. *Proceedings of the National Academy of Sciences of the United States of America* 101: 15867–15869.

Marton FC and Cohen RE (1994) Prediction of a high phase transition in Al_2O_3. *American Mineralogist* 79: 789–792.

Marton FC and Cohen RE (2002) Constraints on lower mantle composition from molecular dynamics of $MgSiO_3$ perovskite. *Physics of the Earth and Planetary Interiors* 134: 239–252.

Marton FC, Ita J, and Cohen RE (2001) Pressure–volume–temperature equation of state of $MgSiO_3$ perovskite from molecular dynamics and constraints on lower mantle composition. *Journal of Geophysical Research B* 106: 8615–8627.

Mary TA, Evans JSO, Vogt T, and Sleight AW (1996) Negative thermal expansion from 0.3 to 1050 Kelvin in ZrW_2O_6. *Science* 272: 735–737.

Matsui M (1988) Molecular dynamics study of $MgSiO_3$ perovskite. *Physics and Chemistry of Minerals* 16: 234–238.

Matsui M, Akaogi M, and Matsumoto T (1987) Computational model of the structural and elastic properties of the ilmenite and perovskite phases of $MgSiO_3$. *Physics and Chemistry of Minerals* 14: 101–106.

Matsui M and Busing WR (1984) Computational modelling of the structural and elastic constants of the olivine and spinel forms of Mg_2SiO_4. *Physics and Chemistry of Minerals* 11: 55–59.

Matsui Y and Tsuneyiki S (1992) Molecular dynamics study of rutile–$CaCl_2$-type phase transition of SiO_2. In: Syono Y and Manghnani MH (eds.) *High-Pressure Research: Application to Earth and Planetary Sciences*, pp. 433–439. Washington DC: Terra Scientific Publishing Co., Tokyo/American Geophysical Union.

Mazin II, Fei Y, Downs RT, and Cohen RE (1998) Possible polytypism in FeO at high pressures. *American Mineralogist* 83: 451–457.

McCammon C (2001) Deep diamond mysteries. *Science* 293: 813–814.

McCammon C, Hutchison M, and Harris J (1997) Ferric iron content of mineral inclusions in diamonds from São Luiz: A view into the lower mantle. *Science* 278: 434–436.

McDonough WF and Rudnick RL (1998) Mineralogy and composition of the upper mantle. In: Hemley RJ (ed.) *Ultra-High Pressure Mineralogy: Physics and Chemistry of the Earth's*

Deep Interior, Vol. 37, pp. 139–164. Washington DC: Mineralogical Society of America.

McMillan P (1985) Vibrational spectroscopy in the mineral sciences. In: Kieffer SW and Navrotsky A (eds.) *Microscopic to Macroscopic. Atomic Environments to Mineral Thermodynamics*, vol. 14, pp. 9–63. Washington DC: Mineralogical Society of America.

McMillan P (1989) Raman spectroscopy in mineralogy and geochemistry. *Annual Review of Earth and Planetary Science* 17: 255–283.

McMillan P and Akaogi M (1987) The Raman spectra of β-Mg_2SiO_4 (modified spinel) and γ-Mg_2SiO_4 (spinel). *American Mineralogist* 72: 361–364.

McMillan P, Akaogi M, Ohtani E, Williams Q, Nieman R, and Sato R (1989) Cation disorder in garnets along the $Mg_3Al_2Si_3O_{12}$–$Mg_4Si_4O_{12}$ join: An infrared, Raman and NMR study. *Physics and Chemistry of Minerals* 16: 428–435.

McMillan P and Hess AC (1988) Symmetry, group theory and quantum mechanics. In: Hawthorne F (ed.) *Spectroscopic Methods in Mineralogy and Geology*, vol. 18, pp. 11–61. Washington DC: Mineralogical Society of America.

McMillan P and Hofmeister A (1988) Infrared and Raman spectroscopy. In: Hawthorne F (ed.) *Spectroscopic Methods in Mineralogy and Geology*, vol. 18, pp. 99–159. Washington DC: Mineralogical Society of America.

McMillan PF (1994) Water solubility and speciation models. In: Carroll M and Holloway JR (ed.) *Volatiles in Magmas*, vol. 30, pp. 131–156. Washington DC: Mineralogical Society of America.

McMillan PF, Dubessy J, and Hemley RJ (1996a) Applications to Earth science and environment. In: Turrell G and Corset J (eds.) *Raman Microscopy. Developments and Applications*, pp. 289–365. New York: Academic Press.

McMillan PF, Gillet P, and Hemley RJ (1996b) Vibrational spectroscopy of mantle minerals. In: Dyar MD, McCammon C, and Schaefer MW (eds.) *Mineral Spectroscopy: A Tribute to Roger G. Burns*, pp. 175–213. Houston, TX: Geochemical Society Geochem Soc Spec Pub N°5.

McMillan PF and Hess AC (1990) *Ab initio* valence force field calculations for quartz. *Physics and Chemistry of Minerals* 17: 97–107.

McMillan PF and Ross NL (1987) Heat capacity calculations for Al_2O_3 corundum and $MgSiO_3$ ilmenite. *Physics and Chemistry of Minerals* 16: 225–234.

McMillan PF and Wolf GH (1995) Vibrational spectroscopy of silicate liquids. In: Stebbins JF, McMillan PF, and Dingwell DB (eds.) *Structure, Dynamics and Properties of Silicate Melts*, vol. 32, pp. 247–315. Washington DC: Mineralogical Society of America.

Meade C, Reffner JA, and Ito E (1994) Synchrotron infrared absorbance of hydrogen in $MgSiO_3$ perovskite. *Science* 264: 1558–1560.

Mehl MJ, Hemley RJ, and Boyd FR (1986) Potential induced breathing model for the elastic moduli and high-pressure behavior of the cubic alkaline-Earth oxides. *Physical Review B* 33: 8685–8696.

Mills IM (1963) Force constant calculations for small molecules. In: Davies M (ed.) *Infrared Spectroscopy and Molecular Structure*, pp. 166–198. New York: Elsevier.

Mittal R and Chaplot SL (1999) Lattice dynamical calculation of isotropic negative thermal expansion in ZrW_2O_6 over 0–1050 K. *Physical Review B* 60: 7234–7237.

Mittal R, Chaplot SL, and Choudhury N (2001) Lattice dynamics calculations of the phonon spectra and thermodynamic properties of the aluminosilicate garnets pyrope, grossular and spessartine $M_3Al_2Si_3O_{12}$ (M = Mg, Ca and Mn). *Physical, Review B* 64: 94320-1–94320-9.

Mittal R, Chaplot SL, Choudhury N, and Loong C-K (2000) Inelastic neutron scattering and lattice-dynamics studies of

almandine $Fe_3Al_2Si_3O_{12}$. *Physical Review B* 61: 94320-1-94320-9.

Moore RO and Gurney JJ (1985) Pyroxene solid solution in garnets included in diamond. *Nature* 318: 553–555.

Muinov M, Kanda H, and Stishov SM (1994) Raman scattering in diamond at high pressure: Isotopic effects. *Physical Review B* 50: 13860–13862.

Murakami M, Hirose K, One S, and Ohishi Y (2003) Stability of $CaCl_2$-type and ζ-PbO_2-type SiO_2 at high pressure and temperature determined by *in situ* X-ray measurement. *Geophysical Research Letters* 30: doi:10.1029/2002GL016722.

Murakami M, Hirose K, Ono S, Tsuchiya M, Isshiki M, and Watanuki T (2004b) High pressure and high temperature phase transitions of FeO. *Physics of the Earth and Planetary Interiors* 146: 273–282.

Murakami M, Hirose K, Kawamura K, Sata N, and Ohishi Y (2004a) Post-perovskite phase transition in $MgSiO_3$. *Science* 304: 855–858.

Nagel L and O'Keeffe M (1971) Pressure, and stress induced polymorphism of compounds with rutile structure. *Mat. Res. Bull* 6: 1317–1320.

Nakagawa T and Tackley PJ (2004) Effects of a perovskite-post perovskite phase change near coremantle boundary in compressible mantle convection. *Geophysical Research Letters* 31: L16611–L16614.

Nakamoto (1978) *Infrared and Raman Spectra of Inorganic and Coordination Compounds*. New York: Wiley.

Navon O (1991) High internal-pressures in diamond fluid inclusions determined by infrared-absorption. *Nature* 353: 746–748.

Navon O and Hutcheon ID (1988) Mantle-derived fluids in diamond micro-inclusions. *Nature* 335: 784–789.

Navrotsky A (1980) Lower mantle phase transitions may generaly have negative pressure–temperature slopes. *Geophysical Research Letters* 7: 709–711.

Navrotsky A (1989) Thermochemistry of perovskites. In: Navrotsky A and Weidner DJ (ed.) *Perovskite: A Structure of Great Interest to Geophysics and Materials Science*, pp. 67–80. Washington, DC: American Geophysical Union.

Navrotsky A (1994) *Physics and Chemistry of Earth Materials*. Cambridge: Cambridge University Press.

Navrotsky A and Weidner DJ (ed.) (1989) *Perovskite: A Structure of Great Interest to Geophysics and Materials Science*. Washington, DC: American Geophysical Union.

Nixon PH (1987) *Mantle Xenoliths*. New York: Wiley.

Oganov AR (2004) Theory of minerals at high and ultrahigh pressures: Structure, properties, dynamics, and phase transitions. In: Katrusiak A and McMillan PF (ed.) *High-Pressure Crystallography*, pp. 199–215. Dordrecht: Kluwer Academic Publishers.

Oganov AR, Brodholt JP, and Price GD (2000) Comparative study of quasiharmonic lattice dynamics, molecular dynamics and Debye model in application to $MgSiO_3$ perovskite. *Physics of the Earth and Planetary Interiors* 122: 277–288.

Oganov AR, Brodholt JP, and Price GD (2001) The elastic constants of $MgSiO_3$ perovskite at pressures and temperatures of the Earth's mantle. *Nature* 411: 934–937.

Oganov AR, Brodholt JP, and Price GD (2002) *Ab initio* theory of phase transitions and thermoelasticity of minerals. In: Gramaccioli CM (ed.) *Energy Modelling in Minerals*, Vol.4, pp. 83–170. Budapest: Eötvös University Press.

Oganov AR, Gillan MJ, and Price GD (2005a) Structural stability of silica at high pressures and temperatures. *Physical Review B* 71: 64104.

Oganov AR and Ono S (2004) Theoretical and experimental evidence for a post-perovskite phase of $MgSiO_3$ in Earth's D″ layer. *Nature* 430: 445–449.

Oganov AR, Price GD, and Scandolo S (2005b) *Ab initio* theory of planetary materials. *Zeitschrift fur Kristallographic* 220: 531–548.

Ono S, Hirose K, Murakami M, and Isshiki M (2002) High-pressure form of pyrite-type germanium dioxide. *Physical Review B* 68: 014103.

Pacalo REG and Gasparik T (1990) Reversals of the orthoenstatite–clinoenstatite transition at high pressures and high temperatures. *Journal of Geophysical Research B* 95: 15853–15858.

Parlinski K and Kawazoe Y (2000) *Ab initio* study of phonons and structural stabilities of the perovskite-type MgSiO$_3$. *Eur. Phys. J. B* 16: 49–58.

Parr RG and Yang W (1989) *Density-functional theory of atoms and molecules*. Oxford: Oxford University Press.

Pauling L and Wilson EB (1963) *Introduction to Quantum Mechanics with Applications to Chemistry*. New York: McGraw-Hill (re-published 1985 by Dover Press).

Pavese A (2002) Vibrational symmetry and spectroscopy. In: Gramaccioli CM (ed.) *Energy Modelling in Minerals*, vol. 4, pp. 171–192. Budapest: Eötvös University Press.

Pavese A, Artioli G, and Moze O (1998) Inelastic neutron scattering from pyrope powder: Experimental data and theoretical calculations. *European Journal of Mineralogy* 10: 59–69.

Pawley AR, McMillan PF, and Holloway JR (1993) Hydrogen in stishovite, with implications for mantle water content. *Science* 261: 1024–1026.

Peercy PS and Morosin B (1973) Pressure and temperature dependence of the Raman active phonons in SnO$_2$. *Physical Review B* 7: 2779–2786.

Perdew JP and Burke K (1996) Comparison shopping for a gradient-corrected density functional. *International Journal of Quantum Chemistry* 57: 309–319.

Perdew JP, Burke K, and Ernzerhof M (1996) Generalized gradient approximation made simple. *Physical Review Letters* 77: 3865–3868.

Perdew JP and Wang K (1992) Accurate and simple analytic representation of the electron-gas correlation energy. *Physical Review B* 45: 13244–13249.

Perdew JP and Zunger A (1981) Self-interaction correction to density-functional approximations for many-electron systems. *Physical Review B* 23: 5048–5079.

Perry CH, Khanna BN, and Rupprecht G (1964a) Dielectric dispersion of some perovskite zirconates. *Physical Review A* 138: 1537–1538.

Perry CH, Khanna BN, and Rupprecht G (1964b) Infrared studies of perovskite titanates. *Physical Review A* 135: 408–412.

Perry CH and Young EF (1967) Infrared studies of some perovskite fluorides. I: Fundamental lattice vibrations. *Journal of Applied Physics* 38: 4616–4628.

Piriou B and Cabannes F (1968) Validité de la méthode de Kramers-Kronig et application à la dispersion infrarouge de la magnésie. *Optica Acta* 15: 271–286.

Piriou B and McMillan P (1983) The high-frequency vibrational spectra of vitreous and crystalline orthosilicates. *American Mineralogist* 68: 426–443.

Pisani C (ed.) (1996) *Quantum-Mechanical Ab Initio Calculation of the Properties of Crystalline Materials. Lecture Notes in Chemistry*, vol. 48. Berlin: Springer.

Prewitt CT and Downs RT (1998) High-pressure crystal chemistry. In: Hemley RJ (ed.) *Ultra-high Pressure Mineralogy: Physics and Chemistry of the Earth's Deep Interior*, Vol. 37, pp. 283–317. Washington DC: Mineralogical Society of America.

Prewitt CT and Finger LW (1992) Crystal chemistry of high-pressure hydrous magnesium silicates. In: Syono Y and Manghani MH (eds.) *High-Pressure Research: Application to Earth and Planetary Sciences*, pp. 269–274. Washington DC:

Terra Scientific Publishing Co., Tokyo/American Geophysical Union.

Price DL, Ghose S, Choudhury N, Chaplot SL, and Rao KR (1991) Phonon density of states in fayalite, Fe$_2$SiO$_4$. *Physica B* 174: 87–90.

Price GD, Parker SC, and Leslie M (1987) The lattice dynamics and thermodynamics of the Mg$_2$SiO$_4$ polymorphs. *Physics and Chemistry of Minerals* 15: 181–190.

Price GD, Wall A, and Patrker SA (1989) The properties and behaviour of mantle minerals: A computer-simulation approach. *Philosophical Transaction of Royal Society of A* 328: 391–407.

Raman CV and Nedungadi TMK (1940) The α–β transformation of quartz. *Nature* 145: 147.

Rao KR, Chaplot SL, Choudhury N, *et al.* (1988) Lattice dynamics and inelastic neutron scattering from forsterite, Mg$_2$SiO$_4$: Phonon dispersion relation, density of states and specific heat. *Physics and Chemistry Minerals* 16: 83–97.

Rao MN, Chaplot SL, Choudhury N, *et al.* (1999) Lattice dynamics and inelastic neutron scattering from sillimanite and kyanite, Al$_2$SiO$_5$. *Physical Review B* 60: 12061–12068.

Rao MN, Goel P, Choudhury N, Chaplot SL, and Ghose S (2002) Lattice dynamics and inelastic neutron scattering experiments on andalusite, Al$_2$SiO$_5$. *Solid State Communications* 121: 333–338.

Redfern SAT (2002) Neutron powder diffraction of minerals at high pressures and temperatures: Some recent technical developments and scientific applications. *European Journal of Mineralogy* 142: 251–262.

Reissland JA (1973) *The Physics of Phonons*. New York: Wiley.

Remler DK and Madden PA (1990) Molecular dynamics without effective potentials via the Car-Parrinello approach. *Molecular Physics* 70: 921–966.

Reynard B and Guyot F (1994) High temperature properties of geikelite (MgTiO$_3$–ilmenite) from high-temperature high-pressure Raman spectroscopy – Some implications for MgSiO$_3$ ilmenite. *Physics and Chemistry of Minerals* 21: 441–450.

Reynard B, Price GD, and Gillet P (1992) Thermodynamic and anharmonic properties of forsterite, α-Mg$_2$SiO$_4$: Computer modelling versus high-pressure and high-temperature measurements. *Journal of Geophysical Research B* 97: 19791–19801.

Reynard B and Rubie D (1996) High-pressure, high-temperature Raman spectroscopic study of ilmenite-type MgSiO$_3$. *American Mineralogist* 81: 1092–1096.

Reynard B, Takir F, Guyot F, Gwanmesia GD, Liebermann RC, and Gillet P (1996) High temperature Raman spectroscopic and x-ray diffraction study of β-Mg$_2$SiO$_4$: Some insights on its high temperature thermodynamic properties and on the $\circledR\square\langle$ phase transformation mechanism and kinetics. *American Mineralogist* 81: 585–594.

Rigden SM, Gwanmesia GD, Fitzgerald JD, Jackson I, and Liebermann RC (1991) Spinel elasticity and seismic structure of the transition zone of the mantle. *Nature* 354: 143–145.

Ringwood AE (1962a) Phase transformations and their bearing on the constitution and dynamics of the mantle. *Geochimica et Cosmochimica Acta* 55: 2083–2110.

Ringwood AE (1962b) Mineralogical constitution of the deep mantle. *Journal of Geophysical Research* 67: 4005–4010.

Ringwood AE (1967) The pyroxene–garnet transformation in the Earth's mantle. *Earth and Planetary Science Letters* 2: 255–263.

Ringwood AE (1975) *Composition and Petrology of the Earth's Mantle*. New York: McGraw-Hill.

Ross NL and McMillan PF (1984) The Raman spectrum of MgSiO$_3$ ilmenite. *American Mineralogist* 69: 719–721.

Rossman GR (1996) Studies of OH in nominally anhdrous minerals. *Physics and Chemistry of Minerals* 23: 299–304.

Salje E (1990) *Phase Transitions in Ferroelastic and Co-Elastic Crystals*. Cambridge: Cambridge University Press.

Samara GA and Peercy PS (1973) Pressure and temperature dependence of the static dielectric constants and Raman spectra of TiO_2 (rutile). *Physical Review B* 7: 1131–1141.

Sauer J (1989) Molecular models in *ab initio* studies of solids and surfaces: From ionic crystals and semiconductors to catalysts. *Chemical Reviews* 89: 199–255.

Sautter V and Gillet P (1994) Les diamants, messagers de la Terre profonde. *La Recherche* 271: 1238–1245.

Sautter V, Haggerty SE, and Field S (1991) Ultradeep (>300 kilometers) ultramafic xenoliths: Petrological evidence from the transition zone. *Science* 252: 827–830.

Schrader B (ed.) (1994) *Infrared and Raman Spectroscopy. Methods and Applications*. New York: VCH.

Schwoerer-Böhning M, Macrander AT, and Arms DA (1998) Phonon dispersion of diamond measured by inelastic X-ray scattering. *Physical Review Letters* 80: 5572–5575.

Scott JF (1968) Evidence of coupling between one- and two-phonon excitation in quartz. *Physics Review Letters* 21: 907–910.

Scott JF (1974) Soft-mode spectroscopy: Experimental studies of structural phase transitions. *Review of Modern Physics* 46: 83–128.

Serghiou G, Zerr A, and Boehler R (1998) Mg, $(Fe)SiO_3$–perovskite stability under lower mantle conditions. *Science* 280: 2093–2095.

Servoin JL and Piriou B (1973) Infrared reflectivity and Raman scattering of Mg_2SiO_4 single crystal. *Physical Status Solidi (B)* 55: 677–686.

Shapiro SM, O'Shea DC, and Cummins HE (1967) Raman scattering study of the alpha–beta transition in quartz. *Physical Review Letters* 19: 361–364.

Sharp TS, El Goresy A, Wopenka B, and Chen M (1999) A post-stishovite SiO_2 polymorph in the meteorite Shergotty: Implications for impact events. *Science* 284: 1511–1513.

Shen G and Heinz DL (1998) High-pressure melting of deep mantle and core materials. In: Hemley RJ (ed.) *Ultra-High Pressure Mineralogy: Physics and Chemistry of the Earth's Deep Interior*, vol. 37, pp. 369–396. Washington DC: Mineralogical Society of America.

Shieh SR, Duffy TS, and Shen G (2005) X-ray diffraction study of phase stability in SiO_2 at deep mantle conditions. *Earth and Planetary Science Letters* 235: 273–282.

Shim S-H, Duffy TS, Jeanloz R, and Shen G (2004) Stability and crystal structure of $MgSiO_3$ perovskite to the core–mantle boundary. *Geophysical Research Letters* 31: L10603.

Shim S-H, Duffy TS, and Shen G (2001) Stability and structure of $MgSiO_3$ perovskite to 2300-kilometer depth in Earth's mantle. *Science* 293: 2437–2440.

Shimanouchi T, Tsuboi M, and Miyazawa T (1961) Optically active lattice vibrations as treated by the GF-matrix method. *Journal of Chemical Physics* 35: 1597–1612.

Shimanouchi T (1970) The molecular force field. In: Henderson D (ed.) *Molecular Properties*, vol. IV, pp. 233–306. New York: Academic Press.

Smith JV and Mason B (1970) Pyroxene–garnet transformation in Coorara meteorite. *Science* 168: 832–833.

Smyth JR (1987) β-Mg_2SiO_4–A potential host for water in the mantle?. *American Mineralogist* 72: 1051–1055.

Spitzer WG, Miller RC, Kleinman A, and Howarth LE (1962) Far infrared dielectric dispersion in $BaTiO_3$, $SrTiO_3$, and TiO_2. *Physical Review* 126: 1710–1721.

Stishov SM and Popova SV (1961) A new dense modification of silica. *Geochemistry International* 10: 923–926.

Stixrude L, Cohen RE, and Hemley RJ (1998) Theory of minerals at high pressure. In: Hemley RJ (ed.) *Ultra-High Pressure Mineralogy: Physics and Chemistry of the Earth's Deep Interior*, vol. 37, pp. 639–671. Washington DC: Mineralogical Society of America.

Stixrude L, Cohen RE, Yu R, and Krakauer H (1996) Prediction of phase transition in $CaSiO_3$ perovskite and implications for lower mantle structure. *American Mineralogist* 81: 1293–1296.

Stixrude L and Lithgow-Bertelloni C (2005) Thermodynamics of mantle minerals. I: Physical properties. *Geophysical Journal International* 162: 610–632.

Stoneham AM (1992) Diamond: Recent advances in theory. In: Field JE (ed.) *The Properties of Natural and Synthetic Diamond*, pp. 3–34. London: Academic Press.

Stowasser R and Hoffman R (1999) What do the Kohn–Sham orbitals and eigenvalues mean? *Journal of the American Chemical Society* 121: 3414–3420.

Teter DM, Hemley RJ, Kresse G, and Hafner J (1998) High-pressure polymorphism in silica. *Physical Review Letters* 80: 2145–2148.

Thompson AB (1992) Water in the Earth's upper mantle. *Nature* 358: 295–302.

Tomioka N and Fujino K (1999) Akimotoite, $(Mg, Fe)SiO_3$, a new silicate mineral of the ilmenite group in the Tenham chondrite. *American Mineralogist* 84: 267–271.

Tossell JA and Vaughan DJ (1992) *Theoretical Geochemistry: Application of Quantum Mechanics in the Earth and Mineral Sciences*. New York: Oxford University Press.

Trave AP, Tangney S, Scandolo S, Pasquarello A, and Car R (2002) Pressure-induced structural changes in liquid SiO_2 from *ab initio* simulations. *Physical Review Letters* 245504: 8924.

Tse J (2004) Computational high pressure science. In: Katrusiak A and McMillan PF (eds.) *High-Pressure Crystallography*, pp. 179–215. Dordrecht: Kluwer Academic.

Tsuchiya J, Tsuchiya T, and Wentzcovich RM (2005) Vibrational and thermodynamic properties of $MgSiO_3$ postperovskite. *Journal of Geophysical Research* 110: B02204–B02210.

Tsuchiya R, Caracas R, and Tsuchiya J (2004a) First principles determination of the phase boundaries of high-pressure polymorphs of silica. *Geophysical Research Letters* doi:10, 1029/2004GL019649.

Tsuneyuki S, Aoki H, Tsukuda M, and Matsui Y (1989) New pressure-induced structural transformations in silica obtained by computer simulation. *Nature* 339: 209–211.

Tsuneyuki S, Tsukuda M, Aoki H, and Matsui Y (1988) First-principles interatomic potential of silica applied to molecular dynamics. *Physical Review Letters* 61: 869–872.

Tsuchiya T, Tsuchiya J, Umemoto K, and Wentzcovich RM (2004b) Phase transition in $MgSiO_3$ perovskite in the earth's lower mantle. *Earth and Planetary Science Letters* 224: 241–248.

Tsuchida Y and Yagi T (1989) A new, post-stishovite high-pressure polymorph of silica. *Nature* 340: 217–220.

Tsuchida Y and Yagi T (1990) New pressure-induced transformations of silica at room temperature. *Nature* 347: 267–269.

van Beest BWH, Kramer GJ, and van Santen RA (1994) Force fields of silicas and aluminophosphates based on *ab initio* calculations. *Physical Review Letters* 64: 1955–1958.

Vohra YK and Weir ST (2002) Designer diamond anvils in high pressure research: Recent results and future opportunities. In: Hemley RJ, Chiarotti GL, Bernasconi M, and Ulivi L (eds.) *High Pressure Phenomena*, Course CXLVII, pp. 87–99. Amsterdam: IOS Press.

Wall A and Price GD (1988) Computer simulation of the structure, lattice dynamics, and thermodynamics of ilmenite-type $MgSiO_3$. *American Mineralogist* 73: 224–231.

Wallace DC (1972) *Thermodynamics of Crystals*. New York: Wiley.

Wang K and Perdew JP (1991) Correlation hole of the spin-polarized electron gas, with exact small-vector and high-density scaling. *Physical Review B* 44: 13298–13307.

Wang SY, Sharma SK, and Cooney TF (1993) Micro-Raman and infrared spectral study of forsterite under high pressure. *American Mineralogist* 78: 469–476.

Wang Y, Weidner DJ, Liebermann RC, and Zhao Y (1994) P–V–T equation of state of (Mg, Fe)SiO$_3$ perovskite: Constraints on the composition of the lower mantle. *Physics of the Earth and Planetary Interiors* 83: 13–40.

Warren JL, Yarnell JL, Dolling G, and Cowley RA (1967) Lattice dynamics of diamond. *Physical Review* 158: 805–808.

Watson GW, Wall A, and Parker SC (2000) Atomistic simulation of the effect of temperature and pressure on point defect formation in MgSiO$_3$ perovskite and the stability of CaSiO$_3$ perovskite. *Journal of Physics: Condensed Mater* 12: 8427–8438.

Weidner DJ (1975) Elasticity of microcrystals. *Geophysical Research Letters* 2: 189–192.

Weidner DJ (1987) Elastic properties of rocks and minerals. In: (ed.) *Methods of Experimental Physics*, vol. 24A, pp. 1–30. New York: Academic Press.

weidner DJ (1998) Rheological studies at high pressure. In: Hemley RJ (ed.) *Ultra-High Pressure Mineralogy: Physics and Chemistry of the Earth's Deep Interior*, vol. 37, pp. 493–524. Washington DC: Mineralogical Society of America.

Weidner DJ and Ito E (1985) Elasticity of MgSiO$_3$ in the ilmenite phase. *Physics of the Earth and Planetary Interiors* 40: 65–70.

Weng K, Xu J, Mao H-K, and Bell PM (1983) Preliminary Fourier-transform infrared spectral data on the SiO$_6{}^{8-}$ octahedral group in silicate perovskite. *Carnegie Institution of Washington Yearbook* 82: 355–356.

Wentzcovich RM, da Silva C, Chelikowsky JR, and Binggeli N (1998) A new phase and pressure induced amorphization in silica. *Physical Review Letters* 80: 2149–2152.

Wentzcovich RM, Martins JL, and Price DD (1993) Ab initio molecular dynamics with variable cell shape: Application to MgSiO$_3$. *Physical Review Letters* 70: 3947–3950.

Wentzcovich RM, Karki BB, Cococcioni M, and de Gironcoli S (2004b) Thermoelastic properties of MgSiO$_3$–perovskite: Insights on the nature of the Earth's lower mantle. *Physical Review Letters* 92: 018501–1018501–4.

Wentzcovich RM, Stixrude L, Karki BB, and Kiefer B (2004a) Akimotoite to pervoskite phase transition in MgSiO$_3$. *Geophysical Research Letters* 31: L10611 doi: 10.1029/2004GL019704.

Williams Q, Jeanloz R, and Akaogi M (1986) Infrared vibrational spectra of beta-phase Mg$_2$SiO$_4$ and Co$_2$SiO$_4$ to pressures of 27 GPa. *Physics and Chemistry of Minerals* 13: 141–145.

Williams Q, Hemley RJ, Kruger MB, and Jeanloz R (1993) High pressure vibrational spectra of quartz, coesite, stishovite and amorphous silica. *Journal of Geophysical Research B* 98: 22157–22170.

Williams Q, Jeanloz R, and Mcmillan PF (1987) The vibrational spectrum of MgSiO$_3$–perovskite: Zero pressure Raman and mid-infrared spectra to 27 GPa. *Journal of Geophysical Research B* 92: 8116–8128.

Wilson EB, Decius JC, and Cross PC (1955) *Molecular Vibrations*. New York: McGraw-Hill, (republished in 1980 by Dover Press).

Winkler B (2002) Neutron sources and instrumentation. *European Journal of Mineralogy* 142: 225–232.

Winkler B (2004) Introduction to high pressure computational crystallography. In: Katrusiak A and McMillan PF (eds.) *High-Pressure Crystallography*, pp. 159–177. Dordrecht: Kluwer Academic.

Winkler B and Dove MT (1992) Thermodynamic properties of MgSiO$_3$ perovskite derived from large scale molecular dynamics simulations. *Physics and Chemistry of Minerals* 18: 407–415.

Wolf GH and Bukowinski MST (1987) Theoretical study of the structural properties and equations of state of MgSiO$_3$ and CaSiO$_3$ perovskites: Implications for lower mantle composition. In: Manghnani MH and Syono Y (eds.) *High Pressure Research in Mineral Physics*, pp. 313–331. Washington DC: American Geophysical Union.

Wolf GH and Bukowinski MST (1988) Variational stabilization of the ionic charge densities in the electron-gas theory of crystals: Applications to MgO and CaO. *Physics and Chemistry of Minerals* 15: 209–220.

Wolf GH and Jeanloz R (1984) Lindemann melting law: Anharmonic correction and test of its validity for minerals. *Journal of Geophysical Research B* 89: 7821–7835.

Wolf GH and Jeanloz R (1985) Lattice dynamics and structural distortions of CaSiO$_3$ and MgSiO$_3$ perovskites. *Geophysical Research Letters* 12: 413–416.

Wolf GH and McMillan PF (1995) Pressure effects on silicate melt structure and properties. In: Stebbins JF, McMillan PF, and Dingwell DB (eds.) *Structure, Dynamics and Properties of Silicate Melts*, vol. 32, pp. 505–561. Washington DC: Mineralogical Society of America.

Wood BJ and Rubie DC (1996) The effect of alumina on phase transformations at the 660-kilometer discontinuity from Fe–Mg partitioning experiments. *Science* 273: 1522–1524.

Wu TC, Bassett WA, Burnley PC, and Weathers MS (1993) Shear-promoted phase transitions in Fe$_2$SiO$_4$ and Mg$_2$SiO$_4$ and the mechanism of deep earthquakes. *Journal of Geophysical Research B* 98: 19767–19776.

Yamaguchi Y, Osamura Y, Goddard JD, and Schaefer HFI (1994) *A New Dimension to Quantum Chemistry. Analytic Derivative Methods in Ab Initio Molecular Electronic Structure Theory*. Oxford: Oxford University Press.

Yan C-S, Vohra YK, Mao H-K, and Hemley RJ (2002) Very high growth rate chemical vapor deposition of single-crystal diamond. *Proceedings of the National Academy of Sciences of the United States of America* 99: 12523–12525.

Yang H and Ghose S (1994) In-situ Fe–Mg order–disorder studies and thermodynamic properties of orthopyroxene (Mg, Fe)2Si$_2$O$_6$. *American Mineralogist* 79: 633–643.

Yang H and Ghose S (1995) A transitional structural state and anomalous Fe–Mg order–disorder in Mg-rich orthopyroxene, (Mg0.75Fe0.25)Si$_2$O$_6$. *American Mineralogist* 80: 9–20.

Young TE, Green HW, Hofmeister AM, and Walker D (1993) Infrared spectroscopic investigation of hydroxyl in ®□Mg$_2$SiO$_4$ and coexisting olivine: Implications for mantle evolution and dynamics. *Physics and Chemistry of Minerals* 19: 409–422.

Yusa H, Akaogi M, and Ito E (1993) Calorimetric study of MgSiO$_3$ garnet and pyroxene: Heat capacities, transition enthalpies, and equilibrium phase relations in MgSiO$_3$ at high pressures and temperatures. *Journal of Geophysical Research B* 98: 6453–6460.

Zerr A and Boehler R (1993) Melting of (Mg, Fe)SiO$_3$–perovskite to 625 kilobars: Indication of a high melting temperature in the lower mantle. *Science* 262: 553–555.

Zhao J, Sturhahn W, Lin J-F, Shen G, Alp EE, and Mao H-K (2004) Nuclear resonant scattering at high pressure and high temperature. *High Pressure Research* 24: 447–457.

Zouboulis ES and Grimsditch M (1991) Raman scattering in diamond up to 1900 K. *Physical Review B* 43: 12490–12493.

Zulumyan NO, Mirgorodskii AP, Pavinich VF, and Lazarev AN (1976) Study of calculation of the vibrational spectrum of a crystal with complex polyatomic anions. Diopside CaMgSi$_2$O$_6$. *Optics and Spectroscopy* 41: 622–627.

8 Theory and Practice – Multianvil Cells and High-Pressure Experimental Methods

E. Ito, Okayama University, Misasa, Japan

8.1	Introduction	198
8.1.1	Beginning of High-Pressure Earth Science	198
8.1.2	Opposed-Type High-Pressure Apparatus and Principles of High-Pressure Design	199
8.1.2.1	Precompression	199
8.1.2.2	Massive support	200
8.1.2.3	Lateral support	201
8.1.3	Conceptual Advantages of the MAAs over the Opposed-Type Apparatus	201
8.2	Multianvil Apparatus	202
8.2.1	Types of MAA	203
8.2.1.1	Hall's tetrahedral apparatus	203
8.2.1.2	Cubic anvil apparatus	203
8.2.1.3	Octahedral anvil apparatus	204
8.2.1.4	Combination of multianvil systems	205
8.2.2	Anvil Materials	206
8.2.3	Pressure Media and Gasket Materials	208
8.2.4	Fixed Points for Pressure Calibration	209
8.3	High-Pressure and High-Temperature Experiments	210
8.3.1	Heating Materials	210
8.3.2	Heating Assemblies and Temperature Measurement	210
8.3.3	Quench Experiment	212
8.3.4	Crystal Growth at High Pressure and High Temperature	213
8.3.5	Electrical Conductivity Measurement	214
8.4	*In Situ* X-Ray Observations Using SR	216
8.4.1	Methods of *In Situ* X-Ray Diffraction Using MAAs	216
8.4.2	Multianvil System Interfaced with SR	216
8.4.3	Experimental Procedures for *In Situ* X-Ray Diffraction Study	218
8.4.4	Pressure Determination	219
8.4.4.1	Pressure standard materials and the pressure scales	219
8.4.4.2	Comparison of the pressure scales	220
8.4.5	Application to Phase Equilibrium Studies	221
8.4.6	Viscosity Measurement by X-Ray Radiography	223
8.4.7	Pressure Generation Using Sintered Diamond Anvil	224
8.5	New Applications and Future Perspectives	225
References		227

Glossary

synchrotron radiation Electromagnetic waves emitted from electrons moving on a circular orbit at velocities close to the velocity of light. In a synchrotron facility, accelerated electrons are kept on moving in a storage ring by magnets stationed along the ring, and intense light is radiated in the direction of a tangent line of the orbit. Synchrotron radiation is polychromatic, extending over energy range from vacuum ultraviolet to hard X-ray and has sharp directivity. Many storage rings for the light source have been constructed, and synchrotron radiation has been widely used in physics, chemistry, biology, and electronic engineering

fields through the structural analysis of solids, liquids, biopolymers, etc. and X-ray lithography. Originally the storage rings used in high-energy physics experiment were employed as the sources of the synchrotron radiation in the 1960s and the 1970s. In the 1980s, exclusive storage rings, the second-generation facilities (e.g., SRS, UK, 1980, 2 GeV; PF, Japan, 1982, 2.5 GeV; NSLF, USA, 1985, 2.5 GeV), were constructed. In the 1990s, the synchrotron facilities entered the third generation, which produce lights of much higher brilliance (e.g., ESRF, France, 1991, 6 GeV; APS, USA, 1996, 7 GeV; SPring-8, Japan, 1997, 8 GeV).

8.1 Introduction

8.1.1 Beginning of High-Pressure Earth Science

Since the Greek philosopher Aristotle pointed out in the fourth century BC that the Earth has a spherical shape, the state of the inside of the Earth has been a fantastic and baffling object for human beings for a long time. In the early twentieth century, the Earth's interior became an object of scientific research due to the information provided by seismology. Jeffreys (1939) and Gutenberg (1948, 1951) constructed the models for distribution of seismic velocities over the entire solid Earth from the surface to the center. The solid Earth was divided into the crust, the mantle, and the core with increasing depth, based on marked changes in seismic velocities. These works made it possible to model the density and thereby the pressure within the Earth as functions of depth (Bullen, 1937). It was shown that the conditions of pressure up to 360 GPa and temperatures up to presumably several thousands of degrees kelvin extended into the Earth's interior. In order to decipher such Earth models in terms of material science, therefore, high-pressure and high-temperature experiments on deep-Earth materials should be carried out to reveal what structures and properties they possess under the corresponding conditions. Research processes to reveal the constitution and state of the Earth's interior by combining the seismology-based knowledge and the results of high-pressure studies may be referred to as the high-pressure Earth sciences.

The embryo of high-pressure Earth sciences would ascend to Bernal's (1936) prophesy that discontinuous increases in seismic velocities around a

400 km depth found by Jeffreys would be caused by the transformation of common olivine to a new phase possessing the spinel structure. It is interesting to note that the crystallographer Bernal knew of the olivine–spinel transformation in Mg_2GeO_4 which occurs at atmospheric pressure. Later, Birch (1952) examined homogeneity of the Earth's interior by analyzing the seismic models of Jeffreys (1939) and Gutenberg (1948, 1951) using the equation of state based on the finite strain theory. Birch concluded that the layer B (Bullen's nomenclature, depths of 33–413 km in the mantle) was composed of olivine, pyroxenes, and garnet, whereas the layer D (depths 984–2898 km) would be composed of close-packed oxides of magnesium, silicon, and iron, similar to corundum, rutile, or spinel in structure. The intermediate layer C was characterized as a layer of transition from familiar silicate minerals in layer B to the high-pressure phases in layer D. Birch also pointed out that the outer core was interpreted as liquid iron alloyed with a small fraction of lighter elements, and the inner core as crystalline iron. Birch's argument was so concrete and convincing that many readers must have dreamed to substantiate his conclusions by conducting high-pressure experiments.

Birch's prediction was soon supported by the high-pressure syntheses of the spinel type of Fe_2SiO_4 by Ringwood (1958) and the rutile type of SiO_2 (stishovite) by Stishov and Popova (1962). The special importance of the latter work should be noted because it was first verified that Si^{++} ion could be accommodated in the sixfold coordinated site with respect to oxide ion at high pressure. It was certainly remarked that the most important effort in high-pressure Earth science in the 1960s was the construction of the phase diagram of the $Mg_2SiO_4 - Fe_2SiO_4$ system, focusing on the clarification of the 410 km seismic discontinuity. This work was carried out actively by both Ringwood's group of ANU (Ringwood and Major, 1970) and Akimoto's group (Akimoto and Fujisawa, 1968) of ISSP, University of Tokyo, somewhat competitively. Reviewing the process of this research is very instructive: the stability field of the spinel (γ) phase was expanded to be richer in the Mg_2SiO_4 component according to the improvement of the high-pressure devices. High-pressure phase relations of many analogous compounds, such as germanates, titanates, and stanates, were widely studied with crystal chemical implications. In the course of the research, it was found that a new phase called modified spinel (β) emerged in compositions close to the Mg_2SiO_4 end member, and has a stability field between those of olivine (α) and spinel (γ). Based on the phase diagram of the $Mg_2SiO_4 - Fe_2SiO_4$ system, it was proposed that the 410 km discontinuity would be responsible for a series of transformation, for example, $\alpha \rightarrow (\alpha + \gamma) \rightarrow (\alpha + \beta) \rightarrow \beta$, of the mantle olivine with a composition close to $(Mg_{0.9}Fe_{0.1})_2SiO_4$ (Ringwood, 1970; Akimoto et al., 1976).

In spite of such epoch-making achievements by both groups, the binary loop of $(\beta + \gamma)$ did not reach to the end-member Mg_2SiO_4 because of limitations of the attainable pressures of the high-pressure apparatus adopted (the Bridgman anvil by the ANU group and the tetrahedral anvil press by the ISSP group). Nevertheless, it was strongly recognized that the performance of the adopted apparatus was crucial to acquire excellent results and thus to promote our knowledge about the Earth's interior. Therefore, many attempts to make novel high-pressure apparatus and also to improve the existing apparatus were vigorously tried in the late 1960s to 1970s. Major goals of the decade were the synthesis of γ-Mg_2SiO_4 and further understanding of the post-spinel transformation, the transformation of the spinel phase at still higher pressures. As a consequence, two devices possessing characteristics complementary to each other, the multianvil apparatus (MAA) and the diamond-anvil cell (DAC), were developed in this period and have been fundamental in the progress of the high-pressure Earth sciences in subsequent years (*see* Chapter 9). Actually, γ-Mg_2SiO_4 was successfully synthesized by Suito (1972) using the MAA, and the decomposition of γ-Fe_2SiO_4 into an oxide mixture of FeO wüstite and SiO_2 stishovite was found by Mao and Bell (1971) by adopting the DAC.

Purposes of this article are to explain the mechanical structure of multianvil types of high-pressure apparatus, together with brief reviews of their development, and to describe their applications to the high-pressure Earth sciences. The multianvil apparatus which have been in practical applications so far are classified into the tetrahedral, cubic, and octahedral types, corresponding to the number of anvils (4, 6, or 8), and the corresponding shapes of sample chamber (tetrahedron, cube, and octahedron, respectively). As the octahedral type has been the most widely adopted in the last few decades, description will be focused on this type of apparatus.

8.1.2 Opposed-Type High-Pressure Apparatus and Principles of High-Pressure Design

As the MAA was developed based on the summarization of faults and limitations of piston–cylinder and other uniaxial opposed-type apparatuses that were already developed, a brief review of these classical types of apparatus may be useful to understand the structural concept of the MAA. In this context, Bundy (1962) described general principles of high-pressure apparatus design, by reviewing the various types of apparatus. By applying these principles, it become possible to generate pressure higher than the yielding and/or fracture strength of materials which are used to build up the high-pressure chamber. The important principles are as follows.

8.1.2.1 Precompression
In the piston–cylinder apparatus, a cylindrical sample assembly with diameter of 10–20 mm is set in a bore space of a thick-walled cylinder, and compressed by inserting a piston with the aid of a uniaxial press as schematically shown in **Figure 1**. The ratio of the inner diameter to the outer diameter of the cylinder is usually around 10. On loading pressure, the maximum hoop tensile stress emerges at the inner wall of the bore, which depends on the ratio but is definitely higher than the applied bore pressure within the elastic limit. Therefore, the attainable pressure never reaches the value of tensile strength of the bore material. In order to overcome this limitation, the central core is kept in a high compression state at zero loading so that the hoop tension at the bore wall is still less than the tensile strength of the material when the bore

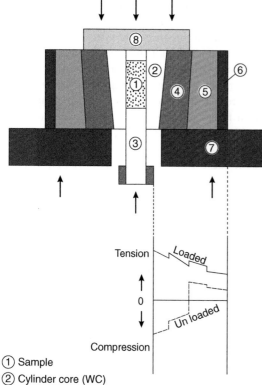

① Sample
② Cylinder core (WC)
③ Piston (WC)
④ Compression ring (hardened steel)
⑤ Compression ring (hardened steel)
⑥ Outer ring (steel)
⑦ Support
⑧ Support

Figure 1 A cross section of the piston–cylinder apparatus. The inner core of WC and the middle rings of hardened steel are in precompression, so that the tensile stress inside the core may not exceed the tensile strength of WC on loading.

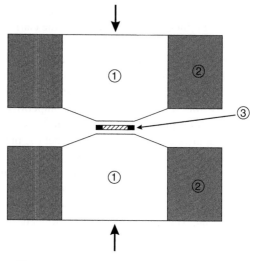

① Sintered carbide anvil
② Steel binding ring
③ Thin disk sample surrounded by pyrophyllite gasket

Figure 2 A cross section of the Bridgman anvil (see text). Ringwood's research on the olivine–spinel transformation was carried out by using this apparatus (cf. Ringwood and Major, 1970).

pressure is applied. This situation is usually realized by thrusting a few coaxial cylinders whose outer diameter is finished to be slightly larger than the inner diameter of the next outer cylinder into a single cylinder. By this method, the maximum loading pressure could go up close to the sum of the compressive and tensile strengths of the bore material. The piston–cylinder apparatus is now widely adopted in the experiments up to pressures c. 5 GPa because of the large sample volume and the performance of precise temperature control.

8.1.2.2 Massive support

One of the simplest high-pressure apparatus is an array of truncated conical pistons with flat faces at the top as shown in **Figure 2**, which was developed by Bridgman in 1952 and is called the Bridgman anvil. Bundy regarded this device as a MAA with two pistons. A thin disk sample surrounded by a pyrophyllite gasket is compressed between the flat faces of the pistons, and the pressure is sustained by the pinched gasket. The tungsten carbide pistons are gripped by hardened steel rings under high hoop tension so that the pistons are under compression and slight elongation. As the top surface area of the piston is around a tenth of the cross-sectional area of the body of the piston, the high compressive stress of the surface top is dispersed backward rapidly without breakage occurring. Apparatus of this type works easily up to higher than 10 GPa which is more than twice higher than the compressive strength of tungsten carbide (WC). This design concept is called the principle of massive support, and has been an important criterion in the design of MAA. Ringwood (1970) carried out high-temperature experiments by inserting a thin resistance heating element with the sample between the anvil faces. It should be mentioned that the DAC is regarded as an evolved apparatus from the Bridgman anvil by adopting the hardest material, diamond.

8.1.2.3 *Lateral support*

Drickamer and Balchan (1962) extensively raised the attainable pressure of the Bridgman anvil-type apparatus by supporting the tapered lateral area of the piston with a pyrophyllite pellet, as illustrated in **Figure** 3. This apparatus, named the Drickamer cell, was mostly used for measuring electrical resistance of metal at room temperature or low temperatures. By using work-hardened tungsten carbide as the piston, the maximum pressure reached nominally higher than 50 GPa. An important role of the lateral support in high-pressure generation is demonstrated in the belt apparatus (Hall, 1960) whose cross section is schematically drawn in **Figure** 4. Two opposed pistons compress a sample retained in a strongly reinforced multilayered cylinder, or belt, in a uniaxial press. The principle of massive support is utilized in both the conically shaped pistons and the cylinder. In addition, both the piston and the cylinder are circumferentially supported by the compressible conical gasket made of laminated pyrophyllite-and-metal. The gasket not only prevents extrusion of the materials inside the chamber but also makes the pressure gradient along the flanks of both the piston and cylinder gentle, preventing them from breaking. Therefore the belt apparatus and similar apparatus can produce pressures higher than 10 GPa over a large volume and have been used most popularly in the industrial field such as diamond synthesis.

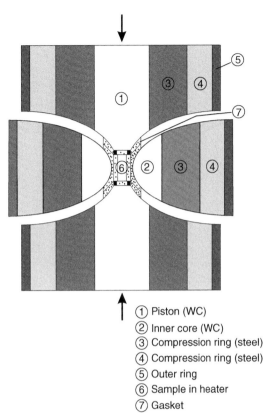

1. Piston (WC)
2. Inner core (WC)
3. Compression ring (steel)
4. Compression ring (steel)
5. Outer ring
6. Sample in heater
7. Gasket

Figure 4   A cross section of the belt apparatus (see text).

8.1.3   Conceptual Advantages of the MAAs over the Opposed-Type Apparatus

MAAs with anvils of 4 and greater were invented in a period from the late 1950s to the 1960s. It is absolutely true that a volume of the three-dimensional sample chamber is intrinsically larger than those of the opposed anvil apparatuses with a two-dimensional sample space such as the Bridgman anvil and the Drickamer cell. Another merit of the MAAs is that the stress states in the samples are much closer to hydrostatic despite the usage of a solid pressure medium due to compression from multi-directions, as schematically shown in **Figure 5**.

When all the anvils are put together, anvil tops form a polyhedral space at the center. A polyhedral pressure medium is set at the central hollow space with or without a preset gasket. As an anvil of the MAA has the shape of a truncated pyramid, it is clear that the anvil is in massive support when the truncated top face compresses the central polyhedral pressure medium. The generated pressure is sustained by friction of the gasket which is caused by strong compression between flanks of the

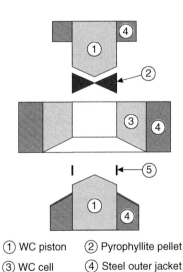

1. WC piston   2. Pyrophyllite pellet
3. WC cell   4. Steel outer jacket
5. Mica insulating sleeve

Figure 3   A cross section of the Drickamar cell (see text).

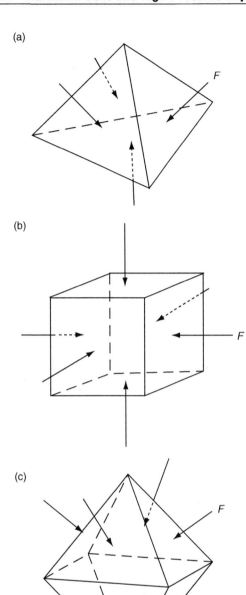

Figure 5 Conceptual drawings for compression of the polyhedral pressure media. All surfaces of the regular polyhedra are thrusted by equivalent normal forces. (a), (b), and (c) show the tetrahedral, cubic, and octahedral compression, respectively.

neighboring anvils. In other words, the anvil is under the lateral support. It is important to point out that all the forces applied to the anvil are compression, which suggests potential of the MAA to generate high pressures. Actually, the MAAs equipped with WC anvil

can generate pressures higher than the piston–cylinder and belt apparatus equipped also with WC. This is because the hoop tensile stress at the inside walls of the bore is the critical factor to determine the maximum attainable pressure in the latter apparatus.

8.2 Multianvil Apparatus

As mentioned above, the MAAs in practical applications have been the tetrahedral, cubic, and octahedral types, corresponding to the number of anvils 4, 6, and 8, respectively. A polyhedral pressure medium larger than the space is usually set with compressive and insulating gaskets in gaps between flanks of adjacent anvils. On thrusting the anvils to the center, the central pressure medium is squeezed to generate high pressure there. The generated pressure is confined by the compressed gasket, which is formed of the preset gasket and the pressure medium extruded into the gaps. It is true that, in many MAAs, more than 90% of the applied force is balanced with the compression of the gasket.

For further increased surface numbers, the regular polyhedral sample chamber changes to a dodecahedron with 12 anvils and to an icosahedron with 20 anvils. With increasing numbers of anvils, the ratio of the stroke of the anvil to shrinkage of the anvil gap increases and thus the sample is more effectively compressed and so the hydrostacity of the solid pressure medium becomes higher. However, the solid angle of the anvil from the apex becomes smaller, which makes the massive support less efficient. A trial of the icosahedral compression carried out at Kawai's laboratory of Osaka University in the late 1960s resulted in the breakage of many anvils. Another difficulty of such MAAs is an inherent difficulty in aligning the anvils at the initial setting, and also during compression. Within my knowledge, the octahedral compression is of the largest number of anvils in practice and is at the best compromising point between the number of anvils and the massive support.

As the pressure in the MAA is confined by the compressed gasket between flanks of adjacent anvils, the generated pressure is sustained on subtle balance with the friction of the gaskets. Therefore, the slight unbalance due to an abrupt change of thrusting force or flow-out of the medium causes an explosive extrusion of a large amount of the pressure medium into the gaps with a big characteristic sound. This phenomenon has been called blow-out, which frequently brings serious damage to the anvils. There is a large hysteresis between compression and decompression trajectories in the relationship of generated pressure versus applied load. The blow-out occurs with high possibility in the later stage of decompression where the pressure decreases the most rapidly with decreasing load. It is quite certain that the higher is the internal friction of pressure medium, the higher probability of the occurrence of blow out.

Three types of methods are adopted to drive anvils of the MAAs: (1) each anvil is driven by an individual ram which is fixed to a solid frame; (2) in order to ensure uniform compression the whole anvil system is immersed in a fluid (oil or water) chamber and is thrust with the aid of hydrostatic pressure into the chamber; and (3) an assembly of anvils is squeezed between a pair of solid guide blocks which is driven by a uniaxial hydraulic press.

In this section, the tetrahedral, cubic, and octahedral apparatus are reviewed successively.

8.2.1 Types of MAA

8.2.1.1 Hall's tetrahedral apparatus
The first MAA was Hall's (1958) tetrahedral apparatus in which four faces of tetrahedral pressure medium are simultaneously compressed by four tungsten carbide anvils with equilateral triangular faces. Pyrophyllite was mostly adopted as the pressure medium. Each anvil is driven by a hydraulic ram which is fixed at a corner of the rigid tetrahedral frame inward as shown in **Figure 6**. The edge length of the pressure medium is made larger than the edge length of the anvil faces. Therefore, on compression, part of pyrophyllite medium is extruded into the gap between the side faces of the anvils to form the gasket which confines the

generated pressure. The maximum attainable pressure was *c.* 12 GPa at room temperature.

It is also interesting to note that the tetrahedral apparatus was interfaced with the X-ray diffraction system by Hall *et al.* (1964) (Section 8.4.1) The tetrahedral apparatus seems to have not been used for the last three decades. However, this apparatus initiated the development of MAA and their resulting versatile abilities for material synthesis and *in situ* X-ray observation.

8.2.1.2 Cubic anvil apparatus
The von Platen's cubic apparatus was probably the first apparatus to compress a cubic pressure medium by six identical anvils along three opposite directions perpendicular to each other. The six anvils have a shape of a split sphere with a truncated square surface at the top and a spherical back surface. The six-anvil assembly, covered by a spherical copper sheath, was put into a cylindrical water reservoir and the cubic sample assembly was compressed with the aid of water pressure in the reservoir. The whole cross section of the apparatus is diagrammatically shown in figure 11 in Bundy (1962). The apparatus was mainly employed to make artificial diamond.

A cubic apparatus in which six anvils are driven by a uniaxial press by means of a set of rigid upper and lower guide blocks was first developed in collaboration between Kyoto University and Kobe Steel Co. in the middle of the 1960s (Osugi *et al.*, 1964). The cross section of the apparatus designed recently is schematically drawn in **Figure 7**, in which the central cubic

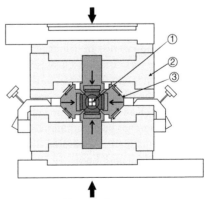

① The Kawai cell

② The DIA-type guide block

③ Sliding surface

Figure 7 The DIA-type guide block compressing the Kawai cell (Ito *et al.*, 2005). There is friction on the sliding interfaces between the inner surfaces of the guide block and the outer surfaces of the anvil support.

Figure 6 Photo of the Hall's tetrahedral anvil apparatus. This apparatus was used in the research of the olivine–spinel transformation in 1960s by Akimoto's group (cf. Akimoto and Fujisawa, 1968).

space is filled with the Kawai cell (see next section). The cubic pressure medium is compressed synchronously by the six tungsten carbide anvils. The upper and lower surfaces of the cube are thrust by the upper and lower anvils fixed to the guide blocks, and the four side surfaces are thrust by the anvils that are forced to proceed toward the center by sliding between the back surfaces of the anvil supports and the inner surfaces of the guide blocks. By utilizing a 500 ton (1 ton = 10^{-2} MN) press, pressures up to 10 GPa can be generated over a volume of 1 cm^3. This apparatus is called the DIA-type apparatus, because the cross section of the anvil support resembles that of Brillouin-cut diamond. It should be noted that *in situ* X-ray observation at high pressures, and simultaneously at high temperature, was first carried out employing this apparatus (Inoue and Asada, 1973).

Cubic anvil apparatuses in which each anvil is backed up by a hydraulic ram like in the Hall's tetrahedral apparatus were also developed in several laboratories (e.g., Shimada, 2000). Six independently moving rams of several meganewtons force are fixed to a steel frame along the three principal directions. The movement of each ram is measured by the linear variable differential transformer (LVDT). Advances of the six rams are regulated by an oil pressure in the pumping system which is semi-manually controlled. As the ram can be driven in a mutually independent manner, this machine has mainly been adopted in the study of the deformation of rocks under the confining pressure (Shimada, 2000).

8.2.1.3 *Octahedral anvil apparatus*

Development of an octahedral anvil apparatus, called the Kawai-type apparatus, in which an octahedral pressure medium was compressed by eight equivalent tungsten carbide anvils, started in the middle of the 1960s under the direction of the late N. Kawai at Osaka University (Kawai, 1966). Some of the technological development has been summarized by Ito *et al.* (2005) and schematically illustrated in **Figure 8** (Kawai, 1966; Kawai and Endo, 1970). A tungsten carbide sphere was split into eight identical blocks by three planes mutually perpendicular and crossing at the center. After truncation of the top to form a regular triangle, the eight spilt sphere anvils were reconstructed with an octahedral pressure medium and adjacent gaskets at the center, and covered with a rubber capsule to be compressed in a cylindrical oil chamber (**Figure 8(a)**) (Kawai, 1966). Therefore the concept is similar to von Platen's cubic apparatus but

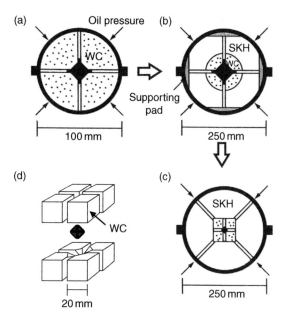

Figure 8 Evolution of the Kawai-type high-pressure apparatus. (A) An eight split-sphere assembly of WC covered by a rubber capsule that is compressed in an oil chamber. (B) On enlargement, six supporting pads are placed at junctions of the assembled split spheres to keep the whole sphere in the perfect shape under compression. (C) The supporting pads turn into six outer split-sphere assembly, and the inner split blocks to inner cubic anvils. (C) The Kawai cell composed of eight cubic anvils and an octahedral pressure medium.

the apparatus was greatly simplified without a cooling system.

According to enlargement to generate higher pressure, the system evolved to the double-staged split-sphere system (Kawai and Endo, 1970) as shown in **Figure 8(c)** via an intermediate stage of **Figure 8(b)**. The assembly of eight cubic anvils of tungsten carbide with the octahedral pressure medium and gaskets, the Kawai cell, is squeezed by the six first-stage anvils made of hardened steel with the square top surface and the cubic split-sphere-shaped back surface. The whole assemblage of the first and the second anvils is compressed by the oil pressure outside the rubber capsule. It should be noted that the versatility of the apparatus was remarkably increased by adopting a disposable cubic tungsten carbide anvil that is much easier to machine than a split sphere. For example, a WC cube can be used for multiple purposes by truncating the different corners in different sizes: for example, the smaller for higher pressure and the larger for a bigger sample volume experiment.

Three of the six first-stage anvils were later glued together in a hemispheric cavity of a steel vessel to form a guide block, and the two guide blocks were aligned in a hydraulic press so that the cubic Kawai cell was compressed by the uniaxial force along the [111] direction (Kawai *et al.*, 1973). The guide block and the 5000 ton press, USSA 5000, installed at the Institute Study of the Earth's Interior (ISEI), Okayama University, are shown in **Figure 9**. In this system, a good match of the spherical surfaces of the anvil back and the cavity of the vessel is very crucial to squeeze the Kawai cell to high tonnages. Ohtani *et al.* (1987) avoided the situation by adopting a set of split-cylinder wedges which was fixed to a bolster plate and supported by a massive cylinder of hardened steel. The split-cylinder system was modified by Walker *et al.* (1990) to an assembly of an unanchored cylindrical cluster of removable wedges set in loose supporting rings, which accommodate appreciable strain under loading. The height to diameter of the cylindrical cluster wedges was chosen so that the

cubic cavity formed by them does not have its angular relationships distorted under appreciable strain.

8.2.1.4 *Combination of multianvil systems*
A double-stage split-sphere apparatus to compress a cubic or tetragonal specimen was developed at the Institute of Mineralogy and Petrology, Siberian Branch of the Russian Academy of Science, Novosibirsk, Russia, under the direction of the late I. Y. Malinovskiy. A photo of the apparatus is shown in **Figure 10**. An assembly of eight steel anvils with split-sphere shape is held in a spherical cavity of the high-pressure vessel and divided into upper and lower parts which is supported by half couplings. The top of the split sphere anvil is truncated so that an octahedral space is formed at the center when the eight are assembled as the Kawai-type apparatus. Six tungsten carbide wedges are put with cubic or tetragonal pressure medium into the octahedral space, and the split sphere anvil assembly is compressed by hydraulic pressure from the outside of the rubber

(b)

(a)

Figure 9 Photos of Kawai-type pressure apparatus installed at ISEI, Okayama University. (a) Lower half part of the guide block make up of three bound split spheres. The Kawai cell covered by mica sheets is set at the center. (b) A 5000 ton press to drive the split sphere guide blocks.

Figure 10 Photo of the double-stage split-sphere apparatus developed at Institute of Mineralogy and Petrology, Siberian Branch of the Russian Academy of Science, Novosibirsk, Russia (see text).

jacket surrounding the split sphere. The apparatus has actively been used in studies on the formation and growth of diamond (Pal'yanov *et al.*, 2002 and references listed herein).

Recently, the Kawai cell has commonly been set in a cubic sample space of the DIA apparatus and squeezed by a uniaxial press. As the combination of both devices is the most feasible for X-ray optics, the double-staged system has been set up in many facilities for high-pressure X-ray experiments using the synchrotron radiation (SR). In this case, however, force applied to the square surface of the Kawai cell is 1/3 times of the applied press load, compared with $1/\sqrt{3}$ times in the case of compression by the uniaxial force in the [111] direction. It is known that friction on the sliding surfaces of the anvil supports and the guide blocks in the DIA apparatus usually loses 10–20% of the applied load (see **Figure 7**). Moreover, a discontinuous slip happens there instead of smooth sliding, which frequently causes the blow-out of the sample in the Kawai cell. In the DIA apparatus the elastic deformation of the guide block is not uniform in the axial and lateral directions, which does not keep the sample chamber exactly cubic. All these shortcomings of the DAI-type apparatus on squeezing the Kawai cell could in principle be overcome by driving the six anvils with independently controlled rams (e.g., Shimada, 2000). Although all such cubic apparatus developed so far have not been operated with the required precision, it is possible to adjust each anvil load to within 10 kN and to synchronize the anvil advances to within a few micrometers by adopting recent high-tech

servomechanism. Construction of such octahedral apparatus with enough force is a task for the future.

8.2.2 Anvil Materials

Let us consider a homogeneous solid material in the stress field that is specified by principal stresses σ_1, σ_2, and σ_3. According to von Mises' theory of the maximum shear stress, the critical condition for the material to commence plastic flow is given as

$$(\sigma_1 - \sigma_2)^2 + (\sigma_2 - \sigma_3)^2 + (\sigma_3 - \sigma_1)^2 = 2\sigma_e^2 \quad [1]$$

where σ_e is the uniaxial compressive strength of the material. Equation [1] indicates that, in order to generate high pressure without plastic deformation or fracture of an anvil, minimization of the left-hand side and/or maximization of the right-hand side are prime requirements. The former requirement demands to avoid the stress concentration on the anvil by selecting the pressure medium and the gasket of appropriate materials together with their suitable sizes. The latter, on the other hand, simply demands the use of a hard material for the anvil.

WC has been the most commonly used for anvils of the MAA. WC as a superhard alloy is a sintered material of the fine-grained intermetallic compound WC with a small amount of Co as a binding agency. Mechanical properties of WC alloy strongly depend on the amount of Co binder as typically shown in **Figure 11**. Compressive strength increases from 350 to 600 kgf mm^{-2} with a decrease in Co content from 25 to 3 wt.%, whereas tensile strength decreases 150 to less than 50 kg mm^{-2} in the same variation of the Co content. Grain size of the WC particle is also an important factor to control toughness of the WC alloy. Recently, WC alloys possessing a compressive strength larger than 700 kgf mm^{-2} have been developed, whose particle sizes are less than 5 μm. The maximum attainable pressure in the Kawai-type apparatus using WC has been limited to *c.* 28 GPa (Kubo and Akaogi, 2000) over volume of *ca.* 2 mm^3, but quite recently pressures up to 41 GPa have been generated by utilizing a newly invented WC alloy (Katsura *et al.*, 2005, unpublished data).

The generated pressure using the Kawai-type apparatus has largely been extended by employing sintered diamond (SD) for the anvil material since the late 1980s (e.g., Ohtani *et al.*, 1989; Kondo *et al.*, 1993) The SD is a composite of diamond and Co metal sintered in the similar mechanism to that for the WC alloy. Mechanical properties of SD also depend

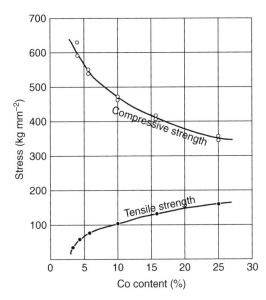

Mechanical properties of superhard alloy in the WC–Co system

Figure 11 Mechanical properties of WC. Compressive strength (upper) and tensile strength (lower) are shown as functions of Co content.

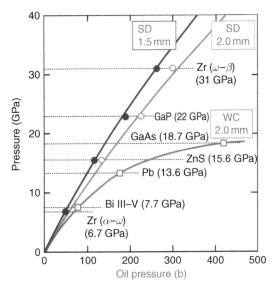

Figure 12 Pressure calibration based on the pressure fixed points. Curves of SD are obtained using cubes with 14 mm edge length and 1.5 and 2.0 mm truncations and that of WC is for 32 mm edge length and 2.0 mm truncation (see text).

on the grain size of the sintered diamond particle. Although detailed information is not given, a lump of SD with a grain size typically less than 10 μm seems to be tougher than that of a larger grain size and is suitable for the anvil material. As the sintering conditions of SD are in the stability field of diamond (e.g., 4.5 GPa and 1500°C), however, available size of the cubic anvil of SD is at present limited to an edge length of 20 mm. The high hardness of SD, on the other hand, makes it difficult to machine SD cubes by polishing. These situations considerably raise the costs of SD cubic anvils compared to WC: the finished 20 mm SD cube costs *c.* US$ 8000. Consequently, the 14 mm cube, which still costs *c.* US$ 2000, has been the biggest one for practical use. It is not possible to directly measure compressive strength of super hard materials such as WC and SD, but the hardness is a good measure to assess it. The Noop hardness of SD is *c.* 49 GPa which is about twice larger than that of WC (Sung and Sung, 1996). In this context, eqn [1] suggests that a much higher pressure can be generated by using SD anvils.

Pressure calibration curves constructed based on the pressure fixed points (see Section 8.2.4) are comparatively reproduced for the cases using SD and WC cubic anvils in the Kawai-type apparatus in **Figure 12**. The curves of SD were obtained using cubes with 14 mm edge lengths and 1.5 and 2.0 mm truncations, and that of WC was for cubes with a 2.0 mm truncation. Difference in pressure efficiency between SD and WD is dramatic. It is suggested the pressure rises almost linearly against load even at higher than 30 Pa in the SD anvil, whereas generated pressure is saturated at around 20 GPa in the WC anvil.

Another type of diamond composite was invented by a research group of ANU (Ringwood *et al.*, 1989). The composite is produced at pressures slightly lower than the diamond stability field using an SiC binder. Recently a similar diamond/SiC composite has been synthesized in a cubic shape by hot isostatic pressing (HIP) processing (Shimono *et al.*, 2003). It is remarkable that the HIP conditions are at 1450°C and 200 MPa for 30 min which are by far easier to produce than those produced within or close to the stability field of diamond. Recently Ohtaka *et al.* (2004) have proved the HIP-processed diamond composite to be hard enough for generation of pressures up to 30 GPa using 10 mm cubes with a truncated edge length of 1.5 mm. It is certain that the HIP method supplies the composite at a much lower price than the composites sintered at high pressure. Hardness of the diamond/SiC composite is believed to be slightly lower than that of the diamond/Co composite. However, an important characteristic of the former is its low absorbency of X-rays, which makes it possible to use the anvils as the windows for versatile *in situ* X-ray observations including radiography.

Boron nitrite (BN) exhibits very similar properties to carbon. Cubic BN (cBN) possessing the zinc blende structure is stabilized under the almost identical $P-T$ conditions where diamond is stable. In cBN, B and N is bound to each other by the covalent bond just like diamond, and the hardness of cBN is the second next to diamond. Therefore, sintered cBN is a suitable material for anvils of MAAs. A remarkable feature of cBN is also the low absorbency for X-rays. Recently, cubes of cBN with edge lengths up to 20 mm have become commercially available.

8.2.3   Pressure Media and Gasket Materials

The following properties are required for the pressure medium of the MAA: (1) low internal friction to keep the sample in a high hydrostatic state, (2) low compressibility to raise pressure effectively, (3) low thermal conductivity, (4) very low electrical conductivity, (5) high melting point which increases with pressure, (6) chemical inertness, and (7) stability at high temperature and high pressure. In addition to the above requirements, in a specified experiment such as *in situ* X-ray observation, high transparency of X-rays is crucial. As some of these requirements contradict each other, the material should properly be selected according to the purpose of experiment.

Pyrophyllite $(Al_4Si_8O_{20}(OH)_4)$ was most widely used as a pressure medium for MAAs until the middle of the 1970s. At temperatures higher than 600°C, pyrophyllite decomposes successively into assemblages of stishovite $(SiO_2) +$ an unknown hydrous aluminum silicate and stishovite $+$ $Al_5Si_5O_{17}(OH)$ at pressures above 10 GPa, and finally into stishovite $+$ corundum (Al_2O_3) at pressures higher than 20 GPa accompanying large volume reduction (Suito, 1986). Therefore, a phase boundary determined by the quenching method using the pyrophyllite medium was generally located at higher pressure than the real one because substantial pressure reduction was accompanied by the decompositions. Instead, semi-sintered MgO with porosity around 35% has been adopted. A special problem of the MgO medium is its high thermal conductivity compared to pyrophyllite, and it is generally difficult to heat the sample to higher than 1000°C. Therefore,

setting a sleeve of low thermal conductivity such as ZrO_2 and $LaCrO_3$ just outside the heater is necessary to reduce heat loss and generate high temperatures. Sometimes CaO-doped ZrO_2 is used as a pressure medium, because no thermal insulator is needed. However, ZrO_2 medium seems to bring blow out more frequently than MgO. In-pressure calibration at room temperature using metallic fixed points (Section 8.2.4), it is recommended to contain the calibrant with soft material such as AgCl in the MgO and ZrO_2 pressure media.

In MAA, a part of the pressure medium extrudes into the anvil gap and forms the gasket according to compression. In order to get better lateral support and also to prevent the blow-out, however, gaskets are usually set at the top of the anvil gap in advance (pre-gasket). One criterion to select material for pre-gasket is to be softer or more compressive than the pressure medium. For the MgO pressure medium, pyrophyllite is adopted as the pre-gasket material. In the Kawai cell, an initial size of octahedron of MgO is substantially larger than the octahedral space formed by truncated triangles of the cubic anvils, and the combination of them both is marked by the code such as 14/8 which means that an octahedron of 14 mm edge length is adopted for the anvils of 8 mm truncation. The thickness of the pyrophyllite gasket is determined by the code.

There is a definite tradeoff on selecting the material and size of the gasket between higher efficiency of pressure generation and higher risk of the blow-out. Yoneda *et al.* (1984) systematically analyzed the stress states of the gasket and the anvil, and proposed a numerical method to select the best configuration of pre-gasket for effective and safe pressure generation. However, the best choice has been determined empirically considering difference in mechanical properties between the anvil material and the pressure medium, and this is the most difficult but interesting issue of the experiment using the MAA. Actually, each laboratory or even each researcher has developed their own assembly and gasket (e.g., Walker *et al.*, 1990). In the case of SD anvil experiments, the baked pyrophyllite gasket, heated at 800–900°C for a few tens of minutes in air, works well for confining high pressures (Ito *et al.*, 2005). Under these conditions, pyrophyllite dehydrates and forms a sintered assemblage of silimanite (Al_2SiO_5) and an aluminous silicate phase $Al_2Si_4O_{10}$, and the assemblage becomes harder with increasing heating duration.

8.2.4 Fixed Points for Pressure Calibration

Determination of the pressure value is of essential importance in applying the experimental results to the Earth's interior, because pressure inside the Earth as a function of depth is believed to be the most trustworthy parameter. Pressure calibration of the MAA has been carried out by detecting pressure-induced phase transitions in specified materials against applied load or pressure (see **Figure 12**). The pressure values of onset of the transformations, the fixed points, were first empirically determined (e.g., Bundy, 1962) and have been revised by the international agreement in the conferences owing to the development of high-pressure technology and research (e.g., Bean *et al.*, 1986). Recently the *in situ* X-ray observation in both the MAA and the DAC has expanded their versatility over wider pressure and temperature ranges by using SR. Pressure values determined from the measured volume of the pressure standard material via the

thermal equation of state (EoS) is believed to be the most reliable because the physical basis for the procedure is clear (see Section 8.4). Nevertheless, pressure fixed points preserve their necessity for an individual laboratory to build up its own calibration and for different laboratories to exchange information to each other in a more orderly manner.

A set of pressure fixed points (PIPS-97), which contained pressure calibration phase transitions at room temperature, equations for well-characterized phase transitions at high pressures and high temperature, was proposed as the consensus in the First International Pressure Calibration Workshop (IPCW) held at Misasa, Japan, in 1997 (Ito and Presnall, 1998). The PIPS-97 had to be published soon under authorship of all those who wished to participate. As the opportunity for publication has been missed for many years, however, on behalf of the IPCW, the author presents the PIPS-97 with several additional data in **Table 1**. The PIPS-97 should

Table 1 Some phase transitions as pressure calibrants (PIPS-97)

P (GPa)	T (°C)	System	Transition	Method	References
2.55	25	Bi	I–II	Electrical resistance	Bean *et al.* (1986)
3.68	25	Tl	II–III	Electrical resistance	Bean *et al.* (1986)
5.5	25	Ba	I–II	Electrical resistance	Bean *et al.* (1986)
7.7	25	Bi	III–V	Electrical resistance	Bean *et al.* (1986)
9.4	25	Sn	I–II	Electrical resistance	Bean *et al.* (1986)
12.3	25	Ba	(Upper)	Electrical resistance	Bean *et al.* (1986)
13.4	25	Pb	I–II	Electrical resistance	Bean *et al.* (1986)
15.6	25	ZnS	Semi/metal	Electrical resistance	Block, (1978)
18.3	25	GaAs	Semi/metal	Electrical resistance	Suzuki *et al.* (1981)
22	25	GaP	Semi/metal	Electrical resistance	Piermarini and Block, (1975)
33	25	Zr	ω–β	Electrical resistance	Akahama *et al.* (1992)
33.2 ± 0.2[a]	25	Zr	ω–β	Electrical resistance/*in situ* X-ray	Tange, (2005)
54 ± 1[a]	25	Fe_2O_3		Electrical resistance/*in situ* X-ray	Ito *et al.* (2005) (unpublished data)
3.2	1200	SiO_2	Qz/Coe	Quench/reversed	Bose and Ganguly, (1995)
4.8→5.8	800→1200	Fe_2SiO_4	α/γ	*In situ* X-ray	Yagi *et al.* (1987)
6.2→5.9	900→1200	$CaGeO_3$	Gt/Pv	*In situ* X-ray	Susaki *et al.* (1985)
8.7→10.1	1000→1530	SiO_2	Coe/St	*In situ* X-ray/reversed	Zhang *et al.* (1996)
12.2→14.3	1000→1400	Mg_2SiO_4	α/β	*In situ* X-ray	Morishima *et al.* (1994)
14.2→15.5	1300→1600	Mg_2SiO_4	α/β	*In situ* X-ray	Katsura *et al.* (2004)
15.1→16.0	500→750	$FeTiO_3$	Il/Pv	*In situ* X-ray	Ming *et al.* (2005)
16	1600	$MgAl_2O_4$	Sp/Per + Cor	Quench/reversed	Akaogi (this meeting)
16.5	1400	$MgSiO_3$	CEn/β + St	Quench	Gasparik, (1989)
16.5	2150	$MgSiO_3$	CEn/Maj	Quench	Presnall and Gasparik, (1990)
15.7→17.4	800→1000	Mg_2SiO_4	β/γ	*In situ* X-ray/reversed	Suzuki *et al.* (2000)
19→20.8	1200→1600	Mg_2SiO_4	β/γ	Quench	Katsura and Ito, (1989)
24.8→23.1	1000→1600	Mg_2SiO_4	γ/Pv + Per	Quench	Ito and Takahashi, (1989)

[a]Added by the author.

semi, semiconductor; Qz, quartz; Coe, coesite; St, stishovite; α, olivine; β, modified spinel; γ, spinel; Gt, garnet; Il, ilmenite; Pv, perovskite; Sp, spinel; Per, periclase; Cor, corundum; CEn, clonoeustatite.

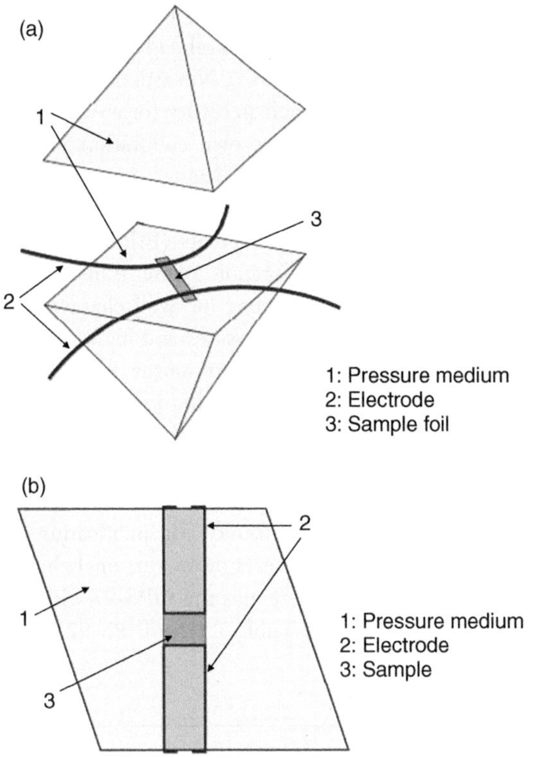

(a)

1: Pressure medium
2: Electrode
3: Sample foil

(b)

1: Pressure medium
2: Electrode
3: Sample

Figure 13 Sample arrangements of octahedral pressure medium for pressure calibration based on the fixed points: (a) a metallic calibrant and (b) semiconductor calibrant (see text).

be improved in accuracy and be expanded to cover wider pressure and temperature conditions in future.

Figure 13 shows arrangements of electrode and calibrant in an octahedral pressure medium for calibration at 300 K. **Figure 13(a)** is for a metallic calibrant such as Sn and Zr, in which the electric resistance is measured by the four-wire method. **Figure 13(b)** is for detecting semiconductor–metal transition displayed by ZnS and GaP, etc. All the electrodes are taken out to the first or the second anvil surfaces which serve as electrical leads. It is noted that, as the transition kinetics are generally low, the load should be raised slowly around the transition point to detect it accurately.

8.3   High-Pressure and High-Temperature Experiments

8.3.1   Heating Materials

In order to perform synthesis of high-pressure materials and to determine the phase equilibrium relations needed in elucidation of the Earth's interior, it

is necessary to heat the sample to at least 1200 K under the prescribed pressure. Sometimes temperatures up to *c.* 3000 K are needed in melting experiments on mantle materials under lower-mantle conditions (e.g., Ito *et al.*, 2004). In MAAs, heating the sample is carried out by the internal heating method in which a small heater and electrodes are set in the pressure medium and the electric power is supplied to the heater via the anvils. The sample is loaded inside a cylindrical heater or between a pair of platy heaters. Widely used heating materials are graphite, $LaCrO_3$, a composite of TiC + diamond, and metals of high melting point such as Pt, Ta, and Re. Among them, the graphite heater works excellently to higher than 2500 K at pressures up to *c.* 11 GPa, but, at still higher pressures, partial transformation of graphite into diamond prevents to serve as a heating material. Maximum attainable temperatures by using various heating materials depend on shape, size, and configuration of the thermal insulator in the sample assembly, and those are typically 1900, 2000, 2700, and 2900 K for Pt, Ta, Re, and $LaCrO_3$, respectively. In the cases of metal heaters, dissolution of small amounts of hydrogen from residual water in the sample assembly into the metal greatly lowers the melting temperature of the metal. The relatively low attainable temperature of the Ta heater compared to its high melting point would be due to this effect. It is noted that graphite is a typical reducing agent whereas $LaCrO_3$ is a strong oxidizing agent especially at high temperature. It should also be noted that direct contact of $LaCrO_3$ with silicates causes a reaction between them. Metal and graphite heaters, on the other hand, can serve as sample containers as well. For the MgO pressure medium, setting a thermal insulator composed of lower thermal conductivity such as $LaCrO_3$ or ZrO_2 outside the heater is essential to generate high temperatures, especially in the metal heater.

8.3.2   Heating Assemblies and Temperature Measurement

An internal heating system in the MAAs was first designed for the tetrahedral press (Bundy, 1962), which is shown in **Figure 14**. A cylindrical graphite heater is put on the line connecting midpoints of two opposite edges of the tetrahedral pyrophyllite, and the electric power to the heater is supplied via the anvils and metal strips which lie across notches worn on the edges of the tetrahedron. The sample,

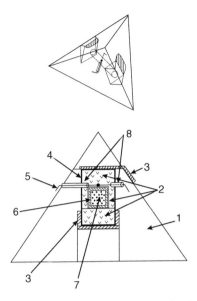

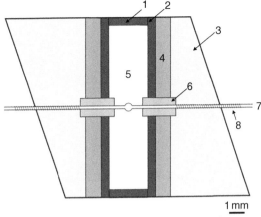

1: Lid; 2: Heater; 3: Pressure medium;
4: Thermal insulator; 5: Sample;
6: Electric insulator; 7: Thermocouple;
8: Coil

Figure 15 Cross section of a basic heating assembly for an octahedral pressure medium (see text).

1: Pyrophyllite;   5: Thermocouple wire;

2: BN;   6: Sample container;

3: Metal electrode; 7: Sample;

4: Graphite heater; 8: Insulating tube

Figure 14 Heating assembly for the Hall's tetrahedral apparatus (see text).

encapsulated within the appropriate material, is put into the heater with a spacer. A thermocouple to measure the temperature passes through the cylindrical heater, and the junction is in contact with the capsule. Terminals of the thermocouple wires are in contact with top surfaces of the remaining anvils or taken out through the gasket to outside of the apparatus. Such configuration of the parts can be recognized as the archetype of the heating assembly in the MAAs and has been modified for the cubic and octahedral apparatus.

A basic assembly for an octahedral pressure medium is schematically illustrated in **Figure 15**. A cylindrical heater and thermal insulator is placed along the central line between the opposite triangles of an MgO octahedron. Lids at the top and bottom of the heater keep the heater in good contact with the anvils of electrodes. A thermocouple penetrates the pressure medium, thermal insulator, and heater at the level of midpoint, which is electrically insulated from the heater by small tubes made of Al_2O_3 or MgO. Outside the tubes the wires pass through small coils made of the same kind of wires which protect the wires from breaking due to flow of the pressure medium and gasket. The coils are effective for

ductile thermocouples such as Pt/Rh wires. This assembly is useful in material synthesis and in determination of the subsolidus phase equilibrium by putting the sample directly into the graphite or metal heater. The straight cylinder heater produces a rather steep temperature gradient in the sample along both the longitudinal and radial directions. In a phase equilibrium study, therefore, it is important to adopt the results analyzed only on the portion adjacent to the thermocouple junction in the quenched sample (e.g., Ito and Takahashi, 1989).

As an outer diameter of the electric insulating tubes is at least 0.6 mm, two holes made at the center of the cylindrical heater for the tubes increases the resistance near the center, and the resultant local heating breaks the heater, especially in the case of a small metal heater. This difficulty can be avoided by packing sample powder into spaces between the thermocouple wire and the small hole on the heater instead of inserting the sleeves, but this method does not always guarantee the insulation at high pressures (Ito and Takahashi, 1989). Moreover, the basic assembly cannot be used in the melting experiment. This problem was avoided by moving the holes close to the cold end of the heater and inserting the thermocouple axially to the sample, as shown in **Figure 16** (Gasparik, 1989). The thermocouple wires are in contact with the sample holder of the metal (Re) to get the junction. The wires cross the Mo ring and exit the assembly near the apexes of the triangular faces. The wires are insulated from

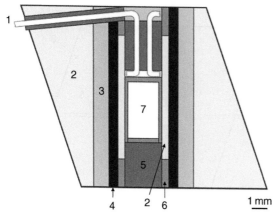

1: Thermocouple wire; 2: Magnesium oxide;
3: Zirconia; 4: Lanthanum chromite;
5: Alumina ceramic; 6: Molybdenum; 7: Sample

Figure 16 Cross section of a heating assembly. The thermocouple is inserted axially to the rhenium sample capsule (see text). Adapted from Gasparik T (1989) Transformation of enstatite–diopside–jadeite pyroxenes to garnet. *Contributions to the Mineralogy and Petrology* 102: 389–505.

the heater, the Mo ring, and the WC anvil with alumina ceramic tubes placed in gutters carved out in one of the triangular surfaces of the octahedron.

The temperature in a cylindrical heater of the small size is strongly inhomogeneous in both the axial and radial directions. Kawashima and Yagi (1988) calculated temperature distribution within a cylindrical heater set in a cubic pressure medium with an edge length of 8 mm using the difference method. Based on the results, they (Kawashima *et al.*, 1990) invented a multistepped graphite heater in which the temperature was almost uniform over 2 mm length in the axial direction. Assessment and improvement of the temperature distribution through the sample were first carried out by Takahashi *et al.* (1982) for the octahedral assembly based on the solubility of enstatite in the diopsidic pyroxene coexisting with the enstatitic pyroxene (Lindsley and Dixon, 1976). Using an assembly similar in size to that of **Figure 15**, they demonstrated that the temperature gradient in a graphite heater tapered by 5°C was reduced to one-third of that in the straight heater at 1500°C. Recently, van Westrenen *et al.* (2003) used the kinetics in $MgAl_2O_4$ spinel growth at the interface between MgO and Al_2O_3 to assess temperature gradients in cylindrical heaters in 18/11 and 8/3 assemblies. They linked the measured thickness of the spinel layer to temperature via an appropriate kinetic equation and showed the lengths of the central uniform temperature

region for both assemblies. It is remarked that, in the small metal heater adopted for experiments at pressures higher than 25 GPa, a steep temperature gradient of 1–2°C μm^{-1} was estimated (Ito and Takahashi, 1989; Ito *et al.*, 2004). A temperature gradient in the $LaCrO_3$ heater is much more moderate compared to the metal and graphite heaters because $LaCrO_3$ is a semiconductor and the relatively low electrical resistance of the hot region lowers the temperature there.

Sample temperature is measured by a thermocouple whose junction is located in or close to the sample (see **Figures 15** and **16**). If the junction were not in the sample, the temperature distribution in the heater should properly be assessed. A difficult experimental problem is that the pressure effect on the emf's of thermocouples is poorly known. Hanneman and Strong (1965) measured the effect for several thermocouples up to 1300°C and 5 GPa, and showed that the absolute temperature correction is proportional to the pressure. They indicated corrections of +40°C for Pt/Pt10% and +10°C for chromel–alumel thermocouples at 5 GPa and 1000°C, respectively. Later, however, these values were substantially reduced by Getting and Kennedy (1970). As to the W/WRe thermocouple, which has recently been most frequently used in the multianvil cell, Li *et al.* (2003) reported that the reading of the W5%Re/W26%Re varied up to 35°C compared with that of the Pt/Pt10% at 15 GPa and at temperatures up to 1800°C. The thermocouple wires are taken out from the apparatus and connected to the reference junction kept at 0°C. Sometimes, however, the wire ends are conveniently connected to the top of the second stage anvils. In this case, the temperature there is the reference of the emf, which should have been measured independently for correction. The anvil top temperature depends on the output of the heater, the size of the second-stage anvil, its thermal insulation, etc., but quickly reaches equilibrium and typically does not exceed 130°C at the sample temperature of 1500°C.

The highest temperature that can be measured with a W/WRe thermocouple with a reliable accuracy is 2300°C. Higher temperatures are estimated by extrapolating a relationship between temperature and applied power obtained at lower temperatures.

8.3.3 Quench Experiment

It is fortunate that most of the high-pressure phases concerned with the Earth's interior can be recovered to the ambient conditions as metastable phases like diamond. Starting material is first pressurized to prescribed pressure and then is heated at the desired temperature by supplying electric power to the heater for certain duration. Then temperature quenching of the sample is performed by shutting off the electric supply. Sample temperature generally goes down to a temperature lower than 100°C from higher than 2000°C within a few seconds because of the small heater. After slow decompression, the sample is taken out from the apparatus. The first-order transformation accompanying reconstruction of the atomic arrangement is kinetically prevented within such a short period of cooling, and the phase stabilized at the experimental conditions is preserved at high pressure and room temperature. We can obtain plenty of information on the phase such as the structure, stability, and physicochemical properties by analyzing the recovered sample in various methods at ambient conditions.

Unquenchable high-pressure phases have been discovered according to the development of high-pressure X-ray diffraction methods. The most typical example is the perovskite type of $CaSiO_3$. Liu and Ringwood (1975) first confirmed the presence of the cubic perovskite-type of $CaSiO_3$ by *in situ* X-ray diffraction on the sample in the DAC at 16 GPa and room temperature after heating it with a laser to about 1500°C. They also observed the retrogressive transformation of the perovskite to glass and ε-$CaSiO_3$ on release of pressure, which was consistent with results of the previous quench experiments (Ringwood and Major, 1971). The retrogressive transformation on pressure release is known for several high-pressure-type oxides; for example, rocksalt-type ZnO (Ito and Matsui, 1974; Inoue, 1975), the fluorite-related types of TiO_2 (Liu, 1978), SnO_2 (Haines and Leger, 1997). The low-pressure phases converted from these compounds after quenching show characteristically broad X-ray diffraction peaks (Ito and Matsui, 1974, 1979). Similarly, recently found post-perovskite phases of $MgSiO_3$ (Murakami *et al.*, 2004; Oganov and Ono, 2004), $MgGeO_3$ (Hirose *et al.*, 2005), and $MnGeO_3$ (Tateno *et al.*, 2006) with the $CaIrO_3$ structure are all unquenchable. The high-pressure phases formed via the second-order transformation such as $CaCl_2$ types of SiO_2 (Tsuchida and Yagi, 1989), GeO_2 and PbO_2 (Haines *et al.*, 1996), and SnO_2 (Haines and Leger, 1997) are also unquenchable.

Any information on unquenchable high-pressure phases can be obtained only by means of *in situ*

measurement. The *in situ* X-ray observation at high pressure and high temperature is the most important methodology in modern high-pressure Earth science.

8.3.4 Crystal Growth at High Pressure and High Temperature

Growth of a single crystal of a high-pressure phase is essential to determine its crystal structure. In addition, a single crystal with enough size makes it possible to carry out many kinds of measurements, for example, transport properties such as atomic diffusion, thermal conductivity and electrical conductivity, creep experiments, elastic resonance spectroscopy, Raman and Brillouin scatterings, and, more recently, X-ray inelastic spectroscopy. Information collected by these measurements provides important clues to understand the constitution and dynamics of the Earth's interior. Therefore, crystal growth of high-pressure phases has been an important role of the MMAs.

According to the classical homogeneous nucleation theory (Turnbull, 1956), a frequency for occurrence of nucleus embryos of a new phase in the mother phase is controlled by the free energy difference ΔG between the new phase and the mother phase at the nucleation site and temperature: for example, both the smaller ΔG and the lower temperature are responsible for lowering the frequency. Growth of crystals, on the other hand, is enhanced by high temperature. Therefore, the favorite conditions for growth of large single crystals would be at relatively high temperature, but should be very close to the phase boundary, which means a small excess of pressure over the phase boundary for the growth of the high-pressure phase. This view was supported through the systematic experiments on the growth of γ-Ni_2SiO_4 spinel in olivine crystals by Hamaya and Akimoto (1982). Sawamoto (1986) succeeded in growing crystals in both the β- and γ-phases of compositions $(Mg,Fe)_2SiO_4$ with sizes larger than 200 μm at temperatures 1700–2200°C and pressures within ±5% of the γ–β phase boundary. Pressure and temperature conditions, at which $MgSiO_3$ garnet (Angel *et al.*, 1989) and ilmenite (Horiuchi *et al.*, 1982) single crystals were synthesized, were also around the phase boundary between the both phases.

For high-pressure phases difficult to grow, appropriate flux could enhance the crystal growth at relatively low temperatures. Sinclair and Ringwood (1978) synthesized stishovite crystals in size to

230 μm by keeping silicic acid with 40% of water at 9 GPa and 700°C for 45 min. Endo *et al.* (1986) obtained stishovite crystals elongated to several hundreds of micrometers at 12 GPa and 1300°C by using Li_2WO_4 as flux. Ito and Weidner (1986) grew $MgSiO_3$ perovskite crystals with sizes to 200 μm by cooling the $MgSiO_3$ melt saturated with H_2O from 1830°C slowly at 27 GPa. Recently, single crystals of perovskite with $MgSiO_3$, and (Mg,Fe,Al) $(Si,Al)O_3$ compositions larger than 300 μm, have been synthesized in NaCl melt by cycling the temperature in a sample chamber (Dobson and Jacobsen, 2004). Quite recently, Anton *et al.* (2007) have succeeded in growing $MgSiO_3$ perovskite crystals with a size exceeding 1 mm by adopting water as a flux in a sophisticated sample capsule. Potential materials as flux for the crystal growth under the high-pressure and high-temperature conditions may be water and some sorts of borates, carbonates, alkali halides, etc.

It is no wonder that a crystal bigger than the sample capsule cannot be grown. Therefore big multianvil presses recently installed (e.g., Frost *et al.*, 2004) must be very advantageous to crystal growth because these presses can compress a larger octahedron of medium to higher pressures.

8.3.5 Electrical Conductivity Measurement

Electrical conductivity within the mantle is explored by examining the electrical current induced by change in the external magnetic field (MT probing) and by analyzing the transient and secular variations of the geomagnetic field originating from the core (GDS probing). Although proposed distributions of electrical conductivity versus depth in the mantle are model dependent, the following characteristics can be noted on the conductivity–depth profile: conductivity increases in the mantle from the order of 10^{-3} S m^{-1} at the top to *c.* 10^0 S m^{-1} at around 800 km depth (e.g., Utada *et al.*, 2003) in a manner reflecting variation of the tectonic setting (e.g., Neal *et al.*, 2000). In the lower mantle, the conductivity is almost constant to 2000 km depth, with values of $3–10$ S m^{-1} (Olsen, 1999) with less lateral heterogeneity compared to the upper mantle.

Almost all mantle minerals are regarded as ionic insulators with large energy gaps, for instance, 6.4 eV (forsterite) and 8 eV (quartz). The mechanism of creating an electrical current is through thermally activated diffusion of the whole ion through the crystal (ionic conduction) or thermally activated

charge transfer between neighboring ions of different valence (hopping conduction). So temperature is the most dominant quantity to control the electrical conductivity. At high pressure, on the other hand, the latter mechanism would become dominant, and the conductivity increases with increasing pressure because the jump distance between sites decreases. The conductivity of the mineral also strongly depends on chemical composition, especially iron content and the Fe^{3+}/Fe^{2+} ratio (and thus on the oxygen fugacity). The introduction of a small amount of water into nominally anhydrous minerals substantially increases the conductivities due to the migration of the proton (e.g., Yoshino *et al.*, 2006).

By measuring the electrical conductivity of mantle minerals over wide pressure and temperature conditions, one can evaluate their activation energy ΔE and activation volume ΔV for conduction depending on the conduction mechanism. These quantities make it possible to extrapolate the measured conductivity to the conditions corresponding to the deep mantle. Comparison of the estimated conductivities with the observed profile leads to an inferred temperature profile if the mineralogy and chemistry there is constrained.

Electrical conductivity at high pressure and temperature has been measured using both the DAC and the Kawai-type apparatus. Although experimental pressure using the DAC reaches 80 GPa (Li and Jeanloz, 1987), temperature control in a laser-heated sample is very difficult with large temperature gradients, and, in addition, experimental temperature is limited to 400°C by adoption of a external heater (e.g., Shankland *et al.*, 1993). Methods for electrical conductivity measurement in the Kawai apparatus have also been developed, in which measurement is carried out up to 25 GPa and at well-defined temperatures to 1700°C under the controlled oxygen fugacity. The thermal state of the lower mantle has been discussed based on electrical conductivities of silicate perovskites (Katsura *et al.*, 1998a; Xu *et al.*, 1998a).

A cross section of the assembly for measuring electrical conductivity in the Kawai apparatus is schematically shown in **Figure 17** (cf. Katsura *et al.*, 1998b). The assembly is for the 8.0/3.0 combination (see Section 8.2.3). The pressure medium is composed of primary MgO. A cylindrical heater is put in a thermal insulator sleeve. The sample is embedded in an Al_2O_3 container which is put between thin Fe disks. The Fe disks are used to keep the oxidation state of the sample below the Fe–FeO buffer, and also to work

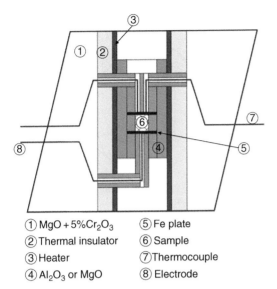

(1) MgO + 5% Cr_2O_3
(2) Thermal insulator
(3) Heater
(4) Al_2O_3 or MgO
(5) Fe plate
(6) Sample
(7) Thermocouple
(8) Electrode

Figure 17 A schematic drawing of the sample assembly of 8/3 cell for electrical conductivity measurement.

as electrodes for the electrical conductivity measurement. The thermocouple is mechanically connected to one Fe disk to measure the sample temperature and also to serve as electrodes for electrical conductivity measurement. The thermocouple is insulated from the heater by Al_2O_3 sleeves. The other Fe disk is connected to a metallic lead for an electrode of electrical conductivity measurement.

The electrical conductivity is measured according to the electric circuit shown in **Figure 18**. The sample

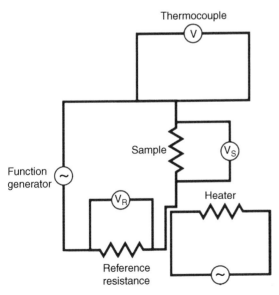

Figure 18 Circuit for electrical conductivity measurement.

is connected to a reference resistance, variable in a range 50–1 $M\Omega$, in a series. A sinusoidal alternating voltage (0.01–0.1 Hz and 0.1–10$V_{\text{peak-to-peak}}$) is applied to this circuit by a function generator. The voltages applied on the sample (V_S) and reference resistance (V_R) are monitored by two digital multimeters. The sinusoidal signal from the function generator is applied to the sample through one side of the thermocouple and the lower guide block. The voltage applied to the sample is measured using the other side of the thermocouple and the lower guide block.

The obtained data points are fitted to the function $V(t) = a + bt + A\sin[\omega(t - \alpha)]$, where t is time and others are fitting parameters, in order to obtain the complex voltages applied on the sample and reference (V_S) and (V_R). The impedance of the sample (Z_S) is obtained from the reference resistance (R_R) and the ratio of the voltages of the sample and reference, namely,

$$Z_S = R_R(V_S/V_R) \qquad [2]$$

Assuming that the impedance of the sample is composed of the resistance of the sample (R_S) and one capacitance (C) in parallel with the sample, the resistance of the sample is obtained from the following equation:

$$1/Z_S = 1/R_S + j\omega C \qquad [3]$$

After decompression and recovery of the assembly, a cross section of the assembly is made along the heater axis, and the thickness and diameter of the sample is measured. The electrical conductivity of the sample is calculated from the resistance and the dimensions of the sample. Xu *et al.* (1998b) also developed a similar assembly to measure electrical conductivity by means of impedance spectroscopy. They set a cylindrical metal shield connected to the ground which is set inside the heater and envelops the sample and the wires. The advantages of the shield are avoiding electrical disturbance from the heater, minimizing the temperature gradient, and reducing leakage current through the pressure medium and lead wires.

Recently, an accessible pressure range for electrical conductivity measurement in the Kawai apparatus has been extended to 35 GPa by adopting SD anvils (see Section 8.4.7) instead of WC anvils (Katsura *et al.*, in preparation).

8.4 *In Situ* X-Ray Observations Using SR

8.4.1 Methods of *In Situ* X-Ray Diffraction Using MAAs

It is desirable to directly observe the sample under the high-pressure and high-temperature conditions. The desire was met in the DAC via the *in situ* optical observation and the X-ray diffraction, because diamond was highly transparent for electromagnetic waves of a broad range of frequency. An X-ray diffraction study in the tetrahedral anvil apparatus was also developed by Barnett and Hall (1964) and Hall *et al.* (1964). The MoKα radiation collimated to *c.* 500 μm was introduced into the tetrahedral pressure medium composed of composites LiF, B, and plastic through the anvil gap, which then was irradiated to the sample packed in NaCl at the center of the tetrahedral pressure medium. The diffracted X-ray was taken out from the opposite anvil gap over the diffraction angles of *c.* ±50°. The diffraction peaks of both the sample and NaCl were recorded on a film or by a counter. The state of the sample such as volume and structure are continuously monitored as a function of pressure which was derived from the volume of NaCl via the EoS for NaCl proposed by Decker (1965). Hall *et al.* (1964) observed *in situ* structural transitions of Cs and the change in electric resistance to 8 GPa simultaneously.

Inoue and Asada (1973) first attached an X-ray diffraction system to the DIA-type apparatus with a pressure medium of a composite of boron and epoxy. They carried out an *in situ* high-pressure and high-temperature X-ray diffraction study up to 13 GPa and 1000°C by setting a paired graphite heater in the pressure medium. Later Inoue (1975) successfully interfaced the energy dispersive system with the DIA apparatus as schematically shown in **Figure 19**. He used a polychromatic X-ray from the W target and the solid-state detector (SSD) for counting the diffracted X-ray. Employing the revised Decker's (1971) NaCl pressure scale, Inoue revealed a shocking fact that pressure values in the quench experiments had substantially been overestimated: that is, the sample pressure fell substantially by heating; 7.5 GPa at room temperature went down to 5 GPa at 1400°C under the fixed press load. Moreover, Inoue demonstrated that the phase boundary for the wurtzite–rocksalt transition in ZnO determined by means of *in situ* X-ray diffraction was dramatically different from that determined by the quench method (Bates *et al.*, 1962).

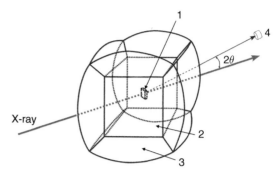

1: Sample; 2: Cubic pressure medium;
3: Gasket; 4: Counter

Figure 19　X-ray diffraction geometry in the energy dispersive method by using the DIA-type apparatus. Polychromatic X-ray is introduced to the sample through the gasket and pressure medium of boron and epoxy composite. The diffracted X-ray is counted by the SSD set at angle 2θ from the direction of the incident X-ray. Adapted from figure 2 of part I in Inoue K (1975) *Development of High Temperature and High Pressure X-Ray Diffraction Apparatus with Energy Dispersive Technique and Its Geophysical Applications.* PhD Thesis, The University of Tokyo.

The energy dispersive method has marked advantages over the angle dispersive method in the X-ray diffraction study using the MAAs. First, information on a wide range of *d*-spacing is collected at once from the intensity/energy profile of the diffracted X-ray collected at the fixed diffraction angle 2θ by SSD. Second, by using a high-energy X-ray (usually 20 to *c.* 120 keV), absorption through the pressure medium and gasket is much reduced compared to usage of the characteristic X-ray such as MoKα radiation. And third, the diffraction area and location in the sample is unmoved due to the fixed diffraction angle. This is specifically important for the high-temperature X-ray diffraction study using the MAA because the sample in a small heater is usually under a steep temperature gradient as mentioned already.

8.4.2 Multianvil System Interfaced with SR

The technical development in the *in situ* X-ray observation recently expanded the versatility of the MAAs. Inoue's (1975) work pointed out that the phase boundaries determined by quench method might contain considerable uncertainty and should be checked by the high *P–T in situ* observation. Initially the rotated anode X-ray source was exclusively employed with the DIA-type apparatus in experiments for both determination of phase boundaries (e.g., Yagi and Akimoto, 1976) and acquisition of compression data on geophysically important minerals (e.g., Sato, 1977). In these works, it took a rather long duration to collect reliable diffraction patterns because the intensity of the X-ray was not sufficiently strong.

In the early 1980s, however, SR X-ray sources became available for high-pressure experiments using the MAAs at Photon Factory (PF), National Laboratory for High Energy Physics, Japan. A new DIA-type apparatus named MAX 80 (Shimomura *et al.*, 1985) was installed at PF. The SR emits an X-ray beam with a high flux, high localizability, low beam divergence, and high flux distribution in a hard X-ray range. These characteristics are quite appropriate to obtain clear diffraction patterns of the sample under compression within a short period. In particular, the high brilliant beam made it possible to study reaction kinetics of high-pressure transformation in silicate (e.g., Rubie *et al.*, 1990) and to measure the viscosity of silicate melt by means of X-ray radiography (Kanzaki *et al.*, 1987).

The great success of MAX 80 promoted the installations of the DIA-type apparatus (SAM 85) interfaced with the SR facility at the National Light Source, Brookhaven National Laboratory, USA (Weidner *et al.*, 1992). The second DIA apparatus (MAX 90) at PF was equipped with SD anvils with a top square edge of 4 mm and the experimental conditions up to 15 GPa and 1500°C was realized (Shimomura *et al.*, 1992). In order to extend *in situ* X-ray observations to higher pressures, the system was transferred into the double-staged system by putting the Kawai cell into the cubic space of the DIA apparatus (see **Figure 7**). An octahedral sample assembly was adopted instead of the cubic one in the single-staged DIA system. Funamori *et al.* (1996) employed sintered diamond cubes with 10 mm edge length for the Kawai cell anvils, and determined the thermoelastic properties of $MgSiO_3$ perovskite up to 30 GPa and 2000 K.

Based on the accomplishments of the double-staged multianvil system at PF, a combination of the DIA press and the Kawai cell (SPEED-1500) was constructed on the beam line BL04B1 at SPring-8, a third-generation SR facility in Hyogo Prefecture in Japan, in 1997 (Utsumi *et al.*, 1998). Recently, a sister apparatus SPEED-Mk. II (Katsura *et al.*, 2004a) was installed on the same beam line in series with SPEED-1500.

X-ray optics adopted for the double-staged multi-anvil system is shown in **Figure 20**. An incident white X-ray beam is introduced to the sample in the octahedral pressure medium through anvil gaps of the first stage (the DIA apparatus) and the second stage (the Kawai cell) anvils. A diffracted X-ray is taken out also between the anvil gaps horizontally at the fixed angle of 2θ and collected by the Ge SSD through the collimator and the receiving slits, which ensure the fixed angle of 2θ. The X-ray that directly passes through the pressure medium and gaskets enters the charge coupled device (CCD) camera to make an X-ray image of the sample, which is a great help to determine the shooting point in the sample assembly. Both SPEED-1500 and SPEED-Mk. II have a compressive force of 15 MN. Advantage of the latter over the former, however, is that the whole apparatus of SPEED-Mk. II can be rotated by $\pm 9°$ around the diffraction center which is acheived by

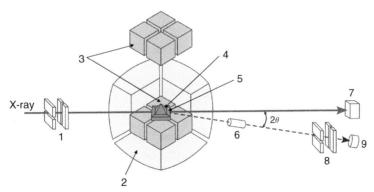

1: Incident slits; 2: 1st-stage anvils; 3: 2nd-stage anvils;
4: Pressure medium; 5: Gasket; 6: Scattering slit (collimator);
7: CCD camera; 8: Receiving slits; 9: Solid-state detector

Figure 20 Schematic drawing of the X-ray optics adopted in double-staged system combining the DIA-type guide block and the Kawai cell (see text).

adjusting the apparatus position precisely. This function is quite effective in obtaining proper diffraction profiles for samples of progressed grain growth or that with the preferred orientation of crystal axes.

8.4.3   Experimental Procedures for *In Situ* X-Ray Diffraction Study

In the energy dispersive diffraction method, the diffracted X-ray is collected by SSD, and analyzed by a multichannel analyzer. In the multichannel analyzer for SPEED-1500 and SPEED-Mk. II, 4096 channels are allotted to energy ranges from 8 to 120 kev with a energy width of 0.027 keV/channel. The interplanar spacing d that causes diffraction in crystal is given by

$$d = ch/2E \sin \theta \qquad [4]$$

,where c, h, E, and θ are the speed of light, the Planck's constant, the energy of X-ray, and the diffraction angle, respectively. The multichannel analyzer is calibrated by detecting the characteristic radiations of several metals such as Cu, Ag, Mo, Ta, Pt, Au, and Pt on the assumption that the energy of each channel is a linear function of the channel number. The 2θ value is calibrated by collecting the diffraction patterns of the standard materials such as Si and MgO, adopting the d-value of each peak calculated using eqn [4].

Figure 21 shows our sample assemblies for *in situ* X-ray observations using SD cubes with an edge length of 14 mm and a truncated corner of 1.5 mm. **Figures 21(a)** and **21(b)** are for electric resistance measurement at room temperature and for examination of phase equilibria at high pressure and temperature, respectively. In both the assemblies, a powdered sample is packed in the central portion of the MgO capsule with a length of 1.5 mm. Both sides of the sample are loaded with fine powdered diamond. In high-temperature experiments, the MgO capsule is put inside a set of cylindrical metal heaters and an $LaCrO_3$ sleeve.

A polychromatic X-ray beam typically collimated to 100 μm vertically and 50 μm horizontally is introduced to the sample and the MgO capsule which serves as the pressure standard, independently. The length of the packed sample is limited to 0.5 mm considering the temperature distribution in the cylindrical heater. The 2θ is typically 6°, and the length of the diffraction area with a parallel-piped shape is about 1.3 mm for the X-ray beam of horizontal width of 50 μm and the collimator with 50 μm diameter. Therefore, in addition to the diffraction from the sample, the diffraction from the diamond is also collected. However, the latter does not seriously interrupt the analysis of the former because diamond has the lowest number of diffraction lines, the lowest thermal expansion, and the lowest

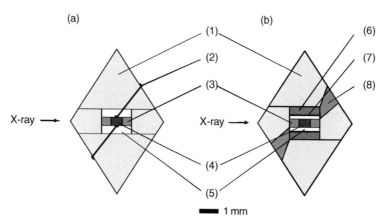

(1): Pressure medium (MgO + 5 wt.% Cr_2O_3); (2): Electrode (Pt); (3): Diamond powder; (4): Sample; (5): Sample capsule (pressure standard) (MgO); (6): Thermal insulation sleeve ($LaCrO_3$); (7): Cylindrical heater (Nichrome); (8): Electrode (sintered TiC).

Figure 21   Cross sections of the octahedral sample assemblies for the *in situ* X-ray observations using SD cubic anvil with 1.5 mm truncation (Ito *et al.*, 2005). (a) and (b) are for room temperature and high temperature, respectively. In (a), electric resistance measurement is conducted using electrodes. Thermocouple in (b) is perpendicular to the paper and in contact with outer surface of the heater.

compressibility. The pressure is determined from the measured unit cell volume of the MgO capsule and the experimental temperature using the EoS of MgO (e.g., Matsui *et al.*, 2000). The cylindrical LaCrO$_3$ heater in which the sample capsule is coaxially loaded can be set perpendicular to the incident X-ray. In this case, parts of the cylindrical LaCrO$_3$ and ZrO$_2$ are removed, and replaced by low atomic number materials such as MgO or diamond to secure the paths of incident and diffracted X-ray (e.g., Ono *et al.*, 2000). The X-ray pass can be secured by using two platy heaters, in which a configuration of the sample and the pressure marker is placed between (e.g., Irifune *et al.*, 1998).

8.4.4 Pressure Determination

One of the primary advantages of the *in situ* X-ray diffraction experiment is that pressure can be monitored continuously in real time. As mentioned in Section 8.2.4, the accuracy of pressure determination is of great importance in experiments concerned with the Earth's interior. For example, global topographies of the 410 km and the 660 km seismic discontinuities are mapped with a resolution of 1–5 km (Flanagan and Shearer, 1998). Pressure increases in the mantle by a ratio of 0.04–0.05 GPa km^{-1}. If one tries to assign the seismic discontinuities to the phase transformations, therefore, the phase boundaries should be determined to at least an accuracy of 0.25 GPa.

8.4.4.1 *Pressure standard materials and the pressure scales*

Pressure determination of the *in situ* X-ray observation using the MAAs is carried out based on the P–V–T EoS of the internal standard material. First the unit cell volume is measured at the prescribed press load and temperature and then it is converted to pressure via the EoS. Therefore the most basic requirement for the pressure standard material is that its EoS is reliably established over wide pressure and temperature ranges. Additional requirements for pressure standard materials, some of which being mutually contradictory, are relatively low absorption for X-rays, simple diffraction profile, low bulk modulus, low yielding strength, chemical inertness, high melting temperature, no phase transition, low grain growth rate, and so on, under high pressure and temperature. Pressure standards which have been most frequently used in the *in situ* experiments for high-pressure Earth sciences are NaCl, MgO, and Au.

NaCl as a pressure standard has been used for 40 years since Hall *et al.* (1964). Decker (1965, 1971) constructed the EoS by combining the lattice energy of the ionic interaction potential and the thermal energy derived from the Debye model and proposed it as a practical pressure scale. Later, Brown (1999) revised the EoS of NaCl to account for more recent pressure–volume data and the shock-wave experiments. Brown's (1999) scale gives slightly lower pressure values than Decker's (1971), for example, by 0.2 GPa at 20 GPa and 1100 K. Advantages of NaCl as a pressure scale are its low bulk modulus (25 GPa under ambient conditions) which defines pressure in high precision, for example, volume determination to within 0.1% precision is converted to a precision of 0.14 GPa at around 20 GPa. Due to its low yielding strength, the deviatoric stress in NaCl is easily released on heating to 1000 K. Nevertheless, its relatively low melting temperature hinders it from usage for experiments at high temperatures due to the pronounced anharmonic and nonharmonic contributions to the EoS and noticeable grain growth. The decisive disadvantage of NaCl as pressure standard is the transformation from B1 to B2 structures, which occurs at 22–20 GPa and at 1600–2000 K (Nishiyama *et al.*, 2003). Therefore NaCl cannot be used in the experiments relevant to the lower mantle at present.

Some of the disadvantages of NaCl can be avoided by using MgO, which possesses a larger bulk modulus of 164 GPa at ambient conditions (Isaak *et al.*, 1989), a higher melting temperature (3100 K at 0 GPa), a relatively low grain growth rate, and no phase transformation is known up to a few hundreds of GPa. The thermal EoS for MgO has been extensively studied based on shock and static compression experiments (e.g., Jamieson *et al.*, 1982; Speziale *et al.*, 2001), on thermodynamic calculation (e.g., Hama and Suito, 1999), and on molecular dynamic simulation (Matsui *et al.*, 2000). Among these four scales, those of Matsui *et al.* (2000) and Speziale *et al.* (2001) are fairly consistent at pressures from 5 to 35 GPa and up to 2000 K with a discrepancy less than 0.5 GPa. Hama and Suito's (1999) and Jamieson *et al.*'s (1982), on the other hand, show large deviation from the former two scales. In particular, Jamieson *et al.*'s (1982) shows characteristically lower deviation up to 1.5 GPa. Matsui *et al.* (2000) incorporated the breathing mode to their molecular dynamic simulation to treat high temperature behavior of MgO, and successfully reproduced the thermal expansivity and the temperature derivatives of elastic constants.

Speziale *et al.* (2001) constructed a Birch–Murnaghan–Debye thermal EoS for MgO by introducing the variable *q* parameter (logarithmic volume derivative of the Grüneisen parameter) to reproduce a wide range of compression data obtained in both static and shock experiments. At present, either Matsui *et al.*'s (2000) or Speziale *et al.*'s (2001) MgO scale may be recommended as standard usage. But it should be noted that the pressures in both scales cross over at around 25 GPa at 2000 K; the pressure determined from Speziale *et al.*'s (2001) is lower than that from Matsui *et al.*'s (2000) at pressures lower than 25 GPa and vice versa at pressures higher than 25 GPa, resulting in a total discrepancy of 0.5 GPa over a pressure range from 20 to 30 GPa. These features of both the scales can be significant if one discusses a slope of the phase boundary. The relatively high bulk modulus of MgO lowers the resolution of pressure determination compared to the NaCl scale. However, pressure is defined with an error less than 0.1 GPa by collecting at least five diffraction lines.

Au has been widely used as a pressure standard material. The bulk modulus of Au at ambient conditions, 171 GPa, is a moderate value as a pressure standard. Due to its high atomic number, however, Au is usually mixed with MgO or silicate sample with a ratio of 1/5 to 1/20 (in weight) when served as the pressure standard. The pressure standard in a composite indicates somewhat erroneous pressure values until the stress homogenization in the composite is achieved by heating. In this respect, a low rigidity of Au, 27.6 GPa, is favorable. The composite is also effective to prevent the grain growth of Au at high temperature. Several EoS's for Au have been constructed for usage as the internal pressure standard; (e.g., Jamieson *et al.*, 1982; Anderson *et al.*, 1989; Shim *et al.*, 2002; Tsuchiya, 2003). The models and derivation procedures for the various equation states for Au were briefly summarized and appraised by Tsuchiya (2003). Among above cited scales, both Jamieson *et al.*'s (1982) and Tsuchiya's (2003) show systematically higher pressures than Shim *et al.*'s (2002), for example, by *c.* 0.8 GPa at 30 GPa and 1500 K, and the discrepancies become larger with increasing pressure and temperature. Anderson, *et al.*'s (1989), on the other hand, indicates lower pressure than Shim *et al.*'s (2002) at high temperature, e.g., by *c.* 0.9 GPa at 30 GPa and 1500 K, whereas it indicates evidently higher values at 300 K. At present, therefore, the pressure scales of Au do not converge, and we may not have any sound reason to select a specified scale.

8.4.4.2 *Comparison of the pressure scales*

Matsui and Nishiyama (2002) compared Matsui *et al.*'s (2000) MgO scale with Anderson *et al.*'s (1989) Au scale based on the simultaneous pressure measurements using both the MgO and Au standards at 1873 K, and concluded that Anderson *et al.*'s (1989) Au scale underestimated pressure by about 1.4 GPa over a pressure range of 20–23 GPa in comparison to Matsui *et al.*'s (2000) MgO scale. Recently, Fei *et al.* (2004a) critically evaluated pressures calculated from different pressure scales of MgO, Au, and Pt based on the X-ray diffraction data collected at simultaneous high pressures and temperatures using multiple internal pressure standards. They demonstrated large discrepancies in pressure determination using different pressure standards or different EoS for the same standard. At temperatures in a range from 1473 to 2173 K and at around 23 GPa, Jamieson *et al.*'s (1982) Au scale indicates 2.5–3.0 GPa higher values than Anderson *et al.*'s (1989) whereas Shim *et al.*'s (2002) scale does almost intermediate values between them, and MgO scale of Jamieson *et al.* signifies 1.5–2.0 GPa lower values that that of Speziale *et al.* (2001) (cf. figure 1 of Fei *et al.*, 2004a). They also showed that the discrepancy between the Au scale of Shim *et al.* (2002) and the MgO scale of Speziale *et al.* (2001) is much smaller, both signifying almost the same values at 2173 K.

Under these circumstances, Fei *et al.* (2004a) modified the Au scale of Shim *et al.* (2002) so as to obtain an Au scale mutually consistent with the MgO scale of Speziale *et al.* (2001). This study makes it possible to compare high pressure/temperature data obtained using MgO and Au pressure scales. As the practical pressure scales, however, the MgO scale should be recommended from the point of view that its EoS is free from the free electron contribution to the thermal pressure and thereby is the least controversial. Two MgO scales derived from different models, Matsui *et al.*'s (2000) and Speziale *et al.*'s (2001), signify fairly close pressures to each other over the wide pressure and temperature range as mentioned in the last section.

In order to construct an absolute pressure scale, simultaneous determination of volume and elasticity at high pressure and high temperature should be achieved. Along this line, Zha *et al.* (2000) proposed a primary pressure scale of MgO at 300 K by integrating single-crystal velocity data from Brillouin scattering measurements and density data from polycrystalline X-ray diffraction. It should be also noted that the state-of-art techniques have been developed for studies of elastic properties of polycrystalline and single-crystal materials using simultaneous ultrasonic and X-ray diffraction techniques at high pressures and temperatures in the DIA-type apparatus (e.g., Li *et al.*, 2004). As in this system, not only the elastic constants but also their pressure and temperature derivatives are determined together with the volume; precise EoS's for the pressure standard materials are constructed. However, the accessible *P–T* range is so far limited up to 13 GPa and 1300 K. The limitation has been overcome by adoption of the double-staged MAA system by Higo *et al* (2006).

Shock-wave Hugoniot data are usually adopted as the fixed points at high pressure and high temperature to constrain the EoS's for standard materials. As most of such data were collected more than 30 years ago, however, reinvestigation using modern techniques may be required to improve the pressure scales.

8.4.5 Application to Phase Equilibrium Studies

Although the pressure scales are not perfect at present and contain some uncertainties, especially at high temperature, it is certain that the *in situ* X-ray diffraction method is more reliable than the quench method in determining pressure values of the MAA. Therefore, important phase boundaries in high-pressure Earth sciences defined by the quench experiments have thoroughly been reinvestigated by means of *in situ* X-ray diffraction as summarized in **Table 2** (see a review by Katsura (2007) for detail). For the transformations that occur at pressures less than 20 GPa, both the quench and the *in situ* X-ray methods have offered generally consistent boundaries. In particular, extrapolation of the phase boundaries for the α–β and β–γ transformations in Mg_2SiO_4 determined *in situ* to higher temperature suggests their locations are very close to those determined by the quench method as shown in **Figure 22**. These situations may indicate the carefulness of the quench experiments and the reliability of the NaCl pressure scale.

However, surprisingly variable results have been reported on the phase boundary for the dissociation of Mg_2SiO_4 (γ) into $MgSiO_3$ (perovskite) and MgO (periclase) as illustrated in **Figure 23**. Location of the boundary has been the most controversial issue relevant to interpretation of the 660 km

Table 2 Phase boundaries, $P(GPa) = aT(K) + b$, for geophysically important reactions determined by means of *in situ* X-ray diffraction and the quench experiment in the MAA

Compound	Reaction[a]	a (GPa K^{-1})	b (GPa)	T-range (K)	P-marker[b]	Reference
SiO2	Coe/St	0.0012	7.7	770–1370	NaCl(B)	Yagi and Akimoto (1976)
		0.002	6.2	137–1800	NaCl(B)	Zhang *et al.* (1996)
		0.002	6.8	870–1470	Q	Suito (1977)
Fe2SiO4	β/γ	0.0019	2.9	1073–1473	NaCl(B)	Yagi *et al.* (1987)
		0.0044	0	973–1473	Q	Akimoto *et al.* (1965)
Mg2SiO4	α/β	0.0032	8.9	1013–1673	NaCl(B)	Morishima *et al.* (1994)
		0.004	7.8	1600–1900	NaCl(B)	Katsura *et al.* (2004b)
		0.002	11.1	1473–1873	Q	Katsura and Ito (1989)
	β/γ	0.0069	8.4	927–1327	NaCl(B)	Suzuki *et al.* (2000)
		0.004	13.1	1473–1873	Q	Katsura and Ito (1989)
	γ/Pv + Pe	−0.003	26.7	1673–2073	Au(A)	Irifume *et al.* (1998)
		−0.0004 ~ −0.0017	22.7 ~ 25.5	1500–2000	Au(A)	Katsura *et al.* (2003)
		−0.0008	23.2	1673–2173	Au(A), Q	Fei *et al.* (2004b)
		−0.0013	25.5	1673–2173	MgO(S), Q	Fei *et al.* (2004b)
		−0.0028	28.3	1373–1873	Q	Ito and Takahashi (1989)
MgSiO3	Il/Pv	−0.0023	26.6	1073–1573	NaCl(B)	Kato *et al.* (1995)
		−0.0035	27.3	1300–1600	Au(A)	Ono *et al.* (2001)
		−0.0029	28.4	1300–1600	Au(J)	Ono *et al.* (2001)
		−0.0027	25.09	1673–2273	Au(A)	Hirose *et al.* (2001)
		−0.0025	26.8	1273–1873	Q	Ito and Takahashi (1989)

[a]Coe: coesite, St:stishovite, α: olivine, β: modified spinel, γ: spinel, Pv: perovskite, Pe: periclase, Il: ilmenite.
[b]NaCl(B): Brown's (1999) NaCl sale, Au(A): Anderson *et al.*'s (1989) Au scale, Au(J): Jamieson *et al.*'s (1982) Au scale, MgO(S): Spaziale *et al.*'s (2001) MgO scale, Q: quench experiment.

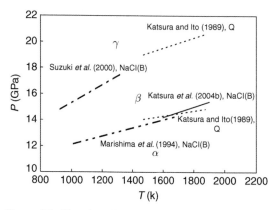

Figure 22 The phase boundaries for the α–β and β–γ transformations in Mg_2SiO_4 determined *in situ* X-ray diffraction method (Morishima *et al.*, 1994; Suzuki *et al.*, 2000; Katsura *et al.*, 2004b) together with those determined by quench method (Katsura and Ito, 1989). All the phase boundaries are in general consistency.

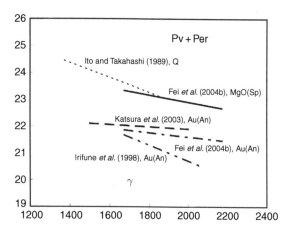

Figure 23 Phase boundaries for the dissociation of Mg_2SiO_4 (γ) into $MgSiO_3$ (perovskite) and MgO (periclase) determined by *in situ* X-ray diffraction method (Irifune *et al.*, 1998; Katsura *et al.*, 2004b; Fei *et al.*, 2004b) together with that determined by quench method (Ito and Takahashi, 1989). Phase boundaries determined based on the Anderson *et al.*'s (1989) Au scale are located at *c.* 2 GPa lower than (Fei *et al.*, 2004b) determined based on Speziale *et al.*'s (2001) MgO scale.

discontinuity. Phase boundaries determined based on the Anderson *et al.*'s (1989) Au scale are located *c.* 2 GPa lower than that determined based on Speziale *et al.*'s (2001) MgO scale (Fei *et al.*, 2004b), in harmony with discussion by Fei *et al.* (2004a). However, there are marked differences in slope for the phase boundaries between those by Katsura *et al.* (2003) and Fei *et al.* (2004b) and that by Irifune *et al.* (1998), although pressure determination in these three works were

carried out based on the Anderson *et al.*'s (1989) Au scale. Fei *et al.*'s (2004b) boundary based on Speziale *et al.*'s (2001) MgO scale is closely located to that of Ito and Takahashi (1989) determined by the quench experiment and supports the conventional interpretation that the 660 km discontinuity would be caused by the dissociation of ringwoodite.

In addition to the mutual inconsistency among different pressure scales, there are at least two issues to make the determination of the phase boundary ambiguous, the uncertainty of temperature determination and the reaction kinetics of phase change.

As pressure scale is a function of temperature as well as the volume of the standard material, precise temperature determination is of essential importance to define pressure with high accuracy. For example, a temperature variation of 100 K causes that of 0.5 and 0.7 GPa at *c.* 1800 K in the MgO and the Au scales, respectively. In the experiments at pressures higher than 20 GPa, the heater in the sample assemblage is generally small (see **Figure 18**), and a steep temperature gradient is inevitably produced in the assemblage. Therefore temperatures of the sample and the standard material differ from that indicated by the thermocouple. It should be also noted that the diffraction area unavoidably covers a region of varying temperature due to the steep temperature gradient through the sample. To minimize this effect, the sample length is usually limited up to 0.5 mm (cf. **Figure 18**). Another difficult problem is that the pressure effect on the emf of the thermocouple is not quantitatively known at high P–T conditions (Section 8.3.2).

In the course of progress of the *in situ* X-ray diffraction study, it has become clear that the reaction kinetics plays an important role in determining the phase relations. It is frequently observed that the starting material of the low-pressure phase rapidly transforms into the intermediate high-pressure phase at relatively low temperature in the course of the first heating (e.g., Yagi and Akimoto, 1976; Irifune *et al.*, 1998; Katsura *et al.*, 2003) and the new phase is so refractory that substantial excess pressure is required to promote further transformation (Ono *et al.*, 2001; Katsura *et al.*, 2003). These features are interpreted as follows. The finely powdered starting material is further pulverized by compression at room temperature as seen from the associated broadening of the diffraction peaks. The fractured grain boundaries and defects formed inside the grains make the starting material very reactive, providing nucleation sites for the new phase. The resultant new phase, on the other hand, is in a well-defined crystalline state with sharp diffraction peaks. It is difficult to increase the reactivity of the new high-pressure phase once formed. This situation hinders observing the reverse reaction that is required to define the precise phase boundary.

8.4.6 Viscosity Measurement by X-Ray Radiography

There are two types of liquids in the 'solid Earth', silicate melt (magma) in the crust and the mantle and molten iron alloy forming the outer core. The viscosity of both the liquids at high pressure is the most crucial property for understanding various magmatic processes and dynamic structure of the outer core. The characteristics of X-rays emitted from SR, high flux and low beam divergence, are appropriate to measure viscosity at high pressure by the *in situ* falling sphere method in the MAA. Kanzaki *et al.* (1987) first measured viscosities of $NaAlSi_3O_8$ (albeit) and $Na_2O. 2SiO_2$ melts up to 3 GPa using the DIA-type apparatus at PF. The shadow images of the falling spheres of Pt and/or Mo (100–300 μm radius) in the melt were clearly observed on the TV, and the viscosity was derived from the falling velocity via Stokes' law.

Now the *in situ* falling sphere viscometry is mostly carried out in the double-staged system (the Kawai cell in the DIA apparatus) to extend experimental pressure. A whole geometry of the X-ray radiography system is shown in **Figure 24**. An X-ray is introduced to the octahedral cell assembly through the gaps between the first- and second-stage anvils, and an X-ray once past the cell illuminates a fluorescent YAG crystal. The visible light generated by the YAG crystal is transferred into the CCD camera where an image of the cell is captured. The system recently installed at beam line BL04B1 of SPring-8 can store 125 frames per second with a resolution of 4 μm per pixel. As the anvil gap of the Kawai cell become narrow at high pressure, typically 1mm at 10 GPa, fast acquisition of the images is essential to obtain accurate viscosity values, especially for a less viscous liquid such as an Fe-alloy.

A typical falling sphere cell assembly (Terasaki *et al.*, 2001) is reproduced in **Figure 25**. A small sphere (100–150 μm diameter) of Pt, W, or Au as a viscosity marker is put at the top of the sample. It should be noted that all the parts, except for the falling spheres and the sample, are made of matters of high X-ray transparency. After compression to the desired pressure, the sample is heated gradually to high temperature below the solidus, and then heated rapidly to the desired temperature above the liquidus. The cell assembly shown in **Figure 25** was designed for viscosity measurement of Fe–FeS melts, in which two spheres are set in the sample capsule of h-BN, the first one at the top of the Fe–FeS powder and the second in the thin silicate layer set above the sample. The silicate possessing a melting temperature slightly higher than the sample is chosen

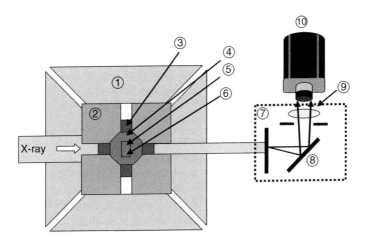

① 1st-stage anvil ② 2nd-stage anvil ③ Gasket ④ Pressure medium ⑤ Falling sphere

⑥ Sample ⑦ Fluorescent YAG ⑧ Mirror ⑨ Lens ⑩ CCD camera

Figure 24 A whole geometry of the X-ray radiography system (see text).

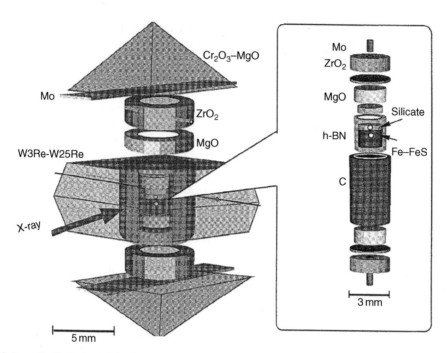

Figure 25 Schematic illustration of the falling sphere assembly (see Terasaki *et al.*, 2001 for detail).

so as to keep the second sphere in the silicate layer when the sample begins to melt. Therefore viscosity data are collected at two different temperatures, corresponding to meting temperatures of the sample and the silicate, in one experimental run (Terasaki *et al.*, 2001). Similar methods have been adopted in viscosity measurements of molten peridotite which has a relatively wide melting interval between solidus and liquidus and is sometimes difficult to define the 'settling velocity' to be transferred to the viscosity coefficient due to the coexistence of solid phases in the melt (Liebske *et al.*, 2005).

Recently, a wide range of viscosities have been measured, for example, from orders of 10^2 Pa s (for alkaline silicates) to $10^{-2}-10^{-3}$ Pa s (for Fe-alloy), over pressures up to 16 GPa (Terasaki *et al.*, 2006). In order to extend experimental pressure, however, usage of diamond/SiC composite as the anvil material is essential. This, on the other hand, will expand the versatility of the X-ray radiography by combining simultaneous X-ray diffraction.

8.4.7 Pressure Generation Using Sintered Diamond Anvil

As mentioned in Section 8.2.2, the capability of high-pressure generation using the Kawai-type apparatus has been extended by adopting SD cubes instead of WC ones. *In situ* X-ray observation using the Kawai cell equipped with SD anvils has widely been carried out to determine the phase equilibrium and EoS of the high-pressure materials up to higher than 50 GPa and temperatures to 2200 K (e.g., Ono *et al.*, 2000; Irifune *et al.*, 2002; Kubo *et al.* 2003; Ito *et al.*, 2005; Tange, 2006). As there has been no fixed point to calibrate pressure higher than 33 GPa (Zr point), pressure calibration in the experiments at higher pressures inevitably relies on the *in situ* X-ray diffraction using the pressure standard material. Results of recent four pressure calibration runs carried out at 300 K are summarized in **Figure 26** (Ito *et al.*, 2005). SD cubes with 14 mm edge length and 1.5 mm truncation were employed with an octahedral $MgO + 5\%$ Cr_2O_3 pressure medium of 5.0 mm edge. The sample assembly was similar to those shown in **Figure 20** and pressure was determined based on Jamieson *et al.*'s (1982) MgO scale. It is remarkable that pressure is generated in a repeatable accuracy of about 2% up to *c.* 60 GPa over the four runs, the maximum attainable pressure reaching 63 GPa. Quite recently, the maximum pressure has been extended to higher than 72 GPa based on Anderson *et al.*'s Au scale (Ito *et al.*, 2006, unpublished data).

Pressure calibration of 1.5 mm TEL based on the MgO scale at room temperature

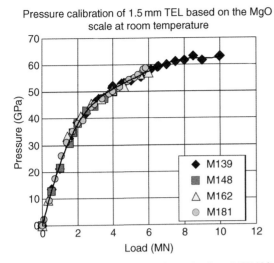

Figure 26 Results of pressure determination at 300 K for the Kawai cell equipped with SD cubes with 14 mm edge length and 1.5 mm truncation (Ito *et al.*, 2005). An octahedral pressure medium of MgO + 5%CrO₃ with 5.0 mm edge length was employed. Pressure determination was based on the Jamieson *et al.*'s (1982) MgO scale. Note a repeatable accuracy of about 2% up to *c.* 60 GPa and the maximum attained pressure of 63 GPa.

8.5 New Applications and Future Perspectives

Two newly developed applications of the Kawai-type apparatus are briefly surveyed here. The first is the determination of rheological properties of mantle minerals at high pressure and temperature which has been developed at CHiPR Stony Brook, USA (e.g., Chen *et al.*, 2004). The stress and strain of the sample under nonhydrostatic conditions in the Kawai cell are measured *in situ* using synchrotron X-ray. Schematic sample assembly is shown in **Figure 27**. Cylindrical sample is placed between

the end plugs of hard corundum, and the sample length is measured from the radiographic images of the gold foils attached at the upper and lower ends of the sample. By using a classic image-processing algorithm (Li *et al.*, 2003), a precision of the strain measurement is increased to 10^{-4} which makes a strain rate accurate to $10^{-7}-10^{-6}$ depending on the time interval between two measurements.

The stress is measured by simultaneously collecting energy dispersive X-ray diffraction patterns of the sample in two perpendicular diffraction planes according to the analytical method of lattice strain $\varepsilon(hkl)$ developed by Singh (1993) and Singh *et al.* (1998). In order to gain accessibility for diffracted X-rays, six of the eight WC cubes are replaced by sintered cBN as shown in **Figure 28**. The Kawai cell is compressed in a set of wedges of six split cylinders. Conical notches are made in the wedges to ensure the path for diffracted X-rays. The whole system is named Tcup. Deformation experiments are carried out up to 10 GPa and higher than 1000°C.

Another interesting new application of the Kawai apparatus is a combination with the acoustic emission (AE) technique performed by Dobson *et al.* (2004) to study the physical processes involved in seismogenesis. Their experimental setup is reproduced in **Figure 29**. AE transducers are mounted on truncations at the back of the furnace anvils. Acoustic waveguides (polycrystalline Al_2O_3 or Fe) are inserted between sample ends and the top truncations along the furnace axis to couple the sample to the furnace anvils. They carefully optimized the following factors essential to data acquisition: choice of waveguide material and appropriate length so that the sample is not crushed or faulted, electrical insulation of the AE transducers from the furnace anvils, and elimination of unwanted noise from sources outside the sample. As a proof of concept, Dobson *et al.*

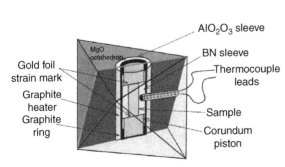

Figure 27 Cell assembly for high-pressure deformation experiments in the Kawai cell (Chen *et al.*, 2004).

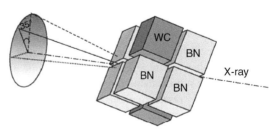

Figure 28 Diffraction geometry in Tcup module (see text) (Chen *et al.*, 2004). Solid line indicates the access of diffracted X-ray if all eight cubes are made of WC, and broken and dotted lines indicate the diffraction in vertical and horizontal planes penetrating through cBN side cubes, respectively.

(a) (b)

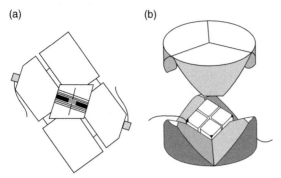

Figure 29 Schematic drawing of setup for AE measurement in the Kawai cell (Dobson *et al.*, 2004). (a) Section of cell and anvil set showing acoustic waveguide (black) connecting the sample to the anvils with transducers mounted on the back. (b) Anvil set in the first-stage wedges.

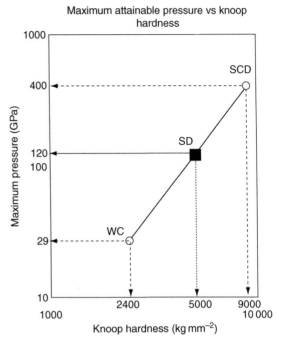

Figure 30 The log–log plot of the maximum attainable pressure by using WC and SCD vs Knoop hardness (Sung and Sung, 1996). The linear relationship indicates that generation of 120 GPa is possible by using SD anvils.

(2004) showed results from AE studies on antigorite dehydration to 8 GPa and the α–β transformation in San Carlos olivine at 16 GPa. Association of this method with *in situ* stress measurement using SR mentioned above will make it possible to determine the exact P–T–σ regime of brittle failure under extreme conditions (Dobson *et al.*, 2004).

Future development of the MAAs will typically be in two directions: the generation of moderate pressures to 30 GPa over a large volume to *c.* 1 cm³ (e.g., Frost *et al.*, 2004) and the generation of higher pressures to over 100 GPa using sintered diamond anvils (e.g., Ito *et al.*, 2005). The first is required to synthesize large single crystals of materials composing the deep interior of the Earth. As already mentioned (Section 8.3.4), the crystals can be used to various kinds of measurements required to understanding the state of the mantle. Large sample volume, in turn, makes it possible to carry out these measurements under conditions deeper in the mantle. It is a great achievement to be able to determine qualitatively the states of the mantle and core materials under conditions up to the core–mantle boundary by extending accessible pressure.

It should be emphasized that *in situ* X-ray observation using SR has become an indispensable tool in the whole high-pressure research. This technique not only offers the most reliable pressure values but also has developed research fields such as reaction kinetics of phase transformation, viscosity, and deformation measurements by means of X-ray radiography.

As mentioned in Section 8.2.2, usage of harder materials for an anvil is essential to extend the maximum attainable pressure in the MAA. Recent improvement of WC has made it possible to generate 40 GPa in the Kawai-type apparatus using the new WC (Katsura *et al.* 2005, unpublished data). However, the most promising anvil material at present is sintered diamond (SD). In **Figure 30**, the maximum pressures attained so far are plotted against Knoop hardness (Sung and Sung, 1996) for WC (in the Kawai type), and single-crystal diamond (SCD in DAC). According to the general trend, the Knoop hardness of SD suggests that pressures up to 120 GPa can be generated by adopting SD. In this connection, Irifune *et al.* (2004) recently synthesized new polycrystalline diamond pellets with grain sizes of 10–20 nm which is binderless and transparent. It is remarkable that Knoop hardness of the pellet is up to 140 GPa which is comparable or even higher than those of natural SCDs. Although the size of the pellet is 1.5 mm diameter and 0.5 mm thickness at present, generation up to multi-megabars may be possible in near future using this material as the anvil. The performance of diamond/SiC composite and sintered cBN is important for *in situ* X-ray observation because cubic anvil of these materials can be used as the windows for the X-ray radiography, the angle dispersive diffraction study, and *in situ* stress measurement.

A serious problem in high-pressure experiments using SD cubes is the high frequency of blow-out which sometimes causes damage of anvils and frequently loss of the sample. One of the main reasons of the blow-out is nonuniform compression of the cubic cell due to uneven deformation of the solid guide block parts. The stick-slipping could be also a major factor of the blow-out. All these shortcomings of the DAI-type apparatus for squeezing the Kawai cell could in principle be overcome by driving the six anvils with independently controlled rams. Although all such cubic apparatus developed so far have not been operated with the required precision, it will be possible to adjust each anvil load to within 10 kN and to synchronize the anvil advances within a few micrometers level by adopting recent high-tech servomechanisms. Construction of such double-staged apparatus with enough force is a task for future consideration.

References

Akimoto S and Fujisawa H (1968) Olivine-spinel solid solution equilibria in the system $Mg_2SiO_4-Fe_2SiO_4$ at $800°C$. *Earth and Plantary Science Letters* 1: 237–240.

Akimoto S, Matsui Y, and Syono Y (1976) High-pressure crystal chemistry of orthosilicates and the formation of the mantle transition zone. In: Strens RGJ (ed.) *Physics and Chemistry of Minerals and Rocks*, pp. 327–363. London: Wiley Interscience.

Anderson OL, Isaak DG, and Yamamoto S (1989) Anharmonicity and the equation of state for gold. *Journal of Applied Physics* 65: 1534–1543.

Angel RJ, Finger IW, Hazen RM, *et al.* (1989) Structure and twinning of single-crystal $MgSiO_3$ garnet synthesized at 17 GPa and $1800°C$. *American Mineralogist* 74: 509–512.

Anton S, Fukui H, Matsuzaki T, *et al.* (2007) Growth of large (1 mm) $MgSiO_3$ perovskite single crystles: A thermal gradient method at ultrahigh pressure. *American Mineralogist* (in press).

Bates CH, White WB, and Roy R (1962) New high-pressure polymorph of zinc oxide. *Science* 137: 993.

Bean VE, Akimoto S, Bell PM, *et al.* (1986) Another step toward an international practical pressure scale. *Physica* 139: 52–54.

Bernal JD (1936) Discussion. *Observatory* 59: 268.

Barnett JD and Hall HT (1964) High pressure–High temperature X-ray diffraction apparatus. *Review of Scientific Instruments* 35: 175–182.

Birch F (1952) Elasticity and constitution of the earth's interior. *Journal of Geophysical Research* 57: 227–286.

Block S (1978) Round-robin study of the pressure phase transition in ZnS. *Acta Crystallographica* (Supplement 316) 34:.

Brown JM (1999) The NaCl pressure standard. *Journal of Applied Physics* 86: 5801–5808.

Bullen KE (1937) Note on the density and pressure inside the Earth. *Transitions of the Royal Society of New Zealand* 67: 122–124.

Bundy FP (1962) General principles of high pressure apparatus design. In: Wentorf RH (ed.) *Modern Very High Pressure Techniques*, pp. 1–24. London: Butterworths.

Chen J, Li L, Weidner D, and Vaughn M (2004) Deformation experiments using synchrotron X-rays: *In situ* stress and strain measurements at high pressure and temperature. *Physics of Earth and Planetary Interreriors* 143-144: 347–356.

Decker DL (1965) Equation of state of NaCl and its use as a pressure gauge in high-pressure research. *Journal Applied Physics* 36: 157–161.

Decker DL (1971) High-pressure equation of state for NaCl, KCl and CsCl. *Journal of Applied Physics* 42: 32393244.

Dobson DP and Jacobsen SD (2004) The flux growth of magnesium silicate perovskite single crystals. *American Mineralogist* 89: 807–811.

Dobson DP, Meredith PG, and Boon SA (2004) Detection and analysis of microseismicity in multi anvil experiments. *Physics of Earth and Planetary Interiors* 143-144: 337–346.

Drickamer HG and Balchan AS (1962) High pressure optical and electrical measurements. In: Wentorf RH (ed.) *Modern Very High Pressure Techniques*, pp. 25–50. London: Butterworths.

Endo S, Akai T, Akahama Y, *et al.* (1986) High temperature X-ray study of single crystal stishovite synthesized with Li_2WO_4. *Physics and Chemistry of Minerals* 13: 146–151.

Fei Y, Li J, Hirose K, *et al.* (2004a) A critical evaluation of pressure scales at high temperatures by *in situ* diffraction measurements. *Physics of Earth and Planetary Interiors* 143-144: 515–526.

Fei Y, Von Orman J, van Westrenen W, *et al.* (2004b) Experimentally determined postspinel transformation boundary in Mg2SiO4 using MgO as an internal standard and its geophysical implications. *Journal of Geophysical Research* 109, doi:10.1029/2003JB002562.

Flanagan MP and Shearer PM (1998) Global mapping of topography on thransition zone velocity discontinuities by stacking SS precursors. *Journal of Geophysical Research* 103: 2673–2692.

Frost DJ, Poe BT, Tronnes RG, Liebske C, Duba A, and Rubie DC (2004) A new large-volume multianvil system. *Physics of Earth and Planetary Interiors* 143-144: 507–514.

Funamori N, Yagi T, Utsumi W, Kondo T, Uchida Y, and Funamori M (1996) Thermoelastic properties of $MgSiO_3$ perovskite determined by *in situ* X-ray observations uo tp 30 GPa and 2000 K. *Journal of Geophysical Reserach* 101: 8257–8269.

Gasparik T (1989) Transformation of enstatite–diopside–jadeite pyroxenes to garnet. *Contributions to the Mineralogy and Petrology* 102: 389–505.

Getting IC and Kennedy GC (1970) Effect of pressure on the emf of chromel–alumel anf platinum-platinum10% rhodium thermocouples. *Journal of Applied Physics* 41: 4552–4562.

Gutenberg B (1948) On the layer of relatively low wave velocity at a depth of about 80 kilometers. *Bulletin of the Seisemological Society of America* 38: 121–148.

Gutenberg B (1951) *PKKP, P'P',* and the Earth's core. *Transitions of the American Geophysical Union* 32: 373–390.

Hall HT (1958) Some high-pressure, high-temperature apparatus design considerations: Equipment for use at 1000,000 atm and $3000°C$. *Review of Scientific Instruments* 29: 267–275.

Hall HT (1960) Ultra-high pressure, high-temperature apparatus: The 'Belt'. *Review of Scientific Instruments* 31: 125–131.

Hall HT, Merrill L, and Barnett JD (1964) High pressure polymorphism in cesium. *Science* 146: 1297–1299.

Haines J and Leger JM (1997) X-ray diiraction study of the phase transitions and structural evolution of tin dioxide at high pressure: Relationships between structural types and implications for other rutile-type dioxides. *Physical Review* B 55: 11144–11154.

Haines J, Leger JM, and Schulte O (1996) The high-pressure phase transition sequence from the rutile-type trough to the cotunite-type structure in PbO_2. *Journal of Physics: Condensed Matter* 8: 1631–1646.

Hama J and Suito K (1999) Thermoelastic properties of periclase and magesiowüstite under high pressure and high temperature. *Physics of Earth and Plantary Intereriors* 114: 165–679.

Hamaya N and Akimoto S (1982) Experimental investigation on the mechanism of olivine spinel transformation: Growth of single crystal from single crystal olivine in Ni_2SiO_4. In: Akimoto S and Manghnani MH (eds.) *Advances in Earth and Planetary Sciences, vol. 12*, pp. 373–389. Dordrecht: D. Reidel.

Hanneman RE and Strong HM (1965) Pressure dependence of the emf of thermocouples to 1300°C and 50 kbar. *Journal ofApplied Physics* 36: 523–528.

Higo Y, Inoue T, Kono Y, and Irifune T (2006) Elastic wave velocity measurements of high pressure phase of mantle. *Prog. Abst. 47th High Press. Conf. Jpn.* 229 (in Japanese).

Hirose K, Kawamira K, Ohishi Y, Tateno S, and Sata N (2005) Stability and equation of state of $MgGeO_3$ post-perovskite phase. *American Mineralogist* 90: 262–265.

Hirose K, Komabayashi T, Murakami M, and Funakoshi K (2001) *In situ* measurements of the majorite–akimotoite–perovskite phase transition boundaries in $MgSiO_3$. *Geophysical Research Letters* 28: 4351–4354.

Horiuchi H, Hirano M, Ito E, and Matsui Y (1982) $MgSiO_3$ (ilmenite-type): Single crystal X-ray diffraction study. *American Mineralogist* 67: 788–793.

Inoue K (1975) *Development of High Temperature and High Pressure X-Ray Diffraction Apparatus with Energy Dispersive Technique and Its Geophysical Applications*. PhD Thesis, The University of Tokyo.

Inoue K and Asada T (1973) Cubic anvil X-ray diffraction press up to 100 kbar and 1000°C. *Japanese Journal of Applied Physics* 12: 1786–1793.

Irifune T, Kurio A, Sakamoto S, Inoue T, Sumiya H, and Funakoshi H (2004) Formation of pure polycrystalline diamond by direct conversion of graphite at high pressure and high temperature. *Physics of Earth and Planetary Interiorers* 143-144: 593–600.

Irifune T, Naka H, Sanehira T, Inoue T, and Funakoshi K (2002) *In situ* X-ray observations of phase transition in $MgAl_2O_4$ spinel to 40 GPa using multianvil apparatus with sintered diamond anvils. *Physics and Chemistry of Minerals* 29: 645–654.

Irifune T, Nishiyama N, Kuroda K, et al. (1998) The postspinel phase boundary in Mg_2SiO_4 determined by *in situ* X-ray diffraction. *Science* 279: 1698–1700.

Isaak DG, Anderson OL, and Goto T (1989) Measured elastic moduli of single-crystal MgO up to 1800 K. Physics and Chemistry of Minerals 32: 721–725; 16: 703–704.

Ito E, Katsura T, Aizawa Y, et al. (2005) High-pressure generation in the Kawai-type apparatus equipped with sintered diamond anvils: Application to the wurtzite–rocksalt transformation in GaN. In: Chen J, Wang Y, Duffy TS, Shen G, and Dobrhinetskaya LF (eds.) *Advances in High-Pressure Technology for Geophysical Applications*, pp. 451–460. Amsterdam: Elsevier B. V.

Ito E, Kubo A, Katsura T, and Walter MJ (2004) Melting experiments of mantle materials under lower mantle conditions with implications for magma ocean differentiation. *Physics of Earth and Planetary Interiors* 143-144: 397–406.

Ito E and Matsui Y (1974) High-pressure synthesis of $ZnSiO_3$ ilmenite. *Physics of Earth and Planetary Interiors* 9: 344–352.

Ito E and Matsui Y (1979) High-pressure transformation in silicates, germinates, and titanates with ABO_3 stoichiometry. *Physics and Chemistry of Minerals* 4: 265–273.

Ito E and Presnall DC (1998) Report on the first international pressure calibration workshop. *The Review of High Pressure Science and Technology* 7: 151–153.

Ito E and Takahashi E (1989) Post-spinel transformations in the system Mg_2SiO_4–Fe_2SiO_4 and some geophysical implications. *Journal of Geophysical Research* 94: 10637–10646.

Ito E and Weidner DJ (1986) Crystal growth of $MgSiO_3$ perovskite. *Geophysical Research Letters* 13: 464–466.

Jamieson JC, Fritz JN, and Manghnani MH (1982) Pressure measurement at high temperature in X-ray diffraction studies: Gold as a primary standard. In: Akimoto S, and Manghnani MH (eds.) *High-Pressure Research in Geophysics.*, pp. 27–48. Tokyo: Center For Academic Publications.

Jeffreys H (1939) Times of P and S, and SKS, and the velocities of P and S. *Monthly Notices of the Royal Astronomical Society. Geophysical Supplement* 4: 498–533.

Kanzaki M, Kurita K, Fujii T, Kato T, Shimomura O, and Akimoto S (1987) A new technique to measure the viscosity and density of silicate melts at high pressure. In: Manghnani MH and Syono Y (eds.) *Geophysics Monograph 39: High-Pressure Research in Mineral Physics*, pp. 195–200. Washington, DC: AGU.

Kato T, Ohtani E, Morishima H, et al. (1995) *In situ* X-ray observation of high-pressure phase transitions of $MgSiO3$ and thermal expansion of $MgSiO_3$ perovskite at 15 GPa by double-stage multi-anvil system. *Journal of Geophysical Research* 100: 20475–20481.

Katsura T (2007) Phase relation studies of mantle minerals by in situ X-ray diffraction using a multi-anvil apparatus. In: ohtani E (ed.) *Advances in High-Pressure Mineralogy: Geological Society of America Special Paper 421*, doi:10.1130/2007.2421(11) (in press), American Mineralogical Society.

Katsura T, Funakoshi K, Kubo A, et al. (2004a) A large-volume high P–T apparatus for in situ X-ray observation 'SPEED-mkII'. *Physica of Earth and Planetary Interiors* 143-144: 497–506.

Katsura T and Ito E (1989) The system Mg_2SiO_4–Fe_2SiO_4 at high pressure and temperatures: Precise determination of stability of olivine, modified spinel. *Journal of Geophysical Research* 94: 15663–15670.

Katsura T, Sato K, and Ito E (1998a) Electrical conductivity of silicate perovskite at lower-mantle conditions. *Nature* 395: 493–495.

Katsura T, Sato K, and Ito E (1998b) Electrical conductivity measurements of minerals at high pressures and temperatures. *The Review of High Pressure Science and Technology* 7: 18–21.

Katsura T, Yamada H, Nishikawa O, et al. (2004b) Olivine–wadsleyite transformation in the system $(Mg,Fe)_2SiO_4$. *Journal of Geophysical Research* 109, doi:10,029/2003JB002438 B02209.

Katsura T, Yamada H, Shinmei T, et al. (2003) Post-spinel transition in Mg_2SiO_4 determined by high P–T in situ X-ray diffractometry. *Physics of Earth and Planetary Interiors* 136: 11–24.

Kawai N (1966) A static high pressure apparatus with tapered multi-piston formed a sphere I. *Proceedings of the Japan Academy* 42: 285–288.

Kawai N and Endo S (1970) The generation of ultrahigh hydrostatic pressure by a split sphere apparatus. *Review of Scientific Instruments* 41: 425–428.

Kawai N, Togaya M, and Onodera A (1973) A new device for high pressure vessels. *Proceedings of the Japan Academy* 49: 623–626.

Kawashima Y and Yagi T (1988) Temperature distribution in a cylindrical furnace for high-pressure use. *Review of Scientific Instruments* 59: 1186–1188.

Kawashima Y, Tsuchida Y, Utsumi W, and Yagi T (1990) A cylindrical furnace with homogeneous temperature distribution for use in a cubic high-pressure press. *Review of Scientific Instruments* 61: 830–833.

Kondo T, Sawamoto H, Yoneda A, Kato A, Matsumuro A, and Yagi T (1993) Ultrahigh-pressure and high temperature generation by use of the MA8 system with sintered diamond anvils. *High Temperatures-High Pressure* 25: 105–112.

Kubo A and Akaogi M (2000) Post-garnet transitions in the system $Mg_4Si_4O_{12}$–$Mg_3Al_2Si_3O_{12}$ up to 28 GPa: Phase relations of garnet, ilemenite and perovskite. *Physics Earth Planetary Interior* 125: 85–102.

Kubo A, Ito E, Katsura T, et al. (2003) In situ X-ray observation of iron using Kawai-type apparatus equipped with sintered diamond: Absence of β-phase up to 44 GPa and 2100 K. *Geophysical Research Letters* 30, doi:10.1029/2002GL016394.

Li B, Kung J, and Liebermann RC (2004) Modern techniques in measuring eleasticity of Earth materials at high pressure and high temperature using ultrasonic interferometry in conjunction with synchrotron X-radiation in multi-anvil apparatus. *Physics of Earth and Planetary Interiors* 143-144: 559–574.

Li J, Hadidiacos C, Mao H-K, Fei Y, and Hemley RJ (2003) Behavior of thermocouples under high pressure in a multi-anvil apparatus. *High Pressure Research* 23: 389–401.

Li Li, Raterron P, Weidner D, and Chen J (2003) Olivine flow mechanisms at 8 GPa. *Physics of Earth and Planetary Interiors* 138: 113–129.

Li X and Jeanloz R (1987) Electrical conductivity of $(Mg,Fe)SiO3$ perovskite a perovskite- dominated assembly at lower mantle conditions. *Geophysical Research Letters* 14: 1075–1078.

Liebske C, Sckmickerler B, Terasaki H, et al. (2005) Viscosity of peridotite liquid up to 13 GPa: Implications for magma ocean viscosities. *Earth and Planetary Science Letters* 240: 589–604.

Lindsley DH and Dixon SA (1976) Diopside–enstatite equi8libria at 850° to 1400°C, 5–35 kb. *American Journal of Science* 276: 1285–12301.

Liu L-G (1978) A fluorite isotype of SnO_2 and a new modification of TiO_2: Implications for the Earth's lower mantle. *Science* 199: 422–425.

Liu L-G and Ringwood AE (1975) Synthesis of a perovskite polymorph of $CaSiO_3$. *Earth and Planetary Science Letters* 28: 209–211.

Mao H-K and Bell PM (1971) High pressure decomposition of spinel (Fe_2SiO_4). *Carnegie Institute Year Book* 70: 176–177.

Matsui M and Nishiyama N (2002) Comparison between the Au and MgO pressure calibration atandards at high temperature. *Geophysical Research Letters* 29, doi:10.1029/2001GL014161.

Matsui M, Parker SC, and Leslie M (2000) The MD simulation of the equation of state of MgO: Applications as a pressure standard at high temperature and high pressure. *American Mineralogist* 85: 312–316.

Ming LC, Kim Y-H, Uchida T, Wang Y, and Rivers M (2006) In situ X-ray diffraction study of phase transitions of $FeTiO_3$ at high pressures and temperatures using a large-volume press and synchrotron radiation. *American Mineralogist* 91: 121–126.

Morishima H, Kato T, Suto M, et al. (1994) The phase boundary between α- and β-Mg_2SiO_4 determined by in situ X-ray observation. *Science* 265: 1202–1203.

Murakami M, Hirose K, Kawamura K, Sata N, and Ohishi Y (2004) Post-perovskite phase transition in $MgSiO_3$. *Science* 304: 855–858.

Neal S, Mackie RL, Larsen JC, and Schultz A (2000) Variations in the electrical conductivity of the upper mantle beneath North America and the Pacific Ocean. *Journal of Geophysical Resaerch* 105: 8229–8242.

Nishiyama N, Katsura T, Funakoshi K, et al. (2003) Determination of the phase boundary between the B1 and B2 phases in NaCl by in situ X-ray diffraction. *Physical Review B* 68: 134109.

Oganov AR and Ono S (2004) Theoretical and experimental evidence for a post-perovskite phase of MgSiO3 in Earth's D'' layer. *Nature* 430: 445–458.

Ohtaka O, Shimono M, Ohnishi N, et al. (2004) HIP production of a diamond/SiC composite and application to high-pressure generation. *Physics of Earth and Planetary Interiors* 1434-144: 587–591.

Ohtani E, Okada Y, Kagawa N, and Nagata Y (1987) Development of a new guide-block system and high pressure and high temperature generation. *Prog. Abst. 28th High Press. Conf. Jpn.* 222-3 (in Japanese).

Ohtani E, Kagawa K, Shimomura O, et al. (1989) High-pressure generation by a multiple anvil system with sintered diamond anvils. *Review of Scientific Instruments* 60: 922–925.

Olsen N (1999) Long-period (30 days-1 year) electromagnetic sounding and the electrical conductivity of the lower mantle beneath Europe. *Geophysical Journal International* 138: 179–187.

Ono S, Ito E, Katsura T, et al. (2000) Thermoelastic properties of the high-pressure phase of SnO_2 determined by in situ X-ray observation up to 30 GPa and 1400 K. *Physics and Chemistry of Minerals* 27: 618–622.

Ono S, Katsura T, Ito E, et al. (2001) In situ observation of ilmenite-perovskite phase transition in $MgSiO_3$ using synchrotron radiation. *Geophysical Research Letters* 28: 835–838.

Osugi J, Shimizu K, Inoue K, and Yasunami K (1964) A compact cubic anvil high pressure apparatus. *Review of Physical Chemistry of Japan* 34: 1–6.

Pal'yanov YN, Sokol AG, Borzdov YM, Khokhryakov AF, and Sobolev NV (2002) Diamond formation through carbonate-silicate interaction. *American Mineralogist* 87: 1009–1013.

Presnall DC and Gasparik T (1990) Melting of enstatite $(MgSiO_3)$ from10 to 16.5 GPa and the forsterite (Mg_2SiO_4)-majorite $(MgSiO_3)$ eutectic at 16.5 GPa Implications for the origin of the mantle. *Journal of Geophysical Research* 95: 15771–15777.

Ringwood AE (1958) The constitution of the mantle IL: Further data on the olivine-spinel transition. *Geochimica et Cosmochimica Acta* 15: 18–29.

Ringwood AE (1970) Phase transformations and the constitution of the mantle. *Physics of Earth and Planetary Interiors* 3: 108–155.

Ringwood AE, Major A, Willis P, and Hibberson WO (1989) Diamond composite tools. In: *Annual Report Research School of Earth Science, ANU*, pp. 38–43.

Ringwood AE and Major A (1970) The system Mg_2SiO_4–Fe_2SiO_4 at high pressures and temperatures. *Physics of Earth and Planetary Interiors* 3: 89–108.

Ringwood AE and Major A (1971) Synthesis of majorite and other high pressure garnets and perovskites. *Physics of Earth and Planetary Interiors* 11: 411–418.

Rubie DC, Tsuchida Y, Yagi T, et al. (1990) An in situ X-ray diffraction study of the kinetics of the Ni_2SiO_4 olivine–spinel transformation. *Journal of Geophysical Research* 95: 15829-15844 .

Sato Y (1977) Pressure–volume relationship of stishovite under hydrostatic compression. *Earth and Planetary Science Letters* 34: 307–312.

Sawamoto H (1986) Single crystal growth of the modifes spinel (β) and spinel (γ) phases of $(Mg, Fe)2SiO_4$ and some geophysical implications. *Physics and Chemistry of Mineralals* 13: 1–10.

Shankland TJ, Peyronneau J, and Poirier JP (1993) Electrical conductivity of the Earth's lower mantle. *Nature* 366: 453–455.

Shimada M (2000) *Mechanical Behavior of Rocks Under High Pressure Conditions*, pp. 1–177. Rotterdam: A. A. Balkema.

Shim S-H, Duffy TS, and Takemura K (2002) Equation of state of gold and its application to the phase boundary near 660 km depth in Earth's mantle. *Earth and Planetary Science Letters* 203: 729–739.

Shimomura O, Utsumi W, Taniguchi T, Kikegawa T, and Nagashima T (1992) A nw high pressure and high temperature apparatus with sintered diamond anvils for synchrotron radiation use. In: Syono Y and Manghnani MH (eds.) *Geophysics Monograph 67: High-Pressure Research: Earth and Planetary Sciences*, pp. 3–11. Washington, DC: AGU.

Shimomura O, Yamaoka S, Yagi T, *et al.* (1985) Multi-anvil type high pressure apparatus for synchrotron radiation, solid state physics under pressure. In: Minomura S (ed.) *Recent Advance with Anvil Devices*, pp. 351–356. Dortrecht: Reidel.

Shimono M, Wada T, Kume S, and Ohtaka O (2003) HIP production of diamond compact. In: Vincenzini P (ed.) CIMTEC 2002, *Proceedings of the 10th International Ceramics Congress Advances in Sciences and Technology* 32, Part C. Techno Srl, pp. 525–532.

Sinclair W and Ringwood AE (1978) Single crystal analysis of the structure of stishovite. *Nature* 272: 714–715.

Singh AK (1993) The lattice strain in a specimen (cubic system) compressed nonhydrostatically in an opposed anvil device. *Journal of Applied Physics* 73: 4278–4286.

Singh AK, Balasingh C, Mao HK, Hemley RJ, and Shu J (1998) Analysis of lattice strain measured under nonhydrostatic pressure. *Journal of Applied Physics* 83: 7567–7575.

Speziale S, Zha C, Duffy TS, Hemley RJ, and Mao HK (2001) Quasi-hydrostatic compression of magnesium oxide to 52 GPa: Implications for pressure–volume–temperature equation of state. *Journal of Geophysical Research* 106: 515–528.

Stishov SM and Popova SV (1962) A new modification of silica. *Geokhimiya* 10: 837–839 (English translation, *Geochemistry*, no.10, 923–926).

Suito K (1972) Phase transformations of pure Mg_2SiO_4 into a spinel structure under high pressures and temperatures. *Journal of Physics of the Earth* 20: 225–243.

Suito K (1986) Disproportionation of pyrophyllite into new phases at high pressure and temperature. *Physica* 139 &140B: 146–150.

Sung C-M and Sung M (1996) Carbon nitrite and other speculative super hard materials. *Materials Chemistry and Physics* 43: 1–18.

Suzuki A, Ohtani E, Morishima H, *et al.* (2000) In situ determination of the phase boundary between wadsleyite and ringwoodite in Mg_2SiO_4. *Geophysical Research Letters* 27: 803–805.

Takahashi E, Yamada H, and Ito E (1982) An ultrahigh pressure assembly to 100 kbar and 1500°C with minimum temperature uncertainty. *Geophysical Research Letters* 9: 805–807.

Tange Y (2006) The system $MgO–FeO–SiO_2$ up to 50 GPa and 2000°C: *An Application of Newly Developed Techniques Using Sintered Diamond Anvils in Kawai-Type High-Pressure Apparatus*. PhD Thesis, Tokyo Institute of Technology.

Tateno S, Hirose K, Sata N, and Ohishi Y (2006) High-pressure behavior of $MnGeO_3$ and $CdGeO_3$ perovskites and the post-perovskite phase transition. *Physics and Chemistry of Minerals* 32: 721–725.

Terasaki H, Kato T, Urakawa S, *et al.* (2001) The effect of temperature, pressure, and sulfur content on viscosity of the Fe–FeS melt. *Earth and Planetary Science Letters* 190: 93–101.

Terasaki H, Suzuki A, Ohtani E, Nishida K, Sakamaki T, and Funakoshi K (2006) Effect of pressure on the viscosity of Fe–S and Fe–C liquids up to 16 GPa. *Geophysical Research Letters* 33 doi:10.1029/2006GL027147.

Tsuchida Y and Yagi T (1989) A new, post-stishovite high-pressure polymorph of silica. *Nature* 340: 217–220.

Tsuchiya T (2003) First-principles prediction of the P–V–T equation of state of gold and the 660-km discontinuity in Earth's mantle. *Journal of Geophysical Research* 108, doi:10.1029/2003JB002446.

Turnbull D (1956) Phase changes. In: Seitz F and Turnbull D (eds.) *Solid State Physics, vol. 3*, pp. 225–306. New York: Academic Press.

Utada H, Koyama T, Shimizu H, and Chave AD (2003) Ab semi-global reference model for electrical conductivity in the mid-mantle beneath the north Pacific region. *Geophysical Research Letters* 30: 2003.

Utsumi W, Funakoshi K, Urakawa S, *et al.* (1998) Spring-8 beamlines for high pressure science with multi-anvil apparatus. *The Review of High Pressure Science and Technology* 7: 1484–1486.

van Westrenen W, van Orman JA, Watson H, Fei Y, and Watson EB (2003) Assessment of temperature gradients in multianvil assemblies using spinel layer growth kinetics. *Geochemistry Geophysics Geosystems* 1036, doi:10.1029/2002GC000474.

von Platen B (1962) A multiple piston, high pressure, high temperature apparatus. In: Wentorf RH (ed.) *Modern very High Pressure Techniques*, pp. 118–136. London: Butterworths.

Walker D, Carpenter MA, and Hitch CM (1990) Some simplifications to multianvil devices for high pressure experiments. *American Mineralogist* 75: 1020–1028.

Weidner DJ, Vaughan MT, Ko J, *et al.* (1992) Characterization of stress, pressure, and temperature in SAM85, A diatype high pressure apparatus. In: Syono Y and Manghnani MH (eds.) *Geophysics Monograph 67: High-Pressure Research: Earth and Planetary Sciences*, pp. 13–17. Tokyo/Washington: AGU.

Xu Y, McCammon C, and Poe BT (1998a) The effect of almina on the electrical conductivity of silicate perovskite. *Science* 282: 922–924.

Xu Y, Poe BT, and Rubie DC (1998b) In-situ electrical conductivity measurements up to 20 GPa in the muti-anvil apparatus. *The Review of High Pressure Science and Technology* 7: 1526–1528.

Yagi T, Akaogi M, Shimomura O, Suzuki T, and Akimoto S (1987) In situ observation of the olivine-spinel transformation in Fe_2SiO_4 using synchrotron radiation. *Journal of Geophysical Research* 92: 6207–6213.

Yagi T and Akimoto S (1976) Direct determination of cosite–stishovite transition by in-situ X-ray measurements. *Tectonophysics* 35: 259–270.

Yoneda A, Yamamoto S, Kato M, Sawamoto H, and Kumazawa M (1984) The use of composite material gaskets to improve pressure generation in multiple anvil devices. *High Temperature–High Pressure* 16: 637–656.

Yoshino T, Matsuzaki T, Yamashita S, and Katsura T (2006) Hydrous olivine unable to account for conductivity anomaly at the top of the asthenoshpere. *Nature* 443: 973–976.

Zhang JZ, Li B, Utsumi W, and Liebermann RC (1966) In situ X-ray observations of the coesite–stishovite transition: Revised phase boundary and kinetics. *Physics and Chemistry of Minerals* 23: 1–10.

Zha C-S, Mao H-k, and Hemley RJ (2000) Elasticity of MgO and a primary pressure scale to 55 GPa. *Proceedings of National Academy of Sciences* 97: 13495–13499.

Relevant Website

http://www.rocklandresearch.com – Rockland Research Corporation.

9 Theory and Practice – Diamond-Anvil Cells and Probes for High *P–T* Mineral Physics Studies

H.-K. Mao, Carnegie Institution of Washington, Washington, DC, USA

W. L. Mao, Los Alamos National Laboratory, Los Alamos, NM, USA

9.1	Diamond-Anvil Cell as a Window to the Earth's Interior	231
9.1.1	Deep Earth Geophysics and Geochemistry	232
9.1.2	Comprehensive High *P–T* Mineral Properties	232
9.1.3	Multiple Radiation Probes	234
9.2	Generation and Characterization of High Pressures	234
9.2.1	Anvils and Axial Windows	234
9.2.2	Gasket and Side Window	236
9.2.3	Samples and Pressure Media	237
9.2.4	DAC Body	238
9.2.5	Compression Mechanism	239
9.2.6	Pressure Calibration	240
9.3	Temperature	241
9.3.1	Resistive Heating	241
9.3.2	Laser Heating	242
9.3.3	Cryogenic	243
9.3.4	Temperature Measurement	244
9.4	Optical Probes	245
9.4.1	Optical Absorption	246
9.4.2	IR Spectroscopy	247
9.4.3	Raman Spectroscopy	247
9.4.4	Brillouin Spectroscopy and Impulsive Stimulated Light Scattering Spectroscopy	248
9.4.5	Fluorescence Spectroscopy	249
9.5	X-Ray Probes	250
9.5.1	Axial XRD	250
9.5.2	Radial XRD	251
9.5.3	Single-Crystal XRD	253
9.5.4	X-Ray Absorption Spectroscopy (XAS)	253
9.5.5	X-Ray Emission Spectroscopy (XES)	254
9.5.6	Inelastic X-Ray Scattering – Near-Edge Spectroscopy	256
9.5.7	Resonance X-Ray Inelastic Spectroscopy	257
9.5.8	Nuclear Resonant X-Ray Spectroscopy	257
9.5.9	Nonresonant Phonon Inelastic X-Ray Scattering	259
References		260

9.1 Diamond-Anvil Cell as a Window to the Earth's Interior

With the rapid emergence of diamond-anvil cell (DAC) research as a major dimension in physical sciences, the literature is growing at an explosive rate. A comprehensive discussion of any one of the sections could easily fill a whole book. Therefore, our strategy is to introduce the essential concept for each subject, lay out the key background, then jump to the latest activities on technical developments, mineral physical applications, current problems, and future prospects. Ample references are provided for readers who seek a deeper understanding of these topics.

9.1.1 Deep Earth Geophysics and Geochemistry

By far the major fraction of minerals in the solid Earth is hidden at great depth under high pressures (P) and temperatures (T). These minerals dictate the formation, evolution, present state, and destination of the planet. Recent geophysical and geochemical studies of the Earth's deep interior present us with a rich array of large-scale processes and phenomena that are not fully understood. These range from the fate of deeply subducting slabs, the origin of plumes, the nature of the core–mantle boundary layer, the differentiation of elements to form the present day crust, mantle, and core, the distribution of trace elements, and the uptake and recycling of volatiles over the course of the Earth's history. Resolving these questions requires a detailed understanding of the chemistry of the relevant materials at high $P–T$. Under these extreme conditions, the physical and chemical behavior of elements can be profoundly altered, causing new and unforeseen reactions to occur and giving rise to structural, elastic, electronic, and magnetic transitions not observed in mineral systems in the near-surface environment. This problem precludes attempts to understand the properties of materials at high $P–T$ using our knowledge of minerals at ordinary conditions.

DAC provides the enabling tool for studying Earth materials at extreme conditions. Following its invention in 1959 (Weir *et al.*, 1959), decades of development have extended the $P–T$ range of the DAC so that it now covers conditions of the entire region spanned by the geotherms of Earth and other terrestrial planets, and significant portions of the Jovian planetary interiors (**Figure 1**). In addition, the excellent transparency of single-crystal diamond windows to a wide range of electromagnetic radiation has allowed the development of numerous analytical probes for comprehensive, *in-situ* characterization of high $P–T$ mineral properties.

9.1.2 Comprehensive High *P–T* Mineral Properties

Knowledge of the key high $P–T$ mineral properties in the following nonexhaustive list is required for understanding deep-Earth geophysical and geochemical phenomena: composition; phase relation; melting curve; crystallographic structure; amorphous or liquid structure; crystallographic site occupancy, order–disorder, and defects; element partitioning; oxidation state; hydration/dehydration; density (ρ); aggregate bulk modulus (K); shear modulus (G); thermal expansivity (α); thermodynamic parameters – heat capacity at constant pressure or volume (C_P, C_V), Debye temperature (Θ_D), Grüneisen parameter (γ), entropy (S), energy (E), etc.; single-crystal elasticity (c_{ij} or s_{ij}); aggregate compressional wave velocity (V_P) and shear wave velocity (V_S); single-crystal V_P and V_S as a function of crystallographic orientation; phonon density of state (DOS); phonon dispersion; preferred orientation under stress; shear strength and hardness; viscosity; diffusivity; creep law and deformation mechanism; thermal conductivity; radiative heat transfer; optical absorption; index of refraction; electric conductivity; dielectric parameter; and magnetism.

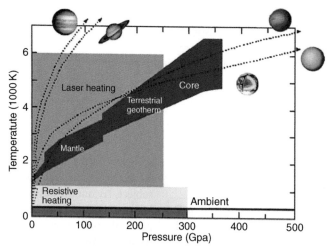

Figure 1 *P–T* range of DAC techniques in comparison to conditions of planetary interiors.

Owing to the often Herculean efforts in improving and developing experimental capabilities all of these properties can now be investigated by DAC probes. Although researchers normally focus on at most a few properties as part of a well-defined study, they still need to maintain a comprehensive view of the entire suite of interrelated properties. Each geophysical problem may require knowledge of a host of properties, and each property may be applicable to a host of problems. Similarly on the experimental side, each technique may provide information on a host of properties, and each property may be determined by a host of complementary techniques. Finally, the integrated high-pressure research impacts on multidisciplinary science and is shown in **Figure 2**.

The following example illustrates the integrated, interactive process that is at the heart of mineral physics, and which shapes the development of experimental techniques for studying key materials properties that apply to Earth problems. The Earth's core plays a central role in the evolution and dynamic processes within the planet. However, answers to some of the most fundamental properties of core remain elusive (e.g., its temperature, chemical composition, crystalline phases, elasticity, and magnetism). As the major constituents of the core, iron and its alloys hold the key to answering these problems, and understanding their high *P–T* behavior has been a focus of mineral physics since the seminal 1952 work of Francis Birch (Birch, 1952). In 1964 Taro Takahashi and William A. Bassett used

X-ray diffraction (XRD) in the DAC to determine the high-pressure crystal structure of iron, and discovered the hexagonal close packed (h.c.p.) polymorph (Takahashi and Bassett, 1964). In the process they also helped initiate what has become the extremely fruitful application of XRD in a DAC for mineral physics studies. Pioneering investigations of the iron melting curve with laser-heated DACs (Boehler, 1993; Saxena *et al.*, 1993; Williams *et al.*, 1987) and the high *P–T* iron phase diagram with simultaneous laser-heating and synchrotron X-ray diffraction (Andrault *et al.*, 1997; Saxena *et al.*, 1995; Yoo *et al.*, 1995) have generated a great deal of excitement, as well as controversy (Anderson, 1997). This in turn has become a main driving force for advancement and improvement of these techniques. Indeed, the 'mission to the Earth's core' (Stevenson, 2003) has united observational, theoretical, and experimental geophysics (Hemley and Mao, 2001a; Scandolo and Jeanloz, 2003), and enriched each discipline through interactions and feedback. Geophysical observations uncovered surprising inner-core properties such as seismic anisotropy, super-rotation, and magnetism (Glatzmaier and Roberts, 1996; Niu and Wen, 2001; Romanowicz *et al.*, 1996; Song and Helmberger, 1998; Song and Richards, 1996; Su *et al.*, 1996). Geodynamicists proposed theories to interpret these observations (Buffett, 2000; Buffett and Wenk, 2001; Karato, 1999; Olson and Aurnou, 1999). *Ab-initio* theoretical calculations (Alfé *et al.*, 1999, 2000; Belonoshko *et al.*, 2003; Laio *et al.*, 2000;

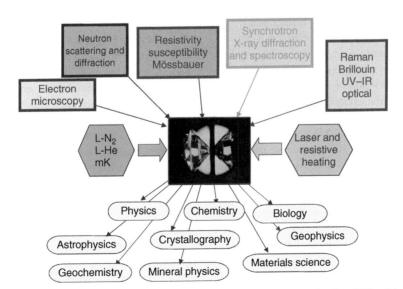

Figure 2   Integration of various analytical techniques for probing high-pressure samples in a DAC subjected to simultaneous high and low temperature. The results impact on multidisciplinary scientific research.

Steinle-Neumann *et al.*, 2001; Stixrude and Cohen, 1995; Vocadlo *et al.*, 2003) were applied to predict the melting, phase stability, elastic anisotropy, and magnetism of iron beyond experimental capabilities. Novel DAC techniques have been developed to overcome previous limitations and provide direct, experimental answers on the properties of high *P–T* iron and iron alloys (Dubrovinsky *et al.*, 2000; Fiquet *et al.*, 2001; Lin *et al.*, 2002; Mao *et al.*, 1998b; Mao *et al.*, 2001; Merkel *et al.*, 2000; Wenk *et al.*, 2000).

9.1.3   Multiple Radiation Probes

Unique advantages of the DAC as a static high-pressure apparatus include the maximum pressure attainable, as well as pressure accuracy, and most importantly the multitude of *in-situ* analytical probes that can be used to study the DAC sample environment. Diamond is transparent to electromagnetic radiation from microwaves, the far infrared (IR), the visible (VIS), and the ultraviolet (UV) up to the diamond bandgap at 5 eV. Diamond is opaque between 5 eV and soft X-rays up to 5 keV, and is then transparent again to X-rays and γ-rays above 5 keV. Diamond is transparent to neutrons but opaque to other particles which require vacuum, such as electrons, ions, and protons. Nonpenetrating radiation between 5 eV and 5 keV which requires vacuum and is therefore incompatible with the high-pressure environment is restricted to *ex situ* studies of samples quenched from high *P–T* environment, or have to be replaced with equivalent probes, such as inelastic scattering.

When monochromatic optical (from IR to UV) or X-ray photons with frequency ν_L, is scattered by the sample, most of the scattered photons keep the original frequency (elastic scattering), but some show peaks at frequencies shifted (inelastic scattering) by the addition or subtraction of ΔE which is the energy difference between the excited and ground states of a specific vibrational or electronic transition of the sample, that is, $\nu = \nu_L - \Delta E$ (Stokes) or $\nu = \nu_L + \Delta E$ (anti-Stokes). Elastic scattering is used in X-ray diffraction and optical imaging, reflection, and absorption spectroscopies. Inelastic scattering includes optical Raman and Brillouin spectroscopies and various phonon and electron X-ray spectroscopies. Together these types of interactions between incident photons and the sample provide a broad range of tools for materials characterization.

Optical, X-ray, and neutron radiation provide complementary probes. For instance, inelastic scattering of light (Brillouin spectroscopy), X-rays, and neutrons can all be used to study phonon dynamics; each technique has its unique merits and limitations. Optical DAC studies can be conducted with laboratory-scaled light sources, lasers, and spectrometers; state-of-the-art X-ray or neutron studies generally require synchrotron or neutron sources at large user facilities. Both optical and neutron radiations are low in energy ($\sim$eV) and have high-energy resolution; X-rays are high in energy (>5 keV) and have low-energy resolution. X-rays and neutrons have short wavelengths comparable to unit cell dimensions and thus access a large region of reciprocal space ($q = 4\pi\sin\theta/\lambda$, where θ is the scattering angle, and λ is the wavelength) to the Brillouin zone edge and beyond; optical radiation has much longer wavelengths, and is thus limited to a very small range near $q = 0$ at the Brillouin zone center. Optical radiation requires clear, colorless windows, and can easily be scattered, refracted, or blocked by interfaces and minor amounts of dark impurities; X-rays can penetrate light element (low Z number) anvils and gaskets regardless of interfaces and additional materials; neutrons have great penetration power through most materials. Optical and X-ray beams can be focused down to micrometer-sized spots for probing minute samples at ultrahigh pressures; neutrons lag behind in microbeam development and are limited to >100 μm-sized samples.

9.2   Generation and Characterization of High Pressures

9.2.1   Anvils and Axial Windows

Anvil devices (**Figure 3**) are used for reaching pressures of the mantle transition zone (>13 GPa) and higher. The anvil is a rigid body consisting of a small pressure-bearing tip on one end that expands to a large base at the other end. The force applied to the low-pressure base (table facet) of opposing anvils is transmitted directed to the small tip (culet facet) where the pressure intensification is inversely proportional to the culet/table area ratio, because pressure is defined as force per unit area. The maximum achievable pressures in an anvil device depend upon the anvil material and the anvil geometry. Colorless, defect-free gemstones are ideal anvil materials owing to their exceptional strength and window qualities. Maximum pressures of 16.7, 25.8, 52, and 300–550 GPa have been reported for cubic zirconia (Xu *et al.*, 1996a), sapphire (Xu *et al.*, 1996b), moissanite (**Figure 3**) (Xu and Mao, 2000), and diamond

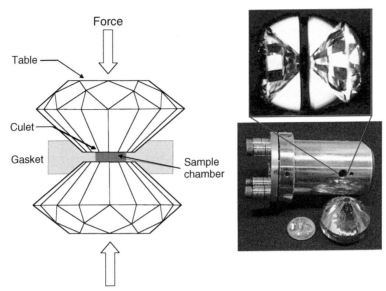

Figure 3   The principle of DAC. Bottom right: A large moissanite anvil is compared with the DAC.

anvils (Xu *et al.*, 1986), respectively in an opposing anvil geometry where the anvils can also act as axial windows (i.e., parallel to the compression axis).

If the anvils were truly rigid bodies, they would be capable of sample compression up to the fracture limit of the anvil. In practice however, all anvils, including diamond, deform elastically and develop a cup (**Figure 4**) resembling the geometry of a 'toroidal cell' (Khvostantsev *et al.*, 1977). The cupping increases with pressure; while this has the desirable effect of providing a larger space for the sample and a

flatter pressure gradient, that also has an adverse effect in that eventually the rims of the two anvils touch, and the pressure ceases to rise with further increases of force. Despite this effect, flat-culet anvils have reached megabar pressures (100 GPa) in 1976 (Mao and Bell, 1976) and a maximum reported pressure of 140 GPa (Yagi and Akimoto, 1982). In order to extend the pressure limit, bevels or convex shapes are added to avoid the touching rim. Beveled anvils with the optimum center flat: outer circle ratio of 1:3–1:10, and a bevel angle of 8.5° can reach

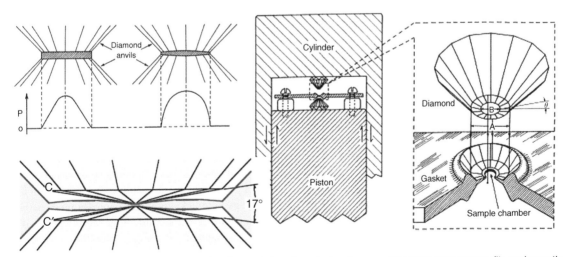

Figure 4   Top left, two flat culets compressing the gasket and producing the smoothly rising pressure profile underneath. Top second left, culets cupping under high pressures and producing sharply rising pressure gradient at rim. Bottom left, culets with 17° bevel angle flattened elastically when central pressure reached 300 GPa (Hemley *et al.*, 1997). Center, diamonds aligned on piston and cylinder which guide the compression on the gasket. Far right, details of the beveled diamond and indented gasket.

200–300 GPa before the bevel flattens as shown by X-ray imaging experiments (Hemley *et al.*, 1997) and finite-element calculations (Merkel *et al.*, 1999).

In addition, the potential of the many exceptional properties of diamond is only partially expressed in natural specimens, but can be optimized in synthetic processes. By varying chemical dopants in the laboratory, the electronic behavior of diamond can be altered from a wide-gap insulator to a *p*- or *n*-type semiconductor (Haenen *et al.*, 2004; Suzuki *et al.*, 2004; Tajani *et al.*, 2004) to a superconductor (Ekimov *et al.*, 2004). Ultrahigh-purity synthetic diamonds open clear optical windows to radiation from the UV to far IR for transmission and inelastic scattering studies with exceedingly low background. Isotopically pure synthetic diamonds provide the highest known thermal conductivity. Defect-free, synthetic, single-crystal diamonds (Sumiya *et al.*, 2005) can achieve a level of perfection unmatched by natural diamonds rivaling perfect silicon single crystals. For diamond anvil application, the apparent hardness and toughness limits the maximum pressures attainable and is therefore more important than the absolute hardness which in practice is unreachable due to the weak cleavage (Brazhkin *et al.*, 2004). Synthetic single-crystal (Yan *et al.*, 2004) or polycrystalline diamonds (Irifune *et al.*, 2003) with properly randomized defects that hamper the propagation of cleavage show greater apparent hardness and toughness than diamonds in which the strength is dictated by cleavage. Exploration of these properties may unleash the full power of diamond anvils and revolutionize the field of high-pressure experimental geophysics and geochemistry.

This exciting prospect has only been limited by our ability to synthesize sizable single-crystal diamond at affordable cost and within reasonable time. The cost and availability of natural single-crystal diamonds limits their anvil size to typically 0.25–0.3 ct (1 ct = 0.2 gm) although the largest diamond used as an anvil was 4 ct (Glazkov *et al.*, 1988). Two existing methods, high-pressure/high-temperature (HPHT) and chemical vapor deposition (CVD), have been used to grow single-crystal diamond. The HPHT method is very expensive (more than natural single-crystal diamond) and slow, and is practically limited to sizes of 2.5–3 ct. The CVD epitaxial process has been successfully used for overlaying circuits on anvil tips for electromagnetic measurements (Jackson *et al.*, 2003, 2005a) and depositing isotopically pure ^{13}C as a pressure sensor (Qiu *et al.*, 2006). The very rapid growth CVD process (Yan *et al.*, 2002)

has produced diamond anvils that reached 200 GPa (Mao *et al.*, 2003b) and a 10 ct size single-crystal diamond (Ho *et al.*, 2006) that could be used for two orders of magnitude larger samples than presently available at megabar pressures.

9.2.2   Gasket and Side Window

Gaskets in anvil devices serve four critical functions: (1) providing lateral support to protect anvil tips, (2) building a gradient from ambient to the peak pressure, (3) encapsulating the sample, and (4) providing an alternative window for X-ray and neutron probes. Outside the diamond culet area, the gasket forms a thick ring that supports the anvils like a belt, without which the anvils would not survive above 30–40 GPa. The effectiveness of the support depends on the tensile strength and thickness of the gasket. Between two parallel culets the thin, flat portion of the gasket sustains a large pressure gradient from the minimum stress at the edge of the culet to the maximum stress at the center. The gasket thickness, *t*, does not depend upon its initial thickness but is rather a function of the shear strength of the material, σ, and the pressure gradient along the radius *r*, $\partial P/\partial r$ (Sung *et al.*, 1977).

$$t = 2\sigma/(\partial P/\partial r) \qquad [1]$$

Different materials are used as gaskets to optimize specific functions and experimental goals. Hardened steel is used as an all-purpose gasket material. High-strength rhenium is used for experiments requiring large thickness or high temperature. Composite gaskets can be constructed to optimize different functions at different parts of the gasket, for example, insulating inserts (MgO or Al_2O_3) are added to metallic gaskets for introduction of electrical leads into the high-pressure region. Using a composite gasket which includes diamond powders or a diamond coating on the central flat region of the gasket greatly increases the shear strength and can increase the gasket thickness by 2–3 times (all other factors remaining constant) (Zou *et al.*, 2001). Since the sample volume in a DAC is defined by the gasket thickness and the hole (sample chamber) diameter in the gasket, doubling the thickness of the sample chamber effectively doubles the sample volume. To eliminate the grain boundary fracture and to avoid interference of XRD patterns from high-Z crystalline gasket materials, high-strength bulk metallic glass has been used as a gasket to above 100 GPa (He *et al.*, 2003). Sample reaction must also be

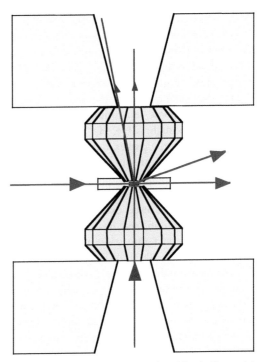

Figure 5 Panoramic DAC with Be gasket. The radial path has larger opening and is much less absorbing to the probing X- radiations than the axial path because of the low absorption coefficient of Be.

carefully considered when choosing gaskets especially with highly reactive samples like hydrogen peroxide (Cynn *et al.*, 1999).

X-ray and neutron radiation can penetrate appropriate gasket materials and probe DAC samples in an equatorial or radial geometry (i.e., perpendicular to the compression axis) (Mao *et al.*, 2001). The gasket materials are chosen for mechanical strength and transparency and low scattering background (**Figure 5**). For X-ray studies, high-strength beryllium (Mao *et al.*, 1998b) (ultimate tensile strength, 517 MPa; micro-yield strength 68.9 MPa), amorphous boron epoxy inserts (Lin *et al.*, 2003b) with kapton belts (Merkel and Yagi, 2005) and compressed, superhard graphite (Mao *et al.*, 2003a) inserts with beryllium belts (Mao *et al.*, 2006c) have been used to reach megabar pressures. With mutually canceling neutron scattering factors and high strength, the Ti_{52}/Zr_{48} zero scattering alloy is used as a DAC gasket for neutron diffraction (Ding *et al.*, 2005).

9.2.3 Samples and Pressure Media

When a solid sample is surrounded by a fluid medium during compression, it is subjected to a hydrostatic pressure, that is, the stress is uniform in all directions. At sufficiently high pressures, all fluids solidify with liquid helium solidifying at the highest pressure of 11 GPa (Besson and Pinceaux, 1979). Pressure is then transmitted through solid media. Solids have finite strength and cause pressure anisotropy or nonhydrostaticity, which refers to directional deviatoric stress at a point, and pressure gradients or inhomogeneity, which refers to the variation of stress conditions at different points in the sample. It is often desirable to eliminate or to reduce the pressure anisotropy and inhomogeneity, and pressure media are chosen for their low strength and chemical inertness in relation to the sample. A mixture of methanol–ethanol is commonly used in the DAC as a fluid medium to 10 GPa (Piermarini *et al.*, 1973) above which its strength rises sharply. Inert-gas solids are commonly used as pressure media to produce quasihydrostatic conditions up to 8 GPa in argon, 20 GPa in neon (Bell and Mao, 1981) and over 100 GPa in helium (Dewaele and Loubeyre, 2005; Loubeyre *et al.*, 1996; Takemura and Singh, 2006).

Gas pressure media can be loaded by placing the entire DAC in a large gas pressure vessel with the sample in a slightly opened gasket hole. The gas is pumped into the vessel to a nominal pressure of

200 MPa which fills the DAC sample chamber, as well as the surroundings. A feed-through mechanism is then used to close the DAC sample chamber and seal the sample in the gas medium inside the gasket. The high-pressure gas-loading method (Mills *et al.*, 1980) has the advantage of being able to load gas mixtures without phase separation and low melting temperature gases such as He, H_2, and Ne, without trapping bubbles in the gasket, but uses a significant quantity of gas at high pressures which requires rigorous safety precautions. A simple alternative is to use liquid nitrogen cooling to liquefy gases inside the DAC. This method is used extensively for loading argon as a gas medium, and has the advantage for loading expensive and rare gases in 'ml' quantities (e.g., [83]Kr (Zhao *et al.*, 2002)). When properly quantified, experiments under deviatoric stress can provide rich additional information about strength, elasticity, and rheology that are unavailable with hydrostatic experiments. The uniaxial compression of a DAC is ideal for quantitative study of deviatoric stress at ultrahigh pressures (Singh *et al.*, 1998a; 1998b). With radial XRD the incident X-rays are directed perpendicular to the compression axis (i.e., through the gasket), allowing for measurement of how the *d*-spacings vary with ϕ, the angle between scattering direction and the compression axis. The difference in *d*-spacings obtained from the $\phi = 0°$ and $\phi = 90°$ gives the deviatoric strain, $Q = (d_{0°} - d_{90°})/3d_\mathrm{P}$, where $d_\mathrm{P} = (d_{0°} + 2d_{90°})/3$ is the *d*-spacing under hydrostatic pressure $\sigma_\mathrm{P} = (\sigma_3 + 2\sigma_1)/3$, where σ_3 and σ_1 are axial and radial stresses, respectively. With the hydrostatic equation of state (EOS), these data can be used to determine the deviatoric stress $t = \sigma_1 - \sigma_3$. The maximum uniaxial stress *t* supported by a material is determined by its strength; that is, $t = \sigma_\mathrm{y} = 2\tau$, where σ_y and τ are the yield and shear strengths of the material, respectively. It is often assumed in high-pressure experiments that $t = \sigma_\mathrm{y}$; however, in general *t* varies with sample environment and the equality (von Mises condition) is true only if the sample is observed to plastically deform under pressure. In **Figure 6** using this method, we demonstrated that the strength of argon could reach 2.7 GPa at 55 GPa (Mao *et al.*, 2006a).

9.2.4 DAC Body

The main functions of the DAC body are to advance the anvils toward one another while allowing sample access to analytical probes. Dozens of DAC designs

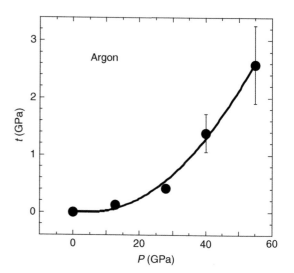

Figure 6 Deviatoric stress, which is the lower bound of the strength, of Ar as a function of pressure (from Mao *et al*, 2006a). The vertical bars indicate the range at different crystallographic orientation (*hkl*). The true error bars are significantly smaller.

with different materials (most commonly hardened steel), sizes, shapes, aspect ratios, and openings of the DAC body have been devised to optimize for specific measurements at varying *P–T* conditions. For instance, cobalt-free steel is used to reduce neutron activation of the DAC, Be seats are used for single-crystal XRD, wide axial openings are used for Brillouin spectroscopy, and panoramic, radial openings are used for radial XRD, neutron diffraction, and X-ray spectroscopy (**Figure 7**). High- (Bassett, 2003) or low-temperature (Kawamura *et al.*, 1985; Pravica and Remmers, 2003; Silvera and Wijngaarden, 1985) alloys are used for DACs operated at extreme temperature conditions. Special nonmagnetic materials, including Be–Cu alloys (Okuchi, 2004; Timofeev *et al.*, 2002), plastics (Matsuda *et al.*, 2004), and ceramics (Yamamoto *et al.*, 1991), are used for making DACs for magnetic measurements. Miniature DACs less than 1 cm in size (Eremets and Timofeev, 1992; Mito *et al.*, 2001; Sterer *et al.*, 1990; Tozer, 1993) have been developed to fit in the limited spaces defined by analytical probes, for example, in the 1 cm bore of the spindel for magic-angle-spin nuclear magnetic resonance (NMR) studies (Tozer, 2002).

For operation in a moderate pressure range <10 GPa, the diamond alignment requirements are not critical, and the DAC design can be focused on ease of operation with maximum access for *in situ* probes (e.g., Merrill and Bassett, 1974; Oger *et al.*,

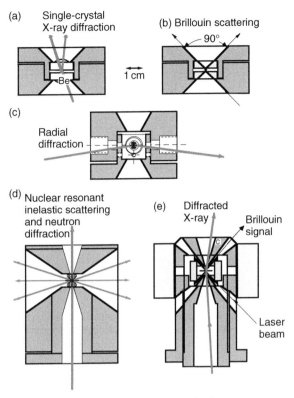

Figure 7 Various types of DAC optimized for specific functions as marked.

2006). For ultrahigh-pressure operation >30 GPa, preservation of the perfect alignment of the anvils during the compression process is absolutely crucial. To push the anvils together requires rigid, frictionless, linear displacement without wobbling or lateral shifting. This is accomplished by close fitting of the hardened steel piston and cylinder to μm accuracy (Mao and Bell, 1976). The same design principles are retained to keep the perfect alignment while maximizing optical and X-ray access through the axis in symmetric DACs (Mao *et al.*, 1998a) or through the radial directions in panoramic DACs (Mao *et al.*, 2001).

The highest stress point of the body is at the table side of the two anvils which are supported by high-strength seats made of tungsten carbide or cubic boron nitride (c-BN). Opposing diamond anvil culets must be aligned to a lateral accuracy of 1–2% of the culet diameter and parallellism within 0.5–1 mrad. This alignment is achieved by two half-cylindrical seats (rockers) perpendicular to each other (Mao and Bell, 1976), or alternatively hemispherical and flat disc seats, to provide two-dimensional tilt and translation alignment capability. More recent designs require manufacturing anvils, seats, and the DAC body to a parallelism of ± 0.5–1 mrad. Then tilt or rocking alignment is no longer necessary, and the rockers are simplified to two flat disks with only translational alignment (Mao *et al.*, 1994). Seats made of another single-crystal diamond backing plate (Yamanaka *et al.*, 2001), conical seats that are supported at the conical side surrounding the table (Boehler and De Hantsetters, 2004), and a combination of the conical seat and diamond backing plate (Krauss *et al.*, 2005) have been used successfully to reach high pressures while maintaining very large axial openings.

9.2.5 Compression Mechanism

Diamond anvils are typically compressed mechanically by screws and springs directly or with an additional lever arm arrangement (Mao *et al.*, 1994), or pneumatically by a diaphragm (Daniels and Ryschkewitsch, 1983) or membrane (LeToullec *et al.*, 1992). Relative to a pneumatic system, the mechanical screw-spring system has the advantages of: (1) stability (a cell has been kept at constant

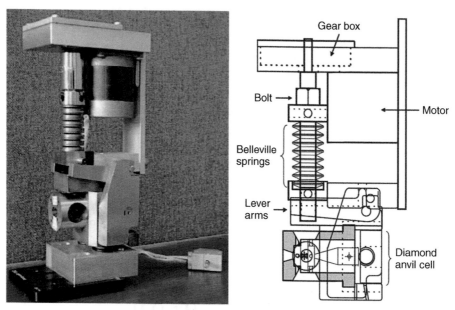

Figure 8 Motorized lever-arm DAC assembly. Left, picture. Right, mechanical drawing. From Mao WL and Mao HK (2006) Ultrahigh-pressure experiment with a motor-driven diamond anvil cell. *Journal of Physics: Condensed Matter* 18: S1069–S1073.

pressure at 200 GPa since 1986) without the concern of gas leak, (2) ease of transportation under high pressures, (3) low-temperature operation unlimited by the freezing point of the pneumatic gas, (4) accommodating large displacement without rupture of membrane, and (5) simplicity and low cost without the attachment of gas bottle and control. A pneumatically controlled membrane, on the other hand, has the advantages of remote and quantitative control of force increment without having to dismount or disturb the DAC.

Another option for remote pressure increment is the attachment of a motor-driven gearbox to a mechanical screw-spring system (**Figure 8**). A motor-driven DAC was used to increase pressure remotely up to 230 GPa while taking XRD patterns in 1–2 GPa steps. 142 XRD patterns were collected over 3 h demonstrating not only the efficient pressure change possible, but also the potential for fine pressure and temporal resolution (Mao and Mao, 2006). Besides providing detailed *P–V* EOS measurements, this type motor-driven DAC could find application for time-dependent strain measurements and for experiments which require precise pressure resolution.

9.2.6   Pressure Calibration

For pressures above several GPa, the only viable primary pressure standard is the pressure-volume (*P–V*) EOS which can be derived from independent

determination of a pair of EOS parameters. In shock-wave measurements, the pair of variables are particle velocity (U_P) and shock velocity (U_S), from which the pressure–density relations are calculated. Such direct determinations of pressure are called primary standards. Once calibrated against a primary standard, any other pressure-dependent variable can be used as a secondary standard for pressure determination. Secondary standards are chosen for their accessibility and resolution. For the widely used 'ruby scale', tiny ruby grains are added in the sample chamber (Piermarini *et al.*, 1975), and the pressure-shift of the ruby fluorescence wavelength can be easily probed with a laser beam through the diamond windows. The ruby scale has been calibrated against shock wave primary standards of Ag, Cu, Mo, and Pd without a pressure medium up to 100 GPa (Mao *et al.*, 1978), and against Ag and Cu in an argon pressure medium up to 80 GPa with ± 6% uncertainty in pressure (Mao *et al.*, 1986). With the improvement of XRD accuracy, the precision of pressure determination from each of the six primary standards, that is, Al, Au, Cu, Pt, Ta, and W, was greatly improved (Chijioke *et al.*, 2005; Dewaele *et al.*, 2004) but the spread of pressures determined from different standard is still as large as 6%, reflecting the need for improving the primary scale.

With improvements in spectroscopic techniques, the precision of ruby measurements and other secondary scales can now resolve 0.2–1% in pressure, but the shock-wave primary calibration that they rely on still carry a 5% uncertainty in pressure which becomes the limiting factor for accuracy. Recently, a new pair of variables has been chosen to improve the accuracy of pressure calibration for anvil devices. They are the density (ρ) measured with XRD and the acoustic velocity (V_ϕ) measured with the ultrasonic method or Brillouin scattering on the same sample under the same compression conditions (Mao and Hemley, 1996). Pressure is derived directly by

$$P = \int V_\phi^2 \mathrm{d}\rho \qquad [2]$$

The resultant P–ρ relation is a primary pressure standard. Such measurements have been carried out on the MgO pressure standard in a helium pressure medium as a primary standard (**Figure 9(a)**) which was used for calibration the following ruby scale with 1% pressure accuracy up to 55 GPa at ambient temperature (Zha *et al.*, 2000, 2004)

$$P(\mathrm{GPa}) = (A/B)\left\{[1 + (\Delta\lambda/\lambda_0)]^B - 1\right\} \qquad [3]$$

where $A = 1904$ GPa, $B = 7.715$, and $\Delta\lambda$ is the ruby fluorescence wavelength shift from its ambient pressure value λ_0 (**Figure 9(b)**).

In summary, the present pressure calibration is robust to ± 6–10% accuracy. A high-pressure community effort is underway to improve the primary calibration by improving the shock-wave equations of state, *ab initio* calculations, or simultaneous

measurements of ρ and V_ϕ at higher pressures and temperatures. In DAC, ρ can be determined by X-ray diffraction to an accuracy of 0.2%, and V_ϕ can be determined by single-crystal Brillouin spectroscopy (Section 9.4.4), impulsive stimulated light scattering (Section 9.4.4), or phonon inelastic X-ray spectroscopy (Section 9.5.9) to an accuracy of 1%. Secondary scales based on X-ray diffraction and optical spectroscopy (e.g., Raman shift of cBN (Goncharov *et al.*, 2005)) are pursued for the goal of 1% accuracy and precision in pressure with the propagation of errors from various sources.

9.3 Temperature

9.3.1 Resistive Heating

XRD up to 1100 K and 100 GPa has been measured using electrical resistance coils external to the diamond anvils (Fei and Mao, 1994) (**Figure 10**). The P–T conditions with external resistance heating are well defined. The singularly high thermal conductivity of diamond results in temperature variations within the sample chamber that are typically less than ± 1 K. However stress-bearing components, including the sample gasket, diamonds, diamond seats, and a portion of the DAC body are also heated which limits the maximum temperature for resistive heating to 1500 K due to softening of these components at high temperatures. Inert and reducing gases,

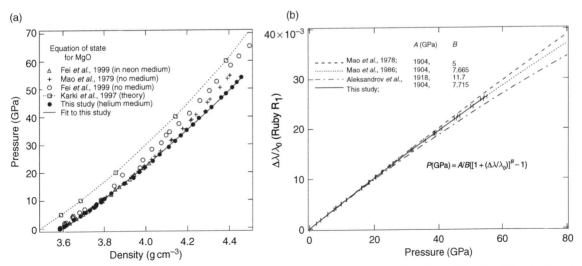

Figure 9 (a) the primary pressure standard of P–ρ equation of state of MgO obtained from X-ray diffraction determination of ρ and single-crystal Brillouin spectroscopy measurements of V_ϕ. (b) the secondary ruby scale based on the MgO standard. From Zha C-S, Mao HK, and Hemley RJ (2000) Elasticity of MgO and a primary pressure scale to 55 GPa. *Proceedings of the National Academy of Sciences* 97: 13494–13499.

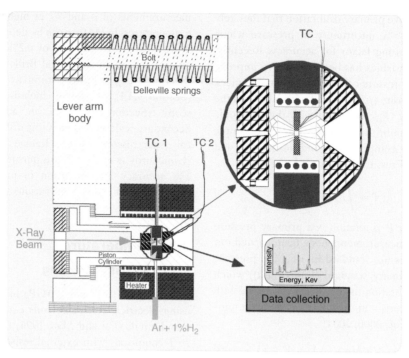

Figure 10 External resistive heating of a DAC with lever arms.

such as argon with 5% H_2, are used to avoid oxidization of the diamonds and metallic gasket. Systematic studies of $P–V–T$ EOS, phase transitions of mantle minerals have been carried out with external resistive heating and simultaneous XRD over a wide $P–T$ range. External resistive heating techniques have been also applied to DAC research in physics, chemistry, and materials science, for example, Degtyareva *et al.* (2005) and Gregoryanz *et al.* (2005). Over a lower pressure range, external resistive heating of the DAC has been used extensively for a full range of optical and X-ray diffraction and spectroscopy studies of hydrothermal fluids and solid (Anderson *et al.*, 2002; Bassett *et al.*, 1993; Bassett, 2003).

To minimize heating of DAC components and concentrate heating closer to the sample, the gasket (Dubrovinsky *et al.*, 1997, 2000) or the diamonds (Burchard *et al.*, 2003) can be used as the resistive heaters. Even closer to the sample, internal resistive heaters have been developed inside the DAC gasket chamber. For electrically conductive samples, the heater can be the sample itself. For instance, iron wire or foil of 5–20 μm size has been heated by passing electric currents up to its melting temperatures for determination of melting and phase transitions (Liu and Bassett, 1975; Mao *et al.*, 1987a). For insulating samples, a metallic wire placed inside the gasket has been successfully used as an internal

heater for heating the silica sample in the tiny hole in the wire for Raman spectroscopic studies up to 3000 K at 10 GPa (Zha and Bassett, 2003).

9.3.2   Laser Heating

Temperatures in excess of 5000 K (Weathers and Bassett, 1987) can be achieved for samples under pressure in DACs by heating with high-power IR lasers (Bassett, 2001; Ming and Bassett, 1974). Due to the transient nature of laser heating (LH) and the steep temperature gradients (from the maximum temperature to ambient over 20 μm radial distance and even steeper axial gradients), in the past contradictory results have been reported by different laboratories, causing confusion about the LH technique. Recently significant progress in shaping and defining the temperature distribution in LHDACs has been made. With a 'double hot-plate' system, which exploits optimal laser characteristics, sample configuration, and optical arrangement, the LH method is approaching the accuracy of other high $P–T$ techniques using internal heaters. In this method, a multimode yttrium–aluminum garnet (YAG) laser, typically 20–100 W total power, provides a flat-top power distribution at the focal spot (Mao *et al.*, 1998a; Shen *et al.*, 1996). The method is further improved by the mixing of two yttrium lithium fluoride (YLF) lasers in TEM_{00}

and two in TEM_{01} modes, to create a flat temperature distribution better than 1%. A hot-plate configuration is created when the heat generation and temperature measurement are concentrated at the planar interface of an opaque sample layer and transparent medium. The heating laser is split into two beams that pass through the opposing diamond anvils to heat the high-pressure sample simultaneously from both sides. The axial temperature gradient in the sample layer in the DAC is eliminated within the cavity of the two parallel hot plates. Variations of double-side LH systems have been integrated with most synchrotron DAC beamlines (Liu *et al.*, 2005; Schultz *et al.*, 2005; Shen *et al.*, 2001; Watanuki *et al.*, 2001) for *in situ* X-ray studies at simultaneous high *P–T* (**Figure 11**). These systems have led to major discoveries of new lower-mantle phases (Kuwayama *et al.*, 2005; Mao *et al.*, 2006c; Murakami *et al.*, 2004).

The LHDAC technique requires an opaque, laser-absorbing sample and does not couple with transparent samples. This problem has been overcome for solid samples by mixing inert laser absorbers, or for fluid samples by using a metallic foil with a high melting point (Pt, Re, or W of 5–15 μm in thickness) as the laser absorber which in effect turns into an internal furnace (Lin *et al.*, 2004b). A small 10–20 μm diameter hole is drilled into the foil which is placed inside a gasket chamber filled with the transparent samples and then compressed to high pressures

(**Figure 2**). A YLF laser beam of 30 μm in diameter is centered at the hole and heats the peripheral metal foil around the small hole; the heated part forms a donut-shaped furnace effectively heating the transparent sample inside. Transparent samples such as CO_2 have been heated up to 1600 K and 65 GPa, indicating the high heating efficiency of the internal donut furnace method.

The progress of LHDAC has been marked by waves of breakthroughs and debates and is a good example of the general progress of most DAC techniques. The first application of Ming–Bassett LH technique for determination of Fe melting temperature at the core conditions (Williams *et al.*, 1987) generated a great deal of excitements. It was followed by the well-known debate that (Boehler, 1993) reported much lower melting temperatures with LHDAC. In 1995, the Royal Society of London Workshop was organized to discuss detailed problems on temperature gradients, temperature measurements, optical aberrations, and melting criteria that resulted in a comprehensive symposium volume (see Jephcoat and Besedin, 1996). The problems were mostly eliminated by subsequent development of new techniques. For instance, double-side LH with flat-top power profile was developed to eliminate the temperature gradient (Shen *et al.*, 1996), and *in-situ* X-ray diffraction was developed for positive identification of phases (Saxena *et al.*, 1995; Yoo *et al.*, 1995). These techniques were further refined, and in 2000, another community workshop was organized to discuss these advances in LHDAC (see Shen *et al.*, 2001). General consensus for iron melting has been reached to at least 100 GPa (Ma *et al.*, 2004). However, many readers may still only be aware of the debate a decade ago. Most recently, high-power fiber lasers with much higher power (100–1500 W), and superior stability, collimation, and flexibility have become available. This new capability will certainly revolutionize the LHDAC technique by lifting it to the next level of accuracy and performance.

9.3.3 Cryogenic

Experiments at simultaneous high-pressure and low-temperature conditions have less direct relevance for deep-Earth behavior but are important for understanding icy bodies. Fundamental understanding of high-pressure mineral physics, as well as constraints for theoretical studies require cryogenic studies of low-temperature phenomena. In principle, the use of high-pressure instrumentation does not affect the

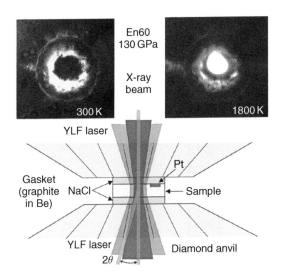

Figure 11 Left bottom, schematic of a laser-heated DAC. Top left, photomicrograph showing the $Mg_{0.6}Fe_{0.4}SiO_3$ orthopyroxene sample (En60) in a transparent graphite ring in Be gasket at 130 GPa before heating; top right, during laser heating to 1800 K.

lowest temperatures possible for simultaneous high-pressure, low-temperature studies.

Cryogenic temperature, high-pressure conditions are created by cooling the entire DAC in a liquid nitrogen or helium flow cryostat which is specially designed for pressure changes through long-stem wrenches for lever-arm driven DAC or gas-line, feedthrough for membrane DAC. The cryostat is designed with feedthrough circuits for electromagnetic measurements and windows for IR, optical Raman, and X-ray probes (**Figure 12**). Temperatures down to 4.2 K have been routinely reached at multimegabar pressures (Eremets *et al.*, 2001b; Goncharov *et al.*, 2001; Struzhkin *et al.*, 2002), with a record low temperature of 27 mK being reached in a dilution refrigerator (Eremets *et al.*, 2000).

The cryogenic DAC has been used extensively for simulating interior conditions of icy satellites in studies of multiple phases of crystalline and amorphous ices (Cai *et al.*, 2005; Yoshimura *et al.*, 2006a, 2006b),

clathrates (Mao *et al.*, 2002), and simple molecular compounds (Mao *et al.*, 2005b). The high-pressure, low-temperature study of Earth and planetary metals (Shimizu *et al.*, 2001), gases (Eremets *et al.*, 2001a; Goncharov *et al.*, 2001; Loubeyre *et al.*, 1993), and ices also provides the fundamental mineral physics basis for understanding theories and predictions of their behaviors at extreme high *P–T* conditions. In addition, the cryogenic DAC is an important tool for low-temperature physics especially for novel superconductivities (Eremets *et al.*, 2001b; Hemley and Mao, 2001b; Shimizu *et al.*, 2002; Struzhkin *et al.*, 1997; Struzhkin *et al.*, 2002).

9.3.4 Temperature Measurement

For external resistive-heating or cryogenic experiments, a large portion of the DAC is heated or cooled to a uniform temperature that can be simply measured by attaching thermocouples at strategic

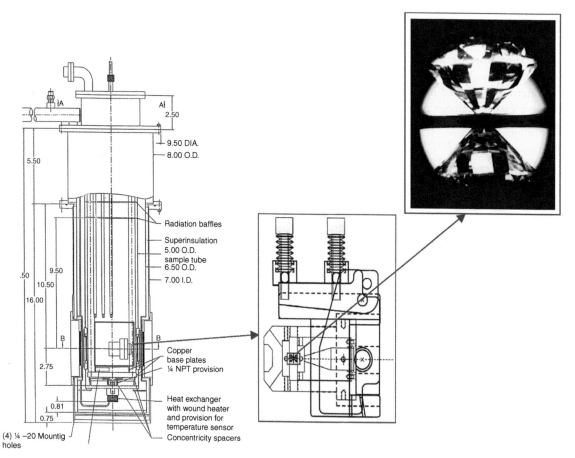

Figure 12 Liquid helium cryostat for high-pressure X-ray diffraction, X-ray Raman, and optical spectroscopic studies of gases, ices and clathrates down to 4 K.

spots on or near the diamonds. For internal resistive-heating or laser-heating experiments, the temperature gradient can be very steep, and the peak temperatures are determined by noncontact, optical pyrometers with μm resolution. Optical probes through the strained, highly dispersive diamond windows often suffer from chromatic aberration, spherical aberration, and astigmatism that introduce errors and controversies in determination of temperature on the steep temperature gradient slope. Fortunately the problems diminish in regions where the gradients are small. Therefore the double-side laser-heating method with flat-top power distribution (from either multimode or controlled mixture of TEM_{00} and TEM_{01} donut modes) has been developed to create a uniformly heated area at the peak temperature.

The accuracy and precision of temperature measurements in the laser-heated DAC have improved significantly with the use of spectral radiometry system (Shen *et al.*, 2001). The thermal radiation spectrum is fit to the Planck radiation function to determine the temperature:

$$I_\lambda = c_1 \varepsilon_\lambda \lambda^{-5} [\exp(1 - c_2/\lambda T)] \qquad [4]$$

where I_λ is spectral intensity, λ wavelength, ε_λ emissivity, and two constants $c_1 = 2\pi hc^2 = 3.7418 \times 10^{-16}$ W m^2 and $c_2 = hc/k = 0.014388$ mK, where h is Planck constant, c speed of light, and k Boltzmann constant. The system response is calibrated by a tungsten ribbon lamp with a known relationship between radiance and power setting based on National Institute for Standards and Technology (NIST) standards. Temperatures on two sides are measured separately with an imaging spectrograph and CCD and equalized by controlling the ratio of beam splitting. Above 1500 K, the use of blackbody spectral radiometry for primary temperature calibration is intrinsically superior to the use of a thermocouple which is a secondary calibration and is affected by the stress effect on the EMF and contaminants (Mao *et al.*, 1971). In addition, the spectral radiometry measurements provide a real-time nonintrusive probe of continuous temperature profile while thermocouples provide an intrusive point measurement. Uniform temperatures of 3000 K (±20) K have been achieved in a high-pressure sample of 15 μm diameter × 10 μm thickness. A significantly finer X-ray microprobe beam has been used for *in situ* characterization of the sample at these uniform *P–T* conditions. For example, phase relations and melting of iron and FeO have been studied at high *P–T* conditions of

the Earth's geotherm (Lin *et al.*, 2003a). Measurements of FeO to 2200 K and 90 GPa show that the (subsolidus) boundary between the B1 (NaCl structure) and the metallic B8 (NiAs structure) phases has a significant positive slope.

A major source of error in spectral radiometry is the uncertainty of the emissivity which is sample dependent and has rarely been characterized as a function of wavelength and pressure (Benedetti and Loubeyre, 2004). The problem is especially severe for transparent samples which have very low emissivities and are far from being ideal black bodies. Another optical probe has been developed for temperature determination of transparent samples from the intensity ratios of the anti-Stokes/Stokes excitation pairs of Raman spectra (Lin *et al.*, 2004b; Santoro *et al.*, 2004). The Stokes and anti-Stokes peak intensities are proportional to the populations in the ground and excited states, respectively, and thus depend on the temperature according to the Boltzmann distribution:

$$I_A/I_S = (\nu_A/\nu_S)^4 [\exp(\Delta\nu/kT)] \qquad [5]$$

where I_A and I_S are the measured intensities of the anti-Stokes and Stokes peaks, respectively, ν_A and ν_S are the frequencies of the anti-Stokes and Stokes peaks, respectively, and $\Delta\nu$ is the relative Raman shift of the Stokes peak from the excitation laser frequency. *In-situ* Raman spectroscopy by laser-heating represents a powerful technique for characterizing high *P–T* properties of molecular compounds present in planetary interiors (Lin *et al.*, 2004b; Santoro *et al.*, 2004).

The principle of using the Boltzmann distribution and anti-Stokes/Stokes ratio for temperature determination works regardless of whether the excitation photon is in the sub-eV optical range or in the 10 keV X-ray range. The anti-Stokes and Stokes sides of the phonon excitation spectra of a laser-heated ^{57}Fe sample in a laser-heated DAC were obtained with nuclear resonant inelastic X-ray scattering spectroscopy yielding accurate, independent, temperature determination up to 1700 K at 73 GPa (Lin *et al.*, 2004c).

9.4 Optical Probes

Through focusing microscope lens, the clear, colorless, single-crystal diamond windows allow many optical measurements with different geometry.

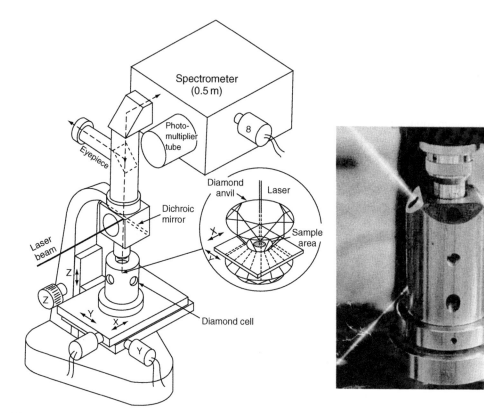

Figure 13   Optical system for DAC.

Figure 13 shows a generic high-pressure microspectrometer: transmission (or absorption) spectroscopy is conducted by focusing the light beam from bottom through the sample and received by the top microscope lens; reflective, Raman and fluorescence spectroscopy, with incident beam through the top lens and scattered beam back through the same lens; Brillouin spectroscopy, with the two beams at a fixed angle. The scattered beam is collected and analyzed by the corresponding spectrometers. Although the details vary slightly, this universal approach has been adopted for many decades in DAC research.

9.4.1   Optical Absorption

Between the UV to IR regions of the electromagnetic spectrum, the color of minerals reveals a wealth of physical and chemical information. The optical appearance of 'colorful' minerals changes as the underlying properties are altered under pressure. These changes can be monitored through a modified transmission and reflectance polarizing microscope and analyzed by spectrometers. High-pressure melting, crystallization, and phase transitions can often be identified visually by movements of grain boundaries and changes of reflectivity, textural, morphology, and refractive indices (Bassett and Takahashi, 1965; Mao *et al.*, 2006b; Takahashi and Bassett, 1964). High-pressure dielectric properties are determined from measurements of refractive indices with optical interferometry (Hemley *et al.*, 1991). High-pressure changes of single-crystal symmetry are revealed by birefringence under crossed polarization or distortion of shape and interfacial angles (Yagi *et al.*, 1979). High-pressure metal-insulator transitions also show conductivity changes in optical range, for example, at the aforementioned pressure-induced transition of graphite to an insulating state, the graphite sample becomes transparent (Hanfland and Syassen, 1989; Utsumi and Yagi, 1991). Pressure-induced changes of optical absorption spectra of transition elements in minerals reveal the change of crystal-field stabilization energy, oxidation states, site occupancy, *d*- or *f*-electron spin state, or charge-transfer process and affect radiative heat transfer, electrical conductivity, and oxidation–reduction processes in the Earth's deep interior (Goncharov *et al.*, 2006; Mao and Bell, 1972). Optical absorption spectra have been measured at pressures up to 80 GPa for the lower-mantle oxide magnesiowüstite. At a pressure of about 60 GPa which corresponds to the high-spin to low-spin transition of Fe^{2+}, absorption in the mid- and near-IR spectral range was enhanced, whereas absorption in the visible UV was reduced. These results indicate that (Mg,Fe)O with low-spin iron will exhibit lower radiative thermal conductivity than (Mg,Fe)O with high-spin iron, which is an important consideration for geodynamic models of convection and plume stabilization in the lower mantle (Goncharov *et al.*, 2006).

9.4.2 IR Spectroscopy

Type II nitrogen-free diamond windows are transparent to IR radiation. Vibrational IR spectroscopy provides detailed information on bonding properties of crystals, glasses, and melts, thereby yielding a microscopic description of thermochemical properties. IR measurements also provide information on electronic excitations, including crystal-field, charge-transfer, and excitonic spectra of insulators and semiconductors, interband and intraband transitions in metals, and novel transitions such as pressure-induced metallization of insulators. The advent of high-pressure synchrotron radiation IR spectroscopy has revolutionized such studies. IR radiation at the VUV ring of the National Synchrotron Light Source (NSLS), for example, has up to $\sim 10^4$ times the brightness of a conventional thermal (lamp) source. The significant enhancement in our ability to probe microscopic samples provided by this source makes it ideally suited to studies of materials under extreme pressures. When coupled with a Fourier transform infrared (FT-IR) interferometer and special microscopes for high-pressure cells recent work has demonstrated a gain of up to five orders of magnitude in sensitivity relative to a grating system commonly used for high-pressure IR measurements. Whereas enhanced flux can also be achieved with new IR laser techniques (e.g., optical parametric oscillators and diode lasers (Cui *et al.*, 1995)), the very broad spectral distribution, ease of interfacing with a conventional FT-IR instrument, and the picosecond time structure provide important advantages over such techniques.

New synchrotron IR experiments have led to a series of discoveries into the nature of Earth and planetary materials at ultrahigh pressures (Hemley *et al.*, 1998). Study of dense hydrogen in a DAC led to the discovery of a number of unexpected phenomena,

including striking intensity enhancements of vibrational modes (Hanfland *et al.*, 1993; Hemley *et al.*, 1994), an unusually complex phase diagram (Goncharov *et al.*, 1995; Mao and Hemley, 1994; Mazin *et al.*, 1997), and new classes of excitations in the solid (Soos *et al.*, 1995). This system is exceedingly rich in physical phenomena and continues to be a crucial testing ground for theories of condensed-matter physics (Mao *et al.*, 1994) and for planetary modeling (Hemley and Mao, 1998). The behavior of H_2O-ice under pressure is important not only for planetary science but is of broad fundamental interest as well. It was predicted 35 years ago (Holzapfel, 1972) that under pressure ice should transform to a very different structure in which conventional hydrogen bonding is lost and the hydrogen atoms are symmetrically disposed between the oxygens. Synchrotron IR measurements have found that ice is indeed profoundly altered at very high pressures. At 60 GPa, synchrotron IR reflectivity measurements show that the material transforms to a new structure that has the spectroscopic signature of the symmetric state (Goncharov *et al.*, 1996). The new form of ice is stable to record pressure of 210 GPa. Under these conditions, ice is no longer a molecular solid but is best described as a dense ionic oxide.

9.4.3 Raman Spectroscopy

Raman spectroscopy is widely used for characterizing structural, vibrational, and electronic properties of minerals at high pressures (McMillan *et al.*, 1996), and can provide finger-printing diagnoses of high *P–T* phases and melting. The appearance and intensity of Raman peaks follow selection rules which are governed by the symmetry of the transition, and the shift of Raman peaks indicates a change in bonding or electronic energy levels.

Developments in Raman spectroscopy technology have taken quantum leaps in the past three decades: from the scanning double-monochromator system in the 1970s to the subtractive-grating filter stage and dispersive spectrograph with a diode array detector in the 1980s, to the single-grating imaging spectrometers with notch filters and CCD detector in the 1990s. The Raman spectrometer has been integrated with a cross-beam (Mao *et al.*, 1987b; Xu *et al.*, 1986) or confocal (Gillet, 1999) microscope to achieve three-dimensional spatial resolution. New Raman spectrometers are much more sensitive and compact, and thus easily coupled DACs for measurement at ultrahigh pressures and varying temperature (Goncharov and Struzhkin, 2003).

Micro-Raman (Gillet, 1999; Mao *et al.*, 1987b) was the first technique that enabled *in situ* characterization of samples at high *P–T* with μm spatial resolution. With ΔE in the range of several cm^{-1} ($1\ cm^{-1} = 8\ meV$) to several thousand cm^{-1}, Raman peaks are largely due to molecular or lattice vibrations, but also cover some electronic and magnetic excitations (Goncharov and Struzhkin, 2003). Micro-Raman spectroscopy has been used extensively as a structural probe for distinguishing high-pressure polymorphs of the same composition, for example, the identification of ikaite (Shahar *et al.*, 2005) and various crystalline and amorphous phases of silica (Hemley *et al.*, 1986) including the soft mode transition from stishovite to the $CaCl_2$ structure (Kingma *et al.*, 1995). Elastic properties can be deduced from Raman measurement of lattice phonons as demonstrated by the estimation of c_{44} modulus of h.c.p.-iron from the Raman E_{2g} phonon (Merkel *et al.*, 2000) and an intermolecular potential study based on the H_2 Raman lattice phonon (Hemley *et al.*, 1990). Raman spectroscopy is the most definitive diagnostic tool for the identification of specific molecules or molecular species. For instance, it was used as the definitive evidence for the X-ray induced breakdown of H_2O into H_2 and O_2 at high pressure (Mao *et al.*, 2006b) (**Figure 14**). The multiplicity and shift of molecular Raman peaks have also been used as diagnostic evidence for phase transitions in simple molecular compounds (Mao *et al.*, 2002; Santoro *et al.*, 2006; Somayazulu *et al.*, 1996).

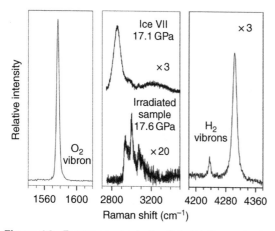

Figure 14 Raman spectra for irradiated H_2O sample at 17.6 GPa showing O_2 and H_2 vibrons which indicates cleaving of the OH bonds and formation of O–O and H–H bonds. Raman spectra of the an unirradiated ice VII sample at 17.1 GPa is shown in the second panel for comparison. From Mao WL, Mao HK, Sturhahn W, *et al.* (2006c) Iron-rich post-perovskite and the origin of ultralow-velocity zones. *Science* 312: 564–565.

9.4.4 Brillouin Spectroscopy and Impulsive Stimulated Light Scattering Spectroscopy

Much of our information on the detailed structure of the deep Earth has been derived from seismic observations in the form of wave velocities (i.e., in the laboratory, velocities of acoustic waves in a crystal can be obtained at high pressure with Brillouin spectroscopy (Bassett and Brody, 1977)), in which a laser beam is scattered by the moving acoustic waves in the sample and the Doppler shift of the laser frequency reveals the wave velocity. Unlike Raman shifts which are usually tens to thousands of wave numbers (cm^{-1}) from the elastic line, Brillouin shifts are typically less than $1\ cm^{-1}$. A multiple (3-6) pass Fabry–Perot interferometer is typically used for gaining the extremely high resolution and rejecting the background caused by Raleigh scattering of the laser line. By measuring velocities of a primary (longitudinal) wave, V_P, and two secondary (shear) wave, V_S, as functions of orientation in a high-pressure single crystal, the elasticity tensor (s_{ij} or c_{ij}) can be completely characterized. Refractive indices and phase transitions at high pressure can also be investigated with Brillouin spectroscopy as by-products (Brody *et al.*, 1981).

Brillouin spectroscopy has been used extensively for measuring elasticity tensor of major mantle minerals and glasses to lower-mantle pressures. The large database includes single-crystal forsterite (Duffy *et al.*, 1995b; Zha *et al.*, 1996), $(Mg,Fe)_2SiO_4$ olivine (Zha *et al.*, 1998), wadsleyite (Zha *et al.*, 1997), hydrous ringwoodite (Wang *et al.*, 2006) pyrope, grossular, andradite (Conrad *et al.*, 1999), silicate perovskite (Jackson *et al.*, 2004b, 2005c; Sinogeikin *et al.*, 2004), calcium oxide (Speziale *et al.*, 2006), magnesium oxide (Zha *et al.*, 2000), silica glass (Zha *et al.*, 1994), silicate glass (Tkachev *et al.*, 2005a, 2005b), and superhard nitrides (Tkachev *et al.*, 2003). High-pressure Brillouin spectroscopy yielded crucial information of single-crystal elasticity of hydrogen (Zha *et al.*, 1993) and helium (Zha *et al.*, 2004), the major component in the solar system, and provided experimental constraints to interior models of Jupiter (Duffy *et al.*, 1994). It is a powerful tool for studying intriguing mineral physics phenomena. For instance single-crystal elasticity of α-quartz was studied in a helium medium to 22 GPa, and the Brillouin measurements revealed a ferroelastic transition nature driven by softening of C_{44} through one of the Born stability criteria.

The extension of Brillouin system capabilities to elevated temperatures with resistive heating has enabled studies of the sound velocity of supercritical fluid hydrogen at high pressures which has application to understanding gas giants (Matsuishi *et al.*, 2002, 2003). Measurement of Brillouin spectra at simulated geotherm *P–T* conditions has been a long-sought after goal for experimental geophysics, but requires being able to reach significantly higher temperatures. Recently, the donut-absorber, double-sided, laser-heating technique has been developed for obtaining Brillouin spectra of H_2O up to 21.8 GPa and 1170 K (**Figure 15**) (Li *et al.*, 2006), opening many possibilities for high *P–T* Brillouin studies (Kuwayama *et al.*, 2005; Mao *et al.*, 2006c; Murakami *et al.*, 2004).

Single-crystal wave velocities have also been obtained up to 20 GPa with a related laser technique, the impulsive stimulated light scattering (ISLS) (Crowhurst *et al.*, 2004). In ISLS, a laser beam is split into two beams and recombined to form a standing wave grating; a second laser is scattered from the grating. The data are collected in the time domain, rather than the frequency domain as in the case of

Brillouin scattering. In addition to acoustic velocities, high-pressure thermal diffusivity can be obtained with ISLS (Chai *et al.*, 1996). Recent developments have combined these two techniques for simultaneous measurement in both the frequency and time domains (Crowhurst *et al.*, 2004).

9.4.5 Fluorescence Spectroscopy

In fluorescence spectroscopy, electron in the sample is first excited by incident photon to an excited energy state, and then the characteristic fluorescence photon is emitted when the electron falls back to the ground state. Energy levels in the sample are revealed as fluorescence photon energy equal to the energy difference of the two states (ΔE). Unlike Raman peaks which move with the excitation laser frequency ($\nu = \nu_{\text{laser}} - \Delta E$), fluorescence peaks appear at frequencies $\nu = \Delta E$ independent of the excitation laser frequency. Another major difference is that the Raman transition is instantaneous with <fs lifetime, whereas the fluorescence spectra have a wide range of lifetimes that can range from picoseconds to milliseconds and even up to seconds.

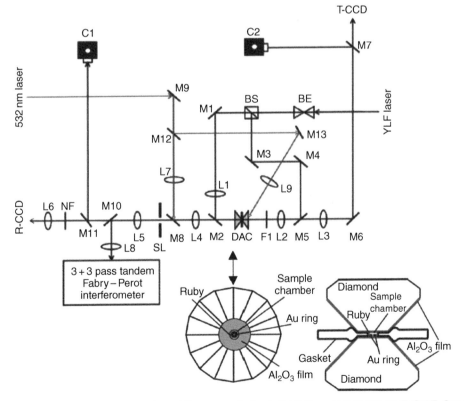

Figure 15 Laser heating system for Brillouin and Raman scattering in DAC. From Li F, Cui Q, He Z, Cui T, Gao C, Zhou Q, and Zou G (2006) Brillouin scattering spectroscopy for a laser heated diamond anvil cell. *Applied Physics Letters* 88: 203507.

The long lifetime characteristics of the fluorescence peaks are often exploited for distinguishing peaks with different origins (Eggert *et al.*, 1989).

Electronic states can be tuned continuously with pressure, or change abruptly at transitions. Monitoring the frequency and lifetime of the fluorescence as a function of pressure reveals detailed changes in electronic states, providing a valuable tool for mineral physics. Minerals containing transition or rare earth elements often emit rich fluorescence corresponding to their complicated *d* and *f* electron states. These minerals are valued for their wide range of luminescence, laser materials, and gemological applications. Fluorescence can also be caused by other nontransition element origins. For instance, under strain the diamond anvils can develop intense fluorescence (Mao and Hemley, 1991).

The pressure tuning of fluorescence spectra provides a scale for calibrating the pressure itself (Barnett *et al.*, 1973; Picard *et al.*, 2006). Unlike Raman peaks which only broaden slightly at high temperature, the commonly used ruby fluorescence peaks become too broad to be useful above 400°C (Sung, 1976). Fluorescence peaks originating from $4f$ electrons often remain sharper at high temperatures than the $3d$ electron Cr^{3+} fluorescence peaks of ruby, and rare earth doped YAG crystals can be used as a high *P–T* calibrant substitute for ruby. For instance, Sm:YAG (Hess and Schiferl, 1992; Liu and Vohra, 1994) and other rare earth doped oxide garnets (Hua *et al.*, 1996; Hua and Vohra, 1997) have been used for this purpose. The spectrometer is gated to optimize the characteristic lifetime of the fluorescence of calibrants while minimizing the thermal black-body radiations which has an infinite lifetime.

9.5 X-Ray Probes

X-rays are a versatile probe for reaching samples in the DAC and obtaining information by various diffraction and spectroscopic techniques. XRD has long been the bread-and-butter probe for studying structure *in situ* at high pressure. XRD is governed by the Bragg equation

$$\lambda = 2d \sin \theta \qquad [6]$$

where λ is the X-ray wavelength, d is the crystallographic interplanar spacing, and θ is the diffraction angle. The *d*-spacings of different lattice planes (*hkl*) can be determined by angular dispersive X-ray diffraction (ADXD) which uses monochromatic

X-radiation with a fixed λ and records the diffraction rings on two-dimensional detectors from which different θ are measured. Alternatively, *d*-spacings can also be determined by energy dispersive X-ray diffraction (EDXD) using polychromatic (white) X-radiation at a fixed θ and recording diffraction peaks at different energies with a solid-state, intrinsic germanium detector, where the X-ray energy (*E*) is related to the wavelength by

$$E\,(keV)^* \,\lambda\left(\overset{\circ}{A}\right) = 12.3986\left(keV^*\overset{\circ}{A}\right) \qquad [7]$$

Before 2000, most X-ray DAC experiments were conducted with XRD; except for a small number of X-ray absorption studies. Recent development at third generation synchrotron X-ray facilities has resulted in an enormous increase in the use of X-ray spectroscopy (XRS) that is revolutionizing DAC studies.

9.5.1 Axial XRD

With an incident X-ray beam passing through one anvil along the DAC compression axis, impinging upon the sample, and diffraction rings exiting through the second anvil, axial XRD has been long established as the standard method for studying high pressure crystallography, phase transitions, *P–V–T* EOS (Mao *et al.*, 1967; Mao *et al.*, 1990), lattice strain (Duffy *et al.*, 1995a), melting (Saxena *et al.*, 1995; Yoo *et al.*, 1995), and pressure-induced amorphization (Hemley *et al.*, 1988), as well as radial distribution functions of amorphous materials (Meade *et al.*, 1992). Ultrahigh *P–T* conditions are achieved at the expense of diminishing sample size leading to increasingly weaker XRD signals that are often overwhelmed by the background signals caused by the surrounding area which usually consists of high Z materials (such as the Re gasket). Successful experiments critically depend upon the ability to focus more photons into the minute sample region which is at the maximum *P–T* and cutting off the 'tail' of the focused beam outside of the maximum *P–T* region. Examples are too numerous to mention, but a few recent ones demonstrate what is possible at the current forefront of XRD DAC mineral physics research. In an *in-situ* XRD study of alumina up to 136 GPa and 2350 K, a phase transformation was discovered at 96 GPa. Rietveld full-profile refinements demonstrated that the high-pressure phase has the Rh_2O_3 (II) (*Pbcn*) structure. This phase is structurally related to corundum, but the AlO_6 polyhedra are highly distorted, with the interatomic bond lengths ranging from 1.690 to 1.847 Å at 113 GPa (Lin *et al.*, 2004a).

The Earth's D″ layer represents the lowermost 130–300 km of the silicate mantle just above the liquid outer core. In 2004, the post-perovskite (ppv) silicate phase was synthesized at core–mantle boundary conditions and characterized experimentally using XRD (Murakami *et al.*, 2004; Oganov and Ono, 2004). This discovery has generated considerable interest since ppv may be the most abundant mineral in the D″ layer (Helmberger *et al.*, 2005), and it may be able to explain a number of the complex seismic features of this region. It is critical to understand its crystal physics and to determine accurate values for the physical and chemical properties of this phase (Shieh *et al.*, 2006) as they provide important input and constraints for deep-Earth models and calculations. Unlike previously known lower-mantle silicate phases, ppv can accommodate a significant amount of iron (i.e., Fe/(Fe + Mg) > 0.6) (Mao *et al.*, 2005a) (**Figure 16**). The ultimate relevance of iron-rich ppv to the Earth's D″ layer depends on the Fe-Mg partitioning between the magnesiowüstite, ferromagnesian perovskite, and ppv silicate phases in the presence of a large reservoir of liquid iron (Kobayashi *et al.*, 2005; Murakami *et al.*, 2005).

9.5.2 Radial XRD

Contrary to the hydrostatic condition that is often viewed as the most desirable, the nonhydrostatic stress in samples compressed in solid medium was generally considered as a nuisance that degrades the experiments. The study of iron in a Be gasket (Mao *et al.*, 1998b) and wüstite in an amorphous boron gasket (Mao *et al.*, 1996) with radial XRD demonstrated that the nonhydrostatic stress, if properly characterized, contained a wealth of elastic and rheological information far more than the information from hydrostatic compression alone, and that hydrostatic and nonhydrostatic measurements are complementary.

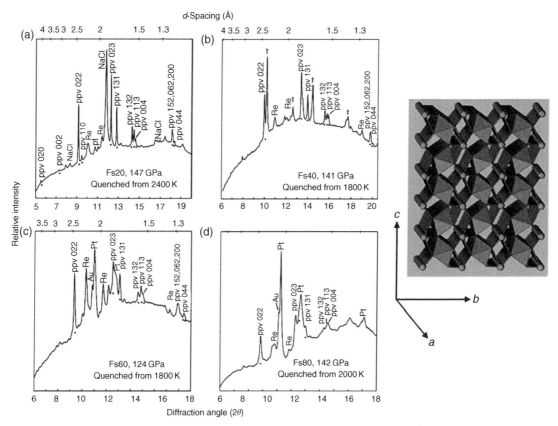

Figure 16 X-ray diffraction patterns for (a) Fs20 ($Mg_{0.8}Fe_{0.2}SiO_3$) at 147 GPa ($\lambda = 0.3888$ Å); (b) Fs40 at 141 GPa ($\lambda = 0.4233$ Å); (c) Fs60 at 124 GPa ($\lambda = 0.4008$ Å); and (d) Fs80 at 142 GPa ($\lambda = 0.4057$ Å). ppv peaks are additionally marked with *. Re oxide peaks are marked with †. The crystal structure of ppv is shown on the right where silicon octahedral are shown in blue and iron and magnesium cations are shown in orange.

Elasticity information can be obtained from probing stress–strain relations in a uniaxially compressed sample. Superimposing a uniaxial compression onto a polycrystalline aggregate already under a confining pressure (P) produces a deviatoric strain (ε_ψ) that is a function of the angle (ψ) between the direction of the strain and the compression axis,

$$\varepsilon_\psi(hkl) = \frac{d_\psi(hkl) - d_\mathrm{P}(hkl)}{d_\mathrm{P}(hkl)} = \left(1 - 3\cos^2\psi\right) \cdot Q\,(hkl) \quad [8]$$

where $d_\psi(hkl)$ is the lattice spacing measured at ψ, and $d_\mathrm{P}(hkl)$ is the lattice spacing under hydrostatic pressure. The slope $Q(hkl)$ is a constant for elastically isotropic materials, but varies with hkl for elastically anisotropic materials, and is an excellent quantitative measure of the direction and degree of elasticity anisotropy. Singh *et al.* (1998a) further developed the formulism for different crystal systems for calculating the full elastic tensor (c_{ij}'s) and velocity anisotropy based on $Q(hkl)$ and additional information from the axial compressibilities (χ_a, χ_b) and the aggregate shear modulus (G) which come from the EOS and NRIXS data, respectively.

To measure strain as a function of ψ, (Kinsland and Bassett, 1976) developed a DAC that had a large side opening which could access the full range of ψ by directing X-rays perpendicular to the diamond axis (radial direction) onto the sample. Mao *et al* (1998b) made modifications and developed a panoramic DAC which extended radial XRD capabilities to megabar pressures by use of an X-ray transparent Be gasket, and analyzed the elasticity of h.c.p.-iron to 220 GPa using the Singh formulism. This technique has also been used successfully for studying elasticity and strength of mantle silicates and oxides below 100 GPa (Kavner, 2003; Shieh *et al.*, 2002; Shieh *et al.*, 2004).

There have been concerns surrounding assumptions related with inverting the polycrystalline XRD data to calculate the single-crystal elastic tensor. A more recent ambient temperature radial XRD study to lower pressure (30 GPa) of the single-crystal elasticity of h.c.p.-iron demonstrated progress in constraining uncertainties in the radial XRD technique (Merkel *et al.*, 2005). Besides improvements that come with using large area detectors which enable precise measurements of the variation of *d*-spacing with orientation, as well as texture analysis, the inversion of elastic moduli was more sophisticated, including effects of lattice preferred orientation in the analysis (Matthies *et al.*, 2001), and using the

constraint of the c_{44} elastic modulus measured using Raman spectroscopy (Merkel *et al.*, 2000). For this lower pressure study, the shape of the anisotropy that was obtained was the same as previous work by Mao *et al.*, 1998b (i.e., in the body diagonal between the *a* and *c*-axes), but the overall magnitude was reduced.

In a very recent study of an analog material h.c.p.-cobalt (Merkel *et al.*, 2006b), further demonstrated that calculations need to include the orientation dependence of differential stress resulting from plastic deformation. Studies using inelastic X-ray scattering spectroscopy (see Section 9.5.9) on single-crystal h.c.p.-cobalt indicate that the fast direction is in the *c*-axis (Antonangeli *et al.*, 2005) which matches well with theory whereas radial XRD results on polycrystalline h.c.p.-cobalt show a fast body diagonal. The error from radial XRD comes from the large shear strength anisotropy in this system and may be a significant problem for using radial XRD studies on polycrystalline h.c.p.-iron to interpret the source of inner-core anisotropy.

The shear strength, elastic moduli, elastic anisotropy, and deformation mechanisms of the two main end-member phases of the lower-mantle MgO and $(Mg_{0.9}Fe_{0.1})SiO_3$-perovskite have been studied using radial XRD up to 47 GPa and 32 GPa at ambient temperature (Merkel *et al.*, 2002; Merkel *et al.*, 2003). It was demonstrated that the elastic moduli of MgO obtained with the radial diffraction method are in agreement with the more precise Brillouin spectroscopy studies. The uniaxial stress component in the polycrystalline MgO sample is found to increase rapidly to 8.5 GPa at a pressure of 10 GPa in all experiments. The uniaxial stress supported by the perovskite aggregate is found to increase continuously with pressure up to 10.9 GPa at 32 GPa. A comparison between the experimental textures and results from polycrystal plasticity gives information regarding the preferred orientation and deformation mechanism of MgO under very high confining pressure at room temperature. Under axial compression, a strong cubic texture developed in MgO could be the source of seismic anisotropy. On the other hand, the measurements displayed no development of significant lattice preferred orientations in the perovskite sample, indicating that deformation by dislocation glide is not the dominant deformation mechanism under these conditions. Assuming that the underlying cause for seismic anisotropy in the deep Earth is elastic anisotropy combined with lattice preferred orientation, these results indicate that silicate perovskite deformed under the conditions of this experiment would not be the source of seismic anisotropy.ᵀ

The radial XRD technique has more recently been implemented with CCD detectors, allowing collection of two-dimensional (δ- and *d*-spacing) diffraction with a monochromatic X-ray beam. The sample is compressed in a wide-angle panoramic DAC and heated with the double-sided laser-heating technique. A study which monitored the development of lattice preferred orientation in a radial XRD experiment on $MgGeO_3$ ppv to above 100 GPa, found that the (100) and (110) slip systems dominate the plastic deformation in this germinate analog. The results provide a new mechanism for development of D" seismic anisotropy (Merkel *et al.*, 2006a).

Both elasticity and rheology can be highly temperature dependent. The next frontier to be explored is radial XRD study *in situ* at high *P–T* conditions analogous to those in the Earth's deep interior. A study of b.c.c.- and h.c.p.-iron with angle-dispersive radial X-ray diffraction measurements under nonhydrostatic conditions up to 30 GPa and 1000 K in laser-heated DAC found that b.c.c.-iron developed preferred orientation compatible with observations under ambient conditions. The preferred orientation of the b.c.c.-phase is inherited by the h.c.p.-phase in accordance with the Burgers orientation relationship, consistent with a martensitic nature of the phase transition. A comparison between the observed texture in h.c.p.-iron with results from polycrystal plasticity modeling suggests that the predominant deformation mechanisms are basal slip and prismatic slip presumably associated with minor mechanical twinning rather than the slip systems that were proposed based on analogy to Cr–Ni alloys (Mao *et al.*, 1998b; Merkel *et al.*, 2004; Merkel *et al.*, 2005).

9.5.3 Single-Crystal XRD

Single-crystal XRD experiments are unique and important sources of structural information crucial for understanding the microscopic mechanisms of high-pressure phenomena. High-pressure single-crystal XRD burgeoned with the invention of 'Merrill–Bassett' DAC (Merrill and Bassett, 1974) that provided a simple, robust hardware, and the development of data collection and processing software by Larry Finger (Hazen and Finger, 1982). A large quantity of high-precision, high-pressure, crystallography data have been collected on single-crystal mineral samples, including troilite (King and

Prewitt, 1978), coesite (Levien and Prewitt, 1981), albite (Benusa *et al.*, 2005), wüstite (Jacobsen *et al.*, 2005), $MgSiO_3$-perovskite (Ross and Hazen, 1989; Vanpeteghem *et al.*, 2006), sillimanite, andalusite (Burt *et al.*, 2006), bernalite (Welch *et al.*, 2005), and mica (Smyth *et al.*, 2000). Subtle, displacive, high-pressure, phase transitions were discovered in single-crystal pyroxnes (Arlt *et al.*, 2000; Jackson *et al.*, 2004a). New rules governing the compression of ionic radii (Shannon and Prewitt, 1970) and tilting of polyhedra were proposed (Angel *et al.*, 2005; Hazen and Finger, 1982). Single-crystal data are used to resolve some very difficult crystallographic problems, for example, the recent solution of the O8 chain structure (Lundegaard *et al.*, 2006).

Compared to other ultrahigh-pressure DAC methods, which typically cover pressure ranges as high as 300 GPa, the pressure range for single-crystal XRD lags far behind. Reports of single-crystal XRD structural analysis is limited to 30 GPa, above which there are only a handful of studies which only provide determination of the orientation matrix and unit cell parameters (e.g., hydrogen with the simple h.c.p. structure to 119 GPa (Loubeyre *et al.*, 1996)), and not full structure refinement. A high-pressure workshop was organized to address the problem of single-crystal X-ray diffraction structure refinement at megabar pressures (Dera *et al.*, 2005), and many different approaches were proposed. The problem will be solvable with sufficient effort. A solution to the problem is emerging with the recent development of the panoramic DAC (Mao *et al.*, 2001) which allows nearly full access around the equatorial plane, as well as both poles of the DAC, and the micro-single-crystal XRD method (Ice *et al.*, 2005; Larson *et al.*, 2002; Tamura *et al.*, 2002), which combines microfocusing, EDXD, and ADXD for single-crystal XRD with sub-μm resolution (**Figure 17**).

9.5.4 X-Ray Absorption Spectroscopy (XAS)

In XAS, the incident X-ray is absorbed when its energy exceeds the excitation energy of deep-core electrons of a specific element in the sample, causing an edge-like absorption spectrum. The X-ray absorption near-edge structure (XANES) within tens of eV to the edge provides information on the symmetry-projected conduction band DOS that is related to the electronic properties including oxidation state, crystal-field splitting, and magnetic spin pairing (Ablett *et al.*, 2003; Jackson *et al.*, 1993). The extended

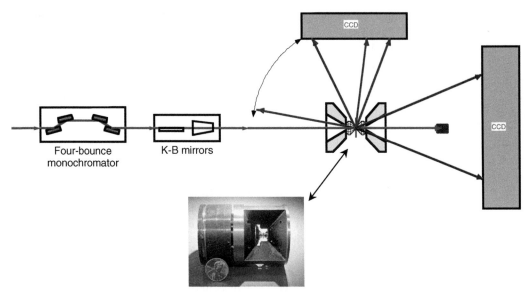

Figure 17 Integrated high-pressure micro-single-crystal XRD and XRS with panoramic DAC for both forward and 90° Laue diffraction and spectroscopy.

X-ray absorption fine structure (EXAFS) to several hundreds of eV above the edge provides local structure information that is particularly valuable for amorphous materials or in low concentrations that XRD method is inapplicable (Anderson *et al.*, 2002; Bassett *et al.*, 2000; Mayanovic *et al.*, 2003). These powerful, element-specific probes reveal the electronic and structural changes taking place at high pressure (Itie *et al.*, 1992).

XAS was one of the first XRS techniques applied to high-pressure synchrotron radiation research, but was previously limited to higher energies due to absorption by the diamond anvils. In addition, Laue diffraction of single-crystal diamonds causes sharp absorption peaks that are detrimental to EXAFS spectroscopy. To greatly reduce the amount of diamond in the beam path, holes are drilled in diamond anvils in the newly developed Bassett-type hydrothermal DAC for studying L-edge of rare earth ions and K-edge of transition element ions in supercritical aqueous solutions at moderate *P–T* conditions (**Figure 18**) (Anderson *et al.*, 2002). The results reveal structural relaxations of ligand-anion complex at increasing *P–T* in mineralizing hydrothermal fluid.

Aternative beam paths through high-strength beryllium and boron gaskets are used in the new panoramic DAC at higher pressures (Mao *et al.*, 2001). Both techniques extend XAS down to 4 keV which covers the K-edges of first-row transition elements and the L-edges of rare earth elements. Specific electronic, magnetic, and optical properties of these elements are crucial for a wide range of applications in materials sciences and technology.

9.5.5 X-Ray Emission Spectroscopy (XES)

In XES, core electrons in the sample are excited by X-rays and subsequently produce fluorescence photons. The energies of the fluorescence photons are analyzed with sub-eV energy resolution of the emission spectral lineshape to provide valuable information on the filled electronic states of the sample. With the panoramic DAC which extends the low end of the high-pressure energy window down to 4 keV, deep-core electron elements heavier than calcium can now be studied at high pressure by a suitable choice of analyzer crystals. Transition-element ions, with their variable valence and magnetic states, control major geochemical processes and geophysical behavior, such as oxidation and reduction, chemical differentiation, elasticity, geomagnetism, conductivity, and radiative heat transfer. Magnetic collapse in transition-metal monoxides has been predicted from first-principles computations at the pressures of the lower mantle and core (Cohen *et al.*, 1997). XES provides a unique probe for the diagnosis of pressure-induced magnetic spin collapse in transition elements. The ultrahigh pressures in the lower mantle and core squeeze atoms and electrons so closely together that they interact differently than under normal conditions, even forcing electrons to pair.

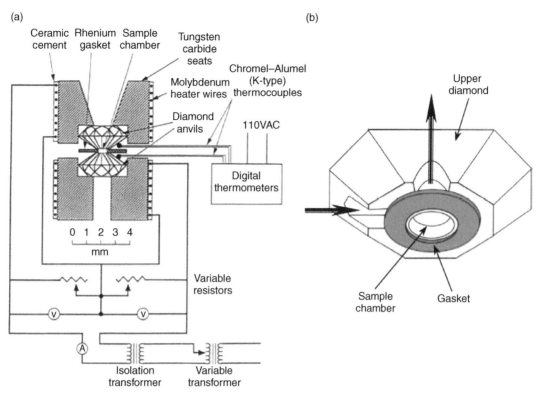

Figure 18 (a) A schematic diagram of a Bassett-type hydrothermal diamond anvil cell (HDAC). (b) The schematic diagram of the upper diamond of the modified HDAC showing the entrance and exit grooves for the incident and fluorescence X-rays, respectively. The laser drilled sample chamber is located in the center of the upper diamond anvil face above the hole in the gasket. From Anderson AJ, Jayanetti S, Mayanovic RA, Bassett WA, and Chou IM (2002) X-ray spectroscopic investigations of fluids in the hydrothermal diamond anvil cell: The hydration structure of aqueous La^{3+} up to 300°C and 1600 bars. *American Mineralogist* 87: 262–268.

In high pressure XES experiments, the incident beam energy is fixed, the emission X-radiation can be collected at any direction, and high-resolution emission spectra are obtained by a synchronized θ-2θ scan of the analyzer and the detector. The excitation X-ray source only needs to have higher energy than that of the fluorescent photons; energy bandwidth (resolution) of the X-ray source is inconsequential. In principle, white (a broad, smooth continuum of energy), pink (an unfiltered undulator beam with energy band path of hundreds eV), and monochromatic (a narrow, eV band path of photons) X-rays can all be used as the source. In practice, if sufficient intensity is available, the monochromatic source is preferred because white and pink X-radiations often cause higher backgrounds and damage the diamond anvils.

The first high-pressure XES study was conducted by collecting the iron K$_\beta$ emission in troilite (FeS) using the white X-ray source the National Synchrotron Light Source, and a high-spin to low-spin transition at 4 GPa coinciding with the FeS I-II structural transition was successfully observed (Rueff *et al.*, 1999). Subsequently with higher source intensity, this technique has been applied with pink source and monochromatic sources at the European Synchrotron Radiation Facility and the Advanced Photon Source for studies of high-spin low-spin transitions in key mantle and core materials and have been extended to ultrahigh pressures above 100 GPa (Badro *et al.*, 1999, 2002, 2003, 2004; Li *et al.*, 2004; Lin *et al.*, 2005a). The results like the reported transitions of iron in ferromagnesian silicate perovskite, the most abundant mineral in the Earth from a high-spin to an intermediate-spin and low-spin state (Badro *et al.*, 2004; Li *et al.*, 2004) and electronic transitions in magnesiowüstite, the second most abundant mineral from a high-spin to a low-spin state (Badro *et al.*, 2003; Lin *et al.*, 2005a) has profound implications on the properties of the lower mantle and place important constraints on models of the deep mantles and cores of terrestrial planets.

9.5.6 Inelastic X-Ray Scattering – Near-Edge Spectroscopy

Near core-electron absorption edge features measured by soft X-ray absorption (XANES) or electron energy loss spectroscopy (EELS) reveal rich information about the nature of chemical bonding. Such information is particularly pronounced and important for light elements, but has been inaccessible for high-pressure studies as the pressure vessel completely blocks the soft X-ray and electron beams. This problem has recently been overcome by applying the X-ray inelastic near-edge spectroscopy (XINES) technique to high pressure. In high-pressure XINES, the high-energy incident X-ray penetrates the pressure vessel and reaches the sample. The scattered photon loses a portion of energy corresponding to the *K*-edge of the low-*Z* sample, but can still exit the vessel to be registered on the analyzer-detector system (**Figure 3**). Inelastic *K*-edge scattering spectra of second-row elements from Li (56 eV) to O (543 eV) at high pressures have been successfully observed at high pressures. This opened a vast, new field of near *K*-edge spectroscopy of the second-row elements.

In XINES experiments, single-crystal analyzers collect scattered X-radiation, and focus it to the detector in a nearly backscattering geometry, thus fixing the energy at the elastic line. The incident X-ray energy is scanned relative to the elastic line to determine the inelastic (Raman) shift. XINES features are relatively insensitive to *q*, and thus the 2θ angle is set at an angle to optimize the intensity, and multiple analyzers can be used to increase the counting rate without concern for their differences in *q*. XINES studies of bonding changes in graphite (Mao *et al.*, 2003a) (**Figure 19**), hexagonal boron nitride (Meng *et al.*, 2004), and H_2O (Cai *et al.*, 2005; Mao *et al.*, 2006b) demonstrate the great promise of this technique and can be easily extended to studies of bonding changes in Earth materials. For example, a more recent study of the near K-edge structure of

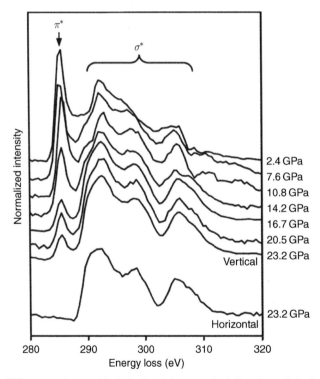

Figure 19 High-pressure IXS spectra for graphite in horizontal and vertical directions plotted as normalized scattered intensity versus energy loss (incident energy–analyzer energy). The scattered intensity is normalized to the incoming intensity. The lower-energy peak, labeled π, corresponds to $1s - \pi^*$ transitions and the higher energy portion, labeled σ, corresponds to $1s - \sigma^*$ transitions. The bottom spectra, taken in the horizontal direction probes bonds in the a-plane and does not show any π-bonding before and after the high-pressure transition. The top seven spectra, taken in the vertical direction, probe the c-plane. After the transition, the σ bonds increase at the expense of the π bonds. From Mao WL, Mao HK, Eng P, *et al.* (2003a) Bonding changes in compressed superhard graphite. *Science* 302: 425–427.

boron and oxygen in B_2O_3 glass to 22.5 GPa, revealed pressure-induced coordination changes which have clear relevance for understanding the bonding in melts and glass in the deep Earth (Lee *et al.*, 2005).

9.5.7 Resonance X-Ray Inelastic Spectroscopy

In resonant inelastic X-ray scattering (RIXS) experiments, both the incident and scattered X-ray energies need to be scanned. The incident X-ray energy is scanned across the core absorption, similar to the procedures for XAS. For each monochromatic incident X-ray energy, fluorescence spectra are collected by scanning analyzers, similar to the procedures of XES spectroscopy. There have been a growing number of RIXS studies of core excitations (Kotani and Shin, 2001) where the final states of the inelastic scattering process in these studies are localized shallow core excitations. These shallow core excitations are the same excitations routinely probed by soft XAS, and are rich in multiplet structures for these highly correlated electronic systems. In fact, the high-energy scale parameters of the highly correlated systems are routinely derived from fitting these spectra with model calculations. For magnetic samples, excitation with circularly polarized X-rays can also provide information on spin-resolved electronic structure, for example, the L_3 edge of Gd (DeGroot *et al.*, 1997; Krisch *et al.*, 1996).

One can obtain selected information connected directly with a specific intermediate state to which the incident photon energy is tuned. RIXS studies of transition-metal and rare-earth systems give us important information on the electronic states, such as the intra-atomic multiplet coupling, electron correlation, and interatomic hybridization. Unusual phase transitions driven by electron correlation effects occur in many transition metals and transition metal compounds at high pressures, some accompanied with large volume collapses (5–17% in rare earth metals and 10–15% in magnetic $3d$ transition metal oxides, for example). However, the exact nature of these transitions is not well understood, including the relationships between the crystal and electronic structures and the role of magnetic moment and order. A RIXS and XES study of Gd metal to 113 GPa suggests Kondo-like aspects in the delocalization of $4f$ electrons (Maddox *et al.*, 2006). Analysis of the RIXS data reveals a prolonged and continuous delocalization with volume throughout

the entire pressure range, so that the volume-collapse transition at 59 GPa is only part of the phenomenon.

Since both incident and scattered photons are hard X-rays that are capable of entering and exiting the DAC, RIXS enable high-pressure *in-situ* spectroscopic studies of rich shallow levels which were previously restricted to soft X-ray, vacuum environment. Multiple electronic levels in transition-metal and rare-earth ions (Hill *et al.*, 1998; Kao *et al.*, 1996), complexes, and clusters in minerals or solutions can be investigated in detail in DAC. For instance, the central question about the redox conditions (Frost *et al.*, 2004; McCammon, 2005) of deep-mantle geochemistry relies on experimental determination of Fe^{3+}/Fe^{2+} ratio of minerals synthesized and quenched from high *P-T* conditions, and analyzed in vacuum with iron *L2,3*-edge ELNES spectroscope (Garvie *et al.*, 2004; Sinmyo *et al.*, 2006; van Aken and Liebscher, 2002). Caliebe *et al.* (1998) demonstrate that for hematite, the $1s2p$ RIXS process can be exploited to obtain spectroscopic information similar to that obtained in *L2,3*-absorption spectroscopy, thus opening the possibility of direct, *in-situ* study of redox at ultrahigh pressure and temperature.

9.5.8 Nuclear Resonant X-Ray Spectroscopy

Important dynamic, thermodynamic, and elastic information on a material, including vibrational kinetic energy, zero-point vibrational energy, vibrational entropy, vibrational heat capacity, Debye temperature, Grüneisen parameter, thermal expansivity, longitudinal velocity, shear velocities, bulk modulus, and shear modulus can be derived from knowledge of the phonon DOS. Studies of the phonon DOS are greatly simplified if the element involved has an isotope which exhibits a nuclear resonant Mössbauer effect. In nuclear resonant X-ray spectroscopy (NRXS), an in-line high resolution monochromator is used to narrow down the photon energy to meV resolution and fine-tune the monochromatic X-ray near the exceedingly narrow nuclear resonant (elastic) line. Avalanche photodiodes (APD) are used to collect only the signal from nuclear resonance absorption, and reject nonresonantly scattered radiation. The APD directly downstream of the sample collects nuclear resonant forward scattering (NRFS) spectra which can provide a precise determination of the hyperfine interaction parameter and Lamb-Mössbauer factor (Jackson *et al.*, 2005b). The APDs surrounding the sample radially collect the nuclear resonant inelastic X-ray scattering (NRIXS) signal which is a result of creation (Stokes) or annihilation (anti-Stokes) of phonons as the incident X-ray beam is scanned over a small range (approximately ±100 meV) around the resonant energy. This phonon excitation spectrum can then be used to determine and calculate phonon DOS (Sturhahn *et al.*, 1995). Although NRIXS measures the phonon DOS rather than the dispersion curve, the DOS is still rich in vibrational and dynamic information. The initial slope gives V_D which is related to V_P and V_S by

$$\frac{3}{V_D^3} = \frac{1}{V_P^3} + \frac{2}{V_S^3} \qquad [9]$$

which is related to the bulk modulus (K) and density (ρ) from X-ray diffraction data by

$$V_\phi^2 = V_P^2 - (4/3)V_S^2 = K/\rho \qquad [10]$$

These two equations can be used to calculate V_P and V_S. It should be noted that unlike the bulk sound speed (V_ϕ) which is more heavily dependent on V_P, V_D is more heavily dependent on V_S, so NRIXS data is particularly useful for constraining the shear

properties. The integration of the whole DOS provides values for key thermodynamic properties necessary for interpreting seismological observations. Theory can be used to deconvolute the dispersion curve from the DOS.

NRIXS has been developed for the Mössbauer isotope ^{57}Fe and can be used to study samples with a minor amount of iron (Hu *et al.*, 2003). After demonstration of DOS measurements on b.c.c.-iron at ambient conditions (Lübbers *et al.*, 2000), NRIXS measurements have been extended to high pressure to obtain the phonon DOS of h.c.p.-iron up to 153 GPa (Mao *et al.*, 2001) and iron-related partial phonon DOS for iron alloys (Lin *et al.*, 2004c; Lin *et al.*, 2003c; Mao *et al.*, 2004) (**Figure 20**), and have attracted particular attention among physicists and geophysicists because iron is an archetypal transition element and is a dominant component in the cores of the Earth and other terrestrial planets (Mars, Venus, and Mercury). Using this method also carries the assumption that the sample contains a single, ^{57}Fe-bearing phase in which iron is evenly distributed which needs to be verified using XRD and microscopic techniques on quenched samples.

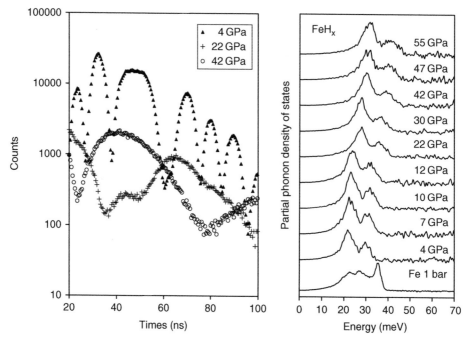

Figure 20 Left, NRFS spectra for FeH$_x$. Data below 22 GPa show fast oscillations diagnostic of a magnetic sample. At 22 GPa the magnetic field becomes very small and above 22 GPa it is negligible. Right, NRIXS spectra showing the partial phonon density of states for FeH$_x$ at high pressure. For comparison, the density of states of pure iron at ambient pressure is shown. From Mao WL, Strurhahn W, Heinz DL, Mao HK, Shu J, and Hemley RJ (2004) Nucler resonant X-ray scattering of iron hydride at high pressure. *Geophysical Research Letters* 31: L15618 (doi:10.1029/2004GL020541).

Separation of iron into different phases would skew the V_D toward the velocity of the slower phase and give a misleading result.

More recently the sound velocities of h.c.p.-iron were determined at high *P–T* by NRIXS in a laser-heated diamond anvil cell at pressures up to 73 GPa and at temperatures up to 1700 K (Lin *et al.*, 2005b). The compressional wave velocities and shear wave velocities of h.c.p.-iron decreased significantly with increasing temperature, and thus cannot be fit to a linear relation, Birch's law, which has guided geophysicists in making correlations of seismic (sound) velocity and density for the past half-century. This result may indicate means that there are more light elements in Earth's core than have been inferred from a linear extrapolation of data collected at ambient temperature.

Experiments simulating the high *P–T* conditions of the core-mantle boundary observed that a large amount of iron incorporated into the ppv silicate phase can significantly change its properties relative to the pure Mg end-member (Mao *et al.*, 2005a). The determined aggregate compressional and shear wave velocities of this iron-rich silicate at high pressure show that a silicate with up to 40% iron end-member can match many of the seismic features observed in ultralow velocity zones above the core-mantle boundary. Mantle dynamics may lead to accumulation of an iron-rich silicate reaction veneer between the iron-poor lower mantle and the iron-rich outer core into ultralow velocity patches observable by seismology (Mao *et al.*, 2006c).

NRIXS is a very promising technique and is currently the only experimental method for constraining shear properties, but there are limitations. The NRIXS process is very inefficient. To obtain a spectra with reasonable counting statistics at an optimized third-generation synchrotron source requires a combination of long counting times (often on the order of hours), maximized sample thickness, and compositions with higher Mössbauer isotope content. This becomes increasingly difficult as researchers try to conduct experiments at higher pressures (which reduces sample size), high temperatures (which limits counting time since the sample must be stably heated over the counting interval and sample thickness since thermal insulating layers must be added), and mantle compositions (which have minor amounts of iron in their compositions). Future improvements with this technique include upgrades at the synchrotron beamline (e.g., increasing flux by adding undulators, improving focus, increasing detector efficiency,

etc.), modifications in the DAC design (to allow additional increased access to detectors), and use of superhard gasket inserts to increase sample thickness (Section 9.2.2).

9.5.9 Nonresonant Phonon Inelastic X-Ray Scattering

Detailed study of phonon dynamics in highly compressed materials is necessary for understanding vibrational thermodynamic properties, elasticity, and phase transition mechanisms, but until very recently has eluded high-pressure experimental investigation. Nonresonant phonon inelastic X-ray scattering (PIXS) is a powerful method which holds great promise for studying phonon dynamics of Earth's materials at high pressure, but this technique is still beset with many challenges. For studying phonons, the energy transfers involved are in the meV range (six orders of magnitude smaller than the incident X-rays) which makes the detection of the energy shifts due to phonon creation and annihilation challenging (Burkel, 2000). Also, the double differential scattering cross-section of PIXS is very small, making such an experiment very demanding even with the brilliance of third-generation synchrotron undulator sources. In addition, there are limitations in sample size and the requirement of navigating the full beam through the high-pressure vessel and separating the very weak sample signal from the strong background signal of the vessel.

Ideally one would want to measure the full phonon dispersion curve which contains all the elasticity information and vibrational thermodynamic properties. Unlike optical techniques which are limited to phonons at or very near the zone center, the advantage for IXS phonon studies is access to the full range of momentum transfers (q), allowing measurement of the full dispersion curve. Single-crystal PIXS can map out the entire phonon dispersion curve (Schwoerer-Böhning and Macrander, 1998). Single-crystal PIXS studies are potentially providing the same amount of information as inelastic neutron scattering (Klotz and Braden, 2000), and have recently been expended to high pressures (Antonangeli *et al.*, 2005; Farber *et al.*, 2006; Occelli *et al.*, 2001; Ruf *et al.*, 2001).

It has been established that the Earth's inner core exhibits elastic anisotropy with compressional waves traveling approximately 3% faster in the radial versus equatorial directions (e.g., (Morelli *et al.*, 1986; Song and Richards, 1996; Tromp, 2001)) which makes understanding the elasticity of iron at inner-core conditions critical for interpreting seismic observations. Currently, static measurements have not reached inner-core *P–T*, so one role of experimental results is the verification of theory which is not limited in *P–T* range but have produced conflicting results for the elastic anisotropy of h.c.p.-iron. High-pressure PIXS studies of polycrystalline iron (since large single-crystal samples of h.c.p.-iron cannot be synthesized) which measure phonon excitations corresponding to scattering of the incident X-ray photons by longitudinal acoustic (LA) phonons have been conducted to over $100\,GPa$ (Antonangeli *et al.*, 2004; Fiquet *et al.*, 2001). These studies show great promise for studying elasticity, but they were conducted axially which limits the access of the uniaxial strain field that develops in the DAC, and a preferred orientation of the sample is assumed rather than measured. Also, thus far, V_S has never been detected in IXSS studies of h.c.p.-iron, and the measurement is limited to V_P. While the study by Antonangeli *et al.* (2004) shows variations in V_P for 50° and 90° relative to the DAC compression axis, multiple points over the full angular range are needed to constrain the shape of the velocity anisotropy. A future advance would be the use of an X-ray transparent gasket which would allow access to the full angular range and determination of the preferred orientation of the sample.

References

Ablett JM, Kao CC, Shieh SR, Mao HK, Croft M, and Tyson TA (2003) High-pressure X-ray near-edge absorption study of thallium rhenium oxide up to 10.86 GPa. *High Pressure Research* 23: 471–476.

Alfé D, Gillan MJ, and Price GD (1999) The melting curve of iron at pressures of the Earth's core from *ab initio* calculations. *Nature* 401: 462–463.

Alfé D, Gillan MJ, and Price GD (2000) Constraints on the composition of the Earth's core from *ab initio* calculations. *Nature* 405: 172–175.

Anderson AJ, Jayanetti S, Mayanovic RA, Bassett WA, and Chou IM (2002) X-ray spectroscopic investigations of fluids in the hydrothermal diamond anvil cell: The hydration structure of aqueous La³⁺ up to 300°C and 1600 bars. *American Mineralogist* 87: 262–268.

Anderson OL (1997) Iron: Beta phase frays. *Science* 278: 821–822.

Andrault D, Fiquet G, Kunz M, Visocekas F, and Häusermann D (1997) The orthorhombic structure of iron: An *in situ* study at high-temperature and high-pressure. *Science* 278: 831–834.

Angel RJ, Zhao J, and Ross NL (2005) General rules for predicting phase transitions in perovskites due to octahedral tilting. *Physical Review Letters* 95: 025503.

Antonangeli D, Krisch M, Fiquet G, Badro J, Farber DL, Bossak A, and Merkel S (2005) Aggregate and single-

crystalline elasticity of hcp cobalt at high pressure. *Physical Review B* 72: 134303.

Antonangeli D, Occelli F, Requardt H, Badro J, Fiquet G, and Krisch M (2004) Elastic anisotropy in textured hcp-iron to 112 GPa from sound wave propagation measurements. *Earth and Planetary Science Letters* 225: 243–251.

Arlt T, Kunz M, Stolz J, Armbruster T, and Angel RJ (2000) P-T-X data on P2₁/c clinopyroxenes and their displacive phase transitions. *Contributions to Mineralogy and Petrology* 138: 35–46.

Badro J, Fiquet G, Guyot F, *et al.* (2003) Iron partitioning in Earth's mantle: Toward a deep lower mantle discontinuity. *Science* 300: 789–791.

Badro J, Fiquet G, Struzhkin VV, *et al.* (2002) Nature of the high-pressure transition in Fe₂O₃ hematite. *Physical Review Letters* 89: 205504.

Badro J, Rueff J-P, Vanko G, Monaco G, Fiquet G, and Guyot F (2004) Electronic transitions in perovskite: Possible nonconvecting layers in the lower mantle. *Science* 305: 383–385.

Badro J, Struzhkin VV, Shu J, *et al.* (1999) Magnetism in FeO at megabar pressures from X-ray emission spectroscopy. *Physical Review Letters* 83: 4101–4104.

Barnett JD, Block S, and Piermarini GJ (1973) An optical fluorescence system for quantitative pressure measurement in the diamond-anvil cell. *Review of Scientific Instruments* 44: 1–9.

Bassett WA (2001) The birth and development of laser heating in diamond anvil cells. *Review of Scientific Instruments* 72: 1270–1272.

Bassett WA (2003) High pressure–temperature aqueous systems in the hydrothermal diamond anvil cell (HDAC). *European Journal of Mineralogy* 15: 773–780.

Bassett WA, Anderson AJ, Mayanovic RA, and Chou IM (2000) Hydrothermal diamond anvil cell for XAFS studies of first-row transition elements in aqueous solution up to supercritical conditions. *Chemical Geology* 167: 3–10.

Bassett WA and Brody EM (1977) Brillouin scattering: A new way to measure elastic moduli at high pressures. In: Manghnani M and Akimoto S (eds.) *High Pressure Research – Applications in Geophysics*, pp. 519–532. New York: Academic Press.

Bassett WA, Shen AH, Bucknum M, and Chou I-M (1993) A new diamond anvil cell for hydrothermal studies to 2.5 GPa and from -190 to 1200°C. *Review of Scientific Instruments* 64: 2340–2345.

Bassett WA and Takahashi T (1965) Silver iodide polymorphs. *American Mineralogist* 50: 1576–1594.

Bell PM and Mao HK (1981) Degree of hydrostaticity in He, Ne and Ar pressure-transmitting media. *Carnegie Institution of Washington, Year Book* 80: 404–406.

Belonoshko AB, Ahuja R, and Johansson B (2003) Stability of the body-centred-cubic phase of iron in the Earth's inner core. *Nature* 424: 1032–1034.

Benedetti LR and Loubeyre P (2004) Temperature gradients, wavelength-dependent emissivity, and accuracy of high and very-high temperatures measured in the laser-heated diamond cell. *High Pressure Research* 24: 423–445.

Benusa MD, Angel RJ, and Ross NL (2005) Compression of albite, NaAlSi₃O₈. *American Mineralogist* 90: 1115–1120.

Besson JM and Pinceaux JP (1979) Melting of helium at room temperature and high pressure. *Science* 206: 1073–1075.

Birch F (1952) Elasticity and constitution of the Earth's interior. *Journal of Geophysical Research* 57: 227–286.

Boehler R (1993) Temperatures in the Earth's core from melting-point measurements of iron at high static pressures. *Nature* 363: 534–536.

Boehler R and De Hantsetters K (2004) New anvil designs in diamond-cells. *High Pressure Research* 24: 391–396.

Brazhkin V, Dubrovinskaia N, Nicol M, *et al.* (2004) What does 'harder than diamond' mean? *Nature Materials* 3: 576–577.

Brody EM, Shimizu H, Mao HK, Bell PM, and Bassett WA (1981) Acoustic velocity and refractive index of fluid hydrogen and deuterium at high pressures. *Journal of Applied Physics* 52: 3583–3585.

Buffett BA (2000) Earth's core and the geodynamo. *Science* 288: 2007–2012.

Buffett BA and Wenk H-R (2001) Texturing of the Earth's inner core by Maxwell stresses. *Nature* 413: 60–62.

Burchard M, Zaitsev AM, and Maresch WV (2003) Extending the pressure and temperature limits of hydrothermal diamond anvil cells. *Review of Scientific Instruments* 74: 1263–1266.

Burkel E (2000) Phonon spectroscopy by inelastic X-ray scattering. *Reports on Progress in Physics* 63: 171–232.

Burt JB, Ross NL, Angel RJ, and Koch M (2006) Equations of state and structures of andalusite to 9.8 GPa and sillimanite to 8.5 GPa. *American Mineralogist* 91: 319–326.

Cai YQ, Mao HK, Chow PC, *et al.* (2005) Ordering of hydrogen bonds in high-pressure low-temperature H_2O. *Physical Review Letters* 94: 025502.

Caliebe WA, Kao C-C, Hastings JB, *et al.* (1998) 1s2p resonant inelastic X-ray scattering in α-Fe_2O_3. *Physical Review B* 58: 13452–13458.

Chai M, Brown JM, and Slutsky LJ (1996) Thermal diffusivity of mantle minerals. *Physics and Chemistry of Minerals* 23: 470–475.

Chijioke AD, Nellis WJ, Soldatov A, and Silvera IF (2005) The ruby pressure standard to 150 GPa. *Journal of Applied Physics* 98: 114905.

Cohen RE, Mazin, II, and Isaak DE (1997) Magnetic collapse in transition metal oxides at high pressure: Implications for the Earth. *Science* 275: 654–657.

Conrad PG, Zha CS, Mao HK, and Hemley RJ (1999) The high pressure, single-crystal elasticity of pyrope, grossular, and andradite. *American Mineralogist* 84: 374–383.

Crowhurst JC, Goncharov AF, and Zaug JM (2004) Impulsive stimulated light scattering from opaque materials at high pressure. *Journal of Physics: Condensed Matter* 16: S1137–S1142.

Cui L, Chen NH, and Silvera IF (1995) Infrared properties of ortho and mixed crystals of solid deuterium at megabar pressures and the question of metallization in the hydrogen. *Physical Review Letters* 74: 4011–4014.

Cynn H, Yoo CS, and Sheffield SA (1999) Phase transition and decomposition of 90% hydrogen peroxide at high pressures. *Journal of Chemical Physics* 110: 6836–6843.

Daniels WB and Ryschkewitsch MG (1983) Simple double diaphragm press for diamond anvil cells at low temperatures. *Review of Scientific Instruments* 54: 115–116.

DeGroot FMF, Nakazawa M, Kotani A, Krisch MH, and Sette F (1997) Theoretical analysis of the magnetic circular dichroism in the 2p3d and 2p4d X-ray emission of Gd. *Physical Review B* B56: 7285.

Degtyareva O, Gregoryanz E, Somayazulu M, Dera P, Mao HK, and Hemley RJ (2005) Novel chain structures in group VI elements. *Nature Materials* 4: 152–155.

Dera P, Prewitt CT, and Jacobsen SD (2005) Structure determination by single-crystal X-ray diffraction (SXD) at megabar pressures. *Journal of Synchrotron Radiation* 12: 547–548.

Dewaele A and Loubeyre P (2005) Mechanical properties of tantalum under high pressure. *Physical Review B* 72: 134106-1–9.

Dewaele A, Loubeyre P, and Mezouar M (2004) Equations of state of six metals above 94 GPa. *Physical Review B* 70: 094112-1-8.

Ding Y, Liu H, Xu J, Prewitt CT, Hemley RJ, and Mao HK (2005) Zone-axis diffraction study of pressure induced inhomogeneity in single-crystal $Fe_{1-x}O$. *Applied Physics Letters* 87: 041912.

Dubrovinsky LS, Saxena SK, and Lazor P (1997) X-ray study of iron with *in-situ* heating at ultrahigh pressure. *Geophysical Research Letters* 24: 1835–1838.

Dubrovinsky LS, Saxena SK, Tutti F, and Rekhi S (2000) *In situ* X-ray study of thermal expansion and phase transition of iron at multimegabar pressure. *Physical Review Letters* 84: 1720–1723.

Duffy TS, Hemley RJ, and Mao HK (1995a) Equation of state and shear strength at multimegabar pressures: Magnesium oxide to 227 GPa. *Physical Review Letters* 74: 1371–1374.

Duffy TS, Vos WL, Zha CS, Hemley RJ, and Mao HK (1994) Sound velocities in dense hydrogen and the interior of Jupiter. *Science* 263: 1590–1593.

Duffy TS, Zha CS, Downs RT, Mao HK, and Hemley RJ (1995b) Elastic constants of forsterite Mg_2SiO_4 to 16 GPa. *Nature* 378: 170–173.

Eggert JH, Goettel KA, and Silvera IF (1989) Ruby at high pressure. II. Fluorescence lifetime of the R line to 130 GPa. *Physical Review B* 40: 5733–5738.

Ekimov EA, Sidorov VA, Bauer ED, *et al.* (2004) Superconductivity in diamond. *Nature* 428: 542–545.

Eremets MI, Gregoryanz E, Struzhkin VV, *et al.* (2000) Electrical conductivity of xenon at megabar pressures. *Physical Review Letters* 85: 2797–2800.

Eremets MI, Hemley RJ, Mao HK, and Gregoryanz E (2001a) Semiconducting non-molecular nitrogen up to 240 GPa and its low-pressure stability. *Nature* 411: 170–174.

Eremets MI, Struzhkin VV, Hemley RJ, and Mao HK (2001b) Superconductivity in boron. *Science* 293: 272–274.

Eremets MI and Timofeev YA (1992) Miniature diamond anvil cell: Incorporating a new design for anvil alignment. *Review of Scientific Instruments* 63: 3123–3126.

Farber DL, Krisch M, Antonangeli D, *et al.* (2006) Lattice dynamics of molybdenum at high pressure. *Physical Review Letters* 96: 115502-1-4.

Fei Y (1999) Effects of temperature and composition on the bulk modulus of (Mg,Fe)O. *American Mineralogist* 84: 272–276.

Fei Y and Mao HK (1994) *In situ* determination of the NiAs phase of FeO at high pressure and high temperature. *Science* 266: 1678–1680.

Fei Y, Mao HK, Shu JF, Parthasarathy G, and Bassett WA (1992) Simultaneous high P–T X-ray diffraction study of b-$(Mg,Fe)_2SiO_4$ to 26 GPa and 900 K. *Journal of Geophysical Research* 97: 4489–4495.

Fiquet G, Badro J, Guyot F, Requardt H, and Krisch M (2001) Sound velocities in iron to 110 gigapascals. *Science* 291: 468–471.

Frank MR, Fei Y, and Hu J (2004) Constraining the equation of state of fluid H_2O to 80 GPa using the melting curve, bulk modulus and thermal expansivity of Ice VII. *Geochimica et Cosmochimica Acta* 68: 2781–2790.

Frost DJ, Liebske C, Langenhorst F, *et al.* (2004) Experimental evidence for the existence of iron-rich metal in the Earth's lower mantle. *Nature* 428: 409–412.

Garvie LAJ, Zega TJ, Rez P, and Buseck PR (2004) Nanometer-scale measurements of $Fe^{3+}/\Sigma Fe$ by electron energy-loss spectroscopy: A cautionary note. *American Mineralogist* 89: 1610–1616.

Gillet P (1999) Introduaction to Raman spectroscopy at extreme pressure and temperature aconditions. In: Wright K and Catlow R (eds.) *Microscopic Properties and Processes in Materials*, pp. 43–69. Dordecht, The Netherlands: Kluwer Academic Publishers.

Glatzmaier GA and Roberts PH (1996) Rotation and magnetism of Earth's inner core. *Science* 274: 1887–1891.

Glazkov VP, Besedin SP, Goncharenko IN, *et al.* (1988) Investigation of the equation of state of molecular deuterium at high pressure using neutron diffraction. *JETP Letters* 47: 661–664.

Goncharov AF, Crowhurst JC, Dewhurst JK, and Sharma S (2005) Raman spectroscopy of cubic boron nitride under extreme conditions of high pressure and temperature. *Physical Review B* 72: 100104 R.

Goncharov AF, Gregoryanz E, Hemley RJ, and Mao HK (2001) Spectroscopic studies of the vibrational and electronic properties of solid hydrogen to 285 GPa. *Proceedings of the National Academy of Sciences* 98: 14234–14237.

Goncharov AF, Mazin, II, Eggert JH, Hemley RJ, and Mao HK (1995) Invariant points and phase transitions in deuterium at megabar pressures. *Physical Review Letters* 75: 2514–2517.

Goncharov AF and Struzhkin VV (2003) Raman spectroscopy of metals, high-temperature superconductors and related materials under high pressure. *Journal of Raman Spectroscopy* 34: 532–548.

Goncharov AF, Struzhkin VV, and Jacobsen SD (2006) Reduced radiative conductivity of low spin (Mg,Fe)O in the lower mantle. *Science* 312: 1205–1208.

Goncharov AF, Struzhkin VV, Somayazulu M, Hemley RJ, and Mao HK (1996) Compression of ice to 210 GPa: Evidence for a symmetric hydrogen bonded phase. *Science* 273: 218–220.

Gregoryanz E, Degtyareva O, Somayazulu M, Hemley RJ, and Mao HK (2005) Melting of dense sodium. *Physical Review Letters* 94: 185502-1–185502-4.

Haenen K, Nesladek M, De Schepper L, Kravets R, Vanecek M, and Koizumi S (2004) The phosphorous level fine structure in homoepitaxial and polycrystalline n-type CVD diamond. *Diamond and Related Materials* 13: 2041–2045.

Hanfland M, Hemley RJ, and Mao HK (1993) Novel infrared vibron absorption in solid hydrogen at megabar pressures. *Physical Review Letters* 70: 3760–3763.

Hanfland M and Syassen K (1989) Optical reflectivity of graphite under pressure. *Physical Review B* 40: 1951–1954.

Hazen RM and Finger LW (1982) *Comparative Crystal Chemistry: Temperature, Pressure, Composition, and the Variation of Crystal Structure*, 231 pp. New York: Wiley.

He D, Zhao Y, Sheng TD, Schwartz RB, *et al.* (2003) Bulk metallic glass gasket for high pressure, *in situ* X-ray diffraction. *Review of Scientific Instruments* 74: 3012–3016.

Helmberger D, Lay T, Ni S, and Gurnis M (2005) Deep mantle structure and the postperovskite phase transition. *Proceedings of the National Academy of Sciences* 102: 17257–17263.

Hemley RJ, Goncharov AF, Lu R, Struzhkin VV, Li M, and Mao HK (1998) High-pressure synchrotron infrared spectroscopy at the National Synchrotron Light Source. *Nuovo Cimento* 20: 539–551.

Hemley RJ, Hanfland M, and Mao HK (1991) High-pressure dielectric measurements of hydrogen to 170 GPa. *Nature* 350: 488–491.

Hemley RJ, Jephcoat AP, Mao HK, Ming LC, and Manghnani M (1988) Pressure-induced amorphization of crystalline silica. *Nature* 334: 52–54.

Hemley RJ and Mao HK (1998) Static compression experiments on low-Z planetary materials. In: Manghnani MH and Yagi T (eds.) *Geophysical Monograph Vol. 101: Properties of the Earth and Planetary Materials at High Pressure and Temperature*, pp. 173–183. Washington, DC: American Geophysical Union.

Hemley RJ and Mao HK (2001a) *In-situ* studies of iron under pressure: New windows on the Earth's core. *International Geology Review* 43: 1–30.

Hemley RJ and Mao HK (2001b) Progress in cryocrystals at megabar pressures. *Journal of Low Temperature Physics* 122: 331–344.

Hemley RJ, Mao HK, Bell PM, and Mysen BO (1986) Raman spectroscopy of SiO$_2$ glass at high pressure. *Physical Review Letters* 57: 747–750.

Hemley RJ, Mao HK, Shen G, *et al.* (1997) X-ray imaging of stress and strain of diamond, iron and tungsten at megabar pressures. *Science* 276: 1242–1245.

Hemley RJ, Mao HK, and Shu JF (1990) Low-frequency vibrational dynamics of hydrogen at ultrahigh pressures. *Physical Review Letters* 65: 2670–2673.

Hemley RJ, Soos ZG, Hanfland M, and Mao HK (1994) Charge-transfer states in dense hydrogen. *Nature* 369: 384–387.

Hess NJ and Schiferl D (1992) Comparison of the pressure-induced frequency shift of Sm:YAG to the ruby and nitrogen vibron pressure scales from 6 to 820 K and 0 to 25 GPa and suggestions for use as a high-temperature pressure calibrant. *Journal of Applied Physics* 71: 2082–2086.

Hill JP, Kao C-C, Caliebe WA, *et al.* (1998) Resonant inelastic X-ray scattering in Nd$_2$CuO$_4$. *Physical Review Letters* 80: 4967–4970.

Ho SS, Yan CS, Liu Z, Mao HK, and Hemley RJ (2006) Prospects for large single crystal CVD diamond. *Industrial Diamond Review* 1/06: 28–32.

Holzapfel WB (1972) On the symmetry of the hydrogen bonds in ice VII. *Journal of Chemical Physics* 56: 712–715.

Hu MY, Sturhahn W, Toellner TS, *et al.* (2003) Measuring velocity of sound with nuclear resonant inelastic X-ray scattering. *Physical Review B* 67: 094304-1–094304-4.

Hua H, Mirov S, and Vohra YK (1996) High-pressure and high-temperature studies on oxide garnets. *Physical Review B* 54: 6200–6209.

Hua H and Vohra YK (1997) Pressure-induced blueshift of Nd^{3+} fluorescence emission in YAlO$_3$: Near infrared pressure sensor. *Applied Physics Letters* 71: 2602–2604.

Ice GE, Dera P, Liu W, and Mao HK (2005) Adapting polychromatic X-ray microdiffraction techniques to high-pressure research: energy scan approach. *Journal of Synchrotron Radiation* 12: 608–617.

Irifune T, Kurio A, Sakamoto S, Inoue T, and Sumiya H (2003) Ultrahard polycrystalline diamond from graphite. *Nature* 421: 599–600.

Itie J, Baudelet F, Dartyge E, Fontaine A, Tolentino H, and San-Miguel A (1992) X-ray absorption spectroscopy and high pressure. *High Pressure Research* 8: 697–702.

Jackson DD, Aracne-Ruddle C, Malba V, Weir ST, Catledge SA, and Vohra YK (2003) Magnetic susceptibility measurements at high pressure using designer diamond anvils. *Review of Scientific Instruments* 74: 2467–2471.

Jackson DD, Malba V, Weir ST, Baker PA, and Vohra YK (2005a) High-pressure magnetic susceptibility experiments on the heavy lanthanides Gd, Tb, Dy, Ho, Er and Tm. *Physical Review B* 71: 184416-1-7.

Jackson JM, Sinogeikin SV, Carpentei MA, and Bass JD (2004a) Novel phase transition in orthoenstatite. *American Mineralogist* 89: 239–244.

Jackson JM, Sturhahn W, Shen GY, *et al.* (2005b) A synchrotron mossbauer spectroscopy study of (Mg,Fe)SiO$_3$ perovskite up to 120 GPa. *American Mineralogist* 90: 199–205.

Jackson JM, Zhang J, and Bass JD (2004b) Sound velocities and elasticity of aluminous MgSiO3 perovskite: Implications for aluminum heterogeneity in Earth's lower mantle. *Geophysical Research Letters* 31: 10614.

Jackson JM, Zhang J, Shu J, Sinogeikin SV, and Bass JD (2005c) High-pressure sound velocities and elasticity of aluminous MgSiO$_3$ perovskite to 45 GPa: Implications for lateral heterogeneity in Earth's lower mantle. *Geophysical Research Letters* 32: 21305.

Jackson WE, Deleon JM, Brown GE, and Waychunas GA (1993) High-temperature XAS study of Fe_2SiO_4 liquid: Reduced coordination of ferrous iron. *Science* 262: 229–233.

Jacobsen SD, Lin JF, Angel RJ, *et al.* (2005) Single-crystal synchrotron X-ray diffraction study of wustite and magnesiowustite at lower-mantle pressures. *Journal of Synchrotron Radiation* 12: 577–583.

Jephcoat AP and Besedin SP (1996) Temperature measurement and melting determination in laser-heated diamond-anvil cells. *Phil. Trans. R. Soc. Lond. A* 354: 1333–1360.

Kao C-C, Caliebe WA, Hastings JB, and Gillet J-M (1996) Resonant Raman scattering in NiO: Resonant enhancement of the charge transfer excitation. *Physical Review B* 54: 16361–16364.

Karato S (1999) Seismic anisotropy of the Earth's inner core resulting from flow induced by Maxwell stresses. *Nature* 402: 871–873.

Kavner A (2003) Elasticity and strength of hydrous ringwoodite at high pressure. *Earth and Planetary Science Letters* 214: 645–654.

Kawamura H, Tachikawa K, and Shimomura O (1985) Diamond anvil cell for cryogenic temperature with optical measurement system. *Review of Scientific Instruments* 56: 1903–1906.

Khvostantsev LG, Vereshchagin LF, and Novikov AP (1977) Device of toroid type for high pressure generation. *High Temperatures – High Pressures* 9: 637–639.

King HE and Prewitt CT (1978) FeS phase-transitions at high-pressures and temperatures. *Physics and Chemistry of Minerals* 3: 72–73.

Kingma KJ, Cohen RE, Hemley RJ, and Mao HK (1995) Transformation of stishovite to a denser phase at lower-mantle pressures. *Nature* 374: 243–245.

Kinsland GL and Bassett WA (1976) Modification of the diamond cell for measuring strain and the strength of materials at pressures up to 300 kilobar. *Review of Scientific Instruments* 47: 130–132.

Klotz S and Braden M (2000) Phonon dispersion of bcc iron to 10 GPa. *Physical Review Letters* 85: 3209–3212.

Kobayashi Y, Kondo T, Ohtani E, *et al.* (2005) Fe–Mg partitioning between (Mg, Fe)SiO3 post-perovskite, perovskite and magnesiowüstite in the Earth's lower mantle. *Geophysical Research Letters* 32: L19301–1–4.

Kotani A and Shin S (2001) Resonant inelastic X-ray scattering spectra for electrons in solids. *Reviews of Modern Physics* 73: 203–246.

Krauss G, Reifler H, and Steurer W (2005) Conically shaped single-crystalline diamond backing plates for a diamond anvil cell. *Review of Scientific Instruments* 76: 105104–105105.

Krisch MH, Sette F, Bergmann U, *et al.* (1996) Observation of magnetic circular dichroism in resonant inelastic X-ray scattering at the L3 edge of gadolinium metal. *Physical Review B* 54: R12673–R12676.

Kuwayama Y, Hirose K, Sata N, and Ohishi Y (2005) The pyrite-type high-pressure form of silica. *Science* 309: 923–925.

Laio A, Bernard S, Chiarotti GL, Scandolo S, and Tosatti E (2000) Physics of iron at earth's core conditions. *Science* 287: 1027–1030.

Larson BC, Yang W, Ice GE, Budal JD, and Tischler JZ (2002) Three-dimensional X-ray structural microscopy with submicrometre resolution. *Nature* 415: 887–890.

Lee SK, Eng PJ, Mao HK, Meng Y, Newville M, Hu MY, and Shu J (2005) Probing of bonding changes in B_2O_3 glasses at high pressure with inelastic X-ray scattering. *Nature Materials* 4: 851–854.

LeToullec R, Loubeyre P, Pinceaux JP, Mao HK, and Hu J (1992) Single crystal X-ray diffraction with a synchrotron source in a MDAC at low temperature. *High Pressure Research* 8: 691–696.

Levien L and Prewitt CT (1981) High-pressure crystal structure and compressibility of coesite. *American Mineralogist* 66: 324–333.

Li F, Cui Q, He Z, Cui T, Gao C, Zhou Q, and Zou G (2006) Brillouin scattering spectroscopy for a laser heated diamond anvil cell. *Applied Physics Letters* 88: 203507.

Li J, Struzhkin VV, Mao HK, *et al.* (2004) Electronic spin state of iron in the Earth's lower mantle. *Proceedings of the National Academy of Sciences* 101: 14027–14030.

Lin J-F, Degtyareva O, Prewitt CT, *et al.* (2004a) Crystal structure of a high-pressure/high-temperature phase of alumina by in situ X-ray diffraction. *Nature Materials* 3: 389–393.

Lin J-F, Heinz DL, Campbell AJ, Devine JM, and Shen G (2002) Iron-silicon alloy in Earth's core? *Science* 395: 313–315.

Lin JF, Heinz DL, Mao HK, *et al.* (2003a) Stability of magnesiowüstite in Earth's lower mantle. *Proceedings of the National Academy of Sciences* 100: 4405–4408.

Lin J-F, Santoro M, Struzhkin VV, Mao HK, and Hemley RJ (2004b) In situ high pressure–temperature Raman spectroscopy technique with laser-heated diamond anvil cells. *Review of Scientific Instruments* 75: 3302–3306.

Lin JF, Shu J, Mao HK, Hemley RJ, and Shen G (2003b) Amorphous boron gasket in diamond anvil cell research. *Review of Scientific Instruments* 74: 4732–4736.

Lin J-F, Struzhkin VV, Jacobsen SD, *et al.* (2005a) Spin transition of iron in magnesiowüstite in the Earth's lower mantle. *Nature* 436: 377–380.

Lin JF, Struzhkin VV, Sturhahn W, *et al.* (2003c) Sound velocities of iron–nickel and iron–silicon alloys at high pressures. *Geophysical Research Letters* 30: doi: 10.1029/2003GL018405.

Lin J-F, Sturhahn W, Zhao J, Shen G, Mao HK, and Hemley RJ (2004c) Absolute temperature measurement in a laser-heated diamond anvil cell. *Geophysical Research Letters* 31: doi: 10.1029/2004GL020599.

Lin J-F, Sturhahn W, Zhao J, Shen G, Mao HK, and Hemley RJ (2005b) Sound velocities of hot dense iron: Birch's law revisited. *Science* 308: 1892–1894.

Liu J and Vohra YK (1994) Sm: YAG optical pressure sensor to 180 GPa: Calibration and structural disorder. *Applied Physics Letters* 64: 3386–3388.

Liu J, Xiao WS, and Li XD (2005) Laser-heated diamond anvil cell technique combined with synchrotron X-ray light source and its applications to Earth's interior material research. *Earth Science Frontiers* 12: 93–101.

Liu LG and Bassett WA (1975) The melting of iron up to 200 kilobars. *Journal of Geophysical Research* 80: 3777–3782.

Loubeyre P, LeToullec R, Häusermann D, *et al.* (1996) X-ray diffraction and equation of state of hydrogen at megabar pressures. *Nature* 383: 702–704.

Loubeyre P, Letoullec R, Pinceaux JP, Mao HK, Hu J, and Hemley RJ (1993) Equation of state and phase diagram of solid ^4He from single-crystal X-ray diffraction over a large P-T domain. *Physical Review Letters* 71: 2272–2275.

Lübbers R, Grünsteudel HF, Chumakov AI, and Wortmann G (2000) Density of phonon states in iron at high pressure. *Science* 287: 1250–1253.

Lundegaard LF, Weck G, McMahon MI, Desgreniers S, and Loubeyre P (2006) Observation of an O_8 molecular lattice in the ε phase of solid oxygen. *Nature* 443: 201–204.

Ma Y, Somayazulu M, Shen G, Mao HK, Shu J, and Hemley RJ (2004) In situ X-ray diffraction studies of iron to Earth-core conditions. *Physics of the Earth and Planetary Interiors* 143–144c: 455–467.

Maddox BR, Lazicki A, Yoo C-S, *et al.* (2006) 4f delocalization in Gd: Inelastic X-ray scattering at ultrahigh pressure. *Physical Review Letters* 96: 215701–1–4.

Mao HK, Badro J, Shu J, Hemley RJ, and Singh AK (2006a) Strength, anisotropy and preferred orientation of solid argon at high pressures. *Journal of Physics: Condensed Matter* 18: S963–S968.

Mao HK, Bassett WA, and Takahashi T (1967) Effect of pressure on crystal structure and lattice parameters of iron up to 300 kilobars. *Journal of Applied Physics* 38: 272–276.

Mao HK and Bell PM (1972) Electrical conductivity and the red shift of absorption in olivine and spinel at high pressure. *Science* 176: 403–406.

Mao HK and Bell PM (1976) High-pressure physics: The 1-megabar mark on the ruby R_1 static pressure scale. *Science* 191: 851–852.

Mao HK, Bell PM, and England JL (1971) Tensional errors and drift of thermocouple electromotive force in the single-stage, piston-cylinder apparatus. *Carnegie Institution of Washington, Year Book* 70: 281–287.

Mao HK, Bell PM, and Hadidiacos C (1987a) Experimental phase relations of iron to 360 kbar, 1400° C determined in an internally heated diamond-anvil apparatus. In: Manghnani MH and Syono Y (eds.) *High-Pressure Research in Mineral Physics*, pp. 135–138. Washington, DC: Terra Scientific.

Mao HK, Bell PM, Shaner J, and Steinberg D (1978) Specific volume measurements of Cu, Mo, Pd and Ag and calibration of the ruby R_1 fluorescence pressure gauge from 0.06 to 1 Mbar. *Journal of Applied Physics* 49: 3276–3283.

Mao HK and Hemley RJ (1991) New optical transitions in diamond at ultrahigh pressures. *Nature* 351: 721–724.

Mao HK and Hemley RJ (1994) Ultrahigh-pressure transitions in solid hydrogen. *Reviews of Modern Physics* 66: 671–692.

Mao HK and Hemley RJ (1996) Experimental studies of the Earth's deep interior: Accuracy and versatility of diamond cells. *Philosophical Transactions of the Royal Society of London A* 354: 1315–1333.

Mao HK, Hemley RJ, and Chao ECT (1987b) The application of micro-Raman spectroscopy to analysis and identification of minerals in thin section. *Scanning Microscopy* 1: 495–501.

Mao HK, Hemley RJ, and Mao AL (1994) Recent design of ultrahigh-pressure diamond cell. In: Schmidt SC, Shaner JW, Samara GA, and Ross M (eds.) *High Pressure Science and Technology – 1993*, vol. 2, pp. 1613–1616. New York: AIP Press.

Mao HK, Shen G, Hemley RJ, and Duffy TS (1998a) X-ray diffraction with a double hot-plate laser-heated diamond cell. In: Manghnani and Yagi T (eds.) *Geophysical Monograph Vol. 101: Properties of the Earth and Planetary Materials at High Pressure and Temperature*, pp. 27–34. Washington, DC: American Geophysical Union.

Mao HK, Shu J, Fei Y, Hu J, and Hemley RJ (1996) The wüstite enigma. *Physics of the Earth and Planetary Interiors* 96: 135–145.

Mao HK, Shu J, Shen G, Hemley RJ, Li B, and Singh AK (1998b) Elasticity and rheology of iron above 220 GPa and the nature of the Earth's inner core. *Nature* 396: 741–743.

Mao HK, Wu Y, Chen LC, Shu JF, and Jephcoat AP (1990) Static compression of iron to 300 GPa and $Fe_{0.8}Ni_{0.2}$ alloy to 260 GPa: Implications for composition of the core. *Journal of Geophysical Research* 95: 21737–21742.

Mao HK, Xu J, and Bell PM (1986) Calibration of the ruby pressure gauge to 800 kbar under quasihydrostatic conditions. *Journal of Geophysical Research* 91(B5): 4673–4676.

Mao HK, Xu J, Struzhkin VV, Shu J, et al. (2001) Phonon density of states of iron up to 153 GPa. *Science* 292: 914–916.

Mao WL and Mao HK (2006) Ultrahigh-pressure experiment with a motor-driven diamond anvil cell. *Journal of Physics: Condensed Matter* 18: S1069–S1073.

Mao WL, Mao HK, Eng P, et al. (2003a) Bonding changes in compressed superhard graphite. *Science* 302: 425–427.

Mao WL, Mao HK, Goncharov AF, et al. (2002) Hydrogen clusters in clathrate hydrate. *Science* 297.

Mao WL, Mao HK, Meng Y, et al. (2006b) X-ray-induced dissociation of H_2O and formation of an O_2–H_2 alloy at high pressure. *Science* 314: 636.

Mao WL, Mao HK, Sturhahn W, et al. (2006c) Iron-rich post-perovskite and the origin of ultralow-velocity zones. *Science* 312: 564–565.

Mao WL, Mao HK, Yan CS, Shu J, Hu JZ, and Hemley RJ (2003b) Generation of ultrahigh pressures using single-crystal chemical vapor deposition diamond anvils. *Applied Physics Letters* 83: 5190–5192.

Mao WL, Meng Y, Shen G, et al. (2005a) Iron-rich silicates in the Earth's D'' layer. *Proceedings of the National Academy of Sciences* 102: 9751–9753.

Mao WL, Sturhahn W, Heinz DL, Mao HK, Shu J, and Hemley RJ (2004) Nuclear resonant X-ray scattering of iron hydride at high pressure. *Geophysical Research Letters* 31: L15618 (doi:10.1029/2004GL020541).

Mao WL, Struzhkin VV, Mao HK, and Hemley RJ (2005b) Pressure–temperature stability of the van der Waals compound $(H_2)_4CH_4$. *Chemical Physical Letters* 242: 66–70.

Matsuda YH, Uchida K, Ono K, Ji Z, and Takeyama S (2004) Development of a plastic diamond anvil cell for high pressure magneto-photoluminescence in pulsed high magnetic fields. *International Journal of Modern Physics B* 18: 3843–3846.

Matsuishi K, Gregoryanz E, Mao HK, and Hemley RJ (2002) Brillouin and Raman scattering of fluid and solid hydrogen at high pressures and temperatures. *Journal of Physics: Condensed Matter* 14: 10631–10636.

Matsuishi K, Gregoryanz E, Mao HK, and Hemley RJ (2003) Equation of state and intermolecular interactions in fluid hydrogen from Brillouin scattering at high pressures and temperatures. *Journal of Chemical Physics* 118: 10683–10695.

Matthies S, Merkel S, Wenk HR, Hemley RJ, and Mao HK (2001) Effect of texture on the determination of elasticity of polycrystalline ε-iron from diffraction measurements. *Earth and Planetary Science Letters* 194: 201–212.

Mayanovic RA, Jayanetti S, Anderson AJ, Bassett WA, and Chou I-M (2003) Relaxation of the structure of simple metal ion complexes in aqueous solutions at up to supercritical conditions. *Journal of Chemical Physics* 118: 719–727.

Mazin, II, Hemley RJ, Goncharov AF, Hanfland M, and Mao HK (1997) Quantum and classical orientational ordering in solid hydrogen. *Physical Review Letters* 78: 1066–1069.

McCammon C (2005) The paradox of mantle redox. *Science* 308: 807–808.

McMillan PF, Hemley RJ, and Gillet P (1996) Vibrational spectroscopy of mantle minerals. In: Dyar MD, McCammon C, and Schaefer MW (eds.) *Mineral Spectroscopy: A Tribute to Roger Burns*, pp. 175–213. Houston: Geochemical Society.

Meade C, Hemley RJ, and Mao HK (1992) High pressure X-ray diffraction of SiO_2 glass. *Physical Review Letters* 69: 1387–1390.

Meng Y, Mao HK, Eng P, et al. (2004) BN under compression: The formation of sp3 bonding. *Nature Materials* 3: 111–114.

Merkel S, Goncharov AF, Mao HK, Gillet P, and Hemley RJ (2000) Raman spectroscopy of iron to 152 gigapascals: Implications for Earth's inner core. *Science* 288: 1626–1629.

Merkel S, Hemley RJ, and Mao HK (1999) Finite element modeling of diamond deformation at multimegabar pressures. *Applied Physics Letters* 74: 656–658.

Merkel S, Kubo A, Miyagi L, et al. (2006a) Plastic deformation of $MgGeO_3$ post-perovskite at lower mantle pressures. *Science* 311: 644–646.

Merkel S, Miyajima N, Antonangeli D, Fiquet G, and Yagi T (2006b) Lattice preferred orientation and stress in polycrystalline hcp-Co plastically deformed under high pressure. *Journal of Applied Physics* 100: 23510.

Merkel S, Shu J, Gillet P, Mao HK, and Hemley RJ (2005) X-ray diffraction study of the single crystal elastic moduli of ε-Fe up to 30 GPa. *Journal of Geophysical Research* 110: B025201–B025212.

Merkel S, Wenk HR, Badro J, Montagnac G, Gillet P, Mao HK, and Hemley RJ (2003) Deformation of $(Mg_{0.9}Fe_{0.1})SiO_3$ perovskite aggregates up to 32 GPa. *Earth and Planetary Science Letters* 209: 351–360.

Merkel S, Wenk H-R, Gillet P, Mao HK, and Hemley RJ (2004) Deformation of polycrystalline iron up to 30 GPa and 1000 K. *Physics of the Earth and Planetary Interiors* 145: 239–251.

Merkel S, Wenk HR, Shu J, *et al.* (2002) Deformation of polycrystalline MgO at pressures of the lower mantle. *Journal of Geophysical Research* 107: doi: 10.1029/2001JB000920.

Merkel S and Yagi T (2005) X-ray transparent gasket for diamond anvil cell high pressure experiments. *Review of Scientific Instruments* 76: 046109–046113.

Merrill L and Bassett WA (1974) Miniature diamond-anvil cell for X-ray diffraction studies. *Review of Scientific Instruments* 45: 290–294.

Mills RL, Liebenberg DH, Bronson JC, and Schmidt LC (1980) Procedure for loading diamond cells with high-pressure gas. *Review of Scientific Instruments* 51: 891–895.

Ming LC and Bassett WA (1974) Laser heating in the diamond anvil press up to 2000°C sustained and 3000°C pulsed at pressures up to 260 kilobars. *Review of Scientific Instruments* 45: 1115–1118.

Mito M, Hitaka M, Kawae T, Takeda K, Kitai T, and Toyoshima N (2001) Development of miniature diamond anvil cell for the superconducting quantum interference device magnetometer. *Proceedinga. of the Symposium on Ultrasonic Electronics.* 40: 641–6644.

Morelli A, Dziewonski AM, and Woodhouse JH (1986) Anisotropy of the inner core inferred from PKIKP travel times. *Geophysical Research Letters* 13: 1545–1548.

Murakami M, Hirose K, Kawamura K, Sata N, and Ohishi Y (2004) Post-perovskite phase transition in $MgSiO_3$. *Science* 304: 855–858.

Murakami M, Hirose K, Sata N, and Ohishi Y (2005) Post-perovskite phase transition and mineral chemistry in the pyrolitic lowermost mantle. *Geophysical Research Letters* 32: L03304.

Niu F and Wen L (2001) Hemispherical variations in seismic velocity at the top of the Earth's inner core. *Nature* 410: 1081–1084.

Occelli F, Krisch M, Loubeyre P, *et al.* (2001) Phonon dispersion curves in an argon single crystal at high pressure by inelastic X-ray scattering. *Physical Review B* 63: 224306.

Oganov AR and Ono S (2004) Theoretical and experimental evidence for a post-perovskite phase of $MgSiO_3$ in Earth's D'' layer. *Nature* 430: 445–448.

Oger PM, Daniel I, and Picard A (2006) Development of a low-pressure diamond anvil cell and analytical tools to monitor microbial activities *in situ* under controlled P and T. *Biochimica et Biophysica Acta* 1764: 434–442.

Okuchi T (2004) A new type of nonmagnetic diamond anvil cell for nuclear magnetic resonance spectroscopy. *Physics of the Earth and Planetary Interiors* 143: 611.

Olson P and Aurnou J (1999) A polar vortex in the Earth's core. *Nature* 402: 170–173.

Picard A, Oger PM, Daniel I, Cardon H, Montagnac G, and Chervin J-C (2006) A sensitive pressure sensor for diamond anvil cell experiments up to 2GPa: FluoSpheres(R). *Journal of Applied Physics* 100: 34915.

Piermarini GJ, Block S, and Barnett JD (1973) Hydrostatic limits in liquids and solids to 100 kbar. *Journal of Applied Physics* 44: 5377–5382.

Piermarini GJ, Block S, Barnett JD, and Forman RA (1975) Calibration of pressure dependence of the R_1 ruby fluorescence line to 195 kbar. *Journal of Applied Physics* 46: 2774–2780.

Pravica M and Remmers B (2003) A simple and efficient cryogenic loading technique for diamond anvil cells. *Review of Scientific Instruments* 74: 2782–2783.

Qiu W, Baker PA, Velisavljevic N, Vohra YK, and Weir ST (2006) Calibration of an isotopically enriched carbon-13 layer pressure sensor to 156 GPa in a diamond anvil cell. *Journal of Applied Physics* 99: 064906.

Romanowicz B, Li XD, and Durek J (1996) Anisotropy in the inner core: Could it be due to low-order convection? *Science* 274: 963–966.

Ross NL and Hazen RM (1989) Single crystal X-ray diffraction study of MgSiO3 perovskite from 77 to 400 K. *Physics and Chemistry of Minerals* 16: 415–420.

Rueff J-P, Kao CC, Struzhkin VV, *et al.* (1999) Pressure induced high-spin to low-spin transition in FeS evidenced by X-ray emission spectroscopy. *Physical Review Letters* 82: 3284–3287.

Ruf T, Serrano J, Cardona M, Pavone P, *et al.* (2001) Phonon dispersion curves in wurtzite-structure GaN determined by inelastic X-ray scattering. *Physical Review Letters* 86: 906–909.

Santoro M, Gorelli FA, Bini R, Ruocco G, Scandolo S, and Crichton WA (2006) Amorphous silica-like carbon dioxide. *Nature* 441: 857–860.

Santoro M, Lin J-F, Mao HK, and Hemley RJ (2004) *In situ* high P–T Raman spectroscopy and laser heating of carbon dioxide. *Journal of Chemical Physics* 121: 2780–2787.

Saxena SK, Dubrovinsky LS, Haggkvist P, Cerenius Y, Shen G, and Mao HK (1995) Synchrotron X-ray study of iron at high pressure and temperature. *Science* 269: 1703–1704.

Saxena SK, Shen G, and Lazor P (1993) Experimental evidence for a new iron phase and implications for Earth's core. *Science* 260: 1312–1314.

Scandolo S and Jeanloz R (2003) The centers of planets. *American Scientist* 91: 516–525.

Schultz E, Mezouar M, Crichton W, *et al.* (2005) Double-sided laser heating system for *in situ* high pressure-high temperature monochromatic X-ray diffraction at the ESRF. *High Pressure Research* 25: 71–83.

Schwoerer-Böhning M and Macrander AT (1998) Phonon dispersion of diamond measured by inelastic X-ray scattering. *Physical Review Letters* 80: 5572–5575.

Shahar A, Bassett WA, Mao HK, Chou I-M, and Mao WL (2005) The stability and Raman spectra of ikaite, $CaCO_3.6H_2O$, at high pressure and temperature. *American Mineralogist* 90: 1835–1839.

Shannon RD and Prewitt CT (1970) Revised values of effective ionic radii. Acta Crystallographica *B* 26: 1046–1048.

Shen G, Mao HK, and Hemley RJ (1996) Laser-heating diamond-anvil cell technique: Double-sided heating with multimode Nd:YAG laser. In: Akaishi M, Arima M, and Irifune T (eds.) *Advanced Materials 96 – New Trends in High Pressure Research*, pp. 149–152. Tsukuba, Japan: National Institute for Research on Inorganic Materials.

Shen G, Rivers ML, Wang Y, and Sutton SR (2001) Laser heated diamond cell system at the advanced photon source for *in situ* X-ray measurements at high pressure and temperature. *Review of Scientific Instruments* 72: 1273–1282.

Shieh SR, Duffy TS, Kubo A, *et al.* (2006) Equation of state of the postperovskite phase synthesized from a natural (Mg,Fe)SiO_3 orthopyroxene. *Proceedings of the National Academy of Sciences* 103: 3039–3043.

Shieh SR, Duffy TS, and Li B (2002) Strength and elasticity of SiO_2 across the stishovite- $CaCl_2$-type structural phase boundary. *Physical Review Letters* 89: 255507.

Shieh SR, Duffy T, and Shen G (2004) Elasticity and strength of calcium silicate perovskite at lower mantle pressures. *Physics of the Earth and Planetary Interiors* 143: 93–106.

Shimizu K, Ishikawa H, Takao D, Yagi T, and Amaya K (2002) Superconductivity in compressed lithium at 20 K. *Nature* 419: 597–599.

Shimizu K, Kimura T, Furomoto S, Takeda K, Kontani K, and Amaya K (2001) Superconductivity in the nonmagnetic state of iron under pressure. *Nature* 412: 316–318.

Silvera IF and Wijngaarden RJ (1985) Diamond anvil cell and cryostat for low-temperature optical studies. *Review of Scientific Instruments* 56: 121–124.

Singh AK, Balasingh C, Mao HK, Hemley RJ, and Shu J (1998a) Analysis of lattice strains measured under non-hydrostatic pressure. *Journal of Applied Physics* 83: 7567–7575.

Singh AK, Mao HK, Shu J, and Hemley RJ (1998b) Estimation of single-crystal elastic moduli from polycrystalline X-ray diffraction at high pressure: Applications to FeO and iron. *Physical Review Letters* 80: 2157–2160.

Sinmyo R, Hirose K, O'Neill HSC, and Okunishi E (2006) Ferric iron in Al-bearing post-perovskite. *Geophysical Research Letters* 33: doi: 10.1029/2006GL025858.

Sinogeikin SV, Zhang J, and Bass JD (2004) Elasticity of single crystal and polycrystalline $MgSiO_3$ perovskite by Brillouin spectroscopy. *Geophysical Research Letters* 31: 6620.

Smyth JR, Jacobsen SD, Swope RJ, *et al.* (2000) Crystal structures and compressibilities of synthetic 2M(1) and 3T phengite micas. *European Journal of Mineralogy* 12: 955–963.

Somayazulu MS, Finger LW, Hemley RJ, and Mao HK (1996) New high-pressure compounds in methane–hydrogen mixtures. *Science* 271: 1400–1402.

Song X and Helmberger DV (1998) Seismic evidence for an inner core transition zone. *Science* 282: 924–927.

Song XD and Richards PG (1996) Observational evidence for differential rotation of the Earth's inner core. *Nature* 382: 221–224.

Soos ZG, Eggert JH, Hemley RJ, Hanfland M, and Mao HK (1995) Charge transfer and electron–vibron coupling in dense solid hydrogen. *Chemical Physics* 200: 23–39.

Speziale S, Shieh SR, and Duffy T (2006) High-pressure elasticity of calcium oxide: A comparison between Brillouin spectroscopy and radial X-ray diffraction. *Journal of Geophysical Research* 111: 2203.

Steinle-Neumann G, Stixrude L, Cohen RE, and Gülseren O (2001) Elasticity of iron at the temperature of the Earth's inner core. *Nature* 413: 57–60.

Sterer E, Pasternak MP, and Taylor RD (1990) A multipurpose miniature diamond anvil cell. *Review of Scientific Instruments* 61: 1117–1119.

Stevenson DJ (2003) Mission to Earth's core – A modest proposal. *Nature* 423: 239–240.

Stixrude L and Cohen RE (1995) High-pressure elasticity of iron and anisotropy of Earth's inner core. *Science* 267: 1972–1975.

Struzhkin VV, Eremets MI, Gan W, Mao HK, and Hemley RJ (2002) Superconductivity in dense lithium. *Science* 298: 1213–1215.

Struzhkin VV, Hemley RJ, Mao HK, and Timofeev YA (1997) Superconductivity at 10 to 17 K in compressed sulfur. *Nature* 390: 382–384.

Sturhahn W, Toellner TS, Alp EE, *et al.* (1995) Phonon density of states measured by inelastic nuclear resonant scattering. *Physical Review Letters* 74: 3832–3835.

Su W, Dziewonski AM, and Jeanloz R (1996) Planet within a planet: Rotation of the inner core of Earth. *Science* 274: 1883–1887.

Sumiya H, Toda N, and Satoh S (2005) Development of high-quality large-size synthetic diamond crystals. *SEI Technical Review* 60: 10–16.

Sung CM (1976) New modification of the diamond anvil press: A versatile apparatus for research at high pressure and high temperature. *Review of Scientific Instruments* 47: 1343–1346.

Sung C-M, Goetze C, and Mao HK (1977) Pressure distribution in the diamond anvil press and the shear strength of fayalite. *Review of Scientific Instruments* 48: 1386–1391.

Suzuki M, Yoshida H, Sakuma N, Ono T, Sakai T, and Koizumi S (2004) Electrical characterization of phosphorus-doped n-type homoepitaxial diamond layers by Schottky barrier diodes. *Applied Physics Letters* 84: 2349–2351.

Tajani A, Tavares C, Wade M, *et al.* (2004) Homoepitaxial {111}-oriented diamond p–n junctions grown on B-doped Ib synthetic diamond. *Physica Status Solidi A* 201: 2462–2466.

Takahashi T and Bassett WA (1964) A high pressure polymorph of iron. *Science* 145: 483–486.

Takemura K and Singh AK (2006) High-pressure equation of state for Nb with a helium-pressure medium: Powder X-ray diffraction experiments. *Physical Review B* 73: 224119.

Tamura N, Celestre RS, MacDowell AA, *et al.* (2002) Submicron X-ray diffraction and its applications to problems in materials and environmental science. *Review of Scientific Instruments* 73: 1369–1372.

Timofeev YA, Struzhkin VV, Hemley RJ, Mao HK, and Gregoryanz EA (2002) Improved techniques for measurement of superconductivity in diamond anvil cells by magnetic susceptibility. *Review of Scientific Instruments* 73: 371–377.

Tkachev SN, Manghnani MH, and Williams Q (2005a) *In situ* Brillouin spectroscopy of a pressure-induced apparent second-order transition in a silicate glass. *Physical Review Letters* 95: 057402-1-4.

Tkachev SN, Manghnani MH, Williams Q, and Ming LC (2005b) Compressibility of hydrated and anhydrous Na_2O–$2SiO_2$ liquid and also glass to 8 GPa using Brillouin scattering. *Journal of Geophysical Research* 110: B07201-1–B07201-12.

Tkachev SN, Solozhenko VL, Zinin PV, Manghnani MH, and Ming LC (2003) Elastic moduli of the superhard cubic BC_2N phase by Brillouin scattering. *Physical Review B* 68: 52104–52106.

Tozer SW (1993) Miniature diamond-anvil cell for electrical transport measurements in high magnetic fields. *Review of Scientific Instruments* 64: 2607–2611.

Tozer SW (2002) High pressure techniques for low temperature studies in DC and pulsed magnetic fields. *International Journal of Modern Physics B* 16: 3395.

Tromp J (2001) Inner-core anisoropy and rotation. *Annual Review of Earth and Planetary Sciences* 29: 47–69.

Utsumi W and Yagi T (1991) Light transparent phase from room temperature compression of graphite. *Science* 252: 1542–1544.

van Aken P and Liebscher B (2002) Quantification of ferrous/ferric ratios in minerals: New evaluation schemes of Fe L_{23} electron energy-loss near-edge spectra. *Physics and Chemistry of Minerals* 29: 188–200.

Vanpeteghem CB, Angel RJ, Ross NL, *et al.* (2006) Al, Fe substitution in the $MgSiO_3$ perovskite structure: A single-crystal X-ray diffraction study. *Physics of the Earth and Planetary Interiors* 155: 96–103.

Vocadlo L, Alfe D, Gillan MJ, Wood IG, Brodholt JP, and Price GD (2003) Possible thermal and chemical stabilization of body-centred-cubic iron in the Earth's core. *Nature* 424: 536–539.

Wang JY, Sinogeikin SV, Inoue T, and Bass JD (2006) Elastic properties of hydrous ringwoodite at high-pressure conditions. *Geophysical Research Letters* 33: 14308.

Watanuki T, Shimomura O, Yagi T, Kondo T, and Isshiki M (2001) Construction of laser-heated diamond anvil cell system for *in situ* X-ray diffraction study at SPring-8. *Review of Scientific Instruments* 72: 1289–1292.

Weathers MS and Bassett WA (1987) Melting of carbon at 50 to 300 kbar. *Physics and Chemistry of Minerals* 15: 105–112.

Weir CE, Lippincott ER, VanValkenburg A, and Bunting EN (1959) Infrared studies in the 1- to 15-micro region to 30,000 atmospheres. *Journal od Research NBS* 63A: 55–62.

Welch MD, Crichton WA, and Ross NL (2005) Compression of the perovskite-related mineral bernalite Fe(OH) (3) to 9 GPa and a reappraisal of its structure. *Mineralogical Magazine* 69: 309–315.

Wenk H-R, Matthies S, Hemley RJ, Mao HK, and Shu J (2000) The plastic deformation of iron at pressures of the Earth's inner core. *Nature* 405: 1044–1047.

Williams Q, Jeanloz R, Bass J, Svendsen B, and Ahrens TJ (1987) The melting curve of iron to 250 Gigapascals: a constraint on the temperature at Earth's center. *Science* 236: 181–182.

Xu J, Mao HK, and Bell PM (1986) High pressure ruby and diamond fluorescence: Observations at 0.21 to 0.55 terapascal. *Science* 232: 1404–1406.

Xu J, Yeh S, Yen J, and Huang E (1996a) Raman study on D_2O up to 16.7 GPa in the cubic zirconia anvil cell. *Journal of Raman Spectroscopy* 27: 823–827.

Xu J, Yen J, Wang Y, and Huang E (1996b) Ultrahigh pressure in gem anvil cell. *High Pressure Research* 15: 127–134.

Xu J-A and Mao HK (2000) Moissanite: A new window for high-pressure experiments. *Science* 290: 783–785.

Yagi T and Akimoto S (1982) Rapid X-ray measurements to 100 GPa range and static compression of α-Fe_2O_3. In: Akimoto S and Manghnani MH (eds.) *High-Pressure Research in Geophysics*, vol. 12, pp. 81–90. Tokyo, Japan: Center for Academic Publications.

Yagi T, Jamieson JC, and Moore PB (1979) Polymorphism in MnF_2 (rutile type) at high pressures. *Journal of Geophysical Research* 84: 1113–1115.

Yamamoto K, Endo S, Yamagishi A, Mikami H, Hori H, and Date M (1991) A ceramic-type diamond anvil cell for optical measurements at high pressure in pulsed high magnetic fields. *Review of Scientific Instruments* 62: 2988–2990.

Yamanaka T, Fukuda T, Hattori T, and Sumiya H (2001) New diamond anvil cell for single-crystal analysis. *Review of Scientific Instruments* 72: 1458–1462.

Yan C-S, Mao HK, Li W, Qian J, Zhao Y, and Hemley RJ (2004) Ultrahard diamond single crystals from chemical vapor deposition. *Physica Status Solidi A* 201: R25–R27.

Yan C-S, Vohra YK, Mao HK, and Hemley RJ (2002) Very high growth rate chemical vapor deposition of single-crystal

diamond. *Proceedings of the National Academy of Sciences* 99: 12523–12525.

Yoo CS, Akella J, Campbell AJ, Mao HK, and Hemley RJ (1995) Phase diagram of iron by *in situ* X-ray diffraction: Implications for the Earth's core. *Science* 270: 1473–1475.

Yoshimura Y, Mao HK, and Hemley RJ (2006a) Direct transformation of ice VII' to low-density amorphous ice. *Chemical Physical Letters* 420: 503–506.

Yoshimura Y, Stewart ST, Maddury Somayazulu, Mao HK, and Hemley RJ (2006b) High-pressure X-ray diffraction and Raman spectroscopy of ice VIII. *Journal of Chemical Physics* 124: 024502-1-7.

Zha C-S and Bassett WA (2003) Internal resistive heating in diamond anvil cell for *in situ* X-ray diffraction and Raman scattering. *Review of Scientific Instruments* 74: 1255–1262.

Zha CS, Duffy TS, Downs RT, Mao HK, and Hemley RJ (1996) Sound velocity and elasticity of single-crystal forsterite to 16 GPa. *Journal of Geophysical Research* 101: 17535–17545.

Zha CS, Duffy TS, Downs RT, Mao HK, and Hemley RJ (1998) Brillouin scattering and X-ray diffraction of San Carlos olivine: Direct pressure determination to 32 GPa. *Earth and Planetary Science Letters* 159: 25–34.

Zha CS, Duffy TS, Mao HK, Downs RT, Hemley RJ, and Weidner DJ (1997) Single-crystal elasticity of β-Mg_2SiO_4 to the pressure of the 410-km seismic discontinuity in the Earth's mantle. *Physics of the Earth and Planetary Interiors* 147: E9–E15.

Zha CS, Duffy TS, Mao HK, and Hemley RJ (1993) Elasticity of hydrogen to 24 GPa from single-crystal Brillouin scattering and synchrotron X-ray diffraction. *Physical Review B* 48: 9246–9255.

Zha CS, Hemley RJ, Mao HK, Duffy TS, and Meade C (1994) Brillouin scattering of silica glass to 57.5 GPa. In: Schmidt SC, Shaner JW, Samara GA, and Ross M (eds.) *High Pressure Science and Technology – 1993*, vol. 1, pp. 93–96. AIP Press.

Zha C-S, Mao HK, and Hemley RJ (2000) Elasticity of MgO and a primary pressure scale to 55 GPa. *Proceedings of the National Academy of Sciences* 97: 13494–13499.

Zha C-S, Mao HK, and Hemley RJ (2004) Elasticity of dense helium. *Physical Review B* 70: 174107–174108.

Zhao JY, Toellner TS, Hu MY, Sturhahn W, Alp EE, Shen GY, and Mao HK (2002) High-energy-resolution monochromator for [83]Kr nuclear resonant scattering. *Review of Scientific Instruments* 73: 1608–1610.

Zou GT, Ma Y, Mao HK, Hemley RJ, and Gramsch S (2001) A diamond gasket for the laser-heated diamond anvil cell. *Review of Scientific Instruments* 72: 1298–1301.

10 Theory and Practice – Techniques for Measuring High P/T Elasticity

J. D. Bass, University of Illinois at Urbana-Champaign, Urbana, IL, USA

10.1	Introduction	269
10.2	Static Compression	270
10.2.1	Piston Cylinder Apparatus	271
10.2.2	Diamond Anvil Cell	271
10.2.3	Multianvil Devices	274
10.3	Ultrasonic Methods	274
10.3.1	Ultrasonic Wave Transmission Techniques	275
10.3.1.1	Pulse transmission	276
10.3.1.2	Ultrasonic echo methods	276
10.3.2	High-Pressure Ultrasonics	278
10.3.3	Vibrational Resonance	280
10.4	Light Scattering Techniques	281
10.4.1	Brillouin Scattering	281
10.4.2	Impulsive Stimulated Scattering	283
10.5	Inelastic X-Ray Scattering	284
10.5.1	Nuclear Resonant Inelastic X-Ray Scattering (NRIXS)	284
10.5.2	Inelastic X-Ray Scattering from Phonons	285
10.6	Shock Waves	286
10.7	Other Techniques	287
References		288

10.1 Introduction

Since the early part of the twentieth century, it has been recognized that elastic modulus measurements are essential for understanding the composition and structure of the Earth, especially deep portions of the Earth that are not sampled directly. This connection between elasticity and the properties of the deep Earth arises naturally, because the field of seismology gives us the acoustic wave velocities throughout the Earth, with ever-increasing detail, whereas the acoustic velocities of minerals are in turn determined by their elastic moduli. Therefore, if we are to utilize the rich source of information on velocity variations in Earth's interior to infer basic properties such as composition and temperature, this requires knowledge of the elastic properties of Earth materials. At perhaps an even more fundamental level, the bulk modulus $K_T = \rho(\partial P/\partial \rho)_T$ (the incompressibility) gives the variation of density with pressure (or depth). Thus, Williamson and Adams (1923) used the few available measurements of elastic wave velocities and the bulk moduli of some rocks and iron metal to determine a density and pressure distribution in the Earth. Much later, Birch (1952) used a broader database on the elastic properties of materials and more modern seismic velocity profiles to make improved inferences about the composition and temperature at depth. These classic studies illustrate the long-standing interest in experimental determination of the elastic properties of Earth materials.

In principle, the isothermal bulk modulus is readily measured by imposing a known pressure on a sample, and measuring the resulting change in volume, or density. These measurements can be done in a variety of ways, some of which will be described in the following sections. One of the challenges in geophysical studies is that many minerals are relatively incompressible, especially high-pressure phases. Moreover, to exploit the information provided by seismological studies, it is necessary to consider the adiabatic elastic properties that govern the propagation of longitudinal and transverse elastic waves in a material, v_P and v_S, respectively. For an isotropic material, the appropriate relations are

$$v_P = \sqrt{\frac{K_S + (4/3)\mu}{\rho}}, \quad v_S = \sqrt{\frac{\mu}{\rho}} \qquad [1]$$

$$K_S = \rho\left(v_P^2 - (4/3)v_S^2\right), \quad \mu = \rho v_S^2 \qquad [2]$$

where ρ is the density, K_S is the adiabatic bulk modulus, and μ is the (adiabatic) shear modulus. The determination of these quantities for Earth materials, using both experimental and theoretical approaches, continues to be one of the main thrusts in experimental geophysics. Clearly, measurements at elevated pressures and temperatures are important for applications to the Earth's interior.

We may separate measurements of elastic properties (both elastic moduli and elastic wave velocities) into two broad classes: those on polycrystalline materials, and single-crystal measurements. Polycrystalline materials include natural rocks or other polyphase assemblages, and aggregates of a single phase. If the samples have randomly oriented grains, then the aggregate velocities and moduli of eqn [1] are obtained directly. Care must be taken to minimize or eliminate the effects of any cracks, pores, preferred orientations of grains (fabric). The effects of such imperfections or microstructures are sometimes difficult to identify or account for. An alternative approach is to perform measurements on individual single crystals. In general, single crystals are acoustically anisotropic (even those belonging to the cubic crystal system), and measurements must be made in several crystallographic directions to completely specify the elastic behavior.

Hooke's law for a crystal of arbitrary symmetry is given as

$$\sigma_{ij} = c_{ijkl}\varepsilon_{kl} \qquad [3]$$

where σ_{ij} is the generalized stress tensor, ε_{kl} is the strain tensor, and c_{ijkl} are the elastic stiffness constants, which we refer to here as the single-crystal elastic moduli (Nye, 1985). The Einstein summation notation is assumed, with the indices $i, j, k, l = 1 \rightarrow 3$. Alternatively, the relation between elastic strain and stress can be expressed as

$$\varepsilon_{ij} = s_{ijkl}\sigma_{kl} \qquad [4]$$

where s_{ijkl} are the elastic compliances. In practice, it is the elastic stiffness, c, that is most often measured. The equations of motion for propagating elastic waves lead to the Christoffel equation:

$$\det\left|c_{ijkl}n_j n_l - \rho v^2 \delta_{ik}\right| = 0 \qquad [5]$$

where n_i are components of the wave normals and δ_{ik} is the Kronecker delta (Musgrave, 1970). Equation [5] shows that the single-crystal elastic moduli can be obtained by performing measurements of sound velocities, v, with specific polarizations of particle motion and propagation directions. Depending on the specific measurement technique and quality of samples, single-crystal moduli can be determined with the highest degree of accuracy of all elastic properties. Knowledge of the c_{ij}'s also gives the elastic anisotropy of crystals and allows one to calculate the acoustic anisotropy of a polycrystal with oriented grains. However, the single-crystal moduli do not yield unique values for the aggregate elastic moduli, except for the single case of the bulk modulus of a single-phase aggregate of grains with cubic symmetry. Theoretical models of the deformation of aggregates, such as the commonly used Voigt and Reuss or Hashin and Shtrickman bounds, can only provide upper and lower bounds on the aggregate elastic moduli (e.g., Watt *et al.*, 1976). Depending on the anisotropy of a material, or the contrast in properties among the constituents of a polyphase aggregate, the uncertainties associated with calculating the properties of an aggregate can be an appreciable part of the total uncertainty in the properties of a mineral assemblage.

Of the various techniques used to measure single-crystal elastic properties, the most familiar of those used in geophysics fall into one of two categories. In an ultrasonics experiment acoustic excitations in the sample are externally generated, usually by means of transducers. The other type of technique is scattering experiments, where one measures the scattering of photons from acoustic phonons that are present in a sample at temperatures well above absolute zero temperature. The probe in this latter type of measurements can be a monochromatic laser beam or synchrotron X-ray beam. With the development of highly monochromatic and energetic X-ray sources at third-generation synchrotrons, along with high-energy resolution detection, synchrotron X-rays are being more widely utilized as a means of measuring the elastic properties of Earth materials at high pressures and temperatures (e.g., Burkel, 2000).

10.2   Static Compression

The most direct and oldest method of measuring an elastic property is by static compression. If a material is compressed hydrostatically and the volume change is measured, then the isothermal bulk modulus, K_T, is determined by the definition given above. A variety of methods have been used to generate pressure and to measure the volume, some of the more common of which are described below. When used with a heater of some sort, compression experiments give the P–V–T equation of state of a material.

In an isothermal static compression experiment, the P–V data are fitted with an equation of state, and the results are commonly given in terms of the bulk modulus evaluated at zero pressure, K_{0T}, and its pressure derivative (also evaluated at zero pressure) $K'_{0T} = (\partial K_{0T}/\partial P)_T$. Perhaps the most commonly used equation for fitting isothermal P–V measurements is the Birch–Murnaghan Eularian finite-strain equation of state (Birch, 1978):

$$P = 3f(1 + 2f)^{5/2}K_{0T}(1 + x_1 f + x_2 f^2 + \cdots) \quad [6]$$

where $f = (1/2)((V/V_0)^{2/3} - 1)$ is the Eularian finite strain parameter, and the coefficients are $x_1 = 3/2(K'_{0T} - 4)$, and $x_2 = (3/2)[K_{0T}K''_{0T} + K'_{0T}(K'_{0T} - 7) + 143/9]$. In practice, K''_{0T} and higher-order derivatives typically cannot be resolved, either because the P–V data are not sufficiently precise, or because they are not obtained over a large enough pressure range. This makes it reasonable to truncate the series in eqn [6] after the term $x_1 f$. In addition, it is common to assume that $K'_{0T} = 4$, further simplifying eqn [6] and data analysis. This assumption is somewhat justified by the observation that for many materials K'_{0T} is approximately 4, but this is far from the case for many important minerals, such as garnets, rutile-structured oxides, hydrous silicates, and enstatite (see Bass (1995)). When one further considers the strong tradeoff of covariance between K_{0T} and K'_{0T} (Bass *et al.*, 1981; Angel, 2000), the approximation $K'_{0T} = 4$ can often lead to highly misleading results. Another equation of state that has become commonly used in recent years is the Vinet 'universal' equation of state (Vinet *et al.*, 1987):

$$P = 3K_{0T}(V_0/V)^{2/3}[1 - (V/V_0)^{1/3} \\ \times \exp\{3/2(K'_{0T} - 1)[1 - (V/V_0)(1/3)]\}$$

Discussions of various aspects of equations of state in geophysics are given by Angel (2000), Poirier (2000), and Stacey (2005). A comprehensive review of static compression results obtained before 1995 is given by Knittle (1995). We now turn our attention to some of the methods of static compression.

10.2.1 Piston Cylinder Apparatus

In its simplest form, a sample is placed in a steel or WC cylinder, and is compressed by pistons that advance into the two ends of the cylinder. The sample volume at high pressure is measured by the displacement of the piston, and the force applied to the piston is used to calculate the pressure. Friction between the piston and cylinder, as well as other frictional effects, need to be accounted for in calculating an accurate pressure (Getting, 1998). Since confined uniaxial compression of a solid (i.e., the pistons directly compressing the sample) is not hydrostatic, the sample is surrounded by a fluid pressure medium (a pentane–isopentane mixture is often used or, more recently, argon; **Figure 1**) for truly hydrostatic workup to ~3 GPa. Above this pressure, a soft solid medium is used to provide 'quasi-hydrostatic' conditions around the sample. Soft metals like Pb or, more commonly, weak minerals like talc or pyrophyllite are used as pressure media. The piston–cylinder method was used extensively in the early days of high-pressure research, and the compression of a wide variety of elements and chemically compex materials were studied with this method (most notably by P.W. Bridgman; see Bridgman (1964). Because much higher pressures are attainable using other instruments, the piston–cylinder technique has been used less for P–V studies in recent years.

10.2.2 Diamond Anvil Cell

In the last few decades, perhaps the most commonly used instrument for measuring high-pressure equations of state has been the diamond anvil cell (DAC). This device is capable of producing the highest pressures of any static compression device, reaching the pressures found deep within Earth's core (e.g., Mao *et al.*, 1990). The diamond anvil cell is also an incredibly versatile instrument. With the DAC, pressure is exerted on a sample by the small culet surfaces of two gem quality diamonds (**Figure 2**). In a vast majority of equation-of-state experiments, the volume is obtained by X-ray diffraction (XRD) measurements of the distance, d, between lattice planes, using Bragg's law: $n\lambda = 2d\sin\theta$. Both angle-dispersive XRD (using either an imaging plate or charged-coupled device (CCD) detector; **Figure 2**), and energy dispersive XRD are used for volume measurements. By exerting a moderate force over the small area of the diamond tips, pressures up to several hundred GPa can be attained (greater than the

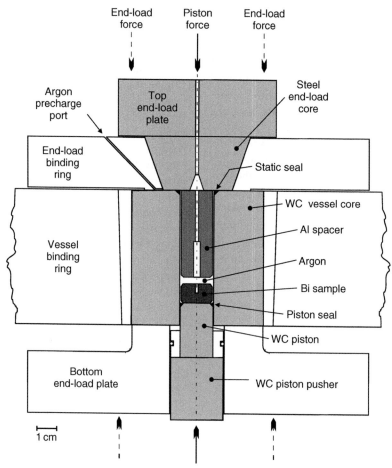

Figure 1 A modern piston cylinder apparatus. In this particular device, hydrostatic pressure is transmitted to the sample by argon gas. Courtesy of I. Getting.

pressure at Earth's center of ~364 GPa). The diameter of the culet surface is typically from 0.1 to 1 mm or more, depending on the desired pressure range. When the diamonds squeeze on the sample directly, for example, compressing a metal foil, extremely large deviatoric stresses are exerted on the sample. To help reduce this effect, the diamonds can instead compress a sample chamber consisting of a hole in a thin metal gasket, transmitting pressure to the sample via a soft pressure-transmitting medium. The best pressure-transmitting media are those with the least strength (He or Ne loaded as pressurized gasses). The choice of pressure medium is important because presence of shear stresses and stress gradients in the sample chamber can lead to systematic errors in measurements of pressure and volume. A detailed review of diamond cell technology and techniques is given in the article by Jayaraman (1983).

Pressure is most often measured indirectly via the fluorescence wavelengths of ruby, which are strongly pressure dependent (Mao *et al.*, 1986; Holzapfel, 2003; Chijioke *et al.*, 2005). Pressure scales based of the fluorescence of other materials are also used (e.g., Datchi *et al.*, 1997), especially at elevated temperatures. One can also use the lattice parameters or unit cell volumes of a material that is unlikely to react with a sample. The most commonly used internal standards for this purpose are MgO (e.g., Speziale *et al.*, 2001), Au (Anderson *et al.*, 1989; Jamieson *et al.*, 1982), or Pt (Holmes *et al.*, 1989) (see also Dewaele *et al.* (2004)). At pressures above 10 or 15 GPa, uncertainties in pressure may be on the order of 10% (Fei *et al.*, 2004; Li *et al.*, 2005). At pressures above 30 GPa, uncertainties in pressure are likely the largest source of error in *P–V* equations of state, especially using a pressure-transmitting medium with significant shear strength (or worse yet, if no pressure medium is used at all).

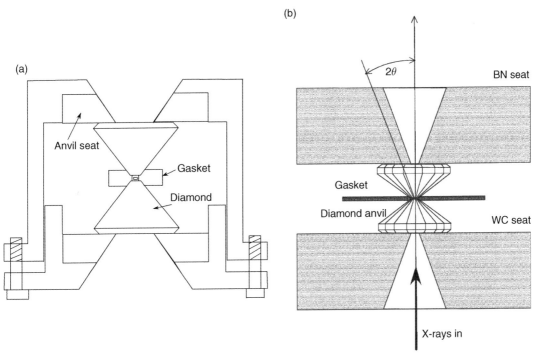

Figure 2   (a) Schematic of a piston-cylinder diamond anvil cell. The sample is between the diamonds, within a hole in the gasket (sample chamber). Screws are used to drive the diamond anvils toward each other to produce high pressure. (b) The basis for X-ray diffraction with DAC. An X-ray beam hits the sample and diffraction is observed at various angles of 2θ. The BN seat is transparent to X-rays, allowing them to be recorded on a 2-D detector. These X-ray measurements yield the unit cell sample volume and density of the sample. Courtesy of Guoyin Shen.

The diamond anvil cell can be used to obtain the full P–V–T equation of state when coupled with a heater. Resistance heaters (e.g., Bassett *et al.*, 1993; Fei, 1999) provide a relatively homogenous environment in which temperature can be accurately measured with thermocouples, but are usually limited in temperature to less than about 1000°C. A resistance heater inside the sample chamber of the diamond cell has been devised for much higher-temperature measurements to 3000 K (Zha and Bassett, 2003). The alternative method is to use laser heating, usually with a yttrium aluminum garnet (YAG) (wavelength $\lambda = 1.06\,\mu\text{m}$) or CO_2 ($\lambda \sim 10\,\mu\text{m}$, e.g., Yagi and Susaki, 1992) laser. A YAG, yttrium lithium floride(YLF), or similar laser is used for heating Fe-bearing silicates or other dark, absorbing samples (e.g., metal). Transparent samples can be heated with a YAG laser by mixing them with an inert absorber, such as graphite or platinum black. An advantage of CO_2 laser heating is that the radiation from a CO_2 laser is absorbed by most transparent oxides and silicates (e.g., Yagi and Susaki, 1992; Fiquet *et al.*, 1996). A modern double-sided

laser-heating system is shown in **Figure 3** (Shen *et al.*, 2001; Meng *et al.*, 2006). The greatest advantage of laser heating is that it allows temperatures of several thousands of kelvin to be obtained while at high pressure. However, the thermal pressure in the DAC is substantial but difficult to characterize (Fiquet *et al.*, 1996, 1998; Andrault *et al.*, 1998). Temperatures are measured using the thermal 'gray-body' radiation from the sample to obtain a color temperature via Planck's radiation function (e.g., Sweeney and Heinz, 1993). Although the accurate measurement of pressure and temperature remains a significant challenge, the laser-heated DAC offers a way to maintain extreme static pressure and temperature conditions so that a variety of types of elasticity measurements can be made. Descriptions of modern beam lines developed specifically for high-pressure–temperature research are described by Mezouar *et al.* (2005), and Meng *et al.* (2006).

The simplicity of the diamond-anvil pressure cell makes it a highly versatile device for a wide variety of spectroscopic studies, not only of the bulk modulus but also phase relations and a host of other properties.

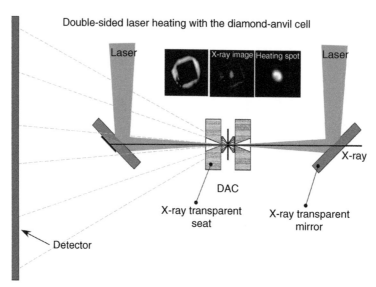

Double-sided laser heating with the diamond-anvil cell

Figure 3 Laser heating with the DAC and X-ray diffraction. Lasers irradiate the sample from both sides to obtain more uniform heating of the sample and smaller thermal gradients. The photographic images show the position of the focused X-ray beam and the laser-heated spot. In this way, the *P–V–T* equation of state of a sample can be determined. Courtesy of Guoyin Shen.

In this section we have described its application to X-ray diffraction studies of the *P–V–T* equation of state. Some other applications of the diamond cell for measurement of elastic properties by spectroscopic methods (e.g., Brillouin scattering, inelastic X-ray scattering) will be mentioned in the following sections.

10.2.3 Multianvil Devices

Although the DAC can attain the highest static pressures, other devices have been developed for compression studies using greater quantities of sample. One class of devices that have found wide use for this purpose is the multianvil presses. These devices can hold much larger sample volumes, approximately 1 mm^3 or larger, can provide relatively uniform heating of a sample, and can accommodate a theromocouple inside the sample chamber. **Figure 4** shows a diagram of a single-stage pressure device with six anvils compressing the sample assembly along the directions of the faces on a cube (a 'cube-anvil' or DIA apparatus). The sample is situated at the center of a cube-shaped pressure-transmitting assembly that may contain a heater and a soft medium immediately around the sample (Wang *et al.*, 1996). Most equation-of-state studies with single-stage devices like the DIA apparatus have been carried out at pressures of about 10–15 GPa, some of them carried out with simultaneous high temperature.

Higher pressures and temperatures can be obtained using any of several variants of the two-stage multianvil device designed by Kawai and Endo (1970). An outer first stage of six steel anvils advance eight tungsten carbide or sintered diamond inner anvils which in turn compress the sample assembly (see Section 10.3). The inner corners of the eight second-stage anvils are truncated to form an octahedral cavity in which the sample assembly is situated and compressed. As with the DIA apparatus, X-rays are introduced into the sample area via gaps between all the anvils. Diffracted X-rays are collected in energy-dispersive mode, giving the information needed to determine the volume of the sample at pressure via an internal calibration standard. The pressure achievable with this apparatus depends on the truncation area of the inner-stage cubes, with smaller truncations yielding greater pressures. The use of sintered diamond anvils can also extend the upper pressure limit. For equation-of-state measurements, pressures of about 30 GPa or even greater can be attained (Irifune, 2002).

10.3 Ultrasonic Methods

Ultrasonic methods are those in which a transducer is used to produce either an acoustic wave that propagates through the sample, or to set the sample into vibration at its resonant frequencies. In their various

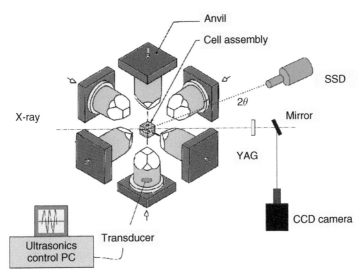

Figure 4 A single-stage six-anvil or cubic anvil apparatus used for *P–V* studies with synchrotron radiation and ultrasonic interferometry. X-rays are introduced through the small spaces between the anvils, and diffracted signals can be measured through these gaps as well. The sample volume is measured by energy dispersive X-ray diffraction recorded on a solid-state detector (SSD). A second measurement of the sample length (and hence volume) is made using X-radiographic images (shadowgraphs) using a fluorescent YAG crystal and CCD camera. Modified from Vaughan *et al.* (1998). Courtesy of B. Li.

forms, both methods have been used since the 1940s for measurements of the adiabatic elastic properties of both single crystals and polycrystalline material at elevated pressures and temperatures. These are mature experimental techniques that are in use in many laboratories around the world.

10.3.1 Ultrasonic Wave Transmission Techniques

Experimental methods that utilize transmitted acoustic pulses are the most precise available. The basis for all of the methods outlined below is measurement of the traveltime of an ultrasonic plane wave through a sample. Knowing the length of the sample and the traveltime, the phase velocity of the ultrasonic wave is defined. Precision on the order of several parts in 10^4 or better can be achieved in favorable cases. When applied to single crystals, ultrasonic transmission techniques are the most accurate and precise of all methods to measure elastic moduli. However, with imperfect specimens, especially polycrystalline samples or with very small samples, the accuracy of the method is not as high. The frequencies used in ultrasonic measurements typically range from hundreds of kHz to GHz. In general, the sample should be as large as possible to minimize possible systematic errors and to increase precision. Working with smaller samples, which is clearly a priority for

geophysical research at high pressures, requires that the frequency of the ultrasonic probe be increased (thereby decreasing the wavelength).

Studies of single crystals by ultrasonic interferometry, as well as other acoustical techniques such as Brillouin scattering, require that sound velocities be measured in several crystallographic directions in order to determine all of the single-crystal elastic moduli. The number of velocity measurements needed depends on the symmetry of the sample, with lower-symmetry materials requiring measurements in a greater number of crystallographic directions (Nye, 1985; Musgrave, 1970). For example, cubic crystals have only three independent and nonzero elastic moduli, whereas monoclinic crystals have 13. In addition, separate transducers may be required for measurements of longitudinal and transverse waves. For each direction in which a velocity measurement is required, parallel surfaces with near perfect flatness must be prepared. Measurements on isotropic aggregates of randomly oriented crystallites require only one measurement each of longitudinal and transverse wave velocities, and yield the aggregate elastic properties directly. However, these advantages are offset by decreased accuracy due to the imperfections often present in polycrystalline materials (e.g., cracks, pores, preferred orientations of grains). Hot pressing and sintering of polycrystalline materials under elevated pressure and temperatures can greatly minimize these effects by producing polycrystalline materials with few imperfections and of near theoretical density (Gwanmesia *et al.*, 1993). One of the powerful advantages of ultrasonic methods is that they can be used with almost any type of sample.

10.3.1.1   *Pulse transmission*

The pulse transmission technique is the simplest for measuring elastic properties by ultrasonics. A transducer at one end of a sample converts an electrical impulse from a pulse generator into a mechanical, or acoustic pulse in the sample. The essential elements of the experimental setup are shown in **Figure 5**. A transducer at the opposite end of the sample senses the transmitted acoustic wave and converts it into an electrical signal that can be recorded and timed. The traveltime through the specimen is thus determined and, with the known length of the sample, a velocity can be calculated.

Although of lower precision then some other ultrasonics methods, the pulse transmission technique is highly versatile, simple, and especially useful for characterizing the properties of rocks and highly attenuating samples. Low frequencies in the range a few kHz to MHz allow one to perform measurements on coarse-grained rocks and other types of samples for which scattering of high-frequency waves would be problematic. This technique was used by Birch (1960) and Simmons (1964), in their classic studies of the P and S velocities in a variety of rocks to 10 kbar. A comprehensive summary of measurements performed on a variety of geologic materials, including sediments, dry and fluid-bearing rocks, and other aggregates, is given by Christensen (1982). This method has continued to find use in characterizing the velocities, elastic moduli, and attenuation in a wide range of geologic materials (e.g., Prasad and Manghnani, 1997; Vanorio *et al.*, 2003).

10.3.1.2   *Ultrasonic echo methods*

In cases where high-quality single crystals or fine-grained polycrystalline samples are available, highly precise measurements of elastic properties can be made by using multiple reflections of acoustic waves, or echoes, within a sample. The various methods developed around this concept involve measuring the time interval between successive echoes, or the phase delay between two echoes of a monochromatic acoustic wave train. Three methods commonly used are the pulse-echo overlap, pulse-superposition, and phase comparison methods. These methods are more accurate and precise than the pulse transmission method and, in cases where high-quality samples are available, pulse-echo methods are often viewed as the gold standard for elastic wave velocity and elastic modulus measurements. Very often 'buffer rods' are placed on either end of the sample, separating the sample and transducer. This can serve to simply remove the transducer from the sample environment, which is necessary in most high-pressure or -temperature experiments. In addition, the use of buffer rods can minimize errors introduced by changes in phase of the ultrasonic waves as they reflect from the ends of the sample (McSkimin, 1950; Davies and O'Connell, 1977; Jackson *et al.*, 1981). A WC buffer rod, arranged as it is used in a modern high-pressure experiment, is shown in **Figure 6**.

In the pulse-echo overlap technique (Papadakis, 1967, 1990), individual pulses are excited in a sample and the same transducer is used to detect reflections. The timing of the pulses is long enough that each pulse undergoes multiple reflections and completely attenuates before a new pulse is introduced. The relative phases of two successive echoes are adjusted so that they align, or overlap, via the triggering rate of an oscilloscope. The triggering frequency needed to overlap echoes is equal to the reciprocal of the

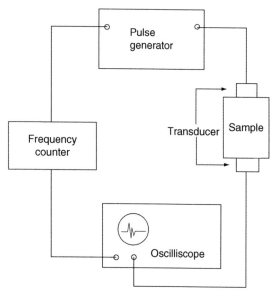

Figure 5   Schematic of a pulse transmission experiments. The transducer attached to the top of the sample produces an ultrasonic pulse, and the bottom transducer receives the transmitted signal.

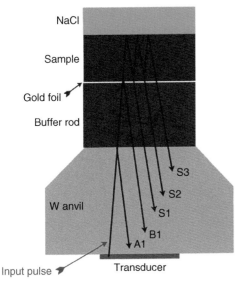

Figure 6 Ultrasonic pulses and echos in a modern high-presssure experiment. The transducer is mounted on a pressure-generating anvil. A buffer rod separates the transducer from the sample embedded in a pressurized medium. Reflections or echos from ends of the sample are indicated. Courtesy of W. Liu and R. C. Liebermann.

traveltime in the sample. The measured travel time, T, is then equal to

$$T = pT_0 - \frac{p\gamma}{2\pi f} + \frac{n}{f} \qquad [6']$$

where T_0 is the intrinsic round-trip traveltime in the sample, p is the number of round trips between reflections, γ is the transducer bond phase shift, f is the frequency of the ultrasonic wave, and n is an integer related to the number of cycles of mismatch in the overlap of a wave train (Papadakis, 1967). Assuming that the phase shift γ is independent of frequency, it can be determined through measurements at slightly different frequencies (McSkimin, 1961; Papadakis, 1967).

With the pulse superposition technique (McSkimin, 1961), the repetition rate of the radio-frequency (RF) pulse generator is varied such that it coincides with the arrival of an echo, or some integral number (p) of echo intervals (i.e., every second echo, or every third echo, and so on). When the timing of the pulses is precisely equal to the time between echoes (i.e., the round-trip traveltime through the specimen) then the amplitude of echoes viewed on an oscilloscope will be at a maximum. Equation [6'] can then be used to determine the true traveltime in the specimen. Clearly, it is advantageous to use large samples because this makes T_0 larger relative to the

other terms in [6'], minimizing errors due to the uncertainties in measurements of γ, T, and the length of the sample.

With the phase comparison method (McSkimin, 1950; Niesler and Jackson, 1989; Jackson and Niesler, 1982), the main experimental variable is the frequency of the ultrasonic wave in the sample. Coherent sinusoidal pulses of frequency f are excited in the sample. The frequency of the ultrasonic waves is varied so that successive echoes are in phase and display a constructive interference maximum or, alternately, destructively interfere to yield a minimum in amplitude. The velocity of the propagating ultrasonic wave is then

$$V_0 = \frac{2lf_0}{n + \gamma/2\pi} \qquad [7]$$

where l is the sample thickness, f_0 is the frequency for maximum constructive interference, n is an integer equal to the number of cycles in the sample at f_0, and γ is the phase angle defined previously. The integer n can be determined by sweeping the frequency through consecutive maxima (or minima, which may be detected with more certainty). The phase comparison method has been used extensively in high-pressure studies of minerals (e.g., Jackson and Niesler, 1982; Niesler and Jackson, 1989), making it a valuable tool in geophysical research. When used with single crystals, it is considered to be one of the most accurate techniques, if not the most accurate, for obtaining elastic properties at high pressure.

The time required to perform a pulse-echo ultra-sonics experiment has been greatly reduced through development of the 'transfer function' method (Li *et al.*, 2002, 2004). Instead of sweeping through a range of frequencies, as in the pulse-echo overlap or phase comparison methods, a broadband pulse is introduced to the buffer rod and sample, and the received signal *y(t)* is recorded digitally. Knowing the input pulse, *x(t)*, and the received signal, the system response, *h(t)*, of the transducer-buffer rod-sample assembly is therefore defined at all frequencies simultaneously from the relation $y(t) = x(t) * h(t)$. The Fourier transform of the input and output signals can be expressed as $Y(f) = X(f)^* H(f)$, where $H(f) = Y(f)/X(f)$ is the transfer function. In essence, the transfer function contains information on the response of the system over a broad range of frequencies (typically several tens of MHz), and it is acquired in a single, fast recording. From the transfer function one can reproduce the received signal of a monochromatic pulse of any given frequency within the frequency band of the source. Thus, one can reproduce the monochromatic pulse-echo overlap or phase comparison experiments, and calculate traveltimes accordingly. With this technique, the frequency response of the system can be obtained in seconds, as opposed to the several minutes typically required to scan in frequency using, for example, the conventional phase comparison method. The speed of the transfer method makes it possible to study kinetics and other time-dependent phenomena.

10.3.2 High-Pressure Ultrasonics

Ultrasonic echo methods have been used extensively for measuring elastic moduli at high pressures and temperatures, employing a variety of high-pressure devices. As noted above, phase comparison has been one of the more commonly used methods for determinations of elastic properties at high pressure. Jackson and Niesler (1982) developed a method whereby a transducer is bonded directly to a sample, and the entire assembly is compressed in a piston–cylinder apparatus. Hydrostatic pressure is transmitted to the sample via a liquid (a pentane–isopentane mixture is commonly used), thus eliminating errors due to deviatoric stresses on the sample. Pressure is measured via the electrical resistance of a manganin wire. While this is a relatively accurate method for high-pressure elastic modulus measurements, it is limited to pressures of about 3 GPa. The pressure limitations of the piston–cylinder apparatus

and fluid pressure media are overcome by using a solid pressure medium and a multianvil apparatus. By using a soft material such as NaCl for transmitting pressure to the sample and annealing the sample and pressure medium assembly to relax deviatoric stresses, quasi-hydrostatic conditions can be attained on samples at high pressure (Weidner *et al.*, 1992). The two types of multianvil apparatuses being used for high-pressure ultrasonics are a single-stage cubic-anvil (DIA) system (**Figure 4**; Weidner *et al.*, 1992; Li *et al.*, 2004), and the two-stage Kawai-type multi-anvil apparatus (**Figure 7**; Kawai and Endo, 1970; Uchida *et al.*, 2002). In the case of the two-stage apparatus, which can reach higher pressures, eight WC anvils with truncated corners surround an octahedral sample chamber containing the sample assembly (**Figure 8**). The transducer is attached to the outer truncated corner of an anvil, which thus serves as a buffer rod. Ultrasonic pulses are then transmitted to the sample by another buffer rod between the WC anvil and the sample (**Figure 8**). When the WC cubes are driven together by six outer or 'first-stage' anvils (not shown in **Figure 7**; see Uchida *et al.*, (2002)), pressures up to >20 GPa can be attained. A heater in the sample assembly allows simultaneous temperatures of >1200°C (Higo *et al.*, 2006). Changes in the sample length, required for accurate velocity determinations, are measured directly using synchrotron X-radiography. A synchrotron X-ray beam is directed through the space between the WC anvils and the transmitted beam is recorded via a YAG crystal (serving as a fluorescent

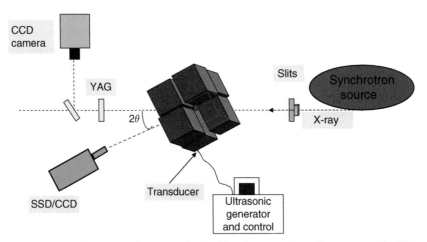

Figure 7 Schematic diagram of the set-up for ultrasonics in a Kawai-type multi-anvil apparatus. The YAG crystal and CCD camera are for x-radiography of the sample to determine it's length in-situ. The solid state detector is used for energy dispersive X-ray diffraction. The transducer that produces an ultrasonic signal is attached to one of the WC cubes which acts as a primary buffer rod. Modified from Vaughan *et al.* (1998) and Li *et al.* (2005).

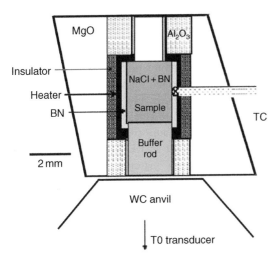

Figure 8 Sample assembly used for ultrasonic velocity measurement experiments at high pressure in a Kawai-type multi-anvil apparatus. Modified from Kung *et al.* (2004) and Li *et al.* (2005).

screen) and detector. Due to their high absorption, the images of metal markers on either side of the sample are apparent in this X-radiograph and give the sample length directly. Because this device allows velocity, sample length, and, hence, volume to be measured in a single experiment, it is possible to obtain the pressure on the sample directly (Ruoff *et al.*, 1973). This remarkable apparatus illustrates the flexibility in adapting ultrasonic methods to a variety of high-pressure devices, due in part to the ability to introduce the signal to the sample via buffer rods.

Ultrasonic measurements of velocities have also been carried out in a torroidal anvil apparatus (Khvostantsev *et al.*, 2004), a device in which pressure is generated by two opposed curved anvils. This device can generate pressures to ~15 GPa on relatively large sample volumes, and high temperatures. Ultrasonic measurements with the toroidal cell date back to the work of FF Voronov in the mid-1970s (see the review by Khvostantsev *et al.*, (2004)). This technology continues to be developed, and has more recently combined synchrotron XRD with ultrasonics capabilities (Lheureux *et al.*, 2000).

In an attempt to perform ultrasonic velocity measurements to even higher pressures, an ultrasonic interferometer has been devised for use with the DACs. The challenge in this experiment is the small sample thickness in a DAC (typically 50 μm or less, in comparison with ~1 mm in a multianvil device). Such thin samples require higher-frequency ultrasound waves. This emerging technique has proved to be successful in measuring both longitudinal and shear elastic wave propagation times in samples compressed in a DAC. Producing shear waves at the requisite GHz frequencies appropriate for samples in a DAC is problematic. An ingenious method of converting longitudinal GHz acoustic waves to transverse waves (P-to-S conversion) is achieved via reflection at a critical angle on the buffer rod (see **Figure 9**) (Jacobsen *et al.*, 2002). Some of the recent technical advances and results obtained with this promising new method are given by Jacobsen *et al.* (2004), Kantor *et al.* (2004), and Reichmann and Jacobsen (2006).

An important application of high-temperature ultrasonic studies has been to the properties of silicate melts. Measurements on dry silicate melts are challenging because of the rather high temperatures

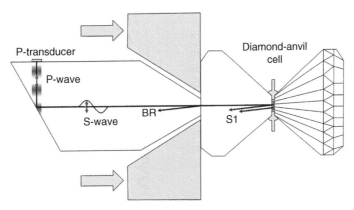

Figure 9 GHz ultrasonics with the diamond anvil cell. Ultrasonic pulses are transmitted to the sample via a buffer rod and the diamond anvil on the left. The production of a shear wave is illustrated via a longitudinal–shear (P-to-S) conversion upon reflection from a polished surface on the buffer rod. Courtesy of Steve D. Jacobsen.

involved (>1000°C), and their high viscosity. The elastic properties of numerous silicate melts, spanning a broad range of compositions, were measured by Rivers and Carmichael (1987). These authors successfully employed both the pulse-echo-overlap technique and an interferometric technique in their experiments. With either method, the ultrasonic waves are introduced into the high-temperature melt via a refractory buffer rod (molybdenum in this case). The sample is contained in a crucible with a flat bottom to reflect ultrasonic waves propagating through the sample. Maxima and minima in the amplitude of the ultrasonic echoes are obtained by moving the buffer rod to vary the path length in the molten sample. The distance between successive amplitude maxima or minima yields the wavelength and the velocity of sound in the liquid. The reader is referred to the earlier work of Baidov and Kunin (1968), Murase *et al.* (1977), and Katahara *et al.* (1981), which document primary developments in this important area of research.

10.3.3   Vibrational Resonance

The resonant vibrational modes of a body are a function of its elastic moduli and external dimensions. Measurements of the vibrational resonance frequencies can therefore be used to obtain all of the elastic properties of a sample in a single experiment. In contrast to the ultrasonic wave transmission methods, where individual velocities are usually measured on different single-crystal samples with specific crystallographic orientations, the vibrational resonance spectrum (amplitude vs. frequency of excitation) yields all of the single-crystal elastic moduli from one experiment. An additional advantage of this technique is that it allows the use of smaller samples for a given ultrasonic frequency range or, alternatively, the use of lower frequencies for a given size of sample (Ohno, 1976). Internal friction or attenuation (Q^{-1}) can also be measured. The technique was applied to measurement of the bulk and shear moduli of elastically isotropic spheres by Fraser and LeCraw (1964) and Soga and Anderson (1967). Interpreting the results for anisotropic crystals, even those of cubic symmetry, is not straightforward. Since the early measurements on elastically isotropic materials, both the experimental and theoretical aspects of resonant ultrasound spectroscopy (RUS) measurements have been refined and applied primarily to high-temperature elasticity studies, including the properties of high-pressure phases at elevated temperatures.

Using RUS methods to determine the single-crystal elastic moduli of anisotropic materials is far more complex than for isotropic substances. The difficulty is mainly in the forward problem of calculating the RUS spectrum expected for an anisotropic crystal of a given shape. This difficulty exists even for crystals of the cubic crystal system, which can display considerable elastic anisotropy. The complexity (number of peaks) of an RUS spectrum increases as the symmetry of a material decreases, and also depends on the detailed shape of a crystal. A major advance in this field was made by Demarest (1971), who solved the problem of calculating the resonant spectrum for crystals with cubic crystallographic symmetry, in the shape of a cube. Demarest's work was extended to the case of orthorhombic crystals by Ohno (1976) and more recently to monoclinic crystals (Isaak *et al.*, 2006). The RUS spectral data are sufficiently complex that assignment of the peaks to specific vibrational modes can be difficult, and any misidentification of the peaks will result in erroneous elastic moduli. Having a good starting model for the elastic moduli greatly assists in evaluation of the data.

In most recent versions of the RUS method, two transducers are attached to opposite corners of a cube or rectangular parallelepiped (**Figure 10**). One of the transducers is used to vibrate the sample and the other acts as a receiver. An experimental concern is the effect the transducers have on the observed resonant frequencies. The theory for analyzing RUS

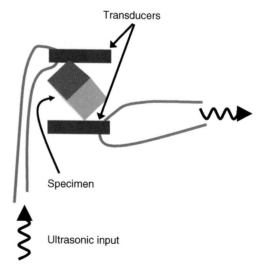

Figure 10   Schematic of a resonant ultrasound experiment. The transducers excite different resonant vibrations as the ultrasonic input is scanned in frequency. Courtesy of Donald G. Isaak.

spectra assumes stress-free boundaries on the sample, and the effect of the force exerted by contact between the transducers and the samples cannot be accounted for theoretically. In practice, the experiment can now be done with sufficiently low stresses exerted on the sample that the effects are thought to be small relative to other sources of error. The high-temperature elastic properties of numerous materials have been measured by RUS, at temperatures as high as 1800 K (e.g., Isaak *et al.*, 1989), and a review of much of those results has been presented by Anderson and Isaak (1995). The technique has recently been applied to measurement of the properties of high-pressure mantle phases at high temperatures (Mayama *et al.*, 2005).

10.4 Light Scattering Techniques

10.4.1 Brillouin Scattering

Over the past few decades, Brillouin scattering has been perhaps the most frequently used technique for measurements of the single-crystal elastic properties of materials of geophysical interest, especially for phases that are stable only at high pressures. This is largely due to the fact that it is an optical, noncontact technique in which no coupling of transducers or other devices to the sample are needed to produce an acoustic excitation. As a result, Brillouin scattering is intrinsically suitable for measurements on small transparent or translucent samples of less than 100 μm in lateral dimensions. The ability to accurately measure the elastic properties of very small samples, either single crystals or well-sintered polycrystalline samples, without physical contact, is a strength of the Brillouin scattering technique. Brillouin scattering can be performed on samples at high pressure in a diamond anvil cell and/or at high temperature. Weidner *et al.* (1975) first applied the Brillouin scattering technique to measurements on a small single-crystal sample of geological importance. Since these first experiments, the technique has continued to be developed and refined for geoscience research.

Brillouin scattering is the inelastic scattering of light (photons) by thermally generated acoustic vibrations (phonons). That is, incident light is scattered from acoustic vibrations that result from thermal motion of atoms in a material. This scattering changes the frequency of the scattered light by an amount that depends on the phase velocity of the acoustic wave. One way to view the process is to consider a sound wave in a material as producing a periodic modulation of the refractive index, which can scatter light. Because

the modulation is propagating with the speed of sound, the scattered light is Doppler-shifted in frequency. Alternatively, the inelastic scattering process may be viewed as the creation or absorption of a phonon by a photon.

With either model of the process, a small portion of the light scattered is shifted in frequency due to the photon–phonon interaction. If we consider incident light of wave vector k_i being scattered into a photon of wave vector k_s through interaction with a phonon of wave vector q, these wave vectors are related through conservation of momentum for the scattering process:

$$k_S - k_i = \pm q \qquad [8]$$

where the '+' sign indicates phonon annihilation (anti-Stokes scattering) and the '−' sign to phonon creation (Stokes scattering). From conservation of energy, the frequencies in the scattering process are related by

$$\omega_s - \omega_i = \pm \Omega \qquad [9]$$

where Ω is the frequency of the phonon. Because $|k_s| \cong |k_i|$, then for an optically isotropic sample and a scattering angle θ between the incident and scattered wave vectors:

$$2|k_s| \sin \theta/2 = q \qquad [10]$$

Note that the scattering angle θ refers to the angle between the incident and scattered light inside the sample, after any refraction at the sample boundaries. Because many minerals are of low symmetry and optically anisotropic, it is useful to consider the more general relation for the velocity (v) of the acoustic wave, that follows from eqns. [8] and [9]:

$$v = \frac{c(\omega_i - \omega_S)}{\omega_i (n_i^2 + n_s^2 - 2n_i n_s \cos \theta)^{1/2}} \qquad [11]$$

where n_i and n_s are the refractive indices for the incident and scattered light, respectively, and c is the speed of light. A special case of particular interest for high-pressure measurements using the diamond anvil cell is one where light enters and exits the sample at the same angle, or symmetrically, through parallel faces of a plate-shaped sample (**Figure 11**). In this case, if the sample is optically isotropic, the expression for the sound velocity is

$$v = \frac{\Delta \omega \cdot \lambda}{2 \sin (\theta^*/2)} \qquad [12]$$

(Whitfield *et al.*, 1976), where θ^* is the external angle between the incoming and the scattered light

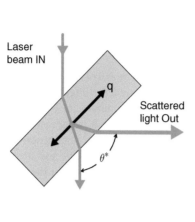
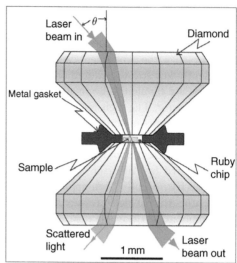

Figure 11 The left-hand panel shows the geometry of a Brillouin scattering experiment with a platelet sample in a symmetric geometry. The external scattering angle in eqn [12] is shown. The geometry is used in high-pressure Brillouin experiments with the diamond anvil cell (right panel). Courtesy of Stanislav V Sinogeikin.

directions outside of the sample (**Figure 11**). This scattering geometry is very convenient and commonly used because the velocities are determined independent of the refractive index of the sample, and because it is compatible with diamond cell measurements. Note that eqn [12] strictly holds only for isotropic materials or when $n_i = n_s$. A backscattering geometry ($\theta = 180°$) is also commonly used, and in this geometry the observed Brillouin shift $\Delta\omega \propto nv$, thus requiring knowledge of the the refractive index to obtain velocities and elastic moduli. In addition, the amplitude of the shear modes goes to zero in a backscattering geometry. An extensive review of Brillouin scattering is given by Sandercock (1982) and references therein.

A schematic of a Brillouin spectrometer is shown in **Figure 12**. The main component is a six-pass Fabry–Perot interferometer that is capable of resolving the very small frequency shift, $\Delta\omega$, of the Brillouin scattered light and the elastically scattered Raleigh light (Sandercock, 1982; Bass, 1989) (**Figure 13**). This requires an instrument with very high finesse. In addition, the Fabry–Perot must have a very high contrast (signal-to-noise ratio) for low-intensity Brillouin peaks to be above any background. The frequencies of the acoustic waves probed by Brillouin scattering are typically in the range of tens of GHz. In a single-crystal Brillouin experiment, the orientation of the sample is changed so that velocities with many values of **q** are probed. In this way, sufficient data may be obtained to constrain all of the single-crystal elastic

moduli. The intrinsic precision of Brillouin measurements is on the order of 0.1–0.5%, depending upon the sample. By collecting large amounts of data (velocity vs orientation), the uncertainties in the experimental results can be reduced considerably. A generalized inversion technique can be used to obtain the single-crystal elastic modulus tensor, c_{ij}, from the velocities and phonon directions (Weidner and Carleton, 1977).

The basic strategy for carrying out a high-pressure Brillouin experiment using the diamond cell is described by Whitfield *et al.* (1976) (**Figure 11**). This work has been followed up by high-pressure Brillouin studies on a variety of materials (Brody *et al.*, 1981; Polian and Grimsditch, 1984; Duffy *et al.*, 1995; Sinogeikin and Bass, 1999), and high-pressure Brillouin studies are now performed in a number of laboratories around the world. It is now possible to measure sound velocities to pressures of 100 GPa and above (e.g., Murakami *et al.*, 2007). High-temperature measurements can be performed using resistance heaters to ~1700 K (Sinogeikin *et al.*, 2000; Jackson *et al.*, 2004) or with CO_2 laser heating to temperatures approaching 3000 K (Sinogeikin *et al.*, 2004). Brillouin scattering can be used to measure the surface wave velocities of metal samples (Sandercock, 1982; Crowhurst *et al.*, 1999).

A new application of Brillouin scattering is its interfacing with synchrotron radiation for simultaneous measurements of sound velocities and volume (density). This capability has recently been developed at the GSECARS (Sector 13) beam line of the advanced

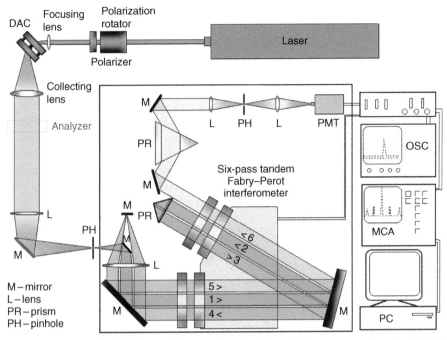

Figure 12 Schematic diagram of a Brillouin spectrometer. An orientation device such as a three-circle Eularian cradle can be at the sample position (where a diamond cell, DAC, is shown here), so that the sample orientation can be changed to access different phonon directions. Light scattered by the sample is analyzed by an electrically scanning Fabry–Perot interferometer. Courtesy of S. V. Sinogeikin.

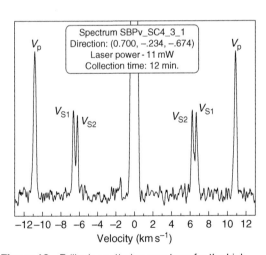

Figure 13 Brillouin scattering spectrum for the high-pressure phase $MgSiO_3$ perovskite. A longitudinal (vP) and two polarizations of transverse waves (vS1 and vS2) are shown symmetrically on either side of the strong elastic peak at zero velocity in the center of the spectrum. The distance of the sharp Brillouin peaks from elastic peak is proportional to the velocity. Adopted from Sinogeikin SV, Lakshtanov DL, Nicolas J, and Bass JD (2004) Sound velocity measurements on laser-heated MgO and Al_2O_3. *Physics of the Earth and Planetary Interiors* 143–144: 575–586.

photon source (Sinogeikin *et al.*, 2006), and should allow the development of improved pressure scales for high-pressure research (Ruoff *et al.*, 1973).

10.4.2 Impulsive Stimulated Scattering

Impulsive stimulated scattering (ISS) is another laser-based technique for measuring the single-crystal elastic moduli, or the bulk elastic properties from polycrystal-line samples. Unlike Brillouin scattering, where the scattering is from intrinsic thermal motion of the atoms, the ISS acoustic signal is produced via the interaction of interfering lasers in the sample. Thus, this method is sometimes referred to as stimulated Brillouin scattering, or laser-induced phonon spectro-scopy. An advantage of the techniques is that the scattered signal is much stronger than with Brillouin scattering, but the nature of the experiment data (acoustic velocities as a function of crystallographic direction) is essentially the same as with Brillouin or ultrasonic interferometry. The ISS technique has addi-tionally been used to measure thermal diffusivity and acoustic attenuation. The theory and applications of the ISS technique have been discussed in review papers by Abramson *et al.* (1999) and Fayer (1982).

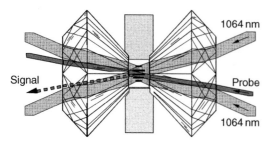

Figure 14 The impulsively stimulated scattering technique. Two laser beams enter (from the right) through the back of a diamond and cross in the sample. The transient grating produced in the sample will diffract a probe beam brought in at the Bragg angle. Courtesy of J.M. Brown.

In outlining the basis for the ISS technique, we follow the descriptions given by Brown *et al.* (1988), Abramson *et al.* (1999), and Zaug *et al.* (1992). Acoustic excitations are produced in a sample by the interference of two pulses from the output of a *Q*-switched, mode-locked laser that are combined in a sample at an angle 2θ (**Figure 14**). The interference of these laser beams produces a periodic variation of intensity which, for absorbing samples, produces a concomitant variation in temperature of the sample. The rapid appearance of this thermal grating and the associated thermal pressure excites a quasi-longitudinal and two quasi-transverse acoustic waves. The wavelengths, λ_A, of the thermal fluctuations and acoustic waves are equal and given by

$$d = \lambda_A = \frac{\lambda}{2\sin\theta} \qquad [13]$$

where λ is the wavelength of the laser light (1064 nm in this case). A third beam from the *Q*-switched laser is frequency doubled, time delayed, and scattered from the thermal grating at the Bragg angle to serve as a probe of the acoustic waves. The intensity of the probe as a function of delay time yields the frequency of the acoustic wave, v_A, and hence its velocity. The raw data are in the time domain, and Fourier transformed to obtain the acoustic frequencies. Note that where a sample is elastically isotropic, only a longitudinal wave is excited. In cases where the sample does not absorb at the 1064 nm frequency of the interfering lasers, electrostriction may be sufficient for the excitation of acoustic phonons.

Like Brillouin scattering, ISS may be used to investigate the properties of surface waves on metals, including at high pressure (Crowhurst *et al.*, 2003) and the elastic moduli of minerals with low symmetry (e.g., Brown *et al.*, 2006).

10.5 Inelastic X-Ray Scattering

The development of intense third-generation synchrotron X-ray sources and the possibility of achieving extremely high-energy resolution, on the order of 1 meV, has opened the opportunity to determine sound velocities through inelastic X-ray scattering (IXS) (Burkel, 2000). Here we mention two types of IXS techniques that have recently been applied to the measurement of velocities and elastic moduli for materials of relevance to geophysics, and which have been developed for high-pressure experiments with the diamond cell.

10.5.1 Nuclear Resonant Inelastic X-Ray Scattering (NRIXS)

This technique provides a way of probing the phonon density of states (PDOSs) by excitation of nuclear resonances for specific isotopes. From the density of states, sound velocities and thermodynamic properties can then be calculated. Intense synchrotron X-rays with a tightly defined energy bandwidth are used to excite the nuclear resonances. One of the optimal isotopes for NRIXS studies is ^{57}Fe, which is fortuitous given the abundance of Fe and Fe-bearing compounds in the Earth.

The configuration of a sample for a high-pressure NRIXS experiment in a diamond cell is shown in **Figure 15**. In an NRIXS experiment, short pulses of synchrotron X-rays excite fluorescent radiation from the Fe sublattice, and this fluorescence is recorded in a short time window between pulses. It is important to note that the experiment selectively probes the Fe atoms, and other elements in the sample do not contribute to the recorded signal. The energy of the impinging X-rays is scanned, typically by less than ±100 meV in steps of ~0.25 meV about the nuclear transition energy of 14.4125 keV for ^{57}Fe. Three avalanche photodiodes (APDs) around the sample record a spectrum of intensity versus energy of the incident radiation, with the zero point in the spectrum corresponding to 14.4125 keV (**Figure 16**). From the NRIXS spectrum, the phonon density of states can be calculated (Sturhahn *et al.*, 1995). An example of the PDOS is given in **Figure 17** for Fe at three different temperatures and high pressure.

The low-frequency part of the PDOS is related to the Debye sound velocity (Hu *et al.*, 2003):

$$\frac{3}{v_D^3} = \frac{1}{v_P^3} + \frac{2}{v_S^3} \qquad [14]$$

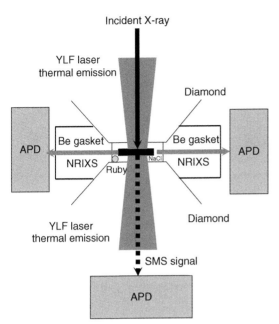

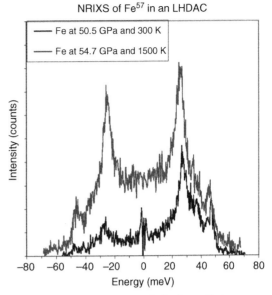

NRIXS of Fe⁵⁷ in an LHDAC

Figure 16 A nuclear resonant inelastic scattering spectrum for ^{57}Fe. Zero energy corresponds to the 14.4125 keV nuclear transition energy of ^{57}Fe. Note the difference in intensity of Stokes and anti-Stokes peaks (symmetrically right and left of zero-energy, respectively), which is an indicator of temperature. Modified from Lin JF, Sturhahn W, Zhao J, Shen G, Mao HK, Hemley RJ (2004) Absolute temperature measurement in a laser-heated diamond anvil cell. *Geophysical Research Letters* 31: L14611 (doi:10.1029/2004GL020599).

Figure 15 Experimental configuration for a nuclear resonant inelastic X-ray scattering experiment at high pressure in the diamond anvil cell. The ^{57}Fe-enriched sample is surrounded by an NaCl pressure medium within the sample chamber of an X-ray transparent Be gasket. Avalanche photodiodes (APDs) surround the sample to detect the nuclear resonant signal. Modified from Lin JF, Sturhahn W, Zhao J, Shen G, Mao HK, Hemley RJ (2005b) Nuclear resonant inelastic X-ray scattering and synchrotron Mössbauer spectroscopy with laser-heated diamond anvil cells. In: Chen J, Wang Y, Duffy TS, Shen G, Dobrzhinetskaya LF (eds.) *Advances in High-Pressure Technology for Geophysical Applications*, pp.397–412. Amsterdam Elsevier B.V.

Parabolic fits to the low-frequency portion of the PDOS gives v_D. If the bulk modulus and density of the sample at high pressure are known, for example, through static compression experiments, then eqns[1], [2], and [14] allow one to solve for the aggregate sound velocities v_P and v_S. The Gruneisen parameter and thermal expansion must also be known to convert isothermal to adiabatic quantities. This method has been used to calculate the velocities of Fe at high pressures and temperatures (Mao *et al.*, 2001; Lin *et al.*, 2005a).

Further details on nuclear resonant inelastic X-ray scattering, and its applications to samples of geophysical interest, can be found in the review articles by Sturhahn (2004) and Lin *et al.* (2005b).

10.5.2 Inelastic X-Ray Scattering from Phonons

IXS by phonons is analogous to Brillouin scattering with X-rays instead of with visible light. The frequency of X-ray photons is changed due to interaction with thermally generated phonons in a material. Only the methods of detecting the frequency shift (energy transfer) of the X-ray photons are different. Like Brillouin scattering, this technique allows the direct measurement of sound velocities in single crystals and polycrystalline samples. A review of IXS is given by Burkel (2000), and its application to materials of geophysical interest and high-pressure research is given by Fiquet *et al.* (2004).

A main advantages of the IXS technique over visible-light scattering methods, such as Brillouin scattering and ISS, is that X-rays penetrate all materials. This makes it possible to determine the body wave sound velocities (v_p and v_s) on optically opaque substances. One can therefore directly measure the elastic moduli and sound velocities of materials relevant to the Earth's core, such as Fe and its alloys (Fiquet *et al.*, 2001, 2004). There are no refractive index effects.

A diagram of the IXS facility at beam line ID28 of the ESRF synchrotron is shown in **Figure 18** to illustrate the measurement techniques. In this experiment the momentum transfer is chosen at desired

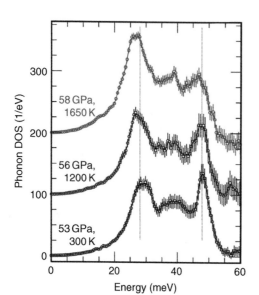

Figure 17 Phonon density of states of Fe at high pressure and tempearatures, determined by NRIXS experiments in the diamond cell. Courtesy of Jung-Fu Lin.

values of $2|k_s| \sin \theta/2 = q$, and the frequency of the input radiation is scanned to determine the energy transfer. Five analyzers are mounted 6.5 m from the sample on a Rowland arm that rotates about a vertical axis through the sample, allowing one to access different values of θ, q. In this way, the phonon dispersion curves for each acoustic mode can be obtained, and are usually fit with a sine function to obtain the velocities corresponding to the Brillouin zone center. The energy is scanned by changing the temperature on the final backscattering Si monochromer. An example of the raw data, showing intensity of scattered X-rays as a function of the

energy transfer of the input X-rays, is shown in **Figure 19**. From the velocities as a function of crystallographic direction, the elastic moduli can be obtained exactly as for other single-crystal techniques described above.

10.6   Shock Waves

Shock wave results on rocks and minerals have had a major influence on our understanding of the formation and evolution of planetary bodies. Dynamic shock compression studies allow one to access pressure and temperature regimes (hundreds of GPa and over 10 000 K) that are inaccessible to other techniques. Shock techniques have been developed to measure a wide variety of properties and processes, and a complete survey of this field cannot be presented here. Rather, we mention only the shock wave equation of state experiments that are the basis for much of what is inferred about the deeper parts of Earth and other planets. Excellent reviews of shock wave theory and experimental methods are given by Rice *et al.* (1958), Al'tschuler (1965), Duvall and Fowles (1963), and McQueen *et al.* (1970).

In a shock wave experiment, a sample is impacted at high velocity by a flyer plate, producing a high-density state that propagates through the sample with shock velocity U_s (**Figure 20**). The locus of all shock states that can be achieved by a sample in given initial state is known as the Hugoniot. The relationship between the Hugoniot and an isotherm or isentrope is shown in **Figure 20**. It is important to realize that the temperature increases rapidly along

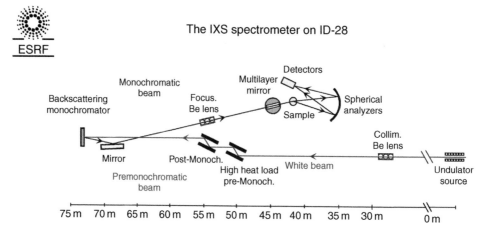

Figure 18 Momentum-resolved inelastic X-ray scattering setup at beam line ID-28 of the ESRF synchrotron in Grenoble, France. Courtesy of M. Krisch.

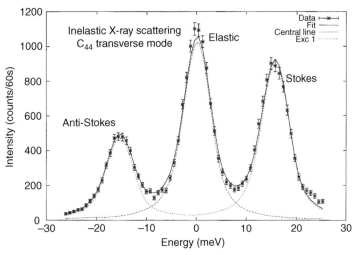

Figure 19 Inelastic x-ray scattering spectrum for Fe and Al-bearing Mg-silicate perovskite. The central peak at zero energy transfer is from elastic scattering. The peaks at $\sim\pm15\,\text{meV}$ are scattering due to the pure shear mode c_{44}.

the Hugoniot, and it is hotter than the isentrope (along which temperature increases adiabatically). Therefore, the pressure corresponding to a given compressed volume is higher along the Hugoniot. In a standard shock wave equation-of-state experiment, the pressure and density (or specific volume) are determined, but the temperature is generally unknown. It is worth noting that the pressures in shock experiments are known with relatively high accuracy, on the order of 1%. This makes shock wave experiments important in the determination of pressure scales for high-pressure research.

From conservation of mass, momentum, and energy across the shock front, one obtains the Rankine–Hugoniot relations:

$$\rho_0 U_S = \rho(U_S - u_P)$$
$$P - P_0 = \rho_0 U_S u_P \qquad [15]$$
$$E - E_0 = (P + P_0)(V_0 - V)/2$$

where U_S and u_P are shock velocity and particle velocity, respectively, and P, V, and E are the pressure, volume, and internal energy in the shocked Hugoniot state. In a shock equation-of-state experiment, the shock velocity and particle velocity are measured. U_s is the transit time of the shock front through the sample. The particle velocity can be obtained by measuring the velocity of the impactor (flyer plate) through an impedance matching method (McQueen *et al.*, 1970). With knowledge of the initial pressure and density of the material, the density, pressure, and change in internal energy in the shocked state are thus known. It is empirically determined that in the absence of phase changes, U_s and u_P

form a linear trend, and the results of shock wave experiments are often represented in this way (Ahrens and Johnson, 1995).

The results of shock equation-of-state experiments have been critical to our understanding of Earth's deep interior. For example, experiments on Fe provide compelling evidence that the density of the outer core is too low for it to be composed of pure Fe, and therefore contains light elements. However, it should be recognized that many other types of sophisticated shock wave experiments are being performed, of which we will mention two. Hugoniot temperatures have been inferred by analysis of the thermal radiation emitted by shocked samples (Lyzenga and Ahrens, 1980; Bass *et al.*, 1990). Measurements of sound velocities can be made along the Hugoniot and can provide evidence for phase transitions that are not readily apparent by other types of data (Brown and McQueen, 1986; Nguyen and Holmes, 2004). New technologies such as the ability to produce shock waves by laser ablation may open up yet further opportunities to investigate the properties of deep Earth materials (Swift *et al.*, 2004).

10.7 Other Techniques

Elasticity measurements have a long history due to their importance in many areas of physics, chemistry, and materials sciences. This review is not exhaustive, and a number of measurement techniques have necessarily not been discussed. Perhaps most noteworthy is inelastic neutron scattering, which is arguably the

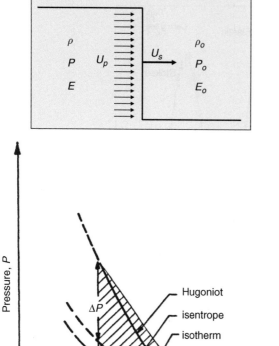

Figure 20 Top: Schematic illustration of the states behind and in front of an advancing shock wave. The bottom panel shows the relationship of the Hugoniot to an isentrope and an isotherm. Top: Courtesy of Paul Asimow.

optimal way to investigate phonons (e.g., Shirane, 1974). This method has found limited applications in geophysics, because of the large sample size requirements for neutron studies. However, significant progress is being made in the use of neutron scattering for investigations at high pressures (Besson *et al.*, 1992). Another technique that been used to provide constraints on the strength and elasticity of materials is radial X-ray diffraction in the diamond cell (e.g., Merkel *et al.*, 2002; Speziale *et al.*, 2006).

Acknowledgments

The author gratefully acknowledge those who have generously provided figures for this paper, including P D Asimow, J M Brown, T Duffy, I Getting, J Hu, D Isaak, S. Jacobsen, M Krisch, B Li, R C Lieberman, A Lin, G Shen, S V Sinogeikin, and W Sturhahn. This work was supported by the National Science Foundation.

References

Abramson EH, Brown JM, and Slutsky LJ (1999) Applications of impulsive stimulated scattering in the Earth and planetary sciences. *Annual Review of Physical Chemistry* 50: 279–313.

Al'tschuler LV (1965) Use of shock waves in high-pressure physics. *Soviet Physics Uspekhi* 8: 52–91.

Ahrens TJ and Johnson ML (1995) Shock wave data for minerals. In: Ahrens TJ (ed.) *Mineral Physics and Crystallography: A Handbook of Physical Constants*, pp. 143–184. Washington, DC: American Geophysical Union.

Anderson OL, Isaak DG, and Yamamoto S (1989) Anharmonicity and the equation of state for gold. *Journal of Applied Physics* 65: 1534–1543.

Anderson OL and Isaak DG (1995) Elastic constants of mantle minerals at high temperature. In: Ahrens TJ (ed.) *Mineral Physics and Crystallography: A Handbook of Physical Constants*, pp. 64–97. Washington, DC: American Geophysical Union.

Andrault D, Fiquet G, Itie JP, et al. (1998) Thermal pressure in the laser-heated diamond-anvil cell: An X-ray diffraction study. *European Journal of Mineralogy* 10: 931–940.

Angel RJ (2000) Equations of state. In: Hazen RM and Downs RT (eds.) *Reviews in Mineralogy and Geochemistry High-Temperature and High-Pressure Crystal Chemistry*, pp. 35–59. Washington, DC: Mineralogical Society of America.

Baidov V V and Kunin LL (1968) Speed of ultrasound and compressibility of molten silicates. *Soviet Physics Uspekhi* 13: 64–65 (English Translation).

Bass JD (1989) Elasticity of grossular and spessartite garnets by Brillouin spectroscopy. *Journal of Geophysical Research* 84: 7621–7628.

Bass JD (1995) Elasticity of minerals, glasses, and melts. In: Ahrens TJ (ed.) *Mineral Physics and Crystallography: A Handbook of Physical Constants*, pp. 45–63. Washington, DC: American Geophysical Union.

Bass JD, Ahrens TJ, Abelson JR, and Tan H (1990) Shock temperature measurements in metals: New results for an Fe alloy. *Journal of Geophysical Research* 95: 21767–21776.

Bass JD, Liebermann RC, Weidner DJ, and Finch S (1981) Elastic properties from acoustic and volume compression experiments. *Physics of the Earth and Planetary Interiors* 25: 140–158.

Bassett WA, Shen AH, Bucknum M, and Chou I-M (1993) A new diamond anvil cell for hydrothermal studies to 2.5 GPa and from 190 to 1200°C. *Review of Scientific Instrument* 64: 2340–2345.

Besson JM, Weill G, Hamel G, Nelmes RJ, Loveday JS, and Hull S (1992) Equation of state of lithium deuteride from neutron diffraction under high pressure. *Physical Review B* 45: 2613–2619.

Birch F (1952) Elasticity and constitution of the Earth's interior. *Journal of Geophysical Research* 57: 227–286.

Birch F (1960) The velocity of compressional waves in rocks to 10 kilobars. Part 1. *Journal of Geophysical Research* 65: 1083–1102.

Birch F (1978) Finite strain isotherm and velocities for single crystal and polycrystalline NaCl at high pressures and 300 K. *Journal of Geophysical Research* 83: 1257–1268.

Bridgman PW (1964) *Collected Experimental Papers of P. W. Bridgeman.* Cambridge MA: Harvard University Press.

Brody EM, Shimizu H, Mao HK, Bell PM, and Bassett WA (1981) Acoustic velocity and refractive index of fluid hydrogen and deuterium at high pressures. *Journal of Applied Physics* 52: 3583–3585.

Brown JM, Abramson EH, and Angel RJ (2006) Triclinic elastic constants for low albite. *Physics and Chemistry of Minerals* 33: 256–265.

Brown JM and McQueen RG (1986) Phase transitions, Gruneisen parameter, and elasticity for shocked iron between 77 GPa and 400 GPa. *Journal of Geophysical Research* 91: 7485–7494.

Brown JM, Slutsky LJ, Nelson KA, and Cheng L-T (1988) Velocity of sound and equations of state for methanol and ethanol to 6.8 GPa in a diamond anvil cell. *Science* 241: 65–67.

Burkel E (2000) Phonon spectroscopy by inelastic X-ray scattering. *Reports on Progress in Physics* 63: 171–232.

Chijioke AD, Nellis WJ, Soldatov A, and Silvera IF (2005) The ruby pressure standard to 150 GPa. *Journal of Applied Physics* 98: 114905.

Christensen NI (1982) Seismic velocities. In: Carmichael RS (ed.) *Handbook of Physical Properties of Rocks*, vol. II, ch. 1, pp. 1–228. Boca Raton Fl: CRC Press.

Crowhurst JC, Hearne GR, Comins JD, Every AG, and Stoddart PR (1999) Surface Brillouin scattering at high pressure - Application to a thin supported gold film. *Physical Review B* 60(R14): 990–993.

Crowhurst JC, Zaug JM, Abramson EH, Brown JM, and Ahre DW (2003) Impulsive stimulated light scattering at high pressure - Precise determination of elastic constants of opaque materials. *High Pressure Research* 23: 373–377.

Datchi F, LeToullec R, and Loubeyre P (1997) Improved calibration of the $SrB_4O_7:Sm^{2+}$ optical pressure gauge: Advantages at very high pressures and high temperatures. *Journal of Applied Physics* 81: 3333–3339.

Davies GF and O'Connell RJ (1977) Transducer and bond phase shifts in ultrasonics and their effects on measured pressure derivatives of elastic moduli. In: Manghnani MH and Akimoto S-I (eds.) *High-Pressure Research, Applications in Geophysics*, pp. 533–562. New York: Academic Press.

Davison L and Graham RA (1979) Shock compression of solics. *Physics Reports* 55: 255–379.

Demarest H (1971) Cube resonance method to determine the elastic constants of solids. *Journal of the Acoustical Society of America* 49: 768–775.

Dewaele A, Loubeyre P, and Mezouar M (2004) Equations of state of six metals above 94 GPa. *Physical Review B* 70: 094112.

Duffy TS, Zha CS, Downs RT, Mao HK, and Hemley RJ (1995) Sound velocity and elasticity of single crystal forsterite to 16 GPa and the composition of the upper mantle. *Nature* 378: 170–173.

Duvall GE and Fowles GR (1963) Shock waves. In: Bradley RS (ed.) *High Pressure Physics and Chemistry*, vol 2, 209 pp. New York: Academic Press.

Fayer MD (1982) Dynamics of molecules in condensed phases – Picosecond holographc grating experiments. *Annual Review of Physical Chemistry* 33: 63–87.

Fei Y (1999) Effects of temperature and composition on the bulk modulus of (Mg,Fe)O. *American Mineralogist* 84: 272–276.

Fei Y, Li J, Hirose K, et al. (2004) A critical evaluation of pressure scales at high temperatures by *in situ* X-ray diffraction measurements. *Physics of the Earth and Planetary Interiors* 143–144: 515–526.

Fiquet G, Andrault D, Dewaele A, Charpin T, Kunz M, and Hausermann D (1998) P–V–T equation of state of $MgSiO_3$ perovskite. *Physics of the Earth and Planetary Interiors* 105: 21–31.

Fiquet G, Andrault D, Itie JP, Gillet P, and Richet P (1996) X-ray diffraction of periclase in a laser-heated diamond-anvil cell. *Physics of the Earth and Planetary Interiors* 95: 1–17.

Fiquet G, Badro J, Guyot F, et al. (2004) Application of inelastic X-ray scattering to the measurements of acoustic wave velocities in geophysical materials at very high pressure. *Physics of the Earth and Planetary Interiors* 143–144: 5–18.

Fiquet G, Badro J, Guyot F, Requardt H, and Krisch M (2001) Sound velocities in iron to 110 GPa. *Science* 291: 468–471.

Fraser DB and LeCraw RC (1964) Novel method of measuring elastic and anelastic properties of solids. *Review of Scientific Instrument* 35: 1113–1115.

Getting IC (1998) New determination of the bismuth I-II equilibrium pressure: A proposed modification to the practical pressure scale. *Metrologia* 35: 119–132.

Gwanmesia GD, Li B, and Liebermann RC (1993) Hot pressing of polycrystals of high-pressure phases of mantle minerals in multi-anvil apparatus. *Pure and Applied Geophysics* 141: 467–484.

Higo Y, Inoue T, Li B, Irifune T, and Liebermann RC (2006) The effect of iron on the elastic properties of ringwoodite at high pressure. *Physics of the Earth and Planetary Interiors* 159: 276–285.

Holmes NC, Moriarty JA, Gathers GR, and Nellis WJ (1989) The equation of state of platinum to 660 GPa (6.6 Mbar). *Journal of Applied Physics* 66: 2962–2967.

Holzapfel WB (2003) Refinement of the ruby luminescence pressure scale. *Journal of Applied Physics* 93: 1813–1818.

Hu M, Sturhahn W, Toellner TS, et al. (2003) Measuring velocity of sound with nuclear resonant inelastic X-ray scattering. *Physical Review B* 67: 094304.

Irifune T (2002) Application of synchrotron radiation and Kawai-type apparatus to various studies in high-pressure mineral physics. *Mineralogical Magazine* 66: 769–790.

Isaak DG, Anderson OL, and Goto T (1989) Measured elastic moduli of single-crystal MgO up to 1800 K. *Physics and Chemistry of Minerals* 16: 704–713.

Isaak DG, Ohno I, and Lee PC (2006) The elastic constants of monoclinic single-crystal chrome-diopside to 1,300 K. *Physics and Chemistry of Minerals* 32: 691–699.

Jackson I and Niesler H (1982) The elasticity of periclase to 3 GPa and some geophysical implications. In: Akimoto S-I and Manghnani MH (eds.) *High-Pressure Research in Geophysics*, pp. 93–113. Tokyo: Center for Academic Publications.

Jackson I, Niesler H, and Weidner DJ (1981) Explicit correction of ultrasonically determined elastic wave velocities for transducer-bond phase shifts. *Journal of Geophysical Research* 86: 3736–3748.

Jackson JM, Sinogeikin SV, Carpenter MA, and Bass JD (2004) Novel phase transition in orthoenstatite. *American Mineralogist* 89: 239–245.

Jacobsen SD, Spetzler HA, Reichmann H-J, et al. (2002) Gigahertz ultrasonic interferometry at high P and T: New tools for obtaining thermodynamic equation of state. *Journal of Physics: Condensed Matter* 14: 11525–11530.

Jacobsen SD, Spetzler H, Reichmann H-J, and Smyth JR (2004) Shear waves in the diamond-anvil cell reveal pressure-induced instability in (Mg,Fe)O. *Proceedings of the National Academy of Sciences* 101: 5867–5871 (doi:10.1073/pnas.0401564101).

Jamieson JC, Fritz JN, and Manghnani MH (1982) Pressure measurement at high temperature in X-ray diffraction studies: Gold as a primary standard. In: Akimoto S and Manghnani MH (eds.) *High Pressure Research in Geophysics*, pp. 27–47. Tokyo: Center for Academic Publications.

Jayaraman A (1983) Diamond anvil cell and high-pressure physical investigations. *Reviews of Modern Physics* 55: 65–108.

Kantor AP, Jacobsen SD, Kantor IY, *et al.* (2004) Pressure-induced magnetization in FeO: Evidence from elasticity and Mössbauer spectroscopy. *Physical Review Letters* 93: 215502-1–215502-4.

Katahara KW, Rai CS, Manghnani MH, and Balogh J (1981) An interferometric technique for measuring velocity and attenuation in molten rocks. *Journal of Geophysical Research* 86: 11779–11786.

Kawai N and Endo S (1970) The generation of ultrahigh hydrostatic pressure by a split sphere apparatus. *Review of Scientific Instrument* 41: 425–428.

Khvostantsev LG, Slesarev VN, and Brazhkin VV (2004) Toroid type high-pressure device: History and prospects. *High Pressure Research* 24: 371–383.

Knittle E (1995) Static compression measurements of equations of state. In: Ahrens TJ (ed.) *Mineral Physics and Crystallography: A Handbook of Physical Constants*, pp. 98–142. Washington, DC: American Geophysical Union.

Kung J, Li B, Uchida T, Wang Y, Neuville D, and Liebermann RC (2004) *In situ* measurements of sound velocities and densities across the orthopyroxene-high-pressure clinopyroxene transition in MgSiO3 at high pressure. *Physics of the Earth and Planetary Interiors* 147: 27–44.

Li B, Chen K, Kung J, Liebermann RC, and Weidner DJ (2002) Sound velocity measurement using transfer function method. *Journal of Physics: Condensed Matter* 14: 11337–11342.

Li B, Kung J, and Liebermann RC (2004) Modern techniques in measuring elasticity of Earth materials at high pressure and high temperature using ultrasonic interferometry in conjunction with synchrotron X-radiation in multi-anvil apparatus. *Physics of the Earth and Planetary Interiors* 143-144: 559–574.

Li B, Kung J, Uchida T, and Wang Y (2005) Pressure calibration to 20 GPa by simultaneous use of ultrasonic and X-ray techniques. *Journal of Applied Physics* 98: 013521-1–013521-5.

Lin JF, Sturhahn W, Zhao J, Shen G, Mao HK, and Hemley RJ (2004) Absolute temperature measurement in a laser-heated diamond anvil cell. *Geophysical Research Letters* 31: L14611 (doi:10.1029/2004GL020599).

Lin JF, Sturhahn W, Zhao J, Shen G, Mao HK, and Hemley RJ (2005a) Sound velocities of hot dense iron: Birch's Law revisited. *Science* 308: 1892–1895.

Lin JF, Sturhahn W, Zhao J, Shen G, Mao HK, and Hemley RJ (2005b) Nuclear resonant inelastic X-ray scattering and synchrotron Mössbauer spectroscopy with laser-heated diamond anvil cells. In: Chen J, Wang Y, Duffy TS, Shen G, and Dobrzhinetskaya LF (eds.) *Advances in High-Pressure Technology for Geophysical Applications*, pp. 397–412. Amsterdam: Elsevier B.V.

Lheureux D, Decremps F, Fischer M, *et al.* (2000) Ultrasonics and X-ray diffraction under pressure in the Paris-Edinburgh cell. *Ultrasonics* 38: 247–251.

Lyzenga GA and Ahrens TJ (1980) Shock temperature measurements in Mg2SiO4 and SiO2 at high pressure. *Geophysical Research Letters* 7: 141–144.

Mao HK, Wu Y, Chen LC, Shu JF, and Jephcoat AP (1990) Static compression of Iron to 300 GPa and Fe$_{0.8}$Ni$_{0.2}$ alloy to 260 GPa – Implications for composition of the core. *Journal of Geophysical Research* 95: 21737–21742.

Mao HK, Xu J, and Bell PM (1986) Calibration of the ruby pressure gauge to 800 kbar under quasihydrostatic conditions. *Journal of Geophysical Research* 91: 4673–4676.

Mao HK, Xu J, Struzhkin VV, *et al.* (2001) Phonon density of states of iron up to 153 gigapascals. *Science* 292: 914–916.

Mayama N, Suzuki I, Saito T, Ohno I, Katsura T, and Yoneda A (2005) Temperature dependence of the elastic moduli of ringwoodite. *Physics of the Earth and Planetary Interiors* 148: 353–359.

McQueen RG, Marsh SP, Taylor JW, Fritz JN, and Carter WJ (1970) The equation of state of solids from shock wave studies. In: Kinslow R (ed.) *High-Velocity Impact Phenomena*, pp. 249. New York: Academic Press.

McSkimin HJ (1950) Ultrasonic measurement techniques applied to small solid specimens. *Journal of the Acoustical Society of America* 22: 413–418.

McSkimin HJ (1961) Pulse superposition method for measuring ultrasonic wave velocities in solids. *Journal of the Acaustical Society of America* 33: 12–16.

Meng Y, Shen G, and Mao HK (2006) Double sided laser heating system at HPCAT for *in-situ* X-ray diffraction at high pressure and high temperatures. *Journal of Physics: Condensed Matter* 18: S1097–S1103.

Merkel S, Wenk HR, Shu J, *et al.* (2002) Deformation of polycrystalline MgO at pressures of the lower mantle. *Journal of Geophysical Research* 107: 2271 (doi:10.1029/2001JB000920).

Mezouar M, Crichton WA, and Bauchau S (2005) Development of a new state-of-the-art beamline optimized for monochromatic single-crystal and powder X-ray diffraction under extreme conditions at the ESRF. *Journal of Synchrotron Radiation* 12: 659–664.

Murakami M, Sinogeikin SV, Hellwig H, and Bass JD (2007) Sound velocity of MgSiO$_3$ perovskite to megabar pressure and mineralogy of Earth's lower mantle. *Earth and Planetary Science Letters* 256: 47–54.

Murase T, Kushiro I, and Fujii T (1977) Compressional wave velocity in partially molten peridotite. *Year Book Carnegie Institution Washington* 76: 414–416.

Musgrave MJP (1970) *Crystal Acoustics*, 289 pp. San Francisco, CA: Holden-Day.

Nguyen JH and Holmes NC (2004) Melting of iron at the physical conditions of the Earth's core. *Science* 427: 339–342.

Niesler H and Jackson I (1989) Pressure derivatives of elastic wave velocities from ultrasonic interferometric measurements on jacketed polycrystals. *Journal of the Acoustical Society of America* 86: 1573–1585.

Nye JF (1985) *Physical Properties of Crystals*, 2nd edn. 329 pp. New York: Oxford University Press.

Ohno I (1976) Free vibration of a rectangular parallelepiped crystal and its application to determination of elastic constants of orthorhombic crystals. *Journal of Physics of the Earth* 24: 355–379.

Papadakis EP (1967) Ultrasonic phase velocity by the pulse-echo-overlap method incorporating diffraction phase corrections. *Journal of the Acoustical Society of America* 42: 1045–1051.

Papadakis EP (1990) The measurement of ultrasonic velocity. In: Thurston RN and Pierce AD (eds.) *Ultrasonic Measurement Methods. Physical Acoustics, volume XIX*, ch. 2, pp. 81–107. San Diego, CA: Academic Press.

Poirier JP (2000) *Introduction to the Physics of the Earth's Interior*, 2nd edn. 312 pp. Cambridge: Cambridge University Press.

Polian A and Grimsditch M (1984) New high-pressure pahse of H2O: Ice X. *Physical Review Letters* 52: 1312–1314.

Prasad M and Manghnani MH (1997) Effects of pore and differential pressure on compressional wave velocity and quality factor in Berea and Michigan sandstones. *Geophysics* 62: 1163–1176.

Reichmann HJ and Jacobsen SD (2006) Sound velocities and elastic constants of ZnAl$_2$O$_4$ spinel and implications for spinel-elasticity systematics. *American Mineralogist* 91: 1049–1054.

Rice MH, McQueen RG, and Walsh JM (1958) Compression of solids by strong shock waves. Seitz F and Turnbull D (eds.) *Solid State Physics*, vol. 6, pp. 1–63. New York: Academic Press.

Rivers ML and Carmichael ISE (1987) Ultrasonic studies of silicate melts. *Journal of Geophysical Research* 92: 9247–9270.

Ruoff AL, Lincoln RC, and Chen YC (1973) A new method of absolute high pressure determination. *Journal of Physics D: Applied Physics* 6: 1295–1306.

Sandercock JR (1982) Trends in Brillouin scattering: Studies of opaque materials, supported films, and central modes. In: Cardona M and Guntherodt G (eds.) *Topics in Applied Physics vol. 51: Light Scattering in Solids III, Recent Results*, pp. 173–206. Berlin: Springer-Verlag.

Shen G, Rivers ML, Wang Y, and Sutton SR (2001) Laser heated diamond cell system at the advanced photon source for *in situ* X-ray measurements at high pressure and temperature. *Review of Scientific Instrument* 72: 1273–1282.

Shirane G (1974) Neutron scattering studies of structural phase transitions at Brookhaven. *Reviews of Modern Physics* 46: 437–449.

Simmons G (1964) Velocity of shear waves in rocks to 10 kilobars. Part 1. *Journal of Geophysical Research* 69: 1123–1130.

Sinogeikin SV and Bass JD (1999) Single-crystal elasticity of MgO at high pressure. *Physical Review* B59: R14141–R14144.

Sinogeikin SV, Bass JD, Prakapenka V, *et al.* (2006) A Brillouin spectrometer interfaced with synchrotron X-radiation for simultaneous X-ray density and acoustic velocity measurements. *Review of Scientific Instrument* 77: 103905.

Sinogeikin SV, Jackson JM, O'Neill B, Palko JW, and Bass JD (2000) Compact high-temperature cell for Brillouin scattering measurements. *Review of Scientific Instrument* 71: 201–220.

Sinogeikin SV, Lakshtanov DL, Nicolas J, and Bass JD (2004) Sound velocity measurements on laser-heated MgO and Al_2O_3. *Physics of the Earth and Planetary Interiors* 143–144: 575–586.

Soga N and Anderson OL (1967) Elastic properties of tektites measured by resonant sphere technique. *Journal of Geophysical Research* 72: 1733–1739.

Speziale S, Zha CS, Duffy TS, Hemley RJ, and Mao HK (2001) Quasi-hydrostatic compression of magnesium oxide to 52 GPa: Implications for the pressure–volume–temperature equation of state. *Journal of Geophysical Research* 106: 515–528.

Speziale S, Shieh SR, and Duffy TS (2006) High-pressure elasticity of calcium oxide: A comparison between Brillouin spectroscopy and radial X-ray diffraction. *Review of Scientific Instrument* 111: (doi:10.1029/2005JB003823).

Stacey FD (2005) High pressure equations of state and planetary interiors. *Reports on Progress in Physics* 68: 341–383.

Sturhahn W (2004) Nuclear resonant spectroscopy. *Journal of Physics: Condensed Matter* 16: S497–S530.

Sturhahn W, Toellner TS, Alp EE, *et al.* (1995) Phonon density of states measured by inelastic nuclear resonant scattering. *Physical Review Letters* 74: 3832–3835.

Sweeney JS and Heinz DL (1993) Thermal analysis in the laser-heated diamond anvil cell. *Pure and Applied Geophysics* 141: 497–507.

Swift DC, Tierrney TE, Kopp RA, and Gammer JT (2004) Shock pressures induced in condensed matter by laser ablation. *Physical Review* E69: 036406.

Uchida T, Wang Y, and Rivers ML (2002) A large-volume press facility at the advanced photon source: Diffraction and imaging studies on materials relevant to the cores of planetary bodies. *Journal of Physics: Condensed Matter* 14: 11517–11523.

Vanorio T, Prasad M, and Nur A (2003) Elastic properties of dry clay mineral aggregates, suspensions and sandstones. *Geophysical Journal Internation* 155: 319–326.

Vinet P, Ferrante J, Rose JH, and Smith JR (1987) Compressibility of solids. *Journal of Geophysical Research* 92: 9319–9325.

Wang Y, Weidner DJ, and Guyot F (1996) Thermal equation of state of CaSiO3 perovskite. *Journal of Geophysical Research* 101: 661–672.

Watt JP, Davies GF, and O'Connell RJ (1976) The elastic properties of composite materials. *Review of Geophysics and Space Physics* 14: 541–563.

Weidner DJ and Carleton HR (1977) Elasticity of coesite. *Journal of Geophysical Research* 82: 1334–1346.

Weidner DJ, Swyler K, and Carleton HR (1975) Elasticity of microcrystals. *Geophysical Research Letters* 2: 189–192.

Weidner DJ, Vaughan MT, Ko J, *et al.* (1992) Characterization of stress, pressure, and temperature in SAM85, a DIA type high pressure apparatus. In: Syono Y and Manghnani MH (eds.) *High Pressure Research: Application to Earth and Planetary Sciences*, pp. 13–17. Washington, DC: American Geophysical Union.

Whitfield CH, Brody EM, and Bassett WA (1976) Elastic moduli of NaCl by Brillouin scattering at high pressure in a diamond anvil cell. *Review of Scientific Instrument* 47: 942–947.

Williamson ED and Adams LH (1923) Density distributions in the Earth. *Journal of the Washington Academy of Sciences* 13: 413–428.

Yagi T and Susaki JI (1992) A laser heating system for diamond anvil using CO_2 laser. In: Syono Y and Manghnani MH (eds.) *High-Pressure Research: Application to Earth and Planetary Sciences*, pp. 51–54. Washington, DC: American Geophysical Union.

Zaug J, Abramson E, Brown JM, and Slutsky LJ (1992) Elastic constants, equations of state and themal diffusivity at high pressure. In: Syono Y and Manghnani MH (eds.) *High-Pressure Research: Application to Earth and Planetary Sciences*, pp. 425–431. Washington, DC: American Geophysical Union.

Zha CS and Bassett WA (2003) Internal resistive heating in diamond anvil cell for *in-situ* X-ray diffraction and Raman scattering. *Review of Scientific Instrument* 74: 1255–1262.

11 Theory and Practice – Measuring High-Pressure Electronic and Magnetic Properties

R. J. Hemley, V. V. Struzhkin, and R. E. Cohen, Carnegie Institution of Washington, Washington, DC, USA

11.1	Introduction	296
11.2	Overview of Fundamentals	296
11.2.1	Bonding in Deep-Earth Materials	296
11.2.1.1	Conventional bonding classification	296
11.2.1.2	Bonding under high $P-T$ conditions	297
11.2.2	Electronic Structure	298
11.2.3	Magnetic Properties	299
11.3	Electronic and Magnetic Excitations	300
11.3.1	Dielectric Functions	300
11.3.1.1	Oscillator fits	301
11.3.1.2	KK analysis	301
11.3.2	Electronic Excitations	302
11.3.2.1	Extended excitations	302
11.3.2.2	Excitons	302
11.3.2.3	Crystal field transitions	302
11.3.2.4	Charge transfer	303
11.3.2.5	Excitations from defects	303
11.3.2.6	Electrical conductivity	303
11.3.3	Magnetic Excitations	303
11.4	Overview of Experimental Techniques	303
11.4.1	Optical Spectroscopies	304
11.4.1.1	Optical absorption and reflectivity	304
11.4.1.2	Luminescence	306
11.4.1.3	IR spectroscopy	306
11.4.1.4	Raman scattering	308
11.4.1.5	Brillouin and Rayleigh scattering	309
11.4.1.6	Nonlinear optical methods	310
11.4.2	Mössbauer Spectroscopy	311
11.4.3	X-Ray and Neutron Diffraction	312
11.4.3.1	X-ray diffraction	312
11.4.3.2	Neutron diffraction	312
11.4.4	Inelastic X-Ray Scattering and Spectroscopy	313
11.4.4.1	X-ray absorption spectroscopy	313
11.4.4.2	X-ray emission spectroscopy	314
11.4.4.3	X-ray inelastic near-edge spectroscopy	315
11.4.4.4	X-ray magnetic circular dichroism	316
11.4.4.5	Electronic inelastic X-ray scattering	317
11.4.4.6	Resonant inelastic X-ray spectroscopy	317
11.4.4.7	Inelastic X-ray scattering spectroscopy	318
11.4.4.8	Compton scattering	318
11.4.4.9	Nuclear resonance forward scattering	318
11.4.4.10	Nuclear resonant inelastic X-ray scattering	319
11.4.4.11	Phonon inelastic X-ray scattering	319

11.4.5	Transport Measurements	320
11.4.5.1	Electrical conductivity	320
11.4.5.2	Magnetic susceptibility	321
11.4.6	Resonance Methods	322
11.4.6.1	Electron paramagnetic and electron spin resonance	322
11.4.6.2	Nuclear magnetic resonance	322
11.4.6.3	de Haas–van Alphen	323
11.5	**Selected Examples**	324
11.5.1	Olivine	324
11.5.2	Magnesiowüstite	324
11.5.3	Silicate Perovskite and Post-Perovskite	326
11.5.4	Volatiles	327
11.5.5	Iron and Iron Alloys	328
11.6	**Conclusions**	329
References		329

Glossary

bounding in deep Earth materials – Description of the binding between atoms and/or molecules in substances that comprise the interior of the planet, typically the mantle and core.

bonding under high *P-T* conditions – Description of the binding between atoms and/or molecules at high pressures and temperatures such as those that characterize the Earth and other planets.

Brillouin and Rayleigh scattering – Inelastic light-scattering technique in which the light is scattered off acoustic waves in a material.

charge transfer – Excitations or bonding states derived by promotion of an electron from one atom or molecule to another atom or molecule.

Compton scattering (CS) – Inelastic scattering of x-ray radiation typically used to that probe electron momentum distributions.

conventional bonding classification – The description of binding forces in atoms and molecules found in gases, solids, and liquids under ambient or near ambient conditions, specifically covalent, ionic, metallic, van der Waals, and hydrogen-bonding.

crystal field transitions – Excitations between quantum states of an atom, ion, or molecule inside a crystal where the states are perturbed (e.g., degeneracies lifted) by the symmetry of the site of the crystal.

defect excitations – Transitions or spectral features arising from defects in the material, including color centers.

de Haas-van Alphen – Technique for measuring Fermi surfaces of metals in a magnetic field.

dielectric functions – The response of a material to an applied electric field.

elastic neutron scattering – Scattering of neutron without a change in particle energy, as in typical neutron diffraction;

elastic X-ray scattering – Scattering of x-rays without changes in photon energy as in typical x-ray diffraction, may be either coherent or incoherent.

electrical conductivity – Movement of charge in a system in response to an applied electric field (static or frequency dependent), and can be intrinsic or extrinsic, electronic or ionic

electronic inelastic X-ray scattering (EIXS) – Technique to probe electronic excitations such as core-level, plasmons, and electronic bands, using x-rays, including momentum q dependence.

electronic structure – Quantum description of electrons in an atom, molecule, or condensed phase (i.e., crystal, liquid, fluid, or glass)

electron paramagnetic resonance (EPR) – Technique which involves resonant microwave absorption between electronic energy levels split in a magnetic field (the Zeeman effect).

excitons – Well defined quantum excitations or quasi-particles such as bound electron-hole pairs in crystals.

extended excitations – Spectral transitions that involve elecvtronic states that are coherent over long distances, as in a metal.

high _P-T_ CARS – Coherent Anti-Stokes Raman Scattering, a nonlinear optical technique whereby a pump beam ω_p and a Stokes beam ω_S are mixed via third order susceptibility, typically used to study vibrational transitions.

inelastic X-ray scattering spectroscopy (IXSS) – General class of techniques measure excitation spectra, probes the dynamical structure factor.

infrared spectroscopy – Class of techniques that use radiation from $10\text{-}10000\,\text{cm}^{-1}$ in absorption, emission, or reflectivity, and used to study vibrational, electronic and magnetic excitations.

iron and iron alloys – Soft, silver-colored metal with an atomic number of 26; the most abundant transition element in the solar system; iron alloys are a mixture of iron and other elements.

Kramers–Kronig analysis – Mathematical treatment that uses causality to determine the full set of optical constants of a material as a function of frequency to its complete spectrum (e.g., absorption and reflectivity).

luminescence – Light emitted by materials; includes both fluorescence (short-lived) and phosphorescence (long-lived).

magnetic excitations – Transitions between magnetic states in materials, such as magnons.

magnetic properties – Properties arising from the presence of unpaired electrons and giving rise to a magnetic field or the response of a material to an external magnetic field.

nonlinear optical methods – Optical techniques in which the response of the material under study involves the non-linear response (susceptibility), or in other words, more than photon.

nuclear magnetic resonance (NMR) – Technique that uses radio frequency radiation to probe transitions associated with nuclei that split in a magnetic field.

nuclear resonance forward scattering (NRFS) – Time-domain technique that uses synchrotron radiation to probe the Mössbauer effect in materials.

nuclear resonant inelastic X-ray scattering (NRIXS) – Synchrotron radiation technique that probes phonon spectra associated with the excitation of the Mössbauer-active nuclei.

olivine – $(Mg,Fe)_2SiO_4$, is a magnesium iron silicate, one of the most common minerals in the Earth's upper mantle.

optical absorption and reflectivity – Techniques used to study transitions in the infrared-visible-ultraviolet spectral range from transmission of light through the material (absorption) or from the light reflected from the surface.

oscillator fits – Analyses of spectra that employ one or more excitations to comprise the spectra profile, often used for modeling dielectric function.

phonon inelastic X-ray scattering (PIXS) – Technique to probe vibrational excitations, including their momentum q dependence, using x-rays.

Raman scattering – Inelastic light scattering technique that measures transitions to an excited state from the ground state (Stokes scattering) and to the ground state from or from a thermally excited state (anti-Stokes scattering).

Rayleigh scattering – Quasi-elastic scattering spectroscopy technique, typically used to explore relaxation phenomena in materials.

resonance methods – Techniques that probe materials by measuring the response through a characteristic frequency or frequencies.

resonant inelastic X-ray spectroscopy (RIXS) – Technique to probe electronic excitations using x-rays scattered at energies that different from the incident x-rays and where the incident radiation is tuned through core-level excitations.

silicate perovskite and post-perovskite – High-pressure silicate phases with the stoichiometry $ASiO_3$, where A is a metal ion (e.g., Mg, Fe, Ca). Silicate perovskite represents the dominant class of phase of the Earth's lower mantle.

silicate post-perovskite – Dense pressure phase of the same stoichiometry as silicate perovskite having the $CaIO_3$ structure and stable at P-T conditions of the base of the mantle.

transition metal oxides – Compounds with the general formula $M_a^i M_b^i O_z$, where M_a^i is a transition metal in a particular oxidation and O is oxygen.

transport measurements – Experiments that quantify the transport of charge (electrons, protons, or ions) and of heat (thermal conductivity).

volatiles – low boiling point compounds usually associated with a planet's atmosphere or crust, including hydrogen, H_2O, CO_2, methane, and ammonia; also the components of these materials that can be bound in high-pressure phases.

X-ray absorption spectroscopy (XAS) – Technique in which incident x-rays are absorbed when energy exceeds the excitation energy of core electrons of a specific element, typically causing an atomic edge-like spectrum.

X-ray emission spectroscopy (XES) – Technique in which core electrons are excited by x-ray and then the core-holes decay through a radiative processe.
X-ray inelastic near-edge spectroscopy (XINES) – Technique in which an incident x-ray is scattered at a lower energy of a core excitation

(e.g., *K*-edge) of a material; typically used to probe structure containing information on bonding states.
X-ray magnetic circular dichroism (XMCD) – Technique to measure the spin polarization of an electronic excitation of materials by the use of circularly polarized x-rays.

11.1 Introduction

Electronic and magnetic properties directly influence large-scale global phenomena, ranging from the initial differentiation of the planet, the formation and transmission of the Earth's magnetic field, the propagation of seismic waves, and the upwelling and downwelling of mass through the mantle. The properties of deep-Earth materials are not readily familiar to us, owing to their distance, both physically and conceptually, from the near-surface environment of the planet. Over the range of pressures encountered with the Earth, rock-forming oxides and silicates compress by factors of 2–3. Even the many incompressible materials are compressed by 50%. Volatiles such as rare gases and molecular species can be compressed by over an order of magnitude. With the large reduction in interatomic distances produced by these conditions, dramatic alterations of electronic and magnetic properties give rise to myriad transformations. Discrete electronic and magnetic transformations can occur, such as metallization of insulators and magnetic collapse. More commonly, these changes are associated with, and drive, other transformations such as structural phase transitions, as well as controlling chemical reactions. Materials considered volatiles under near-surface conditions can be structurally bound in dense, high-pressure phases (e.g., hydrogen in ice, mantle silicates, and ferrous alloys). Gases and H_2O become refractory (high melting point) materials at high pressures (Lin *et al.*, 2005a). The heavy rare gases metallize (Goettel *et al.*, 1989; Reichlin *et al.*, 1989; Eremets *et al.*, 2000). Ions such as Fe^{2+} and Mg^{2+}, which almost always can substitute completely for each other at low pressures, partition into separate phases at depth (Mao *et al.*, 1997), whereas incompatible elements such as Fe and K may form alloys (Parker and Badding, 1996). Noble gases form new classes of compounds (Vos *et al.*, 1992; Sanloup *et al.*, 2002b).

This chapter constitutes an overview of measurements of high-pressure electronic and magnetic properties of Earth materials. The chapter draws largely on a pedagogical review written in 1998 (Hemley *et al.*, 1998b), including numerous updates as a result of considerable advances in the field, particularly in the area of synchrotron techniques (*see* Chapter 1). The article focuses on *in situ* high *P–T* techniques and selected applications in geophysics. We begin with a short review of fundamental properties, with emphasis on how these are altered under very high pressures and temperatures of the Earth's deep interior. This is followed by a discussion of different classes of electronic and magnetic excitations. We then provide a short description of various techniques that are currently used for high-pressure studies. We then provide selected examples, showing how the use of a combination of different techniques reveals insight into high *P–T* phenomena.

11.2 Overview of Fundamentals

11.2.1 Bonding in Deep-Earth Materials

11.2.1.1 Conventional bonding classification

The binding forces that hold the atoms together are generally electrostatic in origin. The forces that keep the atoms from collapsing into each other arise from increased kinetic energy as atoms are brought closer together, the Pauli exclusion principle that keeps electrons apart, the electrostatic repulsion between electrons, and ultimately (as the atoms are brought closer and closer together) the electrostatic repulsion of the nuclei. Crystals can be characterized by the primary source of their binding energy. Following the conventional classifications set out by Pauling (1960), we describe for the purposes of the present chapter the following bonding types – ionic, covalent, metallic, van der Waals, and hydrogen bonding. As pressures increase, the problem changes from understanding the cohesion of crystals, to understanding how electrons and nuclei rearrange in order to minimize repulsive forces.

In ionic crystals, the primary source of binding is from the electrostatic attraction among ions. Prototypical examples are NaCl (Na^+Cl^-) and MgO ($Mg^{2+}O^{2-}$). This ionicity is driven by the increased stability of an ion when it has a filled shell of electrons. For example, oxygen has a nuclear charge $Z = 8$, which means that it would have two 1s-, two 2s-, and four 2p-electrons, but a filled p-shell has six electrons. In a covalent bond, charge concentrates in the bonding region, increasing the potential and kinetic energy of interaction between electrons, but reducing the energy through the electron–nuclear interactions (since each electron is now on average close to both nuclei). Covalent bonds tend to be directional since they are formed from linear combinations of directional orbitals on the two atoms. It is straightforward to develop models for such covalent materials, and such models have ranged from models with empirical parameters to those with parameters obtained from first principles. The high-pressure behavior of metals, and in particular iron and iron alloys, is of paramount importance for understanding the nature of the Earth's core (*see* Chapter 5). The bonding in metals is often thought of as a positive ion core embedded in a sea of electrons, the negative electrons holding the ions together. Under pressure, the conductivity may increase, or decrease: in some cases (e.g., carbon) pressure may drive a transition from a metallic phase (graphite) to an insulator (diamond).

One must also consider weaker bonding interactions. Dispersion, or van der Waals forces, arises from fluctuating dipoles on separated atoms or molecules. Separated, nonoverlapping charge densities have such an attractive force between them, and it varies as $1/r^6$ and higher-order terms at large distances. Even at low pressures, van der Waals forces are largely quenched in solids, and become completely insignificant at high pressures. Hydrogen bonds are essentially ionic bonds involving H and O; a hydrogen atom is covalently bonded to an oxygen atom and attracted electrostatically to a neighboring oxygen ion. As such, the three atoms form a linear linkage with a hydrogen atom asymmetrically disposed between the oxygens, a configuration that is ubiquitous in hydrous materials. A range of degrees of hydrogen bonding at ambient pressure is evident in dense hydrous silicates that are candidate mantle phases. In such structures the application of moderate pressures causes a reduction in oxygen–oxygen distances, which generally enhances bonding between the hydrogen and the more distant oxygen atom. At deep mantle and core pressures, conventional hydrogen bonding is lost, and hydrogens might be symmetrically situated between oxygens.

Many high-pressure minerals exhibit combinations of the types of interactions outlined above. Most common are ionic and covalent bonding, both within the same bond and different degrees of each in different bonds present in the material. Silicates are considered about half-ionic and half-covalent. Transition metals such as Fe are both metallic and covalent. A molecular group that contains strong covalent bonds can also be ionically bound to other ions in the crystal. An example is $CaCO_3$, where the C–O bonds are strong and directional, forming a planar triangle in the CO_3^{2-} carbonate group, which is bound ionically to Ca^{2+} ions.

11.2.1.2 Bonding under high P–T conditions

The bonding in Earth materials characterized under ambient conditions typically does not reflect their state under conditions of the deep Earth. The contributions from each type are in general strongly pressure dependent. High pressure favors high density and high temperatures favor phases with higher entropy S. At lower-mantle and core conditions the density terms dominate, with generally small Claperyon slopes, dP/dT, found for solid-phase transitions. Structures tend to adopt more close-packed configurations with increasing compression. At high pressures, the attractive terms in molecular solids become less and less important, and one could rather think of the atoms as soft spheres. Much of the chemistry and phase diagrams of such materials can be understood as the packing of spheres of various sizes. However, as pressure is increased, the electronic structure also changes (*see* Chapter 6). This happens first in the crystals of heavier rare gases such as Xe, which transforms from the f.c.c. to the h.c.p. structure (Jephcoat *et al.*, 1987) prior to metallization (Goettel *et al.*, 1989; Reichlin *et al.*, 1989; Eremets *et al.*, 2000). On the other hand, the tendency toward close packing is not observed in all cases. There have been a few theoretical treatments of these phenomena (Neaton and Ashcroft, 2001, 1999). In fact, surprisingly complex structures can be adopted, as has been seen in some pure elements (McMahon and Nelmes, 2004). Oxides can also adopt complex structures, driven by subtle pressure-induced changes in orbital hybridization. This is particularly evident in some ferroelectric materials (Wu and Cohen, 2005). In addition, there can be effects of magnetism, or the tendency of electron spins to align: theory and experiment indicate that magnetism tends to decrease and disappear with increasing pressure (Pasternak *et al.*, 1997b; Cohen *et al.*, 2002; Cohen and Mukherjee, 2004; Pasternak and Taylor, 2001; Steinle-Neumann *et al.*, 2004a, 2004b; Steinle-Neumann *et al.*, 2003; Stixrude, 2001).

Changes in bonding also apply to liquids. For the deep Earth, the important systems are silicate and oxide melts as well as liquid iron alloys of the outer core. There is evidence that silicate melts undergo similar structural transformations such as coordination changes, although not abrupt, and spread out over a range of pressures. Detailed information on the electronic properties of geophysically relevant liquids under extreme conditions is lacking.

11.2.2 Electronic Structure

Insight into these phenomena must come from examination of the changes in electronic structure induced by materials subjected to extreme $P-T$ conditions. Because of the presence of long-range interactions in condensed phases, the bonding must in general be viewed from an extended point of view, although a local picture is often employed to simplify the problem. For crystals, the eigenstates can be characterized by the wave number k (Kittel, 1996). The interaction between orbitals on different atoms gives rise to energy bands, which are dispersive as functions of k, and the band energies versus k form the band structure. The core states, or deep levels for each atom, remain sharp delta function-like states, which may be raised or lowered in energy relative to their positions in isolated atoms. These core-level shifts are due largely to the screened Coulomb potential from the rest of the atoms in the crystal. The occupied valence and empty conduction states no longer look like atomic states, but are broadened into energy bands. There are often intermediate states between the core levels and the valence states called semi-core states which are slightly broadened at low pressures, but which become broader and more different from atomic states with increasing pressure. The formation of ions is driven by the increased stability of an ion when it has a filled shell of electrons. The strong attractive electrostatic,

or Madelung, interaction between the ions greatly enhances the crystal stability.

This picture now becomes a means for classifying different materials. Metals have partially occupied states at the Fermi level, the highest occupied energy level in a crystal. One can understand the insulating behavior of materials with fully filled bands by considering what happens when one applies an electric field. An electric field raises the potential at one part of the sample relative to another, and one would think that electrons would then flow down that potential gradient. However, in an insulator, the bands are filled, and due to the Pauli exclusion principle nothing can happen without exciting electrons to states above the gap. This requires a large energy, so there is no current flow for small fields, and thus the materials has a finite susceptibility and the material is insulating. In a metal, the partially filled states at the Fermi level mean that current will flow for any applied field, and the susceptibility is infinite. Actually, the distinction between metallic and insulating systems involves some subtle physics (Resta, 2002), but in general the band picture suffices to understand metals and insulators well within their stability fields. Insulator-to-metal transitions, however, are expected to show complex behavior in some materials (Huscroft *et al.*, 1999). Most transitions found to date seem to involve structural transitions at the same pressure as the metal–insulator transition (e.g., FeO from the rhombohedrally strained rock salt structure to the anti-B8 (inverse NiAs) structure) (Fei, 1996; Mazin *et al.*, 1998; Gramsch *et al.*, 2003; Murakami *et al.*, 2004a), but an isostructural transition has been claimed in MnO (Yoo *et al.*, 2005).

In an insulator, the energy of the highest occupied levels is the valence band, designated E_v and that at the bottom of the conduction band E_c. The difference is the band gap, $E_g = E_c - E_v$. Crystals with band gaps between occupied and unoccupied states should be insulators, and those with partially filled bands should be metals. Localized states may exist at energies between the valence and conduction bands, and these can have major effects on optical and transport properties. If the gap is small or if the material has such intermediate states (e.g., by chemical doping such that electrons can be excited into the conduction bands or holes in the valence bands), the crystal is considered a semiconductor. In a nonmagnetic system, each band holds two electrons, and thus a crystal with an odd number of electrons in the unit cell should be a metal since it will have at least one partially filled band. Magnetic crystals that are insulators by virtue of local magnetic moments are known as Mott insulators (Mott, 1990). These materials include transition metal-containing oxides present in the mantle (and considered core components); such materials exhibit intriguing behavior as a function of pressure and temperature.

We should also consider the electronic structure of noncrystalline materials, including liquids. The existence of a band gap in crystals can be described in terms of the Brillouin zone and Bragg-like reflection of electrons. But a gap does not depend upon the periodicity of the atoms on a lattice. The electrons in the lowest energy states in the conduction band of a noncrystalline material (or disordered crystal) can be localized (Anderson, 1958; Mott, 1990). The difference between the localized and delocalized states is called the mobility edge. These may be important in describing the electronic properties of liquids and amorphous materials formed from deep-Earth crystalline materials (e.g., metastably compressed crystalline minerals and dense melts).

11.2.3 Magnetic Properties

Magnetism arises from the presence of unpaired electrons. Electrons have magnetic moments, but if they are paired up in states as up and down pairs, there will be no net magnetism in the absence of a magnetic field. However, in open-shelled atoms, there may be a net moment. According to Hund's rules, atoms (or isolated ions) will maximize their net magnetic moment, which lowers their total energy due to a decrease in electrostatic energy. In a crystal, interactions with other atoms and formation of energy bands (hybrid crystalline electronic states) may lead to intermediate- or low-spin magnetic structures.

The behavior of iron (Fe^{3+}, Fe^{2+}, and Fe^0) at high pressure is of central importance. Fe^{2+} has six d-electrons, and two end-member situations can be considered. In a site with octahedral symmetry and limited ligand–field interactions, the six d-electrons can be paired to fill three t_{2g} states, forming low-spin, ferrous iron, and there is no net magnetic moment. In high-spin ferrous iron, five up d-states are split in energy from five down d-states by the exchange energy, and five d-electrons go into the three t_{2g} and two e_g lower-energy, majority spin states, and one electron into a high-energy t_{2g} state. This gives a net magnetic moment component of $4\mu_B$ (Bohr magnetons). The fully spin-polarized magnetic

moment component of Fe^{3+} is $5\mu_B$. In crystals, the 4s-states usually dip down in energy with respect to k and are partially occupied, leading to a smaller d-occupation and smaller net moment.

Such open-shell systems have often been discussed in terms of crystal field theory, which has been very successful in rationalizing optical spectra, crystal chemistry, and thermodynamic and magnetic properties of minerals, including their pressure dependencies. The principal assumption of crystal field theory is the existence of localized atomic-like d-orbitals. It has provided a way to rationalize the changes in energy levels in terms of local interactions of the transition metal cations with the coordinating atoms. The symmetry of the coordination polyhedron results in a lifting of the otherwise degenerate d-orbitals. The primary success of crystal field theory arose from symmetry analysis of the splitting of the atomic-like states. These symmetry arguments are rigorous, but the origin of the splitting may not arise from a potential field represented as a point charge lattice, but rather to bonding hybridization. Furthermore, d-states are not pure atomic-like states but are dispersed across the Brillouin zone (energy varies with k).

As shown by Mattheiss (1972), the origin of the crystal field splitting is due largely to hybridization. In the language of quantum mechanics, the e_g–t_{2g} splitting is due to off-diagonal interactions, not shifts in the diagonal elements, as would be the case for conventional electrostatic crystal–field interaction. The splitting can be considered a ligand–field effect, and does not arise from changes in the electrostatic field at an atomic site. Rather, the splitting is due to d–d-interactions between next-nearest neighbors and due to p–d- and s–d-interactions between neighboring oxygen ions. The d–d-interactions lead to splittings that vary as $1/r^5$; although the p–d- and s–d-interactions follow a $1/r^7$ dependence (Harrison, 1980), the metal–metal d–d-splittings dominate. Thus, the same $1/r^5$ dependence of the observed splittings is predicted by electrostatic crystal field model (Drickamer and Frank, 1973) and by hybridization (Mattheiss, 1972).

In the low-temperature ground states, moments on individual ions or atoms can be oriented, giving rise to ferromagnetism if they are lined up in the same direction, or antiferromagnetism if they are oppositely aligned, which depends on the sign of the magnetic coupling $\mathcal{J}$ (which depends on hybridization with other atoms in the crystal). At high temperatures, the moments will disorder, but there are still local moments present. This paramagnetic state is distinguished from the paramagnetic state that arises from loss of local moments (magnetic collapse, or a high-to-low spin transition), although the two states merge continuously into each other with increasing pressure. Ferrimagnetism results when the magnetic moments are unequally distributed over different sublattices. A net spontaneous magnetic moment arises from the incomplete cancellation of aligned spins. An example is magnetite, Fe_3O_4, which consists of a sublattice of Fe^{2+} and one of Fe^{2+} and Fe^{3+}. In some materials, such as f.c.c. and possibly iron (Cohen and Mukherjee, 2004), the moments are non-collinear. For ferromagnetic (FM) and antiferromagnetic (AFM) phases, the magnetic moments are aligned below the Curie and Néel temperatures, respectively. The decrease in magnetism with pressure can be understood qualitatively from the increase in electronic band widths, which eventually become greater than the exchange splitting. This can be understood more quantitatively with the Stoner and extended Stoner models (Cohen *et al.*, 1997).

11.3 Electronic and Magnetic Excitations

Excitations in a system are induced by external (e.g., electronic or magnetic) or internal (e.g., temperature) fields. The coupling of the response to the field is given by the frequency-dependent susceptibility. This can be characterized as the dielectric function $\varepsilon(\omega,k)$, where ω is the frequency and k is the wave vector (the dielectric function is also known as the relative permittivity). The dielectric function is complex, $\varepsilon = \varepsilon_2 + i\varepsilon_2$ where ε_2 contains the contribution from absorption. In general, the response of a crystal to an electric field is dependent on its orientation relative to the crystal axes (which gives rise to properties such as birefringence and optical activity), that is, the dielectric function is a tensorial function. The dielectric tensor is also a function of applied electric field, giving rise to nonlinear optical response (e.g., multiphoton excitations) as described below. Electrons can be excited into extended, or itinerant, states (i.e., across the band gap), or the excitations may be local (i.e., forming a localized electron–hole pair, or exciton).

11.3.1 Dielectric Functions

The index of refraction is related to the dielectric function ε, as $n^2 = \varepsilon$, which can strongly depend on pressure. In general, there are several different contributions to the dielectric function. In materials that contain polar molecules, the molecules can align. There are three main contributions to the dielectric response in insulators: ionic polarization (displacement of ions) in crystals containing nonpolar atoms or ions, electronic polarization (from the deformation of atomic charge distributions), and charge transfer from atom to atom. The electronic contribution to the dielectric function of a material is formally determined by the sum of its electronic excitations (Ashcroft and Mermin, 1976). Hence, measurement of the index of refraction of transparent minerals (i.e., below band gap or absorption edge) can be used to constrain the frequency of higher-energy electronic excitations (e.g., in the ultraviolet (UV) spectrum), (e.g., Eremets, 1996). Optical spectra of oxides and silicates from the near-infrared (IR) to the vacuum UV are determined by electronic transitions involving the valence (bonding) band, the conduction (antibonding) band, and d-electron levels. Piezoelectric crystals become polarized when subjected to a mechanical stress (compressive or tensile). Quartz is a common example. In ferroelectric phases, the crystals are spontaneously polarized and have very high dielectric constants.

This approach has been useful in high-pressure studies when the window of the high-pressure cell precludes measurements above a critical photon energy (Hemley *et al.*, 1991). For example, diamond anvils at zero pressure have an intrinsic absorption threshold of 5.2 eV (i.e., for pure type IIa diamonds). Moreover, type I diamonds contain nitrogen impurities that lower the effective threshold at zero pressure to below 4 eV (depending on the impurities). Although the band gap of diamond increases under pressure, the absorption threshold of diamond (and most likely the gap) decreases in energy under the stresses due to loading in the cell. Full quantitative information on the dielectric function in the diamond transparency window is hard to obtain, although there exist few applications facilitating data analysis. Two mostly used approaches are oscillator model and Kramers–Kronig (KK) analysis of the reflectivity data. Several experimental setups have been reported for reflectivity measurements (Syassen and Sonneschein, 1982; Goncharov *et al.*, 2000); however, only relatively few studies have been performed which yield full dielectric function information under pressure (e.g., Goncharov *et al.*, 1996).

11.3.1.1 Oscillator fits

Classical oscillator fits are based on the following expression for the dielectric constant

$$\epsilon(\omega) = \epsilon_1(\omega) + i\epsilon_2(\omega) = \epsilon_\infty + \sum_{j=1}^{m} \frac{\omega_{pj}^2}{\omega_j^2 - \omega^2 - i\omega\gamma_j} \quad [1]$$

where ω_j and γ_j is the frequency and line width, respectively, of oscillator j. The real and imaginary part of the refractive index (omitting the ω dependence) $N = n + ik$ are related to the dielectric constant by

$$n = \left(\frac{(\epsilon_1^2 + \epsilon_2^2)^{\frac{1}{2}} - \epsilon_1}{2} \right)^{\frac{1}{2}} \quad [2]$$

$$k = \left(\frac{(\epsilon_1^2 + \epsilon_2^2)^{\frac{1}{2}} + \epsilon_1}{2} \right)^{\frac{1}{2}} \quad [3]$$

The following discussion is restricted to analysis of the normal incidence of light at a sample-diamond interface, which is a good approximation to many experimental conditions. The reflectivity from the sample between diamond anvils having a thickness D is given in this case by the following equation, which is equivalent to the formula from (Born and Wolf, 1980), the only difference being that it is written in complex numbers:

$$R = \left| \frac{(N - N_D)^2 (1 - \chi)}{(N + N_D)^2 - \chi(N - N_D)^2} \right|^2 \quad [4]$$

Here, N_D is the refractive index of diamond, and

$$\chi = \exp(i2N\omega D) = \exp(i2n\omega D - 2k\omega D) \quad [5]$$

For the case of a strong absorption of the sample when the reflectivity from the second sample-diamond interface is negligible, one obtains well-known expression

$$R = |r|^2 = \left| \frac{(N - N_D)}{(N + N_D)} \right|^2 = \frac{(n - N_D)^2 + k^2}{(n + N_D)^2 + k^2} \quad [6]$$

11.3.1.2 KK analysis

In general, the response function r (eqn [6]) can be determined from the reflectivity R using the KK procedure (Bassani and Altarelli, 1983). One defines

$$\theta(\omega) = -\frac{2\omega}{\pi} P \int_0^\infty d\omega' \frac{\ln|r(\omega')|}{\omega'^2 - \omega^2} \quad [7]$$

where P is the principal part of the integral. From eqns [7] and [5], it is straightforward to calculate

$$n = \frac{N_D(1 - R)}{1 + R - 2\sqrt{R}\cos(\theta)} \quad [8]$$

$$k = \frac{2N_D(1 - R)\sin(\theta)}{1 + R - 2\sqrt{R}\cos(\theta)} \quad [9]$$

The dielectric constants are given then by

$$\epsilon_1 = n^2 - k^2 \quad [10]$$

$$\epsilon_2 = 2nk \quad [11]$$

This procedure works only in the region of strong absorption, where we can use eqn [4] instead of eqn [3]. Although, all weak IR bands can often be observed in absorption, it is much more convenient to use Fourier transform techniques to carry out the KK transformation, instead of directly using eqn [7]. The following discussion is based on technique outlined in Titchmarsh (1937). The KK transformation is a transformation between the functions connected in the following way:

$$f(x) = -\frac{1}{\pi} P \int_{-\infty}^\infty \frac{g(t)}{t - x} dt, \ g(x) = \frac{1}{\pi} P \int_{-\infty}^\infty \frac{f(t)}{t - x} dt \quad [12]$$

For the present case, $g(x)$ and $f(x)$ can correspond to dielectric functions $\varepsilon_1(\infty) - \varepsilon_\infty$ and $\varepsilon_2(\omega)$. When calculating the phase of the reflectivity, the matching pair would be $\ln|r(\omega)|$ and $\theta(\omega)$. Functions $g(x)$ and $f(x)$ are called Hilbert transforms, and their Fourier transforms $F(t)$ and $G(t)$ are coupled by the following relation (Titchmarsh, 1937):

$$G(t) = -iF(t)\text{sgn}(t) \quad [13]$$

where $\text{sgn}(x) = 0$ when $x < 0$, and $\text{sgn}(x) = 1$, when $x \geq 0$. Thus, taking the Fourier transform of $g(x)$ $(G(t))$, we can find the transform of $f(x)$ $(F(t))$ using eqn [13]; the inverse transform of $F(t)$ gives $f(x)$, thereby giving the KK transform of the function $g(x)$. Fast Fourier transform (FFT) techniques permit the calculation of KK transforms easily and much faster than direct integration of eqn [7]. Formally, the transform should be performed over the frequency range $0 < \omega < \infty$, but experimental data are always limited to a specific frequency range. Procedures are therefore needed to account for missing data, as discussed later in the experimental section. When calculating the KK transform of $\ln|r(\omega)|$ at sample–diamond interface, there may be an additional contribution to the calculated phase shift, which is known as Blashke product (Stern, 1963). The phase shift is given by

$$\theta(\omega) = \theta_c(\omega) + \zeta(\omega) \qquad [14]$$

where

$$\zeta(\omega) = -\mathrm{i}lnB(\omega) = 2\sum_n \tan^{-1}\left[(\omega - \mu_n^r)/\mu_n^i\right] \qquad [15]$$

Here the summation runs over all n, r and i stand for the real and imaginary parts of μ_n, and μ_n^r and μ_n^i are determined from the condition,

$$r(\mu_n) = 0, \quad (\mu_n^r + \mathrm{i}\mu_n^i = \mu_n) \qquad [16]$$

$\theta_c(\omega)$ is a canonical phase shift that results from the KK procedure. The physical reason for occurrence of the Blashke product $\zeta(\omega)$ is a zero in reflectivity $r(\omega)$ in the upper half of the complex ω-plane (eqn [16]), which introduces a singular point in $\ln|r(\omega)|$ and makes the application of the KK relation unjustified in that case. However, one can define the canonical reflectivity,

$$r_c(\omega) = r(\omega)/[(\omega - \mu_n)(-\omega + \mu_n^*)] \qquad [17]$$

which does not have zeros in the upper half of the complex ω-plane. Moreover, on the real ω_r axis, $r(\omega_r) = r_c(\omega_r)$. Thus, the KK relation can be applied to the canonical reflectivity, and the resulting formula for the phase shift will be given exactly by eqn [14]. The Blashke root μ_n is pure imaginary in this case ($\mu_n^r = 0$), because eqn [16] is satisfied when $n(\omega) + \mathrm{i}k(\omega) - N_D = 0$ is fulfilled (which requires $k = 0$). The refractive index $N(\omega)$ is a complex number everywhere except on the imaginary ω_i axis. On that axis, we have

$$n(\mathrm{i}\omega_i) = (\epsilon(\mathrm{i}\omega_i))^{1/2} = \left(1 + \frac{2}{\pi}\int_0^\infty \frac{x\epsilon_2(x)\mathrm{d}x}{x^2}\right)^{1/2} \qquad [18]$$

$n(\mathrm{i}\omega_i)$ decreases monotonically from $n(0)$ to 1 when ω_i increases from 0 to infinity (Landau and Lifshitz, 1977). The condition $n(\mathrm{i}\omega_i) - N_D = 0$ will be satisfied somewhere on the imaginary axis, if $n(0) > N_D$. As shown later in the experimental section, this condition is satisfied for ices at high pressures. Modifications needed to apply the KK procedure correctly will be discussed below. It is useful to point out here that a very simple way to account for the Blashke product is to solve for the condition $n(\mathrm{i}\mu_n) - N_D = 0$ numerically (using eqn [18]), and dielectric function from the oscillator fit (eqn [1]). Another possibility is to apply the KK procedure self-consistently, by iteratively adjusting the Blashke root μ_n.

11.3.2 Electronic Excitations

11.3.2.1 Extended excitations

Interband transitions are electronic excitations between bands, whereas intraband transitions are those within a band. An example of the latter is the damped frequency dependence of electronic excitations associated with itinerant electrons in a metal (which by definition has a partially filled band). The collective longitudinal excitations of the conducting electrons are known as plasma oscillations. Plasma oscillations (plasmons) can be excited by inelastic scattering techniques at high pressures. Plasmons can also be excited in materials with a band gap (i.e., dielectrics), typically at higher energies. In olivines, the band gap varies from 7.8 eV in forsterite to 8.8 eV in fayalite (Nitsan and Shankland, 1976).

11.3.2.2 Excitons

Electronic excitations may occur within the band gap due to the creation of a bound electron-hole pair, called an exciton. The pair is characterized by a binding energy E_b, which is given by the $E_g - E_{ex}$, where E_{ex} is the excitation energy. In some respects, these excitations are the equivalent of localized transitions in molecular spectroscopy, but they can propagate in a crystal and are characterized by a wave vector k. Continuing with the example of olivines at zero pressure, a prominent excitonic absorption is observed near the band gap (Nitsan and Shankland, 1976). The absorption characteristics in this region in dense ferromagnesian silicates and oxides are important for assessing the heat transfer in the Earth, which is predominantly radiative at deep-mantle temperatures.

11.3.2.3 Crystal field transitions

The color of many minerals arises from substitution of transition metals in the crystal structure. The substitution of transition metals in olivines at zero pressure results in a strong absorption beginning 1–2 eV below the band gap, which is ascribed to transitions between wide bands and the d-levels in the system (Nitsan and Shankland, 1976). The development and application of crystal field theory for understanding behavior of open-shell system ions, such as partially filled d-shell of transition elements, in minerals has a long history. As discussed above, a more fundamental quantum mechanical picture shows that orbital hybridization of the central ion and its neighbors (e.g., oxygen atoms) is significant and that the orbital splitting is not a purely electrostatic effect, but has a similar dependence on interatomic distances.

11.3.2.4 Charge transfer

Charge transfer represents another important class of excitations in these materials. Many of these transitions in minerals occur in the UV region (Burns, 1993). Intervalence charge transfer (e.g., metal–metal charge transfer) may occur between ions of different valence state, such as Fe^{+2}–Fe^{3+} or Fe^{2+} and Ti^{4+}. These transitions often give rise to bands in the visible spectrum which show marked shifts with pressure (Mao, 1976).

11.3.2.5 Excitations from defects

Many defects are also associated with electronic properties or are electronic in origin. Electron–hole centers or defects can be produced by deformation or radiation, either naturally or in the laboratory. Examples of naturally occurring systems at ambient pressure that exhibit these properties are smoky quartz and numerous colored diamonds (see Rossman, 1988). Defect luminescence has been observed to be associated with pressure-induced, and it may be connected with emission features observed in shock-wave experiments (see Hemley *et al.*, 1998b).

11.3.2.6 Electrical conductivity

Electrical conductivity is another measure of the response of the system to an applied electric field (static or frequency dependent). Electrical conductivity may be electronic or ionic, and can be intrinsic or extrinsic. If there is a gap between the valence and conduction bands, the electronic conductivity can be thermally activated at finite temperature. This can also give rise to coupled distortions of the lattice as the electrons move through it, called small polarons (see Kittel, 1996). Naturally, electrical conductivity measurements provide a means for the direct identification of the insulator–metal transitions, induced by either pressure or temperature.

In general, electrical properties are also very sensitive to minor chemical impurities and defects in a given specimen (Tyburczy and Fisler, 1995), such as Fe content and volatile content. The conductivity can be written as a complex quantity, giving real and an imaginary part (loss factor); this complex function is measured by impedance spectroscopy. Grain boundary conduction, for example, is observed to decrease under pressure as grain–grain contacts increase. Ionic conductivity may be especially

noticeable at higher temperatures (e.g., premelting or sublattice melting). Such behavior has been proposed for silicate perovskites by analogy to other perovskites as well as theoretical calculations (Hemley and Cohen, 1992).

11.3.3 Magnetic Excitations

Magnons are excited magnetic states which can be excited in light scattering via direct magnetic dipole coupling or an indirect, electric dipole coupling together with spin orbit interaction (Fleury and Loudon, 1968). High-pressure measurements on mineral analogs have been performed to determine the metal–anion distance dependence of the superexchange interaction $\mathcal{J}$ (Struzhkin *et al.*, 1993b). This interaction parameter is a measure of the coupling of metal anions through a (normally) diamagnetic anion, such as the O between two Fe atoms. Magnon excitations may also be observed by IR absorption.

11.4 Overview of Experimental Techniques

An array of ambient-pressure techniques can now be used to probe electronic and magnetic excitations in minerals. Band structure can be probed by photoemission, which involves interaction of electrons with the sample. Other such spectroscopies involving electron interactions include electron energy loss and auger (Cox, 1987), each of which have been used for studying high-pressure phases quenched to ambient conditions. Here, we focus on techniques that have been applied to deep-Earth materials *in situ* at high pressure. Pressure generation techniques will not be discussed. A review of these methods, and specifically diamond-anvil cells, is given elsewhere in this treatise. It is useful to point out briefly that new techniques, such as those based on chemical vapor-deposited diamond, are helping to advance these measurements (Mao *et al.*, 2003b; Yan *et al.*, 2002; Hemley *et al.*, 2005). Cubic zirconia (Xu *et al.*, 1996), moissanite (Xu and Mao, 2000; Xu *et al.*, 2004), and other materials can be used as high P–T anvils. Variations on a number of these techniques can also be used for *in situ* measurements on dynamic compression (e.g., shock-wave) experiments (Ahrens, 1987).

11.4.1 Optical Spectroscopies

11.4.1.1 Optical absorption and reflectivity

Optical spectroscopies have been one of the principle techniques used for *in situ* high-pressure investigations of minerals with high-pressure devices having optical access (e.g., diamond cells). A number of different types of absorption spectrometer systems, including double-beam instruments, have been designed and built for high-pressure mineral studies. In optical experiments on samples in high-pressure cells using refractive optics (lenses), there is a need in general to correct for the index of refraction of the diamond because of chromatic aberrations (wavelength dependence of the focusing). This problem

can also be overcome by the use of reflecting optics (e.g., mirror objectives). Versatile UV–visible–near-IR absorption and reflectivity systems using conventional continuum light sources have been designed for high-pressure applications (Syassen and Sonneschein, 1982; Hemley *et al.*, 1998a) (**Figure 1**). Standard continuum sources and wide variety of laser sources that span the UV–visible and much of the IR regions are now available. An example of a series of optical absorption measurements on magnesiowüstite is shown in **Figure 2(a)**.

The most straightforward is the measurement of the real part of the dielectric function (i.e., off-resonance), corresponding to the refractive index in the visible spectrum of minerals that are transparent

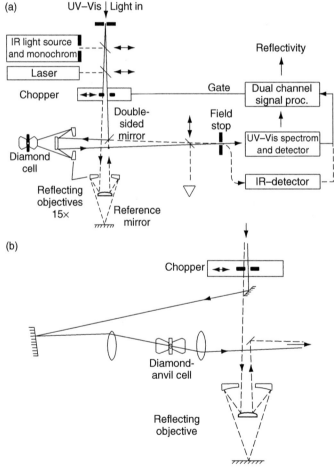

Figure 1   Example of a high-pressure UV–visible–near-IR optical setup. (a) Double-beam instrument for reflectivity measurements. Reflecting microscope optics are used to remove chromatic aberrations along with a grating spectrometer. (b) Modification of the double-beam system for absorption measurements. One of the reflecting objects is replaced by an achromatic lens. The system is capable of measurements from 0.5 to 5 eV; similar systems are in use in a number of high-pressure laboratories. Adapted from Syassen K and Sonneschein R (1982) Microoptical double beam system for reflectance and absorption measurements at high pressure. *Review of Scientific Instruments* 53: 644–650.

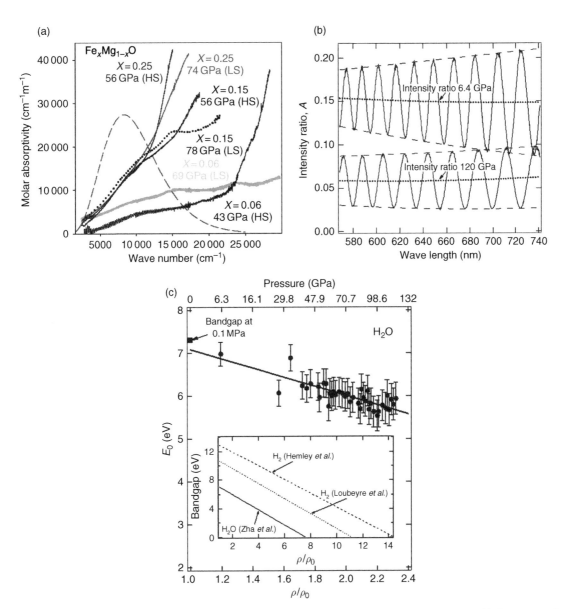

Figure 2 (a) Molar absorptivity (ε) for the high- and low-spin magnesiowüstite with different iron content (x). The values of ε are calculated from Beer's law $A = \varepsilon \times x \times d$, where A is absorbance, d is the sample thickness. $x = 0.25$: blue thin line – 56 GPa, pink thin line – 74 GPa; $x = 0.15$: black thin line – 56 GPa, dotted line – 78 GPa; $x = 0.06$: grey line – 69 GPa, black line, 43 GPa. The dashed (red) line represents the calculated black-body radiation (arbitrary units) at 2400 K (b). Pressure dependence of the reflectivity of H_2O ice relative to diamond. (c) Density dependence of the single oscillator energy E_0 determined from the dispersion of the refractive index of H_2O ice. The band gap for H_2O ice at room pressure is close that determined by the fitted data. Inset: comparison of results for the single oscillator model H_2O and H_2 at higher compression. The density dependence of the effective band gap is nearly parallel for these two materials. (a) Modified from Goncharov AF, Struzhkin VV, Jacobsen SD (2006) Reduced radiative conductivity of low-spin (Mg,Fe)O in the lower mantle. *Science* 312: 205–1208. (b and c) Modified from Zha CS, Hemley RJ, Gramsch SA, Mao HK, and Bassett WA (2007) Optical study of H_2O ice to 120 GPa: dielectric function, molecular polarizability, and equation of state. *Journal of Chemical Physics* 126: 074506.

wide band-gap insulators. Absorption (transmission) or reflectivity measurements with broadband light of thin samples of comparable thickness to the wavelength of light give rise to patterns of constructive and destructive interference in the spectrum. These fringes are straightforward to measure in a diamond cell as shown in the recent measurements for H_2O ice (Zha *et al.*, 2007) (**Figure 2(b)**). The spacing is determined by the refractive index and the thickness of the plate-like sample; each may be determined independently if the order of the fringe is known, and this can be found in several ways. Moreover, the wavelength dependence gives the dispersion of the index, which can be related to the higher-energy electronic excitations, as described above.

Results for ice are shown in **Figure 2(c)**, along with a comparison of the density dependence of the calculated oscillator frequency of H_2O and H_2 (Zha *et al.*, 2007; Hemley *et al.*, 1991; Loubeyre *et al.*, 2002). The results for H_2 are broadly compatible with direct observation of optical absorption of H_2 at pressures of >250 GPa (Mao and Hemley, 1989, 1994; Loubeyre *et al.*, 2002). MgO and diamond both show a decrease in refractive index with pressure. From this, we infer that the band gap of MgO increases with pressure, in agreement with theory (Mehl *et al.*, 1988). Similar behavior is observed for hydrostatically compressed diamond (Eremets *et al.*, 1992; Surh *et al.*, 1992). In contrast, the band gaps of many other materials decrease under pressure (Hemley and Mao, 1997). Also, both theory and experiment indicate that the diamond index can increase with nonhydrostatic stress. Brillouin scattering provides an independent measurements of the refractive index; results for MgO are in good agreement with the interference fringe measurements (Zha *et al.*, 1997).

11.4.1.2 Luminescence

High-pressure luminescence measurements give in principle the same information as does absorption and reflectivity. Luminescence includes fluorescence and phosphorescence, which correspond to shorter- and longer-lived excited states (generally defined as less than or greater than a nanosecond, respectively). The transitions are typically governed by dipole selection rules. For the case of atomic-like transitions, such as localized excitations of ions in crystal fields, this can be understood in terms of the Laporte selection rules (i.e., $\Delta l = \pm 1$, where l is the angular momentum quantum number). An early and particularly important application in mineralogy is the use of this technique in constraining the proportion of Fe^{2+} and Fe^{3+} in various sites in silicates from d-orbital excitations (Rossman, 1988). Charge transfer applications are described above. Other applications include excitations near and across band gaps, and excitations associated with defects.

Luminescence measurements are of obvious importance in the calibration of pressure. This includes excitation of Cr^{3+} in ruby by UV–visible lasers (e.g., He–Cd, Ar^+ ion, doubled Nd-YAG) and measurement of the calibrated shift of the R_1 line in luminescence, currently calibrated to 180 GPa (Mao *et al.*, 1978, 1986; Bell *et al.*, 1986). The recent use of tunable lasers (e.g., titanium-sapphire) and pulsed excitation techniques have allowed the extension of the measurements of ruby luminescence to above 250 GPa (Goncharov *et al.*, 1998). Other luminescent materials, such as Sm-doped yttrium aluminum garnet (Sm-YAG), have been calibrated for this pressure range (Liu and Vohra, 1996) as well as for high temperatures at lower pressures (Hess and Schiferl, 1992). Other applications include the study of vibronic spectra, which are combined (coupled) to electronic and vibrational excitations. Measurements at high pressure reveal the pressure dependence of both the electronic and vibrational levels in the system, which can be used to determine acoustic velocities, (iven assumptions about the dispersion of the modes and correct assignment of typically complex spectra (e.g., Zhang and Chopelas, 1994). There has been continued work on pressure calibration with optical spectroscopic methods (Goncharov *et al.*, 2005b).

11.4.1.3 IR spectroscopy

IR spectroscopy is an extension of optical spectroscopy that typically involves different techniques (e.g., Fourier transform spectroscopy). In addition, the use of synchrotron radiation for IR spectroscopy has been shown to be particularly useful for small samples such as those in high-pressure cells (**Figure 3(a)**). The technique can be used to study electronic and magnetic excitations, insulator–metal transitions, crystal field and charge transfer transitions, and, for metals, interband and intraband transitions (Hemley *et al.*, 1998a). Optical studies under pressure, especially reflectivity and absorbance, provide unique information important for understanding fundamental changes in the electronic structure of these systems close to the insulator–metal transition because of the broad spectral range that extends to long wavelengths (low wave numbers). It can also provide unique information on magnetic excitations (Struzhkin *et al.*, 2000). This study showed how the pressure dependence of the IR and Raman modes can be used to distinguish between magnon and electronic excitations. In addition, IR vibrational spectral profiles contain information on the electronic properties, that is, through the KK transformation. An example of

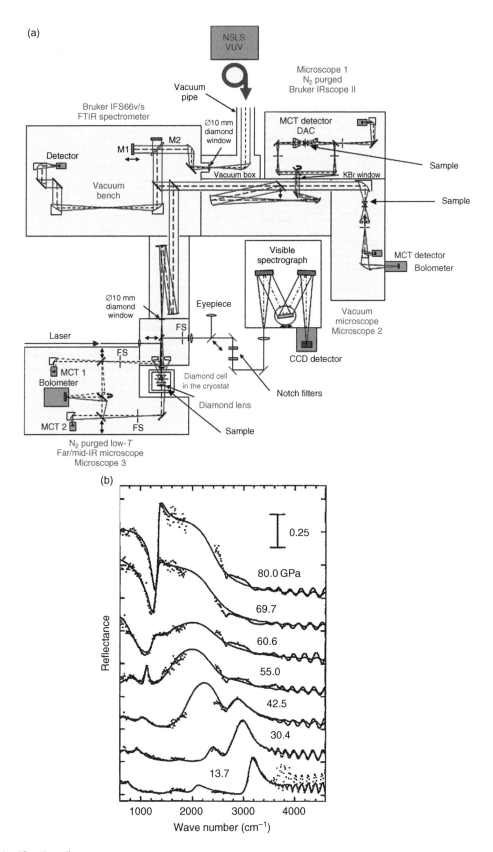

Figure 3 (*Continued*)

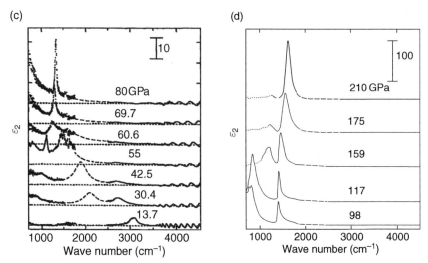

Figure 3 (a) Schematic of the high-pressure synchrotron IR spectroscopy facility at the National Synchrotron Light Source (Beamline U2A). Three different microscopes are optically interfaced with the spectrometers. The sample is located within the high-pressure cell designated by diamond cell or DAC. (b) Representative reflectivity spectra: points, experimental data; lines, oscillator model. Spectra are shifted with 0.25-unit increments in the vertical direction. There are no data in the range of the strong diamond absorption (1800–2400 cm^{-1}). Spurious oscillations at high frequencies originate from Fabry–Perot interference between the diamond anvils. The spectral region around 2600 cm cm^{-1} is complicated by the absorption of diamonds. Peculiarities in the spectra are due to incomplete cancellations of sharp features in reference and sample spectra. (c). Imaginary part of the dielectric constant ε_2 of H_2O obtained from a KK analysis of reflectivity spectra: points, KK transformation of esperimental data; dashed lines, oscillator fit. The curves are offset in the vertical direction for clarity. (d) ε_2 spectra of the H_2O (295 K) showing the positive pressure shift of the stretching mode to v_s 210 GPa. (b-d) Modified from Goncharov AF, Struzhkin VV, Somayazulu M, Hemley RJ, and Mao HK (1996) Compression of ice to 210 gigapascals: Infrared evidence for a symmetric hydrogen-bonded phase. *Science* 273: 218–220.

reflectivity spectra and the transformed data is shown in **Figures 3(b)** and **3(c)**.

11.4.1.4 Raman scattering

Although most commonly used to study vibrational dynamics, laser light scattering spectroscopies can also be used to investigate high-pressure electronic and magnetic properties (**Figure 4**). Raman and Brillouin are inelastic light-scattering techniques that measure transitions to an excited state from the ground state (Stokes scattering) and to the ground state from a thermally excited state (anti-Stokes scattering). Low-frequency crystal field excitations of the appropriate symmetry can be probed, and the pressure dependence of such transitions has been studied in a few cases (Goncharov *et al.*, 1994). The conduction electrons in metals can also be probed by Raman scattering, which in general gives rise to weaker broader features in the light-scattering spectrum. High-pressure studies are complicated by the need the discriminate between the spurious background contaminations from the true electronic contribution. Superconductivity in metals gives rise to structure in

the electronic Raman spectrum, such as gap-like features at low frequency (Zhou *et al.*, 1996).

Such studies include magnetic transitions in AFM materials such as transition metal oxides predicted to undergo high-pressure insulator–metal transitions. These materials have magnetic anomalies that may be difficult to study directly at high pressure, but these anomalies may be associated with pronounced changes in magnetic light-scattering spectra. These excitations have a strong temperature dependence because of the large temperature dependence of magnetic ordering (Struzhkin *et al.*, 1993a). A large number of excitations are observed, including several very strong bands, all of which can be described by the appropriate Hamiltonian. Measurements of the two-magnon excitations in Fe_2O_3 and NiO by high-pressure Raman scattering constrain the magnetic properties of these materials at high density (Massey *et al.*, 1990a, 1990b). The pressure dependence of higher-order effects such as the coupling of the phonons with electron spin-pair excitations and the electronic structure of the material can also be probed (Massey *et al.*, 1992). Although the higher-temperature behavior of such materials is of direct

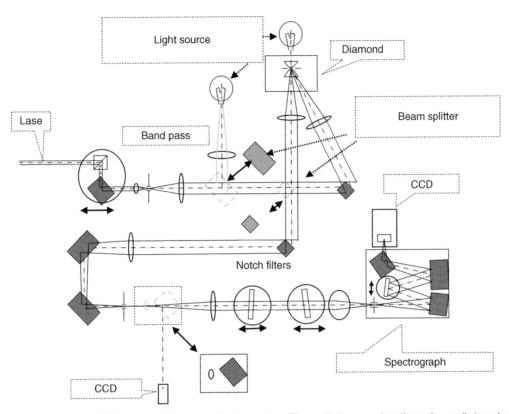

Figure 4 Schematic of a high-pressure Raman scattering system. The excitation wavelength can be easily tuned without changing the holographic optics and consequent realignment of the whole system; use of holographic transmission gratings (notch filters and band-pass filters) at wavelengths other than nominal does not typically result in a deterioration of the system's performance. Rapid changes in both the excitation wavelength and the spectral range are essential for high-pressure spectroscopy. The second important distinction of the system is that it allows rapid modifications of the scattering geometry to be made. In the case of a transparent sample, or if the sample is heterogeneous or very small (e.g., at multimegabar pressures), the backscattering geometry in combination with a confocal spatial filter is probably the best choice. For nontransparent and highly reflective samples, a quasi-backscattering (or angular) geometry may be employed. This arrangement has the advantage that the specular reflection (or direct laser beam in the case of forward scattering) is directed away from the spectrometer. Compared with the backscattering geometry, this expedient reduces the overall background, and allows the observation of lower-frequency excitations. In case of metals, use of the angular excitation geometry can be crucial. To be compatible with this geometry, it was necessary to modify the DAC to allow off-axis entry of the incident light. Specially designed tungsten carbide seats having angular conical holes have been used for this purpose. Use of double spatial filtering (one for the laser and one for the signal) effectively suppresses laser plasma lines and unwanted Raman/fluorescence signals. In this configuration, the laser spot can be made very small (e.g., 2 μm), and the depth of focus substantially reduced. Thus, the Raman signal from the diamond anvil is suppressed. Also, the effects of pressure inhomogeneity can also be substantially obviated. Use of synthetic ultrapure anvils can further reduce diamond fluorescence. This is crucial for studies of very weak scatters and/or materials at very high pressures (above 200 GPa), where strong stress-induced fluorescence of the diamond anvils can be a major obstacle to the acquisition of Raman or fluorescence spectra.

concern for the deep Earth (i.e., along the geotherm), detailed characterization of the low-temperature ground state is essential for understanding higher-temperature properties. The extremely low background signal in light-scattering experiments using ultrapure synthetic single-crystal diamond anvils (Goncharov *et al.*, 1998) is enabling the extension of these light-scattering studies of electronic and magnetic excitations to a wider class of materials, including opaque minerals and metals. This is illustrated by recent measurements on FeO (**Figure 5**; Goncharov and Struzhkin).

11.4.1.5 *Brillouin and Rayleigh scattering*

In principle, other optical light-scattering techniques can be used to investigate electronic and magnetic phenomena at high pressure. Brillouin involves the measurement of acoustic phonons; in metals, the

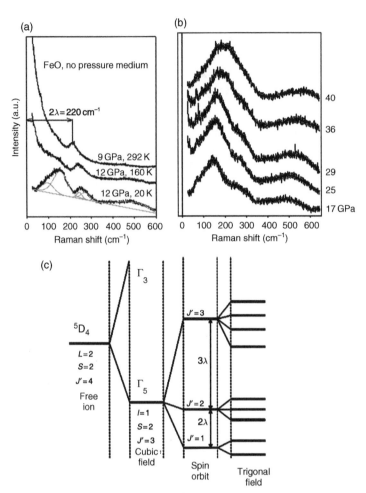

Figure 5 Raman spectra of magnetic excitations of iron oxide under pressure. (a) Temperature dependence; (b) pressure variation of the Raman spectra at low temperature. (c) Energy levels of magnetic excitations in iron oxide; the effect of exchange-driven splitting is not shown. Modified from Goncharov AF and Struzhkin VV (2003) Raman spectroscopy of metals, high-temperature superconductors and related materials under high pressure. *Journal of Raman Spectroscopy* 34: 532–548.

measurements correspond to surface modes or plasmons. Brillouin scattering can also be used to determine the pressure dependence of the refractive index (dielectric function), usually in transparent materials. Rayleigh scattering is a quasi-elastic scattering spectroscopy; it has been used little in high-pressure studies of materials because of the difficulty of obtaining accurate measurements due to typically strong background scattering near the excitation wavelength (laser line) by the anvils in the high-pressure cell.

11.4.1.6 Nonlinear optical methods
One of the limitations of conventional one-photon absorption, reflectivity, and luminescence measurements is the fact that the absorption threshold of the

anvils of the high-pressure cell precludes measurements on samples at higher energies (e.g., in the UV range), as mentioned above. Nonlinear optical techniques can be used to access this region, however. For example, one can perform three-photon excitation of valence electrons to the conduction band or excitonic states in the vicinity of the conduction band, which is above the one-photon threshold of the anvil; this has been achieved in compressed salts in sapphire anvil cells (Lipp and Daniels, 1991). It is now possible to apply and extend these techniques to study high-pressure minerals. Another category of techniques in this class are four-wave mixing experiments, such as coherent anti-Stokes Raman scattering. These rely on the third-order susceptibility and are enhanced when there are two-photon electronic resonances.

Coherent anti-Stokes Raman scattering (CARS) is a nonlinear optical technique whereby a pump beam ω_p and a Stokes beam ω_S are mixed via third-order susceptibility to generate enhanced anti-Stokes emission. Enhanced anti-Stokes emission is observed when the narrowband pump and Stokes beams are detuned such that their difference frequency ($\omega_V = \omega_P - \omega_S$) corresponds to vibrational sideband. A resonantly enhanced signal is emitted in the presence of the second pump photon at the frequency $\omega_{CARS} = 2\omega_P - \omega_S$, subject to momentum conservation (phase matching) of the four-wave mixing process. Signal generation in CARS is coherent, with an intensity that grows quadratically with interaction length, unlike incoherent spontaneous Raman, where intensity grows linearly with interaction length. In order to achieve this enhancement, the pump and Stokes pulses must spatially and temporally overlap at the sample. Moreover, anti-Stokes emission is directional due to phase matching. The Stokes beam may be narrowband, which gives an anti-Stokes response at the difference between the pump and Stokes beams (which may be scanned), or broadband, where a broadband anti-Stokes spectrum may be detected in a spectrometer. The above considerations are important for choosing a correct experimental strategy in the case of the DAC. CARS spectroscopy has been used to study molecular materials under extreme conditions generated by shock waves. High-quality spectra have been obtained using a single (combined) laser pulse of several nanoseconds duration. As described above, in this experiment, the entire CARS spectrum is measured simultaneously using a broadband Stokes pulse. Similar techniques have been used in DAC studies. The overall CARS cross section is determined by the $\chi^{(3)}$ susceptibility, a nonlinear electronic property.

11.4.2 Mössbauer Spectroscopy

Mössbauer spectroscopy probes the recoilless emission and resonant absorption of γ-rays by nuclei (Mössbauer, 1958). The energy levels of the nuclei are sensitive to the electrostatic and magnetic fields present at the nuclei, and thus to changes in chemical bounding, valence (i.e., the oxidation state of an atom, ferrous vs ferric), and magnetic ordering. Mössbauer spectroscopy reveals information on the hyperfine interactions and the local electronic and magnetic field at a nucleus. Only a few nuclei exhibit the

Mössbauer effect (e.g., ^{57}Fe, ^{119}Sn, ^{121}Sb, ^{153}Eu, ^{197}Au), and the great majority of applications in Earth science (and in general) are with ^{57}Fe. The natural abundance of ^{57}Fe is 2.1%, so some studies require enrichment of ^{57}Fe to obtain a stronger signal. The technique has been used extensively in the mineralogy of deep-Earth materials, including in situ high-pressure measurements. A typical probe for mineral systems is based on the reaction, ^{57}Co $+ {}^0\beta^{57} \rightarrow$ Fe $+ \gamma$. In a conventional Mössbauer experiment, one modulates the energy of the γ-rays by continuously vibrating the parent source to introduce a Doppler shift of the radiation; see Dickson and Berry (1986) for an introduction to Mössbauer spectroscopy, and McCammon (1995) for a review of applications to minerals.

The isomer shift (IS) is the shift in the nuclear energy level, and corresponds to the source velocity at which maximum absorption appears. The electric quadrupole splitting (QS) arises from the interaction between the nuclear and quadrupolar moment and the local electric field gradient at the nucleus. This occurs for nuclei with spin quantum number $I > 1$ (i.e., for ^{57}Fe, $I = 3/2$). The hyperfine field is due primarily to the contact interaction between electrons at the nucleus and the nucleus, rather than to the macroscopic magnetic field in a material. A hyperfine field arises from different densities of spin up and down electrons at the nucleus.

Applications include determination of Curie and Néel transition temperatures, valence state, site occupancy, and local order. In studies of deep-Earth materials at high pressure, each of these can be probed as a function of pressure (and temperature), including the measurement of changes across various transformations (crystallographic, electronic, and magnetic). Instrumentation consists of a radioactive source, vibrating drive, and detector (scintillator or proportional counter). Examples of high-pressure studies include measurements on a series of pyroxenes to 10 GPa (Zhang and Hafner, 1992), a phase transition associated with a change in spin state. The Mössbauer technique has been extended into the megabar pressure range as discussed below for FeO (Pasternak et al., 1997b). Mössbauer studies on FeO along with acoustic sound velocity measurements reveal the interplay between magnetic ordering and elasticity (Kantor et al., 2004b). Recent technique developments include the use of perforated anvils to maximize signal (**Figure 6**).

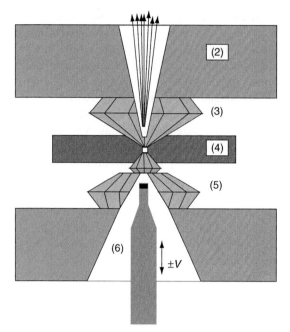

Figure 6 A sketch depicting the setup for high-pressure Mössbauer studies using perforated diamond anvils, Modified from Dadashev A, Pasternak MP, Rozenberg GK, and Taylor RD (2001) Prediction of a new phase transition in Al$_2$O$_3$ at high pressures. *Review of Scientific Instruments* 72: 2633–2637.

11.4.3 X-Ray and Neutron Diffraction

11.4.3.1 X-ray diffraction

Crystal structures are determined by the system optimizing the interactions of electrons and nuclei, and thus by details of bonding interactions between atoms; thus, crystallography is an essential means for studying bonding and evolution of atomic interactions with pressure (Wyckoff, 1923). X-ray diffraction relies on the interaction between X-rays and electron distributions in materials; so, in principle, it can give the real space charge density, but this requires very high quality crystals, corrections for absorption, and measurements to high q. The charge density includes smearing from atomic motions, which must be modeled and subtracted to obtain useful information. Very few experimental charge densities have thus been obtained; one successful study was on stishovite (Spackman *et al.*, 1987). Changes in electron density associated with crossing insulator–metal transition can be identified (Fujii, 1996). Recent work has included studies of oxides by Yamanaka *et al.* (to be published).

The advent of third-generation synchrotron radiation sources and modern area detectors are potentially important for extending these measurements to deep-Earth materials at very high pressures. As discussed above, pressure can induce complex crystal structures in simple elements, in principle yielding detailed information about atomic interactions. Because this structural complexity is driven in part by bonding and electronic structure, detailed understanding of this real space atomic structure of these phases provides insight into the electronic structure. Complementary to real-space distribution of the electrons is the momentum distribution which can be obtained by high-pressure Compton scattering, as described below. The coupling of phonons with the high-spin/low-spin transitions can be observed by X-ray techniques (Schwoerer-Bohning *et al.*, 1996).

11.4.3.2 Neutron diffraction

Neutron scattering is complementary to X-ray scattering, and it has the advantages relative to X-ray scattering by virtue of its sensitivity to low-Z elements, angle-independent form factor, high penetrating power, and sensitivity to magnetic order. Because the scattering amplitude of neutrons by an atom depends on the direction of its magnetic moment, neutron diffraction can be used to map out the orientation of the magnetic moments in the unit cell of a crystal. There has been growing number of *in situ* high-pressure measurements on minerals (Parise, 2006). Advances have been made possible by the development of new classes of high-pressure devices, notably the Paris–Edinburgh cell and various types of gem anvil cells (**Figure 7**). An example is the high-pressure polymorphism of FeS in which the transformation to FeS-II at 3.4 GPa was found to be associated with a spin reorientation transition (Marshal *et al.*, 2000). A second example is Fe$_2$O$_3$, for which the AFM ordering can be tuned by pressure (Goncharenko *et al.*, 1995; Parise *et al.*, 2006). High-pressure neutron scattering of FeO carried out with gem anvil cells revealed a surprising absence of long-range magnetic order to 20 GPa (Ding *et al.*, 2005).

The coupling of phonons with the high-spin/low-spin transitions can be observed by X-ray techniques (Schwoerer-Bohning *et al.*, 1996). Inelastic neutron scattering can also be used to excite magnetic transitions; it requires high energies and the measurement of low momentum transfer. Measurements of crystal field transitions at zero pressure have been reported (Winkler *et al.*, 1997). Given the intensity of current neutron sources, large samples are required.

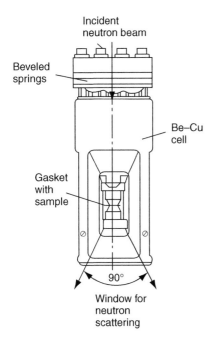

Incident
neutron beam

Beveled
springs

Be–Cu
cell

Gasket
with
sample

90°

Window for
neutron
scattering

View along neutron beam
Variants

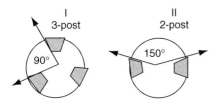

I
3-post

II
2-post

90°

150°

Figure 7 A variation of the 'gem cell'. Modified from Ivanov AS, Goncharenko IN, Somenkov VA, and Braden M (1995) Phonon dispersion in graphite under hydrostatic pressure up to 60 kbar using a sapphire-anvil technique. *Physica B* 213: 1031–1033.

11.4.4 Inelastic X-Ray Scattering and Spectroscopy

Advances in synchrotron radiation techniques have resulted in significant developments in X-ray spectroscopy, as well as diffraction, of minerals and related materials under pressure. X-ray spectroscopies can be used as a probe of electronic structure; these techniques thus constitute an alternative to methods requiring use or measurement of charged particles interacting with a sample (e.g., electrons in photoemission). As such, X-ray spectroscopies can be used for *in situ* study of properties such as band structures and local electronic structure.

Applications to mineralogy and geochemistry of Earth materials in the near-surface environment have

been described previously (Brown *et al.*, 1988). The applications at high pressure through 1998 have been reviewed (Hemley *et al.*, 1998b). Prior to this, very little X-ray spectroscopy had been conducted under pressure, and most of it has been limited to X-ray absorption spectroscopy. With new developments, the following methods are being used and/or developed for high-pressure studies of geophysically important materials: X-ray emission spectroscopy (XES); X-ray absorption spectroscopy (XAS); X-ray magnetic circular dichroism (XMCD); X-ray inelastic near-edge spectroscopy (XINES); electronic inelastic X-ray scattering (EIXS); resonant inelastic X-ray scattering (RIXS); Compton scattering (CS); nuclear resonant forward scattering (NRFS); nuclear resonant inelastic X-ray scattering (NRXS); and phonon inelastic X-ray scattering (PIXS).

These X-ray intensity-limited techniques have become practical with the arrival of the third-generation high-energy synchrotron sources. Since higher pressures are reached at the expense of diminishing sample volume, the pressure limit of these experiments depends upon the ability of passing the maximum number of photons into a microscopic sample volume. The development of high-pressure X-ray spectroscopy has also been hindered by the opaqueness of high-pressure vessels. Recently, third-generation sources have greatly boosted the intensities, and new high-pressure techniques such as the development of high-strength, but X-ray-transparent Be gaskets have extended the window for measurements from 12 keV down to 4 keV. This extended X-ray energy range allows the study of pressure-induced effects on atomic coordination, crystal structures, and electronic properties of a wider class of minerals using a variety of X-ray spectroscopies. The region below 4 keV (e.g., soft X-ray and vacuum UV) is still opaque in all kinds of high-pressure devices; however, inelastic scattering techniques provide a means to access states that cannot be probed by direct (one-photon) processes due to the opaqueness of the anvil and gasket (e.g., in the soft X-ray region).

11.4.4.1 X-ray absorption spectroscopy
In XAS, incident X-ray is absorbed when its energy exceeds the excitation energy of deep-core electrons of a specific element in the sample, causing an atomic edge-like absorption spectrum. Near-edge X-ray absorption fine structure (XANES) (Ablett *et al.*, 2003) provides information on symmetry-projected conduction band density of states (DOS), while the extended X-ray absorption fine structure (EXAFS)

(Buontempo *et al.*, 1998; Pascarelli *et al.*, 2002) provides local structure information. XAS is typically monitored by broadband fluorescence. Different spectral regions in the vicinity of a core-level absorption edge are typically defined on the basis of the different information they contain. EXAFS arises from multiple scattering of electrons in the environment of an atom, from which local structural information (interatomic distances and coordination) is obtained. Spin-dependent X-ray absorption spectroscopy can reveal magnetic properties even in the absence of long-range magnetic order (Hämäläinen *et al.*, 1992).

These element-specific probes reveal electronic and structural changes at high pressure (Itie *et al.*, 1992). XAS was one of the first X-ray spectroscopy techniques applied at high pressure, but was previously limited to higher energies due to the absorption of diamond anvils. Both techniques extend XAS down to 4 keV that covers K edges of first-row transition elements and L edges of rare earth elements. XAS instrumentation is among the simplest in high-pressure X-ray experiments. Only the monochromator needs to be scanned across the specific energy edge of the element of interest. The monochromatic X-rays are focused by mirrors (e.g., meterlong K–B type) upon the sample in the cell. The absorption is measured by monitoring the intensity change of either the transmission or the fluorescence signals due to the absorption/re-emission process, and is normalized relative to a reference intensity measured before the sample. The large, high-demagnification, K–B mirrors provide sufficient intensity, with the beam diameter matching the small sample size at ultrahigh pressure.

High-pressure XAS has been used successfully for transition elements at moderate pressures with BC anvils (Wang *et al.*, 1998), and for higher-Z materials (Itie *et al.*, 1992; Pasternak *et al.*, 1997a). EXAFS is now well developed for high-pressure studies (Itie *et al.*, 1997), and can be measured by absorption or fluorescence. In addition, Laue diffraction of single-crystal diamonds causes sharp absorption peaks that are detrimental to the EXAFS spectroscopy. To greatly reduce the diamond in the beam path, holes can be drilled in diamond anvils, as used in Bassett-type hydrothermal diamond cell for studying ions in aqueous solutions at moderate P–T conditions (Bassett *et al.*, 2000; Anderson *et al.*, 2002), and an alternative beam path through high-strength beryllium and boron gaskets is used in megabar panoramic diamond cells (Mao *et al.*, 2003a). Early high-pressure

work with diamond anvils was limited by the Bragg diffraction from the diamond and by the absorption edge of the diamond and the gasket. Because of the low energy of the Si edge, high-pressure studies of germanate analogs have been carried out (Itié *et al.*, 1989). The development of more transparent gaskets has helped to overcome these problems. EXAFS is particularly useful for study of pressure effects on the structures of liquids (Katayama *et al.*, 1998) and amorphous solids (Mobilio and Meneghini, 1998). A notable example is the observation of the coordination change in crystalline and amorphous GeO_2 (Itie *et al.*, 1989). Mobilio and Meneghini (1998) provide a review of the variety of synchrotron-based X-ray techniques for studying amorphous materials, including techniques such as anomalous scattering that can be used in conjunction with EXAFS.

11.4.4.2 X-ray emission spectroscopy

In this technique, deep-core electrons in the sample are excited by X-rays. The core-holes then decay through either radiative or nonradiative processes. For deep-core holes, the dominant decay channels are radiative processes, producing fluorescence, which is analyzed to provide information on the filled electronic states of the sample. The information provided by XES is complementary to that provided by XAS. Moreover, the final state of the fluorescent process is a one-hole state, similar to the final state of a photoemission process. Thus, the most important information provided by photoelectron spectroscopy, namely large chemical shifts in the core-level binding energies and the valence band density of states, is also available in XES.

The energies of the fluorescent photons are analyzed with sub-electronvolt energy resolution of the emission spectral line shape to provide information on the filled electronic states of the sample. As for XAS, the development of panoramic diamond cells has extended the low end of the high-pressure energy window down to 4 keV (Mao *et al.*, 1998b; Lin *et al.*, 2003a). Deep-core electrons of all elements above Ca can now be studied at high pressure by a suitable choice of analyzer crystals. For instance, transition element ions, with their variable valence and magnetic states, control major geochemical processes and geophysical behaviors, such as oxidation, reduction, chemical differentiation, elasticity, geomagnetism, conductivity, and radiation heat transfer.

In XES experiments, the incident beam energy is fixed, the emission X-radiation can be collected at any direction, and high-resolution emission spectra are obtained by a synchronized θ–2θ scan of the analyzer and the detector. The excitation X-ray source only needs to have higher energy than that of the fluorescent photons; energy band width (resolution) of the X-ray source is inconsequential. In principle, white (a broad, smooth continuum of energy), pink (an unfiltered undulator beam with energy band path of hundreds of electronvolts), and monochromatic (a narrow, electronvolt band path of photon) X-rays can all be used as the source.

Examples of high-pressure XES include study of predicted high-spin/low-spin transitions in iron oxides and sulfides. A high-resolution Rowland circle spectrometer with a double-crystal monochromator is used. A schematic of the K_β emission process in Fe^{2+} is shown in **Figures 8** and **9** (Struzhkin *et al.*, 2004) In an atomic picture, the final state of the K_β is a whole state with $1s^2 3p^5 3d^5$. The measurement also reveals the XANES portion of the spectra, that is, including the pre-edge peak associated with transitions to unoccupied d-states (i.e., 3d in Fe). By measuring the absorption through selective monitoring of the fluorescence peaks, information on the contribution of different spins to the spectrum can be obtained.

XES of the Fe K_β emission was first measured with a white X-ray source at a second generation facility (NSLS); the high-spin–low-spin transition in troilite (FeS), coinciding with its I–II structural transition, was observed (Rueff *et al.*, 1999). Subsequently, with higher source intensity, the technique has been applied at APS and ESRF for studies of high-spin–low-spin transitions in wüstite (Badro *et al.*, 1999), hematite (Badro *et al.*, 2002), magnesiowüstite (Badro *et al.*, 2003), and perovskite (Li *et al.*,

2004) to 100 GPa pressures. Recently, the technique has been extended to high P–T conditions in contribution with laser heating techniques (**Figure 10**).

There have been significant developments in the *ab initio* treatment of such spectra, in particular, the electron–hole attraction, which requires going beyond a simple mean- field approach such as the local density approximation (LDA) (Shirley, 1998). This is needed for XAS and RIXS. Recent measurements of the effect of pressure on plasmon excitations in metals (e.g., the high-pressure phase of Ge) suggest the potential for investigating the electronic structure of core materials at megabar pressures.

11.4.4.3 X-ray inelastic near-edge spectroscopy

Near core-electron absorption edge features measured by soft x-ray absorption (XANES) or electron energy loss spectroscopy (EELS) reveal information on chemical bonding. Such information is particularly pronounced and important for light elements, but has been inaccessible for high-pressure studies as the pressure vessel completely blocks the soft X-ray and electron beams. With XINES, the high-energy incident X-ray penetrates the pressure vessel and reaches the sample. The scattered photon loses a portion of energy corresponding to the K-edge of the low-Z sample, but can still exit the vessel to be registered on the analyzer–detector system. Inelastic K-edge scattering spectra of second-row elements from Li (56 eV) to O (543 eV) at high pressures opened a wide new field of near-K-edge spectroscopy of the second-row elements, as in carbon in graphite (Mao *et al.*, 2003a).

In XINES experiments, spherically bent single-crystal analyzers collect scattered X-radiation, and focus it to the detector at nearly backscattering geometry, thus fixing the energy at the elastic line. The incident X-ray energy is scanned relative to the elastic line to determine the inelastic (Raman) shift. Since XINES probes the core-electron energy level at hundreds of electronvolts, the undulator gap needs be changed to follow the energy scan of the double monochromator. For most high-pressure XINES measurements, the intrinsic peak width is of the order of electronvolts, and the direct beam without a high-resolution monochromator and 1 m analyzer-to-detector distance would usually suffice. The angle 2θ between the incident beam and the scattered X-radiation defines the momentum transfer ($q = 4\pi E \sin \theta$, where E is the scattered X-ray energy). XINES features are relatively insensitive to q, except for the extremely large and small q (Krisch *et al.*, 1997). The 2θ angle for

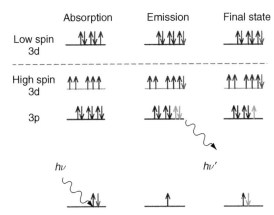

Figure 8 The schematic diagram of the K_β emission process.

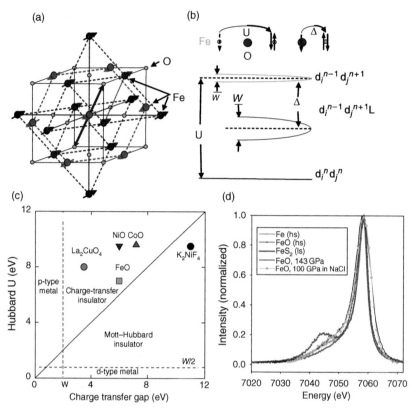

Figure 9 (a) NaCl-type crystal structure of the FeO sample below 17 GPa at room temperature. The structure has a small rhombohedral distortion (elongation along a body diagonal) in the antiferromagnetic phase above 17 GPa. (b) Electronic states in the material. It is energetically unfavorable for the electron to move between Fe atoms – such a transition costs an energy U (Hubbard energy). The transition of the electron from oxygen atom to a neighboring Fe atom is also energetically unfavorable; energy cost for such transition is called charge transfer energy Δ. (c) Zaanen–Sawatzky–Allen diagram, illustrating the balance between Hubbard energy U and charge transfer energy Δ in several strongly correlated materials. (d) X-ray emission sectra of high-spin FeO at ambient pressure, high-spin Fe at ambient pressure, and metallic FeO at 100 GPa. Low-spin XES of FeS_2 are shown for comparison. (a-d) Modified from Struzhkin VV, Hemley RJ, Mao HK, (2004) New condensed matter probes for diamond anvil cell technology. *Journal of Physics: Condensed Matter* 16: 1–16.

high-pressure XINES study can thus be set at an angle (e.g., 30°) to optimize the intensity, and multiple analyzers used to increase the count rate without discriminating q. Data collection efficiency can be increased by the use of an array of analyzers developed; such techniques were used for measurements for graphite (Mao *et al.*, 2003a) and boron nitride (Meng *et al.*, 2004), and H_2O (Cai *et al.*, 2005). Recently, the bonding changes associated with increasing Si coordination from 4 to 6 in silica glass, first measured by X-ray diffraction (1), has been measured using this technique (Lin *et al.*, 2007).

11.4.4.4 X-ray magnetic circular dichroism

This technique measures the spin polarization of an electronic excitation on FM materials by the use of circularly polarized X-rays. It is therefore in principle identical to conventional optical MDC with visible light (magneto-optical Kerr effect). In synchrotron radiation experiments, an X-ray phase plate changes the X-rays from linearly to circularly polarized. XMCD can be observed in both XANES and EXAFS. XMCD in XANES can measure spin-resolved conduction band densities of states, whereas XMCD in EXAFS provides local magnetic structural information. The sign of the dichroism gives the FM coupling between atoms in a metal. High-pressure MCD measurements have been performed (Baudelet *et al.*, 1997). For magnetic samples, magnetic circular dichroism is also observable in XES by exciting core holes with circularly polarized X-rays. With the low end of the high-energy window extended down to 4 keV by the high-strength Be gasket, all elements above Ca ($Z = 20$) in principle can be studied at high pressure with a suitable choice of analyzer crystals.

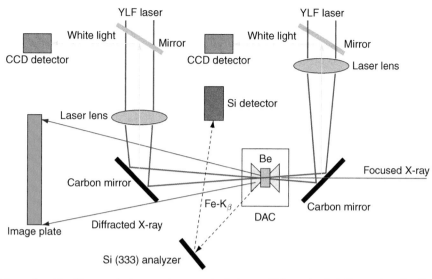

Figure 10 Schematic of the X-ray emission spectroscopy technique with laser-heated diamond anvil cells. The YLF laser beam is focused to 25 μm onto both sides of the sample. The diameter of the focused X-ray beam at 14 keV is less than 10 μm (FWHM); the small beam size insures that the X-ray emission signal from the sample is measured only within the laser-heated spot. The technique has been used to study single-crystal magnesiowüstite (($Mg_{0.76}$,$Fe_{0.24}$)O) loaded in a cell with diamonds having culet sizes of 300 μm. The Fe-K_β fluorescence lines were collected through the Be gasket by a 1 m Rowland circle spectrometer in the vertical scattering geometry. An image plate (MAR345) was used to collect diffracted X-ray in the forward direction.

Crystal analyzers have been developed for inelastic X-ray scattering (IXS) with sub-electronvolt energy resolution of emission spectral line shapes for studies of oxidation states, electronic energy level, spin states, and trace element analysis. In particular, the important 3d and 4f emission bands can be studied to very high pressures.

11.4.4.5 Electronic inelastic X-ray scattering

Pressure has dramatic effects on the energy and dispersion of all electronic bands. Many electronic levels, including some of the most intriguing pressure-induced changes, occur above the intrinsic, 5 eV, band gap of the diamond window, and are thus inaccessible by the standard vacuum techniques such as electron and UV spectroscopy probes. Using 10 keV X-ray beam for excitation and 0.3–1 eV resolution, IXS can in principle probe the full range of high-energy electronic levels, that is, the electronic band structure (Caliebe *et al.*, 2000), Fermi surface (Huotari *et al.*, 2000), excitons (Arms *et al.*, 2001), plasmons (Hill *et al.*, 1996), and their dispersion at high pressure. The instrumentation of EIXS is similar to that of XINES, except that both energy and q need to be scanned to obtain the dielectric function $\varepsilon(E, q)$ and the dynamic structure factor $S(E, q)$. EIXS probes

valence and conduction electronic structures at low energies up to a few tens of electronvolts from the elastic line. Many high-pressure EIXS can use the same 1 eV resolution setup as high-pressure XINES.

11.4.4.6 Resonant inelastic X-ray spectroscopy

Shallow core excitations are the same excitations probed by soft XAS, and are rich in multiplet structures for these highly correlated electronic systems. There have been a growing number of RIXS studies of core excitations (Krisch *et al.*, 1995; Kao *et al.*, 1996a; Bartolome *et al.*, 1997; Hill *et al.*, 1998; Kotani and Shin, 2001), that is, the final states of the inelastic scattering process in these studies are localized shallow core excitations. For magnetic samples, excitation with circularly polarized X-rays can also provide information on spin-resolved electronic structure (Krisch *et al.*, 1996; de Groot *et al.*, 1997).

In RIXS experiments, both incident and scattered X-ray energies are scanned. The incident X-ray energy is scanned across the core absorption similar to the procedures for XAS. For each monochromatic incident X-ray energy, fluorescence spectra are collected by scanning analyzers similar to the procedures of XES spectroscopy. One can obtain selected information connected directly with a specific intermediate state to which the incident photon energy is tuned. RIXS in transition metal and rare earth systems gives us important information on the electronic states, such as the intra-atomic multiplet coupling, electron correlation, and interatomic hybridization. Unusual phase transitions driven by electron correlation effects occur in many transition metals and transition metal compounds at high pressures, some accompanied with large volume collapses (e.g., 5–17% in rare earth metals and 10–15% in magnetic 3d transition metal oxides). The nature of these transitions, including the relationships between the crystal and electronic structures and the role of magnetic moment and order, is an area of active study.

11.4.4.7 Inelastic X-ray scattering spectroscopy

IXSS measures the dynamical structure factor, and, as such, it is similar to inelastic neutron scattering (Ghose, 1988). Formally, $S(q, \omega)$ is related to the space–time Fourier transform of the density correlation function, which provides information on the electronic band structure and elementary excitations (such as phonons and plasmons) and in turn thermal, optical, magnetic, and transport properties of the material (Hill et al., 1996). IXSS has a number of advantages in comparison with other scattering probes, such as those that use (optical) light, electrons, and neutrons (Kao et al., 1996b; Krisch et al., 1995). For example, light scattering can only probe zero-momentum transfer transitions; electron scattering suffers from multiple scattering effects and can only be used in high vacuum conditions. Neutron scattering needs very large samples. On the other hand, IXSS covers wide length (momentum) as well as temporal (energy) scales that are potentially important in high-pressure mineral studies. Low-energy excitations have been studied with ultra-high-resolution ($\sim$1.5 meV or $12\,\mathrm{cm}^{-1}$) IXSS techniques (Ruocco et al., 1996; Sampoli et al., 1997). The resolution that can be achieved is as good as or better than what has been achieved by the best angle-resolved photoemission spectrometer.

Another variation on these techniques is high-resolution resonant X-ray Raman scattering, which is in principle like the more familiar Raman scattering at optical wavelengths. It can be viewed as an instantaneous absorption and emission process where the initial and final states are electronic or magnetic (rather than vibrational) states. Using such resonant IXS, one detects electronic excitations of $\omega = \omega_1 - \omega_2$

and monitor momentum transfer $q = q_1 - q_2$. Examples include the study of the metal–AFM insulator transitions (Isaacs et al., 1996), but the technique has not yet been used for high-pressure minerals. Studies at ambient pressure have shown that it is possible to measuring EXAFS-like spectra for K-edges of low-Z elements by X-ray Raman scattering; similar studies in high-pressure cells could provide probes of local structure (e.g., Si) which cannot be studied by direct EXAFS techniques.

11.4.4.8 Compton scattering

CS is an inelastic scattering at high momentum transfer that probes electron momentum distributions. The details of the momentum distribution reveal information about bonding type (e.g., covalency) and band structure, particularly in combination with electronic structure theory (Isaacs et al., 1999). The availability of intense synchrotron radiation sources, particularly the high brightness at very high energies ($\sim$50 keV), together with the development of high-resolution spectrometers, are opening up new classes of CS experiments on small samples and at high pressures. The method has been applied to elemental solids such as Na and Si in Paris–Edinburgh cells, but not yet to geophysical materials (Tse et al., 2005).

11.4.4.9 Nuclear resonance forward scattering

As described above, Mössbauer spectroscopy has been used extensively in high-pressure mineralogy in laboratory studies with a radioactive parent source. Despite recent advances, high-pressure studies using a conventional Mössbauer source suffer from limited intensity for measurements on small samples, absorption by anvils, and background scattering. The temporal structure of synchrotron radiation can be exploited to perform nuclear resonance spectroscopy in the time domain (Hastings et al., 1991). With this technique, highly monochromatized X-rays from the synchrotron are used to excite narrow nuclear resonances, and the delayed photons are detected (**Figure 11**). Hyperfine splittings are reconstructed from the time-dependent intensity. The phonon densities of states associated with the resonant nuclei can also be measured. The technique has been used successfully to study Fe and Eu at second-generation synchrotron sources (Takano et al., 1991; Nasu, 1996; Chefki et al., 1998). The nuclear resonance forward scattering (NRFS) technique has been extended to probe magnetism up to megabar pressures (**Figure 12**).

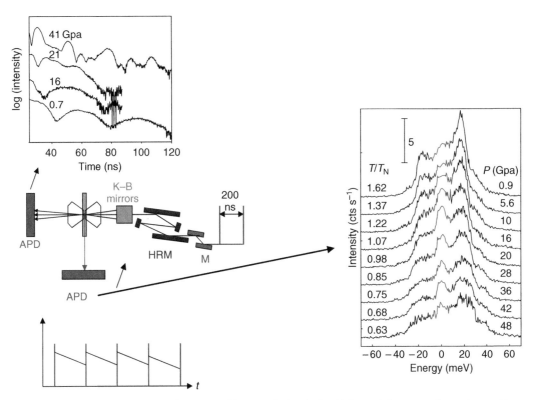

Figure 11 High-pressure nuclear inelastic resonant X-ray scattering. In-line high-resolution monochromator used to narrow down the photon energy to millielectronvolt resolution and fine-tune the monochromatic X-ray near the narrow nuclear resonant (elastic) line. Avalanche photodiodes (APDs) are used to collect only signals from nuclear resonance absorption, and reject all nonresonantly scattered radiation. An APD directly downstream of the sample collects nuclear forward scattering and the APDs surrounding the sample collect NRIXS spectroscopy (Hu *et al*., 2003; Sturhahn *et al*., 1995). Nuclear forward scattering spectra of FeO (top) and nuclear resonant inelastic x-ray scattering spectra of FeO (right) through the pressure-driven transition from the paramagnetic (NaCl structure) to the antiferromagnetic (rhombohedral structure) phase above 17 GPa. Modified from Struzhkin VV, Hemley RJ, Mao HK (2004) New condensed matter probes for diamond anvil cell technology. *Journal of Physics: Condensed Matter* 16: 1–16.

11.4.4.10 Nuclear resonant inelastic X-ray scattering

Nuclear resonant scattering also yields information on the phonon DOS through an inelastic scattering process (**Figure 11**). In principle, the DOS provides constraints on dynamic, thermodynamic, and elastic information of a material, including vibrational kinetic energy, zero-point vibrational energy, vibrational entropy, vibrational heat capacity, Debye temperature, Grüneisen parameter, thermal expansivity, longitudinal velocity, shear velocities, bulk modulus, and shear modulus. Measurements were first carried out on ^{57}Fe in its b.c.c. metallic iron at ambient conditions (Seto *et al*., 1995; Sturhahn *et al*., 1995). The nuclear resonant inelastic X-ray scattering (NRIXS) technique has been extended to high pressure and obtained the DOS of ε-Fe up to 153 GPa (Mao *et al*., 2001) and as a function of pressure and temperature with laser-heating

techniques (Lin *et al*., 2005b). High-pressure studies have been carried out on FeO (Struzhkin *et al*., 2001), Fe–Ni, and Fe–Si alloys (Lin *et al*., 2003b).

11.4.4.11 Phonon inelastic X-ray scattering

For completeness, we also mention phonon scattering, which is useful for understanding vibrational thermodynamic properties, elasticity, and phase transition mechanisms. Because nonresonant PIXS requires high energy resolution (millielectronvolts), and the double differential scattering cross section of phonon IXS is very small, such an experiment is very demanding even with the brilliance of third-generation synchrotron undulator sources. Additional challenges posed by high-pressure studies are limitations in sample size. Ar (Occelli *et al*., 2001) to 20 GPa and ε-Fe to 120 GPa (Fiquet *et al*., 2001; Antonangeli *et al*., 2004) have been

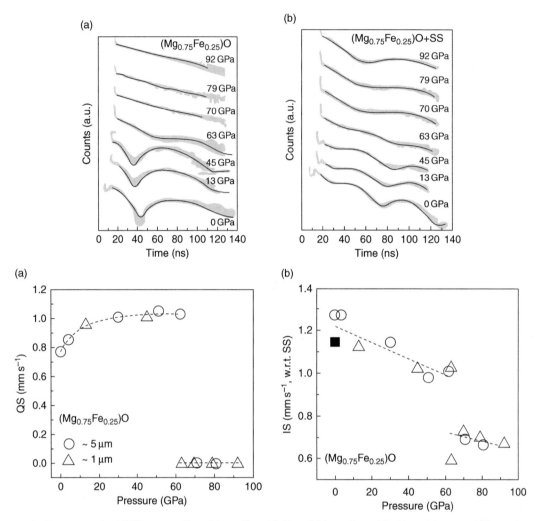

Figure 12 Representative NRFS spectra. Top: (Mg$_{0.75}$,Fe$_{0.25}$)O (a) and (Mg$_{0.75}$,Fe$_{0.25}$,)O with stainless steel (b) as a function of pressure at room temperature. Black line, modeled spectrum with the MOTIF [25] program. The sample thickness was approximately 1 μm, and the stainless steel foil of ∼0.5 μm was used outside the diamond cell to generate quantum beats in the spectra for IS measurements. Evolution of the NRFS spectra of (Mg$_{0.75}$,Fe$_{0.25}$)O with the stainless steel as a reference enables derivation of the IS of the sample as a function of pressure. The quantum beats at 0, 13, and 45 GPa generated from the QS of the high-spin state of iron in the sample, whereas the flat features in the spectra at 70, 79, and 92 GPa indicate disappearance of the QS. The spectrum at 63 GPa is modeled with two states, a state with a QS and a state with zero QS, which may be explained the pressure gradient in the sample chamber. Lower: pressure dependence of (a) QS and in Mg$_{0.75}$Fe$_{0.25}$O as revealed from the modeling of the NRFS spectra. The disappearance of the QS and the significant drop of the IS at above 63 GPa are consistent with the high-spin–low-spin electronic transition of iron in the sample between 51 and 70 GPa. A least-squares fit to the IS (short dash lines) gives d(IS)/dP of –0.0037 (±0.0007) mm s^{-1} GPa^{-1} for the high-spin state and d(IS)/dP of –0.0021 (±0.0010) mm s^{-1} GPa^{-1} for the low-spin state, respectively. Modeling of the NRFS spectrum at 63 GPa indicates coexistence of two states, high-spin state and low-spin state of Fe^{2+}. (a-d) Modified from Lin JF, Gavriliuk AG, Struzhkin VV *et al.* (2006) Pressure-induced electronic spin transition of iron in magnesiowustite-(Mg, Fe)O. *Physical Review B* 73: 113107.

examined. Novel designs of in-line high-resolution monochromators have been adopted for high-pressure geophysics studies. Nonresonant PIXS has been applied to studying the high-pressure elasticity anisotropy of ε-Fe and of d- and f-electron metals can now be studied (Wong *et al.*, 2003; Manley *et al.*, 2003).

11.4.5 Transport Measurements

11.4.5.1 *Electrical conductivity*

Electrical conductivity has been measured under pressure in a wide variety of apparatus since the pioneering work of Bridgman (1949). The principal technique is the four-probe method. The method has been used in multianvil presses for studies of upper- and lower-mantle minerals, including studies of the $(Mg, Fe)SiO_4$ polymorphs of the upper mantle (Xu *et al.*, 1998a) and $(Mg, Fe)SiO_3$ perovskite (Xu *et al.*, 1998b). The method has been also been extensively applied in diamond cells, which provides the opportunity to access higher pressures (Mao and Bell, 1981). A modified, pseudo-four-probe technique has been applied at megabar pressures (see Hemley *et al.*, 2001). The technique has been extended to above 200 GPa at room temperature down to 0.05 K) (Eremets *et al.*, 1998, 2000, 2001) The technique was used early on in conjunction with high *P*–*T* laser heating (e.g., Li and Jeanloz, 1987) and resistive heating (e.g., Peyronneau and Poirier, 1989). With the development of new laser-heating techniques, such as double-sided heating which provide reduced temperature gradients and large heating zones (Mao *et al.*, 1998a), the accuracy of very high *P*–*T* electrical conductivity measurements is expected to be improved.

Microlithographic techniques are allowing still smaller leads to be attached to or actually built into the surface of the anvils for achieving both higher pressure and higher accuracy. The technique can also be coupled with a radiation field for carrying out photoconductivity measurements under pressure. Extensions of these techniques are benefiting from the development of designer anvil methods (Patterson *et al.*, 2000, 2004).

The multianvil experiments mentioned above used a complex impedance method commonly used at ambient pressure (Poe *et al.*, 1998). In this technique, the frequency of the current is varied over a wide range (from 0.01 Hz to 1 MHz); measurements on $(Mg, Fe)_2SiO_4$ polymorphs to 20 GPa as a function of temperature indicate that the conductivity in the high-pressure phases (wadsleyite and ringwoodite) is a factor of 10^2 higher than in olivine (Poe *et al.*, 1998), consistent with previous low-pressure studies. Another useful example is conductivity measured along the Hugoniot in shock-wave experiments (Knittle *et al.*, 1986), including reverberation approaches (Weir *et al.*, 1996).

11.4.5.2 Magnetic susceptibility

The magnetic susceptibility χ describes the response of a system to an applied magnetic field. Diamagnetic materials have negative χ. Paramagnetic and FM materials have positive χ. For paramagnetic materials $\chi \sim 10^{-6}$–10^{-5} m^3 G^{-1}, and for FM materials χ is several orders of magnitude higher. Typically, one reports the differential magnetic susceptibility, $\chi_\alpha =$

$d\chi/dH$. For example, $\chi_\alpha = 1100$ for Fe at external magnetic field $H = 0$. Several classes of high-pressure techniques have been developed. One involves measurement with superconducting quantum interference devices (SQUIDs; Webb *et al.*, 1976). A second is an inductive technique developed by Tissen and Ponyatosvkii (1987) and later extended by Timofeev (1992). Originally applied to study superconductivity under pressure, it has been extended to investigate other pressure-induced magnetic transitions, including those in Fe (Timofeev *et al.*, 2002) (**Figure 13**). A recently developed designer anvil technique has been used to study FM rare earth materials (Jackson *et al.*, 2005). We compared the quality factor (signal/noise) of the designer anvils and conventional coil technique in **Figure 14**.

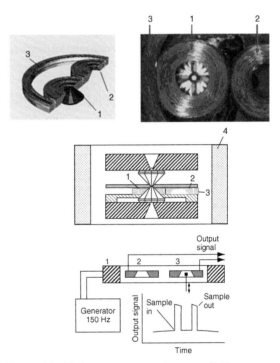

Figure 13 High-pressure magnetic susceptibility technique. Top: double-frequency modulation setup. Coil 4 is used to apply low-frequency magnetic field to modulate the amplitude response from the high-frequency pickup coil 2 due to the superconducting sample. The setup includes two signal generators and two lock-in amplifiers, operating at low (20–40 Hz) and high (155 KHz) frequencies. Lower: schematic representation of the background subtraction principle in magnetic susceptibility measurements: 1 – primary coil; 2 – secondary compensating coil; 3 – secondary signal coil. Removal of the sample from the signal coil produces measurable changes in the total output signal. Modified from Struzhkin VV, Hemley RJ, and Mao HK (2004) New condensed matter probes for diamond anvil cell technology. *Journal of Physics: Condensed Matter* 16: 1–16.

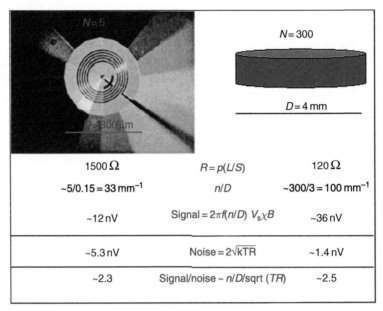

Figure 14   Comparison of signal/noise ratio for designer anvil and standard coil system.

The major issues associated with the use of designer anvils remains their low signal/noise ratio due to high resistance of the coils.

11.4.6   Resonance Methods

11.4.6.1   Electron paramagnetic and electron spin resonance

Various techniques have been developed for electron paramagnetic and electron spin resonance (EPR and ESR), which involves resonant microwave absorption between spin levels split in a magnetic field (Zeeman effect). This requires coupling of the microwave field with the sample, which can be challenging at high pressure because of the small sample size (dimensions less than the wavelength of the radiation) (**Figure 15**). Studies to ~8 GPa have been reported. Measurements on mantle silicates have been carried out at ambient pressures (e.g., Sur and Cooney, 1989).

11.4.6.2   Nuclear magnetic resonance

Nuclear magnetic resonance (NMR) is similar to EPR in that it too involves splitting of magnetic levels (Kirkpatrick, 1988; Stebbins, 1988). Because the magnetic moments of the nuclei are 3 orders of magnitude smaller than that of the electron, the level splitting is much lower in energy (radio frequency range, or ~100 MHz) at typical laboratory magnetic fields (H ~5–10 T). NMR studies of high-pressure phases recovered at zero pressure include ^1H and ^{29}Si

NMR studies of quenched high-pressure hydrous phase (Phillips et al., 1997) and magic angle spinning ^{29}Si relaxation techniques for characterizing naturally shocked samples (e.g., silica phases from Coconino sandstone) (Meyers et al., 1998). The technique has been applied to investigate glasses under pressure (Poe et al., 1993; Yarger et al., 1995). Most recently, two-dimensional NMR experiments applied to pressure quenched glasses reveal details of the changes in bonding associated with pressure densification (Lee et al., 2004, 2005).

There has been important progress in high-pressure NMR techniques (**Figure 16**). Proton NMR measurements have been carried out to 17 GPa (Ulug et al., 1991; Pravica and Silvera, 1998a, 1998b). In situ high-pressure NMR measurements on H_2, H_2O, and H_2O clathrates reveal information on the effect of compression on local bonding properties (e.g., H-bonding) (Okuchi et al., 2005a, 2005b, 2005c). A related measurement is nuclear quadrupolar resonance, which is similar to NMR but consists of measuring nuclear resonances in zero field. Measurements of the quadrupolar resonance ^{63}Cu in Cu_2O have been performed to 6 GPa (Reyes et al., 1992). The numerous double-resonance techniques employed for years in condensed matter and chemical physics generally require larger sample volumes but potentially can be employed at high pressure with continued increases in sample size under pressure.

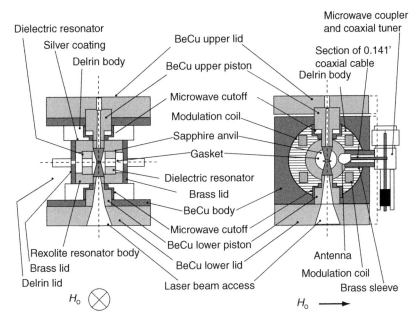

Figure 15 Cross-sectional view of the high-pressure ESR probe including the microwave coupler and coaxial tuner. Modified from Sienkiewicz A, Vileno B, Garaj S, Jaworski M, and Forro L (2005) Dielectric resonator-based resonant structure for sensitive ESR measurements at high hydrostatic pressures. *Journal of Magnetic Resonance* 177: 278–290.

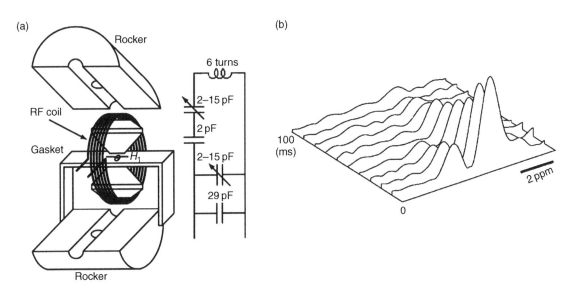

Figure 16 Technique for *in situ* nuclear magnetic resonance (NMR) in diamond-anvil cells. (a) Schematic illustration of the single solenoid RF probe and its circuit. (b) Magnetization of CH_3OH observed at 2.7 GPa. (a,b) Modified from Okuchi T, Cody GD, Mao HK, and Hemley RJ, (2005a) Hydrogen bonding and dynamics of methanol by high-pressure diamond anvil cell NMR. *Journal of Chemical Physics* 122: 244509.

11.4.6.3 de Haas–van Alphen

This is a now standard technique for measuring Fermi surfaces of metals under ambient pressures but it generally requires large, perfect single crystals. The advent of high-field magnets used in conjunction with new classes of diamond cells offers the possibility of extending these measurements to high pressures. Again, such studies of iron alloys under pressure could provide important constraints on the evolution of fundamental electronic properties of core materials with pressures. Other extensions of magnetic susceptibility methods are described above.

11.5 Selected Examples

We now discuss selected examples, focusing on systems where the combination of techniques discussed above have been key to uncovering information on the electronic and magnetic properties of geophysically important materials.

11.5.1 Olivine

The electronic properties of $(Mg, Fe)SiO_4$ olivine and its polymorphs can be considered a textbook problem in high-pressure mineral physics. In olivines, the band gap varies from 7.8 eV in forsterite to 8.8 eV in fayalite (Nitsan and Shankland, 1976). A prominent excitonic absorption is observed near the band gap (Nitsan and Shankland, 1976). Large changes in the optical absorption spectra under pressure were documented in the first *in situ* visible spectroscopy of minerals under pressure (Mao and Bell, 1972). As described above, high-temperature measurements to 20 GPa indicate that the conductivity in the high-pressure phases (wadsleyite and ringwoodite) is a factor of 10^2 higher than in olivine (Xu *et al.*, 1998a; Poe *et al.*, 1998).

More recent work on olivine has included high *P–T* shock compression studies of olivine in which the optical absorption and emission spectra of the material have been used to constrain the emissivity and temperature on shock compression (Luo *et al.*, 2004). The electrical conductivity of olivine containing 0.01–0.08 wt.% water up to 1273 K and 4 GPa indicates that the conductivity is strongly dependent on water content and only modestly dependent on pressure, and can be explained by the motion of free protons (Wang *et al.*, 2006). Recent single crystal measurements of electrical conductivity under pressure reveal a strong dependence on oxygen fugacity (Dai *et al.*, 2006), and the mechanism of hydrogen incorporation and diffusion has been examined (Demouchy and Mackwell, 2006). The radiative heat transfer under pressure has been modeled based on high-pressure vibrational spectra (Hofmeister, 2005). These results have been complemented by subsequent laser flash measurements on olivine at zero pressure (Pertermann and Hofmeister, 2006).

The zero-pressure absorption edge of fayalite (Nitsan and Shankland, 1976) shifts from the UV to the near IR with pressure, such that samples became opaque to visible light above 15–18 GPa; concomitantly, the electrical resistivity decreases by 5 orders of magnitude over this range (Mao and Bell, 1972; Smith and Langer, 1982; Lacam, 1983). A similar decrease in resistivity is also observed on shock-wave compression (Mashimo *et al.*, 1980). The low-temperature antiferromagnetic–paramagnetic transition has been tracked to at least 16 GPa by Mössbauer spectroscopy (Hayashi *et al.*, 1987). First-principles calculations of the electronic structure, and optical and magnetic properties of fayalite are consistent with the redshift of the absorption edge and decrease in electrical resistivity on compression (Jiang and Guo, 2004). Optical, electrical, and X-ray measurements extended to much higher pressures provided evidence that fayalite undergoes amorphization near 40 GPa with a further decrease in the band gap (Williams *et al.*, 1990).

11.5.2 Magnesiowüstite

We consider the high-pressure behavior of magnesiowüstite $(Mg, Fe)O$ as well as its end-member phases MgO and FeO. Several examples of the changes in electronic and magnetic properties of these materials have been discussed already. Periclase (MgO) remains in the rock salt (B1) structure and remains electronically simple to at least 227 GPa (Duffy *et al.*, 1995a), whereas wüstite ($Fe_{1-x}O$) exhibits complex polymorphism under pressure. Wüstite is nonstoichiometric, and contains some ferric iron even in equilibrium with iron metal. The vacancies in wüstite form complex defect clusters (Hazen and Jeanloz, 1984). With increasing pressure, more stoichiometric wüstite can be stabilized. At room temperature and pressure, wüstite has the cubic rock salt (B1) structure. As temperature is lowered, it passes the Néel temperature T_N and becomes antiferromagnetically ordered, and simultaneously assumes a distorted rhombohedral structure. As pressure is increased, T_N increases (Zou *et al.*, 1980), so that pressure promotes the rhombohedral phase. Isaak *et al.* (1993) showed that pressure promotes the rhombohedral distortion even in the absence of magnetism. Theory reveals the origin of the rhombohedral distortion with pressure. Visualization of the charge density as a function of rhombohedral angle (Hemley and Cohen, 1996; Isaak *et al.*, 1993) indicated that Fe–Fe bonding causes the rhombohedral strain, and the increase in Fe–Fe bonding with pressure is associated with the increased angle with pressure.

First-principles density functional theory (DFT) computations predicted that FeO and other transition metal monoxides would undergo magnetic collapse at high pressures (Cohen et al., 1998, 1997). Both LDA and the generalized gradient approximation (GGA) give 100 GPa for a first-order high-spin–low-spin transition in AFM FeO for a cubic lattice with a 7–9% volume collapse, and a continuous transition for FM cubic FeO (Cohen et al., 1998) and for rhombohedrally strained FeO (Cohen et al., 1997). Mössbauer experiments show evidence for a transition at about 90 GPa to a low-spin phase, and also a transition to a nonmagnetic phase with increasing temperature (Pasternak et al., 1997b), consistent with the DFT computations. On the other hand, XES shows no magnetic collapse up to 143 GPa, and reinterprets the 90 GPa transition observed by Mössbauer as a Néel transition (spin disordering) as opposed to magnetic collapse (Badro et al., 1999). LDA+U computations, which model local Coulomb repulsions not included in normal DFT calculations, predict high-spin behavior for AFM FeO to very high pressures (over 300 GPa), consistent with the XES measurements, but show metallization at lower pressures, depending on the value of U (Gramsch et al., 2003). A U of 4.6 eV, found by a self-consistent method (Pickett et al., 1998) and close to another recent estimate of 4.3 eV (Cococcioni and de Gironcoli, 2005), gives a pressure of only 60 GPa for metallization. (Gramsch et al., 2003); the lowest energy structure is predicted to be monoclinic (Gramsch et al., 2003), and was rediscovered by Cococcioni and de Gironcoli (2005). Neutron diffraction shows a monoclinic ground state at 10 K and zero pressure (Fjellvag et al., 2002), but single-crystal X-ray diffraction at high pressures has not resolved the monoclinic distortion (Jacobsen et al., 2005), perhaps because of the lack of high-angle data and/or twinning. LDA+U computations have now been performed for magnesiowüstite (Tsuchiya et al., 2006; Persson et al., 2006), which show reasonable agreement with experiments. An analysis of high-spin–low-spin transitions from a crystal field perspective predicts a continuous magnetic collapse at high temperatures as a function of pressure (Tsuchiya et al., 2006; Sturhahn et al., 2005). However, when the PV term is included in the free energy, it is possible to obtain different behavior

(Persson et al., 2006), including possibly a first-order transition.

At high temperature, the transition originally identified by shock-wave experiments (Jeanloz and Ahrens, 1980) was found to be to a metallic phase (Knittle and Jeanloz, 1986), which was identified as the NiAs (B8) structure by in situ diffraction (Fei and Mao, 1994). The phase diagram indicates that it should occur at low temperatures as well. Subsequent analysis of the diffraction data suggested the formation of a polytype or superlattice between B8 and anti-B8, with Fe in the As-site and O in the Ni-site (Mazin et al., 1998). These B8 and anti-B8 structures can be joined together smoothly, and the boundary between them is the rhombohedrally distorted B1 structure. This phase could form either due to lack of equilibrium, or could even form a unique continuous structure transition between the different phases. This interpretation of the behavior of FeO has been confirmed by recent experiments (Murakami et al., 2004; Kantor et al., 2004a). The phase diagram is shown in **Figure 17**.

Recent studies of (Mg, Fe)O illustrate the importance of using multiple techniques. Theoretical studies indicate that the high-spin–low-spin transitions in a number of structures are dominated by the size of the local coordination polyhedron (Cohen

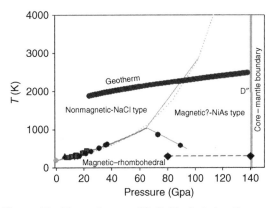

Figure 17 Phase diagram of FeO. Black circles, X-ray diffraction results; red squares: recent nuclear forward scattering results green solid circle: T_N at ambient pressure; blue solid triangle: T_N from high-pressure data at ambient temperature; black solid diamond: beginning and ending of resistive transition. Dotted line is an estimated pressure dependence of TN in rhombohedral phase. Dashed line between 80 and 140 GPa corresponds to the sluggish insulator–metal transition observed in resistivity studies. Modified from Lin JF, Gavriliuk AG, Struzhkin VV, et al. (2006) Pressure-induced electronic spin transition of iron in magnesiowustite-(Mg, Fe)O. *Physical Review B* 73: 113107.

et al., 1997). Thus, a transition metal ion in a smaller site will transform at lower pressures than one in a larger site. Thus, ferric iron in solid solution substituting for Mg^{2+}, which is a smaller ion than Fe^{2+}, transforms at lower pressures than in the pure ferrous iron compound. After being predicted theoretically, this has now been seen in a series of studies. XES shows a high-spin transition at 54–67 GPa in $(Mg_{0.75}Fe_{0.25})O$ and 84–102 GPa in $(Mg_{0.40}Fe_{0.60})O$, again suggesting that the transition in pure FeO is much greater than 100 GPa. Optical absorption spectra have been measured up to 80 GPa for the lower-mantle oxide, (Mg, Fe)O. Upon reaching the high-spin–low-spin transition of Fe^{2+} at about 60 GPa, there is enhanced absorption in the mid- and near-IR spectral range, whereas absorption in the visible–UV is reduced. The observed changes in absorption are attributed to d–d-orbital charge transfer transitions in the Fe^{2+} ion (**Figure 2**). The results indicate that low-spin (Mg, Fe)O will exhibit considerably lower radiative thermal conductivity than the lower-pressure high-spin (Mg, Fe)O (Goncharov *et al.*, 2006). A schematic phase diagram is shown in **Figure 18**.

The Mössbauer data for high-pressure magnesiowüstite are less clear; they show hyperfine field split spectra coexisting with a paramagnetic peak over large pressure ranges (Speziale *et al.*, 2005). The authors interpret the first appearance of a paramagnetic peak at the transition, but the data seem more consistent with coexisting high-spin and low-spin iron, perhaps in difference local environments. In any case, these data are consistent with solid solution in which Mg^{2+} decreases the transition pressure, consistent with earlier predictions. Theory suggests that ferrous iron in the B8 or anti-B8 structures will remain high spin to much higher pressures. This also makes clear the distinction between magnetic collapse and metal–insulator transitions, since normal B8 FeO is predicted to be a high-spin metal, and anti-B8 is predicted to be a high-spin insulator.

11.5.3 Silicate Perovskite and Post-Perovskite

The electronic properties of the silicate perovskite (Mg, Fe)SiO_3 have been the focus of considerable study. UV–visible spectra indicate the presence of crystal field splitting (Shen *et al.*, 1994), although subsequent measurements showed similar results for perovskite and magnesiowüstite (Keppler *et al.*, 1994). A broad feature in the optical spectrum near

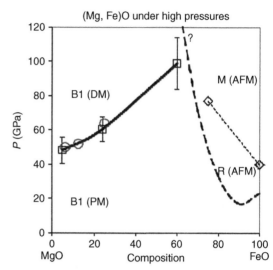

Figure 18 Phase diagram of the magnesiowüstite–(Mg,Fe)O system under high pressures. An isosymmetric transition from the paramagnetic state (PM) to the diamagnetic state (DM) occurs in MgO-rich magnesiowüstite (ferropericlase), and addition of FeO in MgO stabilizes the high-spin state to much higher pressures [8]. On the other hand, addition of MgO in FeO stabilizes the B1 structure relative to the AFM rhombohedral phase to much higher pressures [6]. Potential electronic and structural transitions in FeO-rich region remain to be further understood. Red dashed line represents the calculated high-spin–low-spin transition boundary based on the assumption that the spin transition occurs at the same Fe–O bond length and the iron–iron exchange interaction could be neglected (see Badra *et al.*, 2003). Modified from Lin JF, Gavriliuk AG, Struzhkin VV, *et al.* (2006) Pressure-induced electronic spin transition of iron in magnesiowustite-(Mg, Fe)O. *Physical Review B* 73: 113107.

14 900 cm^{-1} has been assigned to the $Fe^{2+} \rightarrow Fe^{3+}$ charge transfer transition (Keppler *et al.*, 1994). Zhang (1997) measured the Lamb–Mössbauer factor on (Mg, Fe)SiO_3 perovskites and clinoenstatite. The thermally activated electron delocalization found earlier in quenched samples in Mössbauer measurements with a conventional radioactive source (Fei *et al.*, 1994) was suppressed at high pressure. Li and Jeanloz (1987) measured the electrical conductivity of (Mg, Fe)SiO_3 perovskite at high pressures and temperatures by laser heating; subsequent measurements were carried out at lower maximum temperatures by Peyronneau and Poirier (1989) using a resistively heated cell. The latter results fit a hopping conductivity model. Katsura *et al.* (1998) reported that the temperature dependence of the conductivity differed significantly between samples measured at high pressure in its stability field and quenched to ambient pressure.

The strong partitioning of Fe^{2+} in magnesiowüstite relative to ferromagnesium silicate perovskite has been attributed to crystal field stabilization; that is, it arises from the stabilization of Fe^{2+} in the octahedral site of the oxide as compared to the (pseudo) dodecahedral site of the perovskite (Yagi *et al.*, 1979; Burns, 1993). The apparent crystal field stabilization has been used to estimate (or rationalize) partitioning between the two phases. Malavergne *et al.* (1997) find that partitioning results are consistent with observations for inclusions thought to have originated in the lower mantle and proposed to be representative of a pyrolite composition (Kesson and FitzGerald, 1992). McCammon used Mössbauer measurements to show that a significant fraction of the iron in $(Mg, Fe, Al)SiO_3$ perovskite produced in multianvil experiments is Fe^{3+} (McCammon, 1997). The incorporation of Fe^{3+} is strongly coupled to the Al^{3+}; this result does not, however, mean that that the lower mantle is oxidized as pressure stabilized the Fe^{3+} ion (Frost *et al.*, 2004; McCammon, 2005).

Both crystallographic and Mössbauer studies indicate that Fe^{2+} in silicate perovskite resides in the octahedral site (see Hemley and Cohen, 1992). Attempts have been made to determine the site occupancy of Fe^{2+} and Fe^{3+} from Mössbauer spectroscopy (McCammon, 1998). Spectra of $(Mg_{0.95}Fe_{0.95})SiO_3$ reportedly synthesized at low f_{O2} indicates that the Fe^{3+} goes in the octahedral site, whereas higher f_{O2} conditions result in Fe^{3+} on both sites. First-principles theoretical computations show that ferrous iron remains high spin in perovskite and is on the A-site (Caracas and Cohen, 2005a), but ferric iron undergoes a high-spin–low-spin transition in the pressure range of 100–125 GPa (Li *et al.*, 2005).

A revolution in our understanding of the deepest mantle and high-pressure mineral physics was the discovery of the post-perovskite phase in $MgSiO_3$ (Oganov and Ono, 2004; Murakami, *et al.*, 2004; Tsuchiya, 2004), which appears to be stable for a wide range of compositions at pressures in the megabar range (Caracas and Cohen, 2005a, 2005b). Application of the above electronic and magnetic techniques to silicate post-perovskite is challenging because of the need to carry out the experimental *in situ* at very high pressures (e.g., $\sim$100 GPa and above), and much input on the properties of post-perovskite have come from theoretical computations. Nevertheless, a number of measurements are being carried out, including NRIXS and XES, which are providing acoustic velocities and identification of the iron as low spin (**Figure 19**). A growing number of experimental studies are being performed in spite of the challenges (Mao *et al.*, 2006b). Experiments show that most iron in Al-bearing post-perovskite is ferric iron (Sinmyo *et al.*, 2006), consistent with first-principles theory (Li *et al.*, 2005). First-principles DFT calculations show ferrous iron to be unstable to disproportionation to ferric iron (in post-perovskite) and metallic iron in an h.c.p. Fe phase (Zhang and Oganov, 2006), consistent with experiment (McCammon, 2005). Theory predicts ferrous iron to be high spin to pressures well above the highest pressures in the mantle in both perovskite and post-perovskite (Caracas and Cohen, 2005a; Stackhouse *et al.*, 2006).

11.5.4 Volatiles

Volatile uptake and recycling in high-pressure phases provides important constraints on Earth history (Kerrick and Connolly, 2001; Sanloup *et al.*, 2002b, 2002c; Bercovici and Karato, 2003; Brooker *et al.*, 2003). Under pressure, volatiles can become structurally bound components, either as stoichiometric compounds or as soluble components in nominally anhydrous phases (Williams and Hemley, 2001). Dense hydrous silicates present different degrees of hydrogen bonding under ambient conditions. The prototype system is H_2O, which is discussed further below.

Pressure-induced disordering of crystals may be intimately associated with the behavior of the hydrogen through sublattice amorphization or melting (e.g., Duffy *et al.*, 1995b; Parise *et al.*, 1998; Nguyen *et al.*). The pressure dependence of the OH stretching modes show a tendency toward increased hydrogen bonding, but decreased hydrogen bonding is also observed (Faust and Williams, 1996; Hemley *et al.*, 1998a). These results, together with the evidence for disordering, point toward the importance of hydrogen–hydrogen repulsions. Finally, an example of the change in bonding affinities is the formation of iron hydride at high pressure. It produces a d.h.c.p. structure at 3.5 GPa, which is stable to at least 60 GPa (Badding *et al.*, 1992).

There is growing evidence for pressure-induced chemical interactions in rare gases. The low abundance of xenon in the atmosphere (relative to cosmochemical abundances) is a long-standing problem in geochemistry and has given rise to the proposal that the element may be sequestered at depth within the Earth. Near-IR spectroscopy and electrical conductivity measurements show that solid

(a) $$\frac{3}{V_D^3} = \frac{1}{V_P^3} + \frac{2}{V_S^3}$$

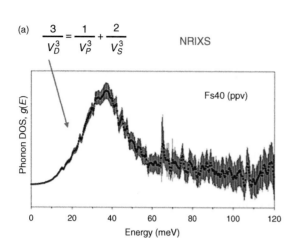

(b)

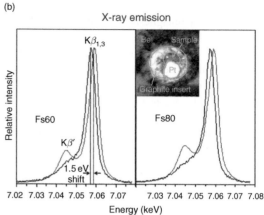

Figure 19 (a) NRIXS spectra showing the phonon DOS for post perovskite phase of Fs40 at 130 GPa after temperature quench from 2000 K. (b) X-ray emission spectra (XES) of the material measured *in situ* at high pressure. (a,b) Modified from Mao WL, Mao HK, Sturhahn W, *et al.* (2006b) Iron-rich post-perovskite and the origin of ultralow-velocity zones. *Science* 312: 564–565.

Xe becomes metallic at 130–150 GPa (Goettel *et al.*, 1989; Reichlin *et al.*, 1989; Eremets *et al.*, 2000). Prior to metallization, the material transforms from the f.c.c. to the h.c.p. structure (Jephcoat *et al.*, 1987). Recent studies have shown that the equilibrium transition pressure is as low as 21 GPa (Caldwell *et al.*, 1997). Experiments to 30 GPa and theoretical calculations to inner-core pressures reported in this study show (surprisingly) no evidence for chemical affinity for Xe and Fe (Caldwell *et al.*, 1997). The melting points of Xe and Ar increase initially with pressure (Jephcoat, 1998) but then flatten out; this has been interpreted as arising from s–p–d-hybridization (Ross *et al.*, 2005).

We also briefly mention planetary gases and ices at high pressure. Dense hydrogen is the most abundant element in the solar system. Accurate determinations of its electronic and magnetic properties to multimegabar pressures (>300 GPa) over a wide temperature range, and ultimately to the high-density plasma, is crucial for planetary geophysics. Both optical spectra and direct measurements of the electrical conductivity show that solid hydrogen remains insulating to at least 230 GPa at low temperatures (<200 K) (Hemley *et al.*, 2002). On the other hand, the dense hot fluid exhibits metallic conductivity at 140–200 GPa (Collins *et al.*, 2001; Nellis *et al.*, 1996). Recent developments have made it possible to contain hydrogen at 1000 K at static megabar pressures, allowing measurements to be performed on the dense hot fluid over this *P–T* range (Gregoryanz *et al.*, 2003).

The changes in bonding and electronic structure of various ices are particularly important for planetary science. The symmetric hydrogen-bond state of ice at 60 GPa reveal intriguing quantum mechanical tunneling effects associated with the transition (Goncharov *et al.*, 1996). There is a steep increase in melting temperature of the ionic symmetric ice phase (Lin *et al.*, 2005a; Goncharov *et al.*, 2005a). Sublattice melting has also been explored (Cavazzoni *et al.*, 1999; Katoh *et al.*, 2002). There are predictions to other high-pressure forms above 0.3 TPa (Benoit *et al.*, 1996; Cavazzoni *et al.*, 1999). Most recently, it has been found that the combination of X-ray irradiation and high pressure can break down H_2O to form an alloy H_2 and O_2 at pressures up to at least ~50 GPa and stable relative to H_2O to ~500 K (Mao *et al.*, 2006a).

11.5.5 Iron and Iron Alloys

The Earth's core plays a central role in the evolution and dynamic processes within the planet. As the major constituents of the core, iron and its alloys hold the key to understanding the nature of this most enigmatic region of the planet (Hemley and Mao, 2001). Geophysical observations have uncovered surprising inner-core properties, such as seismic anisotropy, super-rotation, and magnetism (Song and Richards, 1996; Glatzmaier and Roberts, 1996; Romanowicz *et al.*, 1996; Su *et al.*, 1996; Song and Helmberger, 1998; Tromp, 2001; Niu and Wen, 2001). These observations are supplemented by geodynamic simulations (Karato, 1999; Olson and Aurnou, 1999; Buffett, 2000, 2003; Buffett and Wenk, 2001). *Ab initio* theoretical calculations have been applied to examine and predict melting, phase stabilities, elastic anisotropy, and magnetism of iron beyond experimental capabilities (Stixrude and Cohen, 1995; Alfé *et al.*, 1999, 2000; Laio *et al.*, 2000; Steinle-Neumann *et al.*, 2001; Belonoshko *et al.*, 2003; Vocadlo *et al.*, 2003; *see* Chapter 13). Theory shows that the b.c.c. phase is stabilized by magnetism. There had been much discussion of b.c.c. as the possible structure for iron in the Earth's inner core, but calculations showed that b.c.c. iron is mechanically unstable at high pressures due to the loss of magnetism with pressure (Stixrude and Cohen, 1995). On the other hand, the hexagonal phase (ε-Fe) is nonmagnetic. The reflectivity of iron decreases markedly across the b.c.c.–h.c.p. transition; measurements to 300 GPa showed that this low reflectivity continues to much higher pressure (Reichlin, Hemley, and Mao, unpublished).

Pressure effects on the valence band densities of states and magnetic properties of Fe are being measured with the new synchrotron X-ray techniques described above. Large differences in density of states are predicted between b.c.c. Fe and the two closed-packed phases (h.c.p. and f.c.c.). Spin-dependent K_β emission fine structure can be used to probe localized magnetic properties with XMCD. The element-specific nature of XES and XMCD will be particularly important in the study of transition metal and rare earth alloys. High *P–T* XAS and XRD are providing electronic and structure information for iron melt as well as crystals (Jackson *et al.*, 1993; Sanloup *et al.*, 2002a). High *P–T* NRFS provides information on Mössbauer effect and magnetism (Jackson *et al.*, to be published), and high *P–T* NRIXS coupled with hydrostatic equation of state data yields phonon densities of state, bulk longitudinal- and shear-wave velocities, heat capacity, entropy, Debye temperature, and Grüneisen parameter (Lübbers *et al.*, 2000; Mao *et al.*, 2001; Struzhkin *et al.*, 2001).

11.6 Conclusions

A range of high-pressure techniques are now available to investigate changes in bonding, electronic, and magnetic structure induced by pressure in Earth and planetary materials. In general, recent work has shown that a variety of techniques are required to understand the evolution of these complex systems under pressure. This includes the use of newly developed theoretical methods, which are providing increasingly accurate predictions for energetic properties of these materials under extreme conditions. There is much to be learned about the origin of the behavior of transition metal compounds and solid solutions at high pressures. The study of transitions in Mott insulators is a particularly important current problem, with implications for condensed-matter theory as well. Future work should also focus on both defect properties and polyphase aggregates at high *P–T* conditions: rocks are composite material, yet most mineral physics studies assume that the electronic and magnetic properties of the rock can be determined by adding up the contributions from the component mineral phases. An important question is the extent to which the electronic properties of the composite need to be considered (i.e., from interfacial, nanophase properties (Maxwell-Garnet, 1904)). Effects of such large changes in pressure on chemical properties are established. In general, the large perturbation of pressure on the electronic structure of materials suggests that the partitioning among different phases may be difficult to predict from ambient pressure measurements.

Acknowledgments

This work was supported by NSF-EAR, DOE-BES, and DOE-NNSA (CDAC). We especially acknowledge NSF grants EAR-0310139, EAR-0550040, and EAR-0409321.

References

Ablett JM, Kao CC, Shieh SR, Mao HK, Croft M, and Tyson TA (2003) High-pressure x-ray near-edge absorption study of thallium rhenium oxide up to 10.86 GPa. *High Pressure Research* 23: 471–476.

Ahrens TJ (1987) Shock wave techniques for geophysics and planetary physics. In: Sammis CG and Henyey TL (eds.) *Methods of Experimental Physics*, vol. 24, pp. 185–235. San Diego: Academic Press.

Alfé D, Gillan MJ, and Price GD (1999) The melting curve of iron at the pressures of the Earth's core from *ab initio* calculations. *Nature* 401: 462–463.

Alfé D, Gillan MJ, and Price GD (2000) Constraints on the composition of the Earth's core from *ab initio* calculations. *Nature* 405: 172–175.

Anderson AJ, Jayanetti S, Mayanovic RA, Bassett WA, and Chou I-M (2002) X-ray spectroscopic investigations of fluids in the hydrothermal diamond anvil cell: The hydration structure of aqueous La^{3+} up to 300°C and 1600 bars. *American Mineralogist* 87: 262–268.

Anderson PW (1958) Absence of diffusion in certain random lattices. *Physical Review* 109: 1492–1505.

Antonangeli D, Occelli F, Requardt H, Badro J, Fiquet G, and Krisch M (2004) Elastic anisotropy in textured hcp-iron to 112 GPa from sound wave propagation measurements. *Earth and Planetary Science Letters* 225: 243–251.

Arms DA, Simmons RO, Schwoerer-Bohning M, Macrander AT, and Graber TJ (2001) Exciton dispersion and electronic excitations in hcp ^{4}He. *Physical Review Letters* 87: 156402.

Ashcroft NW and Mermin ND (1976) *Solid State Physics*. New York: Holt Rinehart and Winston.

Badding JV, Mao HK, and Hemley RJ (1992) High-pressure crystal structure and equation of state of iron hydride: Implications for the Earth's Core. In: Syono Y and Manghnani MH (eds.) *High Pressure Research in Mineral Physics: Application to Earth and Planetary Sciences*, pp. 363–372. Washington, DC: Terra Scientific Publishing Co (TERRAPUB).

Badro J, Fiquet G, Guyot F, et al. (2003) Iron partitioning in Earth's mantle: Toward a deep lower mantle discontinuity. *Science* 300: 789–791.

Badro J, Fiquet G, Struzhkin VV, et al. (2002) Nature of the high-pressure transition in Fe_2O_3 hematite. *Physical Review Letters* 89: 205504.

Badro J, Struzhkin VV, Shu J, et al. (1999) Magnetism in FeO at megabar pressures from X-ray emission spectroscopy. *Physical Review Letters* 83: 4101–4104.

Bartolome F, Tonnerre JM, Seve L, et al. (1997) Identification of quadrupolar excitation channels at the L_3 edge of rare-earth compounds. *Physical Review Letters* 79: 3775–3778.

Bassani F and Altarelli M (1983) Interaction of radiation with condensed matter. In: Koch EE (ed.) *Handbook on Synchrotron Radiation*, vol. 1, pp. 463–605. Amsterdam: North-Holland Publishing Company.

Bassett WA, Anderson AJ, Mayanovic RA, and Chou IM (2000) Hydrothermal diamond anvil cell for XAFS studies of first-row transition elements in aqueous solution up to supercritical conditions. *Chemical Geology* 167: 3–10.

Baudelet F, Odin S, Itie JP, et al. (1997) In: *Crystallography at High Pressure Using Synchrotron Radiation: The Next Steps*. Grenoble, France: ESRF.

Baudelet F, Pascarelli S, Mathon O, et al. (2005) X-ray absorption spectroscopy and X-ray magnetic circular dichroism simultaneous measurements under high pressure: The iron bcc–hcp transition case. *Journal of Physics: Condensed Matter* 17: S957–S966.

Bell PM, Xu J, and Mao HK (1986) Static compression of gold and copper and calibration of the ruby pressure scale to pressures to 1.8 megabars. In: Gupta Y (ed.) *Shock Waves in Condensed Matter*, pp. 125–130. New York: Plenum.

Belonoshko AB, Ahuja R, and Johansson B (2003) Stability of the body-centred-cubic phase of iron in the Earth's inner core. *Nature* 424: 1032–1034.

Benoit M, Bernasconi M, Focher P, and Parrinello M (1996) New high-pressure phase of ice. *Physical Review Letters* 76: 2934–2936.

Bercovici D and Karato S-I (2003) Whole mantle convection and the transition-zone water filter. *Nature* 425: 39–44.

Born M and Wolf E (1980) *Principles of Optics*. Oxford: Pergamon Press.

Bridgman PW (1949) *The Physics of High Pressure*. London: G. Bell and Sons.

Brooker RA, Du Z, Blundy JD, et al. (2003) The 'zero charge' partitioning behaviour of noble gases during mantle melting. *Nature* 423: 738–741.

Brown GE, Calas G, Waychunas GA, and Petiau J (1988) X-ray absorption spectroscopy and its applications in mineralogy and geochemistry. In: Hawthorne FC (ed.) *Spectroscopic Methods in Mineralogy and Geology, Reviews in Mineralogy*, vol. 18, pp. 431–512. Washington, DC: Mineralogical Society of America.

Buffett BA (2000) Earth's core and the geodynamo. *Science* 288: 2007–2012.

Buffett BA (2003) The thermal state of Earth's core. *Science* 299: 1675–1677.

Buffett BA and Wenk H-R (2001) Texturing of the Earth's inner core by Maxwell stresses. *Nature* 413: 60–62.

Buontempo U, Filipponi A, Martinez-Garcia D, Postorino P, Mezouar M, and Itie JP (1998) Anomalous bond length expansion in liquid iodine at high pressure. *Physical Review Letters* 80: 1912–1915.

Burns RG (1993) *Mineralogical Applications of Crystal Field Theory*. Cambridge, UK: Cambridge University Press.

Cai YQ, Mao HK, Chow PC, et al. (2005) Ordering of hydrogen bonds in high-pressure low-temperature H_2O. *Physical Review Letters* 94: 025502.

Caldwell WA, Nguyen JH, Pfrommer BG, Mauri F, Louie SG, and Jeanloz R (1997) Structure, bonding, and geochemistry of xenon at high pressures. *Science* 277: 930–933.

Caliebe WA, Soininen JA, Shirley E, Kao C-C, and Hämäläinen K (2000) Dynamic structure factor of diamond and LiF measured using inelastic X-ray scattering. *Physical Review Letters* 84: 3907–3910.

Caracas R and Cohen RE (2005a) Effect of chemistry on the stability and elasticity of the perovskite and post-perovskite phases in the $MgSiO_3$–$FeSO_3$–Al_2O_3 system and implications for the lowermost mantle. *Geophysical Research Letters* 32: L16310.

Caracas R and Cohen RE (2005b) Prediction of a new phase transition in Al_2O_3 at high pressures. *Geophysical Research Letters* 32: L06303.

Cavazzoni C, Chiarotti GL, Scandolo S, Tosatti E, Bernasconi M, and Parrinello M (1999) Superionic and metallic states of water and ammonia at giant planet conditions. *Science* 283: 44–46.

Chefki M, Abd-Elmeguid MM, Micklitz H, Huhnt C, and Schlabitz W (1998) Pressure-induced transition of the sublattice magnetization in $EuCo_2P_2$: Change from local moment Eu(4f)-to Itinerant Co(3d)-Magnetism. *Physical Review Letters* 80: 802–805.

Cococcioni M and de Gironcoli S (2005) Linear response approach to the calculation of the effective interaction parameters in the LDA + U method. *Physical Review B* 71: 035105.

Cohen RE, Fei Y, Downs R, Mazin II, and Isaak DG (1998) Magnetic collapse and the behavior of transition metal oxides: FeO at high pressures. In: Wentzcovitch R, Hemley RJ, Nellis WJ, and Yu P (eds.) *High-Pressure Materials Research*, pp. 27–37. Pittsburgh, PA: Materials Research Society.

Cohen RE, Gramsch S, Mukherjee S, Steinle-Neumann G, and Stixrude L (2002) Importance of magnetism in phase stability, equations of state, and elasticity. In: Hemley RJ, Bernasconi M, Ulivi L, and Chiarotti G (eds.) *Proceedings of the International School of Physics "Enrico Fermi", Vol. CXLVII: High-Pressure Phenomena*, pp. 215–238. Washington, DC: IOS Press.

Cohen RE, Mazin II, and Isaak DG (1997) Magnetic collapse in transition metal oxides at high pressure: Implications for the Earth. *Science* 275: 654–657.

Cohen RE and Mukherjee S (2004) Non-collinear magnetism in iron at high pressures. *Physics of Earth and Planetary Interiors* 143–144: 445–453.

Collins GW, Celliers PM, DaSilva, et al. (2001) Temperature measurements of shock compressed liquid deuterium up to 230 GPa. *Physical Review Letters* 87: 165504.

Cox PE (1987) *The Electronic Structure and Chemistry of Solids*. Oxford: Oxford Science Publications.

Dadashev A, Pasternak MP, Rozenberg GK, and Taylor RD (2001) Prediction of a new phase transition in Al_2O_3 at high pressures. *Review of Scientific Instruments* 72: 2633–2637.

Dai L, Li H, Liu C, *et al.* (2006) Experimental measurement of the electrical conductivity of single crystal olivine at high temperature and high pressure under different oxygen fugacities, *Progress in Natural Science* 16(4): 387–393.

de Groot FMF, Nakazawa M, Kotani A, Krisch MH, and Sette F (1997) Teoretical analysis of the magnetic circulat or dichroism in the $2p4d$ and $2p3d$ X-ray emission of Gd. *Physical Review B* 56: 7285–7292.

Demouchy S and Mackwell S (2006) Mechanisms of hydrogen incorporation and diffusion in iron-bearing olivine. *Physics and Chemistry of Minerals* 33: 347–355.

Dickson DPE and Berry FJ (eds.) (1986) *Mossbauer Spectroscopy*. New York: Cambridge University Press.

Ding Y, Xu J, Prewitt CT, *et al.* (2005) Variable P–T neutron diffraction of wüstite ($Fe_{1-x}O$): Absence of long-range magnetic order to 20 GPa. *Applied Physics Letters* 86: 052505.

Dobson DP, Richmond NC, and Brodholt JP (1997) A high-temperature electrical conduction mechanism in the lower mantle phase $(Mg,Fe)_1$. *Science* 275: 1779–1781.

Drickamer HG and Frank CW (1973) *Electronic Transition and the High Pressure Chemistry and Physics of Solids*. London: Chapman and Hall.

Duffy TS, Hemley RJ, and Mao HK (1995a) Equation of state and shear strength at multimegabar pressures: Magnesium oxide to 227 GPa. *Physical Review Letters* 74: 1371–1374.

Duffy TS, Meade C, Fei Y, Mao HK, and Hemley RJ (1995b) High-pressure phase transition in brucite $Mg(OH)_2$. *American Mineralogist* 80: 222–230.

Eremets M (1996) *High Pressure Experimental Methods*. New York: Oxford University Press.

Eremets MI, Gregoryanz E, Mao HK, Hemley RJ, Mulders N, and Zimmerman NM (2000) Electrical conductivity of Xe at megabar pressures. *Physical Review Letters* 83: 2797–2800.

Eremets MI, Shimizu K, Kobayashi TC, and Amaya K (1998) Metallic CsI at pressures of up to 220 gigapascals. *Science* 281: 1333–1335.

Eremets MI, Struzhkin VV, Timofeev JA, Utjuzh AN, and Shirokov AM (1992) In: Singh AK (ed.) *Recent Trends in High Pressure Research*, pp 362–364.New Delhi: Oxford & IBH.

Eremets MI, Struzhkin VV, Timofeev JA, Utjuzh AN, and Shirokov AM (1992) Refractive index of diamond under pressure. In: Singh AK (ed.) *Recent Trends in High Pressure Research*, pp. 362–364. New Delhi: Oxford & IBH.

Faust J and Williams Q (1996) Infrared spectroscopy of phase B at high pressures: Hydroxyl bonding under compression. *Geophysical Research Letters* 23: 427–430.

Fei Y (1996) In: Dyar MD, McCammon C, and Shaefer MW (eds.) *The Geochemical Society, Special Publication No. 5: Mineral Spectroscopy: A Tribute to Roger G. Burns*, pp 243–254. Houston: The Geochemical Society.

Fei Y and Mao HK (1994) *In situ* determination of the NiAs pbase of FeO at bighpressure and temperature. *Science* 266: 1668–1680.

Fei Y, Virgo D, Mysen BO, Wang Y, and Mao HK (1994) Temperature-dependent electron delocalization in $(Mg,Fe)SiO3$ perovskite. *American Mineralogist* 79: 826–837.

Fiquet G, Badro J, Guyot F, Requardt H, and Krisch M (2001) Sound velocities in iron to 110 gigapascals. *Science* 291: 468–471.

Fjellvag H, Hauback BC, Vogt T, and Stolen S (2002) Monoclinic nearly stoichiometric wüstite at low temperatures. *American Mineralogist* 87: 347–349.

Fleury PA and Loudon R (1968) Scattering of light by one- and two-magnon excitations. *Physical Review* 166: 514–530.

Frost DJ, Liebske C, Langenhorst F, McCammon CA, Trønnes RG, and Rubie DC (2004) Experimental evidence for the existence of iron-rich metal in the Earth's lower mantle. *Nature* 428(6981): 409.

Fujii Y (1996) *International Union of Crystallography. XVII Congress and General Assembly*, Seattle, Washington, C-1.

Ghose S (1988) In: Hawthorne FC (ed.) *Spectroscopic Methods in Mineralogy and Geology, Reviews in Mineralogy*, vol 18, pp. 161–192. Washington, DC: Mineralogical Society of America.

Gilder S and Glen J (1998) Magnetic properties of hexagonal closed-packed iron deduced from direct observations in a diamond anvil cell. *Science* 279: 72–74.

Glatzmaier GA and Roberts PH (1996) Rotation and magnetism of Earth's inner core. *Science* 274: 1887–1891.

Goettel KA, Eggert JH, Silvera IF, and Moss WC (1989) Optical evidence for the metallization of xenon at 132(5) GPa. *Physical Review B* 62: 665–668.

Goncharenko IN, Mignot JM, Andre G, Larova OA, Mirebeau I, and Somenkov VA (1995) Neutron diffraction studies of magnetic structure and phase transitions at very high pressures. *High Pressure Research* 14: 41–53.

Goncharov AF, Goldman N, Fried LE, *et al.* (2005a) Dynamic ionization of water under extreme conditions. *Physical Review Letters* 94: 125508.

Goncharov AF, Hemley RJ, Mao HK, and Shu J (1998) New high-pressure excitations in para-hydrogen. *Physical Review Letters* 80: 101–104.

Goncharov AF and Struzhkin VV (2003) Raman spectroscopy of metals, high-temperature superconductors and related materials under high pressure. *Journal of Raman Spectroscopy* 34: 532–548.

Goncharov AF, Struzhkin VV, Hemley RJ, Mao HK, and Liu Z (2000) New techniques for optical spectroscopy at ultrahigh pressures. In: Manghnani MH, Nellis WJ, and Nicol M (eds.) *Science and Technology of High Pressure*, pp. 90–95. Hyderabad, India: Universities Press.

Goncharov AF, Struzhkin VV, and Jacobsen SD (2006) Reduced radiative conductivity of low-spin (Mg,Fe)O in the lower mantle. *Science* 312: 1205–1208.

Goncharov AF, Struzhkin VV, Ruf T, and Syassen K (1994) High-pressure Raman study of the coupling of crystal-field excitations to phonons in Nd-containing cuprates. *Physical Review B* 50: 13841–13844.

Goncharov AF, Struzhkin VV, Somayazulu M, Hemley RJ, and Mao HK (1996) Compression of ice to 210 gigapascals: Infrared evidence for a symmetric hydrogen-bonded phase. *Science* 273: 218–220.

Goncharov AF, Zaug JM, Crowhurst JC, and Gregoryanz E (2005b) Optical calibration of pressure sensors for high pressure and temperatures. *Journal of Applied Physics* 97: 094917.

Gramsch SA, Cohen RE, and Savrasov SY (2003) Structure, metal-insulator transitions, and magnetic properties of FeO at high pressures. *American Mineralogist* 88: 257–261.

Gregoryanz E, Goncharov AF, Matsuishi K, Mao HK, and Hemley RJ (2003) Raman spectroscopy of hot dense hydrogen. *Physical Review Letters* 90: 175701.

Hämäläinen K, Kao CC, Hastings JB, *et al.* (1992) Spin-dependent X-ray absorption of MnO and MnF_2. *Physical Review B* 46: 14274–14277.

Harrison WA (1980) *Electronic Structure and the Properties of Solids: The Physics of the Chemical Bond*. San Francisco, CA: W.h. Freeman and Company.

Hastings JB, Siddons DP, van Bürck U, Hollatz R, and Bergmann U (1991) Mössbauer spectroscopy using synchrotron radiation. *Physical Review Letters* 66: 770–773.

Hayashi M, Tamura I, Shimomura O, Sawamoto H, and Kawamura H (1987) Antiferromagnetic transition of fayalite under high pressure studied by Mössbauer spectroscopy. *Physics and Chemistry of Minerals* 14: 341–344.

Hazen RM and Jeanloz R (1984) $Fe_{1-x}O$: A review of its defect structure and physical properties. *Reviews of Geophysics and Space Physics* 22: 37–46.

Hellwig H, Daniels WB, Hemley RJ, Mao HK, Gregoryanz E, and Yu Z (2001) Coherent anti-stokes Raman scattering spectroscopy of solid nitrogen to 22 GPa. *Journal of Chemical Physics* 115: 10876–10882.

Hemley RJ, Chen YC, and Yan CS (2005) Growing diamond crystals by chemical vapor deposition. *Elements* 1: 39–43.

Hemley RJ and Cohen RE (1992) Silicate perovskite. *Annual Review of Earth and Planetary Sciences* 20: 553–600.

Hemley RJ and Cohen RE (1996) Structure and bonding in the deep mantle and core. *Philosophical Transactions of the Royal Society of London A* 354: 1461–1479.

Hemley RJ, Eremets MI, and Mao HK (2002) Progress in experimental studies of insulator–metal transitions at multimegabar pressures. In: Hochheimer HD, Kuchta B, Dorhout PK, and Yarger JL (eds.) *Frontiers of High Pressure Research, Vol. II*, pp. 201–216. Amsterdam, The Netherlands: Kluwer.

Hemley RJ, Goncharov AF, Lu R, Li M, Struzhkin VV, and Mao HK (1998a) High-pressure synchrotron infrared spectroscopy at the national synchrotron light source II. *Nuovo Cimento D* 20: 539–551.

Hemley RJ, Hanfland M, and Mao HK (1991) High-pressure dielectric measurements of solid hydrogen to 170 GPa. *Nature* 350: 488–491.

Hemley RJ and Mao HK (1997) Static high-pressure effects in solids. In: Trigg GL (ed.) *Encyclopedia of Applied Physics*, vol. 18, pp. 555–572. New York: VCH Publishers.

Hemley RJ and Mao HK (2001) *In-situ* studies of iron under pressure: New windows on the Earth's core. *International Geological Review* 43: 1–30.

Hemley RJ, Mao HK, and Cohen RE (1998b) High-pressure electronic and magnetic properties. In: Hemley RJ (ed.) *Review in Minerology, Vol. 37: Ultrahigh-Pressure Minerology*, pp. 591–638. Washington, DC: Mineralogical Society of America.

Hess NJ and Schiferl D (1992) Comparison of the pressure-induced frequency shift of Sm:YAG to the ruby and nitrogen vibron pressure scales from 6 to 820 K and 0 to 25 GPa and suggestions for use as a high-temperature pressure calibran. *Journal of Applied Physics* 71: 2082–2085.

Hill JP, Kao CC, Caliebe WA, et al. (1998) Resonant inelastic X-ray scattering in Nd_2CuO_4. *Physical Review Letters* 80: 4967–4970.

Hill JP, Kao CC, Caliebe WAC, Gibbs D, and Hastings JB (1996) Inelastic X-ray scattering study of solid and liquid Li and Na. *Physical Review Letters* 77: 3665–3668.

Hofmeister AM (2005) Dependence of diffusive radiative transfer on grain-size, temperature, and Fe-content: Implications for mantle processes. *Journal of Geodynamics* 40: 51–72.

Hu MY, Sturhahn W, Toellner TS, et al. (2003) Measuring velocity of sound with nuclear resonant inelastic X-ray scattering. *Physical Review B* 67: 094304.

Huotari S, Hämäläinen K, Laukkanen J, et al. (2000) High-pressure compton scattering. In: Manghnani MH, Nellis WJ, and Nicol MF (eds.) *Proceeding of AIRAPT 17*, vol. 2, pp. 1017–1020. Honolulu, HI: Universities Press (India) Limited.

Huscroft C, McMahan AK, and Scalettar RT (1999) Condensed Matter: Electronic Properties, etc – Magnetic and thermodynamic properties of the three-dimensional periodic Anderson Hamiltonian. *Physical Review Letters* 82: 2342–2345.

Isaacs ED, Platzman PM, and Honig JM (1996) Inelastic X-ray scattering study of the metal–antiferromagnetic insulator transition in V_2O_3. *Physical Review Letters* 76: 4211–4214.

Isaacs ED, Shukla A, Platzman PM, Hamann DR, Barbiellini B, and Tulk CA (1999) Covalency of the hydrogen bond in ice: A direct X-ray measurement. *Physical Review Letters* 83: 4445.

Isaak DG, Cohen RE, Mehl MJ, and Singh DJ (1993) Phase stability of wüstite at high pressure from first-principles linearized augmented plane-wave calculations. *Physical Review B* 47: 7720.

Itié J, Baudelet F, Dartyge E, Fontaine A, Tolentino H, and San-Miguel A (1992) X-ray absorption spectroscopy and high pressure. *High Pressure Research* 8: 697–702.

Itié J, Polian A, Calas G, Petiau J, Fontaine A, and Tolentino H (1989) Pressure-induced coordination changes in crystalline and vitreous GeO_2. *Physical Review Letters* 63: 398–401.

Itié JP, Polian A, Martinez D, et al. (1997) *Journal of Physics IV (France)* 7: 31–38.

Ivanov AS, Goncharenko IN, Somenkov VA, and Braden M (1995) Phonon dispersion in graphite under hydrostatic pressure up to 60 kbar using a sapphire-anvil technique. *Physica B* 213: 1031–1033.

Jackson DD, Malba V, Weir ST, Baker PA, and Vohra YK (2005) High-pressure magnetic susceptibility experiments on the heavy lanthanides Gd, Tb, Dy, Ho, Er, and Tm. *Physical Review B* 71: 184416.

Jackson WE, Deleon JM, Brown GE, and Waychunas GA (1993) High-temperature XAS study of Fe_2SiO_4 liquid : Reduced coordination of ferrous iron. *Science* 262: 229–233.

Jacobsen SD, Lin JF, Dera P, et al. (2005) Single-crystal synchrotron X-ray diffraction study of wüstite and magnesiowüstite at lower-mantle pressures. *Journal of Synchroton Radiation* 12: 577.

Jeanloz R and Ahrens TJ (1980) Equations of state of FeO and CaO. *Geophysical Journal of the Royal Astronomical Society* 62: 505–528.

Jephcoat AP (1998) Rare-gas solids in the Earth's deep interior. *Nature* 393: 355–358.

Jephcoat AP, Mao HK, Finger LW, Cox DE, Hemley RJ, and Zha CS (1987) Pressure-induced structural phase transitions in solid xenon. *Physical Review Letters* 59: 2670–2673.

Jiang X and Guo GY (2004) Electronic structure, magnetism, and optical properties of Fe_2SiO_4 fayalite at ambient and high pressures: A GGA+U study. *Physical Review B* 69: 155108.

Kantor AP, Jacobsen SD, Kantor IY, et al. (2004a) Pressure-induced magnetization in FeO: Evidence from elasticity and Mössbauer spectroscopy. *Physical Review Letters* 93: 215502.

Kantor IY, McCammon CA, and Dubrovinsky LS (2004b) Mössbauer spectroscopic study of pressure-induced magnetization in wüstite (FeO). *Journal of Alloys and Compounds* 376: 5.

Kao CC, Caliebe WA, Hastings JB, and Gillet J-M (1996a) X-ray resonant Raman scattering in NiO: Resonant enhancement of the charge-transfer excitations. *Physical Review B* 54: 16361–16364.

Kao CC, Caliebe WA, Hastings JB, Hämäläinen L, and Krisch MH (1996b) *Reviews of Scientific Instruments* 67: 1–5.

Karato S (1999) Seismic anisotropy of the Earth's inner core resulting from flow induced by Maxwell stress. *Nature* 402: 871–873.

Katayama Y, Tsujii K, Oyanagi H, and Shimomura O (1998) Extended x-ray absorption fine structure study on liquid selenium under pressure. *Journal of Non-Crystalline Solids* 232–234: 93–98.

Katoh E, Yamawaki H, Fujihisa H, Sakashita M, and Aoki A (2002) Protonic diffusion in high-pressure ice VII. *Science* 295: 1264–1266.

Katsura T, Sato K, and Ito E (1998) Electrical conductivity of silicate perovskite at lower-mantle conditions. *Nature* 395: 493–495.

Keppler H, McCammon CA, and Rubie DC (1994) Crystal-field and charge-transfer spectra of (Mg,Fe)SiO$_3$ perovskite. *American Mineralogist* 20: 478–482.

Kerrick DM and Connolly JAD (2001) Metamorphic devolatilization of subducted marine sediments and the transport of volatiles into the Earth's mantle. *Nature* 411: 293–296.

Kesson SE and FitzGerald JD (1992) Partitioning of MgO, FeO, NiO, MnO and Cr2O$_3$ between magnesian silicate perovskite and magnesiowüstite: Implications for the origin of inclusions in diamond and the composition of the lower mantle. *Earth and Planetary Science Letters* 111: 229–240.

Kirkpatrick RJ (1988) MAS NMR spectroscopy of minerals and glasses. In: Hawthorne FC (ed.) *Reviews in Mineralogy, Vol 18: Spectroscopic Methods in Mineralogy and Geology*, pp. 341–403. Washington, DC: Mineralogical Society of America.

Kittel C (1996) *Introduction to Solid State Physics*. New York: Wiley.

Knittle E and Jeanloz R (1986) High-pressure metallization of FeO and implications for the Earth's core. *Geophysical Research Letters* 13: 1541–1544.

Knittle E, Jeanloz R, Mitchell AC, and Nellis WJ (1986) Metallization of Fe0.94O at elevated pressure and temperatures observed by shock-wave electrical resistivity measurements. *Solid State Communications* 59: 513–515.

Kotani A and Shin S (2001) Resonant inelastic X-ray scattering spectra in solids. *Reviews of Modern Physics* 73: 203–246.

Krisch M, Sette F, Masciovecchio C, and Verbeni R (1997) Momentum transfer dependence of inelastic X-ray scattering from the LI K edge. *Physical Review Letters* 78: 2843–2846.

Krisch MH, Kao CC, Sette F, Caliebe WA, Hämäläinen L, and Hastings JB (1995) Evidence for a quadrupolar excitation channel at the LIII edge of gadolinium by resonant inelastic X-ray scattering. *Physical Review Letters* 74: 4931–4934.

Krisch MH, Sette F, Bergmann U, et al. (1996) Observation of magnetic circular dichroism in resonant inelastic X-ray scattering at the L$_3$ edge of gadolinium metal (Rapid Communications). *Physical Review B* 54: R12673–R12676.

Lacam A (1983) Pressure and composition dependence of the electrical conductivity of iron-rich synthetic olivines to 200kbar. *Physics and Chemistry of Minerals* 9: 127–132.

Laio A, Bernard S, Chiarotti G, Scandolo S, and Tosatti E (2000) Physics of iron at Earth's core conditions. *Science* 287: 1027–1030.

Landau LD and Lifshitz IM (1977) *Quantum Mechanics: Non-Relativistic Theory*. Oxford: Pergamon Press.

Lee SK, Cody GD, Fei Y, and Mysen BO (2004) Nature of polymerization and properties of silicate melts and glasses at high pressure. *Geochimica et Cosmochimca Acta* 68: 4189–4200.

Lee SK, Mibe K, Fei Y, Cody GD, and Mysen BO (2005) Structure of B$_2$O$_3$ glass at high pressure: A ^{11}B solid-state NMR study. *Physical Review Letters* 94: 165507.

Li X and Jeanloz R (1987) Electrical conductivity of (Mg,Fe)SiO$_3$ perovskite and a perovskite-dominated assemblage at lower mantle conditions. *Geophysical Research Letters* 14: 1075–1078.

Li X and Jeanloz R (1990) Laboratory studies of the electrical conductivity of silicate perovskites at high pressures and temperatures. *Journal of Geophysical Research* 95: 5067–5078.

Li X and Jeanloz R (1987) Electrical conductivity of (Mg,Fe)SiO$_3$ perovskite and a perovskite-dominated assemblage at lower mantle conditions. *Geophysical Research Letters* 14: 1075–1078.

Li XY and Jeanloz R (1990) Laboratory studies of the electrical conductivity of silicate pervoskites at high pressures and temperatures. *Journal of Geophysical Research* 95: 2067–5078.

Lin JF, Gregoryanz E, Struzhkin VV, Somayazulu M, Mao HK, and Hemley RJ (2005a) Melting behavior of H$_2$O at high pressures and temperatures. *Geophysical Research Letters* 32: L11306.

Lin JF, Shu J, Mao HK, Hemley RJ, and Shen G (2003a) Amorphous boron gasket in diamond anvil cell research. *Review of Scientific Instruments* 74: 4732–4736.

Lin JF, Struzhkin VV, Sturhahn W, et al. (2003b) Sound velocities of iron–nickel and iron–silicon alloys at high pressures. *Geophysical Research Letters* 30: 2112–2115.

Lin JF, Sturhahn W, Zhao J, Shen G, Mao HK, and Hemley RJ (2005b) Sound velocities of hot dense iron: Birch's law revisited. *Science* 308: 1892–1894.

Lin JF, Gavriliuk AG, Struzhkin VV, et al. (2006) Pressure-induced electronic spin transition of iron in magnesiowustite-(Mg, Fe)O. *Physical Review B* 73: 113107.

Lin JF, Fukui H, Prendergast D, et al. (2007) Electronic bonding transition in compressed SiO$_2$ glass. *Physical Review B* 75: 012201.

Lipp M and Daniels WB (1991) Electronic structure measurements in KI at high pressure using three-photon spectroscopy. *Physical Review Letters* 67: 2810–2813.

Liu J and Vohra YK (1996) Photoluminescence and X-ray diffraction studies on-Sm doped yttrium aluminum garnet to ultrahigh pressures of 338 GPa. *Journal of Applied Physics* 79: 7978–7982.

Loubeyre P, Ocelli F, and LeToullec R (2002) Optical studies of solid hydrogen to 320 GPa and evidence for black hydrogen. *Nature* 416: 613–617.

Lübbers R, Grünsteudel HF, Chumakov AI, and Wortmann G (2000) Density of phonon states in iron at high pressure. *Science* 287: 1250–1253.

Luo SN, Akins JA, Ahrens TJ, and Asimow PD (2004) Shock-compressed MgSiO$_3$ glass, enstatite, olivine, and quartz: Optical emission, temperatures, and melting. *Journal of Geophysical Research B: Solid Earth* 109: B05205.

Maddox BR, Lazicki A, Yoo CS, et al. (2006) 4f Delocalization in Gd: Inelastic X-Ray scattering at ultrahigh pressure. *Physical Review Letters* 96: 215701.

Malavergne V, Guyot F, Wang Y, and Martinez I (1997) Partitioning of nickel, cobalt and manganese between silicate perovskite and periclase: A test of crystal field theory at high pressure. *Earth and Planetary Science Letters* 146: 499–509.

Manley ME, Lander GH, Sinn H, et al. (2003) Phonon dispersion in uranium measured using inelastic X-ray scattering. *Physical Review B* 67: 052302.

Mao HK (1976) Charge transfer processes at high pressure. In: Strens RGJ (ed.) *The Physics and Chemistry of Rocks and Minerals*, pp. 573–581. New York: John Wiley and Sons.

Mao HK and Bell PM (1972) Electrical conductivity and the red shift of absorption in olivine and spinel at high pressure. *Science* 176: 403–406.

Mao HK and Bell PM (1981) Electrical resistivity measurements of conductors in the diamond-window, high-pressure cell. *Review of Scientific Instruments* 52: 615–616.

Mao HK, Bell PM, Shaner JW, and Steinberg DJ (1978) Specific volume measurements of Cu, Mo, Pd, and Ag and calibration of the ruby R$_1$ fluorescence pressure gauge from 0.06 to 1 Mbar. *Journal of Applied Physics* 49: 3276–3283.

Mao HK and Hemley RJ (1989) Optical studies of hydrogen above 200 gigapascals: Evidence for metallization by band overlap. *Science* 244: 1462–1465.

Mao HK and Hemley RJ (1994) Ultrahigh-pressure transitions in solid hydrogen. *Reviews of Modern Sciences* 66: 671–692.

Mao HK, Shen G, and Hemley RJ (1997) Multivariable dependence of Fe–Mg partitioning in the lower mantle. *Science* 278: 2098–2100.

Mao HK, Shen G, Hemley RJ, and Duffy TS (1998a) X-ray diffraction with a double hot-plate laser-heated diamond cell. In: Manghnani MH and Yagi T (eds.) *Properties of Earth and Planetary Materials at High Pressure and Temperature*, pp. 27–34. Washington DC: American Geophysical Union.

Mao HK, Shu J, Shen G, Hemley RJ, Li B, and Singh AK (1998b) Elasticity and rheology of iron above 220 GPa and the nature of the Earth's inner core. *Nature* 396: 741–743.

Mao HK, Xu J, and Bell PM (1986) Calibration of the ruby pressure gauge to 800 kbar under quasihydrostatic conditions. *Journal of Geophysical Research* 91: 4673–4676.

Mao HK, Xu J, Struzhkin VV, et al. (2001) Phonon density of states of iron up to 153 gigapascals. *Science* 292: 914–916.

Mao WL, Mao HK, Eng P, et al. (2003a) Bonding changes in compressed superhard graphite. *Science* 302: 425–427.

Mao WL, Mao HK, Sturhahn W, et al. (2006) Iron-rich post-perovskite and the origin of ultralow-velocity zones. *Science* 312: 564–565.

Mao WL, Mao HK, Yan CS, Shu J, Hu J, and Hemley RJ (2003b) Generation of ultrahigh pressure using single-crystal, chemical-vapor-deposition diamond anvils. *Applied Physics Letters* 83: 5190–5192.

Mao WL, Mao HK, Meng Y, et al. (2006a) X-ray-induced dissociation of H_2O and formation of an O_2–H_2 alloy at high. *Science* 314: 636–638.

Mao WL, Mao HK, Sturhahn W, et al. (2006b) Iron-rich post-perovskite and the origin of ultralow-velocity zones. *Science* 312: 564–565.

Marshall WG, Nelmes RJ, Loveday JS, et al. (2000) High pressure neutron diffraction study of FeS. *Physical Review B* 61: 11201–11204.

Mashimo T, Kondo KI, Sawaoka A, Syono Y, Takei H, and Ahrens TJ (1980) Electrical conductivity measurements of fayalite under shock compression up to 550 kbar. *Journal of Geophysical Research* 85: 1876–1881.

Massey MJ, Baier U, Merlin R, and Weber WH (1990a) Effects of pressure and isotopic substitution on the Raman spectrum of alpha-Fe_2O_3: Identification of two-magnon scattering. *Physical Review B* 41: 7822–7827.

Massey MJ, Chen NH, Allen JW, and Merlin R (1990b) Pressure dependence of two-magnon Raman scattering in NiO. *Physical Review B* 42: 8776–8779.

Massey MJ, Merlin R, and Girvin SM (1992) Raman Scattering in $FeBO_3$ at high pressures: phonon coupled to spin-pair fluctuations and magneto-deformation potentials. *Physical Review Letters* 69: 2299–2302.

Mattheiss LF (1972) Electronic Structure of the 3*d* Transition-Metal Monoxides. I. Energy-band results. *Physical Review B* 5: 290–306.

Maxwell-Garnet JC (1904) Colours in metal glasses and in metallic films. *Philosophical Transactions of the Royal Society of London* 203: 385.

Mazin II, Fei Y, Downs R, and Cohen RE (1998) Possible polytypism in FeO at high pressures. *American Mineralogist* 83: 451–457.

McCammon CA (1995) Mossbauer spectroscopy of minerals. In: Ahrens TJ (ed.) *Mineral Physics and Crystallography, AGU Reference Shelf 2*, pp. 332–347. Washington: AGU.

McCammon CA (1997) Perovskite as a possible sink for ferric iron in the lower mantle. *Nature* 387: 694–696.

McCammon CA (1998) The crystal chemistry of ferric iron in Mg0.95Fe0.05SiO3 perovskite as determined by Mössbauer spectroscopy in the temperature range 80–293 K. *Physics and Chemistry of Minerals* 25: 292–300.

McCammon CA (2005) The Paradox of mantle redox. *Science* 308: 807–808.

McMahon MI and Nelmes RJ (2004) *Zeitschrift fur Kristallografiya* 219: 742–748.

Meade C, Hemley RJ, and Mao HK (1992) High-pressure X-ray diffraction of SiO_2 glass. *Physical Review Letters* 69: 1387–1390.

Mehl MJ, Cohen RE, and Krakauer H (1988) Lincarized augmented. Plane wave electronic structure calculations for MgO and CaO. *Journal of Geophysical Research* 93: 8009–8022.

Meng Y, Mao HK, Eng PJ, et al. (2004) The formation of sp^3 bonding in compressed BN. *Nature of Materials* 3: 111–114.

Meyers SA, Cygan RT, Assink RA, and Boslough MB (1998) ^{29}Si MAS NMR relaxation study of shocked Coconino sandstone from Meteor Crater, Arizona. *Physics and Chemistry of Minerals* 25: 313–317.

Mobilio S and Meneghini C (1998) Synchrotron radiation and the structure of amorphous materials. *Journal of Non-Crystalline Solids* 232–234: 25–37.

Mössbauer RL (1958) Kernreuonanzfluoreszenz von gammas-trahlung in Ir^{191}. *Zeitschrift fur Physik* 151: 124.

Mott NF (1990) *Met al*–Insulator Transitions. New York: Taylor & Francis.

Murakami M, Hirose K, Ono S, Tsuchiya T, Isshiki M, and Watanuki T (2004a) High pressure and high temperature phase transitions of FeO. *Physics of Earth and Planetary Interiors* 146: 273–282.

Murakami M, Hirose K, Kawamura K, Sata N, and Ohishi Y (2004b) Post-perovskite phase transition in $MgSiO_3$. *Science* 304: 855–858.

Nasu S (1996) High pressure Mossbauer spectroscopy with nuclear forward scattering of synchrotron radiation. *High Pressure Research* 14: 405–412.

Neaton JB and Ashcroft NW (1999) Pairing in dense lithium. *Nature* 400: 141.

Neaton JB and Ashcroft NW (2001) On the constitution of Sodium at higher densities. *Physical Review Letters* 86: 2830.

Nellis WJ, Weir ST, and Mitchell AC (1996) Metallization and electrical conductivity of hydrogen in Jupiter. *Science* 273: 936–938.

Nguyen J, Kruger MB, and Jeanloz R (1997) Evidence for 'partial' (sublattice) amorphization in $Co(OH)_2$. *Physical Review Letters* 78: 1936–1939.

Nitsan U and Shankland TJ (1976) Optical properties and electronic structure of mantle silicates. *Geophysical Journal of the Royal Astronomical Society* 45: 59–87.

Niu F and Wen L (2001) Hemispherical variations in seismic velocity at the top of the Earth's inner-core. *Nature* 410: 1081–1084.

Occelli F, Krisch M, Loubeyre P, et al. (2001) Phonon dispersion curves in an argon single crystal at high pressure by inelastic X-ray scattering. *Physical Review B* 63: 224306.

Oganov AR and Ono S (2004) Theoretical and experimental evidence for a post-perovskite phase of MgSiO3 in Earth's D layer. *Nature* 430: 445.

Okuchi T, Cody GD, Mao HK, and Hemley RJ (2005a) Hydrogen bonding and dynamics of methanol by high-pressure diamond anvil cell NMR. *Journal of Chemical Physics* 122: 244509.

Okuchi T, Hemley RJ, and Mao HK (2005b) Radio frequency probe with improved sensitivity for diamond anvil cell nuclear magnetic resonance. *Review of Scientific Instruments* 76: 026111.

Okuchi T, Mao HK, and Hemley RJ (2005c) A new gasket material for higher resolution NMR in diamond anvil cells. In: Chen J, Wang Y, Duffy T, Shen G, and Dobrzhinetskaya L (eds.) *Advances in High-Pressure Technology for Geophysical Application*, pp. 503–509. Amsterdam, The Netherlands: Elsevier.

Olson P and Aurnou J (1999) A polar vortex in the Earth's core. *Nature* 402: 170–173.

Parise JB, Theroux B, Li R, Loveday JS, Marshall WG, and Klotz S (1998) Pressure dependence of hydrogen bonding in metal deuteroxides: A neutron powder diffraction study of $Mn(OD)_2$ and β-$Co(OD)_2$. *Physics and Chemistry of Minerals* 25: 130–137.

Parise JB (2006) High Pressure Studies. In: Wenk HR (ed.) *Reviews in Mineralogy and Geochemistry: Neutron Scattering in Earth Sciences*, vol. 63, pp. 205–231. Washington, DC: Minerological Society of America.

Parise JB, Locke DR, Tulk CA, Swainson I, and Cranswick L (in press) *Physica B.*

Parker L and Badding JV (1996) Transition element-like chemistry for potassium under pressure. *Science* 273: 95–97.

Pascarelli S, Aquilanti G, Crichton W, et al. (2002) High pressure X-ray absorption and diffraction study of In As. *High Pressure Research* 22: 331–335.

Pasternak MP, Rozenberg GK, Milner AP, et al. (1997a) Pressure-induced concurrent transformation to an amorphous and crystalline phase in berlinite-type $FePO_4$. *Physical Review Letters* 79: 4409–4412.

Pasternak MP and Taylor RD (2001) Pressure-induced metallization and electronic–magnetic properties of some Mott insulators. *Physica Status Solidi B* 223: 65–74.

Pasternak MP, Taylor RD, Jeanloz R, Li X, Nguyen JH, and McCammon C (1997b) High pressure collapse of magnetism in $Fe_{0.94}O$: Mössbauer spectroscopy beyond 100 GPa. *Physical Review Letters* 79: 5046–5049.

Patterson JR, Aracne CM, Jackson DD, et al. (2004) Pressure-induced metallization of the Mott insulator MnO. *Physical Review Letters* 69: 220101.

Patterson JR, Catledge SA, Vohra YK, Akella J, and Weir ST (2000) Electrical and mechanical properties of C_{70} fullerene and graphite under high pressures studied using designer diamond anvils. *Physical Review Letters* 85: 5364–5367.

Pauling L (1960) *The Nature of the Chemical Bond*. Ithaca, NY: Cornell University Press.

Persson K, Ceder G, Bengtson A, and Morgan D (2006) *Ab initio* study of the composition dependence of the pressure-induced spin transition in the $(Mg_{1-x}, Fex)O$ system. *Geophysical Research Letters* 33: L16306.

Pertermann M and Hofmeister AM (2006) Thermal diffusivity of olivine-group minerals at high temperature. *American Mineralogist* 91: 1747–1760.

Peyronneau J and Poirier JP (1989) Electrical conductivity of the Earth's lower mantle. *Nature* 342: 537–539.

Peyronneau J and Poirier JP (1998) In: Manghani M and Yagi T (eds.) *Properties of Earth and Planetary Materials at High Pressure and Temperature*, pp. 77–87. Washington, DC: American Geophysical Union.

Phillips BL, Burnley PC, Worminghaus K, and Navrotsky A (1997) ^{29}Si and ^1H NMR spectroscopy of high pressure hydrous magnesium silicates. *Physics and Chemistry of Minerals* 24: 179–190.

Pickett WE, Erwin SC, and Ethridge EC (1998) Reformulation of the LDA+U method for a local-orbital basis. *Physical Review Letters* 58: 1201–1209.

Poe BT, Xu Y, and Rubie DC (1998) Electrical conductivities of mantle minerals: in-situ high-pressure high-temperature complex impedance spectroscopy. In: Nakahara M (ed.) *Review of High-Pressure Science and Technology, vol 7*, pp. 22–24. Kyoto: Japan Society High-Pressure Science and Technology.

Poe BT, McMillan PF, Coté B, Massiot D, and Coutures JP (1993) Magnesium and calcium aluminate liquids: *In Situ* high-temperature ^{27}Al NMR Spectroscopy. *Science* 259: 786–788.

Pravica MG and Silvera IF (1998a) NMR in a diamond anvil cell at very high pressures. *Review of Scientific Instruments* 69: 479–484.

Pravica MG and Silvera IF (1998b) NMR Study of ortho-para conversion at high pressure in hydrogen. *Physical Review Letters* 81: 4181–4183.

Reichlin R, Brister K, McMahan AK, et al. (1989) Evidence for the insulator-metal transition in xenon from optical, X-Ray, and band-structure studies to 170 GPa. *Physical Review Letters* 62: 669–672.

Resta R (2002) Why are insulators insulating and metals conducting?. *Journal of Physics: Condensed Matter* 14: R625–R656.

Reyes AP, Ahrens ET, Heffner RH, Hammel PC, and Thompson JD (1992) Cuprous oxide manometer for high-pressure magnetic resonance experiments. *Review of Scientific Instruments* 63: 3120–3122.

Romanowicz B, Li XD, and Durek J (1996) Anisotropy in the inner core: Could it be due to low-order convection? *Science* 274: 963–966.

Ross M, Boehler R, and Söderlind P (2005) Xenon melting curve to 80 GPa and 5p–d hybridization. *Physics Review Letters* 95: 257801.

Rossman GR (1988) Optical spectroscopy. In: Hawthorne FC (ed.) *Reviews in Mineralogy, Vol. 18: Spectroscopic Methods in Mineralogy and Geology*, pp. 207–254. Washington, DC: Mineralogical Society of America.

Rueff JP, Kao CC, Struzhkin VV, et al. (1999) Pressure-induced high-spin to low-spin transition in FeS evidenced by X-ray emission spectroscopy (vol 82, pg 3284, 1999). *Physical Review Letters* 83: 3343.

Ruocco G, Sette F, Bergmann U, et al. (1996) Equivalence of the sound velocity in water and ice at mesoscopic wavelengths. *Nature* 379: 521–523.

Sampoli M, Ruocco G, and Sette F (1997) Mixing of longitudinal and transverse dynamics in liquid water. *Physical Review Letters* 79: 1678–1681.

Sanloup C, Guyot F, Gillet P, and Fei Y (2002a) Physical properties of liquid Fe alloys at high pressure and their bearings on the nature of metallic planetary cores. *Journal of Geophysical Research* 107: doi:10.1029/2001JB000808.

Sanloup C, Mao HK, and Hemley RJ (2002b) Evidence for xenon silicates at high pressure and temperature. *Geophysical Research Letters* 29: doi:10.1029/2002GL014973.

Sanloup C, Mao HK, and Hemley RJ (2002c) High-pressure transformations in xenon hydrates. *Proceedings of the National Academy of Sciences* 99: 25–28.

Schwoerer-Bohning M, Klotz S, Besson JM, Burkel E, Braden M, and Pintschovius L (1996) The pressure dependence of the TA1[110] phonon frequencies in the ordered invar alloy Fe_3Pt at pressures up to 7 GPa. *Europhysics Letters* 33: 679–682.

Seto M, Yoda Y, Kikuta S, Zhang XW, and Ando M (1995) Observation of nuclear resonant scattering accompanied by phonon excitation using synchrotron radiation. *Physical Review Letters* 74: 3828–3831.

Shankland TJ, Peyronneau J, and Poirier JP (1993) Electrical conductivity of the Earth's lower mantle. *Nature* 366: 453–455.

Shen G, Fei Y, Halenius U, and Wang Y (1994) Optical absorption spectra of (Mg,Fe)SiO_3 silicate perovskites. *Physics and Chemistry of Minerals* 20: 478–482.

Shirley EL (1998) *Ab Initio* inclusion of electron-hole attraction: Application to X-ray absorption and resonant inelastic X-Ray. *Physical Review Letters* 80: 794–797.

Sienkiewicz A, Vileno B, Garaj S, Jaworski M, and Forro L (2005) Dielectric resonator-based resonant structure for sensitive ESR measurements at high hydrostatic pressures. *Journal of Magnetic Resonance* 177: 278–290.

Smith HG and Langer K (1982) Single crystal spectra of olivines in the range 40,000–50,000 cm^{-1} at pressures up to 200 kbar. *American Mineralogist* 67: 343–348.

Sinmyo R, Hirose K, O'Neill HSC, and Okunishi E (2006) Ferric iron in Al-bearing post-perovskite. *Geophysical Research Letters* 33: L12S13.

Sinogeikin S, Bass J, Prakapenka V, *et al.* (2006) Brillouin spectrometer interfaced with synchrotron radiation for simultaneous X-ray density and acoustic velocity measurement. *Review of Scientific Instruments* 77: 103905.

Song X and Helmberger DV (1998) Seismic evidence for an inner core transition zone. *Science* 282: 924–927.

Song XD and Richards PG (1996) Seismological evidence for differential rotation of the Earth's inner core. *Nature* 382: 221–224.

Spackman MA, Hill RJ, and Gibbs GV (1987) Exploration of structure and bonding in stishovite with Fourier and pseudoatom refinement methods using single-crystal and powder X-ray diffraction data. *Physics and Chemistry of Minerals* 14: 139–150.

Speziale S, Lee VE, Jeanloz R, Milner A, Pasternak MP, and Clark SM (2005) Iron spin transition in Earth's mantle. *Proceedings of the National Academy of Sciences* 102: 17918.

Stackhouse S, Brodholt JP, Dobson DP, and Price GD (2006) Electronic spin transitions and the seismic properties of ferrous ironbearing MgSiO3 post-perovskite. *Geophysical Research Letters* 33: L12S03.

Stebbins JF (1988) NMR spectroscopy and dynamic processes in mineralogy and geochemistry. In: Hawthorne FC (ed.) *Reviews in Mineralogy, Vol. 18: Spectroscopic Methods in Mineralogy and Geology*, pp. 405–429. Washington, DC: Mineralogical Society of America.

Steinle-Neumann G, Cohen RE, and Stixrude L (2004a) Magnetism in iron as a function of pressure. *Journal of Physics: Condensed Matter* 16: S1109–S1119.

Steinle-Neumann G, Stixrude L, and Cohen RE (2003) Physical properties of iron in the inner core. In: Dehant V, Creger K, Karato S-I, and Zatman S (eds.) *Geodynamics Series, Vol. 31: Earth's Core Dynamics, Structure, Rotation*, pp. 137–162. Washington, DC: American Geophysical Union.

Steinle-Neumann G, Stixrude L, and Cohen RE (2004b) Magnetism in dense hexagonal iron. *Proceedings of the National Academy of Sciences* 101: 33–36.

Steinle-Neumann G, Stixrude L, Cohen RE, and Gülseren O (2001) Elasticity of iron at the temperature of the Earth's inner core. *Nature* 413: 57–60.

Stern F (1963) Elementary theory of the optical properties of solids. In: Seitz F and Turnbull D (eds.) *Solid State Physics*, pp. 299–408. New York and London: Academic Press.

Stixrude L (2001) First principles theory of mantle and core phases. *Reviews in Mineralogy and Geochemistry* 42: 319–343.

Stixrude L and Cohen RE (1995) High-pressure elasticity of iron and anisotropy of Earth's inner core. *Science* 267: 1972–1975.

Struzhkin VV, Goncharov AF, Hemley RJ, *et al.* (2000) Coupled magnon–phonon excitations in Sr2CuCl2O2 at high pressure. *Physical Review B* 62: 3895–3899.

Struzhkin VV, Goncharov AF, and Syassen K (1993) Effect of pressure on magnetic excitations in CoO. *Materials Science and Engineering* A168: 107–110.

Struzhkin VV, Mao HK, Hu J, *et al.* (2001) Nuclear Inelastic X-Ray Scattering of FeO to 48 GPa. *Physical Review Letters* 87: 255501.

Struzhkin VV, Goncharov AF, and Syassen K (1993) Effect of pressure on magnetic excitations in CoO. *Materials Science and Engineering* A168: 107–110.

Struzhkin VV, Hemley RJ, and Mao HK (2004) New condensed matter probes for diamond anvil cell technology. *Journal of Physics: Condensed Matter* 16: 1–16.

Sturhahn W, Toellner TS, Alp EE, *et al.* (1995) Phonon density of states measured by inelastic nuclear resonant scattering. *Physical Review Letters* 74: 3832–3835.

Sturhahn W, Jackson JM, and Lin JF (2005) The spin state of iron in Earth's lower mantle minerals. *Geophysical Research Letters* 32: L12307.

Su W, Dziewonski AM, and Jeanloz R (1996) Planet within a planet: Rotation of the inner core of Earth. *Science* 274: 1883–1887.

Sur S and Cooney TS (1989) Electron paramagnetic resonance study of iron (III) and manganese (III) in the glassy and crystalline environments of synthetic fayalite and tephroite. *Physics and Chemistry of Minerals* 16: 693–696.

Surh MP, Louie SG, and Cohen ML (1992) Band gaps of diamond under anisotropic stress. *Physical Review B* 45: 8239–8247.

Syassen K and Sonneschein R (1982) Microoptical double beam system for reflectance and absorption measurements at high pressure. *Review of Scientific Instruments* 53: 644–650.

Takano M, Nasu S, Abe T, *et al.* (1991) Pressure-induced high-spin to low-spin transition in CaFeODN3. *Physical Review Letters* 67: 3267–3270.

Timofeev YA (1992) Detection of superconductivity in high-pressure diamond anvil cell by magnetic susceptibility technique. *Pribory/Tekhnik Eksperiment* 5: 186–189.

Timofeev YA, Struzhkin VV, Hemley RJ, Mao HK, and Gregoryanz EA (2002) Improved techniques for measurement of superconductivity in diamond anvil cells by magnetic susceptibility. *Review of Scientific Instruments* 73: 371–377.

Tissen VG and Ponyatovskii EG (1987) Behavior of Curie temperature of EuO at pressures up to 20 GPa. *JETP Letters* 46: 287–289 (1987).

Titchmarsh EC (1937) *Introduction to the Theory of Fourier Integrals*. Oxford: Clarendon Press.

Tromp J (2001) Inner-core anisotropy and rotation. *Annual Review of Earth and Planetary Sciences* 29: 47–69.

Tse JS, Klug DD, Jiang DT, *et al.* (2005) Compton scattering of elemental Si at high pressure. *Applied Physics Letters* 87: 191905.

Tsuchiya T, Tsuchiya J, Umemoto K, and Wentzcovitch RM (2004) Phase transition in MgSiO3 perovskite in the earth's lower mantle. *Earth and Planetary Science Letters* 224: 241–248.

Tsuchiya T, Wentzcovitch RM, da Silva CRS, and de Gironcoli S (2006) Spin transition in magnesiowüstite in earth's lower mantle. *Physical Review Letters* 96: 198501.

Tyburczy JA and Fisler DK (1995) Electrical properties of minerals and melts. In: Ahrens TJ (ed.) *Mineral Physics and Crystallography: A Handbook of Physical Constants*, vol. pp. 185–208. Washington, DC: American Geophysical Union.

Ulug AM, Conradi MS, and Norberg RE (1991) High pressure NMR: Hydrogen at low temperatures. In: Hochheimer HD and Etters RD (eds.) *Frontiers of High Pressure Research*, pp. 131–141. New York: Plenum.

Ulug AM, Conradi MS, and Norberg RE (1992) High Pressure NMR: Hydrogen at low temperatures. In: Hochheimer HD and Etters RD (eds.) *Frontiers of High Pressure Research*, pp. 131–141. New York: Plenum.

Vočadlo L, Alfè D, Gillan MJ, Wood IG, Brodholt JP, and Price GD (2003) Possible thermal and chemical stabilisation of body-centred-cubic iron in the Earth's core. *Nature* 424: 536–539.

Vos WL, Finger LW, Hemley RJ, Hu JZ, Mao HK, and Schouten JA (1992) Possible thermal and chemical stabilisation of body-centred-cubic iron in the Earth's core. *Nature* 358: 46–48.

Wang D, Mookherjee M, Xu Y, and Karato SI (2006) The effect of water on the electrical conductivity of olivine. *Nature* 443: 977–980.

Wang Y, Weidner DJ, and Meng Y (1998) Advances in equation of state measurements in SAM-85. In: Manghnani MH and Yagi T (eds.) *Properties of the Earth and Planetary Materials at High Pressure and Temperature*, pp. 365–372. Washington, DC: AGU.

Webb AW, Gubser DU, and Towle LC (1976) Cryostate for generating pressures to 100 kbar and temperatures to 0.03 K. *The Review of Scientific Instruments* 47: 59–62.

Weir S, Mitchell AC, and Nellis WJ (1996) Metallization of fluid molecular hydrogen at 140 GPa (1.4 Mbar). *Physical Review Letters* 76: 1860–1863.

Williams Q, Knittle E, Reichlin R, Martin S, and Jeanloz R (1990) Structural and electronic properties of Fe_2SiO_4-fayalite at ultra-high pressures: Amorphization and gap closure. *Journal of Geophysical Research* 95: 21549–21563.

Williams Q and Hemley RJ (2001) Hydrogen in the deep Earth. *Annual Review of Earth and Planetary Sciences* 29: 365–418.

Winkler B, Harris MJ, Eccleston RS, and Knorr K (1997) Crystal field transitions in $Co_2[Al_4Si_5]O_{18}$ cordierite and $CoAl2O_4$ spinel determined by neutron spectroscopy. *Physics and Chemistry of Minerals* 25: 79–82.

Wong J, Krisch M, Farber DL, et al. (2003) Phonon dispersions of fcc δ-plutonium–gallium by inelastic X-ray scattering. *Science* 301: 1078–1080.

Wu Z and Cohen RE (2005) Pressure-induced anomalous phase transitions and colossal enhancement of piezoelectricity in $PbTiO_3$. *Physical. Review Letters* 95: 037601.

Wyckoff RWG (1923) On the hypothesis of constant atomic radii. *Proceedings of the National Academy of Sciences* 9: 33–38.

Xu Y, McCammon CA, and Poe BT (1998) The effect of alumina on the electrical conductivity of silicate perovskite. *Science* 282: 922–924.

Xu J and Mao HK (2000) Moissanite: A window for high-pressure experiments. *Science* 290: 783–785.

Xu J, Mao HK, Hemley RJ, and Hines E (2004) Large volume high pressure cell with support moissanite anvils. *Review of Scientific Instruments* 75: 006404RSI.

Xu Y, Poe BT, Shankland TJ, and Rubie DC (1998) Electrical conductivity of olivine, wadsleyite, and ringwoodite under upper-mantle conditions. *Science* 280: 1415–1418.

Xu J, Yen J, Wang Y, and Huang E (1996) Ultrahigh pressure in gem anvil cells. *High Pressure Research* 15: 127–134.

Yagi T, Mao HK, and Bell PM (1979) Lattice parameters and specific volume for the perovskite phase of orthopyroxene composition, $(Mg,Fe)SiO_3$. *Carnegie Institution of Washington Yearbook* 78: 612–614.

Yan CS, Vohra YK, Mao HK, and Hemley RJ (2002) Very high growth rate chemical vapor deposition of single-crystal diamond. *Proceedings of the National Academy of Sciences* 99: 12523–12525.

Yarger JL, Smith KH, Nieman RA, et al. (1995) Al Coordination changes in high-pressure aluminosilicate liquids. *Science* 270: 1964–1967.

Yoo CS, Maddox B, Klepeis JHP, et al. (2005) First-order isostructural Mott transition in highly compressed MnO. *Eos Transactions of the American Geophysical Union* 94: 115502.

Zha CS, Mao HK, Hemley RJ, and Duffy TS (1997) Elasticity measurement and equation of state of MgO to 60 GPa. *EOS Transactions of the American Geophysical Union* 78: F752.

Zha CS, Hemley RJ, Gramsch SA, Mao HK, and Bassett WA (2007) Optical study of H_2O ice to 120 GPa: dielectric function, molecular polarizability, and equation of state. *Journal of Chemical Physics* 126: 074506.

Zhang F and Oganov AR (2006) Valence and spin states of iron impurities in mantle-forming silicates. *Earth and Planetary Science Letters* 249: 436–443.

Zhang L (1997) *Crystallography at High Pressure Using Synchrotron Radiation: The Next Steps*. Grenoble, France: ESRF.

Zhang L and Chopelas A (1994) Sound velocity of Al_2O_3 to 616 kbar. *Physics of the Earth and Planetary Interiors* 87: 77–83.

Zhang L and Hafner SS (1992) Compressibility of grunerite. *American Mineralogist* 77: 462–473.

Zhou T, Syassen K, Cardona M, Karpinski J, and Kaldis E (1996) Electronic Raman scattering in $YBa_2Cu_4O_8$ at high pressure. *Solid State Communications* 99: 669–673.

Zou G, Mao HK, Bell PM, and Virgo D (1980) High pressure experiments on the iron oxide wüstite ($Fe_{1-x}O$). *Carnegie Institution of Washington Yearbook* 79: 374–376.

12 Theory and Practice – Methods for the Study of High P/T Deformation and Rheology

D. J. Weidner and L. Li, State University of New York (SUNY), Stony Brook, NY, USA

12.1	Introduction	339
12.1.1	Importance of High-Pressure Rheology Measurements	340
12.1.1.1	Lateral variation of asthenosphere viscosity	341
12.1.1.2	Earthquakes	342
12.2	High-Pressure Tools	342
12.2.1	Stress Measurement	342
12.2.2	Strain-Rate Measurement	345
12.2.3	New High-Pressure Deformation Devices	345
12.2.4	The Instruments of Flow	347
12.2.5	Dislocations, Slip Systems, and Texture	348
12.3	New Insights	350
12.3.1	Polycrystalline Materials	350
12.3.2	Activation Volume of Olivine	353
12.3.3	High-Pressure Phases	354
12.4	Conclusion	355
References		355

12.1 Introduction

This section is devoted to recent studies that work to define rheological properties of Earth materials at high pressure and high temperature. The frontier has been defined by the bringing together of exciting new tools for studying these properties. Multianvil high-pressure apparatus, which are capable of reaching into the lower mantle with hydrostatic pressure, are being altered to provide a deformation environment. Samples recovered from these systems deliver information about the deformation process. Synchrotron X-rays have opened new vistas by their ability to see into the pressure chamber and yield stress and strain of the sample *in situ*. Around these metrics, new tools are being developed that can control sample environment. The next phase of these studies will probably see us creating conditions of the deep lower mantle with well-characterized sample environments and precise measures of the quantitative flow process. Along the way, we will gain insights into aggregates and how stress is distributed among the grains. Information on texture, elastic anisotropy, and stress as a function of pressure, temperature, and environment will be derivable from this information.

The dynamic nature of the Earth, including phenomena from earthquakes and volcanic eruptions, to plate tectonics and mantle convection are responding to the plastic nature of the materials of the Earth's mantle. An improved understanding of these dynamic features of the Earth will follow insights of the flow properties of the rocks that make up the Earth. This fundamental principle has defined a field within the Earth sciences that we call 'rheological studies'.

Rheological studies are among the most challenging of the experimental fields that strive to inform us about the Earth. Plastic properties are not state variables. That is, this property does not depend solely on the classical thermodynamic state of the material. In addition to pressure, temperature, and chemical potentials, the plastic properties depend on the history of the sample. History produces the sample with a dislocation density, a grain size, and a texture. History adds the dimension of time. In the Earth, timescales span millions if not billions of years, with strain rates slower than $10^{-15}\,\text{s}^{-1}$, while in the lab, experiments that last for a week are considered long and strain rates slower than $10^{-7}\,\text{s}^{-1}$ imperceptibly slow. If we wish to work in the laboratory, at geological strain rates, then a 1-m-long sample will change its length by $30\,\mu\text{m}$ in 100 years.

Plastic flow belongs in the realm of irreversible thermodynamics, where the concept of steady state replaces that of equilibrium in reversible thermodynamics. Steady state is then governed by an extreme in the rate of entropy production rather than an extreme in the entropy. So too in rheological studies, steady state becomes a goal of experimental probing. Steady state on a macroscopic scale is indicated by the time independence of both stress and strain rate. On a microscopic scale, the defect structure and the grain size must become time independent. Here at least, the role of history is minimized in that history is no longer altering the properties of the sample, and, hopefully, the history prior to the experiment has been removed by the lab-induced flow process. Yet, while steady state is a necessary condition for a well-defined experiment, it is not a sufficient condition. For example, a difference in grain size between two samples can lead to very different flow characteristics. Furthermore, the confirmation of steady state is virtually impossible to do. In this manner, again, steady state is quite similar to the state of equilibrium which is also difficult to verify.

Indeed, natural processes are not restricted to steady-state flow. Phenomena including earthquakes, formation of ductile shear zones, and even glacial rebound occur outside of the steady-state regime. To provide the necessary information to describe these non-steady-state process, an even more complex description is required.

Another complicating feature of rheological processes is their dependence on small amounts of some atomic species. The flow process generally involves movement of atoms. This movement is often greatly influenced by point defects in the solid, indeed by participants that may have a very small concentration in the material. In geological systems, water has extreme effects on the rheological properties of some minerals (quartz and olivine are good examples). Thus, a well-characterized rheological experiment will demonstrate steady state and will characterize the sample and sample environment in terms of grain size, texture, dislocation type, and density, as well as define the chemical potentials of all active species.

With all of these support characterizations in place, the quantitative flow law is then defined as the relation between the stress, σ, and strain rate, $\dot{\varepsilon}$, where stress is a measure of the deviatoric stress, often a differential stress, and strain rate indicates the rate of deformation. A typical high-temperature semi-empirical steady-state relation between stress and strain rate is given by

$$\dot{\varepsilon} = A(\sigma/\mu)^n (b/d)^m \exp\left[-\left(\frac{E^* + PV^*}{RT}\right)\right] \quad [1]$$

where b and μ are the Burger's vector for the active dislocation and the shear modulus of the material, both intrinsic properties of the material. d is the grain size, R is the rare gas constant, P is pressure, and T is temperature. The other parameters A, n, m, E^*, and V^* are flow parameters that are specific to the particular flow mechanism and generally need to be determined empirically. The variables are arranged so that the values within parentheses are dimensionless. Flow by diffusion is generally associated with $n = 1$ and m in the range of 2–5, while dislocation-controlled flow obtains with $m = 0$ (no grain size dependence) and n in the range of 2–5. In order to define all of the parameters in this relation, the experimental protocol needs to be defined to measure strain rate as a function of stress (to define n), grain size (to define m), temperature (to define E^*), and pressure (to define V^*). These measurements should be made on a sample that has achieved steady-state flow (unless one wishes to characterize transient creep) in a well-defined environment.

12.1.1 Importance of High-Pressure Rheology Measurements

Laboratory data do not give us a simple estimate of viscosity in the Earth. First of all, viscosity, the ratio of stress to strain rate, depends on the level of stress as seen in eqn [1]. Thus, to estimate viscosity, we must know the amount of stress. Second, extrapolation over several orders of magnitude of strain rate can lead to large errors. Rheological properties of the Earth's mantle can be inferred from analyses of geodynamic data such as postglacial crustal movement and gravity anomalies (Mitrovica and Forte, 1997; Peltier, 1998). The measurements are enabled by both a large length scale and a long timescale. However, even these data are limited in their ability to define Earth rheology. The depth resolution of postglacial rebound data is limited to $\sim$1000 km (Mitrovica and Peltier, 1991). Furthermore, amount of strain required for postglacial rebound is much smaller than strain encountered in convection. In so much as the rheology maintains a memory of previous deformation, small strain rheology may be different than large strain rheology (Karato, 1998).

Gravity data (e.g., the geoid) have better sensitivity to radial variation in viscosity in the deep mantle (Hager, 1984), but they suffer from nonuniqueness (King, 1995; Thoraval and Richards, 1997). Laboratory studies of rheology can provide information about mechanism, stress and strain dependence, and other variables such as chemical or phase dependence. Therefore, experimental studies on mineral rheology remain a vital component in inferring mantle rheology.

12.1.1.1 Lateral variation of asthenosphere viscosity

As an example, here we illustrate how the activation volume helps constrain the lateral variations in asthenosphere viscosity. We define the asthenosphere viscosity as the lowest viscosity in the upper mantle in a specific region. Generally, the temperature profile is the dominant variable to define regional viscosity. We approximate the mantle temperature profile as adiabatic with a thermal boundary layer at the top. The details of the thermal boundary layer depends on the age of the crust in the region. Older regions have deeper cooling. **Figure 1** illustrates three temperature profiles that could be representative of 20-million-year-old oceans, 80-million-year-old oceans, and very old continental cratons. The geotherm reaches the adiabat at different depths in these models owing to the time that conductive cooling is operating from the surface. The lowest viscosity will not occur at least until the depth

that the temperature is on the adiabat. Since pressure increases with depth, the variation in magnitude of the lowest viscosity from region to region will depend on the activation volume of the flow process. Estimates of olivine's activation volume for power-law creep range from 0 to $30 \, cm^3 \, mol^{-1}$. **Figure 2** illustrates the depth dependence of viscosity as a function of activation volume for a fixed stress along these model geotherms. An activation volume of $20 \, cm^3 \, mol^{-1}$ would render the minimum viscosity under the Canadian Shields 6–7 orders of magnitude greater than the minimum viscosity under 20-million-year-old oceans for the same shear stress, while a 0 activation volume would have no difference in minimum viscosity.

These conclusions are based on maintaining the same dominant slip system as pressure is increased. Karato and Wu (1993) argue that this large pressure-dependent process will force a change in the deformation mechanism to diffusion by the middle of the upper mantle. While diffusion flow is much slower than dislocation flow at atmospheric pressure, the presumed low activation volume for diffusion suggests that pressure will not significantly slow down diffusion flow. Nonetheless, we see the manner that laboratory studies enters the Earth dynamic issues.

The main experimental challenge is to measure the stress and strain rate at mantle pressure and temperature conditions of a specimen that has reached steady-state flow. Low-pressure tools have been applied to precisely apply a force or a strain rate

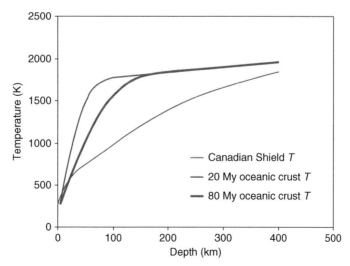

Figure 1 Radial temperature profiles for different geologic settings. Ocean profiles are calculated assuming a cooling lid on a mantle adiabatic temperature gradient. The Canadian Shield is calculated assuming a conduction gradient with localization of radiogenic elements in the crust. From Li L, Raterron P, Weidner D, and Chen J (2003) Olivine flow mechanisms at 8 GPa. *Physics of the Earth and Planetary Interiors* 138: 113–129.

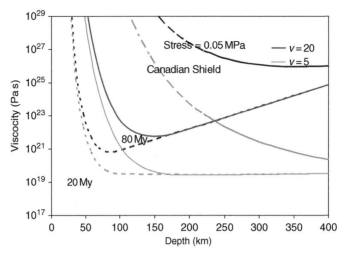

Figure 2 Viscosity vs depth for different geotherms and different activation volumes. The temperatures were calculated from standard cooling profiles for the different regions. From Li L (2003) *Rheology of Olivine at High Temperature and High Pressure*. PhD Thesis, Stony Brook University.

to define the flow of olivine (Chopra, 1997; Durham and Goetze, 1977; Green and Borch, 1987; Hirth and Kohlstedt, 1995; Karato *et al.*, 1986; Kohlstedt *et al.*, 1984; Mei and Kohlstedt, 2000; Paterson, 1990), although at pressure lower than 0.3 GPa (compared to 13 GPa at the olivine stability limit). Torsion apparatus has been applied in studying the high shear strain deformation with pressure up to 0.3 GPa (Bystricky *et al.*, 2000). The force–displacement tools have been pushed to higher pressure using the Griggs-type solid-medium apparatus (up to 2.5 GPa) (Green and Borch, 1987; Karato and Jung, 2003; Ross *et al.*, 1979). Molten salt cell assemblies help to maintain a quantitative measure of the stress at pressures up to 2.2 GPa (Borch and Green, 1989; Green and Borch, 1989). Development of the large volume multianvil press (e.g., (Bussod *et al.*, 1993)) allowed the synthesis of deformed samples at upper-mantle pressure for microstructure examination (Ando *et al.*, 1997; Cordier *et al.*, 2002; Karato, 1989; Karato *et al.*, 1993; Rubie *et al.*, 1993). Results from these higher-pressure experiments have generally been accepted as qualitative.

12.1.1.2 Earthquakes

A second example of the importance of laboratory data concerns the origin of deep-focus earthquakes. The standard friction-mitigated processes should not be operative at depths greater than about 20 km, yet earthquakes persist to depths of 700 km. The process that is responsible for the space–time localization of the stress drop is still unclear. Several alternative

mechanisms for the stress release instability have been suggested over the years, including a thermally induced plastic instability (Hobbs and Ord, 1988; Ogawa, 1987), instabilities accompanying recrystallization (Post, 1977), dehydration embrittlement (Jiao *et al.*, 2000), and instabilities associated with polymorphic phase transformations (Green and Borch, 1989; Kirby *et al.*, 1991; see also reviews by Green and Houston (1995) and Kirby *et al.* (1996)). Again, the main reason uncertainty still exists is the poor knowledge of material properties at high pressures. The rheological properties of deep-Earth materials have been missing so far (Hobbs and Ord, 1988; Karato *et al.*, 2001) and thus impede modeling of the processes. Similarly, greater understanding of possible shearing instabilities and kinetics associated with the phase transformations of olivine to wadsleyite and ringwoodite (Burnley *et al.*, 1991; Green *et al.*, 1990) and the possibility of faulting associated with dehydration of the dense hydrous magnesium silicates (Jung and Green, 2004) need well-controlled deformation experiments under deep-mantle conditions.

12.2 High-Pressure Tools

12.2.1 Stress Measurement

Measurement of deviatoric stress is key to quantifying the rheological properties of a material. Stress provides information about strength relative to the laboratory timescale. Stress, measured in conjunction with ductile strain, can yield flow laws within the context of the state of the sample. The crucial ingredient to facilitate such studies is the development of a stress meter.

Experiments operating below 5 GPa apply the deforming force directly to the sample and design various strategies to measure the magnitude of this deforming force. Stress is the ratio of force to sample area. Typically, a load cell placed in series with the deforming piston defines this force along the deforming column. For systems operating above 1 GPa, confining pressure, combinations of friction between the force gauge and the sample, and forces that support the piston, such as in the multianvil system, continually degrade the ability to relate a measured force to sample stress. These problems become progressively more acute as the pressure of the sample is higher.

A revolutionary technique for stress and strain measurements under high-pressure (and -temperature) conditions *in situ* has now become feasible using high-energy X-rays generated by a synchrotron radiation facility. These new tools are just now being explored and their limitations defined. The exceptional quality of these tools rests in the fact that they are directly monitoring the sample. Stresses are measured in the sample. Different positions in the sample can be isolated to test for stress uniformity. Measurements of stress and strain can be time-resolved with a precision of about 1 min. Strain is also obtained by images of the sample. Distribution of strain with position and time can be defined – all of this done with a current accuracy of a few tens of megapascals in differential stress and 10^{-4} in strain.

A uniaxial stress will introduce elastic strains in the sample. The strain parallel to the axis of compression will generally be larger than the strain perpendicular. The material's elastic moduli, through Hooke's law, quantitatively define the relationship between these strains and the imposed stresses. X-rays sample the distance between lattice planes (called d-spacing) whose normals are parallel to the diffraction vector, which is the bisector between the incident X-ray and the detector. The lattice spacings reflect elastic strain and are insensitive to plastic strain. **Figure 3** illustrates a Debye ring that would be observed from a powder sample in a stress field using a monochromatic X-ray beam and an area detector. The lattice spacing is related to the distance from the center of the image to the Debye ring through Bragg's law. For the angle Ψ of 0, the lattice spacing reflects that measured parallel to the x_1-

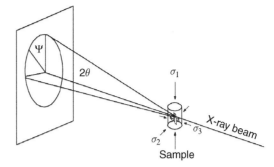

Figure 3   X-ray diffraction geometry. The stress field of the sample is illustrated by σ. In the DDIA configuration, $\sigma_2 = \sigma_3$.

direction. For Ψ of 90°, the X-rays are sampling grains aligned to parallel to the x_2-direction. The strains deduced from these measurements, compared with ambient conditions, are related to the stresses by

$$\varepsilon_{ij} = \sum_{k=1}^{3} \sum_{l=1}^{3} \mathbf{s}_{ijkl}\sigma_{kl} \qquad [2]$$

where $\mathbf{s}$ is the elastic compliance tensor.

A strategy for measuring stress comes from mapping out the d-spacing as a function of Ψ. The values of the elastic moduli are still required, but the details of anisotropy are less critical to yielding precise measurements. With a precision of 10^{-4} for the lattice spacing and a typical elastic modulus of 200 GPa, it should be possible to resolve differential stresses of ~20 MPa. Singh has reduced the elasticity equations to a closed form solution for several crystal symmetries and a concise summary is presented in Singh *et al.* (1998).

In the actual case, more than one set of diffraction planes will produce Debye rings similar to the one illustrated here. Each will be circular in shape for a hydrostatic stress, but distorted by nonhydrostatic stress. The amount of distortion may be different if the elastic compliance tensor is not isotropic. To evaluate this in detail, we need to be careful to account for the crystal orientation when defining the compliance tensor. Thus, in a nonhydrostatic stress field, crystallographic planes with different orientations change their spacing by different amounts. For cubic crystals, the relationship between the Ψ dependence of strain and the differential stress, $\Delta\sigma$, is given by (Singh, 1993)

$$\varepsilon_{hkl}(\Psi) = \Delta\sigma/3 \left\{1 - 3\cos^2(\Psi)\right\} \left\{(S_{11} - S_{12})\left(1 - \Gamma(hkl)\right)\right.$$
$$\left. + S_{44}\,\Gamma(hkl)/2\right\} \qquad [3]$$

Each diffraction peak corresponds to a set of lattice planes defined by the Miller index, _hkl_. In an anisotropic crystal, the elastic moduli will depend on the particular set of planes that are considered. The function, $\Gamma(hkl)$, varies from 0 to 1, and reflects the elastic moduli for the orientation of the particular diffraction peak. Thus, for a particular diffraction plane, _hkl_, the right-hand bracket contains a constant value. Then ε_{hkl} varies in a fixed manner with the angle, Ψ. ε_{hkl} is the deviatoric strain and is defined as the strain relative to the hydrostatic state. Measuring the lattice spacing for a particular _hkl_ at $\Psi = 0°$ and $\Psi = 90°$ is enough to define the differential stress for. This relation for cubic symmetry is generalized by Singh _et al._ (1998) for all crystal symmetries. In the case where the stress field is not uniaxial, but more complicated, this analysis yields the stress field in the plane perpendicular to the X-ray beam, but does not recover the entire stress tensor.

This methodology was first applied in a diamond-anvil cell by Kinsland and Bassett (1977) by passing the X-ray beam through a Be gasket, perpendicular to the axis of the diamonds. The multianvil system has not been accessible by this technique since the anvils themselves cast a shadow and limit the range of Ψ where the diffracted signal can be observed. However, both cubic BN and sintered diamond anvils are both hard and X-ray transparent. Thus, using these anvils, it is now possible to observe diffraction spectra in any plane relative to the incident beam in a multianvil device.

This tool can be used with monochromatic X-rays for angle-dispersive measurements and for energy-dispersive methods with white X-rays, both with specific advantageous and disadvantages. With white X-rays, spectra need to be collected for different values of Ψ through collimators that fix the 2θ value. A conical slit system (Li _et al._, 2004a) can be used for this purpose, as illustrated in **Figure 4**. The slit itself is created by two concentric cones, whose angle is the desired diffraction angle for white, energy-dispersive diffraction. Solid-state detectors with energy discrimination are placed at specific values of Ψ behind the slit system. Each detector is calibrated independently. Lattice spacings are used as illustrated above to define the differential stress. The advantage of this system is that the diffracted signal is collimated allowing diffraction from the sample to be isolated from that of the pressure medium. Monochromatic X-rays are more versatile in defining the orientation of the deformation ellipse and are quite useful when the orientation of the stress field is not well defined by the experimental geometry.

Each diffraction peak represents a different sub-population of grains within the sample. Generally, grains that are oriented so as to contribute to one diffraction line are not oriented to contribute to another. If all of the diffraction lines define the same values of stress, then the stress field is fairly homogeneous within the sample with no distinctions on the basis of the orientation of the crystallites.

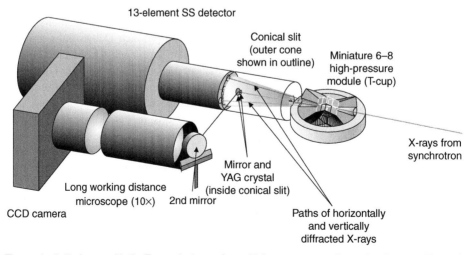

Figure 4   The conical slit shown with the T-cup device and a multielement energy-dispersive detector. The diffracted X-ray from the sample is collimated by a conical slit system which allows diffraction at a fixed 2θ angle. The multiple detector is placed behind the slit system to capture the diffracted beam for several values of ψ.

However, it is quite common for the different peaks to provide different values of the differential stress. This indicates that these different subpopulations of grains experience different stresses. The average stress in the sample, that is the average of force per unit area, will simply be some average of the subpopulation stresses. Later, we will discuss the origin of this stress heterogeneity.

12.2.2 Strain-Rate Measurement

Synchrotron X-rays can also be used to measure sample length. Direct images of the sample can be obtained using an incident beam whose dimensions are larger than the sample. Platinum or gold foils above and below the sample are easily viewed. The X-ray image is obtained by projecting the X-ray onto a fluorescent screen that is viewed with a charge-coupled device (CCD) camera though a magnifying system. A typical image is shown in **Figure 5**. The horizontal lines in this image are gold disks placed at the ends of the sample. The black edges are the shadows of the opaque tungsten carbide anvils. Cubic boron nitride anvils are transparent and can be used in order to view the entire sample since they do not cast such a shadow. **Figure 4** illustrates the position of the microscope for these measurements. The fluorescent screen is located on-axis inside the conical slit system. We determine that, by comparing

Figure 5 Shadowgraph image of a sample at high pressure and temperature. The black boundary is defined by tungsten carbide anvils; the dark lines are from metal foils within the sample.

two images, strains of 10^{-4} can be measured. For images taken 100 s apart, this allows resolution of strain rates of $10^{-6}\,s^{-1}$.

With these tools of stress and strain rate, quantitative flow laws defined by relations such as eqn [1] can be defined. Inspired by the new insights into the high-pressure apparatus, development was then made in the high-pressure instrumentation that could provide a steady-state flow for relatively large strains in samples at high pressure and high temperature.

12.2.3 New High-Pressure Deformation Devices

The use of X-rays to measure both stress and strain directly in the sample, without a proxy such as a force gauge, frees the conceptual design of the deformation equipment. When using a force gauge, one must measure both the force and displacement of a piston and that piston must only push on the sample. Friction on the piston, deformation of the piston, elastic and thermal responses of the piston all must be quantified in order to define the force/displacement of the sample. The piston must push on the sample and only on the sample. With stress and strain of the sample measured directly, these issues are no longer important. The entire anvil can serve as the deforming unit. Indeed, two new systems have emerged that take advantage of the large-volume high-pressure apparatus and use the loading anvils as the instrument to create a deviatoric stress.

The first of these devices is called the DDIA (deformation DIA; Durham *et al.*, 2002; Wang *et al.*, 2003). This system is derived from the DIA apparatus which has been extensively used on synchrotron X-ray sources (Shimomura *et al.*, 1985). This apparatus forces six anvils to synchronously advance on a cubic pressure medium (**Figure 6**). The DDIA modification is to add hydraulic jacks to push on the upper and lower anvils. This allows the application of a nonhydrostatic stress to the entire cubic cell assembly, while at the same time, the main ram can be adjusted so as to maintain a constant pressure. A typical assembly is illustrated in **Figure 7**. The cylindrical geometry of the sample assembly includes hard pistons above and below the sample chamber that transmit the stress produced by the loading anvil to the sample. The independent upper and lower anvils can be driven to produce a constant strain rate in the sample for very large values of strain (in excess of 50%) as the side anvils are withdrawn

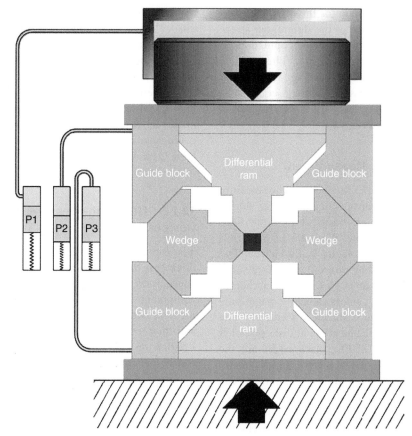

Figure 6   Illustration of DDIA. Four of the six anvils are illustrated (the front and back are not illustrated). P1 is a hydraulic pump that pushes the main jack in the directions of the arrows. This jack drives all six anvils into the sample. Pumps P2 and P3 drive the upper and lower anvils into (out of) the sample independently from the main jack. Figure created by Sabrina Fletcher of LLNL.

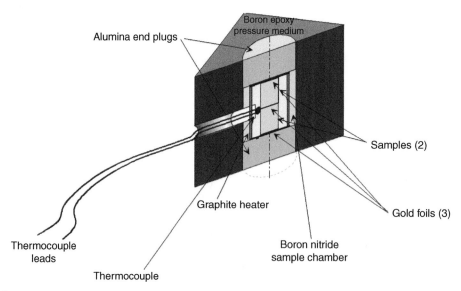

Figure 7   Typical cell assembly. The alumina end plugs serve as deformation anvils at high pressure and temperature.

to maintain a constant volume. The side anvils need to be X-ray transparent in order to measure X-ray diffraction in the horizontal plane. Cubic boron nitride and sintered diamond have both been used for this purpose. The final system allows most samples to achieve a steady-state condition in which stress and strain rate can be measured *in situ*.

The second system that has been recently developed is called the rotational Drickamer cell (Yamazaki and Karato, 2001). This system is patterned after the Drickamer cell (Perez-Albuerne *et al.*, 1964) that has been used on synchrotron sources. This cell is pressurized by two opposed anvils with a containment ring (**Figure 8**). The deformation is obtained by rotating one of the two opposed anvils. This rotation must be done while the anvils are being forced into the sample. A series of hard pistons and furnace elements transmit the rotation from the anvil into the sample (**Figure 8**). The rotation produces a shear strain and stress that varies from the center to the outer edge of the disk-shaped sample. Strain markers that can be imaged with the synchrotron X-ray source, used in conjunction with diffraction-derived stress, define the pertinent variables for these experiments. The most significant of rotational strain systems is their ability to induce very large amounts of strain in the sample. Strains of a few hundred percent are possible. These systems are well suited for evaluating texture development during deformation at high levels of deformation.

12.2.4   The Instruments of Flow

Laboratory determinations of flow laws in the laboratory obtain relevance to flow in the Earth by connecting the microscopic process that enables flow in both arenas. Several types of mechanisms that allow plastic flow are all operating when a solid is experiencing a deviatoric stress. However, the most successful mechanism will dominate the flow. As categorized in Ashby (1972) and presented for many materials in Frost and Ashby (1982), flow regimes are mapped out for materials including minerals with respect to stress and temperature. These maps, much like phase diagrams, indicate the rate-controlling flow process for the given environment. **Figure 9** illustrates a deformation map for MgO. Mechanisms such as dislocation glide-controlled plasticity, power-law creep, and diffusion creep govern different regions. Each regime is enumerated by lines representing strain rates ranging from $1\,s^{-1}$ to $10^{-10}\,s^{-1}$. These lines indicate the strain rate for the particular temperature–stress state. One also sees the manner for the extrapolation of experimental high-strain-rate data (10^{-6}) to the Earth conditions (10^{-15}) as long as the deforming process remains the same. At high stress the flow is controlled by glide where the periodic atomic lattice forces resist dislocation motion. This is often referred to as the Peierls stress region. Here, stress is high enough to overcome obstacles and the activation energy includes a term involving the deviatoric stress. As stress decreases,

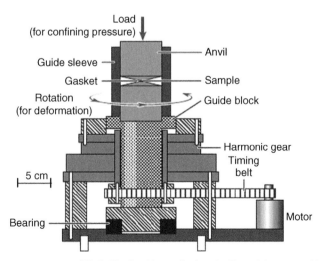

Figure 8   Rotational Drickamer apparatus (RDA). The load is applied vertically and the top and bottom anvils rotate relative to each other by virtue of the motor and the bearing assembly. The sample is sandwiched between two opposing anvils. Reused from Yamazaki D and Karato S (2001) High-pressure rotational deformation apparatus to 15 GPa. *Review of Scientific Instruments* 72: 4207.

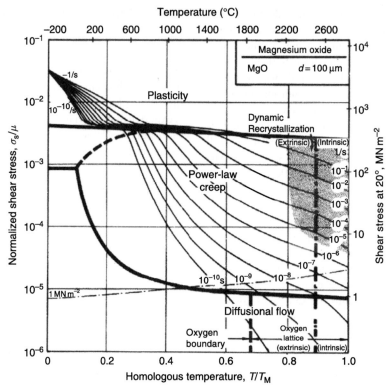

Figure 9 Deformation map of MgO. Dominant deformation mechanisms define different fields in this stress-vs-temperature space. The region labeled 'plasticity' is generally considered to be a region where dislocation glide is governed by the Peierls stress. Lines are drawn to represent fixed strain rates. Reproduced from Frost HJ and Ashby MF (1982) *Deformation-Mechanism Maps*, 166pp. Oxford: Pergamon, with permission from Elsevier.

the obstacles begin to dominate, creating a region where the flow stress is relatively insensitive to either temperature or strain rate. Power-law dislocation creep dominates where T is high enough for recovery processes to overcome the obstacles. At low stress and high temperature, diffusion comes into play. In the diffusion creep region, plasticity is accomplished by movement of atoms, either through the grain or along grain boundaries.

Another mechanism, dynamic recrystallization, appears at the high-temperature end of the map and embedded in the dislocation creep regime. Here flow is controlled by the growth of stress-free grains over stressed grains. Dynamic recrystallization may occur throughout the deformation map, but is not the dominant process in most of this region. It can assist other mechanisms by removing the high-stressed grains or the dislocation-riddled grains. It has been long thought as active in the Earth, based on the observation from the natural rocks. Dynamic recrystallization is unique in that it may alter the grain size thus moving a stress–T point from a field dominated

by power-law creep to one dominated by diffusion. de Bresser *et al.* (1998, 2001) postulate that dynamic recrystallization will evolve the grain size, for a given stress–temperature, so that the boundary between diffusion creep and power-law creep lies on the stress–temperature point.

12.2.5 Dislocations, Slip Systems, and Texture

The flow mechanism is defined by the behavior of the defects. It becomes expressed in the stress–strain-rate relationship. A typical microscopic equation defining the steady-state flow process is: flow rate = concentration of carrier × strength × velocity (Poirier, 1985). Dislocations, defined by their slip vectors, are controlled by the atomic-scale forces. The manifestation of dislocation structures under the transmission electron microscope (TEM) reveals effects of the history of the mobile lattice defects. It further reflects effects of the deforming conditions such as the pressure and temperature. The similarity between the dislocations of natural rocks and those of the experimentally deformed minerals provide sound arguments for the legitimacy of studying mantle rheology in the laboratory.

The study of dislocations of minerals using the TEM for samples that have been deformed at a known condition provides insights into the rheological behavior of the material (Cordier, 2002). Attention on mantle minerals recovered from high-pressure experiments have been a recent research focus including olivine (Couvy *et al.*, 2004a; Li *et al.*, 2006b), wadsleyite (Thurel *et al.*, 2003), garnet (Cordier *et al.*, 1996; Li *et al.*, 2006a; Voegele *et al.*, 1998, 2000, 1999), stishovite (Cordier and Sharp, 1998), quartz (Cordier and Doukhan, 1995), perovskite (Cordier *et al.*, 2004). The slip systems of these minerals are characterized from the quenched samples which have been deformed at mantle pressure and temperature. These studies have enabled different views of the slip systems of minerals, for example, the discovery of the change in dominant high-temperature slip systems of olivine with pressure (Couvy *et al.*, 2004a; Li *et al.*, 2006b)

Even though TEM has been widely used in analyzing the dislocation structure of the minerals, the examination of high-pressure phases is often challenging because some high-pressure phases are sensitive to electron beam damage. Recently, a new approach called large-angle convergent beam electron diffraction (LACBED) has been employed to overcome this problem (Cordier, 2002). LACBED is a defocused diffraction method; it is friendly for beam-sensitive materials and more reliable for providing information on Burgers vectors than conventional TEM. LACBED has been applied in analyzing defects in many of the rock-forming minerals (Cordier, 2002).

While many insights relevant to the Earth come from the laboratory studies, at conditions far from those in nature, confirmation and predictive quantitative results await a theoretical basis. Theoretical approaches based on quantum mechanical models are being pioneered to simulate the dislocations in the minerals (Cordier *et al.*, 2005; Carrez *et al.*, 2006; Durinck *et al.*, 2005a, 2005b). One application is to calculate the energy of a crystal after a generalized stacking fault has been introduced by a rigid shift of a layer of atoms in the crystal along a shear plane. This approach provides insights into the effect of pressure on the plastic strain anisotropy which is difficult to measure in the laboratory.

Dislocation flow can slowly cause the individual grains to rotate with the result that the polycrystalline sample attains a preferred orientation (Wenk, 2002). When seismic waves travel through the regions of the Earth with preferred orientations, velocity anisotropies are often observed (Silver, 1996; Tommasi, 1998) since minerals, such as olivine and pyroxene, are elastically anisotropic crystals. The relation between the fastest direction of P wave and the flow direction has gradually been unveiled as our understanding of the slip systems of Earth minerals progresses (Mainprice and Humbert, 1994). Indeed, the pressure-induced transition of slip systems in olivine noted above has been suggested as a cause of depth-dependent seismic anisotropy (Mainprice *et al.*, 2005).

The preferred orientations (also called texture) of most mantle minerals have been investigated in the laboratory (Bascou *et al.*, 2002; Couvy *et al.*, 2004a, 2004b; Merkel *et al.*, 2003, 2004, 2002; Schafer *et al.*, 1992; Wenk *et al.*, 2005, 2004). Furthermore, a great deal of information from the texture obtained from samples which have experienced deformation under high *P–T* have been used to infer the rheological properties. A common method for analyzing the texture in a recovered sample is to collect electron backscatter diffraction (EBSD) pattern for a sample with low dislocation density (Wright and Adams, 1992) using scanning electron microscope (SEM).

In situ measurements using synchrotron X-ray and neutron sources have great potential for revealing time-dependent development of texture. Both neutron and X-ray diffraction measure a bulk sample rather than surfaces and the measurement is carried out at high *P–T* and controlled strain rate. Synchrotron X-ray diffraction yields the quickest time resolution and also has the smallest scattering volume. Neutron diffraction collects data for more orientations and thus has more access to reciprocal space. For both X-ray and neutron diffraction, a density distribution of crystal orientations relative to the sample coordinates is then mapped out for samples which have been deformed under high pressures.

In parallel with the experimental work, theoretical approaches are also popular in modeling the texture in polycrystals formed during plastic deformation processes (Dawson and Wenk, 2000; Madi *et al.*, 2005; Mainprice *et al.*, 2005; Tommasi *et al.*, 2000; Wenk *et al.*, 1991, 2006). Models have been widely used include viscoplastic self-consistent (VPSC) models (Clausen *et al.*, 1998), finite element model (Dawson, 2002; Dawson *et al.*, 2001), the Taylor model (Taylor, 1938). All models are based on the knowledge of the slip systems and flow equation of the crystals, as well as the link between each individual crystal with the aggregate polycrystal.

12.3 New Insights

12.3.1 Polycrystalline Materials

Rocks are invariably polycrystalline and normally polyphase. Understanding the rheological properties of the constituents does not assure that one can predict the properties of the aggregate. Tullis (2002) describes the complexity of the properties of crustal rocks and features attributable to the polycrystalline character. In general, the reported experimental data at crustal conditions on two-phase rocks with high-strength contrasts support the model in which the strength of the mixture falls between that of the end-member phases (Ji *et al.*, 2004, 2003, 2001). Such models generally assume that the rheological properties of each phase in the mixture are the same as when it is a single phase deformed at the same condition (Takeda, 1998; Tullis *et al.*, 1991). Bruhn *et al.* (1999) reported a weaker mixture than the two-end members (calcite and anhydrite) which have similar strength, in which interphase boundary diffusion was reported as an import agent. Ford and Wheeler (2004) also demonstrated a theoretical model on interface diffusion creep in which the behavior of a two-phase composite is completely different from that of the single-phase aggregates. Dresen *et al.* (1998) reported a stronger mixture which consists of 80% marble and 20% quartz and supported the model which predicts higher strength with lower porosity.

Even with the experimental observations described above, difficulties still remain in quantifying the microstructure, stress, and strain within each phase in the composites. Various analytical models (Ji *et al.*, 2004, 2003, 2001) and finite element models (Bao *et al.*, 1991; Madi *et al.*, 2005; Tullis *et al.*, 1991) have been developed and successfully predict the bulk strength of the composite from its end members for different compositions and microstructures. However, due to the difficulty of detecting the behavior of the individual phases in the composite, these models are somewhat unconstrained. Furthermore, there are fewer experimental constraints on the rheology of aggregates at mantle conditions.

Diffraction-based determinations of stress usually yield multiple estimates of stress for each diffraction pattern; one for each diffraction line. Each diffraction line comes from a different population of grains within the sample. If there is more than one material that makes up the polycrystals, then each material usually contributes diffraction lines and hence measures of stress. Each subpopulation of grains that make up a diffraction line share a common orientation relationship within the sample, and thus a relationship of orientation relative to the applied stress field. This situation provides an unprecedented opportunity to investigate the heterogeneity of stress within the sample that is associated with these subpopulations. Previous measurements of force applied to an area of the polycrystalline sample yields only an average of the stress supported by the constituents. The relationship between these two types of measurements follows from some simple principles:

$$F_i = \int \sigma_{ij} \mathrm{d}A_j \qquad [4]$$

where F_i is the vector component of force, σ_{ij} is a component of the stress tensor, and A_j is a properly oriented area. Thus, the stress deduced from F/A may not be the stress at any point in the solid. The stress variation must satisfy the basic equilibrium equation,

$$\sum_{j=1}^{3} \frac{\partial \sigma_{ij}}{\partial x_j} = 0 \qquad [5]$$

and maintain continuity of displacement. If the stress field within any grain is relatively constant, then eqn [2] becomes a sum over grains,

$$F_i = \sum \sigma_{ij} A_j \qquad [6]$$

where the sum is over grains that define a continuous plane within the sample. In this view, the different subpopulations of grains may take on independent, and different, values of stress. The variation of stress required by eqn [3] would be represented as a broadening of the diffraction peaks, but the stress levels would reflect the properties of the grain. Grain-to-grain variation of stress may occur if the different subpopulations have different strengths. Within the deformation field, no grain will support a greater stress than its strength. Thus, a composite of grains with different strengths could be expected to have subpopulations with different stresses. Here, we give two examples of this effect. The first is for a two-phase aggregate, as reported by (Weidner and Li, 2006).

A mixture of spinel ($MgAl_2O_4$) and MgO is demonstrated. We focus on data obtained during plastic flow of composites with volume distributions of 100% MgO (mgo100), 75% MgO/25% spinel (mgo75sp25), 25% MgO/75% spinel (mgo25sp75), and 100% spinel (sp100). The composites were studied in two experimental deformation runs using a DDIA, the first two samples in one run and the second two samples in the other run. As illustrated in **Figure 10**, two cylindrical samples are stacked along the unique stress axis with metal foils separating the samples and also between the sample and the corundum end piston.

Differential stress is measured from the energy-dispersive diffraction patterns recorded at 0° and 90° relative to the uniaxial compression direction (Li *et al.*, 2004a). Plastic strain as a function of time is defined from sample images (Li *et al.*, 2003; Vaughan *et al.*, 2000). Samples were initially compressed to 5 GPa, then annealed at 1000°C for tens of minutes, then deformed at a strain rate of $10^{-5}\,s^{-1}$ for tens of percent strain, temperature was then lowered to 800°C where deformation continued. Finally, the direction of the deformation was reversed, the sample being lengthened along the unique stress axis. **Figures 11** and **12** illustrate the stress versus strain for the samples. The arrow labeled as 1 indicates the point where the temperature was lowered to 800°C and the arrow labeled as 2 indicates the reversal of the piston loading direction. Stresses at 1000°C are near the uncertainty level of the measurement. Here we focus on the 800°C data as it is robust and well resolved. **Figure 13** summarizes the average loading stress for each material. The error bars indicate the spread of stress measurement based on different diffraction peaks. For spinel, the [400], [311], and [110] diffraction lines are used, while only the [200] could be used for MgO due to the overlap of other lines with some of the spinel lines. As seen in **Figure 13**, the stress supported by spinel is nearly the same for

the 100% spinel and the 75% spinel sample but much less for the 25% spinel sample. The stress in the MgO portion of the sample is nearly the same in all mixtures. **Figure 11** illustrates that the 100% spinel and the 75% spinel, which were run in the same experiment, suffered a factor-of-2 difference in strain even though the stress in the spinel was nearly the same. Thus, the addition of 25% MgO enabled considerable more deformation.

This combination of elastic stresses and strains can guide us to an improved understanding of the range of stress–strain properties that are likely. The large difference in stress between the spinel grains and the MgO grains serves to illustrate that strong and weak grains can self-organize to allow the strong grains to support more stress. Even in the case of 25% spinel, we find that the stress in the spinel is about twice that of the MgO on the loading cycle. For the 75% spinel, the stress in the spinel is essentially the same as in the 100% spinel, thus equal to the strength of the spinel sample. Sill, the small amount of MgO present in this sample is responsible for the doubling of the plastic strain. In the 75% spinel sample, the MgO stress reversed sign during the extension portion of the experiment while the spinel was still under compression. This illustrates that there would remain a significant residual stress in the sample if the loading force were simply removed at this time. However, on further extension, the spinel stress continued to grow to a similar magnitude, but reversed sign as on loading. In the sample with only 25% spinel, the stress in the spinel did not grow to its compressive magnitude during the extension phase of the experiment. Rather, it attained only the stress value of the MgO portion of the sample. This indicates that upon loading, a fabric developed that enabled the strong phase to interact and support more of the stress. On the reversal of stress sign, this fabric did not allow the spinel to continue to support other spinel grains, and the stress in all grains was limited by the strength of the MgO network. Thus, significant fabric develops during the long loading cycle.

A second example of stress variations among subgrain populations comes from a monomineralic phase, MgO, in which different orientations have radically different strengths owing to the orientations of the weak slip planes (Li *et al.*, 2004b; Uchida *et al.*, 2004). Li *et al.* (2004b) report measurements of stress in an MgO sample at 500°C. **Figure 14** illustrates the measured stress using the [111] and [200] diffraction lines with several diffraction patterns as the sample is maintained at constant stress by the DDIA pressure

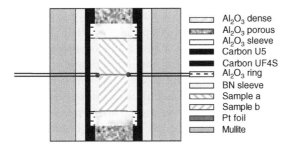

	Al₂O₃ dense
	Al₂O₃ porous
	Al₂O₃ sleeve
	Carbon U5
	Carbon UF4S
	Al₂O₃ ring
	BN sleeve
	Sample a
	Sample b
	Pt foil
	Mullite

Figure 10 Illustration of sample cell for two samples (a and b). This cell allows comparative studies of two materials. The one with the larger strain rate is generally the weakest, as the stresses will be quite similar for the two samples.

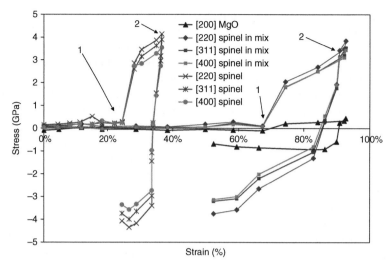

Figure 11 Stress as a function of plastic strain for two samples in the same cell. The set of curves to the right are the 75% spinel sample and deformed approximately twice the pure spinel sample. The arrow labeled 1 indicates the lowering of the temperature from 1000°C to 800°C, while the arrow labeled 2 indicates the reversal of the loading force. From Weidner DJ and Li L (2006) Measurement of stress using synchrotron X-rays. Paper presented at *Journal of Physics*: *Condensed Matter*.

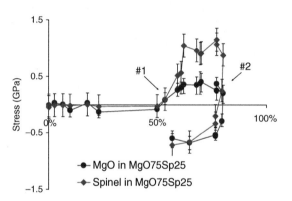

Figure 12 Stress as a function of plastic strain for the 25% spinel sample. The arrow labeled 1 indicates the lowering of the temperature from 1000°C to 800°C, the arrow labeled 2 indicates the reversal of the loading force. From Weidner DJ and Li L (2006) Measurement of stress using synchrotron X-rays. Paper presented at *Journal of Physics*: *Condensed Matter*.

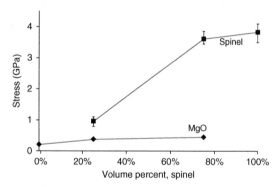

Figure 13 Average stress during loading at 800°C for the two grain populations in a two phase mixture as a function of volume per cent of the spinel phase. The error bars indicate the range of stresses determined from the spinel diffraction peaks, [400], [311], and [220]. Only the MgO [200] peak could be used. From Weidner DJ and Li L (2006) Measurement of stress using synchrotron X-rays. Paper presented at *Journal of Physics*: *Condensed Matter*.

system. One sample that was stressed at about two-thirds of the yield point exhibited a stress state that was the same for all diffraction lines. The sample that was stressed into yield experienced a 50% greater stress measured by the [111] diffraction lines than from the [200] diffraction lines. The difference in stresses between [111] and [200] is explained by plastic flow of MgO, for which the active slip systems required different critical resolved shear stresses and these systems are differently accessible to the different orientations of grains. The plastic strength parallel to [111] can be over an order of magnitude

larger than the strength parallel to [100] (Paterson and Weaver, 1970). Thus, different populations of grains, that depend on the orientation in the polycrystals, support quite different levels of stress.

Different elastic moduli may also be responsible for generating different stresses in different subpopulations. By the same token, such elastic heterogeneity (or anisotropy for a monomineralic assemblage) may generate different strains within the different subpopulations. Such elastic strain anisotropy has been used to predict elastic modulus anisotropy in many studies using X-rays (Duffy *et al.*, 1999; Kavner, 2003;

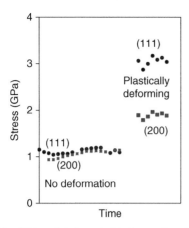

Figure 14 Differential stress in MgO samples at 500°C during uniaxial loading in DDIA. The stress measured by two different diffraction peaks is illustrated by different colors. The higher-stressed sample plastically deformed, while the low-stress sample did not. The different points represent different diffraction spectra collected during the stress cycle. From Li L, Weidner DJ, Chen J, Vaughan MT, Davis M, and Durham WB (2004b) X-ray strain analysis at high pressure: Effect of plastic deformation in MgO. *Journal of Applied Physics* 95(12): 8357–8365.

Kavner and Duffy, 2001; Mao *et al.*, 1998; Merkel *et al.*, 2002). However, care must be exercised in these circumstances. As illustrated in the MgO example, strain anisotropy in a plastically deforming material likely reflects the plastic strength anisotropy of MgO rather than the elastic anisotropy.

12.3.2 Activation Volume of Olivine

The quantification of the rheological properties of mantle rocks at high pressure has been a task for scientists for more than a half century. The purpose is to provide physical basis for understanding the dynamics of the mantle flow. Olivine, the major constituent of the upper mantle for the pyrolite composition, has remained the central focus because of its abundance and in part due to the limitations of the deformation apparatus and the measuring tools. It is one of the most studied minerals, has been reported with a vast amount of information, from single crystal to polycrystal; from forstertite to fayalite; from superdry to superwet samples; from grain size less than 0.5 μm to that of a few millimeters. With such intense attention, the results have converged. Yet, there are a few key variables that remain illusive.

As an illustration, we explore the issue regarding the activation volume (V^*) for which values from $<5\,cm^3\,mol^{-1}$ to larger than $27\,cm^3\,mol^{-1}$ have

been reported from experimental work including those measuring stress–strain-rate relationship (Borch and Green, 1987; Green and Borch, 1987; Karato and Jung, 2003) and those measuring diffusivity of individual cations (Béjina *et al.*, 1999). A small V^* and large V^* will lead to a different Earth dynamic model as illustrated earlier, since a large V^* will magnify the effect of pressure on the strength of mantle exponentially, with significant effects on mantle flow as well as earthquakes (Brodholt and Stein, 1988). A large activation volume for dislocation flow in the presence of a small activation volume for diffusion will lead to a change in flow mechanism with depth (Karato and Wu, 1993), which may provide insight into why seismic anisotropy appears/disappears in the middle of the upper mantle.

The key to defining the effect of pressure on the flow of olivine is to determine the activation volume, V^*, by measuring the flow parameters over a large pressure range. The most advanced tools, at present, for measuring the stress–strain-rate relation at pressure up to 15 GPa is the DDIA. Deforming olivine in the DDIA, we can reach pressure and temperature equivalent to the depth of 410 km in the Earth even though we require a fast deforming process (strain rate of $10^{-5}\,s^{-1}$), as a constant stress is applied. As mentioned earlier, we must rely on a scaling law to correct to Earth strain rates.

In defining the high-pressure flow of olivine, one needs first to identify the flow process that is active at the particular conditions. To map to mantle conditions, power-law creep is the expected process. Power-law creep is characterized by thermal activation of processes that overcome obstacles of dislocation motion. The exponent, *n*, of eqn [1] is in the range of 3–5 while *m* is 0. That is, there is no grain size dependence. Competing processes include dislocation glide at lower temperature and diffusion for higher temperature and smaller grain size. Dislocation glide produces a much stronger dependence of strain rate on stress (an effective *n* would be greater than 10). Dislocations will image as straight lines in TEM images. Diffusion will exhibit a positive value of *m*, thus expressing a grain size-sensitive flow process.

In a series of papers, Li *et al.* (2003, 2004a, 2006b) explore the high-pressure flow of olivine. In relaxation experiments, the flow strength of 0.5 and 5.0 μm samples were found to be identical (Li *et al.*, 2003) in the temperature range of 700–1200°C, indicating that diffusion was not a significant process even for such small grains. Furthermore, the grain size of the two

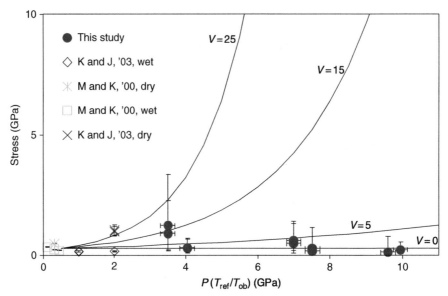

Figure 15 Flow stress for olivine samples corrected to a reference temperature of 1473 K and a strain rate of $10^{-4}\,s^{-1}$. The solid symbols are from Li *et al.* (2006b). The closed/open symbols are for wet samples, while the cross and plus symbols are for dry samples. M&K, '00 is data of Mei and Kohlstedt (2000), K&J, '03 is the data of Karato and Jung (2003). The solid lines illustrate the slopes of flow laws corresponding to several activation volumes. The numbers are in units of $cm^3\,mol^{-1}$. From Li L, Weidner DJ, Raterron P, *et al.* (2006b) Deformation of olivine at mantle pressure using D-DIA. *European Journal of Mineralogy* 18: 7–19.

samples changed during the experiments and tended to converge, indicating the presence of dynamic recrystallization as a significant contributor to the process. TEM revealed sinuous dislocations for samples deformed above 1000°C and quite straight dislocation for samples from experiments below this temperature. This indicates that power-law creep is active at the higher temperature, but dislocation glide is the dominant process below this temperature.

Li *et al.* (2006b) report stress with defined strain rates for steady-state experiments using the DDIA. Their results are illustrated in **Figure 15**. Here, flow stress is plotted as a function of pressure for the specified conditions of temperature and strain rate. The original data are corrected to these conditions using the inferred flow law. The conditions were close enough to those of this plot that the adjustments were fairly minor. The low-pressure data of Karato and Jung (2003) and Mei and Kohlstedt (2000) are also illustrated. Mei and Kohlstedt (2000) studied both wet (water-saturated) and dry samples as indicated here. The samples of Li *et al.* (2006b) were reported to represent a range of water contents, but none of them were dry. There was no correlation of the stress with the water content for these samples. They speculate that a small amount of water will

have a large effect, but the effect may not continue to increase with increasing water solubility.

The data of Li *et al.* (2006b) are consistent with the low-pressure data, further indicating that the controlling process of the flow mechanism was similar in the two cases. Plotted in **Figure 15** are the values of stress corresponding to different activation volumes. The interpretation that is most consistent with all of these observations is that the above 1000°C flow is dominated by a similar power-law creep process throughout the pressure range. Furthermore, this process has a very small pressure dependence, that is $V^* = 0 \pm 5\,cm^3\,mol^{-1}$.

12.3.3 High-Pressure Phases

High-pressure deformation experiments enable rheology measurements on high-pressure phases. Many meaningful state variable experiments can be carried out on metastable phases. Elasticity measurements sample the free-energy function for small atom displacements of a phase with little effect of whether or not there is a more stable phase if the atoms are allowed to move relatively large distances. Flow, however, involves large displacements of the atoms, and the high temperature required to overcome obstacles will generally increase the kinetics to enable a phase transformation to the stable phase. Thus, high-temperature flow for metastable phases is virtually unexplored.

Pyrope garnet is not stable at pressure below ~2 GPa. As a result, there are very few constraints on the flow properties of this material despite its importance as a rock-forming mineral. Karato *et al.* (1995) estimated garnet rheology based on analog studies. Measurements of the flow of other materials in the garnet structure, such as rare earth garnets, taken with a scaling relation based on homologous temperature, can provide a informed estimate of flow properties. The primary assumption is that activation energy scales as T/T_m, where T_m is the melting temperature, among different compounds with the same crystal structure. This scaling equation leads to a picture of garnet that is significantly stronger than olivine.

Li *et al.* (2006a) used the DDIA to investigate the flow properties of pyrope garnet and a pyrope-rich natural garnet. These measurements were made with a comparison to an olivine that would be chemically equilibrated with the garnet. Forsterite was found to be stronger than pyrope at 1200°C, while San Carlos olivine was weaker than the natural garnet with about 30% iron. Thus, iron is more significant in altering the rheological properties of olivine than it is for garnet. Furthermore, garnet is not significantly stronger than olivine under mantle conditions.

12.4 Conclusion

Recent breakthroughs driven by *in situ* characterization of high-pressure/high-temperature samples with synchrotron X-rays are changing the laboratory contribution to our understanding of the flow properties of the Earth's mantle. Right now, we are still on a strong growth curve for these capabilities. These new systems are now shedding light on a wide range of issues. The support of deviatoric stress in a polycrystalline material can be investigated with a new perception. The stress on subpopulations of grains can be directly observed. Deformation at high pressure is now observable. This leads to an understanding of the depth-dependent flow process, both in terms of the effect of pressure of individual phases, and also of the flow process in high-pressure, deep phases.

Future work promises new technical developments accompanied with significant advances in the field. Currently, the resolution of stress is about an order of magnitude lower than for low-pressure deformation experiments. Resolution can be improved with new detector/slit designs. A target of 10 MPa stress resolution at high pressure and temperature with data-gathering times of a few minutes is realizable. Higher-pressure systems are surely coming. The initial stages of DDIA and rotational Drickamer apparatus (RDA) have been proved extremely successful at generating stress and pressure. Following the design criteria, it should be possible to increase pressure to that of the top of the lower mantle. Opposed anvil devices stimulate a new range of possible systems that have the potential to generate conditions found in the middle of the lower mantle. In addition to quantitative flow laws, studies of flow-induced elastic anisotropy will be possible on all of the mantle phases. The connection between geodynamics and seismology will be placed on a firm laboratory basis.

References

Ando J, Irifune T, Takeshita T, and Fujino K (1997) Evaluation of the non-hydrostatic stress produced in a multi-anvil high pressure apparatus. *Physics and Chemistry of Minerals* 24: 139–148.

Ashby MF (1972) A first report on deformation-mechanism maps. *Acta Metallurgica* 20: 887–897.

Bao G, Hutchinson JW, and McMeeking RM (1991) Particle reinforcement of ductile matrices against plastic flow and creep. *Acta Metallurgica Et Materialia* 39: 1871–1882.

Bascou J, Tommasi A, and Mainprice D (2002) Plastic deformation and development of clinopyroxene lattice preferred orientations in eclogites. *Journal of Structural Geology* 24: 1357–1368.

Béjina F, Jaoul O, and Liebermann RC (1999) Activation volume of Si diffusion in San Carlos olivine: Implications for upper mantle rheology. *Journal of Geophysical Research* 104: 25529–25542.

Borch RS and Green HW, II (1987) Dependence of creep in olivine on homologous temperature and its implications for flow in the mantle. *Nature (London)* 330: 345–348.

Borch RS and Green HW, II (1989) Deformation of peridotite at high pressure in a new molten salt cell; comparison of traditional and homologous temperature treatments, physical properties of solid and molten Earth materials. *Physics of the Earth and Planetary Interiors* 55: 269–276.

Brodholt J and Stein S (1988) Rheological control of Wadati-Benioff zone seismicity. *Geophysical Research Letters* 15: 1081–1084.

Bruhn DF, Olgaard DL, and Dell'Angelo LN (1999) Evidence for enhanced deformation in two-phase rocks; experiments on the rheology of calcite-anhydrite aggregates. *Journal of Geophysical Research-B: Solid Earth and Planets* 104: 707–724.

Burnley PC, Green HW, and Prior DJ (1991) Faulting associated with the olivine to Spinel transformation in Mg_2GeO_4 and its implications for deep-focus earthquakes. *Journal of Geophysical Research-B: Solid Earth and Planets* 96: 425–443.

Bussod GY, Katsura T, and Rubie DC (1993) The large volume multi-anvil press as a high P–T deformation apparatus. *Pure and Applied Geophysics* 141: 579–599.

Bystricky M, Kunze K, Burlini L, and Burg JP (2000) High shear strain of olivine aggregates; rheological and seismic consequences. *Science* 290: 1564–1567.

Carrez P, Cordier P, Mainprice D, and Tommasi A (2006) Slip systems and plastic shear anisotropy in Mg_2SiO_4 ringwoodite: Insights from numerical modelling. *European Journal of Mineralogy* 18: 149–160.

Chopra PN (1997) High-temperature transient creep in olivine rocks. *Tectonophysics* 279: 93–111.

Clausen B, Lorentzen T, and Leffers T (1998) Self-consistent modelling of the plastic deformation of FCC polycrystals and its implications for diffraction measurements of internal stresses. *Acta Materialia* 46: 3087–3098.

Cordier P (2002) Dislocations and slip Systems of mantle minerals. Paper presented at *Plastic Deformation of Minerals and Rocks*. Mineralogical Society of America.

Cordier P, Barbe F, Durinck J, Tommasi A, and Walker AM (2005) Plastic deformation of minerals at high pressure: Multiscale numerical modelling. Paper presented at *Mineral Behaviour at Extreme Conditions*. European Mineralogical Union.

Cordier P and Doukhan JC (1995) Plasticity and dissociation of dislocations in water-poor quartz. *Philosophical Magazine A, Physics of Condensed Matter, Structure Defects and Mechanical Properties* 72: 497–514.

Cordier P, Raterron P, and Wang Y (1996) TEM investigation of dislocation microstructure of experimentally deformed silicate garnet. *Physics of the Earth and Planetary Interiors* 97: 121–131.

Cordier P and Sharp TG (1998) Large angle convergent beam electron diffraction determinations of dislocation Burgers vectors in synthetic stishovite. *Physics and Chemistry of Minerals* 25: 548–555.

Cordier P, Thurel E, and Rabier J (2002) Stress determination in multianvil deformation experiments based on dislocation curvatures measurements; application to wadsleyite and ringwoodite. *Geophysical Research Letters* 29: 4.

Cordier P, Ungar T, Zsoldos T, and Tichy G (2004) Dislocation creep in $MgSiO_3$ perovskite at conditions of the Earth's uppermost lower mantle. *Nature* 428: 837–840.

Couvy H, Frost DJ, Heidelbach F, *et al.* (2004a) Shear deformation experiments of forsterite at 11Gpa–1400 degrees C in the multianvil apparatus. *European Journal of Mineralogy* 16: 877–889.

Couvy H, Tommasi A, Mainprice D, Cordier P, Frost D, and Langenhorst F (2004b) Deformation texture in wadsleyite and ringwoodite: Implications for the seismic anisotropy of the transition zone. *Lithos* 73: S20.

Dawson P, Boyce D, MacEwen S, and Rogge R (2001) On the influence of crystal elastic moduli on computed lattice strains in AA-5182 following plastic straining. *Materials Science and Engineering A, Structural Materials Properties Microstructure and Processing* 313: 123–144.

Dawson PR (2002) Modelling deformation of polycrystalline Rocks. Paper presented at *Plastic Deformation of Minerals and Rocks*. Mineralogical Society of America.

Dawson PR and Wenk HR (2000) Texturing of the upper mantle during convection. *Philosophical Magazine A, Physics of Condensed Matter Structure Defects and Mechanical Properties* 80: 573–598.

de Bresser JHP, Peach CJ, Reijs JPJ, and Spiers CJ (1998) On dynamic recrystallization during solid state flow; effects of stress and temperature. *Geophysical Research Letters* 25: 3457–3460.

de Bresser JHP, Ter Heege JH, Spiers CJ, Dresen GE, and Handy ME (2001) Grain size reduction by dynamic recrystallization; can it result in major rheological weakening? Deformation mechanisms, rheology and microstructures. *International Conference on Deformation Mechanisms, Rheology and Microstructures* 90: 28–45.

Dresen G, Evans B, and Olgaard DL (1998) Effect of quartz inclusions on plastic flow in marble. *Geophysical Research Letters* 25: 1245–1248.

Duffy TS, Shen GY, Heinz DL, *et al.* (1999) Lattice strains in gold and rhenium under nonhydrostatic compression to 37 GPa. *Physical Review B* 60: 15063–15073.

Durham WB and Goetze C (1977) Plastic flow of oriented single crystals of olivine; 1, mechanical data. *Journal of Geophysical Research* 82: 5737–5753.

Durham WB, Weidner DJ, Karato S-I, and Wang Y (2002) New developments in deformation experiments at high pressure, paper. Presented at *Plastic Deformation of Minerals and Rocks*. Mineralogical Society of America.

Durinck J, Legris A, and Cordier P (2005a) Influence of crystal chemistry on ideal plastic shear anisotropy in forsterite: First principle calculations. *American Mineralogist* 90: 1072–1077.

Durinck J, Legris A, and Cordier P (2005b) Pressure sensitivity of olivine slip systems: First-principle calculations of generalised stacking faults. *Physics and Chemistry of Minerals* 32: 646–654.

Ford JM and Wheeler J (2004) Modelling interface diffusion creep in two-phase materials. *Acta Materialia* 52: 2365–2376.

Frost HJ and Ashby MF (1982) *Deformation-Mechanism Maps*, 166pp. Oxford: Pergamon Press.

Green HW and Borch RS (1987) The pressure dependence of creep. *Acta Metallurgica* 35: 1301–1315.

Green HW and Borch RS (1989) A new molten-salt cell for precision stress measurement at high-pressure. *European Journal of Mineralogy* 1: 213–219.

Green HW and Houston H (1995) The mechanics of deep earthquakes. *Annual Review of Earth and Planetary Sciences* 23: 169–213.

Green HW, Young TE, Walker D, and Scholz CH (1990) Anticrack-associated faulting at very high-pressure in natural olivine. *Nature* 348: 720–722.

Hager BH (1984) Subducted slabs and the geoid: Constraints on mantle rheology and flow. *Journal of Geophysical Research* 89: 6003–6015.

Hirth G and Kohlstedt KL (1995) Experimental constraints on the dynamics of the partially molten upper mantle; deformation in the diffusion creep regime. *Journal of Geophysical Research-B: Solid Earth and Planets* 100: 1981–2001.

Hobbs BE and Ord A (1988) Plastic instabilities: Implications for the origin of intermediate and deep earthquakes. *Journal of Geophysical Research* 93: 10521–10540.

Ji S, Wang Q, Xia B, and Marcotte D (2004) Mechanical properties of multiphase materials and rocks: A phenomenological approach using generalized means. *Journal of Structural Geology* 26: 1377–1390.

Ji S, Zhao P, and Xia B (2003) Flow laws of multiphase materials and rocks from end-member flow laws. *Tectonophysics* 370: 129–145.

Ji S, Zichao W, and Wirth R (2001) Bulk flow strength of forsterite–enstatite composite as a function of forsterite content. *Tectonophysics* 341: 69–93.

Jiao WJ, Silver PG, Fei YW, and Prewitt CT (2000) Do intermediate- and deep-focus earthquakes occur on preexisting weak zones? An examination of the Tonga subduction zone. *Journal of Geophysical Research, Solid Earth* 105: 28125–28138.

Jung H and Green HW (2004) Experimental faulting of serpentinite during dehydration: Implications for earthquakes, seismic low-velocity zones, and anomalous hypocenter distributions in subduction zones. *International Geology Review* 46: 1089–1102.

Karato S, Riedel MR, and Yuen DA (2001) Rheological structure and deformation of subducted slabs in the mantle transition zone: Implications for mantle circulation and deep earthquakes. *Physics of the Earth and Planetary Interiors* 127: 83–108.

Karato S-I (1989) Defects and plastic deformation in olivine. In: Karato S-I and Toriumi M (eds.) *Rheology of Solids and of the Earth*, pp. 176–208. Oxford: Oxford University Press.

Karato S-I (1998) Micro-physics of post glacial rebound. *GeoResearch Forum* 3–4: 351–364.

Karato S-I and Jung H (2003) Effects of pressure on high-temperature dislocation creep in olivine. *Philosophical Magazine* 83: 401–414.

Karato S-I, Paterson MS, and Fitzgerald JD (1986) Rheology of synthetic olivine aggregates; influence of grain size and water. *Journal of Geophysical Research B* 91: 8151–8176.

Karato S-I, Rubie DC, and Yan H (1993) Dislocation recovery in olivine under deep upper mantle conditions; implications for creep and diffusion. *Journal of Geophysical Research-B: Solid Earth and Planets* 98: 9761–9768.

Karato S-I, Wang Z, Liu B, and Fujino K (1995) Plastic deformation of garnets; systematics and implications for the rheology of the mantle transition zone. *Earth and Planetary Science Letters* 130: 13–30.

Karato S-I and Wu P (1993) Rheology of the upper mantle; a synthesis. *Science* 260: 771–778.

Kavner A (2003) Elasticity and strength of hydrous ringwoodite at high pressure. *Earth and Planetary Science Letters* 214: 645–654.

Kavner A and Duffy TS (2001) Strength and elasticity of ringwoodite at upper mantle pressures. *Geophysical Research Letters* 28: 2691–2694.

King SD (1995) Radial models of mantle viscosity: Results from a genetic algorithm. *Geophysical Journal International* 122: 725–734.

Kinsland GL and Bassett W (1977) Strength of MgO and NaCl polycrystals to confining pressures of 250 Kbar at 25 C. *Journal of Applied Physics* 48: 978–985.

Kirby SH, Durham WB, and Stern LA (1991) Mantle phase-changes and deep-earthquake faulting in subducting lithosphere. *Science* 252: 216–225.

Kirby SH, Stein S, Okal EA, and Rubie DC (1996) Metastable mantle phase transformations and deep earthquakes in subducting oceanic lithosphere. *Reviews of Geophysics* 34: 261–306.

Kohlstedt DL, Ricoult DL, Tressler RE, and Bradt RC (1984) High temperature creep of silicate olivines. *International Symposium on Plastic Deformation of Ceramic Materials, Deformation of Ceramic Materials II* 18: 251–280.

Li L (2003) *Rheology of Olivine at High Temperature and High Pressure*. PhD Thesis, Stony Brook University.

Li L, Long H, Weidner DJ, and Raterron P (2006a) Plastic flow of pyrope at mantle pressure and temperature. Paper presented at *American Mineralogist*.

Li L, Raterron P, Weidner D, and Chen J (2003) Olivine flow mechanisms at 8 GPa. *Physics of the Earth and Planetary Interiors* 138: 113–129.

Li L, Weidner D, Raterron P, Chen J, and Vaughan M (2004a) Stress measurements of deforming olivine at high pressure. *Physics of the Earth and Planetary Interiors* 143–144: 357–367.

Li L, Weidner DJ, Chen J, Vaughan MT, Davis M, and Durham WB (2004b) X-ray strain analysis at high pressure: Effect of plastic deformation in MgO. *Journal of Applied Physics* 95(12): 8357–8365.

Li L, Weidner DJ, Raterron P, *et al.* (2006b) Deformation of olivine at mantle pressure using D-DIA. *European Journal of Mineralogy* 18: 7–19.

Madi K, Forest S, Cordier P, and Boussuge M (2005) Numerical study of creep in two-phase aggregates with a large rheology contrast: Implications for the lower mantle. *Earth and Planetary Science Letters* 237: 223–238.

Mainprice D and Humbert M (1994) Methods of calculating petrophysical properties from lattice preferred orientation data. *Surveys in Geophysics* 15: 575–592.

Mainprice D, Tommasi A, Couvy H, Cordier P, and Frost DJ (2005) Pressure sensitivity of olivine slip systems and seismic anisotropy of Earth's upper mantle. *Nature* 433: 731–733.

Mao H-k, Shu J, Shen G, Hemley RJ, Li B, and Singh AK (1998) Elasticity and rheology of iron above 220 GPa and the nature of the Earth's inner core. *Nature* 296: 741–743.

Mei S and Kohlstedt DL (2000) Influence of water on plastic deformation of olivine aggregates. 2: Dislocation creep regime. *Journal of Geophysical Research-B: Solid Earth and Planets* 105: 21471–21481.

Merkel S, Wenk HR, Badro J, *et al.* (2003) Deformation of (Mg-0.9,Fe-0.1)SiO3 perovskite aggregates up to 32 GPa. *Earth and Planetary Science Letters* 209: 351–360.

Merkel S, Wenk HR, Gillet P, Mao HK, and Hemley RJ (2004) Deformation of polycrystalline iron up to 30 GPa and 1000 K. *Physics of the Earth and Planetary Interiors* 145: 239–251.

Merkel S, Wenk HR, Shu J, *et al.* (2002) Deformation of polycrystalline MgO at pressures of the lower mantle. *Journal of Geophysical Research* 107: 2271.

Mitrovica JX and Forte AM (1997) Radial profile of mantle viscosity; results from the joint inversion of convection and postglacial rebound observables. *Journal of Geophysical Research-B: Solid Earth and Planets* 102: 2751–2769.

Mitrovica JX and Peltier WR (1991) A complete formalism for the inversion of postglacial rebound data – Resolving power analysis. *Geophysical Journal International* 104: 267–288.

Ogawa M (1987) Shear instability in a viscoelastic material as the cause of deep focus earthquakes. *Journal of Geophysical Research* 92: 13801–13810.

Paterson MS (1990) Rock deformation experimentation. In: Duba AG, Durham W, Handin J, and Wang H (eds.) *The Brittle–Ductile Transition in Rocks, the Heard Volume*, pp. 187–194. Washington, DC: AGU.

Paterson MS and Weaver CW (1970) Deformation of polycrystalline MgO under pressure. *Journal of the American Ceramic Society* 53: 463–472.

Peltier WR (1998) Postglacial variations in the level of the sea: Implications for climate dynamics and solid-earth geophysics. *Reviews of Geophysics* 36: 603–689.

Perez-Albuerne E, Forsgren K, and Drickamer H (1964) Apparatus for X-ray measurements at very high pressures. *Review of Scientific Instruments* 35: 29–33.

Poirier JP (1985) *Creep of Crystals, High-Temperature Deformation Processed in Metals, Ceramics and Minerals*. Cambridge: Cambridge University Press.

Post RL (1977) High-temperature creep of Mt Burnet-Dunite. *Tectonophysics* 42: 75–110.

Ross JV, Ave Lallemant HG, and Carter NL (1979) Activation volume for creep in the upper mantle. *Science* 203: 261–263.

Rubie DC, Karato S, Yan H, and O'Neill HSC (1993) Low differential stress and controlled chemical environment in multianvil high-pressure experiments. *Physics and Chemistry of Minerals* 20: 315–322.

Schafer W, Jansen E, Merz P, Will G, and Wenk HR (1992) Neutron-diffraction texture investigation on deformed quartzites. *Physica B* 180: 1035–1038.

Shimomura O, Yamaoka S, Yagi T, *et al.* (1985) Solid state physics under pressure. In: Minomura S (ed.) *Multi-Anvil*

Type X-ray System for Synchrotron Radiation, pp. 351–356. Tokyo: Terra Scientific Publishing.

Silver PG (1996) Seismic anisotropy beneath the continents: Probing the depths of geology. *Annuel Review of Earth Planetary Science* 24: 385–432.

Singh AK (1993) The lattice strains in a specimen (cubic symmetry) compressed nonhydrostatically in an opposed anvil device. *Journal of Applied Physics* 73: 4278–4286.

Singh AK, Balasingh C, Mao HK, Hemley R, and Shu J (1998) Analysis of lattice strains measured under nonhydrostatic pressure. *Journal of Applied Physics* 83: 7567–7578.

Takeda Y-T (1998) Flow in rocks modelled as multiphase continua; application to polymineralic rocks. *Journal of Structural Geology* 20: 1569–1578.

Taylor GI (1938) Plastic Strain in Metals. *Journal of the Institute of Metals* 62: 307–315.

Thoraval C and Richards MA (1997) The geoid constraint in global geodynamics: Viscosity structure, mantle heterogeneity models and boundary conditions. *Geophysical Journal International* 131: 1–8.

Thurel E, Douin J, and Cordier P (2003) Plastic deformation of wadsleyite. III: Interpretation of dislocations and slip systems. *Physics and Chemistry of Minerals* 30: 271–279.

Tommasi A (1998) Forward modeling of the development of seismic anisotropy in the upper mantle. *Earth and Planetary Science Letters* 160: 1–13.

Tommasi A, Mainprice D, Canova G, and Chastel Y (2000) Viscoplastic self-consistent and equilibrium-based modeling of olivine lattice preferred orientations: Implications for the upper mantle seismic anisotropy. *Journal of Geophysical Research, Solid Earth* 105: 7893–7908.

Tullis J (2002) Deformation of granitic rocks: Experimental studies and natural examples. Paper presented at *Plastic Deformation of Minerals and Rocks*. Mineralogical Society of America.

Tullis TE, Horowitz FG, and Tullis J (1991) Flow laws of polyphase aggregates from end-member flow laws. *Journal of Geophysical Research-B: Solid Earth and Planets* 96: 8081–8096.

Uchida T, Wang Y, Rivers ML, and Sutton SR (2004) Yield strength and strain hardening of MgO up to 8 GPa measured in the deformation-DIA with monochromatic X-ray diffraction. *Earth and Planetary Science Letters* 226: 117–126.

Vaughan M, Chen J, Li L, Weidner D, and Li B (2000) Use of X-ray imaging techniques at high-pressure and temperature for strain measurements. Paper presented at *AIRAPT-17*. Hyderabad, India: Universities Press.

Voegele V, Ando JI, Cordier P, and Liebermann RC (1998) Plastic deformation of silicate garnets. I: High-pressure experiments. *Physics of the Earth and Planetary Interiors* 108: 305–318.

Voegele V, Cordier P, Langenhorst F, and Heinemann S (2000) Dislocations in meteoritic and synthetic majorite garnets. *European Journal of Mineralogy* 12: 695–702.

Voegele V, Liu B, Cordier P, *et al.* (1999) High temperature creep in a 2-3-4 garnet: $Ca_3Ga_2Ge_3O_{12}$. *Journal of Materials Science* 34: 4783–4791.

Wang Y, Durham W, Getting IC, and Weidner D (2003) The deformation-DIA: A new apparatus for high temperature triaxial deformation to pressure up to 15 GPa. *Review of Scientific Instruments* 74: 3002–3011.

Weidner DJ and Li L (2006) Measurement of stress using synchrotron X-rays. Paper presented at *Journal of Physics: Condensed Matter*.

Wenk H-R (2002) Texture and anisotropy. Paper presented at *Plastic Deformation of Minerals and Rocks*. Mineralogical Society of America.

Wenk HR, Bennett K, Canova GR, and Molinari A (1991) Modeling plastic-deformation of peridotite with the self-consistent theory. *Journal of Geophysical Research-B: Solid Earth and Planets* 96: 8337–8349.

Wenk HR, Ischia G, Nishiyama N, Wang Y, and Uchida T (2005) Texture development and deformation mechanisms in ringwoodite. *Physics of the Earth and Planetary Interiors* 152: 191–199.

Wenk HR, Lonardelli I, Pehl J, *et al.* (2004) *In situ* observation of texture development in olivine, ringwoodite, magnesiowustite and silicate perovskite at high pressure. *Earth and Planetary Science Letters* 226: 507–519.

Wenk HR, Speziale S, McNamara AK, and Gamero EJ (2006) Modeling lower mantle anisotropy development in a subducting slab. *Earth and Planetary Science Letters* 245: 302–314.

Wright SI and Adams BL (1992) Automatic analysis of electron backscatter diffraction patterns. *Metallurgical and Materials Transactions* A23: 759–767.

Yamazaki D and Karato S (2001) High-pressure rotational deformation apparatus to 15 GPa. *Review of Scientific Instruments* 72: 4207–4211.

13   Theory and Practice – The *Ab Initio* Treatment of High-Pressure and -Temperature Mineral Properties and Behavior

D. Alfè, University College London, London, UK

13.1	Introduction	359
13.2	First-Principles Techniques	361
13.2.1	Density Functional Theory	362
13.2.1.1	XC functionals	363
13.2.1.2	PPs and basis sets	363
13.2.1.3	Ultrasoft (Vanderbilt) PPs	364
13.2.1.4	The projector augmented wave method	365
13.2.2	Beyond DFT	365
13.2.2.1	QMC methods	365
13.3	Mineral Properties and Behavior	367
13.3.1	Static Properties	367
13.3.1.1	Crystal structures and phase transitions	367
13.3.1.2	Elastic constants	368
13.3.2	Finite Temperature	369
13.3.2.1	Molecular dynamics	370
13.3.3	Thermodynamic Properties	371
13.3.3.1	The Helmholtz free energy: Low-temperature and the quasi-harmonic approximation	372
13.3.3.2	Calculation of phonon frequencies	373
13.3.3.3	The Helmholtz free energy: High-temperature and thermodynamic integration	375
13.3.3.4	Melting	377
13.3.3.5	Solutions	380
13.3.3.6	First-principles calculations of chemical potentials	381
13.3.3.7	Volume of mixing	382
13.3.3.8	Solid–liquid equilibrium	382
13.3.3.9	Shift of freezing point	383
13.4	Conclusions	384
References		385

13.1   Introduction

The purpose of this article is to provide some examples about recent developments in the calculation of the high-pressure and -temperature properties of materials using *ab initio*-based techniques, with the description limited to minerals under the relatively benign conditions of the center of the Earth. This article is not intended to review comprehensively all high pressure and temperature *ab initio* calculations on minerals performed to date, nor has the presumption to include the most significant ones. The few examples provided will only serve the purpose of showing what current *ab initio* techniques are capable of predicting.

With the words *ab initio* we refer here to those calculations based on the very basic laws of nature, in which no empirical adjustable parameter is used. Only fundamental constants of physics are allowed. Specifically, since we are interested in the properties of matter, the relevant basic law of physics are those describing the interactions between nuclei and electrons, that is, those of quantum mechanics. We shall see that approximations to exact quantum mechanics are necessary to provide tools that can be used in practice; however, as long as these approximations do not involve the introduction of empirical parameters, we still regard those techniques as *ab initio*.

We start the discussion by recalling the structure of the Earth, which can be broadly described in terms of three main shells. The outermost is the crust, with a thickness of only a few tens of kilometers, mainly formed by silicates. Below the crust we find the mantle, which is customarily divided into an upper mantle and a lower mantle, separated by a transition zone. The mantle makes up most of the volume of the Earth, extending to a depth of 2891 km, almost half way toward the center, and like the crust is also mainly formed by silicates, and in particular by $Mg(Fe)SiO_3$ with some significant fraction of $Mg(Fe)O$ and SiO_2. Below the mantle we find the core, which is divided into an outer liquid core extending from 2891 to 5150 km depths and an inner solid core below that, down to the center of the Earth at 6371 km depth. It is widely accepted that the core is mainly formed by iron, possibly with some 5–10% of nickel, plus a fraction of unknown light impurities which reduce the density by 2–3% in the solid and 6–7% in the liquid with respect to the density of pure iron under the same pressure–temperature conditions.

Studying the high-pressure and -temperature properties of core- and mantle-forming materials is of fundamental importance to the understanding of the formation and evolution of our planet. In particular, knowledge of the thermal structure of the Earth and the thermoelastic properties of Earth-forming minerals will help us to interpret and hopefully predict the behavior of the dynamical processes that happen inside our planet, including the generation of the Earth's magnetic field through the geodynamo, and the convective processes in the mantle which are ultimately responsible for plate tectonics, earthquakes, and volcanic eruptions.

The development of theoretical methods based on the very basic laws of nature of quantum mechanics (developed 80 years ago), coupled with the recent staggering increase of computer power ($\sim$500-fold in the past 10 years), has made it possible to approach the problem from a theoretical–computational point of view. When high-level first-principles methods are used, the results are often comparable in quality with experiments, sometimes even providing informations in regions of the pressure–temperature space inaccessible to experiments.

The exact quantum mechanical treatment of a system containing a large number of atoms is a formidable task. The starting point of nearly every quantum mechanical calculation available is the so-called adiabatic approximation, which exploits the large difference of mass between the nuclei and the electrons. Since the electrons are much lighter, they move so much faster that on the timescale of their movement the nuclei can be considered as fixed. Therefore, one solves only the electronic problem in which the nuclei are fixed and act as an external potential for the electrons. The energy of the electrons, plus the Coulomb repulsion of the nuclei, is therefore a function of the position of the nuclei, and can act as a potential energy for the nuclei. This can be mapped in configuration space to create a potential energy surface, which can later be used to study the motion of the nuclei. Alternatively, forces can be calculated as the derivatives of the potential energy with respect to the position of the nuclei, and these can be used to move the atoms around, relax the system, solve the Newton's equations of motion and perform molecular dynamics (MD) simulations, or calculate harmonic vibrational properties like phonons. The potential energy can also be differentiated with respect to the simulation cell parameters, which provides information on the stress tensor. The solution of the electronic problem also provides insights into the electronic structure of the system, which can be examined to study physical properties like bonding, charge distributions, magnetic densities, polarizabilities, etc.

Most first-principles studies of the high-pressure and -temperature properties of Earth's forming materials are based on the implementation of quantum mechanics known as density functional theory (DFT). This is a technique that was introduced about 40 years ago by Hohenberg and Kohn (HK) (1964), and Kohn and Sham (KS) (1965) in an attempt to simplify the calculation of the ground-state properties of materials (in fact, later shown to be useful also for finite temperature properties (Mermin, 1965)). The basic HK idea was to substitute the cumbersome many-body wavefunction of a system containing N particles, which is a function of $3N$ variables, with the particle density, which is only a function of three variables. The price to pay for this enormous simplification is a modification of the basic equations of quantum mechanics with the introduction of new terms, one of which, called exchange correlation (XC) energy, is unfortunately unknown. However, KS proposed a simple form for the XC functional, known as the local density approximation (LDA) (Kohn and Sham, 1965), that would prove later as the insight that has made DFT so successful and so widespread today. More sophisticated XC functionals were developed in the following decades, and are still being developed today, making DFT an evolving technique with increasingly higher accuracy. One additional attractive feature of DFT is the favorable scaling of computational effort with the size of the system. Traditional DFT techniques scale as N^3, where N is the number of electrons in the system, but large effort is being put into so-called $o(N)$ techniques, which for some materials already provide a scaling which is only directly proportional to the size of the system (Bowler *et al.*, 2002; Soler *et al.*, 2002).

A wide range of properties of minerals have been predicted using DFT techniques, like structural and electronic properties, phase diagrams, thermoelastic properties, speed of sound, transport properties, melting, solutions and partitioninig. In this chapter, we will focus only on a very limited number of applications.

The limitations in accuracy due to the current state of the art of DFT are expected to be progressively removed, either through the formulation of new XC functionals, or with the development of alternative techniques. Among these dynamical mean field theory (Savrasov and Kotliar, 2003) and quantum Monte Carlo (Foulkes *et al.*, 2001) are probably the most promising on a timescale of 5–10 years.

This article is divided into two main sections. In the first section, the main ideas behind first-principles simulations, and in particular DFT and the pseudopotential (PP) approximation are briefly reviewed. Quantum Monte Carlo techniques are also mentioned as promising advances beyond DFT. The second section deals with the properties of materials, first at zero temperature and then at finite temperature. Techniques for the calculation of free energies will be presented, and this will be done by separating the low-temperature regime, where solids can be described within the quasi-harmonic approximation, from the high-temperature regime, where the technique of thermodynamics integration is introduced to calculate free energies. The calculation of melting curves and the thermodynamics of solutions are presented as applications of these techniques.

13.2   First-Principles Techniques

I begin this section by recalling the basic equation of quantum mechanics, the Schrödinger equation, which in the time independent form is

$$\hat{H}\Psi = E\Psi \qquad [1]$$

where $\hat{H}$ is the Hamiltonian of the system, E the total energy, and Ψ the many-body wave function, which is a function of the coordinates of the M nuclei $\{\mathbf{R_i}\}$ and the N electrons $\{\mathbf{r_i}\}$: $\Psi = \Psi(\mathbf{R}_1, \mathbf{R}_2, \ldots, \mathbf{R_M}; \mathbf{r}_1, \mathbf{r}_2, \ldots, \mathbf{r}_N)$. If the system is isolated, in the non-relativistic approximation, the Hamiltonian is given by

$$
\begin{aligned}
&\hat{H}(\mathbf{R}_1, \mathbf{R}_2, \ldots, \mathbf{R_M}; \mathbf{r}_1, \mathbf{r}_2, \ldots, \mathbf{r}_N) \\
&= -\sum_{i=1}^{M} \frac{\hbar^2}{2M_i} \nabla^2_{\mathbf{R}_i} - \sum_{i=1}^{N} \frac{\hbar^2}{2m} \nabla^2_{\mathbf{r}_i} \\
&\quad + \frac{1}{2}\sum_{i,j=1; i\neq j}^{M} \frac{Z_i Z_j e^2}{4\pi\epsilon_0 |\mathbf{R}_i - \mathbf{R}_j|} \\
&\quad + \frac{1}{2}\sum_{i,j=1; i\neq j}^{N} \frac{e^2}{4\pi\epsilon_0 |\mathbf{r}_i - \mathbf{r}_j|} - \sum_{i=1}^{M}\sum_{j=1}^{N} \frac{Z_i e^2}{4\pi\epsilon_0 |\mathbf{R}_i - \mathbf{r}_j|} \quad [2]
\end{aligned}
$$

where the first two terms are the kinetic energy operators for the nuclei and the electrons, respectively, with $\hbar$ the Planck's constant h divided by 2π, and M_i and m the masses of the nuclei and the electrons, respectively. The third and the fourth terms in the equation represent the Coulomb repulsive energy between the nuclei and between the electrons, respectively, with Z_i the charges of the nuclei in units of e, the charge of the electron, and ϵ_0 is the dielectric constant of the vacuum. The last term of the equation represents the electrostatic interaction between the electrons and the nuclei.

The only experimental input in the Schrödinger equation in the nonrelativistic approximation are four fundamental constants: the Plank's constant h, the charge of the electron e, the mass of the electron m, and the dielectric constant of the vacuum ϵ_0, which are the same for every system, plus the mass of the nuclei.

As mentioned in section 13.1, solving the Schrödinger equation is essentially impossible for any real system more complicated than the hydrogen atom, or more generally any system which contains two or more electrons, and therefore approximations are needed to make the problem manageable. The first approximation that can be brought in is the Born–Oppenheimer approximation, also called the adiabatic approximation. Here, one recognizes that the masses of the nuclei M_i are much larger than the mass of the electron m (the lightest possible atom is the hydrogen atom, which is almost 2000 times heavier than the electron), which therefore move on a much faster timescale. This means that, without much loss of accuracy for most systems, one can separate the electronic problem from that of the nuclei, or in other words solve the Schrödinger equation for the electrons only, with the nuclei positions kept fixed. Therefore, we can rewrite eqn [1] thus:

$$\hat{H}(\mathbf{r}_1, \mathbf{r}_2, \ldots, \mathbf{r}_M; \{\mathbf{R}_i\})\Psi(\mathbf{r}_1, \mathbf{r}_2, \ldots, \mathbf{r}_M; \{\mathbf{R}_i\}) = E\{\mathbf{R}_i\}\Psi(\mathbf{r}_1, \mathbf{r}_2, \ldots, \mathbf{r}_M; \{\mathbf{R}_i\}) \quad [3]$$

where now the Hamiltonian depends only parametrically from the positions of the nuclei $\{\mathbf{R}_i\}$, and so do the wave function Ψ and the energy E. Once eqn [3] is solved, the energy $E\{\mathbf{R}_i\}$ can be interpreted as a potential energy for the motion of nuclei. At high temperature (above the Debye temperature), the quantum nature of the nuclei becomes negligible, and with essentially no loss of accuracy one can treat their motion as they were classical particles. This allows to perform MD simulations, in which the Newton equations of motion for the nuclei are solved using the quantum mechanical forces evaluated from the derivative of $E\{\mathbf{R}_i\}$ with respect to the positions $\{\mathbf{R}_i\}$.

13.2.1 Density Functional Theory

The introduction of DFT in 1964 by Hohenberg and Kohn tackled the many-body problem using a completely new approach. Here the main ideas of DFT are briefly outlined. For an in-depth description of DFT, the reader may consult the original papers or the excellent books by Parr and Yang (1989) or Gross and Dreizler (1990), or the recent book by Martin (2004). A simplified (almost) nonmathematical explanation of DFT has been given by Gillan (1997).

Hohenberg and Kohn (1964) proved that the external potential V_{ext} acting on the electrons is uniquely determined (up to a trivial additive constant) by the electron ground-state density $n(\mathbf{r}) = \langle \Psi | \hat{n}(\mathbf{r}) | \Psi \rangle = \int d\mathbf{r}_2 \ldots d\mathbf{r}_N |\Psi(\mathbf{r}, \mathbf{r}_2, \ldots, \mathbf{r}_N)|^2$, where Ψ is the ground-state wavefunction of the system and $\hat{n}(\mathbf{r})$ is the density operator. Here we have omitted the dependence of Ψ from the positions of the nuclei for simplicity. Since $n(\mathbf{r})$ also determines the number of electrons N, and since V_{ext} and N fix the Hamiltonian of the system, it turns out that the electron density completely determines all the electronic ground-state properties of the system, and in fact, as shown later by Mermin (1965), also the finite temperature properties.

One important property of the system is the energy, which can be written as

$$E[n] = F_{\text{HK}}[n] + \int V_{\text{ext}}(\mathbf{r})n(\mathbf{r})d\mathbf{r} \quad [4]$$

with

$$F_{\text{HK}}[n] = \langle \Psi[n] | \hat{T} + \hat{V}_{\text{ee}} | \Psi[n] \rangle \quad [5]$$

where $\hat{T}$ and $\hat{V}_{\text{ee}}$ are respectively the kinetic energy and the electron–electron interaction operators, and $\Psi[n]$ is the ground-state wave function of the system. Note that $F_{\text{HK}}[n]$ does not depend on the external potential and therefore it is a universal functional. This is the crucial result of DFT. Using the variational principle, HK also proved that the ground-state density of the system is the one that minimizes $E[n]$, and the minimum of $E[n]$ is equal to the ground-state energy E_0. The importance of these two results is clear; the only quantity that is needed is the electron density, no matter how many electrons are present in the system.

One year after the publication of the HK paper, Kohn and Sham (KS) (1965) invented an indirect method to solve the problem. The idea is to write the energy functional as an easy part plus a difficult part:

$$F[n] = T_0[n] + E_H[n] + E_{\text{xc}}[n] \quad [6]$$

where $T_0[n]$ is the ground-state kinetic energy of an auxiliary noninteracting system whose density is the same as the one of the real system, $E_H[n]$ is the repulsive electrostatic energy of the classical charge distribution $n(\mathbf{r})$, and $E_{\text{xc}}[n]$ is the XC energy defined through eqn [6].

Minimizing the total energy $E[n]$ under the constraints of orthonormality for the one-particle orbitals of the auxiliary system, $\int \psi_i^*(\mathbf{r})\psi_j(\mathbf{r})d\mathbf{r} = \delta_{ij}$, one finds a set of one-particle Schrödinger-like equations:

$$\left[-\frac{\hbar^2}{2m}\nabla^2 + V_{\text{KS}}(\mathbf{r}) \right]\psi_i(\mathbf{r}) = \epsilon_i \psi_i(\mathbf{r}) \quad [7]$$

where the KS potential is

$$V_{\text{KS}}(\mathbf{r}) = V_{\text{ext}}(\mathbf{r}) + \int \frac{n(\mathbf{r}')}{|\mathbf{r}-\mathbf{r}'|}d\mathbf{r}' + V_{\text{xc}}(\mathbf{r}); \\ V_{\text{xc}}(\mathbf{r}) = \frac{\delta E_{\text{xc}}[n]}{\delta n(\mathbf{r})} \quad [8]$$

and

$$n(\mathbf{r}) = \sum_i f(\epsilon_i - \epsilon_{\text{F}})|\psi_i(\mathbf{r})|^2 \quad [9]$$

with $f(x)$ the Fermi–Dirac distribution and ϵ_{F} the Fermi energy fixed by the condition

$$\int n(\mathbf{r})d\mathbf{r} = N \quad [10]$$

These are the famous KS equations; they must be solved self-consistently because V_{KS} is a functional of the orbitals itself. The generalization to finite temperature is obtained by replacing E with the electronic free energy $U = E - TS$, where S is the electronic entropy, given by the independent-electron formula $S = -k_B T \Sigma_i [f_i \ln f_i + (1 - f_i) \ln(1 - f_i)]$, with f_i the thermal (Fermi–Dirac) occupation number of orbital i.

It is tempting to identify the single-particle eigenvalues ϵ_i with the energy of quasi-particles, and therefore their distribution with the electronic density of states of the system. This would be conceptually wrong, as the KS eigenvalues are only an artificial mathematical tool to arrive at the ground-state density of the system. Nevertheless, it turns out that these DFT density of states often resemble very accurately the real density of states of systems, and they are therefore often used to analyze their electronic structure. However, it is important to remember that even if the exact XC functional $E_{xc}[n]$ were known, one should not expect the DFT density of states to be an exact representation of the real density of states of the system.

When self-consistency is achieved, the electronic free energy of the system is

$$U = \sum_{i=1}^{N} f(\epsilon_i - \epsilon_F)\epsilon_i - \frac{1}{2} \int \frac{n(\mathbf{r})n(\mathbf{r'})}{|\mathbf{r} - \mathbf{r'}|} d\mathbf{r} + E_{xc}[n] - \int V_{xc}(\mathbf{r})n(\mathbf{r})d\mathbf{r} + E^{ion} - TS \quad [11]$$

where E^{ion} is the ionic electrostatic repulsion term. This would be the exact electronic free energy of the system if we knew $E_{xc}[n]$ (which also depends on temperature, though very little is known about this dependence). Unfortunately, the exact form of the XC (free) energy is not (yet) known.

13.2.1.1 XC functionals

Kohn and Sham (1965) also provided an approximate expression for the XC functional, called the LDA. In the LDA, the dependence of functional on the density has the form

$$E_{xc}^{LDA}[n] = \int n(\mathbf{r})\epsilon_{xc}(n(\mathbf{r}))d\mathbf{r} \quad [12]$$

and $\epsilon_{xc}(n)$ is taken to be the XC energy per particle of a uniform electron gas whose density is $n(\mathbf{r})$. This has been accurately calculated using Monte Carlo simulations (Ceperley and Alder, 1980) and parametrized

in order to be given in an analytic form (Perdew and Zunger, 1981).

By construction, this approximation yields exact results if the density of the system is uniform, and should not be very accurate for those systems whose density is highly dishomogeneous, as for example atoms and molecules. However, it turns out to work better then expected for a wide range of materials. In molecules, for example, the LDA usually overestimates the binding energies, but it yields in general good results for equilibrium distances and vibrational frequencies. It was the evidence of the very high quality of the LDA that has been the main factor responsible for the tremendous success of DFT.

Nowadays, a number of sophisticated functionals have become available, like the so-called generalized gradient approximation (GGA) (e.g. Wang and Perdew, 1991), or the recently developed meta-GGA (Tao *et al.*, 2003; Staroverov *et al.*, 2004), and hybrid functionals which contain a certain fraction of exact exchange according to various recipies (e.g. the B3LYP functional (Becke, 1993)), but it is not obvious which one to prefer in general, with the good old LDA itself being competitive in accuracy in a variety of cases. It is also worth mentioning that, when used in combination with plane-waves methods (see next section), XC hybrid functionals usually require a computational effort that is several orders of magnitudes higher than what is required by local XC functionals like the LDA or the GGAs.

Whatever functional is used, these type of calculations all go under the classification of *ab initio*, in the sense that no experimental input is allowed, apart from the four fundamental constants mentioned above. Of course, it would be desirable to have a unique functional with the highest possible accuracy for any system, but this stage has not been reached yet (at the time of writing).

13.2.1.2 PPs and basis sets

In practical cases, it is often necessary to introduce one additional approximation in order to speed up the calculations, known as the PP approximation. In essence, this is a way to freeze the electrons of the core of the atoms, and remove them from the calculations. The justification for doing this is that the core electrons are so tightly bound to the nuclei that they are essentially undisturbed by the chemical bonding, or, conversely, the chemistry of materials is unaffected by the behavior of the core electrons. This implies a saving in complexity and computer time which is proportional to the number of electrons that have been frozen, but, as discussed shortly, the saving becomes enormous in the most widely diffused computer codes which are based on plane-wave expansions of the single-particle KS orbitals.

In order to solve the KS equations, it is necessary to expand the auxiliary KS orbitals in terms of some known basis functions. A variety of possible choices are available. Traditional quantum chemistry codes often use Gaussians, which are quite well-suited for very localized orbitals, and in the course of the years a large amount of expertise has accumulated to create high-quality basis sets for a wide range of materials. The drawback of Gaussians is that the quality of the basis set depends on the choice of the user, and transferability can be an issue when different systems are compared. An alternative set of functions which are totally unbiased and systematically improvable is plane waves. They also have the additional advantage of adapting naturally to calculations in which periodically boundary conditions are employed, which is a very useful set-up even in systems that have no periodicity, in order to reduce finite size effects. Plane-wave calculations are relatively simple, and the evaluation of forces and the stress tensor are not much more difficult than the evaluation of the total energy. A drawback of plane waves is that a large number of them may be needed for describing rapidly varying functions, like the very localized core orbitals, or the valence wave functions in the core region, which need to oscillate widely in order to be orthogonal to the core orbitals. For this reason, plane-wave calculations are almost always associated with the use of PPs.

The first aim of PPs is to eliminate the core electrons from the explicit calculations because they do not participate in the chemical properties of matter, at least until their binding energy is much higher than the energy involved in the chemical properties one wants to study. So one freezes them around the nuclei and redefines the system as it was formed by ions plus valence electrons. We are left now with the problem of dealing with the oscillations in the core region of the valence wave functions, due to the orthogonalization to the core wave functions. The solution to this is the introduction of a PP, which substitutes the ionic Coulomb potential in such a way that the valence pseudo-eigenvalues are the same as the all-electron (AE) ones on some reference configuration in the atom. The pseudo-wave functions coincide with the AE ones from a fixed core radius one, and are as smooth as possible below the core radius, with the only constraint to be normalized (norm-conserving (NC) PPs). To satisfy these

requirements the PP usually must be angular momentum dependent, that is, pseudo-wave functions corresponding to different angular momenta are eigenfunctions of different potentials. However, the long-range behavior of these different potentials must resemble the true one, because above the core radius the pseudo-wave functions are identical to the AE ones. This means that the difference must be confined in the core region and then the PP can be written in the following form (Hamann *et al.*, 1979; Kerker, 1980; Bachelet *et al*, 1982):

$$V_a(\mathbf{r}, \mathbf{r}') = V_a^{\mathrm{loc}}(r)\delta(\mathbf{r} - \mathbf{r}')$$
$$+ \sum_{l=0}^{l_{\max}} V_{a,l}(r) P_l(\hat{\mathbf{r}}, \hat{\mathbf{r}}')\delta(r - r') \qquad [13]$$

where $V_a^{\mathrm{loc}}(r)$ is the local long-range (spherical) part and approaches the AE potential above a cutoff radius r_c^{loc}, and $V_{a,l}(r)$ is the short-range angular-momentum-dependent part, the index a indentifies the atom, and P_l is the projector onto the angular momentum l,

$$P_l(\hat{\mathbf{r}}, \hat{\mathbf{r}}') = \sum_{m=-l}^{l} Y_{l,m}(\theta, \phi) Y_{l,m}^*(\theta', \phi') \qquad [14]$$

with $Y_{l,m}$ being the spherical harmonics. The quality of the PP depends on its transferability properties, that is, the capability to reproduce the AE results over a wide range of electronic configurations, and of course for the atom in different environments.

13.2.1.3 Ultrasoft (Vanderbilt) PPs

The requirement of NC for the pseudo-wave functions can be a limiting factor for numerical calculations when the valence electrons are also very localized around their nuclei. This is a particularly serious problem for first-row elements, such as carbon, and more so for nitrogen, oxygen, and for transition metals, where the *d*-electrons are as localized as shallow core states but have an extraction energy which is not much larger than valence energies, and for this reason cannot be excluded from the calculations. If this is the case, the utilization of NC PPs requires huge plane wave (PW) basis sets to achieve an acceptable accuracy. In a work published in 1990, Vanderbilt showed that, introducing a generalized formalism, the norm conservation constraint could be removed. In this way, one can construct much smoother pseudo-wave functions, with the only constraint of matching the AE ones at and above a fixed core radius (see **Figure 1**). The price to pay for having such smooth

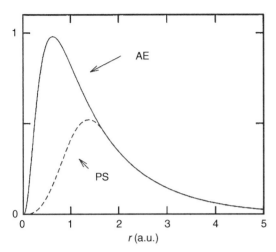

Figure 1 All-electron (solid) and ultrasoft pseudo-(dashed) radial wave functions of the 3*d*-orbital of nickel. $r_c = 1.75$ a.u.

pseudo-wave functions is that, due to the fact that the pseudo-wave functions are not normalized anymore, the charge density has to be restored by adding an 'augmentation' part.

$$n(\mathbf{r}) = \sum_i |\phi_i(\mathbf{r})|^2 + n_{\mathrm{aug}}(\mathbf{r}) \qquad [15]$$

and the KS equations take the generalized form

$$H_{\mathrm{KS}}|\phi_i\rangle = \epsilon_i S|\phi_i\rangle \qquad [16]$$

where S is a nonlocal overlap operator.

13.2.1.4 The projector augmented wave method

In 1994, Blöchl invented a method to reconstruct the AE wave function inside the core region (Blöchl, 1994). The method, called projected augmented wave (PAW), is closely related to the Vanderbilt's PP method, as shown by Kresse and Joubert (1999), but has has been shown to be capable of reproducing essentially the same results of AE calculations, effectively removing the PP approximation. The PAW method is still only available in a handful of computer codes, but as evidences of its advantages accumulate we believe that it will become a standard method of DFT–PW calculations.

13.2.2 Beyond DFT

As the search for the 'divine' functional in DFT goes on, alternative methods to solve the many-body problem are being developed, like dynamical mean field

theory (Savrasov and Kotliar, 2003), and quantum Monte Carlo (QMC) (Foulkes *et al.*, 2001). The QMC method is particularly attractive because it adapts very well to parallel computers. As the speed of single-processor units approaches physical limits, it is conceivable that the direction of the future increase of computer power will be the increase of the number of processors, and therefore Monte Carlo methods will develop naturally on these machines.

QMC methods have been amply described in reviews (Foulkes *et al.*, 2001), including detailed descriptions of how to implement them in practice (Foulkes *et al.*, 2001; Umrigar *et al.*, 1993). The next section presents a brief introduction to the method; the reader is recommended to these reviews for more details.

13.2.2.1 QMC methods

QMC methods encompass a number of different techniques to solve the many-body problem of a system of N interacting quantum particles. In what follows, a brief mention is made of the variational Monte Carlo (VMC) and the diffusion Monte Carlo (DMC) methods.

The VMC method gives an upper bound to the exact ground-state energy E_0. Given a normalized trial wave function $\Psi_T(\mathbf{R})$, where $\mathbf{R} = (\mathbf{r}_1, \mathbf{r}_2, \ldots, \mathbf{r}_N)$ is a 3N-dimensional vector representing the positions of N electrons, and denoting by $\hat{H}$ the many-electron Hamiltonian, the variational energy $E_v \equiv \langle \Psi_T | \hat{H} | \Psi_T \rangle \geq E_0$ is estimated by sampling the value of the local energy $E_L(\mathbf{R}) \equiv \Psi_T^{-1}(\mathbf{R})\hat{H}\Psi_T(\mathbf{R})$ with configurations $\mathbf{R}$, distributed according to the probability density $\Psi_T(\mathbf{R})^2$. Common trial wave functions are of the Slater–Jastrow type:

$$\Psi_T(\mathbf{R}) = D^\uparrow D^\downarrow e^J \qquad [17]$$

where $D^\uparrow$ and $D^\downarrow$ are Slater determinants of up- and down-spin single-electron orbitals, and e^J is the so-called Jastrow factor, which is the exponential of a sum of one-body and two-body terms, with the latter being a parametrized function of electron separation, designed to satisfy the cusp condition, that is, to counterbalance the divergence in the potential when two electrons become very close, with a corresponding divergence in the kinetic energy of opposite sign. The parameters in the Jastrow factor are varied to minimize the variance of the local energy E_L, which also results in a minimization of the local energy itself. Different trial wave functions have also been used, like a more general form of that defined in eqn [17], in which one uses a linear combination of more than one Slater determinants, or antisymmetrized geminals (Casula and Sorella, 2003).

Because of the variational principle, VMC results are an upper bound of true energy, which is not available in VMC (unless, of course, the trial wave function is the exact many-body wave function).

The DMC method is designed to obtain the true ground-state properties of the system. The basic idea is to compute the evolution of the many-body wave function Φ by the time-dependent Schrödinger equation in imaginary time $-\partial\Phi/\partial t = (\hat{H} - E_T)\Phi$, where E_T is an energy offset. The equivalence of this to a diffusion equation allows Φ to be regarded as a probability distribution represented by a population of diffusing walkers. It can be shown that, by adjusting the energy offset E_T appropriately, in the limit of imaginary time going to infinity, the distribution of walkers converges to the ground-state many-body wavefunction, and therefore, in principle, the DMC scheme yields the exact ground-state energy. However, for fermion systems, there is a fundamental problem, viz. Φ changes sign as $\mathbf{R}$ varies, so that it can only be treated as a probability in regions of $\mathbf{R}$-space where it does not change sign. These regions, and the nodal surfaces that defines their boundaries, have to be fixed by the introduction again of some trial wave function Ψ_T. The consequence is that the energy given by DMC is not the true ground-state energy but again an upper bound because of the constraint that the nodal surface is that of Ψ_T. This gives rise to the so-called fixed-node error. The form of Ψ_T is usually the same as that used in VMC, and it is used to provide a starting point for DMC simulations.

An important issue in QMC simulations is the representation of the single-particle orbitals that make up Ψ_T. Much as in DFT, a number of different choices are possible, including Gaussians, plane waves, or B-splines (Alfe and Gillan, 2004). The latter have the attractive feature of being intimately related to plane waves, sharing the properties of being unbiased and systematically improvable, but also of being localized, a crucial property in QMC simulations.

It has also been shown recently that the evaluation of the trial wave function Ψ_T, often the most expensive part of the calculations, can be performed such that its cost scales linearly with the size of the system (Williamson *et al.*, 2001; Manten and Lüchow, 2003; Alfè and Gillan, 2004; Roboredo and Williamson, 2005), making the so-called $o(N)$–QMC techniques promising candidates for calculations on very large systems in the future.

The fixed-node error is the only uncontrollable source of error in DMC calculations; however, for systems containing atoms heavier than those of the first row, it is necessary to use PPs. The reason is that, for a trial wave function of fixed quality, the variance of the energy is proportional to the energy itself. Since core electrons have large negative energies, the variance associated to the evaluation of this component of the energy would mask completely the relatively small energies due to the valence electrons, which are those involved in the chemical bonding. For this reason, it is customary to use PPs also in QMC calculations. The nonlocality that is essential in these PPs gives rise to unavoidable errors in DMC. The reason is that the diffusion equation with a nonlocal Hamiltonian contains a term that can change its sign as time evolves, and therefore presents the same difficulties as the fermion sign problem. To avoid this difficulty one introduces the so-called localization approximation, in which this problematic term is simply neglected. If the trial wave function Ψ_T is close to the true (fixed-node) ground-state wavefunction Ψ, then this approximation introduces an error which is small and proportional to $(\Psi_T - \Psi)^2$. This error, however, is nonvariational, so it can decrease as well as increase the total energy. Some recent promising work has gone toward addressing this problem (Casula *et al.*, 2005).

To summarize, provided all sources of technical errors are kept under control, the only two sources of error in QMC calculations are the fixed-node and the locality approximation. It has been shown that DMC techniques can deliver much higher accuracy than DFT techniques (Filippi *et al.*, 2002; Leung *et al.*, 1999; Grossman *et al.*, 1995; Alfe and Gillan, to be published), hinting that the above two approximations are less serious than those involved in the formulation of XC functionals in DFT calculations (at least for the systems considered). However, DMC total energy calculations are typically 2–3 order of magnitude more expensive than traditional DFT techniques, and for this reason the range of materials studied so far is rather limited, and therefore so is the evidence which would favor DMC compared to DFT. As mentioned already above, however, QMC techniques adapt naturally to large parallel computers, while traditional DFT techniques can take less advantage of this kind of architecture. It is expected therefore that the development and application of QMC techniques will be boosted in the future.

13.3 Mineral Properties and Behavior

Having introduced the computational tools, the discussion must turn to the main purpose of this article, namely how these tools can be applied to the study of the high-pressure and -temperature properties of materials. We will focus in particular on structure and elastic properties, phase diagrams, phase transitions, and the thermodynamics of solutions.

To simulate materials under pressure is not much more difficult than to perform calculations at zero pressure; all that needs to be done is to change the volume of the simulation cell appropriately. In doing so, one possible problem can be the shortening of the nearest-neighbors distance among the atoms, which if drops below the sum of the core radii of the PPs employed (or PAW potentials), may affect the quality of the results. Therefore, some care is necessary in designing the potentials appropriate for the conditions where they need to be used. However, apart from this possible shortcoming, it is often the case that simulations at high pressure are even more accurate than those at low pressure. The reason is that as the pressure increases the charge density becomes more homogeneous, which helps the XC functionals in their work.

By contrast, high-temperature simulations are much more demanding than zero-temperature ones. The reason is that at high temperature it is the free energy that plays the essential role, and an accurate calculation of this requires expensive sampling of the phase space. For solids at not too high temperature, it is often accurate enough to use the quasi-harmonic approximation, in which the potential energy of the crystal is expanded to second order in the displacement of the atoms from their equilibrium positions. This quasi-harmonic potential usually provides a very accurate description of the dynamical properties of the system at low temperature, and gives easy access to the free energy of the system, which can be calculated analytically as a function of temperature. The prefix quasi- is there to indicate that this quasi-harmonic potential depends on the volume of the system. In practice, the quality of the thermodynamics obtained within the quasi-harmonic approximation is often preserved also to temperatures not far from the melting temperature, although at such high temperatures a full account of anharmonic effects becomes necessary, at least to assess the validity of the quasi-harmonic approximation. For highly anharmonic solid and for liquids, one has to resort to MD or Monte Carlo techniques to

sample the phase space. MD simulations are particularly attractive because they also provide dynamical informations like diffusion, or autocorrelation properties which can be used together with the fluctuation–dissipation theorem to evaluate a number of physical properties of the system, as seen below. The next section presents the zero-temperature properties of materials, and the following section discusses finite temperature.

13.3.1 Static Properties

13.3.1.1 Crystal structures and phase transitions

At zero temperature, the main thermodynamic variable is the internal energy of the system E. The simplest possible first-principles calculation one can do is the evaluation of the total energy of a system containing a certain number of atoms at fixed lattice sites. To find the most stable configuration of the atoms, one simply minimizes the total energy with respect to the atomic positions. This is usually done by evaluating forces, which are then used to move the atoms toward their equilibrium positions. For simple crystal structures, this may not be necessary, as the positions may be constrained by symmetry. An example of this is the Earth's mantle mineral MgO (periclase), which at zero pressure has a face centered cubic crystal structure, with the Mg and the O atoms in the primitive cell sitting at a corner and at the center of the conventional cubic cell. In **Figure 2** we show a comparison of pressures as function of volume

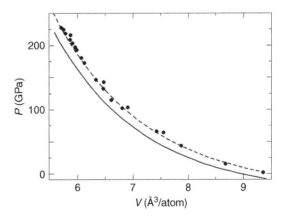

Figure 2 Pressure as function of atomic volume for rocksalt MgO calculated with DFT–LDA (solid line) and DFT–GGA (dashed line), compared with experiments (stars) (Duffy *et al.*, 1995). Calculations do not include zero-point or thermal effects. Reprinted from Alfè, D (2005) *Physical Review Letters* 94: 235701.

between first-principles calculations and experiments. The calculations have been performed with DFT and both the LDA and the GGA approximations known as PW91 (Wang and Perdew, 1991) for the XC. The GGA results appear to agree very well with the experimental data, while the LDA ones underestimate the pressure slightly. However, these particular calculations do not include zero-point motion and room-temperature thermal expansion, which are instead present in the experiments. With those effects included (Karki *et al.*, 2000), the experimental data would fall roughly in the middle of the two theoretical curves.

As the pressure is increased, the crystal structure of the material may change. For MgO this does not happen up to at least 227 GPa (Duffy *et al.*, 1995), but it is believed that as the pressure is increased still further MgO would transform to the structure of CsCl, which is simple cubic with the Mg and the O atoms in the primitive cell still sitting at the corner and the center of the cube, respectively. The pressure at which the phase transformation occurs is defined by the point where the enthalpies of the two crystal structures cross. These enthalpies can be computed using *ab initio* techniques, by computing the energy E and the pressure p of the crystal as function of volume V, and then construct the enthalpy $H = E + pV$. For MgO, this has been done using a number of different techniques in the past, including various flavors of DFT (Chang and Cohen, 1984; Mehl *et al.*, 1988; Karki *et al.*, 1997; Jaffe *et al.*, 2000; Drummond and Ackland, 2002; Oganov and Dorokupets, 2003; Oganov *et al.*, 2003) and, very recently, QMC (Alfe *et al.*, 2005). The latest DFT results point toward a transition at ~500 GPa, and the QMC results show that the transition is at 600 ± 30 GPa.

Another example of phase transition between different crystal structures is the transformation of iron from the zero-pressure magnetic body-centered-cubic (b.c.c) to the hexagonal-close-packed (h.c.p) structure at a pressure between 10 and 15 GPa (Jephcoat *et al.*, 1986). Like the previous MgO case, the cubic structure is symmetric and it is necessary only to evalute the energy of the crystal as a function of volume. In contrast to b.c.c, however, the h.c.p crystal structure has an additional degree of freedom, arising from the lack of a symmetry relating the hexagonal plane and the direction perpendicular to the plane. This additional degree of freedom, known as the c/a ratio, needs to be optimized for every volume V. Once this is done, an enthalpy curve can be constructed and compared with that obtained

from the b.c.c structure. Calculations using DFT with the LDA or various GGAs have been performed, and it has been shown that in this particular case the LDA gives poor agreement with the experiments, even failing to predict the correct zero-pressure crystal structure (Cho and Scheffler, 1996; Zhu *et al.*, 1992; Leung *et al.*, 1991; Körling and Häglund, 1992; Singh *et al.*, 1991), while PW91, for example, predicts the transition between 10 and 13 GPa (depending on the exact details of the pseudo- or PAW potential) (Alfe *et al.*, 2000), in good agreement with the experimental value which is in the range 10–15 GPa (Jephcoat *et al.*, 1986).

Finally, as a last example, we mention the Earth's mantle mineral $MgSiO_3$ perovskite, which at mantle pressures has an orthorombic crystal structure with the atoms in the cell not constrained by symmetry operations (Wentzcovitch *et al.*, 1993; Stixrude and Cohen, 1993; D'Arco *et al.*, 1993). It follows that at any fixed volume one needs to optimize not only the lattice vectors, but also the positions of the atoms in the cell. Very recently, this mineral has been found to display a phase transition to a new phase, named post-perovskite (Murakami *et al.*, 2004; Oganov and Ono, 2004). This transition has also been found by *ab initio* calculations (Tsuchiya *et al.*, 2004; Iitaka *et al.*, 2004; Oganov and Ono, 2004). The transition pressure from the static first-principles calculations appears to be ~100 GPa, which is below the core–mantle pressure. However, Oganov and Ono (2004) and Tsuchiya *et al.* (2004) showed that high-temperature harmonic effects are responsible for an increase in the transition pressure, which is therefore predicted to be close to that at the top of D″ zone at the bottom of the mantle.

13.3.1.2   Elastic constants

Most of what we know about the interior of our planet comes from seismology, and therefore from the elastic behavior of the minerals inside the Earth. The theory of elasticity of crystals can be found in standard books (Wallace, 1998), and therefore we will not dwell on it for too long. Briefly, if a crystal is subjected to an infinitesimal stress $d\sigma_{ij}$, with i and j running through the three Cartesian directions in space, then it will deform according to the strain matrix $d\epsilon_{ij}$:

$$d\sigma_{ij} = \sum_{k,l} c_{ijkl} d\epsilon_{kl}   [18]$$

The constant of proportionality between stress and strain, c_{ijkl}, is a fourth-rank tensor of elastic constants.

With no loss of generality we can assume $d\sigma_{ij}$ and $d\epsilon_{ij}$ to be symmetric ($d\sigma_{ij} \neq d\sigma_{ji}$ would imply a nonzero torque on the crystal, which would simply impose an angular acceleration and not a deformation), and therefore the elastic constant tensor is also symmetric. It is therefore possible to rewrite the second-rank tensors $d\sigma_{ij}$ and $d\epsilon_{ij}$ as six components (vectors), in the Voigt notation, with the index pairs 11, 22, 33, 23, 31, and 12 represented by the six symbols 1, 2, 3, 4, 5, and 6, respectively. In this notation, the stress–strain relation appears as

$$d\sigma_i = \sum_j C_{ij} d\epsilon_j \qquad [19]$$

with i and j going from 1 to 6. Elastic constants are given as the coefficients C_{ij} in this notation. The matrix C_{ij} is symmetric, so that the maximum number of independent elastic constants of a crystal is 21. Because of crystal symmetries, the number of independent constants is usually much smaller. For example, in cubic crystals, there are only three elastic constants, in h.c.p Fe there are five, and in orthorhombic $MgSiO_3$ perovskite there are nine.

Equation [19] provides the route to the calculation of the elastic properties of materials, and it can be applied both at zero and high temperatures. At zero temperature, the components of stress tensor can be calculated as (minus) the partial derivative of the internal energy with respect to the components of the strain:

$$\sigma_{ij} = -\partial E/\partial \epsilon_{ij}|\epsilon \qquad [20]$$

Examples of zero-temperature calculations of elastic constants include the DFT calculations of Stixrude and Cohen (1995) on the h.c.p crystal structure of iron at Earth's inner core conditions, which suggested a possible mechanism based on the partial alignment of h.c.p crystallites to explain the seimic anysotropy of the Earth's inner core. Another example is the recent GGA and LDA calculations of the elastic constant of the recently discovered post-perovskite phase by three groups (Oganov and Ono, 2004; Oganov *et al.*, 2005; Tsuchiya *et al.*, 2004; Iitaka *et al.*, 2004), which showed that this phase is elastically very anisotropic, and that with a proper alignment it is possible to explain the observed seismic anisotropy of the D″ region.

13.3.2   Finite Temperature

The extension to finite temperature properties of materials could simply be obtained by substituting the internal energy E with the Helmholtz free energy F. The ij component of the stress tensor σ_{ij} is (minus) the partial derivative of F with respect to strain ϵ_{ij}, taken at constant T, and holding constant all the other components of the strain tensor:

$$\sigma_{ij} = -\partial F/\partial \epsilon_{ij}|_{\epsilon,T} \qquad [21]$$

Similarly, the pressure p is obtained as (minus) the partial derivative of F with respect to volume, taken at constant temperature:

$$p = -\partial F/\partial V|_T \qquad [22]$$

If the system of interest is at sufficiently high temperature (above the Debye temperature), the nuclei can be treated as classical particles, and the expression of the Helmholtz free energy F for a system of N identical particles enclosed in a volume V, and in thermal equilibrium at temperature T is (Frenkel and Smit, 1996)

$$F = -k_B T \ln \left\{ \frac{1}{N!\Lambda^{3N}} \int_V d\mathbf{R}_1 \cdots d\mathbf{R}_N \, e^{-\beta U(\mathbf{R}_1, \ldots, \mathbf{R}_N; T)} \right\} \qquad [23]$$

where $\Lambda = h/(2\pi M k_B T)^{1/2}$ is the thermal wavelength, with M the mass of the particles, h the Planck's constant, $\beta = 1/k_B T$, k_B is the Boltzmann constant, and $U(\mathbf{R}_1, \ldots \mathbf{R}_N; T)$ the potential energy function, which depends on the positions of the N particles in the system, and possibly on temperature, in which case U is the electronic free energy. The multidimensional integral extends over the total volume of the system V.

By taking minus the derivative of F with respect to volume at constant temperature we obtain

$$p(V,T) = \frac{N k_B T}{V}$$
$$+ \frac{\int_V d\mathbf{R}_1 \cdots d\mathbf{R}_N e^{-\beta U(\mathbf{R}_1, \ldots, \mathbf{R}_N; T)} \left(\frac{-\partial U(\mathbf{R}_1, \ldots, \mathbf{R}_N; T)}{\partial V} \right)_T}{\int_V d\mathbf{R}_1 \cdots d\mathbf{R}_N e^{-\beta U(\mathbf{R}_1, \ldots, \mathbf{R}_N; T)}}$$
$$= \frac{N k_B T}{V} + \left\langle \left(\frac{-\partial U(\mathbf{R}_1, \ldots, \mathbf{R}_N; T)}{\partial V} \right)_T \right\rangle \qquad [24]$$

The first term is the kinetic pressure and is present also in a system with no interactions between the particles (the ideal gas). The appearance of this term can be also understood by realizing that the integral appearing in eqn [23] is proportional to V^N. The second term is the canonical thermal average of the derivative of the potential (free) energy function with respect to volume, taken at constant temperature.

A similar expression holds for the stress tensor, which at finite temperature is evaluated as the thermal average of the derivative of F with respect to the strain components, plus the kinetic term.

Analogously, the internal energy $E = (\partial(F/T)/\partial(1/T))_V$ is given by

$$E(V,T) = \frac{3}{2}Nk_B T + \langle U(\mathbf{R}_1 \cdots \mathbf{R}_N; T)\rangle \qquad [25]$$

This shows that a number of finite properties can simply be calculated by taking the thermal average of the corresponding quantity evaluated at $T = 0$, plus a trivial kinetic term.

13.3.2.1 *Molecular dynamics*

If the system is ergodic, thermal averages (represented in the equations above as $\langle \cdot \rangle$) can be calculated as time averages along an MD simulation (Alder and Wainwright, 1959; Gibson *et al.*, 1960). The main idea here is to move the ions according to the Newton's equations of motion. This is achieved in practice by discretizing the equation of motion $m_i \partial v_i/\partial t = \mathbf{F}_i$ (m_i is the mass of atom i, $\mathbf{v}_i$ is its velocity, and $\mathbf{F}_i$ is the force acting on it). This is done by dividing time into time steps Δt, and approximate the solution of the equation of motion, for example, as proposed by Verlet (1967) (*see also* Allen and Tildesley (1987), and Frenkel and Smit (1996)):

$$\begin{aligned}\mathbf{r}_i(t + \Delta t) &= \mathbf{r}_i(t) + \mathbf{v}_i(t)\Delta t + \frac{1}{2m_i}\mathbf{F}_i(t)\Delta t^2 \\ \mathbf{v}_i(t + \Delta t) &= \mathbf{v}_i(t) + \frac{1}{2m_i}(\mathbf{F}_i(t) + \mathbf{F}_i(t + \Delta t))\Delta t\end{aligned} \qquad [26]$$

If the time step is small enough (usually less than 1/20 of a typical vibrational period), the Verlet algorithm conserves well the total energy of the system, both on short and long timescales. If the volume of the simulation cell V and the number of atoms N are kept constant, this so-called (N, V, E) simulation generates configurations in phase space that are distributed according to the microcanonical ensemble.

It is a standard result of statistical mechanics that thermal averages evaluated either in the micro- or in the canonical ensemble are equivalent if the system is sufficiently large, but it is useful to be able to perform simulations in ensembles other than the microcanonical one. For example, in order to obtain thermal averages in the canonical ensemble (i.e. constant N, V, and T), like the pressure in eqn [24], one can couple

the system with an external heat bath, following the prescription of Andersen (1980) or Nosé (1984) and Hoover (1985). When combined with the Parrinello–Rahman constant-stress technique (Parrinello and Rahman, 1980), this also allows simulations to be performed at constant T and $\sigma_{\alpha\beta}$ (see e.g. Wentzcoritch *et al.*, 1993).

The forces F_i can be calculated within the framework of DFT, in which case the method is sometimes called first-principles molecular dynamics (FPMD). As mentioned earlier, for the method to be applicable in practice, one makes the fundamental approximation that the dynamics of the ions is decoupled from that of the electrons. This is usually justified due to the large difference in masses between the two sets of particles.

The first FPMD simulation was performed by Car and Parrinello (CP) (1985), who proposed an elegant method to keep the electrons on the ground state along the MD trajectory. This was done by including the electronic degrees of freedom into a generalized Lagrangian, by assigning a ficticious mass to the single-particle wave functions and treat them as dynamical variables, like the positions of the ions. With a judicious choice of this ficticious mass, the electronic degrees of freedom would remain decoupled from the ionic ones, following them adiabatically while remaining in the their ground state (at least for nonmetallic systems). An alternative method is to bring the electrons to the ground state at each time step (Kresse and Furthmuller, 1996), which is usually more costly than the CP scheme, but this is compensated by the possibility of making longer time steps. This method can be also easily applied to metallic systems, and can become very efficient if combined with the extrapolation of the single-particle wave functions (Arias *et al.*, 1992; Mead, 1992) and the electronic charge density (Alfè, 1999).

As some examples of the use of FPMD to compute thermal averages, and study the high-temperature properties of solids, we mention the work of Oganov *et al.* (2001), who studied the high-temperature elastic constants of $MgSiO_3$ perovskite, and more recently that of Wookey *et al.* (2005), who studied the high temperature elastic constants of $MgSiO_3$ post-perovskite, and Gannarelli *et al.* (2005), who studied the elastic behavior of h.c.p iron at inner core p and T conditions.

Using MD simulations, it is also possible to study the properties of liquids. For example, in a liquid, the atoms are free to diffuse throughout the whole volume, and this behavior can be characterized by diffusion coefficients D_α, where α runs over different species in the system. These D_α are straightforwardly related to the mean square displacement of the atoms through the Einstein relation (Allen and Tildesley, 1987):

$$\frac{1}{N_\alpha}\left\langle \sum_{i=1}^{N_\alpha} |\mathbf{r}_{\alpha i}(t_0 + t) - \mathbf{r}_{\alpha i}(t_0)|^2 \right\rangle \to 6D_\alpha t, \text{as } t \to \infty \quad [27]$$

where $\mathbf{r}_{i\alpha}(t)$ is the vector position at time t of the ith atom of species α, N_α is the number of atoms of species α in the cell, and $\langle\rangle$ means time average over t_0. The diffusion coefficient can also be used to obtain a rough estimate of the viscosity η of the liquid, by using the relation between the two stated by the Stokes–Einstein relation:

$$D_\eta = \frac{k_B T}{2\pi a} \quad [28]$$

This technique was used by de Wijs *et al.* (1998) to estimate the viscosity of liquid iron at Earth's core conditions. The Stokes–Einstein relation (eqn [28]) is exact for the Brownian motion of a macroscopic particle of diameter a in a liquid of viscosity η. The relation is only approximate when applied to atoms; however, if a is chosen to be the nearest-neighbors distance of the atoms in the solid, eqn [28] provide results that agree within 40% for a wide range of liquid metals. To calculate the viscosity rigorously, it is possible to use the Green–Kubo relation

$$\eta = \frac{V}{k_B T} \int_0^\infty dt \langle \sigma_{xy}(t)\sigma_{xy}(0) \rangle \quad [29]$$

where σ_{xy} is the off-diagonal component of the stress tensor $\sigma_{\alpha\beta}$ (α and β are Cartesian components). This relation was used in the context of first-principles calculations for the first time by Alfè and Gillan (1998b), who first calculated the viscosity of liquid aluminum at ambient pressure and a temperature of 1000 K, showing that the method provided results in good agreement with the experiments, and then applied the method to the calculation of the viscosity of a liquid mixture of iron and sulfur under Earth core conditions. In **Figure 3**, we show the integral in eqn [29] calculated from 0 to time t for this iron–sulfur liquid mixture. In principle, this has to be computed from zero to infinity, as stated in eqn [29]; however, in this particular case, there is nothing to be gained by extending the integral beyond about 0.2 ps, after which the integrand has decayed to zero and it is dominated by statical noise. The figure also shows the

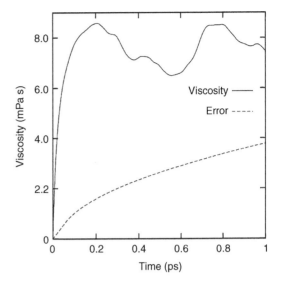

Figure 3 Viscosity integral of the average stress autocorrelation function and its statistical error as a function of time for liquid Fe/S under Earth's core conditions. Reprinted from Alfè and Gillan (1998b).

computed statistical error on the integral, and from this it was possible to infer the value for the viscosity $\eta = 9 \pm 2$ mPa s, in good agreement with that obtained from the diffusion coefficient via the Einstein relation [28], calculated to be $\eta \sim 13$ mPa s in a previous paper (Alfè and Gillan, 1998a).

13.3.3 Thermodynamic Properties

The phase stability of a system is determined by the minimum of its Gibbs free energy $G = F + pV$. Since $p = -\partial F/\partial V|_T$, also knowledge of F as function of V and T allows the computation of G.

More generally, equilibrium in a multispecies system is determined by the chemical potentials μ_i, with i running over the different species, which represents the constant of proportionality between the energy of the system and the amount of the specie i (Wannier, 1966):

$$\mu_i = \left(\frac{\partial E}{\partial N_i}\right)_{S,V} \quad [30]$$

where S is the entropy, and N_i is the number of particles of the species i. Alternative equivalent definitions of the chemical potential are (Wannier, 1966; Mandl, 1997):

$$\mu_i = \left(\frac{\partial F}{\partial N_i}\right)_{T,V} = \left(\frac{\partial G}{\partial N_i}\right)_{T,p} = -T\left(\frac{\partial S}{\partial N_i}\right)_{E,V} \quad [31]$$

Equilibrium between two phases is determined by the condition of equality of the chemical potential of each individual species in the two phases. The next section starts by considering a single-component system. The extension to multicomponent systems will be considered in Section 13.3.3.5.

13.3.3.1 The Helmholtz free energy: Low-temperature and the quasi-harmonic approximation

For a solid at low temperature, F can be easily accessed by treating the system in the quasi-harmonic approximation. This is obtained by expanding the potential (free) energy function U around the equilibrium positions of the nuclei. The first term of the expansion is simply the energy of the system calculated with the ions in their equilibrium positions, $E_{perf}(V, T)$ (this is a free energy at finite temperature, and therefore depends both on V and T). If the crystal is in its minimum energy configuration, the linear term of the expansion is zero, and by neglecting terms of order three and above in the atomic displacements we have the quasi-harmonic potential as

$$U_{harm} = E_{perf} + \frac{1}{2} \sum_{ls\alpha, l't\beta} \Phi_{ls\alpha, l't\beta} u_{ls\alpha} u_{l't\beta} \qquad [32]$$

where $\mathbf{u}_{ls}$ denotes the displacement of atom s in unit cell l, α and β are Cartesian components, and $\Phi_{ls\alpha, l't\beta}$ is the force constant matrix, given by the double derivative $\partial^2 U / \partial u_{ls\alpha} \partial u_{l't\beta}$ evaluated with all atoms at their equilibrium positions. This force constant matrix gives the relation between the forces $\mathbf{F}_{ls}$ and the displacements $\mathbf{u}_{l't}$, as can be seen by differentiating eqn [32] and ignoring the higher-order anharmonic terms:

$$F_{ls\alpha} = -\partial U / \partial u_{ls\alpha} = -\sum_{l't\beta} \Phi_{ls\alpha, l't\beta} u_{l't\beta} \qquad [33]$$

Within the quasiharmonic approximation, the potential energy function U_{harm} completely determines the physical properties of the system, and in particular the free energy, which takes the form

$$F(V, T) = E_{perf}(V, T) + F_{harm}(V, T) \qquad [34]$$

where the quasiharmonic component of the free energy is

$$F_{harm} = k_B T \sum_n \ln(2 \sinh(\hbar \omega_n / 2 k_B T)) \qquad [35]$$

with ω_n the frequency of the nth vibrational mode of the crystal. In a periodic crystal, the vibrational modes can be characterized by a wave vector $\mathbf{k}$, and for each such wave vector there are three vibrational modes for every atom in the primitive cell. If the frequency of the sth mode at wave vector $\mathbf{k}$ is denoted by $\omega_{\mathbf{k}s}$, then the vibrational free energy is

$$F_{harm} = k_B T \sum_{\mathbf{k}s} \ln(2 \sinh(\hbar \omega_{\mathbf{k}s} / 2 k_B T)) \qquad [36]$$

The vibrational frequencies $\omega_{\mathbf{k}s}$ can be calculated from first principles, and we shall see below how this can be done.

Once this quasi-harmonic free energy is known, all the thermodynamical properties of the system can be calculated. In particular, the pressure is given by

$$p = -\partial F / \partial V |_T = -\partial E_{perf} / \partial V |_T - \partial F_{harm} / \partial V |_T \qquad [37]$$

The last term in the equation above is the ionic component of the thermal pressure, and it is different from zero because the vibrational frequencies $\omega_{\mathbf{k}s}$ depend on the volume of the crystal. In fact, it is easy to see from eqn [36] that even at zero temperature there is a finite contribution to the quasi-harmonic free energy, given by

$$F_{harm}(V, 0) = \sum_{\mathbf{k}s} \frac{\hbar \omega_{\mathbf{k}s}}{2} \qquad [38]$$

This zero-point energy contribution to the harmonic free energy is also responsible for a contribution to the pressure. Since usually the vibrational frequencies $\omega_{\mathbf{k}s}$ increase with decreasing volume, these contributions are positive, and are responsible for the phenomenon of thermal expansion in solids.

The dependence of $E_{perf}(V, T)$ on T also means that there is an electronic contribution to the thermal pressure, which is also positive, and in some cases (i.e., iron at Earth's core conditions) can be a significant fraction of the thermal pressure, and a non-negligible fraction of the total pressure (Alfè *et al.*, 2001).

As an example of a calculation of the thermal expansivity of minerals using the quasi-harmonic approximation, in **Figure 4**, the temperature dependence of the thermal expansivity α of MgO at pressures up to 200 GPa, compared with experimental results, is shown. At ambient pressure, the first-principles $\alpha(T)$ agrees very closely with experiment

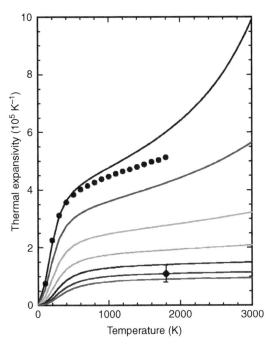

Figure 4   Temperature dependence of thermal expansivity α of MgO on isobars at 0, 10, 30, 60, 100, 150, and 200 GPa (curves from top to bottom). Experimental data at zero pressure are indicated by filled circles. Average value of α over temperature range 300–3300 K and pressure range 169–196 GPa derived from shock experiments is indicated by the diamond.

up to ~1000 K (about one-third of the melting temperature), which is a result of the good agreement of the calculated phonons with the experimental ones. The increasingly poor agreement at high T is due to anharmonic effects coming into play. These can be taken into account by avoiding the quasi-harmonic approximation and evaluating the full free energy of the system, as we shall see in Section 13.3.3.3.

13.3.3.2   Calculation of phonon frequencies

There are two different first-principles strategies for calculating phonon frequencies. The method that is easier to understand starts from the fact that the force constant matrix expresses the proportionality between displacements and forces, when the displacements are small enough for this relationship to be linear. All that has to be done in principle is to displace a single atom t in cell l' in Cartesian direction β, all other atoms being held fixed at their equilibrium positions; the forces $F_{ls\alpha}$ on all the atoms then give directly the elements of the force constant

matrix $\Phi_{ls\alpha,l't\beta}$ for the given $(l't\beta)$. If this procedure is repeated for all other $(l't\beta)$, all the elements of the force constant matrix can be obtained. Translational invariance implies that the number of separate calculations required to do this is at most 3 times the number of atoms in the primitive cell, but for most materials symmetry relations can be used to reduce this number substantially. This strategy, sometimes called the small displacement method (Kresse *et al.*, 1995), is implemented, for example, in the PHON code (Alfè, 1998). Although the small displacement method is widely used, and can be very accurate, a word of caution is in order. Since DFT calculations on condensed matter always use periodic boundary conditions, the repeating cell must be large enough so that the elements $\Phi_{ls\alpha,l't\beta}$ have all fallen off to negligible values at the boundary of the repeating cell. This is readily achieved for some materials, particularly metals. However, in ionic materials, the force constant elements fall off only as r^{-3}, and convergence can be slow. Moreover, in polar materials, Coulomb forces produce a macroscopic electric field in the limit of zero wave vector. This electric field is responsible for a splitting in the frequencies of the vibrational modes parallel and perpendicular to the electric field (the so-called LO–TO splitting). This effect can be taken into account by adding a nonanalytic contribution to the dynamical matrix at wave vector **k** which has the form (Giannozzi, 1991)

$$D_{s\alpha,t\beta}^{\text{na}} = \left(m_s m_t\right)^{-1/2} \frac{4\pi e^2}{\Omega} \frac{(\mathbf{k} \cdot \mathbf{Z}_s^*)_\alpha (\mathbf{k} \cdot \mathbf{Z}_t^*)_\beta}{\mathbf{k} \cdot \epsilon^\infty \cdot \mathbf{k}} \quad [39]$$

where $\mathbf{Z}_s^*$ is the Born effective charge tensor for atom s, ϵ^∞ the high-frequency static dielectric tensor, and m_s, m_t the mass of the atoms. These two quantities can be calculated in the framework of density functional perturbation theory (DFPT) (Baroni *et al.*, 1987, 2001; Giannozzi *et al.*, 1991), which also provides a second elegant strategy for the calculation of phonons in crystals. The main idea in DFPT, pioneered by Baroni *et al.* (1987), is to exploit the Hellmann–Feynman theorem to show that a linear-order variation in the electron density upon application of a perturbation to the crystal is responsible for a variation in the energy up to second (in fact, third (Gonze and Vigneron, 1989)) order of the perturbation. Using standard perturbation theory, this linear order variation of the electronic charge density can be calculated using only unperturbed wave functions, which therefore only require calculations on the ground-state crystal. If the perturbation is a phonon wave with wave vector **k**, calculation of the density change to linear order in the perturbation can be used to determine the force constant matrix at wave vector **k**. This can be done for any arbitrary wave vector, without the need for the construction of a supercell. The implementation of the method is by no means straightforward, and for further details the reader should consult the original papers (Baroni *et al.*, 1987; Giannozzi *et al.*, 1991).

As an example of first-principles calculations of phonon frequencies using the small displacement method (Alfè, 1998) in **Figure 5** the phonon dispersion relations for b.c.c iron under ambient conditions, compared with experimental data is shown. We see that the agreement between theory and experiments is very good almost everywhere in the Brillouin zone, with discrepancies being at worst ~3%.

Phonons can also be calculated at high pressure, and as an illustration of this, in **Figure 6** we show a comparison between DFT–GGA calculated phonons, using the small displacement method (Alfè, 1998), and nuclear resonant inelastic X-ray scattering (NRIXS) (Seto *et al.*, 1995; Sturhahn *et al.*, 1995) experiments, of phonon density of states of b.c.c and h.c.p iron from 0 to 153 GPa (Mao *et al.*, 2001). The agreement between theory and experiments is good in the whole pressure region, being slightly better at high pressure.

The NRIXS technique has also been recently used to measure the partial density of states of FeS as a function of pressure (Kobayashi *et al.*, 2004). Measurements were taken at pressures of 1.5, 4.0, and 9.5 GPa, in the troilite, MnP-type, and monoclinic crystal structures of FeS, respectively, and were compared with first-principles calculations based on DFT–GGA. The agreement between the calculations and the experiments was reasonably good, although the FeS in the troilite structure was found to be unstable. The calculations were also used to provide the total density of states, which provided thermodynamic quantities such as the entropy and the specific heat.

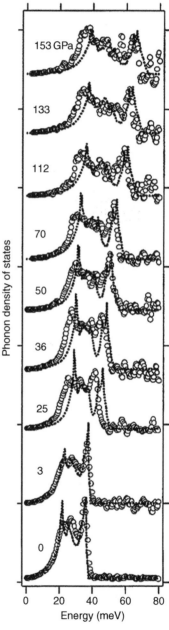

Figure 6 Phonon density of states of b.c.c Fe (pressure $p = 0$ and 3 GPa) and h.c.p Fe (p from 25 to 153 GPa). Dotted curves and open circles show first-principles theory and experiment respectively.

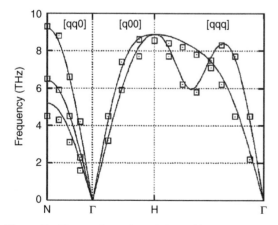

Figure 5 Phonon dispersion relations of ferromagnetic bcc Fe. Lines and open squares show first-principles theory and experiment, respectively. Reproduced from Alfè D, Kresse G, and Gillan MJ (2000) *Physical Review B* 61: 132.

13.3.3.3 The Helmholtz free energy: High-temperature and thermodynamic integration

At high temperature, anharmonic effects in solids may start to play an important role, and the quasi-harmonic approximation may not be accurate enough. Moreover, if the system of interest is a liquid, the quasi-harmonic approximation is of no use. This section describes a method to calculate the free energy of solids and liquid in the high-temperature limit, provided that the temperature is high enough that the quantum nature of the nuclei can be neglected. If this is the case, the Helmholtz free energy F is defined as in eqn [23].

Performing the integral in eqn [23] to calculate F is extremely difficult. However, it is less difficult to calculate changes in F as some specific variables are changed in the system. For example, we have seen that by taking the derivative of F in eqn [23] with respect to volume at constant T we obtain (minus) the pressure. Therefore, the difference of F between two volumes can be obtained by integrating the pressure p, which can be calculated using an MD simulation. Similarly, by integrating the internal energy E, one obtains differences in F/T.

It is equally possible to calculate differences in free energy between two systems having the same number of atoms N, the same volume V, but two different potential energy functions U_0 and U_1. This can be done by introducing an intermediate potential energy function U_λ such that for $\lambda = 0$, $U_\lambda = U_0$; and for $\lambda = 1$, $U_\lambda = U_1$; and such that for any value of $0 < \lambda < 1$, U_λ is a continuous and differentiable function of λ. For example, a convenient form is

$$U_\lambda = (1 - f(\lambda))U_0 + f(\lambda)U \qquad [40]$$

where $f(\lambda)$ is an arbitrary continuous and differentiable function of λ in the interval $0 \leq \lambda \leq 1$, with the property $f(0) = 0$ and $f(1) = 1$. According to eqn [23], the Helmholtz free energy of this intermediate system is

$$F_\lambda = -k_B T \ln \left\{ \frac{1}{N!\Lambda^{3N}} \int_V d\mathbf{R}_1 \cdots d\mathbf{R}_N \, e^{-\beta U_\lambda(\mathbf{R}_1, \ldots, \mathbf{R}_N; T)} \right\} \qquad [41]$$

Differentiating this with respect to λ gives

$$\frac{dF_\lambda}{d\lambda} = \frac{\int_V d\mathbf{R}_1 \cdots d\mathbf{R}_N e^{-\beta U_\lambda(\mathbf{R}_1, \ldots, \mathbf{R}_N, T)} \left(\frac{\partial U_\lambda}{\partial \lambda} \right)}{\int_V d\mathbf{R}_1 \cdots d\mathbf{R}_N e^{-\beta U_\lambda(\mathbf{R}_1, \ldots, \mathbf{R}_N, T)}}$$

$$= \left\langle \frac{\partial U_\lambda}{\partial \lambda} \right\rangle_\lambda \qquad [42]$$

and therefore by integrating $dF_\lambda/d\lambda$ one obtains

$$\Delta F = F_1 - F_0 = \int_0^1 d\lambda \left\langle \frac{\partial U_\lambda}{\partial \lambda} \right\rangle_\lambda \qquad [43]$$

This also represents the reversible work done on the system as the potential energy function is switched from U_0 to U_1. In most cases, a suitable choice for the function that mixes U_0 and U_1 is simply $f(\lambda) = \lambda$, and the thermodynamic formula [43] takes the simple form

$$\Delta F = F_1 - F_0 = \int_0^1 d\lambda \langle U_1 - U_0 \rangle_\lambda \qquad [44]$$

This way to calculate free energy differences between two systems is called thermodynamic integration (Frenkel and Smit, 1996). The usefulness of the thermodynamic integration formula expressed in eqn [43] becomes clear when one identifies U_1 with the DFT potential (free) energy function, and with U_0 some classical model potential for which the free energy is easily calculated, to be taken as a reference system. Then eqn [43] can be used to calculate the DFT free energy of the system by evaluating the integrand $\langle U_1 - U_0 \rangle_\lambda$ using FPMD simulations at a sufficiently large number of values of λ and calculating the integral numerically. Alternatively, one can adopt the dynamical method described by Watanabe and Reinhardt (1990). In this approach, the parameter λ depends on time, and is slowly (adiabatically) switched from 0 to 1 during a single simulation. The switching rate has to be slow enough so that the system remains in thermodynamic equilibrium, and adiabatically transforms from the reference to the *ab initio* system. The change in free energy is then given by

$$\Delta F = \int_0^{T_{sim}} dt \frac{d\lambda}{dt} (U_1 - U_0) \qquad [45]$$

where T_{sim} is the total simulation time, $\lambda(t)$ is an arbitrary function of t with the property of being continuous and differentiable for $0 \leq t \leq 1$, $\lambda(0) = 0$, and $\lambda(T_{sim}) = 1$.

Thermodynamic integration can be used to calculate the free energies of both solids and liquids. It is clear from eqn [43] that the choice of the reference system is almost completely irrelevant (of course, the stable phase of the system cannot change as λ is switched from 0 to 1), provided that ΔF can be calculated in practice. So, if the goal is to obtain *ab initio* free energies, it is essential to minimize the amount of _ab initio_ work in order to make the calculations feasible. This is achieved by requiring that (1) the integrand in eqn [43] is a smooth function of λ, (2) the thermal averages $\langle U_1 - U_0 \rangle_\lambda$ can be computed within the required accuracy on the timescales accessible to FPMD, and (3) the convergence of ΔF as function of the number of atoms N in the system is again achieved with N accessible to first-principles calculations. All points (1), (2), and (3) could obviously be satisfied by a perfect reference system, that is, a system which differs from the _ab initio_ system only by an arbitrary constant. In this trivial case, the integrand in eqn [43] would be a constant, and thermal averages could be calculated on just one configuration and with cells containing an arbitrary small number of atoms. The next thing close to a constant is a slowly varying object, and this therefore provides the recipe for the choice of a good reference system, which has to be constructed in such a way that the fluctuations in $U_1 - U_0$ are as small as possible. If this is the case, thermal averages of $U_1 - U_0$ are readily calculated on short simulations. Moreover, $\langle U_1 - U_0 \rangle_\lambda$ is a smooth function of λ, so a very limited number of simulations for different values of λ are needed and, finally, convergence of $\langle U_1 - U_0 \rangle_\lambda$ with respect to the size of the system is also quick. In fact, if the fluctuations in $U_1 - U_0$ are small enough, one can simply write $F_1 - F_0 \simeq \langle U_1 - U_0 \rangle_0$, with the average taken in the reference system ensemble. If this is not good enough, the next approximation is readily shown to be

$$F_1 - F_0 \simeq \langle U_1 - U_0 \rangle_0 - \frac{1}{2k_B T} \langle [U_1 - U_0 - \langle U_1 - U_0 \rangle_0]^2 \rangle_0 \quad [46]$$

This form is particularly convenient since one only needs to sample the phase space with the reference system, and perform a number of _ab initio_ calculations on statistically independent configurations extracted from a long classical simulation.

Once the Helmholtz free energy of the system is known, it can be used to derive its thermodynamical properties. For example, it is possible to calculate properties on the so-called Hugoniot line, and compare the results with those obtained in shock-wave experiments. The data that emerge most directly from shock experiments consist of a relation between the pressure p_H and the molar volume V_H on the Hugoniot line, which is the set of thermodynamic states given by the Rankine–Hugoniot formula (Poirier, 1991)

$$\frac{1}{2} p_H (V_0 - V_H) = E_H - E_0 \quad [47]$$

where E_H is the molar internal energy behind the shock front, and E_0 and V_0 are the molar internal energy and volume in the zero-pressure state ahead of the front. These experiments are particularly useful in identifying the melting transition. This is done by monitoring the speed of sound, which shows discontinuities at two characteristic pressures p_s and p_l, which are the points where the solid and liquid Hugoniots meet the melting curve. Below p_s, the material behind the shock front is entirely solid, while above p_l it is entirely liquid; between p_s and p_l, the material is a two-phase mixture. To illustrate an example of the quality of the DFT–GGA predictions of the Hugoniot line, we show in **Figure 7** the calculations of the $p(V)$ relation on the Hugoniot by Alfè _et al._ (2002a) for solid and liquid iron, compared with the experimental data obtained by Brown and McQueen (1986). We can see that the agreement between the theory and experiments is extremely good. The two theoretical curves come from raw and free energy-corrected calculations (see below). In **Figure 8** we show a comparison of the calculated speed of sounds of the liquid with those obtained in the shock experiments. Again, the agreement between the two sets of data is extremely good.

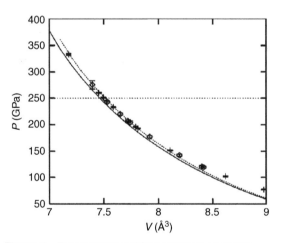

Figure 7 Experimental and first-principles Hugoniot pressure p of Fe as a function of atomic volume V. Symbols show the measurements of Brown and McQueen (1986). Solid curve is first-principles pressure obtained when calculated equilibrium volume of b.c.c. Fe is used in the Hugoniot–Rankine equation; dotted curve is the same, but with experimental equilibrium volume of b.c.c. Fe. The comparison is meaningful only up to a pressure of c. 250 GPa (horizontal dotted line), at which point the experiments indicate melting. Reprinted from Alfè D, Gillan MJ, and Price GD (2002a) _Physical Review B_ 65: 165118.

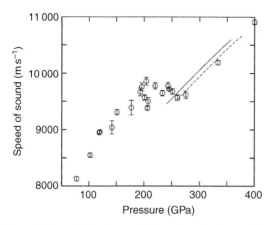

Figure 8 Longitudinal speed of sound on the Hugoniot. Circles: experimental values from Brown and McQueen (1986); continuous and dashed curves: present *ab initio* values without and with free energy correction (see text). Reprinted from Alfè D, Gillan MJ, and Price GD (2002a) *Physical Review B* 65: 165118.

13.3.3.4 Melting

The first to use thermodynamic integration in the context of first-principles calculations were Sugino and Car (SC) (1995), who calculated the Gibbs free energies G of solid and liquid silicon at ambient pressure to obtain the DFT–LDA melting temperature T_m, given by the condition $G^l(T_m) = G^s(T_m)$, with superscript l and s indicating liquid and solid, respectively. In their work, SC used a Stillinger–Weber potential (Stillinger and Weber, 1985) as reference system for both solid and liquid, coupled with the scheme described in eqn [45] to compute free energy differences. They found that using a switching time T_m of less than 1 ps was already sufficient to obtain a statistical accuracy on the free energies capable of predicting the melting temperature with an error of the order of 50 K. The importance of the SC work was that it showed that with a judicious choice of the reference system these kinds of calculations are entirely accessible to first-principles techniques. The calculated melting LDA temperature of Si was 1350 K, about 20% lower than the experimental value (1680 K). Subsequent work showed that by using the GGA approximation the calculated zero-pressure melting point was $\sim$1500 K (Alfè and Gillan, 2003), and that by using the recently developed metaGGA (Tao *et al.*, 2003; Staroverov *et al.*, 2004) the results were very close to the experimental value (Wang *et al.*, unpublished). Shortly after the work of SC, de Wijs *et al.* (1998) used DFT–LDA in combination with thermodynamic integration to

calculated the zero-pressure melting point of Al. They found the value of 890 K, in good agreement with the experimental value 933 K.

Encouraged by these early successes, DFT–GGA and thermodynamic integration were used by Alfè (1999) and Alfè *et al.* (2001, 2002) to calculate the free energies of solid and liquid iron under Earth's core conditions, which they used to obtain a number of thermodynamic properties, including the whole melting curve in the region $\sim$50–400 K. They discovered that a simple sum of inverse power pair-potentials of the form $U_{IP}(r) = B/r^{\alpha}$, where r is the distance between two ions and B and α are two adjustable parameters, did an excellent job in describing the energetics of the liquid and the high-temperature solid, provided B and α were appropriately adjusted. As mentioned in the previoous section, an additional crucial advantage of having a good reference system is that convergence of $F_1 - F_0$ with respect to the size of the system is very rapid, and in fact for both solid and liquid iron Alfè *et al.* already found that with 64-atom systems $F_1 - F_0$ was converged to within better than 10 meV/atom, which in turns implied melting temperature converged to better than 100 K with respect to this single technical point. Their best estimate for the melting point at the inner–outer core boundary pressure of 330 GPa was $T_m = 6350 \pm 300$ K, where the error quoted is the result of the combined statistical errors in the free energies of solid and liquid. Systematic errors due to the approximations of DFT are more difficult to estimate, and a definite word can only be pronounced after the problem is explored with a more accurate implementation of quantum mechanics. We hope that QMC techniques may serve this scope in the near future.

At the present state of knowledge, the experimental understanding of the melting point of Fe under this conditions is still scarce, as experiments based on diamond-anvil cells (DACs) cannot reach these pressures. Moreover, even in the region of the phase space where DAC experiments are possible, there is still considerable disagreement between different groups (Boehler, 1993; Ma *et al.*, 2004), and between DAC and shock-wave experiments (Brown and McQueen, 1986; Nguyen and Holmes, 2004).

On the theoretical side, we also mention the work of Laio *et al.* (2000) and Belonoshko *et al.* (2000), who performed DFT-based simulations to calculate the melting curve of Fe under Earth's core conditions, although their approach was rather different from that of Alfè (1999) and Alfè *et al.* (2001, 2002a). Instead of calculating free energies, Laio *et al.* (2000) and Belonoshko *et al.* (2000) fitted a classical model potential to their first-principles calculations, and then used the classical potential to compute the melting curve. To do so, they used the coexistence method, in which solid and liquid are simulated in contact in a box. This method is an alternative route to the calculation of melting curves, and therefore equivalent to the free energy approach. However, Laio *et al.* and Belonoshko *et al.* found that iron melted at 5400 and 7000 K, respectively, at the pressure of 330 GPa. These large differences, and the difference with the value 6350 K reported by Alfè *et al.*, are due to the quality of the classical potentials employed, and in particular the free energy differences between these classical potentials and the DFT system. This was later investigated by Alfè *et al.* (2002b), who showed that it is possible to assess the differences in free energies between the classical potential and the DFT one, and correct for it. In particular, it was shown that at a fixed pressure p, a first approximation of the difference T' in the melting temperature between the classical potential and the *ab initio* system is given by

$$T' = \Delta G^{ls}(T_{mod})/S^{ls}_{mod} \quad [48]$$

where S^{ls}_{mod} is the entropy of fusion of the model potential, T_{mod} its melting temperature, and $\Delta G^{ls} = (G^l_{ab} - G^l_{mod}) - (G^s_{ab} - G^s_{mod})$, where G is the Gibbs free energy, the subscripts ab and mod indicate the *ab initio* and the model system, respectively, and the superscripts l and s indicate liquid and solid, respectively. These differences of Gibbs free energies can be calculated using thermodynamic integration, which if the model potential is not too different fro the *ab initio* one, can be calculated using the perturbative approach outlined in eqn [46] above. The relation between ΔG, evaluated at constant p, and ΔF, calculated at constant V, is readily shown to be

$$\Delta G = \Delta F - \frac{1}{2}V\kappa_T \Delta p^2 + o(\Delta p^3) \quad [49]$$

where κ_T is the isothermal compressibility and Δp is the change of pressure when U_{mod} is replaced by U_{ab} at constant V and T. Once this corrections were applied, the results of Belonoshko *et al.* (2000) came in perfect agreement with those of Alfè *et al.* (2002a). These results are all displayed in **Figure 9**, together with a number of experimental data.

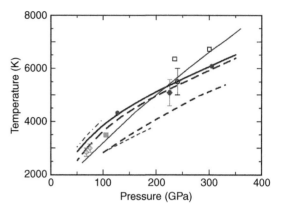

Figure 9 Comparison of melting curve of Fe from DFT calculations and experimental data: black solid and dashed curves: first-principles results of Alfè *et al.*, (2002a) without and with free-energy correction (see text); red filled circles: corrected coexistence results from (Alfè *et al.*, 2004); blue dashed curve: first-principles based results of Laio *et al.* (2000); purple curve: first-principles based results of Belonoshko *et al.* (2000); black chained and maroon dashed curves: DAC measurements of Williams *et al.* (1987) and Boehler (1993); green diamonds and green filled square: DAC measurements of Shen *et al.* (1998) and Ma *et al.* (2004); black open squares, black open circle, and magenta diamond: shock experiments of Yoo *et al.* (1993), Brown and McQueen (1986), and Nguyen and Holmes (2004). Error bars are those quoted in original references.

The coexistence method mentioned above is an alternative route to the calculation of melting properties and as such delivers the same results if applied consistently. For its very nature, the method is intrinsically very expensive, because it requires simulations on systems containing large number of atoms, typically many hundreds or even thousands. For this reason, until very recently, it had been only applied to calculations employing classical potentials. However, it has been recently shown that the method can in fact be applied also in the context of first-principles calculations. As computers become faster and faster, this method will become more and more common also within the context of first-principles calculations.

In **Figure 10**, I show the low-pressure melting curve of aluminum, as obtained by Vocado and Alfè (2002) with the free energy approach and the GGA XC potential and, later, with the coexistence approach (Alfè, 2003), using the very same electronic structure techniques. This was the first time that the coexistence method was being applied using first-principles techniques. The simulations were performed on systems containing up to 1728 atoms, and the results showed that already with modest

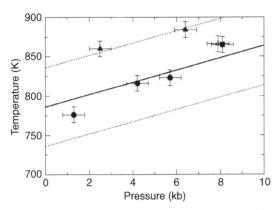

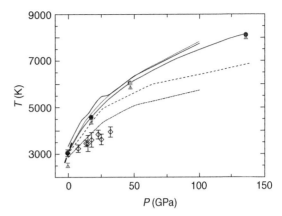

Figure 10 Temperatures and pressures at which liquid and solid Al coexist in first-principles simulations containing 1728 atoms (filled square), 1000 atoms (filled circles), and 512 atoms (filled triangles) performed using Γ-point sampling only. Open triangle is the result from a 2 × 2 × 1 **k**-point grid on a 512-atom system. The solid line is the lower end of the melting curve calculated using the free energy approach in Vočadlo and Alfè (2002); light dashed lines represent error bars. Results from simulations using Γ-point sampling only for 1000- and 512-atom systems were reported in Alfè (2003).

Figure 11 Melting curve of MgO obtained with DFT–LDA coexistence simulations performed on 432-atom cells (blue dots and heavy solid line), 1024-atom cell (green square), and DFT–GGA results (red triangles) (Alfè, 2005) compared with experiments (open diamonds) (Zerr and Boehler, 1994). Other curves show results of earlier modeling work based on interaction models. Reproduced from (Alfè, 2005).

system sizes of 512 atoms the points calculated with the coexistence approach differed from those obtained with the free energy approach by no more than 50 K. In fact, this difference was reduced even further if the 512-atom simulation was carried out with a denser **k**-point grid sampling, showing that genuine size effects are very small already in systems containing 512 atoms. Note that the zero-pressure melting temperature of 786 K obtained with the GGA is not in very good agreement with the experimental value 933 K. Vocadlo and Alfè argued that the main cause of this discrepancy was a deficiency of the GGA of predicting the correct zero-pressure density, which comes out too large, and this would effectively correspond to calculating the melting temperature at a negative pressure. By applying a correction to the Helmholtz free energy in order to obtain the zero-pressure density for the solid, and assuming that the same correction also applied to the liquid, they found that this corrected free energy predicted a zero-pressure melting temperature of 912 K, in excellent agreement with the experimental datum. The same argument was also applied by Alfè *et al.* (2002a) to the melting curve of iron. In that case, the GGA equation of state predicts a pressure which is between 2.5% and 10% lower than experiments, in the high- and low-pressure region of the phase diagram, respectively. Here, adding a correction to the free energy

in order to match the experimental pressure lowered the melting curve, so that the corrected melting temperature at 330 GPa was 6200 K.

A short time after the coexistence work on aluminum, a second calculation using the same technique to calculate the melting curve of LiH appeared (Ogitsu *et al.*, 2003), and more recently for the calculation of the melting curve of hydrogen up to 200 GPa (Bonev *et al.*, 2004), and the melting curve of MgO in the pressure range 0–135 GPa (Alfè, 2005), reported in **Figure 11**. This last work was performed with both the LDA and the GGA XC functionals, and in spite of the results at high pressure being very similar, at zero pressure the two calculations differed by about 20%, with the LDA being in good agreement with the experimental datum, and the GGA predicting a lower melting temperature. This work also pointed out that the contribution of the electronic entropy to the free energy was non-negligible, and very different between solid and liquid, such that it is responsible for a lowering of about 130(750) K of the melting temperature in the low(high) pressure region of the phase diagram. This may be surprising, as solid MgO is a large band-gap insulator, and indeed a significant band gap remains in the high-temperature solid. However, the absence of local order in the liquid is such that the band gap is greatly reduced, with a correspondingly significant increase of electronic entropy. This emphasizes the need to accurately include electronic effects in the calculation of melting properties.

13.3.3.5 *Solutions*

This section presents the discussion on systems formed by more than one species of atoms. For example, consider two different substances, mix them together, and in general they will form a solution, like sugar and coffee. The solvent is the substance present in the largest quantity (milk), and solute the other (sugar). In general, solutions may have more than one solute and/or more than one solvent, but for simplicity we will focus here only on binary mixtures.

This section discusses how first-principles methods can be used to study the thermodynamical properties of solutions. As an example of the methods to be discussed, calculations on iron alloyed with either sulfur, or silicon or oxygen, at the conditions of pressure and temperature of the Earth's inner core boundary (ICB), and how these calculations have been used to estimate the composition of the core, are presented. The techniques, however, are completely general, and can be applied also to other systems under different thermodynamic conditions.

As mentioned above, the behavior of solutions can be understood in terms of the chemical potential μ_i, defined in eqns [30] and [31]. Consider now a solution with N_A particles of solvent A and N_X particles of solute X, with $N = N_A + N_X$. In the high-temperature limit, the Helmholtz free energy of this system is

$$F = -k_B T \ln \frac{1}{\Lambda_A^{3N_A} \Lambda_X^{3N_X} N_A! N_X!}$$

$$\times \int_V d\mathbf{R}_1 \cdots d\mathbf{R}_N e^{-U(\mathbf{R}_1,\ldots,\mathbf{R}_N;T)/k_B T} \quad [50]$$

According to eqn [31], we have

$$\mu_X = \left(\frac{\partial F}{\partial N_X}\right)_{T,V} = F(N_A, N_X + 1) - F(N_A, N_X) \quad [51]$$

which can be evaluated using eqn [50]:

$$\mu_X = -k_B T \ln \frac{1}{\Lambda_X^3 (N_X + 1)}$$

$$\times \frac{\int_V d\mathbf{R}_1 \cdots d\mathbf{R}_N d\mathbf{R}_{N+1} e^{-U(\mathbf{R}_1,\ldots,\mathbf{R}_N,\mathbf{R}_{N+1};T)/k_B T}}{\int_V d\mathbf{R}_1 \cdots d\mathbf{R}_N e^{-U(\mathbf{R}_1,\ldots,\mathbf{R}_N;T)/k_B T}} \quad [52]$$

The ratio of the two integrals is an extensive quantity, but μ_X is an intensive quantity; therefore, it is useful to rewrite the expression as follows:

$$\mu_X = -k_B T \ln \frac{N}{(N_X + 1)}$$

$$\times \left\{ \frac{1}{N\Lambda_X^3} \frac{\int_V d\mathbf{R}_1 \cdots d\mathbf{R}_N d\mathbf{R}_{N+1} e^{-U(\mathbf{R}_1,\ldots,\mathbf{R}_N,\mathbf{R}_{N+1};T)/k_B T}}{\int_V d\mathbf{R}_1 \cdots d\mathbf{R}_N e^{-U(\mathbf{R}_1,\ldots,\mathbf{R}_N;T)/k_B T}} \right\} \quad [53]$$

so that the value in curly brackets is now independent of the system size. By setting $c_X = N_X/N$ (which in the limit of large N and N_X is the same as $(N_X + 1)/N$) and

$$\bar{\mu}_X = -k_B T$$

$$\times \ln \left\{ \frac{1}{N\Lambda_X^3} \frac{\int_V d\mathbf{R}_1 \cdots d\mathbf{R}_N d\mathbf{R}_{N+1} e^{-U(\mathbf{R}_1,\ldots,\mathbf{R}_N,\mathbf{R}_{N+1};T)/k_B T}}{\int_V d\mathbf{R}_1 \cdots d\mathbf{R}_N e^{-U(\mathbf{R}_1,\ldots,\mathbf{R}_N;T)/k_B T}} \right\} \quad [54]$$

we can rewrite the chemical potential in our final expression:

$$\mu_X(p, T, c_X) = k_B T \ln c_X + \bar{\mu}_X(p, T, c_X) \quad [55]$$

The first term of eqn [55] depends only on the number of particles of solute present in the solution, while the second term is also responsible for all possible chemical interactions.

For small concentration of solute we can make a Taylor expansion of $\bar{\mu}_X$:

$$\bar{\mu}_X = \mu_X^0 + \lambda c_X + o(c_X^2) \quad [56]$$

where $\lambda = (\partial \bar{\mu}_X / \partial c_X)_{p,T}$. If the solution is so dilute that the particles of the solute do not interact with each other, we can stop the expansion to the first term, and we have

$$\mu_X(p, T, c_X) = k_B T \ln c_X + \mu_X^0(p, T) \quad [57]$$

A system in which eqn [57] is strictly satisfied is called an ideal solution.

To find an expression for the chemical potential of the solvent we employ the Gibbs–Duhem equation, which for a system at constant pressure and constant temperature reads (Wannier, 1966)

$$\sum_i N_i d\mu_i = 0 \quad [58]$$

In particular, in our two-component system, the Gibbs–Duhem equation implies

$$c_A d\mu_A + c_X d\mu_X = 0 \qquad [59]$$

which gives (Alfè *et al.*, 2002c)

$$\mu_A(p, T, c_X) = \mu_A^0(p, T) + (k_B T + \lambda_X(p, T))\ln(1 - c_X) \\ + \lambda_X(p, T)c_X + O(c_X^2) \qquad [60]$$

where μ_A^0 is the chemical potential of the pure solvent. To linear order in c_X, this gives

$$\mu_A(p, T, c_X) = \mu_A^0(p, T) - k_B T c_X + O(c_X^2) \qquad [61]$$

Though μ_A^0 is the chemical potential of the pure solvent, note that μ_X^0 is not the chemical potential of the pure solute, unless the validity of eqn [57] extends all the way up to $c_X = 1$.

13.3.3.6 First-principles calculations of chemical potentials

To calculate μ_X, it is useful to consider the difference in chemical potential between the solute and the solvent $\mu_{XA} = \mu_X - \mu_A$, which is equal to the change of Helmholtz free energy of the system as one atom of solvent is transmuted into an atom of solute at constant volume V and constant temperature p. This transmutation does not obviously correspond to a real physical process, but provides a perfectly rigorous way of calculating the difference of chemical potentials:

$$\mu_{XA} = k_B T \ln\frac{c_X}{1 - c_X} + 3k_B T \ln(\Lambda_X/\Lambda_A) + m(c_X) \qquad [62]$$

where Λ_X and Λ_A are the thermal wavelengths of solute and solvent, and

$$m(c_X) = -k_B T \ln\frac{\int_V d\mathbf{R} e^{-\beta U(N_A-1, N_X+1; \mathbf{R})}}{\int_V d\mathbf{R} e^{-\beta U(N_A, N_X; \mathbf{R})}} \qquad [63]$$

with $U(N_A, N_X; \mathbf{R})$ the potential energy of the system with N_A atoms of solvent and N_X atoms of solute, and $U(N_A - 1, N_X + 1; \mathbf{R})$ the one for the system in which one of the atoms of solvent has been transmuted into solute.

The thermodynamic integration technique described in Section 13.3.3.3 can now be used to compute $m(c_X)$ in the liquid state. This is done by defining an intermediate potential $U_\lambda = \lambda U(N_A - 1, N_X + 1; \mathbf{R}) + (1 - \lambda)U(N_A, N_X; \mathbf{R})$, so that $m(c_X)$ can be expressed as

$$m(c_X) = \int_0^1 d\lambda \langle U(N_A - 1, N_X + 1) - U(N_A, N_X) \rangle_\lambda \qquad [64]$$

In practice, the calculation of $m(c_X)$ is done by performing two separate simulations, one with N_A atoms of solvent and N_X of solute and the other with $N_A - 1$ atoms of solvent and $N_X + 1$ atoms of solute. At the end of each time step, forces are computed in both systems, and their linear combination $f_\lambda = \lambda f(N_A - 1, N_X + 1) + (1 - \lambda)f(N_A, N_X)$ is used to evolve the system in time in order to compute the thermal average $\langle U(N_A - 1, N_X + 1) - U(N_A, N_X) \rangle_\lambda$. This is repeated at a number of different values of λ and the integral is performed numerically. Alternatively, an approach similar to the one described in eqn [45] is also possible, in which λ is slowly varied from 0 to 1 in the course of the simulation.

To improve statistics, it is useful to transmute many atoms of solvent into solute. In this case, one does not obtain directly μ_{XA} at a chosen concentration, but an integral of this over a range of concentrations. However, by repeating the calculations transmuting a different number of atoms at a time, it is possible to extract informations about the value of μ_{XA} in a whole range of concentration, as described in Alfè *et al.* (2002c). To illustrate the feasibility of these kind of calculations, in **Figure 12** we show the value of the integrand in eqn 64 as function of λ for an iron–oxygen liquid mixture at a pressure of 370 GPa and a temperature of 7000 K. The error in calculating this integral is of the order of 0.1 eV, which is very small for the purposes of evaluating the partitioning of oxygen between solid and liquid iron, and the depression of melting point resulting from this partitioning (see below). This pressure is somewhat higher than the ICB pressure of 330 GPa. The temperature is also

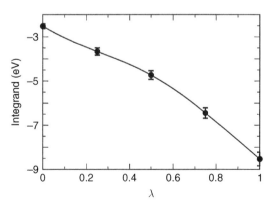

Figure 12 The integrand $\langle U_1 - U_0 \rangle_\lambda$ (eV units) appearing in the thermodynamic integration formula in eqn [64]. Results shown refer to oxygen solute for $N_X = 12$ and $N = 64$. Filled circles show values computed from *ab initio* MD simulations, with bars indicating statistical errors. Curve is a polynomial fit to the computed values.

higher than the melting temperature of iron; however, it was argued in Alfè *et al.* (2002c) that this would not change the chemical potential of O by more than 0.3 eV, which again does not have any effect on the conclusions to be presented below.

In the solid state, thermodynamic integration is not the most appropriate way of calculating the chemical potential difference μ_{XA}. This is clear, because in the zero-temperature limit, at infinite dilution $m(c_X \rightarrow 0)$ is simply the change in internal energy when one atom in the perfect lattice of solvent is replaced by a solute atom, the impurity system being relaxed to equilibrium. This requires only a static calculation of the type described in Section 13.3.1. At finite temperatures in the infinite dilution limit, $m(c_X \rightarrow 0)$ can be obtained from the quasi-harmonic vibrational frequencies of the pure A system and the system containing a single X impurity. If anharmonic effects are significant, as they are in the case of O substituted in h.c.p Fe (Alfè *et al.*, 2000), thermodynamic integration can be be used to estimate the anharmonic effects. These methods can also be generalized to include the variation of $m(c_X)$ with c_X to linear order in c_X (Alfè *et al.*, 2002c).

The evaluation of $m(c_X)$ only gives access to the difference between the chemical potential of the solute and that of the solvent. Therefore, in order to to obtain μ_X, a preliminary calculation of μ_A is necessary. This can be done on the pure solvent system using the techniques described in Section 13.3.3.3.

13.3.3.7 Volume of mixing

It is often interesting to study the change of volume of a solution as a function of the concentration of solute. To this end, it is useful to express the volume of the system as the partial derivative of the Gibbs free energy with respect to pressure, taken at constant temperature and number of particles:

$$V = (\partial G / \partial p)_{T, N_X, N_A} \qquad [65]$$

If we now add to the system one particle of solvent at constant pressure, the total volume changes by v_A, and becomes $V + v_A$. We call v_A the partial molar volume of the solvent. The total Gibbs free energy is $G + \mu_A$, so that according to eqn [65], $v_A = (\partial \mu_A / \partial p)_{T, N_X, N_A} = (\partial \mu_A / \partial p)_{T, c_X}$, where the last equality stems from the fact that μ_A only depends on N_X and N_A through the molar fraction c_X (we assume here that c_X does not change when we add one particle of solvent to the system; this is obviously true if the number of atoms of solvent N_A is already very large).

The partial volume in general depends on c_X, p, and T, but under the assumption of ideality $v_A = (\partial \mu_A^0 / \partial p)_{T, c_X}$, and it depends only on p and T. In an ideal solution, v_A is the same as in the pure solvent.

Similarly, the partial molar volume of the solute is $v_X = (\partial \mu_X / \partial p)_{T, c_X}$, which becomes independent on c_X under the assumption of ideality. Notice that this is not in general equal to the partial volume of the pure solute. As an example of the calculation of partial molar volumes, we mention the partial molar volumes of iron and X, with X being either sulfur, or silicon, or oxygen, obtained by Alfè *et al.* (2002d) in a binary mixture of iron at the conditions mentioned above of $p = 370$ GPa and $T = 7000$ K. These were $v_{Fe} = 6.97$, $v_{Si} = 6.65$, $v_S = 6.65$, $v_O = 4.25$ Å^3. Although ideality was not assumed in these calculations, the partial molar volumes of Si, S, and O were found to be rather independent from their concentration in liquid Fe. It is interesting to note that both sulfur and silicon have a partial molar volume which is very similar to that of iron, while oxygen is significantly smaller. This is the main reason for the large difference in the behavior of sulfur and silicon on one side, and oxygen on the other, and the resulting partitioning of oxygen between solid and liquid iron (see below).

13.3.3.8 Solid–liquid equilibrium

We want to study now the conditions that determine equilibrium between solid and liquid, and in particular how the solute partitions between the two phases and how this partitioning affects the melting properties of the solution. Thermodynamic equilibrium is reached when the Gibbs free energy of the system is at its minimum, and therefore $0 = dG = d(G^l + G^s)$, where superscripts s and l indicate quantities in the solid and in the liquid, respectively. In a multicomponent system, the Gibbs free energy can be expressed in terms of the chemical potentials of the species present in the system (Wannier, 1966; Mandl, 1997):

$$G = \sum_i N_i \mu_i \qquad [66]$$

Using eqn [66] and the Gibbs–Duhem eqn [58], we obtain:

$$dG = \sum_i \mu_i \, dN_i \qquad [67]$$

If the system is isolated, particles can only flow between the solid and the liquid, and we have $dN_i^s = -dN_i^l$, which implies:

$$dG = \sum_i dN_i (\mu_i^l - \mu_i^s) \qquad [68]$$

If $\mu_i^l < \mu_i^s$ there will be a flow of particles from the solid to the liquid region ($dN_i > 0$), so that the Gibbs free energy of the system is lowered. The opposite will happen if $\mu_i^l > \mu_i^s$. The flow stops at equilibrium, which is therefore reached when $\mu_i^l = \mu_i^s$. In particular, in our two-component system, equilibrium between solid and liquid implies that the chemical potentials of both solvent and solute are equal in the solid and liquid phases:

$$\mu_X^s(p, T_m, c_X^s) = \mu_X^l(p, T_m, c_X^l)$$
$$\mu_A^s(p, T_m, c_X^s) = \mu_A^l(p, T_m, c_X^l)$$
[69]

where T_m is the melting temperature of the solution at pressure p. Using eqn [55], we can rewrite the first of the two equations above as

$$\bar{\mu}_X^s(p, T_m) + k_B T_m \ln c_X^s = \bar{\mu}_X^l(p, T_m) + k_B T_m \ln c_X^l \quad [70]$$

from which we obtain an expression for the ratio of concentrations of solute between the solid and the liquid,

$$c_X^s / c_X^l = \exp\left\{ [\bar{\mu}_X^l(p, T_m) - \bar{\mu}_X^s(p, T_m)] / k_B T_m \right\} \quad [71]$$

In general, $\bar{\mu}_X^l < \bar{\mu}_X^s$, because the greater mobility of the liquid can usually better accomodate particles of solute, and therefore their energy (chemical potential) is lower. This means that the concentration of the solute is usually smaller in the solid.

Equation [71] was used by Alfè *et al.* (2000, 2002b, 2002c) to put constraints on the composition of the Earth's core. The constraints came from a comparison of the calculated density contrast at ICB, and that obtained from seismology, which is between $4.5 \pm 0.5\%$ (Master and Shearer, 1990) and $6.7 \pm 1.5\%$ (Masters and Gubbins, 2003). This density contrast is significantly higher than that due to the crystallization of pure iron, and therefore must be due to the partitioning of light elements between solid and liquid. Alfè *et al.* (2000, 2002a, 2002c) considered sulfur, silicon, and oxygen as possible impurities, and using first principles obtained the partitions for each impurity. The calculations showed that for both sulfur and silicon $\bar{\mu}_X^l$ and $\bar{\mu}_X^s$ are very similar, which means that c_X^s and c_X^l are also very similar, according to eqn [71]. As a result, the density contrast of an Fe/S or an Fe/Si system is not much different from that of pure Fe, and still too low when compared with the seismological data. By contrast, for oxygen, $\bar{\mu}_o^l$ and $\bar{\mu}_o^s$ are very different, and the partitioning between solid and liquid is very large. This results in a much too large density contrast, which also does not agree with the sesmological data. These results are summarized in **Figure 13**. The conclusion from

these calculations was that none of these binary mixtures can be viable for the core. The density contrast can of course be explained by ternary or quaternary mixtures. Assuming no cross-correlated effects between the chemical potentials of different impurities, and based on the seismological density contrast of $4.5 \pm 0.5\%$ (Masters and Shearer, 1990), Alfè *et al.* (2002a, 2002c) proposed an inner core containing about 8.5% of sulfur and/or silicon and almost no oxygen, an outer core containing about 10% of sulfur and/or Si, and an additional 8% of oxygen. The more recent seismological datum of $6.7 \pm 1.5\%$ (Masters and Gubbins, 2003) would change the estimate of the core composition to an inner core containing about 7% of sulfur and/or Si and still almost no oxygen, and an outer core containing about 8% of sulfur and/or Si, and an additional 13% of oxygen. One consequence of the large partitioning of oxygen between solid and liquid is that as the solid core grows it expels oxygen in the liquid, which by converting its gravitational energy helps driving the convective motions that are responsible for the generation of the Earth's magnetic field (Gubbins *et al.*, 2004).

13.3.3.9 Shift of freezing point

The partitioning of the solute between the solid and the liquid is generally responsible for a change in the melting temperature of the mixture with respect to that of the pure solvent. To evaluate this, we expand the chemical potential of the solvent around the melting temperature of the pure system, T_m^0:

$$\mu_A(p, T_m, c_X) = \mu_A(p, T_m^0, c_X) - s_A^0 \delta T + \cdots \quad [72]$$

where $\delta T = (T_m - T_m^0)$ and $s_A^0 = -\left(\frac{\partial \mu_A}{\partial T}\right)_{T=T_m^0}$ is the entropy of the pure solvent at T_m^0. We now impose continuity across the solid/liquid boundary:

$$\mu_A^{0s}(p, T_m^0) - s_A^{0s}\delta T - k_B T_m c_X^s = \mu_A^{0l}(p, T_m^0)$$
$$- s_A^{0l}\delta T - k_B T_m c_X^l \quad [73]$$

where we have considered only the linear dependence of μ_A on c_X (See eqn [61]). Noting that $\mu_A^{0s}(p, T_m^0) = \mu_A^{0l}(p, T_m^0)$ we have

$$\delta T = \frac{k T_m}{s_A^{0l} - s_A^{0s}}(c_X^s - c_X^l) \quad [74]$$

Since usually $c_X^s < c_X^l$, there is generally a depression of the freezing point of the solution.

Using eqn [74], and the composition estimated from the density contrast of 4.5%, Alfè *et al.* (2002c, 2002d) estimated a depression of about 600–700 K of the

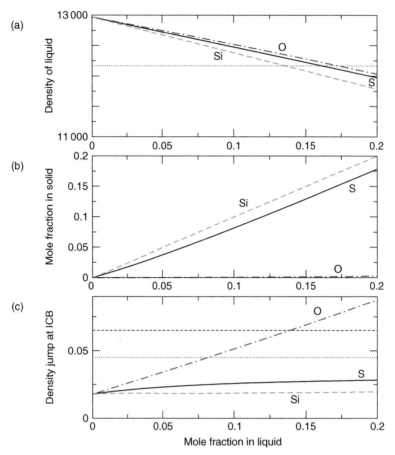

Figure 13   Liquid and solid impurity mole fractions c_X^l and c_X^s of impurities $X = $ S, Si, and O, and resulting densities of the inner and outer core of the Earth predicted by first-principles simulations. Solid, dashed, and chain curves represent S, Si, and O, respectively. (a) liquid density ρ^l (kg m^{-3} units); horizontal dotted line shows density from seismic data (Dziewonski and Anderson, 1981; Masters and Shearer, 1990); (b) mole fractions in solid resulting from equality of chemical potentials in solid and liquid; (c) relative density discontinuity $\delta\rho/\rho^l$ at the ICB; horizontal dotted and dashed lines are values from free oscillation data (Masters and Shearer, 1990; Masters and Gubbins, 2003). Adapted from Alfè *et al.* (2002d).

melting temperature of the core mixture with respect to the melting temperature of pure Fe, and they suggested an ICB temperature of about 5600 K. This estimate would go down by a further 300 K if the latest estimate of 6.7% for the density constrast at ICB was used (Masters and Gubbins, 2003), which would therefore result in a temperature at the ICB of about 5300 K.

13.4   Conclusions

Some *ab intio* techniques, based on the fundamental laws of quantum mechanics, that are currently being used to study the properties of high pressure and temperature minerals, are described in this article. Focus has been on the formulation of quantum mechanics known as DFT, with various approximations for the XC functional, and it is shown that this

technique can predict very reliably a number of static and elastic properties, like low-temperature equation of states, phase transition, and elastic constants. DFT has also been successfully used to predict the vibrational properties of a large number of materials, and the examples provided here show that these predictions are in good agreement with the experimental data, at both zero and high pressure. To study the high-temperature properties of solids and liquids, *ab initio* techniques coupled with MD have proved to be a powerful and reliable tool for both solids and liquids. This has been illustrated by mentioning the calculation of the high-temperature elastic constants of the mantle-forming minerals MgSiO$_3$ in both the perovskite and the recently discovered post-perovskite phases, and the elastic behavior of h.c.p iron under the conditions of the Earth's inner core, and the dynamical properties of liquids like diffusion and viscosity.

A major contribution to the calculation of the *ab initio* high-pressure and -temperature properties of minerals has been given by the possibility of calculating free energies, and it has been described how this can be done for both liquids and solids, using the technique of thermodynamic integration. The access to the free energies of materials has allowed the calculation of a number of thermodynamical properties. Here, I have described the calculation of melting properties, and in particular the melting curve of iron under Earth's core conditions. Also how *ab initio* techniques coupled with thermodynamic integration can be used to calculate chemical potentials in binary solutions is shown, and the method is illustrated by showing the calculation of the chemical potentials of oxygen, sulfur, and silicon in solid and liquid iron at Earth's core conditions. These calculations were used in conjunction with seismological data to estimate the probable composition of the Earth's inner and outer core.

In the description of melting properties, it has been pointed out that current XC functionals do not always provide perfect agreement with the experiments. It is shown how the application of empirical corrections can improve these results, but this is not a completely satisfactory solution, as it introduces some arbitrariness in the calculations, which do not conform to our original definition of *ab initio* anymore. Therefore, it is important to be able to go beyond DFT, and for this QMC techniques have been mentioned, which are sufficiently accurate and general that they will become a major player in the field of first-principles calculations of material properties. The development of these techniques will be helped by the availability of ever more powerful computers.

Also mentioned is the technique of the coexistence of phases to calculate melting curves. This technique is intrinsically very expensive, as it needs calculations performed on systems containing hundreds or even thousands of atoms. However, recently it has just become possible to apply this technique in the context of *ab initio* calculations, with simulations reported on up to 1728 atom cells. This method has been illustrated with the recent calculation of the melting curve of MgO under Earth's mantle conditions. The availability of faster computers will help making these coexistence calculations more routinely applied, and possibly will also allow simulation of the eutectic behavior of solutions, in which a liquid is in coexistence with more than one solid phase.

Acknowledgments

This work has bee supported by the Royal Society of Great Britain and by the Engineering and Physical Sciences Research Council in the UK.

References

Alfè D (1998) Program available at http://chianti.geol.ucl.ac.uk/~dario (accessed Oct 2006).
Andersen HC (1980) *Journal of Chemical Physics* 72: 2384.
Alder BJ and Wainwright TE (1959) *Journal of Chemical Physics* 31: 459.
Alfè D (1999) *Computer Physics Communications* 118: 31.
Alfè D, Price GD, and Gillan MJ (2001) *Physical Review B* 64: 045123.
Alfè D (2003) *Physical Review B* 68: 064423.
Alfè D (2005) *Physical Review Letters* 94: 235701.
Alfè D, Alfredsson M, Brodholt J, Gillan MJ, Towler MD, and Needs RJ (2005) *Physical Review B* 72: 014114.
Alfè D and Gillan MJ (1998a) *Physical Review B* 58: 8248.
Alfè D and Gillan MJ (1998b) *Physical Review Letters* 81: 5161.
Alfè D, Gillan MJ, and Price GD (2002a) *Physical Review B* 65: 165118.
Alfè D, Gillan MJ, and Price GD (2002d) *Earth and Planetary Science Letters* 195: 91.
Alfè D and Gillan MJ (2003) *Physical Review B* 68: 205212.
Alfè D and Gillan MJ (2004) *Physical Review B* 70: 161101.
Alfè D and Gillan M (2004b) *Journal of Physics: Condensed Matter* 16: L305.
Alfè D and Gillan MJ (2006) The energetics of oxide surfaces by quantum Monte Carlo. *Journal of Physics: Condensed Matter* 18: L435.
Alfè D, Kresse G, and Gillan MJ (2000) *Physical Review B* 61: 132.
Alfè D, Price GD, and Gillan MJ (2000a) *Nature* 405: 172.
Alfè D, Price GD, and Gillan MJ (2000b) *Geophysical Research Letters* 27: 2417.
Alfè D, Price GD, and Gillan MJ (2002b) *Journal of Chemical Physics* 116: 6170.
Alfè D, Price GD, and Gillan MJ (2002c) *Journal of Chemical Physics* 116: 7127–7136.
Alfè D, Vočadlo L, Price GD, and Gillan MJ (2004) *Journal of Physics Condensed Matter* 16: S973.
Allen MP and Tildesley DJ (1987) *Computer Simulation of Liquids.* New York: Oxford Science Publications.
Arias TA, Payne MC, and Joannopoulos JD (1992) *Physical Review B* 45: 1538 (1992) *Physical Review Letters* 69: 1077.
Bachelet GB, Hamann DR, and Schlüter M (1982) *Physical Review B* 26: 4199.
Baroni S, Gianozzi P, and Testa A (1987) *Physical Review Letters* 58: 1861.
Baroni S, de Gironcoli S, Dal Corso A, and Gianozzi P (2001) *Review Modern Physics* 73: 515.
Becke AD (1993) *Journal of Chemical Physics* 98: 5648.
Belonoshko AB, Ahuja R, and Johansson B (2000) *Physical Review Letters* 84: 3638.
Blöchl PE (1994) *Physical Review B* 50: 17953.
Boehler R (1993) *Nature* 363: 534.
Bonev SA, Schwegler F, Ogitsu T, and Galli G (2004) *Nature* 431: 669.
Bowler DR, Miyazaki T, and Gillan MJ (2002) *Journal of Physics: Condensed Matter* 14: 2781.

Brown JM and McQueen RG (1986) *Journal of Geophysical Research* 91: 7485.

Car R and Parrinello M (1985) *Physical Review Letters* 55: 2471.

Casula M, Filippi C, and Sorella S (2005) *Physical Review Letters* 95: 100201.

Casula M and Sorella S (2003) *Journal of Chemical Physics* 119: 6500.

Ceperley D and Alder B (1980) *Physical Review Letters* 45: 566.

Chang KJ and Cohen ML (1984) *Physical Review B* 30: 4774.

Cho JH and Scheffler M (1996) *Physical Review B* 53: 10685.

D'Arco Ph, Sandrone G, Dovesi R, Orlando R, and Saunders VR (1993) *Physics and Chemisty of Minerals* 20: 407.

de Wijs GA, Kresse G, and Gillan MJ (1998) *Physical Review B* 57: 8223.

Dreizler RM and Gross EKU (1990) *Density Functional Theory*. Berlin: Springer.

Drummond ND and Ackland GJ (2002) *Physical Review B* 65: 184104.

Duffy TS, Hemley RJ, and Mao H-K (1995) *Physical Review Letters* 74: 1371.

Dziewonski AM and Anderson DL (1981) *Physics of the Earth and Planetary Interiors* 25: 297.

Filippi C, Healy SB, Kratzer P, Pehlke E, and Scheffler M (2002) *Physical Review Letters* 89: 166102.

Foulkes WMC, Mitaš L, Needs RJ, and Rajagopal G (2001) *Reviews of Modern Physics* 73: 33.

Frenkel D and Smit B (1996) *Understanding Molecular Simulation*. San Diego: Academic Press.

Gannarelli CMS, Alfè D, and Gillan MJ (2005) *Physics of the Earth and Planetary Interiors* 152: 67.

Gibson JB, Goland AN, Milgram M, and Vineyard MGH (1960) *Physical Review* 120: 1229.

Giannozzi P, de Gironcoli S, Pavone P, and Baroni S (1991) *Physical Review B* 43: 7231.

Gillan MJ (1997) *Contemporary Physics* 38: 115.

Gonze X and Vigneron J-P (1989) *Physics Review B* 39: 13120.

Grossman JC, Mitas L, and Raghavachari K (1995) *Physical Review Letters* 75: 3870; erratum: *ibid* 76: 1006.

Gubbins D, Alfè D, Masters G, Price D, and Gillan MJ (2004) *Geophysical Journal International* 157: 1407.

Hamann DR, Schlüter M, and Chiang C (1979) *Physical Review Letters* 43: 1494.

Hohenberg P and Kohn W (1964) *Physical Review* 136: B864.

Hoover WG (1985) *Physical Review A* 31: 1695.

Iitaka T, Hirose K, Kawamura K, and Murakami M (2004) *Nature* 430: 442.

Jaffe JE, Snyder JA, Lin Z, and Hess AC (2000) *Physical Review B* 62: 1660.

Jephcoat AP, Mao HK, and Bell PM (1986) *Journal of Geophysical Research* 91: 4677.

Karki BB, Stixrude L, Clark SJ, Warren MC, Ackland GJ, and Crain J (1997) *American Mineralogist* 82: 51.

Karki BB, Wentzcovitch RM, de Gironcoli S, and Baroni S (2000) *Physical Review B* 61: 8793.

Karki BB, Wentzcovitch RM, de Gironcoli S, and Baroni S (1999) *Science* 286: 1705.

Kerker GP (1980) *Journal of Physics C* 13: L189.

Kobayashi H, Kamimura T, Alfè D, Sturhahan W, Zhao J, and Alp EE (2004) *Physical Review Letters* 93: 195503.

Kohn W and Sham L (1965) *Physical Review A* 140: 1133.

Körling M and Häglund J (1992) *Physical Review B* 45: 13293.

Kresse G, Furthmüller J, and Hafner J (1995) *Europhysics Letters* 32: 729.

Kresse G and Furthmuller J (1996) *Computational Materials Sciences* 6: 15; (1996) *Physical Review B* 54: 11169.

Kresse G and Joubert D (1999) *Physical Review B* 59: 1758.

Laio A, Bernard S, Chiarotti GL, Scandolo S, and Tosatti E (2000) *Science* 287: 1027.

Leung TC, Chan CT, and Harmon BN (1991) *Physical Review B* 44: 2923.

Leung W-K, Needs RJ, Rajagopal G, Itoh S, and Ihara S (1999) *Physical Review Letters* 83: 2351.

Ma Y, Somayazulu M, Shen G, Mao HK, Shu J, and Hemley RJ (2004) *Physics of the Earth and Planetary Interiors* 143–144: 455.

Mandl F (1997) *'Statistical Physics'* 2nd ed., New York: Wiley.

Manten S and Lüchow A (2003) *Journal of Chemical Physics* 119: 1307.

Mao HK, Xu J, Struzhkin VV, *et al.* (2001) *Science* 292: 914.

Martin RM (2004) *Electronic Structure*. Cambridge: Cambridge University Press.

Masters TG and Gubbins D (2003) *Physics of the Earth and Planetary Interiors* 140: 159.

Masters TG and Shearer PM (1990) *Journal of Geophysics Research* 95: 21691.

Mead CA (1992) *Review of Modern Physics* 64: 51.

Mehl MJ, Cohen RE, and Krakauer H (1988) *Journal of Geophysical Research* 93: 8009.

Mermin ND (1965) *Physical Review* 6 137: A1441.

Murakami M, Hirose K, Kawamura K, Sata N, and Ohishi Y (2004) *Science* 304: 855.

Nguyen JH and Holmes NC (2004) *Nature* 427: 339.

Nosé S (1984) *Molecular Physiology* 52: 255; (1984) *Journal of Chemical Physics* 81: 511.

Oganov AR, Brodholt JP, and Price GD (2001) *Nature* 411: 934.

Oganov AR and Dorokupets PI (2003) *Physical Review B* 67: 224110.

Oganov AR, Gillan MJ, and Price GD (2003) *Journal of Chemical Physics* 67: 224110.

Oganov AR, Martoňák R, Laio A, Raiteri P, and Parrinello M (2005) *Nature* 438: 1142.

Oganov AR and Ono S (2004) *Nature* 430: 445.

Ogitsu T, Schwegler F, Gygi F, and Galli G (2003) *Physical Review Letters* 91: 175502.

Parr RG and Yang W (1989) *Density-Functional Theory of Atoms and Molecules*. New York: Oxford Science Publications.

Parrinello M and Rahman A (1980) *Physical Review Letters* 45: 1196.

Perdew JP and Zunger A (1981) *Physical Review B* 23: 5040.

Poirier J-P (1991) *'Introduction to the Physics of the Earth's Interior'*. Cambridge: Cambridge University Press.

Roboredo FA and Williamson AJ (2005) *Physical Review B* 71: 121105.

Savrasov PY and Kotliar G (2003) *Physical Review Letters* 90: 056401.

Seto M, Yoda Y, Kikuta S, and Zhang XW (1995) *Physical Review Letters* 74: 3828.

Shen G, Mao H, Hemley RJ, Duffy TS, and Rivers ML (1998) *Geophysical Research Letters* 25: 373.

Singh DJ, Pickett WE, and Krakauer H (1991) *Physical Review B* 43: 11638.

Soler JM, Artacho E, Gale JD, *et al.* (2002) *Journal of Physics: Condensed Matter* 14: 2745.

Staroverov VN, Scuseria GE, Tao J, and Perdew J (2004) *Physical Review B* 69: 075102.

Stillinger FH and Weber TA (1985) *Physical Review B* 31: 5262.

Stixrude L and Cohen RE (1993) *Nature* 364: 613.

Stixrude L and Cohen RE (1995) *Science* 267: 1972.

Sturhahn W, Toellner TS, Alp EE, *et al.* (1995) *Physics Review Letters* 74: 3832.

Sugino O and Car R (1995) *Physical Review Letters* 74: 1823.

Tao J, Perdew JP, Staroverov VN, and Scuseria GE (2003) *Physical Review Letters* 91: 146401.

Tsuchiya T, Tsuchiya J, Umemoto K, and Wentzcovitch R (2004) *Earth and Planetary Science Letters* 224: 241.

Umrigar CJ, Nightingale MP, and Runge KJ (1993) *Journal of Chemical Physics* 99: 2865.

Vanderbilt D (1990) *Physical Review B* 41: 7892.

Verlet L (1967) *Physical Review* 159: 98.

Vočadlo L and Alfè D (2002) *Physical Review B* 65: 214105.

Wallace DC (1998) lsquo. *Thermodynamics of Crystals*. New York: Dover Publications.

Wang Y and Perdew J (1991) *Physical Review B* 44: 13298.

Wang W, Scandolo S, and Car R (unpublished).

Wannier GH (1966) *Statistical Physics*. New York: Dover Publications, Inc.

Watanabe M and Reinhardt WP (1990) *Physics Review Letters* 65: 3301.

Wentzcovitch RM, Martins JL, and Price GD (1993) *Physical Review Letters* 70: 3947.

Williams Q, Jeanloz R, Bass JD, Svendesen B, and Ahrens TJ (1987) *Science* 286: 181.

Williamson AJ, Hood RQ, and Grossman JC (2001) *Physical Review Letters* 87: 246406.

Wookey J, Stackhouse S, Kendall JM, Brodholt J, and Price GD (2005) *Nature* 438: 1004.

Yoo CS, Holmes NC, Ross M, Webb DJ, and Pike C (1993) *Physical Review Letters* 70: 3931.

Zerr A and Boehler R (1994) *Nature* 371: 506.

Zhu J, Wang XW, and Louie SG (1992) *Physical Review B* 45: 8887.

14 Properties of Rocks and Minerals – Constitutive Equations, Rheological Behavior, and Viscosity of Rocks

D. L. Kohlstedt, University of Minnesota, Minneapolis, MN, USA

14.1	Introduction	389
14.2	Role of Lattice Defects in Deformation	390
14.2.1	Point Defects – Thermodynamics and Kinetics	390
14.2.1.1	Thermodynamics	390
14.2.1.2	Kinetics	392
14.2.2	Line Defects – Structure and Dynamics	394
14.2.3	Planar Defects – Structure and Energy	396
14.3	Mechanisms of Deformation and Constitutive Equations	398
14.3.1	Constitutive Equations	399
14.3.1.1	Diffusion creep	399
14.3.1.2	Deformation involving dislocations	400
14.3.2	Role of Fluids in Rock Deformation	405
14.3.2.1	Role of melt in rock deformation	405
14.3.2.2	Role of water in rock deformation	407
14.4	Upper-Mantle Viscosity	410
14.4.1	Upper-Mantle Viscosity Profile	411
14.4.2	Western US Mantle Viscosity Profile	411
14.5	Concluding Remarks	412
References		413

14.1 Introduction

The rheological behavior of mantle rocks determines the mechanical response of lithospheric plates as well as the nature of convective flow in the deeper mantle. Geodynamic models of these large-scale processes require constitutive equations that describe rheological properties of the appropriate mantle rocks. The forms of such constitutive equations (flow laws or combinations of flow laws) are motivated by micromechanical analyses of plastic deformation phenomena based on physical processes such as ionic diffusion, dislocation migration, and grain-boundary sliding. Critical parameters in flow laws appropriate for rock deformation such as the dependence of strain rate on stress, grain size, temperature dependence (activation energy), and pressure dependence (activation volume) must be determined from careful, well-designed laboratory experiments. Geophysical observations of the mechanical behavior of Earth's mantle in response to changes in the stress environment due for example to an earthquake or the retreat of a glacier provide tests of the applicability of the resulting constitutive equations to deformation occurring under geological conditions.

Confident extrapolation of flow laws determined under laboratory conditions requires an understanding of the physical processes involved in deformation. Minerals and rocks deform plastically by a number of different mechanisms, each of which requires defects in the crystalline structure. Diffusion creep involves the movement of atoms or ions and thus of point defects; dislocation creep entails the glide and climb of line defects; grain-boundary sliding necessitates motion along planar defects. Within any deformation regime, more than one deformation process may be important. For example, diffusion creep can be divided into a regime dominated by diffusion along grain boundaries and another dominated by diffusion through grain interiors (the so-called grain matrix). Grain size, temperature, and pressure dictate which of these two processes is more important. Likewise, in dislocation creep, deformation can be divided into low-temperature and high-temperature regimes, with glide controlling the rate of deformation in the former and climb in the latter. As temperature decreases, differential stress necessarily increases if deformation is to continue at a given rate.

Plastic deformation is a thermally activated, kinetic, irreversible process. Experimentalists frame laboratory investigations of plastic deformation in terms of strain rate as a function of differential stress, temperature, pressure, grain size, and other thermomechanical and structural parameters. In contrast, geodynamicists cast plasticity in terms of viscosity, the ratio of stress to strain rate. In both cases, flow laws or constitutive equations describe the relationships between these parameters, thus providing the basis for extrapolating from laboratory to Earth conditions. Tests of the appropriateness of laboratory-derived flow laws to processes occurring in Earth at much slower strain rates (higher viscosities) – and thus necessarily much lower stresses and/or much lower temperatures and/or much coarser grain sizes – include comparisons of microstructures observed in experimentally deformed samples with those in naturally deformed rocks and of viscosities derived from laboratory experiments with those determined from geophysical observations.

The present chapter, therefore, starts with an examination of the role of defects in plastic deformation, including a discussion of the influence of water-derived point defects on kinetic processes. Next, a section on some of the important mechanisms of deformation introduces constitutive equations that describe plastic flow by linking thermomechanical parameters such as strain rate, differential stress, temperature, pressure, and water fugacity with structural parameters such as dislocation density, grain size, and melt fraction. Finally, profiles of viscosity versus depth in the upper mantle are used to test the applicability of flow laws determined from micromechanical models and laboratory experiments to plastic flow taking place deep beneath Earth's surface.

14.2 Role of Lattice Defects in Deformation

Defects in rocks are essential for plastic deformation to proceed. Zero-dimensional or point defects (specifically vacancies and interstitials) allow flow by diffusive transport of ions, one-dimensional (1-D) or line defects (dislocations) permit deformation by glide and climb, and 2-D or planar defect (grain–grain interfaces) facilitate deformation by grain-boundary sliding and migration. More than one type of defect may be

simultaneously involved in deformation. For example, diffusion creep involves not only diffusion but also grain-boundary sliding (Raj and Ashby, 1971), dislocation processes also frequently couple with grain-boundary sliding (Langdon, 1994), and dislocation creep necessitates glide on several slip systems often combined with climb (von Mises, 1928; Groves and Kelly, 1969; Kelly and Groves, 1969; Paterson, 1969). In addition, dynamic recrystallization frequently acts as a recovery mechanism, producing a new generation of dislocation-free grains that are more easily deformed than the parent grains (e.g., Tullis and Yund, 1985; de Bresser et al., 2001; Drury, 2005). In this section, some fundamental aspects of point defects, dislocations, and grain boundaries are introduced as background for discussing the rheological behavior of rocks and the associated constitutive equations.

14.2.1 Point Defects – Thermodynamics and Kinetics

14.2.1.1 Thermodynamics

At temperatures above absolute zero, in thermodynamic equilibrium crystalline grains contain finite populations of vacancies and self-interstitials. Although the enthalpy of a crystal increases linearly with the addition of vacancies or self-interstitials, the entropy decreases nonlinearly. Hence, the Gibbs free energy of a crystal is minimized not when the crystal is perfect (i.e., defect free) but rather when the crystal contains a finite concentration of vacancies and self-interstitials (i.e., when the crystal is imperfect) (see e.g., Devereux, 1983, pp. 296–300). In metals, the concentrations of these point defects in thermodynamic equilibrium are determined by temperature and pressure through their Gibbs energies of formation (Schmalzried, 1981, pp. 37–38). In ionic materials, the situation is more complicated because the concentrations of the different types of point defects are coupled through the necessity for electroneutrality, which requires that the sum of the concentrations of the various types of positively charged point defects equals the sum of the concentrations of all of the negatively charged point defects (Schmalzried, 1981, pp. 38–42). The concentration of one type of positively charged point defect often greatly exceeds the concentrations of the other positively charged point defects; a similar situation will exist for negatively charged defects. This combination of positive and negative defects defines the majority point defects or the defect type. The charge neutrality condition is then approximated by equating the concentration of the positively charged to the concentration of the negatively charged majority point defects. An example serves to illustrate the salient elements of point defect thermodynamics.

Consider the case of a simple metal oxide, MeO, where Me is an ion such as Mg, Ni, Co, Mn, or Fe. Structural elements for this system include $Me_{Me}^{\times}$, $O_O^{\times}$, $Me_{Me}^{\cdot}$, $Me_i^{\cdot\cdot}$, $V_O^{\cdot\cdot}$, V_{Me}'', and O_i''. The first two entries are termed 'regular structural elements' and the last five are called 'irregular structural elements' or point defects in the ideal crystal lattice. The Kröger–Vink (1956) notation is used to indicate the atomic species, A, occupying a specific crystallographic site, S, and having an effective charge, C, relative to the ideal crystal lattice: A_S^C. In this nomenclature, the symbols ×, ·, and ' indicate neutral, one positive, and one negative effective charges. Vacant sites are denoted by a V, and ions located at interstitial sites are indicated by a subscript i. The defect $Me_{Me}^{\cdot}$ corresponds to a 3+ ion sitting at a site normally occupied by a 2+ ion, such as a ferric iron in a ferrous iron site. In a strict sense, charge neutrality is given by

$$[Me_{Me}^{\cdot}] + 2[V_O^{\cdot\cdot}] + 2[Me_i^{\cdot\cdot}] = 2[V_{Me}''] + 2[O_i''] \quad [1a]$$

where the square brackets, [ ], indicate molar concentration. In a Mg-rich system, one possible disorder type is the combination of $Me_i^{\cdot\cdot}$ and V_{Me}'' (Catlow *et al.*, 1994), such that the charge neutrality condition is approximated as

$$[Me_i^{\cdot\cdot}] = [V_{Me}''] \quad [1b]$$

In contrast, in an Fe-rich quasi-binary material such as (Mg, Fe)O, likely majority point defects are $Fe_{Me}^{\cdot}$ and V_{Me}'' (Hilbrandt and Martin, 1998), yielding the charge neutrality condition

$$[Fe_{Me}^{\cdot}] = 2[V_{Me}''] \quad [1c]$$

where $Fe_{Me}^{\cdot}$ is often written as $h^{\cdot}$, indicating a highly mobile electron hole localized at a Fe^{2+} site.

To obtain expressions that give the dependences of point defect concentrations on thermodynamic variables such as temperature, pressure, and component activities, reaction equations are needed. The first reaction equation should involve the majority point defects. In the case of the disorder type given by eqn [1b], the appropriate reaction starts with a regular structural element resulting in

$$Me_{Me}^{\times} \rightleftarrows Me_i^{\cdot\cdot} + V_{Me}'' \quad [2]$$

for which the law of mass action yields

$$a_{Me_i^{\cdot\cdot}} a_{V_{Me}''} = a_{Me_{Me}^{\times}} K_2 \quad [3a]$$

where K_2 is the reaction constant for the reaction in eqn [2]. Since the activity of $Me_{Me}^{\times}$ is little affected by the presence of a small concentration of point defects (typically $<10^{-3}$), $a_{Me_{Me}^{\times}} \approx [Me_{Me}^{\times}] \approx 1$. Likewise, for ideally dilute solutions of point defects, the activities of the point defects can be replaced by their concentrations such that eqn [3a] becomes

$$[Me_i^{\cdot\cdot}][V_{Me}''] = K_2 = \exp\frac{-G_F}{RT} \quad [3b]$$

where G_F is the Gibbs energy for the reaction in eqn [2] and RT has the usual meaning. This reaction describes the formation of the so-called Frenkel point defects (a vacancy–interstitial pair). If the charge neutrality condition in eqn [1b] and the mass action equation in eqn [3b] are combined, the dependence on temperature and pressure of the concentrations of the Frenkel defects is

$$[Me_i^{\cdot\cdot}] = [V_{Me}''] = \sqrt{K_2} = \exp\frac{-G_F}{2RT} \quad [4]$$

Frenkel majority point defects are an example of a thermal disorder type, that is, the concentrations of these point defects depend on temperature and pressure but not on the activities of the components. A second example of thermal disorder occurs if Schottky point defects, V_{Me}'' and $V_O^{\cdot\cdot}$, are the majority point defects.

For the disorder type given by eqn [1c], the appropriate reaction involves a regular structural element plus a neutral crystal component to yield

$$\frac{1}{2}O_2(srg) + 2Fe_{Me}^{\times} + Me_{Me}^{\times} \rightleftarrows 2Fe_{Me}^{\cdot} + V_{Me}'' + MeO(srg)$$
$$[5]$$

where the oxygen is provided from a site of repeatable growth (srg) such as a dislocation, grain boundary, or crystal–gas interface (surface). The law of mass action yields

$$[Fe_{Me}^{\cdot}]^2[V_{Me}'']a_{MeO} = [Fe_{Me}^{\times}]^2[Me_{Me}^{\times}]f_{O_2}^{1/2}K_5 \quad [6]$$

where f_{O_2} is the oxygen fugacity. If the charge neutrality condition in eqn [1c] is now substituted into eqn [6] and the approximation $[Me_{Me}^{\times}] \approx a_{MeO} = 1$ is made, then

$$\begin{aligned}[Fe_{Me}^{\cdot}] &= 2[V_{Me}''] \\ &= \sqrt[3]{2}[Fe_{Me}^{\times}]^{2/3}f_{O_2}^{1/6}K_5^{1/3} \\ &= \sqrt[3]{2}[Fe_{Me}^{\times}]^{2/3}f_{O_2}^{1/6}\exp\frac{-G_5}{3RT}\end{aligned} \quad [7]$$

Thus, if a neutral crystal component is added from a site of repeatable growth, the concentrations of the majority point defects will depend not only on pressure and temperature but also on component activities, often referred to as an activity-dependent disorder type. In this case, it should be noted that the material will be non-stoichiometric with typically $[V''_{Me}] \gg [V^{\cdot\cdot}_O]$. Also, a comparison of eqn [4] with eqn [7] demonstrates that the form as well as the magnitude of the concentration of V''_{Me} depends on the charge neutrality condition.

Since $[V^{\cdot\cdot}_O]$ is generally several orders of magnitude smaller than $[V''_{Me}]$, O diffuses much more slowly than Me, at least through the interiors of grains. Consequently, as discussed below, the rate of diffusion creep (specifically, Nabarro–Herring creep) and of dislocation climb are limited by the rate of O diffusion. Hence, it is important to examine the dependence of the concentration of $V^{\cdot\cdot}_O$ on temperature, pressure, and oxygen fugacity. Since the dependence of the concentration of V''_{Me} on these parameters has been determined above, formulation of a reaction involving $V^{\cdot\cdot}_O$ and V''_{Me} provides a good starting point. The point defects $V^{\cdot\cdot}_O$ and V''_{Me} are related via the Schottky formation reaction

$$Me^{\times}_{Me} + O^{\times}_O \rightleftarrows V''_{Me} + V^{\cdot\cdot}_O + MeO \qquad [8]$$

It should be emphasized that it is essential in reaction equations such as eqn [8] that the number and identity of the atomic species, the crystallographic sites, and the effective charges (recall A^C_S) be the same on the two sides of the reaction equation. The law of mass action for eqn [8] reads

$$[V''_{Me}][V^{\cdot\cdot}_O] = K_8 = \exp\frac{-G_8}{RT} \qquad [9]$$

again using the approximations $a_{Me^{\times}_{Me}} \approx a_{O^{\times}_O} \approx 1$ with a_{MeO} equal to unity since MeO is present. For the Frenkel thermal disorder type given by eqn [1b], $[Me^{\cdot\cdot}_i] = [V''_{Me}]$, if eqn [9] is now combined with eqn [4], then

$$[V^{\cdot\cdot}_O] = [V''_{Me}]^{-1}K_8 = \frac{K_8}{\sqrt{K_2}} = \exp\frac{-(G_8 - G_F/2)}{RT} \qquad [10]$$

For the activity-dependent disorder type given by eqn [1c], $[Fe^{\cdot}_{Me}] = 2[V''_{Me}]$, if eqn [9] is combined with eqn [7], then

$$\begin{aligned}[V^{\cdot\cdot}_O] &= [V''_{Me}]^{-1}K_8 \\ &= \sqrt[3]{4}[Fe^{\times}_{Me}]^{-2/3}f_{O_2}^{-1/6}\frac{K_8}{K_5^{1/3}} \\ &= \sqrt[3]{4}[Fe^{\times}_{Me}]^{-2/3}f_{O_2}^{-1/6}\exp\frac{-(G_8 - G_5/3)}{RT} \qquad [11]\end{aligned}$$

Under hydrous conditions, additional point defects must be considered. Hydrogen ions, that is protons, $\dot{p}$ occupy interstitial sites, $H^{\cdot}_i$, near oxygen ions thus corresponding to the point defects $(OH)^{\cdot}_O$. This water-derived point defect can be introduced through dissociation of a water molecule by the reaction

$$H_2O(srg) + 2O^{\times}_O + Me^{\times}_{Me} \rightleftarrows 2(OH)^{\cdot}_O + V''_{Me} \\ + MeO(srg) \qquad [12]$$

In addition, point defect associates such as those produced between $(OH)^{\cdot}_O$ and V''_{Me} can form

$$(OH)^{\cdot}_O + V''_{Me} \rightleftarrows \{(OH)^{\cdot}_O - V''_{Me}\}' \qquad [13]$$

where the curly brackets { } indicate a defect associate. This and other defect associates as well as $(OH)^{\cdot}_O$ must now be included in the charge neutrality equation, eqn [1]. New disorder types such as

$$[(OH)^{\cdot}_O] = \left[\{(OH)^{\cdot}_O - V''_{Me}\}'\right] \qquad [14]$$

and

$$[Fe^{\cdot}_{Me}] = \left[\{(OH)^{\cdot}_O - V''_{Me}\}'\right] \qquad [15]$$

must also be considered. A summary of the dependencies of the concentrations of various point defects on oxygen fugacity and water fugacity for the system MeO is provided in **Table 1** for a range of charge neutrality conditions.

14.2.1.2 Kinetics

Diffusion of atoms and ions through crystalline solids takes place by movement of point defects, namely, vacancies and self-interstitials. Here the discussion focuses on vacancies; a parallel analysis applies for self-interstitials. Since the fraction of vacant sites, X_V, on any specific sublattice is small, typically $<10^{-3}$ to $\ll 10^{-3}$, vacancies diffuse much more rapidly than the ions on that sublattice. From the point of view of a vacancy, all of the neighboring sites are available, while from the perspective of an ion, possibly one but more likely no sites are available at any particular moment. This connection between the movement of ions and that of vacancies leads to the following relationship between the diffusion coefficient for ions, D_{ion}, and that for vacancies, D_V:

$$X_{ion}D_{ion} = X_V D_V \qquad [16a]$$

where X_{ion} is the fraction of sites on a given sublattice occupied by the ions associated with that sublattice.

Table 1 Dependence of point defect concentrations on oxygen fugacity and water fugacity, expressed as the exponents p and q in the relationship $[\ ] \propto f_{O_2}^p f_{H_2O}^q$, for several charge neutrality conditions for MeO

Charge neutrality	$Me_{Me}^{\cdot}$	V_{Me}''	$V_O^{\cdot\cdot}$	$\{Me_{Me}^{\cdot} - V_{Me}''\}'$	$(OH)_O^{\cdot}$	$\{(OH)_O^{\cdot} - V_{Me}''\}'$
$[Me_{Me}^{\cdot}] = 2[V_{Me}'']$	1/6, 0	1/6, 0	−1/6, 0	1/3, 0	−1/12, 1/2	1/12, 1/2
$[(OH)_O^{\cdot}] = 2[V_{Me}'']$	1/4, −1/6	0, 1/3	0, −1/3	1/4, 1/6	0, 1/3	0, 2/3
$[Me_{Me}^{\cdot}] = \left[\{(OH)_O^{\cdot} - V_{Me}''\}'\right]$	1/8, 1/4	1/4, −1/2	−1/4, 1/2	3/8, −1/4	−1/8, 3/4	1/8, 1/4
$[(OH)_O^{\cdot}] = \left[\{(OH)_O^{\cdot} - V_{Me}''\}'\right]$	1/4, 0	0, 0	0, 0	1/4, 0	0, 1/2	0, 1/2

Since $X_{ion} = (1 - X_V) \approx 1$, eqn [16a] is usually approximated as

$$D_{ion} \approx X_V D_V \qquad [16b]$$

such that

$$D_{ion} \ll D_V \qquad [16c]$$

that is, vacancies diffuse orders of magnitude faster than ions because $X_V \ll 1$.

One consequence of the rapid diffusion of vacancies (as well as of electron holes) is that equilibrium vacancy concentrations are attained relatively rapidly in response to changes in oxygen partial pressure or oxide activity, typically in much less than an hour under laboratory conditions for which diffusion distances are on the scale of the sample size (Mackwell *et al.*, 1988; Wanamaker, 1994). Point defect concentrations respond even more quickly to changes in temperature and pressure for which the appropriate diffusion length is the distance between sites of repeatable growth, which serve as sources and sinks for point defects.

The concentration of vacancies in a mineral will differ from one sublattice to the next. In a system composed of Me = (Mg, Fe), Si, and O, the concentration of vacancies on the Me sublattice is generally several orders of magnitude larger than the concentrations on the Si and O sublattices, that is, $[V_{Me}''] >> [V_O^{\cdot\cdot}], [V_{Si}'''']$ (e.g., Nakamura and Schmalzried, 1983, 1984; Tsai and Dieckmann, 2002). As a result, $D_{Me} >> D_O, D_{Si}$.

For deformation by diffusion creep or dislocation climb processes, the flux of ions directly enters expressions for the strain rate. The deformation rate is calculated from analyses of the flux of ions between regions experiencing different stress states, such as between grain boundaries in a rock that are oriented normal to and those that are oriented parallel to the maximum principal stress. In an ionic solid, the fluxes of all of the constituent ions must be coupled in order to maintain stoichiometry (Ruoff, 1965; Readey,

1966; Gordon, 1973; Dimos *et al.*, 1988; Jaoul, 1990; Schmalzried, 1995, pp. 345–346). As a consequence, the creep rate of a monominerallic rock is controlled by the rate of diffusion of the slowest diffusing ion.

In the case of olivine, Me_2SiO_4, for diffusion along parallel paths (i.e., 1-D diffusion), the fluxes of Me^{2+}, Si^{4+}, and O^{2-} are given by (e.g., Schmalzried, 1981, p. 63; Schmalzried, 1995, pp. 78–82)

$$j_{Me^{2+}} = -\frac{D_{Me}C_{Me}}{RT}\frac{\partial\eta_{Me^{2+}}}{\partial\xi}$$
$$= -\frac{D_{Me}C_{Me}}{RT}\left(\frac{\partial\mu_{Me^{2+}}}{\partial\xi} + 2F\frac{\partial\varphi}{\partial\xi}\right) \qquad [17a]$$

$$j_{Si^{4+}} = -\frac{D_{Si}C_{Si}}{RT}\frac{\partial\eta_{Si^{4+}}}{\partial\xi}$$
$$= -\frac{D_{Si}C_{Si}}{RT}\left(\frac{\partial\mu_{Si^{4+}}}{\partial\xi} + 4F\frac{\partial\varphi}{\partial\xi}\right) \qquad [17b]$$

and

$$j_{O^{2-}} = -\frac{D_O C_O}{RT}\frac{\partial\eta_{O^{2-}}}{\partial\xi}$$
$$= -\frac{D_O C_O}{RT}\left(\frac{\partial\mu_{O^{2-}}}{\partial\xi} - 2F\frac{\partial\varphi}{\partial\xi}\right) \qquad [17c]$$

where j is the flux, D the self-diffusivity, C the concentration, and η the electrochemical potential of the appropriate ionic species. The electrochemical potentials for the various ions are written in terms of the chemical potential, μ, and the electrical potential, φ, as (e.g., Schmalzried, 1981, p. 63)

$$\eta = \mu + zF\varphi \qquad [18]$$

where z is the charge on the ion and F is the Faraday constant. To maintain stoichiometry during diffusion-controlled creep in which transport of entire lattice molecules must occur between sites of repeatable growth such as dislocations and grain boundaries, the flux coupling condition is

$$\frac{j_{Me^{2+}}}{C_{Me}} = \frac{j_{Si^{4+}}}{C_{Si}} = \frac{j_{O^{2-}}}{C_O} \qquad [19a]$$

Equation [19a] applies specifically to the case in which the all of the ions diffuse along a single path or parallel paths; the more general case in which diffusion occurs along multiple paths (e.g., along grain boundaries and through grain interiors) is treated in Dimos *et al.* (1988; see also Kohlstedt, 2006).

The flux coupling equation provides a means of eliminating the $\partial\varphi/\partial\xi$ term from the flux equations. The flux coupling condition for olivine in eqn [19a] can be rewritten as

$$2j_{Me^{2+}} = 4j_{Si^{4+}} = j_{O^{2-}} \qquad [19b]$$

Then, since $D_{Me} >> D_O > D_{Si}$, the approximation

$$\frac{\partial\eta_{Me^{2+}}}{\partial\xi} = \frac{\partial\mu_{Me^{2+}}}{\partial\xi} + 2F\frac{\partial\varphi}{\partial\xi} \approx 0 \qquad [20]$$

must hold in order to satisfy the flux coupling condition given by eqn [19b]. Using the implicit assumption of local thermodynamic equilibrium, the chemical potentials of the ions can be written in terms of the chemical potentials of the oxides as

$$\mu_{Me^{2+}} + \mu_{O^{2-}} = \mu_{MeO} \qquad [21a]$$

and

$$\mu_{Si^{4+}} + 2\mu_{O^{2-}} = \mu_{SiO_2} \qquad [21b]$$

Thus, the flux coupling condition in eqn [19b] leads to the relation

$$D_{Si}\frac{\partial\mu_{SiO_2}}{\partial\xi} = (D_O + 2D_{Si})\frac{\partial\mu_{MeO}}{\partial\xi} \qquad [22]$$

The flux of Si^{4+}, the slowest ionic species, then becomes

$$j_{Si^{4+}} = -\frac{C_{Si}D_{Si}}{RT}\left(\frac{\partial\mu_{SiO_2}}{\partial\xi} - 2\frac{\partial\mu_{MeO}}{\partial\xi}\right) \qquad [23a]$$

Since the chemical potential of olivine is

$$\mu_{Me_2SiO_4} = 2\mu_{MeO} + \mu_{SiO_2}$$

the flux of Si^{4+} is

$$j_{Si^{4+}} = -\frac{C_{Si}D_{Si}}{RT}\frac{D_O}{D_O + 4D_{Si}}\frac{\partial\mu_{Me_2SiO_4}}{\partial\xi} \qquad [23b]$$

The chemical potential of olivine contains an activity, a, and a stress, σ, term such that

$$\frac{\partial\mu_{Me_2SiO_4}}{\partial\xi} = RT\frac{\partial\ln a_{Me_2SiO_4}}{\partial\xi} - V_{Me_2SiO_4}\frac{\partial\sigma}{\partial\xi} \qquad [24]$$

where $V_{Me_2SiO_4}$ is the molar volume of olivine (e.g., Ready, 1966; Gordon, 1973; Dimos *et al.*, 1988; Schmalzried, 1995, p. 334). The first term on the

right-hand side of eqn [24] can be expressed in terms of the gradient in the concentration of vacancies on the Me sublattice (Dimos *et al.*, 1988; Jaoul, 1990) as

$$RT\frac{\partial\ln a_{Me_2SiO_4}}{\partial\xi} \approx -2RT\frac{\partial[V''_{Me}]}{\partial\xi}$$

$$\approx -2[(V''_{Me})^o]V_{Me_2SiO_4}\frac{\partial\sigma}{\partial\xi} \qquad [25]$$

where $[(V''_{Me})^o]$ is the concentration of Me vacancies under hydrostatic stress conditions. The gradient in the concentration in metal vacancies results from the gradient in normal stress between sources for vacancies (regions of minimum compressive stress) and sinks for vacancies (regions of maximum compressive stress). Since $[(V''_{Me})^o]$ is small, $<10^{-3}$, the second term in eqn [24] dominates such that

$$\frac{\partial\mu_{Me_2SiO_4}}{\partial\xi} \approx -V_{Me_2SiO_4}\frac{\partial\sigma}{\partial\xi} \qquad [26]$$

Therefore, if $D_O >> D_{Si}$

$$j_{Si^{4+}} = \frac{C_{Si}D_{Si}}{RT}V_{Me_2SiO_4}\frac{\partial\sigma}{\partial\xi} \qquad [27]$$

14.2.2 Line Defects – Structure and Dynamics

Dislocations are line defects that separate regions of a crystal that have slipped from those that have not slipped. Their motion by glide and climb is central to deformation of most crystalline solids, much like point defects are essential for diffusion. Unlike point defects, dislocations are not equilibrium defects. While the change in enthalpy associated with the formation of a dislocation is quite large, the change in entropy is relatively small. In a simplistic analogy, the former might be considered equal to the change in enthalpy due to the formation of of N vacancies, where the vacancies form a row of the same length as the dislocation line. In contrast, the change in entropy resulting from the formation of N vacancies constrained to lie along a line will be significantly less than that associated with the formation of N vacancies randomly distributed in a crystal. The result is that, at equilibrium, a crystal is expected to be dislocation free (Devereux, 1983, pp. 368–372).

A dislocation is characterized by two vector quantities, its line direction, ℓ, which defines the direction of the dislocation at each point along its length, and a displacement or Burgers vector, **b**, which defines the displacement of the lattice produced as the dislocation moves through a crystal. For a dislocation loop such as illustrated in **Figure 1**, the tangential line

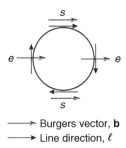

→ Burgers vector, **b**
→ Line direction, ℓ

Figure 1 Sketch of dislocation loop indicating the direction of the Burgers vector, **b**, and of the dislocation line, ℓ, at each point along the loop. The edge and screw segments of the dislocation loop are denoted by e and s, respectively.

direction changes from one point to the next along the loop, while the Burgers vector is the same at every point. Regions along the dislocation loop for which **b** is perpendicular to ℓ are termed edge segments, while regions for which **b** is parallel to ℓ are termed screw segments. Regions of a dislocation for which **b** and ℓ are neither perpendicular nor parallel are referred to as mixed segments.

Movement of a dislocation can take place by glide and climb. An edge dislocation can move in one of two ways. If an edge dislocation moves in the glide plane defined by $\mathbf{b} \times \ell$, its motion is conservative. In contrast, if an edge dislocation moves normal to its glide plane, the process is nonconservative in that lattice molecules are added to or taken away from the dislocation line by diffusion such that the dislocation climbs out of its glide plane. A screw dislocation does not have a specific glide plane since **b** and ℓ are parallel to each other. Hence, a screw dislocation generally glides on the plane that offers least resistance to its motion. If it encounters an obstacle to its movement, a screw dislocation can cross slip off of its original glide plane onto a parallel glide plane in order to continue moving. A dislocation must

always move in a direction perpendicular to its line direction. The displacement resulting from glide of an edge dislocation is thus parallel to the direction of dislocation glide, while the displacement resulting from movement of a screw dislocation is perpendicular to its direction of motion defined by the velocity vector, **v**, as illustrated in **Figure 2**. Finally, a slip system is defined for a dislocation by the combination of a unit vector normal to the slip plane, **n**, and **b**. As an example, important slip systems in clinopyroxene include $\{110\}$ $\langle 1\bar{1}0 \rangle$, $\{110\}[001]$, and $(100)[001]$ (Bystricky and Mackwell, 2001). Slip systems in olivine include $(010)[100]$, $(001)[100]$, $(100)[001]$, and $(010)[001]$ with the dominate slip system determined by stress, water concentration, as well as pressure and temperature conditions (Carter and Avé Lallemant, 1970; Jung and Karato, 2001; Mainprice *et al.*, 2005).

To first order, dislocations with short displacement vectors are preferred over those with longer displacement vectors since the energy per unit length of a dislocation, E_{disl}, is proportional to the square of the Burgers vector. For an edge dislocation in an elastically isotropic material (Weertman and Weertman, 1992, pp. 45–52)

$$E_{disl} = \frac{Gb^2}{4\pi(1-v)} \ln\frac{r}{r_c} \qquad [28]$$

where G is the shear modulus, r is the mean spacing between dislocations, and r_c is the radius of the dislocation core. In eqn [28], the contribution of the core of the dislocation to the elastic strain energy has been neglected. For elastically anisotropic materials, the full matrix of elastic constants must be considered, as discussed in detail by Hirth and Lothe (1968, pp. 398–440).

Slip generally occurs on the closest packed plane that contains **b**, that is, the most widely separated

(a)   (b)   (c)

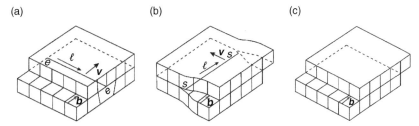

Figure 2 Sketch illustrating that (a) the movement of an edge dislocation, e, and (b) the movement of a screw dislocation, s, with the same Burgers vector (c) produce the same displacement of the upper half of a crystal relative to the lower half. Adapted from Kelly A and Groves GW (1970) *Crystallography and Crystal Defects*, 428 pp. Reading, MA: Addison-Wesley Publishing Company.

planes and thus, in general, the most weakly bonded planes. In olivine, $|[100]| < |[001]| < |[010]|$ suggesting that $\mathbf{b} = [100]$ should dominate. However, plastic deformation of a polycrystalline material cannot proceed without opening up void space if only one or two independent slip systems operate. Homogeneous plastic deformation of a rock requires dislocation glide on five independent slip systems (von Mises, 1928). This constraint is relaxed to glide on four independent slip systems if deformation is allowed to be heterogeneous (Hutchinson, 1976). If dislocations both glide and climb, homogeneous deformation necessitates the operation of just three independent slip systems (Kelly and Groves, 1969). Processes such as grain-boundary sliding, ionic diffusion, and twinning can also relax the von Mises criterion.

Glide and climb of dislocations involve steps along the dislocation line. Steps that lie in the glide plane, kinks, facilitate glide. Steps out of the glide plane, jogs, facilitate climb.

Dislocations interact with one another and with an applied stress field through the stress field that arises due to the elastic distortion caused by the presence of each dislocation. The stress fields associated with edge and screw dislocations in an elastically isotropic and elastically anisotropic media are derived in Hirth and Lothe (1968, pp. 398–440) and in Weertman and Weertman (1992, pp. 22–41). The force on a dislocation is then calculated using the Peach–Koehler equation (Peach and Koehler, 1950). In its simplest form, the force per unit length, f, on a dislocation is

$$f = \sigma b \qquad [29]$$

where the stress σ arises, for example, from a neighboring dislocation or an external force.

14.2.3 Planar Defects – Structure and Energy

Planar defects include twin boundaries, stacking faults, grain boundaries, and interphase boundaries. The emphasis in this chapter is on grain boundaries and interphase boundaries, because of the important role of these two types of interfaces in plastic deformation.

If the misorientation across a grain boundary is not too large, say, $<15°$ (i.e., a low-angle boundary), the interface can be constructed from a periodic array of dislocations. A low-angle tilt boundary consists of

a series of parallel edge dislocations, as illustrated schematically in **Figure 3** and imaged in **Figure 4**. In the case of a low-angle tilt boundary, the spacing between dislocations, h, can be expressed in terms of the misorientation angle, ϑ, across the boundary as

$$h = \frac{b}{2\sin(\vartheta/2)} \qquad [30a]$$

For a small misorientation, the misorientation angle can be approximated by

$$\vartheta \approx \frac{b}{h} \qquad [30b]$$

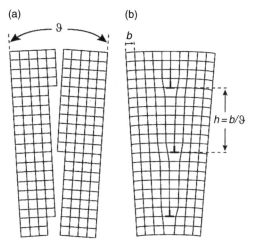

Figure 3 Sketch of a low-angle tilt boundary formed by a series of periodically spaced edge dislocations, which are denoted by the inverted 'T' symbols. Adapted from Read WT Jr (1953) *Dislocation in Crystals*, 175pp. New York: McGraw-Hill Book Company.

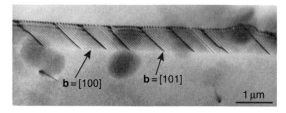

Figure 4 Bright-field transmission electron micrograph of a low-angle tilt boundary in olivine. Two sets of dislocations, one with $\mathbf{b} = [100]$ and the other with $\mathbf{b} = [101]$, are present in this boundary. The darker $\mathbf{b} = [101]$ dislocations are more widely spaced than the lighter, periodically spaced $\mathbf{b} = [100]$ dislocations. This image was taken with the diffraction condition $\mathbf{g} = (002)$, hence the $\mathbf{b} = [100]$ dislocations exhibit only residual contrast. Adapted from Goetze C and Kohlstedt DL (1973) Laboratory study of dislocation climb and diffusion in olivine. *Journal of Geophysical Research* 78: 5961–5971.

Starting with eqn [28], the elastic strain energy per unit area of a low-angle tilt boundary can then be derived (Read and Shockley, 1950)

$$E_{\text{tilt}} = \frac{Gb^2}{4\pi(1-v)} \frac{\vartheta}{b} (A - \ln \vartheta) \qquad [31]$$

where the parameter A accounts for the fraction of the elastic strain energy associated with the cores of the dislocations. Once the cores of the dislocations in a tilt boundary begin to overlap, this simple description of a grain boundary breaks down.

A low-angle twist boundary requires two, generally orthogonal, sets of screw dislocations. The elastic energy per unit area for a low-angle twist boundary is similar to that for a low-angle tilt boundary. A sketch of the atomic structure of a low-angle twist boundary formed by rotating the lattices of two grains relative to each through an angle ϑ, thus forming orthogonal sets of dislocations, is presented in **Figure 5**. A transmission electron micrograph of a twist boundary is shown in **Figure 6**.

As illustrated in **Figure 7**, grain-boundary energy increases rapidly with increasing ϑ, as described by eqn [31], up to $\vartheta \approx 20°$. At larger values of ϑ, grain-boundary energy is approximately constant except for a few minima that occur at specific orientations. Other than twin boundaries, these minima occur at orientations for which a significant fraction of the lattice points in the two neighboring grains are nearly co-incident (Chan and Balluffi, 1985, 1986). A simple example of this coincidence structure illustrating the type of periodicity that can be present in specific grain boundaries is shown in **Figure 8** for a twist boundary.

Grain and interphase boundaries influence the physical properties of rocks in several important ways. First, grain-boundary sliding provides a mechanism for producing strain in response to an applied differential stress. Grain-boundary sliding takes place by the movement of grain-boundary dislocations. As discussed in the next section, grain-boundary sliding is a critical part of diffusion creep (Raj and Ahsby, 1971). Grain-boundary sliding can also be important during deformation in which dislocation glide and climb operate (e.g., Langdon, 1994). In this case, lattice dislocations can dissociate and enter grain boundaries resulting in enhanced or stimulated grain-boundary sliding (Pshenichnyuk et al., 1998). As discussed below, grain-boundary sliding often provides an additional mechanism of deformation and can be particularly important in rocks composed of minerals with fewer than five independent slip systems.

(a)

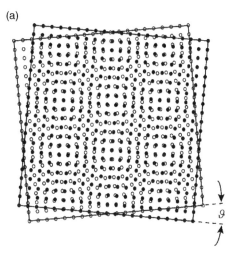

(b)

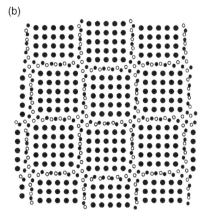

Figure 5 Sketch of atomic configuration for a [001] low-angle twist boundary in a simple cubic lattice with the boundary parallel to the plane of the figure. In this example, $\vartheta \approx 10°$. (a) In the unrelaxed atomic configuration, the open circles represent atoms in the plane just above the boundary, while the closed circles represent the atoms in the plane just below the boundary. (b) In the relaxed atomic configuration, the grains join in regions in which the atomic match is good with two orthogonal sets of screw dislocations located between these regions. Adapted from Schmalzried H (1995) *Chemical Kinetics of Solids*, 433pp. New York: VCH Publishers.

Second, grain boundaries migrate in response to a difference in dislocation density between neighboring grains. As expressed in eqn [28], each dislocation introduces elastic strain energy into a crystalline grain. If during deformation the dislocation density becomes higher in one grain than in the next grain, a thermodynamic driving force exists that can cause the grain boundary to migrate toward the region of higher dislocation density, absorbing lattice dislocations into the grain boundary thus reducing the

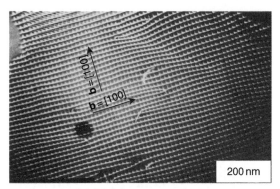

Figure 6 Dark-field transmission electron micrograph of two orthogonal sets of screw dislocations, one with **b** = [100] and the other with **b** = [001], forming a low-angle twist boundary in the (010) plane of olivine. The boundary is very nearly parallel to the plane of the figure. Adapted from Ricoult DL and Kohlstedt DL (1983a) Structural width of low-angle grain boundaries in olivine. *Physics and Chemistry of Minerals* 9: 133–138. Ricoult DL and Kohlstedt DL (1983b) Low-angle grain boundaries in olivine. In: Yan MF and Heuer AH (eds.) *Advances in Ceramics, Character of Grain Boundaries*, vol. 6, pp. 56–72. Columbus: American Ceramic Society.

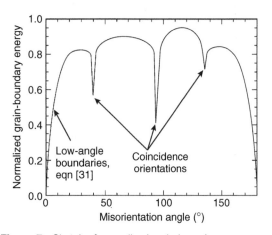

Figure 7 Sketch of normalized grain-boundary energy vs misorientation angle illustrating the energy minima that arise when the two neighboring grains are oriented such that a significant fraction of lattice sites are coincident.

stored strain energy. A dislocation-free region is left in the wake of the migrating boundary.

Third, grain boundaries are short-circuit paths for diffusion. The rate of diffusion along a grain boundary is in general several orders of magnitude faster than through a grain interior (Shemon, 1983, pp. 164–175). Thus, even though the cross-sectional area of a grain boundary is small relative to that of the grain itself, the flux of atoms passing through a grain boundary can be significant. The result is that, in the diffusion creep regime, grain-boundary diffusion is often the primary

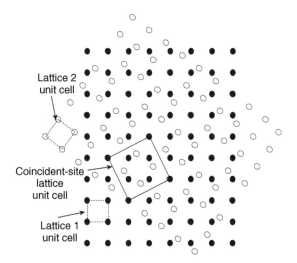

Figure 8 Sketch of lattice sites in two simple cubic grains that have been rotated relative to each other by ~53°. At this orientation, a fraction of the lattice sites from the two grains coincide forming a coincident-site lattice with a unit cell that is larger than those of the two adjoining grains.

mechanism for deformation. It should also be noted that grain boundaries provide storage sites for elements that, due to size or charge, are incompatible in the grain interior (Hiraga *et al.*, 2003, 2004, 2007; Hiraga and Kohlstedt, 2007). The presence of incompatible elements at grain and interphase boundaries perturbs the structure and the local charge state of the boundary, which in turn can affect diffusion kinetics (Mishin and Herzi, 1999).

14.3 Mechanisms of Deformation and Constitutive Equations

The mechanism of deformation that dominates the plastic deformation of a rock depends both on the properties of the rock such as grain size, d, and phase content, Φ, and on thermodynamic parameters such as temperature, pressure, P, water fugacity, and component (oxide) activity, a_{ox}. The dependence of strain rate, $\dot{\varepsilon}$, on these quantities can be expressed in a general constitutive equation or flow law as

$$\dot{\varepsilon} = \dot{\varepsilon}(d,\ \Phi,\ T,\ P,\ f_{H_2O},\ a_{ox},\ \ldots) \qquad [32]$$

Diffusion-controlled creep tends to be important at low differential stresses in fine-grained rocks. In contrast, dislocation-dominated processes govern flow at higher stresses in coarser-grained rocks. In this section, some of the constitutive equations used to describe flow in the diffusion and dislocation deformation regimes are first introduced. Then, the influence on rheological behavior of two important fluids, water and melt, are discussed.

14.3.1 Constitutive Equations

14.3.1.1 Diffusion creep

Diffusion creep can be divided into two regimes, one in which diffusion through the interiors of grains controls the rate of deformation (Nabarro, 1948; Herring, 1950) and the other in which diffusion along grain boundaries governs the rate of deformation (Coble, 1963). The relative importance of these two mechanisms depends primarily on grain size and temperature.

14.3.1.1.(i) Grain matrix diffusion

The rate of deformation in the Nabarro–Herring creep regime in which diffusion through grain interiors limits the rate of creep is described by the relation (Nabarro, 1948; Herring, 1950)

$$\dot{\varepsilon}_{NH} = \alpha_{NH} \frac{\sigma V_m}{RT} \frac{D_{gm}}{d^2} \quad [33]$$

where α_{NH} is a geometrical term and D_{gm} is the diffusion coefficient for the slowest species diffusing through the grain matrix (gm), that is, the grain interior. The important points to note are that strain rate is linearly proportional to the differential stress, inversely proportional to the square of the grain size, and exponentially dependent on temperature and pressure through the diffusion coefficient since

$$D_{gm} = D_{gm}^o \exp\left(-\frac{\Delta E_{gm} + P\Delta V_{gm}}{RT}\right)$$
$$= D_{gm}^o \exp\left(-\frac{\Delta H_{gm}}{RT}\right) \quad [34]$$

where D_{gm}^o is a material-dependent parameter and ΔE_{gm}, ΔV_{gm}, and ΔH_{gm} are the activation energy, activation volume, and activation enthalpy for grain matrix diffusion, respectively.

14.3.1.1.(ii) Grain-boundary diffusion

A similar expression applies in the Coble creep regime in which the creep rate is limited by diffusion through grain boundaries (Coble, 1963):

$$\dot{\varepsilon}_C = \alpha_C \frac{\sigma V_m}{RT} \frac{\delta D_{gb}}{d^3} \quad [35]$$

where α_C is a geometrical term, δ is the diffusion width of the grain boundary, which is approximately equal to the structural width of ~ 1 nm (e.g., Atkinson, 1985; Carter and Sass, 1981; Ricoult and Kohlstedt,

1983a, 1983b), and D_{gb} is the diffusion coefficient for the slowest species diffusing along the boundaries. For grain boundaries

$$D_{gb} = D_{gb}^o \exp\left(-\frac{\Delta E_{gb} + P\Delta V_{gb}}{RT}\right)$$
$$= D_{gb}^o \exp\left(-\frac{\Delta H_{gb}}{RT}\right) \quad [36]$$

where D_{gb}^o is a material-dependent parameter and ΔE_{gb}, ΔV_{gb}, and ΔH_{gb} are the activation energy, activation volume, and activation enthalpy for grain-boundary diffusion, respectively. Again, the strain rate increases linearly with increasing differential stress and increases exponentially with inverse temperature while decreasing exponentially with increasing pressure. However, in contrast to the Nabarro–Herring case, strain rate for Coble creep varies inversely as the cube, rather than the square, of the grain size.

Comparison of eqn [33] with eqn [35] reveals the following points: (1) Both creep mechanisms give rise to Newtonian viscous behavior ($\dot{\varepsilon} \propto \sigma^1$) with viscosity, $\eta \equiv \sigma/\dot{\varepsilon}$, independent of stress. (2) As long as the grain size is small enough that diffusion creep dominates over dislocation creep, Nabarro–Herring creep ($\dot{\varepsilon} \propto 1/d^2$) is more important than Coble creep ($\dot{\varepsilon} \propto 1/d^3$) at larger grain sizes. (3) Nabarro–Herring creep dominates at higher temperatures and Coble creep at lower temperatures because $\Delta H_{gb} < \Delta H_{gm}$.

The analyses of Nabarro (1948) and Herring (1950) and of Coble (1963) strictly apply to deformation of a single spherical grain. Subsequent analyses pointed out the necessity of grain-boundary sliding in diffusion creep of polycrystalline materials (Liftshitz, 1963; Raj and Ashby, 1971). The fact that grains tend to be equiaxed in samples deformed in the diffusion creep field indicates that grain-boundary sliding and grain rotation are crucial. The creep process is essentially that of grain-boundary sliding accommodated by diffusion both through grain interiors and along grain boundaries, if grain boundaries are too weak to a support shear stress. For this case, $\alpha_{NH} = 14$ and $\alpha_C = 14\pi$. Since grain-boundary and grain-matrix diffusion are independent/parallel processes, eqn [33] and [35] can be combined to give

$$\dot{\varepsilon}_{diff} = 14\left(\frac{\sigma V_m}{RT}\right)\left(D_{gm} + \frac{\pi\delta D_{gb}}{d}\right)\left(\frac{1}{d^2}\right) \quad [37]$$

In general, this deformation mechanism is referred to simply as diffusion creep.

14.3.1.2 Deformation involving dislocations

As differential stress and/or grain size increases, a transition occurs from diffusion creep to dislocation-dominated deformation. One estimate of the stress required to move from the diffusion creep regime to the dislocation creep regime is based on a calculation of the stress required to operate a Frank–Reed dislocation source, σ_{FR}, (Frank and Read, 1950) and thus generate the dislocations necessary to sustain deformation:

$$\sigma_{FR} \approx \frac{2Gb}{L} \quad [38]$$

where L is the length of the dislocation segment operating as the source of new dislocations. If, the length of a dislocation line is limited by the grain size, then for $G = 70\,GPa$, $b = 0.5\,nm$, and $L = d = 10\,\mu m$, $\sigma_{FR} \approx 7\,MPa$

The strain produced by a single dislocation moving across a grain is very small, thus many dislocations must be generated and move to accomplish a significant amount of deformation. The strain, ε, produced in a cubic grain of dimension d by one dislocation moving through the grain is

$$\varepsilon = \frac{b}{d} \quad [39]$$

Thus, if $d = 1\,mm$, $\varepsilon \approx 5 \times 10^{-7}$. Greater strain is achieved by moving a larger number of dislocations through the grain. The density of dislocations, ρ, scales with applied differential stress as

$$\rho \approx \left(\frac{\sigma}{Gb}\right)^2 \quad [40]$$

such that $\rho \approx 10^9\,m^{-2}$ for a differential stress of 1 MPa. The strain produced by this density of dislocation moving a distance x is

$$\varepsilon = \rho b x \quad [41]$$

For $x = d$, $\varepsilon \approx 5 \times 10^{-4}$ (still a small value). To produce geologically significant strains, dislocations must be generated, move through crystalline grains, and then be removed so that new dislocations can be generated to maintain an approximately constant density of dislocations and keep the deformation process going. If steady-state deformation is attained, then a balance must be achieved, and one step in this series of steps will limit the rate of deformation.

A number of models have been proposed to describe the rate of deformation in terms of applied stress, temperature, pressure, and other thermodynamic and structural variables (e.g., Poirier, 1985, pp. 94–144; Evans and Kohlstedt, 1995). In the present chapter, characteristic elements of these models are reviewed in order to provide a framework for examining laboratory-determined deformation data and extrapolating to geological conditions.

The dependence of the rate of deformation accomplished by dislocation processes on differential stress can be examined through the Orowan equation, which is obtained by differentiating eqn [41] with respect to time and noting that in steady state the dislocation density is independent of time, $\partial\rho/\partial t = 0$ (Orowan, 1940; Poirier, 1985, pp. 62–63):

$$\dot{\varepsilon} = \rho b \bar{v} \quad [42]$$

where $\bar{v}$ is the average dislocation velocity. It should be noted, however, that the $\partial\rho/\partial t$ term is critical when discussing transient deformation (e.g., deformation immediately following a change in stress). As discussed in reference to eqn [40], the dislocation density is proportional to approximately the square of the stress, such that $\dot{\varepsilon} \propto \rho \propto \sigma^2$. Additional dependence of strain rate on differential stress enters through the dislocation velocity term.

In this discussion, dislocation creep is divided into three regimes. As with diffusion creep, the main parameters determining the relative importance of the various dislocation creep mechanisms are differential stress, temperature, pressure, and grain size. First, at high temperature (low stress) and relatively large grain size, all of the strain can be accomplished by glide and climb of dislocations. These conditions define the dislocation creep regime. Second, at high temperature but smaller grain size, grain-boundary sliding operates in conjunction with dislocation processes to produce strain. These conditions define a regime in which dislocation processes are accommodated by grain-boundary sliding, simply referred to here as the grain-boundary sliding regime. Third, at low temperatures (high stresses), deformation takes place by dislocation glide limited by the intrinsic resistance of the lattice. These conditions delineate the low-temperature plasticity regime.

14.3.1.2.(i) Dislocation creep

To date, by far the majority of analyses of plastic deformation in geological materials have used a power-law equation to describe experimental results. A general form of this flow law is

$$\dot{\varepsilon} = A \frac{\sigma^n}{d^m} f_{O_2}^p f_{H_2O}^q \exp - \frac{Q_{cr}}{RT} \qquad [43]$$

where A is a material-dependent parameter and Q_{cr} is the activation energy (strictly, enthalpy) for creep. The power-law form of the creep equation arises by considering the role of dislocation climb and grain-boundary sliding in the deformation process.

The potential importance of dislocation climb as the rate-controlling step of high-temperature, steady-state creep was suggested in the early 1950s by Mott (1951, 1953, 1956). Subsequently, several climb-controlled creep models were developed that relate strain rate to differential stress and temperature, the latter though the self-diffusivity (for reviews see, e.g., Weertman (1978), Poirier (1985, pp. 94–144), Cannon and Langdon (1988), Evans and Kohlstedt (1995), and Weertman (1999)). The rate of climb of edge dislocations depends directly on diffusive fluxes of ions (e.g., Hirth and Lothe, 1968, pp. 506–519; Poirier, 1985, pp. 58–62). The justification for emphasizing the importance of climb in high-temperature $(T > (2/3)T_m$, where T_m is the melting temperature) deformation of crystalline materials is the observed one-to-one correlation between the activation energy for creep and the activation energy for self-diffusion of the slowest ionic species. This correlation has been observed for a large number of metallic and ceramic materials (e.g., Dorn, 1956; Sherby and Burke, 1967; Mukerjee *et al.*, 1969; Takeuchi and Argon, 1976; Evans and Knowles, 1978) as well as for olivine (Dohmen *et al.*, 2002; Chakraborty and Costa, 2004; Kohlstedt, 2006).

Two climb-based models of dislocation creep merit particular attention because of their pioneering contributions in this area and because they incorporate most of the elements found in subsequent models. Weertman (1955, 1957a) developed a flow law for steady-state dislocation creep in which dislocation glide produces the strain but dislocation climb controls the stain rate. In this model, a dislocation source generates a dislocation loop that expands by gliding until it encounters an obstacle such as a dislocation loop that was generated on a parallel glide plane. The two dislocations interact to form a dipole that prevents further glide until the two dislocations comprising the dipole climb to annihilate one another. Once the dipole is annihilated, the dislocation sources produce new dislocation loops (i.e., dislocation multiplication) to continue the deformation process. If glide is rapid and climb is slow then the average dislocation velocity is

$$\bar{v} = \frac{\ell_g + \ell_c}{t_g + t_c} \approx \frac{\ell_g}{\ell_c} v_c \qquad [44]$$

where ℓ_g is the glide distance, ℓ_c is the climb distance, and v_c is the climb velocity (Poirier, 1985, p. 110; Weertman, 1999). The climb velocity, which is determined by diffusion of the slowest atomic/ionic species, is given by the expression (Hirth and Lothe, 1968, pp. 506–519)

$$v_c = 2\pi \frac{\sigma V_m}{RT} \frac{D}{b} \frac{1}{\ln(R_o/r_c)} \qquad [45]$$

where the average spacing between dislocations, R_o, is usually written in terms of the dislocation density as $R_o \approx 1/\sqrt{\rho}$, and the inner cutoff (core) radius, r_c, is generally set at $r_c \approx b$. A flow law is then obtained by inserting eqn [40] and [45] into the Orowan equation, eqn [42], to yield (Weertman, 1999)

$$\dot{\varepsilon} = 2\pi \frac{GV_m}{RT} \left(\frac{\sigma}{G}\right)^3 \frac{D}{b^2} \frac{1}{\ln(G/\sigma)} \frac{\ell_g}{\ell_c} \qquad [46]$$

Nabarro (1967) formulated a model for steady-state dislocation creep based solely on dislocation climb. Bardeen–Herring sources (Bardeen and Herring, 1952) generate dislocations that form a network and continuously climb. Dislocation multiplication occurs by operation of dislocation sources thus increasing their density, while climb of dislocations of opposite sign toward one another results in annihilation thus decreasing their density. A balance between multiplication and annihilation results in steady-state creep described by the flow law (Nabarro, 1967; Nix *et al.*, 1971)

$$\dot{\varepsilon} = 2 \frac{GV_m}{RT} \left(\frac{\sigma}{G}\right)^3 \frac{D}{b^2} \frac{1}{\ln(4G/\pi\sigma)} \qquad [47]$$

Although the steady-state strain rates obtained from the models developed by Weertman and Nabarro, eqns [47] and [46], respectively, differ in magnitude, these flow laws share fundamental elements. Both yield a cubic dependence of strain rate on differential stress and a linear dependence of strain rate on diffusivity. That is, both result in power-law equations similar to that given in eqn [43] with $m = 0$. The exponential dependence of strain rate on temperature enters through D as does at least part of the dependence of strain rate on oxygen and/or water fugacity. In detail, the dependence of diffusivity on fugacity enters through the concentration of the point defects (e.g., vacancies) that enable diffusion of ions, as expressed in eqn [16].

As discussed by Hirth and Lothe (1968, pp. 506–529) and Poirier (1985, pp. 58–62), the climb velocity is directly proportional to the concentration of jogs, c_j, times the migration velocity of jogs, v_j:

$$v_c = c_j v_j \qquad [48]$$

The jog migration velocity is directly proportional to the flux of ions to or from a dislocation and thus directly proportional to the self-diffusion coefficient of the slowest ion. It is usually assumed that dislocations are fully saturated with jogs (i.e., $c_j = 1$) such that the concentration of vacancies remains at local equilibrium along the dislocation line. If the concentration of jogs is significantly below this level (i.e., $c_j << 1$), then the jog concentration introduces an extra energy term into the climb velocity, such that the activation energy for climb will be larger than that for self-diffusion of the slowest ion. In addition, as discussed below, if the jog concentration depends on the concentration of other point defects, then c_j may also be a function of thermodynamic parameters such as oxygen fugacity and water fugacity (e.g., Hobbs, 1981, 1983, 1984).

14.3.1.2.(ii) Grain-boundary sliding and migration

Boundaries or interfaces separating neighboring grains often contribute substantially to plastic deformation of rocks. In the diffusion creep regime, sliding along grain and interphase boundaries is an essential aspect of deformation (Raj and Ashby, 1971). Thus, as discussed above, diffusion creep might be more descriptively named diffusion-accommodated grain-boundary sliding.

Grain-boundary sliding (gbs) is also important during deformation in which dislocations dominate flow. Grain-boundary sliding has received a great deal of attention in the literature on superplastic deformation, a deformation process in which a solid is deformed (in tension) to very large strains (~1000% elongation) without failing by necking and subsequent fracture (see for example, Ridley, 1995). Based on microstructural analyses, Boullier and Gueguen (1975) argued that superplastic flow is responsible for deformation in at least some mylonites. Schmid *et al.* (1977) and Goldsby and Kohlstedt (2001) discussed the possibility of superplastic flow in their deformation experiments on limestone and ice,

respectively. However, as Poirier (1985, p. 204) pointed out "superplasticity is a behavior, not a definite phenomenon." Thus, in this paper, emphasis is given to the coupling between grain-boundary sliding and dislocation activity in the deformation process without appealing specifically to the term superplasticity.

In this regime, strain rate varies with differential stress according to a power-law relationship such as the one presented in eqn [43] with $n > 1$. This observation immediately implies that grain-boundary sliding cannot be accommodated simply by diffusion, which would lead to a linear dependence of strain rate on stress (e.g., Ashby and Verrall, 1973). A number of models examine deformation processes involving grain-boundary sliding coupled with dislocation motion. A few of these studies are mentioned here to bring out the key elements of this process, with more extensive lists of references available in Poirier (1985, section 7.4) and elsewhere (Kaibyshev, 1992; Langdon, 1994, 1995).

Ball and Hutchinson (1969), Mukherjee (1971), and Gifkins (1976, 1978) considered the motion of grain-boundary dislocations impeded by obstacles in the grain boundaries. Stress concentrations that develop at ledges or triple junctions are relieved by the generation of dislocations that then move through the adjoining grains. These dislocations, piled up against neighboring grain boundaries, are removed by climb into the grain boundaries. As these extrinsic dislocations move along grain boundaries, they contribute to grain-boundary sliding. Kaibyshev (2002) in particular emphasized the importance of this excess population of dislocation in the grain boundaries due to interaction of the grain boundaries with lattice dislocations. This approach leads to the flow law

$$\dot{\varepsilon} = A_{gbs} \frac{D_{gb}}{d^2} \frac{G V_m}{RT} \left(\frac{\sigma}{G}\right)^2 \qquad [49]$$

where the value of the parameter A_{gbs}, which depends on the details of the grain-boundary sliding process and the dislocation recovery mechanism (see Langdon, 1994), has a value of order 10. This type of analysis is applied to the situation in which a subgrain structure does not build up within the individual grains so that dislocations move relatively freely from one grain boundary, across the grain, to the opposite grain boundary.

If grains contain subgrains, then dislocations that are generated by sliding along a grain-boundary glide across the adjoining grain until they encounter a subgrain boundary. The dislocations then climb into the subgrain boundary, interact with dislocations in the subgrain wall, and are annihilated (Langdon, 1994). This type of analysis leads to a flow law of the form

$$\dot{\varepsilon} = B_{\mathrm{gbs}} \frac{D_{\mathrm{gm}}}{d^1} \frac{GV_{\mathrm{m}}}{RT} \left(\frac{\sigma}{G}\right)^3 \qquad [50]$$

with $B_{\mathrm{gbs}} \approx 1000$. Note that eqns [49] and [50] differ in their dependence on stress (σ^2 vs σ^3), on grain size ($1/d^2$ vs $1/d$), and on temperature (Q_{gb} vs Q_{gm}).

In silicates, evidence for creep involving grain-boundary sliding and dislocations motion has been identified in deformation data for dry olivine (Kohlstedt and Wang, 2001; Hirth and Kohlstedt, 2003) and is apparent in the results for dry clino-pyroxene (Bystricky and Mackwell, 2001) for which a coarse-grain rock is a factor of $\sim$10 stronger than a fine-grained sample fabricated from powders pre-pared from the coarse-grain rock. Interestingly, this grain-boundary sliding regime appears to be absent in olivine deformed under hydrous conditions, pos-sibly because an enhancement of dislocation climb provides the extra deformation mechanism required to fulfill the von Mises criterion (Mei *et al.*, 2002; Hirth and Kohlstedt, 2003).

In addition to their contribution to deformation by producing strain as they slide, grain boundaries facil-itate deformation by migrating in response to the strain energy associated with high densities of dislocations within the grains. This dynamic recrys-tallization process does not itself produce strain; however, it does remove dislocations within grains, resulting in a population of small grains that are, at least initially, free of dislocations. Dislocations gen-erated at grain boundaries can more easily glide across these dislocation-free grains, unimpeded by interactions with other dislocations, than through larger grains containing high densities of dislocations. In addition, since the contribution of grain-boundary sliding to deformation increases with decreasing grain size as indicated in eqns [49] and [50], grain-boundary sliding can contribute to deformation in regions of small grains, thus also enhancing the rate of deformation. Hence, dynamic recrystallization often results in 'strain softening', a reduction in creep strength with progressive deformation (i.e., increasing strain) of a rock (e.g., de Bresser *et al.*, 2001; Drury, 2005). Dynamic recrystallization can also occur by subgrain rotation by which dislocations are removed from the interiors of grains by addition to low-angle boundaries (e.g., Poirier, 1985,

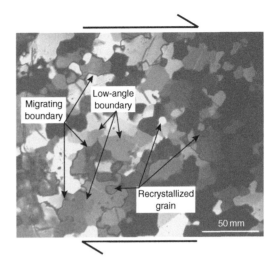

Figure 9 Transmitted-light optical micrograph with polarizers crossed of thin section of $(Fe_{0.5}Mg_{0.5})_2SiO_4$ illustrating the process of dynamic recrystallization occurring both by subgrain rotation and grain-boundary migration. Sample was deformed in torsion to a shear strain of $\gamma = 4$. Courtesy of Y-H Zhao.

pp. 169–177). As dislocations are added to these (low-angle) subgrain boundaries, the misorientation angle increases until distinct new grains (i.e., high-angle grain boundaries) form. An example of dynamic recrystallization involving grain-boundary migration and simultaneous subgrain rotation is pre-sented in **Figure 9**. Again, the removal of dislocations from within the grains reduces temporarily the num-ber of obstacles to dislocation glide, and the formation of smaller grains makes the operation of grain-size-sensitive deformation processes (diffusion creep and grain-boundary sliding) more favorable.

14.3.1.2.(iii) Low-temperature plasticity At lower temperatures and hence higher stresses, power-law formulations underpredict the observed increase in strain rate with increasing stress. As tem-perature decreases, diffusion becomes too slow to permit a significant contribution of dislocation climb to the deformation process. The rate of defor-mation then becomes limited by the ability of dislocations to glide past obstacles. In silicate miner-als with a significant component of covalent bonding, dislocations must overcome the resistance imposed by the lattice itself, often referred to as the Peierls barrier (e.g., Frost and Ashby, 1982, pp. 6–9). Motion of dislocations over the Peierls barrier requires the nucleation and migration of kinks (steps along dis-location lines that lie in the glide plane), with the

nucleation rate of pairs of kinks generally assumed to limit the dislocation velocity (Frost and Ashby, 1982, p. 8). The dislocation velocity is determined by the glide velocity, v_g, and consequently by the kink velocity, v_k:

$$v \approx v_g = c_k v_k \quad [51]$$

where c_k is the concentration of kinks. The activation enthalpy for the nucleation and migration of double kinks, $\Delta H_k(\sigma)$, is a function of differential stress. A detailed analysis for $\Delta H_k(\sigma)$ yields (Kocks *et al.*, 1975, p. 243; Frost and Ashby, 1982, p. 8)

$$\Delta H_k(\sigma) = \Delta H_k^o \left[1 - \left(\frac{\sigma}{\sigma_P}\right)^r\right]^s \quad [52]$$

where ΔH_k^o is the Helmholtz free energy of an isolated pair of kinks, σ_P is the Peierls stress, and r and s are model-dependent parameters (Frost and Ashby, 1982, p. 9). From Orowan's equation, eqn [42], the flow law then becomes

$$\dot{\varepsilon} = \dot{\varepsilon}_P \left(\frac{\sigma}{G}\right)^2 \exp - \left\{\frac{\Delta H_k^o}{RT}\left[1 - \left(\frac{\sigma}{\sigma_P}\right)^r\right]^s\right\} \quad [53]$$

where $\dot{\varepsilon}_P$ is a material-dependent parameter.

An illustrative method for displaying the conditions under which a given deformation mechanism dominates flow is the deformation mechanism map (Frost and Ashby, 1982). At present, olivine is the only mineral for which data exist in the diffusion, dislocation, grain-boundary sliding, and low-temperature plasticity regimes. Four parameters are usually considered when constructing a deformation mechanism map: strain rate, stress, grain size, and temperature. Examination of eqn [32] indicates that other parameters could be included (e.g., pressure, water fugacity, and melt fraction) depending on the application. Here, the classical form is first used in which stress is plotted as a function of temperature for a fixed grain size (and fixed pressure, water fugacity, etc.) with strain rate shown parametrically. In **Figures 10(a)** and **10(b)**, differential stress is plotted as a function of temperature for dunite rocks composed of 10-μm grains and 1-mm grains, respectively. Flow parameters for the diffusion, dislocation, and grain-boundary sliding regimes are taken from Hirth and Kohlstedt (2003), while those for the low-temperature plasticity regime are from Goetze (1978) and Evans and Goetze (1979). For a fine-grained rock with $d = 10\,\mu m$ such as might be found in a shear zone (**Figure 10(a)**), diffusion creep dominates at geological strain rates, 10^{-15}–$10^{-10}\,s^{-1}$ for stresses below ~300 MPa. At higher stresses, grain-boundary sliding becomes important, and at still higher stresses, low-temperature plasticity dominates. For a coarser-grained rock with $d = 1\,mm$ such as might be expected in the upper mantle (**Figure 10(b)**), deformation at geological strain rates and stresses appropriate for the asthenosphere occurs near the boundary between the diffusion and dislocation creep regimes, while deformation in the lithosphere requires dislocation creep or low-temperature

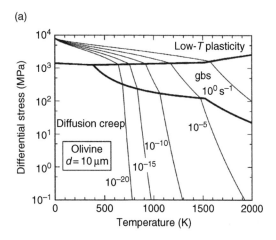
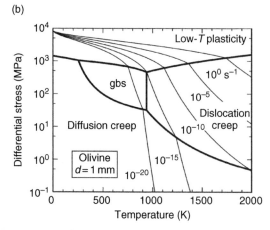

Figure 10 Deformation mechanism maps plotted as differential stress as a function of temperature for dunite with grain sizes of (a) 10 μm and (b) 1 mm. (a) At the smaller grain size for a strain rate of $10^{-15}\,s^{-1}$, deformation occurs by diffusion creep for stresses up to ~300 MPa above which deformation is dominated by grain-boundary sliding and then by low-temperature plasticity. (b) At the larger grain size for a strain rate of $10^{-15}\,s^{-1}$, with increasing stress deformation moves from the diffusion creep regime to the dislocation creep regime to the low-temperature plasticity regime.

plasticity. Strong seismic anisotropy in the mantle at depths less than ~250 km indicates that deformation occurs by dislocation processes (e.g., Montagner, 1985; Nishimura and Forsyth, 1989; Gaherty and Jordan, 1995). In contrast, the apparent lack of seismic anisotropy at greater depths, below the Lehmann discontinuity (Lehmann, 1959, 1961), has been interpreted as an indicator of diffusion creep (Karato, 1992), although more recent observations indicate that a weak anisotropy exists in this deeper region, characteristic of dislocation activity on the $(hk0)[001]$ slip system (Mainprice *et al.*, 2005).

An alternative form of the deformation mechanism map that provides insight into mantle deformation process is obtained by plotting stress as a function of grain size with strain rate again shown parametrically, as illustrated in **Figure 11**. The greater importance of grain-boundary sliding at the lower temperature (800°C vs 1300°C) is clear in this comparison. This behavior reflects the fact that the activation energy for grain-boundary sliding below ~1250°C appears to be smaller than that for dislocation creep, ~400 kJ mol^{-1} versus ~500 kJ mol^{-1}(Hirth and Kohlstedt, 2003) although this point remains to be confirmed experimentally (e.g., Mei and Kohlstedt, 2000b). Also, the contribution of low-temperature plasticity is, as expected, greater at 800°C. Finally, at 1300°C for geological stresses, strain rates, and asthenospheric grain sizes, deformation occurs close to the boundary between diffusion and dislocation creep.

14.3.2 Role of Fluids in Rock Deformation

Two important, often-present components that can result in a substantial weakening of rocks are melt and water. If either is present, even in small concentration, it can profoundly reduce the viscosity of a rock.

14.3.2.1 Role of melt in rock deformation

The influence of melt on rock viscosity enters constitutive equations in two ways (Kohlstedt, 2002; Xu *et al.*, 2004). First, melt provides a path along which ions can diffuse more rapidly than they can either through grain interiors or along grain boundaries. Second, since melt does not support shear stresses, for a given applied macroscopic stress, the stress at the grain scale will be locally increased if melt is present.

Both of these contributions to the high-temperature strength of a rock were included in the analysis by Cooper *et al.* (1989) of the effect of melt on diffusion creep of an aggregate composed of a fluid plus a solid phase. Such analyses depend critically on the distribution of the melt phase. In this particular model, the melt distribution was assumed to be that dictated by surface tension for an isotropic system under a hydrostatic state of stress for which the dihedral angle, θ, is determined by the relative values of the solid–liquid

(a)

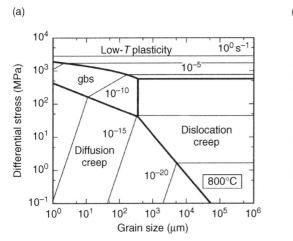

(b)

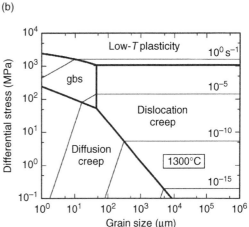

Figure 11 Deformation mechanism maps plotted as differential stress as a function of grain size for dunite at temperatures of (a) 800°C and (b) 1300°C. (a) At the lower temperature for a strain rate of 10^{-15} s^{-1}, deformation occurs by diffusion creep for grain sizes less than ~200 μm and by dislocation creep at larger grain sizes. At higher strain rates, grain-boundary sliding contributes at grain sizes smaller than ~50 μm such as might be present in shear zones. (b) At the higher temperature, deformation is dominated by diffusion creep at finer grain sizes and by dislocation creep at coarser grain sizes.

and solid–solid interfacial energies, γ_{sl} and γ_{ss}, respectively:

$$\cos\left(\frac{\theta}{2}\right) = \frac{\gamma_{ss}}{2\gamma_{sl}} \qquad [54]$$

For the olivine + mid-ocean ridge basalt (MORB) case, the average dihedral angle is $\sim 40°$ (i.e., $0 < \theta < 60°$) such that, ideally, melt forms an interconnected network along three-grain junctions and through four-grain corners, such as illustrated in **Figure 12**. Average values for dihedral angle of $<60°$ are observed for many rocks composed of silicate minerals plus a silicate melt. However, as illustrated in **Figure 13**, for melt fractions greater than ~ 0.05 the model of Cooper and Kohlstedt (Cooper *et al.*, 1989) significantly underpredicts the decrease in viscosity measured in experiments on partially molten rocks in the diffusion creep regime (Hirth and Kohlstedt, 1995a; Zimmerman and Kohlstedt, 2004).

One important aspect not taken into account in the above analysis is the fact that the wetting behavior of most partially molten systems is not isotropic (e.g., Cooper and Kohlstedt, 1982). As illustrated in **Figure 14**, microstructural observations demonstrate that melt wets a fraction of the grain boundaries, behavior not predicted for an isotropic system with $\theta > 0°$. Furthermore, the percentage of grain boundaries wetted by melt increases with increasing melt fraction, ϕ (Hirth and Kohlstedt, 1995b). At the time of writing of this chapter, a model that appears to deal adequately with the effect of this microstructural distribution of melt on diffusion creep behavior is being developed (Takei and Holtzman, 2006). This model expresses the viscosity of a partially molten rock in terms of grain-boundary contiguity (the ratio of grain-boundary area to grain-boundary area plus grain-melt interfacial area). This model extends that of Cooper and Kohlstedt (Cooper *et al.*, 1989) from two to three dimensions and includes the observed anisotropic wetting of grains. One important result of

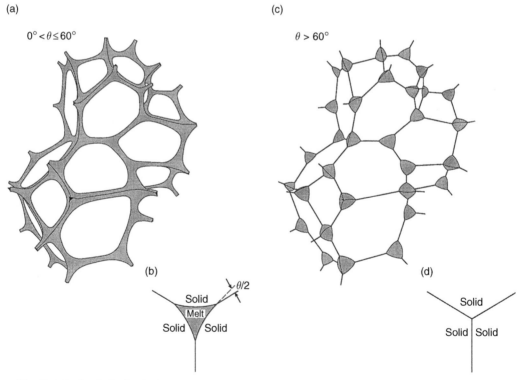

(a) $0° < \theta \leq 60°$

(c) $\theta > 60°$

(b)

Solid $\theta/2$

Melt

Solid   Solid

(d)

Solid

Solid | Solid

Figure 12 Sketch of melt distribution in a partially molten rock with isotropic interfacial energies. In (a) with $0° < \theta \leq 60°$, melt wets all of the triple junctions and four-grain junctions; (b) cross-section through a triple junction midway along a grain edge illustrates the presence of melt. In (c) with $\theta > 60°$, melt is confined to four-grain junctions; (d) cross-section through a triple junction midway along a grain edge illustrates the absence of melt. Adapted from Riley GN, Jr and Kohlstedt DL (1990) An experimental study of melt migration in an olivine–melt system. In: Ryan MP (ed.) *Magma Transport and Storage*, pp. 77–86. New York: John Wiley & Sons.

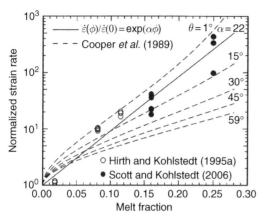

Figure 13 Plot of normalized strain rate, $\dot{\varepsilon}(\phi)/\dot{\varepsilon}(0)$, vs melt fraction. Circles are the experimentally determined values. Dashed curves are based on model of Cooper *et al.* (1989) for $1° \leq \theta \leq 59°$. Solid line is a fit of the experimental data to the empirical relationship $\dot{\varepsilon}(\phi)/\dot{\varepsilon}(0) = \exp(\alpha\phi)$, yielding $\alpha = 22$. For $\phi < 0.15$, data are from Hirth and Kohlstedt (1995a). For $\phi > 0.15$, data are from Scott and Kohlstedt (2006).

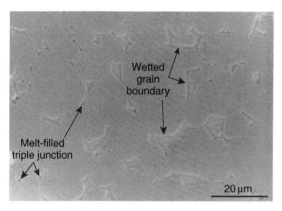

Figure 14 Backscattered scanning electron micrograph of sample of olivine +5% basalt. Both melt-filled triple junctions and wetted grain boundaries are identified. The presence of wetted grain boundaries demonstrates that the interfacial energies are anisotropic in this system, since melt would be confined to triple junctions in an isotropic system with a similar dihedral angle of $\theta \approx 40°$.

this model for the diffusion creep regime is a factor-of-five decrease in viscosity caused by increasing ϕ from 0 to 0.001 (i.e., 0.1% melt). This model yields a good fit to the experimental data for $\phi > 0$. The predicted factor-of-five decrease in viscosity in going from $\phi = 0$ to $\phi = 0.001$ has not been observed experimentally, possibly because most of the samples fabricated for these experiments contain a trace amount of melt due to impurities introduced during

the crushing and grinding process or produced from inclusions within the olivine crystals used to prepare fine-grained powders from which samples were synthesized (Faul and Jackson, 2006).

Experimental determinations of strain rate as a function of melt fraction in both the diffusion creep regime and the dislocation creep regime have been empirically fit to the relation

$$\dot{\varepsilon}(\phi) = \dot{\varepsilon}(0)\exp(\alpha\phi^r) \qquad [55]$$

for melt fractions from near zero to the rheologically critical melt fraction (RCMF) with $r = 1$ for the olivine + basalt system (Kelemen *et al.*, 1997; Mei *et al.*, 2002; Scott and Kohlstedt, 2006) as well as for an organic crystal plus melt system (Takei, 2005) and $r = 3$ for a partially molten granitic rock (Rutter and Neuman, 1995, Rutter *et al.*, 2006). The RCMF is the melt fraction at which, with increasing melt fraction, viscosity of a melt-solid system moves from that of a solid framework with interstitial melt to that of a melt containing grains in suspension.

Melt has a greater effect on viscosity in the dislocation creep regime than in the diffusion creep regime, at least in the olivine + basalt and partially molten lherzolite systems (Mei *et al.*, 2002; Zimmerman and Kohlstedt, 2004; Scott and Kohlstedt, 2006). This point is illustrated in **Figure 15** with $\alpha = 21$ in the diffusion creep regime and $\alpha = 32$ in the dislocation creep regime. At melt fractions greater than the RCMF, a partially molten rock behaves as a fluid (melt) containing suspended particles. The Einstein–Roscoe equation (Roscoe, 1952),

$$\eta_{ER}(\phi) = \frac{\mu}{(1.35\phi - 0.35)^{2.5}} \qquad [56]$$

which reasonably well describes the viscosity of a system with $\phi > 0.25$–0.30, is also shown in **Figure 15**. In eqn [56], μ is melt viscosity. Finally, it should be noted that the onset of melting will affect grain-boundary chemistry, which will indirectly affect rock viscosity through its influence on the rates of ionic diffusion along grain boundaries (Hiraga *et al.*, 2007).

14.3.2.2 Role of water in rock deformation

Water weakening of nominally anhydrous silicate minerals was first observed in the mid-1960s in an experimental study of the strength of quartz (Griggs and Blacic, 1965). In these solid-medium deformation experiments, samples deformed with a hydrous

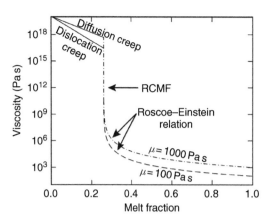

Figure 15 Plot of viscosity of a partially molten rock versus melt fraction illustrating the transition between flow controlled by a framework of solid particles with interstitial melt to flow dominated by melt with suspended particles. At low melt fractions, dislocation creep and diffusion creep are described by eqn [55] with $r = 1$ and $\alpha = 21$ and 32, respectively. At the RMCF, which marks the transition from solid-like to fluid-like behavior, viscosity plummets rapidly with increasing melt fraction. At higher melt fractions, $\phi > 0.25$–0.30, the Roscoe–Einstein relation, eqn [56], describes deformation of a mixture of melt with suspended particles. Adapted from Scott T and Kohlstedt DL (2006) The effect of large melt fraction on the deformation behavior of peridotite. *Earth and Planetary Science Letters* 246: 177–187.

confining medium (talc) were weaker than samples deformed with an anhydrous confining medium. Subsequently, water weakening has been reported for other nominally anhydrous minerals (NAMs) including olivine (e.g., Avé Lallemant and Carter, 1970), pyroxene (Avé Lallemant, 1978), and feldspar (e.g., Tullis and Yund, 1980). These first studies of water weakening emphasized the phenomenon. A second generation of experiments focused on the mechanism of water weakening by studying the dependence of viscosity on water fugacity, that is, on the concentration of water-derived point defects (Kronenberg and Tullis, 1984; Kohlstedt et al., 1995; Post et al., 1996; Mei and Kohlstedt, 2000a, 2000b; Karato and Jung, 2003; Chen et al., 2006).

The first model of water or hydrolytic weakening was built on a mechanism in which water hydrolyzes strong Si–O bonds forming weaker Si–OH$\cdots$OH–Si bonds (Griggs, 1967). Thus, dislocation glide is easier under hydrous conditions than under anhydrous conditions; effectively, the Peierls stress/barrier decreases as Si–O–Si bridges become hydrolyzed.

In this analysis, the rate of dislocation glide is limited by the propagation of kinks along dislocations, a process facilitated by diffusion of water along dislocation cores. The dislocation velocity is then taken to be proportional to the concentration of water (Griggs, 1974).

More recent models of water weakening have emphasized the role of water-derived point defects, particularly hydrogen ions (protons) (Hobbs, 1981, 1983, 1984; Poumellec and Jaoul, 1984; Mackwell et al., 1985). Protons diffuse rapidly in NAMs (for a review, see Ingrin and Skogby, 2000) such as quartz (Kronenberg et al., 1986), olivine (Mackwell and Kohlstedt, 1990; Kohlstedt and Mackwell, 1998, 1999; Demouchy and Mackwell, 2003), and pyroxene (Ingrin et al., 1995; Hercule and Ingrin, 1999; Carpenter Woods et al., 2000; Stalder and Skogby, 2003). Thus, initially dry, millimeter-size samples can be hydrated in high-temperature, high-pressure experiments lasting a few hours or less. The presence of protons in a NAM affects the velocity of a dislocation in two possible ways. First, since protons are charged, a change in the concentration of protons will result in a change in the concentration of all other charged point defects. Hence, the introduction of protons into a NAM will directly affect the concentrations of vacancies and self-interstitials on each ionic sublattice and consequently the rates of diffusion of the constituent ions and the rate of dislocation climb. In addition, extrinsic point defects, specifically water-derived point defects, can affect the concentrations of kinks and jogs along dislocation lines and thus dislocation velocity as expressed in eqns [48] and [51] (Hirsch, 1979, 1981; Hobbs, 1981, 1984). The effect of water-derived point defects on the concentrations of kinks and jogs can enter in two ways. First, for minerals containing a transition metal such as Fe (i.e., semiconducting silicates), the presence of water-derived point defects will affect the concentration of electron holes, h$\cdot$, which in turn can affect the concentrations of kinks and jogs by ionizing initially neutral jogs (Hirsch, 1979, 1981; Hobbs, 1984). For the case of kinks, a positively charged kink, k$\cdot$, can be produced from a neutral kink by the reaction

$$k^\times + h\cdot \rightleftarrows k\cdot \tag{57}$$

for which the law of mass action yields

$$[k\cdot] = K_{57}[k^\times][h\cdot] \tag{58}$$

The total concentration of kinks then becomes

$$[k^{tot}] = [k^\times] + [k^\cdot] = [k^\times](1 + K_{57}[h^\cdot]) \qquad [59]$$

where the second term can be much larger than unity and dominate the kink population. The dependence of kink concentration on water (proton) concentration or water fugacity then enters through concentration of positively charged kinks, thus enhancing the dislocation glide velocity. A similar argument can be used for increasing the concentration of jogs, and thus the dislocation climb velocity. While this approach provides one mechanism for increasing dislocation velocity in NAMs, it is not easily extended to minerals such as quartz that do not contain significant concentrations of transition metals. Therefore, a more direct influence of water-derived point defects on kink and jog concentrations needs to be considered. In point defect notation, the addition of protons, $\dot{p}$, to a nominally anhydrous silicate will affect the concentration of kinks through a reaction such as

$$k^\times + \dot{p}\{k^\times - \dot{p}\}^\cdot \qquad [60]$$

where the curly brackets { } indicate the formation of a neutral kink associated with a proton. Application of the law of mass action yields

$$[\{k^\times - \dot{p}\}^\cdot] = K_{60}[k^\times][\dot{p}] \qquad [61]$$

The total concentration of kinks is then

$$[k^{tot}] = [k^\times] + [\{k^\times - \dot{p}\}^\cdot] = [k^\times]\left(1 + K_{60}[\dot{p}]\right) \qquad [62]$$

Again, a similar equation applies for jogs. The dependence of kink and jog concentrations on water concentration/fugacity then comes directly through the dependence of proton concentration on water fugacity.

The high-temperature deformation behaviors of anorthite, clinopyroxene, and olivine deformed in the dislocation creep regime under anhydrous and ander hydrous conditions are compared in **Figure 16**. A similar comparison is made for the diffusion creep regime by Hier-Majumder *et al.* (2005a). In the dislocation creep regime, the dependence of strain rate on water fugacity has been quantified for olivine (Mei *et al.*, 2002; Hirth and Kohlstedt, 2003; Karato and Jung, 2003) and for clinopyroxene (Chen *et al.*, 2006). In the case of olivine, strain rate increases as water fugacity to the ~1st power, while in the case of clinopyroxene, strain rate increases as water fugacity to the ~3rd power. In the diffusion creep regime, creep rate increases as water fugacity to the ~1st

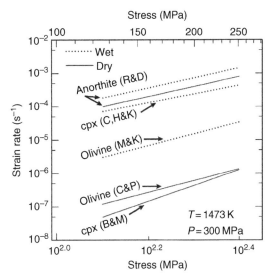

Figure 16 Strain rate vs stress for samples of anorthite, clinopyroxene, and olivine deformed in the dislocation creep regime under anhydrous and hydrous conditions. In each case, samples deformed under hydrous conditions are weaker than those deformed under anhydrous conditions. The water fugacity for samples deformed under hydrous conditions at a confining pressure of 300 MPa and 1473 K is ~300 MPa. Results are taken from R&D, Rybacki and Dresen (2000); C,H,&K, Chen *et al.* (2006); M&K, Mei and Kohlstedt (2000a); C&P, Chopra and Paterson (1984); and B&M, Bystricky and Mackwell (2001).

power for olivine (Mei and Kohlstedt, 2000a) and to the ~1.4 power for clinopyroxene (Hier-Majumder *et al.*, 2005a). To assess possible explanations for these observed dependences of strain rate on water fugacity, **Table 2** summarizes the relationships between the concentration of various point defects and water fugacity for several charge neutrality conditions for a transition-metal silicate. (1) For olivine under both anhydrous and hydrous conditions, the activation energies for dislocation creep (Hirth and Kohlstedt, 2003) and those for Si self-diffusion (Dohmen *et al.*, 2002; Chakraborty and Costa, 2004) agree within experimental error (Kohlstedt, 2006). This observation suggests that the high-temperature creep of olivine is controlled by dislocation climb limited by diffusion of the slowest ionic species, Si, under both dry and wet conditions. In this case, the jog concentration is unity, that is, the dislocations are fully saturated with jogs (see eqn [48] and the discussion that follows). Under hydrous conditions, the water fugacity exponent, q, in eqn [43] of ~1 is consistent with Si diffusing by a vacancy mechanism with the major Si vacancy being part of the point defect

Table 2 Dependence of point defect concentrations on water fugacity, expressed as the exponent q in the relationship $[\ ] \propto f_{H_2O}^q$, for several charge neutrality conditions for a transition-metal silicate

	$[Fe^{\cdot}_{Me}]$ $[h^{\cdot}]$	$[Fe'_{Si}]$	$[V''_{Me}]$	$[(OH)^{\cdot}_{O}]$ $[p^{\cdot}]$	$[H'_{Me}]$	$[(2H)^{x}_{Me}]$	$[O'_i]$	$[V''''_{Si}]$	$[H'''_{Si}]$	$[(2H)''_{Si}]$	$[(3H)'_{Si}]$	$[(4H)^{x}_{Si}]$
$[h^{\cdot}]=2[V''_{Me}]$	0	0	0	1/2	1/2	1	0	0	1/2	1	3/2	2
$[p^{\cdot}]=2[V''_{Me}]$	−1/6	1/6	1/3	1/3	2/3	1	1/3	2/3	1	4/3	5/3	2
$[p^{\cdot}]=[Fe'_{Si}]$	−1/4	1/4	1/2	1/4	3/4	1	1/2	1	5/4	3/2	7/4	2
$[h^{\cdot}]=[H'_{Me}]$	1/4	−1/4	−1/2	3/4	1/4	1	−1/2	−1	−1/4	1/2	5/4	2
$[p^{\cdot}]=[H'_{Me}]$	0	0	0	1/2	1/2	1	0	0	1/2	1	3/2	2

associate $\{2(OH)^{\cdot}_{O} - V''''_{Si}\}'' \equiv (2H)''_{Si}$ and charge neutrality given by $[(OH)^{\cdot}_{O}] = [H'_{Me}]$ (see **Table 2**). Diffusion of Me ions also varies as water fugacity to the 1st power (Hier-Majumder *et al.*, 2005b), indicating that $(2H)^{x}_{Me}$ is the primary point defect on the Me sublattice responsible for diffusion (see **Table 2**). (2) The situation appears to be more complex for clinopyroxene. The stress exponent is $n = 2.7 \pm 0.3$, consistent with the climb-controlled creep models discussed above. Unfortunately, diffusion data are lacking for comparison with the creep data. The relatively strong dependence of strain rate on water fugacity, $q = 3.0 \pm 0.6$, in the dislocation creep regime suggests that a simple climb-controlled mechanism may not apply for high-temperature deformation of clinopyroxene (see **Table 2**). One possible way to reconcile this large value for the water fugacity exponent is as follows: (1) Creep is climb controlled; (2) charge neutrality condition is given by $[h^{\cdot}] = [H'_{Me}]$ or by $[p^{\cdot}] = [H'_{Me}]$; (3) strain rate is limited by Si diffusion by a vacancy mechanism involving the defect associate $\{4(OH)^{\cdot}_{O} - V''''_{Si}\}^{x} \equiv (4H)^{x}_{Si}$; (4) dislocations are undersaturated with jogs with the jog concentration dominated by neutral jogs associated with protons, $\{k^{x} - p^{\cdot}\}'$, as in eqns [61] and [62]. These scenarios yield a water fugacity exponent for strain rate of 2¾ and 2½, both within the experimental uncertainty of the measured value. Clearly, diffusion data are critical in order to develop this type of argument more completely.

14.4 Upper-Mantle Viscosity

In this section, viscosity profiles for the upper mantle constrained by geophysical observations and calculated from laboratory-derived flow laws are compared. Two situations are considered: first, global average values determined from analyses of glacial isostatic adjustment are examined relative to those calculated for dunite containing $1000 \, H/10^6 Si$. Second, the viscosity of the upper-mantle region beneath the western US is compared to the viscosity predicted for dunite deforming under water-saturated conditions.

Laboratory results for dunite are used because it is the only rock for which data are available over a wide range of upper-mantle conditions. Justification for this choice is based on the observations that olivine is the major mineral in the upper mantle and that, under conditions for which data are available, the flow behavior of dunite is nearly identical to that of lherzolite (Zimmerman and Kohlstedt, 2004). Furthermore, flow laws for melt-free dunite are used because, at small melt fractions, homogeneously distributed melt has a relatively small effect on viscosity. For $\phi = 0.02$, viscosity is a factor of <2 smaller than the viscosity of a melt-free rock (**Figures 13** and **15**); larger melt fractions would be difficult to trap in the mantle over long periods of time. By comparison, the viscosity of dunite containing 1000 ppm water is a factor of >100 smaller than the viscosity of dry dunite (**Figure 16**). Thus, the effect of water rather than melt is emphasized here. It should be noted that melt might have a larger effect on viscosity if it segregates into melt-rich bands as observed in large-strain experiments on partially molten peridotites (Holtzman *et al.*, 2003a, 2003b, 2005); however, this area remains to be fully explored.

Finally, viscosity profiles are calculated for dislocation creep rather then diffusion creep. Three points justify this choice. First, strong seismic anisotropy at depths of <250 km (e.g., Nishimura and Forsyth, 1989) and weak but persistent seismic anisotropy at great depths (Mainprice *et al.*, 2005) indicate that deformation in the upper mantle is dominated by dislocation processes. Second, analysis of the influence of rheological properties on the evolution of the tip of a subducting slab indicates that non-Newtonian behavior (i.e., dislocation creep) facilitates the initiation of subduction (Billen and Hirth, 2005). Third, as illustrated in **Figures 10** and **11** as well as in Hirth (2002) and Hirth and Kohlstedt (2003), deformation of dunite at temperatures above ~1000°C occurs primarily by dislocation glide and climb with contributions from grain-boundary sliding.

14.4.1 Upper-Mantle Viscosity Profile

Recent papers by Kaufmann and Lambeck (2002) and Hirth and Kohlstedt (2003) have investigated upper-mantle viscosity profiles based on analyses of glacial isostatic adjustment and laboratory-derived flow laws, respectively (Dixon *et al.*, 2004). Kaufmann and Lambeck (2002) examined changes of Earth's surface associated with the retreat of Late Pleistocene ice sheets to obtain global average models of viscosity profiles. They obtained a rather narrow range of values for upper-mantle viscosity of between 3×10^{20} and 3×10^{21} Pa s for the depths of 60–410 km with a best fitting result of 3×10^{20} to 5×10^{20} Pa s for 120–410 km, as illustrated in **Figure 17**, for which a rigid, elastic lithosphere overlies the viscoelastic upper mantle.

Hirth and Kohlstedt (2003) critically reviewed published experimental data from laboratory experiments on dunite rocks and olivine aggregates. Best-fit

values for the flow-law parameters in eqn [43], summarized in their table 1, were used in the present analysis. The activation volume as a function of depth was calculated using their equation 3, which is based on a calculation of the elastic strain necessary to create and move point defects. The temperature as a function of depth was determined based on an adiabatic geotherm with a potential temperature, T_p, of 1350°C. A water concentration of 1000 H/10^6Si was employed following the study of Hirth and Kohlstedt (1996). Two viscosity profiles are presented in **Figure 17**, one assuming constant stress ($\sigma =$ constant) and the other assuming constant rate of viscous energy dissipation, that is, $\sigma\dot{\varepsilon} =$ constant (e.g., Christensen, 1989). For additional details, the reader is referred to Hirth and Kohlstedt (2003). The viscosity profile with $\sigma\dot{\varepsilon}$ constant, which changes by a factor of three in viscosity in going from 60 to 410 km, is much steeper than the viscosity profile with σ constant, which changes by a factor of 120 in viscosity over the same depth range. Hence, it is difficult to reconcile profiles obtained with σ constant with the constraints obtained from glacial isostatic adjustment results. In contrast, profiles produced from laboratory data with $\sigma\dot{\varepsilon}$ constant are in reasonably good agreement in terms of both the narrow width and the magnitude reported from the glacial isostatic adjustment analyses.

14.4.2 Western US Mantle Viscosity Profile

In their analysis of the role of water on the lateral variation in upper-mantle viscosity, Dixon *et al.* (2004) tabulated constraints on the viscosity structure of the upper mantle beneath the western US (their table 1). Values of viscosity were based on the investigations of Bills *et al.* (1994), Kaufmann and Amelung (2000), Nishimura and Thatcher (2003), and Pollitz *et al.* (2000, 2003). These studies examined the response of Earth's surface to smaller, more local loads (earthquakes and lake level changes) than analyzed in the glacial isostatic adjustment work. Thus, their resolution is best in the shallower portion of the upper mantle. Dixon *et al.* (2004) emphasized the distinctly lower values of mantle viscosity observed in the western US (10^{18}–10^{19} Pa s) compared with the global average values (10^{20}–10^{21} Pa s). These authors make a compelling case that this difference is due to enrichment in water of the upper mantle beneath the western US associated with subduction of the Farallon Plate, which hydrates and weakens the mantle. In **Figure 18**, the range of viscosity obtained for the western US is compared with viscosity profiles

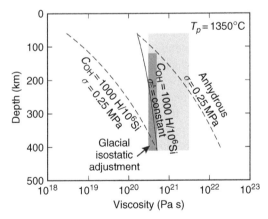

Figure 17 Viscosity profiles for anhydrous dunite with $\sigma = 0.25$ MPa, for hydrous dunite with $C_{OH} = 1000$ H/10^6Si with $\sigma = 0.25$ MPa, and for hydrous dunite with $C_{OH} = 1000$ H/10^6Si with $\sigma\dot{\varepsilon} =$ constant. Flow-law parameters are from Hirth and Kohlstedt (2003). The light gray box defines the range of global average values reported by Kaufmann and Lambeck (2002) based on analyses of glacial isostatic adjustment. The dark gray box delineates the range of global average values reported by Kaufmann and Lambeck (2002) for their best-fitting result from analyses of glacial isostatic adjustment.

calculated based on the flow laws determined by Hirth and Kohlstedt (2003). Viscosity profiles are again calculated for σ constant and for $\sigma\dot{\varepsilon}$ constant. The geotherm was taken from figure 1 of Dixon *et al.* (2004), which was obtained from P-wave tomographic inversion for the western US (Goes and van der Lee, 2002). Water content was taken to be saturated following the solubility law determined by Kohlstedt *et al.* (1996) and Zhao *et al.* (2004). As illustrated in **Figure 18**, the flow law for dislocation creep of dunite determined from laboratory experiments yields a viscosity profile that falls within the range of values obtained from measurements of Earth's response to changes in more local surface loads. The profile with $\sigma\dot{\varepsilon}$ constant yields a relatively small decrease in viscosity with increasing depth, while the profile with σ constant produces a much larger decrease in viscosity over the same depth range. In either case, the good agreement between the viscosity profiles determined from field observation and those calculated based on experimentally determined flow laws supports extrapolation of the laboratory-derived constitutive equations to problems of geophysical interest.

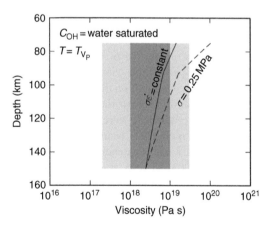

Figure 18 Viscosity profiles for water-saturated dunite with $\sigma = 0.25$ MPa and with $\sigma\dot{\varepsilon} =$ constant using flow-law parameters from Hirth and Kohlstedt (2003). Temperature profile is taken from figure 1 of Dixon *et al.* (2004), which comes from the P-wave tomography inversion of Goes and van der Lee (2002). Water concentration was taken to be saturated and was determined from the studies of Kohlstedt *et al.* (1996) and Zhao *et al.* (2004). The light gray box delineates the range of values summarized by Dixon *et al.* (2004) for the western US based on the work of Bills *et al.* (1994), Kaufmann and Amelung (2000), Nishimura and Thatcher (2003), Politz *et al.* (2000), and Politz (2003). The dark gray box defines the viscosity bounds at 10^{18}–10^{19} Pa s.

14.5 Concluding Remarks

While the good agreement between laboratory and field observations for the viscosity structure of the upper mantle is encouraging, even exciting, several important areas remain that require experimental investigation of the rheological properties of mantle rocks. Possibly the most obvious void is the paucity of deformation studies on transition zone and lower-mantle minerals and rocks. Progress in this field has been slow, largely due to a lack of apparatuses capable of carrying out experiments under well-controlled thermomechanical conditions at very high pressures. However, recent developments combining a new generation of deformation apparatuses such as the deformation DIA (D-DIA) and the rotational Drickamer apparatus (RDA) with high-intensity, synchrotron-produced X-rays hold promise for significant progress in this area (e.g., Durham *et al.*, 2002; Thurel *et al.*, 2003; Xu *et al.*, 2003, 2005).

For the upper mantle, further work is needed to quantify the deformation behavior in the low-temperature plasticity regime, particularly under hydrous conditions, for application to studies of deformation of the lithosphere. Furthermore, laboratory (e.g., Jung and Karato, 2001) and seismic (e.g., Mainprice *et al.*, 2005) observations indicate that a transition in slip system occurs below a depth of $\sim$250 km in Earth's upper mantle, related to changes in water, stress, and/or pressure conditions. At present, flow laws for the upper mantle are primarily based on experiments carried out at relatively modest pressures for which measurements of lattice-preferred orientations indicate that deformation in olivine is dominated by dislocations with [100] Burgers vectors. If dislocations with [001] Burgers vectors become important in deeper portions of the upper mantle, the flow law determined at very high pressures may deviate from that extrapolated from lower pressure conditions (Couvy *et al.*, 2004).

Finally, to date, most studies of plastic deformation of rocks have concentrated on steady-state creep with little attention given to transient deformation. While a steady-state approach may work reasonably well for describing convective flow in the mantle, it cannot provide an adequate description of deformation that occurs in an environment for which thermomechanical conditions are changing and microstructures are simultaneously evolving (e.g., Evans, 2005). This point has been appreciated by a number of researchers, some of whom have taken a state-variable or mechanical-equation-of-state approach (e.g., Lerner and Kohlstedt, 1981; Covey-Crump, 1994, 1998; Rutter, 1999; Stone *et al.*, 2004). Such approaches will be critical in the development of the next generation of constitutive equations, which must be capable of describing flow in regions such as shear zones and metamorphic belts where changes in thermal and chemical environment including fluid chemistry and evolution of materials parameters including dislocation density, grain size, and lattice-preferred orientation occur continuously.

Acknowledgments

This work was supported by NSF grants OCE-0327143 and 0648020 (deformation of partially molten rocks), EAR- 0439747 (deformation of nominally anhydrous minerals), and EAR-0409719 (grain boundaries in rocks).

References

Ashby MF and Verrall RA (1973) Micromechanisms of flow and fracture and their relevance to the rheology of the upper mantle. *Philosophical Transactions of the Royal Society of London* A 288: 59–95.

Atkinson A (1985) Grain boundary diffusion – Structural effects and mechanisms. *Journal de Physique* 46: 379–391.

Avé Lallemant HG (1978) Experimental deformation of diopside and websterite. *Tectonophysics* 48: 1–27.

Avé Lallemant HG and Carter NL (1970) Syntectonic recrystallization of olivine and modes of flow in the upper mantle. *Geological Society of American Bulletin* 81: 2203–2220.

Ball A and Hutchinson MM (1969) Superplasticity in the aluminium–zinc eutectoid. *Metal Science Journal* 3: 1–7.

Bardeen J and Herring C (1952) Diffusion in alloys and the Kirkendall effect. In: Shockley (ed.) *Imperfections in Nearly Perfect Crystals*, pp. 261–288. New York: Wiley.

Billen MI and Hirth G (2005) Newtonian versus non-Newtonian upper mantle viscosity: Implications for subduction initiation. *Geophysical Research Letters* 32: L19304 (doi:10.1029/2005GL023457).

Bills BG, Currey DR, and Marshall GA (1994) Viscosity estimates for the crust and upper mantle from patterns of lacustrine shoreline deformation in the eastern Great Basin. *Journal of Geophysical Research* 99: 22059–22096.

Boullier AM and Gueguen Y (1975) Origin of some mylonites by superplastic flow. *Contributions to Mineralogy and Petrology* 50: 93–104.

Bystricky M and Mackwell SJ (2001) Creep of dry clinopyroxene aggregates. *Journal of Geophysical Research* 106: 13443–13454.

Cannon WR and Langdon TG (1988) Review: Creep of ceramics. Part 2: An examination of flow mechanism. *Journal of Materials Science* 23: 1–20.

Carpenter Woods S, Mackwell SJ, and Dyar D (2000) Hydrogen in diopside: Diffusion profiles. *The American Mineralogist* 85: 480–487.

Carter CB and Sass SL (1981) Electron diffraction and microscopy techniques for studying grain boundary structure. *Journal of the American Ceramic Society* 64(6): 335–345.

Carter NL and Avé Lallemant HG (1970) High temperature deformation of dunite and peridotite. *Bulletin of the Geological Society of America* 81: 2181–2202.

Catlow CRA, Bell RG, and Gale JD (1994) Computer modeling as a technique in materials chemistry. *Journal of Materials Chemistry* 4: 781–792 (doi:10.1039/JM9940400781).

Chakraborty S and Costa F (2004) Fast diffusion of Si and O in San Carlos olivine under hydrous conditions. *Goldschmidt Conference*, Copenhagen, Denmark.

Chan S-W and Balluffi RW (1985) Study of energy vs. misorientation for grain boundaries in gold by crystallite rotation method. I: (001) Twist boundaries. *Acta Metallurgica* 33: 1113–1119.

Chan S-W and Balluffi RW (1986) Study of energy vs. misorientation for grain boundaries in gold by crystallite rotation method. Part II: Tilt boundaries and mixed boundaries. *Acta Metallurgica* 34: 2191–2199.

Chen S, Hiraga T, and Kohlstedt DL (2006) Water weakening of clinopyroxene in the dislocation creep regime. *Journal of Geophysical Research* 111: B08203 (doi:10.1029/2005JB003885).

Chopra PN and Paterson MS (1984) The role of water in the deformation of dunite. *Journal of Geophysical Research* 89: 7681–7876.

Christensen UR (1989) Mantle rheology, constitution, and convection. In: Peltier WR (ed.) *Mantle Convection: Plate Tectonics and Global Dynamics, The Fluid Mechanics of Astrophysics and Geophysics*, pp. 595–655. New York: Gordon and Breach Science Publishers.

Coble R (1963) A model for boundary diffusion controlled creep in polycrystalline materials. *Journal of Applied Physics* 34: 1679–1682.

Cooper RF and Kohlstedt DL (1982) Interfacial energies in the olivine-basalt system. In: Akimota S and Manghnani MH (eds.) *High-Pressure Research in Geophysics, Advances in Earth and Planetary Sciences*, vol. 12, pp. 217–228. Tokyo: Center for Academic Publications Japan.

Cooper RF, Kohlstedt DL, and Chyung K (1989) Solution-precipitation enhanced creep in solid–liquid aggregates which display a non-zero dihedral angle. *Acta Metallurgica* 37: 1759–1771.

Couvy H, Frost D, Heidelbach F, *et al.* (2004) Shear deformation experiments of forsterite at 11 GPa – 1400°C in the multianvil apparatus. *European Journal of Mineralogy* 16: 877–889.

Covey-Crump SJ (1994) The application of Hart's state variable description of inelastic deformation to Carrara marble at *T* < 450°C. *Journal of Geophysical Research* 99: 19793–19808.

Covey-Crump SJ (1998) Evolution of mechanical state in Carrara marble during deformation at 400°C to 700°C. *Journal of Geophysical Research* 103: 29781–29794.

de Bresser JHP, Ter Heege JH, and Spiers CJ (2001) Grain size reduction by dynamic recrystallization: Can it result in major rheological weakening? *International Journal of Earth Sciences* 90: 28–45.

Demouchy S and Mackwell SJ (2003) Water diffusion in synthetic iron-free forsterite. *Physics and Chemistry of Minerals* 30: 486–494 (doi: 10.1007/s00269-003-0342-2).

Devereux OF (1983) *Topics in Metallurgical Thermodynamics*, 494pp. New York: Wiley.

Dimos D, Wolfenstine J, and Kohlstedt DL (1988) Kinetic demixing and decomposition of multicomponent oxides due to a nonhydrostatic stress. *Acta Metallurgica* 36: 1543–1552.

Dixon JE, Dixon TH, Bell DR, and Malservisi R (2004) Lateral variation in upper mantle viscosity: Role of water. *Earth and Planetary Science Letters* 222: 451–467.

Dohmen R, Chakraborty S, and Becker H-W (2002) Si and O diffusion in olivine and implications for characterizing plastic flow in the mantle. *Journal of Geophysical Research* 29: 2030 (doi:10.1029/2002GL015480).

Dorn JE (1956) Some fundamental experiments on high temperature creep. In: *Creep and Fracture of Metals at High Temperatures*, pp. 89–132. London: Her Majesty's Stationary Office.

Drury MR (2005) Dynamic recrystallization and strain softening of olivine aggregates in the laboratory and the lithosphere. In: Gapais D, Brun JP, and Cobbold PR (eds.) Geological Society, Special Publications: *Deformation Mechanisms, Rheology and Tectonics: From Minerals to the Lithosphere,* vol. 243, pp. 143–158. London: Geological Society.

Durham WB, Weidner DJ, Karato S-I, and Wang Y (2002) New developments in deformation experiments at high pressure (review). In: Karato S-I and Wenk H-R (eds.) *Plastic Deformation of Minerals and Rocks, Reviews in Mineralogy,* vol. 5, pp. 21–49. Washington DC: Mineralogical Society of America.

Evans B (2005) Creep constitutive laws for rocks with evolving structure. In: Bruhn D and Burlini L (eds.) *Geophysical Society, Speical Publications: High Strain Zones: Structure and Physical Properties,* vol. 245, pp. 329–346. London: Geological Society.

Evans B and Goetze C (1979) The temperature variation of hardness of olivine and its implications for polycrystalline yield stress. *Journal of Geophysical Research* 84: 5505–5524.

Evans B and Kohlstedt DL (1995) Rheology of rocks. In: Ahrens (ed.) *Rock Physics and Phase Relations: A Handbook of Physical Constants*, pp. 48–165. Washington: American Geophysical Union.

Evans HE and Knowles G (1978) Dislocation creep in non-metallic materials. *Acta Metallurgica* 26: 141–145.

Faul U and Jackson I (2006) The effect of melt on the creep strength of polycrystalline olivine. *EOS Transactions of the American Geophysical Union.* Fall Meeting Supplement 87(52): MR11B-0129.

Frank FC and Read WT (1950) Multiplication processes for slow moving dislocations. *Physical Review* 70: 722–723.

Frost HJ and Ashby MF (1982) *Deformation-Mechanism Maps: The Plasticity and Creep of Metals and Ceramics*, 166pp. Oxford: Pergamon Press.

Gaherty JB and Jordan TH (1995) Lehmann discontinuity at the base of an anisotropic layer beneath continents. *Science* 268: 1468–1471.

Gifkins RC (1976) Grain boundary sliding and its accommodation during creep and superplasticity. *Metallurgical Transactions* 7A: 1225–1232.

Gifkins RC (1978) Grain rearrangements during superplastic deformation. *Journal of Material Science* 13: 1924–1936.

Goes S and van der Lee S (2002) Thermal structure of the North American uppermost mantle inferred from seismic tomography. *Journal of Geophysical Research* 107 (doi:10.1029/2000JB000049).

Goetze C (1978) The mechanisms of creep in olivine. *Philosophical Transactions of the Royal Society of London A* 288: 99–119.

Goetze C and Kohlstedt DL (1973) Laboratory study of dislocation climb and diffusion in olivine. *Journal of Geophysical Research* 78: 5961–5971.

Goldsby DL and Kohlstedt DL (2001) Superplastic deformation of ice: Experimental observations. *Journal of Geophysical Research* 106: 11017–11030.

Gordon RS (1973) Mass transport in the diffusional creep of ionic solids. *Journal of the American Ceramic Society* 56: 147–152.

Griggs DT (1967) Hydrolytic weakening of quartz and other silicates. *Geophysical Journal of the Royal Astronomical Society* 14: 19–31.

Griggs DT (1974) A model of hydrolytic weakening in quartz. *Journal of Geophysical Research* 79: 1653–1661.

Griggs DT and Blacic JD (1965) Quartz: Anomalous weakness of synthetic crystals. *Science* 147: 292–295.

Groves GW and Kelly A (1969) Change of shape due to dislocation climb. *Philosophical Magazine* 19: 977–986.

Hercule S and Ingrin J (1999) Hydrogen in diopside: Diffusion, kinetics of extraction-incorporation, and solubility. *The American Mineralogist* 84: 1577–1587.

Herring C (1950) Diffusional viscosity of a polycrystalline solid. *Journal of Applied Physics* 21: 437–445.

Hier-Majumder S, Anderson IM, and Kohlstedt DL (2005b) Influence of protons on Fe–Mg interdiffusion in olivine. *Journal of Geophysical Research* 110: B02202 (doi:10.1029/2004JB003292).

Hier-Majumder S, Mei S, and Kohlstedt DL (2005a) Water weakening of clinopyroxene in diffusion creep. *Journal of Geophysical Research* 110: B07406 (doi:10.1029/2004JB003414).

Hilbrandt N and Martin M (1998) High-temperature point defect equilibria in iron-doped MgO: An *in situ* Fe-K XAFS study on the valence and site distribution of iron in $(Mg_{1-x}Fe_x)O$. *Berichte der Bunsengesellschaft fuer Physikalische Chemie* 102: 1747–1759.

Hiraga T, Anderson IM, and Kohlstedt DL (2003) Chemistry of grain boundaries in mantle rocks. *The American Mineralogist* 88: 1015–1019.

Hiraga T, Anderson IM, and Kohlstedt DL (2004) Partitioning in mantle rocks: Grain boundaries as reservoirs of incompatible elements. *Nature* 427: 699–703.

Hiraga T, Hirschmann MM, and Kohlstedt DL (2007) Equilibrium interface segregation in the diopside–forsterite system. II: Applications of interface enrichment to mantle geochemistry. *Geochimica et Cosmochimica Acta* 71: 1281–1289 (doi:10.1016/j.gca.2006.11.020).

Hiraga T and Kohlstedt DL (2007) Equilibrium interface segregation in the diopside–forsterite system. I: Analytical techniques, thermodynamics, and segregation characteristics. *Geochimica et Cosmochimica Acta* 71: 266–280 (doi:10.1016/j.gca.2006.11.019).

Hirsch PB (1979) A mechanism for the effect of doping on dislocation mobility. *Journal de Physique* 40(C6): 117–121.

Hirsch PB (1981) Plastic deformation and electronic mechanisms in semiconductors and insulators. *Journal of Physical Colloquium* 42(C3): 149–159.

Hirth G (2002) Laboratory constraints on the rheology of the upper mantle. In: Karato S-I and Wenk H-R (eds.) *Plastic Deformation in Minerals and Rocks, Reviews in Mineralogy and Geochemistry*, vol. 51, pp. 97–120. Washington, DC: Mineralogical Society of America.

Hirth G and Kohlstedt DL (1995a) Experimental constraints on the dynamics of the partially molten upper mantle. Deformation in the diffusion creep regime. *Journal of Geophysical Research* 100: 1981–2001.

Hirth G and Kohlstedt DL (1995b) Experimental constraints on the dynamics of the partially molten upper mantle. 2: Deformation in the dislocation creep regime. *Journal of Geophysical Research* 100: 15441–15449.

Hirth G and Kohlstedt DL (1996) Water in the oceanic upper mantle: Implications for rheology, melt extraction and the evolution of the lithosphere. *Earth and Planetary Science Letters* 144: 93–108.

Hirth G and Kohlstedt DL (2003) Rheology of the upper mantle and the mantle wedge: A view from the experimentalists. In: Eiler J (ed.) *Geophysical Monograph, Vol.138: Inside the*

Subduction Factory, pp. 83–105. Washington, DC: American Geophysical Union.

Hirth JP and Lothe J (1968) *Theory of Dislocations,* 780pp. New York: McGraw-Hill Book Company.

Hobbs BE (1981) The influence of metamorphic environment upon the deformation of materials. *Tectonophysics* 78: 335–383.

Hobbs BE (1983) Constraints on the mechanism of deformation of olivine imposed by defect chemistry. *Tectonophysics* 78: 335–383.

Hobbs BE (1984) Point defect chemistry of minerals under hydrothermal environment. *Journal of Geophysical Research* 89: 4026–4038.

Holtzman BK, Groebner NH, Zimmerman ME, Ginsberg SB, and Kohlstedt DL (2003a) Deformation-driven melt segregation in partially molten rocks. *Geochemistry Geophysics Geosystems* 4: 8607 (doi:10.1029/2001GC000258).

Holtzman BK, Kohlstedt DL, and Phipps Morgan J (2005) Viscous energy dissipation and strain partitioning in partially molten rocks. *Journal of Petrology* 46: 2569–2592 (doi:10.1093/petrology/egi065).

Holtzman BK, Kohlstedt DL, Zimmerman ME, Heidelbach F, Hiraga T, and Hustoft J (2003b) Melt segregation and strain partitioning: Implications for seismic anisotropy and mantle flow. *Science* 301: 1227–1230.

Hutchinson JW (1976) Bounds and self-consistent estimates for creep of polycrystalline materials. *Proceedings of the Royal Society of London* A 348: 101–127.

Ingrin J, Hercule S, and Charton T (1995) Diffusion of hydrogen in diopside: Results of dehydration experiments. *Journal of Geophysical Research* 100: 15489–15499.

Ingrin J and Skogby H (2000) Hydrogen in nominally anhydrous upper-mantle minerals: Concentration levels and implications. *European Journal of Mineralogy* 12: 543–570.

Jaoul O (1990) Multicomponent diffusion and creep in olivine. *Journal of Geophysical Research* 95: 17631–17642.

Jung H and Karato S-I (2001) Water-induced fabric transitions in olivine. *Science* 293: 1460–1463.

Kaibyshev O (1992) *Superplasticity of Alloys, Intermetallides, and Ceramics.* Berlin/New York: Springer.

Kaibyshev OA (2002) Fundamental aspects of superplastic deformation. *Materials Science and Engineering* 324: 96–102.

Karato S-I (1992) On the Lehmann discontinuity. *Geophysical Research Letters* 19: 2255–2258.

Karato S-I and Jung H (2003) Effects of pressure on high-temperature dislocation creep in olivine. *Philosophical Magazine* 83: 401–414.

Kaufmann G and Amelung F (2000) Reservoir-induced deformation and continental rheology in vicinity of Lake Mead, Nevada. *Journal of Geophysical Research* 105: 16341–16358.

Kaufmann G and Lambeck K (2002) Glacial isostatic adjustment and the radial viscosity profile from inverse modeling. *Journal of Geophysical Research* 107: 2280 (doi:10.1029/2001JB000941).

Kelemen PB, Hirth G, Shimizu N, Spiegelman M, and Dick HJB (1997) A review of melt migration processes in the adiabatically upwelling mantle beneath spreading ridges. *Philosophical Transactions of the Royal Society of London A* 355: 283–318.

Kelly A and Groves GW (1969) Change in shape due to dislocation climb. *Philosophical Magazine* 19: 977–986.

Kelly A and Groves GW (1970) *Crystallography and Crystal Defects,* 428pp. Reading, MA: Addison-Wesley Publishing Company.

Kocks UF, Argon AS, and Ashby MF (1975) Thermodynamics and kinetics of slip. In: *Progress in Materials Sciences,* vol. 19, 288pp. New York: Pergamon Press.

Kohlstedt DL (1992) Structure, rheology and permeability of partially molten rocks at low melt fractions. In: Phipps-Morgan J, Blackman DK, and Sinton JM (eds.) *Geophysical Monographs, Vol 71: Mantle Flow and Melt Generation at Mid-Ocean Ridges,* pp. 103–121. Washington: American Geophysical Union.

Kohlstedt DL (2002) Partial melting and deformation. In: Karato S-I and Wenk HR (eds.) *Plastic Deformation in Minerals and Rocks, Reviews in Mineralogy and Geochemistry,* vol. 51, pp. 105–125. Washington, DC: Mineralogical Society of America.

Kohlstedt DL (2006) The role of water in high-temperature rock deformation. In: Keppler H and Smyth J (eds.) *Water in Nominally Anhydrous Minerals, Reviews in Mineralogy and Geochemistry,* vol. 62, pp. 377–396. Washington, DC: Mineralogical Society of America.

Kohlstedt DL, Evans B, and Mackwell SJ (1995) Strength of the lithosphere: Constraints imposed by laboratory experiments. *Journal of Geophysical Research* 100: 17587–17602.

Kohlstedt DL, Keppler H, and Rubie DC (1996) Solubility of water in the α, β and γ phases of $(Mg,Fe)_2SiO_4$. *Contributions to Mineralogy and Petrology* 123: 345–357.

Kohlstedt DL and Mackwell SJ (1998) Diffusion of hydrogen and intrinsic point defects in olivine. *Zeitschrift fur Physikalische Chemie* 207: 147–162.

Kohlstedt DL and Mackwell SJ (1999) Solubility and diffusion of 'water' in silicate minerals. In: Wright K and Catlow R (eds.) *Microscopic Properties and Processes in Minerals,* pp. 539–559. Netherlands: Kluwer Academic Publishers.

Kohlstedt DL and Wang Z (2001) Grain-boundary sliding accommodated dislocation creep in dunite. Fall Meeting Supplement Abstract 82(47): T21C-01.

Kröger FA and Vink HJ (1956) Relation between the concentrations of imperfections in crystalline solids. In: Seitz F and Turnball D (eds.) *Solid State Physics,* vol. 3, pp. 307–435. San Diego CA: Academic Press.

Kronenberg AK, Kirby SH, Aines RD, and Rossman GR (1986) Solubility and diffusional uptake of hydrogen in quartz at high water pressures: Implications for hydrolytic weakening. *Journal of Geophysical Research* 91: 12723–12744.

Kronenberg AK and Tullis J (1984) Flow strengths of quartz aggregates: Grain size and pressure effects due to hydrolytic weakening. *Journal of Geophysical Research* 89: 4281–4297.

Langdon TG (1994) A unified approach to grain boundary sliding in creep and superplasticity. *Acta Metallurgica et Materialia* 42: 2437–2443.

Langdon TG (1995) Mechanisms of superplastic flow. In: Ridley N (ed.) *Superplasticity: 60 Years after Pearson,* pp. 9–24. London: Institute of Materials.

Lehmann I (1959) Velocities of longitudinal waves in the upper part of the Earth's mantle. *Annales de Geophysique* 15: 93–118.

Lehmann I (1961) S and the structure of the upper mantle. *Geophysical Journal of the Royal Astronomical Society* 4: 124–138.

Lerner I and Kohlstedt DL (1981) Effect of γ radiation on plastic flow of NaCl. *Journal of the American Ceramic Society* 64: 105–108.

Liftshitz IM (1963) On the theory of diffusion-viscous flow of polycrystalline bodies. *Soviet Physics JETP* 17: 909–920.

Mackwell SJ, Dimos D, and Kohlstedt DL (1988) Transient creep of olivine: Point defect relaxation times. *Philosophical Magazine A* 57: 779–789.

Mackwell SJ and Kohlstedt DL (1990) Diffusion of hydrogen in olivine: Implications for water in the mantle. *Journal of Geophysical Research* 95: 5079–5088.

Mackwell SJ, Kohlstedt DL, and Paterson MS (1985) The role of water in the deformation of olivine single crystals. *Journal of Geophysical Research* 90: 11319–11333.

Mainprice D, Tommasi A, Couvy H, Cordier P, and Frost DJ (2005) Pressure sensitivity of olivine slip systems and seismic anisotropy of Earth's upper mantle. *Nature* 433: 731–733.

Mei S, Bai W, Hiraga T, and Kohlstedt DL (2002) Influence of water on plastic deformation of olivine–basalt aggregates. *Earth and Planetary Science Letters* 201: 491–507.

Mei S and Kohlstedt DL (2000a) Influence of water on plastic deformation of olivine aggregates. Part 2: Dislocation creep regime. *Journal of Geophysical Research* 105: 21471–21481.

Mei S and Kohlstedt DL (2000b) Influence of water on plastic deformation of olivine aggregates. 1: Diffusion creep regime. *Journal of Geophysical Research* 105: 21457–21469.

Mishin Y and Herzig C (1999) Grain boundary diffusion: Recent progress and future research. *Materials Science and Engineering* A260: 55–71.

Montagner J-P (1985) Seismic anisotropy of the Pacific Ocean inferred from long-period surface waves dispersion. *Physics of the Earth and Planetary Interiors* 38: 28–50.

Mott NF (1951) The mechanical properties of metals. *Proceedings of the Physical Society* 64: 729–742.

Mott NF (1953) A theory of work hardening of metals. II: Flow without slip lines, recovery and creep. *Philosophical Magazine* 44: 742.

Mott NF (1956) A discussion of some models of the rate-determining process in creep. In: *Creep and Fracture of Metals at High Temperatures*, pp. 21–24. London: Her Majesty's Stationary Office.

Mukherjee AK (1971) The rate controlling mechanism in superplasticity. *Materials Science and Engineering* 8: 83–89.

Mukerjee AK, Bird JE, and Dorn JE (1969) Experimental correlations for high-temperature creep. *Transactions of ASME* 62: 155–179.

Nabarro F (1948) Deformation of crystals by the motion of single ions. In: *Report on a Conference on the Strength of Solids*, pp. 75–90. London: Physical Society.

Nabarro FRN (1967) Steady-state diffusional creep. *Philosophical Magazine* 16: 231–237.

Nakamura A and Schmalzried H (1983) On the nonstoichiometry and point defects of olivine. *Physics and Chemistry of Minerals* 10: 27–37.

Nakamura A and Schmalzried H (1984) On the Fe^{2+}–Mg^{2+} interdiffusion in olivine. *Berichte der Bunsengesellschaft fuer Physikalische Chemie* 88: 140–145.

Nishimura CE and Forsyth DW (1989) The anisotropic structure of the upper mantle in the Pacific. *Geophysical Journal* 96: 203–230.

Nishimura T and Thatcher W (2003) Rheology of the lithosphere inferred from post-seismic uplift following the 1959 Hebgen Lake earthquake. *Journal of Geophysical Research* 108: 2389 (doi:10.1029/2002JB002191).

Nix WD, Gasca-Neri R, and Hirth JP (1971) A contribution to the theory of dislocation climb. *Philosophical Magazine* 23: 1339–1349.

Orowan E (1940) Problems of plastic gliding. *Proceedings of the Physical Society* 52: 8–22.

Paterson MS (1969) The ductility of rocks. In: Argon AS (ed.) *Physics of Strength and Plasticity*, pp. 377–392. Cambridge, MA: MIT Press.

Peach MO and Koehler JS (1950) The forces exerted on dislocations and the stress fields produced by them. *Physical Review* 80: 436–439.

Poirier JP (1985) *Creep of Crystals: High-temperature Deformation Processes in Metals, Ceramics and Minerals*, 260pp. Cambridge: Cambridge University Press.

Pollitz FF (2003) Transient rheology of the uppermost mantle beneath the Mojave Desert, California. *Earth and Planetary Science Letters* 215: 89–104.

Pollitz FF, Peltzer G, and Bürgmann R (2000) Mobility of continental mantle; evidence from postseismic geodetic observations following the 1992 Landers earthquake. *Journal of Geophysical Research* 105: 8035–8054.

Post AD, Tullis J, and Yund RA (1996) Effects of chemical environment on dislocation creep of quartzite. *Journal of Geophysical Research* 101: 22143–22155.

Poumellec B and Jaoul O (1984) Influence of pO_2 and pH_2O on the high temperature plasticity of olivine. In: Tressler RE and Bradt RC (eds.) *Deformation of Ceramic Materials II. Materials Science Research*, vol. 18, pp. 281–305. New York: Plenum Press.

Pshenichnyuk AI, Astanin VV, and Kaibyshev OA (1998) The model of grain-boundary sliding stimulated by intragranular slip. *Philosophical Magazine* 77: 1093–1106.

Raj R and Ashby MF (1971) On grain boundary sliding and diffusional creep. *Metallurgical Transactions* 2: 1113–1127.

Read WT Jr (1953) *Dislocations in Crystals*, 175pp. New York: McGraw-Hill Book Company.

Read WT and Shockley W (1950) Dislocation models of crystal grain boundaries. *Physical Review* 78: 275–289.

Ready DW (1966) Chemical potentials and initial sintering in pure metals and ionic compounds. *Journal of Applied Physics* 37: 2309–2312.

Ricoult DL and Kohlstedt DL (1983a) Structural width of low-angle grain boundaries in olivine. *Physics and Chemistry of Minerals* 9: 133–138.

Ricoult DL and Kohlstedt DL (1983b) Low-angle grain boundaries in olivine. In: Yan MF and Heuer AH (eds.) *Advances in Ceramics, Character of Grain Boundaries*, vol. 6, pp. 56–72. Columbus: American Ceramic Society.

Ridley N (1995) C.E. Pearson and his observations of superplasticity. In: Ridley N (ed.) *Superplasticity: 60 Years after Pearson*, pp. 1–5. London: Institute of Materials.

Riley GN Jr and Kohlstedt DL (1990) An experimental study of melt migration in an olivine-melt system. In: Ryan MP (ed.) *Magma Transport and Storage*, pp. 77–86. New York: John Wiley & Sons.

Roscoe R (1952) The viscosity of suspensions of rigid spheres. *British Journal of Applied Physics* 3: 267–269.

Ruoff AL (1965) Mass transfer problems in ionic crystals with charge neutrality. *Journal of Applied Physics* 36: 2903–2907.

Rutter EH (1999) On the relationship between the formation of shear zones and the form of the flow law for rocks undergoing dynamic recrystallization. *Tectonophysics* 303: 147–158.

Rutter EH, Brodie KH, and Irving DH (2006) *Journal of Geophysical Research* 111: B06407 (doi:10.1029/2005JB004257).

Rutter EH and Neumann DHK (1995) Experimental deformation of partially molten Westerly granite, with implications for the extraction of granitic magmas. *Journal of Geophysical Research* 100: 15697–15715.

Rybacki E and Dresen G (2000) Dislocation and diffusion creep of synthetic anorthite aggregates. *Journal of Geophysical Research* 105: 26017–26036.

Schmalzried H (1981) *Solid State Reactions*, 2nd edn., 254pp. Weinheim: Verlag Chemie.

Schmalzried H (1995) *Chemical Kinetics of Solids*, 433pp. New York: VCH Publishers.

Schmid S, Boland JN, and Paterson MS (1977) Superplastic flow in fine grained limestone. *Tectonophysics* 43: 257–291.

Scott T and Kohlstedt DL (2006) The effect of large melt fraction on the deformation behavior of peridotite. *Earth and Planetary Science Letters* 246: 177–187.

Sherby OD and Burke PM (1967) Mechanical behavior of crystalline solids at elevated temperature. In: Chalmers B and Hume-Rothery W (eds.) *Progress in Materials Science*. New York: Pergamon Press.

Shewmon PG (1983) *Diffusion in Solids*, 203pp. Jenks, OK: J.Williams Book Company.

Stalder R and Skogby H (2003) Hydrogen diffusion in natural and synthetic orthopyroxene. *Physics and Chemistry of Minerals* 30: 12–19.

Stone DS, Plookphol T, and Cooper RF (2004) Similarity and scaling in creep and load relaxation of single-crystal halite (NaCl). *Journal of Geophysical Research* 109: doi:10.1029/2004JB003064.

Takei Y (2005) Deformation-induced grain boundary wetting and its effects on the acoustic and rheological properties of partially molten rock analogue. *Journal of Geophysical Research* 110: B12203 (doi:10.1029/2005JB003801).

Takei Y and Holtzman BK (2006) Viscous constitutive relations of partially molten rocks in terms of grain-boundary contiguity. *EOS Transactions of the American Geophysical Union* Fall Meeting Supplement 87(52): V23D-0654.

Takeuchi S and Argon AS (1976) Review: Steady-state creep of single-phase crystalline matter at high temperature. *Journal of Material Science* 11: 1542–1566.

Thurel E, Cordier P, Frost D, and Karato S-I (2003) Plastic deformation of wadsleyite. II: High-pressure deformation in shear. *Physics and Chemistry of Minerals* 30: 267–270 (doi: 10.1007/s00269-003-0313-7).

Tsai T-L and Dieckmann R (2002) Variations in the oxygen content and point defects in olivines, $(Fe_x, Mg_{1-x})2SiO_4$, $0.2 \leq x \leq 1.0$. *Physics and Chemistry of Minerals* 29: 680–694.

Tullis J and Yund RA (1980) Hydrolytic weakening of experimentally deformed Westerly granite and Hale albite rock. *Journal of Structural Geology* 2: 439–451.

Tullis J and Yund RA (1985) Dynamic recrystallization of feldspar: A mechanism for ductile shear zone formation. *Geology* 13: 238–241.

von Mises R (1928) Mechanik der plastischen Formänderung von Kristallen. *Zietschrift Angewandten Mathematik und Mechanik* 8: 161–185.

Wanamaker BJ (1994) Point defect diffusivities in San Carlos olivine derived from reequilibration of electrical conductivity following charges in oxygen fugacity. *Geophysical Research Letters* 21: 21–24.

Weertman J (1955) Theory of steady-state creep based on dislocation climb. *Journal of Applied Physics* 26: 1213–1217.

Weertman J (1957a) Steady-state creep through dislocation climb. *Journal of Applied Physics* 28: 362–364.

Weertman J (1957b) Steady-state creep of crystals. *Journal of Applied Physics* 28: 1185–1189.

Weertman J (1978) Creep laws for the mantle of the Earth. *Philosophical Transactions of the Royal Society of London* A 288: 9–26.

Weertman J (1999) Microstructural mechanisms of creep. In: Meyers MA, Armstrong RW, and Kirschner H (eds.) *Mechanics and Materials: Fundamentals and Linkages*, pp. 451–488. New York: Wiley.

Weertman J and Weertman JR (1992) *Elementary Dislocation Theory*, 213pp. Oxford: Oxford University Press.

Xu Y, Nishihara Y, and Karato S-I (2005) Development of a rotational Drickamer apparatus for large-strain high-pressure deformation experiments. In: Chen J, Wang Y, Duffy TS, Shen G, and Dobrzhinetskaya LF (eds.) *Frontier of High-Pressure Research: Applications to Geophysics*, pp. 167–182. Amsterdam: Elsevier.

Xu Y, Weidner DJ, Chen J, Vaughan MT, Wang Y, and Uchida T (2003) Flow-law for ringwoodite in the subduction zone. *Physics of the Earth and Planetary Interiors* 136: 3–9.

Xu Y, Zimmerman ME, and Kohlstedt DL (2004) Deformation behavior of partially molten mantle rocks. In: Karner GD, Driscoll NW, Taylor B, and Kohlstedt DL (eds.) *Rheology and Deformation of the Lithosphere at Continental Margins: MARGINS Theoretical and Experimental Earth Science Series*, vol. 1, pp. 284–310. New York: Columbia University Press.

Zhao Y-H, Ginsberg SG, and Kohlstedt DL (2004) Solubility of hydrogen in olivine: Effects of temperature and Fe content. *Contributions to Mineralogy and Petrology* 147: 155–161 (doi:10.1007/s00410-003-0524-4).

Zimmerman ME and Kohlstedt DL (2004) Rheological properties of partially polten lherzolite. *Journal of Petrology* 45: 275–298.

15 Properties of Rocks and Minerals – Diffusion, Viscosity, and Flow of Melts

D. B. Dingwell, University of Munich, Munich, Germany

15.1	Mass and Momentum Transport in Melts: Georelevance	419
15.1.1	Role of Melts in Earth Differentiation	419
15.1.2	Melt-Bearing Reaction Kinetics	420
15.1.3	Magma Transport at Depth	420
15.1.4	Magma Mixing, Mingling, and Unmixing	420
15.1.5	Volcanic Ascent and Eruption	420
15.1.6	Emplacement of Natural Glass	421
15.2	Dynamics and Relaxation in Melts	421
15.2.1	A Note on the Definition of Terms	421
15.2.2	Structural Relaxation	421
15.2.3	Spectroscopic Observation of Structural Relaxation	422
15.2.4	Phenomenology of the Glass Transition	422
15.2.5	Atomistic Origin of Relaxation	423
15.3	Diffusion in Melts	423
15.3.1	Fundamental Concepts	423
15.3.2	Methods of Investigation	423
15.3.3	Tracer Diffusivities: Self-Diffusivities and Chemical Diffusion	424
15.3.4	Temperature and Pressure Dependence of Diffusion	425
15.3.5	Concentration-Dependent Diffusion	425
15.3.6	Intrinsic versus Extrinsic Diffusivities	425
15.4	Melt Rheology	426
15.4.1	Methods of Investigation	426
15.5	Viscosity of Liquids	427
15.5.1	Newtonian Rheology in Silicate Melts	427
15.5.2	Temperature Dependence of Newtonian Viscosity	427
15.5.3	Pressure Dependence of Newtonian Viscosity	428
15.5.4	Compositional Dependence of Newtonian Viscosity	428
15.5.5	Non-Newtonian Rheology in Silicate Melts	430
15.5.6	Isostructural Viscosity	431
15.5.7	Failure of Melts	431
15.6	Concluding Statements	431
References		434

15.1 Mass and Momentum Transport in Melts: Georelevance

15.1.1 Role of Melts in Earth Differentiation

The differentiation of the Earth is generally thought to have been accomplished in the presence of the molten phase (Wood *et al.*, 2006). This picture extends from the scale of global differentiation into core, mantle, and crust, through the differentiation into upper and lower sections of each, and down to the scale of local differentiation providing chemical specialization of great extremes within the crust; the latter with their global significant economic importance. Nothing is more central to this picture than the transport properties of magma. With lower viscosities and possessing higher diffusivities than their neighboring solid phases, the melts involved in these processes are the material conduit for chemical transport from one reservoir to another. One needs but to pose the question of differentiation efficiency on a cool subsolidus body to appreciate the meaning of this attribute of the terrestrial planets.

15.1.2   Melt-Bearing Reaction Kinetics

Equally significant in the relevance of melt transport properties to Earth processes are the observations that have been made regarding the relationship between melt viscosity and the diffusivities of melt components (Hofmann, 1980; Shimizu and Kushiro, 1984; Dingwell, 1990; Mungall *et al.*, 1999; Mungall, 2002). The kinetics of reactions within the Earth's interior may be substantially controlled by the diffusivities of chemical components present in the reacting system. In general, diffusivities in the molten state are higher than in the solid-state equivalent phases. Reactions in the presence of a melt phase may thus be strongly influenced by the diffusivities of components in the melt phase itself. For this reason, the experimental investigation and the systemization of results of chemical and trace diffusivities in silicate melts have been intensively pursued in the past decades. Systematization of the results in terms of the relationship to viscosity has revealed some interesting and useful approximations, which are discussed in this chapter.

15.1.3   Magma Transport at Depth

The transport of magmas at depth is essentially controlled by the presence of stress gradients, vertical or otherwise. As such, the rheology of the magma is the coefficient of material response that scales the deformation of the magma body and its transport relative to neighboring rock masses. Experimental recognition of the wide range of magnitude of rheology of magma – together with its variable dependence on temperature, chemical composition, physical state, pressure, and (variably) strain rate – has generated a rich literature on the deformation of magma in the deep Earth. Central to all of these considerations has been the premise that, during magma deformation and transport, conditions obtain that permit the plastic deformational response of the magma. Under such circumstances, the rheological literatures of liquid silicates, crystal and/or bubble suspensions, and partially molten rock, all play major roles in filling out our physical picture of magma rheology at depth (Dingwell *et al.*, 1993).

15.1.4   Magma Mixing, Mingling, and Unmixing

Multicomponent silicate melts are stable over very wide ranges of temperature, pressure, and composition. Further, within the Earth's mantle and crust, there is ample evidence, both textural, chemical and by geophysical inference, for the temporal coexistence and juxtaposition/interference of distinct melt batches within regional magmatic terrains. Thus, the phenomena of magma mingling (physical mixing) and mixing (chemical mixing) are now thought to be common in magmatic systems, often serving to precondition subvolcanic systems for subsequent eruptions. Magma mingling and mixing occurs between magmas that are, often via extensive differentiation, very different in composition (e.g., basalt-rhyolite). Magma mingling and mixing also occurs between chemically distinct but quite similar compositions in subvolcanic environments. Evidence of both is provided most dramatically in coerupted mingling lava products. The kinetics of magma mingling and mixing are controlled by the relative viscosities and the chemical diffusivities of the melts, respectively. Here the question of turbulence tehaotic laminar flow in magma mingling remains largely uninvestigated.

Finally, unmixing of magma into two coexisting liquids also occurs in nature. Liquid immiscibility is a phenomenon whose kinetics are very poorly understood. The experimental investigation of liquid immiscibility is advancing rapidly and unmixing kinetics are clearly a key factor in its investigation (Gurenko *et al.*, 2005; Vekster *et al.*, 1998, 2002, 2005, 2006).

15.1.5   Volcanic Ascent and Eruption

There is hardly a more spectacular demonstration of the influence of the transport properties on melt transport than the eruption of a volcano. The range of eruptive styles of volcanic eruptions, from highly explosive Plinian eruptions, through the slow creep of rheomorphic flows, to the ready flow of rivers of lava over tens of kilometers is determined by a series of conditions that are either directly or indirectly linked to the viscosity of the melt (Dingwell, 1996, 1998a, 1998b). The rapidity of volcanic ascent and eruption lead to a number of consequences for the flow of magma. The dominant role of solid–liquid reactions in determining the differentiation of liquids and their transport at depth gives way to the relatively rapid generation of thermal and rheological consequences of effects such as the exsolution of volatile phases (Bagdassarov and Dingwell, 1992, 1993a, 1993b), the strain rate-driven onset of shear thinning (Dingwell and Webb, 1989, 1990; Webb and Dingwell, 1990a, 1990b), and even the development of viscous dissipation (Costa and Macedonio, 2003; Rosi *et al.*, 2004; Vedeneeva *et al.*, 2005). Numerical simulations, of increasingly critical importance in testing the consequences of mechanistic hypotheses for transport during eruptions, are critically dependent on an adequate characterization of melt viscosity (Dingwell, 1998b; Papale, 1999).

15.1.6 Emplacement of Natural Glass

Viscosity also serves to quantify the conditions for the transformation from the liquid to the glassy state. Volcanism, at least on the Earth and Moon, commonly generates volcanic glass. The observation of abundant survival of this unstable thermodynamic and fluid dynamic state is a tribute to the sluggishness of crystal nucleation and growth kinetics in silicates, combined with the rapid cooling history provided in volcanic settings. The disequilibrium nature of the glassy state means that the properties of the glass phase are path dependent (Tool and Eichlin, 1931; Tool, 1946; Moynihan *et al.*, 1974, 1976a, 1976b). This, in turn, leads to the conclusion that the glass properties can be used to obtain information on the conditions of their emplacement (Stevenson *et al.*, 1995; Gottsmann *et al.*, 2002; Giordano *et al.*, 2005). This possibility has been exploited in recent years using calorimetric techniques to determine the effective viscosity at the glass transition and thus constrain the cooling history of natural glass (Wilding *et al.*, 1995, 1996a, 1996b, 2000; Gottsmann and Dingwell, 2001a, 2001b, 2002; Gottsmann *et al.*, 2004). The striking conclusion is that hyaline volcanic facies can endure a very wide range of effective cooling histories and can suffer several transects from liquid to glassy behavior during their emplacement.

15.2 Dynamics and Relaxation in Melts

15.2.1 A Note on the Definition of Terms

In the context of the subject matter of this chapter, it is important to note here that the term 'melt' is commonly very loosely defined. The term implies perhaps 'something which was created by melting'.

This definition alone, however, says nothing about the fundamental nature of the properties of the system. Yet it is the liquid-like values of the transport properties of the 'melt' phase that lend it its greatest significance with respect to the Earth sciences. Here we will follow the usage that the melt phase describes an amorphous, typically silicate, phase whose properties may be either liquid-like or solid-like, depending on the environment and the conditions of the natural transport process, the experiment, or the simulation in which the melt exists. In this sense, the phenomenology of the melt's response will define its behavior as a liquid or as a solid, or, in the terms to be employed here, as a liquid or a glass (Dingwell and Webb, 1990).

15.2.2 Structural Relaxation

The nature and phenomenology of structural relaxation in silicate melts, well appreciated in the glass and ceramic literature as well as in modern chemistry (e.g., Angell *et al.*, 2000), has also been described in the context of mineral physics, petrology, and volcanology in a way which brings its relevance to the Earth sciences to the fore (Dingwell and Webb, 1989; Dingwell, 1995a, 1995b). The fundamental picture of liquid structure upon which the atomistic interpretation of its relaxation ultimately rests has been intensively investigated using a wide range of spectroscopic techniques. It is not the task of this chapter to review those structural studies for which ample summaries already exist (Mysen, 1988; Stebbins *et al.*, 1995; Mysen and Richet, 2005). Rather, it should be noted here that almost all of the fundamental systematics and phenomenology of structural relaxation have been mapped out prior to the modern spectroscopic analysis of melt structure. This was made possible by using physical properties of the melt phase as proxies for its structural state. Arguably, the first attempt to do so is provided by the definition of a state parameter of glass, the so-called 'fictive temperature' commonly attributed to Tool (1946). The variation of the structure of the melt was obtained via the measurement of a precisely determined physical property of the glassy state (density, refractive index) in the quenched state, as well as during the thermal recyling (heat capacity) (e.g., Narayanaswamy, 1971, 1988). A graphic illustration of the variation of fictive temperature with heating and cooling through the glass transition is provided in **Figure 1**.

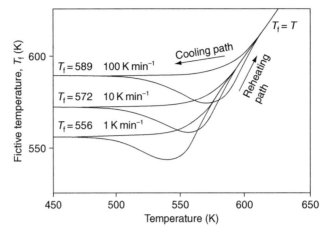

Figure 1   The variation of fictive temperature with temperature during the cooling of a supercooled liquid into the glassy state and during the heating of a glass into the supercooled liquid state. The fictive temperature is a proxy for all structure-dependent properties of the melt. Fictive temperature equals temperature in the liquid state, is constant in the glassy state, and exhibits a hysteresis in the glass transition region that depends on heating/cooling rate and the temperature dependence of viscosity and structure. Reproduced from Narayanaswamy OS (1988) Thermorheological simplicity in the glass transition. *Journal of American Ceramic Society* 71: 900–904.

15.2.3   Spectroscopic Observation of Structural Relaxation

Spectroscopic investigations of silicate melts have been highly successful in the elucidation of various aspects of the coordination of individual cations and anions in the silicate melt structure as well as information on the degree of connectivity of tetrahedral structural units (Q species) in silicate melts (Stebbins *et al.*, 1995; Mysen and Richet, 2005). The success of such studies raises the question of whether structural relaxation can be investigated directly by using spectroscopic species data. There are several possible experimental protocols (Dingwell, 1995a). First, the structure of carefully quenched silicate glasses can be treated via fictive temperature analysis in order to obtain the temperature dependence of species concentration involved in homogeneous equilibria (Liu *et al.*, 1987, 1988; Dingwell and Webb, 1990). Second, high-temperature, high-pressure spectroscopic investigations may directly record the temperature dependence of speciation and contain thereby the observation of the inflection in temperature dependence of speciation marking the glass transition temperature (Nowak and Behrens, 1995; Shen and Keppler, 1995). Third, the high-temperature equilibration of species may be monitored in so-called time series relaxation experiments where the quenched products are analyzed for the approach to equilibrium recorded by spectroscopic analysis of their species concentrations (Zhang *et al.*, 1995, 1997). Fourth, the spectroscopic investigation of

silicate melts at high temperatures has resulted in a series of estimates of the timescales for relaxation through self-diffusion of the fundamental structural units being spectroscopically resolved. Nuclear magnetic resonance studies of the mobility of network-forming components in silicate melts have linked the relaxation of Si–O bonds between structural units containing variable numbers of nonbridging oxygens closely to the timescale of viscous flow (Stebbins, 1995). This close link between the relaxation of structural units as small as individual bridging bonds and the macroscopic relaxation of shear stress is strong evidence for the exclusion of large flow units involving some sort of polyanionic units (polymers) during the flow of silicate melts. In fact, there is no clear evidence for a polymeric nature of the structure of silicate melts based on physical properties. Nevertheless, polymer-based models for the activity and solubility of chemical components appear to be capable of predicting the chemical properties of melts (Moretti, 2005).

15.2.4   Phenomenology of the Glass Transition

The phenomenology of the glass transition is an ancient observation associated with glassmaking in prehistoric times. The fundamental observation is the transformation of the properties of a melt from those of a liquid to those of a solid, or vice versa. For certain properties, the transformation is qualitative in nature, for example, the development of rigidity in

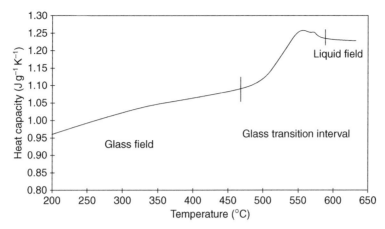

Figure 2 Phenomenology of the glass transition. The heat capacity of a silicate melt expresses a temperature-dependent glassy segment at lower temperature, a higher segment in the liquid state at higher temperature, and a hysteresis in the glass transition interval whose path links the two states – glassy and liquid. The increase in heat capacity at the glass transition results from the self-diffusion of atoms in the structure which enables temperature-dependent configurational changes in the structure to follow their temperature-dependent equilibria. In the glass, such reactions are largely frozen out.

the glassy state. For most properties however, the transition is somewhat more subtle. Yet quantitative determination of the effect of the glass transition on temperature-dependent properties such as density, compressibility, electrical conductivity, and heat capacity are a major source of quantitative information on the properties of silicate melts in general. The fundamental phenomenology of the glass transition can be illustrated with the example of the temperature- and state-dependent properties, thermal expansivity and heat capacity (**Figure 2**).

15.2.5 Atomistic Origin of Relaxation

The advent of spectroscopic investigations of silicate melts at elevated temperatures has now aided considerably in placing the relaxation behavior of silicate melts on an atomistic basis. Simple observations have, for some time, implied a relationship between atomic mobility, diffusion, and viscosity. Experimental advances in the obtainment of diffusivity and viscosity data have led to the somewhat paradoxical confirmation of Stokes–Einstein or Eyring formulations of viscosity–diffusivity relationships for certain chemical components of silicate melts (Shimizu and Kushiro, 1984) under certain conditions yet rejection for others (e.g., Hofmann, 1980; Jambon, 1982). We now know that the picture of atomistic origins of relaxation clearly must distinguish between individual chemical components of the melt phase such that some components appear to control the

relaxation responsible for determining the rheology of the melt (Si, O) and others do not (Li, Na). These relations are possibly best illustrated with the aid of a so-called 'relaxation map' of all available atomistic mobility information for a particular melt composition (**Figure 3**).

15.3 Diffusion in Melts

15.3.1 Fundamental Concepts

Diffusion reflects temperature and establishes chemical equilibrium. It is the fundamental diffusive mobility of atoms in materials at temperature that is ultimately responsible for the achievement of equilibrium in chemical reactions, for the re-equilibrative response by relaxation of a system or phase to a perturbation of its thermodynamic state, and for the maintenance through time of the state of equilibrium by dynamic diffusive exchange of atoms/bonds. As such, diffusion is at the heart of the establishment and maintenance of equilibrium.

15.3.2 Methods of Investigation

Information on the magnitude of diffusivities in silicate melts comes from a range of experimental approaches, both direct and indirect (Chakraborty, 1995).

Direct determinations of diffusivity in silicate melts have been dominated by studies involving the microanalytical determination of experimentally

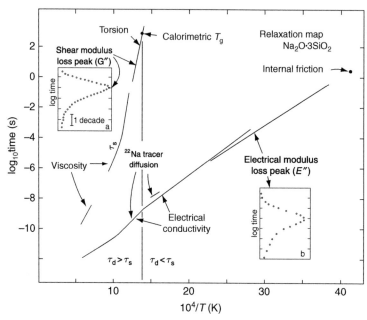

Figure 3 A relaxation map for a sodium silicate melt. Various experimental data sources on differing melt properties (mechanical, calorimetric, electric) are combined and compared with each other in order to obtain an overview of the relaxational landscape and locations of relaxational modes, within the silicate melt structure. Two fundamental relaxation modes are seen here – a slow one related to Si mobility and a faster one related to Na mobility. Reproduced from Dingwell DB (1990) Effects of structural relaxation on cationic tracer diffusion in silicate melts. *Chemical Geology* 82: 209–216.

generated gradients in either isotopic or chemical concentration. A range of experimental geometries have been employed including cylindrical half space couples, sphere absorption/desorption, thin source, and others. All of these techniques rely upon the spatial resolution of chemical or isotopic gradients which have usually been induced via annealing of an initial step function concentration gradient over time. As such, time series form the strategy surrounding the central experimental variable of time. Combined with variations in pressure and temperature, the diffusivity, its temperature and pressure dependence, are mapped out for the silicate melt. Variation of further field variables, such as electric field, has been pursued in the materials literature but has played no significant role in the investigations within the geoscientific community.

Direct derivation of self-diffusivities via relaxation techniques in spectroscopy has also played a role in our understanding of silicate mobility. Motional averaging of individual species, for example, has yielded diffusivity estimates for cations in silicate melts.

As discussed above, viscosity–diffusivity relationships have been elaborated for silicate melts with the result that diffusivities can be inferred from viscosities for selected components in silicate melts. Although the viscosity database far outweighs the diffusivity database for which those relationships apply, this method of indirect determination of diffusivity via viscosity determination or relaxation behavior has not been exploited significantly in the geoscientific literature.

15.3.3 Tracer Diffusivities: Self-Diffusivities and Chemical Diffusion

It is essential, at the outset of thinking about diffusion in melts, to distinguish between self-diffusion on the one hand and chemical diffusion on the other. The self-diffusion of a component occurs in the absence of chemical gradients. The chemical diffusion of a component occurs in the presence of and in response to a gradient in its chemical potential expressed as a concentration gradient in the host phase. The simplest consequence is that chemical diffusion is directed, whereas self-diffusion is not. The most common approach to the determination of self-diffusion in silicate melts has been via isotopic doping in diffusion couples. Tracer diffusion typically describes diffusion of a species in the presence of a minimal activity gradient. Although, in principle, anything detectable can be considered as a tracer in such a system, the phrase tracer diffusion usually refers to a trace element concentration level doping of a system and the subsequent spatial resolution of its chemical diffusion. At very low levels of concentration, the inference is sometimes made that tracer diffusion can approach the value of self-diffusion in the system.

In general, self-diffusivities and chemical diffusivities are different in magnitude and cannot be compared for the purpose of deriving relative structural information about the diffusion matrix, the silicate melt. Self-diffusivities contain fundamental information on the silicate melt structure and/or the structural role of the diffusing species. Chemical diffusivities contain information on the activity–composition relationships of the diffusing species in the host melts.

15.3.4 Temperature and Pressure Dependence of Diffusion

The temperature dependence of diffusivities in silicate melts at high temperature is most commonly described via an Arrhenian relationship (i.e., a logarithmic dependence of a property upon the reciprocal of absolute temperature). This situation is certainly valid for the typically restricted ranges of temperature involved in high-temperature determinations of diffusivities in low-viscosity liquids. It is probably not valid for some species when observed over a larger temperature range (Dingwell, 1990). Arrhenian activation energies of diffusion range widely as a function of ionic properties, melt composition, and absolute temperature range. Below, the potential complexity of the temperature dependence of diffusivities is discussed within the context of their relationship to viscosity.

The pressure dependence of diffusion is highly variable. At a given temperature and pressure range, it can be either positive or negative, depending on the identity of the ion. As a function of pressure, the pressure dependence of diffusivity may also switch from a positive to a negative pressure dependence (Poe *et al.*, 1997; Poe and Rubie, 1998).

The structural basis for the interpretation of the temperature and, especially, the pressure dependence of diffusivities is far from complete. General relationships of the temperature dependence are discussed in more detail below. General aspects of the pressure dependence are given to date in very broad terms in the literature, usually without direct

structural data to confirm them. Progress is badly needed here if we are to develop truly generalizable relationships.

15.3.5 Concentration-Dependent Diffusion

Concentration dependence of diffusivity in silicate melts has been demonstrated for several components. Systems exhibiting concentration-dependent diffusivity possess, by definition, significant activity and, thus, concentration gradients. Diffusion in the presence of such chemical concentration gradients is chemical diffusion. Thus, concentration dependence of diffusivities can be viewed as the case where the compositional dependence of the chemical diffusivity of the diffusing speciation is approximated by the incorporation of a single compositional gradient, that of the diffusing species itself. One of the best-established cases of the approximation of concentration-dependent diffusivity is that of H_2O in rhyolitic melts (Behrens *et al.*, 2004).

15.3.6 Intrinsic versus Extrinsic Diffusivities

The relationship between viscosity and diffusivity may be illustrated well using the abundant data now available on the tracer and/or self-diffusivity of various cations in multicomponent silicate melts for which the viscosity may be calculated. This is attempted in **Figures 4** and **5**. Cationic diffusivities are plotted versus melt viscosity for a range of

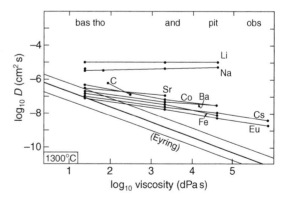

Figure 4 Tracer diffusion data plotted vs the estimated viscosity of the host melt. With the exception of the fast cations, Na and Li, the diffusivities of all others cluster near the calculated Si diffusivity. These are defined as extrinsic diffusivities. Reproduced from Dingwell DB (1990) Effects of structural relaxation on cationic tracer diffusion in silicate melts. *Chemical Geology* 82: 209–216.

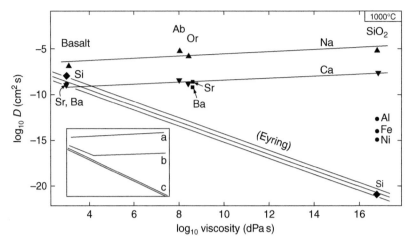

Figure 5 An expanded version of **Figure 4** where additional data on the diffusivity of cations are plotted vs viscosity of the host melt. Here, intrinsic diffusivities can be distinguished clearly from extrinsic diffusivities. Perhaps the simplest aspect of the distinction is the opposing viscosity dependence of intrinsic (positive) vs extrinsic (negative) diffusivities. Reproduced from Dingwell DB (1990) Effects of structural relaxation on cationic tracer diffusion in silicate melts. *Chemical Geology* 82: 209–216.

multicomponent silicate melts. The diffusivities of a structurally highly diverse range of cations can be seen to exhibit very similar values of diffusivity and its temperature dependence. Two cations (Na, Li) appear to form exceptions to the rule, exhibiting distinctly higher diffusivities. It is important to note that the viscosity dependence, and through it the composition dependence, of the cationic diffusivities is negative for the large grouping of cations clustered at lower diffusivity values but positive for the cations of higher diffusivity. Without any further constraints, diffusivity, as an expression of structure in these melts, appears paradoxical. The cations of lower diffusivity appear to behave independently of their cationic properties. C and Cs, two cations of extremely different cationic properties, have virtually identical diffusivities. There is however a further constraint, that is, the diffusivity of Si. It may be reliably estimated through the Stokes–Einstein relationship to lie, as represented in **Figures 4** and **5**, just at the base of the cluster of slower cationic diffusivities. In fact, the relative location of the Si diffusivity led Dingwell (1990) to suggest that the values of diffusivity being expressed by most of the cations in **Figure 4** are controlled by the viscosity of the melt itself. These were termed then 'extrinsic diffusivities'. They are subequal for all such affected cations, there being only one value of viscosity at a given composition and temperature. Extrinsic diffusivities have, in addition, relatively high activation energies, and they are necessarily

sensitive to any change in melt viscosity induced in the system.

The Na and Li data are, if joined by further data from melts of higher viscosity (**Figure 5**), shown to be a second class of diffusivities with respect to structural relaxation, such that cations exhibiting extrinsic diffusivities in very low viscosity melts exhibit the alternative behavior of Na and Li in more viscous melts. These diffusivities, characterized by low activation energies, higher diffusivities than Si, and an insensitivity to melt viscosities, were termed 'intrinsic diffusivities'. Intrinsic diffusivities were so named as their striking variation in magnitude and activation energy appears to reflect their individual cationic properties quite rationally.

15.4   Melt Rheology

15.4.1   Methods of Investigation

The experimental measurement of silicate melt viscosity has a history going back a century. Studies directly dedicated to the Earth sciences have been performed for almost as long (e.g., Kozu and Kani, 1935). Inferences were drawn from such studies from the very start regarding their geological and geophysical application (Bowen, 1934). Elevated pressures and temperatures were employed very early (see Dingwell, 1998c) in such studies and fundamental constraints on volcanism and magmatism were derived. This long history, together with the observation that melt viscosity ranges over more than 10

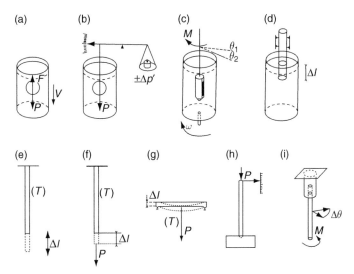

Figure 6 Geometric arrangements for experimental strategies of the determination of viscosity. Viscosity experiments may be grouped into two classes – those in which the stress is controlled and the strain rate monitored, and those in which the strain rate is controlled and the resultant stress is monitored. Reproduced from Zarzycki J (1991) *Glasses and the Vitreous State*, 505pp. New York: Cambridge University Press.

orders of magnitude, has resulted in the application of a remarkable range of experimental methods for the investigation of melt viscosity (**Figure 6**). In principle, viscosity is determined either by setting the rate of deformation (strain rate) and obtaining the stress sustained by the liquid against its deformation or by setting the applied stress and obtaining the deformation rate of the sample.

15.5 Viscosity of Liquids

15.5.1 Newtonian Rheology in Silicate Melts

The Newtonian nature of the viscosity of silicate liquids has been extensively investigated. It is standard practice in the most common methods of viscometry employed for silicate liquids to systematically vary the strain rate during measurement in order to test for non-Newtonian behavior. It is safe to say that on the basis of the considerable viscosity data set available today, the vast majority of melt deformation and transport processes within the deep Earth are likely to occur in the presence of a melt phase exhibiting purely Newtonian behavior. The issue of the boundaries of Newtonian behavior in the laboratory has been investigated extensively. Toward the low strain rate side, the boundaries have not been found. We infer that they do not exist. Thus, there is no experimental evidence of yield strengths for

liquid silicates. Toward the high strain rate side, the boundary of Newtonian behavior, or the so-called onset of non-Newtonian behavior, has been found, as described in detail below.

15.5.2 Temperature Dependence of Newtonian Viscosity

The temperature dependence of the viscosity of silicate melts may be expressed over limited temperature intervals as an exponential function of inverse absolute temperature.

In this so-called Arrhenian description, the temperature dependence is reflected in the magnitude of a coefficient commonly referred to as the activation energy of viscous flow. By analogy to rate theory, the activation energy was traditionally interpreted as reflecting an average energy barrier to flow. The temperature dependence of viscosity of several melts is illustrated in **Figure 7**, where the variation in activation energy, at a given temperature, is reflected in the range of slope values. The activation energy of geologic melts varies, at superliquidus temperatures, from ~30 to 70 kcal mol⁻¹.

Inspection of **Figure 7** readily reveals that the temperature dependence of the viscosity of silicate melts is, in general, non-Arrhenian. The temperature dependence increases with decreasing temperature, to a greater or lesser extent, for all melt compositions.

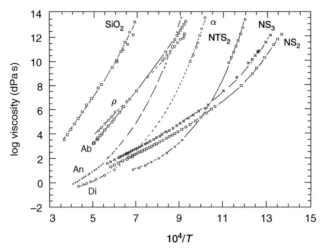

Figure 7 The non-Arrhenian nature of the temperature dependence of the Newtonian viscosity of silicate melts as exhibited by several analog compositions of relevance to geological processes. The general trend toward more fragile behavior at lower absolute temperatures for a constant viscosity is evident. Ab, albite melt; An, anorthite; Di, diopside melt; NTS_2, sodium titanosilicate melt; NS_3, sodium trisilicate; NS_2, sodium disilicate. Reproduced from Richet, P and Bottinga Y (1995) Rheology and configurational entropy of silicate melts. In: Stebbins JF, Dingwell DB, and McMillan PF (eds.) (1995) *Reviews in Mineralogy, Vol. 32: Structure, Dynamics and Properties of Silicate Melts*, pp. 21–66. Washington, DC: Mineralogical Society of America.

This aspect of the temperature dependence of viscosity is most commonly expressed in terms of the addition of a term to the temperature dependence that effectively reduces the reciprocal absolute temperature. This empirical description of temperature-dependent melt viscosities using the Tammann–Vogel–Fulcher formulation, despite providing a good description of the data, lacks a theoretical basis.

A theoretical basis underlies the Adam–Gibbs (1965) formulation of viscous flow (based on cooperative relaxation) by Richet (1984). Here the viscosity is expressed as an inverse function of the configurational entropy of the silicate melt. Adam–Gibbs theory holds the promise of a link between external thermodynamic properties of the silicate melt phase and its flow behavior. As such, it has been analyzed experimentally by comparison of calorimetric and rheological data for model silicates. The results for diopside liquid are presented in **Figure 8**, where the linerization of the temperature dependence of melt viscosity occurs via a normalization to the temperature-dependent value of the configurational entropy (Richet *et al.*, 1986).

15.5.3 Pressure Dependence of Newtonian Viscosity

The pressure dependence of the viscosity of silicate melts has been investigated predominantly in the low-viscosity regime at relatively high temperatures, typically above the high-pressure liquidus of the melts (Kushiro, 1976; Kushiro *et al.*, 1976; Scarfe *et al.*, 1987; Wolf and McMillan, 1995). Such investigations have been based chiefly on falling sphere viscometry using Stokesian principles (see review by Dingwell, 1998c). **Figure 9** contains a collection of pressure-dependent viscosity curves. The complexity of these trends implies that we do not yet have the proper structural information necessary for describing the pressure-induced change in flow mechanisms in silicate melts at high temperature in a systematized fashion. Undoubtedly, some surprises lie ahead.

15.5.4 Compositional Dependence of Newtonian Viscosity

The database on compositional dependence of Newtonian viscosity is vast. The salient features are summarized here. **Figure 10** illustrates the relative viscosities of alkali and alkaline earth aluminosilicate melts over a wide range of silica contents. It can be seen that the relative viscosities of these systems do not change and that the viscosity increases with decreasing field strength of the network-stabilizing cation. This has been interpreted as a consequence of the relative mean bond strengths in these tectosilicate liquids. Further, the fundamental feature of nonlinear

(a)

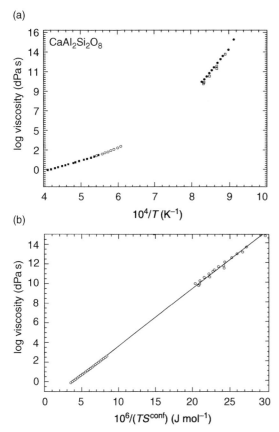

(b)

Figure 8 The normalization of the non-Arrhenian temperature dependence of the viscosity of diopside melt using calorimetrically derived data for the temperature dependence of the configurational entropy of the glass and liquid. (a) Viscosity vs. reciprocal absolute T and (b) viscosity vs. reciprocal absolute T normalized to S^{conf}. Reproduced from Richet P, Robie RA, and Hemmingway BS (1986) Low temperature heat capacity of diopside glass ($CaMgSi_2O_6$) a calorimetric test of the configurational entropy theory applied to the viscosity of silicate liquids. *Geochimica et Cosmochimica Acta* 50: 1521–1533.

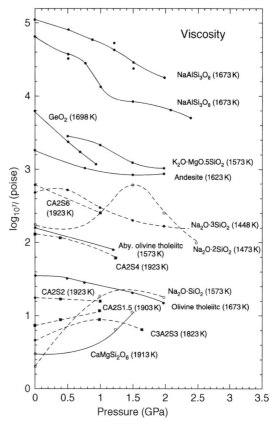

Figure 9 The pressure dependence of the Newtonian viscosity of silicate melts. The data are largely derived from Stokesian settling experiments. Reproduced from Scarfe CM, Mysen BO, and Virgo DL (1987) Pressure dependence of the viscosity of silicate melts. In: Mysen BO (ed.) *Magmatic processes: Physiochemical principles*, Geochemical Society Special Publication 1, pp. 59–67. University Park, PA: Geochemical Society and Wolf GH and McMillan PF (1995) Pressure effects on silicate melt properties. *Reviews in Mineralogy* 32: 505–561.

compositional variation of viscosity is underlined by the curvature of the viscosity trends with decreasing silica content.

Coming from this relatively simple and ordered picture of the composition dependence of viscosity, we can obtain an impression of the variation in viscosity generated by melt composition from **Figure 11** where the individual influences of a large number of components on the viscosity of a haplogranitic melt are illustrated. The influence of a further variable in melt composition, oxidation state, on melt viscosity is summarized for several ternary silicate melts in **Figure 12**. Reduction of Fe-bearing melts results in a nonlinear decrease in viscosity. The viscosities of

highly oxidized melts are very sensitive to oxidation state whereas moderately reduced melts are relatively insensitive.

The first non-Arrhenian model for geologically revelant melt compositions was generated for the system calcalkaline rhyolite (metaluminous haplogranite)–H_2O (Hess and Dingwell, 1996). Its parametrization illustrates all of the essential features of the addition of water to a multicomponent silicate melt (**Figure 13**). Sufficient low-temperature data now exist to begin the effective modeling of non-Arrhenian melt viscosity in multicomponent compositional space (Giordano and Dingwell, 2003; Giordano *et al.*, 2006). **Figure 14(a)** illustrates an example of the compositional range of natural multicomponent melts investigated and their measured

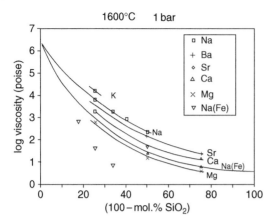

Figure 10 Systematics of the composition dependence of high-temperature melt viscosity in the alkali and alkaline earth aluminosilicate joins of nominal tectosilicate stoichiometry. Reproduced from Dingwell DB (1989) Shear viscosities of ferrosilicate liquids. *American Mineralogist* 74: 1038–1044.

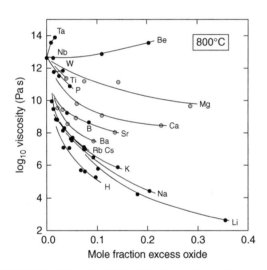

Figure 11 The comparative influences of over 20 components on the viscosity of a leucogranitic melt. These data are derived from new dilatometer-based techniques for the determination of high-viscosity melt properties at low temperatures. Reproduced from Dingwell DB (1997) The brittle–ductile transition in high-level granites: Material constraints. *Journal of Petrology* 38: 1635–1644.

viscosities. The quality of multicomponent fitting is illustrated in **Figure 14(b)**, where the measured versus predicted viscosities are compared over 10 log units of the viscosity. **Figure 15** illustrates the massive influence of water on the glass transition temperature of granitic melts. The next challenge in multicomponent modeling thus clearly involves the incorporation of volatile components in a non-Arrhenian model (Gordano *et al.*, 2007).

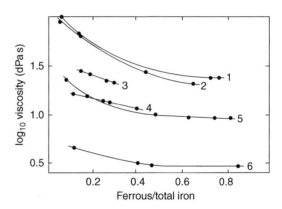

Figure 12 The influence of the redox state of iron on the viscosity of several ternary melts. A clear parametrization of the influence of redox on multicomponent melt viscosities remains outstanding at present. Reproduced from Dingwell DB (1991) Redox viscometry of some iron-bearing silicate liquids. *American Mineralogist* 76: 1560–1562.

15.5.5 Non-Newtonian Rheology in Silicate Melts

The onset of non-Newtonian viscosity has been accessed experimentally using longitudinal compression, simple shear, and elongation (Simmons *et al.*, 1982, 1988; Hessenkemper and Brueckner, 1988; Manns and Brueckner, 1988; Simmons and Simmons, 1989; Webb and Dingwell, 1990a, 1990b). The elongational results for natural nephelinitic, basaltic, andesitic, and rhyolitic melts are presented in **Figure 16** as a function of strain rate. Also included is the estimated relaxational strain rate from the Maxwell relation. The offset between the onset of noticeable strain rate dependence of the viscosity and the relaxational relaxation strain rate is constant and independent of composition. Thus, normalization of the results to the relaxational strain rate generates a single master curve for the onset of non-Newtonian viscosity at high strain rates, independent of composition. Further, the results from compressive longitudinal, simple shear, and these elongational studies also adhere to a single master curve. Thus, the tensorial properties of the strain do not influence the non-Newtonian onset significantly. In this context, it is noteworthy that the volume and shear viscosities of silicate melts are equal. In **Figure 17**, the assumption of an equivalence of the volume and shear viscosities is demonstrated to hold up well in the comparison of ultrasonically derived longitudinal viscosities and concentric cylinder-derived shear viscosities.

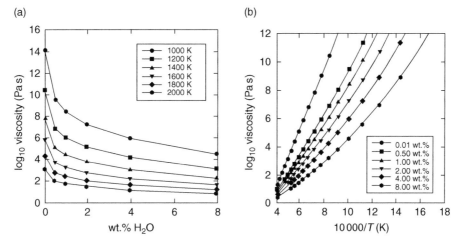

Figure 13 The temperature and water content dependence of the shear viscosity of haplogranitic melts. This parametrization was compiled using spectroscopic, volume, and shear stress relaxation data. 1, $KFeSi_2O_x$ at $1380°C$; 2, $RbFeSi_2O_x$ at $1470°C$; 3, $Ba_{0.5}FeSi_2O_x$ at $1345°C$; 4, $Sr_{0.5}FeSi_2O_x$ at $1350°C$; 5, $NaFeSi_2O_x$ at $1430°C$; 6, $CaFeSi_2O_x$ at $1400°C$.Reproduced from Hess K-U and Dingwell DB (1996) Viscosities of hydrous leucogranitic melts: A non-Arrhenian model. *American Mineralogist* 81: 1297–1300.

15.5.6 Isostructural Viscosity

Inherent in the temperature dependence of the viscosity of melts is that the thermal energy in the system as well as the configurational state of the liquid are both varying with temperature. Thus, the temperature dependence of the so-called 'equilibrium viscosity' is a combination of effects. To obtain the temperature dependence of a single structured liquid, special loading–unloading experiments have been devised and conducted with the result that the activation energy sinks to about half its equilibrium value (**Figure 18**). This implies that about half of the temperature dependence of equilibrium viscosity is due to configurational changes in the structure of the melt and the other half is due to thermal energy increase in the individual bonds.

15.5.7 Failure of Melts

Incursions into the non-Newtonian flow of silicate melts are commonly terminated by melt failure. The non-Newtonian flow of the melt implies that the thermodynamic equilibrium structure of the melt cannot be maintained during viscous flow, generating a runaway process of structure collapse in the melt. The consequence of the collapse is shear thinning followed by melt failure. The stresses for these failures have been experimentally constrained and compared with estimates from other sources in the literature (Romano *et al.*, 1996). The experimental observation of melt failure, together with its obvious significance for explosive volcanic eruptions, has in the meantime spawned a whole new field of investigation: experimental volcanology (Alidibirov and Dingwell, 1996a, 1996b, 2000). To date, the essential character of melt failure and fragmentation has yielded quantifications of the fragmentation threshold (**Figure 19**; Spieler *et al.*, 2004b), fragmentation speed (Spieler *et al.*, 2004a; Scheu *et al.*, 2006), fragmentation efficiency (Spieler *et al.*, 2003, Kueppers *et al.*, 2006a, 2006b), and the influence of permeability on these parameters (Mueller *et al.*, 2005). The transition from flow to failure/fragmentation, which lies at the heart of major explosive eruptions (Dingwell, 1996), may even be a multiple, cyclic phenomenon on extruding high-viscosity lavas (Tuffen *et al.*, 2003; Gonnerman and Manga, 2003; Tuffen and Dingwell, 2004) and thereby a central feature of magma ascent with significant implications for the kinematics of erupted magmas (Marti *et al.*, 1999; Kennedy *et al.* 2005).

15.6 Concluding Statements

Rheology, viscosity, and transport lie at the heart of the significance of melt generation and transport in the evolution and behavior of our planet. Revised understanding of these transport properties in the

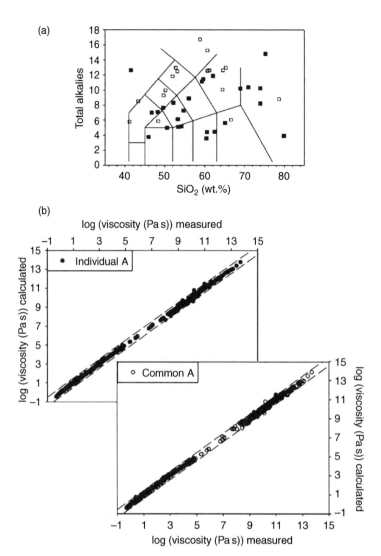

Figure 14 (a) The compositional basis of multicomponent modeling of non-Arrhenian viscosity of silicate melts employed by Giordano and co-workers. (b) The quality of fit of the SM approach of Giordano and co-workers to the multicomponent melt viscosity data set for the compositions illustrated in (a). The model is fitted over 10 log units of viscosity. Reproduced from Giordano D and Dingwell DB (2003) Non-Arrhenian multicomponent melt viscosity: A model. *Earth and Planetary Science Letters* 208: 337–349 and Giordano D, Mangiacapra A, Potuzak M, *et al.* (2006) An expanded non-Arrhenian model for silicate melt viscosity: A treatment for metaluminous, peraluminous and peralkaline liquids. *Chemical Geology* 229: 42–56.

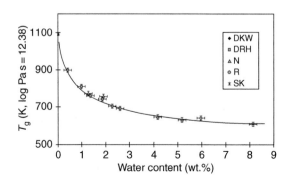

Figure 15 The variation of the glass transition temperature of hydrous granitic melts with water content as obtained by volume, shear stress, and structural relaxation studies. Reproduced from Dingwell DB (1998d) The glass transition in hydrous granitic melts. *Physics of the Earth and Planetary Interiors* 107: 1–8, where original data sources are listed.

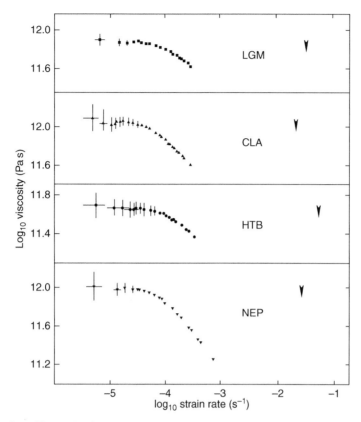

Figure 16 The onset of non-Newtonian rheology in geological melts as determined using fiber elongation viscometry. LGM, Little Glass Mountain rhyolite; CLA, Crater Lake andesite; HTB, Hawaiian tholeiitic basalt; NEP, nephelinite. The deviation from Newtonian behavior occurs ~3 log units of strain rate below the relaxational strain rate (signified by the arrowheads). Reproduced from Webb SL and Dingwell DB (1990a) Non-Newtonian rheology of igneous melts at high stresses and strain rates: Experimental results for rhyolite, andesite, basalt and nephelinite. *Journal of Geophysical Research* 95: 15695–15701.

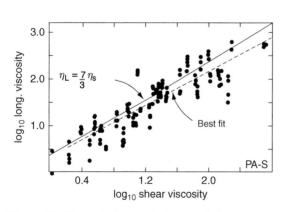

Figure 17 The equivalence of volume and shear viscosities of silicate melts demonstrated by the comparison of ultrasonically derived longitudinal viscosities with concentric cylinder-derived shear viscosities. Reproduced from Dingwell DB and Webb SL (1989) Structural relaxation in silicate melts and non-Newtonian melt rheology in igneous processes. *Physics and Chemistry of Minerals* 16: 508–516.

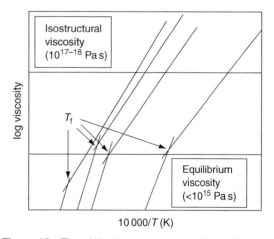

Figure 18 The relative temperature dependence of the equilibrium viscosity vs the isostructural viscosity. The temperature dependence of the equilibrium viscosity can be seen to be composed thereby from two contributions, one structural and one thermal, which are similar in magnitude. Reproduced from Dingwell DB (1995a) Relaxation in silicate melts: Some applications in petrology. In: Stebbins JF, Dingwell DB, and McMillan PF (eds.) *Reviews in Mineralogy, Vol. 32: Structure, Dynamics and Properties of Silicate Melts*, pp. 21–66. Washington, DC: Mineralogical Society of America.

Threshold (=lowest ΔP necessary to initiate fragmentation)

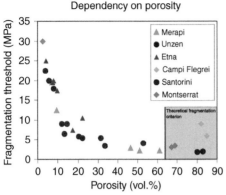

Figure 19 The fragmentation threshold of vesicular magma as a function of porosity. The fragmentation threshold is defined as the overpressure at which the magma will fail by fragmentation during rapid decompression. Reproduced from Spieler O, Kennedy B, Kueppers U, Dingwell DB, Scheu B, and Taddeucci J (2004b) A fragmentation threshold for the initiation and cessation of explosive eruptions. *Earth and Planetary Science Letters* 226: 139–148.

past decade have led to fundamental corrections in our understanding of the liquid state and its behavior in natural processes. Equally, entirely new experimental branches have opened up due to the questions posed by this revised appreciation of diffusion, viscosity, and flow of melts.

References

Adam G and Gibbs JH (1965) On the temperature dependence of cooperative relaxation properties in glass-forming liquids. *Journal of Chemical Physics* 43: 139–146.

Alidibirov M and Dingwell DB (1996a) High temperature fragmentation of magma by rapid decompression. *Nature* 380: 146–149.

Alidibirov M and Dingwell DB (1996b) An experimental facility for the investigation of high temperature magma fragmentation during rapid decompression. *Bulletin of Volcanology* 58: 411–416.

Alidibirov M and Dingwell DB (2000) Three fragmentation mechanisms for highly viscous magma under rapid decompression. *Journal of Volcanology and Geothermal Research* 100: 413–421.

Angell CA, Ngai KL, McKenna GB, McMillan PF, and Martin SW (2000) Relaxation in glassforming liquids and amorphous solids. *Journal of Applied Physics* 88(6): 3113–3157.

Bagdassarov N and Dingwell DB (1992) A rheological investigation of vesicular rhyolite. *Journal of Volcanological and Geothermal Research* 50: 307–322.

Bagdassarov N and Dingwell DB (1993a) Frequency-dependent rheology of vesicular rhyolite. *Journal of Geophysical Research* 98: 6477–6487.

Bagdassarov N and Dingwell DB (1993b) Deformation of foamed rhyolites under internal and external stresses. *Bulletin of Volcanology* 55: 147–154.

Behrens H, Zhang YX, and Xu ZG (2004) H_2O diffusion in dacitic and andesitic melts. *Geochimica Cosmochim Acta* 68: 5139–5150.

Bowen NL (1934) Viscosity data for silicate melts transactions. *American Geophysical Union* 15: 249–255.

Chakraborty S (1995) Diffusion in silicate melts. In: Stebbins JF, Dingwell DB, and McMillan PF (eds.) *Reviews in Mineralogy, Vol. 32: Structure, Dynamics and Properties of Silicate Melts*, pp. 411–503. Washington, DC: Mineralogical Society of America.

Costa A and Macedonio G (2003) Viscous heating effects influence with temperature-dependent viscosity: Triggering of secondary flows. *Journal of Fluid Mechanics* 540: 21–38.

Dingwell DB (1989) Shear viscosities of ferrosilicate liquids. *American Mineralogist* 74: 1038–1044.

Dingwell DB (1990) Effects of structural relaxation on cationic tracer diffusion in silicate melts. *Chemical Geology* 82: 209–216.

Dingwell DB (1991) Redox viscometry of some iron-bearing silicate liquids. *American Mineralogist* 76: 1560–1562.

Dingwell DB (1995a) Relaxation in silicate melts: Some applications in petrology. In: Stebbins JF, Dingwell DB, and McMillan PF (eds.) *Reviews in Mineralogy, Vol. 32: Structure, Dynamics and Properties of Silicate Melts*, pp. 21–66. Washington, DC: Mineralogical Society of America.

Dingwell DB (1995b) Viscosity and anelasticity of melts and glasses. In: Ahrens T. (ed.) *Mineral Physics and Crystallography: A Handbook of Physical Constants (AGU Reference Shelf 2)*, pp. 209–217. Washington, DC: American Geophysical Union.

Dingwell DB (1996) Volcanic dilemma: Flow or blow? *Science* 273: 1054–1055.

Dingwell DB (1997) The brittle–ductile transition in high-level granites: Material constraints. *Journal of Petrology* 38: 1635–1644.

Dingwell DB (1998a) Magma degassing and fragmentation. Recent experimental advances. In: Freundt A and Rosi M (eds.) *Explosive Volcanism: A Physical Description*, pp. 318. Berlin: Springer-Verlag.

Dingwell DB (1998b) A physical description of magma relevant to explosive silicic volcanism. In: Gilbert JS and Sparks RSJ (eds.) *Physics of Explosive Volcanic Eruptions*, Geological Society of London Special publication 145, pp. 9–26. Bath: The Geological Society.

Dingwell DB (1998c) Melt viscosity and diffusion under elevated pressures. *Reviews in Mineralogy* 37: 397–424.

Dingwell DB (1998d) The glass transition in hydrous granitic melts. *Physics of the Earth and Planetary Interiors* 107: 1–8.

Dingwell DB, Bagdassarov N, Bussod G, and Webb SL (1993) Magma rheology. In: Luth RW (ed.) Mineralogical Association of Canada Short Course on Experiments at High Pressure and Applications to the Earth's Mantle, pp. 131–196. Québec: Mineralogical Association of Canada.

Dingwell DB, Courtial P, Giordano D, and Nichols A (2004) Viscosity of peridotite liquid. *Earth and Planetary Science Letters* 226: 127–138.

Dingwell DB and Webb SL (1989) Structural relaxation in silicate melts and non-Newtonian melt rheology in igneous processes. *Physics and Chemistry of Minerals* 16: 508–516.

Dingwell DB and Webb SL (1990) Relaxation in silicate melts. *European Journal of Mineralogy* 2: 427–449.

Giordano D and Dingwell DB (2003) Non-Arrhenian multicomponent melt viscosity: A model. *Earth and Planetary Science Letters* 208: 337–349.

Giordano D, Mangiacapra A, Potuzak M, et al. (2006) An expanded non-Arrhenian model for silicate melt viscosity: A treatment for metaluminous, peraluminous and peralkaline liquids. *Chemical Geology* 229: 42–56.

Giordano D, Nichols ARL, and Dingwell DB (2005) Glass transition temperatures of natural hydrous melts: A relationship with shear viscosity and implications for the welding process. *Journal of Volcanological and Geothermal Research* 142: 105–118.

Gordano D, Russell JK, and Dingwell DB (2007) Viscosity of magmatic liquids: An empirical model. *EOS* .

Gonnermann H and Manga M (2003) Explosive volcanism may not be an inevitable consequence of magma fragmentation. *Nature* 426: 432–435.

Gottsmann J and Dingwell DB (2001a) Cooling dynamics of phonolitic rheomorphic fall-out deposits on Tenerife, Canary Islands. *Journal of Volcanological and Geothermal Research* 105: 323–342.

Gottsmann J and Dingwell DB (2001b) The cooling of frontal flow ramps: A calorimetric study of the Rocche Rosse rhyolite flow, Lipari, Aeolian Islands, Italy. *Terra Nova* 13: 157–164.

Gottsmann J and Dingwell DB (2002) The thermal history of a rheomorphic air-fall deposit: The 8 ka pantellerite flow of Mayor Island, New Zealand. *Bulletin of Volcanology* 64: 410–422.

Gottsmann J, Giordano D, and Dingwell DB (2002) Predicting shear viscosity at the glass transition during volcanic processes: A calorimetric calibration. *Earth and Planetary Science Letters* 198: 417–427.

Gottsmann J, Harris AJL, and Dingwell DB (2004) Thermal history of Hawaiian Páhoehoe lava crusts at the glass transition: Implications for flow rheology and flow emplacement. *Earth and Planetary Science Letters* 228(3–4): 343–353.

Gurenko AA, Vekster, IV, Meixner A, Thomas R, Dorfman AM, and Dingwell DB (2005) Matrix effects and partitioning of boron isotopes between immiscible Si-rich and B-rich liquids in the Si–Al–B–Na–Ce–D system: a SIMS study of glasses quenched from centrifuge experiments. *Chemistry Geology* 222: 268–280.

Hess K-U and Dingwell DB (1996) Viscosities of hydrous leucogranitic melts: A non-Arrhenian model. *American Mineralogist* 81: 1297–1300.

Hessenkemper H and Brückner R (1988) Load-dependent flow behavior of silicate melts. *Glastechnische Berichte* 61: 312–320.

Hofmann A (1980) Diffusion in natural silicate melts: A critical review. In: Hargreaves RB (ed.) *Physics of Magmatic Systems*, pp. 385–417. Princeton, NJ: Princeton University Press.

Jambon A (1982) Tracer diffusion in granitic melts – Experimental results for Na, K, RB, Cs, Ca, Sr, Ba, Ce, Eu to $1300°C$ and a model of calculation. *Journal of Geophysical Research* 87: 797–810.

Kennedy B, Spieler O, Scheu B, Kueppers U, Taddeucci J, and Dingwell DB (2005) Conduit implosion during vulcanian eruptions. *Geology* 33: 581–584.

Kozu S and Kani K (1935) Viscosity determinations in the ternary system diopside–albite–anorthite at high temperatures. *Proceedings of the Imperial Academy of Japan* 11: 383–385.

Kueppers U, Schen B, Spieler O, and Dingwell DB (2006a) Fragmentation efficiency of explosive volcanic eruptions: A study of experimentally generated pyroclasts. *JVGR* 153: 125–135.

Kueppers U, Derugini D, and Dingwell DB (2006b) 'Explosive energy' during volcanic eruptions from fractal analysis of products. *EPSL* 248(3–4): 800–807.

Kushiro I (1976) Changes in viscosity and structure of melt of $NaAlSi_2O_6$ composition at high pressures. *Journal of Geophyical Research* 81: 6347–6350.

Kushiro I, Yoder HS, and Mysen BO (1976) Viscosities of basalt and andesite melts at high pressures. *Journal of Geophyical Research* 81: 6351–6356.

Liu SB, Pine A, Brandriss M, and Stebbins JF (1987) Relaxation mechanisms and effects of motion in albite ($NaAlSi_3O_8$) liquid and glass: A high temperature NMR study. *Physics and Chemistry of Minerals* 15: 155–162.

Liu SB, Stebbins JF, Schneider E, and Pines A (1988) Diffusive motion in alkali silicate melts: An NMR study at high temperature. *Geochimica et Cosmochimica Acta* 52(2): 527–538.

Manns P and Brückner R (1988) Non-Newtonian flow behavior of a soda-lime silicate glass at high rates of deformation. *Glastechnische Berichte* 61: 46–56.

Marti J, Soriano C, and Dingwell B (1999) Tube pumice: Strain marker of the ductile–brittle transition in explosive volcanism. *Nature* 402: 650–653.

Moretti R (2005) Polymerisation, basicity, oxidation state and their role in ionic modelling of silicate melts. *Annales Geophysicae* 48: 583–608.

Moynihan CT, Easteal AJ, DeBolt MA, and Tucker J (1976a) Dependence of fictive temperature of glass on cooling rate. *Journal of American Ceramic Society* 59: 12–16.

Moynihan CT, Easteal AJ, Tran DC, Wilder JA, and Donovan EP (1976b) Heat capacity and structural relaxation of mixed alkali glasses. *Journal of American Ceramic Society* 59: 137–140.

Moynihan CT, Easteal AJ, and Wilder J (1974) Dependence of the glass transition temperature on heating and cooling rate. *Journal of Physical Chemistry* 78: 2673–2677.

Mueller S, Spieler O, Scheu B, and Dingwell DB (2005) Permeability and degassing of dome lavas undergoing rapid decompression: An experimental determination. *Bulletin of Volcanology* 67: 526–538.

Mungall J (2002) Empirical models relating viscosity and tracer diffusion in magmatic silicate melts. *Geochimica et Cosmochimica Acta* 66: 125–143.

Mungall J, Dingwell DB, and Chaussidon M (1999) Chemical diffusivities of 18 trace elements in granitoid melts. *Geochimica et Cosmochimica Acta* 63: 2599–2610.

Mysen BO (1988) *Structure and Properties of Silicate Melts*. Amsterdam: Elsevier, 354 pp.

Mysen BO and Richet P (2005) *Structure and Properties of Silicate Melts,* 2nd edn., 560 pp. Amsterdam: Elsevier.

Narayanaswamy OS (1971) A model of structural relaxation in glass. *Journal of American Ceramic Society* 54: 491–498.

Narayanaswamy OS (1988) Thermorheological simplicity in the glass transition. *Journal of American Ceramic Society* 71: 900–904.

Nowak M and Behrens H (1995) The speciation of water in haplogranitic glasses and melts determined by *in situ* near infrared spectroscopy. *Geochimica et Cosmochimica Acta* 59: 3445–3450.

Papale P (1999) Strain-induced magma fragmentation in explosive eruptions. *Nature* 397: 425–428.

Poe B, McMillan PF, Rubie DC, Chakraborty S, Yarger J, and Diefenbacher J (1997) Silicon and oxygen selfdiffusivities in silicate liquids measured to 15 gigapascals and 2800 kelvin. *Science* 276: 1245–1248.

Poe B and Rubie D (1998) Transport properties of silicate melts at high pressure. In: Aoki H, Syono Y, and Hemley RJ (eds.) *Physics Meets Mineralogy*, pp. 340–353. Cambridge: Cambridge University Press.

Richet P (1984) Viscosity and configurational entropy of silicate melts. *Geochimica et Cosmochimica Acta* 48: 471–484.

Richet P and Bottinga Y (1995) Rheology and configurational entropy of silicate melts. In: Stebbins JF, Dingwell DB, and McMillan PF (eds.) (1995) Reviews in Mineralogy, Vol. 32:

Structure, Dynamics and Properties of Silicate Melts, pp. 21–66. Washington, DC: Mineralogical Society of America.

Richet P, Robie RA, and Hemmingway BS (1986) Low temperature heat capacity of diopside glass ($CaMgSi_2O_6$) a calorimetric test of the configurational entropy theory applied to the viscosity of silicate liquids. *Geochimica et Cosmochimica Acta* 50: 1521–1533.

Romano C, Dingwell DB, and Sterner SM (1994) Kinetics of quenching of hydrous feldspathic melts: Quantification using synthetic fluid inclusions. *American Mineralogist* 79: 1125–1134.

Romano C, Mungall J, Sharp T, and Dingwell DB (1996) Tensile strengths of hydrous vesicular glasses: An experimental study. *American Mineralogist* 81: 1148–1154.

Rosi M, Landi P, Polacci M, DiMuro A, and Zandomeneghi D (2004) Role of conduit shear on ascent of the crystal-rich magma feeding the 800-year-BP Plinian eruption of Quilotoa Volcano (Ecuador). *Bulletin of Volcanology* 66: 307–321.

Scarfe CM, Mysen BO, and Virgo DL (1987) Pressure dependence of the viscosity of silicate melts. In: Mysen BO (ed.) *Magmatic processes: Physiochemical principles*, Geological Society Special publication 1, pp. 59–67. University Park, PA: Geochemical Society.

Scheu B, Spieler O, and Dingwell DB (2005) Dynamics of explosive volcanism at Unzen: An experimental contribution. *Bulletin of Volcanology* 69: 175–187.

Shen A and Keppler H (1995) Infrared spectroscopy of hydrous silicate melts to 1000 degrees C and 10 kbar: Direct observation of H_2O speciation in a diamond-anvil cell. *American. Mineralogist* 80: 1335–1338.

Shimizu N and Kushiro I (1984) Diffusivity of oxygen in jadeite and diopside melts at high pressures. . *Geochimica et Cosmochimica Acta* 48: 1295–1303.

Simmons J, Mohr R, and Montrose C (1982) Non-Newtonian viscous flow in glass. *Journal of Applied Physics* 53: 4075–4080.

Simmons J, Ochoa R, Simmons K, and Mills J (1988) Non-Newtonian viscous flow in soda-lime-silica glass at forming and annealing conditions. *Journal of the American Ceramic Society* 76: 904.

Simmons J and Simmons C (1989) Nonlinear viscous flow in glass forming. *Bulletin of the Ceramic Society of America* 11: 1949–1955.

Spieler O, Alidibirov M, and Dingwell DB (2003) Grain-size characteristics of experimental pyroclasts of 1980 Mount St. Helens cryptodome dacite: Effects of pressure drop and temperature. *Bulletin of Volcanology* 65: 90–104.

Spieler O, Dingwell DB, and Alidibirov A (2004a) Magma fragmentation speed: An experimental determination. *Journal of Geothermal and Volcanological Research* 129: 109–123.

Spieler O, Kennedy B, Kueppers U, Dingwell DB, Scheu B, and Taddeucci J (2004b) A fragmentation threshold for the initiation and cessation of explosive eruptions. *Earth and Planetary Science Letters* 226: 139–148.

Stebbins JF (1995) Dynamics and structure of silicate and oxide melts: Nuclear magnetic resonance studies. *Reviews in Mineralogy* 32: 191–246.

Stebbins JF, Dingwell DB, and McMillan PF (eds.) (1995) *Reviews in Mineralogy, Vol. 32: Structure, Dynamics and Properties of Silicate Melts*. Washington, DC: Mineralogical Society of America, 616pp.

Stevenson R, Dingwell DB, Webb SL, and Bagdassarov NL (1995) The equivalence of enthalpy and shear relaxation in rhyolitic obsidians and quantification of the liquid-glass transition in volcanic processes. *Journal of Volcanology and Geothermal Research* 68: 297–306.

Tool AQ (1946) Relation between inelastic deformability and thermal expansion of glass in its annealing range. *Journal of American Ceramic Society* 29: 240–253.

Tool AQ and Eichlin CG (1931) Variations caused in the heating curves of glass by heat treatment. *Journal of American Ceramic Society* 14: 276–308.

Tuffen H and Dingwell DB (2004) Fault textures in volcanic conduits: Evidence for seismic trigger mechanisms during silicic eruptions. *Bulletin of Volcanology* 67: 370–387.

Tuffen H, Dingwell DB, and Pinkerton H (2003) Repetitive fracture and healing in silicic magmas: A link between flow banding and fossil earthquakes? *Geology* 31: 1089–1092.

Vedeneeva EA, Melnik OE, Barmin AA, and Sparks RSJ (2005) Viscous dissipation in explosive volcanic flows. *Geophysical Research Letters.* 32(5): L05303.

Vekster IV, Pettibon C, Senner GA, Dorfman AM, and Dingwell DB (1998) Trace element partitioning immiscible silicate–carbonate liquid systems: An initial experimental study using a centrifuge autoclave. *Journal of Petrology* 39: 2015–2031.

Vekster IV, Dorfman AM, Dingwell DB, and Zotov N (2002) Element partitioning between immiscible borosilicate liquid: A high temperature centrifuge study. *Geochimica et Cosmochimica Acta* 66: 2603–2614.

Vekster IV, Dorfman AM, Kamenetskey M, Dulski D, and Dingwell DB (2005) Partitioning of lanthanides and Y between immiscible silicate and fluoride melts. Fluorite and cryolite and the origin of the lanthanide tetrad effect in igneous rocks. *GCA* 69: 2847–2860.

Vekster IV, Dorfman AM, Danyushevsky LV, Jakobsen JK, and Dingwell DB (2006) Immiscible silicate liquid partion coefficients: Implication for crystal-melt element partitioning and basalt pefrogenesis. *Contributions to Mineralogy and Petrology* 152: 685–702.

Webb SL and Dingwell DB (1990a) Non-Newtonian rheology of igneous melts at high stresses and strain rates: Experimental results for rhyolite, andesite, basalt and nephelinite. *Journal of Geophysical Research* 95: 15695–15701.

Webb SL and Dingwell DB (1990b) The onset of non-Newtonian rheology in silicate melts. *Physics and Chemistry of Minerals* 17: 125–132.

Wilding M, Dingwell DB, Batiza R, and Wilson L (2000) The cooling rates of hyaloclastites: Applications of relaxation geospeedometry to undersea volcanics. *Bulletin of Volcanology* 61: 527–536.

Wilding M, Webb SL, and Dingwell DB (1995) Evaluation of a relaxation geothermometer for volcanic glasses. *Chemical Geology* 125: 137–148.

Wilding M, Webb SL, and Dingwell DB (1996a) Tektite cooling rates: Calorimetric geospeedometry applied to a natural glass. *Geochimica et Cosmochimica Acta* 60: 1099–1103.

Wilding M, Webb SL, Dingwell DB, Ablay G, and Marti J (1996b) The variation of cooling rates within volcanic facies from Tenerife, Canary Islands. *Contributions to Mineralogy and Petrology* 125: 151–160.

Wolf GH and McMillan PF (1995) Pressure effects on silicate melt properties. *Reviews in Mineralogy* 32: 505–561.

Wood BJ, Walter MJ, and Wade J (2006) Accretion of the Earth and segregation of its core. *Nature* 441: 825–833.

Zarzycki J (1991) *Glasses and the Vitreous State*, 505 pp. New York: Cambridge University Press.

Zhang Y, Stolper EM, and Ihinger PD (1995) Kinetics of the reaction $H_2O + O = 2OH$ in rhyolitic and albitic glasses: Preliminary results. *American Mineralogist* 80: 593–612.

Zhang Y, Jenkins J, and Xu Z (1997) Kinetics of the reaction $H_2O + O \leftrightarrow 2OH$ in rhyolitic glasses upon cooling: Geospeedometry and comparison with glass transition. *Geochimica et Cosmochimica Acta* 61: 2167.

16 Seismic Anisotropy of the Deep Earth from a Mineral and Rock Physics Perspective

D. Mainprice, Université Montpellier II, Montpellier, France

16.1	Introduction	437
16.2	Mineral Physics	443
16.2.1	Elasticity and Hooke's Law	443
16.2.2	Plane Waves and Christoffel's Equation	448
16.2.3	Measurement of Elastic Constants	453
16.2.4	Effective Elastic Constants for Crystalline Aggregates	456
16.2.5	Seismic Properties of Polycrystalline Aggregates at High Pressure and Temperature	458
16.2.6	Anisotropy Minerals in the Earth's Mantle and Core	461
16.2.6.1	Upper mantle	461
16.2.6.2	Transition zone	463
16.2.6.3	Lower mantle	465
16.2.6.4	Inner core	468
16.3	Rock Physics	472
16.3.1	Introduction	472
16.3.2	Olivine the Most Studied Mineral: State of the Art – Temperature, Pressure, Water, Melt, ETC	472
16.3.3	Seismic Anisotropy and Melt	477
16.4	Conclusions	482
References		482

16.1 Introduction

Seismic anisotropy is commonly defined as the direction-dependent nature of the propagation velocities of seismic waves. However, this definition does not cover all the seismic manifestations of seismic anisotropy. In addition to direction-dependent velocity, there is direction-dependent polarization of P- and S-waves, and anisotropy can contribute to the splitting of normal modes. Seismic anisotropy is a characteristic feature of the Earth, with anisotropy being present near the surface due to aligned cracks (e.g., Crampin, 1984), in the lower crust, upper and lower mantle due to mineral-preferred orientation (e.g., Karato, 1998; Mainprice *et al.*, 2000). At the bottom of the lower mantle (D″layer, e.g., Kendall and Silver (1998)) and in the solid inner core (e.g., Ishii *et al.*, 2002a) the causes of anisotropy are still controversial (**Figure 1**). In some cases multiple physical factors could be contributing to the measured anisotropy, for example, mineral-preferred orientation and alignment of melt inclusions at mid-ocean ridge systems (e.g., Mainprice, 1997). In the upper mantle, the

pioneering work of Hess (1964) and Raitt *et al.* (1969) from Pn velocity measurements in the shallow mantle of the ocean basins showed azimuthal anisotropy. Long-period surface waves studies (e.g., Nataf *et al.*, 1984; Montagner and Tanimoto, 1990) have since confirmed that azimuthal and SH/SV polarization anisotropy are global phenomena in the Earth's upper mantle, particularly in the top 200 km of the upper mantle. Anisotropic global tomography, based on surface and body wave data, has shown that anisotropy is very strong in the subcontinental mantle and present generally in the upper mantle, but significantly weaker at greater depths (e.g., Beghein *et al.*, 2006; Panning and Romanowicz, 2006). The long wavelengths used in long-period surface wave studies means that such methods are insensitive to heterogeneity less than the wavelength of about 1000 km. In an effort to address the problem of regional variations of anisotropy, the splitting of SKS teleseismic shear waves that propagate vertically have been extensively used. At continental stations SKS studies show that the azimuth of the fast polarization direction is parallel to the trend of mountain belts (Kind *et al.*, 1985; Silver and Chan 1988,

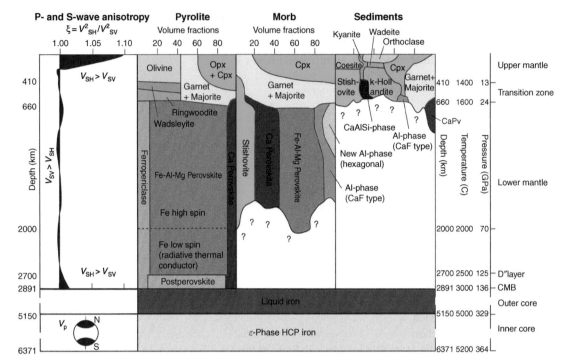

Figure 1 The simplified petrology and seismic anisotropy of the Earth's mantle and core. The radially (transverse isotropic) anisotropic model of S-wave anisotropy of the mantle is taken from Panning and Romanowicz (2006). The icon at the inner core represents the fast P-wave velocities parallel to the rotation axis of the Earth. The petrology of mantle is taken from Ono and Oganov (2005) for pyrolite, Perrillat *et al.* (2006) for the transformed MORB and based on Irifune *et al.* (1994) as modified by Poli and Schmidt (2002) for the transformed argillaceous sediments.

1991; Vinnik *et al.*, 1989; Silver, 1996; Fouch and Rondenay, 2006). From the earliest observations it was clear that the anisotropy in the upper mantle was caused by the preferred orientation of olivine crystals induced by plastic deformation related to mantle flow processes at the geodynamic or plate tectonic scale.

The major cause of seismic anisotropy in the upper mantle is the crystal-preferred orientation (CPO) caused by plastic deformation. Knowledge of the CPO and its evolution requires well-characterized naturally deformed samples, experimentally deformed samples and numerical simulation for more complex deformation histories of geodynamic interest. The CPO not only causes seismic anisotropy, but also records some aspects of the deformation history. Samples of the Earth's mantle are readily found on the surface in the form of ultramafic massifs, xenoliths in basaltic or kimberlitic volcanics and as inclusions in diamonds. However, samples from depths greater than 220 km are extremely rare. Upper-mantle samples large enough for the measurement of CPO have been recovered from kimberlitic volcanics in South Africa to a depth of about 220 km established by geobarometry (e.g., Boyd, 1973).

Kimberlite mantle xenoliths of deeper origin (>300 km) with evidence for equilibrated majorite garnet which is now preserved as pyrope garnet with exsolved pyroxene have been reported (Haggerty and Sautter, 1990; Sautter *et al.*, 1991). The Alpe Arami peridotite garnet lherzolite is proposed to have been exhumed from a minimum depth of 250 km based on clinoenstatite exsolution lamellae present in diopside grains (Bozhilov *et al.*, 1999). Samples of even deeper origin are preserved as inclusions in diamonds. Although most diamonds crystallize at depths of 150–200 km, some diamonds contain inclusions of majorite (Moore and Gurney, 1985), enstatite and ferripericlase (Scott-Smith *et al.*, 1984), and $CaSiO_3 + (Fe, Mg)SiO_3 + SiO_2$ (Harte and Harris, 1993). The mineral associations imply transition zone (410–660 km) and lower-mantle origins for these diamond inclusions (Kesson and Fitz, 1991). Although these samples help to constrain mantle petrology, they are too small to provide information about CPO. Hence, knowledge of CPO in the transition zone, lower mantle, and inner core will be derived from deformation experiments at high pressure and temperatue (e.g., olivine – Couvy *et al.*, 2004; ringwoodite – Karato *et al.*, 1998; perovskite – Cordier *et al.*, 2004b; MgGeO$_3$ postperovskite – Merkel *et al.*, 2006a; ε-phase iron – Merkel *et al.*, 2005).

It has been accepted since the PREM seismic model (Dziewoński and Anderson, 1981) that the top 200 km of the Earth's mantle is anisotropic on a global scale (**Figure 1**). However, there are exceptions, for example, under the Baltic shield the anisotropy increases below 200 km (Pedersen *et al.*, 2006). The seismic discontinuity at about 200 km was first reported by the Danish seismologist Inge Lehmann (1959, 1961), that now bears her name. However, the discontinuity is not always present at the same depth. Anderson (1979) interpreted the discontinuity as the petrological change of garnet lherzolite to eclogite. More recently, interpretations have favored an anisotropy discontinuity, although even this is controversial (see Vinnik *et al.* (2005)), proposed interpretations include: a local anisotropic decoupling shear zone marking the base of the lithosphere (Leven *et al.*, 1981), a transition from an anisotropic mantle deforming by dislocation creep to isotropic mantle undergoing diffusion creep (Karato, 1992), simply the base of an anisotropic layer beneath continents (e.g., Gaherty and Jordan, 1995), or the transition from [100] to [001] direction slip in olivine (Mainprice *et al.*, 2005). Global tomography studies show that the base of the anisotropic subcontinental mantle may vary in depth from 100 to 450 km (e.g., Polet and Anderson, 1995), but most global studies favor an anisotropy discontinuity for S-waves at around 200–250 km, which is stronger and deeper (300 km) beneath continents (e.g., Deuss and Woodhouse, 2002; Ritsema *et al.*, 2004; Panning and Romanowicz, 2006) and weaker and shallower (200 km) beneath the oceans. There is also evidence for weak seismic discontinuities at 260 and 310 km which has been reported in subduction zones by Deuss and Woodhouse (2002).

A major seismic discontinuity at 410 km is due to the transformation of olivine to wadsleyite (e.g., Helffrich and Wood, 1996) with a shear wave impedance contrast of 6.7% (e.g., Shearer, 1996). The 410 km discontinuity has topography within 5 km of the global average. The olivine to wadsleyite transformation will result in the lowering of anisotropy with depth. Global tomography models (e.g., Montagner, 1994a, 1994b; Montagner and Kennett, 1996; Beghein *et al.*, 2006; Panning and Romanowicz, 2006) indicate that the strength of

anisotropy is less in the transition zone (410–660 km) than in the upper mantle (**Figure 1**). A global study of the anisotropy of transion zone by Trampert and van Heijst (2002) has detected a weak anisotropy shear wave of about 1–2%. The surface wave overtone technique used by Trampert and van Heijst (2002) cannot localize the anisotropy within the 410–660 km depth range; however the only mineral with a strong anisotropy and significant volume fraction in the transition zone is wadsleyite occurring between 410 and 520 km. Between 520 and 660 km there is an increase the very weakly anisotropic phases, such as garnet, majorite, and ringwoodite in the transition zone (**Figure 1**). Tommasi *et al.* (2004) have shown that the CPO predicted by a plastic flow model using the experimentally observed slip systems of wadsleyite can reproduce the weak anisotropy observed by Trampert and van Heijst (2002). A weaker discontinuity at 520 km, with a shear wave impedance contrast of 2.9%, has been reported by Shearer and co-workers (e.g., Shearer, 1996; Flanagan and Shearer, 1998). The discontinuity at 520 km depth has been attributed to the wadsleyite to ringwoodite transformation by Shearer (1996). Deuss and Woodhouse (2001) have reported the 'splitting' of the 520 km discontinuity into two discontinuities at 500 and 560 km, they interpret the variations of depth of 520 km, and presence of two discontinuities at 500 and 560 km in certain regions can only be explained by variations in temperature and composition (e.g Mg/Mg + Fe ratio), which effect the phase transition Clapeyron. Regional seismic studies by Vinnik and Montagner (1996) and Vinnik *et al.* (1997) show evidence for a weakly anisotropic (1.5%) layer for S-waves at the bottom 40 km of the transition zone (620–660 km). The some global tomography models (e.g., Montagner and Kennett, 1996; Montagner, 1998) also show significant transverse isotropic anisotropy in the transition zone with $V_{SH} > V_{SV}$ and $V_{PH} > V_{PV}$. Given the low intrinsic anisotropy of most of the minerals in the lower part of the transition zone, Karato (1998) suggested that this anisotropy is due to petrological layering caused by garnet and ringwoodite rich layers of transformed subducted oceanic crustal material. Such transversely isotropic medium with a vertical symmetry axis would not cause any splitting for vertically propagating S-waves and would not produce the azimuthal anisotropy observed by Trampert and van Heijst (2002), but would produce the difference between horizontal and vertical velocities seen by global tomography. A global study supports this suggestion, as high-velocity slabs of former oceanic lithosphere are conspicuous structures just above the 660 km discontinuity in the circum-Pacific subuction zones (Ritsema *et al.*, 2004). A regional study by Wookey *et al.* (2002) also finds significant shear wave splitting associated with horizontally travelling S-waves, which is compatiable with a layered structure in the vicinity of the 660 km discontinuity. However, recent anisotropic global tomography models do not show significant anisotropy in this depth range (Beghein *et al.*, 2006; Panning and Romanowicz, 2006).

The strongest seismic discontinuity at 660 km is due to the dissociation of ringwoodite to perovskite and ferropericlase (**Figure 1**) with a shear wave impedance contrast of 9.9% (e.g., Shearer, 1996). The 660 km discontinuity has an important topography with local depressions of up to 60 km from the global average in subduction zones (e.g., Flanagan and Shearer, 1998). From 660 to 1000 km a weak anisotropy is observed in the top of the lower mantle with $V_{SH} < V_{SV}$ and $V_{PH} < V_{PV}$ (e.g., Montagner and Kennett, 1996; Montagner, 1998). Karato (1998) attributed the anisotropy to the CPO of perovskite and possibly ferropericlase caused by plastic deformation in the convective boundary layer at the top of the lower mantle. In this depth range, Kawakatsu and Niu (1994) have identified a flat seismic discontinuity at 920 km with S to P converted waves with an S-wave velocity change of 2.4% in Tonga, Japan Sea, and Flores Sea subuction zones. They suggested that this feature is thermodynamically controlled by some sort of phase transformation or alternatively we may suggest it marks the bottom of the anisotropic boundary layer proposed by Montagner (1998) and Karato (1998). Reflectors in lower mantle have been reported by Deuss and Woodhouse (2001) at 800 km depth under North America and at 1050 and 1150 km beneath Indonesia; they only considered the 800 km reflector to be a robust result. Karki *et al.* (1997b) have suggested that the transformation of the highly anisotropic SiO_2 polymorphs stishovite to $CaCl_2$ structure at 50 ± 3 GPa at room temperature may be the possible explanation of reflectivity in the top of the lower mantle. However, according to Kingma *et al.* (1995) the transformation would take place at 60 GPa at lower mantle temperatures in the range 2000–2500 K, corresponding to depth of 1200–1500 km, that is, several hundred kilometers below

the 920 km discontinuity. It is highly speculative to suggest that free silica is responsible for the 920 km discontinuity as a global feature as proposed by Kawakatsu and Niu (1994). Ringwood (1991) suggested that 10% stishovite would be present from 350 to 660 km in subducted oceanic crust and this would increase to about 16% at 730 km. Hence, in the subduction zones studied by Kawakatsu and Niu (1994), it is quite possible that significant stishovite could be present to 1200 km and may be a contributing factor to the seismic anisotropy of the top of the lower mantle. From 1000 to 2700 km the lower mantle is isotropic for body waves or free oscillations (e.g., Montagner and Kennett, 1996; Beghein *et al.*, 2006; Panning and Romanowicz, 2006). Karato *et al.* (1995) have suggested by comparison with deformation experiments of fine-grained analogue oxide perovskite that the seismically isotropic lower mantle is undergoing deformation by superplasticity or diffusive creep which does not produce a crystal-preferred orientation. In the bottom of the lower mantle the D″ layer (100–300 km thick) appears to be transversely isotropic with a vertical symmetry axis characterized by $V_{SH} > V_{SV}$ (**Figure 1**) (e.g., Kendall and Silver, 1996, 1998), which may be caused by CPO of the constituent minerals, shape-preferred orientation of horizontally aligned inclusions, possibly melt (e.g., Williams and Garnero, 1996; Berryman, 2000) or core material. It has been suggested that the melt fraction of D″ may be as high as 30% (Lay *et al.*, 2004). Seismology has shown that D″ is extremely heterogeneous as shown by globally high fluctuations of shear (2–3%) and compressional (1%) wave velocities (e.g., Mégnin and Romanowcz, 2000; Ritsema and van Heijst, 2001; Lay *et al.*, 2004), a variation of the thickness of D″ layer between 60 and 300 km (e.g., Sidorin *et al.*, 1999a), P- and S-wave velocities variations are some times correlated and sometimes anticorrelated (thermal, chemical, and melting effects?) (e.g., Lay *et al.*, 2004), ultralow velocity zones (ULVZs) at the base of D″ with V_p 10% slower and V_s 30% slower than surrounding material (e.g., Garnero *et al.*, 1998), regions with horizonal (e.g., Kendall and Silver, 1998; Kendall, 2000) or inclined anisotropy (e.g., McNamara *et al.*, 2003; Garnero *et al.*, 2004; Maupin *et al.*, 2005; Wookey *et al.*, 2005a) in the range 0.5–1.5% and isotropic regions, localized patches of shear velocity discontinuity, even predicted the possibility of a globally extensive phase transformation (Nataf and Houard, 1993) and its Claperon slope (Sidorin *et al.*, 1999b). Until recently, the candidate phase for this transition was SiO_2. However, the mineralogical picture of the

D″ layer has been completely changed with the discovery of postperovskite by Murakami *et al.* (2004), which is produced by the transformation of Mg-perovskite in the laboratory at pressures greater than 125 GPa at high temperature. Seismic modeling of the D″ layer using the new phase diagram and elastic properties of perovskite and postperovskite can explain many features mentioned above near the core–mantle boundary (Wookey *et al.*, 2005b).

The inner solid core was the last elastic shell of the Earth to be identified by the Danish seismologist Inge Lehmann in a paper published in 1936 with the short title P′. She identified P-waves that traveled through the core region (PKP, where K stands for core) at epicentral distances of 105–142° in contradiction to the expected travel times for a single core model. She proposed a two-shell model for the core with a uniform velocity of about 10 km s^{-1} with a small velocity discontinuity between each shell and a inner shell radius of 1400 km, close to the actually accepted value of 1221.5 km from PREM (Dziewoński and Anderson, 1981). The liquid nature of the outer core was first proposed by Jeffreys (1926) based on shear wave arrival times and the solid nature of inner core was first proposed by Birch (1940) based on the compressibility of iron at high pressure. Given the great depth (5149.5 km) and the number of layers a seismic wave has to traverse to reach the inner core and return to the surface it is not surprising the first report of anisotropy of the inner core was inferred fifty years after the discovery of the inner core. Poupinet *et al.* (1983) were the first to observe that PKIKP (where K now stands for the outer core and I is for inner core) P-waves travel about 2 s faster parallel to the Earth's rotation axis than waves traveling the equatorial plane. They interpreted their observations in terms of a possible heterogeneity of the inner core. Shortly afterwards, a PKIKP travel time study by Morelli *et al.* (1986) and normal modes (free oscillations) by Woodhouse *et al.* (1986) reported new observations and interpreted the results in terms of anisotropy. However, the interpretation of PKIKP body wave travel times in terms of anisotropy remained controversial, with an alternative interpretation being that the inner core had a nonspherical structure (e.g., Widmer *et al.*, 1992). Finally, the observation of large differential travel times for PKIKP for paths from the South Sandwich Islands to Alaska by Creager (1992), Song and Helmberger (1993), and the interpretation of higher-quality free

oscillation data by Tromp (1993, 1994) and Durek and Romanowicz (1999) gave further strong support for the homogenous transverse anisotropy interpretation. The general consensus became that the inner core is strongly anisotropic, with a P-wave anisotropy of about 3–4% with the fast velocity direction parallel to the Earth's rotation axis (see reviews by Creager (2000), Song (1997), and Tromp (2001). However, many studies have suggested variations to this simple anisotropy model of the inner core. It has been suggested that the symmetry axis of the anisotropy is tilted from the Earth's rotation axis (Shearer and Toy, 1991; Creager, 1992; Su and Dziewoński, 1995) by 5–10°. A significant difference in the anisotropy between Eastern and Western Hemispheres of the inner core has been reported by Tanaka and Hamaguchi (1997) and Creager (1999) with the Western Hemisphere having significantly stronger anisotropy than the Eastern Hemisphere that is nearly isotropic. Several recent studies concur that the outer part (100–200 km) of the inner core is isotropic and inner part is anisotropic (e.g., Song and Helmberger, 1998; Garcia and Souriau, 2000, 2001; Song and Xu, 2002; Garcia, 2002). It has also been suggested that there is a small innermost inner core with radius of about 300 km with distinct transverse isotropy relative to the outermost inner core by Ishii and Dziewoński (2002) **Figure 2(a)**. The innermost core has the slowest P-wave velocity at 45° to the east–west direction and the outer part has a weaker anisotropy with slowest P-wave velocity parallel to the east–west direction. Using split normal mode constraints, Beghein and Trampert (2003) also showed that there is a change in velocity structure with radius in the inner core; however, their model (**Figure 2(b)**) shows that the symmetry of the P- and S-wave changes at about 400 km radius, suggesting a radical change, such as a phase transition of iron. Much of the complexity of the observations seems to be station and method dependent (see Ishii *et al.* (2002a, 2002b). In a detailed study Ishii *et al.* (2002a, 2002b) derive a model that simultaneously satisfies normal mode, absolute travel time, and differential travel time data. It has allowed them to separate a mantle signature and regional structure from global anisotropy of the inner core. Their preferred model of homogeneous transverse isotropy with a symmetry axis aligned with the rotation axis contradicts many of models proposed above, but is similar to

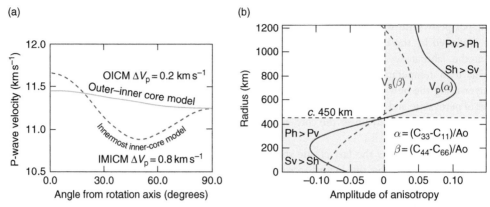

Figure 2   (a) The outer–inner core and innermost inner-core P-wave anisotropy models of Ishii and Dziewoński (2002). Note that variation of V_p is only $0.2\,km\,s^{-1}$ for outer–inner core model (300–1221 km radius) and $0.8\,km\,s^{-1}$ for innermost inner core model (0–300 km radius). (b) P- and S-wave inner-core models of Beghein and Trampert (2003). Note the veriation of V_p in the outermost and innermost core is between about 5 and $10\,km\,s^{-1}$, but the symmetry is different. The amplitude of S-wave anisotropy is weaker than P-wave, but there is symmetry change for both at c. 450 km radius.

previous suggestions. In a study of inner-core P-wave anisotropy using both finite-frequency and ray theories, Calvet *et al.* (2006) found that the data can be explained by three families of models which all exhibit anisotropy changes at a radius between 550 and 400 km (compared to 300 km for Ishii and Dziewoński (2002, 2003) and about 400 km for Beghein and Trampert (2003)). The first model has a weak anisotropy with a slow P-wave velocity symmetry axis parallel to Earth's rotation axis. The second model has a nearly isotropic innermost inner core. Lastly, the third model has a strongly anisotropic innermost inner core with a fast symmetry axis parallel to Earth's rotation axis. These models have very different implications for the origin of the anisotropy and the history of Earth's core. These divergences partly reflect the uneven sampling of the inner core by PKP(DF) paths resulting from the spatial distribution of earthquakes and seismographic stations. More data are required to improve the determination of the anisotropic structure of the innermost inner core, as for example, nearly antipodal waves that may provide crucial constraints on the structures at the centre of the Earth.

Theoretical studies of the process of generating the Earth's magnetic field through fluid motion of the outer core have predicted that the electromagnetic torque would force the inner core to rotate relative to the mantle (Steenbeck and Helmis, 1975; Gubbins, 1981; Szeto and Smylie, 1984; Glatzmaier and Roberts, 1995, 1996). Song and Richards (1996) first reported seismic differential

travel time observations based of three decades of data supporting the eastward relative rotation of inner core by about 1° per year faster than the daily rotation of the mantle and crust. Around the equator of the inner core, this rotation rate corresponds to a speed of a few tens of kilometers per year. The interpretation of the observed travel times required using the seismic anisotropy model inner core established by Su and Dziewoński (1995). Clearly to establish such small relative rotation rate a detailed knowledge of anisotropy, heterogeneity, and shape of the inner core is required, as travel times will change with direction and time (Song, 2000). The result was supported by some studies (e.g., Creager, 1997) and challenged by other studies (e.g., Souriau *et al.*, 1997), but all indicated a smaller rotation rate than 1° per year. The most recent study to date by Zhang *et al.* (2005) confirms a rotation rate of 0.3–0.5° using techniques that avoids artifacts of poor event locations and contamination by small-scale heterogeneities. Shear waves are very useful for determining the magnitude and orientation of anisotropy along individual ray paths. Despite several claims to have observed shear waves (e.g., Julian *et al.*, 1972; Okal and Cansi 1998), the analysis of the attenuation (Doornbos, 1974) and frequency range (Deuss *et al.*, 2000) reveals that these claims are unjustified. The analysis and validation of observations by Deuss *et al.* (2000) provide the first reliable observation of long period (20–30 s) shear waves providing new possibilities for exploring the anisotropy of inner core.

The transverse isotropy observed for P-waves traveling through the inner core could be explained by the CPO of hexagonal close packed (h.c.p.) from of iron (ε-phase) crystals or a layered structure. Analysis of the coda of short period inner core boundary reflected P-waves (PKiKP) require only a few percent heterogeneity at length scales of 2 km (Vidale and Earle, 2000) which suggests a relatively homogenous nonstructured inner core; however, this interpretation of coda has recently been questioned by Poupinet and Kennett (2004). The mechanism responsible for the CPO has been the subject of considerable speculation in recent years. The suggested mechanisms include: alignment of crystals in the magnetic field as they solidify from the liquid outer core (Karato, 1993), the alignment of crystals by plastic flow under the action of Maxwell normal stresses caused the magnetic field (Karato, 1999), faster crystal growth in the equatorial region of the inner core (Yoshida *et al.*, 1996), anisotropic growth driven by strain energy, dendric crystal growth aligned with the direction of dominant heat flow (Bergman, 1997), plastic flow in a thermally convective regime (Jeanloz and Wenk, 1988; Wenk *et al.*, 1988; Wenk *et al.*, 2000), and plastic flow under the action of magnetically induced Maxwell shear stresses (Buffett and Wenk, 2001). An alternative explication was proposed by Singh *et al.* (2000) to explain the P-wave anisotropy and the low shear wave velocity of about 3.6 km s^{-1} (Deuss *et al.* 2000) is the presence of a volume fraction of 3–10% liquid iron (or FeS) in the form of oblate spherical inclusions aligned in the equatorial plane in a matrix of iron crystals with their c-axes aligned parallel to the rotation axis as originally proposed by Stixrude and Cohen (1995). The S-wave velocity and attenuation data are mainly from the outer part of the inner core and hence it was suggested that the liquid inclusions are present in this region. Note this is in contradiction with other studies, which suggest that the outer core has a low anisotropy. Several problems posed by all models to different degrees include: whether the inner core is thermally convective (see Weber and Machetel, 1992; Yukutake, 1998); the viscosity of the inner core; the strength of magnetic field and magnitude of the Maxwell stresses necessary to cause crystal alignment; and the presence of liquids or even the ability of the models to correctly predict the magnitude and orientation of the seismic P-wave anisotropy. Given the range of seismic models and the variety of physical phenomena proposed to explain these models, better constraints on the seismic data, probably using better-quality data and a wider

geographical distribution of seismic stations in polar regions are urgently required.

In this chapter, our current knowledge of the seismic anisotropy of the constituent minerals of the Earth's interior and our ability to extrapolate these properties to mantle conditions of temperature and pressure (**Figure 1**) are reviewed. The chapter begins by reviewing the fundamentals of elasticity, plane wave propagation in anisotropic crystals, the measurements of elastic constants, and the effective elastic constants of crystalline aggregates.

16.2 Mineral Physics

16.2.1 Elasticity and Hooke's Law

Robert Hooke's experiments demonstrated that extension of a spring is proportional to the weight hanging from it which was published in *De Potentia Restitutiva* (or of Spring Explaining the Power of Springing Bodies 1678), establishing that in elastic solids there is a simple linear relationship between stress and strain. The relationship is now commonly known as Hooke's law (**Ut tensio, sic vis** – which translated from Latin is 'as is the extension, so is the force' – was the solution to an anagram announced two years early in 'A Description of Helioscopes and some other Instruments 1676', to prevent Hooke's rivals from claiming to have made the discovery themselves!). In the case of small (infinitesimal) deformations a Maclaurin expansion of stress as a function of strain developed to first order correctly describes the elastic behavior of most linear elastic solids:

$$\sigma_{ij}(\varepsilon_{kl}) = \sigma_{ij}(0) + \left(\frac{\partial \sigma_{ij}}{\partial \varepsilon_{kl}}\right)_{\partial \varepsilon_{kl}=0}$$
$$\times \varepsilon_{kl} + \frac{1}{2}\left(\frac{\partial \sigma_{ij}}{\partial \varepsilon_{kl}\partial \varepsilon_{mn}}\right)_{\substack{\partial \varepsilon_{kl}=0 \\ \partial \varepsilon_{mn}=0}} \varepsilon_{kl}\varepsilon_{mn} + \cdots$$

as the elastic deformation is zero at a stress of zero then $\sigma_{ij}(0) = 0$ and restricting our analysis to first order, then we can define the fouth rank elastic tensor c_{ijkl} as

$$c_{ijkl} = \left(\frac{\partial \sigma_{ij}}{\partial \varepsilon_{kl}}\right)_{\partial \varepsilon_{kl}=0}$$

where ε_{kl} and σ_{ij} are, respectively, the stress and infinitesimal strain tensors. Hooke's law can now be expressed in its traditional form as

$$\sigma_{ij} = c_{ijkl}\,\varepsilon_{kl}$$

The coefficients of the elastic fourth-rank tensor c_{ijkl} translate the linear relationship between the second-rank stress and infinitesimal strain tensors. The four indices ($ijkl$) of the elastic tensor have values between 1 and 3, so that there are $3^4 = 81$ coefficients. The stress tensor is symmetric as we assume that stresses acting on opposite faces are equal and opposite; hence, there are no stress couples to produce a net rotation of the elastic material. The infinitesimal strain tensor is also symmetric; because we assume that pure and simple shear quantities are so small that their squares and products can be neglected. Due to the symmetric symmetry of stress and infinitesimal strain tensors, they only have six independent values rather than nine for the nonsymmetric case; hence, the first two (i,j) and second two (k,l) indices of the elastic tensor can be interchanged:

$$c_{ijkl} = c_{jikl} \quad \text{and} \quad c_{ijkl} = c_{ijlk}$$

The permutation of the indices caused by the symmetry of stress and strain tensors reduces the number of independent elastic coefficients to $6^2 = 36$ because the two pairs of indices (i,j) and (k,l) can only have six different values:

$$1 \equiv (1,1)\ 2 \equiv (2,2)\ 3 \equiv (3,3)\ 4 \equiv (2,3) = (3,2)\ 5 \equiv (3,1)$$
$$= (1,3)\ 6 \equiv (1,2) = (2,1).$$

It is practical to write a 6 by 6 table of 36 coefficients with two Voigt indices m and n (c_{mn}) that have values between 1 and 6, whereas the representation of the c_{ijkl} tensor with 81 coefficients would be a printer's nightmare. The relation between the Voigt (mn) and tensor indices ($ijkl$) can be expressed most compactly by

$$m = \delta_{ij}\,i + (1 - \delta_{ij})(9 - i - j) \quad \text{and}$$
$$n = \delta_{kl}\,k + (1 - \delta_{kl})(9 - k - 1)$$

where δ_{ij} is the Kronecker delta ($\delta_{ij} = 1$ when $i = j$ and $\delta_{ij} = 0$ when $i \neq j$).

Combining the first and second laws of thermodynamics for stress–strain variables we can define the variation of the internal energy (dU) per unit volume of a deformed anisotropic elastic body as a function of entropy (dS) and elastic strain ($d\varepsilon_{ij}$) at an absolute temperature (T) as

$$dU = \sigma_{ij}\,d\varepsilon_{ij} + T\,dS$$

U and S are called state functions. From this equation it follows that the stress tensor at constant entropy can be defined as

$$\sigma_{ij} = \left(\frac{\partial U}{\partial \varepsilon_{ij}}\right)_S = c_{ijkl}\varepsilon_{kl}$$

hence

$$c_{ijkl} = \left(\frac{\partial \sigma_{ij}}{\partial \varepsilon_{kl}}\right) \quad \text{and} \quad c_{klij} = \left(\frac{\partial \sigma_{kl}}{\partial \varepsilon_{ij}}\right)$$

and finally we can write the elastic constants in terms of internal energy and strain as

$$c_{ijkl} = \frac{\partial}{\partial \varepsilon_{kl}}\left(\frac{\partial U}{\partial \varepsilon_{ij}}\right)_S$$
$$= \left(\frac{\partial^2 U}{\partial \varepsilon_{ij}\partial \varepsilon_{kl}}\right)_S = \left(\frac{\partial^2 U}{\partial \varepsilon_{kl}\partial \varepsilon_{ij}}\right)_S = c_{klij}$$

The fourth-rank elastic tensors are referred to as second-order elastic constants in thermodynamics, because they are defined as second-order derivatives of a state function (e.g., internal energy $\partial^2 U$ for adiabatic or Helmhotz free energy $\partial^2 F$ for isothermal constants) with respect to strain. It follows from these thermodynamic arguments that we can interchange the first pair of indices (ij) with second (kl):

$$c_{ijkl} = c_{jikl} \quad \text{and} \quad c_{ijkl} = c_{ijlk} \quad \text{and now} \quad c_{ijkl} = c_{klij}$$

The additional symmetry of $c_{ijkl} = c_{klij}$ permutation reduces the number of independent elastic coefficients from 36 to 21 and tensor with two Voigt indices is symmetric, $c_{mn} = c_{nm}$. Although we have illustrated the case of isentropic (constant entropy, equivalent to an adiabatic process for a reversible process such as elasticity) elastic constants which intervene in the propagation of elastic waves whose vibration is too fast for thermal diffusion to establish heat exchange to achieve isothermal conditions, these symmetry relations are also valid for isothermal elastic constants which are used in mechanical problems. The vast majority of elastic constants reported in the literature are determined by the propagation ultrasonic elastic waves and are adiabatic. More recently, elastic constants predicted by atomic modeling for mantle conditions of pressure, and in some cases temperature, are also adiabatic (see review by Karki *et al.* (2001), also *see* Chapter 6).

The elastic constants in the literature are presented in the form of 6×6 tables for the triclinic symmetry with 21 independent values; here the independent values are shown in bold characters in the upper diagonal of c_{mn} with the corresponding c_{ijkl}.

$$
\begin{bmatrix}
\mathbf{c_{11}} & \mathbf{c_{12}} & \mathbf{c_{13}} & \mathbf{c_{14}} & \mathbf{c_{15}} & \mathbf{c_{16}} \\
c_{12} & \mathbf{c_{22}} & \mathbf{c_{23}} & \mathbf{c_{24}} & \mathbf{c_{25}} & \mathbf{c_{26}} \\
c_{13} & c_{23} & \mathbf{c_{33}} & \mathbf{c_{34}} & \mathbf{c_{35}} & \mathbf{c_{36}} \\
c_{14} & c_{24} & c_{34} & \mathbf{c_{44}} & \mathbf{c_{45}} & \mathbf{c_{46}} \\
c_{15} & c_{25} & c_{35} & c_{45} & \mathbf{c_{55}} & \mathbf{c_{56}} \\
c_{16} & c_{26} & c_{36} & c_{46} & c_{56} & \mathbf{c_{66}}
\end{bmatrix}
$$

$$
=
\begin{bmatrix}
\mathbf{c_{1111}} & \mathbf{c_{1122}} & \mathbf{c_{1133}} & \mathbf{c_{1123}} & \mathbf{c_{1113}} & \mathbf{c_{1112}} \\
c_{1122} & \mathbf{c_{2222}} & \mathbf{c_{2233}} & \mathbf{c_{2223}} & \mathbf{c_{2213}} & \mathbf{c_{2212}} \\
c_{1133} & c_{2233} & \mathbf{c_{3333}} & \mathbf{c_{3323}} & \mathbf{c_{3313}} & \mathbf{c_{3312}} \\
c_{1123} & c_{2223} & c_{3323} & \mathbf{c_{2323}} & \mathbf{c_{2313}} & \mathbf{c_{2312}} \\
c_{1113} & c_{2213} & c_{3313} & c_{2313} & \mathbf{c_{1313}} & \mathbf{c_{1312}} \\
c_{1112} & c_{2212} & c_{3312} & c_{2312} & c_{1312} & \mathbf{c_{1212}}
\end{bmatrix}
$$

In the triclinic system there are no special relationships between the constants. On the other extreme is the case of isotropic elastic symmetry defined by just two coefficients. Note this is not the same as cubic symmetry, where there are three coefficients and that a cubic crystal can be elastically anisotropic. The isotropic elastic constants can be expressed in the four-index system as

$$
c_{ijkl} = \lambda\delta_{ij}\delta_{kl} + \mu\left(\delta_{ik}\delta_{jl} + \delta_{il}\delta_{jk}\right)
$$

where λ is Lamé's coefficient and μ is the shear modulus. λ and μ are often referred to as Lamé's constants after the French mathematician Gabriel Lamé, who first published his book *Leçons sur la théorie mathématique de l'élasticité des corps solides* in 1852. In the two-index Voigt system the independent values are

$$
c_{11} = c_{22} = c_{33} = \lambda + 2\mu
$$

$$
c_{12} = c_{23} = c_{13} = \lambda
$$

$$
c_{44} = c_{55} = c_{66} = \frac{1}{2}(c_{11} - c_{12}) = \mu
$$

In matrix form this is written as

$$
\begin{bmatrix}
c_{11} & c_{12} & c_{12} & 0 & 0 & 0 \\
c_{12} & c_{11} & c_{12} & 0 & 0 & 0 \\
c_{12} & c_{12} & c_{11} & 0 & 0 & 0 \\
0 & 0 & 0 & \frac{1}{2}(c_{11}-c_{12}) & 0 & 0 \\
0 & 0 & 0 & 0 & \frac{1}{2}(c_{11}-c_{12}) & 0 \\
0 & 0 & 0 & 0 & 0 & \frac{1}{2}(c_{11}-c_{12})
\end{bmatrix}
$$

where the two independent values are c_{11} and c_{12}. Another symmetry that is very important in seismology is the transverse isotropic medium (or hexagonal crystal symmetry). In many geophysical applications of transverse isotropy, the unique symmetry direction (X3) is vertical and the other perpendicular elastic axes (X1 and X2) are horizontal and share the same elastic properties and velocities. It is very common in seismological papers to use the notation of Love (1927) for the elastic constants of transverse isotropic media where

$$
\mathbf{A} = \mathbf{c_{11}} = c_{22} = \mathbf{c_{1111}} = c_{2222}
$$

$$
\mathbf{C} = \mathbf{c_{33}} = \mathbf{c_{3333}}
$$

$$
\mathbf{F} = \mathbf{c_{13}} = c_{23} = \mathbf{c_{1133}} = c_{2233}
$$

$$
\mathbf{L} = \mathbf{c_{44}} = c_{55} = \mathbf{c_{2323}} = c_{1313}
$$

$$
\mathbf{N} = \mathbf{c_{66}} = \frac{1}{2}(c_{11} - c_{12}) = \mathbf{c_{1212}} = \frac{1}{2}(\mathbf{c_{1111}} - \mathbf{c_{1122}})
$$

and

$$
\mathbf{A} - 2\mathbf{N} = \mathbf{c_{12}} = c_{21} = c_{11} - 2\mathbf{c_{66}} = \mathbf{c_{1212}} = c_{2211}
$$
$$
= c_{1111} - 2\mathbf{c_{1212}}
$$

$$
\begin{bmatrix}
\mathbf{c_{11}} & \mathbf{c_{12}} & \mathbf{c_{13}} & 0 & 0 & 0 \\
c_{12} & \mathbf{c_{11}} & \mathbf{c_{13}} & 0 & 0 & 0 \\
c_{13} & c_{13} & \mathbf{c_{33}} & 0 & 0 & 0 \\
0 & 0 & 0 & \mathbf{c_{44}} & 0 & 0 \\
0 & 0 & 0 & 0 & c_{44} & 0 \\
0 & 0 & 0 & 0 & 0 & \frac{1}{2}(c_{11}-c_{12})
\end{bmatrix}
$$

$$
=
\begin{bmatrix}
\mathbf{A} & \mathbf{A} - 2\mathbf{N} & \mathbf{F} & 0 & 0 & 0 \\
\mathbf{A} - 2\mathbf{N} & \mathbf{A} & \mathbf{F} & 0 & 0 & 0 \\
\mathbf{F} & \mathbf{F} & \mathbf{C} & 0 & 0 & 0 \\
0 & 0 & 0 & \mathbf{L} & 0 & 0 \\
0 & 0 & 0 & 0 & \mathbf{L} & 0 \\
0 & 0 & 0 & 0 & 0 & \mathbf{N}
\end{bmatrix}
$$

and the velocities in orthogonal directions that characterize a transverse isotropic medium are functions of the leading diagonal of the elastic tensor and are given as

$$\mathbf{A} = c_{11} = \rho V^2{}_{PH} \mathbf{C} = c_{33} = \rho V^2{}_{PV} \mathbf{L} = c_{44} = \rho V^2{}_{SV} \mathbf{N}$$
$$= c_{66} = \rho V^2{}_{SH}$$

where ρ is density, V_{PH} and V_{PV} are the velocities of horizontally (X1 or X2) and vertically (X3) propagating P-waves, V_{SH} and V_{SV} are the velocities of horizontally and vertically polarized S-waves propagating horizontally.

Elastic anisotropy can be characterized by taking ratios of the individual elastic coefficients. Thomsen (1978) introduced three parameters to characterize the elastic anisotropy of any degree, not just weak anisotropy, for transverse isotropic medium,

$$\varepsilon = c_{11} - c_{33}/2\, c_{33} = A - C/2C$$

$$\gamma = c_{66} - c_{44}/2\, c_{44} = N - L/2L$$

and

$$\delta* = \frac{1}{2}c^2{}_{33}\left[2(c_{13} + c_{44})^2 - (c_{33} - c_{44})(c_{11} + c_{33} - 2c_{44})\right]$$

$$\delta* = \frac{1}{2}C^2\left[2(F + L)^2 - (C - L)(A + C - 2L)\right]$$

Thomsen also proposed a weak anisotropy version of the $\delta*$ parameter,

$$\delta = \left[(c_{13} + c_{44})^2 - (c_{33} - c_{44})^2\right]/\left[2c_{33}(c_{33} - c_{44})\right]$$
$$= \left[(F + L)^2 - (C - L)^2\right]/\left[2C(C - L)\right]$$

These parameters go to zero in the case of isotropy and have values of much less than 1 (i.e., 10%) in the case of weak anisotropy. The parameter ε describes the P-wave anisotropy and can be defined in terms of the normalized difference of the P-wave velocity in the directions parallel to the symmetry axis (X3, vertical axis) and normal to the symmetry axis (X12, horizontal plane). The parameter γ describes the S-wave anisotropy and can be defined in terms of the normalized difference of the SH-wave velocity in the directions normal to the symmetry axis (X12, horizontal plane) and parallel to the symmetry axis (X3, vertical axis), but also in terms SH and SV, because SH parallel to the symmetry axis has the same velocity as SV normal to the symmetry axis:

$$\varepsilon = V_P(X12) - V_P(X3)/V_P(X3) = V_{PH} - V_{PV}/V_{PV}$$

$$\gamma = V_{SH}(X12) - V_{SH}(X3)/V_{SH}(X3)$$
$$= V_{SH}(X12) - V_{SV}(X12)/V_{SV}(X12) = V_{SH} - V_{SV}/V_{SV}$$

Thomsen (1986) found that the parameter $\delta*$ controls most of the phenomena of importance for exploration geophysics, such as velocities inclined to the symmetry axis (vertical), some of which are non-negliable even when the anisotropy is weak. The parameter $\delta*$ is an awkard combination of elastic parameters, which is totally independent of the velocity in the direction normal to the symmetry axis (X12 horizontal plane) and which may be either postitive or negative. Mensch and Rasolofosaon (1997) have extended the application of Thomsen's parameters to anisotropic media of arbitary symmetery and the associated analysis in terms of the perturbation of a reference model that can exhibit strong S-wave anisotropy.

In the domain of one- or three-dimensional radial anisotropic seismic tomography, it has been the practice to use the parameters φ, ξ, and η to characterize the transverse anisotropy, where

$$\varphi = c_{33}/c_{11} = C/A = V_{PV}^2/V_{PH}^2$$

$$\xi = c_{66}/c_{44} = N/L = V_{SH}^2/V_{SV}^2$$

$$\eta = c_{13}/(c_{11} - 2c_{44}) = F/(A - 2L)$$

For characterizing the anisotropy of the inner core, some authors (e.g., Song, 1997) use a variant of Thomsen's parameters, $\varepsilon'' = (c_{33} - c_{11})/2c_{11} = (C - A)/2A$ (positions of c_{11} and c_{33} reversed from Thomsen, double prime has been added to avoid confusion here with Thomsen's parameter)

$$\gamma = (c_{66} - c_{44})/2c_{44} = (N - L)/2L$$
(same as Thomsen)

$$\sigma = (c_{11} + c_{33} - 4c_{44} - 2c_{13})/2c_{11}$$
$$= (A + C - 4L - 2F)/2A$$
(very different from Thomsen's $\delta*$)

others (e.g., Woodhouse et al., 1986) use

$$\alpha = (c_{33} - c_{11})/A_o = (C - A)/A_o$$

$$\beta = (c_{66} - c_{44})/A_o$$

$$\gamma = (c_{11} - 2c_{44} - c_{13})/A_o = (A - 2N - F)/A_o$$

where $A_o = \rho_o V_{po}^2$ is calculated using the density ρ_o and P-wave velocity V_{po} at the center of the spherically symmetric reference Earth model, PREM (Dziewoński and Anderson, 1981). With at least four different sets of triplets of anisotropy parameters to describe transverse isotropy in various domains of seismology, the situation is complex for a researcher who wants to compare the anisotropy from different published works. Even when comparisons are made, for example, for the inner core (Calvet *et al.*, 2006), drawing conclusions may be difficult as the parameters refect only certain aspects of the anisotropy.

In studying the effect of symmetry of the elastic properties of crystals, one is directly concerned with only the 11 Laue classes and not the 32 point groups, because elasticity is a centrosymmetrical physical property. The velocity of an elastic wave depends on its direction of propagation in an anisotropic crystal, but not the positive or negative sense of the direction. In this chapter we are restricting our study to second-order elastic constants, corresponding to small strains characteristic elastic deformations associated with the propagation of seismic waves. If we wanted to consider larger finite strains or the effect of an externally applied stress, we would need to consider third-order elastic constants as the approximation adopted in limiting the components of the strain tensor to terms of the first degree in the derviatives is no longer justified. For second-order elastic constants the two cubic and two hexagonal Laue classes are not distinct (e.g., Brugger, 1965) and may be replaced by a single cubic and a single hexagonal class, which results in only nine distinct symmetry classes for crystals shown in **Table 1**.

Table 1   Second-order elastic constants of all Laue crystal symmetries

Cubic (3) 23, $m3$, 432, $\bar{4}3m$, $m3m$

$$\begin{bmatrix} c_{11} & c_{12} & c_{12} & 0 & 0 & 0 \\ c_{12} & c_{11} & c_{12} & 0 & 0 & 0 \\ c_{12} & c_{12} & c_{11} & 0 & 0 & 0 \\ 0 & 0 & 0 & c_{44} & 0 & 0 \\ 0 & 0 & 0 & 0 & c_{44} & 0 \\ 0 & 0 & 0 & 0 & 0 & c_{44} \end{bmatrix}$$

Hexagonal (5) 6, $\bar{6}$, 6/m, 622, 6mmm, $\bar{6}$2m, 6/mmm

$$\begin{bmatrix} c_{11} & c_{12} & c_{13} & 0 & 0 & 0 \\ c_{12} & c_{11} & c_{13} & 0 & 0 & 0 \\ c_{13} & c_{13} & c_{33} & 0 & 0 & 0 \\ 0 & 0 & 0 & c_{44} & 0 & 0 \\ 0 & 0 & 0 & 0 & c_{44} & 0 \\ 0 & 0 & 0 & 0 & 0 & \frac{1}{2}(c_{11}-c_{12}) \end{bmatrix}$$

Trigonal (6) 32, 3m, $\bar{3}m$

$$\begin{bmatrix} c_{11} & c_{12} & c_{13} & c_{14} & 0 & 0 \\ c_{12} & c_{11} & c_{13} & -c_{14} & 0 & 0 \\ c_{13} & c_{13} & c_{33} & 0 & 0 & 0 \\ c_{14} & -c_{14} & 0 & c_{44} & 0 & 0 \\ 0 & 0 & 0 & 0 & c_{44} & c_{14} \\ 0 & 0 & 0 & 0 & c_{14} & \frac{1}{2}(c_{11}-c_{12}) \end{bmatrix}$$

Trigonal (7) 3, $\bar{3}$

$$\begin{bmatrix} c_{11} & c_{12} & c_{13} & c_{14} & -c_{25} & 0 \\ c_{12} & c_{11} & c_{13} & -c_{14} & c_{25} & 0 \\ c_{13} & c_{13} & c_{33} & 0 & 0 & 0 \\ c_{14} & -c_{14} & 0 & c_{44} & 0 & c_{25} \\ -c_{25} & c_{25} & 0 & 0 & c_{44} & c_{14} \\ 0 & 0 & 0 & c_{25} & c_{14} & \frac{1}{2}(c_{11}-c_{12}) \end{bmatrix}$$

Tetragonal (6) 422, 4mm, $\bar{4}$2m, 4/mmm

$$\begin{bmatrix} c_{11} & c_{12} & c_{13} & 0 & 0 & 0 \\ c_{12} & c_{11} & c_{13} & 0 & 0 & 0 \\ c_{13} & c_{13} & c_{33} & 0 & 0 & 0 \\ 0 & 0 & 0 & c_{44} & 0 & 0 \\ 0 & 0 & 0 & 0 & c_{44} & 0 \\ 0 & 0 & 0 & 0 & 0 & c_{66} \end{bmatrix}$$

Tetragonal (7), 4, $\bar{4}$ 4/m

$$\begin{bmatrix} c_{11} & c_{12} & c_{13} & 0 & 0 & c_{16} \\ c_{12} & c_{11} & c_{13} & 0 & 0 & -c_{16} \\ c_{13} & c_{13} & c_{33} & 0 & 0 & 0 \\ 0 & 0 & 0 & c_{44} & 0 & 0 \\ 0 & 0 & 0 & 0 & c_{44} & 0 \\ c_{16} & -c_{16} & 0 & 0 & 0 & c_{66} \end{bmatrix}$$

(Continued)

Table 1 (Continued)

Orthorhombic (9) 222, *mm2, mmm*

$$
\begin{bmatrix}
c_{11} & c_{12} & c_{13} & 0 & 0 & 0 \\
c_{12} & c_{22} & c_{23} & 0 & 0 & 0 \\
c_{13} & c_{23} & c_{33} & 0 & 0 & 0 \\
0 & 0 & 0 & c_{44} & 0 & 0 \\
0 & 0 & 0 & 0 & c_{55} & 0 \\
0 & 0 & 0 & 0 & 0 & c_{66}
\end{bmatrix}
$$

Mononclinic (13) 2, *m*, 2/*m*

$$
\begin{bmatrix}
c_{11} & c_{12} & c_{13} & 0 & c_{15} & 0 \\
c_{12} & c_{22} & c_{23} & 0 & c_{25} & 0 \\
c_{13} & c_{23} & c_{33} & 0 & c_{35} & 0 \\
0 & 0 & 0 & c_{44} & 0 & c_{46} \\
c_{14} & c_{25} & c_{35} & 0 & c_{55} & 0 \\
0 & 0 & 0 & c_{46} & 0 & c_{66}
\end{bmatrix}
$$

Triclinic (21) 1, $\bar{1}$

$$
\begin{bmatrix}
c_{11} & c_{12} & c_{13} & c_{14} & c_{15} & c_{16} \\
c_{12} & c_{22} & c_{23} & c_{24} & c_{25} & c_{26} \\
c_{13} & c_{23} & c_{33} & c_{34} & c_{35} & c_{36} \\
c_{14} & c_{24} & c_{34} & c_{44} & c_{45} & c_{46} \\
c_{15} & c_{25} & c_{35} & c_{45} & c_{55} & c_{56} \\
c_{16} & c_{26} & c_{36} & c_{46} & c_{56} & c_{66}
\end{bmatrix}
$$

The number in brackets is the number of independent constants.

16.2.2 Plane Waves and Christoffel's Equation

There are two types of elastic waves, which propagate in an isotropic homogeneous elastic medium, the faster compressional (or longitudinal) wave with displacements parallel to propagation direction and slower shear (or transverse) wave with displacements perpendicular to the propagation direction. In anisotropic elastic media there are three types of waves:one compressional wave and two shear waves with, in general, different velocities. In order to understand the displacements associated with different waves, their relationship to the propagation direction and elastic anisotropy, it is important to consider the equation of propagation of a mechanical disturbance in an elastic medium. If we ignore the effect of gravity, we can write the equation of displacement (u_i) as function of time (t) as

$$
\rho\left(\frac{\partial^2 u_i}{\partial t^2}\right) = \left(\frac{\partial \sigma_{ij}}{\partial x_j}\right)
$$

where ρ is the density and x_j is position. Using Hooke's law, stress can be written as

$$
\sigma_{ij} = c_{ijkl}\left(\frac{\partial u_l}{\partial x_k}\right)
$$

Hence, elasto-dynamical equation describing the inertial forces can be rewritten with one unknown (the displacement) as

$$
\rho\left(\frac{\partial^2 u_i}{\partial t^2}\right) = c_{ijkl}\left(\frac{\partial^2 u_l}{\partial x_j x_k}\right)
$$

The displacement of monochromatic plane wave can be described by any harmonic form as a function of time (e.g. Fedorov, 1968):

$$
u = A \exp i(k.x - \omega t)
$$

where A the amplitude vector, which gives the direction and magnitude of particle motion, t is time, n the propagation direction normal to the plane wavefront, ω the angular frequency, which is related to frequency by $f = \omega/2\pi$, k is the wave vector which is related to the phase velocity (V) by $V = \omega/k$ and the plane wavefront normal (n) by $k = (2\pi/\lambda)n$, where λ is the wavelength. For plane waves the total phase $\phi = (k.x - \omega t)$ is a constant along the wavefront. Hence, the equation for a surface of equal phase at any instant of time (t) is a plane perpendicular to the propagation unit vector (n). Now if we insert the solution for the time-dependant displacement into the elasto-dynamical equation, we find the

Christoffel equation first published by E.B. Christoffel in 1877:

$$C_{ijkl} s_j s_l p_k = \rho V^2 p_i \quad \text{or} \quad C_{ijkl} s_j s_l p_k = \rho p_i$$

$$\left(C_{ijkl} n_j n_l - \rho V^2 \delta_{ik} \right) p_k = 0 \quad \text{or} \quad \left(C_{ijkl} s_j s_l - \rho \delta_{ik} \right) p_k = 0$$

where V are the phase velocities, ρ is density, p_k polarization unit vectors, n_j propagation unit vector, and s_j are the slowness vector of magnitude $1/V$ and the same direction as the propagation direction (n). The polarization unit vectors p_k are obtained as eigenvectors and corresponding eigenvalues of the roots of the equation

$$\det| C_{ijkl} n_j n_l - \rho V^2 \delta_{ik} | = 0 \quad \text{or}$$
$$\det| C_{ijkl} s_j s_l - \rho \delta_{ik} | = 0$$

We can simplify this equation by introducing the Christoffel (Kelvin–Christoffel or acoustic) tensor $T_{ik} = C_{ijkl} n_j n_l$ and three wave moduli $M = \rho V^2$; hence, $\det| T_{ik} - M \delta_{ik} | = 0$. The equation can be written in full as

$$\begin{vmatrix} T_{11} - M & T_{12} & T_{13} \\ T_{21} & T_{22} - M & T_{23} \\ T_{31} & T_{32} & T_{33} - M \end{vmatrix} = 0$$

which upon expansion yields the cubic polynominal in M

$$M^3 - \mathrm{I}_T M^2 + \mathrm{II}_T M - \mathrm{III}_T = 0$$

where $\mathrm{I}_T = T_{ii}$, $\mathrm{II}_T = \frac{1}{2}(T_{ii}T_{jj} - T_{ij}T_{ij})$, and $\mathrm{III}_T = \det| T_{ij} |$ are the first, second, and third invariants of the Christoffel tensor. The three roots of the cubic polynominal in M are the three wave moduli M. The eigenvectors (e_j) associated with each wave moduli can be found by solving $(T_{ij} - M\delta_{ij})e_j = 0$. Analytical solutions for the Christoffel tensor have been proposed in various forms by Červený (1972), Every (1980), Mainprice (1990), Mensch and Rasolofosaon (1997), and probably others.

The Christoffel tensor is symmetric because of symmetry of the elastic constants, and hence

$$T_{ik} = C_{ijkl} n_j n_l = C_{jikl} n_j n_l = C_{ijlk} n_j n_l = C_{klij} n_j n_l = T_{ki}$$

The Christoffel tensor is also invariant upon the change of sign of the propagation direction (n) as the elastic tensor is not sensitive to the presence or absence of a center of symmetry, being a centro-symmetric physical property. Because the elastic strain energy ($(1/2)C_{ijkl}\varepsilon_{ij}\varepsilon_{kl}$) of a stable crystal is always positive and real (e.g., Nye, 1957), the eigenvalues of the 3×3 Christoffel tensor (being a Hermitian matrix) are three positive real values of the wave moduli (M) corresponding to ρV_p^2, ρV_{s1}^2, ρV_{s2}^2 of the plane waves propagating in the direction n. The three eigenvectors of the Christoffel tensor are the polarization directions (also called vibration, particle movement or displacement vectors) of the three waves. As the Christoffel tensor is symmetric, the three eigenvectors (and polarization) vectors are mutually perpendicular. In the most general case, there are no particular angular relationships between polarization directions (p) and the propagation direction (n); however, typically the P-wave polarization direction is nearly parallel and the two S-waves polarizations are nearly perpendicular to the propagation direction and they are termed quasi-P or quasi-S waves. If polarizations of the P-wave and two S-waves are, respectively, parallel and perpendicular to the propagation direction, which may happen along a symmetry direction, then the waves are termed pure P and pure S or pure modes. Only velocities in pure mode directions can be directly related to single elastic constants (Neighbours and Schacher, 1967). In general, the three waves have polarizations that are perpendicular to one another and propagate in the same direction with different velocities, with $V_p > V_{s1} > V_{s2}$. A propagation direction for which two (or all three) of the phase velocities are identical is called an acoustic axis, which occur even in crystals of triclinic symmetry. Commonly, the acoustic axis is associated with the two S-waves having the same velocity. The S-wave may be identified by their relative velocity $V_{s1} > V_{s2}$ or by their polarization being parallel to a symmetry direction or feature, for example, SH and SV, where the polarization is horizontal and vertical to the third axis of reference Cartesian frame of the elastic tensor (X3 in the terminology of Nye (1957), X3 is almost always parallel to the crystal c-axis) in mineral physics and perpendicular to the Earth's surface in seismology.

The velocity at which energy propagates in a homogenous anisotropic elastic medium is defined as the average power flow density divided by average total energy density and can be calculated from the phase velocity using the following relationship given by Fedorov (1968):

$$V_i^e = C_{ijkl} p_j p_l n_k / \rho V$$

The phase and energy velocities are related by a vector equation $V_i^e \cdot n_i = V_i$. It is apparent from this relationship that V^e is not in general parallel to propagation direction (n) and has a magnitude equal or greater to the phase velocity ($V = \omega/k$). The propagation of waves in real materials occurs as packets of waves typically having a finite band of frequencies. The propagation velocity of wave packet is called the group velocity and this is defined for plane waves of given finite frequency range as

$$V_i^g = (\partial\omega/\partial k_i)$$

The group velocity is in general different to the phase velocity except along certain symmetry directions. In lossless anisotropic elastic media the group and energy velocities are identical (e.g., Auld, 1990); hence, it is not necessary to evaluate the differential angular frequency versus wave vector to obtain the group velocity as $V^g = V^e$. The group velocity has direct measurable physical meaning that is not apparent for the energy or phase velocities.

Various types of plots have been used to illustrate the variation of velocity with direction in crystals. Velocities measured by Brillouin spectroscopy are displayed via graphs of velocity as a function of propagation directions used in the experiments. The phase velocities V_p, V_{sh}, and V_{sv} of Stishovite are shown in **Figure 3**, using the elastic constants from Weidner *et al.* (1982); although this type of plot may be useful for displaying the experimental results, it does not convey the symmetry of the crystal. In crystal acoustics, the phase velocity and slowness surfaces have traditionally illustrated the anisotropy

of elastic wave velocity in crystals as a function of the propagation direction (n) and plots of the wavefront (ray or group) surface given by tracing the extremity of the energy velocity vector defined above. The normal to the slowness surface has the special property of being parallel to the energy velocity vector. The normal to the wavefront surface has the special property of being parallel to the propagation (n) and wave vector (k). We can illustrate these polar reciprocal properties using the elastic constants of the h.c.p. ε-phase of iron, which is considered to be the major constituent of the inner core, determined by Mao *et al.* (1998) at high pressure (**Figure 4**). Note that the twofold symmetry along the $\mathbf{a}[2\bar{1}\bar{1}0]$ axis of hexagonal ε-phase is respected by the slowness and wavefront surfaces of the SH-waves. The wavefront surface can be regarded as a recording after 1 s of the propagation from a spherical point source at the center of the diagram. The wavefront is a surface that separates the disturbed regions from the undisturbed ones. Anisotropic media have velocities that vary with direction, and hence the phase velocity and the slowness surfaces with concave and convex undulations in three dimensions. The undulations are not sharp as velocities and slownesses change slowly with orientation. In contrast, the wave surface can have sharp changes in direction, called cusps or folded wave surfaces in crystal physics (e.g., Musgrave, 2003) and triplications or caustics in seismology (e.g., Vavrycuk, 2003), particularly for S-waves, which correspond in orientation to undulations in the phase velocity and slowness surfaces. The high-pressure form of SiO_2 called Stishovite illustrates the various facets of the phase velocity, slowness, and wavefront surfaces in a highly anisotropic mineral

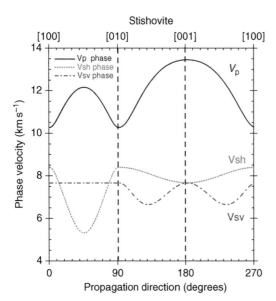

Figure 3 The variation of velocity with direction for tetragonal Stishovite as described by Weidner *et al.* (1982).

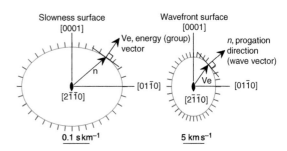

Figure 4 The polar reciprocal relation between the slowness and wavefront surfaces in hexagonal ε-phase iron at 211 GPa using the elastic constants determined by Mao *et al.* (1998). The normal to the slowness surface is the energy vector (brown) and the normal to the wavefront surface is the propagation direction (parallel to the wave vector). Note the twofold symmetry on the surfaces.

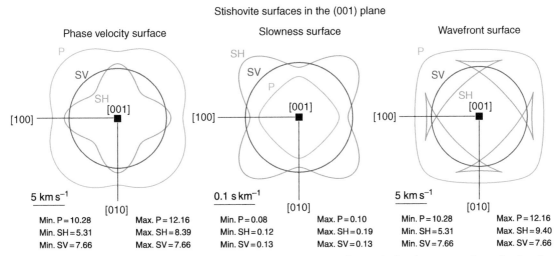

Min. P = 10.28 | Max. P = 12.16
Min. SH = 5.31 | Max. SH = 8.39
Min. SV = 7.66 | Max. SV = 7.66

Min. P = 0.08 | Max. P = 0.10
Min. SH = 0.12 | Max. SH = 0.19
Min. SV = 0.13 | Max. SV = 0.13

Min. P = 10.28 | Max. P = 12.16
Min. SH = 5.31 | Max. SH = 9.40
Min. SV = 7.66 | Max. SV = 7.66

Figure 5   The three surfaces used to characterize acoustic properties, the phase velocity, slowness, and wavefront surfaces for tetragonal Stishovite. Note fourfold symmetry of the surfaces and the cusps on the SH wavefront surface. The elastic constants of Stishovite were measured by Weidner *et al.* (1982) at ambiante conditions.

(**Figure 5**). Velocity clearly varies strongly with propagation direction; in the case of Stishovite, the variation for SH is very important, whereas SV is constant in the (001) plane. Stishovite has tetragonal symmetry; hence, the *c*-axis has fourfold symmetry that can clearly be identified in the various surfaces in (001) plane. There are orientations where the SH and SV surfaces intersect; hence, there is no shear wave splitting (S-wave birefringence) as both S-waves have the same velocity. The phase velocity and slowness surfaces have smooth changes in orientation corresponding to gradual changes in velocity. In contrast, along the **a**[100] and **b**[010] directions, the SH wavefront has sharp variations in orientation called cusps. The cusps on the SH wavefront are shown in more detail in **Figure 6**, where the cusps on the wavefront are clearly related to minima of the slowness (or maxima on the phase velocity) surface. The propagation of SH in the **a**[100] direction is instructive; if one considers seismometer at the point S, then seismometer will record first the arrival of wavefront AA′, then BB′ and finally CC′. The parabolic curved nature of the cusp AA′ is also at the origin of the word caustic to describe this phenomena by analogy with the convergent rays in optics, whereas the word triplication evokes the arrival of the three wavefronts. Although we are dealing with homogeneous anisotropic medium, a single crystal of stishovite, the seismometer will record three arrivals for SH, plus of course SV and P, giving

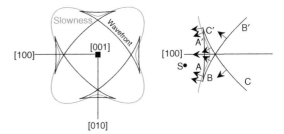

Figure 6   Cusps on the wavefront surface of tetragonal Stishovite and its relation to the slowness surface in the [100] direction. The propagation directions of the wavefront are marked by arrows every 10° (see the text for detailed discussion).

a total of five arrivals for a single mechanical disturbance. Media with tetragonal elastic symmetry are not very common in seismology, whereas media with hexagonal (or transverse isotropic) symmetry are very common and has been postulated, for example, for the D″ layer above the core–mantle boundary (CMB) (e.g., Kendall, 2000). The three surfaces of the hexagonal ε-phase of iron are shown in **Figure 7** in the **a**(2$\bar{1}\bar{1}$0) and **c**(0001) planes, which are respectively the perpendicular and parallel to the elastic symmetry axis (*c*-axis or X3) of transverse isotropic elastic symmetry. First, in this chapter the wave properties in the **a**(2$\bar{1}\bar{1}$0) plane are considered. In this plane the maximum P-wave velocity is 45° from the *c*-axis and minimum parallel to the *a*- and

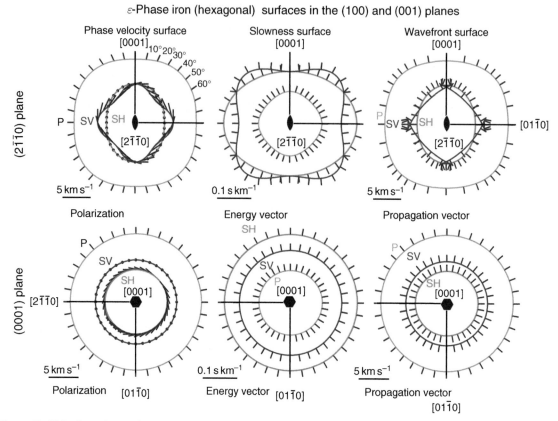

Figure 7   Velocity surfaces of ε-phase of iron in the second-order prism and basal plane illustrating the anisotropy in a hexagonal or transverse isotropic structure. Note the perfectly isotropic (circular) velocity surfaces in the base plane. The polarizations are marked on the phase velocity surfaces and for the basal plane the polarizations for P are normal to the surface and parallel to the propagation direction, where as for S they are normal and vertical for SV and tangential and horizontal for SH as in an isotropic medium. The energy vector and propagation direction are normal to the slowness and wavefront surface in the basal plane.

c-axes. The maximum SH velocity is parallel to the c-axis and the minimum parallel to b-axis, where as the SV velocity has a maximum parallel to the a- and c-axes and a minimum 45° to the c-axis. The polarization direction of P-waves is not perpendicular to the phase velocity surface in general, and not parallel to the propagation direction, except along the symmetry directions **m**$[01\bar{1}0]$ and **c**$[0001]$ axes. The SH polarizations are all normal to the c-axis and hence they appear as points in the **a**$(2\bar{1}\bar{1}0)$ plane. The SV polarizations are inclined to the a-axis; hence, they appear as lines of variable length depending on their orientation. The SH and SV velocity surfaces intersect parallel to the c-axis and at 60° from the c-axis, where they have the same velocity. The minima in the SV slowness surface along the a- and c-axes correspond to the cusps seen on the wavefront surface. The wave properties in the (0001) plane are completely different as both the P- and S-waves display a single velocity, hence the

name transverse isotropy as the velocities do not vary with direction in this plane perpendicular (transverse) to the unique elastic symmetry axis (c-axis or X3). The isotropic nature of this plane is also shown by the polarizations of the P-waves, which are normal to the phase velocity surface, and parallel to the propagation direction, as in the isotropic case. Similarly for the S-waves, the polarization directions for SV are parallel and SH are normal to the symmetry axis, and both are perpendicular to the propagation direction.

The illustrations used so far are only two-dimensional sections of the anisotropic wave properties. Ideally, we would like to see the three-dimensional form of the velocity surfaces and polarizations. A three-dimensional plot of the P, SH, and SV velocities in **Figure 8** shows the geometrical relation of the polarizations to the crystallographic axes, but is too complicated to see the variation in velocities; only a few directions have been plotted for clarity.

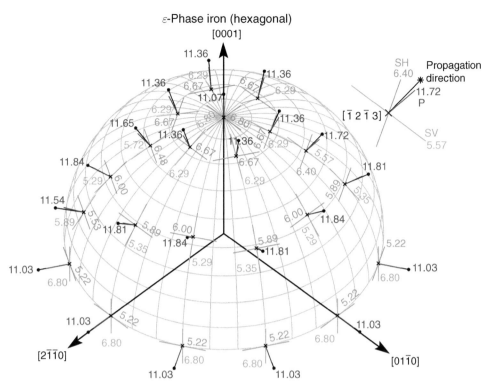

Figure 8 A three-dimensional illustration of the propagation direction (black), P-wave polarization (red), SH-wave (green), and SV-wave (blue) in hexagonal ε-phase iron at 211 GPa. The sphere is marked with grid at 10° intervals. The SH-wave polarizations are organized around the [001] direction in a hexagonal pattern. The P-wave polarization is not in general parallel and hence the S-wave polarizations are not perpendicular to the propagation, as illustrated for the [1213] direction. However, along symmetry directions, such as [21̄1̄0], the P-wave polarization is parallel and hence the S-wave polarizations are perpendicular to the propagation.

A more practical representation that is directly related to spherical plot in **Figure 8** is the pole figure plot of contoured and shaded velocities with polarizations shown in **Figure 9**. The circular nature of the velocity and polarization around the sixfold symmetry axis **c**[0001] is immediately apparent. The maximum shear wave splitting and SV velocity is in the basal (0001) plane. From the plot we can see that ε-phase of iron is very anisotropic at the experimental conditions of Mao *et al.* (1998) with a P-wave anisotropy of 7.1%. The shear wave anisotropy has a maximum of 26.3%, because in this transverse isotropic structure, SH has a minimum and SV has a maximum velocity in the basal plane. At first sight the pole figure plot of polarizations appears complex. To illustrate the representation of S-wave polarizations on a pole figure a single propagation direction in **Figure 10** has been drawn. In this chapter, stishovite has been chosen as an example mineral because it is very anisotropic and hence the angles between the polarizations are clearly not parallel (qP) or pendicular (qS1, qS2) to the propagation direction.

16.2.3 Measurement of Elastic Constants

Elastic properties can be measured by a various methods, including mechanical stress–strain, ultrasonic, resonant ultrasound spectroscopy Brillouin spectroscopy, nonhydrostatic radial X-ray diffraction, X-ray and neutron inelastic scattering, and shock measurements. In addition to physical measurements, atomic scale first-principles methods can predict elastic properties of crystals (see review by Karki *et al.* (2001), also *see* Chapter 6). The classical mechanical stress–strain measurements of elastic constants are no longer used due to the large errors and most compilations of single-crystal elastic constants (e.g., Bass, 1995; Isaak, 2001) are mainly based on ultrasonic measurements, the traditional technique at ambient conditions for large specimens. The measurement of the elastic constants for minerals from the deep Earth using classical techniques requires large (<1 cm^3) gem quality crystals. However, many minerals of the deep Earth are not stable, or meta-stable, at ambient conditions and no large gem quality crystals are available. Furthermore, the main applications of elastic constants

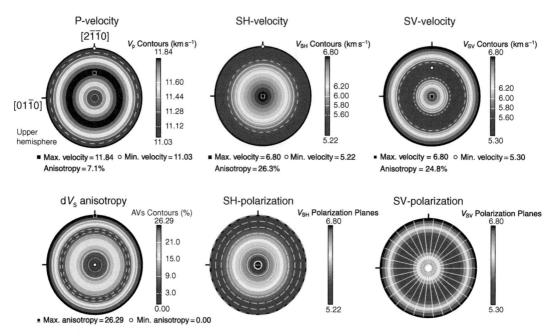

Figure 9 The pole figure of plot for hexagonal ε-phase iron where the circular symmetry around the [0001] axis is clearly visible for velocities, dV_s anisotropy and S-wave polarizations.

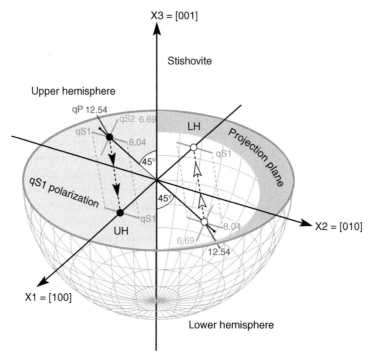

Figure 10 The pole figure of plot for strongly anisotropic stishovite illustrating the projection of the qS1 polarization onto the equatorial plane of projection. On the left-hand side the upper hemisphere projection down on to the equatorial plane is shown. On the right-hand side the lower hemisphere projection up toward the equatorial plane is shown. The velocities and the polarizations for qP, qS1, and qS2 have the same velocities (in km s^{-1}) and orientations, respectively, for the propagation direction in the positive (upper hemisphere) or negative (lower hemisphere) sense due to the centrosymmetic property of elasticity. Note that the change from an upper hemisphere to lower hemisphere projection for a centrosymmetic property is achieved by a 180° rotation in azimuth around X3.

are for the interpretation of seismological data at the high pressure and temperature conditions of the Earth's mantle and core. Although ultrasonic techniques are still widely used, new methods are constantly being developed and refined for measurements at higher pressures and temperatures (Liebermann and Li, 1998; Li et al., 2004; Jacobsen et al., 2005).

Ultrasonic measurements require a mechanical contact between the transducers that produce and detect the ultrasonic signal and the sample. For experiments at high pressure and temperature the contact is generally made via the high-pressure pistons and a high-temperature ceramic buffer rod or directly with the diamond anvils. Various corrections are necessary to take into account the ray paths through the pistons and buffer rods and transducer-bond phase shift effects. The technique most commonly used is based on the ultrasonic interferometry method introduced by Jackson and Niesler (1982) to obtain accurate pressure derivatives of single-crystal MgO to 3 GPa in a piston cylinder apparatus. The use of ultrasonic interferometry in conjunction with synchrotron X-radiation in multianvil devices permits a more accurate measurement of elastic constants at simultaneous high pressure and temperature (Li et al., 2004), as the sample length can be directly measured by X-radiography in situ, thereby reducing uncertainties in velocity measurements. A new gigahertz ultrasonic interferometer has recently been developed for the diamond anvil cell (Jacobsen et al., 2005). The gigahertz frequency reduces the wavelength in minerals to a few micrometers, which allows the determination of velocity and the elastic constants for samples of a few tens of micrometers in thickness. Another ultrasonic technique called resonant ultrasound spectroscopy has been used to study the elastic constants of minerals with high precision to very high temperatures at ambient pressure (e.g., Isaak, 1992; Isaak et al., 2005).

Brillouin scattering spectroscopy has become an established tool for measuring elastic constants since the introduction of laser sources and multipass Fabry–Perot interferometers (e.g., Vacher and Boyer, 1972). Unlike the ultrasonic techniques which usually measure the velocity in relatively few crystallographic directions, and where possible in pure mode directions, using Brillouin scattering the velocities can be measured along many propagation directions and the numerous velocities are inverted to obtain a least-squares determination of the elastic constants (Weidner and Carleton, 1977); hence, it is very suitable for the measurement of low-symmetry crystals. Brillouin scattering has several advantages for measurements of transparent minerals at high pressure and temperature in a diamond anvil cell; it only requires a small sample; no physical contact with the sample is required and the Brillouin peaks increase with temperature (Sinogeikin et al., 2005). The elastic constants of MgO have been determined to high pressure (55 Gpa; Zha et al., 2000) and high temperature (1500 K; Sinogeikin et al., 2005) using Brillouin scattering. A related technique using laser-induced phonon spectroscopy in the form of impulsive scattering in a diamond anvil cell has been used to measure elastic constants of mantle minerals to 20 GPa (e.g., Chai et al., 1977, 1997; Abramson et al., 1997).

Several X-ray and neutron diffraction and scattering techniques (e.g., Fiquet et al., 2004) have recently been developed to explore elastic behavior at extreme pressures (>100 GPa) in diamond anvils. Experiments using X-ray radial diffraction of polycrystalline samples under nonhydrostatic stress have been used to estimated the single-crystal elastic constants by using the measured lattice strains and crystal-preferred orientation combined with model polycrystalline stress distribution (e.g., Voigt or Reuss) (e.g., Mao et al., 1998; Merkel et al., 2005, 2006b). The uncertainties in the absolute value of the elastic constants may be on the order of 10–20% and model-dependent stress or strain distributions may be limited by the presence of elastic and plastic strain in some cases; however, this technique provides valuable information at extreme pressures. Inelastic X-ray scattering has provided volume averaged P-wave velocities of h.c.p. iron as a function of pressure to 110 GPa (Fiquet et al., 2001). Using knowledge of the CPO combined with different scattering geometries, Antonangeli et al. (2004) have determined the P-wave velocities in several directions and C_{11} elastic constant of h.c.p. iron, providing a valuable independent validation of results from radial diffraction. The nuclear-resonant inelastic X-ray scattering technique provides a direct probe of the phonon density of states. By the integration of the measured phonon density of states, the elastic and thermodynamic parameters are obtained; when combined with a thermal equation of state, the P- and S-wave velocities of h.c.p. iron have been determined to 73 GPa and 1700 K in a laser-heated diamond anvil cell by Lin et al. (2005), illustrating the rapid progress in this area.

16.2.4   Effective Elastic Constants for Crystalline Aggregates

The calculation of the physical properties from microstructural information (crystal orientation, volume fraction, grain shape, etc.) is important for uppermantle rocks because it gives insight into the role of microstructure in determining the bulk properties; it is also important for synthetic aggregates experimentally deformed at simulated conditions of the Earth's interior. A calculation can be made for the *in situ* state at high temperature and pressure of the deep Earth for samples where the microstructure has been changed by subsequent chemical alteration (e.g., the transformation olivine to serpentine) or mechanically induced changes (e.g., fractures created by decompression). The *in situ* temperatures and pressures can be simulated using the appropriate single-crystal derivatives. Additional features not necessarily preserved in the recovered microstructure, such as the presence of fluids (e.g., magma) can be modeled (e.g., Blackman and Kendall, 1997; Mainprice, 1997; Williams and Garnero, 1996). Finally, the effect of phase change on the physical properties can also be modeled using these methods (e.g., Mainprice *et al.*, 1990). Modeling is essential for anisotropic properties as experimental measurements in many directions necessary to fully characterize anisotropy are not currently feasible for the majority of the temperature and pressure conditions found in the deep Earth.

In the following, we will only discuss the elastic properties needed for seismic velocities, but the methods apply to all tensorial properties where the bulk property is governed by the volume fraction of the constituent minerals. Many properties of geophysical interest are of this type, for example, thermal conductivity, thermal expansion, elasticity and seismic velocities. However, these methods do not apply to properties determined by the connectivity of a phase, such as the electrical conductivity of rocks with conductive films on the grain boundaries (e.g., carbon). We will assume that the sample may be microscopically heterogeneous due to grain size, shape, orientation, or phase distribution, but will be considered macroscopically uniform. The complete structural details of the sample are in general never known, but a 'statistically uniform' sample contains many regions, which are compositionally and structurally similar, each fairly representative of the entire sample. The local stress and strain fields at every point $\mathbf{r}$ in a linear elastic polycrystal completely determined by Hooke's law are as follows:

$$\sigma_{ij}(\mathbf{r}) = C_{ijkl}(\mathbf{r})\varepsilon_{kl}(\mathbf{r})$$

where $\sigma_{ij}(\mathbf{r})$ is the stress tensor, $C_{ijkl}(\mathbf{r})$ is the elastic stiffness tensor, and $\varepsilon_{kl}(\mathbf{r})$ the strain tensor at point $\mathbf{r}$. The evaluation of the effective constants of a polycrystal would be the summation of all components as a function of position, if we know the spatial functions of stress and strain. The average stress $<\sigma>$ and strain $<\varepsilon>$ of a statistically uniform sample are linked by an effective macroscopic modulus C^* that obeys Hookes's law of linear elasticity,

$$C^* = <\sigma><\varepsilon>^{-1}$$

where

$$<\varepsilon> = \frac{1}{V} \int \varepsilon(r)\, \mathrm{d}r$$

and

$$<\sigma> = \frac{1}{V} \int \sigma(r)\, \mathrm{d}r$$

and V is the volume, the notation $<>$ denotes an ensemble average. The stress $\sigma(\mathbf{r})$ and strain $\varepsilon(\mathbf{r})$ distribution in a real polycrystal varies discontinuously at the surface of grains. By replacing the real polycrystal with a 'statistically uniform' sample, we are assuming that $\sigma(\mathbf{r})$ and strain $\varepsilon(\mathbf{r})$ are varying slowly and continuously with position $\mathbf{r}$.

A number of methods are available for determining the effective macroscopic effective modulus of an aggregate. We will briefly present these methods which try to take into account an increasing amount of microstructural information, which of course results in increasing theoretical complexity, but yields estimates which are closer to experimental values. The methods can be classified by using the concept of the order of the statistical probability functions used to quantitatively describe the microstructure (Kröner, 1978). A zero-order bound is given when one has no statistical information of the microstructure of the polycrystal and for example we do not know the orientation of the component crystals; in this case we have to use the single-crystal properties. The maximum and minimum of the single-crystal property are the zero-order bounds. The simplest and best-known averaging techniques for obtaining estimates of the effective elastic constants of polycrystals are the Voigt (1928) and Reuss (1929) averages. These averages only use the volume fraction of each phase, the orientation and the elastic constants of the single crystals or grains. In terms of statistical probability functions, these are first-order bounds as only the first-order correlation function is used, which is the volume fraction. Note no information about the shape or position of neighboring grains is used. The Voigt average is found by simply assuming that the strain field is everywhere constant (i.e., $\varepsilon(\mathbf{r})$ is independent of $\mathbf{r}$). The strain at every position is set equal to the macroscopic strain of the sample. C^* is then estimated by a volume average of local stiffnesses $C(\mathbf{g}_i)$ with orientation $\mathbf{g}_i$, and volume fraction V_i,

$$C^* \approx C^{\text{Voigt}} = \left[\sum_i V_i C(\mathbf{g}_i) \right]$$

Reuss average is found by assuming that the stress field is everywhere constant. The stress at every position is set equal to the macroscopic stress of the sample. C^* or S^* is then estimated by the volume average of local compliances $S(\mathbf{g}_i)$,

$$C^* \approx C^{\text{Reuss}} = \left[\sum_i V_i S(\mathbf{g}_i) \right]^{-1}$$

$$S^* \approx S^{\text{Reuss}} = \left[\sum_i V_i S(\mathbf{g}_i) \right]$$

$$C^{\text{Voigt}} \neq C^{\text{Reuss}} \quad \text{and} \quad C^{\text{Voigt}} \neq \left[S^{\text{Reuss}} \right]^{-1}$$

These two estimates are not equal for anisotropic solids with the Voigt being an upper bound and the Reuss a lower bound. A physical estimate of the moduli should lie between the Voigt and Reuss average bounds as the stress and strain distributions are expected to be somewhere between uniform strain (Voigt bound) and uniform stress (Reuss bound). Hill (1952) observed that arithmetic mean (and the geometric mean) of the Voigt and Reuss bounds, sometimes called the Hill or Voigt–Reuss–Hill (VRH) average, is often close to experimental values. The VRH average has no theoretical justification. As it is much easier to calculate the arithmetic mean of the Voigt and Reuss elastic tensors, all authors have tended to apply the Hill average as an arithmetic mean. In Earth sciences, the Voigt, Reuss, and Hill averages have been widely used for averages of oriented polyphase rocks (e.g., Crosson and Lin, 1971). Although the Voigt and Reuss bounds are often far apart for anisotropic materials, they still provide the limits within which the experimental data should be found.

Several authors have searched for a geometric mean of oriented polycrystals using the exponent of the average of the natural logarithm of the eigenvalues of the stiffness matrix (Matthies and Humbert, 1993). Their choice of this averaging procedure was guided by the fact that the ensemble average elastic stiffness $<C>$ should equal the inverse of the ensemble average elastic compliances $<S>^{-1}$, which is not true, for example, of the Voigt and Reuss estimates. A method of determining the geometric mean for arbitrary orientation distributions has been developed (Matthies and Humbert, 1993). The method derives from the fact that a stable elastic solid must have an elastic strain energy that is positive. It follows from this that the eigenvalues of the elastic matrix must all be positive. Comparison between Voigt, Reuss, Hill, and self-consistent estimates shows that the geometric mean provides estimates very close to the self-consistent method, but at considerably reduced computational complexity (Matthies and Humbert, 1993). The condition that the macroscopic polycrystal elastic stiffness $<C>$ must equal the inverse of the aggregate elastic compliance $<S>^{-1}$ would appear to be a powerful physical constraint on the averaging method (Matthies and Humbert, 1993). However, the arithmetic (Hill) and geometric means are very similar (Mainprice and Humbert, 1994), which tends to suggest that they are just mean estimates with no additional physical significance.

The second set of methods use additional information on the microstructure to take into account the mechanical interaction between the elastic elements of the microstructure. Mechanical interaction will be very important for rocks containing components of very different elastic moduli, such as solids, liquids, gases, and voids. The most important approach in this area is the 'self-consistent' (SC) method (e.g., Hill, 1965). The SC method was introduced for materials with a high concentration of inclusions where the interaction between inclusions is significant. In the SC method, an initial estimate of the anisotropic homogeneous background medium of the polycrystal is calculated using the traditional volume averaging method (e.g., Voigt). All the elastic elements (grains, voids, etc.) are inserted into the background medium using Eshelby's (1957) solution for a single ellipsoidal inclusion in an infinite matrix. The elastic moduli of the ensemble, inclusion, and background medium are used as the 'new' background medium for the next inclusion. The procedure is repeated for all inclusions and repeated in an iterative manner for the polycrystal until a convergent solution is found. The interaction is notionally taken into account by the evolution of the background medium that contains information about the inclusions, albeit in a homogenous form. As the inclusion can have an ellipsoidal shape an additional microstructural parameter is taken into account by this type of model.

Several scientists (e.g., Bruner, 1976) have remarked that the SC progressively overestimates the interaction with increasing concentration. They proposed an alternative differential effective medium (DEM) method in which the inclusion concentration is increased in small steps with a re-evaluation of the elastic constants of the aggregate at each increment. This scheme allows the potential energy of the medium to vary slowly with inclusion concentration (Bruner, 1976). Since the addition of inclusions to the backgound material is made in very small increments, one can consider the concentration step to be very dilute with respect to the current effective medium. It follows that the effective interaction between inclusions can be considered negligible and we can use the inclusion theory of Eshelby (1957) to take into account the interaction. In contrast, the SC uses Eshelby's theory plus an iterative evaluation of the background medium to take into account the interaction. Mainprice (1997) has compared the results of SC and DEM for anisotropic oceanic crustal and mantle rocks containing melt inclusions and found the results to be very similar for melt fractions of less than 30%. At higher melt fractions the SC exhibits a threshold value around 60% melt, whereas the DEM varies smoothly up a 100% melt. The presence of a threshold in the SC calculations is due to the specific way that the interaction is taken into account. The estimates of both methods are likely to give relatively poor results at high fractions of a phase with strong elastic contrast with the other constituents as other phenomena, such as mechanical localization related to the percolation threshold, are likely to occur.

The third set of methods uses higher-order statistical correlation functions to take into account the first- or higher-order neighbor relations of the various microstructural elements. The factors that need to be statistically described are the elastic constants (determined by composition), orientation, and relative position of an element. If the element is considered to be small relative to grain size, then grain shape and the heterogeneity can be accounted for the relative position correlation function. Nearest neighbors can be taken into account using a two-point correlation function, which is also called an autocorrelation function by some authors. If we use the 'statistically uniform' sample introduced above, we are effectively assuming that all the correlation functions used to describe the microstructure up to order infinity are statistically isotropic; this is clearly

a very strong assumption. In the special case where all the correlation functions up to order infinity are defined, Kröner (1978) has shown that the upper and lower bounds converge for the self-consistent method so that $C^{sc} = (S^{sc})^{-1}$. The statistical continuum approach is the most complete description and has been extensively used for model calculations (e.g., Beran $et\ al.$, 1996; Mason and Adams, 1999). Until recently, it has been considered too involved for practical application. With the advent of automated determination of crystal orientation and positional mapping using electron back-scattered diffraction (EBSD) in the scanning electron microscope (Adams $et\ al.$, 1993), digital microstructural maps are now available for the determination of statistical correlation functions. This approach provides the best possible estimate of the elastic properties but at the expense of considerably increased computational complexity.

The fact that there is a wide separation in the Voigt and Reuss bounds for anisotropic materials is caused by the fact that the microstructure is not fully described by such methods. However, despite the fact that these methods do not take into account such basic information as the position or the shape of the grains, several studies have shown that the Voigt or the Hill average are within 5–10% of experimental values for low porosity rocks free of fluids. For example, Barruol and Kern (1996) showed for several anisotropic lower crust and upper mantle rocks from the Ivrea zone in Italy that the Voigt average is within 5% of the experimentally measured velocity.

16.2.5 Seismic Properties of Polycrystalline Aggregates at High Pressure and Temperature

Orientation of crystals in a polycrystal can be measured by volume diffraction techniques (e.g., X-ray or neutron diffraction) or individual orientation measurements (e.g., U-stage and optical microscope, electron channelling or EBSD). In addition, numerical simulations of polycrystalline plasticity also produce populations of crystal orientations at mantle conditions (e.g., Tommasi $et\ al.$, 2004). An orientation, often given the letter **g**, of a grain or crystal in sample coordinates can be described by the rotation matrix between crystal and sample coordinates. In practice, it is convenient to describe the rotation by a triplet of Euler angles, for example, **g** $= (\varphi 1\ \ \phi\ \ \varphi 2)$ used by Bunge (1982). One should be aware that there are many different definitions of Euler angles that are used in the physical sciences. The orientation distribution function (ODF) $f(\mathbf{g})$ is defined as the volume fraction of orientations with an orientation in the interval between $\mathbf{g}$ and $\mathbf{g} + d\mathbf{g}$ in a space containing all possible orientations given by

$$\Delta V / V = \int f(\mathbf{g})\, d\mathbf{g}$$

where $\Delta V / V$ is the volume fraction of crystals with orientation $\mathbf{g}$, $f(\mathbf{g})$ is the texture function, and $d\mathbf{g} = 1/8\pi^2 \sin\phi\, d\varphi 1\, d\phi\, d\varphi 2$ is the volume of the region of integration in orientation space.

To calculate the seismic properties of a polycrystal, one must evaluate the elastic properties of the aggregate. In the case of an aggregate with a crystallographic fabric, the anisotropy of the elastic properties of the single crystal must be taken into account. For each orientation $\mathbf{g}$ the single-crystal properties have to be rotated into the specimen coordinate frame using the orientation or rotation matrix g_{ij},

$$C_{ijkl}(\mathbf{g}) = g_{ip} \cdot g_{jq} \cdot g_{kr} \cdot g_{lt}\, C_{pqrt}(\mathbf{g}^\circ)$$

where $C_{ijkl}(\mathbf{g})$ is the elastic property in sample coordinates, $g_{ij} = g(\varphi 1\, \phi\, \varphi 2)$ the measured orientation in sample coordinates, and $C_{pqrt}(\mathbf{g}^\circ)$ is the elastic property in crystal coordinates.

The elastic properties of the polycrystal may be calculated by integration over all possible orientations of the ODF. Bunge (1982) has shown that integration is given as

$$<C_{ijkl}>^m = \int C_{ijkl}{}^m(\mathbf{g}) \cdot f(\mathbf{g})\, d\mathbf{g}$$

where $<C_{ijkl}>^m$ is the elastic properties of the aggregate of mineral m. Alternatively, it may be determined by simple summation of individual orientation measurements:

$$<C_{ijkl}>^m = \sum C_{ijkl}{}^m(\mathbf{g}) \cdot v(\mathbf{g})$$

where $v(\mathbf{g})$ is the volume fraction of the grain in orientation $\mathbf{g}$. For example, the Voigt average of the rock for m mineral phases of volume fraction $v(m)$ is given as

$$<C_{ijkl}>^{\text{Voigt}} = \sum v(m) <C_{ijkl}>^m$$

The final step is the calculation of the three seismic phase velocities by the solution of the Christoffel equation, details of which are given above.

To calculate the elastic constants at pressures and temperatures, the single-crystal elastic constants are given at the pressure and temperature of their measurement by using the following relationship:

$$\begin{aligned}
C_{ij}(PT) = {}& C_{ij}(P_o\, T_o) + (dC_{ij}/dP) \cdot \Delta P \\
& + \frac{1}{2}(d^2 C_{ij}/dP^2) \cdot \Delta P^2 + (dC_{ij}/dT) \cdot \Delta T \\
& + (d^2 C_{ij}/dPdT) \cdot \Delta P \cdot \Delta T
\end{aligned}$$

where $C_{ij}(PT)$ are the elastic constants at pressure P and temperature T, $C_{ij}(P_o T_o)$ the elastic constants at a reference pressure P_o (e.g., 0.1 MPa) and temperature T_o (e.g., 25°C), dC_{ij}/dP is the first-order pressure derivative, dC_{ij}/dT is the first-order temperature derivative, $\Delta P = P - P_o$ and $\Delta T = T - T_o$. The equation is a Maclaurin expansion of the elastic tensor as a function of pressure and temperature, which is a special case of a Taylor expansion as the series is developed about the elastic constants at the reference condtion $C_{ij}(P_o T_o)$. The series only represent the variation of the C_{ij} in their intervals of pressure and temperature of convergence; in other words, the pressure and temperature range of the experiments or atomic modeling calculations used to determine the derivatives. Note this equation is not a polynomial and care has to taken when using the results of data fitted to polynomials, as for example the second-order derivatives fitted to a polynomial should be multiplied by 2 for use in the equation above, for example, second-order pressure derivatives for MgO given by Sinogeikin and Bass (2002). Also, note this equation is not a Eulerian finite strain equation of state (e.g., Davies, 1974, see below) and data fit to such an equation will not have derivatives compatible with the equation above, for example, the pressure derivatives of Brucite determined by Jiang et al. (2006). The second-order pressure derivatives $d^2 C_{ij}/dP^2$ are available for an increasing number of mantle minerals (e.g., olivine, orthopyroxene, garnet, MgO) and first-order temperature derivatives seem to adequately describe the temperature dependence of most minerals, although second-order derivatives are also available in a few cases (e.g., garnet, fayalite, fosterite, rutile; see Isaak (2001) for references). Experimental measurements of the cross pressure–temperature derivatives $d^2 C_{ij}/dPdT$ (that is the temperature derivative of the C_{ij}/dP at constant temperature) are still very rare. For example, despite the fact that MgO (periclase) is a well-studied reference material for high-pressure studies, the complete set of single-crystal cross-derivatives were measured for the first

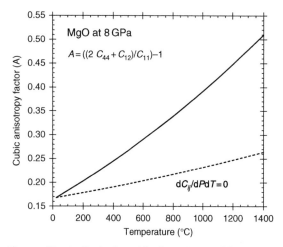

Figure 11 An illustration of the importance of the cross pressure–temperature derivatives for cubic MgO at 8 GPa pressure. With increasing temperature at constant pressure, the anisotropy measured by the cubic anisotropy factor (A) increases by a factor of 2. Data from Chen *et al.* (1998).

time by Chen *et al.* (1998) to 8 GPa and 1600 K. The effect of the cross-derivatives on the V_p and dV_s anisotropy of MgO is dramatic; the anisotropy is increased by a factor of 2 when cross derivatives are used (**Figure 11**). Note when a phase transitions occurs, then the specific changes in elastic constants at pressures near the phase transition will have to be taken into account, for example, the SiO_2 polymorphs (Karki *et al.*, 1997b; Cordier *et al.*, 2004a; Carpenter, 2006). The seismic velocities also depend on the density of the minerals at pressure and temperature which can be calculated using an appropriate equation of state (Knittle, 1995). The Murnaghan equation of state derived from finite strain is sufficiently accurate at moderate compressions (Knittle, 1995) of the upper mantle and leads to the following expression for density as a function of pressure:

$$\rho(P) = \rho_o(1 + (K'/K) \cdot (P - P_o))^{1/K'}$$

where K is bulk modulus, $K' = dK/dP$ the pressure derivative of K, ρ_o is the density at reference pressure P_o and temperature T_o. For temperature the density varies as

$$\rho(T) = \rho_o\left[1 - \int \alpha_v(T) dT\right] \approx \rho_o[1 - \alpha_{av}(T - T_o)]$$

where $\alpha_v(T) = (1/V)(\partial V/\partial T)$ is the volume thermal expansion coefficient as a function of temperature and α_{av} is an average value of thermal expansion which is constant over the temperature range (Fei,

1995). According to Watt (1988) an error of less 0.4% on the P- and S-velocity results from using α_{av} to 1100 K for MgO. For temperatures and pressures of the mantle, the density is described by

$$\rho(P, T) = \rho_o\left\{(1 + (K'/K) \cdot (P - P_o))^{1/K'}[1 - \alpha_{av}(T - T_o)]\right\}$$

An alternative approach for the extrapolation elastic constants to very high pressures is Eulerian finite strain theory (e.g., Davies, 1974). The theory is based on a Maclaurin expansion of the free energy in terms of Eulerian finite volumetric strain. For example, Karki *et al.* (2001) reformulated Davies's equations for the elastic constants in terms of finite volumetric strain (f) as

$$C_{ijkl}(f) = (1 + 2f)^{7/2}$$
$$\times \left[C_{oijkl} + b_1 f + \frac{1}{2}b_2 f + \cdots\right] - P\Delta_{ijkl}$$

where

$$f = \frac{1}{2}\left[(V_o/V)^{2/3} - 1\right]$$

$$b_1 = 3K_o\left(dC_{oijkl}/dP\right) - 5C_{oijkl}$$

$$b_2 = 9K_o^2\left(d^2 C_{oijkl}/dP^2\right)$$
$$+ 3(K_o/dP)\left(b_1 + 7C_{oijkl}\right) - 16b_1 - 49C_{oijkl}$$

$$\Delta_{ijkl} = -\delta_{ij}\delta_{kl} - \delta_{ik}\delta_{jl} - \delta_{il}\delta_{jk}$$

However, with advent of practical computational methods for applying first-principles (*ab initio*) methods to calculation of elastic constants of minerals at extreme pressures reduces the Eulerian finite strain theory to a descriptive tool, if tensors are available at the appropriate pressure (and temperature) conditions. The simple Maclaurin series expansions given above for pressure and temperature are a compact way of describing the variation of the elastic tenors in experimentations at high pressure and temperature. Extrapolation of the simple Maclaurin series expansions outside the range of experimental (or computational) data is not recommended, as the formulation is descriptive. The relative complexity of the Eulerian finite volumetric strain formulation has the merit of a physical basis and hence extrapolation beyond the experimental data range may be undertaken with caution. In practice, applications of the Eulerian finite volumetric strain formulation have been limited to high symmetry crystals (e.g., Li *et al.*, 2006b, cubic Ca-pervoskite, trigonal brucite;

Jiang *et al.*, 2006), a new thermodynamically correct formulation of the problem in terms of pressure and temperature has recently proposed by Stixrude and Lithgow-Bertelloni (2005a, 2005b).

16.2.6 Anisotropy Minerals in the Earth's Mantle and Core

To understand the anisotropic seismic behavior of polyphase rocks in the Earth's mantle, it is instructive to first consider the properties of the component single crystals. In this section emphasis is on the anisotropy of individual minerals rather than the magnitude of velocity. The percentage anisotropy (*A*) is defined here as $A = 200 (V_{maximum} - V_{minimum})/(V_{maximum} + V_{minimum})$, where the maximum and minimum are found by exploring a hemisphere of all possible propagation directions. Note for P-wave velocities the anisotropy is defined by the maximum and minimum velocities in two different propagation directions, for example, the maximum *A* is given by the maximum and minimum V_p in a hemipshire, or for V_p in two specific directions such as the vertical and horizontal can be used. For S-waves in an anisotropic medium there are two orthogonally polarized S-waves with different velocites for each propagation direction; hence, *A* can be defined for each direction. The consideration of the single-crystal properties is particularly important for the transition zone (410–660 km) and lower mantle (below 660 km) as the deformation mechanisms and resulting preferred orientation of these minerals under the extreme conditions of temperature and pressure are very poorly documented by experimental investigations. In choosing the anisotropic single-crystal properties, where possible, the most recent experimental determinations have been included. A major trend in recent years is the use of computational modeling to determine the elastic constants at very high pressures and more recently at high temperatures. The theoretical modeling gives a first estimate of the pressure and temperature derivatives in a range not currently accessible to direct measurement (see review by Karki *et al.* (2001), also *see* Chapter 6). Although there is an increasing amount of single-crystal data available to high temperature or pressure, no data is available for simultaneous high temperature and pressure of the Earth's lower mantle or inner core (see **Figure 1** for the pressure and temperatures).

16.2.6.1 Upper mantle

The upper mantle (down to 410 km) is composed of three anisotropic and volumetrically important phases: olivine, enstatite (orthopyroxene), and diopside (clinopyroxene). The other volumetrically important phase is garnet, which is nearly isotropic and hence not of great significance to our discussion of anisotropy.

Olivine. A certain number of accurate determinations of the elastic constants of olivine are now available which all agree that the anisotropy of V_p is 25% and maximum anistropy of V_s is 18% at ambient conditions for a mantle composition of about Fo90. The first-order temperature derivatives have been determined between 295 and 1500 K for forsterite (Isaak *et al.*, 1989a) and olivine (Isaak, 1992). The first- and second-order pressure derivatives for olivine were first determined to 3 GPa by Webb (1989). However, a determination to 17 GPa by Zaug *et al.* (1993) and Abramson *et al.* (1997) has shown that the second-order derivative is only necessary for elastic stiffness modulus C_{55}. The first-order derivatives are in good agreement between these two studies. The second-order derivative for C_{55} has proved to be controversial. Zaug *et al.* (1993) where first to measure nonlinear variation of C_{55} with pressure, but other studies have not reproduced this behavior (e.g., for olivine: Chen *et al.*, 1996 and Zha *et al.*, 1998 or fosterite Zha *et al.*, 1996). The anisotropy of the olivine single crystal increases slightly with temperature (+ 2%) using the data of Isaak (1992) and reduces slightly with increasing pressure using the data of Abramson *et al.* (1997).

Orthopyroxene. The elastic properties of orthopyroxene (Enstatite or Bronzite) with a magnesium number (Mg/(Mg + Fe)) near the typical upper mantle value of 0.9 has also been extensively studied. The V_p anisotropy varies between 15.1% (En80 Bronzite; Frisillo and Barsch (1972)) and 12.0% (En100 Enstatite; Weidner *et al.* (1978)) and the maximum V_s anisotropy between 15.1% (En80 Bronzite; Webb and Jackson, 1993), and 11.0% (En100 Enstatite; Weidner *et al.*, 1978). Some of the variation in the elastic constants and anisotropy may be related to composition and structure in the orthopyroxenes. The first-order temperature derivatives have been determined over a limited range between 298 and 623 K (Frisillo and Barsch, 1972). The first- and second-order pressure derivatives for Enstatite have been determined up to 12.5 GPa by Chai *et al.* (1997). This study confirms an earlier one of Webb and Jackson to 3 GPa that showed that first- and second-order pressure derivatives are needed to describe the elastic constants at mantle pressures. The anisotropy of V_p and V_s does not vary significantly with pressure using the data of Chai *et al.* (1997) to 12.5 GPa. The anisotropy of V_p and V_s does increase by about 3% when extrapolating to 1000°C using the first-order temperature derivatives of Frisillo and Barsch (1972).

Clinopyroxene. The elastic constants of clinopyroxene (Diopside) of mantle composition have only been experimentally measured at ambient conditions (Levien *et al.*, 1979; Collins and Brown, 1998); both studies show that V_p anisotropy is 29% and V_s anisotropy is between 20% and 24%. There are no measured single-crystal pressure derivatives. In one of the first calculations of the elastic constants of a complex silicate at high pressure, Matsui and Busing (1984) predicted the first-order pressure derivatives of diopside from 0 to 5 GPa. The calculated elastic constants at ambient conditions are in good agreement with the experimental values and the predicted anisotropy for V_p and V_s of 35.4% and 21.0%, respectively, is also in reasonable agreement. The predicted bulk modulus of 105 GPa is close to the experimental

value of 108 GPa given by Levien *et al.* (1979). The pressure derivative of the bulk modulus 6.2 is slightly lower than the value of 7.8 ± 0.6 given by Bass *et al.* (1981). Using the elastic constants of Matsui and Busing (1984), the V_p anisotropy decreases from 35.4% to 27.7% and V_s increases from 21.0% to 25.5% with increasing pressure from ambient to 5 GPa. In the absence of experimental measurements, the author would recommend using the values given by Matsui and Busing (1984). A major problem until recently was the lack of clinopyroxene temperature derivatives. Isaak *et al.* (2005) have mesured the temperature derivatives of chrome diopside to 1300 K at room pressure that are notably smaller than other mantle minerals. The single-crystal seismic properties of olivine, enstatite, and diopside at 220 km depth are illustrated in **Figure 12**. Garnet is nearly isotropic with V_p anisotropy of 0.6% and V_s of 1.3%. With an increasing interest in the water cycle in the deep Earth and subduction processes, the properties of hydrated phases are being studied. Although very little is currently known about the elastic properties of olivine containing water, the single-crystal elastic constants of the A-phase have

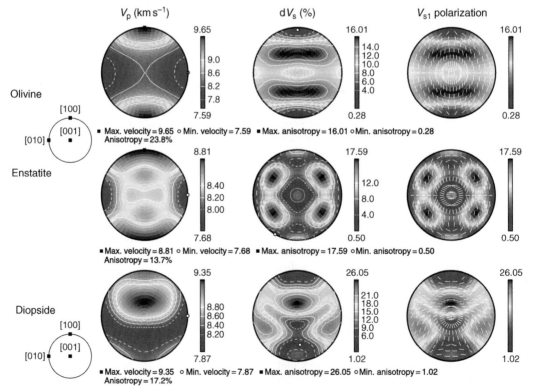

Figure 12 Single-crystal anisotropic seismic properties of upper mantle minerals olivine (orthorhombic), enstatite (orthorhombic), and diopside (monoclinic) at about 220 km (7.1 GPa, 1250°C).

been determined at ambient conditions (Sanchez-Valle *et al.*, 2006). The A-phase is the only dense hydrous magnesium silicate occurring along the fosterite–humite join. The A-phase is hexagonal and can contain up 11.8 wt.% water and is about 8% less dense than olivine. The hydration of forsterite to phase A decreases the bulk and shear moduli by about 18% and 21%, respectively, while both compressional and shear wave velocities decrease by about 7%. These results suggest that water could be identified seismologically, if phase A is present in abundance in cold (below 1000°C at 15 GPa) subducted slabs. The A-phase is anisotropic with V_p anisotropy of 11.6% and dV_s of 15.8%. However, their actuel presence in the mantle is currently considered as speculative (see Jacobsen (2006)).

16.2.6.2 Transition zone

Over the last 20 years a major effort has been made to experimentally determine the phase petrology of the transition zone and lower mantle. Although single crystals of upper-mantle phases are readily available, single crystals of transition zone and lower mantle for elastic constant determination have to be grown at high pressure and high temperature. The petrology of the transition zone is dominated by garnet, majorite, wadsleyite, ringwoodite, calcium-rich perovskite, clinopyroxene, and possibly stishovite.

Majorite. The pure Mg end-member majorite of the majorite–pyrope garnet solid solution has tetragonal symmetry and is weakly anisotropic with 1.8% for V_p and 9.1% for V_s (Pacalo and Weidner, 1997). A study of the majorite–pyrope system by Heinemann *et al.* (1997) shows that tetragonal form of majorite is restricted to a composition of less 20% pyrope and hence is unlikely to exist in the Earth's transition zone. Majorite with cubic symmetry is nearly isotropic with V_p anisotropy of 0.5% and V_s of 1.1%. Pressure derivatives and temperature derivatives for majorite and majorite–pyrope have been determined by Sinogeikin and Bass (2002), respectively. Cubic majorite has very similar properties to pyrope garnet (Chai *et al.*, 1997) as might be expected. The elastic properties of sodium-rich Majorite have studied by Pacalo *et al.* (1992).

Wadsleyite. The elastic constants of Mg_2SiO_4 wadsleyite were first determined by Sawamoto *et al.* (1984) and this early determination was confirmed by Zha *et al.* (1987) with a V_p anisotropy of 16% and V_s of 17%. The $(Mg, Fe)_2SiO_4$ wadsleyite has slightly lower velocities and higher anisotropies (Sinogeikin *et al.*, 1998). The first-order pressure derivatives

determined from the data of Zha *et al.* (1997) to 14 GPa show that the anisotropy of Mg_2SiO_4 wadsleyite decreases slightly with increasing pressure. At pressures corresponding to the 410 km seismic discontinuity (*c.* 13.8 GPa), the V_p anisotropy would be 11.0% and V_s 12.5%.

Ringwoodite. The elastic constants of Mg_2SiO_4 ringwoodite were first measured by Weidner *et al.* (1984) and $(Mg, Fe)_2SiO_4$ ringwoodite by Sinogeikin *et al.* (1998) at ambient conditions with V_p anisotropy of 3.6% and 4.7%, and V_s of 7.9% and 10.3%, respectively. Kiefer *et al.* (1997) have calculated the elastic constants of Mg_2SiO_4 ringwoodite to 30 GPa. Their constants at ambient conditions give a V_p anisotropy of 2.3% and V_s of 4.8% very similar to the experimental results of Weidner *et al.* (1984). There is a significant variation (5–0%) of the anisotropy of ringwoodite with pressure, 15 GPa (*c.* 500 km depth) the V_p anisotropy is 0.4% and V_s is 0.8%; hence, ringwoodite is nearly perfectly isotropic at transition zone pressures. Single-crystal temperature derivatives have been measured for ringwoodite (Jackson *et al.*, 2000), but none are available for wadsleyite. Olivine transforms to wadsleyite at about 410 km, and wadsleyite transforms to ringwoodite at about 500 km, both transformations result in a decrease in anisotropy with depth. The gradual transformation of clinopyroxene to majorite between 400 and 475 km would also result in a decrease in anisotropy with depth. The seismic anisotropy of wadsleyite and ringwoodite is illustrated in **Figure 13**. A recent *ab initio* molecular dynamics study by Li *et al.* (2006a) has shown that ringwoodite is nearly isotropic at transition zone conditions with P- and S-wave anisotropy close to 1%, extrapolated experimental values to a depth of 550 km (19.1 GPa, 1520°C) suggest that the V_p and V_s anisotropy are 3.3% and 8.2%. Both Ringwoodite and Wadsleyite have hydrous forms. Wadsleyite can contain up to 3.3 wt.% water, in fact perfectly anhydrous Wadsleyite is unknown. No single-crystal elastic constants are available for hydrous Wadsleyite. Ringwoodite can contain up to 2.5 wt.% water and the single-crystal elastic properties of pure Mg end member (Inoue *et al.*, 1998; Wang *et al.*, 2003) and Fo90 hydrous ringwoodite (Jacobsen *et al.*, 2004) have be measured at ambiante conditions. The pressure derivatives for Fo90 hydrous ringwoodite have recently been determined to 9 GPa (Jacobsen and Smyth, 2006). Hydrous ringwoodite has the same anisotropy as the anhydrous mineral at ambient pressure. Both forms have an anisotropy that decreases with pressure, the hydrous from would become

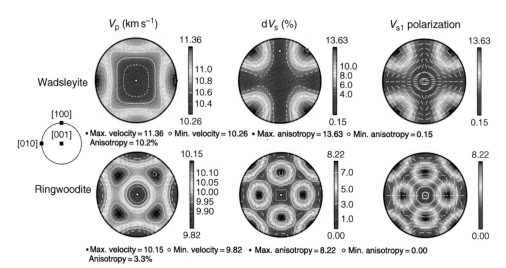

Figure 13 Single-crystal anisotropic seismic properties of transition zone minerals wadsleyite (orthorhombic) at about 450 km (15.2 GPa, 1450°C) and ringwoodite (cubic) at about 550 km (19.1 GPa, 1520°C).

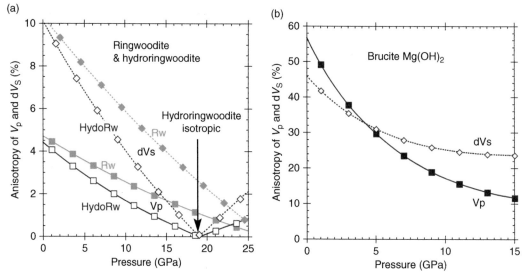

Figure 14 The variation of seismic P and S anisotropy with increasing pressure for hydrous minerals. Note the decrease in anisotropy with increasing pressure. (a) Ringwoodite (cubic) and hydrous ringwoodite (cubic)). (see text for details). (b) Brucite (trigonal). (a) Data from Jacobsen SD and Smyth JR (2006) Effect of water on the sound velocities of ringwoodite in the transition zone. In: Jacobsen SD and van der Lee S (eds.) *Geophysical Monograph Series, 168: Earth's Deep Water Cycle*, pp. 131–145. Washington, DC: American Geophysical Union. (b) Data from Jiang F, Speziale S and Duffy TS (2006) Single-crystal elasticity of brucite, $Mg(OH)_2$, to 15 GPa by Brillouin scattering. *American Mineralogist* 91: 1893–1900.

isotropic at about 17 GPa (**Figure 14(a)**). Comparison with Brucite $Mg(OH)_2$ is interesting as it is a model system for understanding dense hydrous magnesium minerals (alphabet phases) under hydrostatic compression and an important structural unit of many layer silicates, such as chlorite, lizardite and talc. The single-crystal elastic constants of Brucite have recently been measured to a pressure of 15 GPa by Jiang *et al.* (2006). The seismic anisotropy of brucite is

exceptionally high at ambient conditions with P- and S-wave anisotropy of 57% and 46%, respectively. As in hydrous ringwoodite both forms of anisotropy decrease with increasing pressure being 12% and 24% at 15 GPa for P- and S-wave anisotropy of 57% and 46%, respectively (**Figure 14(b)**). The stronger decrease of P-wave anisotropy compared to S-wave is probably related to the very important linear compressibility along the c-axis. The apparent symmetry of the

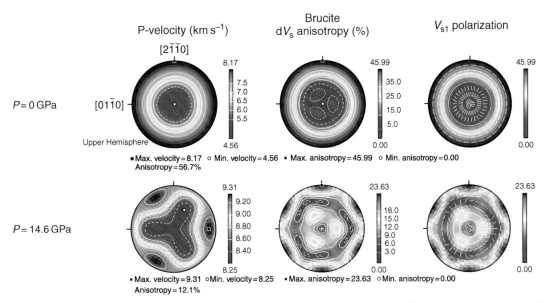

Figure 15 Single-crystal anisotropic seismic properties of brucite (trigonal) at 0 and 14.6 GPa, data from Jiang *et al.* (2006). Note the change in velocity distribution of P-waves, dV$_s$ anisotropy and S1 polarization orientations, which reflect the nearly hexagonal symmetry of Brucite's elastic properties at low pressure and the increasingly trigonal nature (threefold *c*-axis at the center of the pole figure) with increasing pressure.

seismic anisotropy of brucite changes with pressure (**Figure 15**). Brucite is stable to 80 GPa and modest temperatures, so it could be an anisotropic component of cold subducted slabs.

16.2.6.3 Lower mantle

The lower mantle is essentially composed of perovskite, ferripericlase, and possibly minor amount of SiO$_2$ in the form of stishovite in the top part of the lower mantle (e.g., Ringwood, 1991). Ferripericlase is the correct name for (Mg, Fe),O with small percentage of iron, less than 50% in Mg site; previously this mineral was incorrectly called magnesiowustite, which should have more than 50% Fe. It is commonly assumed that there is about 20% Fe in ferripericlase in the lower mantle. MgSiO$_3$ may be in the form of perovskite or possibly ilmenite. The ilmenite structured MgSiO$_3$ is most likely to occur at the bottom of the transition zone and top of the lower mantle. In addition, perovskite transforms to postperovskite in the D″ layer, although extact distribution with depth (or pressure) of the phases will depend on the local temperature and their iron content.

Perovskite (MgSiO$_3$, CaSiO$_3$). The first determination of the elastic constants of pure MgSiO$_3$ perovskite at ambient conditions was given by Yeganeh-Haeri *et al.* (1989). However this determination has been replaced by a more accurate study of a better quality crystal (Yeganeh-Haeri, 1994). The

1994 study gives V$_p$ anisotropy of 13.7% and V$_s$ of 33.0%. The [010] direction has the maximum dV$_s$ anisotropy. A new measurement of the elastic constants of MgSiO$_3$ perovskite at ambient conditions was made by Sinogeikin *et al.* (2004), gives V$_p$ anisotropy of 7.6% and dV$_s$ of 15.4%, which has very similar velocity distribution to the determination of Yeganeh-Haeri (1994), but the anisotropy is reduced by a factor of 2. Karki *et al.* (1997a) calculated the elastic constants of MgSiO$_3$ perovskite 140 GPa at 0 K. The calculated constants are in close agreement with the experimental measurements of Yeganeh-Haeri (1994) and Sinogeikin *et al.* (2004). Karki *et al.* (1997a) found that significant variations in anisotropy occurred with increasing pressure, first decreasing to 6% at 20 GPa for V$_p$ and to 8% at 40 GPa for V$_s$ and then increasing to 12% and 16%, respectively, at 140 GPa. At the 660 km seismic discontinuity (*c.* 23 GPa) the V$_p$ and V$_s$ anisotropy would be 6.5% and 12.5%, respectively. Recent progress in finite temperature first-principles methods for elastic constants has allowed their calculation at lower mantle pressures and temperatures. Oganov *et al.* (2001) calculated the elastic constants of Mg-perovskite at two pressures and three temperatures for the lower mantle (**Figure 16**). More recently, Wentzcovitch *et al.* (2004) have calculated the elastic constants over the complete range of lower mantle conditions and produced pressure and temperature derivatives.

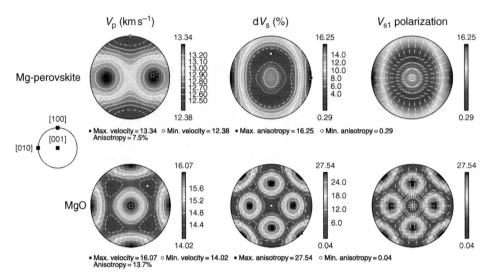

Figure 16   Single-crystal anisotropic seismic properties of lower mantle minerals Mg-perovskite (orthorhombic) and MgO (periclase – cubic) at about 2000 km (88 GPa, 3227°C) using the elastic constants determined at high PT by Oganov *et al.* (2001) and Karki *et al* (2000) respectively.

The results for pure Mg-perovskite from Oganov *et al.* and Wentzcovitch *et al.* agree quite closely for P-wave anisotropy, but Oganov's elastic constants give a higher S-wave anisotropy. At ambient conditions the results from all studies are very similar with V_p anisotropy is 13.7% and the V_s anisotropy is 33.0%. When extrapolated along a geotherm using the pressure and temperature derivatives of Wentzcovitch *et al.* (2004) the P- and S-wave anisotropies about the same at 8% at 1000 km depth and again similar anisotropies of 13% at 2500 km depth. The other perovskite structure present in the lower mantle is $CaSiO_3$ perovskite, recent *ab initio* molecular dynamics study by Li *et al.* (2006b) has shown that this mineral is nearly isotropic at lower mantle conditions with P- and S-wave anisotropy close to 1%.

MgSiO₃ ilmenite. Experimental measurements by Weidner and Ito (1985) have shown that $MgSiO_3$ ilmenite of trigonal symmetry is very anisotropic at ambient conditions with V_p anisotropy of 21.1% and V_s of 36.4%. Pressure derivatives to 30 GPa have been obtained by first-principles calculation (Da Silva *et al.*, 1999) and the anisotropy decreases with increasing pressure to 9.9% for P-waves and 24.8% for S-waves at 30 GPa.

Ferripericlase. The other major phase is ferripericlase $(Mg, Fe)O$, for which the elastic constants have been determined. The elastic constants of the pure end-member periclase MgO of the MgO–FeO solid solution series has been measured to 3 GPa by Jackson and Niesler (1982). Isaak *et al.* (1989a,

1989b) have measured the temperature derivatives for MgO to 1800 K. Both these studies indicate a V_p anisotropy of 11.0% and V_s of 21.5% at ambient conditions. Karki *et al.* (1997c) calculated the elastic constants of MgO to 150 GPa at 0 K. The thermoelasticty of MgO at lower mantle temperatures and pressures has been studied by Isaak *et al.* (1990), Karki *et al.* (1999, 2000) (**Figure 16**), and more recently by Sinogeikin *et al.* (2004, 2005); there is good agreement between these studies for the elastic constants and pressure and temperature derivatives. However, the theoretical studies do not agree with experimentally measured cross pressure–temperature derivatives of Chen *et al.* (1998). At the present time only the theoretical studies permit the exploration of the seismic properties of MgO at lower mantle conditions. They find considerable changes in anisotropy is preserved at high temperature with increasing pressure, along a typical mantle geotherm, MgO is isotropic near the 670 km discontinuity, but the anisotropy of P- and S-waves increases rapidly with depth reaching 17% for V_p and 36% for V_s at the D″ layer. The anisotropy of MgO increases linearly from 11.0% and 21.5% for V_p and V_s, respectively, at ambient conditions to 20% and 42%, respectively, at 1800 K according to the data of Isaak *et al.* (1989b). The effect of temperature on anisotropy is more important a low pressure than at lower mantle pressures, where the effect of pressure dominates according to the results of Karki *et al.* (1999). Furthermore, not only the magnitude of the anisotropy of MgO, but also the orientation of the anisotropy changes with increasing pressure according to the calculations of Karki *et al.* (1999), for example, the fastest V_p is parallel to [111] at ambient pressure and becomes parallel to [100] at 150 GPa pressure and fastest S-wave propagating in the [110] direction has a polarization parallel to [001] at low pressure that changes to [1–10] at high pressure. Ferripericlase–magnesiowustite solid solution series has been studied at ambient conditions by Jacobsen *et al.* (2002), for ferripericlase with 24% Fe the P-wave anisotropy is 10.5% and maximum S-wave is 23.7%, slightly higher than pure MgO at the same conditions. Data are required at lower mantle pressures to evaluate if the presence of iron has a significant effect on anisotropy of ferripericlase of mantle composition.

SiO₂ polymorphs. – The free SiO_2 in the transition zone and the top of the lower mantle (to a depth of 1180 km or 47 GPa) will be in the form of stishovite. The original experimental determination of the single-crystal elastic constants of stishovite by Weidner *et al.* (1982) and the more recent calculated constants of Karki *et al.* (1997b) both indicate a V_p and V_s anisotropy at ambient

conditions of 26.7–23.0% and 35.8–34.4%, respectively, making this a highly anisotropic phase. The calculations of Karki *et al.* (1997a) show that the anisotropy increases dramatically as the phase transition to $CaCl_2$ structured SiO_2 is approached at 47 GPa. The V_p anisotropy increases from 23.0% to 28.9% and V_s from 34.4% to 161.0% with increasing pressure from ambient to 47 GPa. The maximum V_p is parallel to [001] and the minimum parallel to [100]. The maximum dV_s is parallel to [110] and the minimum parallel to [001].

Postperovskite. Finally, this new phase is present in the D″ layer. Discovered and published in May 2004 by Murakami *et al.* (2004), the elastic constants at 0 K were rapidly established at low (0 GPa) and high (120 GPa) pressure by static atomistic calculations (Iitaka *et al.*, 2004; Oganov and Ono, 2004; Tsuchiya *et al.*, 2004). From these first results we can see that Mg-postperovskite is very different to Mg-perovskite as there are substantial changes in the distribution of the velocity anisotropy with increasing pressure (**Figure 17**). At zero pressure the anisotropy is very high, 28% and 47% for

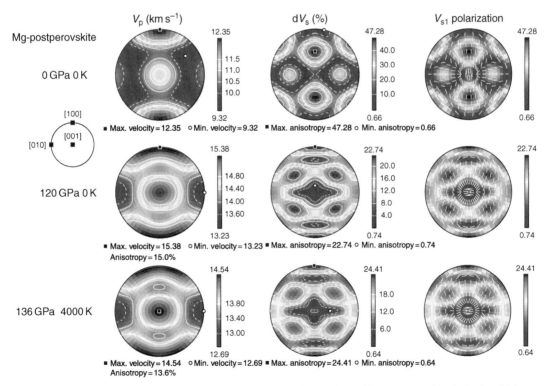

Figure 17 Single-crystal anisotropic seismic properties of the D″ layer mineral Mg-postperovskite (orthorhombic). Increasing the pressure from 0 to 120 GPa at a temperature of 0 K decreases the anisotropy and also changes the distribution the maximum velocities and S-wave polarizations. Elastic constants calculated by Tsuchiya *et al.* (2004). Increasing the pressure to 136 GPa and temperature to 4000 K, there are relatively minor changes in the anisotropy. Elastic constants calculated by Stackhouse *et al.* (2005).

P- and S-waves, respectively. The maximum for V_p is parallel to [100] with small submaxima parallel to [001] and minimum near [111]. The shear wave splitting (dV_s) has maxima parallel to <101> and <110>. At 120 GPa the anisotropy has reduced to 15% and 22% for P- and S-waves, respectively. The distribution of velocities has changed, with the P maximum still parallel to [100] and submaximum parallel to [001] which is now almost the same velocity as parallel to [100], the minimum is now parallel to [010]. The S-wave splitting maxima have also changed and are now parallel to <111>. Apparently, the compression of the postperovskite stucture has caused important elastic changes as in MgO. An *ab initio* molecular dynamics study by Stackhouse *et al.* (2005) at high temperature revealed that the velocity distribution and anisotropy were little affected by inceasing the temperature from 0 to 4000 K when at a pressure of 136 GPa (**Figure 17**). Wentzcovitch *et al.* (2006) produced a more exten-sive set of high-pressure elastic constants and pressure and temperature derivatives, similar P distributions and slightly different shear wave splitting pattern with maximum along the [001] axis which is not present in the Stackhouse *et al.* velocity surfaces or in high pressure 0 K results.

In conclusion for the mantle, we can say that the general trend favors an anisotropy decrease with increasing pressure and increase with increasing temperature; olivine is a good example of this beha-vior for minerals in upper mantle and transition zone. The changes are limited to a few percent in most cases. The primary causes of the anisotropy changes are minor crystal structural rearrangements rather than velocity changes due to density change caused by compressibility with pressure or thermal expansion with temperature. The effect of tempera-ture is almost perfectly linear in many cases; some minor nonlinear effects are seen in diopside, MgO, and SiO_2 polymorphs. Nonlinear effects with increasing pressure on the elastic constants cause the anisotropy of wadsleyite, ringwoodite to first decrease. In the case of the lower mantle minerals Mg-perovskite and MgO, there is a steady increase in the anisotropy in increasing depth; this is a very marked effect for MgO. Stishovite also shows major changes in anisotropy in the pressure range close to the transformation to the $CaCl_2$ structure. The sin-gle-crystal temperature derivatives of wadsleyite, ilmenite $MgSiO_3$, and stishovite are currently unknown which make quantitative seismic

anisotropic modeling of the transition zone and upper part of the lower-mantle speculative. To illustrate the variation of anisotropy as a function of mantle conditions of temperature and pressure, the seismic properties along a mantle geotherm (**Figure 18**) were calculated. The mantle geotherm is based on the PREM model for the pressure scale. The temperature scale is based on the continental geotherm of Mercier (1980) from the surface to 130 km and Ito and Katsura (1989) for the transition zone and Brown and Shankland (1981) for the lower mantle. The upper-mantle minerals olivine (V_p, V_s) and enstatite (V_p) show a slight increase of aniso-tropy in the first 100 km due to the effect of temperature. With increasing depth, the trend is for decreasing anisotropy except for V_s of enstatite and diopside. In the transition zone and lower man-tle, the situation is more complex due to the presence of phase transitions. In the transition zone diopside may be present to about 500 km with an increasing V_s and decreasing V_p anisotropy with depth. Wadsleyite is less anisotropic than olivine at 410 km, but significantly more anisotropic than ringwoodite found below 520 km. Although the lower mantle is known to be seismically isotropic, the constituent minerals are anisotropic. MgO shows important increase in anisotropy with depth (10–30%), at 670 km it is isotropic and 2800 km it is very anisotropic, possibly being candidate mineral to explain anisotropy of the D″ layer. Mg-perovskite is strongly anisotropic (*c.* 10%) throughout the lower mantle. The SiO_2 polymorphs are all strongly anisotropic, particularly for S-waves. If free silica is present in the transition zone or lower mantle, due perhaps to the presence of subducted basalt (e.g., Ringwood, 1991), then even a small volume fraction of the SiO_2 polymorphs could influence the seismic anisotropy of the mantle. However, to do so, the SiO_2 polymorphs would have to be oriented, either due to dislocation glide (plastic flow), oriented grain growth, or anisometric crystal shape (viscous flow) (e.g., Mainprice and Nicolas, 1989). Given that the SiO_2 polymorphs are likely to be 10% less by volume (Ringwood, 1991) and hence would not be the load-bearing framework of the rock, it is more likely that the inequant shape of SiO_2 polymorphs would control their orientation during viscous flow.

16.2.6.4   Inner core

Unlike the mantle the Earth's inner core is composed primarily of iron, with about 5 wt.% nickel and very small amounts of other siderophile elements such as

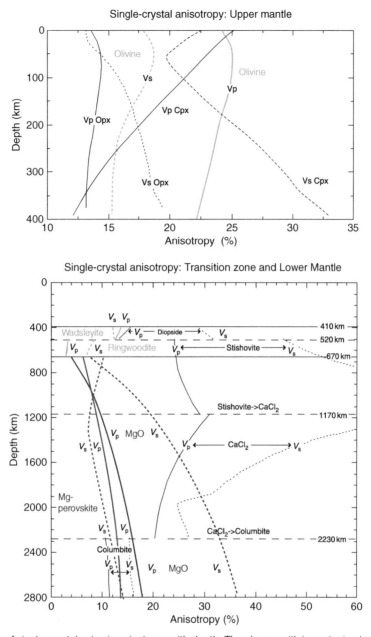

Figure 18 Variation of single-crystal seismic anisotropy with depth. The phases with important volume fractions (see **Figure 1**) (olivine, wadsleyite, ringwoodite, Mg-perovskite, and MgO) are highlighted by a thicker line. P-wave anisotropy is the full line and the S-wave is the dashed line (see text fro details).

chromium, manganese, phosphorus, cobalt, and some light elements such as are oxygen, sulfur, and silicon. The stable structure of iron at ambient conditions is body-centered cubic (b.c.c.). When the pressure is increased above 15 GPa iron transforms to an h.c.p. structure called the ε-phase. At high pressure and temperature iron most likely remains h.c.p. However, there have been experimental observations

of a double hexagonal close-packed structure (d.h.c.p.) (Saxena *et al.*, 1996) and a distorted h.c.p. structure with orthorhombic symmetry (Andrault *et al.*, 1997). Atomic modeling at high pressure and high temperature suggests that the h.c.p. structure is still stable at temperatures above 3500 K in pure iron (Vočadlo *et al.*, 2003a), although the energy differences between the h.c.p. and b.c.c. structures is very small and the authors speculate that the b.c.c. structure may be stabilized by the presence of light elements. A suggestion echoed by the seismic study of Beghein and Trampert (2003). To make a quantitative anisotropic seismic model to compare with observations, one needs either velocity measurements or the elastic constants of single-crystal h.c.p. iron at the conditions of the inner core. The measurement or first-principles calculation of the elastic constants of iron is major challenge for mineral physics. The conditions of the inner core are extreme with pressures from 325 to 360 GPa, and temperatures from 5300 to 5500 K. To date experimental measurements have been using diamond anvil cells to achieve the high pressures on polycrystalline h.c.p. iron. Inelastic X-ray scattering has been used to measure V_p at room temperature and high pressure up to 110 GPa at 298 K (Fiquet *et al.*, 2001), up to 153 GPa (Mao *et al.*, 2001) and the anisotropy of V_p has been characterized in two directions up to 112 GPa (Antonangeli *et al.*, 2004). V_p has been determined at simultaneous high pressure and temperature up to 300 GPa and up to 1200 K from X-ray Debye–Waller temperature factors (Dubrovinsky *et al.*, 2001) and up to 73 GPa and 1700 K using inelastic X-ray scattering by Lin *et al.* (2005). Radial X-ray diffraction has been used to measure the elastic constants of polycrystalline iron with simultaneous measurement of the CPO at room temperature and pressures up to 211 GPa (Singh *et al.*, 1998; Mao *et al.*, 1998 (with corrections, 1999); Merkel *et al.*, 2005), which is still well below

inner-core pressures. The experimental results are plotted in **Figure 19** as V_p, V_{SH}, and V_{SV} of a hexagonal media. Note that the polycrystals have a strong uniaxial (fiber) CPO with high concentration of *c*-axes parallel to compression direction of the diamond anvil cell as the symmetry axis and hence has hexagonal elastic symmetry like the single crystal.

In order to simulate the *in situ* conditions the static (0 K), elastic constants have been calculated at inner-core pressures (Stixrude and Cohen, 1995). The calculated elastic constants predict maximum P-wave maximum velocity parallel to the *c*-axis and the difference in velocity between the *c*- and *a*-axes is quite small (**Figure 20**). The anisotropy of the calculated elastic constants being quite low required that the crystal-preferred orientation is very strong; it was even suggested that inner core could be a single crystal of h.c.p. iron to be compatible with the seismic observations, and that *c*-axis is aligned with Earth's rotation axis. Other studies with calculations in static conditions include Steinle-Neumann *et al.* (2001) and Vočadlo *et al.* (2003b), which indicated a higher anisotropy for P-waves than Stixrude and Cohen (1995), but with same P-wave distribution with a maximum parallel to the *c*-axis. The first attempt to introduce temperature into first-principles methods for iron by Laio *et al.* (2000) produced estimates of the isotropic bulk and shear modulus at inner-core conditions (325 GPa and 5400 K) and single-crystal elastic constants at conditions comparable with the

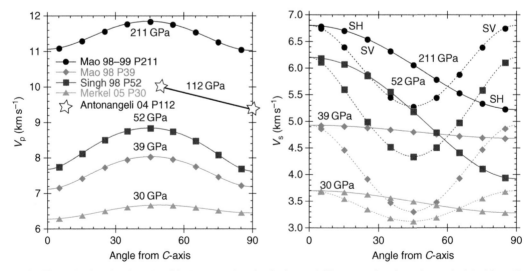

Figure 19 The seismic velocities at ambient temperature in single-crystal hexagonal ε-phase iron calculated from the experimentally determined elastic constants or measured velocity in the case of Antonangeli *et al.* (2004) (see text for discussion).

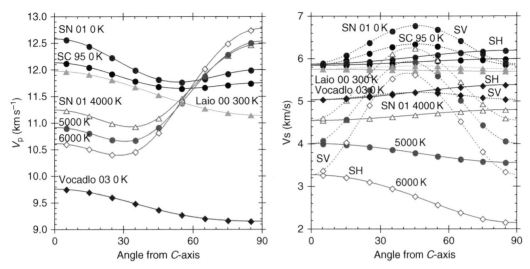

Figure 20 The seismic velocities at 0 K or high temperature in single-crystal hexagonal e-phase iron calculated from the theoretically determined elastic constants (see text for discussion).

experimental study of Mao *et al.* (1998) (210 GPa and 300 K). The first study of simulate inner core temperatures (up to 6000 K) and pressures by Steinle-Neumann *et al.* (2001) produced two unexpected results, first the increase of the unit cell axial c/a ratios by a large amount (10%) with increasing temperature, and second the migration of the P-wave maximum velocity to the basal plane at high temperature. These new high-temperature results required a radical change in the seismic anisotropy model with one-third of c-axes being aligned normal to the Earth's rotation axis giving an excellent agreement with travel time differences. However, recently new calculations (Gannarelli *et al.*, 2003, 2005; Sha and Cohen, 2006) have failed to reproduce the large change in c/a axial ratios with temperature, which casts some doubt on the elastic constants of Steinle-Neumann *et al.* at high temperature. It should be said that high-temperature first-principls calculations represent frontier science in this area. What is perhaps even more troublesome is that there is very poor agreement, or perhaps one should say total disagreement, between experimental results and first principles for P- and S-wave velocity distribution in single-crystal h.c.p. iron at low temperature and high pressure where methods are considered to be well established. Experimental techniques can be criticized as one has to use polycrystalline samples and so the results represent a lower bound on anisotropy of the single-crystal elastic constants. The diamond anvil cells imposed a strong axial strain on the samples that results in a very strong orientation of c-axes

parallel to the compression axis that can be used as reasonable proxy for a single crystal, even before corrections for the presence of CPO are made. The radial X-ray diffraction techniques allow the measurement of the CPO (texture) *in situ*; on the other hand, it requires a model of the microscopic stress–strain distribution to determine the single-crystal elastic constants, where at present simple models like the constant stress (Reuss) are used. The fact that the samples undergo elastic and plastic strain complicates the rigorous interpretation of this type of experiment (Weidner *et al.*, 2004; Merkel *et al.*, 2006b) in terms of elastic constants. One reassuring observation is that different experimental techniques, for example, radial X-ray diffraction and inelastic X-ray scattering, both give a similar P-wave velocity distribution with the maximum velocity at 45° to the c- and a-axes and an apparently increasing anisotropy with increasing pressure. All reports of first-principles calculations at 0 K give similar P-wave velocity with the maximum parallel to the c-axis. However, differences can be seen in the magnitude of the anisotropy and the position of the minimum velocity, minimum at 50° from the c-axis (Stixrude and Cohen, 1995; Steinle-Neumann *et al.*, 2001) or at 90° (Laio *et al.*, 2000; Vočadlo *et al.*, 2003b). The differences between experiment and theory are even more flagrant for S-waves. The experimental data show that SH is greater than SV except at more than 60° from the c-axis, SH has maximum parallel to the c-axis, and SV has minimum velocity at 45° to the c-axis. Results from theory show more variability between authors, SV is greater than SH, and SV has maximum velocity at 45° to the *c*-axis in the work of Stixrude and Cohen (1995) and Steinle-Neumann *et al.* (2001), which is the exact opposite of the experimental results. Laio *et al.* (2000) and Vočadlo *et al.* (2003b) have a much lower S-wave anisotropy than in the experiments. From this brief survey of recent results in this field, it is clear that there is still much to do to unravel the meaning of seismic anisotropy of the inner core and physics of iron at high pressure and temperature in particular. Although the stability of h.c.p. iron at inner-core conditions has been questioned from time to time on experimental or theoretical grounds, that the inner core may not be pure iron (e.g., Poirier, 1994), the major problem at the present time is to get agreement between theory and experiment at the same physical conditions. Interpretation of the mechanisms responsible for inner-core seismic anisotropy is out of the question without a reliable estimate of elastic constants of h.c.p. iron; indeed if the inner core is composed of h.c.p. iron (e.g., Beghein and Trampert, 2003; Vočadlo *et al.*, 2003a), nowhere in the Earth is Francis Birch's 'high pressure language' (positive proof = vague suggestion etc, Birch, 1952) more appropriate.

16.3 Rock Physics

16.3.1 Introduction

In this section the contribution of CPO to seismic anisotropy in the deep Earth with cases of olivine and the role of melt is illustrated. The CPO in rocks of upper-mantle origin is now well established (e.g., Mercier, 1985; Nicolas and Christensen, 1987; Mainprice *et al.*, 2000) as direct samples are readily available from the first 50 km or so, and xenoliths provide further sampling down to depths of about 220 km. Ben Ismaïl and Mainprice (1998) created a database of olivine CPO patterns from a variety of the upper-mantle geodynamic environments (ophiolites, subduction zones, and kimberlites) with a range of microstructures. However, for the deeper mantle (e.g., Wenk *et al.*, 2004) and inner core (e.g., Merkel *et al.*, 2005) we had to rely traditionally on high-pressure and -temperature experiments to characterize the CPO at extreme conditions. In recent years, the introduction of various types of polycrystalline plasticity models to stimulate CPO development for complex strain paths has allowed a high degree of forward modeling using either slip systems

determined from studying experimentally deformed samples using transmission electron microscopy (e.g., wadsleyite – Thurel *et al.*, 2003; ringwoodite – Karato *et al.*, 1998), X-ray diffraction peak broadening analysis for electron radiation sensitive minerals (e.g., Mg-perovskite – Cordier *et al.*, 2004b) or predicted systems from atomic-scale modeling of dislocations (e.g., olivine – Durinck *et al.*, 2005a, 2005b; ringwoodite – Carrez *et al.*, 2006). The polycrystalline plasticity modeling has allowed forward modeling of upper mantle (e.g., Chastel *et al.*, 1993; Blackman *et al.*, 1996; Tommasi, 1998), transition zone (e.g., Tommasi *et al.*, 2004), lower mantle (e.g., Wenk *et al.*, 2006), D″ layer (e.g., Merkel *et al.*, 2006a), and the inner core (e.g., Jeanloz and Wenk, 1988; Wenk *et al.*, 2000).

16.3.2 Olivine the Most Studied Mineral: State of the Art – Temperature, Pressure, Water, Melt, ETC

Until the papers by Jung and Karato (2001), Katayama *et al.* (2004), and Katayama and Karato (2006) were published, the perception of olivine dominated flow in the upper mantle was quite simple with [100]{0kl} slip be universally accepted as the mechanism responsable for plastic flow and the related seismic anisotropy (e.g., Mainprice *et al.*, 2000). The experimental deformation of olivine in hydrous conditions at 2 GPa pressure and high temperature by Karato and co-workers produced a new type of olivine CPO developed at low stress with [001] parallel to the shear direction and (100) in the shear plane, which they called C-type, which is associated with high water content. They introduced a new olivine CPO classification that illustrated the role of stress and water content as the controlling factors for the development of five CPO types (A, B, C, D, and E) (**Figure 21**). The five CPO types are assumed to represent the dominant slip system activity on A ≡ [100](010), B ≡ [001](010), C ≡ [001](100), D ≡ [100]{0kl}, and E ≡ [100](001). The Ben Ismaïl and Mainprice (1998) olivine CPO database with 110 samples has been taken and the percentages for each CPO type and added an additional class called AG-type (or axial b- [010] girdle by Tommasi *et al.*, 2000) have been estimated which is quite common in naturally deformed samples. The CPO types in percentage of the database are A-type (49.5%), D-type (23.8%), AG-type (10.1%), E-type (7.3%), B-type (7.3%), and C-type (1.8%). It is clear that CPO associated with [100] direction

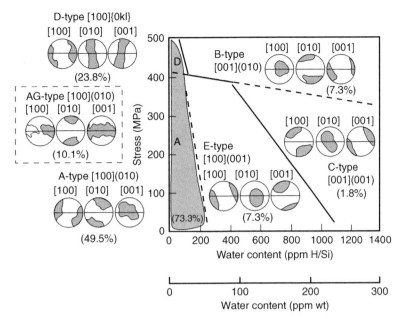

Figure 21 The classification of olivine CPO proposed by Jung and Karato (2001) as a function of stress and water content. The water content scale in ppm H/Si is that originally used by Jung and Karato. The water content scale in ppm wt is more recent calibration used by Bolfan-Casanova (2005). The numbers in brackets are the percentage of samples with the fabric types found in the database of Ben Ismaïl and Mainprice (1998).

slip (A, AG, D, E types) represent 90.8% of the database and therefore only 9.2% are associated with [001] direction slip (B and C types). Natural examples of all CPO types taken from the database are shown in **Figure 22**, with the corresponding seismic properties in **Figure 23**. There is only one unambiguous C-type sample and another with transitional CPO between B and C types. The database contains samples from palaeo-mid ocean ridges (e.g., Oman ophiolite), the circum Pacific suduction zones (e.g., Philippines, New Caledonia, Canada) and subcontinental mantle (e.g., kimberlite xenoliths from South Africa). There have been some recent reports of the new olivine C-type CPO (e.g., Mizukami *et al.*, 2004) associated with high water content and others from ultrahigh pressure (UHP) rocks (e.g., Xu *et al.*, 2006) have relatively low water contents. It is instructive to look at the solubility of water in olivine to understand the potential importance of the C-type CPO. In **Figure 24**, the experimentally determined solubility of water in nominally anhydrous upper-mantle silicates (olivine, cpw, opx, and garnet) in the presence of free water are shown over the upper-mantle pressure range. The values given in the review by Bolfan-Casanova (2005) are in H_2O ppm wt using the calibration of Bell *et al.* (2003), so the values of

Karato *et al.* in $H/10^6$ Si using the infra-red calibration of Paterson (1982) have to be multiplied by 0.22 to obtain H_2O ppm wt. If free water is available, then olivine can incorporate, especially below 70 km depth, many times the concentration necessary for C-type CPO to develop according to the results of Karato and co-workers.

Why is it that the C-type CPO is relatively rare? It is certain that deforming olivine moving slowly toward the surface will lose its water due to the rapid diffusion of hydrogen. For example, even xenoliths transported to the surface in a matter of hours lose a significant fraction of their initial concentration (Demouchy *et al.*, 2006). Hence, it very plausible that in the shallow mantle (less than about 70 km depth), the C-type will not develop because the solubility of water is too low in olivine at equilibrium conditions and that 'wet' olivine upwelling from greater depths and moving toward the surface by slow geodynamic processes will lose their excess water by hydrogen diffusion. In addition, any 'wet' olivine coming into contact with basalt melt tends to 'dehydrate' as the solubility of water in the melt phase is hundreds to thousands of times greater than olivine (e.g., Hirth and Kohlstedt, 1996). The melting will occur in upwelling 'wet' peridotites at a well-defined depth when the solidus is exceeded and the volume fraction

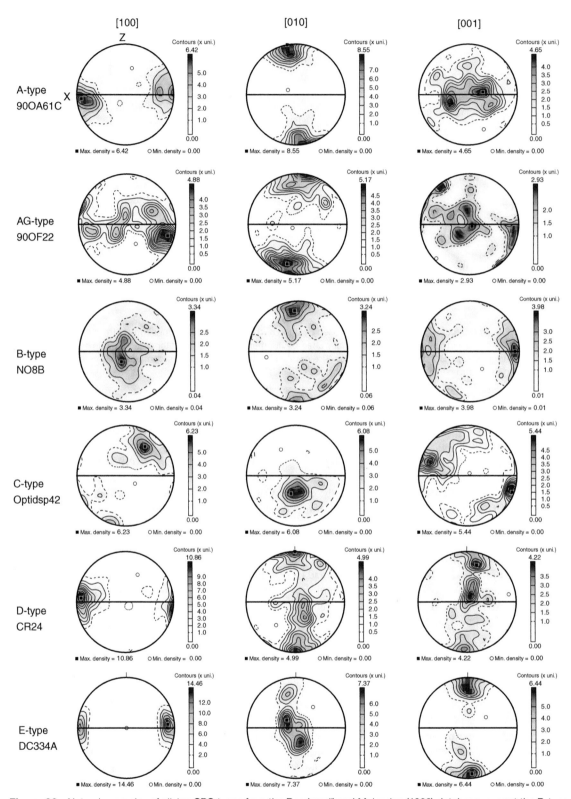

Figure 22 Natural examples of olivine CPO types from the Ben Ismaïl and Mainprice (1998) database, except the B-type sample NO8B from K. Mishibayashi (per.com., 2006). X marks the lineation, the horizontal line is the foliation plane. Contours given in times are uniform.

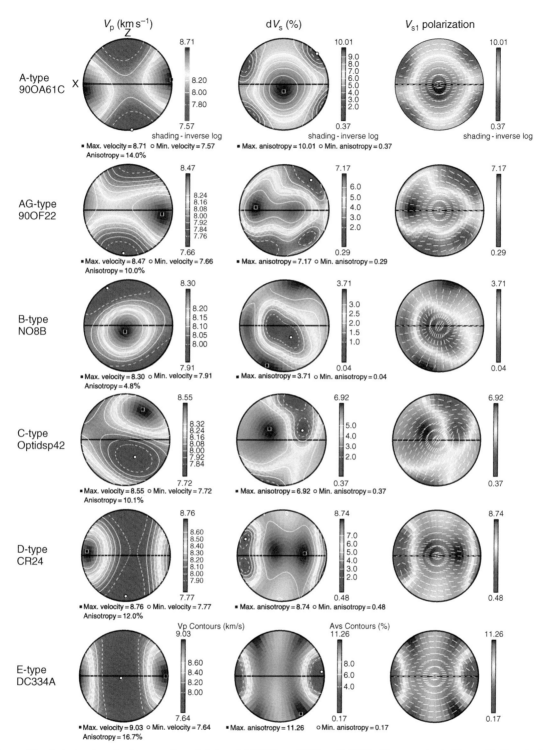

Figure 23   Anisotropic seismic properties of olivine at 1000°C and 3 GPa with CPO of the samples in **Figure 20**. X marks the lineation, the horizontal line is the foliation plane.

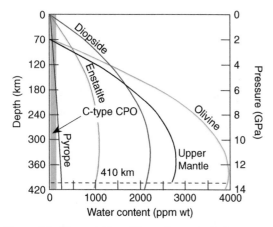

Figure 24 The variation of the water content of upper mantle silicates in the presence of free water from Bolfan-Casanova (2005). The water content of the experimentally produced C-type CPO by Jung and Karato is indicated by the light blue region.

of melt produced is controlled by the amount of water present. Karato and Jung (1998) estimate that melting is initiated at about 160 km in the normal mid-ocean ridge (NMORB) source regions and at greater depth of 250 km in back-arc-type MORB with the production of 0.25–1.00% melt, respectively. Given the small melt fraction the water content olivine is unlikely to be greatly reduced. As more significant melting will of course occur when the 'dry' soldius is exceeded with 0.3% melt per km melting at about 70 km, then melt (3% or more) is quickly produced and the water content of olivine will be greatly reduced. So the action of reduced solubility and selective partitioning of water into the melt phase are likely made the first 70 km of oceanic mantle olivine very dry. The other region where water is certainly present is in subduction zones, where relatively water-rich sediments, oceanic crust, and partly hydrated oceanic lithosphere will add to the hydrous budget of the descending slab. Partial melting is confined in the mantle wedge to the hottest regions, where the temperatures approach the undisturbed asthenospheric conditions at 45–70 km depth, and 5–25 wt.% melt will be produced by a lherzolitic source (Ulmer, 2001). Hence even in the mantle wedge the presence of large volume fractions of melt is likely to reduce the water content of olivine to low levels, possibly below the threshold of the C-type CPO, for depths shallower than 70 km.

Is water the only reason controlling the development of the C-type fabric or [001] direction slip in general? The development of slip in the [001]

direction on (010) and (100) planes at high stresses in olivine has been known experimentally since transmission electron microscopy study of Phakey *et al.* (1972). Recently, Couvy *et al.* (2004) have produced B- and C-type CPO in fosterite at very high pressure (11 GPa) using nominally dry starting materials. However, as pointed out by Karato (2006) the infrared spectrophotometry of the deformed samples revealed the present of water, presumably due to the dehydatation of sample assembly at high pressure. However, concentration of water in the fosterite increases linearly with time at high pressure, whereas the CPO is acquired at the beginning of the experiment in the stress-relaxation tests conducted by Couvy *et al.* (2004), so it is by no means certain that significant water was present at the beginning of the experiment. Mainprice *et al.* (2005) suggested that pressure was the controlling variable, partly inspired by recent atomic modeling of dislocations in fosterite (Durinck *et al.*, 2005b), which shows that the energy barrier for [100] direction slip increases with hydrostatic pressure, whereas for [001] it is constant, which could explain the transition form [100] to [001] with pressure. It remains a possibility that [001] slip occurs at high pressure in dry samples, but experiments with better control of water content are needed to confirm or deny this possibility.

What are the seismic consequences of the recent discovery of C-type CPO in experiments in hydrous conditions? The classic view of mantle flow dominated by [100]{0kl} slip is not challenged by this new discovery as most of the upper mantle is will be dry and at low stress. The CPO associated with [100] slip, that is, 90.8% of the Ben Ismaïl and Mainprice database (1998), produces seismic azimuthal anisotropy for horizontal flow with maximum V_p, polarization of the fastest S-wave parallel to the flow direction and $V_{SH} > V_{SV}$. The seismic properties of all the CPO types are shown in **Figure 23**. It remains to apply the C-, B-, and possibly E-type fabrics to hydrated section of the upper mantle. Katayama and Karato (2006) propose that the mantle wedge above subduction zones are regions where the new CPO types are likely to occur, with the B-type CPO occurring in the low-temperature (high stress and wet) subduction where an old plate is subducted (e.g., NE Japan). The B-type fabric would result anisotropy parallel normal to plate motion (i.e., parallel to the trench). In the back-arc, the A-type (or E-type reported by Michibayashi *et al.* (2006) in the back-arc region of NE Japan) fabric is likely dominant because water content is significantly reduced in this region due to the generation of island arc magma. Changes in the dominant type of olivine fabric can result in complex seismic anisotropy, in which the fast shear-wave polarization direction is parallel to the trench in the fore-arc, but is normal to it in the back-arc (Kneller *et al.*, 2005). In higher-temperture subduction zones (e.g., NW America, Cascades) the C-type CPO will develop in the mantle wedge (low stress and wet), giving rise to anisotropy parallel to plate motion (i.e., normal to the trench). Although these models are attractive, trench parallel flow was first described by Russo and Silver (1994) for flow beneath the Nasca Plate, where there is a considerable path length of anisotropic mantle to generate the observed differential arrival time of the S-waves. On the other hand, a well-exposed peridotite body analog of arc-parallel flow of south central Alaska reveals that horizontal stretching lineations and olivine [100] slip directions are subparallel to the Talkeetna arc for over 200 km, clearly indicating that mantle flow was parallel to the arc axis (Mehl *et al.*, 2003). The measured CPO of olivine shows that the E-type fabric is dominant along the Talkeetna arc; in this case, foliation is parallel to the Moho suggesting arc parallel shear with a horizontal flow plane. Tommasi *et al.* (2006) also report the E-type fabric with trench parallel tectonic context in a highly depleted peridotie massif from the Canadian Cordillera in dunites associated with high degrees of melting, and hence probably dry, whereas harzburgites have an A-type fabric. In limiting the path length to the region of no melting, in coldest part of the mantle wedge, above the plate, the vertical thickness is also constrained to about 45–70 km depth, if one accepts the arguments for melting (Ulmer, 2001). For NE Japan the volcanic front is about 70 km above the top of the slab defined by the hypocenter distribution of intermediate-depth earthquakes (Nakajima and Hasegawa, 2005, their figure1), typical S-wave delay times are 0.17 s (maximum 0.33 s, minimum 0.07 s). The delay times from local slab sources are close to the minimum of 0.07 s for trench parallel fast S-wave polarizations to the east of the volcanic front (i.e., above the coldest part of the mantle wedge) and are over 0.20 s for the trench normal values (i.e., the back-arc side). Are these CPO capable of producing a recordable seismic anisotropy over such a short path length? The vertical S-wave anisotropy can be estimated from the delay time given by Nakajima and Hasegawa (2005) (e.g., 0.17 s), the vertical path length (e.g., 70 km), the average S-wave velocity (e.g., 4.46 km s^{-1}) to give a 1.1% S-wave anisotropy, which is less than

maximum S-wave anisotropy of 1.7% given for a B-type CPO in a vertical direction given by Katayama and Karato (2006) for horizontal flow, so B-type CPO is compatible with seismic delay time, even if we allow some complexity in the flow pattern. The case for C-type in the high-temperature subduction zones is more difficult to test, as the S-wave polarization pattern will be the same for C- and A-types (Katayama and Karato, 2006). The clear seismic observations of fast S-wave polarizations parallel to plate motion (trench normal) given for the Cascadia (Currie *et al.*, 2004) and Tonga (Fischer *et al.*, 1998) subduction zones, which would be compatible with A- or C-types. In general, some care has to taken to separate below slab, slab, and above slab anisotropy components to test mantle wedge anisotropy.

16.3.3 Seismic Anisotropy and Melt

The understanding of the complex interplay between plate separation, mantle convection, adiabatic decompression melting, and associated volcanism at mid-ocean ridges in the upper mantle (e.g., Solomon and Toomey, 1992), presence of melt in the deep mantle in the D'' layer (e.g., Williams and Garnero, 1996) and inner core (Singh *et al.*, 2000) are challenges for seismology and mineral physics. For the upper mantle two contrasting approaches have been used to study mid-ocean ridges, on the one hand marine geophysical (mainly seismic) studies of active ridges and on the other hand geological field studies of ophiolites, which represent 'fossil' mid-ocean ridges. These contrasting methods have yielded very different views about the dimensions of the mid-ocean ridge or axial magma chambers (AMCs). The seismic studies have given us three-dimensional information about seismic velocity and attenuation in the axial region. The critical question is how can this data be interpreted in terms of geological structure and processes? To do so we need data on the seismic properties at seismic frequencies of melt containing rocks, such as hartzburgites, at the appropriate temperature and pressure conditions. Until recently laboratory data for filling these conditions was limited for direct laboratory measurements to isotropic aggregates (e.g. Jackson *et al.*, 2002), but deformation of initially isotropic aggregates with a controlled melt fraction in shear (e.g., Zimmerman *et al.*, 1999; Holtzman *et al.*, 2003) allows simultaneous development of the CPO and anisotropic melt distribution. To obtain information concerning anisotropic rocks, one can use various modeling techniques to estimate the seismic properties of idealized rocks (e.g., Mainprice, 1997; Jousselin and Mainprice, 1998; Taylor and Singh, 2002) or experimentally deformed samples in shear (e.g., Holtzman et al., 2003). This approach has been used in the past for isotropic background media with random orientation distributions of liquid filled inclusions (e.g., Mavko, 1980; Schmeling, 1985; Takei, 2002). However, their direct application to mid-ocean ridge rocks is compromised by two factors. First, field observations on rock samples from ophiolites show that the harzburgites found in the mid-ocean ridges have strong crystal-preferred orientations (e.g., Nicolas and Boudier, 1995), which results in the strong elastic anisotropy of the background medium. For the case of the D″ layer and the inner core the nature of the background media is not well defined and an isotropic medium has been assumed (Williams and Garnero, 1996; Singh et al., 2000). Second, field observations show that melt films tend to be segregated in the foliation or in veins, so that the melt-filled inclusions should be modeled with a preferred shape orientation.

The rock matrix containing melt inclusions is modeled using effective medium theory, to represent the overall elastic behavior of the body. The microstructure of the background medium is represented by the elastic constants of the crystalline rock, including the CPO of the minerals and their volume fractions. Quantitative estimates of how rock properties vary with composition and CPO can be divided into two classes. There are those that take into account only the volume fractions with simple homogenous strain or stress field and upper and lower bounds for anisotropic materials such as Voigt–Reuss bounds, which give unacceptably wide bounds when the elastic contrast between the phases is very strong, such as a solid and a liquid. The other class takes into account some simple aspects of the microstructure, such as inclusion shape and orientation. There are two methods for the implementation of the inclusions in effective medium theory to cover a wide range of concentrations; both methods are based on the analytic solution for the elastic distortion due to the insertion of a single inclusion into an infinite elastic medium given by Eshelby (1957). The uniform elastic strain tensor inside the inclusion (ε_{ij}) is given by

$$\varepsilon_{ij} = \frac{1}{2}\left(G_{ikjl} + G_{jkil}\right) C_{klmn}\, \varepsilon^{*}mn$$

where G_{ikjl} is the tensor Green's function associated with displacement due to a unit force applied in a given direction, C_{klmn} are the components of the

background medium elastic stiffness tensor and ε_{mn}^{*} is the eigenstrain or stress-free strain tensor due to the imaginary removal of the inclusion from the constraining matrix. The symmetrical tensor Green's function G_{ikjl} is given by Mura (1987)

$$G_{ikjl} = \frac{1}{4\pi} \int_0^{\pi} \sin\theta\, \mathrm{d}\theta \int_0^{2\pi} \left(K_{ij}^{-1}(x) x_k x_l\right) \mathrm{d}\phi$$

with $K_{ip}(x) = C_{ijpl} x_j x_l$, the Christoffel stiffness tensor for direction (x), and $x_1 = \sin\theta\,\cos\phi/a_1$, $x_2 = \sin\theta\,\sin\phi/a_2$ and $x_3 = \cos\theta/a_3$.

The angles θ and ϕ are the spherical coordinates that define the vector x with respect to the principal axes of the ellipsoidal inclusion. The semiaxes of the ellipsoid are given by a_1, a_2, and a_3. The integration to obtain the tensor Green's function must be done by numerical methods, as no analytical solutions exist for a general triclinic elastic background medium. Greater numerical efficiency, particularly for inclusions with large axial ratios, is achieved by taking the Fourier transform of G_{ikjl} and using the symmetry of the triaxial ellipsoid to reduce the amount of integration (e.g., Barnett, 1972). The self-consistent (SC) method introduced by Hill (1965) uses the solution for a single inclusion and approximates the interaction of many inclusions by replacing the background medium with the effective medium.

In the formulation of SC scheme by Willis (1977), a ratio of the strain inside the inclusion to the strain in the host medium can be identified as A_i,

$$A_i = [I + G(C_i - C^{\mathrm{scs}})]^{-1}$$

$$\langle \varepsilon^{\mathrm{scs}} \rangle = \sum_{i=1}^{i=n} V_i A_i \langle \sigma^{\mathrm{scs}} \rangle = \sum_{i=1}^{i=n} V_i C_i A_i$$

$$C^{\mathrm{scs}} = \langle \sigma^{\mathrm{scs}} \rangle \langle \varepsilon^{\mathrm{scs}} \rangle^{-1}$$

where I is the symmetric four rank unit tensor $I_{ijkl} = (1/2)\ (\delta_{ik}\ \delta_{jl} + \delta_{il}\ \delta_{jk})$, δ_{ik} is the Kronecker delta, V_i is volume fraction and C_i are the elastic moduli of the ith inclusion. The elastic contants of the SC scheme (C^{scs}) occur on both sides of the equation because of the stain ratio factor (A), so that solution has to be found by iteration. This method is the most widely used in Earth sciences, being relatively simple to compute and well established (e.g., Kendall and Silver, 1996, 1998). Certain consider that when the SC is used for two phases, for example, a melt added to a solid crystalline background matrix, the melt inclusions are isolated (not connected) below 40% fluid content, and the solid and fluid phases can only be considered to be mutually fully interconnected (bi-connected) between 40% and 60%. For our application to magma bodies, one would expect such interconnection at much lower melt fractions. The second method is DEM. This models a two-phase composite by incrementally adding inclusions of melt phase to a crystalline background phase and then recalculating the new effective background material at each increment. McLaughlin (1977) derived the tensorial equations for DEM as follows:

$$\frac{\mathrm{d}C^{\mathrm{DEM}}}{\mathrm{d}V} = \frac{1}{(1-V)}\left(C_i - C^{\mathrm{DEM}}\right)A_i$$

Here again the term A_i is the strain concentration factor coming from the Eshelby formulation of the inclusion problem. To evaluate the elastic moduli (C^{DEM}) at a given volume fraction V, one needs to specify the starting value of C^{DEM} and the component that is the inclusion. Unlike the SC, the DEM is limited to two components A and B. Either A or B can be considered to be the included phase. The initial value of C^{DEM} is clearly defined at 100% of phase A or B. The incremental approach allows the calculations at any composition irrespective of starting concentrations of original phases. This method is also implemented numerically and addresses the drawback of the SC in that either phase can be fully interconnected at any concentration. Taylor and Singh (2002) attempted to take advantages of both of these methods and minimize their shortcomings by using a combined effective medium method, a combination of the SC and DEM theory. Specifically, they used the formulation originally proposed by Hornby *et al.* (1994) for shales, an initial melt-crystalline composite is calculated using the SC with melt fraction in the range 40–60% where they claim that each phase (melt and solid) is connected and then uses the DEM method to incrementally calculate the desired final composition that may be at any concentration with a bi-connected microstructure.

To illustrate the effect on oriented melt inclusions, the data from the study of a harzburgite sample (90OF22) collected from the Moho transition zone of the Oman ophiolite (Mainprice, 1997) will be used. The crystal-preferred orientation and petrology of the sample have been described by Boudier and Nicolas (1995) and CPO of the olivine (AG type) is given in **Figure 22**. The mapping area records a zone of intense melt circulation below a fast spreading palaeo mid-ocean ridge at a level between the asthenospheric mantle and the oceanic crust. The DEM effective

medium method combined with Gassmann's (1951) poro-elastic theory to ensure connectivity of the melt system at low frequency relevant to seismology is used (see Mainprice (1997) for further details and references). The harzburgite (90OF22) has a composition of 71% olivine and 29% opx. The composition combined with CPO of the constituent minerals and elastic constants extrapolated to simulate conditions of 1200°C and 200 MPa where the basalt magma would be liquid predicts the following P-wave velocities in the principal structural direction $X = 7.82$, $Y = 7.69$, $Z = 7.38$ km s^{-1} ($X =$ lineation, $Z =$ normal to foliation, Y is perpendicular to X and Z). The crystalline rock with no melt has essentially an orthorhombic seismic anisotropy. First, by adding the basalt spherical basalt inclusions, the velocities for P- and S-waves decrease and attuation increases (**Figure 25**) with increasing melt fraction and the rock becomes less anisotropic, but preserves its orthorhombic symmetry. When pancake-shaped basalt inclusions with $X:Y:Z = 50:50:1$ are added, to simulate the distribution of melt in the foliation plane observed by Boudier and Nicolas (1995), certain aspects of the original orthorhombic symmetry of the rock are preserved, such as the difference between V_{p} in the X- and Y-directions. However, many velocities and attenuations change illustrating the domination of the transverse isotropic symmetry with Z-direction symmetry axis associated with the pancake-shaped basalt inclusions. The V_{p} in the Y-direction decreases rapidly with increasing melt fraction, causing the seismic anisotropy of P-wave velocities between Z and X or Y to increase. The contrast in behavior for P-wave attenuation is also very strong with attenuation (Q^{-1}) increasing for Z and decreasing for X- and Y-directions. For the S-waves, the effects are even more dramatic and more like a transverse isotropic behavior. The S-waves propagating in X-direction with a Y-polarization, $V_{\mathrm{s}}X(Y)$, and those propagating in the Y-direction with X-polarization, $V_{\mathrm{s}}Y(X)$ (see **Figure 26** for directions), are fast velocities; because they are propagating along the XY (foliation) plane with polarizations in XY plane, we can call these V_{SH} waves for a horizontal foliation. In contrast, all the other S-waves have either their propagation or polarization (or both) direction in the Z-direction and have the same lower velocity, these we can call V_{SV}. Similarly for the S-wave attenuation V_{SH} is less attenuated than V_{SV}. From this study we can see that a few percent of aligned melt inclusions with high axial ratio can change the symmetry and increase the anisotropy of crystalline aggregate (see

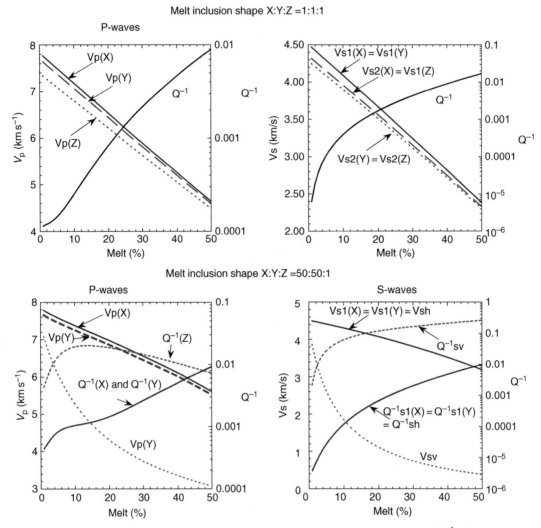

Figure 25 The effect of increasing basalt melt fraction on the seismic velocities and attenuation (Q^{-1}) of a hartzburgite (see text for details).

Figure 26 for summary), completely replacing the anisotropy associated with crystalline background medium in the case of S-waves. Taylor and Singh (2002) come to the same conclusion, that S-wave anisotropy is an important diagnostic tool for the study of magma chambers and regions of partial melting.

One of the most ambitious scientific programs in recent years was the Mantle Electromagnetic and Tomography (MELT) Experiment that was designed to investigate the forces that drive flow in the mantle beneath a mid-ocean ridge, MELT Seismic Team, (1998). Two end-member models often proposed can be classified into two groups; the flow is a passive response to diverging plate motions, or buoyancy forces supplied a plate-independent component variation of density caused by pressure release partial melting of the ascending peridotite. The primary objective in this study was to constrain the seismic structure and geometry of mantle flow and its relationship to melt generation by using teleseismic body waves and surface waves recorded by the MELT seismic array beneath the superfast spreading southern East Pacific Rise (EPR). The observed seismic signal was expected to be the product of elastic anisotropy caused by the alignment of olivine crystals due to mantle flow and the presence of aligned melt channels or pockets of unknown structure at depth (e.g., Kendall 1994; Blackman et al., 1996; Blackman and Kendall, 1997; Mainprice 1997). Observations

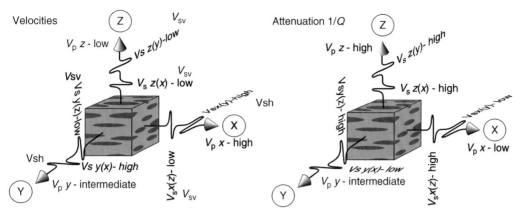

Figure 26 A graphical illustration of the 'pancake'-shaped melt inclusions (red) in the foliation (*XY*) plane (where *X* is the lineation) and the relation between velocity and attenuation (Q^{-1}). The melt is distributed in the foliation plane. The velocities have an initial orthorhombic symmetry with $V_p X > Y > Z$. The directions of high velocity are associated with low attenuation. P-waves normal to the foliation have the lowest velocity. S-waves with polarizations in the foliation (*XY*) plane have the highest velocities (V_{SH}). Diagram inspired a figure in the thesis of Barruol (1993).

revealed that on the Pacific plate (western) side of the EPR had lower seismic velocities (Forsyth *et al.*, 1998; Toomey *et al.*, 1998) and greater shear wave splitting (Wolfe and Solomon, 1998). The shear wave splitting showed that the fast shear polarization was consistently parallel to the speading direction, and at no time parallel to the ridge axis with no null splitting being recorded near the ridge axis. In addition, the delay time between S-wave arrivals on the Pacific plate was twice that of the Nasca plate. P delays decreased within 100 km of the ridge axis (Toomey *et al.*, 1998) and Rayleigh surface waves indicated a decrease in azimuthal anisotropy near the ridge axis. Any model of the EPR must take into account that the average spreading rate at 17° S on the ridge is 73 mm yr^{-1} and the ridge migrates 32 mm yr^{-1} to the west. Anisotropic modeling of the P and S data within 500 km of the ridge axis by Toomey *et al.* (2002) and Hammond and Toomey (2003) showed that a best-fitting finite-strain hexagonal-symmetry two-dimensional flow model had an asymmetric distribution of higher melt fraction and temperature, dipping to the west under the Pacific plate and lower melt fraction and temperature with an essentially horizontal structure under the Nasca plate. Hammond and Toomey (2003) introduce low melt fractions, <2%) in relaxed (connected) cuspate melt pockets (Hammond and Humphreys, 2000) to match the observed velocities. Blackman and Kendall (2002) used a three-dimensional texture flow model to predicted pattern of upper-mantle flow beneath the EPR oceanic spreading center with asymmetric asthenospheric flow

pattern. Blackman and Kendall (2002) explored a series of models for the EPR and found that asymmetric thermal structure proposed by Toomey *et al.* (2002) produced the model in closest agreement with seismic observations. The three-dimensional model shows that shear wave splitting will be lowest at about 50 km to the west of the EPR on the Pacific plate with similar low value at 400 km to the east on the Nasca plate, and not the ridge axis because of the underlying asymmetric mantle structure. The MELT experiment has shown that melt flow beneath a fast spreading ridge is more complicated than originally predicted, with a deep asymmetric structure present to 200 km depth. The influence of the near surface configuration (e.g., ridge migration) was also important in controlling the asthenospheric return flow toward the Pacific superswell in the west (Hammond and Toomey, 2003). The influence of melt geometry appears to be small in the case of the EPR, as the essential anisotropic seismic structure is captured by models that do not have melt geometries with strong shape-preferred orientation. The situation may be different for the oceanic crust at fast oceanic spreading centers. The seismic anisotropy in regions of important melt production, such as Iceland (Bjarnason *et al.*, 2002), does not show the influence of melt geometry on anisotropy, but rather the influence of large-scale mantle flow. In other contexts, such as the rifting, for example, the Red Sea (Vauchez *et al.*, 2000) and East African Rifts (Kendall *et al.*, 2004), the melt geometry does seem to have an important influence of seismic anisotropy.

16.4   Conclusions

In this chapter some aspects of seismology that have a bearing on the geodynamics of the deep interior of the Earth have been reviewed. In particular, the emphasis is on the importance of seismic anisotropy and the variation of anisotropy on a global and regional basis. In the one-dimensional PREM model (Dziewoński and Anderson, 1981), anisotropy was confined to the first 250 km of the upper mantle. Subsequently, other studies of the mantle found various additional forms of global anisotropy associated with the transition zone, the 670 km boundary layer, and the D″ layer (e.g., Montagner, 1994a, 1994b; Montagner and Kennett, 1996; Montagner, 1998). The most recent global studies using more complete data sets and new methods of analysis emphasize the exceptionally strong nature of the upper-mantle anisotropy, the anisotropy of the D″ layer and no significant deviation from the original isotropic PREM for the rest of the mantle (e.g., Beghein *et al.*, 2006; Panning and Romanowicz, 2006). To explain velocity variations that are observed in the mantle studies using probabilistic tomography, the authors put the emphasis on chemical heterogeneity and lateral temperature variations in the mantle (Trampert *et al.*, 2004; Deschamps and Trampert, 2003). Trampert and van der Hilst (2005) argue that spatial variations in bulk major element composition dominates buoyancy in the lowermost mantle, but even at shallower depths, its contribution to buoyancy is comparable to thermal effects. The case of the D″ layer is perhaps even more challenging, as it is clearly a region with strong temperature and compositional gradients (e.g., Lay *et al.*, 2004), but also a regionally varying seismic anisotropy (e.g., Maupin *et al.*, 2005; Wookey *et al.*, 2005a) with an overall global signature (e.g., Montagner and Kennett, 1996; Beghein *et al.*, 2006; Panning and Romanowicz, 2006). The inner core has well-known travel time variations that can be modeled to fit various single or double concentric layered anisotropy scenarios. Recent studies (Ishii and Dziewoński, 2002; Beghein and Trampert, 2003) tend to favor a difference in anisotropy between the outermost and innermost inner core; however, they disagree in the magnitude and symmetry of the anisotropy. The most recent study (Calvet *et al.*, 2006) further suggests that data set is too poor to distinguish between several current models. In the mantle and the inner core, there are often differences between studies at the global and regional scales, and differences between one- and three-dimensional global models.

The seismic sampling over different radial and lateral length scales using surface and body waves of variable frequency has made reference models very important in the reporting and understanding of complex data sets. Kennett (2006) has shown, for example, that it is difficult to achieve comparable P- and S-wave definition for the whole mantle. Mineral physics can play important role because a representation based on elastic moduli rather than wave speeds would provide a better interpretation in terms of composition, temperature, and anisotropy.

In addressing the basics of elasticity, wave propagation in anisotropic crystals and the nature of the anisotropy polycrystalline aggregate with CPO, the author hopes to have provided some of keys necessary for the interpretation of seismic anisotropy. CPO produced by plastic deformation is the link between deformation history and the seismic anisotropy of the Earth's deep interior. We have seen above that many regions of the mantle (e.g., lower mantle) do not have a pronounced seismic anisotropy. However, from mineral physics we have seen that in the upper mantle olivine has a strong elastic anisotropy, in the transition zone wadsleyite is quite anisotropic, in the lower mantle Mg-perovskite and MgO have inceasing anisotropy with depth, in the D″ layer postperovskite is very anisotropic and the inner-core h.c.p. iron is moderately anisotropic. In addition, if we add minerals from the hydrated mantle in suduction regions, such as the A-phase, hydro-ring-woodite, and brucite, these can be quite anisotropic to exceeding anisotropic in the case of brucite. Potentially, the mineralogy suggests that seismic anisotropy could be present if these minerals have a CPO. Aligned melt inclusions and compositional layers can also produce anisotropy. To understand why there are regions in the deep Earth that have no seismic anisotropy is clearly a challenge for mineral physics, seismology, and geodynamics.

References

Abramson EH, Brown JM, Slutsky LJ, and Zaug J (1997) The elastic constants of San Carlos olivine to 17 GPa. *Journal of Geophysical Research* 102: 12253–12263.

Adams BL, Wright SL, and Kunze K (1993) Orientation mapping: The emergence of a new microscopy. *Metallurgical Transactions A* 24: 819–831.

Anderson DL (1979) Deep structure of the continents. *Journal of Geophysical Research* 84: 7555–7560.

Anderson OL and Isaak DG (1995) Elastic constants of mantle minerals at high temperature. In: Ahrens TJ (ed.) *Handbook of Physical Constants*, pp. 64–97. Washington, DC: American Geophysical Union special publication.

Andrault D, Fiquet G, Kunz M, Visocekas F, and Häusermann D (1997) The orthorhombic structure of iron: An *in situ* study at high temperature and high pressure. *Science* 278: 831–834.

Antonangelia D, Occellia F, Requardta H, Badro J, Fiquet G, and Krisch M (2004) Elastic anisotropy in textured hcp-iron to 112 GPa from sound wave propagation measurements. *Earth and Planetery Science Letters* 225: 243–251.

Auld BA (1990) *Acoustic Fields and Waves in Solids*, 2nd edn., vol. 1, 425 pp. Malabar, FL: Krieger Publication.

Barnett DM (1972) The precise evaluation of derivatives of the anisotropic elastic Green's functions. *Physica Status Solidi (b)* 49: 741–748.

Barruol G (1993) *Pétrophysique de la croûte inférieure. Rôle de l'Anisotropies Sismiques sur la Réflectivité et Déphasage des ondes S*. Thesis Université Montpellier II. 272pp.

Barruol G and Kern H (1996) P and S waves velocities and shear wave splitting in the lower crustal/upper mantle transition (Ivrea Zone). Experimental and calculated data. *Physics of the Earth and Planetary Interiors* 95: 175–194.

Bass JD (1995) Elastic properties of minerals, melts, and glasses. In: Ahrens TJ (ed.) *Handbook of Physical Constants*, pp. 45–63. Washington, DC: American Geophysical Union special publication.

Bass JD, Liebermann RC, Weidner DJ, and Finch SJ (1981) Elastic properties from acoustic and volume compression experiments. *Physics of the Earth and Planetary Interiors* 25: 140–158.

Beghein C and Trampert J (2003) Robust normal mode constraints on the inner-core anisotropy from model space search. *Science* 299: 552–555.

Beghein C, Trampert J, and van Heijst HJ (2006) Radial anisotropy in seismic reference models of the mantle. *Journal of Geophysical Research* 111: B02303 (doi:10.1029/2005JB003728).

Bell DR, Rossman GR, Maldener A, Endisch D, and Rauch F (2003) Hydroxide in olivine: A quantitative determination of the absolute amount and calibration of the IR spectrum. *Journal of Geophysical Research* 108 (doi:10.1029/2001JB000679).

Ben Ismail W and Mainprice D (1998) An olivine fabric database: An overview of upper mantle fabrics and seismic anisotropy. *Tectonophysics* 296: 145–157.

Beran MJ, Mason TA, Adams BL, and Olsen T (1996) Bounding elastic constants of an orthotropic polycrystal using mesurements of the microstructure. *Journal of Mechanics Physics Solids* 44: 1543–1563.

Berryman JG (2000) Seismic velocity decrement ratios for regions of partial melt in the lower mantle. *Geophysical Research Letters* 27: 421–424.

Bergman MI (1997) Measurements of elastic anisotropy due to solidification texturing and the implications for the Earth's inner core. *Nature* 389: 60–63.

Bjarnason IT, Silver PG, Rümpker G, and Solomon SC (2002) Shear wave splitting across the Iceland hot spot: Results from the ICEMELT experiment. *Journal of Geophysical Research* 107(B12): 2382 (doi:10.1029/2001JB000916).

Blackman DK and Kendall J-M (1997) Sensitivity of teleseismic body waves to mineral texture and melt in the mantle beneath a mid-ocean ridge. *Philosophical Transactions of the Royal Society of London A* 355: 217–231.

Blackman DK, Kendall J-M, Dawson P, Wenk H-R, Boyce D, and Phipps MJ (1996) Teleseismic imaging of subaxial flow at mid-ocean ridges: Travel-time effects of anisotropic mineral texture in the mantle. *Geophysical Journal International* 127: 415–426.

Blackman DK and Kendall J-M (2002) Seismic anisotropy in the upper mantle, 2, Predictions for current plate boundary flow models. *Geochemistry Geophysics Geosystems* 3(9): 8602 (doi:10.1029/2001GC000247).

Birch F (1940) The alpha–gamma transformation of iron at high pressure, and the problem of the Earth's magnetism. *American Journal of Science* 235: 192–211.

Birch F (1952) Elasticity and constitution of the Earth's interior. *Journal of Geophysical Research* 57: 227–286.

Bolfan-Casanova N (2005) Water in the Eath's mantle. *Mineralogical Magazine* 63(3): 229–257.

Boudier F and Nicolas A (1995) Nature of the Moho transition zone in the Oman ophiolite. *Journal of Petrology* 36: 777–796.

Boyd FR (1973) A pyroxene geotherm. *Geochemica etCosmochimica Acta* 37: 2533–2546.

Bozhilov KN, Green HW, II, and Dobrzhinetskaya L (1999) Clinoenstatite in the Alpe Arami periodotite: Additional evidence for very high-pressure. *Science* 284: 128–132.

Buffett BA and Wenk H-R (2001) Texturing of the Earth's inner core by Maxwell stresses. *Nature* 413: 60–63.

Bunge HJ (1982) *Texture Analysis in Materials Sciences*, 593 pp. London: Buttleworth.

Brown JM and Shankland TJ (1981) Thermodynamic parameters in the Earth as determined from seismic profiles. *Geophysical Journal of the Royal Astronomical Society* 66: 579–596.

Brugger K (1965) Pure modes for elastic waves in crystals. *Journal of Applied Physics* 36: 759–768.

Bruner WM (1976) Comments on Seismic velocities in dry and saturated cracked solids. *Journal of Geophysical Research* 81: 2573–2576.

Calvet M, Chevrot S, and Souriau A (2006) P-wave propagation in transversely isotropic media II. Application to inner core anisotropy: Effects of data averaging, parametrization and *a priori* information. *Physics of the Earth and Planetary Interiors* 156: 21–40.

Červený V (1972) Seismic rays and ray intensities in inhomogeneous anisotropic media. *Geophysical Journal of the Royal Astronomical Society* 29: 1–13.

Carpenter MA (2006) Elastic properties of minerals and the influence of phase transitions. *American Mineralogist* 91: 229–246.

Carrez P, Cordier P, Mainprice D, and Tommasi A (2006) Slip systems and plastic shear anisotropy in Mg2SiO4 Ringwoodite: Insights from numerical modelling. *European Journal of Mineralogy* 18: 149–160.

Chai M, Brown JM, and Slutsky LJ (1977) The elastic constants of a pyrope–grossular–almandine garnet to 20 GPa. *Geophysical Research Letters* 24: 523–526.

Chai M, Brown JM, and Slutsky LJ (1997) The elastic constants of an aluminous orthopyroxene to 12.5 GPa. *Journal of Geophysical Research* 102: 14779–14785.

Chastel YB, Dawson PR, Wenk H-R, and Bennet K (1993) Anisotropic convection with implications for the upper mantle. *Journal of Geophysical Research* 98: 17757–17771.

Chen G, Li B, and Liebermann RC (1996) Selected elastic moduli of single-crystal olivines from ultrasonic experiments to mantle pressures. *Science* 272: 979–980.

Chen G, Liebermann RC, and Weidner DJ (1998) Elasticity of single-crystal MgO to 8 gigapascals and 1600 kelvin. *Science* 280: 1913–1916.

Collins MD and Brown JM (1998) Elasticity of an upper mantle clinopyroxene. *Physics and Chemistry of Minerals* 26: 7–13.

Cordier P, Mainprice D, and Mosenfelder JL (2004a) Mechanical instability near the stishovite - CaCl2 phase transition: Implications for crystal preferred orientations and seismic properties. *European Journal of Mineralogy* 16: 387–399.

Cordier P, Ungàr T, Zsoldos L, and Tichy G (2004b) Dislocation creep in MgSiO3 perovskite at conditions of the Earth's uppermost lower mantle. *Nature* 428: 837–840.

Couvy H, Frost DJ, Heidelbach F, *et al.* (2004) Shear deformation experiments of forsterite at 11 GPa - 1400°C in the multianvil apparatus. *European Journal of Mineralogy* 16: 877–889.

Crampin S (1984) Effective anisotropic elastic constants for wave propagation through cracked solids. *Geophysical Journal of the Royal Astronomical Society* 76: 135–145.

Creager KC (1992) Anisotropy of the inner core from differential travel times of the phases PKP and PKIKP. *Nature* 356: 309–314.

Creager KC (1997) Inner core rotation rate from small-scale heterogeneity and time-varying travel times. *Science* 278: 1284–1288.

Creager KC (1999) Large-scale variations in inner-core anisotropy. *Journal of Geophysical Research* 104: 23127–23139.

Creager KC (2000) Inner core anisotropy and rotation. In: Karato S-I, Stixrude L, Liebermann R, Masters G, and Forte A (eds.) *Geophysical Monograph Series, 117: Mineral Physics and Seismic Tomography from the Atomic to the Global Scale*, pp. 89–114. Washington, DC: American Geophysical Union.

Crosson RS and Lin JW (1971) Voigt and Reuss prediction of anisotropic elasticity of dunite. *Journal of Geophysical Research* 76: 570–578.

Currie CA, Cassidy JF, Hyndman R, and Bostock MG (2004) Shear wave anisotropy beneath the Cascadia subduction zone and Western North America craton. *Geophysical Journal International* 157: 341–353.

Da Silva CRS, Karki BB, Strixrude L, and Wentzcovitch RM (1999) *Ab initio* study of the elastic behavior of MgSiO3 ilmenite at high pressure. *Geophysical Research Letters* 26: 943–946.

Davies GF (1974) Effective elastic moduli under hydrostatic stress I, Quasi-harmonic theory. *Journal of Physics and Chemistry of Solids* 35: 1513–1520.

Demouchy S, Jacobsen SD, Gaillard F, and Stern CR (2006) Rapid magma ascent recorded by water diffusion profiles in mantle olivine. *Geology* 34: 429–432 (doi: 10.1130/G22386.1).

Deschamps F and Trampert J (2003) Mantle tomography and its relation to temperature and composition. *Physics of the Earth and Planetetary Interiors* 140: 277–291.

Deuss A and Woodhouse JH (2001) Seismic observations of splitting of the mid-mantle transition zone discontinuity in Earth's mantle. *Science* 294: 354–357.

Deuss A and Woodhouse JH (2002) A systematic search for mantle discontinuities using SS-precursors. *Geophysical Research Letters* 29(8) (doi:10.1029/2002GL014768).

Deuss A, Woodhouse JH, Paulssen H, and Trampert J (2000) The observation of inner core shear waves. *Geophysical Journal International* 142: 67–73.

Doornbos DJ (1974) The anelasticity of the inner core. *Geophysical Journal of the Royal Astronnomical Society* 38: 397–415.

Dubrovinsky LS, Dubrovinskaia NA, and Le Bihan T (2001) Aggregate sound velocities and acoustic Gruneisen parameter of iron up to 300 GPa and 1,200 K. *Proceedings of the National Academy of Sciences USA* 98: 9484–9489.

Duffy TS and Vaughan MT (1988) Elasticity of enstatite and its relationship to crystal structure. *Journal of Geophysical Research* 93: 383–391.

Duffy TS, Zha CS, Downs RT, Mao HK, and Hemley RJ (1995) Elasticity of forsterite to 16 GPa and the composition of the upper mantle. *Nature* 378: 170–173.

Durek JJ and Romanowicz B (1999) Inner core anisotropy inferred from direct inversion of normal mode spectra. *Geophysical Journal International* 139: 599–622.

Durinck J, Legris A, and Cordier P (2005a) Influence of crystal chemistry on ideal plastic shear anisotropy in forsterite: First

principle calculations. *American Mineralologist* 90: 1072–1077.

Durinck J, Legris A, and Cordier P (2005b) Pressure sensitivity of olivine slip systems: First-principle calculations of generalised stacking faults. *Physics and Chemistry of Minerals* 32: 646–654.

Dziewoński AM and Anderson DL (1981) Preliminary reference Earth model. *Physics of the Earth and Planetary Interiors* 25: 297–356.

Eshelby JD (1957) The determination of the elastic field of a ellipsoidal inclusion, and related problems. *Proceedings of the Royal Society of London, A* 241: 376–396.

Every AG (1980) General closed-form expressions for acoustic waves in elastically anisotropic solids. *Physical Review B* 22: 1746–1760.

Fedorov FI (1968) *Theory of Elastic Waves in Crystals*, 375 pp. New York: Penum Press.

Fei Y (1995) Thermal expansion. In: Ahrens TJ (ed.) *Minerals Physics and Crystallography: A Handbook of Physical Constants*, pp. 29–44. Washington, DC: American Geophysical Union.

Fiquet G, Badro G, Guyot F, Requardt H, and Krisch M (2001) Sound velocities in iron to 110 Gigapascals. *Science* 292: 468–471.

Fiquet G, Badro J, Guyot F, *et al.* (2004) Application of inelastic X-ray scattering to measurements of acoustic wave velocities in geophysical materials at very high pressure. *Physics of the Earth and Planetery Interiors* 143–144: 5–18.

Fischer KM, Fouch MJ, Wiens DA, and Boettcher MS (1998) Anisotropy and flow in Pacific subduction zone back-arcs. *Pure and Applied Geophysics* 151: 463–475.

Flanagan MP and Shearer PM (1998) Global mapping of topography on transition zone velocity discontinuties by stacking SS precursors. *Journal of Geophysical Research* 103: 2673–2692.

Forsyth DW, Webb SC, Dorman LM, and Shen Y (1998) Phase velocities of Rayleigh waves in the MELT experiment of the East Pacific Rise. *Science* 280: 1235–1238.

Fouch MJ and Rondenay S (2006) Seismic anisotropy beneath stable continental interiors. *Physics of the Earth and Planetary Interiors* 158: 292–320.

Frisillo AL and Barsch GR (1972) Measurement of single-crystal elastic constants of bronzite as a function of pressure and temperature. *Journal of Geophysical Research* 77: 6360–6384.

Fukao Y, Widiyantoro S, and Obayashi M (2001) Stagnant slabs in the upper and lower mantle transition region. *Reviews of Geophysics* 39: 291–323.

Garnero EJ, Revenaugh JS, Williams Q, Lay T, and Kellogg LH (1998) Ultralow velocity zone at the core–mantle boundary. In: Gurnis M, Wysession ME, Knittle E, and Buffett BA (eds.) *The Core–Mantle Boundary Region*, pp. 319–334. Washington, DC: American Geophysical Union.

Garnero EJ (2000) Heterogeneity of the lowermost mantle. *Annual Review of Earth and Planetary Sciences* 28: 509–537.

Garnero EJ, Maupin V, Lay T, and Fouch MJ (2004) Variable azimuthal anisotropy in Earth's lowermost mantle. *Science* 306: 259–260.

Gaherty JB and Jordan TH (1995) Lehmann discontinuity as the base of an anisotropic layer beneath continents. *Science* 268: 1468–1471.

Gannarelli CMS, Alfè D, and Gillan MJ (2003) The particle-in-cell model for *ab initio* thermodynamics: Implications for the elastic anisotropy of the Earth's inner core. *Physics of the Earth and Planetary Interiors* 139: 243–253.

Gannarelli CMS, Alfè D, and Gillan MJ (2005) The axial ratio of hcp iron at the conditions of the Earth's inner core. *Physics of the Earth and Planetary Interiors* 152: 67–77.

Garcia R (2002) Seismological and mineralogical constraints on the inner core fabric. *Geophysical Research Letters* 29: 1958 (doi:10.1029/2002GL015268).

Garcia R and Souriau A (2000) Inner core anisotropy and heterogeneity level. *Geophysical Research Letters* 27: 3121–3124.

Garcia R and Souriau A (2001) Correction to: Inner core anisotropy and heterogeneity level. *Geophysical Research Letters* 28: 85–86.

Gassmann F (1951) Uber die elastizitat poroser medien. *Vier. der Natur Gesellschaft* 96: 1–23.

Glatzmaier GA and Roberts PH (1995) A three dimensional convective dynamo solution with rotating and finitely conducting inner core and mantle. *Physics of the Earth and Planetetary Intereriors* 91: 63–75.

Glatzmaier GA and Roberts PH (1996) Rotation and magnetism of Earth's inner core. *Science* 274: 1887–1891.

Gubbins D (1981) Rotation of the inner core. *Journal of Geophysical Research* 86: 11695–11699.

Haggerty SE and Sautter V (1990) Ultra deep (>300 km) ultramafic, upper mantle xenoliths. *Science* 248: 993–996.

Hammond WC and Humphreys ED (2000) Upper mantle seismic wave velocity: Effects of realistic partial melt geometries. *Journal of Geophysical Research* 105: 10975–10986.

Hammond WC and Toomey DR (2003) Seismic velocity anisotropy and heterogeneity beneath the Mantle Electromagnetic and Tomography Experiment (MELT) region of the East Pacific Rise from analysis of P and S body waves. *Journal of Geophysicsl Research* 108(B4): 2176 (doi:10.1029/2002JB001789).

Harte B and Harris JW (1993) Lower mantle inclusions from diamonds. *Terra Nova* 5 (supplement 1): 101.

Heinemann S, Sharp TG, Sharp SF, and Rubie DC (1997) The cubic–teragonal phase transition in the system majorite (Mg4Si4O12) – pyrope (Mg3Al2Si3O12) and garnet symmetry in the Earth's transition zone. *Physics and Chemistry of Minerals* 24: 206–221.

Helffrich GR and Wood BJ (1996) 410 km discontinuity sharpness and the form of the olivine α–β phase diagram: Resolution of apparent seismic contradictions. *Geophysical Journal International* 126: F7–F12.

Hess HH (1964) Seismic anisotropy of the uppermost mantle under oceans. *Nature* 203: 629–631.

Hill R (1952) The elastic behaviour of a crystalline aggregate. *Proceedings of the Physics Society of London Series. A* 65: 349–354.

Hill R (1965) A self consistent mechanics of composite materials. *Journal of Mechanics and Physics of Solids* 13: 213–222.

Hirth G and Kohlstedt DL (1996) Water in the oceanic upper mantle: Implications for rheology, melt extration and the evolution of the lithosphere. *Earth and Planetetary Science Letters* 144: 93–108.

Holtzman B, Kohlstedt DL, Zimmerman ME, Heidelbach F, Hiraga T, and Hustoft J (2003) Melt segregation and strain partitioning: Implications for seismic anisotropy and mantle flow. *Science* 301: 1227–1230.

Hornby BE, Schwartz LM, and Hudson JA (1994) Anisotropic effective-medium modelling of the elastic properties of shales. *Geophysics* 59: 1570–1583.

Iitaka T, Hirose K, Kawamura K, and Murakami M (2004) The elasticity of the MgSiO3 postperovskite phase in the Earth's lowermost mantle. *Nature* 430: 442–445.

Inoue T, Weidner DJ, Northrup PA, and Parise JB (1998) Elastic properties of hydrous ringwoodite (γ-phase) in Mg2SiO4. *Earth and Planetary Science Letters* 160: 107–113.

Irifune T, Ringwood AE, and Hibberson WO (1994) Subduction of continental crust and terrigenous and pelagic sediments: An experimental study. *Earth and Planetary Science Letters* 126: 351–368.

Isaak DG (1992) High-temperature elasticity of iron-bearing olivine. *Journal of Geophysical Research* 97: 1871–1885.

Isaak DG (2001) Elastic properties of minerals and planetary objects. In: Levy M, Bass H, and Stern R (eds.) *Handbook of Elastic Properties of Solids, Liquids, and GasesVolume III: Elastic Properties of Solids: Biological and Organic Materials, Earth and Marine Sciences*, pp. 325–376. New York: Academic Press.

Isaak DG, Anderson OL, Goto T, and Suzuki I (1989a) Elasticity of single-crystal forsterite measured to 1,700 K. *Journal of Geophysical Research* 94: 5895–5906.

Isaak DG, Anderson OL, and Goto T (1989b) Measured elastic moduli of single-crystal MgO up to 1800 K. *Physics and Chemistry of Minerals* 16: 704–713.

Isaak DG, Cohen RE, and Mehl MJ (1990) Calculated elastic and thermal properties of MgO at high pressures and temperatures. *Journal of Geophysical Research* 95: 7055–7067.

Isaak DG and Ohno I (2003) Elastic constants of chrome-diopside: Application of resonant ultrasound spectroscopy to monoclinic single-crystals. *Physics and Chemistry of Minerals* 30: 430–439.

Isaak DG, Ohno I, and Lee PC (2005) The elastic constants of monoclinic single-crystal chrome-diopside to 1,300 K. *Physics and Chemistry of Minerals* 32: (doi:10.1007/s00269-005-0047-9).

Ishii M, Dziewoński AM, Tromp J, and Ekström G (2002b) Joint inversion of normal mode and body wave data for inner core anisotropy: 2. Possible complexities. *Journal of Geophysical Research* 107: (doi:10.1029/2001JB000713).

Ishii M and Dziewoński AM (2002) The innermost inner core of the Earth: Evidence for a change in anisotropic behavior at the radius of about 300 km. *Proceedings of the National Academy of Sciences USA* 99: 14026–14030.

Ishii M and Dziewoński AM (2003) Distinct seismic anisotropy at the centre of the Earth. *Physics of the Earth and Planetary Interiors* 140: 203–217.

Ishii M, Tromp J, Dziewoński AM, and Ekström G (2002a) Joint inversion of normal mode and body wave data for inner core anisotropy: 1. Laterally homogeneous anisotropy. *Journal of Geophysical Research* 107: (doi:10.1029/2001JB000712).

Ito E and Katsura T (1989) A temperature profile of the mantle transition zone. *Geophyical Research Letters* 16: 425–428.

Jackson I and Niesler H (1982) The elasticity of periclase to 3 GPa and some geophysical implications. In: Akimoto S and Manghnani MH (eds.) *High-Pressure Research in Geophysics*, pp. 93–113. Japan: Center for Academic Publications.

Jackson I, Fitz GJD, Faul UH, and Tan BH (2002) Grainsize sensitive seismic wave attenuation in polycrystalline olivine. *Journal of Geophysical Research* 107(B12): 2360 (doi:10.1029/2001JB001225).

Jackson JM, Sinogeikin SV, Bass JD, and Weidner DJ (2000) Sound velocities and elastic properties of γ-Mg2SiO4 to 873K by Brillouin spectroscopy. *American Minerologist* 85: 296–303.

Jacobsen SD, Reichmann HJ, Spetzler HA, et al. (2002) Structure and elasticity of single-crystal (Mg,Fe)O and a new method of generating shear waves for gigahertz ultrasonic interferometry. *Journal of Geophysical Research* 107: 2037 (doi:10.1029/2001JB000490).

Jacobsen SD, Smyth JR, Spetzler H, Holl CM, and Frost DJ (2004) Sound velocities and elastic constants of ironbearing hydrous ringwoodite. *Physics of the Earth and Planetary Interiors* 143–144: 47–56.

Jacobsen SD, Reichmann HJ, Kantor A, and Spetzler HA (2005) A gigahertz ultrasonic interferometer for the diamond anvil cell and high-pressure elasticity of some iron-oxide minerals. In: Chen J, Wang Y, Duffy TS, Shen G, and Dobrzhinetskaya LF (eds.) *Advances in High-Pressure Technology for Geophysical Applications*, pp. 25–48. Amsterdam: Elsevier.

Jacobsen SD (2006) Effect of water on the equation of state of nominally anhydrous minerals. *Reviews in Mineralogy and Geochemistry* 62: 321–342.

Jacobsen SD and Smyth JR (2006) Effect of water on the sound velocities of ringwoodite in the transition zone. In: Jacobsen SD and van der Lee S (eds.) *Geophysical Monograph Series, 168: Earth's Deep Water Cycle*, pp. 131–145. Washington, DC: American Geophysical Union.

Jeanloz R and Wenk H-R (1988) Convection and anisotropy of the inner core. *Geophysical Research Letters* 15: 72–75.

Jeffreys H (1926) The rigidity of the Earth's central core. *Monthly Notices of the Royal Astronomical Society Geophysics* 371–383.

Jiang F, Speziale S, and Duffy TS (2006) Single-crystal elasticity of brucite, Mg(OH)2, to 15 GPa by Brillouin scattering. *American Mineralogist* 91: 1893–1900.

Jousselin D and Mainprice D (1998) Melt topology and seismic anisotropy in mantle peridotites of the Oman ophiolite. *Earth and Planetary Science Letters* 167: 553–568.

Julian BR, Davies D, and Sheppard RM (1972) PKJKP. *Nature* 235: 317–318.

Jung H and Karato S-I (2001) Water-induced fabric transitions in olivine. *Science* 293: 1460–1463.

Karato S-I (1992) On the Lehmann discontinuity. *Geophysical Research Letters* 19: 2255–2258.

Karato S-I (1993) Inner core anisotropy due to magnetic field-induced preferred orientation of iron. *Science* 262: 1708–1711.

Karato S-I (1998) Seismic anisotropy in the deep mantle, boundary layers and the geometry of mantle convection. *Pure and Applied Geophysics* 151: 565–587.

Karato S-I (1999) Seismic anisotropy of Earth's inner core caused by Maxwell stress-induced flow. *Nature* 402: 871–873.

Karato S-I (2006) Remote Sensing of hydrogen in Earth's mantle. *Reviews in Mineralogy and Geochemistry* 62: 343–375.

Karato S-I, Dupas-Bruzek C, and Rubie DC (1998) Plastic deformation of silicate spinel under the transition zone conditions of the Earth's mantle. *Nature* 395: 266–269.

Karato S-I and Jung H (1998) Water, partial melting and the origin of the seismic low velocity and high attenuation zone in the upper mantle. *Earth and Planetetary Science Letters* 157: 193–207.

Karato S-I, Zhang S, and Wenk H-R (1995) Superplastic in Earth's lower mantle: Evidence from seismic anisotropy and rock physics. *Science* 270: 458–461.

Karki BB, Stixrude L, and Crain J (1997b) *Ab initio* elasticity of three high-pressure polymorphs of silica. *Geophysical Research Letters* 24: 3269–3272.

Karki BB, Stixrude L, Clark SJ, Warren MC, Ackland GJ, and Crain J (1997a) Elastic properties of orthorhombic MgSiO3 pervoskite at lower mantle pressures. *American Mineralogist* 82: 635–638.

Karki BB, Stixrude L, Clark SJ, Warren MC, Ackland GJ, and Crain J (1997c) Structure and elasticity of MgO at high pressure. *American Mineralogist* 82: 52–61.

Karki BB, Wentzcovitch RM, de Gironcoli M, and Baroni S (1999) First principles determination elastic anisotropy and wave velocities of MgO at lower mantle conditions. *Science* 286: 1705–1707.

Karki BB, Wentzcovitch RM, de Gironcoli M, and Baroni S (2000) High pressure lattice dynamics and thermoelasticity of MgO. *Physical Review B1* 61: 8793–8800.

Karki BB, Strixrude L, and Wentzocovitch RM (2001) High-pressure elastic properties of major materials of Earth's mantle from from first principles. *Review of Geophysics* 39: 507–534.

Katayama I, Jung H, and Karato S-I (2004) A new type of olivine fabric from deformation experiments at modest water content and low stress. *Geology* 32: 1045–1048.

Katayama I and Karato S-I (2006) Effect of temperature on the B- to C-type olivine fabric transition and implication for flow pattern in subduction zones. *Physics of the Earth and Planetary Interiors* 157: 33–45.

Kawakatsu H and Niu F (1994) Seismic evidence for a 920 km discontinuity in the mantle. *Nature* 371: 301–305.

Kendall J-M (1994) Teleseismic arrivals at a mid-ocean ridge: Effects of mantle melt and anisotropy. *Geophysical Research Letters* 21: 301–304.

Kendall J-M (2000) Seismic anisotropy in the boundary layers of the mantle. In: Karato S, Forte AM, Liebermann RC, Masters G, and Stixrude L (eds.) *Earth's Deep Interior: Mineral Physics and Tomography from the Atomic to the Global Scale*, pp. 133–159. Washington, DC: American Geophysical Union.

Kendall J-M and Silver PG (1996) Constraints from seismic anisotropy on the nature of the lowermost mantle. *Nature* 381: 409–412.

Kendall J-M and Silver PG (1998) Investigating causes of D" anisotropy. In: Gurnis M, Wysession M, Knittle E, and Buffet B (eds.) *The Core–Mantle Boundary Region*, pp. 409–412. Washington, D.C: American Geophysical Union.

Kendall J-M, Stuart GW, Ebinger CJ, Bastow ID, and Keir D (2004) Magma-assisted rifting in Ethiopia. *Nature* 433: 146–148 (doi:10.1038/nature03161).

Kennett BLN (2006) On seismological reference models and the perceived nature of heterogeneity. *Physics of the Earth and Planetary Interiors* 159: 129–139.

Kesson SE and Fitz GJD (1991) Partitioning of MgO, FeO, NiO, MnO and Cr2O3 between magnesian silicate perovskite and magnesiowüstite: Implications for the inclusions in diamond and the composition of the lower mantle. *Earth and Planetary Science Letters* 111: 229–240.

Kiefer B, Stixrude L, and Wentzcovitch RM (1997) Calculated elastic constants and anisotropy of Mg2SiO4 spinel at high pressure. *Geophysical Research Letters* 24: 2841–2844.

Kind R, Kosarev GL, Makeyeva LI, and Vinnik LP (1985) Observation of laterally inhomogeneous anisotropy in the continental lithosphere. *Nature* 318: 358–361.

Kingma KJ, Cohen RE, Hemley RJ, and Mao H-K (1995) Transformations of stishovite to a denser phase at lower-mantle pressures. *Nature* 374: 243–245.

Kneller EA, van Keken PE, Karato S-I, and Park J (2005) B-type fabric in the mantle wedge: Insights from high-resolution non-Newtonian subduction zone models. *Earth and Planetary Science Letters* 237: 781–797.

Knittle E (1995) Static compression measurements of equation of state. In: Ahrens TJ (ed.) *Minerals Physics and Crystallography: A Handbook of Physical Constants*, pp. 98–142. Washington, D.C: American Geophysical Union.

Kröner E (1978) Self-consistent scheme and graded disorder in polycrystal elasticity. *Journal of Physics F: Metal Physics* 8: 2261–2267.

Kumazawa M (1969) The elastic constants of single crystal orthopyroxene. *Journal of Geophysical Research* 74: 5973–5980.

Laio A, Bernard S, Chiarotti GL, Scandolo S, and Tosatti E (2000) Physics of iron at Earth's core conditions. *Science* 287: 1027–1030.

Lay T, Garnero EJ, and Williams Q (2004) Partial melting in a thermo–chemical boundary layer at the base of the mantle. *Physics of the Earth and Planetary Interiors* 146: 441–467.

Lehmann I (1936) P'. *Union Geodesique at Geophysique Internationale, Serie A, Travaux Scientifiques* 14: 87–115.

Lehmann I (1959) Velocites of longitudinal waves in the upper part of the Earth's mantle. *Annals of Geophysics* 15: 93–118.

Lehmann I (1961) S and the structure of the upper mantle. *Geophyical Journal of the Royal Astronomical Society* 4: 124–138.

Leven J, Jackson I, and Ringwood AE (1981) Upper mantle seismic anisotropy and lithosphere decoupling. *Nature* 289: 234–239.

Levien L, Weidner DJ, and Prewitt CT (1979) Elasticity of diopside. *Physics and Chemistry of Minerals* 4: 105–113.

Li B, Kung J, and Liebermann RC (2004) Modern techniques in measuring elasticity of Earth materials at high pressure and high temperature using ultrasonic interferometry in conjunction with synchrotron X-radiation in multi-anvil apparatus. *Physics of the Earth and Planetary Interiors* 143–144: 559–574.

Li L, Weidner DJ, Brodholt J, Alfé D, and Price GD (2006a) Elasticity of Mg_2SiO_4 ringwoodite at mantle conditions. *Physics of the Earth and Planetary Interiors* 157: 181–187.

Li L, Weidner DJ, Brodholt J, Alfé D, Price GD, Caracas R, and Wentzcoitch R (2006b) Elasticity of CaSiO3 perovskite at high pressure and high temperature. *Physics of the Earth and Planetary Interiors* 155: 249–259.

Liebermann RC and Li B (1998) Elasticity at high pressures and temperatures. In: Hemley R (ed.) *Ultrahigh-Pressure Mineralogy: Physics and Chemistry of the Earth's Deep Interior*, pp. 459–492. Washington, DC: Mineralogical Society of America.

Lin J-F, Sturhahn W, Zhao J, Shen G, Mao H-K, and Hemley RJ (2005) Sound Velocities of hot dense iron: Birch's law revisited. *Science* 308: 1892–1894.

Love AEH (1927) *A Treatise on the Mathematical Theory of Elasticity*. New York: Dover Publications.

McLaughlin RA (1977) A study of the differential scheme for composite materials. *International Journal of Engineering Science* 15: 237–244.

McNamara AK, van Keken PE, and Karato S-I (2003) Development of finite strain in the convecting lower mantle and its implications for seismic anisotropy. *Journal of Geophysical Research* 108(B5): 2230 (doi:10.1029/2002JB001970).

Mainprice D (1990) A FORTRAN program to calculate seismic anisotropy from the lattice preferred orientation of minerals. *Computers and Geosciences* 16: 385–393.

Mainprice D (1997) Modelling anisotropic seismic properties of partially molten rocks found at mid-ocean ridges. *Tectonophysics* 279: 161–179.

Mainprice D, Barruol G, and Ben Ismaïl W (2000) The anisotropy of the Earth's mantle: From single crystal to polycrystal. In: Karato S-I, Stixrude L, Liebermann R, Masters G, and Forte A (eds.) *Geophysical Monograph Series, 117: Mineral Physics and Seismic Tomography from the Atomic to the Global Scale*, pp. 237–264. Washington, DC: American Geophysical Union.

Mainprice D and Humbert M (1994) Methods of calculating petrophysical properties from lattice preferred orientation data. *Surveys in Geophysics 15* Special issue 'Seismic properties of crustal and mantle rocks: laboratory measurements and theorietical calculations' 575–592.

Mainprice D, Humbert M, and Wagner F (1990) Phase transformations and inherited lattice preferred orientation: Implications for seismic properties. *Tectonophysics* 180: 213–228.

Mainprice D and Nicolas A (1989) Development of shape and lattice preferred orientations: Application to the seismic anisotropy of the lower crust. *Journal of Structural Geology* 11(1/2): 175–189.

Mainprice D, Tommasi A, Couvy H, Cordier P, and Frost DJ (2005) Pressure sensitivity of olivine slip systems: Implications for the interpretation of seismic anisotropy of the Earth's upper mantle. *Nature* 433: 731–733.

Mao H-K, Shu J, Shen G, Hemley RJ, Li B, and Singh AK (1998) Elasticity and rheology of iron above 220 GPa and the nature of Earth's inner core. *Nature* 396: 741–743 (Correction, 1999 *Nature* 399: 280).

Mao HK, Xu J, Struzhkin VV, *et al.* (2001) Phonon density of states of iron up to 153 gigapascals. *Science* 292: 914–916.

Mason TA and Adams BL (1999) Use of microstructural statistics in predicting polycrystalline material properties. *Metallurgical Transactions A* 30: 969–979.

Masters G and Gilbert F (1981) Structure of the inner core inferred from observations of its spheroidal shear modes. *Geophysical Research Letters* 8: 569–571.

Matsui M and Busing WR (1984) Calculation of the elastic constants and high-pressure properties of diopside, CaMgSi2O6. *American Mineralogist* 69: 1090–1095.

Matthies S and Humbert M (1993) The realization of the concept of a geometric mean for calculating physical constants of polycrystalline materials. *Physica Status Solidi (b)* 177: K47–K50.

Mavko GM (1980) Velocity and attenuation in partially molten rocks. *Journal of Geophysical Research* 85: 5173–5189.

Maupin V, Garnero EJ, Lay T, and Fouch MJ (2005) Azimuthal anisotropy in the D'' layer beneath the Caribbean. *Journal of Geophysical Research* 110: B08301 (doi:10.1029/2004JB003506).

Mégnin C and Romanowicz B (2000) The three-dimensional shear velocity structure of the mantle from the inversion of body, surface, and higher-mode waveforms. *Geophysical Journal International* 143: 709–728.

Mehl L, Hacker BR, Hirth G, and Kelemen PB (2003) Arc-parallel flow within the mantle wedge: Evidence from the accreted Talkeetna arc, South Central Alaska. *Journal of Geophysical Research* 108(B8): 2375 (doi:10.1029/2002JB002233).

MELT Seismic Team (1998) Imaging the deep seismic structure beneath a midocean ridge: The MELT experiment. *Science* 280: 1215–1218.

Mensch T and Rasolofosaon P (1997) Elastic-wave velocities in anisotropic media of arbitary symmetry – generalization of Thomsen's parameters ε, δ and γ. *Geophysical Journal International* 128: 43–64.

Merkel S, Kubo A, Miyagi L, Speziale S, Duffy TS, Mao H-K, and Wenk H-R (2006a) Plastic deformation of $MgGeO_3$ post-perovskite atlower mantle pressures. *Science* 311: 644–646.

Merkel S, Miyajima N, Antonangeli D, Fiquet G, and Yagi T (2006b) Lattice preferred orientation and stress in polycrystalline hcp-Co plastically deformed under high pressure. *Journal of Applied Physics* 100: 023510 (doi:10.1063/1.2214224).

Merkel S, Shu J, Gillet H-K, Mao H, and Hemley RJ (2005) X-ray diffraction study of the single-crystal elastic moduli of ε-Fe up to 30 GPa. *Journal of Geophysical Research* 110: B05201 (doi:10.1029/2004JB003197).

Mercier J-C (1980) Single-pyroxene thermobarometry. *Tectonophysics* 70: 1–37.

Mercier J-C (1985) Olivine and pyroxenes. In: Wenk HR (ed.) *Preferred Orientation in Deformed Metals and Rocks: An Introduction to Modern Texture Analysis*, pp. 407–430. Orlando, FL: Academic Press.

Michibayashi K, Abe N, Okamoto A, Satsukawa T, and Michikura K (2006) Seismic anisotropy in the uppermost mantle, back-arc region of the northeast Japan arc:

Petrophysical analyses of Ichinomegata peridotite xenoliths. *Geophysical Research Letters* 33: L10312 (doi:10.1029/2006GL025812).

Mizukami T, Simon W, and Yamamoto J (2004) Natural examples of olivine lattice preferred orientation patterns with a flow-normal a-axis maximum. *Nature* 427: 432–436.

Montagner J-P (1994a) What can seismology tell us about mantle convection ? *Reviews of Geophysics* 32: 115–137.

Montagner J-P (1994b) Where can seismic anisotropy be detected in the Earth's mantle ? In boundary layers. *Pure and Applied Geophysics* 151: 223–256.

Montagner J-P (1998) Where can seismic anisotropy be detected in the Earth's mantle? In boundary layers.... *Pure and Applied Geophysics* 151: 223–256.

Montagner J-P and Anderson DL (1989) Contraints on elastic combinations inferred from petrological models. *Physics of the Earth and Planetary Interiors* 54: 82–105.

Montagner J-P and Kennett BLN (1996) How to reconcile body-wave and normal-mode reference Earth models. *Geophysical Journal International* 125: 229–248.

Montagner J-P and Tanimoto T (1990) Global anisotropy in the upper mantle inferred from the regionalization of phase velocities. *Journal of Geophysical Research* 95: 4797–4819.

Moore RO and Gurney JJ (1985) Pyroxene solid solution in garnets included in diamond. *Nature* 335: 784–789.

Morelli A, Dziewoński AM, and Woodhouse JH (1986) Anisotropy of the inner core inferred from PKIKP travel times. *Geophysical Research Letters* 13: 1545–1548.

Mura T (1987) *Micromechanics of Defects in Solids*. Dordrecht, The Netherlands: Martinus Nijhoff.

Musgrave MJP (2003) *Crystal Acoustics – Introduction to the Study of Elastic Waves and Vibrations in Crystals*, 621 pp. New York: Acoustical Society of America.

Murakami M, Hirose K, Kawamura K, Sata N, and Ohishi Y (2004) Post-perovskite phase transition in MgSiO3. *Science* 304: 855–858.

Neighbours JR and Schacher GE (1967) Determination of elastic constants from sound-velocity measurements in crystals of general symmetry. *Journal of Applied Physics* 38: 5366–5375.

Nakajima J and Hasegawa A (2005) Shear-wave polarization anisotropy and subduction-induced flow in the mantle wedge of northeastern Japan. *Earth and Planetary Science Letters* 225: 365–377.

Nataf HC, Nakanishi I, and Anderson DL (1984) Anisotropy and shear velocity heterogeneities in the upper mantle. *Geophysical Research Letters* 11(2): 109–112.

Nataf HC, Nakanishi I, and Anderson DL (1986) Measurements of mantle wave velocities and inversion for lateral inhomogeneities and anisotropy. *Journal of Geophysical Research* 91(B7): 7261–7307.

Nataf H and Houard S (1993) Seismic discontinuity at the top of D: A world-wide feature? *Geophysical Research Letters* 20: 2371–2374.

Nicolas A and Boudier F (1995) Mapping mantle diapirs and oceanic crust segments in Oman ophiolites. *Journal of Geophysical Research* 100: 6179–6197.

Nicolas A, Boudier F, and Boullier AM (1973) Mechanism of flow in naturally and experimentally deformed peridotites. *American Journal of Science* 273: 853–876.

Nicolas A and Christensen NI (1987) Formation of anisotropy in upper mantle peridotites; a review. In: Fuchs F and Froidevaux C (eds.) *Geodynamics Series, Vol. 16: Composition, Structure and Dynamics of the Lithosphere–Asthenosphere System*, pp. 111–123. Washington, DC: American Geophysical Union.

Nye JF (1957) *Physical Properties of Crystals – Their Representation by Tensors And Matrices*. Oxford: Oxford University Press.

Oganov AR, Brodholt JP, and Price GD (2001) The elastic constants of MgSiO3 perovskite at pressures and temperatures of the Earth's mantle. *Nature* 411: 934–937.

Oganov AR and Ono S (2004) Theoretical and experimental evidence for a post-perovskite phase of MgSiO3 in Earth's D''layer. *Nature* 430: 445–448.

Okal EA and Cansi Y (1998) Detection of PKJKP at intermediate periods by progressive multichannel correlation. *Earth and Planetary Science Letters* 164: 23–30.

Ono S and Oganov AR (2005) *In situ* observations of phase transition between perovskite and CaIrO3-type phase in MgSiO3 and pyrolitic mantle composition. *Earth and Planetary Science Letters* 236: 914–932.

Pacalo REG and Weidner DJ (1997) Elasticity of majorite, MgSiO3 tetragonal garnet. *Physics of the Earth and Planetary Interiors* 99: 145–154.

Pacalo REG, Weidner DJ, and Gasparik T (1992) Elastic properties of sodium-rich majorite garnet. *Geophysical Research Letteres* 19(18): 1895–1898.

Panning M and Romanowicz B (2006) Three-dimensional radially anisotropic model of shear velocity in the whole mantle. *Geophysical Journal International* 167: 361–379 (doi:10.1111/j.1365-246X.2006.03100.x).

Paterson MS (1982) The determination of hydroxyl by infrared absorption in quartz, silicate glasses and similar materials. *Bulletin of Mineralogy* 105: 20–29.

Pedersen HA, Bruneton M, Maupin V, and the SVKLAPKO Seismic Tomography Working Group (2006) Lithospheric and sublithospheric anisotropy beneath the Baltic shield from surface-wave array analysis. *Earth and Planetary Science Letters* 244: 590–605.

Perrillat J-P, Ricolleau A, Daniel I, et al. (2006) Phase transformations of subducted basaltic crust in the upmost lower mantle. *Physics of the Earth and Planetary Interiors* 157: 139–149.

Phakey P, Dollinger G, and Christie J (1972) Transmission electron microscopy of experimentally deformed olivine crystals. In: Heard HC, Borg IY, Carter NL, and Raleigh CB (eds.) *Geophysics Monograph Series: Flow and Fracture of Rocks*, vol. 16, pp. 117–138. Washington, DC: AGU.

Polet J and Anderson DL (1995) Depth extent of cratons as inferred from tomographic studies. *Geology* 23: 205–208.

Poli S and Schmidt M (2002) Petrology of subducted slabs. *Annual Review of Earth and Planetary Sciences* 30: 207–235.

Poupinet G, Pillet R, and Souriau A (1983) Possible heterogeneity of the Earth's core deduced from PKIKP travel times. *Nature* 305: 204–206.

Poupinet G and Kennett BLN (2004) On the observation of high frequency PKiKP an dits coda in Australia. *Physics of the Earth and Planetary Interiors* 146: 497–511.

Raitt RW, Schor GG, Francis TJG, and Morris GB (1969) Anisotropy of the Pacific upper mantle. *Journal of Geophysical Research* 74: 3095–3109.

Ritsema J and van Heijst H-J (2001) Constraints on the correlation of P and S-wave velocity heterogeneity in the mantle from P, PP, PPP, and PKPab traveltimes. *Geophysical Journal International* 149: 482–489.

Ritsema J, van Heijst H-J, and Woodhouse JH (2004) Global transition zone tomography. *Journal of Geophysical Research* 109: B02302 (doi:10.1029/2003JB002610).

Reuss A (1929) Berechnung der Fliessgrenze von Mischkristallen auf Grund der Plastizitätsbedingung für Einkristalle. *Zeitschrift fur Angewandte Mathematik und Mechanik* 9: 49–58.

Ringwood AE (1991) Phase transitions and their bearing on the constitition and dynamics of the mantle. *Geochimica et Cosmochimica Acta* 55: 2083–2110.

Poirier J-P (1994) Light elements in the Earth's outer core: A critical review. *Physics of the Earth and Planetary Interiors* 85: 319–337.

Russo RM and Silver PG (1994) Trench-parallel flow beneath the Nazca plate from seismic anisotropy. *Science* 263: 1105–1111.

Sanchez-Valle C, Sinogeikin SV, Smyth JR, and Bass JD (2006) Single-crystal elastic properties of dense hydrous magnesium silicate phase A. *Americn Mineralologist* 91: 961–964.

Sautter V, Haggerty SE, and Field S (1991) Ultra deep (>300 km) xenoliths: New petrological evidence from the transition zone. *Science* 252: 827–830.

Sawamoto H, Weidner DJ, Sasaki S, and Kumazawa M (1984) Single-crystal elastic properties of the modified-spinel (beta) phase of magnesium orthosilicate. *Science* 224: 749–751.

Saxena SK, Dubrovinsky LS, and Häggkvist P (1996) X-ray evidence for the new phase of β-iron at high temperature and high pressure. *Geophysical Research Letters* 23: 2441–2444.

Scott-Smith BH, Danchin RV, Harris JW, and Stracke KJ (1984) Kimberlites near Orroroo, South Australia. In: Kornprobst J (ed.) *Kimberlites I: Kimberlites and Related Rocks*, pp. 121–142. Amsterdam: Elsevier.

Schmeling H (1985) Numerical models on the influence of partial melt on elastic, anelastic ane electrical properties of rocks. Part I: elasticity and anelasticity. *Physics of the Earth and Planetary Interiors* 41: 34–57.

Sha X and Cohen RE (2006) Thermal effects on lattice strain in e-Fe under pressure. *Physical Review B* 74: (doi:10.1103/PhysRevB.74.064103).

Sharp TG, Bussod GYA, and Katsura T (1994) Microstructures in b-Mg1.8Fe0.2SiO4 experimentally deformed at transition-zone conditions. *Physics of the Earth and Planetary Interiors* 86: 69–83.

Shearer PM (1996) Transition zone velocity gradients and the 520 km discontinuity. *Journal of Geophysical Research* 101: 3053–3066.

Shearer PM and Toy KM (1991) PKP(BC) versus PKP(DF) differential travel times and aspherical structure in the Earth's inner core. *Journal of Geophysical Research* 96: 2233–2247.

Sidorin I, Gurnis M, and Helmberger DV (1999a) Evidence for a ubiquitous seismic discontinuity at the base of the mantle. *Science* 286: 1326–1331.

Sidorin I, Gurnis M, and Helmberger DV (1999b) Dynamics of a phase change at the base of the mantle consistent with seismological observations. *Journal of Geophysical Research* 104: 15005–15023.

Silver PG (1996) Seismic anisotropy beneath the continents: Probing the depths of geoloy. *Annual Review of Earth and Planetary Sciences* 24: 385–432.

Silver PG and Chan W (1988) Implications for continental structure and evolution from seismic anisotropy. *Nature* 335: 34–39.

Silver PG and Chan WW (1991) Shear wave splitting and subcontinental mantle deformation. *Journal of Geophysical Research* 96: 16429–16454.

Singh AK, Mao H-K, Shu J, and Hemley RJ (1998) Estimation of single crystal elastic moduli from polycrystalline X-ray diffraction at high pressure: Applications to FeO and iron. *Physical Review Letters* 80: 2157–2160.

Singh SC, Taylor MAJ, and Montagner J-P (2000) On the presence of liquid in Earth's inner core. *Science* 287: 2471–2474.

Sinogeikin SV, Katsura T, and Bass JD (1998) Sound velocities and elastic properties of Fe-bearing wadsleyite and ringwoodite. *Journal of Geophysical Research* 103: 20819–20825.

Sinogeikin SV and Bass JD (2002) Elasticity of pyrope and majorite–pyrope solid solutions to high temperatures. *Earth and Planetary Science Letters* 203: 549–555.

Sinogeikin SV, Bass JD, and Katsura T (2001) Single-crystal elasticity of γ-(Mg0.91Fe0.09)2SiO4 to high pressures and to high temperatures. *Geophysical Research Letters* 28: 4335–4338.

Sinogeikin SV, Lakshtanov DL, Nichola JD, and Bass JD (2004) Sound velocity measurements on laser-heated MgO and Al2O3. *Physics of the Earth and Planetary Interiors* 143–144: 575–586.

Sinogeikin SV, Lakshtanov DL, Nichola JD, Jackson JM, and Bass JD (2005) High temperature elasticity measurements on oxides by Brillouin spectroscopy with resistive and IR laser heating. *Journal of European Ceramic Society* 25: 1313–1324.

Solomon SC and Toomey DR (1992) The structure of mid-ocean ridges. *Annual Review of the Earth and Planetary Sciences* 20: 329–364.

Song X (1997) Anisotropy of the Earth's inner core. *Reviews of Geophysics* 35: 297–313.

Song X (2000) Joint inversion for inner core rotation, inner core anisotropy, and mantle heterogeneity. *Journal of Geophysical Research* 105: 7931–7943.

Song X and Helmberger DV (1993) Anisotropy of the Earth's inner core. *Geophysical Research Letters* 20: 2591–2594.

Song X and Helmberger DV (1998) Seismic evidence for an inner core transition zone. *Science* 282: 924–927.

Song X and Richards PG (1996) Seismological evidence for differential rotation of the Earth's inner core. *Nature* 382: 221–224.

Song X and Xu XX (2002) Inner core transition zone and anomalous PKP(DF) waveforms from polar paths. *Geophysical Research Letters* 29(4): (10.1029/2001GL013822).

Souriau A, Roudil P, and Moynot B (1997) Inner core differential rotation: Facts and artefacts. *Geophysical Research Letters* 24: 2103–2106.

Stackhouse S, Brodholt JP, Wookey J, Kendall J-M, and Price GD (2005) The effect of temperature on the seismic anisotropy of the perovskite and post-perovskite polymorphs of MgSiO3. *Earth and Planetary Science Letters* 230: 1–10.

Steenbeck M and Helmis G (1975) Rotation of the Earth's solid core as a possible cause of declination, drift and reversals of the Earth's magnetic field. *Geophysical Journal of the Royal Astronomical Society* 41: 237–244.

Steinle-Neumann G, Stixrude L, Cohen RE, and Gulseren O (2001) Elasticity of iron at the temperature of the Earth's inner core. *Nature* 413: 57–60.

Stevenson DJ (1987) Limits on lateral density and velocity variations in the Earth's outer core. *Geophysical Journal of the Royal Astronomical Society* 88: 311–319.

Stixrude L and Cohen RE (1995) High-pressure elasticity of iron and anisotropy of Earth's inner core. *Science* 267: 1972–1975.

Stixrude L and Lithgow-Bertelloni C (2005a) Mineralogy and elasticity of the oceanic upper mantle: Origin of the low-velocity zone. *Journal of Geophysical Research* 110: B03204 (doi:10.1029/2004JB002965).

Stixrude L and Lithgow-Bertelloni C (2005b) Thermodynamics of mantle minerals – I. Physical properties. *Geophysical Journal International* 162: 610–632.

Su WJ and Dziewoński AM (1995) Inner core anisotropy in three dimensions. *Journal of Geophysical Research* 100: 9831–9852.

Szeto AMK and Smylie DE (1984) The rotation of the Earth's inner core. *Philosophical Transactions of the Royal Society of London, Series A* 313: 171–184.

Takei Y (2002) Effect of pore geometry on Vp/Vs: From equilibrium geometry to crack. *Journal of Geophysical Research* 107(B2): 2043 (doi:10.1029/2001JB000522).

Tanaka S and Hamaguchi H (1997) Degree one heterogeneity and hemispherical variations of anisotropy in the inner core form PKP(BC)–PKP(DF) times. *Journal of Geophysical Research* 102: 2925–2938.

Taylor MAJ and Singh SC (2002) Composition and microstructure of magma bodies from effective medium theory. *Geophysical Journal International* 149: 15–21.

Thomsen L (1986) Weak elastic anisotropy. *Geophysics* 51: 1954–1966.

Thurel E, Douin J, and Cordier P (2003) Plastic deformation of wadsleyite: III. Interpretation of dislocations and slip systems. *Physics Chemistry of Minerals* 30: 271–279.

Tommasi A (1998) Forward modeling of the development of seismic anisotropy in the upper mantle. *Earth and Planetary Science Letters* 160: 1–13.

Tommasi A, Mainprice D, Canova G, and Chastel Y (2000) Viscplastic self-consistent and equilibrium-based modeling of olivine lattice preferred orientations. 1.Implications for the upper mantle seismic anisotropy. *Journal of Geophysical Research* 105: 7893–7908.

Tommasi A, Mainprice D, Cordier P, Thoraval C, and Couvy H (2004) Strain-induced seismic anisotropy of wadsleyite polycrystals: Constraints on flow patterns in the mantle transition zone. *Journal of Geophysical Research* 109(B12405): 1–10.

Tommasi A, Vauchez A, Godard M, and Belley F (2006) Deformation and melt transport in a highly depleted peridotite massif from the Canadian Cordillera: Implications to seismic anisotropy above subduction zones. *Earth and Planetary Science Letters* 252: 245–259.

Toomey DR, Wilcock WSD, Solomon SC, Hammond WC, and Orcutt JA (1998) Mantle seismic structure beneath the MELT region of the East Pacific Rise from P and S tomography. *Science* 280: 1224–1227.

Toomey DR, Wilcock WSD, Conder JA, *et al.* (2002) Asymmetric mantle dynamics in the MELT Region of the East Pacific Rise. *Earth and Planetary Science Letters* 200: 287–295.

Trampert J, Deschamps F, Resovsky J, and Yuen D (2004) Probabilistic tomography maps significant chemical heterogeneities in the lower mantle. *Science* 306: 853–856.

Trampert J and van Heijst HJ (2002) Global azimuthal anisotropy in the transition zone. *Science* 296: 1297–1299.

Trampert J and van der Hilst RD (2005) Towards a quantitative interpretation of global seismic tomography. Earth's deep mantle: Structure, composition, and evolution. *Geophysical Monograph Series* 160: 47–63.

Tromp J (1993) Support for anisotropy of the Earth's inner core. *Nature* 366: 678–681.

Tromp J (1994) Normal mode splitting due to inner core anisotropy. *Geophysical Journal International* 121: 963–968.

Tromp J (2001) Inner-core anisotropy and rotation. *Annual Review of Earth and Planetary Sciences* 29: 47–69.

Tsuchiya T, Tsuchiya J, Umemoto K, and Wentzcovitch RM (2004) Elasticity of MgSiO3-post-perovskite. *Geophysical Research Letters* 31: L14603 (doi:10.1029/2004GL020278).

Ulmer P (2001) Partial melting in the mantle wedge – The role of H_2O on the genesis of mantle-derived 'arc-related' magmas. *Physics of the Earth and Planetary Interiors* 127: 215–232.

Vacher R and Boyer L (1972) A tool for the measurement of elastic and photoelastic constants. *Physical Review B* 6: 639–673.

Vauchez A, Tommasi A, Barruol G, and Maumus J (2000) Upper mantle deformation and seismic anisotropy in continental rifts. *Physics and Chemistry of the Earth, Part A Solid Earth and Geodesy* 25: 111–117.

Vavrycuk V (2003) Parabolic lines and caustics in homogeneous weakly anisotropic solids. *Geophysical Journal International* 152: 318–334.

Vidale JE and Earle PS (2000) Fine-scale heterogeneity in the Earth's inner core. *Nature* 404: 273–275.

Vinnik LP, Chevrot S, and Montagner J-P (1997) Evidence for a stagnant plume in the transition zone? *Geophysical Research Letters* 24: 1007–1010.

Vinnik LP, Kind R, Kosarev GL, and Makeyeva LI (1989) Azimuthal anisotropy in the lithosphere from observations of long-period S-waves. *Geophysical Journal International* 99: 549–559.

Vinnik LP and Montagner JP (1996) Shear wave splitting in the mantle PS phases. *Geophysical Research Letters* 23(18): 2449–2452.

Vinnik LP, Kurnik, and Farra V (2005) Lehmann discontinuity beneath North America: No role for seismic anisotropy. *Geophysical Research Letters* 32: L09306 (doi:10.1029/2004GL022333).

Vočadlo L, Alfe D, Gillan MJ, Wood IG, Brodholt JP, and Price GP (2003a) Possible thermal and chemical stabilization of body-centred-cubic iron in the Earth's core. *Nature* 424: 536–539.

Vočadlo L, Alfè D, Gillan MJ, and Price GD (2003b) The properties of iron under core conditions from first principles calculations. *Physics of the Earth and Planetary Intereriors* 140: 101–125.

Voigt W (1928) *Lerrbuch der Kristallphysik*. Leipzig: Teubner-Verlag.

Wang J, Sinogeikin SV, Inoue T, and Bass JD (2003) Elastic properties of hydrous ringwoodite. *American Mineralogist* 88: 1608–1611.

Watt JP (1988) Elastic properties of polycrystalline materials: Comparison of theory and experiment. *Physics and Chemistry of Minerals* 15: 579–587.

Webb SL (1989) The elasticity of the upper mantle orthosilicates olivine and garnet to 3 GPa. *Physics and Chemistry of Minerals* 516: 684–692.

Webb SL and Jackson I (1993) The pressure dependence of the elastic moduli of single-crystal orthopyroxene $(Mg_{0.8}Fe_{0.2})SiO_3$. *European Journal of Mineralogy* 5: 1111–1119.

Weber P and Machetel P (1992) Convection within the inner core and thermal implications. *Geophysical Research Letters* 19: 2107–2110.

Weidner DJ, Bass JD, Ringwood AE, and Sinclair W (1982) The single-crystal elastic moduli of stishovite. *Journal of Geophysical Research* 87: 4740–4746.

Weidner DJ and Carleton HR (1977) Elasticity of coesite. *Journal of Geophysical Research* 82: 1334–1346.

Weidner DJ and Ito E (1985) Elasticity of $MgSiO_3$ in the ilmenite phase. *Physics of the Earth and Planetary Interiors* 40: 65–70.

Weidner DJ, Sawamoto H, Sasaki S, and Kumazawa M (1984) Single-crystal elastic properties of the spinel phase of Mg_2SiO_4. *Journal of Geophysical Research* 89: 7852–7860.

Weidner DJ, Wang H, and Ito J (1978) Elasticity of orthoenstatite. *Physics of the Earth and Planetary Interiors* 17: P7–P13.

Weidner D, Li L, Davis M, and Chen J (2004) Effect of plasticity on elastic modulus measurements. *Geophysical Research Letters* 31: L06621 (doi:10.1029/2003GL019090).

Wenk H-R, Baumgardner J, Lebensohn R, and Tomé C (2000) A convection model to explain anisotropy of the inner core. *Journal of Geophysical Research* 105: 5663–5677.

Wenk HR, Lonardelli I, Pehl J, *et al.* (2004) *In situ* observation of texture development in olivine, ringwoodite, magnesiowüstite and silicate perovskite. *Earth and Planetary Science Letters* 226: 507–519.

Wenk H-R, Speziale S, McNamara AK, and Garnero EJ (2006) Modeling lower mantle anisotropy development in a subducting slab. *Earth and Planetary Science Letters* 245: 302–314.

Wenk H-R, Takeshita T, and Jeanloz R (1988) Development of texture and elastic anisotropy during deformation of hcp metals. *Geophysical Research Letters* 15: 76–79.

Wentzcovitch RM, Karki BB, Cococcioni M, and de Gironcoli S (2004) Thermoelastic properties of of MgSiO3 perovskite: Insights on the nature of the Earth's lower mantle. *Physical Review Letters* 92: 018501.

Wentzcovitch RM, Tsuchiya T, and Tsuchiya J (2006) MgSiO3-post perovskite at D'' conditions. *Proceedings of the National Academy of Sciences USA* 103: 543–546 (doi:10.1073_pnas.0506879103).

Widmer R, Masters G, and Gilbert F (1992) Observably split multiplets—data analysis and interpretation in terms of large-scale aspherical structure. *Geophysiocal Journal International* 111: 559–576.

Williams Q and Garnero EJ (1996) Seismic evidence for partial melt at the base of the Earth's mantle. *Science* 273: 1528–1530.

Willis JR (1977) Bounds and self-consistent estimates for the overall properties of anisotropic composites. *Journal of the Mechanics and Physics of Solids* 25: 185–202.

Wolfe CJ and Solomon SC (1998) Shear-wave splitting and implications for mantle flow beneath the MELT region of the East Pacific Rise. *Science* 280: 1230–1232.

Woodhouse JH, Giardini D, and Li X-D (1986) Evidence for inner core anisotropy from splitting in free oscillation data. *Geophysical Research Letters* 13: 1549–1552.

Wookey J, Kendall J-M, and Barruol G (2002) Mid-mantle deformation inferred from seismic anisotropy. *Nature* 415: 777–780.

Wookey J, Kendall J-M, and Rumpker G (2005a) Lowermost mantle anisotropy beneath the north Pacific from differential S-ScS splitting. *Geophysical Journal International* 161: 829–838.

Wookey J, Stackhouse S, Kendall J-M, Brodholt J, and Price GD (2005b) Efficacy of the post-perovskite phase as an explanation for lowermost-mantle seismic properties. *Nature* 438: 1004–1007.

Yeganeh-Haeri A (1994) Synthesis and re-investigation of the elastic properties of single-crystal magnesium silicate perovskite. *Physics of the Earth and Planetary Interiors* 87: 111–121.

Yeganeh-Haeri A, Weidner DJ, and Ito E (1989) Elasticity of MgSiO$_3$ in the perovskite structure. *Science* 243: 787–789.

Yoshida SI, Sumita I, and Kumazawa M (1996) Growth model of the inner core coupled with the outer core dynamics and the resulting elastic anisotropy. *Journal of Geophysical Research* 101: 28085–28103.

Yukutake T (1998) Implausibility of thermal convection in the Earth's solid inner core. *Physics of the Earth and Planetary Interiors* 108: 1–13.

Zaug JM, Abramson EH, Brown JM, and Slutsky LJ (1993) Sound velocities in olivine at Earth mantle pressures. *Science* 260: 1487–1489.

Zha C-S, Duffy TS, Downs RT, Mao H-K, and Hemley RJ (1996) Sound velocity and elasticity of single-crystal fosterite to 16 GPa. *Journal of Geophysical Research* 101: 17535–17545.

Zha C-S, Duffy TS, Downs RT, Mao H-K, Hemley RJ, and Weidner DJ (1987) Single-crystal elasticity of β-Mg$_2$SiO$_4$ to the pressure of the 410 km seismic discontinuity in the Eath's mantle. *Physics of the Earth and Planetary Interiors* 147: E9–E15.

Zha C-S, Duffy TS, Downs RT, Mao H-K, and Hemley RJ (1998) Brillouin scattering and X-ray diffraction of San Carlos olivine: Direct pressure determination to 32 GPa. *Earth and Planetary Science Letters* 159: 25–33.

Zha CS, Mao HK, and Hemley RJ (2000) Elasticity of MgO and a primary pressure scale to 55 GPa. *Proceedings of the National Academy of Sciences USA* 97: 13494–13499.

Zhang S and Karato S-I (1995) Lattice preferred orientation of olivine aggregates in simple shear. *Nature* 375: 774–777.

Zhang J, Song X, Li Y, Richards PG, Sun X, and Waldhauser F (2005) Inner core differential motion confirmed by earthquake waveform doublets. *Science* 309: 1357–1360.

Zimmerman ME, Zhang S, Kohlstedt DL, and Karato S-I (1999) Melt distribution in mantle rocks deformed in shear. *Geophysical Research Letters* 26(10): 1505–1508.

Xu Z, Wang Q, Ji S, *et al.* (2006) Petrofabrics and seismic properties of garnet peridotite from the UHP Sulu terrane (China): Implications for olivine deformation mechanism in a cold and dry subducting continental slab. *Tectonophysics* 421: 111–127.

17 Properties of Rocks and Minerals – Physical Origins of Anelasticity and Attenuation in Rock

I. Jackson, Australian National University, Canberra, ACT, Australia

17.1	Introduction	496
17.2	Theoretical Background	496
17.2.1	Phenomenological Description of Viscoelasticity	496
17.2.2	Intragranular Processes of Viscoelastic Relaxation	499
17.2.2.1	Stress-induced rearrangement of point defects	499
17.2.2.2	Stress-induced motion of dislocations	500
17.2.3	Intergranular Relaxation Processes	505
17.2.3.1	Effects of elastic and thermoelastic heterogeneity in polycrystals and composites	505
17.2.3.2	Grain-boundary sliding	506
17.2.4	Relaxation Mechanisms Associated with Phase Transformations	508
17.2.4.1	Stress-induced variation of the proportions of coexisting phases	508
17.2.4.2	Stress-induced migration of transformational twin boundaries	509
17.2.5	Anelastic Relaxation Associated with Stress-Induced Fluid Flow	509
17.3	Insights from Laboratory Studies of Geological Materials	512
17.3.1	Dislocation Relaxation	512
17.3.1.1	Linearity and recoverability	512
17.3.1.2	Laboratory measurements on single crystals and coarse-grained rocks	512
17.3.2	Stress-Induced Migration of Transformational Twin Boundaries in Ferroelastic Perovskites	513
17.3.3	Grain-Boundary Relaxation Processes	513
17.3.3.1	Grain-boundary migration in b.c.c. and f.c.c. Fe?	513
17.3.3.2	Grain-boundary sliding	514
17.3.4	Viscoelastic Relaxation in Cracked and Water-Saturated Crystalline Rocks	516
17.4	Geophysical Implications	517
17.4.1	Dislocation Relaxation	517
17.4.1.1	Vibrating string model	517
17.4.1.2	Kink model	518
17.4.2	Grain-Boundary Processes	519
17.4.3	The Role of Water in Seismic Wave Dispersion and Attenuation	519
17.4.4	Bulk Attenuation in the Earth's Mantle	520
17.4.5	Migration of Transformational Twin Boundaries as a Relaxation Mechanism in the Lower Mantle?	520
17.4.6	Solid-State Viscoelastic Relaxation in the Earth's Inner Core	521
17.5	Summary and Outlook	521
References		522

Nomenclature

a	spacing between adjacent Peierls valleys
a_d	period of kink potential measured parallel to dislocation line
b	Burgers vector
d	grain size; characteristic separation of vacancy sources
f	frequency; the reciprocal of $du/dx = \tan\phi$ (**Figure 4(c)**)
g	dimensionless multiplier of T_m/T in exponent for thermal activation of relaxation time

h	separation of closely spaced dislocations in subgrain wall
h_j	jth coefficient in Fourier series representing grain-boundary topography
k	Boltzmann constant
m_l	effective mass per unit length of dislocation
n	exponent in term βt^n representing transient creep in the Andrade model
p	internal variable providing link between stress and anelastic strain
$p_L(L)$	distribution of dislocation segment lengths
p_s	small fluctuation in pressure P
p_0	stress-dependent equilibrium value
s	Laplace transform variable; distance measured along dislocation line; exponent describing grain-size sensitivity of Q^{-1}
t	time
u	displacement of dislocation segment, or position of grain boundary, in y-direction
u_0	x-independent amplitude of u; position of straight dislocation segment under static stress σ
u_m	maximum displacement for newly formed kink pair
w_k	width of a single geometrical kink
x	coordinate measuring distance parallel to dislocation line or grain boundary
x_1	unstable equilibrium spacing of kinks of opposite sign
y	coordinate measuring distance perpendicular to dislocation line
A	measure of anisotropy or inter-phase variability in elastic moduli or thermal expansivity involved in thermoelastic relaxation strength
B	dislocation drag coefficient
C_{OH}	concentration of hydroxyl ions
D	diffusivity (either of matter or heat)
$D(\tau)$	normalised distribution of anelastic relaxation times
D_b	grain-boundary diffusivity
D_k	kink diffusivity
D_{mc}	multicomponent diffusivity
E	activation energy for relaxation time
E_d	elastic strain energy per unit length of dislocation or line tension

E_P	activation energy describing position (period) of dissipation peak
E_{up}	binding energy of dislocation to pinning point
G	shear modulus
G_m	free energy of kink migration
G_{HS}	mean of the Hashin–Shtrikmann bounds on the shear modulus of a polycrystalline aggregate
G_R	relaxed value of shear modulus (anelastic relaxation only)
G_{RS}	Reuss average of single-crystal elastic constants for shear
G_U	unrelaxed shear modulus
H_{int}	kink interaction energy
H_k	formation energy for a single geometrical kink
H_{kp}	kink-pair formation energy
H_m	enthalpy of kink migration
H_{mc}	activation enthalpy for multicomponent diffusion
$J(t)$	creep function
J_1	real part of J^*
J_2	negative imaginary part of J^*
J_U	unrelaxed compliance (ε/σ)
J^*	dynamic compliance
K_f	bulk modulus of fluid
K_m	bulk modulus of melt
K_s	bulk modulus of solid matrix
$\underline{K}$	bulk modulus of composite of coexisting phases
L	length of dislocation segment between adjacent pinning points
L_c	critical dislocation segment length for dislocation multiplication
$\underline{L}$	subgrain diameter
N	number of terms retained in Fourier series representing grain-boundary topography
N_v	number of dislocation segments of given length L per unit volume of internal variable p
P	
Q^{-1}	inverse quality factor ($=\tan\delta$), a measure of strain energy dissipation
Q_D^{-1}	height of the Debye dissipation peak of the standard anelastic solid
R	radius of curvature of dislocation bowed out under applied stress; gas constant; radius of grain-edge tubule

S_m	entropy of kink migration	ν	Poisson's ratio
T	absolute temperature	ν_{k0}	attempt frequency of dislocation
T_m	melting temperature		vibration
T_o	oscillation period $(= 2\pi/\omega)$	ρ	crystal density
U	equilibrium distance for grain-bound-	ρ_k^-, ρ_k^+	densities of kinks of opposite sign
	ary sliding		(number per unit length of dislocation
$U(u)$	Peierls potential		line)
X	chemical composition	ρ_{k0}^-, ρ_{k0}^+	as above, under conditions of zero
Y	Youngs modulus		stress
α	negative of exponent describing fre-	ρ_k^{eq}	equilibrium kink density: number of
	quency dependence of Q^{-1}		kinks per unit length of dislocation line
α_b	aspect ratio of grain-boundary region	σ	stress, usually shear stress
	$(= \delta/d)$	σ_n	normal stress
α_e	aspect ratio (minimum/maximum	σ_P	Peierls stress
	dimension) of ellipsoidal inclusion	σ_t	tectonic shear stress
α_t	aspect ratio of grain-edge tubule	σ_{up}	unpinning shear stress
	$(= 2R/d)$	σ_0	amplitude of sinusoidally time-varying
β	coefficient of term βt^n representing		stress
	transient creep in the Andrade model	τ	relaxation time
γ	multiplicative constant of order unity	τ_A	anelastic relaxation time
γ_s	multiplicative constant	τ_d	duration of transient diffusional creep
δ	phase lag between stress and strain;	τ_{dr}	timescale for draining of cylindrical
	width of grain-boundary region $(<<d)$		laboratory specimen
δG	anelastic relaxation of the shear	τ_e	relaxation time for elastically accom-
	modulus		modated grain-boundary sliding
δJ	anelastic relaxation of the compliance	τ_M	Maxwell (viscous) relaxation time
δP	width in pressure of binary loop	$\tau_{s,t}$	relaxation time for melt squirt between
δT_0	width of melting interval		grain-edge tubules
δT_s	pressure-induced perturbation to soli-	$\tau_{s,e}$	relaxation time for melt squirt between
	dus and liquidus temperatures		ellipsoidal inclusions
$\delta\phi$	pressure-induced perturbation in melt	ϕ	melt fraction; the angle betweeen a
	fraction		dislocation segment and its Peierls
ε	strain		valley
ε_a	anelastic strain	ω	angular frequency
ε_e	elastic strain	ω_0	(angular) resonance frequency of dis-
ε_m	maximum anelastic strain due to		location segment
	migration of geometrical kinks on a	Γ	Gamma function
	dislocation segment	Δ	anelastic relaxation strength
ε_0	amplitude of sinusoidally time-varying	Λ	dislocation density, that is, dislocation
	strain		length per unit volume $= N_v L$
η	viscosity or effective viscosity	Λ_m	density of mobile dislocations
η_1	viscosity of dashpot in parallel with	Ω	molecular volume of diffusing species
	spring in the anelastic element of the	$(\delta\rho/\rho)_0$	fractional density contrast between
	Burgers model		coexisting polymorphs
η_b	grain-boundary viscosity	$(\delta\rho/\rho)_a$	fractional change in density resulting
η_f	fluid viscosity		from small change in proportions of
η_m	melt viscosity		coexisting phases; anelastic volu-
λ	wavelength of periodic grain-boundary		metric strain
	topography or of seismic wave	$(\delta\rho/\rho)_e$	fractional change in density caused by
μ_k	kink mobility $(= D_k/kT)$		pressure p_s

17.1 Introduction

The Earth and its constituent materials are subjected to naturally applied stresses that vary widely in intensity and timescale. At sufficiently high temperatures, nonelastic behavior will be encountered – even at the low stress amplitudes of seismic wave propagation, tidal forcing, and glacial loading (**Figure 1**). For harmonic loading, nonelastic behavior is manifest in a phase lag between stress and strain giving rise to strain energy dissipation and associated frequency dependence (dispersion) of the relevant modulus or wave speed. In the case of glacial loading, the transient creep of mantle material of sufficiently low viscosity allows time-dependent relaxation of the stresses imposed by the growth and decay of ice sheets with consequent time-dependent surface deformations.

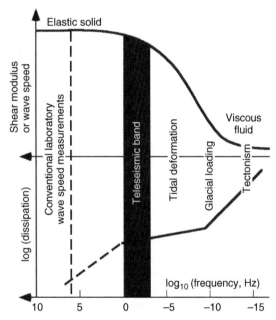

Figure 1 A schematic mechanical relaxation spectrum for the Earth. The mechanical behavior of the Earth's hot interior and its constituent materials change progressively from that of an elastic solid to that of a viscous fluid with decreasing frequency or increasing timescale. This transition involves the thermally activated mobility of crystal defects including vacancies, dislocations, twin domain and grain boundaries, and of phase boundaries and intergranular melt. The resulting viscoelastic behavior is manifest in the dissipation of strain energy and associated frequency dependence of the shear modulus and wavespeed. Adapted from Jackson I, Webb SL, Weston L, and Boness D (2005) Frequency dependence of elastic wave speeds at high-temperature: A direct experimental demonstration. *Physics of the Earth and Planetary Interiors* 148: 85–96.

Departures from elastic behavior have long been studied in metals. Notable early successes came from the application of resonance (pendulum) methods at fixed frequencies near 1 Hz that revealed low-temperature dissipation peaks attributable to various types of point-defect relaxation (e.g., Zener, 1952; Nowick and Berry, 1972). Until recently most of the available information concerning elastic wave speeds in geological materials and their variations with pressure and temperature came from ultrasonic and opto-acoustic techniques at frequencies at least six orders of magnitude higher than those of teleseismic waves (**Figure 1**). However, there is growing recognition of the need for a thorough understanding of viscoelastic behavior in minerals and rocks at low strains and seismic frequencies. It is the purpose of this article to assess the relaxation mechanisms most likely to produce significant departures from elastic behavior under the high-temperature conditions of the Earth's interior. Among these are intragranular mechanisms involving the motion of point defects and dislocations, grain-boundary migration and sliding, the stress-induced redistribution of an intergranular fluid phase, and effects associated with phase transformations.

Section 17.2 focuses on the theoretical framework for the description of viscoelasticity provided by broad phenomenological considerations and models for the operation of specific microphysical relaxation mechanisms. A thorough review of this material in a materials science context is provided by Schaller *et al.* (2001). In Section 17.3, an attempt will be made to integrate the results of the still relatively few seismic-frequency laboratory experiments on geological materials into the theoretical framework. Section 17.4 outlines selected applications of the emerging understanding of viscoelastic relaxation in the interpretation of seismological wave speed and attenuation models.

17.2 Theoretical Background

17.2.1 Phenomenological Description of Viscoelasticity

For sufficiently small stresses, the stress–strain behavior is expected to be linear and the response is represented in the time domain by the 'creep function' $\mathcal{J}(t)$ which is defined as the strain resulting from the application at time $t = 0$ of unit step-function stress, that is,

$$\sigma(t) = 0, \quad t < 0$$
$$= 1, \quad t \geq 0 \quad [1]$$

(e.g., Nowick and Berry, 1972). For the special case of elastic behavior, the strain appears essentially instantaneously (delayed only by the finiteness of the elastic wave speeds), and thereafter remains constant for the duration of stress application. However, at high temperatures and low frequencies, the response $\mathcal{J}(t)$ will usually be more complicated than the elastic ideal, involving in addition to the instantaneous (elastic) component, a time-dependent contribution which may be a mixture of recoverable (anelastic) and irrecoverable (viscous) strains (**Figure 2**). The simplest moderately realistic example of such linear viscoelastic rheology, that can be constructed from (elastic) springs and (viscous) dashpots arranged in series and parallel combinations, is the Burgers model (e.g., Findley *et al.*, 1976; Cooper, 2003) with the creep function

$$\mathcal{J}(t) = \mathcal{J}_U + \delta\mathcal{J}\,[1 - \exp(-t/\tau)] + t/\eta \qquad [2]$$

$\mathcal{J}_U$ and $\delta\mathcal{J}$ are the magnitudes of the instantaneous (elastic) and anelastic (time-dependent but recoverable) contributions, whereas τ and η are the time constant for the development of the anelastic response, and the steady-state Newtonian viscosity, respectively (**Figure 2**). The widely used 'standard anelastic solid' (e.g., Nowick and Berry, 1972) is the Burgers model without the series dashpot; its creep function is accordingly given by the first two terms of eqn [2]. More empirically successful in the description of transient creep, but physically less transparent, is the Andrade model (e.g., Poirier, 1985) for which the creep function is

$$\mathcal{J}(t) = \mathcal{J}_U + \beta t^n + t/\eta, \quad 1/3 < n < 1/2 \qquad [3]$$

The middle term with coefficient β and time t raised to the fractional power n represents transient creep. The relative merits of these alternative models as parametrizations of the high-temperature torsional microcreep behavior of our experimental assemblies have recently been evaluated. It has been found that the βt^n term in the Andrade model typically better fits the early part of a torsional microcreep record than does the Burgers model with its unique anelastic relaxation time τ. However, the Burgers model, generalized to include a suitable distribution $D(\tau)$ of relaxation times,

$$\mathcal{J}(t) = \mathcal{J}_U\left\{1 + \Delta \int_0^\infty D(\tau)[1 - \exp(-t/\tau)]\,d\tau\right\} + t/\eta \qquad [4]$$

provides a versatile alternative to the Andrade model in describing the viscoelastic rheology revealed, for example, by experimental torsional microcreep tests.

In eqn [4], Δ is the fractional increase in compliance associated with complete $(t = \infty)$ anelastic relaxation, known as the relaxation strength.

The strain $\varepsilon(t) = \varepsilon_0 \exp\,i(\omega t - \delta)$ resulting from the application of sinusoidally time-varying stress $\sigma(t) = \sigma_0 \exp(i\omega t)$, can be evaluated from the creep function provided that the behavior is linear (i.e., described by a differential equation which is linear in stress and strain and their respective time derivatives). This is done by superposition of the responses to each of a series of infinitesimal step-function applications of stress, that together represent the history $\sigma(t)$ of stress application (e.g., Nowick and Berry, 1972). Thus, an expression is obtained for the 'dynamic compliance' $\mathcal{J}^*(\omega)$ given by

$$\mathcal{J}^*(\omega) = \varepsilon(t)/\sigma(t) = i\omega \int_0^\infty \mathcal{J}(\xi) \exp(-i\omega\xi)\,d\xi \qquad [5]$$

where $\omega = 2\pi f$ is the angular frequency. This integral is the Laplace transform of $\mathcal{J}(t)$ with transform variable $s = i\omega$, or equivalently, within a multiplicative constant, the Fourier transform of the function which is $\mathcal{J}(\xi)$ for $\xi \geq 0$ and zero elsewhere. Since the Laplace transforms of each of the terms in the Burgers and Andrade creep functions are tabulated in standard compilations (e.g., Abramowitz and Stegun, 1972), analytical expressions for the dynamic compliance are readily derived (e.g., Jackson, 2000).

Thus, the dynamic compliance $\mathcal{J}^*(\omega)$ for the simple Burgers model is

$$\mathcal{J}^*(\omega) = \mathcal{J}_U + \delta\mathcal{J}/(1 + i\omega\tau) - i/\eta\omega \qquad [6]$$

with real and negative imaginary parts

$$\begin{aligned}\mathcal{J}_1(\omega) &= \mathcal{J}_U + \delta\mathcal{J}/(1 + \omega^2\tau^2) \\ \mathcal{J}_2(\omega) &= \omega\tau\delta\mathcal{J}/(1 + \omega^2\tau^2) + 1/\eta\omega\end{aligned} \qquad [7]$$

For the generalized Burgers model the equivalent expressions are

$$\mathcal{J}^*(\omega) = \mathcal{J}_U\left\{1 + \Delta \int_0^\infty D(\tau)\,d\tau/(1 + i\omega\tau)\right\} - i/\eta\omega$$

$$\mathcal{J}_1(\omega) = \mathcal{J}_U\left\{1 + \Delta \int_0^\infty D(\tau)\,d\tau/(1 + \omega^2\tau^2)\right\} \qquad [8]$$

$$\mathcal{J}_2(\omega) = \omega\mathcal{J}_U\Delta \int_0^\infty \tau D(\tau)\,d\tau/(1 + \omega^2\tau^2) + 1/\eta\omega$$

For the Andrade model, the corresponding expressions are

$$\begin{aligned}\mathcal{J}^*(\omega) &= \mathcal{J}_U + \beta\Gamma(1 + n)(i\omega)^{-n} - i/\eta\omega \\ \mathcal{J}_1(\omega) &= \mathcal{J}_U + \beta\Gamma(1 + n)\omega^{-n}\cos(n\pi/2) \\ \mathcal{J}_2(\omega) &= \beta\Gamma(1 + n)\omega^{-n}\sin(n\pi/2) + 1/\eta\omega\end{aligned} \qquad [9]$$

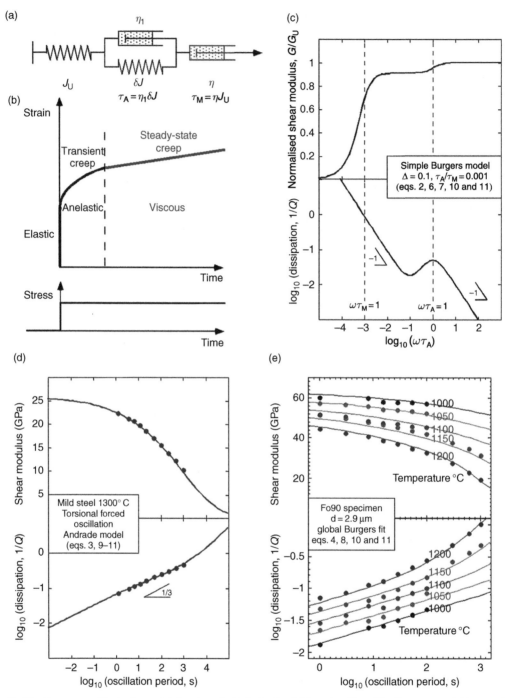

Figure 2 The Burgers and Andrade models for viscoelastic behavior. (a) Construction of the Burgers model from series and parallel combinations of Hookean springs and linear viscous dashpots. (b) The creep function for the Burgers model (eqn [2]). (c) The frequency-dependent shear modulus and dissipation for the Burgers model with $\Delta = 0.1$ and $\tau_M/\tau_A = 10^3$ (eqns [7], [10], and [11]). (d) The variations of shear modulus and dissipation for the Andrade model. The broad absorption band, within which $Q^{-1} \sim T_o^n$ (or ω^{-n}), is the result of the implicit infinitely wide distribution of anelastic relaxation times. This example is the least-squares fit of eqns [3] and [9]–[11] to forced-oscillation data for mild steel at tested at 1300°C and 200 MPa. (e) Another example of a broad absorption band, here for melt-free Fo$_{90}$ olivine. The data indicated by the plotting symbols are compared with the generalized Burgers model (eqns [4], [8], [10], and [11]) fitted to such data for a suite of four genuinely melt-free olivine polycrystals. See Faul and Jackson (2005) for details.

where $\Gamma(1 + n)$ is the Gamma function (Findley *et al.*, 1976; Gribb and Cooper, 1998a).

From these analytical expressions for $\mathcal{J}_1(\omega)$ and $\mathcal{J}_2(\omega)$, the shear modulus

$$G(\omega) = [\mathcal{J}_1^2(\omega) + \mathcal{J}_2^2(\omega)]^{-1/2} \qquad [10]$$

and the associated strain energy dissipation

$$Q^{-1}(\omega) = \mathcal{J}_2(\omega)/\mathcal{J}_1(\omega) \qquad [11]$$

are readily evaluated. It follows from eqns [10] and [11] that forced-oscillation measurements of $G(\omega)$ and $Q^{-1}(\omega)$, like the results of microcreep tests, can be represented by a creep-function model in an appropriately internally consistent manner.

In principle, then, knowledge of the creep function $\mathcal{J}(t)$ for arbitrary t and of the dynamic compliance $\mathcal{J}^*(\omega)$ for arbitrary ω provide equivalent insight into the mechanical behavior of the material. Conversion from one description to the other proceeds through integral transforms such as eqn [4] known as the Kronig–Kramers relations (e.g., Nowick and Berry, 1972, p. 37). These alternative descriptions of the mechanical behavior in the time and frequency domains are naturally associated with microcreep and forced oscillation experiments, respectively. Practical considerations related to the acquisition and processing of forced oscillation and microcreep data transform the nature of the relationship between the two methods from strict equivalence to complementarity. Thus, the intensively sampled relatively short-period (< 1–1000 s) stress- and strain-versus-time sinusoids of forced-oscillation experiments can be filtered very effectively with Fourier techniques to yield precise determinations of the relative amplitudes and phase of the stress and strain and hence $\mathcal{J}^*(\omega)$. For periods longer than 1000 s, however, the acquisition of forced-oscillation records representing a substantial number of oscillation periods becomes prohibitively time consuming. It is here that the microcreep method has an important advantage. In addition, the capacity to test explicitly the extent of the recovery of the nonelastic strain following removal of the steady stress is a major advantage of the microcreep method.

Thermodynamic description of anelasticity is based on the notion that stress σ and strain ε are linked not only directly through the unrelaxed modulus $G_U = \mathcal{J}_U^{-1}$, but also indirectly through a third, internal, variable p (Nowick and Berry, 1972; **Figure 3**). If the following conditions are met, the behavior is that of the standard anelastic solid: (1) there exists an equilibrium value $p_0(\sigma)$ of p that is proportional to σ (in the absence

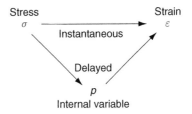

Figure 3 The thermodynamic basis for anelasticity. A third variable p provides an additional indirect and time-dependent link between stress and strain. Redrawn after Nowick and Berry (1972).

of an equilibrium value, the behavior is viscous); (2) $\partial p/\partial t \propto p(t) - p_0$; and (3) the delayed contribution to the strain is proportional to p. There are many different types of internal variable potentially associated with anelastic behavior in geological materials. These include the spatial arrangement of point defects or instantaneous position of a mobile segment of a dislocation line, the position of a mobile twin-domain wall, pressure in an intergranular fluid phase that is spatially variable at the grain or larger scale, and the proportions and chemical compositions of coexisting mineral phases. Many of these possibilities have been considered elsewhere (e.g., Jackson and Anderson, 1970; Karato and Spetzler, 1990). In this review attention will be focused on those mechanisms most likely to exhibit substantial relaxation strength ($\Delta > 0.001$) at teleseismic and lower frequencies (sub-Hz) under the pressure–temperature conditions of the Earth's deep interior.

17.2.2 Intragranular Processes of Viscoelastic Relaxation

The stress-induced migration or redistribution of defects with thermally activated mobility results in viscoelastic behavior even of single crystals when mechanically tested at sufficiently high temperatures. Such defects include vacancies, interstitials, substitutional impurity atoms, and dislocations.

17.2.2.1 Stress-induced rearrangement of point defects

Strain–energy dissipation peaks unambiguously associated with the stress-induced migration of point defects have been documented in metals (e.g., Nowick and Berry, 1972). Examples include the stress-induced reorientation of solvent–solute pairs in face-centered cubic (f.c.c.) and other metallic solid solutions (the Zener relaxation), redistribution of C and other small interstitial atoms in body-centered cubic (b.c.c.) metals such as Fe (the Snoek relaxation) and the stress-induced H diffusion on grain or specimen scale (the Gorsky effect). Such effects might also be expected in mantle minerals – for example, stress-induced repartitioning of Mg and Fe between nonequivalent crystallographic sites. However, relaxation times will generally be very short at mantle temperatures (e.g., <10 ms at 1000°C for Mg/Fe reordering between adjacent M1 and M2 sites in olivine, Aikawa *et al.*, 1985) and relaxation strengths are typically <10^{-3} (Karato and Spetzler, 1990).

17.2.2.2 Stress-induced motion of dislocations

Dislocation glide has long been recognized as an important mechanism of viscoelastic relaxation in crystalline solids. Theoretical models of the anelasticity associated with the reversible stress-induced glide of dislocation segments located between adjacent pinning

points were first developed in the 1950s to explain low-temperature (<0°C) internal friction peaks in deformed f.c.c. metals (Koehler, 1952; Seeger, 1956; Granato and Lücke, 1956). Although there was general agreement on the relatively large relaxation strength for this process, the characteristic timescale for the stress-induced motion proved more difficult to estimate reliably because of its dependence on structure and chemical composition of the dislocation core and the interactions amongst dislocations and between dislocations and impurities.

17.2.2.2.(i) The vibrating-string model of dislocation damping
An applied shear stress σ exerts a force σb per unit length on a suitably oriented dislocation of Burgers vector b, causing an initially straight dislocation segment to bow out in its glide plane (**Figure 4(a)**). This tendency is opposed by the line tension or elastic strain energy per unit length

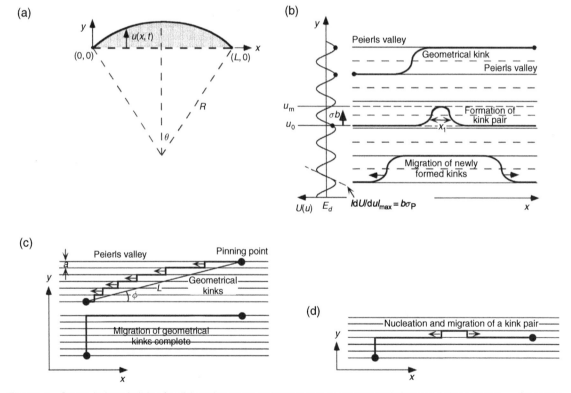

Figure 4 Stress-induced glide of a dislocation segment in response to an externally imposed shear stress σ_{yz} (involving forces parallel to $+y$ $(-y)$ acting on the upper (lower) faces of a cube straddling the glide plane in the page. (a) The force exerted on the dislocation by the applied stress and the dislocation line tension are in balance when the dislocation segment has bowed out into an arc with a radius of curvature inversely proportional to the applied stress (eqn [12]). (b) Modulation of the energy of a straight dislocation by the periodic Peierls potential $U(u)$ and its role in the nucleation and migration of kinks (redrawn after Seeger, 1981). (c) The stress-induced migration from right to left of geometrical kinks contributes a well-defined maximum anelastic strain (eqn [30]). (d) At moderately high temperature, the formation and migration of kink pairs will contribute additional anelastic strain.

$E_d \sim (1/2)Gb^2$, where G is the (unrelaxed) shear modulus. (The consequences of the periodic variation of the elastic strain energy of the dislocation about its average value E_d with position in the crystal lattice will be discussed in the following section.) The balance between the force exerted by the applied stress and the line tension determines the relationship between stress and radius of curvature R of the bowed-out segment as

$$R\sigma = Gb/2 \qquad [12]$$

(e.g., Nowick and Berry, 1972). Dislocation segments with length $L < L_c = 2R(\sigma) = Gb/\sigma$ bow out stably under the influence of a static stress σ and will be the primary focus of attention here.

Longer segments ($L > L_c$) participate in dislocation multiplication by the Frank–Read mechanism, thereby increasing the dislocation density and facilitating the transition to power-law dislocation creep. The grain size d imposes an absolute upper limit on the length of dislocation segments. In order to completely exclude the possibility of such dislocation multiplication, and thus guarantee linear behavior, it is accordingly required that

$$\sigma/G < b/d \qquad [13]$$

This condition is assessed below where the applicability of the theory of dislocation relaxation to laboratory experiments and to the Earth's interior is considered.

The classic vibrating-string dislocation-damping model of Koehler (1952) and Granato and Lücke (1956) has been revisited in simplified form by Nowick and Berry (1972) (see also Simpson and Sosin (1972), Minster and Anderson (1981), and Fantozzi *et al.* (1982)). A dislocation line with equilibrium orientation parallel to x is pinned by bound impurity atoms or intersections with other dislocations at $x = 0$ and $x = L$ (**Figure 4(a)**) allowing displacement $u(x, t)$ parallel to y in response to an externally applied oscillating shear stress $\sigma = \sigma_0 \exp(i\omega t)$. The equation of motion expressing the force balance on unit length of dislocation is

$$m_1 \partial^2 u/\partial t^2 + B\partial u/\partial t - E_d \partial^2 u/\partial x^2 = \sigma b = \sigma_0 b \exp(i\omega t) \qquad [14]$$

(Granato and Lücke, 1956). In this equation, ρ, $m_1 \sim \rho b^2$, B, and $-1/\partial^2 u/\partial x^2$ are respectively the crystal density, the effective mass per unit length of dislocation, the drag coefficient, and the instantaneous radius of curvature of the dislocation segment.

Neglecting high-frequency perturbations to the shape of the oscillating line segment (treated by Granato and Lücke (1956)), Nowick and Berry (1972, pp. 413–417) derived the approximate solution

$$u(x, t) = u_0(Lx - x^2)\exp(i\omega t) \qquad [15]$$

where

$$u_0 = (b\sigma_0/2E_d)/[1 - (\omega/\omega_0)^2 + i\omega\tau] \qquad [16]$$

with

$$\omega_0^2 = 12E_d/m_1 L^2 \qquad [17]$$

and

$$\tau = BL^2/12E_d \qquad [18]$$

For a population of N_v dislocation segments per unit volume of common length L and thus a dislocation density (length per unit volume)

$$\Lambda = N_v L \qquad [19]$$

the instantaneous value of the anelastic strain, calculated from the area swept out by the population of similarly favorably oriented bowing dislocations, is then

$$\varepsilon_a(t) = (\Lambda b/L) \int_0^L u(x, t)\,\mathrm{d}x = (\Lambda L^2/6)bu_0 \exp(i\omega t) \qquad [20]$$

and the resulting anelastic contribution toward the dynamic compliance (eqn [5]) is

$$\delta \mathcal{J}(\omega) = \varepsilon_a(t)/\sigma(t)$$
$$= (\Lambda b^2 L^2/12E_d)/[1 - (\omega/\omega_0)^2 + i\omega\tau] \qquad [21]$$

For frequencies well below the dislocation resonance frequency ω_0 (estimated from eqn [17] as 20 MHz–2 GHz for $L = 1$–$100\ \mu m$ in olivine), the behavior is approximately that of the standard anelastic solid (eqns [7] without the '$1/\eta\omega$' term). For $\Delta << 1$, eqns [10] and [11] thus become

$$Q^{-1}(\omega) = \Delta\omega\tau/(1 + \omega^2\tau^2)$$
$$\delta G/G = \Delta/(1 + \omega^2\tau^2) \qquad [22]$$

with

$$\Delta = \Lambda Gb^2 L^2/12E_d = (1/6)\Lambda L^2$$
$$\tau \sim (1/6)BL^2/Gb^2 \qquad [23]$$

17.2.2.2.(ii) The role of kinks in dislocation mobility: migration of geometrical kinks
However, it is well known that the elastic strain energy of a dislocation is not constant as assumed for the vibrating string model of dislocation motion, but instead varies about the average value E_d with position u within the crystal lattice. The periodic variation

normal to the close-packed direction is known as the Peierls potential $U(u)$ (**Figure 4(b)**). Especially if the Peierls potential is large, as is typical of silicate minerals, it will be energetically favorable for a stationary or moving dislocation line to lie mainly within a series of low-energy Peierls valleys – the individual straight segments being offset a distance a by kinks (**Figure 4(b)**). The kink picture of dislocation mobility dates from the mid-1950s (Seeger, 1956; Brailsford, 1961; Seeger and Wüthrich, 1976) and has been comprehensively reviewed by Seeger (1981).

We first consider abrupt kinks (of zero width), for which the position $u(x)$ of a dislocation segment of length L is given by

$$u(x) = \int_0^L (\rho_k^- - \rho_k^+)\,\mathrm{d}x \qquad [24]$$

where ρ_k^- and ρ_k^+ are the numbers of kinks of opposite sign per unit length of dislocation line. Brailsford (1961) expressed the kink densities as functions of x and t that satisfy continuity equations allowing for the thermal generation and recombination of kinks and for their stress-induced migration and diffusion (with mobility and diffusivity μ_k and D_k, respectively). In this way, he derived a differential equation for the motion of the dislocation segment analogous to eqn [14] with the inertial term neglected

$$\partial u/\partial t - D_k \partial^2 u/\partial x^2 = \sigma b a^2 \mu_k (\rho_k^- + \rho_k^+) \qquad [25]$$

A series solution, valid to first order in stress $\sigma_0 \exp(i\omega t)$, was obtained by fixing ρ_{k0}^- and ρ_{k0}^+ at their zero-stress levels, ρ_{k0}^-, and ρ_{k0}^+ respectively. The dominant leading term is of the form

$$u(x, t) - u_0 = u_1 \sin(\pi x/L)\exp(i\omega t)/(1 + i\omega\tau_L) \quad [26]$$

Substitution into eqn [25] and integration with respect to x allows evaluation of the anelastic relaxation time and relaxation strength associated with the motion of pre-existing (geometrical) kinks as

$$\tau = (L/\pi)^2/D_k \sim (L/\pi a_d)^2$$
$$\times [v_{k0}\exp(S_m/k)]^{-1}\exp(H_m/kT) \qquad [27]$$

and

$$\Delta = (8/\pi^4)\Lambda L^2(ab^2 G/kT)$$
$$\times a(\rho_{k0}^- + \rho_{k0}^+) \sim (4/5\pi^4)\Lambda L^2(ab^2 G/kT) \qquad [28]$$

(Brailsford, 1961; see also Seeger and Wüthrich (1976), Seeger (1981), and Fantozzi et al. (1982)). The kink diffusivity D_k in eqn [27] is given approximately by

$$D_k = a_d^2 v_{k0}\exp(-G_m/kT) \qquad [29]$$

where $G_m = H_m - TS_m$ is the free energy of kink migration, v_{k0} is the attempt frequency of dislocation vibration, and $a_d \sim b$ is the period of the kink potential. Equation [28] gives the relaxation strength for a population of dislocations with dislocation density Λ and uniform segment lengths L with an average value of $a(\rho_{k0}^- + \rho_{k0}^+) \sim \tan \phi$ of $1/10$ (ϕ is the inclination of the dislocation line from the Peierls valley; Brailsford, 1961).

Importantly, this analysis of dislocation relaxation based on migration of geometrical kinks yields the same key scalings, whereby $\Delta \propto \Lambda L^2$ and $\tau \propto L^2$, as the vibrating string model (Fantozzi et al., 1982; eqns [23]). The mild temperature dependence of Δ arises from the ratio of kink mobility to diffusivity.

From **Figure 4(c)**, it is clear that the maximum area swept out by migration of geometrical kinks on a dislocation segment of length L is given by $L^2 \sin \phi \cos \phi/2$. The maximum anelastic strain resulting from the migration of geometrical kinks

$$\varepsilon_m = b\Lambda L \sin \phi \cos \phi/2 \qquad [30]$$

(e.g., Karato, 1998) will be compared with the strains of laboratory experiments and seismic wave propagation in a later section.

17.2.2.2.(iii) The role of kinks in dislocation mobility: formation and migration of kink pairs

The more elaborate analysis of Seeger (1981) includes a quantitative description of the finite width of kinks and the formation and migration of new kink pairs in the stressed crystal. The starting point is the time-independent force balance per unit length of dislocation (c.f. eqn [14]) with the additional restoring force tending to constrain dislocation segments to lie in the Peierls valleys

$$E_d \mathrm{d}^2 u/\mathrm{d}x^2 - \mathrm{d}U/\mathrm{d}u + \sigma b = 0 \qquad [31]$$

Under a static stress σ, a straight dislocation in equilibrium will thus occupy a position u_0 such that

$$(\mathrm{d}U/\mathrm{d}u)|_{u=u_0} = b\sigma \qquad [32]$$

and the maximum value of $|\mathrm{d}U/\mathrm{d}u|$ is thus associated with the Peierls stress σ_P (**Figure 4(b)**).

A more general solution involving kinks is sought by rewriting $\mathrm{d}^2 u/\mathrm{d}x^2$ in eqn [31] as $-(1/f^3)\mathrm{d}f/\mathrm{d}u$, where $f = \mathrm{d}x/\mathrm{d}u$. Integration then yields

$$1/f^2 = (2/E_d)\int_{u_0}^{u_m}[\mathrm{d}U/\mathrm{d}u - \sigma b]\mathrm{d}u$$
$$= (2/E_d)[U(u_m) - U(u_0) - \sigma b(u_m - u_0)] \qquad [33]$$

The requirement that $1/f(u_m) = 0$ means that u_m must satisfy

$$U(u_m) - U(u_0) - \sigma b(u_m - u_0) = 0 \qquad [34]$$

with the understanding that for $\sigma = 0$, $u_0 = 0$, and $u_m = a$.

The shape of either a positive or negative geometrical kink ($\sigma = 0$) or a kink pair ($\sigma \neq 0$) is then obtained by further integration as

$$x - x_0 = \int_{u_0}^{u_m} f(u)\,du = \pm (2/E_d)^{-1/2}$$
$$\int_{u_0}^{u_m} [U(u) - U(u_0) - \sigma b(u - u_0)]^{-1/2}\,du \qquad [35]$$

This powerful general result (Seeger, 1981, eqn [5]) provides the basis for a detailed description of the role of kinks in anelastic relaxation. Both pre-existing (geometrical) kinks and newly formed kink pairs may contribute. The following is a brief outline of the way in which estimates of relaxation strength Δ and relaxation time τ have been derived through considerations based on eqn [35]. In order to illustrate key aspects of the theory, some results are evaluated for a specific (sinusoidal) functional form for the Peierls potential

$$U(u) = U(0) + (a/2\pi)b\sigma_P[1 - \cos(2\pi u/a)] \qquad [36]$$

along with the approximations $a \sim b$ and $E_d \sim (1/2)Gb^2$.

The energy of formation H_k and width w_k of a single geometrical kink are

$$H_k = \int_{-\infty}^{\infty} \{E_d[ds/dx - 1] + [U(u) - U(0)]\}\,dx$$
$$= (2E_d)^{1/2} \int_0^u [U(u) - U(0)]^{1/2}\,du \qquad [37]$$

with $ds = (dx^2 + du^2)^{1/2}$ and

$$w_k = a|dx/du|_{u = a/2} = 2a^2 E_d/\pi H_k \qquad [38]$$

For the sinusoidal Peierls potential (eqn [36]), these expressions can be evaluated as

$$H_k = (2a/\pi)(2ab\sigma_P E_d/\pi)^{1/2}$$
$$\sim (2/\pi^{3/2})Gb^3(\sigma_P/G)^{1/2} \qquad [39]$$

and

$$w_k = (\pi^{1/2}/2)b(\sigma_P/G)^{-1/2} \qquad [40]$$

The functional forms of eqns [39] and [40] illustrate the close relationship between the Peierls stress and the geometry and energetics of kink formation. The higher the Peierls stress, the narrower are the kinks and the greater is their formation energy.

Allowance for the work done against the external stress in forming a kink pair yields

$$H_{kp} = 2(2E_d)^{1/2} \int_{u_0}^{u_m} [U(u) - U(u_0)$$
$$- \sigma b(u - u_0)]^{1/2}\,du < 2H_k \qquad [41]$$

An estimate of the kink-pair formation energy that is more accurate for low stress conditions ($\sigma/\sigma_P << 1$) corresponds to the unstable equilibrium spacing x_1 of kinks of opposite sign at which their interaction energy H_{int} is maximized:

$$H_{kp} = 2H_k - H_{int}(x_1) \qquad [42]$$

Under these conditions, there is a balance between the forces associated with the applied stress and with the long-range elastic interaction seeking respectively to increase and decrease the kink separation (**Figure 4(b)**).

Analysis of the statistical mechanics of the dislocated crystal (with and without kinks) leads to estimates of the entropy of kink formation, the kink-pair formation rate, and hence the equilibrium density of kinks of given sign per unit length of dislocation line, given by

$$\rho_k^{eq} = (1/w_k)(2\pi H_k/kT)^{1/2}\exp(-H_k/kT) \qquad [43]$$

(for the sinusoidal potential). The velocity of a dislocation moving perpendicular to its Peierls valley by the formation and separation of kink pairs is then calculated. Finally, expressions are obtained for the relaxation strength

$$\Delta = \Lambda L^2 b^2 G_u/12E_d \sim (1/6)\Lambda L^2 \qquad [44]$$

and the relaxation time

$$\tau = [kT/(\rho_k^{eq})^2 D_k](L/2a^2 E_d)(1 + \rho_k^{eq}L) \qquad [45]$$

(eqns [45] and [50] of Seeger (1981)). The latter applies to conditions of stress sufficiently low for the temperature-dependent equilibrium kink density to be maintained and for all but very low temperatures (<10 K). It follows from eqns [29], [43], and [45] that different effective activation energies of approximately $2H_k + H_m$ and $H_k + H_m$, respectively, are expected for the low- and high-temperature regimes defined by $\rho_k^{eq}L << 1$ and $\rho_k^{eq}L >> 1$, respectively. The lower activation energy $H_k + H_m$ for the high-temperature regime reflects the greater probability of kink annihilation at high kink densities.

The relaxation strength (eqn [44]) is identical to that for the vibrating string model (eqn [23]) as expected from geometrical considerations. At the high kink densities of the high-temperature regime

($\rho_k^{eq}L \gg 1$), the relaxation time given by eqn [45] is equivalent to eqn [23] for the vibrating string model with the drag coefficient $B = 6kT/\rho_k^{eq}D_k a^2$, reasonably, inversely proportional to the product of the kink density and kink mobility (D_k/kT).

17.2.2.2.(iv) Relaxation strength for dislocation damping

The foregoing analysis (eqns [23] and [44]) yields a substantial relaxation strength $\Delta \sim 0.1-1$ for the stress-induced migration of favourably oriented intragranular dislocation segments with the dislocation density $\Lambda \sim L^{-2}$ of a three-dimensional (3-D) (Frank) network. A relaxation strength of order unity for olivine at upper-mantle temperatures is similarly expected from eqn [28] for relaxation due to the motion of geometrical kinks if the dislocation density approaches that of the Frank network. Somewhat lower values of $\Delta \sim 0.01 - 0.1$ are inferred once allowance is made for more typical geometries. However, for dislocations closely spaced (separation h) in the walls of subgrains of dimension L, the predicted relaxation strength is enhanced relative to that for a Frank network by the factor L/h which can be substantially greater than 10 (Friedel *et al.*, 1955; Nowick and Berry, 1972).

17.2.2.2.(v) Absorption band behavior: distribution of relaxation times

The foregoing theory is readily generalized to accommodate a wide distribution of dislocation segment lengths (e.g., Schoeck, 1963). If $p_L(L)dL$ is the number of dislocation segments per unit volume of length between L and $L + dL$, then the total dislocation length per unit volume is

$$\Lambda = \int_0^\infty p_L(L)L\,dL \qquad [46]$$

The internal friction and frequency-dependent modulus are, respectively,

$$Q^{-1}(\omega) = \Delta \int_0^\infty p_L(L)L^3\omega\tau(L)dL/[1 + \omega^2\tau^2(L)] \quad [47]$$

and

$$G(\omega)/G_u = 1 - \Delta \int_0^\infty p_L(L)L^3 dL/[1 + \omega^2\tau^2(L)] \quad [48]$$

where Δ is now the anelastic relaxation strength for the entire distribution of dislocation segment lengths.

For geological materials tested in the laboratory, the dissipation commonly displays a monotonic variation of Q^{-1} with frequency and temperature represented by

$$Q^{-1} \sim [\omega \exp(E/RT)]^{-\alpha} \qquad [49]$$

with $\alpha \sim 0.3 \pm 0.1$ (Berckhemer *et al.*, 1982; Jackson *et al.*, 2002). Geophysical observations of the attenuation of seismic body and surface waves and free oscillations, and of strain–energy dissipation at the longer periods of earth tides and the Chandler wobble, are broadly consistent with this view of the frequency dependence of Q^{-1} (e.g., Minster and Anderson, 1981; Shito *et al.*, 2004). If eqn [47] with $\tau(L) \sim L^2$ (e.g., eqns [23], [27], and [45]) is to yield $Q^{-1} \sim \omega^{-\alpha}$, it is required that the distribution of dislocation segment lengths adopt the form

$$p_L(L) \sim L^{2\alpha - 4} \qquad [50]$$

strongly skewed toward short dislocation lengths (Minster and Anderson, 1981; Karato, 1998). Alternatively, or additionally, a distribution of activation energies for dislocation migration might contribute to the breadth of an anelastic absorption band (Minster and Anderson, 1981).

17.2.2.2.(vi) Dislocation relaxation: transition from anelastic to viscous behavior

The foregoing analysis is based on the assumption that dislocation segments are firmly pinned by bound impurities and/or interactions with other dislocations. Under these circumstances, the restoring force provided by line tension or recovery of kink-pair formation energy ensures that the nonelastic strain is recoverable upon removal of the applied stress, meaning that the behavior is anelastic. However, at sufficiently high temperatures and/or stress amplitudes, dislocation segments may break free from their pinning points, thereby allowing larger, irrecoverable (viscous) strains. By balancing the work done in moving the dislocation away from the pinning point with the binding energy E_{up} of the dislocation to the pinning point, the critical stress σ_{up} for unpinning (at 0 K) can be estimated as

$$\sigma_{up} = E_{up}/abL \qquad [51]$$

For finite temperature T, thermal activation will allow unpinning at somewhat lower stresses with a probability proportional to $\exp[-(E_{up} - \sigma abL)/kT]$ (Nowick and Berry, 1972, pp. 364–365).

17.2.2.2.(vii) Role of water in nominally anhydrous minerals

The presence of intragranular water (or more accurately hydrogen-related defects) is known to affect both large-strain rheology and electrical conductivity of nominally anhydrous silicate minerals like olivine and wadsleyite. Arguing mainly by analogy with such observations, Karato and Jung (1998) suggested that the concentration C_{OH} of hydroxyl within mineral grains and/or grain boundaries might similarly enhance seismic wave attenuation and dispersion through its influence on the concentrations and mobilities of key defects. For parametrization they suggested a variant of eqn [49]:

$$Q^{-1} \sim [(A + BC_{OH})/\omega \exp(gT_m/T)]^{\alpha} \qquad [52]$$

in which T_m is the relevant melting temperature and g is a dimensionless constant.

17.2.3 Intergranular Relaxation Processes

17.2.3.1 Effects of elastic and thermoelastic heterogeneity in polycrystals and composites

The role of the heterogeneous microscopic stress field that arises in stressed polycrystals as the inevitable consequence of the elastic anisotropy of the component crystallites has been somewhat neglected in discussions of the nature of the high-temperature internal friction background. During testing under conditions of high temperature and low frequency, these intergranular fluctuations in stress may be subject to relaxation by the reversible migration of grain boundaries (Leak, 1961). Such grain-boundary migration is driven by the minimization of the elastic strain energy of the stressed polycrystal (Kamb, 1959). The anelastic relaxation involves enlargement of those grains oriented for high compliance under the prevailing stress relative to neighboring grains oriented for lower compliance (Nowick and Berry, 1972, p. 453; **Figure 5**).

A useful indication of the magnitude of the modulus relaxation (Kumazawa, 1969) derives from consideration of the physical basis for the Voigt and Reuss bounds on the effective elastic moduli of a

polycrystal. In such discussions, the Reuss (lower) bound on the effective moduli assumes special significance because it corresponds to a state of uniform stress throughout the polycrystal. It is just this condition that is approached through relaxation of the initially inhomogeneous internal stress field. Kumazawa therefore concluded that the difference between the Voigt–Reuss–Hill (or the average G_{HS} of the more closely spaced Hashin–Shtrikman bounds) and Reuss averages (G_{RS}) of the single-crystal elastic constants is a measure of the inherent nonelastic behavior of the polycrystal. Thus, the fractional relaxation of the shear modulus is given by

$$\delta G/G = (G_{HS} - G_{RS})/G_{HS} \qquad [53]$$

Shear moduli intermediate between G_{HS} and G_{RS} can accordingly be achieved through the relaxation toward uniformity of the inhomogeneous internal stress field arising from the elastic anisotropy of the individual crystallites.

Elastic heterogeneity in polycrystals and multiphase composites leads also to bulk dissipation Q_K^{-1} arising from the coupling between macroscopic volumetric strain and internal shear strains (Budiansky and O'Connell, 1980). Their estimates of Q_K^{-1}/Q_G^{-1} for representative upper-mantle materials will be considered in Section 17.4. The magnitude of Q_K^{-1} and the extent of the associated relaxation of the bulk modulus depend on the spacing between the HS-average and Reuss bounds on the effective bulk modulus (c.f. eqn [53]). These are typically much more closely spaced than for the shear modulus resulting in more modest anelastic relaxation of the bulk modulus (Section 17.4).

Anisotropy and inter-phase variability (collectively, heterogeneity) of elastic constants and thermal expansivity within polycrystals and composite materials result in spatial heterogeneity of the temperature changes caused by adiabatic compression, that may be relaxed toward isothermal conditions by intergranular or inter-phase heat flow (Zener, 1952; Budiansky et al., 1983). A formal phenomenological theory (Zener, 1952, pp. 84–89) for such thermoelastic behavior, valid at frequencies sufficiently high for the diffusion of heat to be confined to the immediate neighborhood of the grain or inter-phase boundaries, yields

$$Q^{-1} \sim A(D/\omega)^{1/2}/d \qquad [54]$$

(cf. eqn [49]), where D is the thermal diffusivity and d is the grain size. The dissipation scales with the measure A of the heterogeneity of the relevant physical property. The relevance of thermoelastic

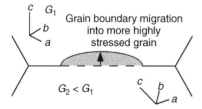

Figure 5 Reversible stress-induced grain-boundary migration as an anelastic relaxation mechanism. The driving force for the (normal) migration of the boundary is the reduction of elastic strain energy. Redrawn after Jackson et al. (2000).

damping to bulk dissipation Q_K^{-1} in the Earth's mantle will be considered briefly in Section 17.4.

17.2.3.2 Grain-boundary sliding

17.2.3.2.(i) Overview Grain-boundary sliding has been widely invoked as a plausible mechanism for the ubiquitous transition from elastic through anelastic to viscous behavior in fine-grained polycrystalline materials. Central to the theory of grain-boundary sliding is the notion of a grain boundary region, of finite width δ, distinguished from the crystalline lattices of the neighboring grains by a higher degree of positional disorder, and consequently, higher diffusivities D_b for the various atomic and molecular species and lower viscosity η_b. At sufficiently low-temperature and/or high-frequency, grain-boundary sliding will be inhibited, and an unrelaxed shear modulus G_U representative of the strictly elastic behavior of the polycrystal is expected (**Figure 6**, uppermost panel). With increase of temperature and/or timescale of the mechanical test beyond an appropriate threshold, it is envisaged that relaxation (to zero) of the distribution of grain-boundary shear stress allows a finite amount of sliding along suitably oriented boundaries accommodated by elastic distortion of the neighboring grains (Ashby, 1972). The modified distribution of normal stress associated with the accommodating elastic distortion provides the restoring force required for macroscopically anelastic (i.e., recoverable) behavior (**Figure 6**, middle panel). At still longer timescales and higher temperatures, it is envisaged that viscous

deformation occurs with grain-boundary sliding accommodated by diffusional transport of matter away from boundary regions of high chemical potential (normal stress) to other parts of the grain boundary at lower chemical potential (Raj and Ashby, 1971). The transition from the anelastic behavior associated with elastically accommodated grain-boundary sliding to steady-state viscous behavior requires an adjustment of the distribution of normal stress that is facilitated by grain-boundary diffusion (**Figure 6**, lowermost panel).

17.2.3.2.(ii) Elastically accommodated grain-boundary sliding The elastic regime prevailing following completion of elastically accommodated sliding was analyzed in the classic work of Raj and Ashby (1971). The component of the normal stress parallel to the mean direction x of the grain boundary and integrated over the wavelength λ of the periodic boundary balances the externally applied shear stress σ. The elastic strains caused by the newly created distribution of normal stress are those required to restore grain shape compatibility across the slipped boundary. With the boundary topography $u(x)$ for a 2-D array of hexagonal grains (of unrelaxed shear modulus G_U and Poisson's ratio ν) represented by a Fourier series

$$u(x) = \sum_{j=1}^{\infty} b_j \cos(2\pi j x/\lambda) \qquad [55]$$

the following expressions were obtained for the normal stress σ_n and the equilibrium distance U of reversible sliding:

$$\sigma_n(x) = -\sigma\lambda \sum_{j=1}^{\infty} j^2 b_j \sin(2\pi j x/\lambda) / \left[\pi \sum_{j=1}^{\infty} j^3 b_j^2\right] \qquad [56]$$

$$U = (1-v)\lambda^3 \sigma / \left[2\pi^3 G_U \sum_{j=1}^{\infty} j^3 b_j^2\right] \qquad [57]$$

Numerical results, apparently based on truncation of the infinite series after $N = 100$ terms, yielded a finite sliding distance and a corresponding large anelastic relaxation strength

$$\Delta = 0.57(1-v) \qquad [58]$$

independent of grain size. For example, for $v = 0.26$, appropriate for olivine at 1000–1300°C (Anderson and Isaak, 1995), $\Delta \sim 0.42$. The relaxed shear modulus G_R and the height Q^{-1}_D of the Debye dissipation peak (as appropriate for the standard anelastic solid) are given by the following expressions:

$$G_R/G_U = 1/(1+\Delta), \quad Q^{-1}_D = (\Delta/2)/(1+\Delta)^{1/2} \qquad [59]$$

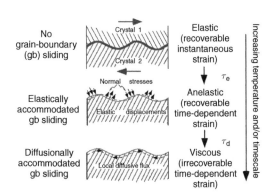

Figure 6 Grain-boundary sliding as a viscoelastic relaxation mechanism. The low effective viscosity of the grain-boundary region facilitates local shear stress relaxation and sliding providing for a seamless transition with increasing temperature and/or timescale from elastic through anelastic to viscous behavior as described in the text. Adapted from Ashby MF (1972) Boundary defects and atomistic aspects of boundary sliding and diffusional creep. *Surface Science* 31: 498–542.

However, because $h_j \sim j^{-2}$ (Raj, 1975), the infinite sum $\Sigma j^3 h_j^2$ in the denominators of eqns [56] and [57] fails to converge – suggesting that sufficiently sharp grain-edge intersections might inhibit elastically accommodated sliding (Faul et al., 2004; Jackson et al., 2006). The higher values of $\Delta \sim 0.8$ obtained in the modeling of sliding between spherical grains (Zener, 1941; Kê, 1947) and the lower values (0.23 for $v = 0.26$) reported from a finite-element study with a fine mesh (Ghahremani, 1980) are consistent with this suggestion.

The relaxation time for elastically accommodated grain-boundary sliding is given by

$$\tau_e = \gamma \eta_b d / G_U \delta \qquad [60]$$

where γ, of order 1, is either a constant or a function of Poisson's ratio v for the unrelaxed material (Kê, 1947; Nowick and Berry, 1972; Mosher and Raj, 1974; O'Connell and Budiansky, 1977; Ghahremani, 1980).

17.2.3.2.(iii) Diffusionally accommodated grain-boundary sliding
The possibilities of elastic and steady-state diffusional accommodation of grain-boundary sliding were examined separately by Raj and Ashby (1971). Grain-boundary sliding with 'sequential' occurrence of elastic and diffusional accommodation was explored by Raj (1975) in his analysis of transient diffusional creep. The transient required to adjust the normal stress distribution from that prevailing on completion of elastically accommodated sliding (eqn [56]) to that required for steady-state diffusional creep is of approximate duration

$$\tau_d = (1 - v) k T d^3 / [40 \pi^3 G_U \delta D_b \Omega] \qquad [61]$$

Ω is the molecular volume of the diffusing species and D_b is the grain-boundary diffusivity. The transient creep rate calculated by Raj is enhanced relative to the corresponding steady-state diffusional creep rate by a numerical factor that varies approximately as $(t/\tau_d)^{-1/2}$, which integrates to a creep function of the Andrade form (eqn [3]) – as recognized and emphasized by Gribb and Cooper (1998a; see also Cooper (2003)). This creep function implicitly involves a monotonic and infinitely wide distribution of anelastic relaxation times (Jackson, 2000) and therefore does not result in a dissipation peak. The diffusional creep transient should instead be responsible for a wide absorption band, within which

$$Q^{-1} \sim T_o^{1/2} d^{-s} \qquad [62]$$

with period $T_o = 2\pi/\omega$ and $1 < s < 3/2$ (Gribb and Cooper, 1998a; Faul et al., 2004).

Through use of the well-known connection between grain-boundary diffusivity and viscosity, it can be shown that $\tau_e/\tau_d \ll 1$. Thus, it is predicted that the dissipation peak associated with the transition between elastic and anelastic behavior and the onset of appreciable distributed dissipation associated with the progressive transition between anelastic and viscous behavior should be widely separated in oscillation period – temperature space (Jackson et al., 2002).

However, a consistent pattern of high-temperature viscoelastic behavior at variance with the classic Raj–Ashby theory is emerging from intensive laboratory studies on fine-grained geological and ceramic materials (Section 17.3). These observations have provided the motivation for a new approach to the micromechanical modeling of grain-boundary sliding free from the principal limitations of Raj–Ashby theory, namely that elastic and diffusional accommodation occur separately and sequentially, and that grain edges are only moderately sharp (Jackson et al., 2006). The boundary-value problem incorporating both sliding and diffusion has been solved with a perturbation approach, valid only for a gently sloping periodic boundary, providing for the first time, the complete relaxation spectrum (Morris and Jackson, 2006). As expected, the boundary viscosity determines the behavior at short periods T_o – with Q^{-1} increasing with increasing T_o toward the elastically accommodated sliding peak of classical theory (Raj and Ashby, 1971). Further increase of T_o leads to a regime beyond the peak where grain-boundary diffusion dominates – its influence extending progressively with increasing period from relaxation of stress concentrations at grain corners, where Q^{-1} varies only mildly with period, ultimately to grain-scale diffusion associated with steady-state creep ($Q^{-1} \sim T_o$). Significantly, the peak height varies inversely with the boundary slope, but a final answer for finite boundary slope with arbitrarily sharp grain-edge intersections awaits the results of numerical analysis in progress.

Finally, in the context of diffusional creep it should be stressed that free surfaces, grain boundaries, polygonized (subgrain) boundaries, and even individual dislocations act as easy sources and sinks of vacancies. Stress-induced migration of vacancies results in diffusional creep with a strain rate

$$d\varepsilon/dt = \gamma D \sigma b^3 / d^2 k T \qquad [63]$$

where d is the characteristic separation of vacancy sources and sinks which can be substantially smaller than the grain size and the dimensionless constant γ is of order unity (Friedel, 1964, pp. 311–314). It was suggested by Friedel that such diffusional creep might be an important contributor to the high-temperature internal friction background. However, as a viscous process it would result in a stronger frequency dependence $Q^{-1} \sim \omega^{-1}$ than is observed experimentally (see below).

17.2.4 Relaxation Mechanisms Associated with Phase Transformations

17.2.4.1 Stress-induced variation of the proportions of coexisting phases

Relaxation associated with stress-induced change of the proportions of coexisting crystalline or solid and liquid phases has been analyzed in detail by Darinskiy and Levin (1968) and Vaišnys (1968; see also Jackson and Anderson (1970)). The following simplified treatment is intended to explain the nature of such relaxation. Consider a region of pressure (P)–temperature (T)–bulk composition (X) space within which two phases such as a pair of low- and high-pressure polymorphs with density contrast $(\delta\rho/\rho)_0$ coexist in thermodynamic equilibrium. A superimposed small fluctuation p_s in pressure (under isothermal conditions), arising for example, from a seismic compressional wave will induce changes in the equilibrium proportions and compositions of the coexisting phases. If the width of the two-phase region is δP (for given T and X), the magnitude of the implied fractional density perturbation is approximately

$$\varepsilon_a = (\delta\rho/\rho)_a = (p_s/\delta P)(\delta\rho/\rho)_0 \qquad [64]$$

The volumetric strain specified by eqn [64] is proportional to the pressure perturbation and the attainment of the perturbed equilibrium will require finite time for diffusional rearrangement of the various chemical species. Moreover, the process is reversible on removal of the pressure perturbation. The conditions for anelastic behavior (identified above) are thus clearly satisfied. The corresponding elastic strain is given by

$$\varepsilon_e = (\delta\rho/\rho)_e = p_s/\underline{K} \qquad [65]$$

where $\underline{K}(X, T)$ is the appropriate average of the unrelaxed bulk moduli for the coexisting phases.

The relaxation strength, given by the ratio $\varepsilon_a/\varepsilon_e$ of the anelastic and elastic strains as in eqn [4], is thus

$$\Delta_v = (\underline{K}/\delta P)(\delta\rho/\rho)_0 \qquad [66]$$

Because $\underline{K}/\delta P$ can be very large, Δ_v can be substantial (>1) even for a modest density contrast of order 0.01 (Vaišnys, 1968). The thermodynamic coexistence of the two phases, especially in subequal proportions near the middle of the two-phase loop, presumably means that nucleation is no barrier to further incremental transformation in either direction. An increase in the volume fraction of whichever phase is favored by the instantaneous value of the fluctuating hydrostatic pressure at the expense of the other phase would presumably be accomplished most readily by the migration normal to itself of the phase boundary across which grains of the two phases are in contact. The adjustment $\sim p_s/\delta P$ to the proportions of the coexisting phases could then be accomplished by multicomponent diffusion with a diffusivity D given approximately by that of the most slowly diffusing major-element species (usually Si in silicate minerals). For equal proportions of the two phases, the necessary change in volume requires the formation of an outer rind of thickness

$$\delta \sim (d/3)(p_s/\delta P) \qquad [67]$$

on an approximately spherical grain of diameter d. If diffusion were the rate-controlling step, the relaxation time would thus be of order

$$\tau \sim \delta^2/D(T) = (d^2/9D_0)(p_s/\delta P)^2 \exp(H/RT) \qquad [68]$$

In this simple analysis several major sources of complication have been ignored, particularly in estimation of the relaxation time. No attempt has been made to model the time-dependent stress at the phase boundary or the transformation kinetics, potentially strongly influenced by the rheology of the surrounding medium (Darinskiy and Levin, 1968; Morris, 2002). Nevertheless, the broad feasibility of such relaxation of the bulk modulus for seismic wave propagation in the transition zone will be explored below.

These ideas are also readily applied to stress-induced variations in the proportion of melt in partially molten material. In this case the pressure perturbation p_s slightly modifies the solidus and liquidus temperatures by an amount

$$\delta T_s \sim (\mathrm{d}T_m/\mathrm{d}P)p_s \qquad [69]$$

If δT_0 is the width of the melting interval, then the change to the melt fraction caused by the pressure fluctuation p_s is

$$\delta\phi \sim \delta T_s/\delta T_0 \qquad [70]$$

which plays an analogous role to $p_s/\delta P$ in the foregoing analysis (Vaišnys, 1968; Jackson and Anderson, 1970; Mavko, 1980).

17.2.4.2 Stress-induced migration of transformational twin boundaries

Another possible mechanism of seismic wave attenuation and dispersion, with particular potential application to the Earth's silicate perovskite-dominated lower mantle (and perhaps parts of the lithosphere), involves the stress-induced migration of boundaries between ferroelastic twins. The ABO_3 perovskite structure relevant to the lower mantle, comprising a 3-D corner-connected framework of cation-centered BO_6 octahedra with larger A cations occupying the cage sites, displays an extraordinary versatility based on the systematic tilting and/or rotation of adjacent octahedra. In this way the structure is not only able to accommodate cations that vary widely in size and charge, but also able to respond flexibly to changing conditions of pressure and temperature. Thus, many perovskites cooled from high temperature undergo a series of displacive phase transformations to structures of progressively lower symmetry, each transition accompanied by the appearance of a spontaneous shear strain. The reduction of symmetry typically results in the formation of a microstructure comprised of transformational twins with different orientations (relative to the lattice of the parent crystal) meeting in domain walls.

The nature of anelastic relaxation associated with the stress-induced motion of such domain walls has been explained by Harrison and Redfern (2002; see also Schaller et al., 2001) as follows. Under an applied stress, the free energy degeneracy between adjacent twin domains is removed resulting in a force acting on the domain wall. Thus, domain walls will tend to migrate so as to enlarge the domains of relatively low free energy at the expense of those of higher energy. The result is an additional nonelastic contribution to the macroscopic strain that is proportional to the product of the spontaneous strain, the density of domain walls, and the distance through which they move. Interactions between domain walls, between domain walls and associated defects, and between domain walls and the surface of a laboratory specimen result in a finite mobility or equivalently a characteristic timescale for their motion that is thermally activated. Moreover, the process of domain wall migration is reversible on removal of the applied stress – meeting the final criterion for anelastic behavior discussed in Section 17.2.1.

17.2.5 Anelastic Relaxation Associated with Stress-Induced Fluid Flow

The presence within a rock of an intergranular fluid phase creates additional opportunities for anelastic relaxation. Well-developed theoretical models predict the dependence of elastic bulk and shear moduli and attenuation upon the volume fraction and viscosity η_f of the fluid, and upon the nature of the grain-scale fluid distribution and the angular frequency ω of the applied stress field (Walsh, 1968, 1969; O'Connell and Budiansky, 1977; Mavko, 1980; Schmeling, 1985; Hammond and Humphreys 2000a, 2000b; see review by Jackson (1991)).

The model microstructure typically comprises a population of fluid-filled inclusions of specified shape (e.g., ellipsoids, grain-boundary films, or grain-edge tubes) and connectivity embedded within a crystalline matrix, subject to an externally imposed stress field. As a consequence of the distortion of the matrix, each fluid inclusion is exposed to a particular state of stress. Depending upon the frequency of the applied stress, any one of four distinct fluid stress regimes may be encountered: glued, saturated-isolated, saturated-isobaric, and drained, listed here in order of decreasing frequency (O'Connell and Budiansky, 1977; **Figure 7**). The response to an externally imposed shear stress will be considered first. At sufficiently high frequencies, within the glued regime, the fluid is able to support a nonhydrostatic stress with an effective shear modulus $|G| = \omega\eta_f$ not substantially less than the matrix rigidity so that relative tangential displacement of neighboring grains is inhibited and the influence of the fluid is minimal. The saturated–isolated regime, where the fluid no longer supports shear stress but fluid pressure varies between (even adjacent) inclusions of different orientation, is encountered at somewhat lower frequencies. For still lower frequencies, fluid pressures are equilibrated by fluid flow between adjacent inclusions ('melt squirt', Mavko and Nur, 1975) in the saturated–isobaric regime. Finally, at very low frequencies, drained conditions will apply for which no stress-induced perturbation

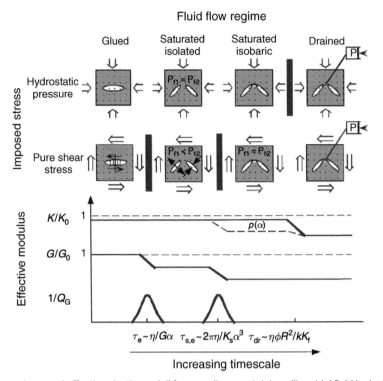

Figure 7 Fluid-flow regimes and effective elastic moduli for a medium containing ellipsoidal fluid inclusions of given aspect ratio α, as envisaged by O'Connell and Budiansky (1977). The upper row of panels shows the response of the medium to hydrostatic compression, whereas the lower row shows the response to shear. The vertical black bars highlight the transitions between fluid-flow regimes that are expected to cause marked modulus relaxation. A distribution $p(\alpha)$ of fluid-inclusion aspect ratios leads to partial relaxation of the bulk modulus in the saturated isobaric regime as indicated by the dashed line within the saturated isobaric regime. P_{f1} and P_{f2} are the pore pressures in representative inclusions of different orientation relative to the applied stress. Adapted from Lu C and Jackson I (2006) Low-frequency seismic properties of thermally cracked and argon saturated granite. *Geophysics* 71: F147–F159.

of pore fluid pressure (relative to that in an external reservoir) can be sustained.

The transition between the glued, saturated–isolated, and saturated–isobaric regimes are possible causes of dispersion and attenuation under the conditions of laboratory petrophysical experimentation and for seismic waves within the Earth's crust and its deeper interior. The transition from the glued regime to the saturated–isolated regime, in which the fluid no longer supports shear stress but undergoes no grain-scale flow, is formally identical to the process of elastically accommodated grain-boundary sliding reviewed above. The grain-boundary fluid inclusion is simply identified with the boundary region of aspect ratio $\alpha_b = \delta/d$ and low viscosity η_b, and the relaxation time is given by eqn [59] rewritten in terms of α_b as

$$\tau_e = \gamma \eta_b / G_u \alpha_b \qquad [71]$$

The relaxation time for fluid squirt between the triple-junction tubules characteristic of a nonwetting melt was estimated by Mavko (1980) as

$$\tau_{s,t} = 160 \, \eta_m / K_m \alpha_t^2 \qquad [72]$$

where $\alpha_t = 2R/d$ is the aspect ratio of the tubules (diameter/length), K_m is the bulk modulus of the melt, and η_m the viscosity of the melt. The relaxation time for melt squirt between ellipsoidal inclusions was given by O'Connell and Budiansky (1977) as

$$\tau_{s,e} = 2\pi \eta_m / K_s \alpha_e^3 \qquad [73]$$

where K_s is the bulk modulus of the solid and α_e the aspect ratio (width/length) of the ellipsoidal melt inclusions.

The characteristic timescale for each of these three processes of anelastic relaxation thus depends on the viscosity of the fluid phase (or grain-boundary material) as well as the aspect ratio of the fluid inclusion or grain-boundary region of relatively low viscosity. The variations of relaxation time for representative ranges of aspect ratio and viscosity for each of the three processes are contoured in **Figure 8**. Each line is the locus of all combinations of aspect ratio and viscosity that yield a particular relaxation time. All combinations (α, η) that plot above (below) a contour corresponding to a given relaxation time τ, are associated with unrelaxed (relaxed) behavior for oscillation period $T_o = \tau$. For a given mechanism to produce a relaxation peak centered within the teleseismic band (1–1000 s), it is therefore required that the relaxation time calculated as a function of aspect ratio and viscosity fall between the 1 and 1000 s contours indicated in **Figure 8**. An interpretation of dissipation data from laboratory experiments on melt-bearing fine-grained olivine polycrystals, based on **Figure 8**, is presented in Section 17.3.3.

The effective elastic moduli, and hence relaxation strengths associated with the transitions between these fluid-flow regimes, depend upon both the volume

fraction ϕ, and aspect ratio, α (minimum/maximum dimension) of the fluid inclusions. However, for relatively low fluid fractions and low aspect ratios (i.e., fluid-filled cracks rather than pores), the behavior is controlled primarily by a single variable known as crack density, ε (O'Connell and Budiansky, 1974). For a population of spheroidal inclusions of common low aspect ratio, ε is given by

$$\varepsilon = 3\phi/4\pi\alpha \qquad [74]$$

which definition can be generalized to accommodate a distribution of crack aspect ratios.

For the transition between the glued and saturated isolated regimes, the relaxation of the shear modulus is given, for low crack density ε, by

$$\delta G \sim 32(1-v)G\varepsilon/[15(2-v)] = (32/35)G\varepsilon \qquad [75]$$
$$(\text{for } v = 1/4)$$

For the transition from the saturated isolated regime to the saturated isobaric regime, the further relaxation of the shear modulus is

$$\delta G \sim 32(1-v)G\varepsilon/45 = (8/15)G\varepsilon \quad (\text{for } v = 1/4) \qquad [76]$$

meaning that fluid squirt is only about half as effective as elastically accommodated grain-boundary sliding as a relaxation mechanism (O'Connell and Budiansky, 1977). A nonwetting fluid, confined to grain-edge tubules, is much less effective in reducing the shear modulus in both saturated isolated and saturated isobaric regimes (Mavko, 1980).

As regards the bulk modulus, the presence of fluid inclusions with a bulk modulus K_f substantially lower than that (K_s) of the crystalline matrix will result in a modest reduction of the modulus for the drained regime given by

$$K \sim K_s[1 - (K_s/K_f)\phi] \qquad [77]$$

(O'Connell and Budiansky, 1974; Hudson, 1981). In response to externally imposed changes in mean stress (**Figure 7**), fluid inclusions of common aspect ratio but different orientation experience the same perturbation in pore pressure – meaning that there is no distinction between the saturated isolated and isobaric regimes. Given the existence of a distribution of fluid-inclusion aspect ratios $p(\alpha)$, however, stress-induced fluid flow will occur, for example, between cracks and pores, with associated partial relaxation of the bulk modulus (Budiansky and O'Connell, 1980; **Figure 7**). In a partially molten medium, the maintenance of equilibrium proportions of the coexisting crystalline and liquid phases results in a lower effective compressibility for the fluid

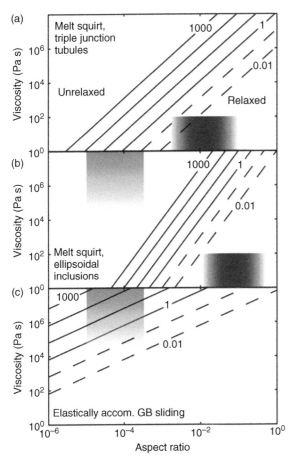

Figure 8 Relaxation times for elastically accommodated grain-boundary sliding and melt squirt as functions of the aspect ratio of the melt inclusion or grain-boundary region and of the melt or grain-boundary viscosity as described in the text. Values of the relevant elastic moduli, viscosities, and aspect ratios are appropriate for olivine-basalt system at 1200–1300°C. After Schmeling (1985), from Faul *et al.* (2004) with permission.

phase on sufficiently long timescales, and a large anelastic reduction in the bulk modulus as discussed in the previous section.

17.3 Insights from Laboratory Studies of Geological Materials

17.3.1 Dislocation Relaxation

17.3.1.1 Linearity and recoverability

Linearity of dislocation-controlled mechanical behavior requires that the dislocation density be independent of the applied stress and therefore that dislocation multiplication be avoided. For synthetic polycrystals of representative grain size $(1–100) \times 10^{-6}$ m and Burgers vector of 5×10^{-10} m, the critical elastic strain amplitudes for the onset of Frank–Read multiplication of the longest possible dislocation segments (with $L = d$), calculated from eqn [13], are 5×10^{-4} to 5×10^{-6}, respectively. These correspond (for $G = 50$ GPa) to stresses of 25–0.25 MPa, respectively. The linear regime is thus experimentally accessible over a wider range of stress and strain amplitudes in more fine-grained polycrystals (e.g., Gribb and Cooper, 1998a). The same argument applied to the larger grain sizes (1–10 mm) of the Earth's upper mantle yields critical elastic strain amplitudes of 5×10^{-7} to 5×10^{-8} for the onset of dislocation multiplication and hence nonlinear behavior. Of course, the presence of impurities and other dislocations means that dislocation segments will generally be substantially smaller than the grain size – expanding the field for linear behaviour to larger strains.

Recoverability of the nonelastic strains associated with dislocation migration requires that dislocations remain firmly pinned by impurities or interactions with other dislocations. For a scenario $(E_{up} \sim 200–400$ kJ mol^{-1}, $a \sim b \sim 5 \times 10^{-10}$ m and $L \sim (3–30) \times 10^{-6}$ m) representative of silicate materials under laboratory or upper-mantle conditions, the critical stress for unpinning σ_{up} is predicted (eqn [51]) to be of order 0.1–1 MPa corresponding to shear strain amplitudes of order $10^{-6}–10^{-5}$. It follows that thermally assisted unpinning is likely to be encountered in laboratory internal friction experiments typically performed at strain amplitudes of $10^{-6}–10^{-4}$ (e.g., Nowick and Berry, 1972; Karato and Spetzler, 1990). Under these circumstances, one would expect more strongly frequency-dependent attenuation ($Q^{-1} \sim \omega^{-1}$ in the viscous regime; **Figure 2**) and possibly nonlinear behavior resulting

from dislocation multiplication in the longer dislocation segments. Unpinning is less likely at the lower stresses and strains of teleseismic wave propagation ($<10^{-6}$, e. g., Shearer, 1999, p.19).

17.3.1.2 Laboratory measurements on single crystals and coarse-grained rocks

Notwithstanding the potential importance of dislocation relaxation mechanisms in seismic wave attenuation, relevant laboratory experiments on geological materials have been few and far between. Gueguen et al. (1989) tested both untreated and predeformed single crystals of forsterite in low-strain ($10^{-6}–10^{-5}$) torsional forced oscillation at periods of $10^{-3.5}–10$ Hz and temperatures as high as 1400°C. The prior deformation was a compressive test at 1600°C to 1% strain with the maximum principal stress (20 MPa) oriented parallel to [111]$_c$, which direction later served also as the orientation of the torsional axis. The effect of the prior deformation was to substantially increase the average dislocation density from $\sim 10^9$ to $\sim 10^{11}$ m^{-2}. For each specimen the observed dissipation was dominated by an intense and broad absorption band (eqn [49]) with an activation energy $E = 440 \pm 50$ kJ mol^{-1} and exponent $\alpha = 0.20 \pm 0.03$. A minor absorption peak superimposed upon the background was observed near 0.1 Hz. The dissipation increased markedly as the result of prior deformation, about fourfold at 1400° and 1000 s period. Gueguen et al. argued plausibly that this observation, along with the similar activation energy for high-temperature dislocation creep in the same material, implicates relaxation associated with the stress-induced migration of the dominant $b = [100]$ edge dislocations. The similarly high activation energies for dislocation damping and creep have subsequently been interpreted by Karato (1998) to suggest a common reliance on the nucleation and migration of kink pairs (see below).

Torsional forced-oscillation measurements, performed at low strain amplitudes ($\sim 10^{-6}$) and high temperatures (to 1500 K) on single-crystal MgO with an average dislocation density of about 5×10^9 m^{-2}, are similarly well described by eqn [49] with $E = 230$ kJ mol^{-1} and $\alpha = 0.30$ (Getting et al., 1997). The much lower activation energy for dislocation damping than for dislocation creep (400 kJ mol^{-1}; Hensler and Cullen, 1968) led Karato (1998) to suggest that geometrical kink migration might account for the dislocation damping with nucleation and migration of kink pairs responsible for the dislocation creep.

Absorption-band behavior probably attributable at least in part to dislocation damping also dominates the high-temperature internal friction measured on relatively coarse-grained rocks (Berckhemer *et al.*, 1982; Kampfmann and Berckhemer, 1985; Jackson *et al.*, 1992). Activation energies for dissipation are generally comparable to those for creep – consistent with Karato's interpretation of the Gueguen *et al.* data for single-crystal forsterite. None of these studies involved a systematic examination of dislocation microstructure, and the shear moduli measured (at ambient pressure) in Berckhemer's laboratory were significantly compromised by thermal microcracking.

17.3.2 Stress-Induced Migration of Transformational Twin Boundaries in Ferroelastic Perovskites

Harrison and his colleagues have reported the results of a thorough study of the anelastic behavior associated with the stress-induced motion of domain walls in single crystals of the ferroelastic perovskite LaAlO$_3$ (Harrison and Redfern, 2002; Harrison *et al.*, 2004). They combined three-point bending experiments performed under combined steady and oscillating load with *in situ* optical observations of the domain wall motions. Essentially elastic behavior was observed for temperatures $T > T_c$ within the stability field of the high-symmetry phase (T_c is the phase transformation temperature). Strongly anelastic behavior, observed within the stability field of the low-symmetry ferroelastic phase, is associated with the thermally activated motion of the domain walls separating transformational twins formed spontaneously on cooling below T_c. At sufficiently low temperatures, domain-wall mobility is reduced to such an extent that the domain microstructure is frozen and elastic behavior with a high unrelaxed modulus prevails. The transition between the frozen and ferroelastic regimes is associated with pronounced relaxation ($\delta Y/Y \sim 0.9$) of Young modulus Y and an associated dissipation peak of amplitude ($\tan \delta = Q^{-1} \sim 1$). The dissipation peak is somewhat broader than the Debye peak of the standard anelastic solid and was modeled by Harrison and Redfern (2002) with a Gaussian distribution of activation energies suggestive of pinning of domain walls by oxygen vacancies. Subsequent studies on a number of additional ferroelastic perovskites have shown that appreciable domain wall mobility and associated anelastic relaxation may be restricted to phases with positive volume strain relative to the cubic parent

and thick domain walls (Daraktchiev *et al.*, 2006). A broad dissipation peak newly recognized in previously published torsional forced-oscillation data for polycrystalline CaTiO$_3$ perovskite, is potentially attributable to the motion of twin domain boundaries. However, the CaTiO$_3$ peak is more probably associated with the rounding of grain edges at triple-junction tubules containing silicate impurity as argued in the following section.

17.3.3 Grain-Boundary Relaxation Processes

17.3.3.1 Grain-boundary migration in b.c.c. and f.c.c. Fe?

Like many other metals, the low-temperature body-centered cubic (b.c.c.) polymorph of iron was studied with resonant (pendulum) techniques in the early days of internal friction studies. More recently, Jackson *et al.* (2000) employed subresonant forced-oscillation methods to probe the viscoelastic behavior of mildly impure polycrystalline iron over an extended temperature range reaching well into the stability field of the high-temperature f.c.c. phase.

Markedly viscoelastic behavior, dominated by the monotonically frequency and temperature-dependent 'high-temperature background', was observed within each of the b.c.c. and f.c.c. stability fields for both mild steel and soft iron compositions. The behavior of the b.c.c. phase is adequately described by eqn [49] with an activation energy of $E = 280(30)$ kJ mol^{-1} not very different from that for lattice diffusion (239 kJ mol^{-1}; Frost and Ashby, 1982) and $\alpha = 0.20(2)$. For the f.c.c. phase, α determined during staged cooling from 1300°C to 800°C typically assumes values of 0.2–0.3 at the highest and lowest temperatures within this range but reaches significantly lower values ~ 0.1 at intermediate temperatures. The result is a poorly defined Q^{-1} plateau at short periods and 900–1100°C, suggestive of a broad Q^{-1} peak superimposed upon the background (Jackson *et al.*, 2000, figure 8b).

Each of these phases displays unusually strong elastic anisotropy at elevated temperatures and ambient pressure: G_{RS}/G_{HS} decreasing from 0.91 to 0.61 between 20°C and 900°C for b.c.c.-Fe (Dever, 1972), whereas the only available single-crystal elasticity data for f.c.c.-Fe yield $G_{RS}/G_{HS} = 0.74(1)$ at 1155°C. Relaxation of the initially inhomogeneous stress field toward the Reuss state of uniform stress by grain-boundary migration, first suggested by Leak (1961), is accordingly a potentially important relaxation mechanism in polycrystalline iron prompting more detailed investigation now in progress.

17.3.3.2 Grain-boundary sliding

The high-temperature behavior of fine-grained synthetic rocks (and analogous ceramic materials) has been the focus of a substantial body of recent work. Fine-grained ultramafic materials with and without a small melt fraction have been made by hot-isostatic pressing of powders prepared by crushing natural dunites or olivine crystal separates, or by a solution–gelation procedure (Tan *et al.*, 1997, 2001; Gribb and Cooper, 1998a, 2000; Jackson *et al.*, 2002, 2004; Xu *et al.*, 2004). These studies have consistently revealed broad absorption-band behavior; superimposed dissipation peaks were reported only in the studies of Jackson *et al.* (2004) and Xu *et al.* (2004) of melt-bearing materials.

Of these materials, only the olivine polycrystals, prepared from hand-picked natural olivine crystals or sol–gel precursors, are pure enough to remain genuinely melt free during testing by high-temperature mechanical spectroscopy (Jackson *et al.*, 2002; see also Jackson *et al.* (2004)). The reconstituted dunite tested by Gribb and Cooper (1998a) is only nominally melt free – the usual inventory of impurities being expected to produce 0.1–1% melt at temperatures of 1200–1300°C. Accordingly, it is the study by Jackson *et al.* (2002, 2004) that defines the base-line behavior of genuinely melt-free polycrystalline olivine. The results depart markedly from the prescriptions of the widely used Raj–Ashby model of elastically and diffusionally accommodated grain-boundary sliding. First, despite intensive sampling of the transition from elastic behavior through essentially anelastic into substantially viscous deformation, dissipation peaks attributable to elastically accommodated grain-boundary sliding are conspicuous by their absence. Instead, the dissipation increases monotonically with increasing oscillation period and temperature (**Figure 9(a)**) in the manner of the high-temperature background. Second, the variation with oscillation period and grain size, given by $Q^{-1} \sim (T_o/d)^{1/4}$, is much milder than predicted for the diffusional creep transient (eqn [62]). For other ceramic systems lacking a widely distributed grain-boundary phase of low viscosity, 'grain-boundary' Q^{-1} peaks are similarly absent or very broad and of subsidiary significance relative to the background dissipation (Pezzotti *et al.*, 1998; Lakki *et al.*, 1999; Webb *et al.*, 1999). Third, complementary microcreep tests on the same olivine materials clearly demonstrate a continuous transition from elastic through anelastic to viscous behavior, implying that the timescales τ_e and τ_d given by eqns

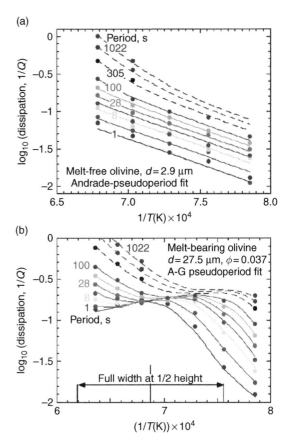

Figure 9 Representative data displaying contrasting variations of dissipation with oscillation period and temperature (a) monotonic absorption-band behavior for a genuinely melt-free polycrystalline olivine, the curves representing an Andrade-pseudoperiod fit to the Q^{-1} data. (b) A broad dissipation peak superimposed upon the monotonic background for a melt-bearing olivine polycrystal, the curves representing an Andrade–Gaussian pseudoperiod fit to Q^{-1} data. (a and b) Adapted from Jackson I, Faul UH, Fitz Gerald J, and Morris SJS (2006) Contrasting viscoelastic behaviour of melt-free and melt-bearing olivine: Implications for the nature of grain-boundary sliding. *Materials Science and Engineering A* 442: 170–174.

[60] and [61] above are not as widely separated as suggested by the Raj–Ashby theory (Jackson *et al.*, 2002).

In the presence of small basaltic melt fractions ranging from 10^{-4} to 4×10^{-2}, qualitatively different behavior has been observed in the form of a broad dissipation peak superimposed upon a dissipation background enhanced relative to that for melt-free material of the same grain size (Jackson *et al.*, 2004; **Figure 9(b)**). The peak width exceeds that of the Debye peak of the standard anelastic solid by about two decades in period and is independent of temperature but varies mildly with the breadth of the grain-size distribution. The peak height is well described by power-law dependence upon maximum melt fraction. The peak position (period) varies exponentially with $1/T$ (with a high activation energy E_P of $720\,kJ\,mol^{-1}$ – possibly reflecting both thermal and compositional influences upon viscosity) and linearly with grain size. This grain-size sensitivity is that expected of elastically accommodated grain-boundary sliding (eqn [60]). These observations have been compared and contrasted with the somewhat different findings of Gribb and Cooper (2000) and Xu *et al.* (2004) by Faul *et al.* (2004).

It has long been understood that the viscoelastic relaxation associated with partial melting should depend upon the extent to which the melt wets the grain boundaries (e.g., Stocker and Gordon, 1975). It is well known that basaltic melt in textural equilibrium does not normally wet olivine grain faces – being confined instead to a network of interconnected grain-edge tubules of cuspate triangular cross-section along with a population of larger localized melt pockets (Faul *et al.*, 1994). Among the melt-bearing olivine polycrystals described by Faul *et al.* (2004), for example, the melt fraction ranges widely from $\sim 10^{-4}$ to 4×10^{-2} with aspect ratios of 3×10^{-3} to 10^{-1} for grain-edge tubules and 10^{-2} to 5×10^{-1} for the isolated melt pockets. For such aspect ratios and the bulk melt viscosities of 1–$100\,Pa\,s$ appropriate for 1200–$1300°C$, melt squirt, whether between adjacent tubules or between adjacent melt pockets is expected at timescales substantially shorter than seismic periods (**Figure 8(a)** and **8(b)**) in accord with the results of previous analyses (Schmeling, 1985; Hammond and Humphreys, 2000a). Moreover, the relaxation strength for squirt flow between melt tubules is low (Mavko, 1980): ~ 0.02 corresponding to a Q^{-1} peak height of only 0.01 for a melt fraction ϕ of 0.04. For these and other reasons given by Faul *et al.* (2004), melt squirt was rejected as a viable explanation for the melt-related dissipation peak. Instead, it was recognized that elastically accommodated grain-boundary sliding involving boundary regions of aspect ratio $\sim 10^{-4}$ has the potential to contribute to attenuation at seismic frequencies – provided that the effective grain-boundary viscosity is, not unreasonably, 10^4–$10^9\,Pa\,s$ for temperatures of 1300–$1000°C$ (**Figure 8(c)**).

It remained to explain the absence of a Q^{-1} peak associated with elastically accommodated grain-boundary sliding in the genuinely melt-free materials. The key microstructural difference that appears to correlate with the presence or absence of a dissipation peak in these materials is the rounding of olivine grain edges at the triple-junction tubules of

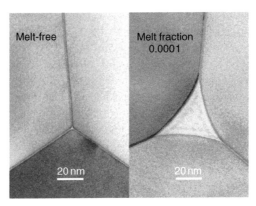

Figure 10 Contrasting grain-edge microstructures in melt-free and melt-bearing olivine polycrystals. For the melt-free material, olivine grain edges are tightly interlocking, whereas the presence of grain-edge melt tubules results in a significant radius on each olivine grain edge. These microstructural differences are considered responsible for the qualitatively different dissipation behaviors highlighted in **Figure 9**. Adapted from Faul UH, Fitz Gerald JD, and Jackson I (2004) Shear-wave attenuation and dispersion in melt-bearing olivine polycrystals. Part II: Microstructural interpretation and seismological implications. *Journal of Geophysical Research* 109: B06202 (doi:10.1029/2003JB002407).

the melt-bearing specimens (**Figure 10**). In contrast, the tight grain-edge interlocking characteristic of melt-free material (**Figure 10**) apparently inhibits elastically accommodated sliding – a conclusion supported by the foregoing discussion (following eqn [59]) of the Raj and Ashby model. A corollary of this interpretation is that the background dissipation in both classes of material would be attributed to diffusionally accommodated grain-boundary sliding. Thus for the melt-bearing materials, it is suggested that grain-boundary sliding occurs with concurrent elastic and diffusional accommodation.

Broadly similar observations have recently been made for polycrystalline MgO. Dense polycrystals of high purity display only a monotonically frequency and temperature-dependent Q^{-1} background and associated dispersion of the shear modulus (Barnhoorn *et al.*, 2006). In marked contrast, a previous study of MgO of lower purity (Webb and Jackson, 2003) revealed a pronounced dissipation peak, superimposed upon the background, and now attributed to elastically accommodated grain-boundary sliding facilitated by the presence of a grain-boundary phase of low viscosity.

Further examples of both types of high-temperature viscoelastic behavior are provided by the fine-grained polycrystalline titanate perovskites studied as analogues for the high-pressure silicate perovskites by

Webb *et al.* (1999). The purest and microstructurally simplest of these specimens was an $SrTiO_3$ specimen of 5 μm grain size. Its grains are untwinned and its essentially impurity-free grain boundaries meet in grain edge triple-junctions invariably less than 20 nm in cross-sectional dimension. The lack of an extensive network of substantial grain-edge triple-junction tubules is reflected in background-only dissipation generally well described for temperatures between 900°C and 1300°C by eqn [49]. On the other hand, the $CaTiO_3$ specimens of 3 and 20 μm grain size are pervasively twinned and contain a significant level of silicate impurity responsible for well-developed networks of grain-edge tubules of triangular-cuspate cross-section. With the benefit of hindsight, deviations from power-law fits to the $Q^{-1}(T_o)$ data at 1000°C and 1050°C (Webb *et al.*, 1999, table 4 and figure 13) for the 20 μm specimen are consistent with the presence of a broad dissipation peak analogous to those characteristic of melt-bearing olivine. For the more fine-grained $CaTiO_3$ specimen, any such peak may be masked by the relatively low-temperature onset of markedly viscous behavior.

The growing body of experimental observations thus indicates that the transition from elastic through anelastic to viscous behavior in fine-grained geological (and ceramic) materials is not satisfactorily described by the classic Raj–Ashby theory. In particular,

1. sufficiently pure materials display no high-temperature dissipation peak attributable to elastically accommodated grain-boundary sliding;
2. low grain-boundary viscosity is a necessary but not sufficient condition for elastically accommodated sliding: rounding of grain edges at triple junctions is also required; and
3. diffusional accommodation of grain-boundary sliding, presumably responsible for the ubiquitous high-temperature background with its mildly and monotonically frequency-dependent dissipation, can occur without or alongside elastic accommodation (Barnhoorn *et al.*, 2006).

These observations have provided the motivation for a new approach to the micromechanical modeling of grain-boundary sliding free from the principal limitations of Raj–Ashby theory as discussed in Section 17.2.

17.3.4 Viscoelastic Relaxation in Cracked and Water-Saturated Crystalline Rocks

The sparse experimental observations concerning the effect of water saturation on the elastic moduli of

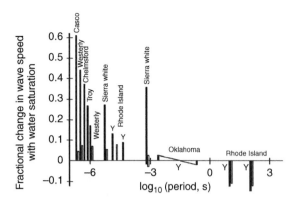

Figure 11 The effect of water saturation of granite (with zero pressure and confining pressure) on the speed of compressional (unlabeled solid bars), bar-mode longitudinal (solid bars labeled 'Y'), and shear waves (unfilled bars). The sources of the data for the various granite specimens (named after their provenance) are given by Lu and Jackson (2006, figure 11) – after which this figure has been redrawn.

cracked crystalline rocks of relatively low-porosity, recently assembled by Lu and Jackson (2006), are reproduced in **Figure 11**. The existence of a mechanical relaxation spectrum first suggested by Gordon (1974) is confirmed by this more recent compilation. Ultrasonic, resonance and forced-oscillation methods together provide access to more than eight decades of period or frequency. The fractional change in elastic wave speed $\delta V_i/V_i$ that results from water saturation (with $P_f = 0$) is plotted for the compressional (i = P), shear (i = S), and extensional (i = Y) bar modes. For μs–ms periods, the wave speeds are consistently increased by water saturation, the effect being more pronounced for compressional and extensional modes than in shear. The sign is reversed for periods greater than ~0.1 s with significant negative perturbations to both extensional and shear wave speeds for the 10–100 s periods of forced-oscillation methods.

Interpretation of the broad trend evident in **Figure 11** requires estimates of the characteristic times for the various fluid-related relaxation mechanisms discussed in Section 17.2, for the special case of water saturation of cracks of aspect ratio ~10^{-3}. For relaxation of shear stress within an individual grain-scale fluid inclusion (eqn [71]), $\tau_e \sim \eta_f/G_u\alpha \sim 10^{-11}$ s; for squirt between adjacent cracks (eqn [73]), $\tau_{s,e} = 2\pi\eta_f/K_s\alpha_e^3 \sim 10^{-4}$ s; and for the draining of a cm-sized laboratory rock specimen of porosity $\phi \sim 0.01$ and permeability ~10^{-18} m^2, $\tau_{dr} = \eta_f\phi R^2/kK_f \sim 10^{-1}$ s (Lu and Jackson, 2006). Accordingly, Lu and Jackson (2006) concluded that the ultrasonic measurements at periods $<10^{-5}$ s probe the saturated isolated regime, whereas near-zero values of $\delta G/G$, and hence $\delta V_S/V_S$, at the ms periods of resonance techniques are characteristic of the saturated isobaric regime. Near-zero values of $\delta K/K$ for periods $>10^{-3}$ s are indicative of the drained regime. Chemical effects of water saturation, such as increased crack density and reduction in the stiffness of grain contacts, are required to explain the negative values of $\delta V/V$ in experimental data obtained at periods greater than 0.01 s (Lu and Jackson, 2006).

17.4 Geophysical Implications

17.4.1 Dislocation Relaxation

17.4.1.1 Vibrating string model

The possibility that viscoelastic relaxation associated with the stress-induced motion of dislocations might explain the attenuation and dispersion of seismic waves, especially in the Earth's upper mantle, has been widely canvassed (Gueguen and Mercier, 1973; Minster and Anderson, 1981; Karato and Spetzler, 1990; Karato, 1998). Gueguen and Mercier (1973) invoked the vibrating string model of dislocation motion as an explanation for the upper-mantle zone of low wave speeds and high attenuation. They associated an intragranular dislocation network with any observable high-temperature relaxation peak and suggested that a grain-boundary network of more closely spaced dislocations might explain the high-temperature background attenuation. It was suggested that impurity control of dislocation mobility might yield relaxation times appropriate for the seismic-frequency band.

A more ambitious attempt to produce a model of dislocation-controlled rheology consistent with both seismic-wave attenuation and the long-term deformation of the mantle was made by Minster and Anderson (1981). They envisaged a microstructure established by dislocation creep in response to the prevailing tectonic stress σ_t and comprising subgrains of dimension $\underline{L}$

$$\underline{L} = \gamma_s Gb/\sigma_t = \gamma_s L_c \qquad [78]$$

with a mobile dislocation density

$$\Lambda_m = \gamma(\sigma_t/Gb)^2 = \gamma L_c^{-2} \qquad [79]$$

where $L_c = Gb/\sigma_t$ is the critical dislocation link length for multiplication by the Frank–Read mechanism (see discussion following eqn [12]). From laboratory experiments on olivine (Durham

et al., 1977), it was estimated that $\gamma_s \sim 20$ for well-annealed mantle olivine; γ is a constant of order unity (Frost and Ashby, 1982). Furthermore, it was argued that the distribution of dislocation segment lengths will be dominated by those of length

$$L \sim (5/3)L_c \qquad [80]$$

for which the dislocation multiplication time, calculated with a migration velocity given by eqn [14], is a minimum. It follows from eqns [78]–[80] that the dislocation density within the subgrains is essentially that (L^{-2}) of the Frank network with dislocation relaxation strength $\Delta \sim 0.1$ (eqn [23]). For $\sigma_t \sim 1$ MPa (strain rate of 10^{-15} s^{-1} with effective viscosity of 10^{21} Pa s), $L_c \sim 30 \times 10^{-6}$ m and $\Lambda_m \sim 10^9$ m^{-2}. It was suggested that the total dislocation density should be much higher – being dominated by dislocations organised into the subgrain walls. The climb-controlled motion of these latter dislocations was invoked to explain deformation at tectonic stresses and timescales (Minster and Anderson, 1981).

Their suggestion of a dominant link length disallows an absorption band based on a wide distribution of segment lengths (eqn [50]). Minster and Anderson (1981) suggested instead that the seismic absorption band reflects a spectrum of activation energies. A frequency-independent upper-mantle Q^{-1} of 0.025 was shown to require a range δE of about 80 kJ mol^{-1} and to produce an absorption band only two to three decades wide.

A unified dislocation-based model for both seismic wave attenuation and the tectonic deformation of the Earth's mantle has been sought more recently by Karato and Spetzler (1990) and Karato (1998). Karato and Spetzler (1990) emphasized the apparent paradox whereby a Maxwell model based on reasonable steady-state viscosities seriously underestimates the level of seismic-wave attenuation (their figure 3). Equally, a transient-creep description of seismic-wave attenuation extrapolated to tectonic timescales (Jeffreys, 1976) seriously overestimates the mantle viscosity. For a unified dislocation-based model, it is thus required that dislocations have much greater mobility (velocity/stress) at seismic frequencies than for tectonic timescales. Karato and Spetzler's assessment of relaxation strengths and relaxation times, based on the vibrating-string model of dislocation motion (eqn [23]) and revisited with a somewhat higher dislocation density of 10^9 m^{-2} (eqn [79] applied to the upper mantle as above) is in general accord with that of Minster and Anderson. In particular, reasonable levels of seismic wave attenuation ($10^{-1} > \Delta > 10^{-3}$) would require segment lengths L of $(1–10) \times 10^{-6}$ m, and a specific range of dislocation mobilities (b/B in eqn [14]) would be needed for relaxation times in the seismic band (Karato and Spetzler, 1990, figure 7). The issue in all such analyses is to relate key parameters controlling Δ, τ, and indeed the steady-state effective viscosity η, to the intrinsic and/or extrinsic properties of dislocations, as appropriate. Karato and Spetzler noted that dislocation mobility is expected to increase with increasing temperature and water content but to decrease with increasing pressure. The segment length L for the vibrating string model was shown to depend in different ways on the tectonic stress according to whether pinning is effected by interaction with impurities or other dislocations.

17.4.1.2 Kink model

The feasibility of seismic-wave attenuation resulting from the stress-induced migration of kinks was also briefly explored by Karato and Spetzler (1990) and analyzed in more detail by Karato (1998). In the latter study, Karato identified three distinct stages of dislocation relaxation to be expected with progressively increasing temperature or stress and/or decreasing frequency. Stage I involves the reversible migration of pre-existing geometrical kinks along Peierls valleys with characteristic timescale and relaxation strength given by eqns [27] and [28] above. The maximum anelastic strain resulting from migration of geometrical kinks (eqn [30]) is of order 10^{-6} for a Frank network of dislocations with $\Lambda \sim L^{-2} \sim 10^{9}\,\text{m}^{-2}$ – consistent with the large relaxation strength for this mechanism (eqn [28]).

Stage II is associated with the spontaneous nucleation of isolated kink pairs followed by the separation of the newly formed kinks of opposite sign. The resulting anelastic relaxation, usually associated with the Bordoni dissipation peak for deformed metals (e.g., Seeger, 1981; Fantozzi et al., 1982), is described in detail by Seeger's theory outlined above and culminating in eqns [43]–[45]. At the still higher temperatures/larger stresses of stage III, it was envisaged by Karato that continuous nucleation and migration of kinks, possibly enhanced by unpinning of dislocations, would allow the arbitrarily large, irrecoverable strains of viscous behavior.

The relaxation times expected for geometrical kink migration and kink-pair formation and migration in the Earth's upper mantle were shown by Karato (1998) to be broadly compatible with strain–energy dissipation at teleseismic and lower frequencies. It was thus

concluded that attenuation and associated dispersion, microcreep, and large-strain creep possibly share a common origin in dislocation glide. The results of such calculations, repeated with eqns [27] and [28] and [43]–[45] above, are presented in **Figure 12**. For this purpose we have used $a_\text{d} \sim b \sim a \sim 5 \times 10^{-10}$ m, $G_\text{u} = 60$ GPa, $v_{\text{k}0}\exp(S_\text{m}/k) \sim 10^{13}$ Hz (Seeger and Wüthrich, 1976), and have chosen a relatively large value of $H_\text{m} = 200\,\text{kJ}\,\text{mol}^{-1}$ intended to reflect kink interaction with impurities and $\sigma_\text{P}/G = 0.01$ which, through eqn [39], fixes H_k at $162\,\text{kJ}\,\text{mol}^{-1}$ (so that $2H_\text{k} + H_\text{m}$ is of order $500\,\text{kJ}\,\text{mol}^{-1}$: c.f. Karato, 1998, figure 3). Substantially larger values of σ_P/G have been suggested (Frost and Ashby, 1982, p. 106; Karato and Spetzler, 1990, p. 407), but when combined with Seeger's theory yield implausibly high kink formation energies. Relaxation times have been calculated for indicative lengths of 1 and $100\,\mu\text{m}$ for dislocation segments. The results of these calculations (**Figure 12**), like those of Karato (1998), indicate that the migration of geometrical kinks could result in anelastic relaxation within the seismic frequency band at typical upper-mantle temperatures. The much higher activation energy for the formation and

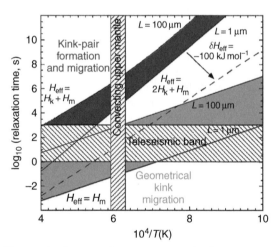

Figure 12 Computed relaxation times for Seeger's (1981) kink model of dislocation mobility. The green band represents relaxation associated with the migration of geometrical kinks with segment lengths of $(1–100) \times 10^{-6}$ m and an activation energy for kink migration $H_\text{m} = 200\,\text{kJ}\,\text{mol}^{-1}$. Relaxation times calculated from eqns [36]–[39] for the formation and migration of kink pairs are indicated by the red band for the same range of segment lengths and an activation energy for the formation of a single kink of $162\,\text{kJ}\,\text{mol}^{-1}$ corresponding through Seeger's (1981) theory with $\sigma_\text{P}/G = 0.01$. The effect of an effective activation energy lower by $100\,\text{kJ}\,\text{mol}^{-1}$ is shown by the broken line.

migration of kink pairs (here $2H_k + H_m = 524\,kJ\,mol^{-1}$) results in much longer relaxation times – ranging from $\sim 10^4$ to $10^{6.5}$ s at the highest plausible upper-mantle temperatures $\sim 1700\,K$. An effective activation energy lower by about $100\,kJ\,mol^{-1}$ would be required to bring the relaxation involving kink-pair formation and migration into substantial overlap with the seismic frequency band for typical temperatures of the convecting upper mantle as indicated by the broken line in **Figure 12**. Alternatively, kink-pair formation and migration might result in relaxation at the day-to-month timescales of Earth tides. Similarly, high activation energies for attenuation and creep in olivine are interpreted by Karato (1998) to imply the need for kink nucleation for both processes, whereas the low reported activation energy for attenuation in single-crystal MgO might reflect a dominant role for the migration of geometrical kinks.

17.4.2 Grain-Boundary Processes

Gribb and Cooper (1998a) explained their observations of high-temperature absorption band behavior (eqn [49] with $\alpha \sim 0.35$) for a reconstituted dunite of $3\,\mu m$ grain size in terms of the transient diffusional creep modeled by Raj (1975). The strong $Q^{-1} \sim d^{-1}$ grain-size sensitivity implied by this model (eqn [62]) extrapolated to mm–cm grain size seriously underestimates upper-mantle seismic wave attenuation. Gribb and Cooper (1998a) sought to resolve this discrepancy by invoking diffusionally accommodated sliding on subgrain boundaries rather than grain boundaries. It was suggested that this mechanism could also reconcile the Gribb and Cooper data with those of Gueguen *et al.* (1989) for pre-deformed single-crystal forsterite. The increased attenuation resulting from prior deformation has been more naturally interpreted by others (including Gueguen *et al.* (1989)) as prima facie evidence of dislocation relaxation as discussed above.

The milder grain-size sensitivity $Q^{-1} \sim d^{-1/4}$ subsequently directly measured by Jackson *et al.* (2002) for melt-free polycrystalline olivine yields extrapolated levels of attenuation for mantle grain sizes more consistent with seismological observations. More recently, Faul and Jackson (2005) have employed a generalized Burgers model (eqns [8]) to provide an internally consistent description of the variations of both shear modulus and dissipation with frequency, temperature, and grain size for melt-free polycrystalline olivine. Extrapolation (principally in grain size) of this model to upper-mantle

conditions provides a satisfactory first-order explanation for much of the variability of seismic wave speeds and attenuation in the upper mantle. The presence of a zone of low shear wave speeds and high attenuation in the oceanic upper mantle and its systematic variation with age of the overlying lithosphere can be explained by lateral temperature variations without recourse to more exotic explanations such as water or partial melting (Anderson and Sammis, 1970; Karato and Jung, 1998). Beneath the continents, lateral variations in seismic structure can be explained by variations in the depth at which the conductive part of the geotherm intersects a common mantle adiabat.

In more active tectonic provinces such as mid-ocean ridges, subduction zones, and back-arc basins, partial melting is expected to influence the wave speeds and attenuation. Superposition of the experimentally observed melt-related dissipation peak upon a melt-enhanced background, extrapolated to upper-mantle conditions, results in generally higher but less markedly frequency-dependent attenuation than for melt-free material under otherwise similar conditions (Faul *et al.*, 2004, figure 13).

17.4.3 The Role of Water in Seismic Wave Dispersion and Attenuation

The crystalline rocks of the Earth's upper crust are commonly pervasively cracked and often water-saturated. The sparse laboratory measurements so far performed on cracked and water-saturated low-porosity rocks are strongly suggestive of intense viscoelastic relaxation (Section 17.3). The inference is that modulus relaxation and strain–energy dissipation, associated with stress-induced fluid flow on various spatial scales, are to be expected for frequencies ranging from ultrasonic (MHz) to teleseismic (mHz–Hz). It follows that conventional (ultrasonic) measurements will tend to overestimate seismic wave speeds (especially V_P) and hence also the ratio V_P/V_S. Accordingly, high-frequency (ultrasonic) laboratory data cannot be applied directly in modeling the wave speeds and attenuation in fluid-saturated crustal rocks. Rather, it would be expected from the analysis of Sections 17.2 and 17.3 that the shear modulus of water-saturated rock at teleseismic frequencies should be comparable with that of the dry rock. Contrasting behavior is expected for the bulk modulus of the cracked rock – which for undrained conditions will be significantly increased by water saturation (eqn [77]).

In the deeper crust and mantle, free water will typically have a more transient existence, but a small amount of water dissolved in a major nominally anhydrous mineral (like olivine or pyroxene) has the potential to significantly enhance solid-state viscoelastic relaxation, for example, by increasing the mobility of dislocations. There are as yet no directly relevant experimental data, but the potential has recently been demonstrated (Shito et al., 2006) for future analyses in which observed variations in V_P, V_S, and Q^{-1} might be combined to map variations in chemical composition, temperature, and the water content of the upper mantle.

17.4.4 Bulk Attenuation in the Earth's Mantle

If pressure-induced changes in the proportions of coexisting crystalline phases were to result in relaxation of the bulk modulus within the seismic frequency band (for which there is little seismological evidence), conditions would be most favorable in the upper mantle and transition zone. It is here that relatively high temperatures facilitating rapid phase transformation are combined with phase transformations involving substantial density contrast. Consider, for example, the olivine–wadsleyite transformation with an *in situ* density contrast of $\sim 6\%$. This is diluted in the pyrolite composition by an almost equal volume of the $(M, Al)(Si, Al)O_3$ ($M = Mg$, Fe) component playing an essentially passive role. However, the presence of the latter component and partitioning between it and the M_2SiO_4 phases reduces the width of the binary loop to $\delta P < 0.3$ GPa (Stixrude, 1997; Irifune and Isshiki, 1998; Weidner and Wang, 2000). Thus, eqn [66] for the relaxation strength yields $\Delta_v \sim (150\,\text{GPa}/0.3\,\text{GPa})(0.03) = 15$, but of course any resulting compressional wave attenuation would be confined to a layer of thickness ~ 10 km comparable with the seismic wavelength. For the wider two-phase loop of the pyroxene–garnet phase transformation, $\Delta_v \sim (150\,\text{GPa}/5\,\text{GPa})(0.03) = 0.9$, so that any seismic wave attenuation, although of lower intensity, would more readily observed. The large relaxation strengths for this process mean that it probably plays an important role somewhere in the Earth's wide mechanical relaxation spectrum. For the olivine–wadsleyite transformation, the relaxation time estimated from eqn [78] with D_{Si} in olivine at 1600 K (Brady, 1995) and $d = 10^{-2}$ m is

$$\tau(s) = 2 \times 10^{17}(p_s/\delta P)^2 = 2 \times 10^{17} \times (K\varepsilon_e/\delta P)^2 \sim 10^{23}\varepsilon_e^2 \qquad [81]$$

or 10^5–10^9 s for a strain ε_e of 10^{-9}–10^{-7} suggesting the possibility of bulk relaxation at tidal and longer periods.

At the shorter periods of teleseismic waves, significant bulk dissipation Q_K^{-1} may arise from the elastic heterogeneity of aggregates of anisotropic mineral grains as explained in Section 17.3. Values of $Q_K^{-1}/Q_G^{-1} \sim 0.01$ were calculated for polycrystalline forsterite and a peridotitic composite, representative of upper-mantle materials by Budiansky and O'Connell (1980). For the mixture of silicate perovskite and magnesiowüstite that dominates the mineralogical makeup of the Earth's lower mantle, the relaxation of the bulk modulus is expected to decrease with increasing pressure along the high-temperature adiabat from $\sim 2\%$ at zero pressure to $< 1\%$ in the deepest mantle (Heinz, et al., 1982; Jackson, 1998). Thermoelastic relaxation (eqn [54]) may enhance the bulk dissipation that is the direct consequence of elastic heterogeneity, to levels sufficient to explain the modest amount of bulk dissipation evident in the damping of the low-order radial modes of free oscillation of the Earth (Budiansky et al., 1983).

17.4.5 Migration of Transformational Twin Boundaries as a Relaxation Mechanism in the Lower Mantle?

It has been suggested by Harrison and Redfern (2002) that the mobile transformational twin boundaries that are clearly responsible for intense dissipation and associated modulus relaxation observed in the laboratory in some pervasively twinned perovskite crystals might also operate in the Earth's silicate perovskite-dominated lower mantle. Pinning of such domain walls by oxygen vacancies as in $LaAlO_3$ perovskite would result in anelastic relaxation times for lower-mantle conditions much shorter than teleseismic periods. Stronger pinning of domain walls by defects, impurities, and grain boundaries would be required for relaxation times within the seismic band (Harrison and Redfern, 2002). Indeed, pinning so effective as to eliminate significant anelastic relaxation has subsequently been observed in orthorhombic $(Ca, Sr)TiO_3$ perovskites (Daraktchiev et al., 2006). More critically, however, substantially different microstructures are to be expected in the silicate perovskites of the lower mantle. The sustained operation of tectonic stress over geological timescales should enlarge the domains of any transformational twins that are favorably oriented with respect to the ambient stress field resulting in a population of mainly single-domain grains and incidentally some degree of lattice preferred orientation. It therefore seems unlikely that transformational twinning will play a major role in seismic wave attenuation in the lower mantle.

17.4.6 Solid-State Viscoelastic Relaxation in the Earth's Inner Core

The combination of low (but finite) rigidity and marked attenuation that characterize the Earth's inner core have long been regarded as highly anomalous and have often been explained in terms of partial melting, as reviewed by Jackson *et al.* (2000; see also Vočadlo (2007)). However, the relatively low value of G/K and hence high Poisson's ratio ($v = 0.44$) are at least in part the expected consequence of the compression of a close-packed crystalline solid (Falzone and Stacey, 1980). First-principles calculations suggest values of v 0.29–0.32 at inner core densities and 0 K (Vočadlo *et al.*, 2003a, table 1) and the possibility of a further substantial increase due to the usual anharmonic influence of temperature (Steinle-Neumann *et al.*, 2003; Vočadlo, 2007). Laboratory measurements reviewed above have revealed pronounced viscoelastic relaxation in both b.c.c. and f.c.c. phases of mildly impure iron at homologous temperatures (T/T_m) reaching 0.6 and 0.9, respectively (Jackson *et al.*, 2000). Given the high homologous temperature ($T/T_m \sim 1$) of the inner core, intense solid-state viscoelastic relaxation is inevitable. The experimental evidence for solid-state viscoelastic relaxation in iron would have direct relevance to the Earth's inner core if the b.c.c. phase were to be stabilized by high temperature and/or the accommodation of light alloying elements (Vočadlo *et al.*, 2003b; Vočadlo, 2007). The inevitable contribution of solid-state relaxation to the observed seismic wave attenuation ($Q^{-1}_G \sim 0.01$–0.02) and to any required viscoelastic relaxation of the shear modulus ($\sim 1.5\%$ per decade of frequency for $Q^{-1} = 0.01$) must therefore be considered before partial melting is invoked (e.g., Vočadlo, 2007).

17.5 Summary and Outlook

This survey of viscoelastic relaxation mechanisms potentially relevant to the Earth's deep interior highlights the disparity in maturity between theory and experimental observations. Long-established theoretical models provide plausible, if highly idealized, descriptions of viscoelastic relaxation associated with dislocation motion and grain-boundary sliding. The stress-induced redistribution of an intergranular fluid is similarly well-described theoretically. That this substantial framework has not yet been more thoroughly tested against experimental observations is attributable in part to limitations in the experimental methods that have been overcome only relatively recently. The pioneering experimental work of the 1950s and 1960s focused on low-temperature internal friction peaks in metals investigated with resonant (i.e., pendulum) techniques. This isochronal approach has the major disadvantage that it provides no opportunity to observe the relaxation spectrum by continuous variation of frequency under isothermal conditions. More recently, there has been growing recognition of the superiority of mechanical spectroscopy methods in which subresonant forced oscillations probe the relaxation spectrum under isothermal high-temperature conditions affording the benefit of a stable microstructure (Woirgard *et al.*, 1981; Berckhemer *et al.*, 1982; Gadaud *et al.*, 1990; Jackson and Paterson, 1993; Getting *et al.*, 1997; Gribb and Cooper, 1998b; Schaller *et al.*, 2001). The latter technique is particularly well suited to investigation of the high-temperature background – the monotonic variation of Q^{-1} with frequency and temperature that typically dominates the dissipation at high temperature and low frequency. In addition, complementary microcreep tests are now being more widely used to distinguish between recoverable (anelastic) and permanent (viscous) strains.

Sustained application of these improved methods to well-characterized specimens of geological and ceramic materials is beginning to elucidate the microscopic processes responsible for their high-temperature viscoelastic relaxation. The consistent picture emerging from studies of fine-grained polycrystalline materials lacking a dispersed secondary phase of low viscosity is that of a broad absorption band (eqn [49]) involving mild frequency dependence of Q^{-1} and an Arrhenian variation with temperature with an activation energy often comparable with that for creep. The lack of a dissipation peak attributable to elastically accommodated grain-boundary sliding is difficult to reconcile with the classic Raj–Ashby model of the breakdown of elastic behavior – suggesting instead that elastically accommodated sliding is precluded by sufficiently tight

(~nm) grain-edge interlocking. In marked contrast, the presence of a widely dispersed grain-boundary phase of relatively low viscosity occupying grain-edge tubules seems to be closely correlated with the observation of a broad dissipation peak superimposed upon the high-temperature background. Further experimental work with carefully prepared and characterized materials is needed to assess the generality of this distinction. This disconnect between the existing theory of grain-boundary sliding and the experimental observations of grain-size sensitive viscoelastic relaxation is providing the motivation to revisit the theory in order to clarify the role of sharp/rounded grain edges and the possibility of sliding with concurrent elastic and diffusional accommodation.

Although so far only rarely determined, the grain-size sensitivity of viscoelastic relaxation is potentially diagnostic of the relative contributions of grain-boundary and intragranular mechanisms. For example, the mild grain-size sensitivity of the dissipation and dispersion measured on fine-grained polycrystalline olivine suggests that such grain-boundary processes contribute significantly – but admits the possibility also of significant intragranular contributions.

Dislocation damping is a likely intragranular contributor to seismic wave attenuation especially in the shallower parts of the upper mantle where seismic anisotropy attests to the development of olivine lattice preferred orientation by dislocation creep. The potential of stress-induced migration of (pre-existing) geometrical kinks (Karato, 1998; **Figure 12**) to cause significant relaxation in the seismic frequency band has been demonstrated. However, unequivocal experimental evidence of dislocation damping in upper-mantle materials is limited to the single study of Gueguen *et al.* (1989) that demonstrated a correlation between dissipation and dislocation density. Extensive systematic studies of untreated and variously pre-deformed specimens will be needed to clarify the nature of dislocation relaxation including the role of pinning and the relative contributions of subgrain and subgrain-boundary dislocations. The possibility that both grain-boundary and dislocation relaxation are enhanced by the presence of small concentrations of water-related defects remains to be explored experimentally.

Both dislocation-related and grain-boundary relaxation processes thus have the potential to make substantial contributions to seismic wave dispersion and attenuation. Dislocation relaxation should

become progressively more important with increasing grain size for two reasons: (1) prior/ongoing tectonic deformation by dislocation rather than diffusional creep is required to generate an appropriate density of mobile dislocations, and (2) low grain-boundary area per unit volume will diminish the relative importance of grain-boundary relaxation. Thermal activation of the mobility of the defects responsible for the dominant high-temperature background means that solid-state viscoelastic relaxation will become progressively more intense as the relevant solidus is approached. Solid-state relaxation mechanisms may thus account for most of the seismologically observed variation of wave speeds and attenuation.

However, relaxation mechanisms associated with partial melting and phase transformations must also contribute at least locally to viscoelastic relaxation in the mantle. Melt squirt is thought to occur at periods shorter than those of teleseismic wave propagation and accordingly might be partly responsible for reduced wave speeds (but not for the observed attenuation) beneath mid-ocean ridges and in subduction-zone and back-arc environments. Facilitation of elastically accommodated grain-boundary sliding by grain-edge melt tubules, manifest as a broad peak superimposed upon the background dissipation, can produce nearly frequency-independent attenuation across wide ranges of conditions (Faul *et al.*, 2004). In parts of the transition zone, stress-induced variations of the proportions of coexisting phases may result in relaxation of the bulk modulus and associated dissipation of strain energy at tidal and longer periods.

References

Abramowitz M and Stegun IA (1972) *Handbook of Mathematical Functions with Formulas, Graphs, and Mathematical Tables.* New York: Dover Publications.

Aikawa N, Kumazawa M, and Tokonami M (1985) Temperature dependence of intersite distribution of Mg and Fe in olivine and the associated change of lattice parameters. *Physics and Chemistry of Minerals* 12: 1–8.

Anderson OL and Isaak DG (1995) Elastic constants of mantle minerals at high temperature. In: Ahrens TJ (ed.) *AGU Reference Shelf, vol. 2: Mineral Physics and Crystallography: A Handbook of Physical Constants*, pp. 64–97. Washington, DC: American Geophysical Union.

Anderson DL and Sammis CG (1970) Partial melting in the upper mantle. *Physics of the Earth and Planetary Interiors* 3: 41–50.

Ashby MF (1972) Boundary defects and atomistic aspects of boundary sliding and diffusional creep. *Surface Science* 31: 498–542.

Barnhoorn A, Jackson I, Fitz Gerald JD, and Aizawa Y (2006) Suppression of elastically accommodated grain-boundary sliding in high-purity MgO. *Journal of the European Ceramic Society* (in press).

Berckhemer H, Kampfmann W, Aulbach E, and Schmeling H (1982) Shear modulus and Q of forsterite and dunite near partial melting from forced oscillation experiments. *Physics of the Earth and Planetary Interiors* 29: 30–41.

Brady JB (1995) Diffusion data for silicate minerals, glasses, and liquids. In: Ahrens TJ (ed.) *AGU Reference Shelf, Vol. 2: Mineral Physics & Crystallography: A Handbook of Physical Constants*, pp. 269–290. Washington, DC: American Geophysical Union.

Brailsford AD (1961) Abrupt-kink model of dislocation motion. *Physical Review* 122: 778–786.

Budiansky B and O'Connell RJ (1980) Bulk dissipation in heterogeneous media. In: Nemat-Nasser S (ed.) *AMD, Vol. 42: Solid Earth Geophysics and Geotechnology*, pp. 1–10. New York: ASME.

Budiansky B, Sumner EE, and O'Connell RJ (1983) Bulk thermoelastic attenuation of composite materials. *Journal of Geophysical Research* 88: 10343–10348.

Cooper RF (2003) Seismic wave attenuation: Energy dissipation in viscoelastic crystalline solids. In: Karato S and Wenk H (eds.) *Reviews in Minerology and Geochemistry: Plastic Deformation in Minerals and Rocks*, pp. 253–290. Washington DC: Mineralogical Society of America.

Daraktchiev M, Harrison RJ, Mountstevens EH, and Redfern SAT (2006) Effect of transformation twins on the anelastic behavior of polycrystalline $Ca_{1-x}Sr_xTiO_3$ and $Sr_xBa_{1-x}SnO_3$ perovskite in relation to the seismic properties of Earth's mantle perovskite. *Materials Sciences and Engineering A* 442: 199–203.

Darinskiy BM and Levin YN (1968) Theory of internal friction due to phase transitions. *Physics of Metals and Metallography – USSR* 26(6): 98–105.

Dever DJ (1972) Temperature dependence of the elastic constants in alpha-iron single crystals: Relationship to spin order and diffusion anomalies. *Journal of Applied Physics* 43: 3293–3300.

Durham WB, Goetze C, and Blake B (1977) Plastic flow of oriented single crystals of olivine. Part II: Observations and interpretations of the dislocation structures. *Journal of Geophysical Research* 82: 5755–5770.

Falzone AJ and Stacey FD (1980) Second-order elasticity theory: Explanation for the high Poisson's ratio of the inner core. *Physics of the Earth and Planetary Interiors* 22: 371–377.

Fantozzi G, Esnouf C, Benoit W, and Ritchie IG (1982) Internal friction and microdeformation due to the intrinsic properties of dislocations: The Bordoni relaxation. *Progress in Materials Science* 27: 311–451.

Faul UH, Fitz Gerald JD, and Jackson I (2004) Shear-wave attenuation and dispersion in melt-bearing olivine polycrystals. Part II: Microstructural interpretation and seismological implications. *Journal of Geophysical Research* 109: B06202 (doi:10.1029/2003JB002407).

Faul UH and Jackson I (2005) The seismological signature of temperature and grain size variations in the upper mantle. *Earth and Planetary Science Letters* 234: 119–134.

Faul UH, Toomey DR, and Waff HS (1994) Intergranular basaltic melt is distributed in thin, elongated inclusions. *Geophysical Research Letters* 20(1): 29–32.

Findley WN, Lai JS, and Onaran K (1976) *Creep and Relaxation of Non-Linear Viscoelastic Materials*. Amsterdam: North-Holland.

Friedel G (1964) *Dislocations*. Oxford: Pergamon.

Friedel J, Boulanger C, and Crussard C (1955) Constantes elastiques et frottement interieur de l'aluminium polygonise. *Acta Mettalurgica* 3: 380–391.

Frost HJ and Ashby MF (1982) *Deformation-Mechanism Maps. The Plasticity and Creep of Metals and Ceramics*. Oxford: Pergamon Press.

Gadaud P, Guisolan B, Kulik A, and Schaller R (1990) Apparatus for high-temperature internal friction differential measurements. *Review of Scientific Instruments* 61: 2671–2675.

Getting IC, Dutton SJ, Burnley PC, Karato S, and Spetzler HA (1997) Shear attenuation and dispersion in MgO. *Physics of the Earth and Planetary Interiors* 99: 249–257.

Ghahremani F (1980) Effect of grain boundary sliding on anelasticity of polycrystals. *International Journal of Solids and Structures* 16: 825–845.

Gordon RB (1974) Mechanical relaxation spectrum of crystalline rock containing water. *Journal of Geophysical Research* 79: 2129–2131.

Granato A and Lücke K (1956) Theory of mechanical damping due to dislocations. *Journal of Applied Physics* 27: 583–593.

Gribb TT and Cooper RF (1998a) Low-frequency shear attenuation in polycrystalline olivine: Grain boundary diffusion and the physical significance of the Andrade model for viscoelastic rheology. *Journal of Geophysical Research* 103(B11): 27267–27279.

Gribb TT and Cooper RF (1998b) A high temperature torsion apparatus for the high-resolution characterization of internal friction and creep in refractory metals and ceramics: Application to the seismic-frequency, dynamic response of Earth's upper mantle. *Review of Scientific Instruments* 69: 559–564.

Gribb TT and Cooper RF (2000) The effect of an equilibrated melt phase on the shear creep and attenuation behaviour of polycrystalline olivine. *Geophysical Research Letters* 27(15): 2341–2344.

Gueguen Y, Darot M, Mazot P, and Woirgard J (1989) Q^{-1} of forsterite single crystals. *Physics of the Earth and Planetary Interiors* 55: 254–258.

Gueguen Y and Mercier JM (1973) High attenuation and the low-velocity zone. *Physics of the Earth and Planetary Interiors* 7: 39–46.

Hammond WC and Humphreys ED (2000a) Upper mantle seismic wave attenuation: Effects of realistic partial melt distribution. *Journal of Geophysical Research* 105(B5): 10987–10999.

Hammond WC and Humphreys ED (2000b) Upper mantle seismic wave velocity: Effects of realistic partial melt geometries. *Journal of Geophysical Research* 105(B5): 10975–10986.

Harrison RJ and Redfern SAT (2002) The influence of transformation twins on the seismic-frequency elastic and anelastic properties of perovskite: Dynamical mechanical analysis of single crystal $LaAlO_3$. *Physics of the Earth and Planetary Interiors* 134: 253–272.

Harrison RJ, Redfern SAT, and Salje EKH (2004) Dynamical excitation and anelastic relaxation of ferroelastic domain walls in $LaAlO_3$. *Physical Review B* 69: 144101.

Heinz DL, Jeanloz R, and O'Connell RJ (1982) Bulk attenuation in a polycrystalline Earth. *Journal of Geophysical Research* 87(B9): 7772–7778.

Hensler JH and Cullen GV (1968) Stress, temperature, and strain rate in creep of magnesium oxide. *Journal of the American Ceramic Society* 51: 557–559.

Hudson JA (1981) Wave speeds and attenuation of elastic waves in material containing cracks. *Geophysical Journal of the Royal Astronomical Society* 64: 133–150.

Irifune T and Isshiki M (1998) Iron partitioning in a pyrolite mantle and the nature of the 410-km seismic discontinuity. *Nature* 392: 702–705.

Jackson I (1991) The petrophysical basis for the interpretation of seismological models for the continental lithosphere.

Geological Society of Australia Special Publication 17: 81–114.

Jackson I (1998) Elasticity, composition and temperature of the Earth's lower mantle: A reappraisal. *Geophysical Journal International* 134: 291–311.

Jackson I (2000) Laboratory measurement of seismic wave dispersion and attenuation: Recent progress. In: Karato S, Forte AM, Liebermann RC, Masters G, and Stixrude L (eds.) *Earth's Deep Interior. Mineral Physics and Tomography from the Atomic to the Glocal Scale*, pp. 265–289. Washington, DC: AGU.

Jackson DD and Anderson DL (1970) Physical mechanisms of seismic wave attenuation. *Reviews of Geophysics and Space Physics* 8: 1–63.

Jackson I, Faul UH, Fitz Gerald J, and Morris SJS (2006) Contrasting viscoelastic behaviour of melt-free and melt-bearing olivine: Implications for the nature of grain-boundary sliding. *Materials Science and Engineering A* 442: 170–174.

Jackson I, Faul UH, Fitz Gerald J, and Tan BH (2004) Shear-wave attenuation and dispersion in melt-bearing olivine polycrystals. Part I: Specimen fabrication and mechanical testing. *Journal of Geophysical Research* 109: B06201 (doi:10.1029/2003JB0002406).

Jackson I, Fitz Gerald JD, Faul UH, and Tan BH (2002) Grainsize sensitive seismic wave attenuation in polycrystalline olivine. *Journal of Geophysical Research* 107:B12 2360 (doi:10.1029/2001JB001225).

Jackson I, Fitz Gerald JD, and Kokkonen H (2000) High-temperature viscoelastic relaxation in iron and its implications for the shear modulus and attenuation of the Earth's inner core. *Journal of Geophysical Research* 105: 23605–23634.

Jackson I and Paterson MS (1993) A high-pressure, high-temperature apparatus for studies of seismic wave dispersion and attenuation. *PAGEOPH* 141: 445–466.

Jackson I, Paterson MS, and Fitz Gerald JD (1992) Seismic wave attenuation in Åheim dunite: An experimental study. *Geophysical Journal International* 108: 517–534.

Jackson I, Webb SL, Weston L, and Boness D (2005) Frequency dependence of elastic wave speeds at high-temperature: A direct experimental demonstration. *Physics of the Earth and Planetary Interiors* 148: 85–96.

Jeffreys H (1976) *The Earth: Its Origin, History and Physical Constitution*, 6th edn., 574 pp. Cambridge: University Printing House.

Kamb WB (1959) Theory of preferred crystal orientation developed by crystallization under stress. *Journal of Geology* 67: 153–170.

Kampfmann W and Berckhemer H (1985) High temperature experiments on the elastic and anelastic behaviour of magmatic rocks. *Physics of the Earth and Planetary Interiors* 40: 223–247.

Karato S (1998) A dislocation model of seismic wave attentuation and micro-creep in the Earth: Harold Jeffreys and the rheology of the solid Earth. *PAGEOPH* 153: 239–256.

Karato S and Jung H (1998) Water, partial melting and the origin of the seismic low velocity and high attenuation zone in the upper mantle. *Earth and Planetary Science Letters* 157: 193–207.

Karato S and Spetzler HA (1990) Defect microdynamics in minerals and solid state mechanisms of seismic wave attenuation and velocity dispersion in the mantle. *Reviews of Geophysics* 28: 399–421.

Koehler JS (1952) The influence of dislocations and impurities on the damping and the elastic constants of metal single crystals. In: Shockley W and Hollomon JH (eds.) *Imperfections in Nearly Perfect Crystals*, pp. 197–216. New York: John Wiley & Sons.

Kumazawa M (1969) The elastic constant of polycrystalline rocks and the non-elastic behavior inherent to them. *Journal of Geophysical Research* 74: 5311–5320.

Kê T (1947) Experimental evidence of the viscous behaviour of grain boundaries in metals. *Physical Review* 71(8): 533–546.

Lakki A, Schaller R, Carry C, and Benoit W (1999) High-temperature anelastic and viscoplastic deformation of fine-grained magnesia- and magnesia/yttria-doped alumina. *Journal of the American Ceramic Society* 82(8): 2181–2187.

Leak GM (1961) Grain boundary damping. Part I: Pure iron. *Proceedings of the Physical Society* 78: 1520–1528.

Lu C and Jackson I (2006) Low-frequency seismic properties of thermally cracked and argon saturated granite. *Geophysics* 71: F147–F159.

Mavko GM (1980) Velocity and attenuation in partially molten rocks. *Journal of Geophysical Research* 85: 5173–5189.

Mavko GM and Nur A (1975) Melt squirt in the asthenosphere. *Journal of Geophysical Research* 80: 1444–1448.

Minster JB and Anderson DL (1981) A model of dislocation-controlled rheology for the mantle. *Philosophical Transactions of the Royal Society of London* 299: 319–356.

Morris SJS (2002) Coupling of inerface kinetics and transformation-induced strain during pressure-induced solid-solid phase changes. *Journal of the Mechanics and Physics of Solids* 50: 1363–1396.

Morris SJS and Jackson I (2006) A micromechanical model of attenuation by diffusionally assisted grain-boundary sliding. *EOS (Transactions American Geophysical Union)* 87(52), Fall Meet. Suppl., Abstract MR21A-0009.

Mosher DR and Raj R (1974) Use of the internal friction technique to measure rates of grain boundary sliding. *Acta Metallurgica* 22: 1469–1474.

Nowick AS and Berry BS (1972) *Anelastic Relaxation in Crystalline Solids*. New York: Academic Press.

O'Connell RJ and Budiansky B (1974) Seismic velocities in dry and saturated cracked solids. *Journal of Geophysical Research* 79: 5412–5426.

O'Connell RJ and Budiansky B (1977) Viscoelastic properties of fluid-saturated cracked solids. *Journal of Geophysical Research* 82: 5719–5735.

Pezzotti G, Kleebe HJ, and Ota K (1998) Grain-boundary viscosity of polycrystalline silicon carbides. *Journal of the American Ceramic Society* 81(12): 3293–3299.

Poirier J-P (1985) *Creep of Crystals. High-Temperature Deformation Processes in Metals, Ceramics and Minerals*. Cambridge UK: Cambridge University Press.

Raj R (1975) Transient behaviour of diffusion-induced creep and creep rupture. *Metallurgical Transactions A* 6A: 1499–1509.

Raj R and Ashby MF (1971) On grain boundary sliding and diffusional creep. *Metallurgical Transactions* 2: 1113–1127.

Schaller R, Fantozzi G, and Gremaud G (2001) (eds.) Mechanical spectroscopy Q^{-1} 2001 with applications to materials science. *Materials Science Forum* 366–368: 683 pp.

Schmeling H (1985) Numerical models on the influence of partial melt on elastic, anelastic and electric properties of rocks. Part I: Elasticity and anelasticity. *Physics of the Earth and Planetary Interiors* 41: 34–57.

Schoeck G (1963) Friccion interna debido a la interaccion entre dislocaciones y atomos solutos. *Acta Metallurgica* 11: 617–622.

Seeger A (1956) On the theory of the low-temperature internal friction peak observed in metals. *Philosophical Magazine* 1: 651–662.

Seeger A (1981) The kink picture of dislocation mobility and dislocation-point-defect interactions. *Journal of Physics IV* 42: 201–228.

Seeger A and Wüthrich C (1976) Dislocation relaxation processes in body-centred cubic metals. *Il Nuovo Cimento* 33: 38–73.

Shearer PM (1999) *Introduction to Seismology*. New York: Cambridge University Press.

Shito A, Karato S, and Park J (2004) Frequency dependence of Q in the Earth's upper mantle inferred from continuous spectra of body waves. *Geophysical Research Letters* 31: L12603.

Shito A, Karato S, Matsukage NK, and Nishihara Y (2006) Toward mapping water content from seismic tomography: Applications to subduction zone upper mantle. In: Jacobsen SD and van der Lee S (eds.) AGU *Monograph V. 168: Earth's Deep Water Cycle*, pp. 225–236. Washington, DC: American Geophysical Union.

Simpson HM and Sosin A (1972) Contribution of defect dragging to dislocation damping. Part I: Theory. *Physical Review B* 5: 1382–1392.

Steinle-Neumann G, Stixrude L, and Cohen RE (2003) Physical properties of iron in the inner core. In: Dehant V, Creager KC, Karato S-I, and Zatman S (eds.) *AGU Geodynamic Series 31: Earth's Core, Dynamics, Structure, Rotation*, pp. 137–161. Washington, DC: AGU.

Stixrude L (1997) Structure and sharpness of phase transitions and mantle discontinuities. *Journal of Geophysical Research* 102(B7): 14835–14852.

Stocker RL and Gordon RB (1975) Velocity and internal friction in partial melts. *Journal of Geophysical Research* 79: 2129–2131.

Tan BH, Jackson I, and Fitz Gerald JD (1997) Shear wave dispersion and attenuation in fine-grained synthetic olivine aggregates: Preliminary results. *Geophysical Research Letters* 24(9): 1055–1058.

Tan BH, Jackson I, and Fitz Gerald JD (2001) High-temperature viscoelasticity of fine-grained polycrystalline olivine. *Physics and Chemistry of Minerals* 28: 641–664.

Vaišnys JR (1968) Propagation of acoustic waves through a system undergoing phase transformations. *Journal of Geophysical Research* 73: 7675–7683.

Vočadlo L, Alfè D, Gillan MJ, and Price GD (2003a) The properties of iron under core conditions from first principles calculations. *Physics of the Earth and Planetary Interiors* 140: 101–125.

Vočadlo L, Alfè D, Gillan MJ, Wood IG, Brodholt JP, and Price GD (2003b) Possible thermal and chemical stabilization of body-centred cubic iron in the Earth's core. *Nature* 424: 536–539.

Vočadlo L (2007) *Ab initio* calculations of the elasticity of iron and iron alloys at inner core conditions: Evidence for a partially molten inner core? *Earth and Planetary Science Letters* 254: 227–232.

Walsh JB (1968) Attenuation in partially melted material. *Journal of Geophysical Research* 73: 2209–2216.

Walsh JB (1969) New analysis of attenuation in partially melted rock. *Journal of Geophysical Research* 74: 4333–4337.

Webb S and Jackson I (2003) Anelasticity and microcreep in polycrystalline MgO at high temperature: An exploratory study. *Physics and Chemistry of Minerals* 30: 157–166.

Webb S, Jackson I, and Fitz Gerald JD (1999) Viscoelastic rheology of the titanate perovskites $CaTiO_3$ and $SrTiO_3$ at high temperature. *Physics of the Earth and Planetary Interiors* 115: 259–291.

Weidner DJ and Wang Y (2000) Phase transformations: Implications for mantle structure. In: Karato S (ed.) *Geophysical Monograph Series, vol. 117. Earth's Deep Interior: Mineral Physics and Seismic Tomography From the Atomic to the Global Scale*, pp. 215–235. Washington, DC: American Geophysical Union.

Woirgard J, Mazot P, and Rivière (1981) Programmable system for the measurement of the shear modulus and internal friction, on small specimens at very low frequencies. *Journal of Physics C* 5: 1135–1140.

Xu Y, Zimmerman ME, and Kohlstedt DL (2004) Deformation behaviour of partially molten mantle rocks. In: Karner G (ed.) *MARGINS Theoretical and Experimental Earth Science*, pp. 284–310. New York: Columbia University Press.

Zener C (1941) Theory of the elasticity of polycrystals with viscous grain boundaries. *Physical Review* 60: 906–908.

Zener C (1952) *Elasticity and Anelasticity of Metals*. Chicago, IL: The University of Chicago Press.

18 Properties of Rocks and Minerals – High-Pressure Melting

R. Boehler, Max Planck Institute für Chemie, Mainz, Germany

M. Ross, University of California, Livermore, CA, USA

18.1	Melting Properties of Lower-Mantle Components	527
18.1.1	Melting of (Mg, Fe)SiO₃-Perovskite	527
18.1.2	Melting of MgO	528
18.1.3	Partial Melting in the Lower Mantle	530
18.2	Iron Phase Diagram	531
18.2.1	f.c.c.–h.c.p. Transition	531
18.2.2	Iron Melting below 100 GPa	532
18.2.3	Iron Melting above 100 GPa	533
18.2.4	Diamond-Cell Measurements of Iron and Transition Metals	533
18.2.5	Local Structures of Icosahedral Short-Range Order in Transition Metal Melts	535
18.2.6	Shock Melting	536
18.2.7	Theoretical Calculations of the Iron Melting Curve	537
18.3	Viscosity of Iron at Earth Core Conditions	538
References		539

18.1 Melting Properties of Lower-Mantle Components

The lower mantle encompasses the pressure range from 24 to 135 GPa at the core–mantle boundary (CMB). The three major components are $(Mg, Fe)SiO_3$-perovskite (PV) and $(Mg, Fe)O$-magnesiowüstite (MW) making up about 90%, and $CaSiO_3$-perovskite. With regards to melting, the perovskite phase drew most of the attention in early theoretical and experimental work. Based on the phase diagram measured in multi-anvil presses at lower pressures (up to about 25 GPa), and the very high melting temperature of MgO at ambient pressure, it was believed that the eutectic composition of the lower mantle lies near the composition of the major component $(Mg, Fe)SiO_3$-perovskite. Therefore, the eutectic melting temperature of the lower mantle would be close to that of the end-member PV. However, more recent measurements made in the laser-heated diamond cell have shown that this was incorrect. The measurements indicate, that at about 60 GPa and 4200 K, the two melting curves cross and perovskite becomes the higher melting phase. Thus, with increasing depth, the eutectic melting composition would have to move away from that of PV and the first partial melt would be rich in $(Mg, Fe)O$. Due to gravitational segregation, this liquid MW-rich phase could therefore be the major phase at the very bottom of the mantle and its physical properties would be more suitable for explaining the unusual seismic signature (Boehler, 2000).

18.1.1 Melting of (Mg, Fe)SiO₃-Perovskite

Both the early theoretical and experimental estimates of the melting temperature of $(Mg, Fe)SiO_3$-perovskite for lower mantle pressures showed extremely large differences. The first molecular dynamics (MD) calculations overestimated the PV melting temperature by 2000 K, even at pressures as low as 30 GPa, and early diamond-cell measurements showed positive and negative slopes and even a combination of both (see references in Zerr and Boehler (1993)). The extremely large variations clearly demonstrated the experimental and computational difficulties to determine melting of minerals at high compression. Recently, improved techniques for using CO_2-lasers to heat minerals directly without metal absorbers and the use of inert gas pressure media, such as argon, allowed more systematic melting measurements of minerals and even multicomponent systems at mantle pressures (Zerr and Boehler, 1993, 1994; Zerr et al., 1997, 1998). It is now clear that the melting temperatures of PV drastically increase with increasing pressures, as evident from **Figure 1**. Similar high melting slopes were measured for Ca-perovskite (Shen and Lazor, 1995; Zerr *et al.*, 1997). There is still some disagreement on the melting temperature at the pressure of the CMB (135 GPa). The melting temperatures of Zerr and Boehler (ZB) and Shen and Lazor (SL) agree well within the experimental uncertainty, which is typically ±200 K. The extrapolated melting curve of ZB agrees well with recent shock measurements on MgSiO$_3$-glass (Akins *et al.*, 2004; Luo *et al.*, 2004) if the last data point of ZB at 62 GPa is omitted, and this would suggest significant flattening in the melting curve.

It is not a straightforward process to determine the exact melting temperatures from these shock experiments because the solid shock conditions often overdrive the melting curve, and the estimated melting temperature at 110 GPa of 5200 K has a reported uncertainty of ±500K. In the same shock study, crystalline enstatite is shown to melt at higher pressures but at slightly lower temperature (5000 K at 170 GPa), indicating a melt maximum at just above 100 GPa. This data point further supports a flat melting curve and even suggests that the melt at these high pressures becomes more dense than the solid, even if a post-perovskite phase is assumed (Murakami *et al.*, 2004).

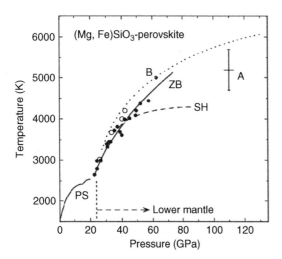

Figure 1 Melting temperatures of Mg–Si-perovskite from different studies. Diamond cell measurements: black dots, ZB (Zerr and Boehler, 1993); open circles, Shen and Lazor (1995); dashed curve, Sweeney and Heinz (1998). Calculations: dotted curve, B (Belonoshko *et al.*, 2005); shock melting point (A), cross by Akins *et al* (2004) and Luo *et al.* (2004). PS is the solidus of upper-mantle peridotite. There is consensus that the melting temperatures of PV rise very steeply, but the amount of flattening at higher pressures is still not clear.

There is agreement between the ZB measurements and recent two-phase quantum MD simulations corrected by simulations employing classical effective potentials (Belonoshko *et al.*, 2005). The simulations are useful in that they predict a significant flattening of the PV melting curve above 62 GPa, the highest pressure attained experimentally. Combining the diamond-cell measurement of ZB and SL, the shock measurement and the latest theoretical work, the best estimate for the melting temperture of PV at the CMB is between 5500 and 6000 K.

Sweeney and Heinz (1998) using conventional YAG laser-heating techniques without an argon pressure medium made measurements to about 80 GPa. At lower pressures, the melting temperatures agree with ZB within their experimental uncertainties, but their melting slope above about 40 GPa is almost zero, which is in disagreement with the other estimates. There are to date no additional diamond-cell data on perovskite melting.

18.1.2 Melting of MgO

At ambient conditions, MgO has the NaCl rock-salt(B1) structure which is known from experiments to be stable up to 227 GPa (Duffy *et al.*, 1995). The room-temperature isotherm and the elasticity have been measured in diamond cells till to 55 GPa (Zha *et al.*, 2000). Due to the high initial melting temperature of MgO (3060 K at 1 b), the only experimental determination of the melting curve has been that using a laser-heated diamond-anvil cell (DAC) to 32 GPa and 4000 ± 300 K (Zerr and Boehler, 1994). The melting curve is rather flat and resembles those of ionic solids at high compression (Boehler *et al.*, 1996, 1997). Since these experiments cover only the first 25% of the mantle pressure range, their melting slope ($dT/dP = 36$ K/GPa) plays an important role in constraining theoretical predictions (Aguado and Madden, 2005; Alfé, 2005; Belonoshko and Dubrovinsky, 1996; Strachan *et al.*, 1999; Vocadlo and Price, 1996). **Figure 2** shows a comparison of the experimental and predicted theoretical melting curves. An extrapolation of the ZB measurements leads to an estimated melting temperature of about 5000 K at the CMB using a Lindemann equation with a Grüneisen parameter of 1.3, determined from this measured curve (Zerr and Boehler, 1994). The measured melting temperatures are substantially lower than all the theoretical results, which among themselves are in poor agreement and exhibit a wide spread.

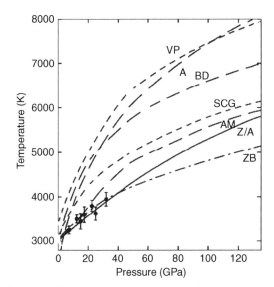

Figure 2 MgO melting: experiment and theory. The only measurements are from ZB up to 32 GPa (Zerr and Boehler, 1994). Theory: AM (Aguado and Madden, 2005) and their new extrapolation Z/A of the ZB data; SCG (Strachan *et al.*, 1999); BD (Belonoshko and Dubrovinsky, 1996); VP (Vocadlo and Price, 1996); A (Alfé, 2005); for description, see text.

Theoretical calculations for the oxides are complicated by the fact that the oxygen anion, O^{2-}, which does not exist in the free state, is stabilized by its local environment (Watson, 1958). Oxygen anions in oxide materials have been characterized as having diffuse electronic charge distributions and large polarizabilities that make quantitative modeling difficult (Vidal-Valet *et al.*, 1978). In MgO, the charge densities are close to the situation Mg^+O^-. For the other alkaline oxides the charge densities imply $X^{2+}O^{2-}$. It was shown (Lacks and Gordon, 1993) that the O^{2-} anion can be distorted in the direction of neighboring cations creating a nonspherical charge distribution that lowers the Coulomb energy. Since the liquid is not locked into a periodic structure, it can be influenced to lower its energy by forming local arrangements differing from the usual hard-sphere-like liquid structures. In order to include possible local structures in their simulations, Strachan *et al.* (1999; SCG), and Aguado and Madden (2005; AM) employed *a priori* methods to obtain transferable potentials. Transferable potentials provide an accurate description of interatomic forces over a range of perfect and defective crystal structures, liquid phases, and low-coordinated sites as those encountered at surfaces and clusters. Transferable potential parameters may be obtained by fitting to interatomic

forces and stress tensor components that are obtained by *ab initio* methods for a large number of atomic configurations. Transferability is essential in the case of melting, because in contrast to the solid the liquid is not restricted to a symmetric periodic structure.

To calculate melting, SCG used the 'transferable' potential for both the solid and liquid phases to calculate the enthalpy and volume changes. The melting curve was obtained by integrating the Clausius–Clapeyron equation. They also predicted a B1–B2 transition at 400 GPa and 0 K and a melting temperature of $\sim$6100 K at the CMB. AM also developed a transferable potential, which they also used to calculate the volume and enthalpy, and locate the melting curve by integrating the Clausius–Clapeyron equation. They found that at 1 bar, MgO melts from a wurtzite structure with a melting slope in good agreement with the slope determined by ZB. Near 10 GPa, the wurtzite structure reconverts to rock salt and lies in agreement with the ZB melting curve to the highest measured pressure of 32 GPa. AM predict the temperature of $\sim$5950 K at the CMB.

Based on an inspection of the DAC data, AM note that the extrapolation procedure of ZB, to predict a melting temperature of $\sim$5150 K at the CMB, has a large uncertainty. Since their calculations are in reasonable agreement with ZB up to about 23 GPa, AM combined their calculations with the experimental data and constructed a revised melting curve. The new melting curve predicts a temperature of 5850 K at CMB, higher than ZB by 700 K, but lower than other theoretical estimates.

Intermolecular potentials fitted to the room-temperature solid properties were used to calculate the melting curve by the method of coexisting phases (Belonoshko and Dubrovinsky, 1996; BD) and (Vocadlo and Price, 1996; VP). By employing a potential determined solely from fits to solid state data, for both the solid and liquid, they assumed implicitly that the potential is transferable from solid to liquid, which is questionable in the case of MgO. In the most recent study, Alfé (2005) calculated an MgO melting curve by carrying out simulations of the coexisting phases, but rather than employing effective interaction potentials, he used first-principles density functional theory (DFT). His temperatures are the highest of those calculations reported. The coexisting phase method employed by BD and Alfé is an approximation that amounts to determining a state mechanical equilibrium between the two phases, $(P, T)_s = (P, T)_l$, but not to thermodynamic equilibrium, $(P, T, G)_s = (P, T, G)_l$. The problem is that the formation of local structures may lower the internal energy, hence the Gibbs energy, without significantly altering the pressure. While the coexisting phase method can be useful in many situations, the method is not rigorous.

The lowest calculated melting slopes, in best agreement with experiment, are those by SCG and AM, who employed transferable potentials for both phases, and employed the rigorous Clausius–Clapeyron equation to calculate the melting curve. Since the number of configurations employed in developing their potentials was necessarily finite, their reported melting pressures should be considered as an upper bound. Consequently, we believe, a best estimate of the melting temperature of MgO at the CMB is in the range 5150–5850 K.

18.1.3 Partial Melting in the Lower Mantle

The lower mantle likely behaves as a eutectic system with the end members MgO (or MW) and PV. Even though the study of the melt properties of these end members is of fundamental importance, their melting temperatures may have little direct relevance to mantle melting. In view of the large discrepancies between the theoretical and experimental melting temperatures, one has to keep in mind that the experiments on PV, MgO, and MW were made using identical experimental setups, temperature measurements, and melt detection, and they were carried out by the same persons. This resulted in two important facts even if one assumes any possible systematic errors: first that the melting slopes between PV and MgO are entirely different, and second that the two melting curves are likely to cross each other between 50 and 60 GPa (see **Figure 3**). From these facts, one must conclude that the melting temperature and chemical composition of a partial melt in the mantle will change with depth. In fact, melting measurements for a material composed of all the major elements in the lower mantle – O, Si, Mg, Fe, Al, Ca, Ti, Cr, Ni, and Na (the sample was a glass with a pyrolite composition) – showed a significant lowering of the melting temperature with respect to PV and MW (LM 'rock' in **Figure 3**). The solidus temperatures in these experiments were bracketed on the basis of textural changes observed in the recovered samples from a series of $P-T$ runs (Zerr *et al.*, 1998). The measured pyrolite solidus for this multicomponent system lies approximately

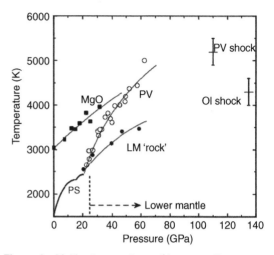

Figure 3 Melting temperatures of lower-mantle components measured in the diamond cell by Zerr and Boehler (1993, 1994; Zerr *et al.* 1998). PS is the solidus of upper-mantle peridotite. The melting curves of MgO and PV are likely to cross at about 60 GPa and their melting temperatures in the deep mantle may be similar if the PV melting continues to flatten (see text). The eutectic temperatures of a pyrolite system (LM rock) is about 1000 K below that of the end members MgO and PV. Shock melting points for $MgSiO_3$-PV (Akins *et al.*, 2004) and Mg_2SiO_4-olivine ($MgO:MgSiO_3 = 1:1$) (Holland and Ahrens, 1997) are also shown.

parallel to the melting curve of MgO, but is offset to lower temperatures by about 1000 K.

The solidus temperatures, extrapolated along the same path as MgO, are in good agreement with the shock melting point of Mg_2SiO_4-olivine at 130 GPa and 4300 K (Holland and Ahrens, 1997). This is not a coincidence, because the olivine composition ($MgO:MgSiO_3 = 1:1$) may well represent the eutectic composition in the MgO–PV system (see **Figure 4**). There exists also a systematic agreement between the DAC and shock measurments in that the difference in the melting temperatures between the eutectic and PV is about 1000 K. We conclude that a reasonable estimate for the solidus temperatures at the CMB (135 GPa) is about 4300 K, definitely much lower than any prediction for MgO or PV. The predicted simplified phase diagram for the lower mantle is shown in **Figure 4**.

Because the solidus temperature is very close to the predicted core temperature at the CMB (Boehler, 2000), partial melting in the lowermost mantle is in agreement with the experimental evidence for melting from both shock and diamond-cell work. The important task for future work will be the investigation of the nature and chemical composition of this melt. This

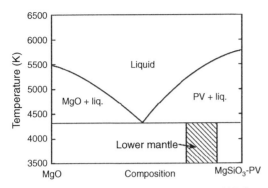

Figure 4 Schematic eutectic phase diagram of MgO and PV with predicted melting temperatures at the CMB (see text). The eutectic temperature is about 1000 K lower than that of the end members PV and MgO. For comparison, the shock temperature of olivine ($MgSiO_3$:MgO = 1:1) at 130 GPa is 4300 ± 300 K (Holland and Ahrens, 1997) and that of $MgSiO_3$ is 5200 ± 500 K at 110 GPa. The composition of the lower mantle is accepted to be closer to PV than to MgO and therefore a partial melt must be MgO rich (Boehler, 2000).

melt is likely enriched in MgO (MW) because the lower mantle composition is likely to be closer to PV than to MgO, as shown shematically in **Figure 4**. These studies will be important not only for understanding the unusual signature of the seismic signal neat the CMB, but also for the understanding of early mantle differentiation, when the mantle was much hotter, and possibly entirely molten.

18.2 Iron Phase Diagram

Understanding the Earth's thermal gradient and dynamics requires an accurate knowledge of the equation of state and melting temperature of Fe at the inner–outer core boundary (IOCB) at 330 GPa. Melting temperatures at megabar pressures (1 Mbar = 100 GPa) can be obtained from diamond cell and shock experiments and from theory. The melting curve was determined up to 200 GPa using the laser-heated DAC (Boehler, 1993), and shock melting has been measured between 220 and 240 GPa (Brown and McQueen, 1986; Nguyen and Holmes, 2004). Theoretical melting calculations have been made to 360 GPa (Alfé *et al.*, 1999, 2004b; Belonoshko *et al.*, 2000; Laio *et al.*, 2000). The extrapolated measurements and the theoretical predictions of the melting temperatures at the IOCB range from 4800 to 6500 K.

18.2.1 f.c.c.–h.c.p. Transition

Most of the experimental work on the iron phase diagram was done below 100 GPa. Although it has little relevance to the state of the core, these data are important for understanding the existing large discrepancies at higher pressures and assuring a smooth continuity of pressure-induced material properties. At ambient temperature, iron has a b.c.c. structure and transforms into h.c.p. at 13 GPa (Boehler *et al.*, 1990). The h.c.p. phase was found to be stable up to about 300 GPa by X-ray diffraction (Mao *et al.*, 1990). The high *P–T* phase diagram is shown in **Figure 5**. The f.c.c.–h.c.p. phase boundary has been measured in the diamond cell and in a multi-anvil press to 40 GPa and 1600 K (Boehler, 1986; Kubo *et al.*, 2003; Mao *et al.*, 1987) (highest data points by Kubo and Mao shown in **Figure 5**), and extended to near the melting curve (Boehler, 1993). These data are in good agreement. Combined with shock melting measurements and new X-ray measurements (see below),

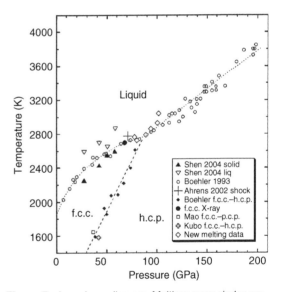

Figure 5 Iron phase diagram. Melting: open circles are older measurements (Boehler, 1993). Triangles are X-ray measurements (Shen *et al.*, 2004). The cross is a shock melting measurement on preheated iron (Ahrens *et al.*, 2002). Open diamonds are new measurements in an argon pressure medium. The f.c.c.–h.c.p. phase boundary: open square and cross are the highest data points by Mao *et al.* and Kubo *et al.*, respectively (Kubo *et al.*, 2003; Mao *et al.*, 1987); diamonds are from Boehler (1993). The solid dot is a new X-ray measurement documenting the thus far highest pressure conditions for f.c.c. iron. Other DAC data up to 140 GPa (Saxena *et al.*, 1993) (not shown in the figure for clarity) were measured with similar techniques as by Boehler (1993) and are nearly identical. Dashed lines are guides to the eye.

the f.c.c.–h.c.p.–liquid triple point has to be above 71 GPa, probably close to 80 GPa.

Several years ago, it was suggested, based on X-ray diffraction measurements in the laser-heated diamond cell, that the phase diagram of iron may be more complicated containing orthorhombic or d.h.c.p. phases. More recently, however, no such transitions were found (Shen *et al.*, 1998, 2004). It is highly likely that both the previously observed distortion from h.c.p. to orthorhombic, and a change in stacking order h.c.p. → d.h.c.p. were due to deviatoric stresses using hard pressure media in the diamond cell.

In order to clarify these earlier discrepancies on the iron phase behavior, we made new measurements at the new laser-heating beamline ID 27 at the ESRF using argon as a pressure medium. Heating technique and temperature measurements were identical to those in our own laboratory (see below). We found that 68 GPa and 2700 K is the highest pressure and temperature where f.c.c. iron could be clearly identified. The actual pressure was probably higher, because it was measured at room temperature and was not corrected for a thermal pressure increase. For the cell configuration used in these measurements, this pressure increase could be of the order of 1 GPa (Chudinovskikh and Boehler, 2003). No other iron phases other than f.c.c. and h.c.p. could be observed. Newer shock velocity measurements on preheated iron (Ahrens *et al.*, 2002; also shown in **Figure 5**) also show no additional transitions along the shock path other than melting at 71 GPa, indicating that iron melted from the f.c.c. phase.

18.2.2   Iron Melting below 100 GPa

Most of the iron melting measurements are at pressures below 100 GPa. The results from different laboratories using six different methods in the laser-heated diamond cell and shock experiments are now in very good agreement within the experimental uncertainty of about ±100 K. Therefore, the iron melting curve to 100 GPa is probably better constrained than that for any other material (Ahrens *et al.*, 2002; Boehler, 1993; Jephcoat and Besedin, 1996; Saxena *et al.*, 1994; Shen *et al.*, 2004; Yoo *et al.*, 1996).

The measurement of pressure and temperature in the laser-heated diamond cell is now straightforward and their uncertainty is much smaller than the existing discrepancies at the highest pressure of 200 GPa. The detection of melting is more challenging, but there are a number of physical property changes between solid and liquid that are measurable in the diamond cell. For iron, discontinuous changes in the resistivity of thin wires (Boehler, 1986) and changes in reflectivity and laser power (Boehler *et al.*, 1990; Boehler, 1993; Saxena *et al.*, 1993) have been measured. The most drastic change between solid and liquid is the loss of strength at melting. This often causes motion on the sample surface that can be made visible with the speckle pattern of an argon-laser beam (Boehler *et al.*, 1996; Jephcoat and Besedin, 1996). These methods were compared in some experiments with the first observation of microscopic melt features on the surface of the quenched sample for iron (Boehler, 1993) and for a number of other metals (Errandonea *et al.*, 2001). Additionally, the observation of motion was successfully tested against known melting temperatures for tungsten in vacuum (Boehler and Errandonea, 2002) and alkali halides at low pressures (Boehler *et al.*, 1996). The results for iron from these different measurements are in very good agreement. Additionally, the effect of different pressure media on the melting temperatures has been thoroughly checked (Boehler, 1993; Boehler *et al.*, 1990). These melting results have been recently confirmed using X-ray diffraction, which probably yields the most conclusive melting criterion. However, X-ray experiments are significantly more challenging than optical measurements because they can only be done with synchrotron radiation and require special conditions for temperature gradients and sample size, which limits the pressure range. Up to 58 GPa, diffusive X-ray scattering from laser-heated iron samples has been documented above the melting temperature (Shen *et al.*, 2004), shown by the open triangles in **Figure 5**. All the measurements without melt features fall below the Boehler melting curve, and all data indicative of melting fall above that curve. This has been used to argue for higher melting temperatures. One has too keep in mind, however, that in order to produce an X-ray pattern that contains diffuse diffractions, a finite amount of material has to be in the molten state. Moreover, the heated surface from which the temperature is measured is always hotter than the bulk material and, thus, the data points indicating melting must lie significantly above the melting curve. This fact may lead to an overestimation of the melting temperature using X-ray diffraction (see also for tantalum (Errandonea *et al.*, 2003). Thus, we see no experimental evidence for higher melting temperatures as often claimed.

Jephcoat and Besedin (1996) compared the melting temperatures of argon and iron and predicted a crossover of the argon and the iron melting curves at 47 GPa and 2750 ± 200 K. From more recent extended melting measurements of argon (Boehler et al., 2001) and Shen et al.'s (2004) new iron melting data, this crossover point is probably better constrained at 50 ± 1 GPa and 2600 ± 50 K, but still in good agreement with the earlier estimate.

Additional evidence for 'low' iron melting temperatures below 100 GPa comes from shock velocity measurements on preheated iron (Ahrens et al., 2002). At $P-T$ conditions shown by the cross in **Figure 5** (71 ± 2 GPa, 2775 ± 160 K), a very sharp drop in the shock velocity was observed which is due to melting. Temperatures in this work were calculated and possible overshoot of the equilibrium melt conditions was not taken into account. An accurate estimate of the uncertainty in this measurement is therefore difficult.

Recently, we collected new data using argon as a pressure medium to much higher pressures (>100 GPa) than in previous work where the pressure medium above 35 GPa was Al_2O_3. We employed several improved experimental techniques. Two new 40 W diode-pumped Nd:vanadate lasers (INAZUMA, Spectra-Physics) with perfect TEM00 mode and perfect power stability resulted in larger hot spots and reduced temperature fluctuations to less than 10 degrees. A new gasket, consisting of tungsten and diamond powder, provided a significantly larger cell volume and thus a more efficient thermal insulation of the sample from the diamonds, leading to a reduction in the temperature gradients. Moreover, we significantly improved the optical resolution by replacing the previous reflecting objective with achromatic lenses. These achromats, however, used in a conventional configuration, cause errors in the temperature measurements from small hot spots due to their finite chromatic aberration (Boehler, 2000). We measured these temperature errors as a function of numerical aperture, which was varied by inserting an iris diaphragm in the parallel light path between the two achromats. This error becomes negligible when the numerical aperture is reduced by a factor of 3. This iris reduces the optical resolution to about 2–3 μm and the light intensity by a factor of 10. For temperature measurement, this is sufficient and does not reduce its accuracy. The iris is opened during visual observation to increase the optical resolution.

Even though argon is solid above 50 GPa along the iron melting curve, it remains very soft during heating and recrystallizes, as evident from the very small pressure gradients measured throughout the

pressure chamber after heating. Below about 70 GPa, melting of iron is readily and reproducibly detected by motion using the well-tested laser-speckle method, which is described elsewhere (Boehler, 2000) and discussed above. Above ~70 GPa, however, motion on the sample surface during melting becomes very sluggish, and at 100 GPa either does not exist or becomes undetectable. This was generally observed for many other materials during a vast amount of measurements in the past few years. A possible cause for this phenomenon may be an increase in the viscosity of the melt at high pressure (discussed in more detail below). The previously reported observation of motion up to 200 GPa (Boehler, 1993) must therefore have been due to a combination of low optical resolution, laser beam instabilities, higher temperature gradients in the Al_2O_3-matrix, and change in reflectivity. Using the new optics, a reproducible sharp increase in the reflectivity upon melting and decrease upon freezing of the sample could be more clearly observed. These new data are shown by the open diamonds in **Figure 5**. It should be noted that relating changes in the reflectivity to melting or freezing have been thoroughly checked against other methods (listed above) at lower pressures.

In summary, from the very good agreement between the visually measured melting temperatures, the new X-ray measurements, and the shock measurements, one has to conclude that at 100 GPa the iron melting temperature must be below 3000 K. The older melting temperatures of over 4000 K (Williams et al., 1987) at this pressure should therefore be discarded.

18.2.3 Iron Melting above 100 GPa

The melting temperature of iron at core pressures remains uncertain as a result of significant differences between the diamond-cell measurements, shock data, and theory. The results of these measurements (**Figure 5**) and theoretical calculations (Alfé et al., 1999; Belonoshko et al., 2000; Laio et al., 2000) are plotted in **Figure 6**. The differences may be resolved, at least partially, by the presence of local structures in the melt that can also account for a high core viscosity.

18.2.4 Diamond-Cell Measurements of Iron and Transition Metals

The only diamond-cell measurements above 100 GPa are from Boehler (1993); up to 200 GPa)

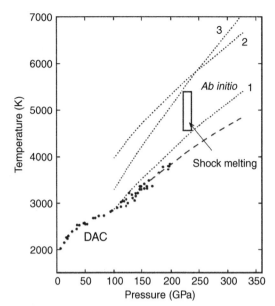

Figure 6 Iron melting curves. Diamond cell (DAC) measurements (Boehler, 1993) with extrapolation up to 330 GPa (IOCB). Shock melting experiments (Nguyen and Holmes, 2004), and from theory 1, (Laio *et al.*, 2000), 2, (Alfé *et al.*, 1999), and 3 (Belonoshko *et al.*, 2000). The differences in the melting temperatures at 3.3 Mbar (inner core boundary) are over 2500 K. The error bar for the shock data is the estimated uncertainty in the temperature calculation.

and Saxena *et al.* (1994; up to 140 GPa). These two sets of measurements using similar methods to detect melting are in good agreement and well within the experimental uncertainty, suggesting a melting temperature at 200 GPa near 4000 K. A generous absolute uncertainty in these measurements is less than ± 200 K at the highest pressures. This includes the uncertainty in the melt detection, the precision of the temperature measurements, and worst-case assumptions on changes in the emissivity-wavelength dependence of iron at these conditions compared to those measured at one atmosphere.

The two largest possible error sources in these measurements are uncertainties in the melting criterion and chemical reactions between molten iron and the Al_2O_3-pressure medium in these experiments. The possibility of lowering the iron melting temperatures due to chemical reactions between molten iron and Al_2O_3 can be excluded because recovered iron samples have since been frequently checked by electron-microprobe analysis yielding negative results. There is also no indication of a reaction between liquid iron and argon from our recent *in situ* synchrotron X-ray measurements up to

100 GPa and to over 3600 K. The visual measurement of melting still needs to be confirmed above 100 GPa using other methods. This is a great experimental challenge, but in view of the good agreement between different methods to detect melting at lower pressures, we are unable to assign additional error sources in these measurements which could drive the iron melting curve to significantly higher temperatures.

The melting temperature of iron at the IOCB (330 GPa) was determined, using the Kraut–Kennedy relationship, by a linear extrapolation of the melting temperature and density measured between 100 and 200 GPa to 330 GPa. K–K is a simplified version of the Lindemann melting relationship. In volume, this extrapolation from 200 to 330 GPa is only 11% of the total compression of iron and leads to a melting temperature of less than 5000 K at the IOCB. The total uncertainty is unlikely to exceed 300 K. These measurements are in serious disagreement with the predictions of Alfé *et al.* and Belonoshko *et al.*, but are in much better agreement with those of Laio *et al.* The agreement with shock data is mixed.

In order to better judge the reasonableness of our DAC Fe melting measurements a systematic study was undertaken for a wide range of other elemental materials with the purpose of acquiring greater insight into the phenomena of melting. **Figure 7** contains a plot of transition metal melting curves and of two well-understood nontransition materials, Ar (Zha *et al.*, 1986) and Al (Boehler and Ross, 1997). All of the experimental data were collected using essentially the same experimental setups. The melting curves of Cu, Ar, and Al have significantly higher melting temperatures and melting slopes than the f.c.c. transition metals (Fe, Co, Ni). The b.c.c. (W, Ta, Mo) melting curves have even smaller, nearly flat melting curves (Errandonea *et al.*, 2001).

To obtain some physical understanding as to why transition metals have relatively low melting slopes, a model (Ross *et al.*, 2004) was developed that explained that for the case of partially filled d-electron bands the loss of structural periodicity associated can lead to a lowering of the liquid energy relative to the solid, thereby lowering the melting temperature. According to this model, Cu, which has a filled d-electron band, should have a higher melting slope than its neighbor Ni (which has a partially filled d-band). To test this model, measurements of the Cu melting curve were extended to near 100 GPa (3650 K), and the Ni melting curve to 84 GPa (2970 K) (Japel *et al.*, 2005). These

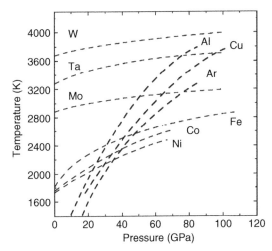

Figure 7 Melting systematics. The transition metals Fe, Co, and Ni, which melt from the f.c.c. phase, have steeper melting curves than the b.c.c. transition metals W, Ta, and Mo (Errandonea *et al.*, 2001). Elements with high d-electron band occupancy (Cu, Al, Ar) have very steep melting curves (Japel *et al.*, 2005).

data (**Figure 7**) show that the melting slope for Cu is about 2.5 times steeper than for Ni (and Fe and Co). In effect, the 'withdrawal' of a single electron from the filled Cu d-shell, to 'create' Ni with a partially filled d-shell, is sufficient to cause a dramatic drop in the melting slope. The model predicts that with increasing pressure the d-electron population rises, due to the s–d-transition, thereby maintaining a low melting slope. The neighboring metals Ni, Co, and Fe have nearly identical melting curves. Overall, we find the Fe melting curve is typical of the f.c.c. transition metals. While it is the main component of the Earth's core, it does not have a special place in the pantheon of metals.

18.2.5 Local Structures of Icosahedral Short-Range Order in Transition Metal Melts

A possible mechanism for lowering the transition metal melting slopes may be the presence in the melt of clusters of icosahedral short-range order (ISRO). In contrast to the solid, the absence of long-range order allows the liquid greater flexibility to lower its energy by forming local structures. ISRO is favored by the fivefold symmetric d-electron character of the charge density. Evidence for the presence of ISRO clustering in stable and under-cooled melts of transition metals at ambient conditions have been reported by Schenk *et al.* (2002) and Lee *et al.* (2004) using neutron scattering and *in situ* X-ray diffraction methods, respectively. Among these metals are Ni,

Fe, Zr (Schenk *et al.*, 2002), Fe, and Ti (Lee *et al.*, 2004). First-principles MD simulations made by Jakse and Pasturel (2004) also found evidence for the existence of short-range order in the stable and undercooled melts of Ni, Zr, and Ta.

Icosahedral clusters consisting of 13 atoms, 12 atoms with fivefold symmetry surrounding a central atom (Lee *et al.*, 2004) and interacting by a Lennard-Jones potential has an energy that is 8.4% lower than a close-packed arrangement (Frank, 1952). Although it is impossible to create a crystalline structure in which each atom has fivefold symmetry, randomly packed ISRO clusters of varying sizes may evolve continuously and be interconnected throughout the liquid (Jónsson and Anderson, 1988). As a result, a fractional presence of ISRO will reduce the average liquid energy and increase the mixing entropy relative to a hard-sphere-like liquid, thereby lowering the melting temperature. The concentration of ISRO clusters in transition metal melts can be expected to increase at elevated pressures due to the s–d electron transfer. Recently, measurements made in a laser-heated DAC were reported that extended the melting curve of Xe to 80 GPa and 3350 K (Ross *et al.*, 2005). The steep lowering of the melting slope (dT/dP) observed near 17 GPa and 2750 K were shown to be the result of hybridization of the p-like valence and d-like conduction states with the formation of ISRO clusters. Falconi *et al.* (2005) reported X-ray diffraction measurements in liquid Cs in which they observed a discontinuity in the liquid density above 3.9 GPa, accompanied by a structural change, "suggesting the existence of pressure-induced dsp^3 electronic hybridization."

The presence of local structures in the iron melt at elevated pressure are likely to increase its viscosity. Qualitative experimental observations we made of the melt motion using the laser-speckle method (see Section 18.2.2) may support this hypothesis. For all materials ranging from noble gases to alkali halides, oxides, and many metals, we observed that the vigor of motion strongly decreased with increasing pressure along, and even far above, the melting curve. Often, the disappearance of motion limited the pressure range of these melting measurements. The exact nature of this observed motion is not understood, but apart from temperature gradients, surface tension, and the strength of the pressure medium, it is likely related to the viscosity of the melt. The direct measurement of melt viscosities at very high pressures is probably impossible, but reporting qualitative observations may be helpful to support this hypothesis. At 'low' pressures up to about 50 GPa, we observe motions in liquid iron comparable to those observed in stirred light oil at ambient conditions (about 1 Pa s). The vigor of motion is certainly significantly less than that measured for water or molten iron at one atmosphere (10^{-3} Pa s) or that observed for molten water at several tens of gigapascal (Schwager *et al.*, 2004). Above 70 GPa, motion in liquid iron slows down significantly and is probably comparable to that of molten lava ($10^3 - 10^4$ Pa s). At 100 GPa, there is no evidence of motion, not even at temperatures exceeding the melting temperatures (measured by the change in reflectivity, see above) up to 1000 K. In the absence of further tests and in view of the complex nature of motion, however, it would be speculative to draw direct conclusions to the pressure dependence of iron melt viscosities.

Shen *et al.* (2004) have recently reported X-ray scattering data for liquid Fe measured in a DAC up to 58 GPa. They conclude that the structure factor preserves the same shape as that of a close-packed hard-sphere liquid over the full pressure range, and that this supports the view that the viscosity of liquid iron does not change from its ambient value. However, an inspection of their data (their **Figure 3**) shows that while the location of the first peak remains unchanged from 27 GPa to 58 GPa, the position of the second and higher peaks shift continuously to Q values lower than those for hard sphere, indicating pressure-induced non-hard-sphere-like liquid structure. This data needs to be further analyzed in terms of an alternative liquid ordering model.

18.2.6   Shock Melting

Phase transitions may be identified in experiments employing shock waves by the measurement of discontinuities; in sound velocities, temperature, and pressure–volume Hugoniot curves. Discontinuities in the $P-V$ Hugoniot, originating from melting, are generally too small to be reliable, except for very accurate data. Because of the loss of shear strength in the liquid, measurements of the sound velocity has proved to be the most useful method for detecting the pressure and density at melting. Since temperature measurements have not proven feasible for opaque materials, such as Fe (Yoo *et al.*, 1993), they need to be calculated. These calculations (Boness and Brown, 1990; Brown and McQueen, 1986) depend on estimates of the specific heat and Grüneisen parameter, which for iron lead to uncertainties of order ±500 K. Each of the diagnostic methods suffer from the possibility that the short

nanosecond timescales of the shock transit may cause overshoot of the equilibrium melt pressure resulting in an overestimate of the melting temperatures.

Brown and McQueen (1986) have reported two discontinuities in the sound velocity from measurements along the iron Hugoniot, one at 200 GPa (~4000 K) and a second at 243 GPa (~5500 K). The first was identified as the onset of a solid–solid transition, and the second as the onset of shock melting. Recently, however, Nguyen and Holmes (2004) have reported sound velocity measurements, which show only a single transition, interpreted as melting, with an onset at 225 GPa (5100 ± 500 K), a pressure lying between the two transitions of BM. These data are plotted in **Figure 8**.

Except for the region between about 200 GPa and 240 GPa, the shock melting data of Brown and McQueen and Nguyen and Holmes are in good agreement. A consistent picture can be drawn in which the solid melts at 200 GPa to a viscous melt, and the normal, less viscous liquid exists above 225–243 GPa. The corresponding calculated temperature at 200 GPa is 4350 ± 500 K (Boness and Brown, 1990), compared to 3850 ± 200 K on Boehler's melting curve. Thus, the two data sets would agree within their experimental uncertainties. We suggest that a plausible explanation

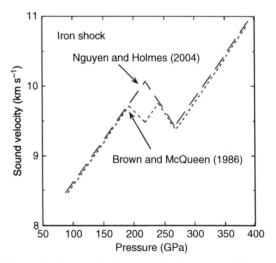

Figure 8 Sound velocities along the iron Hugoniot (Brown and McQueen, 1986; Nguyen and Holmes, 2004). BM (lower dashed line) have reported two discontinuities, one near 200 GPa (~4000 K) and a second at 243 GPa (~5500 K). The first was identified as the onset of solid–solid transition, and the second as the onset of shock melting. NH (upper dashed line) have reported only a single transition, interpreted as melting, with an onset at 225 GPa (5100 ± 500 K).

for the offset is due to an overshoot and superheating in the shock experiment.

In order to superheat a bulk solid, there must be a barrier to the nucleation of the melt. Appreciable bulk superheating of solids with shock waves has been observed in alkali halides (Boness and Brown, 1993; Swenson *et al.*, 1985) and fused quartz (Lyzenga *et al.*, 1983) and discussed by Luo (Luo and Ahrens, 2003) for metals. In the case of alkali halides and fused quartz, the overshoot is due to a need for a structural reorganization from the solid to liquid on a nanosecond timescale. A possible mechanism for an overshoot in iron is the presence of ISRO clusters in the melt providing a viscosity sufficiently high as to delay melting by about 10% in pressure, from 200 to 225 GPa. Brown (2001) has reported that an examination of a definitive set of Los Alamos Hugoniot data (Brown *et al.*, 2000) for iron, in the pressure range to 442 GPa, shows a small density-change discontinuity of about −0.7% at 200 GPa. Brown infers that this supports the conclusion that a phase of iron other than h.c.p. may be stable above 200 GPa.

18.2.7 Theoretical Calculations of the Iron Melting Curve

Calculations for iron melting have been made to 360 GPa (Alfé *et al.*, 2002, 2004a; Belonoshko *et al.*, 2000; Laio *et al.*, 2000). The results of these calculations were shown in **Figure 6**. The experimental DAC Fe melting curve is in agreement with the theoretical predictions of Laio *et al.* (2000), but those of Alfé *et al.* and Belonoshko *et al.* (Alfé *et al.*, 2002; Belonoshko *et al.*, 2000) are considerably higher in temperature and do not converge with the iron melting curve at 100 GPa.

Belonoshko *et al.* based their two-phase equilibrium calculations on model potentials fitted to quantum calculations of the Fe solid phases at 0 K, so there is no assurance that these potentials can reproduce correctly the physical properties of the solid or liquid melt at extreme conditions. A more rigorous method for calculating the properties, *ab initio* density functional theory molecular dynamics (DFT-MD), was employed by Alfé *et al.* (1999, 2002, 2004a) and Laio *et al.*, (2000). In the DFT-MD method, the total energy is calculated quantum mechanically, and the ions are moved in accordance with the calculated forces. An accurate simulation requires a large number of atoms and computer time. Because it is extremely slow and expensive to carry out MD DFT simulations of

melting on large systems, Alfé *et al.* employ a perturbation approximation to calculate free energy of the melt. The calculations start out with atoms interacting by an inverse power potential and switches continuously to the system of interest, in this case iron, over a reversible path. At each step along the trajectory a full simulation is made for the coupled system. Since the free energy of the inverse-power system is well known, the free energy of the system of interest can be determined by adding to it the change in free energy calculated along the path (de Konig *et al.*, 1999). While this method has proved to be very useful for many systems of interest, it raises the question of whether the adiabatic switching method can form ISRO clusters, or any other locally coordinated structure over a continuous reversible path, by starting from a simple inert gas-like liquid. In other words, is the method ergodic? Can it access all thermodynamically available states of the system in allowable time?

In an attempt to address these questions, we have compared the calculated melting curves for Fe (Alfé *et al.*, 2002; Laio *et al.*, 2000) and Cu (Vocadlo *et al.*, 2004) with the experimental melting curves (**Figure 9**). The calculations of Vocadlo *et al.* and Alfé *et al.*, for Cu and Fe, respectively, were made with essentially the same computer program using the adiabatic switching method. The Cu melting calculations of Vocadlo *et al.* (2004) are in agreement with the Cu experimental curve (Japel *et al.*, 2005),

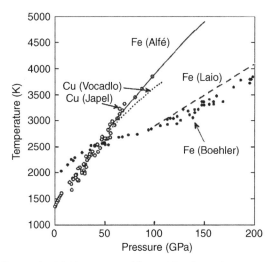

Figure 9 Melting curves of Cu and Fe. Open circles are experimental DAC data for Cu (Japel *et al.*, 2005), and solid dots are for Fe (Boehler, 1993). Also shown are theoretical melting curves for Cu (Vocadlo *et al.*, 2004), and two for Fe (Alfé *et al.*, 1999; Laio *et al.*, 2000).

but the Fe melting curve calculated by Alfé *et al.* also falls on the same Cu experimental curve. The adiabatic switching method predicts melting curves for Cu and Fe that are almost indistinguishable, in drastic contrast to the experimental evidence, using identical techniques for both metals. Since Cu has a closed d-band below an s-like valence band while Fe has a partially filled d-band at the Fermi surface, this suggests a failure of the switching method to properly account for the differentiating role of d-electrons in transition metal melting. It appears that in the case of iron, the final melt structure is biased by the starting configuration. Nevertheless, the adiabatic switching method remains useful for carefully chosen systems.

Laio *et al.*, by taking an approach similar to that of Aguado and Madden, employed a 'force-matching' procedure in which the interactions were calculated by the DFT method at a number of atomic configurations in the solid and liquid at a $P-T$ condition and fitted to a parameterized potential. The potential constructed this way has limited transferability, and will not be transferable to other $P-T$ conditions for which the process needs to be repeated. The melting temperature at a given $P_m - T_m$ point is calculated by a solid–liquid coexistence method. Laio *et al.*'s report is brief, and a full understanding of why the results agree with experiments is lacking. But the fact that the 'force-matching' procedure does address the local interactions suggests that this method may provide an improved approach to transition metals.

Since the solid structure is well understood, it is apparent from our discussion that uncertainties in the liquid modeling remain the more important problem. Given the wide range in the theoretical predictions for the melting curves for any geological material, it might be a useful exercise for theorists to first set their sights on developing improved liquid models before reporting new sets of melting calculations. Traditionally, experimental data have been used to validate theoretical models.

18.3   Viscosity of Iron at Earth Core Conditions

Thirty years ago, Jacobs (1975) made the frequently cited statement: "The viscosity of the Earth's core is one of the least well known physical parameters of the Earth. Estimates in the literature of the kinematic viscosity differ by many orders of magnitude, ranging from 10^{-3} to 10^{11} cm^2/sec." These predicted viscosities range from values measured for liquid iron at

1 atm (similar to water, about 10^{-3} Pa s) to those of glass-like liquids (like pitch, 10^{10} Pa s). There remains even today a broad consensus that the region of the outer core, from 135 to 330 GPa, suffers from a very high unexplained viscosity (Anderson, 1980; Palmer and Smylie, 1999; Stevenson, 1983; Smylie, 1999). In order to reconcile the high core viscosities derived from various seismic observations with low liquid iron viscosities, Stevenson suggested a two-phase suspension model (solid iron or oxide particles in liquid iron) and obtained high viscosities of up to 10^{11} Pa s. Such a view appears consistent with a melt containing randomly packed ISRO clusters in liquid iron. Clearly, more experimental work on liquid structures are necessary.

Theoretical estimates of the viscosity of molten iron at core conditions were done using several unknown thermodynamic parameters such as the activation energy (Poirier, 1988), and the predicted low viscosities have been recently supported by MD simulations (Wijs *et al.*, 1998). There are, however, qualitative trends indicating an increasing viscosity of liquid iron with increasing pressure. At low pressure, there have been a number of attempts, for example, by LeBlanc and Secco (1996), to measure the viscosity of liquid iron–sulfur alloys at high pressure to a few gigapascals. A theoretical analysis of grain-size measurements of quenched iron liquid to 10 GPa is also indicative of a strong pressure dependence of the viscosity (Brazhkin and Lyapin, 2000). These measurements are probably not representative for core conditions, but they support the generally observed increase of viscosity with pressure such as Bridgman's early experiment on organic liquids, measurements for the noble gases (Trappeniers *et al.*, 1980), and our own observations for liquid iron (see above).

The shock melting and laser-heated DAC measurements near 200–243 GPa and 4000–5000 K, while falling short of the conditions of the IOCB, now represent the only experimental results near that boundary. From our qualitative observations of the speckle motion of the iron melt in the laser-heated DAC, one may conclude that the viscosity of molten iron does not resemble that of water or liquid iron at 1 atm (10^{-3} Pa s) commonly used by dynamo theorists, but is orders of magnitude greater. Moreover, a further investigation of the reasons for the discrepancies between the DAC and shock melting temperature, and between the two sets of shock measurements, may have the potential for providing new information for estimating the viscosity of molten iron at core conditions. The very large uncertainty in the prediction of the viscosity of liquid iron at outer-core conditions requires a new understanding of the atomic ordering in the melt.

Acknowledgment

Work by Marvin Ross was partially supported by the University of California under the auspices of the US DOE by the Lawrence Livermore National Laboratory.

References

Aguado A and Madden PA (2005) New insights into the melting behavior of MgO from molecular dynamics simulations: The importance of premelting effects. *Physical Review Letters* 94: 1–4, 068501.

Ahrens T, Holland KG, and Chen G (2002) Phase diagram of iron, revised-core temperatures. *Geophysical Research Letters* 29: 54-1–54-4.

Akins JA, Luo S, Asimov PD, and Ahrens T (2004) Shock-induced melting of MgSiO$_3$ perovskite and implications for melts in the Earth's lowermost mantle. *Geophysical Research Letters* 31: 1–4, L14612.

Alfé D (2005) Melting curve of MgO from first principles calculations. *Physical Review Letters* 94: 1–4, 235701.

Alfé D, Gillan MJ, and Price GD (1999) The melting curve of iron at the pressures of the Earth's core from *ab initio* calculations. *Nature* 401: 462–464.

Alfé D, Price GD, and Gillan MJ (2002) Iron under Earth's core conditions: Liquid-state thermodynamics and high-pressure melting curve from *ab initio* calculations. *Physical Review B* 65: 165118-1/165118-11.

Alfé D, Price GD, and Gillan MJ (2004a) The melting curve of iron from quantum machanics calculations. *Journal of Physics and Chemistry of Solids* 65: 1573–1580.

Alfé D, Vocadlo L, Price GD, and Gillan MJ (2004b) Melting curve of materials: Theory versus experiments. *Journal of Physics: Condensed Matter* 16: S973–S982.

Anderson DL (1980) Bulk attenuation in the Earth and viscosity of the core. *Nature* 285: 204–207.

Belonoshko AB, Ahuja R, and Johansson B (2000) Quasi-*ab initio* molecular dynamic study of Fe melting. *Physical Review Letters* 84: 3638–3641.

Belonoshko AB and Dubrovinsky LS (1996) Molecular dynamics of NaCl (B1 and B2) and MgO (B1) melting: Two-phase simulation. *American Mineralogist* 81: 303–316.

Belonoshko AB, Skorodumova NV, Rosengren A, *et al.* (2005) High-pressure melting of MgSiO$_3$. *Physical Review Letters* 94: 1–4, 195701.

Boehler R (1986) The phase diagram of iron to 430 kbar. *Geophysical Research Letters* 13: 1153–1156.

Boehler R (1993) Temperatures in the Earth's core from melting-point measurements of iron at high static pressures. *Nature* 363: 534–536.

Boehler R (2000) High-pressure experiments and the phase diagram of lower mantle and core materials. *Reviews of Geophysics* 38: 221–245.

Boehler R and Errandonea D (2002) The laser-heated diamond cell: High *P–T* phase diagrams. In: Hemley RJ, Chiarotti GL, Bernasconi M, and Ulivi L (eds.) *Enrico Fermi Course*, pp. 55–72. Varenna, Italy: IOS Press.

Boehler R and Ross M (1997) Melting curve of aluminum in a diamond cell to 0.8 Mbar: Implications for iron. *Earth and Planetary Science Letters* 153: 223–227.

Boehler R, Ross M, and Boercker DB (1996) High-pressure melting curves of alkali halides. *Physical Review B* 53: 556–563.

Boehler R, Ross M, and Boercker DB (1997) Melting of LiF and NaCl to 1 Mbar: Systematics of ionic solids at extreme conditions. *Physical Review Letters* 78: 4589–4592.

Boehler R, Ross M, Söderlind P, and Boercker DB (2001) High pressure melting curves of argon, krypton, and xenon: Deviation from corresponding states theory. *Physical Review Letters* 86: 5731–5734.

Boehler R, von Bargen N, and Chopelas A (1990) Melting, thermal expansion, and phase transitions of iron at high pressures. *Journal of Geophysical Research* 95: 21731–21736.

Boness DA and Brown JM (1990) The electronic band structures of iron, sulfur, and oxygen at high pressure and the Earth's core. *Journal of Geophysical Research* 95: 21721–21730.

Boness DA and Brown JM (1993) Bulk superheating of solid KBr and CsBr with shock waves. *Physical Review Letters* 71: 2931–2934.

Brazhkin VV and Lyapin AG (2000) Universal viscosity growth in metallic melts at megabar pressures: The vitreous state of the Earth's inner core. *Physics-Uspekhi* 43: 493–508.

Brown JM (2001) The equation of state of iron to 450 GPa: Another high pressure solid phase? *Geophysical Research Letters* 28: 4339–4342.

Brown JM, Fritz JN, and Hixson RS (2000) Hugoniot data for iron. *Journal of Applied Physics* 88: 5496–5498.

Brown JM and McQueen RG (1986) Phase transitions, Grüneisen parameter and elasticity for shocked iron between 77 GPa and 400 GPa. *Journal of Geophysical Research* 91: 7485–7494.

Chudinovskikh L and Boehler R (2003) MgSiO$_3$ phase boundaries measured in the laser-heated diamond cell. *Earth and Planetary Science Letters* 219: 285–296.

de Konig M, Antonelli A, and Yip S (1999) Optimized free-energy evaluation using a single reversible-scaling simulation. *Physical Review Letters* 83: 3973–3976.

Duffy TS, Hemley RJ, and Mao HK (1995) Equation of state and shear strength at multimegabar pressures: MgO to 227 GPa. *Physical Review Letters* 74: 1371–1374.

Errandonea D, Schwager B, Ditz RCG, Boehler R, and Ross R (2001) Systematics of transition metal melting. *Physical Review B* 63: 1–4 132104.

Errandonea D, Somayazulu M, Häusermann D, and Mao HK (2003) Melting of tantalum at high pressure determined by angle dispersive X-ray diffraction in a double-sided laser-heated diamond-anvil cell. *Journal of Physics: Condensed Matter* 15: 7635–7649.

Falconi S, Lundegaard LF, Hejny C, and McMahan AK (2005) X-ray diffraction study of liquid Cs up to 9.8 GPa. *Physical Review Letters* 94: 1–4, 125507.

Frank FC (1952) Super cooling of liquids. *Proceedings of the Royal Society of London, Series A* 215: 43–46.

Holland KG and Ahrens TJ (1997) Melting of (Mg, Fe)$_2$SiO$_4$ at the core–mantle boundary of the Earth. *Science* 275: 1623–1625.

Jacobs JA (1975) *The Earth's Core*. London: Academic Press.

Jakse N and Pasturel A (2004) *Ab initio* molecular dynamics simulations of local structure of supercooled Ni. *Journal of Chemical Physics* 120: 6124–6127.

Japel S, Schwager B, Boehler R, and Ross M (2005) Melting of copper and nickel at high pressure; the role of d-electrons. *Physical Review Letters* 95: 167801–167804.

Jephcoat AP and Besedin SP (1996) Temperature measurement and melting determination in the laser-heated diamond-anvil cell. In: Jephcoat AP, Angel RJ, and O'Nions RK (eds.) *Philosophical Transactions: Development in High-Pressure, High-Temperature Research and the Study of the Earth's Deep Interior*, pp. 1333–1360. London: Royal Society of London.

Jónsson H and Anderson HC (1988) Icosahedral ordering in the Lennard-Jones liquid and glass. *Physical Review Letters* 60: 2295–2298.

Kubo T, Ito E, Katsura T, Shinmei T, and Yamada H (2003) *In situ* X-ray observation of iron using Kawai-type apparatus equipped with sintered diamond: Absence of β phase up to 44 GPa and 2100 K. *Geophysical Research Letters* 30(26): 1–4.

Lacks DJ and Gordon RG (1993) Crystal-structure calculations with disordered ions. *Physical Review B* 48: 2889–2908.

Laio A, Bernard GL, Chiarotti GL, Scandolo S, and Tosatti E (2000) Physics of iron at Earth's core conditions. *Science* 287: 1027–1030.

LeBlanc GE and Secco RA (1996) Viscosity of an Fe–S liquid up to 1300°C and 5 GPa. *Geophysical Research Letters* 23: 213–216.

Lee GW, Gangopadhyay AK, Kelton KF, *et al.* (2004) Difference in icosohedral short-range order in early and late transition liquids. *Physical Review Letters* 93: 037802-1–037802-4.

Luo S and Ahrens T (2003) Superheating systematics of crystalline solids. *Applied Physics Letters* 82: 1836–1838.

Luo SN, Akins JA, Ahrens T, and Asimov PD (2004) Shock-compressed MgSiO$_3$ glass, enstatite, olivine, and quartz: Optical emission, temperatures, and melting. *Journal of Geophysical Research* 109: 1–14, B05205.

Lyzenga GA, Ahrens TJ, and Mitchell AC (1983) Shock temperatures of SiO$_2$ and their geophysical implications. *Journal of Geophysical Research* 88: 2431–2444.

Mao HK, Bell PM, and Hadidiacos C (1987) Experimental phase relations of iron to 360 kbar, 1400°C. Determined in an internally heated diamond anvil apparatus. In: Manghnani MH and Syono Y (eds.) *High Pressure Research in Mineral Physics, Geophysical Monograph Series*, pp. 135–138. Washington, DC: AGU.

Mao HK, Wu Y, Chen LC, Shu JF, and Jephcoat AP (1990) Static compression of iron to 300 GPa and Fe$_{0.8}$Ni$_{0.2}$ alloy to 260 GPa: Implications for composition of the core. *Journal of Geophysical Research* 95: 21737–21742.

Murakami MK, Hirose K, and Kavamura K (2004) Post-perovskite phase transition in MgSiO$_3$. *Science* 304: 855–858.

Nguyen JH and Holmes NC (2004) Melting of iron at the physical conditions of the Earth's core. *Nature* 427: 339–342.

Palmer A and Smylie DE (1999) VLBI onservations of free nutations and viscosity at the top of the core. *Physics of the Earth and Planetary Intereriors* 148: 285–301.

Poirier JP (1988) Transport properties of liquid metals and viscosity of the Earth's core. *Geophysical Journal* 92: 99–105.

Ross M, Boehler R, and Söderlind P (2005) Xenon melting curve to 80 GPa and 5p–d hybridization. *Physical Review Letters* 95: 257801–257804.

Ross M, Yang LH, and Boehler R (2004) Melting of aluminum, molybdenum and the light actinides. *Physical Review B* 70: 184112-1–184112-8.

Saxena SK, Shen G, and Lazor P (1993) Experimental evidence for a new iron phase and implications for the Earth's core. *Science* 260: 1312–1314.

Saxena SK, Shen G, and Lazor P (1994) Temperatures in Earth's core based on melting and phase transformation experiments on iron. *Science* 264: 405–407.

Schenk T, Holland-Moritz D, Simonet V, Bellissent R, and Herlach DM (2002) Icosahedral short-range order in deeply undercooled metallic melts. *Physical Review Letters* 89: 075507-1–075507-4.

Schwager B, Chudinovskikh L, Gavriliuk AG, and Boehler R (2004) Melting curve of H$_2$O to 90 GPa measured in a laser-heated diamond cell. *Journal of Physics: Condensed Matter* 16: S1177–S1179.

Shen G and Lazor P (1995) Measurement of melting temperatures of some minerals under lower mantle pressures. *Journal of Geophysical Research* 100: 17699–17713.

Shen GY, Mao HK, Hemley RJ, Duffy TS, and Rivers ML (1998) Melting and crystal structure of iron at high pressures and temperatures. *Geophysical Research Letters* 25: 373–376.

Shen GY, Prakapenka VB, River ML, and Sutton SR (2004) Structure of liquid iron at pressures up to 58 GPa. *Physical Review Letters* 92: 1–4, 185701.

Smylie DE (1999) Viscosity near Earth's inner core. *Science* 284: 461–463.

Stevenson DJ (1983) Anomalous bulk viscosity of two-phase fluids and implications for planetary interiors. *Journal of Geophysical Resarch* 88: 2445–2455.

Strachan T, Cagin T, and Goddard WA, III (1999) Phase diagram of MgO from density-functional theory and molecular dynamics simulations. *Physical Review B* 60: 15084–15093.

Sweeney JS and Heinz DL (1998) Laser-heating through a diamond-anvil cell: Melting at high pressures. In: Manghnani MYT (ed.) *Properties of Earth and Planetary Materials at High Pressure and High Temperature, Geophysical Monograph Series*. Washingron, DC: AGU.

Swenson CA, Shaner J, and Brown JM (1985) Hugoniot overtake sound velocity measurements on CsI. *Physical Review B* 34: 1457–1463.

Trappeniers NJ, Van Der Gulik PS, and Van Den Hooff H (1980) The viscosity of argon at very high pressure, up to the melting line. *Chemical Physics Letters* 70: 438–443.

Vidal-Valet JP, Vidal JP, and Kurki-Suonio K (1978) X-ray study of the atomic charge densities in MgO, CaO and BaO. *Acta Crystallographica A* 34: 594–602.

Vocadlo L, Alfé D, Price GD, and Gillan MJ (2004) *Ab initio* melting curve of copper by the phase coexistence approach. *Journal of Chemical Physics* 120: 2872–2878.

Vocadlo L and Price GD (1996) The melting of MgO – Computer calculations via molecular dynamics. *Physics and Chemistry of Minerals* 23: 42–49.

Watson RE (1958) Analytic Hartree–Fock solutions for O^{-2}. *Physical Review* 111: 1108–1110.

Wijs GA, Kresse G, Vocadlo L, *et al.* (1998) The viscosity of liquid iron at the physical conditions of the Earth's core. *Nature* 23: 805–807.

Williams Q, Jeanloz R, Bass J, Svendsen B, and Ahrens TJ (1987) The melting curve of iron to 2.5 Mbar: A constraint on the temperature of the Earth's core. *Science* 236: 181–182.

Yoo CS, Holmes NC, Ross M, Webb DJ, and Pike C (1993) Shock temperatures and melting of iron at Earth core conditions. *Physical Review Letters* 70: 3931–3934.

Yoo CS, Söderlind P, Moriarty JA, and Cambell AJ (1996) dhcp as a possible new ε' phase of iron at high pressures and temperatures. *Physical Letters A* 214: 65–70.

Zerr A and Boehler R (1993) Melting of (Mg, Fe)SiO$_3$-perovskite to 625 kbar: Indication of a high melting temperature in the lower mantle. *Science* 262: 553–555.

Zerr A and Boehler R (1994) Constraints on the melting temperature of the lower mantle from high pressure experiments on MgO and magnesiowüstite. *Nature* 371: 506–508.

Zerr A, Diegeler A, and Boehler R (1998) Solidus of the Earth's deep mantle. *Science* 281: 243–245.

Zerr A, Serghiou G, and Boehler R (1997) Melting of CaSiO$_3$-perovskite to 430 kbar and first *in-situ* measurements of lower mantle eutectic temperatures. *Geophysical Research Letters* 24: 909–912.

Zha CS, Boehler R, Young DA, and Ross M (1986) The argon melting curve to very high pressures. *Journal of Chemical Physics* 85: 1034–1036.

Zha CS, Mao HK, and Hemley RJ (2000) Elasticity of MgO and a primary pressure scale to 55 GPa. *Proceedings of the National Academy of Sciences* 97: 13494–13499.

19 Properties of Rocks and Minerals – Thermal Conductivity of the Earth

A. M. Hofmeister and J. M. Branlund, Washington University, St. Louis, MO, USA

M. Pertermann, Rice University, Houston, TX, USA

19.1	Introduction	544
19.1.1	The Importance of Thermal Conductivity to Geophysics	544
19.1.2	Types of Thermal Transport and Justification for Omitting Metals	545
19.1.3	History of Mineral Physics Efforts and the Current State of Affairs	546
19.1.4	Scope of the Present Chapter	546
19.2	Theory of Heat Flow in Electrically Insulating Solid Matter	547
19.2.1	Phonon Scattering	547
19.2.1.1	Acoustic models of lattice heat transport	547
19.2.1.2	The damped harmonic oscillator (DHO) – phonon gas model	549
19.2.1.3	Bulk sound model for pressure derivatives	550
19.2.2	Radiative Transport in Partially Transparent Materials (Insulators)	550
19.2.2.1	Distinguishing direct from diffusive radiative transport on the basis of frequency-dependent attenuation	550
19.2.2.2	Spectroscopic models for diffusive radiative transport inside the Earth	552
19.2.2.3	Understanding the general behavior of $k_{rad,dif}$ from asymptotic limits	554
19.2.2.4	An approximate formula that connects k_{rad} with concentration	554
19.3	Experimental Methods for the Lattice Contribution	555
19.3.1	Conventional Techniques Involving Multiple Physical Contacts	555
19.3.2	Methods Using a Single Physical Contact	556
19.3.3	Contact-Free, Laser-Flash Analysis	557
19.3.4	Additional Contact-Free (Optical) Techniques	559
19.4	The Database on Lattice Transport for Mantle Materials	559
19.4.1	Evaluation of Methodologies, Based Primarily on Results for Olivine	559
19.4.1.1	Ambient conditions	559
19.4.1.2	Elevated temperature	564
19.4.1.3	Elevated pressure	566
19.4.2	Laser-Flash Data on Various Minerals	566
19.4.2.1	Effect of chemical composition and hydration on room temperature values	567
19.4.2.2	Comparison of $D(T)$ for dense oxides and silicates	568
19.4.3	Comparison of the Room Temperature Lattice Contribution to Theoretical Models and Estimation of D and k for Some High-Pressure Phases	568
19.4.3.1	Ambient conditions and compositional dependence	569
19.4.3.2	Elevated temperature	570
19.4.3.3	Elevated pressure	571
19.4.4	Lattice Thermal Conductivity and Its Temperature Dependence	571
19.5	Calculation of the Effective Thermal Conductivity for Diffusive Radiative Transfer	572
19.6	Conclusions	573
References		573

Nomenclature

a	lattice constant	PTGS	picosecond transient grating spectroscopy
A	absorption coefficient	q	wave vector
$B, B_1, b, b_1 \ldots$	fitting parameters	Q	contact resistance
c	speed of light	r	radiative contribution
C, c	heat capacity	R	reflectivity
d	thickness or grain-size	s	$hc\nu_{ave}/k_B$
dif,rad	diffusive and radiative	S	source function
D	thermal diffusivity	t	time
DHO	damped harmonic oscillator	T	temperature
f	extinction coefficient	u	group velocity (sound speed)
F	flux	V	volume
FWHM	full width at half maximum	X	concentration
G	shear modulus	Z	formula units in the primitive cell
h	Planck's constant	α	thermal expansivity
H	heat source term	γ	Grüneisen parameter
i,j,k,m	integer indices or exponents	δ	Anderson–Grüneisen parameter
I	intensity	θ	Debye temperature
k	thermal conductivity	ϑ	temperature function
k_B	Boltzmann's constant	λ	wavelength
K	bulk modulus	Λ	mean free path
lat	lattice	ν	frequency
L	grid spacing	ξ	emissivity
LFA	laser flash analysis	ρ	density
M	molar formula weight	σ	Stephan–Boltzmann constant
n	index of refraction	τ	lifetime
n,x,y,z	scalar directions	φ	angle
P	pressure	ω	circular frequency

19.1   Introduction

19.1.1   The Importance of Thermal Conductivity to Geophysics

Fourier (*c.* 1820) recognized that the geothermal gradient could be used to estimate the Earth's age. His equation links thermal conductivity (k) and temperature (T) to the flux:

$$F = -k\,\partial T/\partial n \qquad [1]$$

where the length-scale n is normal to the surface. To solve eqn [1], Kelvin (*c.* 1864) treated the Earth as a one-dimensional (1-D), semi-infinite solid; his age of 100 My for the Earth was accepted until the discovery of radioactivity in the early 1900s. Kelvin's solution is still used to model global power (flux times surface area) from measurements of oceanic heat flux (Pollack *et al.*, 1993; discussed by Hofmeister and Criss 2005a, 2005b, 2006).

The surface flux is one clue to Earth's thermal state. Probing the workings of Earth's interior and its thermal evolution also requires detailed knowledge of k or of thermal diffusivity:

$$D = \frac{k}{\rho C_P} \qquad [2]$$

where ρ is density and C_P is heat capacity. Temperature derivatives of D or k are crucial, which is revealed by examining the equation for heat conduction in an isotropic medium:

$$\rho C_P \frac{\partial T}{\partial t} = k\nabla^2 T + H$$
$$+ \frac{\partial k}{\partial T}\left[\left(\frac{\partial T}{\partial x}\right)^2 + \left(\frac{\partial T}{\partial y}\right)^2 + \left(\frac{\partial T}{\partial z}\right)^2\right] \qquad [3]$$

where t is time and H describes heat sources such as radioactivity (Carslaw and Jaeger, 1959). The specific forms and values of $k(T)$, or $D(T)$, strongly influence the temperature distribution obtained in conductive cooling models due to the nonlinear nature of eqn [3]. Examples include warming of subducting slabs, stability of olivine near the transition zone, occurrence of deep earthquakes (Hauck *et al.*, 1999; Branlund *et al.*, 2000), buckling of the lithosphere (Gerbault, 2000), the geotherm (Hofmeister, 1999), and lithospheric cooling (Honda and Yuen, 2001, 2004).

Thermal transport properties play a crucial role in mantle convection, as this phenomenon results from competition between diffusion of heat (thermal conductivity), resistance to motion (viscosity), and buoyancy forces. Variable thermal conductivity has mostly been neglected in modern geodynamic studies, in view of the much stronger, exponential dependence of viscosity on temperature (e.g., Tackley, 1996), with a few exceptions (e.g., Yuen and Zhang, 1989). However, even regimes considered to be dominated by viscosity, as at high Rayleigh numbers, are impacted by variable k (Dubuffet *et al.*, 2000). Numerical convection models of the Earth have shown that the functional form for $k(T)$ strongly influences the character of the solutions through feedback in the temperature equation. Feedback occurs because k is a coefficient inside the differential operator of the temperature equation, which is nonlinear, much like eqn [3] for conduction. The effect of variable k on mantle convection is not overridden by simultaneously and strongly varying viscosity (van den Berg *et al.*, 2004, 2005; Yanagawa *et al.*, 2005). Small changes in k with T are important because (1) the time-dependence of an infinite Prandtl number fluid such as the mantle is primarily governed by the temperature equation, and (2) the temperature equation functions as a 'master' equation, whereas the momentum equation, which is governed by viscosity, responds instantaneously as the 'slave' equation (Haken, 1977; Hofmeister and Yuen (in press)). Consequently, accurate characterization of k (or D) at temperature is essential in modeling planetary heat flow on any scale (e.g., Starin *et al.*, 2000; Yuen *et al.*, 2000; Schott *et al.*, 2001; Dubuffet *et al.*, 2002).

19.1.2 Types of Thermal Transport and Justification for Omitting Metals

The functional dependence of k (or D) on T is regulated by processes operating on microscopic to mesoscopic scales (**Figure 1**). In the Earth, heat transfer is diffusive, as assumed in geodynamic models. Transport of heat by scattering of 'phonons'

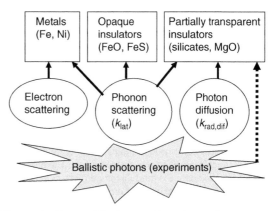

Figure 1 Types of Earth materials and associated mechanisms of heat transport. Ovals indicate mechanisms operating in the mantle. Starburst indicates mechanisms present only in the laboratory.

(quantized lattice vibrations) within each individual mineral grain is termed lattice conductivity (k_{lat}). For electrical insulators such as mantle minerals, heat is also diffused radiatively by 'photons' and represented by an effective thermal conductivity, $k_{rad,dif}$, calculated from spectra. The process involves grain-to-grain progressive absorption and re-emission of photons down a shallow temperature gradient in the medium (**Figure 2(a)**). However, laboratory experiments with steep temperature gradients involve a different kind of radiative transfer, termed boundary-to-boundary, direct, or ballistic (**Figure 2(b)**), wherein light from the source warms the thermocouple with negligible or scant participation of the medium (e.g., Lee and Kingery, 1960; Kanamori *et al.*, 1968; Hofmeister, 2005). Confusion of direct with diffusive processes led to misinterpretation of experimental data on minerals used in geophysical applications, starting with the efforts of Birch and Clark (1940) and continuing today (cf. Gibert *et al.*, 2005; Pertermann and Hofmeister, 2006).

Geophysical studies concerning heat transport in the metallic core utilize k obtained from electrical conductivity (σ) using Drude's classical model (e.g., Anderson, 1998), which assumes that heat is transported through collisions involving electrons. However, the Wiedemann–Franz ratio ($k/\sigma T$) depends on temperature, which is unexpected (Burns, 1990), and measurements of k for Fe and Ni metals in particular do not follow the pressure dependence predicted for electronic transport of heat (e.g., Sundqvist, 1981). To accurately predict heat transport at extremely high P and T of the core requires experimental measurements of $k(P, T)$

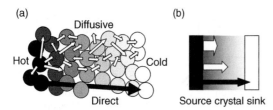

Figure 2 Schematics of photon transport mechanisms. (a) Diffusive vs direct radiative transfer in an internally heated medium comprised of grains. In a low temperature gradient, each grain is effectively isothermal. The shades of the grains indicate the gradual temperature change. Diffusion (white arrows) involves emissions of photons by a hot grain, which are absorbed by nearby warm grains, which emit light in accord with their cooler temperatures. Each grain is both an emitter and a receiver. Black arrows show direct transfer of a photon from a hot to a cold grain; here, no interaction with the intervening grains occurs. (b) Mixed direct and diffusive radiative transfer in laboratory experiments. Shading indicates the temperature gradient across a single crystal or glass. Heat from photons diffuses into the medium because the intensity of the light is increasingly attenuated with distance, indicated by the inverse correlation of arrow width and length. Direct transfer occurs when some amount of light traverses the entire sample, warming the sink (sensor). (a) Reprinted from Hofmeister AM (2005) The dependence of radiative transfer on grain-size, temperature, and pressure: Implications for mantle processes. *Journal of Geodynamics* 40: 51–72.

for diverse metals and improvements in both theoretical models. Therefore, this chapter does not further discuss metals.

19.1.3 History of Mineral Physics Efforts and the Current State of Affairs

Earth materials were probed in the pioneering experiments on the temperature (Eucken, 1911) and pressure response of thermal conductivity (Bridgeman, 1924). The seminal study of Birch and Clark (1940) found that k depends weakly on chemical composition, and increases with T, even though only 400°C was reached. Based on the interpretation that such increases represent strong diffusive radiative transfer, theoretical arguments that $k_{rad,dif}$ goes at T^3 (Kellett, 1952), inference of high interior temperatures for the Earth, and thermal modeling, Lubimova (1958) and MacDonald (1959) concluded that radiative transfer is a key process in Earth's mantle. Interestingly, Preston (1956) pointed out that grain boundaries and ionic Fe content could limit the efficacy of radiative transfer. Evidence for plate tectonics led to the inference that mantle

convection provides an efficient means for expulsing heat from the Earth, which made it unneccessary to call on the presence of strong radiative transfer to prevent an overly hot interior. Consequently, detailed calculations of radiative transfer for mantle minerals were limited to olivine (e.g., Shankland *et al.*, 1979). Only recently have issues of high interior radiative transfer and the effect of grain size been revisited (e.g., Hofmeister, 1999).

Measurement and calculations in geoscience from *c.* 1960–1980 focused on the lattice contribution paralleling theoretical and experimental efforts in materials science (e.g., Klemens, 1958; Slack, 1979). Few studies of deep-Earth materials were performed during 1980–2000 largely due to the view that viscosity has overriding importance in geodynamics (Tozer, 1965). However, lithospheric processes and materials have been a continued focus (summarized by Clauser and Huenges (1995); also see Seipold (1998) and Vosteen and Schellschmidt (2003)).

Mineral physics and geodynamics research into thermal transport have recently been stimulated by the spectroscopic model for both components of $k(T)$ of Hofmeister (1999) and its implementation in geodynamic models mainly by D. A. Yuen and colleagues. Concurrent advances made in the accurate measurement of D using the contact-free, laser-flash technique that dominates research in materials science and engineering (Degiovanni *et al.*, 1994; Mehling *et al.*, 1998) have only recently been transferred to Earth science (Büttner *et al.*, 1998; Hofmeister 2004b, 2006; Pertermann and Hofmeister, 2006). Methods commonly used involve contact with thermocouples (e.g., Xu *et al.*, 2004; Osako *et al.*, 2004; Gibert *et al.*, 2005) and are subject to unwanted and irrelevant direct radiative transfer, as well as to systematic underestimation of k due to contact resistance at interfaces (Hofmeister, 2007). Inconsistencies exist even for mantle olivine, which has been probed more than any other mineral (see Gibert *et al.* (2003)).

19.1.4 Scope of the Present Chapter

To address the inconsistencies in the existing database, this report relies on the materials science literature, and clarifies conditions needed for radiative transfer to be diffusive. A primary goal is to ascertain conditions under which contact measurements of k or D, which constitute the bulk of experimental determinations, are trustworthy. Single crystals are the focus because (1) pressures and temperatures inside the Earth should produce well-annealed grain boundaries, minimizing resistance at interfaces, and thus single-crystal data should represent the mantle, (2) data on polycrystals (rocks) will be affected by grain boundary behavior, for example, the T dependence could be altered by differential thermal expansion, whereas the P dependence could be affected by compression of pore space, (3) understanding heat transport in single crystals is a precursor to deciphering more complicated behavior of agglomerates, and (4) problems associated with single-crystal measurements are likely present in studies of rocks, but are difficult to recognize. The unfortunate conclusion is that results from contact methods for geophysically important silicates and oxides systematically underestimate k or D at low T due to thermal resistance at interfaces, and overestimate k at high T due to spurious radiative transfer. Problems with the measurements have limited understanding of microscopic behavior, necessitating evaluation of theoretical models of k_{lat} and $k_{rad,dif}$. Effects of P and T are the focus because the dependence on chemical composition and phase are implicit in the physical properties which enter into the formulae. Spectral data for lower-mantle phases are insufficient to constrain $k_{rad,dif}$, so estimates are provided.

19.2 Theory of Heat Flow in Electrically Insulating Solid Matter

19.2.1 Phonon Scattering

Debye's (1914) analogy of the scattering of phonons to collisions of molecules in a gas forms the basis of all models of heat transport in insulators. Peierls (1929) and Klemens (1958) related lattice thermal conductivity to summations of the form

$$k_{lat} = \frac{1}{3} \frac{\rho}{ZM} \sum_{j=1}^{3} \sum_{i=1}^{3NZ} c_{ij} u_{ij}^2 \tau_i \qquad [4]$$

where is M is the molar formula weight, Z is the number of formula units in the primitive unit cell, u_{ij} is the group velocity ($= d\omega_i/dq_j$), q_j is the wave vector, $\omega_i = 2\pi\nu_i$ is the circular frequency of a given mode, ν_i is frequency, τ_i is the mean free lifetime, $i = 1$ to $3NZ$ sums the normal modes of a crystal with N atoms in the formula unit, and $j = 1$ to 3 sums the

three orthogonal directions. Also, the Einstein heat capacity of each vibrational mode

$$c_i = \left(\frac{h\nu_i}{k_B T}\right)^2 \exp\left(\frac{h\nu_i}{k_B T}\right) \Big/ \left[1 - \exp\left(\frac{h\nu_i}{k_B T}\right)\right]^2 \qquad [5]$$

is calculated per mole at constant volume, where h is Planck's constant, and k_B is Boltzmann's constant. All such models are approximate, including those obtaining parameters from quantum mechanical formulations, because the underlying, mean free gas theory of Claussius is inexact, for example, even the factor of 3 in eqn [4] is approximate (Reif, 1965).

Given the success of Debye's model for C_V, parameters for acoustic modes have generally been used to calculate k_{lat}, and optic modes were considered to play a minor part (e.g., Slack, 1979). Several problems exist: (1) Near-zero speeds have been the reason provided for neglecting optic modes, but slow speeds exist only at the center and edges of the Brillouin zone and at the barycenter of the peak in frequency space (e.g., Burns, 1990; Hofmeister, 2001). Speeds of acoustic modes are also near zero at the Brillouin zone edge, and thus acoustic modes have similar limitations in transporting heat. (2) Most importantly, lifetimes are roughly estimated and are generally assumed to go as $1/T$ (Ziman, 1962; Roufosse and Klemens, 1973). An alternative model was developed to quantify lifetimes from spectra of optic modes (e.g., Hofmeister, 1999). The following subsections summarize the forms predicted for k_{lat} and D.

19.2.1.1 Acoustic models of lattice heat transport

By assuming three phonon collisions and that the mean free path goes as $1/T$ above the Debye temperature (θ), Peierls (1929) reproduced Eucken's (1911) empirical law:

$$k_{lat} = B/T \quad \text{for} \quad T > \theta \qquad [6]$$

Dugdale and MacDonald (1955) estimated the fitting parameter B in eqn [6] using dimensional analysis:

$$k_{lat} = \frac{C_V}{3\alpha} \frac{u}{\gamma_{th}} \frac{a}{T} = \frac{VK_T}{3} \frac{u}{\gamma_{th}^2} \frac{a}{T} \qquad [6a]$$

where $\gamma_{th} = \alpha V K_T/C_V$ is the thermal Gruneisen parameter, α is thermal expansivity, K_T is the bulk modulus, V is molar volume, u is a sound speed, and a^3 is the volume of the unit cell. Models based on Peierls' work give different values of B. Julian's (1965) correction of Liebfried and Schlömann's (1954) work provides

$$k_{lat} = \frac{24}{20} \frac{4^{1/3}}{\gamma_{th}^2} \left(\frac{k_B \theta}{h}\right)^3 \frac{ZMa}{T} \qquad [6b]$$

which differs by a factor of ~ 7 from that of Roufosse and Klemens (1973):

$$k_{lat} = \frac{1}{2^{1/6} \pi^{1/3}} \frac{3^{1/3}}{\gamma_{th}^2} \left(\frac{k_B \theta}{h}\right)^3 \frac{ZMa}{T} \qquad [6c]$$

The prefactors in eqn [6b] and [6c] pertain to $T > \theta$, which is ~ 700 K for mantle minerals.

Because the mean free path of the phonons cannot become shorter than the unit cell, eqn [6] is no longer valid at some very high temperature (e.g., Ziman, 1962). Roufosse and Klemens (1974) explored the limiting temperature with their relationship

$$k_{lat} = \frac{B}{T} \left[\frac{2}{3} \left(\frac{B_1}{T}\right)^{1/2} + \frac{1}{3} \frac{T}{B_1} \right] \quad \text{for } T > \theta \qquad [7]$$

where the constant B_1 is related to the Debye frequency and the number of atoms in the formula. For simple solids, B_1 is ~ 500 K, and deviations from eqn [6] of $\sim 15\%$ are seen only above twice this temperature. For complex crystal structures, B_1 should be 1000–2000 K, and thus eqn [6] should be valid up to the melting temperatures of minerals (Roufosse and Klemens, 1974). Liebfried and Schlöman in 1960 reached the same conclusion.

For four-phonon events and $T > \theta$, Pomeranchuk (1943) found that $k \sim T^{-5/4}$, whereas Klemens (1969) proposed that

$$k_{lat} \sim \frac{1}{BT + B_1 T^2} \qquad [8]$$

and Ziman (1962) provided:

$$k_{lat} \sim \frac{1}{BT} + \frac{1}{B_1 T^2} \qquad [9]$$

Defects in crystals (e.g., structural incongruities or chemical substitutions) have been considered as the source of departures from Eucken's law (eqn [6]). The form

$$k_{lat} = B/T^m, \quad 0 < m < 1 \qquad [10]$$

can be traced back to experimental work by Lees (1905), but as discussed by Roufosse and Klemens (1973), subsequent measurements on the same material provided the inverse temperature dependence of eqn [6]. Nonetheless, theoretical studies have tried to quantify this effect. By assuming that the frequency

dependence for scattering of 'photons' applies to scattering of 'phonons' from point defects, Klemens (1960) concluded that $m = 1/2$ for temperatures near or above θ, and applied this result to solid solutions. Roufosse and Klemens (1974) state that defect concentrations > 5 atom% are needed for this result to be valid. Scattering from defects and vacancies has also been used to validate a formula that is widely used in analysis of rocks:

$$k_{lat} \sim \frac{1}{B + B_1 T} \quad \text{for } T > \theta \qquad [11]$$

(Madarasz and Klemens, 1987). Because heat capacity is nearly constant for $T > \theta$ and ρ is weakly temperature dependent, the forms for k_{lat} in eqns [6]–[11] apply to D_{lat} as well.

Near 298 K, k is roughly approximated. Ziman (1962) provided

$$\sim \exp(-\theta/T) \quad \text{for } T < \theta \qquad [12]$$

In geoscience, the high-temperature forms (eqns [6]–[11]) are commonly compared to measurements at 298 K.

The pressure dependence of k is inferred from that of the prefactors in the high T limit (e.g., Pierrus and Sigalas, 1985). Debye's model restricts values of mode Grüneisen parameters ($\gamma_i = \partial \ln v_i / \partial \ln V$, where v_i is frequency) by assuming linear dispersion, that is, $u_i \sim v_i a \sim v_i V^{1/3}$ for a cubic lattice. The definition of group velocity, $u_i = dv_i/ds$, where $s = 2\pi/a$ for the cubic lattice, gives $u_i \sim \gamma_i v_i V^{1/3}$. Thus, Debye's model requires that mode Grüneisen parameters are constant and near unity, and their derivatives, $q_i = \partial \ln \gamma_i / \partial \ln V$ are null (Hofmeister, 2007). Using $\theta \propto v$ in eqns [6b] and [6c] leads to

$$\frac{\partial(\ln(k_{lat}))}{\partial P} = \frac{1}{K_T} \left(3\gamma_{ave} + 2q_{ave} - \frac{1}{3} \right)$$

$$\approx \frac{4}{K_T}, \quad T > \theta \text{ and } q_{ave} \sim 0 \qquad [13a]$$

In evaluating eqn [6a], $<u>$ is approximated as $(V K_T)^{1/2}$, the bulk sound speed, giving

$$\frac{\partial(\ln(k_{lat}))}{\partial P} = \frac{1}{K_T} \left(\frac{3}{2} \frac{\partial K_T}{\partial P} + 2q_{th} - \frac{11}{6} \right)$$

$$\approx \frac{6}{K_T} \quad \text{for } T > \theta \qquad [13b]$$

Roughly, $q_{th} = 1 + \delta_T - \partial K_T / \partial P$, where $\delta_T = (1/\alpha K_T)(\partial K_T/\partial T)$. This approximation omits terms on the order of 0.1. Generally, q_{th} is near 1 and $\partial K_T/\partial P = K'$ is near 4 (Anderson and Isaak, 1995; Knittle, 1995; Hofmeister and Mao, 2002).

19.2.1.2 The damped harmonic oscillator (DHO) – phonon gas model

Lifetimes of optic modes obtained through the DHO model of Lorentz pertain to heat transport (Hofmeister, 1999, 2001, 2004b, 2004c, 2006; Giesting and Hofmeister, 2002; Giesting *et al.*, 2004). The lifetime of any mode, τ_i, equals $1/2\pi \mathrm{FWHM}_i$, where FWHM_i is the full width at half maximum from individual peaks in the dielectric functions that are extracted from infrared (IR) reflectivity data (e.g., Spitzer *et al.*, 1962). Raman peaks directly provide FWHM_i, although instrumental broadening may exist. For each mode, FWHM_i describes its interaction with all other modes in a real crystal, and includes effects such as anharmonicity and scattering from defects, and should represent an average across the Brillouin zone.

Because information is lacking on the specific dependence of u_{ij} on wave vector for each mode, average properties are used to describe thermal conductivity. Equation [4] reduces to

$$k_{\mathrm{lat}} = \frac{\rho}{3ZM} C_V \langle u \rangle^2 \tau = \frac{\rho}{3ZM} \frac{C_V \langle u \rangle^2}{2\pi \langle \mathrm{FWHM} \rangle} \quad [14]$$

Velocities of acoustic and optic modes should be roughly similar inside the Brillouin zone, so acoustic velocities are averaged to provide $<u>$. The specific form for $<u>$, for example, a simple versus a weighted average, is determined from comparing calculated to measured k for series of isostructural compounds (Hofmeister, 2001, 2004c).

Models have focused on thermal conductivity for historical reasons. However, the temperature dependence of k_{lat} (specifically, the maximum near 200 K) is complex, due to the 'S'-shaped T-response of heat capacity (Hofmeister, 2006). Because $C_P(T)$ is well-known for minerals (e.g., Berman and Brown, 1985), and $\rho(T)$ is fairly well-constrained due to thermal expansivity being small (e.g., Fei, 1995), the temperature response of k_{lat} is readily derived from eqn [2] once $D(T)$ is known. A simple formula is obtained from eqn [14] by neglecting the $\sim$3% difference between C_P and C_V:

$$D_{\mathrm{lat}}(T) = \frac{\langle u(T) \rangle \Lambda(T)}{3Z} = \frac{\langle u(T) \rangle^2}{6\pi Z \langle \mathrm{FWHM}(T) \rangle} \quad [15]$$

where $\Lambda = <u>\tau$ is the mean free path, and M has been absorbed into the heat capacity. Possible forms for $D(T)$ are inferred from the responses of u and FWHM to T.

Sound velocities of dense minerals depend weakly on temperature; $\partial(\ln u)/\partial T$ is near $-0.004\%\ \mathrm{K}^{-1}$ for grossular and pyrope garnets and is -0.006 to $-0.008\%\ \mathrm{K}^{-1}$ for forsterite and olivine (Anderson and Isaak, 1995). FWHM are obtained from IR reflectivity data., but measurements at temperature are rare. Absorption peak widths cannot be used, as these are a convolution of the longitudinal (LO) and transverse (TO) components. Raman spectra of various garnets show that peak widths increase strongly as temperature increases (Gillet *et al.*, 1992), but this parameter was not quantified. Raman spectra show that forsterite and olivine peak widths increase strongly with T (Kolesov and Geiger, 2004a), such that $\partial(\ln \mathrm{FWHM})/\partial T$ is near $0.14\%\ \mathrm{K}^{-1}$ for three Raman peaks examined in detail. This result compares closely with quantitative analysis of IR data for MgO which provides $\partial(\ln \mathrm{FWHM})/\partial T = 0.25\%\ \mathrm{K}^{-1}$ (Kachare *et al.*, 1972). Thus, the temperature dependence of D is dominated by the response of FWHM. From the above data and eqn [15], $1/D$ should depend directly on T. Observation of finite peak width for MgO at 4 K (e.g., Kachare *et al.*, 1972) and near 5 K for Raman bands of fayalite (Kolesov and Geiger, 2004b) suggests the form

$$\frac{1}{D} = \frac{6\pi Z}{u_0^2}(\mathrm{FWHM}_0 + bT + \cdots)(1 + b_1 T + \cdots)$$
$$\approx B + B_1 T + B_2 T^2 + \cdots \quad [16]$$

where b, B, etc. are constants. The linear coefficient B_1 should dominate and has a positive sign (Hofmeister, 2006). To allow for the dependence of u on T being nonnegligible, Pertermann and Hofmeister (2006) represented thermal diffusivity by

$$D \approx B + \frac{B_1}{T} + \frac{B_2}{T^2} + \cdots \quad [17]$$

The above results hold at all temperatures. However, at very high T, peak widths can become constant, resulting in constant values for either $1/D$ or D. In the kinetic theory of gases, lifetimes are connected with number of molecules per unit volume (e.g., Reif, 1965). At 298 K, the FWHM of diverse crystalline structures appears to be connected with the number of phonons within the primitive unit cell (Hofmeister, 2004c). As temperature increases, overtone-combination modes are excited, but saturation in the number of modes occurs when T is high enough that the continuum dominates the statistics (see Mitra, 1969), and increasing T no longer significantly changes the number of phonons. For silicate spinels, overtone-combinations are not observed above about $\sim$2400 cm^{-1} (Hofmeister and Mao,

2001), indicating that continuum statistics should control vibrational behavior at the energy equivalent temperature of $\sim$1600 K. This example suggests that constant D should occur at $T \sim 2\theta$.

Assuming that the mode Grüneisen parameters, $\gamma_i = (K/v_i)(\partial v_i/\partial P)$ where K pertains to the volume of the vibrating unit are roughly equal, and that a_i^3 on average equals the volume, gives

$$\frac{\partial(\ln(k_{lat}))}{\partial P} = \frac{1}{K_T}\left(m\gamma_{ave} + \frac{1}{3} - 2q_{ave}\right)$$
$$- \frac{\partial\ln <\text{FWHM}>}{\partial P}$$
$$\approx \frac{4}{K_T} \qquad [18]$$

where $m = 2$ for $\theta > \sim$250 K, $m = 3$ for $\theta \sim 400$ K, and $m = 4$ for $\theta > \sim 550$ K (refractory mantle minerals). Previous derivations of $\partial \ln k_{lat}/\partial P$ (Hofmeister, 1999) are missing the term $-2q_{ave}$ because linear dispersion was assumed in error. However, as q_{ave} in eqn [18] averages q_i, this should be very close to 0 (Hofmeister and Mao, 2002). Roughly, γ_{ave} equals γ_{th}.

The pressure dependence of τ or FWHM should be small; this property is closely related to population of vibrational states which are governed by temperature, not pressure. Few relevant spectroscopic measurements exist. Raman measurements above 100 K of ZnO and NaNO$_3$ (Serrano et al., 2003; Jordan et al., 1994) give $\partial\ln(\text{FWHM})/\partial P = 0$. Derivatives below 100 K from other materials are larger, $\sim$1–10% GPa^{-1}, but this trend could be due to the methanol–ethanol medium being stiff at lower T (Hofmeister, 2007). It appears that $\partial \ln(\text{FWHM})/\partial P$ is small above 298 K. Eliminating the small terms in eqn [18], provides the previous relation, $\partial \ln k_{lat}/\partial P = (4\gamma_{th} + 1/3)/K_T$, which sets an upper limit.

19.2.1.3   Bulk sound model for pressure derivatives

From eqn [14] and assuming that $<u>$ is the bulk sound velocity provides

$$\frac{\partial(\ln(k_{lat}))}{\partial P} = \frac{1}{K_T}\frac{\partial K_T}{\partial P} + \frac{1}{C_V}\frac{\partial C_V}{\partial P} + \frac{1}{\tau}\frac{\partial\tau}{\partial P} \cong \frac{K'}{K_T} \approx \frac{4}{K_T} \quad [19]$$

(Hofmeister, 2007). The pressure derivative of C_P is near zero for minerals, as inferred from the thermodynamic identity $\partial C_P/\partial P = -T\partial^2 V/\partial T^2 = -TV(\alpha^2 + \partial\alpha/\partial T)$. Available measurements on MgO (Andersson and Bäckström, 1986), olivine, and garnet (Osako et al., 2004) corroborate this inference, providing $C_P^{-1}\partial C_P/\partial P \sim 0.4\%$ GPa^{-1} which is 1/10th of $k_{lat}^{-1}\partial k_{lat}/\partial P = 5\%$ GPa^{-1}. Given the

uncertainties, slightly positive values from experiments, and slightly negative values from above thermodynamic identiy, $C_P^{-1}\partial C_P/\partial P$ is negligible. As discussed above, the last term is small and negative since compression should provide more frequent collisions and hence shorter lifetimes. Thus, an upper limit is provided by the simplied version of eqn [19]:

$$\frac{\partial(\ln(k_{lat}))}{\partial P} = \frac{K'}{K_T} \qquad [19a]$$

The small size of the C_P derivative and eqn [2] leads to a model-independent relation:

$$\frac{\partial(\ln(k_{lat}))}{\partial P} = \frac{1}{K_T} + \frac{\partial(\ln D)}{\partial P} \qquad [20]$$

19.2.2   Radiative Transport in Partially Transparent Materials (Insulators)

Radiative transfer is conceptually difficult. Not only is this process frequency and temperature dependent, but three different length scales are involved (the size of the particulates involved in physical scattering, the distance over which about half of the photon flux is absorbed, and the distance over which temperature changes significantly). Depending on the relative sizes of these length scales, radiative transfer can occur with and without participation of the intervening medium (**Figures 1** and **2**), that is, as diffusive or direct end-member types. It is commonly assumed that average absorption coefficients represent photon mean free paths (e.g., Ross, 1997). This over simplification has led to misinterpretation of direct radiative processes, which occur under optically thin conditions, as being diffusive, which requires optically thick conditions, and questionable use of laboratory experiments to constrain behavior of photons in the mantle. To ascertain which type of radiative transfer operates we compare the above length scales.

19.2.2.1   Distinguishing direct from diffusive radiative transport on the basis of frequency-dependent attenuation

The average grain size in the mantle is $\sim$1 mm, based on rock samples. If a trivial amount (0.005%) of the photons are back-reflected at grain boundaries due to mismatches between indices of refraction, then scattered light is essentially extinguished over distances of 2–200 m. This extinction length does not include absorption losses, and thus holds for frequency regions with virtually no absorbance. Mantle temperature gradients are below $\sim$10 K km^{-1}. Light emitted from any given grain thus should not reach another grain with a temperature differing by more than 1 K, so direct radiative transfer is impossible. Grain-to-grain diffusion (i.e., absorption and re-emission) of internally generated heat describes the physical process within the mantle at all frequencies, and is calculated from optical spectra (Section 19.2.2.2).

That radiative processes operating in the laboratory differ from those in the Earth is evident in the temperature gradients, which are ~ 1 K mm^{-1} in laser-flash experiments (Parker *et al.*, 1961) and larger in contact measurements, compared to 10^{-5} K mm^{-1} in mantle boundary layers. Direct radiative transfer occurs when a significant proportion of the photon flux crosses the sample (**Figure 2**), and is aided by strong temperature contrasts across a small distance (Lee and Kingery, 1960). The concern in laboratory experiments is that the transparency of the near-IR region (**Figure 3**) allows the graybody emissions of the heater to cross the sample largely unattenuated, warming the thermocouple without participation of the medium.

How much light crosses the sample at any given frequency depends on the true absorption coefficient $[A(v)]$ and thickness (d)

$$A(v)d = -\ln[I_T(v)/I_0(v)] + 2\ln[1 - R(v)] \quad [21]$$

and where I_T is the intensity of light exiting the crystal, and I_0 is intensity entering the crystal. The reflectivity ($R = I_R/I_0$, where I_R is intensity reflected from the surface) term accounts for laboratory measurements being obtained from a sample with flat, parallel surfaces. For weakly absorbing spectral regions, and normal incidence

$$R = (n - 1)^2/(n + 1)^2 \quad [22]$$

where n is the index of refraction. If additional scattering is present, such as from grain boundaries or imperfections, a baseline is subtracted to obtain A. Roughly, a medium is optically thin if $A(v)d < 1$, which allows >36% of the light to pass through. For a medium to be optically thick and the process to be considered diffusive, $A(v)d > 2$, in which case <14% of the light is transmitted (Siegel and Howell, 1972).

Emission characteristics are also germane. Planck's blackbody function is

$$I_{bb}(v) = \frac{2hv^5}{c^3}\left[\frac{1}{\exp(hcv/k_B T) - 1}\right] \quad [23]$$

where the energy density is provided on a per wavelength basis. Blackbody intensity increases about 10-fold between 500 and 900°C. The blackbody peak also shifts rapidly with T:

$$v_{max} = (\text{in cm}^{-1}) = 3.451\,T(\text{in K}) \quad [24]$$

and dominates the near-IR at temperatures of heat transfer experiments (**Figure 3**).

Measuring absolute values of A at high T is difficult (Grzechnik and McMillan, 1998). Relevant near-IR data on minerals concern hydroxyl absorptions. Not all reported A values include a baseline correction, and virtually all of the mineralogical literature uses common logarithms in a format analagous to eqn [21]. We account for these differences, and use olivine to represent minerals with Fe^{2+} in octahedral sites because characteristics of the d-d electronic transitions are largely determined by the site occupied by Fe, whereas characteristics of the overtones of the Si–O tetrahedron vary little among diverse minerals (cf. olivine to quartz near 2000 cm^{-1} in **Figure 3**). Similarly, quartz should represent Fe-free minerals.

For quartz, $A < 0.05$ mm^{-1} for 3700 cm^{-1} < v < 4200 cm^{-1} from $\sim$20 to 579°C, using $n = 1.55$ in eqns [22] and [21] and figure 16 of Aines and Rossman (1985). This weak absorbance is associated with overtones of hydroxyl or water, which are

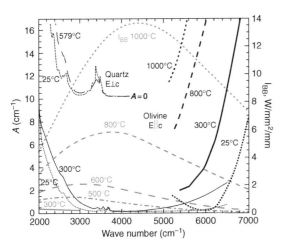

Figure 3 Near-IR spectra of olivine and quartz, and blackbody curves as functions of frequency and temperature. Assorted gray curves and left Y-axis = blackbody function at various temperatures, as labeled (widely spaced dots in the upper left corner only = 1000° C). Olivine (Fo$_{90}$) data = medium black lines from Hofmeister (2005) and thick lines from Ullrich *et al.* (2002). Quartz data = fine lines from Aines and Rossman (1985), offset for clarity: A actually is 0 for the flat section near 4000 cm^{-1}. For temperatures accessed in contact measurements of k or D (up to $\sim$800° C), the blackbody curve occupies the transparent spectral region of Fe-free minerals and those with octahedral Fe^{2+}.

~1/100th the intensity of the fundamentals present near 3500 cm^{-1} (**Figure 3**). The data indicate that low absorbance should persist into the visible region and to higher temperatures. Measurements of microcrystalline and milky quartz up to 700°C (Yamagishi *et al.*, 1997) and fused silica up to ~1400°C (Grzechnik and McMillan, 1998) provide confirmation. Quartz is quite transparent (~80% of the near-IR to visible flux is transmitted across 4 mm), and direct radiative transfer is expected over many centimeters, thicknesses much larger than used in contact experiments (e.g., Kanamori *et al.*, 1968; Gibert *et al.*, 2005).

Olivine spectra (**Figure 3**) show that *A* is essentially negligible up to ~300°C in the transparent near-IR region. The rise at low frequency at 25°C between 5000 and 6000 cm^{-1} seen by Ullrich *et al.* (2002) is not supported by our results or those of Taran and Langer (2001), and may be due to low sensitivity near the limit of their spectral range. For this reason, it is difficult to extrapolate the measurements of Ullrich *et al.* (2002) into the near-IR in order to estimate *A*(*T*) at the most transparent frequencies. We suggest that *A* is ~0.1 mm^{-1} at 4000 cm^{-1} near 800°C, and thus that >50% of the flux in the narrow region from ~3000 to 4500 cm^{-1} is transmitted for 4 mm lengths typical in heat transfer experiments. This inference is corroborated by the results of Gibert *et al.* (2005) wherein direct radiative transport is observed in temperature–time curves of large olivine samples (Section 19.3.2). Most of the blackbody peak at these temperatures occupies this spectral segment (**Figure 3**), and thus radiative transfer for Fe^{2+}-bearing samples is largely direct at the temperatures of up to 800°C probed in conventional methods. Direct radiative transport is unavoidable in experimental measurements of other minerals with octahedral Fe^{2+}: pyroxene, the α-, β-, and γ-polymorphs of olivine, and magnesiowüstite (cf. Burns, 1970; Goto *et al.*, 1980; Ross, 1997; Keppler and Smyth, 2005).

The presence of direct radiative transfer in contact measurements is indicated by an increase in *k* with *T* above ~400°C (**Figure 4**). This change is associated with the strong nonlinear increase of blackbody flux in the transparent near-IR, not with involvement of Fe^{2+} absorptions, as these negligibly overlap with the blackbody curve at such low temperatures (**Figure 3**). That the change in $\partial k/\partial T$ for quartz, which lacks Fe^{2+} bands, occurs at a similar temperature (Kanamori *et al.*, 1968) and shows that

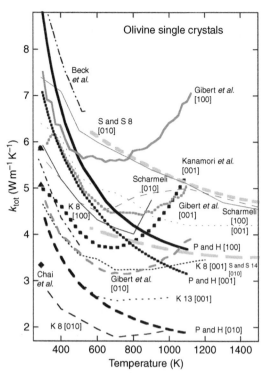

Figure 4 Literature data on the temperature dependence of the measured total thermal conductivity from single-crystal olivines. Black symbols (labeled with orientations) are *Fe*# = Fe/(Fe + Mg) = 0.11 measured using PTGS (Chai *et al.*, 1996). Solid lines = [100]; dotted lines = [001]; dashed lines = [010]. Light gray heavy lines are polynomial fits for *Fe*# = 0.08 and *Fe*# = 0.13 using a laser-wave method (Schatz and Simmons, 1972a). Medium black lines are laser-flash data (Perterman and Hofmeister, 2006). Medium gray lines are for a remote heating thermocouple method (Gibert *et al.*, 2005). All other studies use contact methods. Dot-dashed lines = unknown composition and orientations of Beck *et al.* (1978). Small squares are (001) for *Fe*# = 0.18 (Kanamori *et al.*, 1968). Widely spaced dotted line labeled K13 is *Fe*# = 0.13 (Kobayashi, 1974). Thin lines labeled K8 are the three axes for *Fe*# = 0.08 (Kobayashi, 1974). Very thin lines are three orientations of *Fe*# = 0.08 acquired at 2.5 GPa (Schärmeli, 1982).

direct radiative transfer in the near-IR governs both minerals.

Figure 5 summarizes conditions under which radiative transfer is diffusive or direct. The amount of direct radiative transfer in laboratory experiments depends on boundary conditions.

19.2.2.2   *Spectroscopic models for diffusive radiative transport inside the Earth*

For an optically thick, nonopaque medium in local radiative equilibrium

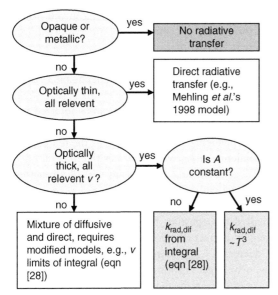

Figure 5 Flowchart for discerning whether radiative transfer is diffusive or direct, with links to appropriate models, discussed in the text. 'All relevant ν' refers to frequencies up to where I_{bb} is negligible at the temperature of interest.

$$k_{\mathrm{rad,dif}}(T) = \frac{4\pi}{3}\int_0^\infty \Lambda(v,T)$$
$$\times \frac{c}{v^2}\frac{\partial\{[n(v,T)]^2 S(v,T)\}}{\partial T}\,\mathrm{d}v \qquad [25]$$

provides energy density on a per wavelength basis, where the source function (S) in geophysical models has generally been assumed to be Planck's blackbody function and Λ was related to absorption characteristics only (Clark, 1957; Shankland *et al.*, 1979). Based on radiative transfer models in other disciplines (e.g., Brewster, 1992; Hapke, 1993; Kaufmann and Freedman, 2002), Hofmeister (2004a, 2005) revised geophysical models to account for grain-size. Wave numbers ($v = 1/\lambda$) are presumed. The convention of Brewster (1992), wherein π occurs in eqn [23] rather than in eqn [25], was used only by Hofmeister (2004b).

Few materials emit as perfect blackbodies. Instead, the intensity produced is $S = \xi I_{bb}$, where emissivity (ξ) is a material property. Because each grain in Earth's mantle serves as a source (**Figure 2**), emissivity is obtained from Kirchhoff's (*c.* 1861) law of radiation:

$$\xi(v) = 1 - \exp[-dA(v)] \qquad [26]$$

The importance of emissivity is confirmed in a thought experiment: removing one grain from the

mantle leaves a cavity that vibrations cannot propogate into. Radiative flux into the cavity is $F_{\mathrm{rad}} = \xi\sigma T^4$, where σ is the Stefan–Boltzmann constant (e.g., Halliday and Resnick, 1966). From this relation and eqn [1], $F_{\mathrm{rad}} = -k_{\mathrm{rad,dif}}\partial T/\partial r$, so $k_{\mathrm{rad,dif}}$ must be proportional to ξ.

The mean free path is controlled by both scattering and attenuation:

$$\frac{1}{\Lambda} = \frac{1}{d} + A(v) \qquad [27]$$

(e.g., Lee and Kingery, 1960). We neglect effects of grain shape and use an average grain-size in eqns [25]–[27], as is assumed when incorporating viscous damping in geodynamic models (e.g., Ranalli, 2001). Since forward scattering dominates, we do not consider a directional dependence nor do we distinguish between grain-size and thickness.

To evaluate eqn [25], an additional physical process in a grainy medium must be accounted for: the amount of light exiting any given grain is reduced from the incident intensity not only by absorption, but also by back reflections at its surfaces (eqns [21] and [22]). This step is necessary because back reflections and high absorbance together produce opaque spectral regions for which $k_{\mathrm{rad,dif}} = 0$. Frequency regions that are opaque are excluded from the integral. Opacity occurs when the product $dA(v)$ reaches a critical value, controlled by R (eqns [21] and [22]), such that the light from any given grain cannot reach its second nearest-neighbor. When $A(v)d$ approaches 7 ($\sim$<0.1% transmission), the sample is effectively opaque. Allowing this limit to vary significantly (from 0.005 to 2%) gives similar results (Hofmeister, 2005).

Given the approximate nature of radiative transfer models, eqn [25] is simplified based on the dependence of material properties on v, T, and P. The near-IR to UV spectral range need only be considered, as the lattice modes in the IR and the metal-oxygen charge transfer bands in the far-UV are too intense to permit radiative transfer. The result (Hofmeister, 2005) is

$$k_{\mathrm{rad,dif}}(T) = \frac{4\pi dn^2}{3}\sum\int_{\mathrm{lower}}^{\mathrm{upper}}\frac{(1 - e^{-dA})}{(1 + dA)v^2}$$
$$\times \frac{\partial[I_{bb}(v,T)]}{\partial T}\,\mathrm{d}v \qquad [28]$$

where the sum allows for possible existence of multiple transparent regions.

The pressure dependence of $k_{\mathrm{rad,dif}}$ is unimportant compared to temperature effects. The relevant quantity, $\partial A/\partial P=(\partial A/\partial v)(\partial v/\partial P)$, has opposite signs above and below the peak center, causing $\partial k_{\mathrm{rad,dif}}/\partial P$ to be highly temperature dependent, due to the peak positions and widths of A and I_{bb} both depending on T (Hofmeister, 2005). At mantle temperatures, the derivative is close to zero. Evaluating $\partial k_{\mathrm{rad,dif}}/\partial P$ relevant to the mantle requires near-IR spectra at simultaneously elevated P and T, and cannot be determined from only $\partial v/\partial P$ at 298 K as previously considered (e.g., Ross, 1997; Keppler and Smyth, 2005).

19.2.2.3 Understanding the general behavior of $k_{\mathrm{rad,dif}}$ from asymptotic limits

For dark, but not opaque materials (dA large but <7), eqn [28] reduces to the result of Shankland *et al.* (1979). If, in addition, A is constant over all frequency, and eqn [28] is integrated from $v=0$ to ∞, Clark's (1957) formula results in

$$k_{\mathrm{rad,dif}} = \frac{16}{3}\sigma T^3/A \quad \text{(for d}A \text{ large and constant)} \quad [29]$$

Although these conditions are rarely met in minerals, eqn [29] describes some key features of radiative diffusion. Note that d is not explicitly contained in the formula. For an opaque medium ($A \to \infty$) and any value of d, the limit ($k_{\mathrm{rad,dif}} \to 0$) is obtained. When d$A$ exceeds a critical value depending on reflectivity and texture (discussed above), $k_{\mathrm{rad,dif}} \to 0$. The results of Shankland *et al.* (1979), derived without considering grain size, are invalid for very weakly absorbing regions unless grain size is very, very large.

For nearly transparent materials and A independent of v and T, eqn [28], integrated from $v=0$ to ∞, provides a very different result:

$$k_{\mathrm{rad,dif}} = \frac{16}{3}\sigma T^3 \mathrm{d}^2 A \quad \text{(for d}A \text{ small and constant)} \quad [30]$$

Thus, as $A \to 0$, $k_{\mathrm{rad,dif}} \to 0$. Diffusive radiative transfer is absent in a fully transparent medium because photons cannot warm the medium unless absorbed. As $d \to 0$, $k_{\mathrm{rad,dif}} \to 0$. This case approximates exceedingly fine-grained ceramics that are effectively opaque due to back reflections. Physical scattering dominates this asymptotic limit.

The form of $k_{\mathrm{rad,dif}}(A)$ can be inferred from eqns [29] and [30]. Because $\partial k_{\mathrm{rad,dif}}/\partial A$ is positive at low dA but negative at high dA, and both limits for k_{rad} are 0, a maximum exists in $k_{\mathrm{rad,dif}}(A)$. The maximum is associated with moderate A and grain size, wherein physical scattering is not overwhelming, significant

light is emitted by individual grains in the mantle, and this light is not completely extincted proximal to where it was emitted.

Radiative transfer is more complex than T^3 law in eqns [29] and [30] because A depends on v and T. However, the trend in $k_{\mathrm{rad,dif}}$ with T that should result from evaluating eqn [25] can be inferred from these equations (Hofmeister and Yuen (in press)). The dependence of A on v weakens the temperature dependence of k_{rad} as it truncates the integral of eqn [28]. For small dA, A increasing with T strengthens the temperature dependence of $k_{\mathrm{rad,dif}}$ (eqn [30]). The effects of $A(v)$ and $A(T)$ roughly cancel at small dA, suggesting for small dA, the T^3 law is roughly correct. But for large dA, A increasing with T decreases $k_{\mathrm{rad,dif}}$ (eqn [29]) Dark samples are opaque over larger portions of the integral, which further decreases k_{rad}. A much weaker dependence than T^3 is expected for $k_{\mathrm{rad,dif}}$ at high T for large dA.

19.2.2.4 An approximate formula that connects k_{rad} with concentration

If the peak is narrow and the product Ad is low enough that the material is partially transparent over most frequencies, eqn [28] can be evaluated using the trapezoidal rule:

$$k_{\mathrm{rad,dif}}(T) = \frac{4dn^2}{3} \frac{[1-\exp(-dA_{\mathrm{ave}})]}{(1+dA_{\mathrm{ave}})} \Delta v \frac{2\pi h^2 c^3 v_{\mathrm{ave}}^4}{k_{\mathrm{B}}T^2}$$
$$\times \frac{\exp(hcv_{\mathrm{ave}}/k_{\mathrm{B}}T)}{[\exp(hcv_{\mathrm{ave}}/k_{\mathrm{B}}T)-1]^2} \quad [31]$$

where v_{ave} and A_{ave} represent the average frequency and average absorption coefficient for a peak that spans the interval Δv (Hofmeister, 2004a). For a single band, A_{ave} equals the half height. For broad or multiple peaks, $k_{\mathrm{rad,dif}}$ could be calculated by summing the results for a series of segments, for example, 1/10th of each peak in the spectrum could be evaluated individually using eqn [31]. This procedure could be applied to the near-IR and UV tails. As a first approximation, we consider a single peak, and neglect the temperature dependence of the spectral parameters.

The simplest version of the Beer–Lambert law (e.g., Rossman, 1988) is used to link concentration (X) to absorption:

$$X = A_{\mathrm{max}}/f' = A_{\mathrm{ave}}/f \quad [32]$$

where f or f' is termed an extinction coefficient, and is obtained by comparing spectra from standards with known concentration.

The variables in eqn [31] are separable. Combining eqns [31] and [32] gives

$$k_{\text{rad,dif}} = \vartheta(T) \frac{8\pi ck_{\text{B}} dn^2}{3} v_{\text{ave}}^2 \Delta v \frac{1 - \exp(-dfX)}{1 + dfX} \quad [33]$$

where the temperature function is given by

$$\vartheta(T) = \left(\frac{s}{T}\right)^2 \frac{\exp(s/T)}{[\exp(s/T) - 1]^2} \quad [34]$$

and $s = hcv_{\text{ave}}/k_{\text{B}}$. Equation [34] is the same as Einstein's heat capacity function (eqn [5]). Clearly, the T^3 law does not describe materials with absorption bands.

19.3 Experimental Methods for the Lattice Contribution

19.3.1 Conventional Techniques Involving Multiple Physical Contacts

Many studies of $k(T)$ or $D(T)$ in Earth science utilize conventional techniques wherein a heat source and thermocouples directly contact the sample. Contact methods are limited to $\sim$1200 K due to limitations of the materials used, long measurement times, and radiative heat losses from the surface (e.g., Parker et al., 1961). To determine k, steady-state methods are generally used which require information on both the flux and the temperature gradient (eqn [1]); both boundary conditions and sample geometry influence the results. Obtaining D involves measurement of the time it takes a thermal disturbance to travel a certain distance. Some techniques provide both variables; the transient hot wire and transient hot strip methods provide k more accurately than D ($\pm 6\%$ and 30%, respectively: Hammerschmidt and Sabuga, 2000). Tye (1969) and Somerton (1992) describe various methods in detail.

A large database on minerals at ambient conditions (Horai, 1971) was created using the needle-probe method, wherein k is determined from a powder (grain size <50 μm) dispersed in water. In contrast, most studies of Earth materials at high T use Ångström's (c. 1861) method or its variants. In brief, a sinusoidal source of heat at frequency ω is applied to one end of a rod-shaped sample, and the phase of the decaying sinusoidal temperature wave which passes into the sample is measured at a distance Z along the rod. Solutions to Fourier's equation in 1-D show that the phase depends on $Z(\omega/2D)^{1/2}$. Correction terms address boundary conditions such as surface losses

via radiation to the surroundings. Thermocouple attachment involves drilling a hole in the center of the sample and inserting a smaller-diameter wire (Xu et al., 2004).

For hard solids, conventional measurements commonly differ by $\pm 20\%$ even at room temperature (Ross et al., 1984). For olivine single-crystals with compositions of $\text{Fe}/(\text{Fe} + \text{Mg}) \sim 0.1$, discrepancies nearing 50% are seen at 298 K (**Figure 4**). Offsets between various data sets are consistent up to 500–700 K, whereupon $\partial k/\partial T$ changes sign. Less regular trends occur at higher temperatures and can be explained by variable amounts of unwanted, direct radiative transfer-associated transparency in the near-IR spectral region (discussed above). Particulars of the experiments (e.g., contact adhesives, sample thickness, chemical composition) likely cause of the variation in the position and degree of the upturns in $k_{\text{tot}}(T)$. Radiative effects are reduced for samples with small grain size or strong near-IR absorptions (e.g., Fe-rich garnet). Some researchers tried to separate phonon from photon components by assuming that $k_{\text{rad,dif}} \sim cT^3$ (e.g., Schatz and Simmons, 1972a, 1972b). Schärmeli (1982) subtracted $k_{\text{rad,dif}}(T)$ calculated after Shankland et al. (1979) from his measurements to obtain k_{lat}. However, all formulae for $k_{\text{rad,dif}}$ assume optically thick conditions (**Figure 5**), which are not met in the near-IR region in the laboratory (**Figures 3 and 5**; Section 19.2.3.1); consequently, this approach is invalid.

Radiative transfer does not explain the observed discrepancies at ambient temperature. Systematic errors in conventional measurements also arise from thermal contact resistance and differential thermal expansion (e.g., Parker et al., 1961). Contact resistance at the interface explains the variation seen in olivine at room temperature (e.g., **Figure 4**). Surface roughness creates 'gaps' through which phonons cannot propagate. Oxidation is another contributing factor (Fried, 1969). This impediment can be described as a thermal resistance, Q, which is the inverse of the thermal conductivity of the interface. Thermal resistances add

$$\frac{1}{k_{\text{meas}}} = \frac{1}{k_{\text{sample}}} + Q \quad [35]$$

(Carslaw and Jaeger, 1959) and thus, contact measurements set a lower limit on k at 298 K, in the absence of other problems. Differential thermal expansion between thermocouples and sample can make Q temperature dependent.

Use of tall cylinders to measure anisotropy is problematic. LO modes are excited indirectly as the phonons propogate along **z** (**Figures 6(a)** and **6(b)**) analogous to IR absorption experiments (e.g., Berman, 1963). As the thickness-to-diameter ratio increases, proportionately more LO modes are present. For noncubic samples, LO modes of the [001] orientation are closely related to one of the two types of TO modes for [100] and [001], thus long cylinders mix polarizations.

Pressure derivatives of k obtained for hard solids using contact methods can be unreliable because deformation alters the geometry and cracking reduces thermal contact; values for $\partial(\ln k)/\partial P$ often vary by a factor of 3 between studies (Ross *et al.*, 1984; Bäckström, 1977; Brydsten *et al.*, 1983; Alm and Bäckström, 1974). Some uncertainties are due to pressure determinations and nonhydrostatic conditions, especially in the older works. Hot wire, static radial flow, and variations of Ångström's method have been used. (*Note:* Radial methods average k over two directions.) Recent measurements used Ångström's method in a multianvil apparatus to attain high pressures (Xu *et al.*, 2004). The two-strip

method (Andersson and Bäckström, 1986; Osako *et al.*, 2004) is the most accurate of the techniques used at pressure for hard solids and single crystals.

19.3.2 Methods Using a Single Physical Contact

Recently, Schilling (1999), Höfer and Schilling (2002), and Gibert *et al.* (2005) utilized an approach in which heat is applied remotely (flux from a filament), but the sample is in contact with thermocouples for determining heat flow. We use the abbreviation RHTM (remote heating thermocouple measurement). Problems associated with multiple contact methods (contact resistance, polarization mixing, and spurious radiative transfer) can affect RHTM experiments.

The raw data consist of the change in temperature as a function of the time subsequent to the applied heat (light) pulse. The resulting temperature–time curves indicate optically thin conditions, but are not analyzed in the same manner as laser-flash data (next section). Finite-difference calculations are used to determine D from the temperature difference across the sample by assuming (1) that a constant describes the heat loss from the surface, and (2) that direct radiative transfer is proportional to the heat from the pulse (specifically it is assumed that the rear surface receives light in proportion to the emitted heat inferred from measurements of temperature on the front surface of the sample). This approach assumes constant, frequency-independent absorbance. The equations assume instantaneous application of heat, but in actuality the pulse width is very large, about one-third the time taken for heat to traverse the sample.

Radiative transfer is incompletely removed by the finite-difference calculations, as concluded by the authors and evidenced in positive $\partial D/\partial T$ above $\sim$600 K. Höfer and Schilling (2002) and Gibert *et al.* (2005) try to separate the components by assuming forms for each of the phonon and photon components (specifically, $D_{\text{lat}} = 1/(B_0 + B_1 T)$ and $D_{\text{rad}} = BT^3$). This approach is invalid for several reasons. (1) The T^3 law (Clark, 1957) is the form derived for k_{rad}, not D_{rad}. Given their relation (eqn [2]), and the temperatures over which fitting is performed, C_P increases nonlinearly with T and ρ decreases with T, such that their responses do not cancel. Moreover, the T^3 form is inappropriate for minerals in general and for Fe^{2+}-bearing olivine in particular (Section 19.2.2). (2) It is not possible to simultaneously have optically thin

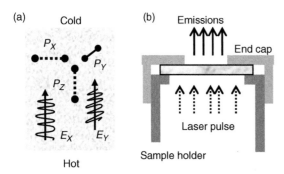

Figure 6 Schematics of geometric effects on heat transfer. (a) Long cylinders used in multiple- and single-contact methods. Phonons (squiggles) scattered upwards (along **z**) have amplitude along **x** and **y**, as do any electromagnetic waves. Inside the sample, atom pairs vibrate in three directions. Those with dipoles oriented along **x** and **y** interact directly with the heat rising from below, because the electric field of these dipoles are in the same direction as that of the heat wave. X and Y dipoles are TO modes for the orientation shown. Indirect coupling stimulates the LO modes (dipoles along **z**), which are TO modes in the perpendicular orientations. (b) Thin plates used in laser-flash analysis. Stippled, disk-shaped sample. Thin samples approximate the ideal situation where only TO modes exist. The endcap (dark gray) allows only the emitted light from the center to reach the detector, reducing edge effects.

and optically thick conditions with constant absorbance, and so this approach is inconsistent. (3) It is highly unlikely that the diffusive type of radiative transfer is present in these experiments at the low temperatures accessed (Section 19.2.3.1). If direct and diffusive radiative transfer do coexist, the processes would occur at different frequencies, requiring an analysis which accounts for the v and T dependence of A.

Uncertainties of 5% for RHTM rest on comparing quartz data (Höfer and Schilling, 2002) with previous measurements obtained using conventional methods (e.g., Kanamori *et al.*, 1968; Beck *et al.*, 1978), which include spurious radiative transfer and contact resistance effects. Also, laser-flash measurements (below) show that D_{lat} of any given quartz sample depends on its impurity content, including hydroxyl (Branlund and Hofmeister, 2004).

The 3ω method (e.g., Cahill *et al.*, 1992) also involves a single contact. Thermal conductivity is determined below $\sim$300 K using a third-harmonic detection scheme from the self-heating of a narrow metal line deposited on the sample. This radial method mixes two polarizations. An important advantage is insensitivity to blackbody radiation. This technique has been used at pressures up to 0.8 GPa (Chen *et al.*, 2004), but not on Earth materials.

19.3.3 Contact-Free, Laser-Flash Analysis

Accurate (nominally $\pm$2%) measurements of k_{lat} are provided by the contact-free, laser-flash technique that was originally developed by Parker *et al.* (1961). This technique is well-established in applied science, but because it is uncommon in Earth science (Holt, 1975; Büttner *et al.*, 1998; Hofmeister, 2006), details are provided.

Main components are a controlled atmosphere furnace, a high-energy pulsed laser, and an IR detector (e.g., Bräuer *et al.*, 1992). A sample in the form of a small slab with parallel faces ($\sim$0.3–3 mm thick by 6–15 mm diameter) is held at temperature in the furnace while emissions from the top of the slab are monitored remotely with the IR detector (**Figures 6** and **7**). Additional heat is supplied remotely to the bottom of the slab by a pulse from an IR laser. As heat from the pulse diffuses from the bottom to the top of the sample, the increase in emissions is recorded by the IR detector (**Figure 7**). Because emissions are directly related

to temperature ($\sim T^4$), the detector response is known as a temperature–time curve. Top and bottom surfaces of the sample are graphite coated (thickness $\sim$1 μm) to absorb laser light, thereby shielding the detector, to enhance intensity of the emissions (e.g., Blumm *et al.*, 1997), and to buffer oxygen fugacity to C–CO, preventing oxidation of Fe^{2+}. Use of metal coatings is discussed below. Neither thermocouples nor heater contact the sample, and neither power input nor temperature need be quantified. Equations used for data analysis require that the pulse width be significantly shorter than the time heat takes to cross the sample, which is met (**Figure 7**). Because the rise in sample temperature associated with the pulse is small, $\sim$4 K, D is approximately constant during data acquisition, and the T dependence of D is determined by varying furnace temperature (Parker *et al.*, 1961). For thin samples (**Figure 6**), the proportion of LO modes and the effect of contact with the holder on heat flow are minimized.

A key advantage of LFA in studying Earth materials is that it permits separation of the lattice component from unwanted radiative transfer. As heat diffuses by phonon collisions from the bottom coat to the top of the sample, the temperature of the top graphite coat gradually rises (**Figure 7(b)**). The top coat reaches a peak emissions at some finite time after the laser pulse, because phonons travel near the speed of sound. In contrast, radiative transfer between the graphite layers is recorded as a virtually instantaneous rise in emissions after the pulse (**Figure 7(c)**), because photons travel near the speed of light. Phonon and photon heat transfer are thus visually discernable in the temperature–time curves (**Figure 7**), and are extracted using the mathematical model of Mehling *et al.* (1998); see also Hofmann *et al.* (1997). Mehling *et al.*'s model was developed for optically thin conditions. The formulation accounts for absorbance being frequency dependent, although values of optical properties are not needed. Blumm *et al.* (1997) established model accuracy by comparing D calculated by applying Mehling *et al.*'s (1998) model to measurements of glass coated with graphite only (which has some radiative transfer) to D calculated by applying Cowan's (1963) model (see below) to measurements of the same glass, for which radiative transfer was suppressed by first applying an Au metal coating and then overcoating with graphite as in **Figure 7(a)**. This double-coating technique of Degiovanni *et al.* (1994) only completely suppresses

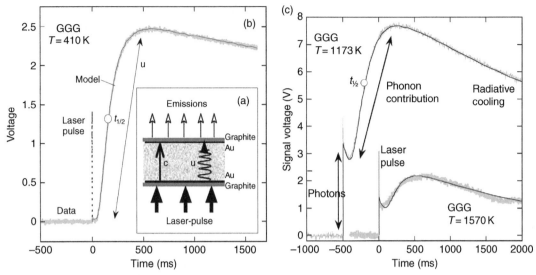

Figure 7 Temperature–time curves from laser-flash analysis and connection with microscopic process. (a) Physical transfer of heat during the experiment. A plate-shaped sample (stipple) is coated top and bottom with first gold and then graphite and held at some given temperature in a furnace. Emissions (light arrows) from the top coat are recorded by an infrared detector, giving a baseline. The laser pulse (heavy arrow) is absorbed by the bottom graphite coat. Heat from the bottom coat then either traverses the sample as photons (medium arrow) near light speed (c), or as phonons (squiggle arrow) near sound speed (u). For pale or transparent samples, radiative transfer is boundary-to-boundary. (b) Record of heat transfer near room temperature. Emissions are measured as signal voltage and set to zero before the laser fires at $t = 0$ (dotted line). Gray curve, data; thin black curve, model. Virtually all heat is carried by the phonons, causing the emissions from the top coat to rise gradually with time. The position for the half rise time (t_{half}), used in analysis (e.g., eqn [36]), is shown by the circle. (c) High-temperature results, symbols as in part (b). Radiative transfer from the bottom coat warms the top coat, giving the sharp rise in emissions almost instantaneously (dotted double arrow). Slower phonon travel occurs afterwards (solid double arrow). Emissions subsequently decrease as the sample re-equilibrates with its surroundings by radiating from its surface. The curves for 1173 K are shifted by −500 ms for clarity. The match between data and model at 1570 K is marginally acceptable: data were not collected beyond this point. Modified from Hofmeister AM (2006) Thermal diffusivity of garnets at high temperature. *Physics and Chemistry of Minerals* 33: 45–62.

boundary-to-boundary radiative transfer at low temperature, so Mehling *et al.*'s (1998) model is essential for geophysical studies.

Sources of experimental uncertainty in laser-flash measurements are implicit in simple models for temperature–time curves arising from phonon transport alone (e.g., opaque samples, or very low temperatures). Fourier's equation for heat flow in 1-D is the basis for analysis. For the simplest case of adiabatic heating, without any radiative heat transfer,

$$D = 0.1388 \, d^2 / t_{half} \qquad [36]$$

where d is sample thickness and t_{half} is half of the time required for temperature to reach the maximum after the laser pulse (Parker *et al.*, 1961). (Note: Equation [36] was applied to olivine by Holt (1975), but is not appropriate due to radiative transfer.) Subsequently, Cowan (1963) accounted for radiative heat losses from the front and back surfaces of the sample to the surroundings. Cape and Lehman (1963) allowed

for the finite width of the laser pulse as well as for additional heat loss through the sides of the sample. Other models (Heckman, 1973; Azumi and Takahashi, 1981) explore effects of idealized laser pulse shapes. These models were improved by Josell *et al.* (1995) and Blumm and Opfermann (2002) who also developed an algorithm to take the measured shape of the laser pulse into account.

Although equations more complicated than eqn [36] are used in LFA, the strong dependence of D on thickness remains still contributes the main source of uncertainty. The reliability of each data point is insured by requiring that calculated and measured temperature–time curves match. The accuracy for the technique is considered to be 2%, determined through benchmarking against metals and graphite (Blumm and Opfermann, 2002). These opaque and soft materials lack radiative transfer and have good thermal contact, allowing calibration against results from conventional methods. Standard reference materials (SRMs) obtained from the National Institute of Standards are used (e.g., Henderson *et al.*, 1998a, 1998b).

19.3.4 Additional Contact-Free (Optical) Techniques

Schatz and Simmons (1972a, 1972b) developed a time-varying, quasi-steady-state technique that is contact-free for measurement of *k(T)*. The sample is maintained at temperature in a furnace, and a CO_2 laser applies an additional oscillating heat input to one side of the sample. Heat emitted from the opposite side is monitored with a near-IR detector. Samples were not coated, so photon processes contribute to *k*. The T^3 formula for k_{rad} is used, which presumes optically thick conditions at all frequencies (**Figure 5**). In contrast, for sample thickness of ~5 mm, optically thin conditions dominate below ~1000 K (**Figure 3**; figure 9 in Schatz and Simmons, 1972b). Above ~1500 K, samples should be optically thick, but uncertainties are large.

Picosecond transient grating spectroscopy (PTGS) has provided thermal diffusivity at ambient conditions and at ~5 GPa (Zaug *et al.*, 1992; Chai *et al.*, 1996). Results at *T* are available in thesis form (Harrell, 2002). Pressure derivatives should be accurate in this contact-free method. In brief, laser pulses with wavelength $\lambda = 532$ nm crossing at an angle 2φ are used to create an interference pattern with grid spacing $L = 0.5\ \lambda/\sin\varphi$. The signal decays as $\exp(-2t/\tau)$ and the diffusivity is obtained from the decay rate (τ) and a radiative component (*r*) using

$$\frac{1}{\tau} = r + \frac{4\pi^2 D}{L^2} \quad [37]$$

The decay is measured by the Bragg diffraction of a third laser pulse. It is argued that *r* can be neglected, and in any case the technique provides a maximum value of *D*. As discussed by Chai *et al.* (1996), a concern is whether long-lived electronic states participate, in which case τ reflects processes in addition to the desired thermal relaxation, and *D* calculated will be less than the true value. The authors claim uncertainties in *D* of 2%, but benchmarking was not performed.

Other all-optical techniques exist, but have not been implemented in Earth science. A contact-free method for pressure studies was developed by Pangilinan *et al.* (2000). Optical heating and magnetization thermometry were developed for low temperatures by Hao *et al.* (2004).

19.4 The Database on Lattice Transport for Mantle Materials

19.4.1 Evaluation of Methodologies, Based Primarily on Results for Olivine

Direct comparison of methods requires single-crystal data to eliminate textural and porosity differences. Only for a few minerals (olivine, quartz, NaCl) have multiple measurements of single crystals been made using diverse methods (**Table 1**). Comparisons of other minerals are precluded due to variations in chemical compositions, imprecise compositions for some samples, nonoverlapping temperature ranges, or varying amounts of disorder.

19.4.1.1 Ambient conditions

For olivine, sample variations are unimportant because (1) San Carlos material was mainly used, and (2) LFA, which is the most accurate method, gave the same *D* values for samples from two different localities. For some orientations, *D* varies up to a factor of 2 among published studies (**Table 1**). Quartz (SiO_2) and synthetic samples of NaCl are essentially pure.

Contact resistance primarily causes the discrepancies as demonstrated by the correlation of room temperature values with the number of thermal contacts (**Figure 8**; Hofmeister, 2007). Lee and Hasselman (1985) reached a similar conclusion by replacing the remote detector in LFA experiments with a thermocouple. For olivine, orientationally averaged values are reduced by 13% per contact. For NaCl and quartz, *D* is reduced from laser-flash results by 4–5% per contact. Less scatter is observed if average values, rather than individual orientations, are compared (**Figure 8**). Averaging also minimizes possible effects of trace impurities. The behavior is consistent with polarization mixing occurring in tall cylinder geometries (**Figure 6**). RHTM measurements of the olivine axis with the highest *D* give values lower than laser-flash results, whereas the axis with lowest *D* has higher values than laser flash results, and both methods provide similar *D* for the intermediate axis. The greatest difference occurring for the tallest cylinder (**Table 1**) supports this deduction. As an additional test, we measured *D* from a single section of olivine as a function of thickness and found that at a height-to-diameter ratio, *b/d*, of 0.1, LFA provides intrinsic values of *D* but polarization mixing clearly occurs when $b/d = 0.2$, elevating *D* by 20%.

Table 1 Summary of k or D measurements of oriented single-crystals at room temperature of geophysically relevant solids

Phase	Face	L (mm)	Method	$D_{R.T.}$ (mm^2 s^{-1})	Upturn T (°C)	Reference, notes
Olivines						
Fo$_{93-94}$ Needles CA	100	1.60	Laser-flash	3.25	—	Pertermann and Hofmeister (2006)
	001	1.88		2.59	—	
	010	1.05		1.66	—	Pertermann and Hofmeister (2006); this work
Fo$_{92}$ Pakistan	010	0.70	Laser-flash	1.65	—	
	010	1.55		2.04	—	
Fo$_{91}$ San Carlos AZ	100	9.1	RHTM	2.71	500	Gibert *et al.* (2005)
	001	6.1, 6.6		2.5	550	
	010	4.5, 5.6		1.77	525	
Fo$_{89}$ San Carlos AZ	100	0.5	PTGS	2.16	R. T. only	Chai *et al.* (1996)
	001	0.5		1.87		
	010	0.5		1.25		
Fo$_{93}$	100	~0.35	Ångström	2.55	Low T	Osako (1997) as cited by Osako *et al.* (2004)
Pakistan	001			2.21		
	010			1.53		
Fo$_{92}$	100	2	Thin wire	2.4$^{c,\,d}$	—	Schärmeli (1982)
San Carlos	001	2		2.1$^{c,\,d}$	—	Extrapolated from high P
	010	2		1.8$^{c,\,d}$	350	
Fo$_{92}$	100	3–5	Modified Ångström	2.18	450	Kobayashi (1974)
Arizona	001	3–5		1.71	525	
	010	3–5		1.07	>550	
Fo$_{88}$ Japan	001	3–5	Modified Ångström	1.85	500	Kanamori *et al.* (1968)
Garnets						
Py$_{71}$Al$_{15}$Gr$_{14}$ Garnet Ridge AZ	Xtl	1.17	Laser-flash	1.27	—	Hofmeister (2006)
~Py$_{60}$Al$_{40}$, Ca?^{a} Delaware, PA	Xtl	3–5	Modified Ångström	1.11	550	Kanamori *et al.* (1968)
~Py$_{55}$Al$_{45}$, Ca? unknown	Xtl	8.6	Steady state	1.54^c	Cryogenic	Slack and Oliver (1971)
Py$_{51}$Al$_{33}$Gr$_{16}$ unknown	Xtl	0.5	PTGS	1.02	R. T. only	Chai *et al.* (1996)
~Py$_{50}$Al$_{50}$, Ca?^{a} Adirondack NY	Xtl	3–5	Modified Ångström	1.09	600	Kanamori *et al.* (1968)
~Py$_{49}$Al$_{51}$, Ca?	Xtl	11.1	Steady state	1.47^c	Cryogenic	Slack and Oliver (1971)
Py$_{43}$Al$_{53}$Gr$_4$ unknown	Xtl	1.03	Laser-flash	1.43	—	Hofmeister (2006)
Py$_{39}$Al$_{50}$Gr$_{11}$	Xtl	1.24	Laser-flash	1.23	—	Hofmeister (2006)

Gore Mt. NY Py$_{35}$Al$_{45}$Gr$_{20}$		1.24	Laser-flash	1.06	—	Hofmeister (2006)
Garnet Ridge AZ Py$_{34}$Al$_{58}$Gr$_{8}$ India		Not reported	Ångström	1.19	Low T	Osako (1997) as cited by Osako et al. (2004)
Py$_{21}$Al$_{74}$Gr$_{5}$ Ft. Wrangle AK		1.61	Laser-flash	1.25	—	Hofmeister (2006)
Py$_{24}$Al$_{74}$Gr$_{1}$ Bahia, Brazil		1.05	Pulse	1.19	Extrap. from P	Osako et al. (2004)
Pyroxene						
Al-rich enstatite	100	0.5	PTGS	1.26	R. T. only	Chai et al. (1996)
Kilbourne Hole NM	010	0.5		1.05		
	001	0.5		1.66		
Mg$_{0.86}$Fe$_{0.14}$SiO$_{3}$	c	3–5	Modified Ångström	1.25	—	Kobayashi (1974)
Unknown	c	3–5		2.35		
CaMg$_{0.97}$Fe$_{0.03}$Si$_{2}$O$_{6}$	001	1.24	Laser-flash	2.54	—	Hofmeister and Yuen (in review)
Dekalb NY	001	1.05		3.84		
CaMg$_{0.95}$Fe$_{0.05}$Si$_{2}$O$_{6}$	100	3–5	Modified Ångström	1.18	250	Kobayashi (1974)
Quebec Canada	010	3–5		1.28	350	
	001	3–5		1.81	250	
Periclase						
MgO synthetic	xtl	0.84	Laser flash	15.22	—	This work
MgO synthetic	xtl	7.5	Two-strip	16.5^c	R. T. only	Andersson and Bäckström (1986)
MgO synthetic	xtl	12.4	Steady state	17.8^c	cryogenic	Slack (1962)
		11.1		16.4^c		
MgO unknown	xtl	2.2	Xenon-flash	19^e	Low T	Makarounis and Jenkins (1962)
Corundum						
α-Al$_2$O$_3$ synthetic, Linde	001	4–10	Steady state	8.1	Low T	McCarthy and Ballard (1951)
	100	4–10		7.4		
Quartz						
SiO$_2$, anhydrous Hot springs AK	001	1.145	Laser-flash	7.17	—	Branlund and Hofmeister (in prep.)
	100	1.198		3.70		
SiO$_2$, 50 H/10^{6}Si synthetic	001	1.740	Laser-flash	6.17	—	Branlund and Hofmeister (in prep.)
	100	1.973		4.45		
SiO$_2$ synthetic	001	8.5–20	RHTM	7.00	Near α–β transition	Höfer and Schilling (2002)
	100	8.5–20		3.60		
SiO$_2$ Minas Gerais Brazil	001	3–5	Modified Ångström	7.14	Near α–β transition	Kanamori et al. (1968)
	100	3–5		3.33		
Spinel						
MgAl$_2$O$_4$	Xtl	1.075	Laser-flash	7.62	—	This work

(Continued)

Table 1 (Continued)

Phase	Face	L (mm)	Method	$D_{R.T.}$ (mm² s⁻¹)	Upturn T (°C)	Reference, notes
Fe + Zn + Cr < 1%, Burma $MgAl_2O_4$	Xtl	1.1	Steady state	7.5	Cryogenic	Slack (1962)
Zn ~ 0.2%, Burma $MgAl_{0.80}Fe_{0.30}Al_{1.90}O_4$ Parker Mine Canada	Xtl	1.78	Laser-flash	2.00	—	This work
$Mg_{0.77}Fe_{0.38}al_{1.84}O_4$ Thailand	Xtl	0.735	Laser-flash	1.75	—	This work
$Mg_{0.75}Fe_{0.40}Al_{1.85}O_4$ Australia	Xtl	1.1	Steady state	1.8^c	Cryogenic	Slack (1962)
Orthoclase						
$K_{0.95}Na_{0.05}AlSi_3O_8$ with 1.2 wt% Fe_2O_3 Itrongay Madagascar	100 010 001	6.1–10 6.1–10 6.1–10	RHTM	0.83 1.11 1.04	~200 ~300 ~300	Höfer and Schilling (2002)
$KAlSi_3O_8$, some Fe^{3+} Fianarantsoa Madagascar	Xtl^b	~2	Radial 3ω	1.13	Cryogenic	Cahill et al. (1992) not chemically analyzed
~$KAlSi_3O_8$ Plenty River, Australia	100 010 001	Not reported	Steady state	1.29 1.43 1.34	R. T. only	Sass (1965) not chemically analyzed

^{a}The stated compositions of Kanamori et al. (1968) are inconsistent with the indices of refraction and lattice constants, indicating that some Ca and/or Mn is present in non-negligible amounts. Slack (1962) did not determine Ca contents, and no cross-check is available.

^{b}Orientation is not reported; the method is radial which collects data over two directions.

^{c}Thermal conductivity was reported. We used C_P and ρ from Anderson and Isaak (1995) or Berman and Brown (1985) to convert k to D.

^{d}Extrapolated from measurements taken at 2.5 GPa using the average value of $D^{-1}\partial D/\partial P = 5\%$ GPa (from **Table 3**, discussed below).

^{e}Radiative transfer effects were not accounted for in the older analyses of laser-flash data, causing overestimation of D values.

Notes: Xtl implies cubic sample, not oriented. Cryogenic implies measurements were acquired below ~298 K; R. T. only implies measurements were acquired near 298 K; low T indicates measurements were obtained below 500° C.

Although no contact with thermocouples exists, PTGS provides very low values for olivine, ~25% below D from laser-flash data (**Figure 8**). Values for garnet and pyroxene seem low as well (**Table 1**). Harrell's (2002) PTGS results for San Carlos olivine are 8% at ~295 K for the same sample studied by Chai *et al.* (1996) using the same technique. Processes in addition to thermal relaxation apparently

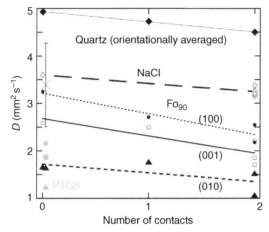

contribute to the rate of decay of the laser pulse (eqn [37]). Coupling of vibronic with electronic transitions may be involved. Absolute values of D from optical method of Pangilinan *et al.* (2000) are also uncertain (**Figure 8**): see their discussion for details.

Data on aggregates are evaluated in view of the above results. The main database of k at ambient conditions (Horai, 1971) made using the needle-point method yields low thermal diffusivity values, compared to all other methods, for a wide range of minerals (**Table 2**). The same conclusion was reached by comparing data on garnets, accounting for variations in chemical composition (Hofmeister, 2006).

In studies of solid aggregates of olivine wherein single crystals were also examined, D at ambient conditions of the dunite is slightly lower than that obtained by averaging the orientations, but larger than $D_{[010]}$ (cf. **Tables 1** and **3**). Lower D for aggregates has been ascribed to preferred orientation of the grains. However, contact measurements for dunites and ceramics provide D on average below laser-flash values by ~40%, whereas contact results for single crystals are low by ~25% (**Figures 8** and **9**). Because ceramics should lack preferred orientation, D of aggregates could be reduced from intrinsic values through contact losses at grain interfaces. Alternatively, small amounts of radiative transfer existing at 298 K for single crystals, but not for aggregates (next section) compensates for the reduction in D due to contact losses.

In short, contact resistance with heaters and thermocouples, and possibly among constituent grains, leads to systematic and substantial underestimation of D and therefore of k_{lat} (~20% for anisotropic samples, somewhat less for cubic symmetry). Measurements using long cylinders also

Figure 8 Thermal diffusivity at 298 K as a function of the number of physical contacts with heaters and/or thermocouples. Lines are least-squares linear fits. Dot, square, and triangles indicate [100], [001], and [010] orientations of olivine, respectively (**Table 1**). Diamonds indicate weighted directional average of quartz (**Table 1**). PTGS measurements of olivine (gray) have problems other than contact, see text. Plus sign represents NaCl (Pierrus and Sigalas, 1985; Håkasson *et al.*, 1988; Håkasson and Andersson, 1986; Brydsten *et al.*, 1983), averaging to $D = 3.25 \, mm^2 \, s^{-1}$. LFA results of $D = 3.6 \, mm^2 \, s^{-1}$ is the average of 10 measurements on a crystal purchased from IR Crystal Laboratories. X indicates NaCl from Pangilinan *et al.* (2000). Modified from Hofmeister AM (in press) Dependence of thermal transport properties on pressure. *Proceedings of the National Academy of Sciences.*

Table 2 Comparison of thermal diffusivity from single-crystals (average of the orientations over the available studies, **Table 1**) to needle-probe measurements on powder in water

Phase	Laser-flash	RHTM	Contact	Needle-probe[d]
Fo$_{93-95}$	2.50	2.32	1.93	1.88[a]
Py$_{39}$Al$_{50}$Gr$_{11}$	1.24			1.44
Mg$_{0.85}$Fe$_{0.15}$SiO$_3$			1.61	1.62
CaMg$_{0.96}$Fe$_{0.02}$Si$_2$O$_6$	2.97		1.42	2.22
SiO$_2$	4.93[b]	4.73	4.6	3.91
Mg$_{0.9}$Fe$_{0.1}$Al$_2$O$_4$[c]	5.4		5.5	3.21

[a]Average of three samples.
[b]Average of two samples.
[c]The density of Horai's (1971) sample suggests the slightly ferrous composition listed here. Linear interpolation of the data in **Table 1** was used to obtain the crystal D-values for this composition.
[d]Horai (1971).

Table 3 Thermal diffusivity and its derivatives at room temperature of forsteritic olivine, dunites, and aggregates

Method	Sample	D (mm^2 s^{-1})	$\partial D/\partial T$	$D^{-1}\partial D/\partial P$ (% GPa^{-1})	P_{max} (GPa)	Reference and Notes
Laser-flash	Crystal	2.50	−0.011	—	—	Pertermann and Hofmeister (2006)
	Dunite	2.31	−0.010	—	—	Pertermann and Hofmeister (2006)
		2.05	−0.0091	—	—	Pertermann and Hofmeister (2006)
RHTM	Crystal	2.32	−0.0072	—	—	Gibert et al. (2005)
	Dunite	2.34	−0.0063	—	—	Gibert et al. (2005)
Contact	Crystal	2.06	−0.0090	3.6	8.3	Osako et al. (2004), extrapolated
	Ceramic	1.3	−0.0052	3.6	10	Xu et al. (2004), extrapolated
	Dunite[a]	1.40	—	11.1	1	Gibert et al. (2003)
		1.38	—	8.1	1	Gibert et al. (2003)
		1.28	—	9.1	1	Gibert et al. (2003)
	Crystal	1.65	−0.004	—	—	Kobayashi (1974)
	Dunite[a]	1.3-1.9	—	—	—	Kobayashi (1974)
	Fragment	2.07	−0.006	5	5.6	Beck et al. (1978)@350 K
	Fragment	3.47	−0.007	7	5.6	Beck et al. (1978)
	Dunite	1.37	−0.0018	5	5.6	Beck et al. (1978)
	Dunite	1.34	−0.0028	4	5.6	Beck et al. (1978)
	Ceramic[b]	1.14	−0.0020	4.7	9	Katsura (1995)@400 K
	Ceramic[c]	1.0	−0.0028	11	0.2	Staudacher (1973)@373 K
	Ceramic	1.92	−0.0030	15	0.5	Fujisawa et al. (1968)@400 K
	Ceramic[d]	1.08	−0.0027	—	—	Fujisawa et al. (1968)@500 K
PTGS	Crystal	1.76	−0.0041	4–6	5	Chai et al. (1996); Harrell (2002)

[a]Different orientations of the same sample were examined.
[b]Contains 5% enstatite, which has a higher-pressure derivative (**Table 4**).
[c]Mg_2SiO_4.
[d]Fe_2SiO_4.
Notes: crystal, average of three orientations. Fragment, unoriented crystal; Dunite, natural rock; characterized to varying degrees. Ceramic, hot pressed, fine-grained, synthetic near Fo$_{90}$ unless noted otherwise.

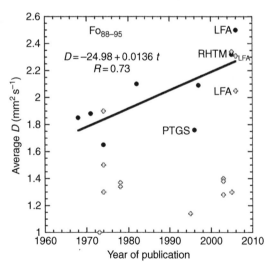

Figure 9 Time evolution of thermal diffusivity for olivine. Data from **Table 1**. Points not labeled involve conventional methods. Circles and line indicate single crystals and cross indicates aggregates.

underestimate anisotropy. Samples with high D values seem to be more accurately measured by contact methods. However, radiative transfer may offset such losses for single crystals, particularly for very transparent materials (spinel gems, MgO, and SiO$_2$). Results obtained by Slack (1962, 1964) are generally high (**Table 1**) which may be due to direct radiative transfer ocurring inside his large gemstones. Insufficient information exists to distinguish between these possibilities, and the proportion of each effect likely depends on individual experiments.

19.4.1.2 Elevated temperature

Thermal conductivity of olivine single-crystals decreases steeply from ambient temperature values (**Figure 4**). Initial slopes ($\partial k/\partial T$ or $\partial D/\partial T$) are nearly parallel, but as temperature increases to about 600 K, slopes from conventional methods and RHTM become less steep compared to laser flash results. At ∼800 K, many studies have positive $\partial k/\partial T$

which is known to originate from unwanted, direct radiative transfer (Section 19.2.2.1). Radiative transfer must be present at low temperature also because (1) gradual changes occur in both the lattice component and in direct radiative transfer with temperature, that is, no abrupt 'onset' exists, (2) a small radiative component is clearly seen by ~350 K (and sometimes at as low as 298 K) in time–temperature curves from LFA experiments, even when samples are coated with metal and/or graphite. For RHTM, initial values of $\partial D/\partial T$ are larger than for conventional methods (**Table 3** and **Figure 4**). We attribute this difference to removal of some, but not all, of direct radiative transfer effects (e.g., Gibert *et al.*, 2005). Direct radiative transfer is not limited to measurements with obvious upturns, but almost certainly occurs in contact measurements with flat or shallowly decreasing $\partial k/\partial T$ as well, for reasons similar to those enumerated above. The amount of direct radiative transfer depends on experimental details such as transparency, sample thickness, iron content, physical scattering, the temperature gradient, and the emissivity of the heater and thermocouples in contact with the sample.

That radiative transfer is lower in aggregates due to physical scattering accounts in part for the discrepancies in D between aggregates and single crystals (**Figure 9**), that is, single-crystal measurements are overall higher because a radiative component offsets contact losses.

Systematic errors likely coexist in $\partial D/\partial T$ due to contact resistance. One possibility is that both contact resistance and radiative transfer contribute to the variability at elevated temperature. These two effects cannot be differentiated, given the available data.

The accuracy of D from silicate spinels by Xu *et al.* (2004) is evaluated by examining their results for olivine obtained under similar conditions. Because their samples are ceramics, radiative transfer is not expected. At high T, $\partial D/\partial T$ is similar to that from laser-flash measurements (**Figure 10**), consistent with fine-grained ceramics suppressing radiative transfer. However, extrapolation to 1 atm. provides D similar to those of the [010] axis with the lowest value. Preferred orientation was not detected, and is not expected for their fine grain-size, so D of the ceramic at 1 atm should instead nearly equal the average of the three orientations. At 400 K, the average of the single-crystal measurements is 1.69 mm^2 s^{-1}, and Xu *et al.*'s (2004) values are below this by 0.61 mm^2 s^{-1} (36%). At 1000 K, the average of the single-crystal measurements is 0.77 mm^2 s^{-1}, and

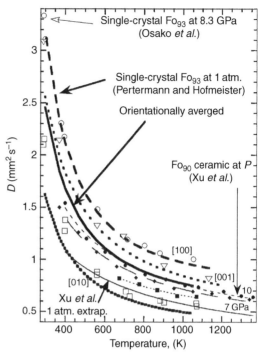

Figure 10 Comparison of D for olivine aggregates (filled symbols: Xu *et al.*, 2004) and single-crystals (open symbols: Osako *et al.*, 2004), both at P and T, to 1 atm. single-crystal data from laser-flash analysis (heavy broken lines: Pertermann and Hofmeister, 2006). Heavy solid line indicates average of LFA oriented values at 1 atm. Open circles indicate [100]; open triangles indicate [001]; open squares indicate [010]. Thin broken lines show fits of Xu *et al.*'s data to $D = a + b/T$ with residual = 0.99. Dot indicates 10 GPa; diamond indicates 7 GPa; square indicates 4 GPa. Thin solid line indicates extrapolation from high-pressure fits of Xu *et al.* (2004) to $D = c/T^n$ to 1 atm. Modified from Hofmeister AM (in press) Dependence of thermal transport properties on pressure. *Proceedings of the National Academy of Sciences*.

Xu *et al.*'s (2004) values are below this by 0.20 mm^2 s^{-1} (33%). One explanation for this trend is that contact resistance is temperature dependent. Alternatively, although radiative transfer is not expected for ceramics, these samples are very thin and it is possible that a small amount exists at elevated T. The data are insufficiently accurate to constrain the room temperature value through fitting, that is, fits to $D = B_0 + B_1/T$ (**Figure 10**) or B/T^m (by the authors) have similar residuals but extrapolate differently. Based on comparing their results for olivine to LFA data, we suggest that Xu *et al.*'s (2004) analysis underestimates D of wadsleyite and ringwoodite by ~33% above 400 K.

For single-crystal olivine, $D(T)$ at 8.3 GPa (Osako *et al.*, 2004) nearly coincides with LFA results at

1 atm. (**Figure 10**), whereas dunite curves at 10 GPa (Xu *et al.*, 2004) match the average of the LFA curves. Slight radiative transfer is evident in [010] above 800 K (**Figure 10**), which suggests that lesser amounts may affect [001] and [100]. Radiative transfer likely compensates for contact losses. Therefore, we did not curve-fit the results of Osako *et al.* (2004).

Similarly, *D(T)* curves from single-crystal almandine at high *P* in the contact experiments of Osako *et al.* (2004) are parallel to LFA results at 1 atm. and low *T*, but some radiative transfer is evident in the trend of *D* with *T* above 800 K (**Figure 10**). Garnets with Fe in dodecahedral coordination have an absorption band near 5000 cm^{-1} (e.g., Manning, 1967) which suppresses direct radiative transport, but the samples of Osako *et al.* (2004) are thin enough (~0.35 mm) to transmit light. Occurrence of both radiative transfer and contact resistance at 298 K are consistent with the small difference of 5% for garnets at ambient conditions (**Table 1**).

Although contact methods do not provide absolute values, *D(T)* curves so obtained at various pressures are roughly parallel to each other and to LFA results at 1 atm., suggesting that the *P* and *T* components of *D* are separable. Thus, LFA data on *D(T)* for dense phases can be extrapolated to mantle conditions, if reliable pressure derivatives for *D* exist (next section).

19.4.1.3 *Elevated pressure*

Pressure derivatives have almost entirely been obtained using contact methods (**Tables 3** and **4**). One concern is the accuracy to which pressure was determined, which is difficult to gauge for older studies and those which use manganin coils. We do not dismiss these studies, but it is necessary to dismiss studies in which the sample cracked, which increases thermal resistance, or is porous, wherein compression serves to reduce porosity. Both cases artificially elevate pressure derivatives. On this basis, we dismiss the results of Yukutake and Shimada (1978) on SiO$_2$. Results for fused silica and quartz by Horai and Sasaki (1989) are discarded because these were obtained at very low pressures and have much higher derivatives than any of the other studies, including those wherein the sample cracked (Kieffer *et al.*, 1976; **Table 4**).

Pressure derivatives for olivine systematically decrease as *P* of the experiments increase (**Table 3**). This behavior is attributed to progressive compression altering the interface, that is, better thermal contact occurs when the sample is under pressure.

Comparing data at two different elevated pressures should provide accurate derivatives, whereas comparing data at ambient and elevated pressure likely overestimates $\partial D / \partial P$. The most reliable values for olivine are thus obtained from the recent single-crystal study at very high pressure (Osako *et al.*, 2004), and are corroborated by the high-pressure study of hot-pressed ceramic by Xu *et al.* (2004). Their values are similar to those obtained using PTGS, which does not involve thermal contact. Chai *et al.* (1996) report that room temperature PTGS measurements of San Carlos olivine at 1 atm. and 4.8 GPa provide $D^{-1} \partial D / \partial P$ of about 4% GPa^{-1}. Harrell (2002) reported a significantly higher derivative for *D* of 6% GPa^{-1} for the same sample with the same technique. Contact measurements of NaCl (**Table 4**) also have $\partial D / \partial P$ depending on *P*. We suggest that the most recent, optical measurement of Pangilinan *et al.* (2000) best constrains the pressure derivative.

High slopes for stiff mantle minerals at low pressure are not intrinsic, and are attributed to compression of the thermal contacts, pore space, or trace interstitial phases. Pressure derivatives from low-*P* studies of granular rocks are unlikely to represent intrinsic behavior for any material. Andersson and Bäckström (1986) and Katsura's (1997) measurements constrain *D(P)* for MgO. Glass and quartz are problematic due to radiative transfer and cracking. The remaining measurements (last six lines in **Table 4**) are uncorroborated, and the pressure derivatives must be considered uncertain.

19.4.2 Laser-Flash Data on Various Minerals

To help constrain *D* of magnesiowüstite and ringwoodite, new LFA data are presented for ceramic MgO, and single crystals of MgO, MgAl$_2$O$_4$, and Mg$_{0.8}$Fe$_{0.3}$Al$_{1.9}$O$_4$. The ceramic has 4.5% porosity. This sample and single-crystal MgO were purchased from Alfa-Aesar. Natural, essentially end-member spinel is nearly gem-quality and pale pink due to trace Cr. High *T* results are shown for another slightly impure gem (Mg$_{0.96}$Fe$_{0.01}$Zn$_{0.02}$Al$_{2.01}$O$_4$). The opaque hercynite has a few fractures. Experimental procedures are as described by Hofmeister (2006).

Figure 11 shows that *D* of MgO and both spinels decrease with temperature such that very flat trends are seen above some high temperature of ~1200–1600 K. The trend below this transition temperature are well-described by eqn [17], but using eqn [16] requires fewer fitting parameters for MgO and both

Table 4 Pressure derivatives of thermal conductivity from mostly contact methods for minerals and insulators other olivine

Sample	Form	T (°C)	P_{max} (GPa)	$k^{-1}\partial k/\partial P$ (% GPa^{-1})	γ_{th}	K_T (GPa)	Reference and Notes
MgO	Polycrystal	100	5	4.0[a]	1.54	160	Katsura (1997)
	Crystal	25	1.2	5.0			Andersson and Bäckström (1986)
	Crystal	~22	4	6.8[b]			Yukutake and Shimada (1978)
	Crystal	20	5	2.0			MacPherson and Schloessin (1982)
NaCl	Crystal	19	5.6	18	1.58	23.8	MacPherson and Schloessin (1982)
	Crystal	69	5	17			Beck et al. (1978)
	Crystal	22	4	31			Yukutake and Shimada (1978)
	Polycrystal	~20	4	32.7			Pierrus and Sigalas (1985)
	Polycrystal	~20	2	31			Håkasson et al. (1988)
	Polycrystal	22	2	30.7			Håkasson and Andersson (1986)
	Polycrystal	40	1.8	36[a]			Kieffer et al. (1976)
	Crystal	~20	1.7	27[c]			Pangilinan et al. (2000)
SiO$_2$ glass	Slab	100	9	−3.7[a]	0.036	36.5	Katsura (1993)
	Slab	~22	1	−4[a]			Andersson and Dzhavadov (1992)
	Slab	40	3.6	+1.8[a,b]			Kieffer et al. (1976)
	Slab	~22	1.0	6			Horai and Sasaki (1989)
quartz ⊥**c**	Crystal	40	3	20[a,b]	0.667	37.5	Kieffer et al. (1976)
	Crystal	64	5.3	2.2[d]			Beck et al. (1978)
	Crystal	~22	1.2	50			Horai and Sasaki (1989)
quartz ‖**c**	Crystal	82	5.3	11[d]			Beck et al. (1978)
quartz, both	Crystal	~22	1.2	29			Horai and Sasaki (1989)
coesite	2% porosity	19	4	3.9	0.41	113	Yukutake and Shimada (1978)
	Polycrystal	76	5.6	1.4–4.4			Beck et al. (1978), two samples
stishovite	1% porosity	25	4	9.0	1.2	306	Yukutake and Shimada (1978)
Py$_{25}$Al$_{74}$Gr$_1$	Crystal	~22	8.3	4.6	~1.1	177	Osako et al. (2004)
Mg$_{0.85}$Fe$_{0.15}$O$_3$	Crystal	40	5.6	7	0.97	107	Schloessin and Dvorak (1972)
~NaAlSi$_2$O$_6$	Unknown	~22	3	4.6	1.06	125	Osako et al. (2004)
sulfur	Polycrystal	27	2	69	1.12	8.8	Nilsson et al. (1982)
CaF$_2$	Crystal	R.T.	1.0	11	1.83	86.3	Andersson and Bäckström (1987)
NaClO$_3$	Polycrystal	22	2	32	1.85	24.6	Franson and Ross (1983)

[a]Calculated from eqn [20].
[b]Cracked during the experiment, values questionable.
[c]Optical technique.
[d]Partial conversion to coesite, values questionable.
Notes: For additional alkali halides with rocksalt or CsI structures see Ross et al. (1984); Hofmeister (2007).

spinels, and is provided in **Table 5**. Ceramic MgO has slightly lower D values, in accord with the porosity (**Figure 11**). The hercynite has much lower D and a flatter trend than essentially end-member spinel.

19.4.2.1 Effect of chemical composition and hydration on room temperature values

For solid solutions of olivine and garnet, the lowest values of D or k_{lat} are measured near the middle of each binary, with steep decreases from end-member values (Hofmeister, 2006; Pertermann and Hofmeister, 2006). Similar, but more irregular, trends were seen in contact measurements of feldspar, olivine, pyroxene, and garnet (Horai, 1971). For low concentrations of impurities in olivine-group minerals (<10 atom% from end-member), D for each

orientation was found to linearly depend on the mass of the formula unit (Pertermann and Hofmeister, 2006).

That impurity ions lower D and k_{lat} in contact measurements is well-known (e.g., Slack, 1964). Laser-flash measurements show that hydration lowers the lattice component of thermal diffusivity for calcic and Fe–Mg garnets, and glasses. Hydration depresses D above the glass transition, suggesting that melts respond likewise. For olivine and quartz, D is unaffected, but the expected response of ρ and C_P to hydration means that k_{lat} decreases with hydration. The best-constrained trends (the garnets) provide

$$\partial(\ln D)/\partial X = -0.0030\%(\text{ppm H}_2\text{O})^{-1} \quad [38]$$

19.4.2.2 Comparison of D(T) for dense oxides and silicates

Similar behavior was observed for the oxides (**Figure 11**), 23 garnets with different chemical compositions (Hofmeister, 2006), 11 olivine-group crystals and 7 polycrystals (Pertermann and Hofmeister, 2006), and 11 samples of quartz (Branlund and Hofmeister (in press)) For garnets, thermal diffusivity is constant once temperature (T) exceeds a critical value (T_{sat}) of ~1100–1500 K. From ~290 K to T_{sat}, the measurements are best represented by eqn [16]. The constants vary little among diverse chemical compositions, suggesting that the oxygen sublattice controls heat transport. Higher-order terms are needed only when T_{sat} is low. The initial slope $\partial D/\partial T$ decreases as D at 298 K decreases. For olivines, saturation was not reached, although the trends became very flat above 800–1400 K, depending on the composition and orientation. Olivine family minerals are better fit by eqn [17].

For quartz, the displacive phase transition near 573 K creates a 'λ'-shaped curve in $1/D$, much like that in C_P, which precludes curve fitting for the α polymorph. For high-temperature β-SiO$_2$, D depends linearly on T over the narrow T range examined, and the trends are very flat, much like the high-T results for the dense minerals shown in **Figure 11**. For a pure, dry β-quartz from Hot Springs, AK,

$$D_{[001]} = 1.3483 - 0.00001\,T \qquad [39]$$

$$D_{[100]} = 1.0522 - 0.00002\,T \qquad [40]$$

19.4.3 Comparison of the Room Temperature Lattice Contribution to Theoretical Models and Estimation of D and k for Some High-Pressure Phases

Ascertaining heat transport at ambient conditions for high-pressure phases and extrapolation to mantle conditions requires a model that has been benchmarked against reliable data. To date, few minerals have been accurately characterized, and only rough estimates can be provided for high-pressure

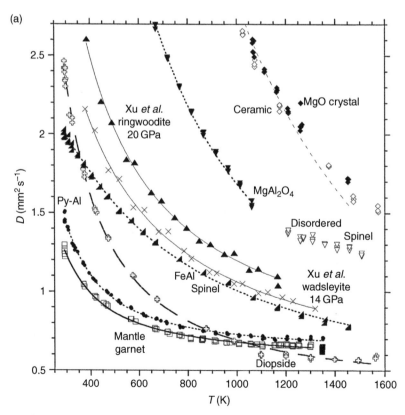

Figure 11 (Continued)

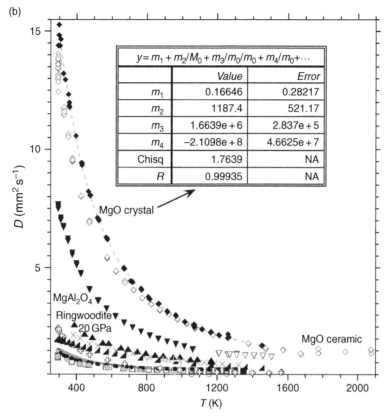

Figure 11 Laser-flash (at 1 atm.) and high-pressure measurements of thermal diffusivity as a function of temperature. Filled diamonds indicate MgO crystal. Open diamonds indicate ceramic MgO. Filled down-pointing triangles indicate near end-member spinel of **Table 1**. Open triangles indicate spinel with ~3% impurities that was disordered during heating. Up-pointing triangles indicate Fe-rich ringwoodite at 20 Gpa from Xu *et al.* (2004). X indicates Fe-rich wadsyleyite at 14 GPa from Xu *et al.* (2004). Right triangle indicates Fe-rich spinel. Plus indicates diopside. Gray line indicates Al-Py at 8.3 GPa from Osako *et al.* (2004). Gray circles indicate same composition of almandine at 1 atm. Open circles indicate almandine near melting. Filled circles indicate Py-Al. Open squares indicate mantle garnet (Ant Hill sample). Filled squares indicate partial melted mantle garnet. Laser-flash garnet data from Hofmeister (2006), see **Tables 1** and **5** for details on other samples. (a) Expanded view at low temperature showing fits to eqn [17]. Note parallel trends for the oxides. (b) Full view of oxides and spinels. MgO requires a high-order fit, given in the inset. The tendency is towards constant *D* at high *T*.

polymorphs. Existing data point to the validity of the DHO model.

19.4.3.1 Ambient conditions and compositional dependence

Even for simple substances (e.g., diatomic solids), agreement of absolute values with the acoustic models is lackluster (see Slack, 1979). Because of this and because acoustic models actually are valid only above the Debye temperature, comparisons are made to T and P derivatives (next sections).

Room temperature contact results for olivine, spinel, and garnet family minerals and calcium aluminates are reasonably well represented by the DHO model, using various schemes for $<u>$ for each mineral family, for example, $<u> = u_p$ for diatomics, but $1/2(u_p + u_s)$ for olivines and spinels (Hofmeister, 2001; Giesting and Hofmeister, 2002; Giesting *et al.*, 2004; Hofmeister, 2004b, 2004c). Unfortuately, the above calibrations were made against systematically low k_{lat} values from contact measurements, so predictions in these references set lower limits. One exception is majoritic garnets, since the contact data for pyrope-almandine garnets are reasonably close to LFA results (~5% in many cases, see Tables). For end-member majorite, $MgSiO_3$, $D = 3.50\,mm^2\,s^{-1}$ at 298 K from Giesting *et al.* (2004). Additional vibrational spectroscopic data and laser-flash measurements for a wide range of structures are needed to utilize this model predictively for other mantle phases. As is the case for contact methods, the model sets a lower limit on D.

Table 5 Laser flash results for lattice heat transport of dense phases and estimates for ringwoodite and perovskite

Sample	Fit of D^{-1} vs T in $mm^{-2}s$ for T in K	Fit range (K)	D_{sat} ($mm^2 s^{-1}$)	$k_{lat}^{-1}\partial k_{lat}/\partial P$ (%GPa^{-1})[a]
MgO	$-0.0033499+0.00029431T+9.4369\times10^{-8}T^2$	290–1500	1.50	4.16
Spinel	$-0.064053+0.00065317T$	290–1060	$\sim$1.4[b]	2.8
Hercynite	$-0.29973+0.00068264T$	290–1450	0.7[b]	$\sim$3
Ringwoodite[c]	$fx\ (0.26533+0.0010477T-2.3724\times10^{-7}T^2)^c$	400–1170	—	2.9
Olivine [100][d]	$-0.19368+0019392T-7.0201\times10^{-7}T^2$	290–1250	—	4.3 (bulk)
[001]	$-0.15084+0.0020126T-6.7467\times10^{-7}T^2$	290–1150	—	—
[010]	$-0.50339+0.0042683-1.7989\times10^{-6}T^2$	290–1030	—	—
$Py_{43}Al_{53}Gr_4$[e]	$-0.097321+0.0034104T-2.6156\times10^{-6}T^2+6.8585\times10^{-10}T^3$	290–1300	0.695	$\sim$3.7
$Py_{35}Al_{45}Gr_{20}$[e]	$0.049403+0.0032299T-2.3992\times10^{-6}T^2+6.0168\times10^{-10}T^3$	290–1105	0.661	$\sim$3.7
Perovskite[f]	$(0.062\pm0.014)+(0.00215\pm0.00005)T$	160–340	—	2.4
Diopside $\|c$[g]	$-0.022932+0.0010246T-2.3197\times10^{-7}T^2$	290–1500	0.58	3.5
$\perp c$[g]	$-0.16466+0.002055T-4.184\times10^{-7}T^2$			
CaF_2[h]	$0.53911-0.0002419T+1.4396\times10^{-6}T^2$	470–830	—	8.9

[a]Value at ambient conditions, calculated from $(4\gamma_{th}+1/3)/K_T$, thermodynamic parameters summarized by Hofmeister and Mao (2003).
[b]$T_{sat}\sim$>1500 K.
[c]$f=2.15$, calculated assuming that γ-Mg_2SiO_4 has the same D at 298 K as $MgAl_2O_4$, and that the dependence of D on Fe content for ringwoodite is the same as that measured for olivine by Pertermann and Hofmeister (2006). Polynomial in T is from fitting the high-pressure contact measurements of Xu et al. (2004).
[d]Perterman and Hofmeister (2006) list fits to eqn [17] for olivine and olivine family minerals, as these better represent the data. For consistency and comparison, fits to D^{-1} are given here.
[e]Hofmeister (2006). Data on 21 other garnet compositions are provided. These garnets should represent majorite-pyrope solid solutions near the middle of the binary (see Giesting et al., 2004).
[f]$MgSiO_3$, room temperature $D=1.72\ mm^2s^{-1}$ and $D(T)$ from contact measurements of Osako and Ito (1991): spectroscopic calculations give similar values (Hofmeister, 2004b). Both provide a lower limit. Using recent values of C_P sets a lower limit on $k_{lat}=5.7\ Wm^{-1}K^{-1}$ at 298 K, see Hofmeister (2004b). These values are 5–25% too low, based on Figures 8–10.
[g]Hofmeister and Yuen (in press).
[h]Hofmann et al. (1997).
Notes: Source for $D(T)$ is this work, unless noted. Residuals for all fits are better than 0.995.

The nonlinear compositional dependence of D is consistent with the DHO model, but not with the acoustic formulations (eqns [6] and [10]). In particular, impurities lower D substantially from end-members (Perterman and Hofmeister, 2006): no minimum is needed, as suggested by acoustic models, and the form of the temperature dependence does not differ between end-members and solid solutions as postulated in acoustic models. For garnets, sound speeds change linearly across various binary solid-solutions, but <FWHM> increases sharply from the end-members, having a maximum near mid-binary, thus eqn [15] reproduces the experimentally determined compositional dependence of D and k_{lat} for garnets (Giesting and Hofmeister, 2002). Cation disorder (and also protonation) causes the nonlinear dependence of the FWHM on composition because structural disorder adds new vibrational peaks (e.g., two-mode behavior, Chang and Mitra, 1969). Because more phonons exist in the disordered (or wet) than in the corresponding ordered (or dry) phase, the number of scattering events per primitive unit cell is larger, decreasing phonon mean free paths and thus D.

19.4.3.2 Elevated temperature

For all materials studied using LFA, $D(T)$ at high temperature has a very low slope, and for some materials, D is constant at very high T. The asymptotic limit (T_{sat}) was reached for many different garnets (Hofmeister, 2006), $MnGeO_4$ olivine (Pertermann and Hofmeister, 2006), diopside (Hofmeister and Yuen, in press), and MgO (**Figure 11**). This observation is not consistent with traditional models based on acoustic modes. For many samples, asymptotic values are reached before melting occurs, in contrast to the deduction of Roufosse and Klemens (1974). Also, the mean free path at saturation (λ_{sat}, computed from eqn [15]) is significantly larger than the primitive lattice parameter (as predicted in the acoustic models), even if the average sound speed is approximated by u_S, which is probably less than <u>. Neither do the forms derived from acoustic models of heat transport (eqns [6]–[12]) describe the data below T_{sat}. The same behavior is seen regardless of whether the chemical composition is end-member, near end-member, or extensive solid solution, and thus disorder and defects do not control the functional dependence of D on temperature, in contrast to the predictions of the acoustic models.

The observed behavior with T is consistent with the damped harmonic oscillator model. Existence of a high-T asymptote for D is derived from statistical thermodynamics. Specifically, D is independent of T at roughly double the Debye temperature, which is consistent with saturation of overtone-combination phonon densities. From ∼290 K to T_{sat}, the measurements are represented by eqn [16] and provide constants varying little among diverse chemical compositions with the garnet structure (Hofmeister, 2006). That the linear coefficient dominates is consistent with the roughly inverse temperature behavior long considered to dominate vibrational transport. That olivine is better represented by eqn [17] is attributed to sound speeds of olivine being less constant than those of garnet; $\partial(\ln u)/\partial T$ is near −0.004% K^{-1} for grossular and pyrope garnets but almost double for forsterite and olivine, −0.006 to −0.008% K^{-1} (Anderson and Isaak, 1995).

Phonon transport is best represented by inverse thermal diffusivity wherein $1/D$ goes as T^n where n is between 1 and 3 up to ∼200 K, depends on a quadratic or cubic polynomial at moderate T (eqn [16]), but approaches a constant above some very high temperature. Because no mechanism for exchanging heat between phonons exists at $T = 0$ K, then $1/D$ should approach 0 as T approaches 0, that is $1/D$ goes as T^m, where $1 < m < 3$ at cryogenic temperatures (see Slack and Oliver, 1971, Cahill *et al.*, 1992). The predicted and observed temperature response of $1/D$ mimics the well-known form for heat capacity, in that acoustic modes control heat transport near cryogenic temperatures, optic phonons dominate above ambient temperature, and a limit analogous to that of Dulong and Petit is reached at high temperature, due to full population of discrete phonon states. The change in $C_P(T)$ results from populations of vibrational levels changing with temperature. The possibility of interaction (phonon–phonon scattering, described by FWHM) concomitantly increases with population of the levels. Therefore, $1/D$, being proportional to FWHM, has a temperature dependence like that of C_P.

19.4.3.3 Elevated pressure

Reliable data on the pressure derivatives $k_{lat}^{-1}\partial k_{lat}/\partial P$ (**Tables 3** and **4**) are best fit to eqn [18], derived

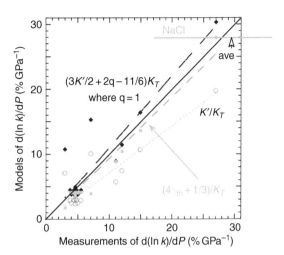

Figure 12 Comparison of calculated to reliable measured pressure derivatives of thermal conductivity. Gray indicates comparison with eqn [18] (DHO model). Diamond and dashed line indicates comparison with eqn [13b] (dimensional analysis). Open circles and dotted line indicates comparison with eqn [19a] (bulk sound model) which used K' from the compilation of Hofmeister and Mao (2003). Solid line indicates 1:1 ideal correspondence. Gray line indicates range of NaCl measurements. Short arrow points to the average.

from the summation using the optic model (**Figure 12**). Our comparison omits the unconfirmed measurements of the soft alkali halides and data on glasses. Equation [13a], obtained from three different acoustic models, significantly overpredicts the measurements and is not shown in **Figure 12**. Dimensional analysis (eqn [13b]) slightly overpredicts the results with significant scatter. A better fit could be obtained with q_{th} slightly less than unity. The bulk sound model (eqn [19a]) underpredicts the data with scatter but is reasonable for hard silicates and oxides. For high-pressure extrapolations, dependence of K_T and γ_{th} on P need be accounted for in applying eqn [18] or [19a]. Different results are obtained from these two formulations because mode velocities depend on mode frequencies.

19.4.4 Lattice Thermal Conductivity and Its Temperature Dependence

Because D has a simpler temperature dependence than k_{lat} (eqn [2]) and the physical properties relating these (ρ and C_P) are fairly well constrained, the focus of this report has been on D. Understanding heat transport has been impeded and seems to include erroneous beliefs because of the historical focus on thermal conductivity.

For all insulators, thermal conductivity peaks below room temperature (Slack and Oliver, 1971). The peak in $k_{lat}(T)$ exists because the product $\rho C_P D$ has a peak near or below room temperature. The behavior has been demonstrated for garnets (Hofmeister, 2006), but should be universally true for electrical insulators, as follows. Because D and ρ are lower order in T at low temperature than is C_P, which goes as T^3, heat capacity increasing as T increases dominates k in the cryogenic regime. Above room temperature, k decreases as T increases and asymptotically approaches a constant value at high temperature because the decrease in D with T is strong; the decreases in ρ with T is weak and are not offset by the weak increase C_P with T. The distinct behavior of k for different temperature regions results from combining $1/D$ and C_P, both of which are complex functions of temperature (e.g., **Figure 11**; Berman and Brown, 1985). Because the behavior of $1/D$ and C_P with T are both responses to changing populations of the various types of vibrational states with temperature, the complex form for k has the same origin. It is unnecessary, and probably incorrect, to describe k at different temperatures in terms of normal versus umklapp processes as commonly assumed (e.g., Ziman, 1962).

19.5 Calculation of the Effective Thermal Conductivity for Diffusive Radiative Transfer

Olivine is only mineral which has sufficient spectroscopic data at temperature (Taran and Langer, 2001; Ullrich *et al.*, 2002) to allow evaluation of $k_{rad,dif}$ from the integral of eqn [28]. Measurements at ~290 K and pressure (e.g., Keppler and Smyth, 2005; Goncharov *et al.*, 2006) do not constrain mantle values, providing only the roughest of estimates. It may be possible to use the pressure dependence to project the room temperature data to high pressures, especially if some temperature measurements are available. One concern is absolute values of A, since measurements are of transmitted light and include a reflection component.

The results for olivine should reasonably approximate other minerals with Fe^{2+} in an octahedral site, that is, the mantle minerals orthopyroxene, clinopyroxene, magnesiowustite, wadsleyite, and ringwoodite. The results (Hofmeister, 2005) depend

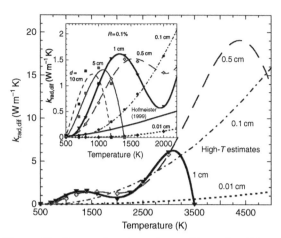

Figure 13 Temperature dependence of $k_{rad,dif}$ for olivine. Symbols are value calculated from eqn [28], assuming that opacity occurs for $dA \geq 7$. Curves are polynomial fits, labled with grain size. Gray curve indicates parametrization of Hofmeister (1999). Inset are low-temperature values. Reprinted from Hofmeister AM (2005) The dependence of radiative transfer on grain-size, temperature, and pressure: Implications for mantle processes. *Journal of Geodynamics* 40: 51–72.

strongly on grain size (**Figure 13**). Interface reflectance increasingly influences the results as d increases above 5 mm, but not for small grain sizes likely in the mantle. For $d < 2$ mm and R ~0.1%

$$k_{rad,dif} = 10d[0.36776 - 0.0010594T + 8.3496T^{-2}] \quad [41]$$

For large grain-size, radiative transfer is complex due to the convolution of the strong d-d absorptions with the blackbody curve.

The T^3 law is obviously not valid for minerals with absorption bands, as confirmed by eqn [34], derived for narrow bands and low dA. Equation [33] confirms the essentially linear dependence of $k_{rad,dif}$ on d found for olivine (eqn [41]). Fitting eqn [33] below 2500 K to a polynomial confirms the format for the T dependence of eqn [41].

From eqn [33] and the information above and in the theory section, $k_{rad,dif}$ depends largely on how dark (i.e., Fe-rich) or how pale (i.e., very Fe-poor) individual grains are, more than on the specific spectrum possessed by a given mineral phase. Because completely opaque minerals do not transfer radiation, very Fe-rich minerals have low $k_{rad,dif}$. Very Fe-poor minerals have low $k_{rad,dif}$ because these are poor emitters, and thus $k_{rad,dif}$ has a maximum at moderate Fe contents. At a given Fe content, the dependence on d is similar to that on X, but it is truely the product dA or dX which controls radiative diffusion. The convolution of grain-size and absorbance (or concentration) and the convolution of Planck's blackbody curve with changes in spectral parameters with temperature makes radiative transfer in the mantle a highly complex and counter-intuitive phenomena.

19.6 Conclusions

The unfortunate conclusion is that methods commonly used in geologic science, which involve contact, fall short of providing accurate and independent measurements of phonon transport, due to both the presence of unwanted direct radiative transfer and resistance at contact interfaces. The laser-flash technique is accurate, benchmarked against soft, opaque materials, but one drawback is that the sample size required is too large for high-pressure synthetics, and another is the cost of the instrumentation. Improving accuracy and precision in determining heat transport properties requires further implementation of existing all-optical techniques to a greater variety of minerals and high-pressure structures, and, possibly, development of additional all-optical techniques with similar or greater accuracy than the laser-flash measurements. The PTGS method is not recommended, as the results reflect processes in addition to thermal relaxation.

Available data indicate that the P and T components of D (or of k_{lat}) can be modeled separately and recombined to provide data at mantle conditions. Acoustic-based models do not reproduce observed behavior of D. Pressure derivatives are predictable from the DHO model, given that thermal Grüneisen parameters and bulk moduli are well-known. Understanding the temperature derivatives requires additional laser-flash experiments and some vibrational spectroscopy at temperature. For the high-pressure silicates, only rough estimates can be provided at present.

A new model for diffusive radiative transport accounts for grain size, which is equally as important as attenuation of light through absorption. Spectroscopic data are insufficient to make full use of this model at present. High temperature, simultaneously elevated P and T, and a wide frequency range need to be investigated in spectroscopic measurements. At present, only approximate results and limited information on the pressure, temperature, composition, and grain-size dependence of $k_{rad,dif}$ are available, except for olivine, wherein results

should be as accurate as possible up to 2000 K, given the approximate nature of all radiative transfer models.

Acknowledgments

This work is supported by NSF grants EAR-0207198 and -0440088.

References

Aines RD and Rossman GR (1985) The high temperature behavior of trace hydrous components in silicate minerals. *American Mineralogist* 70: 1169–1179.

Alm O and Bäckström G (1974) Thermal conductivity of KCl up to 19 kbar. *Journal of Physics and Chemistry of the Solids* 35: 421–424.

Anderson OL (1998) The Gruneisen parameter for iron at outer core conditions and the resulting conductive heat and power in the core. *Physics of the Earth and Planetary Interiors* 109: 179–197.

Anderson OL and Isaak DG (1995) Elastic constants of mantle minerals at high temperature. In: Ahrens TJ (ed.) *Handbook of Physical Constants*, vol. 3, pp. 64–96. Washington, DC: American Geophysical Union.

Andersson S and Bäckström G (1986) Techniques for determining thermal conductivity and heat capacity under hydrostatic pressure. *Reviews in Scientific Instruments* 57: 1633–1639.

Andersson S and Dzhavadov L (1992) Thermal conductivity and heat capacity of amorphous SiO_2: Pressure and volume dependence. *Journal of Physics: Condensed Matter* 4: 6209–6216.

Azumi T and Takahashi Y (1981) Novel finite pulse-width correction in flash thermal diffusivity measurement. *Reviews in Scientific Instruments* 52: 1411–1413.

Bäckström G (1977) Measurement of thermophysical properties of solids under high pressure. In: Cezairliyan A (ed.) *Proceedings of the 7th Symposium on Thermophysical Porperties*, pp. 169–80. New York: ASME.

Beck AE, Darba DM, and Schloessin HH (1978) Lattice conductivities of single-crystal and polycrystalline materials at mantle pressures and temperatures. *Physics of the Earth and Planetary Interiors* 17: 35–53.

Berman RG (1963) Infrared absorption at longitudinal optic frequency in cubic crystal films. *Physical Review* 130: 2193–2198.

Berman RG and Brown TH (1985) Heat capacity of minerals in the system Na_2O–K_2O–CaO–MgO–FeO–Fe_2O_3–Al_2O_3–SiO_2–TiO_2–H_2O–CO_2: Representation, estimation, and high temperature extrapolation. *Contributions to Mineralogy and Petrology* 89: 168–183.

Birch F and Clark H (1940) The thermal conductivity of rocks and its dependence upon temperature and composition. *American Journal of Science* 238: 529–558; 613–635.

Blumm J and Opfermann J (2002) Improvement of the mathematical modeling of flash measurements. *High Temperatures–High Pressures* 34: 515–521.

Blumm J, Henderson JB, Nilson O, and Fricke J (1997) Laser flash measurement of the phononic thermal diffusivity of glasses in the presence of ballistic radiative transfer. *High Temperatures–High Pressures* 34: 555–560.

Bräuer H, Dusza L, and Schulz B (1992) New laser flash equipment LFA 427. *Interceramic* 41: 489–492.

Branlund JM and Hofmeister AM (2004) Effects of hydrogen impurities on the lattice thermal diffusivity of quartz and quartzites up to 1000°C. *EOS Transactions AGU, 85, Fall Meeting Supplement,* Abstract.

Branlund JM and Hofmeister AM (in review) Thermal diffusivity of quartz to 1000°C: Effects of impurities and the α–β phase transition. *Physics and Chemistry of Minerals.*

Branlund JM, Kameyama MC, Yuen DA, and Kaneda Y (2000) Effects of temperature-dependent thermal diffusivity on shear instability in a viscoelastic zone: Implications for faster ductile faulting and earthquakes in the spinel stability field. *Earth and Planetary Science Letters* 182: 171–185.

Bridgeman PW (1924) The thermal conductivity and compressibility of several rocks under high pressures. *American Journal of Science* 7: 81–102.

Brewster MQ (1992) *Thermal Radiative Transfer and Properties.* New York: John Wiley and Sons.

Brydsten U, Gerlich D, and Bäckström G (1983) Thermal conductivity of single-crystal NaCl under uniaxial compression. *Journal of Physics C: Solid State Physics* 16: 143–146.

Büttner R, Zimanowski B, Blumm J, and Hagemann L (1998) Thermal conductivity of a volcanic rock material (olivine–melilitite) in the temperature range between 288 and 1470 K. *Journal of Volcanology and Geothermal Research* 80: 293–302.

Burns G (1990) *Solid State Physics.* San Diego: Academic Press.

Burns RG (1970) *Mineralogical Applications of Crystal Field Theory.* Oxford: Cambridge University Press.

Cahill D, Watson SK, and Pohl RO (1992) Lower limit of thermal conductivity of disordered solids. *Physical Review B* 46: 6131–6140.

Cape JA and Lehman GW (1963) Temperature and finite-pulse effects in the flash method for measuring thermal diffusivity. *Journal of Applied Physics* 34: 1909–1913.

Carslaw HS and Jaeger JC (1959) *Conduction of Heat in Solids,* 2nd edn. New York: Oxford University Press.

Chai M, Brown JM, and Slutsky LJ (1996) Thermal diffusivity of mantle minerals. *Physics and Chemistry of Minerals* 23: 470–475.

Chang IF and Mitra SS (1968) Application of a modified random-element-isodisplacement model to long-wavelength optic phonons of mixed crystals. *Physical Review* 172: 924–933.

Chen F, Shulman J, Xue Y, Chu CW, and Nolas GS (2004) Thermal conductivity measurement under hydrostatic pressure using the 3 omega method. *Reviews in Sciientific Instruments* 75: 4578–4584.

Clark SP, Jr. (1957) Radiative transfer in the Earth's mantle. *Transactions of the American Geophysical Union* 38: 931–938.

Clauser C and Huenges E (1995) Thermal conductivity of rocks and minerals. In: Ahrens TJ (ed.) *Mineral Physics and Crystallography. A Handbook of Physical Constants,* pp. 45–63. Washington, DC: American Geophysical Union.

Cowan DR (1963) Pulse method of measuring thermal diffusivity at high temperatures. *Journal of Applied Physics* 34: 926–927.

Debye P (1914) *Vortrage über die kinetische Theorie der Materie und der Electrizität.* Berlin: B.G. Teuber.

Degiovanni A, Andre S, and Maillet D (1994) Phonic conductivity measurement of a semi-transparent material. In: Tong TW (ed.) *Thermal conductivity 22,* pp. 623–633. Lancaster, PN: Technomic.

Dubuffet F, Yuen DA, and Yanagawa TK (2000) Feedback effects of variable thermal conductivity on cold downwellings in high Rayleigh number convection. *Geophysical Research Letters* 27: 2981–2984.

Dubuffet F, Yuen DA, and Rainey ESG (2002) Controlling thermal chaos in the mantle by positive feedback from radiative thermal conductivity. *Nonlinear Proceedings in Geophysics* 9: 1–13.

Dugdale JS and MacDonald DKC (1955) Lattice thermal conductivity. *Physical Review* 98: 1751–1752.

Eucken A (1911) Über die Temperaturabhängigkeit der Wärmeleitfähigkeit fester Nichtmetalle. *Annales de Physique Leipzig* 34: 186–221.

Fei Y (1995) Thermal expansion. In: Ahrens TJ (ed.) *Mineral Physics and Crystallography. A Handbook of Physical Constants,* pp. 29–44. Washington, DC: American Geophysical Union.

Franson Å and Ross RG (1983) Thermal conductivity, heat capacity and phase stability of solid sodium chlorate ($NaClO_3$) under pressure. *Journal of Physics C: Solid State Physics* 12: 219.

Fried E (1969) Thermal conduction contribution to heat transfer at contacts. In: Tye RP (ed.) *Thermal conductivity,* vol. 2, ch. 5, pp. 253–75. London: Academic Press.

Fujisawa H, Fujii N, Mizutani H, Kanamori H, and Akimoto S (1968) Thermal diffusivity of Mg_2SiO_4, Fe_2SiO_4, and NaCl at High Pressures and Temperatures. *Journal of Geophysical Research* 75: 4727–4733.

Gerbault M (2000) At what stress level is the central Indian Ocean lithosphere buckling? *Earth and Planetary Science Letters* 178: 165–181.

Giesting PA and Hofmeister AM (2002) Thermal conductivity of disordered garnets from infrared spectroscopy. *Physical Review B* 65 (doi 10.1103/PhysRev.B.65.144305).

Giesting PA, Hofmeister AM, Wopenka B, Gwanmesia GD, and Jolliff BL (2004) Thermal conductivity and thermodynamics of majoritic garnets: Implications for the transition zone. *Earth and Planetary Science Letters* 218: 45–56.

Gibert B, Schilling FR, Tommasi A, and Mainprice D (2003) Thermal diffusivity of olivine single-crystals and polycrystalline aggregates at ambient conditions – A comparison. *Geophysical Research Letters* 30 (doi:10.1029/2003GL018459).

Gibert B, Schilling FR, Gratz K, and Tommasi A (2005) Thermal diffusivity of olivine single crystals and a dunite at high temperature: Evidence for heat transfer by radiation in the upper mantle. *Physics of the Earth and Planetary Interiors* 151: 129–141.

Gillet P, Fiquet G, Malezieux JM, and Geiger CA (1992) High-pressure and high-temperature Raman spectroscopy of end-member garnets: Pyrope, grossular and andradite. *European Journal of Mineralogy* 4: 651–664.

Goncharov AF, Viktor V, Struzhkin VV, Steven D, and Jacobsen SD (2006) Reduced radiative conductivity of low-spin (Mg,Fe)O in the lower mantle. *Science* 312: 1205–1207.

Goto T, Ahrens TJ, Rossman GR, and Syono Y (1980) Absorption spectrum of shock-compressed Fe^{2+}-bearing MgO and the radiative conductivity of the lower mantle. *Physics of the Earth and Planetary Interiors* 22: 277–288.

Grzechnik A and McMillan PF (1998) Temperature dependence of the OH^- absorption in the SiO_2 glass and melt to 1975 K. *Ameican Mineralogist* 83: 331–338.

Haken H (1977) *Synergetics.* Berlin: Springer-Verlag.

Håkasson B and Andersson P (1986) Thermal conductivity and heat capacity of solid NaCl and NaI under pressure. *Journal of Physics and Chemistry of the Solids* 47: 355–362.

Håkasson B, Andersson P, and Bäckström G (1988) Improved hot-wire procedure for thermophysical measurements under pressure. *Reviews in Scientific Instruments* 59: 2269–2276.

Halliday D and Resnick R (1966) *Physics*. New York: John Wiley and Sons.

Hammerschmidt U and Sabuga W (2000) Transient hot wire (THW) mehtod: Uncertainty assessment. *International Journal of Thermophysics* 21: 1255–1278.

Hao Y-H, Neuwmann M, Enss C, and Fleischmann (2004) Contactless technique for thermal conductivity measurement at very low temperature. *Reviews in Scientific Instruments* 75: 2718–2725.

Hapke B (1993) *Theory of Reflectance and Emittance Spectroscopy*. Cambridge: Cambridge University Press.

Harrell MD (2002) *Anisotropic Thermal Diffusivity in Olivines and Pyroxenes to High Temperature*. PhD Thesis, University of Washington, Seattle.

Hauck SA, Phillips RJ, and Hofmeister AM (1999) Variable conductivity: Effects on the thermal structure of subducting slabs. *Geophysical Research Letters* 26: 3257–3260.

Heckman RC (1973) Finite pulse-time and heat loss effects in pulse thermal diffusivity measurements. *Journal of Applied Physics* 44: 1455–1460.

Henderson JB, Giblin F, Blumm J, and Hagemann L (1998a) SRM 1460 series as a thermal diffusivity standard for laser flash instruments. *International Journal of Thermophysics* 19: 1647–1656.

Henderson JB, Hagemann L, and Blumm J (1998b) Development of SRM 8420 series electrolytic iron as a thermal diffusivity standard. *Netzsch Applications Laboratory Thermophysical Properties Section Report No. I-9E*.

Höfer M and Schilling FR (2002) Heat transfer in quartz, orthoclase, and sanidine at elevated temperature. *Physics and Chemistry of Minerals* 29: 571–584.

Hofmann R, Hahn O, Raether F, Mehling H, and Fricke J (1997) Determination of thermal diffusivity in diathermic materials by the laser-flash technique. *High Temperatures–High Pressures* 29: 703–710.

Hofmeister AM (1999) Mantle values of thermal conductivity and the geotherm from phonon lifetimes. *Science* 283: 1699–1706.

Hofmeister AM (2001) Thermal conductivity of spinels and olivines from vibrational spectroscopy at ambient conditions. *American Mineralogist* 86: 1188–1208.

Hofmeister AM (2004a) Enhancement of radiative transfer in the mantle by OH⁻ in minerals. *Physics of the Earth and Planetary Interiors* 146: 483–485.

Hofmeister AM (2004b) Thermal conductivity and thermodynamic properties from infrared spectroscopy. In: King P, Ramsey M, and Swayze G (eds.) *Infrared Spectroscopy in Geochemistry, Exploration Geochemistry, and Remote Sensing*, ch. 5, pp. 135–154. Ottawa, ON: Mineralogical Association of Canada.

Hofmeister AM (2004c) Physical properties of calcium aluminates from vibrational spectroscopy. *Geochemica et Cosmochimica Acta* 68: 4721–4726.

Hofmeister AM (2005) The dependence of radiative transfer on grain-size, temperature, and pressure: Implications for mantle processes. *Journal of Geodynamics* 40: 51–72.

Hofmeister AM (2006) Thermal diffusivity of garnets at high temperature. *Physics and Chemistry of Minerals* 33: 45–62.

Hofmeister AM (2007) Dependence of thermal transport properties on pressure. *Proceedings of the National Academy of Sciences*, doi: 10.1073/pnas.0610734104.

Hofmeister AM and Criss RE (2005a) Earth's heat flux revisited and linked to chemistry. *Tectonophysics* 395: 159–177.

Hofmeister AM and Criss RE (2005b) Reply to 'Comments on Earth's heat flux revised and linked to chemistry' by R. Von Herzen, E.E. Davis, A. Fisher, C.A. Stein and H.N. Pollack. *Tectonophysics* 395: 193–198.

Hofmeister AM and Criss RE (2006) Comment on 'Estimates of heat flow from Cenozoic seafloor using global depth and age data' by M. Wei and D. Sandwell. *Tectonophysics* 428: 95–100.

Hofmeister AM and Mao HK (2001) Evaluation of shear moduli and other properties of silicate spinels from IR Spectroscopy. *American Mineralogist* 86: 622–639.

Hofmeister AM and Mao HK (2002) Redefinition of the mode Gruneisen parameter for polyatomic substances and thermodynamic implications. *Proceedings of the National Academy of Sciences* 99: 559–564.

Hofmeister AM and Mao HK (2003) Pressure derivatives of shear and bulk moduli from the thermal Gruneisen parameter and volume–pressure data. *Geochemica et Cosmochimica Acta* 66: 1207–1227.

Hofmeister AM and Yuen DA (in press) The threshold dependencies of thermal conductivity and implications on mantle dynamics. *Journal of Geodynamics* doi:10.1016/j.jog.2007.02.003.

Holt JB (1975) Thermal diffusivity of olivine. *Earth and Planetary Science Letters* 27: 404–408.

Honda S and Yuen DA (2001) Interplay of variable thermal conductivity and expansivity on the thermal structure of oceanic lithosphere. *Geophysical Research Letters* 28: 351–354.

Honda S and Yuen DA (2004) Interplay of variable thermal conductivity and expansivity on the thermal structure of the oceanic lithosphere II. *Earth, Planets and Space* 56: e1–e4.

Horai K (1971) Thermal conductivity of rock-forming minerals. *Journal of Geophysical Research* 76: 1278–1308.

Horai K and Sasaki J (1989) The effect of pressure on the thermal conductivity of silicate rocks up to 12 kbar. *Physics of the Earth and Planetary Interiors* 55: 292–305.

Jordan M, Schuch A, Righini R, Signorini JF, and Jodl H-J (1994) Phonon relaxation processes in crystals (NaNO₃) at high pressure and low temperature. *Journal of Chemical Physics* 101: 3436–3443.

Josell D, Warren K, and Czairliyan A (1995) Correcting an error in Cape and Lehman's analysis for determining thermal diffusivity from thermal pulse experiments. *Journal of Applied Physics* 78: 6867–6869.

Julian CL (1965) Theory of heat conduction in rare-gas crystals. *Physical Review A* 137: 128–137.

Kachare A, Andermann G, and Brantley LR (1972) Reliability of classical dispersion analysis of LiF and MgO reflectance data. *Journal of Physics and Chemistry of the Solids* 33: 467–475.

Kanamori H, Fujii N, and Mizutani H (1968) Thermal diffusivity of rock-forming minerals. *Journal of Geophysical Research* 73: 595–605.

Katsura T (1993) Thermal diffusivity of silica glass at pressures up to 9 GPa. *Physics and Chemistry of Minerals* 20: 201–208.

Katsura T (1995) Thermal diffusivity of olivine under upper mantle conditions. *Geophysical Journal International* 122: 63–69.

Katsura T (1997) Thermal diffusivity of periclase at high temperatures and high pressures. *Physics of the Earth and Planetary Interiors* 101: 73–77.

Kaufmann R and Freedman WJ (2002) *Universe*. New York: W.H. Freeman.

Kellett BS (1952) The steady flow of heat through hot glass. *Optical Society of America, Journal* 42: 339–343.

Keppler H and Smyth JR (2005) Optical and near-infrared spectra of ringwoodite to 21.5 GPa: Implications for radiative heat transport in the mantle. *American Mineralogist* 90: 1209–1212.

Kieffer S, Getting WIC, and Kennedy GC (1976) Experimental determination of the pressure dependence of the thermal diffusivity of teflon, sodium chloride, quartz, and silica. *Journal of Geophysical Research* 81: 3018–3024.

Klemens PG (1958) Thermal conductivity and lattice vibrational modes. *Solid State Physics* 7: 1–98.

Klemens PG (1960) Thermal resistance due to point defects at high temperatures. *Physical Review* 119: 507–509.

Klemens PG (1969) *Thermal Conductivity*. New York: Academic Press.

Knittle E (1995) Static compression measurements of equations of state. In: Ahrens TJ (ed.) *Mineral Physics and Crystallography: A Handbook of Physical Constants*, pp. 98–142. Washington, DC: American Geophysical Union.

Kobayashi Y (1974) Anisotropy of thermal diffusivity in olivine, pyroxene and dunite. *Journal of the Physics of Earth* 22: 359–373.

Kolesov BA and Geiger CA (2004a) A Raman spectroscopic study of Fe–Mg olivines. *Physics and Chemistry of Minerals* 31: 142–154.

Kolesov BA and Geiger CA (2004b) A temperature-dependent single-crystal Raman spectrosocopic study of fayalite: Evidence for phonon–magnetic excitation coupling. *Physics and Chemistry of Minerals* 31: 155–161.

Lee HL and Hasselman DPH (1985) Comparison of data for thermal diffusivity obtained by laser-flash method using thermocouple and photodector. *Journal of American Ceramic Society* 68, C12–C13.

Lee DW and Kingery WD (1960) Radiation energy transfer and thermal conductivity of ceramic oxides. *Journal of American Ceramic Society* 43: 594–607.

Lees CH (1905) Effects of temperature and pressure on the thermal conductivities of solids – Part I. The effect of temperature on the thermal conductivities of some electrical insulators. *Philosophical Transactions of Royal Society of London Series A* 204: 433–466.

Liebfried G and Schlömann E (1954) Warmleitung in elektrische isolierenden Kristallen. *Nachrichten aus Gesundheitswesen Wissenschaften Goettingen Mathematik und Physik* K1: 71–93.

Lubimova H (1958) Thermal history of the Earth with consideration of the variable thermal conductivity of the mantle. *Geophysical Journal of the Royal Astronomical Society* 1: 115–134.

MacDonald GJF (1959) Calculations on the thermal history of the Earth. *Journal of Geophysical Research* 64: 1967–2000.

Madarasz FL and Klemens PG (1987) Reduction of lattice thermal conductivity due to point defects at intermediate temperatures. *International Journal of Thermophysics* 8: 257–262.

Manning PG (1967) The optical absorption spectra of the garnets almandine-pyrope, pyrope and spessartine and some structural interpretations of mineralogical significance. *Canadian Mineralogist* 9: 237–251.

Makarounis O and Jenkins RJ (1962) USNRDL-TR-599, AD 295 887. *Not seen, cited by Touloukian et al. 1977; Andersson and Bäckström (1986)*.

McCarthy KA and Ballard SS (1951) New data on the thermal conductivity of optical crystals. *Journal of Optical Society of America* 41: 1062–1063.

Mehling H, Hautzinger G, Nilsson O, Fricke J, Hofmann R, and Hahn O (1998) Thermal diffusivity of semitransparent materials determined by the laser-flash method applying a new mathematical model. *International Journal of Thermophysics* 19: 941–949.

Mitra SS (1969) Infrared and Raman spectra due to lattice vibrations. In: Nudelman S and Mitra SS (eds.) *Optical Properties of Solids*, pp. 333–452. New York: Plenum Press.

Nilsson O, Sandberg O, and Bäckström G (1982) Thermal properties of sulfur under pressure. *ETPC Proceedings* 8: 159–165.

Osako M (1997) Thermal diffusivity of olivine and garnet single crystals. *Bulletin of National Science Museum of Tokyo Series E* 20: 1–7.

Osako M and Ito E (1991) Thermal diffusivity of MgSiO$_3$ perovskite. *Geophysical Research Letters* 18: 239–242.

Osako M, Ito E, and Yoneda A (2004) Simultaneous measurements of thermal conductivity and thermal diffusivity for garnet and olivine under high pressure. *Physics of the Earth and Planetary Interiors* 143–144: 311–320.

Pangilinan GI, Ladouceur HD, and Russell TP (2000) All-optical technique for measuring thermal properties of materials at static high pressure. *Reviews in Scientific Instruments* 71: 3846–3852.

Parker JW, Jenkins JR, Butler PC, and Abbott GI (1961) Flash method of determining thermal diffusivity, heat capacity, and thermal conductivity. *Journal of Applied Physics* 32: 1679–1684.

Peierls RE (1929) Zur kinetische Theorie der Warmeleitung in Kristallen. *Annales de Physique Leipzig* 3: 1055–1101.

Pierrus J and Sigalas I (1985) Thermophysical properties of the sodium halides under pressure. *Journal of Physics C: Solid State* 19: 1465–1470.

Pertermann M and Hofmeister AM (2006) Thermal diffusivity of olivine-group minerals. *American Mineralogist* 91: 1747–1760.

Pollack HN, Hurter SJ, and Johnson JR (1993) Heat flow from the Earth's interior: Analysis of the global data set. *Reviews of Geophysics* 31: 267–280.

Pomeranchuk I (1943) Heat conductivity of dielectrics at high temperatures. *Journal of Physics (USSR)* 7: 197–201.

Preston FW (1956) Thermal conductivity in the depths of the Earth. *American Journal of Science* 25: 754–757.

Ranalli G (2001) Mantle rheology: Radial and lateral viscosity variations inferred from microphysical creep laws. *Journal of Geodynamics* 32: 425–444.

Reif F (1965) *Fundamentals of Statistical and Thermal Physics*. New York: McGraw Hill.

Ross RG, Andersson P, Sundqvist B, and Bäckström G (1984) Thermal conductivity of solids and liquids under pressure. *Reports on Progress in Physics* 47: 1347–1402.

Ross NL (1997) Optical absorption spectra of transition zone minerals and implications from radiative heat transfer. *Physics and Chemistry of the Earth* 22: 113–118.

Rossman GR (1988) Optical spectroscopy. *Reviews of Mineralogy* 18: 207–254.

Roufosse MC and Klemens PG (1973) Thermal conductivity of complex dielectric crystals. *Physical Review B* 7: 5379–5386.

Roufosse MC and Klemens PG (1974) Lattice thermal conductivity of minerals at high temperatures. *Journal of Geophysical Research* 79: 703–705.

Sass JH (1965) The thermal conductivity of fifteen feldspar specimens. *Journal of Geophysical Research* 70: 4064–4065.

Schärmeli GH (1982) Anisotropy of olivine thermal conductivity at 2.5 GPa up to 1500 K measured on optically non-thick sample. In: Schreyer W (ed.) *High-Pressure Researches in Geoscience*, pp. 349–373. Stuttgart: E Schweizerbartsche Verlag.

Schatz JF and Simmons G (1972a) Thermal conductivity of Earth materials at high temperature. *Journal of Geophysical Research* 77: 6966–6983.

Schatz JF and Simmons G (1972b) Method of simultaneous measurement of radiative and lattice thermal conductivity. *Journal of Applied Physics* 43: 2588–2594.

Schilling FR (1999) A transient technique to measure thermal diffusivity at elevated temperature. *European Journal of Mineralogy* 11: 1115–1124.

Schloessin HH and Dvorak Z (1972) Anisotropic lattice thermal conductivity in enstatite as a function of pressure and temperature. *Geophysical Journal of the Royal Astronomical Society* 27: 499–516.

Schott B, van den Berg AP, and Yuen DA (2001) Focussed time-dependent Martian volcanism from chemical differentiation coupled with variable thermal conductivity. *Geophysical Research Letters* 28: 4271–4274.

Seipold U (1998) Temperature dependence of thermal transport properties of crystalline rocks – A general law. *Tectonophysics* 291: 161–171.

Serrano J, Manjon FJ, Romaro AH, *et al.* (2003) Dispersive phonon linewidths: The E_2 phonons of ZnO. *Physical Review Letters* 90: 055510.

Shankland TJ, Nitsan U, and Duba AG (1979) Optical absorption and radiative heat transport in olivine at high temperature. *Journal of Geophysical Research* 84: 1603–1610.

Siegel R and Howell JR (1972) *Thermal Radiation Heat Transfer*. New York: McGraw-Hill.

Slack G (1962) Thermal conductivity of MgO, Al_2O_3, $MgAl_2O_4$, and Fe_3O_4 crystals from 3° to 300° K. *Physical Review* 126: 427–441.

Slack G (1964) Thermal conductivity of pure and impure silicon, silicon carbide, and diamond. *Journal of Applied Physics* 35: 3460–3466.

Slack G (1979) The thermal conductivity of nonmetallic solids. *Solid State Physics* 34: 1–73.

Slack GA and Oliver DW (1971) Thermal conductivity of garnets and phonon scattering by rare-earth ions. *Physical Review B* 4: 592–609.

Somerton WH (1992) *Thermal Properties and Temperature-Related Behavior of Rock/Fluid Systems*. Amsterdam: Elsevier.

Spitzer WG, Miller RC, Kleinman DA, and Howarth LW (1962) Far-infrared dielectric dispersion in $BaTiO_3$, $SrTiO_3$, and TiO_2. *Physical Review* 126: 1710–1721.

Starin L, Yuen DA, and Bergeron SY (2000) Thermal evolution of sedimentary basin formation with temperature-dependent conductivity. *Geophysical Research Letters* 27: 265–268.

Staudacher W (1973) Die Temperature-Leitfähigkeit von natürlichem Olivin bei hohen Drucken und Temperaturen. *Zeitschift Für Geophysika* 39: 979–988.

Sundqvist B (1981) Thermal conductivity and Lorentz number of nickel under pressure. *Solid State Communications* 37: 289–291.

Tackley PJ (1996) Effects of strongly variable viscosity on three-dimensional compressible convection in planetary mantles. *Journal of Geophysical Research* 101: 3311–3332.

Taran MN and Langer K (2001) Electronic absorption spectra of Fe^{2+} ions in oxygen-based rock-forming minerals at temperatures between 297 and 600 K. *Physics and Chemistry of Minerals* 28: 199–210.

Touloukian YS, Powell RW, Ho CY, and Klemens PG (1970) *Thermal Conductivity of Nonmetallic Solids*. New York: Plenum Press.

Tozer D (1965) Heat transfer and convection currents. *Philosophical Transactions of the Royal Society of London A* 258: 252–271.

Tye RP (1969) *Thermal Conductivity*. London: Academic Press.

Ullrich K, Langer K, and Becker KD (2002) Temperature dependence of the polarized electronic absorption spectra of olivines. Part I – Fayalite. *Physics and Chemistry of Minerals* 29: 409–419.

van den Berg AP, Yuen DA, and Rainey ESG (2004) The influence of variable viscosity on delayed cooling due to variable thermal conductivity. *Physics of the Earth and Planetary Interiors* 142: 283–295.

van den Berg AP, Rainey ESG, and Yuen DA (2005) The combined influences of variable thermal conductivity, temperature- and pressure-dependent viscosity and core–mantle coupling on thermal evolution. *Physics of the Earth and Planetary Interiors* 149: 259–278.

Vosteen H-D and Schellschmidt R (2003) Influence of temperature on thermal conductivity, thermal capacity and thermal diffusivity for different types of rock. *Physics and Chemistry of the Earth* 28: 499–509.

Xu Y, Shankland TJ, Linhardt S, Rubie DC, Langenhorst F, and Klasinski K (2004) Thermal diffusivity and conductivity of olivine, wadsleyite, and ringwoodite to 20 GPa and 1373 K. *Physics of the Earth and Planetary Interiors* 143–144: 321–326.

Yamagishi H, Nakashima S, and Ito Y (1997) High temperature infrared spectra of hydrous microcrystalline quartz. *Physics and Chemistry of Minerals* 24: 66–74.

Yanagawa TKB, Nakada M, and Yuen DA (2005) The influence of lattice thermal conductivity on thermal convection with strongly temperature-dependent viscosity. *Earth, Planets and Space* 57: 15–28.

Yuen DA and Zhang S (1989) Equation of state and rheology in deep mantle convection. In: Navrotsky A and Weidner DJ (eds.) *Perovskites*. pp. 131–146. Washington, DC: American Geophysical Union.

Yuen DA, Vincent AP, Bergeron SY, *et al.* (2000) Crossing of scales and nonlinearities in geophysical processes. In: Boschi E, Ekstrom G, and Morelli A (eds.) *Problems in Geophysics for the New Millenium*, pp. 403–463. Bologna, Italy: Editrice Compositori.

Yukutake H and Shimada M (1978) Thermal conductivity of NaCl, MgO, coesite and stishovite up to 40 kbar. *Physics of the Earth and Planetary Interiors* 17: 193.

Zaug J, Abransom E, Brown JM, and Slutsky LJ (1992) Elastic constants, equations of state and thermal diffusivity at high pressure. In: Syono Y and Manghnani MH (eds.) *High-Pressure*, pp. 157–166. Washington, DC: Terra/AGU.

Ziman JM (1962) *Electrons and Phonons: The Theory of Transport Phenomena in Solids*. Oxford: Clarendon Press.

20 Properties of Rocks and Minerals – Magnetic Properties of Rocks and Minerals

R. J. Harrison, R. E. Dunin-Borkowski, T. Kasama, E. T. Simpson, and J. M. Feinberg,
University of Cambridge, Cambridge, UK

20.1	Introduction	580
20.2	Magnetism at the Atomic Length Scale	581
20.2.1	Exchange Interactions and Magnetic Structure in Fe-Bearing Oxides	581
20.2.2	Atomistic Simulations of Magnetic Ordering	584
20.2.2.1	Theory	584
20.2.2.2	Application to magnetic nanoparticles	586
20.2.2.3	Application to coupled magnetic and chemical ordering in solid solutions	587
20.3	Magnetism at the Nanometer Length Scale	591
20.3.1	Theory of Off-Axis Electron Holography of Magnetic Materials	591
20.3.1.1	Amplitude and phase of a TEM image	591
20.3.1.2	Calculation of the mean inner potential	593
20.3.1.3	Formation of an electron hologram	594
20.3.1.4	Processing of the electron hologram	595
20.3.1.5	Removing the mean inner potential contribution	596
20.3.2	Interpretation of Electron Holographic Phase Images	597
20.3.2.1	Quantification of the magnetic induction	597
20.3.2.2	Visualization of the magnetic induction	598
20.3.3	Experimental Results	598
20.3.3.1	Electron holography of isolated magnetite crystals	598
20.3.3.2	Electron holography of chains of closely spaced magnetite crystals	601
20.3.3.3	Electron holography of two-dimensional magnetite nanoparticle arrays	603
20.3.3.4	Exchange interactions across antiphase boundaries in ilmenite–hematite	606
20.4	Magnetism at the Micrometer Length Scale	608
20.4.1	Theory	609
20.4.1.1	The micromagnetic energy	609
20.4.1.2	Discretization of the micromagnetic energy	610
20.4.1.3	Finite element discretization	611
20.4.2	Applications of Micromagnetic Simulations	612
20.4.2.1	Equilibrium domain states in isolated magnetite particles	612
20.4.2.2	Temperature dependence of domain states in isolated particles	614
20.4.2.3	Field dependence of domain states	616
20.4.2.4	Magnetostatic interactions between particles	617
20.5	Magnetism at the Macroscopic Length Scale	618
20.5.1	Theory	619
20.5.1.1	First-order reversal curves and the FORC distribution	619
20.5.1.2	Interpretation of the FORC diagram	620
20.5.1.3	Extended FORCs and the reversible ridge	620
20.5.2	FORC Diagrams as a Function of Grain Size	621
20.5.2.1	SP particles	621
20.5.2.2	SD particles	622
20.5.2.3	PSD particles	623
20.5.2.4	MD particles	623

20.5.3	Mean-Field Interactions and FORC Diagrams	623
20.5.4	Practical Applications of FORC Diagrams	624
20.6	Summary	625
References		626

20.1 Introduction

Magnetic minerals are pervasive in the natural environment, and are present in all types of rocks, sediments, and soils. These minerals retain a memory of the geomagnetic field that was present during the rock's formation. Paleomagnetic recordings have been exploited for more than 50 years to map the movements of the continental and oceanic plates, and have proved to be one of the most powerful tools for reconstructing the geological history of the Earth and other planets (Connerney *et al.*, 1999, 2004; Acuna *et al.*, 1999). The variation in intensity of the geomagnetic field, as determined from rocks and archeological material, has been used to provide an understanding of the behavior of the geodynamo and to constrain models of fluid motion in the Earth's core (Labrosse and Macouin, 2003; Gallet *et al.*, 2005; Valet *et al.*, 2005). More recently, magnetic mineralogy has been used to trace changes in the climate, as the magnetic minerals that are present in any sample are indicative of the environment in which they are formed (Kumar *et al.*, 2005).

Interpretations of rock magnetic measurements are completely reliant on an accurate understanding of the physical processes by which a material acquires and maintains a faithful record of the geomagnetic field. Since the pioneering work of Néel (1948, 1949), rock magnetists have attempted to develop a quantitative understanding of how assemblages of magnetic minerals in single-domain (SD), pseudo-single-domain (PSD), or multidomain (MD) states acquire and maintain natural remanent magnetization (NRM) (see Dunlop and Özdemir (1997) for a detailed overview). The theories work well in ideal cases, that is, when magnetic grains are homogeneous, defect free, and sufficiently well separated from each other, the magnetic interactions between them can be neglected. They begin to fail, however, when the mineral is heterogeneous at the nanometer scale, as is necessarily the case when the magnetic grains form part of a nanoscale intergrowth. Recent studies have demonstrated that nanoscale microstructures are extremely common in magnetic minerals, and that they have a significant impact on

their macroscopic magnetic properties (Harrison and Becker, 2001; Harrison *et al.*, 2002; McEnroe *et al.*, 2001, 2002; Robinson *et al.*, 2002, 2004, 2006; Harrison *et al.*, 2005; Feinberg *et al.*, 2004, 2005). These microstructures not only determine the intensity and stability of macroscopic magnetism recorded in rocks – thereby controlling the fidelity of paleomagnetic recordings at the global scale – but are extremely important in an industrial context, by proving natural analogs of future generations of high-density magnetic recording media (Skumryev *et al.*, 2003; Puntes *et al.*, 2004).

This review describes the current state of the art in the field of computational and experimental mineral physics, as applied to the study of magnetic minerals. Particular emphasis is placed on the relationship between nanoscale microstructure and macroscopic magnetic properties. For a comprehensive review of the magnetic properties of specific rocks and minerals, the reader is referred to Hunt *et al.* (1995) and Dunlop and Özdemir (1997). Arguably, the most significant recent advance is the application to mineral magnetism of off-axis electron holography, a transmission electron microscopy (TEM) technique that yields a two-dimensional vector map of magnetic induction with nanometer spatial resolution (Harrison *et al.*, 2002). Electron holography is capable of imaging the magnetization states of individual magnetic particles and the magnetostatic interaction fields between neighboring particles: two factors that play a central role in the interplay between magnetism and microstructure. By combining this capability with electron tomography, it is now possible to determine both the micromagnetic structures and the three-dimensional morphologies of nanoscale magnetic particles directly and quantitatively as a function of temperature and applied magnetic field. In tandem with these techniques, advances in the application of atomistic and micromagnetic simulations to the study of magnetic ordering in minerals have opened the way to novel interpretations and modeling of nanoscale magnetic properties (Robinson *et al.*, 2002). Only now are the sizes of systems that are accessible to both experimental and computational studies converging at the nanometer length scale. This convergence provides unique opportunities for tackling problems that lie at the frontiers of rock magnetism.

This review is organized in order of increasing length scale of magnetic interactions. Section 20.2 deals with magnetism at the atomic length scale. It contains a brief description of exchange interactions and magnetic structure in Fe-bearing oxides, and the use of atomistic simulations of magnetic ordering to the study magnetism at surfaces and interfaces. Section 20.3 deals with magnetism at the nanometer length scale. Following a summary of the theory of electron holography, recent applications of holography to the study of magnetic minerals are reviewed. Section 20.4 deals with magnetism at the micrometer length scale, including advances in micromagnetic simulations that allow the magnetic behavior of particles with realistic three-dimensional morphologies to be modeled. In Section 20.5, we move to the macroscopic length scale, with a description of how new approaches for the measurement of macroscopic magnetic properties (i.e., FORC diagrams) are providing quantitative information about the spectrum of coercivities and interaction fields that exist at the microscopic scale.

20.2 Magnetism at the Atomic Length Scale

20.2.1 Exchange Interactions and Magnetic Structure in Fe-Bearing Oxides

The driving force for magnetic ordering in Fe-bearing oxides is the superexchange interaction between neighboring transition metal cations via intermediate oxygen anions (Goodenough, 1966). The magnitudes and signs of superexchange interactions define the magnetic ground state and the magnetic ordering temperature of the mineral, and play a fundamental role in determining its macroscopic magnetic properties. The exchange interaction energy for classical spins can be expressed as

$$E_{mag} = -\sum_{i \neq j} \mathcal{J}_{ij} \mathbf{S}_i \cdot \mathbf{S}_j \qquad [1]$$

where $\mathbf{S}_i$ and $\mathbf{S}_j$ and are the spins on atoms i and j, and $\mathcal{J}_{ij}$ is the corresponding exchange integral. Positive values of $\mathcal{J}_{ij}$ lead to parallel (i.e., ferromagnetic) alignment of spins; negative values lead to antiparallel (i.e., antiferromagnetic) alignment.

Empirical values of $\mathcal{J}_{ij}$ can be obtained from spin-wave dispersion curves measured using inelastic neutron scattering (Samuelsen, 1969; Samuelsen and Shirane, 1970; Brockhouse, 1957; Watanabe and Brockhouse, 1962; Glasser and Milford, 1963; Phillips and Rosenberg, 1966).

Alternatively, theoretical values can be obtained from first-principles calculations (Sandratskii, 1998; Matar, 2003). The simplest theoretical approach involves calculating the total energies of several different collinear arrangements of spins, and then determining values of $\mathcal{J}_{ij}$ directly from eqn [1] (Sandratskii et al., 1996; Rollmann et al., 2004). This approach is limited, however, by the small number of alternative structures that can be generated for a given unit cell. The use of noncollinear magnetic structures (Sandratskii, 1998) provides a more general procedure for calculating exchange integrals out to arbitrary cation–cation separations (Uhl and Siberchicot, 1995). This approach is based on the calculation of the total energies of spin-spiral configurations, over a grid of wave vectors within the Brillouin zone. It is often found that exchange interactions that are determined by spin-wave and first-principles methods overestimate magnetic ordering temperatures significantly (Sandratskii et al., 1996; Uhl and Siberchicot, 1995). If necessary, the calculated values of $\mathcal{J}_{ij}$ can be scaled or refined to provide better agreement with experimental observations (Burton, 1985; Harrison and Becker, 2001; Harrison, 2006).

Exchange integrals for hematite (Fe_2O_3), ilmenite ($FeTiO_3$), and magnetite (Fe_3O_4) are listed in **Table 1** and **Figure 1**. The crystal structures of hematite–ilmenite and magnetite are compared in **Figure 2**. Superexchange interactions are highly sensitive to the relative positions of the two cations and the intermediate oxygen anion, varying in magnitude approximately as $\cos^2 \psi$, where ψ is the cation–oxygen–cation bond angle (Coey and Ghose, 1987). In hematite (**Figure 2(a)**), Fe^{3+} cations occupy two-thirds of the octahedral interstices within a hexagonal close-packed oxygen sublattice, forming symmetrically equivalent A and B layers parallel to the (001) basal plane (space group $R\bar{3}c$). Each octahedron shares a face with an octahedron in the layer above or below, and edges with three octahedral in its own layer. Since both face- and edge-sharing octahedral–octahedral linkages have $\psi \sim 90°$ (**Table 1**), both first- and second-nearest-neighbor interactions are weak. Third- and fourth-nearest-neighbor interactions involve corner-sharing octahedra in adjacent layers ($\psi \sim 120°$ and $132°$, respectively; **Table 1**). These interactions are large and negative, leading to an antiferromagnetic ground state in which

Table 1   Summary of magnetic exchange intergrals for hematite, ilmenite, and magnetite

Nearest neighbor	Cation–cation distance (Å)	Interaction type	Polyhedral sharing	Cation–oxygen–cation bond angle (deg)	Magnetic interaction parameter (K)
Hematite (spin waves; Samuelsen and Shirane, 1970)					
1	2.9	Interlayer	Face	86.5	12
2	2.971	Intralayer	Edge	93.9	3.2
3	3.36	Interlayer	Corner	119.7	−59.4
4	3.71	Interlayer	Corner	131.6	−46.4
5	3.99	Double	None	NA	−2
Ilmenite (spin waves; Ishikawa et al., 1985)					
1	3.03	Intralayer	Edge	89.1	9.8
2	4.074	Double	None	NA	−8.9
3	5.088	Intralayer	None	NA	0.56
4	5.9	Intralayer	None	NA	0.68
Magnetite (first-principles calculations; Uhl and Siberchiot, 1995)					
1	2.97	Oct–oct	Edge	87.8	9.6
2	3.48	Tet–oct	Corner	126.8	−33.4
3	3.635	Tet–tet	None	NA	−2.1
4	5.14	Oct–oct	None	NA	−0.17
5	5.45	Tet–oct	None	NA	0.65
6	5.45	Tet–oct	None	NA	0.3
7	5.936	Tet–tet	None	NA	−0.73
8	6.63	Oct–oct	None	NA	−0.48
9	6.88	Tet–oct	None	NA	0.57
10	6.96	Tet–tet	None	NA	0.67
11	7.852	Oct–oct	None	NA	0.13
12	8.06	Tet–oct	None	NA	−0.26
13	8.06	Tet–oct	None	NA	0.2

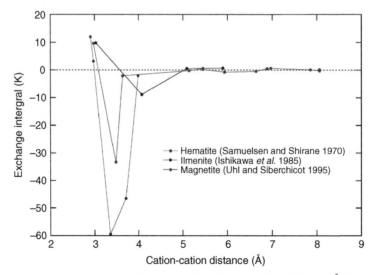

Figure 1   Magnetic superexchange integrals, J_{ij} (K), as a function of cation-cation distance (Å) for hematite (blue), ilmenite (black), and magnetite (red). Data for hematite and ilmenite were measured using inelastic neutron scattering (Samuelsen and Shirane, 1970; Ishikawa *et al.*, 1985). Data for magnetite were calculated using first-principles methods (Uhl and Siberchicot, 1995).

(a)

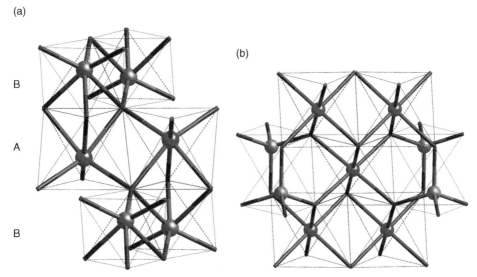

(b)

Figure 2   Comparison of the local structural topology of (a) hematite/ilmenite and (b) magnetite. In ilmenite, A layers (red) are occupied by Fe^{2+} and B layers (blue) are occupied by Ti^{4+} (or vice versa). Hematite has the same structural topology, but all layers are occupied by Fe^{3+}. In magnetite, tetrahedral sites (blue) are occupied by Fe^{3+} cations and octahedral sites (red) are occupied by both Fe^{2+} and Fe^{3+} cations.

A-layer spins are antiparallel to B-layer spins. Above 260 K, spins lie along one of the three <100> crystallographic axes within the basal plane (Besser *et al.*, 1967). The antiferromagnetic sublattices are canted by an angle of $\sim 0.13°$, leading to a weak parasitic moment within the basal plane oriented at 90° to <100> (Dzyaloshinskii, 1958). By considering arbitrary, noncollinear configurations of the atomic magnetic moments, Sandratskii and Kübler (1996) used first-principles calculations to demonstrate that the canted magnetic structure appears as a direct consequence of spin–orbit coupling (**Figure 3**). Below 260 K, the spin alignment switches to [001] (the Morin transition), and the canting is lost.

In ilmenite, Fe^{2+} and Ti^{4+} are ordered onto A and B (or B and A) layers, and the equivalency of the layers is lost (space group $R\bar{3}$). Since one layer is fully occupied by Ti^{4+}, the strong interlayer interactions that were present in hematite are eliminated. Second-nearest-neighbor interactions extend across the intervening Ti^{4+} layers. These weak negative interactions result in antiferromagnetic ordering below 60 K. The spins in one A layer are then aligned antiparallel to those on the adjacent A layers, parallel and antiparallel to [001].

In magnetite (**Figure 2(b)**), cations occupy tetrahedral and octahedral interstices within a cubic closed-packed oxygen sublattice. Octahedra are occupied by Fe^{2+} and Fe^{3+} cations, whereas tetrahedra are occupied

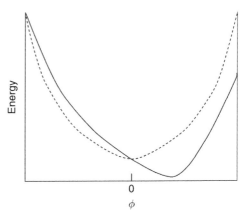

Figure 3   Schematic illustration of the variation in total energy of canted antiferromagnetic hematite with the canting angle, ϕ, as determined by first-principles calculations. Without the inclusion of spin–orbit coupling (dashed line), the energy minimum occurs when the spins are exactly antiparallel to each other ($\phi = 0$). With spin–orbit coupling included (solid line), the energy minimum occurs when the spins are slightly canted with respect to each other ($\phi > 0$). Reproduced from Sandratskii LM and Kübler J (1996) First-principles LSDF study of weak ferromagnetism in Fe_2O_3. *Europhysics Letters* 33: 447–452.

exclusively by Fe^{3+}. Octahedra share edges with adjacent octahedra and corners with adjacent tetrahedra. There are no shared oxygens between adjacent tetrahedra. The first-nearest-neighbor interaction is weak, due to the unfavorable cation–oxygen–cation bond angle (**Table 1**). The dominant negative tetrahedral–octahedral interaction leads to a ferrimagnetic structure, in which spins on the octahedral sites are antiparallel to those on the tetrahedral sites. Exchange interactions are weak when the two cations are not linked directly by a common oxygen, leading to a weak tetrahedral–tetrahedral interaction and a rapid decrease in interaction strength for cation–cation separations that are greater than ~4 Å. Spins point parallel to <111> above 130 K and <100> axes below 130 K. For further details of the crystal and magnetic structures of minerals, the reader is referred to Banerjee (1991), Dunlop and Özdemir (1997), and Harrison (2000).

20.2.2 Atomistic Simulations of Magnetic Ordering

20.2.2.1 Theory

Magnetic ordering in minerals was first described using a mean-field model by Néel (1948). For some applications, for example, for estimating Néel temperatures from $\mathcal{J}_{ij}$, this macroscopic approach remains extremely useful (Stephenson, 1972a, 1972b). When studying materials that are heterogeneous at the nanometer scale, however, the mean-field model is inappropriate, and the atomistic nature of the magnetic interactions must be taken into account.

Atomistic simulations are increasingly used to study magnetism at surfaces and interfaces (Kodama, 1999; Kodama and Berkowitz, 1999; Kachkachi et al., 2000a, 2000b; Dimian and Kachkachi, 2002; Kachkachi and Dimian, 2002; Garanin and Kachkachi, 2003; Kachkachi and Mahboub, 2004; Harrison and Becker, 2001; Robinson et al., 2002; Harrison, 2006; Harrison et al., 2005). It is currently practical to describe systems containing ~10^4 magnetic atoms (Kodama and Berkowitz, 1999). For magnetite, this limitation corresponds to a spherical particle of diameter ~8 nm. If surface properties are not of interest, then an effectively infinite (bulk) system can be simulated by creating a large supercell of the crystal structure and applying periodic boundary conditions (Mazo-Zuluaga and Restrepo, 2004; Harrison, 2006).

The magnetic energy of such a system is a sum of exchange, anisotropy, magnetostatic, and dipole–dipole interaction terms:

$$E_{\mathrm{mag}} = -\sum_{i \neq j} \mathcal{J}_{ij} \mathbf{S}_i \cdot \mathbf{S}_j - \sum_i K_i (\mathbf{S}_i \cdot \mathbf{e}_i)^2$$
$$- (g\mu_B) \sum_i \mathbf{B} \cdot \mathbf{S}_i + E_{\mathrm{d}} \qquad [2]$$

where g is the Landé factor, μ_B is the Bohr magneton, **B** is an externally applied magnetic field, K is a uniaxial anisotropy constant, **e** is the corresponding uniaxial anisotropy axis, and E_{d} is the demagnetizing energy due to dipole–dipole interactions (Kodama and Berkowitz, 1999; Kachkachi et al., 2000a). E_{d} can be expressed in the form

$$E_{\mathrm{d}} = \frac{(g\mu_B)^2}{2} \sum_{i \neq j} \frac{(\mathbf{S}_i \cdot \mathbf{S}_j) \mathbf{R}_{ij}^2 - 3(\mathbf{S}_i \cdot \mathbf{R}_{ij})(\mathbf{R}_{ij} \cdot \mathbf{S}_j)}{R_{ij}^5} \qquad [3]$$

where $\mathbf{R}_{ij}$ is the vector joining atoms i and j. For ellipsoidal particles, eqn [3] simply generates a macroscopic shape anisotropy (Kachkachi et al., 2000a). The large computational overhead involved in summing eqn [3] over all pairs of atoms can be avoided, therefore, by describing this shape anisotropy by a macroscopic approximation of the form

$$E_{\mathrm{d}} = \frac{1}{2V} (D_x M_x^2 + D_y M_y^2 + D_z M_z^2) \qquad [4]$$

where D_x, D_y and D_z are demagnetizing factors and M_x, M_y, and M_z are the components of net magnetization along the principal axes of the ellipsoid (Stacy and Banerjee, 1974). Kachkachi et al. (2000a) found that eqn [4] yielded identical results to eqn [3] for nanoparticles of maghemite (γ-Fe$_2$O$_3$). Different values of K and **e** can be specified for atoms in the core and at the surface of a particle. For a surface atom, **e** is given by the sum of vectors joining the atom to its nearest neighbors, and points approximately perpendicular to the surface (Kodama and Berkowitz, 1999; Garanin and Kachkachi, 2003; Kachkachi and Mahboub, 2004). In this way, the anisotropy is enhanced when the local symmetry is lower than that of the bulk structure. Values for surface anisotropy constants, K, of ~1–4 k_B/cation (where k_B is the Boltzmann constant) are suggested by electron paramagnetic resonance measurements of dilute magnetic cations substituted onto low-symmetry sites in nonmagnetic oxides (Low, 1960). Bulk anisotropies are at least two orders of magnitude smaller.

Monte Carlo methods provide an efficient way of determining the equilibrium spin configuration for a given temperature and applied field (Kachkachi et al., 2000a; Mazo-Zuluaga and Restrepo, 2004; Harrison, 2006). An atom is chosen at random, and its spin direction changed by a random amount. If the resulting energy change, ΔE_{mag}, is negative, then the change is accepted. If ΔE_{mag} is positive, then the change is accepted with a probability of

$\exp(-\Delta E_{mag}/k_B T)$. After a sufficient number of steps, the system reaches equilibrium. The equilibrium configuration is obtained by averaging over a number of steps until the system converges to the desired statistical significance. In many applications (e.g., for the simulation of hysteresis loops), it is not only the equilibrium configuration that is important, but the transitional configurations adopted during the approach to equilibrium. In these cases, a dynamic solution to eqn [3] is required. One approach is to use the Landau–Lifshitz–Gilbert (LLG) equation to calculate the trajectory of each spin (Brown, 1963), in the form

$$\frac{d\mathbf{m}}{dt} = \gamma \mathbf{m} \times \mathbf{H}_e - \lambda \mathbf{m} \times (\mathbf{m} \times \mathbf{H}_e) \qquad [5]$$

where $\mathbf{m}$ is the magnetic moment of a given atom, γ is the gyromagnetic ratio, λ is a damping constant, and $\mathbf{H}_e$ is the effective magnetic field acting on that atom:

$$\mathbf{H}_e = -\frac{dE_{mag}}{d\mathbf{m}} \qquad [6]$$

The first term in eqn [5] describes the precession of the magnetic moment about the effective field direction. The second term decreases the precession angle over time (damping), eventually orienting the magnetic moment along the effective field direction. Although the LLG method has been applied successfully to the study of magnetic nanoparticles (Dimian and Kachkachi, 2002; Kachkachi and Dimian, 2002; Kachkachi and Mahboub, 2004), the method takes many iterations to converge, and can often predict unreasonably large coercivities in atomistic simulations due to the large value of the effective exchange field relative to the applied field. Kodama and Berkowitz (1999) adapted the two-dimensional conjugate direction algorithm of Hughes (1983) to provide a more efficient method of energy minimization for three-dimensional atomistic simulations (achieving convergence in 5–15 iterations). Whichever method is used, finite temperatures can be modeled by applying random rotations to the spins between energy minimization steps. The magnitudes of rotations are adjusted to give $\Delta E_{mag} = N k_B T$, where N is the number of spins in the particle. Random rotations of the individual spins can be combined with random uniform rotations of all of the spins to model collective modes of thermal relaxation, such as superparamagnetism.

The term 'chemical ordering' is used to describe changes in the distribution of magnetic and nonmagnetic atoms in a crystal, which may be brought about by both order–disorder and exsolution processes. In homogeneous systems, the coupling between magnetic and chemical ordering can be described by using established thermodynamic models (Inden, 1981; Kaufman, 1981; Burton and Davidson, 1988; Burton, 1991; Ghiorso, 1997; Harrison and Putnis, 1997, 1999). In heterogeneous systems, however, the presence of internal interfaces and phase boundaries necessitates the use of an atomistic approach. The chemical energy of a given atomic configuration can be written in terms of chemical exchange interaction parameters (Bosenick et al., 2001) in the form

$$E_{chem} = E_0 + \sum_{p,\,q} N_{p,\,q} \mathcal{J}_{p,\,q}^{chem} \qquad [7]$$

where $\mathcal{J}_{p,q}^{chem}$ is the energy associated with placing a pair of unlike cations (labeled p) at a given separation (q) within the structure, and $N_{p,q}$ is the number of times that each type of pair appears in the configuration. E_0 is constant for a fixed bulk composition, and can be neglected. Values for $\mathcal{J}_{p,q}^{chem}$ can be obtained from first-principles or empirical-potential calculations (Becker et al., 2000; Warren et al., 2000a, 2000b; Dove, 2001; Bosenick et al., 2001; Vinograd et al., 2004). Harrison et al. (2000a) used static-lattice calculations to estimate $\mathcal{J}_{p,q}^{chem}$ for the ilmenite–hematite solid solution. These estimates were then refined by fitting the model to cation distribution data obtained using neutron diffraction (Harrison et al., 2000b; Harrison and Redfern, 2001).

In order to simulate coupled magnetic and chemical ordering, a combination of two different Monte Carlo steps must be performed: spin flips and atom swaps. In the spin-flip step, the spin of a randomly chosen atom is changed by a random amount, and the change in magnetic energy, ΔE_{mag}, is used to determine whether this change is accepted or rejected. In the atom-swap step, two atoms are chosen at random and their positions are exchanged. If either atom is magnetic, then the swap will also change the configuration of the spins, and the total energy change, $\Delta E = \Delta E_{chem} + \Delta E_{mag}$, is used to determine whether the swap is accepted or rejected. Atom swaps preserve the net spin of the system. Therefore, after a given number of atom swaps, an equal number of spin flips are performed in order to allow the spin configuration to equilibrate with respect to the new atomic configuration. The alternation of atom swaps and spin flips is repeated until the system reaches a state of global equilibrium with respect to both chemical and magnetic degrees of order.

20.2.2.2   *Application to magnetic nanoparticles*

For magnetic nanoparticles with sizes of 1–10 nm, surface atoms (defined as those having fewer nearest neighbors than the bulk structure) make up at least 25% of the total number of atoms. Finite-size and surface effects give rise to magnetic properties that deviate significantly from those of the bulk material (Kodama, 1999; Kachkachi *et al.*, 2000b). The effect of surface anisotropy and surface roughness on the spin structure of 2.5 nm diameter $NiFe_2O_4$ particles is illustrated in **Figure 4** (Kodama, 1999; Kodama and Berkowitz, 1999). Whereas smooth nanoparticles may adopt uniform spin configurations (**Figures 4(a)** and **4(b)**), rough surfaces (characterized by the presence of surface vacancies and broken exchange interactions) typically display surface spin disorder (**Figures 4(c)** and **4(d)**). A number of different surface spin configurations can be adopted,

depending on the thermal and field history of the particle. Energy barriers between different surface spin configurations can be very high, leading to high-field irreversibility (**Figure 4(c)**). In $NiFe_2O_4$, for example, hysteresis persists in magnetic fields of up to 16 T (Kodama *et al.* 1996), implying effective anisotropy fields for surface spins that are ~400 times larger than the bulk magnetocrystalline anisotropy field. The observation of shifted hysteresis loops during field cooling (Kodama *et al.*, 1997) implies that certain surface configurations freeze in preferentially, and that there is strong exchange coupling between surface and core spins.

The effect of varying surface anisotropy and exchange coupling on the hysteresis properties of SD ferromagnetic particles has been explored by Kachkachi and Dimian (2002). Significant deviations from the classical Stoner-Wohlfarth (1948) model are observed when the surface anisotropy and exchange

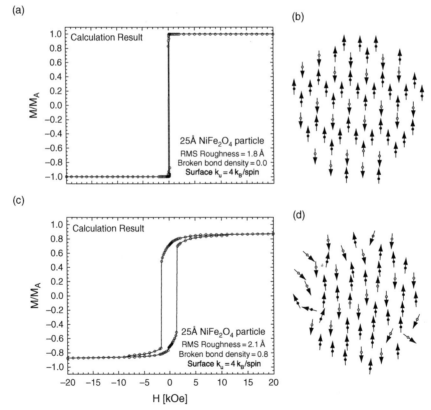

Figure 4   Calculated hysteresis loops and spin configurations of a 2.5 nm diameter spherical particle of $NiFe_2O_4$. Simulations were performed using an atomistic model of magnetic ordering (Kodama, 1999; Kodama and Berkowitz, 1999). Arrows in (b) and (d) show the spins on individual Fe atoms. Light and dark circles correspond to tetrahedral and octahedral sites, respectively. A particle with no broken bonds and low surface roughness has a low coercivity (a) and no surface spin disorder (b). A particle with a higher broken-bond density and surface roughness displays high coercivity and high-field irreversibility (c), resulting from the presence of significant surface spin disorder (d). Reproduced from Kodama RH (1999) Magnetic nanoparticles. *Journal of Magnetism and Magnetic Materials* 200: 359–372.

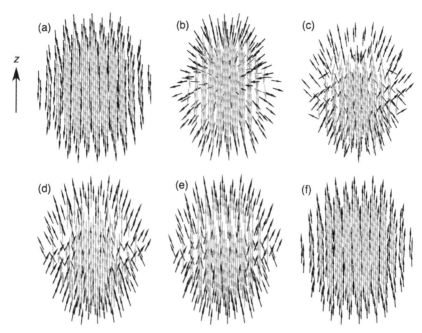

Figure 5 Atomistic simulation of nonuniform magnetic switching in a ferromagnetic nanoparticle with strong surface anisotropy. The spherical particle contains 176 surface spins (dark arrows) and 184 core spins (gray arrows). The easy axis for core spins is parallel to the z-direction (indicated). The easy axis for surface spins is approximately normal to the particle surface. In this example, the core and surface anisotropy constants were chosen to be equal in magnitude and the exchange integral between neighboring spins was one-tenth the magnitude of the core/surface anisotropy constant. Each spin configuration was obtained at a different value of the applied magnetic field, starting with a saturating field in the negative z-direction (a) and ending with a saturating field in the positive z-direction (f). The surface spins are observed to switch magnetization direction before the core spins (c). Core spins are observed to switch magnetization direction in a cluster-like fashion (d, e). Adapted from Kachkachi H., and Dimian M (2002) Hysteretic properties of a magnetic particle with strong surface anisotropy. *PhysicalReview B* 66: 174419.

constants are of similar magnitude (**Figure 5**). These deviations are associated with nonuniform reversal mechanisms, which involve the successive switching of surface and core spins.

The magnetic properties of maghemite (γ-Fe_2O_3) nanoparticles have been investigated using both Monte Carlo and conjugate direction methods (Kachkachi *et al.*, 2000a; Kodama and Berkowitz, 1999). Mössbauer spectroscopy indicates that surface spins in maghemite nanoparticles are highly canted (Coey, 1971). Their magnetic properties are dominated by surface effects, which result in high coercive fields, high-field irreversibility, and shifted hysteresis loops in field-cooled samples (Kachkachi *et al.*, 2000a; Tronc *et al.*, 2000). Particles are not saturated in fields of up to 5.5 T (**Figure 6(a)**), and an anomalous increase in magnetization appears below 70 K (**Figure 6(b)**), which is more pronounced in smaller particles (**Figure 6(c)**). The simulated contribution to the magnetization from core and surface spins is illustrated in **Figure 7** (Kachkachi *et al.*, 2000a). Assuming that exchange interactions between surface atoms are an order of magnitude weaker than interactions between bulk atoms, an anomalous increase in magnetization observed at low temperatures can be attributed to the ordering of surface spins.

20.2.2.3 Application to coupled magnetic and chemical ordering in solid solutions

The magnetic properties of the ilmenite–hematite solid solution are influenced profoundly by nanoscale microstructures resulting from chemical ordering. Slowly cooled rocks that contain finely exsolved hematite–ilmenite have strong and extremely stable magnetic remanence, which may account for some of the magnetic anomalies that are present in the deep crust and on planetary bodies that no longer retain a magnetic field, such as Mars (McEnroe *et al.*, 2001, 2002, 2004a, 2004b, 2004c; Kasama *et al.*, 2003, 2004). This remanence has been attributed to the presence of a stable ferrimagnetic substructure, which is associated with the coherent

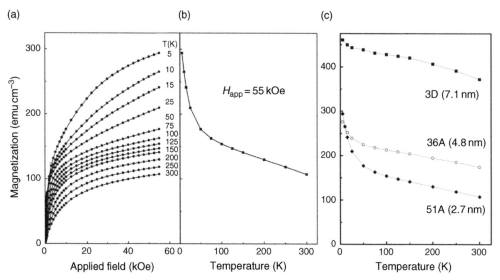

Figure 6 (a) Magnetization of a diluted assembly of γ-Fe$_2$O$_3$ nanoparticles, with a mean diameter of 2.7 nm, as a function of magnetic field at different temperatures. (b) Temperature dependence of magnetization at 55 kOe, extracted from (a), showing an anomalous increase in magnetization below 70 K. (c) Temperature dependence of magnetization in a field of 55 kOe for three samples with different mean diameters (2.7, 4.8, 7.1 nm). The anomalous increase in magnetization at low temperatures is more pronounced in the smaller particles, consistent with a surface effect. Reproduced from Kachkachi H, Ezzir A, Noguès M, and Tronc E (2000a) Surface effects in nanoparticles: Application to maghemite γ-Fe$_2$O$_3$. *European Physical Journal B* 14: 681–689.

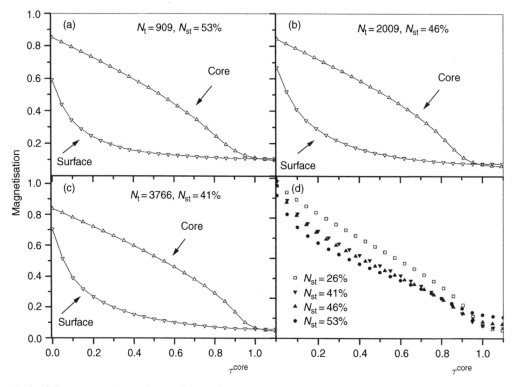

Figure 7 (a–c) Temperature dependence of the surface and core magnetization (per site) and (d) mean magnetization, as obtained from atomistic Monte Carlo simulations of an ellipsoidal maghemite nanoparticle. The exchange interactions on the surface are taken to be 1/10 times those in the core, leading to an increase in the surface magnetization contribution at low temperatures. Temperature is given in reduced units ($\tau_c = T/T_c^{core}$, where T_c^{core} is the critical temperature of the core spins). Reproduced from Kachkachi H, Ezzir A, Noguès M, and Tronc E (2000a) Surface effects in nanoparticles: Application to maghemite γ-Fe$_2$O$_3$. *European Physical Journal B* 14: 681–689.

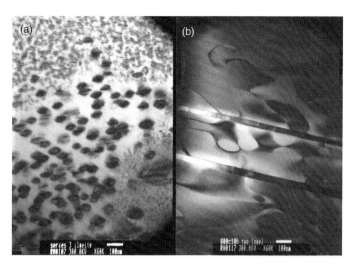

Figure 8   Examples of nanoscale microstructures in the ilmenite–hematite solid solution. (a) A natural hemo-ilmenite containing abundant nanoscale exsolution lamellae of hematite in an ilmenite host (McEnroe *et al.*, 2002). Scale bar = 100 nm. (b) A synthetic sample of the ilmenite–hematite solid solution (ilm70) containing curved antiphase domains (APDs) and antiphase boundaries (APBs), formed after cooling through the $R\bar{3}c$ to $R\bar{3}$ cation ordering phase transition. The APDs are crosscut by two titanomagnetite lamellae, the result of annealing the sample under slightly reducing conditions. Scale bar = 100 nm.

interface between nanoscale ilmenite and hematite exsolution lamellae (the so-called 'lamellar magnetism hypothesis'; Harrison and Becker, 2001; Robinson *et al.*, 2002, 2004; Harrison, 2006; **Figure 8(a)**). Rapidly cooled members of the hematite–ilmenite series, on the other hand, are well known for their ability to acquire self-reversed thermoremanent magnetization (i.e., they acquire a remanent magnetization on cooling that is antiparallel to the applied field direction). This phenomenon is related to the presence of fine-scale twin domains that form on cooling through the $R\bar{3}c$–R $\bar{3}$ cation-ordering phase transition (Ishikawa and Syono, 1963; Nord and Lawson, 1989, 1992; Hoffman, 1992; Bina *et al.*, 1999; Prévot *et al.*, 2001; Lagroix *et al.*, 2005; **Figure 8(b)**).

Harrison (2006) used Monte Carlo simulations to investigate the consequences of coupling between magnetic and chemical ordering in the ilmenite–hematite solid solution (**Figure 9**). Key features of the equilibrium phase diagram are reproduced successfully by the simulations: (1) a paramagnetic (PM) to antiferromagnetic (AF) transition in the hematite-rich, cation-disordered ($R\bar{3}c$) solid solution; (2) a PM $R\bar{3}c$ to PM $R\bar{3}$ cation ordering transition in the ilmenite-rich solid solution; (3) a PM $R\bar{3}c$ + PM $R\bar{3}$ miscibility gap developing below a tricritical point at $x = 0.58 \pm 0.02$, $T = 1050 \pm 25$ K; and (4) an AF $R\bar{3}c$ + PM $R\bar{3}$ miscibility gap developing below a eutectoid point at $x = 0.18 \pm 0.02$, $T = 800 \pm 25$ K.

A snapshot of the simulated cation/spin configuration obtained at 100 K for a bulk composition 30%

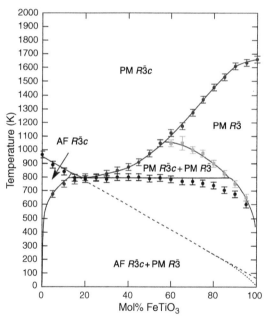

Figure 9   Summary of the equilibrium phase relations in the ilmenite–hematite solid solution determined by Monte Carlo simulation (Harrison, 2006). Dashed and dotted lines show the metastable magnetic ordering temperatures for the cation-disordered and cation-ordered solid solution, respectively.

Fe_2O_3 70% $FeTiO_3$ (ilm70) is shown in **Figure 10**. The supercell contains two nanoscale precipitates of AF $R\bar{3}c$ hematite within a host of PM $R\bar{3}$ ilmenite. The lower precipitate has a thickness of 2 nm (~1.5 unit cells), corresponding to the lower size limit of exsolution lamellae typically observed in natural samples (Robinson *et al.*, 2002). The precipitate contains a total of 11 Fe-bearing cation layers: 9 Fe^{3+} layers, which are bounded by mixed Fe^{3+}–Fe^{2+} 'contact' layers. The natural tendency for hematite lamellae to form with an odd number of Fe-bearing layers results in the formation of a 'defect' moment due to the presence of uncompensated spins (**Figure 10(c)**). Under favorable conditions, this 'lamellar magnetism' far outweighs the spin-canted moment of the hematite phase (Robinson *et al.*, 2004). The upper precipitate has a less well-defined shape and an atomically rough interface with the ilmenite host. It has a thickness of 0.7–1.4 nm (~0.5–1 unit cells), which corresponds to the length scale of the compositional clustering that is commonly observed in natural samples in the vicinity of precipitate-free zones (McEnroe *et al.*, 2002). The more irregular shape and rough interface of the upper precipitate

enhances the spin imbalance, yielding a larger net magnetization (**Figure 10(d)**).

Samples that have cooled rapidly through the $R\bar{3}c$–$R\bar{3}$ transition develop a high degree of short-range cation order, which is characterized by the formation of fine-scale twin domains (Harrison and Redfern, 2001; Nord and Lawson, 1989, 1992; **Figure 8(b)**). Adjacent domains have an antiphase relationship with each other, in terms of the ordering of Fe and Ti layers; an Fe-rich layer becomes a Ti-rich layer on crossing the twin wall and vice versa. In order to highlight this relationship, twin domains and twin walls are often referred to as antiphase domains (APDs) and antiphase domain boundaries (APBs), respectively. **Figure 11** shows the results of Monte Carlo simulations of a 48-layer supercell of ilm70, with APBs at its center and upper/lower boundaries (Harrison, 2006). The degree of cation order is defined by the order parameter:

$$Q = \frac{N_{Ti}^{B} - N_{Ti}^{A}}{N_{Ti}^{A} + N_{Ti}^{B}} \qquad [8]$$

where N_{Ti}^{A} and N_{Ti}^{B} are the number of Ti^{4+} cations on A and B layers, respectively. The cation/spin configuration after annealing the supercell in the simulation at 850 K is shown in **Figures 11(a)** **11(d)**. The APDs are cation ordered ($Q = \pm 1$) and the APBs are cation disordered ($Q = 0$) (**Figure 11(a)**). The APBs are enriched in Fe relative to the APDs, although the magnitude of this enrichment is enhanced by the immiscibility of ilmenite and hematite at this temperature (**Figure 11(b)**). The spin profile at 25 K shows the presence of oppositely magnetized ferrimagnetic domains, a consequence of the switch round of Fe-rich and Ti-rich layers at the APB (**Figure 11(c)**). Experimental confirmation of such negative exchange coupling, obtained using electron holography, is discussed in Section 20.3.3.4. The APB shown in **Figure 11(c)** is characterized by an asymmetric spin profile. Only the very center of the APB can be classed as antiferromagnetic. At 400 K, the majority of the supercell is magnetically disordered, whereas the APBs retain a narrow region of magnetic order (**Figure 11(d)**).

After annealing at 1100 K, the supercell contains a well-ordered domain and a smaller, less well-(anti)-ordered domain (**Figure 11(e)**). This situation is reached as the system attempts to remove one APD and establish an equilibrium state of homogenous long-range order. Fe enrichment now occurs at the APBs and across the less well-(anti)ordered domain.

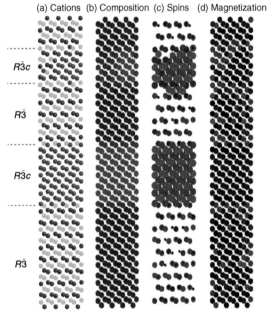

Figure 10 Snapshots of a combined simulation of cation and magnetic ordering in the ilmenite–hematite solid solution (ilm70) at 100 K (Harrison, 2006). (a) Distribution of Fe^{3+}, Fe^{2+}, and Ti (red, green, and blue, respectively). (b) Local chemical composition (red = hematite, brown = ilmenite) calculated by averaging the number of Ti cations within the first four coordination shells around each site. (c) Magnitude and direction of spin on each site (red = negative, blue = positive, symbol size proportional to magnitude of spin). (d) Local ferrimagnetic moment (blue = positive, red = negative), calculated by averaging the spin values within the first four coordination shells around each site. The blue regions highlight the ferrimagnetism associated with local spin imbalance at the interface between ilmenite and hematite precipitates.

The spin profile at 25 K indicates a strong ferrimagnetic moment associated with the ordered domain and a weak ferrimagnetic moment with the anti-ordered domain (**Figure 11(g)**). At 375 K, magnetic order is lost in the ordered domain, whereas weak magnetic order is retained across the anti-ordered domain and boundary regions (**Figure 11(h)**). These properties lead to a self-reversal in the net magnetization on cooling (**Figure 12**). Magnetic ordering in the Fe-enriched anti-ordered domain sets in below 425 K, yielding a weak positive ferrimagnetic moment. Magnetic order spreads to the ordered domain on cooling below 350 K. Below 250 K, the moment of the anti-ordered domain is outweighed by the oppositely oriented moment of the ordered domain, and the net magnetization reverses. In contrast, no net reversal is observed in the 850 K simulation, which contains equally well-ordered and anti-ordered domains, despite the enhanced enrichment of Fe at the APB.

20.3 Magnetism at the Nanometer Length Scale

Off-axis electron holography is an advanced TEM technique that allows a two-dimensional projection of the in-plane component of the magnetic induction in a specimen to be mapped with a spatial resolution approaching the nanometer scale. The high spatial resolution of this technique makes it ideal for the study of magnetic particles that are in the SD to PSD size range, as well as for magnetic minerals that are structurally and/or chemically heterogeneous. Its ability to provide images of stray magnetic fields also makes electron holography an ideal technique for the study of magnetostatic interactions between magnetic nanoparticles. We begin by reviewing theoretical and practical aspects of electron holography. We then describe its recent application to several different magnetic minerals.

20.3.1 Theory of Off-Axis Electron Holography of Magnetic Materials

20.3.1.1 Amplitude and phase of a TEM image

The formation of a TEM image can be described in terms of the electron wave function in the image plane of the microscope:

$$\psi(\mathbf{r}) = A(\mathbf{r})\exp[i\phi(\mathbf{r})] \qquad [9]$$

where A is amplitude, ϕ is phase shift (with respect to a wave that has traveled through vacuum alone), and $\mathbf{r}$ is a vector in the plane of the sample (Cowley, 1995). As an electron passes through the microscope, it experiences a phase shift that is associated with both the electrostatic potential of the sample (**Figure 13(a)**) and the in-plane components of the magnetic induction (**Figure 13(b)**). In a conventional TEM image, only the spatial distribution of the image intensity

$$I(\mathbf{r}) = \psi(\mathbf{r})\psi^*(\mathbf{r}) = A^2(\mathbf{r}) \qquad [10]$$

is recorded, and all information about the phase shift is lost. Electron holography is an interferometric technique that allows phase information to be recovered. After subtraction of the electrostatic contribution to the phase shift (see below), a phase image can be converted directly into a quantitative two-dimensional map of the in-plane magnetic induction in the sample (Tonomura, 1992; Völkl *et al.*, 1998; Dunin-Borkowski *et al.*, 2004; Midgely, 2001).

The phase shift (measured relative to that of an electron that has passed through vacuum alone) is given by the expression

$$\phi(x) = C_E \int V_0(x, z)\mathrm{d}z - \left(\frac{e}{\hbar}\right) \iint B_\perp(x, z)\mathrm{d}x\,\mathrm{d}z \qquad [11]$$

where x is a direction in the plane of the sample, z is the incident electron beam direction, V_0 is the mean inner potential, and $B_\perp$ is the component of magnetic induction perpendicular to both x and z (Reimer, 1991). C_E is a constant that depends on the accelerating voltage of the TEM, in the form

$$C_E = \left(\frac{2\pi}{\lambda}\right)\left(\frac{E + E_0}{E(E + 2E_0)}\right) \qquad [12]$$

where λ is the electron wavelength, and E and E_0 are the kinetic and rest mass energies of the incident electrons, respectively. Values of C_E for a range of accelerating voltages are listed in **Table 2**. If neither V_0 nor $B_\perp$ varies with z inside the specimen, and both parameters are zero outside the specimen, eqn [11] can be expressed more simply in the form

$$\phi(x) = C_E V_0(x)t(x) - \left(\frac{e}{\hbar}\right) \int B_\perp(x)t(x)\,\mathrm{d}x \qquad [13]$$

where t is the thickness of the sample. Differentiating with respect to x then leads to the expression

$$\frac{\mathrm{d}\phi(x)}{\mathrm{d}x} = C_E \frac{\mathrm{d}}{\mathrm{d}x}\{V_0(x)t(x)\} - \left(\frac{e}{\hbar}\right)B_\perp(x)t(x) \qquad [14]$$

For constant V_0 and t, the first term in eqn [14] is zero, and the gradient of the phase shift is proportional to the desired in-plane component of the magnetic induction in the specimen (**Figure 13(b)**). The phase gradient can then be written in the form

$$\frac{\mathrm{d}\phi(x)}{\mathrm{d}x} = -\left(\frac{et}{\hbar}\right) B_\perp(x) \qquad [15]$$

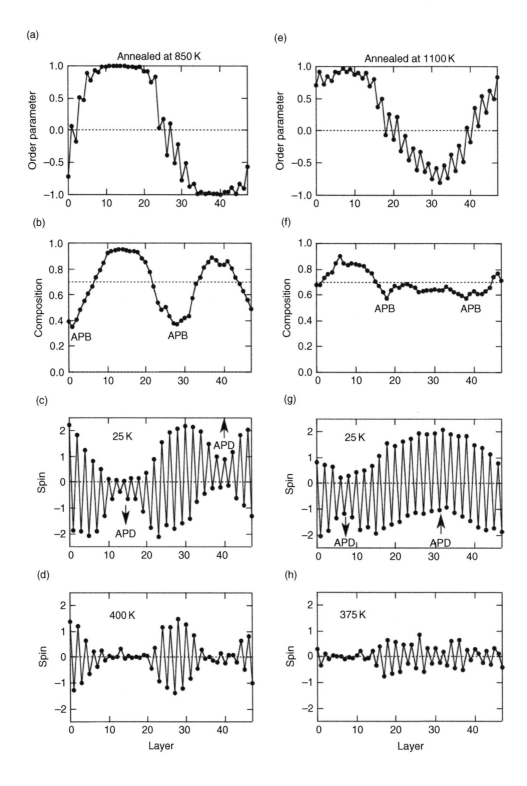

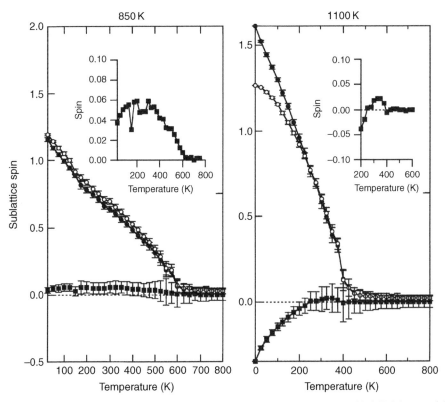

Figure 12 Temperature dependence of sublattice spins S_A and S_B (circles) and net spin $|S_A|-|S_B|$ (squares) for the simulations pre-annealed at (a) 850 K and (b) 1100 K. Insets show expanded view of the variation in net spin. A self-reversal of the net magnetization occurs in (b) due to the negative exchange coupling between poorly ordered and well ordered APDs (see **Figures 11(e)–11(h)**).

Unfortunately, in most cases, both V_0 and t vary across the specimen, and careful separation of the magnetic and mean inner potential contributions to the measured phase shift is required before quantitative analysis of the magnetic induction is possible (see Section 20.3.1.5).

20.3.1.2 Calculation of the mean inner potential

For a specimen that has a single composition and crystallographic orientation, the mean inner potential contribution to the phase shift is proportional to the specimen thickness (**Figure 13(a)**). Direct

Figure 11 Average values of order parameter, composition, and spin on each of the 48 layers of an $8 \times 8 \times 8$ supercell of ilm70, pre-annealed at (a–d) 850 K and (e–h) 1100 K (Harrison, 2006; Harrison et al., 2005). A starting configuration with APBs at the bottom and the center of the supercell was chosen in each case. (a) The order parameter profile at 850 K shows two fully ordered/antiordered APD ($Q = 1$ and $Q = -1$, respectively) separated by APBs ($Q = 0$). (b) The composition profile at 850 K shows that unmixing has taken place within the PM $R\bar{3}c$ + PM $R3$ miscibility gap, with the PM $R3$ phase corresponding to the APDs and the PM $R\bar{3}c$ phase corresponding to the APBs. Dashed line indicates the bulk composition, $x = 0.7$. (c) The spin profile at 25 K shows that the APDs are strongly ferrimagnetic. The APD centered on layer 14 has a net negative spin, whereas the APD centered on layer 40 has a net positive spin (indicated by the arrows). (d) The spin profile at 400 K shows that the Fe-rich APBs remain magnetically ordered, whereas the Fe-poor APDs are magnetically disordered. The APBs are associated with a small net spin (see **Figure 12**). (e) The order parameter profile at 1100 K shows a fully ordered APD ($Q \sim 1$) and a less well-(anti)ordered APD ($Q \sim -0.75$). (f) The composition profile at 1100 K shows that the well-ordered APD has $x > 0.7$, whereas the less well-ordered APD has $x < 0.7$. Evidence for Fe enrichment at the APBs is also seen. (g) The spin profile at 25 K shows that the well-ordered APD is strongly ferrimagnetic, whereas the ferrimagnetic spin of the less well-ordered APD is decreased by the influence of the boundary regions. (h) The spin profile at 375 K shows that the less well-ordered APD and boundary regions are magnetically ordered, whereas the well-ordered APD is magnetically disordered. The magnetically ordered regions carry a small net spin that is opposite to the net spin of the well-ordered APD (see **Figure 12**).

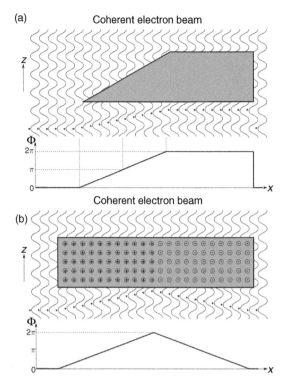

(a) Coherent electron beam

(b) Coherent electron beam

Figure 13 Schematic illustration of the phase shift experienced by electrons passing through a specimen in the TEM. (a) The mean inner potential contribution to the phase shift of electrons passing through a sample of uniform structure and chemical composition reflects changes in the specimen thickness (first term in eqn [13]). (b) The magnetic contribution to the phase shift, given by the second term in eqn [13], reflects the in-plane component of the magnetic induction, integrated along the electron beam direction. For a sample containing two uniformly magnetized domains, one magnetized out of the plane of the diagram (crosses) and one magnetized into the plane of the diagram (dots), the gradient of the phase shift is constant within the domains and changes sign at the domain wall.

Table 2 Values of the constant C_E in eqn [12] as a function of TEM accelerating voltage

Voltage (kV)	$C_E(\times 10^6$ rad/V m)
100	9.24396
200	7.2884
300	6.52616
400	6.12141
500	5.87316
600	5.70724
700	5.58974
800	5.50296
900	5.43679
1000	5.38502

measurement of V_0 using electron holography is possible if an independent measurement of the specimen thickness is available. Such measurements are rare, however, and it is often necessary to calculate theoretical values of V_0. An estimate for V_0 can be obtained by assuming that the specimen can be described as a collection of neutral free atoms (the 'nonbinding' approximation), and by using the expression

$$V_0 = \left(\frac{h^2}{2\pi m e \Omega}\right) \sum_{\Omega} f_{el}(0) \qquad [16]$$

where $f_{el}(0)$ is the electron scattering factor at zero scattering angle (with dimensions of length), Ω is the unit cell volume, and the sum is performed over all atoms in the unit cell (Reimer, 1991). Calculated values for $f_{el}(0)$ have been tabulated by Doyle and Turner (1968) and Rez et al. (1994). Equation [16] leads to an overestimation of V_0 by approximately 10%, because the redistribution of electrons due to bonding (which typically results in a contraction of the electron density around each atom) is neglected. Calculated upper limits of V_0 for common magnetic oxide minerals are listed in **Table 3**.

20.3.1.3 Formation of an electron hologram

The microscope setup for electron holography is illustrated schematically in **Figure 14**. A field-emission gun (FEG) is used to provide a highly coherent source of electrons. For studies of magnetic materials, a Lorentz lens (a high-strength minilens located below the lower objective pole-piece) allows the microscope to be operated at high magnification with the objective lens switched off and the sample in magnetic-field-free conditions. Since a large

Table 3 Calculated V_0 values for a range of magnetic minerals using eqn [16] and electron scattering factors tabulated by Doyle and Turner (1968)

Mineral Name (formula)	V_0(V)
Hematite (Fe_2O_3)	19.3
Ilmenite ($FeTiO_3$)	19.9
Magnetite (Fe_3O_4)	19.1
Ulvöspinel (Fe_2TiO_4)	19.1
Maghemite (γ-Fe_2O_3)	17.8
Pseudobrookite (Fe_2TiO_5)	17.3
Pseudobrookite ($FeTi_2O_5$)	17.9

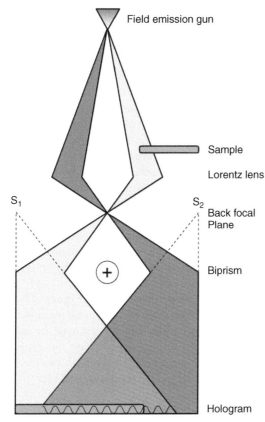

Field emission gun

Sample

Lorentz lens

S_1

S_2

Back focal
Plane

Biprism

Hologram

Figure 14 Schematic illustration of the setup used for generating off-axis electron holograms. The sample occupies approximately half the field of view. Essential components are the field-emission electron gun source, which provides coherent illumination, and the positively charged electrostatic biprism, which overlap of the sample and (vacuum) reference waves. The Lorentz lens allows imaging of magnetic materials in close-to-field-free conditions.

amount of off-line image processing must be carried out to process holograms and in particular to remove the mean inner potential contribution to the phase shift, there are great advantages in recording holograms digitally, using a charge-coupled device (CCD) camera.

For off-axis electron holography, the sample is typically placed half-way across the field of view, so that part of the electron wave passes through the sample (the sample wave) and part passes through vacuum (the reference wave). A voltage of 50–200 V is applied to an electrostatic biprism wire (typically a <1 μm diameter quartz wire coated in Pt or Au) mounted in place of one of the selected area diffraction apertures. This voltage deflects the sample and reference waves, causing them

to overlap. If the electron source is sufficiently coherent, then an interference fringe pattern (an electron hologram) is formed in the overlap region. The sample and reference waves can then be considered as originating from two virtual sources, S_1 and S_2 (**Figure 14**). The angles of the sample and reference waves differ by a small amount, which is proportional to the biprism voltage, and can be described by the wave vector $\mathbf{q}_c$. The intensity in the overlap region is then given by the expression

$$I_{hol}(\mathbf{r}) = |\psi(\mathbf{r}) + \exp[2\pi i \mathbf{q}_c \cdot \mathbf{r}]|^2$$
$$= 1 + A^2(\mathbf{r}) + 2A(\mathbf{r})\cos[2\pi i \mathbf{q}_c \cdot \mathbf{r} + \phi(\mathbf{r})] \quad [17]$$

Hence, an electron hologram consists of a sum of the intensities of the sample and reference waves, onto which is superimposed a set of cosinusoidal fringes with local amplitude A and phase shift ϕ. An example of hologram, acquired from a sample containing maghemite inclusions in a matrix of hematite, is shown in **Figure 15(a)**. Local shifts in the positions of the holographic interference fringes, visible in the inset to **Figure 15(a)**, are directly proportional to the phase shift of the electron wave, which results, in turn, from variations in the thickness, mean inner potential, and magnetic induction of the inclusion and the host. A broader set of Fresnel fringes are also visible at the edges of the hologram in **Figure 15(a)**. These fringes are caused by the edge of the biprism wire.

20.3.1.4 Processing of the electron hologram

The sequence of processing steps that is required to extract a phase map, $\phi(\mathbf{r})$, from an electron hologram is illustrated in **Figure 15**. First a hologram of the region of interest is acquired (**Figure 15(a)**). The sample is then moved away from the field of view and a reference hologram is acquired from a region of vacuum (**Figure 15(b)**). Next, both the sample and the reference holograms are Fourier transformed (**Figure 15(c)**). The Fourier transform of eqn [17] comprises a central peak at $\mathbf{q} = 0$ and two sidebands at $\mathbf{q} = \pm \mathbf{q}_c$:

$$FT[I_{hol}(\mathbf{r})] = \delta(\mathbf{q}) + FT[A^2(\mathbf{r})] + \delta(\mathbf{q} + \mathbf{q}_c)$$
$$\otimes FT[A(\mathbf{r})\exp(i\phi(\mathbf{r})] + \delta(\mathbf{q} - \mathbf{q}_c)$$
$$\otimes FT[A(\mathbf{r})\exp(-i\phi(\mathbf{r})] \quad [18]$$

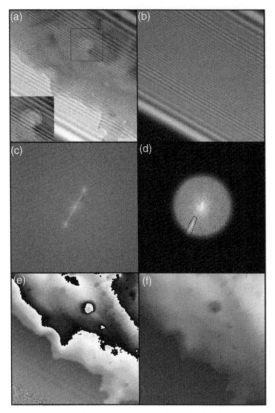

Figure 15 Sequence of image processing steps required to convert an electron hologram into a phase-shift image. (a) Original electron hologram of the region of interest (a natural sample of hematite containing nanoscale inclusions of maghemite). Broad Fresnel fringes, caused by the edges of the biprism wire, are visible in the upper right and lower left. The inset is a magnified image of the outlined region, showing the change in position of the fine-scale holographic fringes as they pass through an inclusion. (b) A reference hologram recorded over a region of vacuum. (c) Fourier transform of the electron hologram shown in (a), comprising a central peak, two side bands, and a diagonal streak due to the Fresnel fringes. (d) A mask is applied to the Fourier transform in (c) in order to isolate one side band. The Fresnel streak is removed by assigning a value of zero to pixels falling inside the region shown by the dashed line. (e) Inverse Fourier transform of (d) yields the complex image wave, which in turn yields a modulo 2π image of the holographic phase shift. (f) Automated phase unwrapping algorithms are used to remove the 2π phase discontinuities from (e) to yield the final phase shift image.

The sidebands contain the Fourier transforms of either the complex image wave or its conjugate. Both amplitude and phase information are recovered by isolating one sideband digitally (**Figure 15(d)**)

and performing an inverse Fourier transform of this part of the Fourier transform alone. The diagonal streak at the lower left of **Figure 15(d)** results from the presence of Fresnel fringes visible in the raw holograms (**Figures 15(a)** and **15(b)**). This streak can lead to artifacts in the reconstructed phase map, and is normally masked out (i.e., replaced by pixels with values of zero) before the inverse Fourier transform is performed. The complex image waves that are derived from the sample and reference holograms are divided by each other to remove phase shifts caused by inhomogeneities in the charge and thickness of the biprism wire, and distortions caused by aberrations of the microscope lenses and the recording system (de Ruijter and Weiss, 1993). The phase shift is then obtained by evaluating the arctangent of the ratio of the imaginary and real components of the corrected complex image wave (**Figure 15(e)**). The initial phase map is presented modulo 2π. The 2π discontinuities can be removed by using one of a number of automated phase unwrapping algorithms (Ghiglia and Pritt, 1998) to produce an 'unwrapped' final phase image (**Figure 15(f)**).

20.3.1.5 Removing the mean inner potential contribution

If the direction of magnetization in the sample can be reversed exactly, for example, by applying a large magnetic field to the specimen, then the magnetic contribution to the phase shift changes sign in eqn [11]. If phase images that have been acquired before and after magnetization reversal are added together digitally, then the magnetic contribution to the phase shift cancels out, leaving twice the mean inner potential contribution. Magnetization reversal can be performed *in situ* in the TEM by using the magnetic field of the TEM objective lens (**Figure 16**). The sample is typically tilted to an angle of $\pm 30°$ to the horizontal. The objective lens is then turned on to provide a chosen vertical magnetic field of up to 2 T. The objective lens is then turned off and the sample tilted back to the horizontal prior to acquisition of the hologram. In practice, the two saturation remanent states may not be exactly equal and opposite to each other. It is then necessary to repeat the switching process several times, so that nonsystematic differences between switched pairs of phase images average out. Systematic differences between switched pairs, which can lead to artifacts in the final magnetic induction map, are often identified by

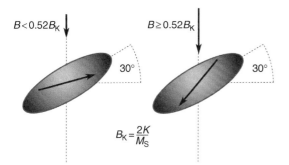

Figure 16 Schematic illustration of magnetic switching in the TEM. A uniaxial particle with anisotropy constant K and saturation magnetision M_s, initially magnetized to the right, is tilted to an angle of 30° to the horizontal. A chosen current is passed through the objective lens of the TEM, exposing the sample to a downward pointing magnetic field of up to 2 T. The direction of magnetization switches when the vertical field reaches $0.52B_K$, where $B_K = 2K/M_s$. The objective lens is then switched off and the sample is tilted back to the horizontal.

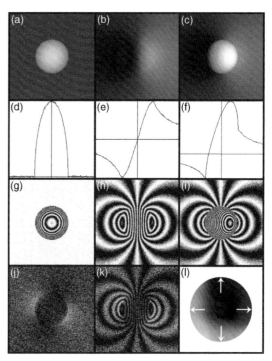

Figure 17 Simulation of the holographic phase shift associated with a 200 nm diameter spherical particle of magnetite. The particle is uniformly magnetized in the vertical direction. The mean inner potential contribution to the phase shift is shown in (a), the magnetic contribution is shown in (b), and the sum of the two is shown in (c). (d–f) Profiles of (a–c), taken horizontally through the center of the particle (i.e., in a direction normal to the magnetization direction). The analytical form of these curves is given by eqns [19] and [20]. (g–i) Cosine of 4 times the phase shift shown in (a–c). (j) Color map derived from the gradient of the magnetic contribution to the phase shift (b). The hue and intensity of the color indicates the direction and magnitude of the integrated in-plane component of magnetic induction, according to the color wheel shown in (l). The color can be combined with the contour map, as shown in (k).

inspection. Once the mean inner potential contribution to the phase shift has been determined in this way, it can be subtracted from each individual phase image of the same region of the sample. By varying the magnitude of the applied field, it is possible to record a series of images that correspond to any desired point on the remanent hysteresis loop.

When interpreting the subsequent remanent hysteresis loop, it is important to remember that holography measures the coercivity of remanence (H_{cr}) rather than the coercivity (H_c) of the sample. For SD particles, however, $H_{cr}/H_c \sim 1$, allowing an estimate of the coercivity to be made. A sample with uniaxial anisotropy constant, K, and saturation magnetization, M_s, switches when the vertical field reaches $0.52B_K$ (assuming a tilt angle of 30° to the horizontal), where $B_K = 2K/M$ is the coercivity for fields applied along the anisotropy axis (Stoner and Wohlfarth, 1948).

20.3.2 Interpretation of Electron Holographic Phase Images

20.3.2.1 Quantification of the magnetic induction

A quantitative measure of $B_\perp$, integrated in the electron beam direction, can be obtained from the gradient of the magnetic contribution to the phase shift (eqn [15]). This measurement includes contributions from the internal magnetization of the sample, the internal demagnetizing field, and stray magnetic fields created by the sample in the vacuum surrounding it. A simulation of the contributions to the phase shift associated with the presence of a uniformly magnetized 200 nm diameter spherical particle of magnetite is shown in **Figure 17**. The total phase shift (**Figure 17(c)**) is the sum of mean inner potential (**Figure 17(a)**) and magnetic (**Figure 17(b)**) contributions. An analytical expression that describes the phase shift shown in **Figure 17(c)**, along a line passing through the center of the particle in a direction perpendicular to $B_\perp$, is

$$\phi(x)|_{x \leq a} = 2C_E V_0 \sqrt{a^2 - x^2} + \left(\frac{e}{\hbar}\right) B_\perp \left[\frac{a^3 - (a^2 - x^2)^{3/2}}{x}\right]$$

[19]

$$\phi(x)|_{x > a} = \left(\frac{e}{\hbar}\right) B_\perp \left(\frac{a^3}{x}\right)$$

[20]

where a is the radius of the particle (de Graef *et al.*, 1999). The mean inner potential and magnetic contributions to this phase profile are shown in **Figure 17(d)–17(f)**. The difference between the minimum and maximum values of the magnetic contribution to the phase shift in **Figure 17(e)** is

$$\Delta\phi_{mag} = 2.044 \left(\frac{e}{\hbar}\right) B_\perp a^2$$

[21]

For a uniformly magnetized cylinder of radius a, the equivalent expression is

$$\Delta\phi_{mag} = \pi \left(\frac{e}{\hbar}\right) B_\perp a^2$$

[22]

To a good approximation, ion-beam thinned TEM specimens of magnetic materials can often be described locally as plates or wedges of semi-infinite extent in the horizontal plane. If such a sample is magnetized uniformly parallel to its edge, then the effect of demagnetizing and stray fields on the measured phase shift may be negligible, and it may then be possible to determine $B_\perp$ directly by using eqn [15]. This approach requires, however, that the local specimen thickness is known. A measure of the sample thickness can be obtained using energy filtered imaging (Egerton, 1996). Two images of the sample are acquired: an unfiltered image (formed using both elastically and inelastically scattered electrons) and a zero-loss energy-filtered image. The log of the ratio between the unfiltered and zero-loss energy-filtered images yields the quantity t/λ_{in}, where λ_{in} is a mean free path for inelastic scattering. Values of λ_{in} can be calculated or measured experimentally (Egerton, 1996; Golla-Schindler *et al.*, 2005). Harrison *et al.* (2002) determined a value for $\lambda_{in} = 170$ nm for magnetite. Care is required when using this approach, as the effective magnetic thickness of a sample may be significantly smaller than its physical thickness, due to the presence of magnetically 'dead' layers on its surfaces, resulting from specimen preparation techniques such as ion-beam thinning. For a sample of known $B_\perp$ and V_0, an estimate of the thickness of the

magnetically dead layers can be obtained by comparing the physical thickness of the specimen derived from the mean inner potential contribution to the phase shift (the first term in eqn [13]) with the magnetic thickness derived from eqn [15]. For an Ar-ion milled synthetic sample of ilm70, assuming a value for $V_0 = 19.6$ V (calculated using eqn [16]) and a saturation induction of 0.225 T, the average difference between the magnetic and physical specimen thickness was found to be ~40 nm in total.

20.3.2.2 Visualization of the magnetic induction

The in-plane component of the integrated magnetic induction can be visualized by adding contours to the magnetic contribution to the phase shift, as shown in **Figure 17(h)** in the form of the cosine of the phase image. The spacing of the contours can be varied by multiplying the phase map by a constant before calculating its cosine (an 'amplification' factor of 4 was used in **Figure 17**). By calculating the horizontal and vertical derivatives of the magnetic contribution to the phase shift ($d\phi_{mag}/dx$ and $d\phi_{mag}/dy$), a vector field can be determined and displayed in the form of either an arrow map or a color map (**Figure 17(j)**), whereby the direction and magnitude of the projected in-plane magnetic induction are represented by the hue and intensity of a color, respectively, according to the color wheel shown in **Figure 17(l)**. Color can also be added to the cosine image if desired (**Figure 17(k)**).

20.3.3 Experimental Results

20.3.3.1 Electron holography of isolated magnetite crystals

An experimental study of isolated magnetic nanoparticles allows the effects of particle size, shape, and magnetocrystalline anisotropy on their magnetic state to be assessed without the complicating influence of magnetostatic interactions. Magnetotactic bacteria provide a convenient source of high-purity, relatively defect-free magnetite crystals with varying morphologies, aspect ratios, and sizes in the range 10–200 nm (Devouard *et al.*, 1998; Bazylinski and Frankel, 2004; Arató *et al.*, 2005). Although magnetotactic bacteria normally grow crystals in closely spaced chains, the preparation of bacteria for TEM examination by air drying inevitably leads to cell damage and a degree of chain breakup. **Figure 18(a)** shows a high-resolution TEM image of an isolated 50 nm magnetite crystal from a bacterial cell. This crystal was separated by at least 500 nm

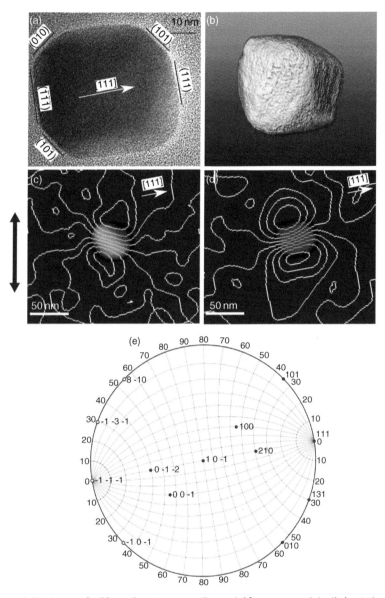

Figure 18   (a) High-resolution image of a 50 nm diameter magnetite crystal from a magnetotactic bacterium (image courtesy of M. Pósfai). (b) Three-dimensional reconstruction of the same particle, obtained using electron tomography (image courtesy of R. Chong). (c,d) Remanent states of the particle at room temperature and 90 K, respectively. The remanent states were obtained after tilting the sample to $\pm 30°$ in the vertical 2 T field of the TEM objective lens. The in-plane component of the applied field was directed along the black double arrow. (e) Stereographic projection showing the crystallographic orientation of the sample. At room temperature, the remanent magnetization direction is close to [131]. At 90 K, the remanent direction is close to either [210] or [012].

from adjacent crystals. The three-dimensional morphology and orientation of the crystal were determined by using electron tomography, from a series of two-dimensional high-angle annular dark-field (HAADF) images taken over an ultrahigh range of tilt angles (**Figure 18(b)**). The tomographic reconstruction reveals that the particle is elongated slightly in the [111] direction in the plane of the specimen (as indicated by the white arrow in **Figure 18(a)**). The

crystallographic orientation of the particle is shown in the form of a stereogram in **Figure 18(e)**.

Electron holography of the magnetite crystal was performed both at room temperature (**Figure 18(c)**) and at 90 K (**Figure 18(d)**). The magnetic contribution to the phase shift was isolated by performing a series of *in situ* magnetization reversal experiments, as described in Section 20.3.1.5. The direction of the in-plane component of the applied field is indicated by the black double arrow. Both images show uniformly magnetized SD states, including the characteristic return flux of an isolated magnetic dipole (**Figure 17(k)**). In both cases, the remanent magnetization direction appears to make a large angle to the applied field direction. At room temperature, the phase contours in the crystal make an angle of ~30° to the [111] elongation direction (**Figure 18(c)**). The contours are parallel to the [111] elongation direction at 90 K (below the Verwey transition; **Figure 18(d)**).

Figure 19(a) shows a profile of the magnetic contribution to the phase image that was used to create **Figure 18(c)**, taken along a line passing through the centre of the crystal in a direction perpendicular to the phase contours. A least-squares fit of the experimental profile to eqns [19] and [20] yielded a value for $B_\perp$ of 0.6 ± 0.12 T. This value is equal to the room temperature saturation induction of magnetite, suggesting that the magnetization direction of the particle lies exactly in the plane of the specimen, close to the [131] crystallographic direction (**Figure 18(e)**). This direction corresponds to the longest diagonal dimension of the particle, which is consistent with shape anisotropy dominating the magnetic state of the crystal at room temperature. The 90 K phase profile (**Figure 19(b)**) yielded a value for $B_\perp$ of 0.46 ± 0.09 T. This value is lower than the saturation induction of magnetite at 90 K, suggesting that, at remanence, the magnetization direction in the crystal is tilted out of the plane by ~40° to the horizontal. This direction is close to either [210] or [012] (**Figure 18(e)**). Below the Verwey transition, the magnetocrystalline anisotropy of

magnetite is known to increase considerably in magnitude (Muxworthy and McClelland, 2000), and to switch from $<111>_{cubic}$ to $[001]_{monoclinic}$. The $[001]_{monoclinic}$ easy axis can lie along any one of the original $<100>_{cubic}$ directions. Both the [100] and [001] directions of the original cubic crystal lie close to the observed remanence direction, suggesting that magnetocrystalline anisotropy has a more significant impact on the remanence direction than shape anisotropy at 90 K. The fact that the remanence direction in **Figure 18(d)** is perpendicular to the applied field direction suggests that this choice may be influenced by the morphology of the crystal.

Theoretical predictions of the effect of particle size and shape on the magnetic state of magnetite are shown in **Figure 20** (Butler and Banerjee, 1975; Muxworthy and Williams, 2006). The upper solid line shows the theoretical boundary between SD and two domain states (Butler and Banerjee, 1975). The dashed line shows the boundary between SD and single vortex (SV) states predicted by micromagnetic simulations (Muxworthy and Williams, 2006; see Section 20.4). For equidimensional particles, the equilibrium SD/SV transition is predicted to occur at a particle size of 70 nm (Fabian et al., 1996; Williams and Wright, 1998) and the transition to a superparamagnetic (SP) state is observed to occur below 25–30 nm (Dunlop and Özdemir, 1997). The observation of a stable SD state for the roughly equidimensional 50 nm crystal in **Figure 18** is in agreement with the expected behavior.

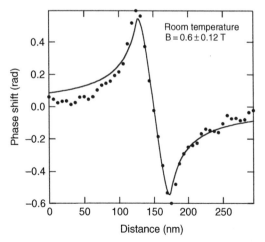
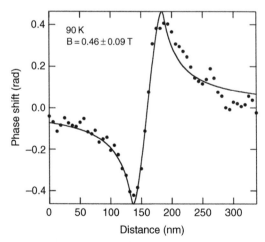

Figure 19   Profiles of the magnetic contribution to the phase shift across the magnetite particle shown in **Figure 18** at (a) room temperature and (b) 90 K (closed circles). Profiles were taken through the center of the particle in a direction normal to the contours shown in **Figures 18(c)** and **18(d)**. Solid lines are least-squares fits to the data using eqns [19] and [20], yielding $B_\perp = 0.6 \pm 0.12$ T at room temperature, and $B_\perp = 0.46 \pm 0.09$ T at 90 K.

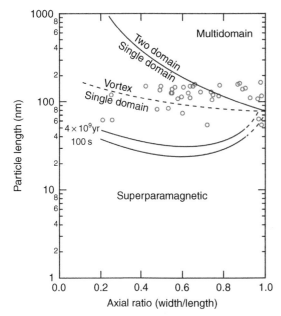

Figure 20 Equilibrium threshold sizes for SP, SD, SV, and two domain magnetic states as a function of particle length and axial ratio. Upper solid line shows the calculated boundary between SD and two domain states (Butler and Banerjee, 1975). Lower solid lines show the sizes for SP behavior with relaxation times of 4×10^9 years and 100 s (Butler and Banerjee, 1975). Dashed line shows the boundary between SD and SV states for uniaxial ellipsoidal particles, calculated using finite-element micromagnetic methods. Open circles show the sizes and aspect ratios of the magnetite blocks from region B in **Figure 24**.

20.3.3.2 Electron holography of chains of closely spaced magnetite crystals

Figures 21(a) and **21**(b) show a bright-field TEM image and a three-dimensional tomographic reconstruction, respectively, of a double chain of magnetite crystals from a magnetotactic bacterial cell (Simpson *et al.*, 2005). Each crystal has its [111] crystallographic axis aligned accurately (to within $4°$) of the chain axis, as shown by the arrows in **Figure 21**(a), but is rotated by a random angle about this axis (like beads on a string). From a magnetic perspective, the alignment of the crystals ensures that their room temperature magnetocrystalline easy axes are closely parallel to the chain axis. The largest crystals are elongated slightly along [111], with lengths of 92–94 nm and widths of 82–88 nm. Isolated crystals of this size would be expected to adopt SV states at equilibrium (**Figure 20**). Electron holography of similar chains revealed, however, that such crystals are magnetized uniformly parallel to each chain axis (**Figure 21**(c)). For such highly aligned chains of crystals, magnetostatic interactions move the

boundary between SD and SV states to larger particle sizes, and promote the stability of SD states. This effect, which is also predicted by micromagnetic simulations (Muxworthy *et al.*, 2003a; see Section 20.4), enables bacteria to grow SD crystals to much larger sizes than would otherwise be possible, thereby optimizing the overall magnetic moment of the chain.

In **Figure 21**, strong magnetostatic interactions between crystals, combined with their high degree of alignment, result in uniform magnetic phase contours that are constrained tightly along the chain axis. In this case, the magnetic induction of the chain can be quantified by assuming that it has approximately a cylindrical geometry (eqn [22]), yielding a value for $B_\perp$ of 0.62 T for one of the central crystals. This value is close to that predicted for magnetite ($B_0 = 0.6$ T), suggesting that the crystals are magnetized parallel to their length and to the chain axis. Interacting chains of closely spaced crystals are always magnetized along the chain axis, thus providing a reliable magnetic moment for magnetotaxis. This is confirmed by hysteresis (Pan *et al.*, 2005; see Section 20.5) and remanence experiments (Hanzlik *et al.*, 2002) on magnetotactic bacteria, which demonstrate that chains of magnetosomes act like a single elongated particle (with uniaxial anisotropy) and switch as a single unit.

When the alignment of crystals in a chain is less than perfect, it is possible to produce a nonuniform magnetization state by the application of a suitably oriented magnetic field (**Figure 22**). **Figure 22**(a) illustrates the magnetization state of two double chains of magnetite crystals (from the same bacterial cell) after the application of a magnetic field with an in-plane component of 1 T parallel to the chain axes. Despite the imperfect alignment of the crystals in the upper double chain, all four chains are magnetized along their length. Some flux divergence is evident where large gaps occur between crystals in the upper double chain, indicating that magnetostatic interactions are weakened slightly at these positions. **Figure 22**(b) illustrates the magnetization state after application of a similar magnetic field perpendicular to the chain axes. Although three out of the four chains are unaffected by the change in the applied field direction, the uppermost chain is split into two halves, each of which has a small component of its magnetization in the direction of the applied field. The two crystals in the center of the chain are now arranged in an energetically unfavorable opposing configuration, and the magnetization of one of the crystals is deflected so that it points at an angle of $\sim 60°$ to the chain axis.

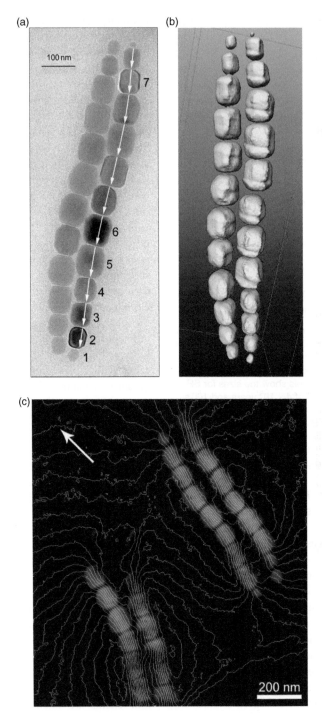

Figure 21   (a) Bright-field TEM image of a double chain of magnetite magnetosomes, acquired at 400 kV using a JEOL 4000EX TEM (image courtesy of M. Pósfai). The white arrows are approximately parallel to [111] in each crystal. (b) Electron tomographic reconstruction of the three-dimensional morphology of the double magnetosome chain shown in (a) (image courtesy of R Chong). (c) Magnetic phase contours measured using electron holography from two pairs of bacterial magnetite chains at 293 K, after magnetizing the sample parallel and antiparallel to the direction of the white arrow. Figs. 21a and c adapted from Simpson ET, Kasama T, Pósfai M, Buseck PR, Harrison RJ, and Dunin-Borkowski RE (2005) Magnetic induction mapping of magnetite chains in magnetotactic bacteria at room temperature and close to the Verwey transition using electron holography. *Journal of Physics: Conference Series* 17: 108–121.

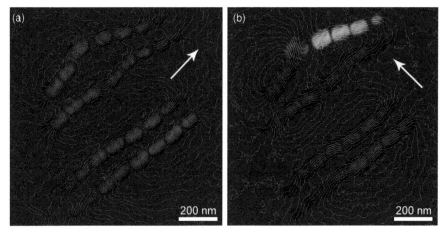

Figure 22 Illustration of the effect of changing the applied magnetic field direction on magnetic induction maps measured from two pairs of magnetite chains at 293 K. The applied field directions are indicated using white arrows. In (a) the chains are magnetized in the same direction. In (b) the topmost chain is partially magnetized antiparallel to the other chains in the figure. Adapted from Simpson ET, Kasama T, Pósfai M, Buseck PR, Harrison RJ, and Dunin-Borkowski RE (2005) Magnetic induction mapping of magnetite chains in magnetotactic bacteria at room temperature and close to the Verwey transition using electron holography. *Journal of Physics: Conference Series* 17: 108–121.

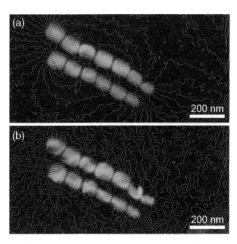

Figure 23 Magnetic induction maps acquired from two pairs of bacterial magnetite chains at (a) 293 K and (b) 116 K. In the room temperature holograms, the contours are parallel to each other within the crystals and only deviate as a result of their morphologies and positions. At 116 K, this regularity is less evident. The field lines undulate to a greater degree within the crystals, as well as at kinks in the chains. The small vortex in the lower chain in (b) is likely to be an artifact resulting from diffraction contrast in this crystal. Adapted from Simpson ET, Kasama T, Pósfai M, Buseck PR, Harrison RJ, and Dunin-Borkowski RE (2005) Magnetic induction mapping of magnetite chains in magnetotactic bacteria at room temperature and close to the Verwey transition using electron holography. *Journal of Physics: Conference Series* 17: 108–121.

Figure 23 shows magnetic induction maps that have been acquired from a similar pair of magnetite chains both at room temperature and at 116 K (close to the Verwey transition). Whereas the contours

are highly parallel to each other and to the chain axes at room temperature (**Figure 23(a)**), their direction is far more variable and irregular at low temperature (**Figure 23(b)**). This behavior is most pronounced in some crystals in **Figure 23(b)** that show S-shaped magnetic configurations. As mentioned above, in magnetite the change in structure from cubic to monoclinic at the Verwey transition is associated with a change in easy axis from <111> to <100>. Although particle interactions and shape anisotropy result in the preservation of the overall magnetic induction direction in **Figure 23(b)** along the chain axis at low temperature, it is likely that the undulation of the contours along the chain axes results from a competition between the effects of magnetocrystalline anisotropy, shape anisotropy and magnetostatic interactions, which are only mutually favorable at room temperature.

20.3.3.3 Electron holography of two-dimensional magnetite nanoparticle arrays

In contrast to the sizes of crystals in magnetotactic bacteria, the grain sizes of primary magnetic minerals in most igneous and metamorphic rocks exceed the MD threshold. Such rocks are less likely to maintain strong and stable natural remanent magnetization (NRM) over geological times than those containing SD grains. It has long been proposed, however, that solid state processes such as sub-solvus exsolution can transform an MD grain into a collection of SD grains, thus increasing the stability of the NRM (Davis and Evans, 1976). This transformation is brought about by the formation of intersecting paramagnetic exsolution lamellae, which divide the host grain into a three-dimensional array of isolated magnetic regions that have SD–PSD sizes.

An excellent example of this phenomenon occurs in the magnetite-ulvöspinel (Fe_3O_4–Fe_2TiO_4) solid solution (Davis and Evans, 1976; Price, 1980, 1981). This system forms a complete solid solution at temperatures above $\sim450°C$ but unmixes at lower temperatures (Ghiorso, 1997). Intermediate bulk compositions exsolve during slow cooling to yield an intergrowth of SD- or PSD-sized magnetite-rich blocks that are separated by nonmagnetic ulvöspinel-rich lamellae. **Figure 24(a)** illustrates the typical microstructure observed in a natural sample of exsolved titanomagnetite (Harrison *et al.*, 2002). This image is a composite chemical map, obtained using energy-filtered TEM imaging, showing the distribution of Fe in blue (magnetite) and Ti in red (ulvöspinel). Ulvöspinel lamellae form

preferentially parallel to {100} planes of the cubic magnetite host lattice. In TEM sections that are oriented parallel to {100}, this symmetry generates a rectangular array of cuboidal magnetite blocks. Profiles of the Fe and Ti distribution along the line marked C (**Figure 24(b)**) demonstrate that the blocks are essentially free of Ti, that is, that they are nearly pure magnetite.

Harrison *et al.* (2002) used electron holography to determine the magnetic remanence states of region B in **Figure 24(a)**. The magnetite blocks were found to be primarily in SD states (**Figure 25**). The dimensions of the blocks, which are plotted on **Figure 20** for reference, indicate that the vast majority would display SV states at remanence if they were isolated and at equilibrium. Micromagnetic simulations of isolated cuboidal particles (Section 20.4) indicate that SD states can exist in metastable form up to a certain size above the equilibrium SD–SV threshold (Fabian *et al.*, 1996; Williams and Wright, 1998; Witt *et al.*, 2005). The majority of the blocks in **Figure 25** fall within the limits of metastability for SD states

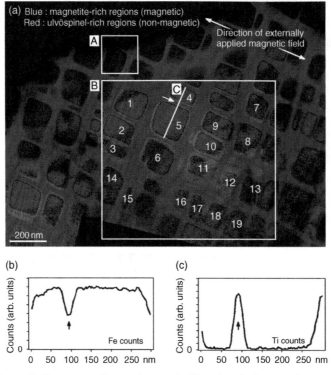

Figure 24 (a) Chemical map of a titanomagnetite sample, acquired by using electron spectroscopic imaging (Harrison *et al.*, 2002). Blue and red correspond to Fe and Ti concentrations, respectively. The blue regions are magnetic and are rich in magnetite (Fe_3O_4), whereas the red regions are nonmagnetic and rich in ulvöspinel (Fe_2TiO_4). The numbers refer to individual magnetite-rich blocks, which are discussed in the text. (b, c) Line profiles obtained from the Fe and Ti chemical maps, respectively, along the line marked C in (a). The short arrows mark the same point in the three pictures.

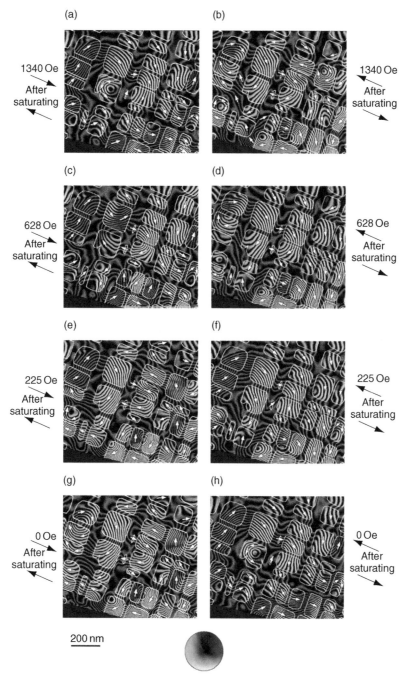

Figure 25 Magnetic microstructure of region B in **Figure 24(a)** measured by using electron holography (Harrison *et al.*, 2002). Each image corresponds to a different magnetic remanent state, acquired with the sample in field-free conditions. The outlines of the magnetite-rich regions are marked in white, while the direction of the measured magnetic induction is indicated both using arrows and according to the color wheel shown at the bottom. Images (a), (c), (e), and (g) were obtained after applying a large field toward the top left of each picture, then the indicated field toward the bottom right, after which the external magnetic field was removed for hologram acquisition. Images (b), (d), (f), and (h) were obtained after applying identical fields in the opposite directions.

calculated by Witt *et al.* (2005). It appears that the presence of magnetostatic interactions favors the adoption of metastable SD states over equilibrium SV states. This behavior results from the fact that the demagnetizing energy – which destabilizes the SD state with respect to the vortex state in isolated particles – is reduced greatly in an array of strongly interacting SD particles.

Transitions between different magnetic states in an individual block can be seen in **Figure 25**. For example, block 8 (labeled in **Figure 24(a)**) in **Figure 25(e)** is magnetized NNW (blue), whereas in **Figure 25(f)** it is magnetized SSE (yellow). It contains an off-centered vortex in **Figure 25(b)**, suggesting that magnetization reversal in this block may occur via the formation, displacement, and subsequent annihilation of a vortex (Enkin and Williams, 1994; Guslienko *et al.*, 2001), rather than by the coherent rotation of the SD moment.

Several blocks are observed to act collectively to form magnetic 'superstates' that would normally be observed in a single, larger magnetized region. One example is where two or more blocks interact to form a single vortex superstate. Two-, three-, and five-block vortex superstates are visible in **Figure 25** (e.g., blocks 1 and 2 in **Figure 25(g)** and blocks 1, 2, 3, 5, and 6 in **Figure 25(e)**. A similar superstate involving three elongated blocks is shown in **Figures 26(a)** and **26(b)** and schematically in **Figure 27(a)**. The absence of closely spaced contours between the superstate and the adjacent single vortex in **Figure 26(b)** shows that stray interaction fields are eliminated in the intervening ulvöspinel. Flux closure is achieved with considerably less curvature of magnetization within the three-component assembly than is required in the adjacent conventional vortex,

reducing the exchange energy penalty associated with the nonuniform magnetization configuration.

A second example of collective behavior involves the interaction of a chain of blocks to form an SD superstate that is magnetized parallel to the chain axis but perpendicular to the easy axes of the individual blocks. This behavior is illustrated schematically in **Figure 27(b)** and can be found in several places in **Figure 25** (e.g., blocks 16, 17, and 18 in **Figures 25(a)**, **25(b)**, **25(d)**, **25(f)**, and **25(h)**). An extreme example of this behavior is shown in **Figure 28**, which shows saturation isothermal remanent states in an exsolved titanomagnetite inclusion within clinopyroxene (Feinberg *et al.*, 2004, 2005). These states were recorded after tilting the sample by angles of $\pm 30°$ and applying a 2 T vertical field (**Figure 16**). The in-plane component of the field was parallel to the elongation direction of the central blocks. Nevertheless, strong interactions between the blocks (which are separated by ~ 15 nm of ulvöspinel) constrain the remanence to lie almost perpendicular to the elongation direction of the individual blocks and to the applied field direction. The expected remanent state of such a system might have been expected to involve adjacent blocks being magnetized in an alternating manner along their elongation directions, as seen in blocks 16, 17, and 18 and blocks 9, 10, and 11 in **Figures 25(c)**, **25(e)**, and **25(g)** and shown schematically in **Figure 27(c)**.

A further example of magnetostatic interactions between blocks is shown in Fig. 26d. The two largest blocks (colored green) are both magnetized in the same direction. The small block between them (colored red) is magnetized in the opposite direction, apparently because it follows the flux return paths of its larger neighbors.

20.3.3.4 Exchange interactions across antiphase boundaries in ilmenite–hematite

Harrison *et al.* (2005) used electron holography to study the nature of the exchange coupling at APBs in ilmenite–hematite. A sample of ilm 70 was synthesized from the oxides under controlled oxygen fugacity at 1573 K, quenched through the cation-ordering phase transition and annealed for 10 h at 1023 K. Representative magnetic induction maps are shown in **Figures 29(a)–29(c)**. Each figure, which is derived from the gradient of the magnetic contribution to the recorded phase shift, shows a magnetic remanent state obtained at a different stage of the switching process. The direction and magnitude of the in-plane magnetic flux are defined by the hue and intensity of the color, respectively.

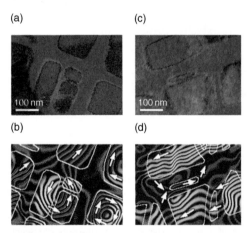

(a)   (c)

(b)   (d)

Figure 26 (a, c) Chemical maps (blue Fe, red Ti) from two regions not shown in **Figure 24**. (b, d) The corresponding magnetic microstructures, in the same format as **Figure 25**. (b) Three adjacent magnetite-rich regions combining to form a single vortex; (d) a small region that is magnetically antiparallel to its larger neighbors.

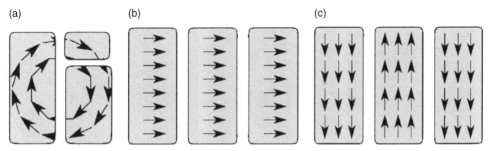

Figure 27 Schematic diagrams showing some of the possible magnetization states of three closely spaced regions of magnetic material.

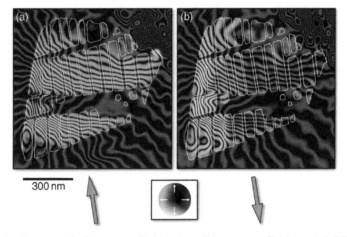

Figure 28 Magnetic induction maps of a titanomagnetite inclusion within pyroxene (Feinberg *et al.*, 2004, 2005). The inclusion is an intergrowth of elongated magnetite blocks (outlined in white) separated by lamellae of ulvöspinel. (a) and (b) correspond to saturation remanant states obtained after tilting the sample to ±30° in the vertical 2 T of the TEM objective lens, such that the in-plane component of the applied field was directed along the gray arrows. Note that the magnetic microstructures in (a) and (b) are the exact reverse of each other, allowing the mean inner potential to be determined using the method described in Section 20.3.1.5.

The magnetization is constrained by shape and magnetocrystalline anisotropy to lie either parallel or antiparallel to the intersection of the specimen plane with the (001) crystallographic plane (indicated by the double black arrow in **Figure 29**). As a result, regions with strong in-plane magnetization appear either blue or green. Regions that have no in-plane magnetization appear as dark bands. Analysis (see **Figure 30**) shows that the dark bands in **Figure 29** are associated with three distinct types of magnetic wall. A finger-like region of reversed magnetization (labeled '1' in **Figure 29(a)**) enlarges by the movement of its left-hand boundary as the applied field is increased (**Figure 29(b)**). This left-hand boundary is a conventional free-standing 180° Bloch wall. In contrast, in regions where a 180° reversal in magnetization direction coincides exactly with the position of an APB (e.g., at regions labeled '2'), the reversal results from

negative exchange coupling across the APB, as predicted by Monte Carlo simulations (**Figure 11**). This type of boundary is referred to as a 180° 'chemical' wall, and occurs without any out-of-plane rotation of the magnetic moments. A third type of magnetic wall appears as thick dark bands, which are also coincident with the positions of APBs (e.g., at regions labeled '3'). Such walls form when the negative exchange coupling between adjacent APDs is overcome at sufficiently large fields, forcing the magnetization direction on either side to point in the same direction. These walls are referred to as 0° walls.

For a 180° Bloch wall, the in-plane component of the magnetic induction is generally described by an expression of the form

$$B_\perp(x) = B_0 \tanh\left(\frac{x}{w}\right) \quad [23]$$

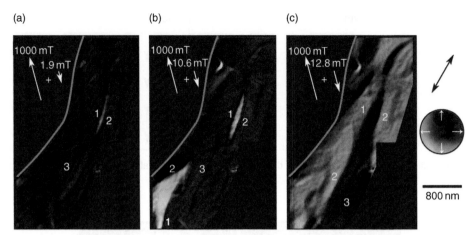

Figure 29 Magnetic microstructure of ilm 70 containing several APDs (Harrison *et al.*, 2005). The sample edge is indicated by the gray line. Prior to each measurement, the sample was exposed to a saturating field with an in-plane/out-of-plane component of +1000/+1732 mT, followed by a smaller field with an in-plane/out-of-plane component of (a) −1.9/+3.3 mT, (b) −10.6/+18.4 mT, (c) −12.8/+22.2 mT. White arrows indicate the direction of the in-plane component of the applied field. The hue and intensity of the color indicates the direction and magnitude of the in-plane component of the magnetic flux in the sample in field-free conditions, as defined by the color wheel on the right. The blue-purple and green-yellow colors correspond to equal and opposite in-plane magnetizations in the direction indicated by the black double arrow. The dark bands indicate regions with weak in-plane magnetization (magnetic domain walls). Dark bands that separate regions of blue and green color correspond to 180° magnetic and chemical walls (e.g., at regions labeled '1' and '2', respectively). Dark bands that are surrounded by regions of the same color correspond to 0° magnetic walls (e.g., at regions labeled '3').

where B_0 is the saturation induction and $2w$ is the wall width. By substituting eqn [23] into eqn [13], the magnetic phase profile across a 180° Bloch wall (assuming a constant thickness, t) is

$$\phi(x) = B_0 tw \ln\left(\cosh\left(\frac{x}{w}\right)\right) \qquad [24]$$

Equation [24] provides an excellent fit to the phase profile of a 180° Bloch wall for $2w = 19$ nm (**Figure 30(a)**). In contrast, the phase profile on either side of a 180° chemical wall is nonlinear (see below), and the reversal in the slope of the phase profile at the center of the wall occurs much more abruptly (**Figure 30(b)**). A fit to the central portion of this wall yields $2w = 7$ nm (dashed line in **Figure 30(b)**). This value is close to the resolution limit of the measurements, and provides an upper limit for the width of the chemical wall. A 0° wall can be considered as the superposition of a 180° Bloch wall and a 180° chemical wall. Assuming that both types of wall can be described by eqn [23] with the same value of w, the in-plane component of magnetic induction is of the form

$$B_\perp(x) = B_0 \tanh^2\left(\frac{x}{w}\right) \qquad [25]$$

By substituting eqn [25] into eqn [13], the magnetic phase profile across a 0° wall is

$$\phi(x) = B_0 t\left[x - 10\tanh\left(\frac{x}{w}\right)\right] \qquad [26]$$

Equation [26] provides an excellent fit to the phase profile of a 0° Bloch wall (**Figure 30(c)**). An average of 13 measurements yielded $2w = (50 \pm 14)$ nm for 0° walls. Previous studies demonstrated that self-reversed thermoremanent magnetization (SR–TRM) was observed only when APDs were below 80–100 nm in size (Nord and Lawson, 1989, 1992). This limit is imposed by the formation of 0° walls, which allow negative exchange coupling between adjacent domains to be overcome when the APDs are much larger than 50 nm in size.

20.4 Magnetism at the Micrometer Length Scale

Due to the large number of atoms and spins involved, atomistic simulations, which describe the discrete arrangement of atoms and spins on a crystalline lattice, are currently unsuitable for systems larger than ∼10 nm (Section 20.2). The magnetization states of larger particles (or collections of particles) are more efficiently described using micromagnetic simulations (Brown 1963). Micromagnetics is the study of magnetization at the nanometer to micrometer

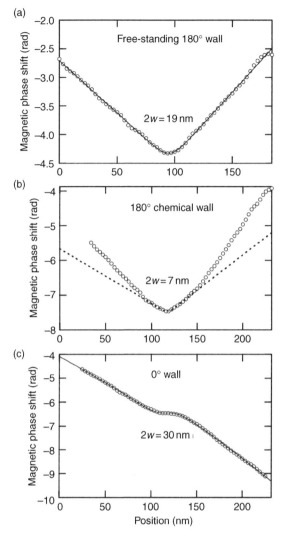

Figure 30 Profiles of the holographic phase shift, ϕ, across three distinct types of magnetic domain wall in ilm 70 (Harrison et al., 2005). The gradient of each profile is proportional to the in-plane component of the magnetic flux (eqn [15]). (a) A free-standing 180° Bloch wall. The solid line is a least-squares fit to the observed profile, obtained by using eqn [24] and yielding a wall width of 19 nm. (b) A 180° 'chemical' wall that is coincident with an APB. The dashed line is a fit to the central portion of the wall, obtained by using eqn [24] and yielding a wall width of 7 nm. (c) A 0° magnetic wall that is coincident with the same APB as in (b). The slope of the phase profile has the same sign on either side of the wall, indicating that the direction of magnetization is the same. The solid line is a fit to the profile, obtained using eqn [26] and yielding a wall width of 30 nm.

length scale (i.e., a length scale that is much larger than that of the crystalline lattice but smaller than that of a magnetic domain). Micromagnetic models treat magnetization as a classical, continuous

vector field in space. The energy of the system is described by a number of macroscopic constants, whose values can be derived from their microscopic equivalents (eqn [2]). Micromagnetic simulations of magnetic minerals have progressed rapidly from one-dimensional (Moon and Merrill, 1984, 1985; Moon, 1991) to two-dimensional (Newell et al., 1993; Xu et al., 1994), and finally three-dimensional (Schabes and Bertram, 1988a, 1988b; Williams and Dunlop, 1989, 1990; Wright et al., 1997; Fabian et al., 1996; Williams and Wright, 1998) models of homogeneous isolated single crystals with simple geometric shapes. The recent application of finite element/boundary element methods (FEM/BEM) to micromagnetic simulations now allows the simulation of heterogeneous, polycrystalline systems that have complex and realistic morphologies (Fidler and Schrefl, 2000). In addition, substantial progress is being made in the application of micromagnetic simulations to the study of magnetostatic and exchange interactions between arrays of closely spaced magnetic particles (Muxworthy et al., 2003a; Muxworthy et al., 2004; Carvallo et al., 2003).

20.4.1 Theory

We begin by summarizing the basic principles of micromagnetics, as applied to rock magnetic problems. For detailed reviews of the technical aspects of micromagnetic simulations, the reader is referred to Brown (1963), Wright et al. (1997), Fabian et al. (1996), and Fidler and Schrefl (2000).

20.4.1.1 *The micromagnetic energy*

Micromagnetism is a continuum approximation, in which the magnetization of a particle is taken to be a continuous function of position:

$$\mathbf{m}(\mathbf{r}) = \frac{\mathbf{M}(\mathbf{r})}{M_s} = \begin{pmatrix} \alpha \\ \beta \\ \gamma \end{pmatrix} = \begin{pmatrix} \cos(\phi)\sin(\theta) \\ \sin(\phi)\sin(\theta) \\ \cos(\theta) \end{pmatrix} \quad [27]$$

where $\mathbf{m}$ is a unit vector parallel to the magnetization direction $\mathbf{M}$ at position $\mathbf{r}$ and M_s is the saturation magnetization. The direction of $\mathbf{m}$ is defined either in terms of direction cosines α, β, and γ or in terms of the polar coordinates ϕ and θ. $\mathbf{M}$ represents the local average of many thousands of individual spins.

Minimization of the microscopic exchange energy (eqn [1]) requires $\mathbf{m}$ to be uniform throughout a grain. Deviations from uniform magnetization at the macroscopic length scale impose deviations on the angles between adjacent spins at the atomic scale. On the assumption that these angular deviations are small, the macroscopic exchange energy can be expressed in the form of a truncated Taylor expansion of eqn [1]:

$$E_{ex} = A \int_V (\nabla \mathbf{m})^2 \, dV \qquad [28]$$

where the exchange constant A is related to the atomistic exchange integrals and V is the volume of the particle. The exchange energy is positive wherever gradients in the macroscopic magnetization occur (e.g., within domain walls) and zero wherever the magnetization is uniform (e.g., within domains).

Since the angular relationship between the atomic spins and the net magnetization is fixed, the macroscopic expression for the magnetocrystalline anisotropy energy is equivalent to that in eqn [2]. For unixial anisotropy this expression is

$$E_a = - \int_V K (\mathbf{m} \cdot \mathbf{e})^2 \, dV \qquad [29]$$

while for cubic magnetocrystalline anisotropy, the expression is

$$E_a = \int_V [K_1 (\alpha^2 \beta^2 + \beta^2 \gamma^2 + \gamma^2 \alpha^2) + K_2 \alpha^2 \beta^2 \gamma^2] \, dV \qquad [30]$$

Similarly, the macroscopic expression for the magnetostatic energy is equivalent to that in eqn [2]:

$$E_h = - \mu_0 M_s \int_V \mathbf{H}_{ext} \cdot \mathbf{m} \, dV \qquad [31]$$

where $\mathbf{H}_{ext}$ is the applied magnetic field.

Calculation of the demagnetizing energy is the most challenging and computationally intensive part of any micromagnetic simulation. A macroscopic expression for the demagnetizing energy (eqn [3]) can be formulated in terms of the demagnetizing field, $\mathbf{H}_d(\mathbf{r})$, which is the sum of the magnetic fields at position $\mathbf{r}$ created by all of the magnetic moments in the particle:

$$E_d = - \frac{1}{2} \mu_0 M_s \int_V \mathbf{m} \cdot \mathbf{H}_d \, dV \qquad [32]$$

A general method for calculating $\mathbf{H}_d$ follows from Maxwell's equations in a current-free region with static electric and magnetic fields:

$$\nabla \times \mathbf{H} = 0 \qquad [33]$$

$$\nabla \cdot \mathbf{B} = \mu_0 \nabla \cdot (\mathbf{H} + \mathbf{M}) = 0 \qquad [34]$$

From eqn [33], it follows that the magnetic field, $\mathbf{H} = \mathbf{H}_{ext} + \mathbf{H}_d$, can be described as the gradient of a magnetic scalar potential, ϕ:

$$\mathbf{H} = -\nabla \phi \qquad [35]$$

By rearranging eqn [34], one obtains Poisson's equation

$$\nabla^2 \phi = \nabla \cdot \mathbf{M} \qquad [36]$$

Outside the particle, $\mathbf{M}$ is zero, and eqn [36] reduces to the Laplace equation

$$\nabla^2 \phi = 0 \qquad [37]$$

The general solution to eqn [36] is of the form

$$\phi(\mathbf{r}) = \frac{1}{4\pi} \left[\int_V \frac{\rho(\mathbf{r}')}{|\mathbf{r} - \mathbf{r}'|} \, dV' + \int_S \frac{\sigma(\mathbf{r}')}{|\mathbf{r} - \mathbf{r}'|} \, dS' \right] \qquad [38]$$

where $\rho(\mathbf{r}) = -\nabla \cdot \mathbf{M}$ is the density of magnetic volume charges due to nonzero divergence of the magnetization within the interior of the particle, and $\sigma(\mathbf{r}) = \mathbf{M} \cdot \hat{n}$ is the density of magnetic surface charges due to the component of magnetization normal to the particle surface ($\hat{n}$).

The total energy of the system is the sum of exchange, anisotropy, magnetostatic, and demagnetizing energies:

$$E_{tot} = E_{ex} + E_a + E_h + E_d \qquad [39]$$

Important energy terms that are missing from eqn [39] include the magnetoelastic energy arising from the stress fields surrounding dislocations and other lattice defects and the magnetostrictive self-energy associated with the elastic strain of an inhomogeneously magnetized particle. For a mineral such as magnetite, magnetostriction can be neglected for particles that are smaller than $\sim 6 \, \mu m$ (Hubert, 1967). The inclusion of magnetostriction into micromagnetic models of titanomagnetite is discussed by Fabian and Heider (1996).

20.4.1.2 Discretization of the micromagnetic energy

Although micromagnetism is a continuum approach, numerical calculation and minimization of the total energy (eqn [39]) requires discretization of the volume that describes the object of interest. The most common approach is to divide the object into a three-dimensional mesh of cubic elements. Each element is assigned a magnetization vector at its center (eqn [27]) and is assumed to be magnetized homogeneously. The elements must be large enough to average out the discrete effects of the crystalline lattice (i.e., they should be significantly larger than the unit cell size), yet small enough that the angular differences between the magnetization directions of adjacent cubes are smaller than $\sim 15°$ (Williams and Wright, 1998). This upper limit is imposed by the use of a truncated Taylor expansion for the exchange energy (eqn [28]), which assumes that the gradient of the magnetization is small. Assuming that the magnetization varies most rapidly at domain walls (which have a width of ~ 100 nm wide in magnetite), and that approximately 4–10 elements are required over this distance to obtain an accurate value for the exchange energy, the maximum element size is of the order 10–25 nm. Larger element sizes can be used in cases where the magnetization remains fairly uniform, and also if the primary interest is in examining the effects of magnetostatic interactions between particles rather than the magnetization states of individual particles (Muxworthy et al., 2003a). A minimum of two elements per exchange length, $l = \sqrt{A/K_d}$, where $K_d = \mu_0 M_S^2/2$, is usually recommended (Rave et al., 1998).

The exchange, anisotropy, and magnetostatic contributions to the total energy (eqn [39]) are functions of the local magnetization and its derivatives. After discretization of the particle volume, these terms are readily calculated using finite difference (FD) methods (Wright et al., 1997).

Calculating the demagnetizing energy, however, involves summing over contributions from all elements in the system. The assumption that each element is magnetized uniformly (i.e., that $\nabla \cdot \mathbf{M} = 0$) eliminates the volume-charge contribution to the magnetic scalar potential (the first term in eqn [38]) and reduces the problem to summing the surface-charge contributions from the faces of each element. The calculation can be simplified further by transforming eqn [38] into a product of spatial terms (i.e., terms that depend only on the geometric relationship between pairs of elements) and angular terms (i.e., terms that depend on the direction of magnetization within each element). The demagnetizing energy can then be expressed in the form

$$E_d = \frac{\mu_0 M_s^2}{8\pi} \sum_{l=1}^{N} \sum_{m=1}^{N} W_{l-m}^{\alpha\beta} \alpha_l \beta_m \qquad [40]$$

where $W_{l-m}^{\alpha\beta}$ are spatial coefficients (evaluated using the method of Rhodes and Rowlands 1954), α and β are angular terms corresponding to the charges of different faces of each element, and N is the number of elements (Wright et al., 1997). The spatial terms

can be evaluated once at the start of the simulation and stored in a look-up table. The summation can be accelerated greatly using fast fourier transform (FFT) methods, whereby eqn [40] is rewritten as a convolution and summed in frequency space (Fabian et al., 1996; Wright et al., 1997).

The total energy must be minimized in order to obtain the equilibrium magnetization state of the object. Dynamic approaches make use of the LLG equation of motion (eqn [5]), and are particularly suitable for the study of magnetization reversal processes. Alternatively, conjugate gradient (Fabian et al., 1996; Wright et al., 1997), Monte Carlo (Kirschner et al., 2005), or simulated annealing (Thomson et al., 1994; Winklhofer et al., 1997) methods may be used. Whereas simulated annealing and Monte Carlo methods are typically used to find the magnetic domain state that corresponds to the absolute energy minimum (AEM) of the object, the use of LLG and conjugate gradient techniques typically results in determination of magnetic domain states that represent local energy minima (LEM). The LEM state that is obtained depends on the initial state of the particle. Simulations typically start with the smallest particle size, which is initialized with a uniform magnetization state in a chosen direction. The final magnetic structure obtained for that particle size serves as the initial guess for the next, slightly larger, particle size. In this way, systematic changes in domain structure as a function of particle size and shape can be determined.

20.4.1.3 Finite element discretization

Most naturally occurring magnetic particles have irregular morphologies. Discretization using a regular array of cubes provides a poor description of noncuboidal grain shapes (**Figure 31**). Curved boundaries are approximated simply by assigning a value of $M = 0$ to certain elements of the regular cubic array ('cell blanking'). The finite-difference discretization of a sphere shown in **Figure 31(a)** has a highly stepped surface, which may result in magnetostatic artifacts that can drastically alter its predicted magnetic domain structure, behavior, and stability. Improvements to this approach can be made by assigning values of M according to the volume fraction of each element that is enclosed by the true particle volume (Witt et al., 2005). In this way, elements that occur entirely within the particle have $M = M_s$, elements entirely outside the particle have $M = 0$, and those at the boundary have $0 < M < M_s$. State-of-the-art micromagnetic simulations involve the use of finite element methods (FEMs) to simulate magnetic

(a)   (b)

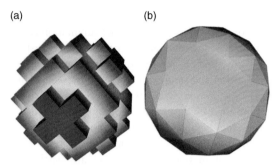

Figure 31   Discretization of a spherical particle using (a) regular array of 343 cubic elements (including blanks) and (b) finite element mesh of 60 tetrahedra (image courtesy of W. Williams).

domain structure in complex geometries (Fidler and Schrefl, 2000). Efficient discretization is then carried out using a combination of triangles, squares, and rectangles (in two dimensions) or tetrahedra, cubes, and hexahedra (in three dimensions). An FEM discretization using 60 tetrahedral elements (**Figure 31(b)**) provides a much better representation of a sphere than the finite difference discretization using 343 cubic elements (**Figure 31(a)**). Finite element models reduce magnetostatic artifacts that originate on grain surfaces drastically. In order to determine the demagnetizing energy when using FEMs, each node of the finite element mesh is associated with a value of the magnetic scalar potential. Values of ϕ are determined by solving Poisson's equation (eqn [36]) inside the particle and Laplace's equation (eqn [37]) outside the particle, subject to the following boundary conditions (Fidler and Schrefl, 2000):

$$\phi_{\text{int}} = \phi_{\text{ext}} \tag{41}$$

$$(\nabla\phi_{\text{int}} - \nabla\phi_{\text{ext}}) \cdot \hat{n} = \mathbf{M} \cdot \hat{n} \tag{42}$$

Because FEMs do not require the use of a regular periodic array of nodes, it is possible to adapt the mesh to better suit a given pattern of nonuniform magnetization. For example, it is more efficient to have a high density of nodes in regions where the magnetization varies rapidly and a low density of nodes in regions where the magnetization remains uniform. Adaptive mesh algorithms actively modify the finite element mesh in response to the changing magnetization state of the system and guarantee that accurate solutions are obtained near magnetic inhomogeneities or domain walls, while keeping the number of elements to a minimum (Fidler and Schrefl, 2000; Scholz et al., 1999).

20.4.2   Applications of Micromagnetic Simulations

20.4.2.1   Equilibrium domain states in isolated magnetite particles

High-resolution micromagnetic studies of isolated cuboidal magnetite particles in the size range from 10 nm to 4 μm have been performed by Williams and Wright (1998), Fabian et al. (1996), and Witt et al. (2005) using FFT-accelerated finite difference methods combined with conjugate gradient energy minimization (Wright et al., 1997). The domain states that are found to be stable in cubic particles in the size range 10–400 nm are (1) a uniformly magnetized SD state; (2) a flower (F) state (**Figure 32(a)**); and (3) an SV state (**Figure 32(b)**). Fabian et al. (1996), Winklhofer et al. (1997), and Witt et al. (2005) demonstrated that a double-vortex (DV) state (**Figure 32(c)**) exists as an LEM state in cubes that are larger than 300 nm, although the appearance of this state appears to be sensitive to the precision used in the simulations.

(a)   (b)   (c)

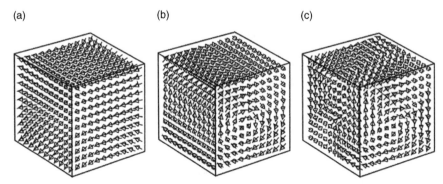

Figure 32   Calculated domain states occurring in cubic grains of magnetite at room temperature for a grain with edge length of 120 nm (a) single domain (flower state), (b) single vortex state, and (c) double vortex state. The [001] axis aligns with the z-axis of the cube. It was necessary to constrain (c) for a 120 nm cube. Reproduced from Muxworthy AR, Dunlop DJ, Williams W (2003b) High-temperature magnetic stability of small magnetite particles. *Journal of Geophysical Research* 108: 2281.

Both F and SV states reduce the component of magnetization normal to the particle surface, thereby reducing the demagnetizing energy (eqn [38]). Although F states are not magnetized uniformly, they still obey the Néel (1949) SD theory of thermoremanent and viscous remanent magnetization. For this reason, SD and F states are often referred to interchangeably.

The variation in the total micromagnetic energy of a magnetite cube with particle size is illustrated in **Figure 33** (Muxworthy *et al.*, 2003b). The starting configuration was a 50 nm cube with uniform magnetization. This SD state relaxes to an F state, which is then used as the starting configuration for the next largest particle. The F state remains (meta)stable up to a particle size of 96 nm. Above this critical size, it relaxes spontaneously to an SV state that has a much lower energy. If the SV state is studied as a function of gradually decreasing particle size, then its energy intersects that of the F state at a particle size of 64 nm. Hence, 64 and 96 nm correspond to the lower and upper limits for the sizes of isolated magnetite cubes that can support metastable F states. Above a particle size of 64 nm, the SV state represents the stable AEM state of the particle. The F state can exist, however, as a metastable LEM state up to a particle size of 96 nm. On decreasing the particle size, the SV to F state transition is continuous, corresponding to a gradual 'unwinding' of the vortex

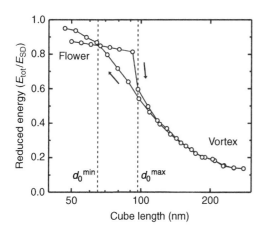

Figure 33 Calculated micromagnetic energy density of a magnetite cube as a function of edge length *d* for an initial SD configuration at room temperature (**Figure 32(a)**). The grain size was gradually increased until the SD structure collapsed to a vortex structure at d_0^{max} = 96 nm. The size was then gradually decreased until a SD state formed at d_0^{min} = 64 nm. Reproduced from Muxworthy AR, Dunlop DJ, and Williams W (2003b) High-temperature magnetic stability of small magnetite particles. *Journal of Geophysical Research* 108: 2281.

(Williams and Wright, 1998). There is close agreement between different micromagnetic studies regarding the lower limit of SV stability (64–70 nm) but significant variation regarding the upper limit (96–220 nm) (Fabian *et al.*, 1996; Williams and Wright, 1998; Muxworthy *et al.*, 2003b; Witt *et al.*, 2005). The lower limit is defined strictly as the particle size at which the absolute energies of the two alternative states become equal, whereas the upper limit is determined by the disappearance of the energy barrier separating two states that have very different energies. The latter transition is sensitive to the precision of the calculation and the method used to determine the energy minimum, and is therefore subject to more variation from study to study.

A gradual transition to classical MD states occurs for particle sizes in the range 1–4 μm. This transition, described in **Figure 34**, is characterized by (Williams and Wright, 1998): (1) an alignment of the near-surface magnetization parallel to the particle surface; (2) an alignment of the magnetization with the magnetocrystalline easy axes (or the projection of the easy axes on the particle surface); (3) an increase in the fraction of the particle volume occupied by regions of uniform magnetization; (4) a decrease in the size, together with a more domain-wall-like appearance, of the nonuniformly magnetized regions; (5) tilting of vortex cores away from [001], allowing larger regions of magnetization to point along the magnetocrystalline easy axes; and (6) vortex cores in larger particles becoming nucleation centers for domain walls.

The magnetic structures of noncuboidal particles have been investigated using a modified version of the cell-blanking technique by Witt *et al.* (2005). The equilibrium SD–SV threshold size in isolated particles with octahedral morphology was found to be *d* = 88 nm, identical to that observed for cubic particles (*d* is defined here as the the diameter of a sphere with the same volume as the particle). The similarity between the equilibrium SD–SV threshold size of cubes and octahedra is not surprising, since they have identical demagnetizing factors. There is a large difference, however, in the upper size limit for metastable SD states (*d* = 320 nm for octahedra versus *d* = 160 nm for cubes). This difference is illustrated further in **Figure 35**, in which the lower and upper limit of stability for F states in cuboid (**Figure 35(a)**) and noncuboid (**Figure 35(b)**) particles are compared. The particle morphologies that were used to produce **Figure 35(b)** are similar to those observed in

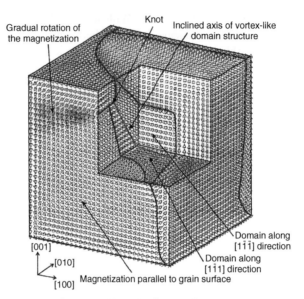

Figure 34 The magnetization structure of a 1 μm cubic grain of magnetite with lines indicating the positions of major domain boundaries. The top surface shows a domain wall which is nucleated on a vortex core (labeled a knot in the diagram). The center of the grain is dominated by domains aligned toward an easy magnetocrystalline anisotropy axis, and the magnetization of the surface lies in the plane of the surface to reduce the magnetostatic free pole energy. Reproduced from Williams W, and Wright TM (1998) High-resolution micromagnetic models of fine grains of magnetite. *Journal of Geophysical Research* 103: 30537–30550.

many magnetotactic bacteria (**Figure 35(c)**), and are elongated along a <111> crystallographic direction. The shaded areas in **Figures 35(a)** and **35(b)** show the range in the size and aspect ratio of magnetite crystals in natural magnetotactic bacteria that have this morphology (Petersen *et al.*, 1989). A significant proportion of these particles lies above the upper limit of stability for F states predicted for cuboidal particles (**Figure 35(a)**). In cuboidal particles, magnetostatic interactions along the bacterial chain would be required to prevent the formation of vortex states. All of the particles, however, lie within the stability limit for F states for the more realistic particle morphologies (**Figure 35(b)**), implying that magnetostatic interactions may not be required to stabilize such states in large naturally occurring bacterial magnetosomes. The stabilization of F states in magnetosomes results in part from the elongation of the particles along <111> (so that magnetocrystalline and shape anisotropies act in unison) and in part from the more rounded ends of the crystals (which inhibit flowering and reduce the tendency to de-nucleate the F state). Nonuniform magnetization states have been observed in large magnetite magnetosomes using electron holography (McCartney *et al.*, 2001).

20.4.2.2 Temperature dependence of domain states in isolated particles

The temperature dependence of magnetic domain states, and the thermal relaxation properties of SD and PSD particles, are of central importance to the theories of thermoremanent and viscous remanent magnetization (Néel, 1949). Most micromagnetic simulations are designed to minimize the internal energy of the system (eqn [39]) rather than its Gibb's free energy. Consequently, the effective temperature of the simulations is 0 K, and the effects of entropy and thermal fluctuations are neglected. There are several ways of incorporating temperature into micromagnetic simulations. A basic approach, which neglects thermal fluctuations, is to use temperature-dependent values of A, K, and M_s in the calculation of the internal energy (Muxworthy and Williams, 1999; Muxworthy *et al.*, 2003b). The temperature dependencies of A, K, and M_s in magnetite are given by Heider and Williams (1988), Fletcher and O'Reilly (1974), Abe *et al.* (1976), Bickford *et al.* (1957), Pauthenet and Bochirol (1951), Belov (1993), and Muxworthy and McClelland (2000). This approach allows the temperature dependence of equilibrium domain states to be determined, but may incorrectly predict

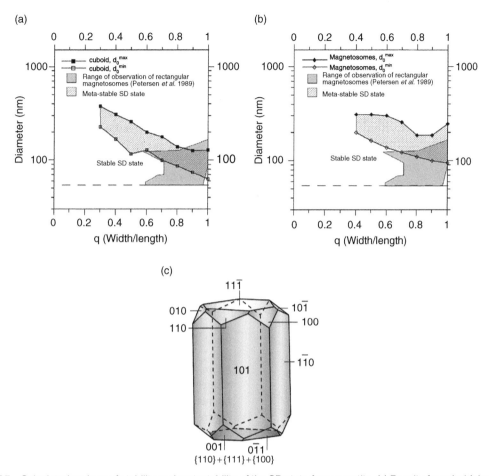

Figure 35 Calculated regions of stability and metastability of the SD state for magnetite. (a) Results for cuboidal particles. (b) Results for particles with morphologies similar to those seen in some strains of magnetotactic bacteria (c). Displayed is the SD–PDS transition as a function of width over length of the respective particles. The shaded area delineates microscopically observed magnetosome shapes (Petersen *et al.*, 1989). The dashed area corresponds to the calculated region where flower states are metastable. Above this area, the SD state is unstable and cannot persist. Parts (a) and (b) reproduced from Witt A, Fabian K, and Bleil U (2005) Three-dimensional micromagnetic calculations for naturally shaped magnetite: Octahedra and magnetosomes. *Earth and Planetary Science Letters* 233: 311–324. Part (c) from Bazylinski DA, and Frankel RB (2004) Magnetosome formation in prokaryotes. *Nature Reviews Microbiology* 2: 217–230.

the existence of LEM states that would be unstable in the presence of thermal fluctuations. Thermal fluctuations can be incorporated into micromagnetic simulations by adding a random thermal field to the effective field (eqn[6]) and then determining the dynamic response of the system using the LLG equation (eqn [5]) (Fidler and Schrefl, 2000; Scholz *et al.*, 2001). Alternatively, Monte Carlo methods can be used (Kirschner *et al.*, 2005). Atomistic Monte Carlo simulations (see Section 20.2.2.1) are first used to determine the equilibrium spin configuration at a given temperature. The value of M_s to be used in the micromagnetic simulations is then obtained by averaging the spin configuration over

the volume of one micromagnetic element. Thereafter, nonatomistic Monte Carlo techniques, analogous to those described in Section 20.2.2.1, are used to determine the equilibrium domain state of the micromagnetic model.

The lower and upper limits for the sizes of metastable F states in cubic magnetite particles at high temperatures are illustrated in **Figure 36** (Muxworthy *et al.*, 2003b). The equilibrium SD–SV threshold size increases from 70 nm at room temperature to approximately 90 nm close to the Néel temperature. The upper SD–SV threshold size increases from 96 to ~200 nm close to the Néel temperature, considerably extending the size range

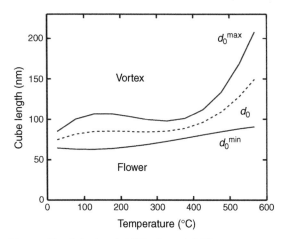

Figure 36 Calculated d_0^{max}, d_0, and d_0^{min} vs. temperature for cubic grains of magnetite. Above d_0^{max} only the vortex state is possible, whereas below d_0^{min}, only the flower or SD state is possible. Between d_0^{max} and d_0^{min} it is possible for the grain to be in either state. Reproduced from Muxworthy AR, Dunlop DJ, and Williams W (2003b) High-temperature magnetic stability of small magnetite particles. *Journal of Geophysical Research* 108: 2281.

over which metastable F states can exist. The SD–SV threshold size at low temperature (below the Verwey transition) has been explored by Muxworthy and Williams (1999). The large increase in the magnitude of the magnetocrystalline anisotropy at the Verwey transition (Muxworthy and McClelland, 2000) stabilizes the SD state with respect to the SV state, increasing the lower SD–SV threshold size to 140 nm at 110 K. However, the increase in magnetocrystalline anisotropy also increases considerably the height of the energy barrier that separates the SD and SV states, increasing the probability of particles becoming trapped in a metastable SV state on cooling below the Verwey transition.

In order to calculate the thermal relaxation properties of SD and PSD particles, it is necessary to determine the magnitudes of the energy barriers that separate different LEM states. These values can be calculated using constrained micromagnetic simulations (Enkin and Williams, 1994; Winklhofer et al., 1997; Muxworthy et al., 2003b). In an unconstrained simulation, the magnetic moments of all of the elements are allowed to vary, so that the system evolves toward the nearest LEM state (**Figure 33**). In a constrained simulation, the system is forced to adopt a non-LEM state by fixing the orientations of the magnetic moments in some of the elements during the simulation (**Figure 37**). For example, by

constraining the moments on opposite faces of a cuboidal particle to be either parallel or antiparallel, it can be forced to adopt an F or an SV state, respectively (**Figure 37(a)**). **Figures 37(b)** and **37(c)** show the calculated energy of a particle with an aspect ratio 1.2 as the moments on opposite faces are rotated independently of each other through 360° (Muxworthy et al., 2003b). At 27°C (**Figure 37(b)**) the SV state is the AEM state, and there are two nondegenerate SD LEM states at 90° to each other. The SD state with the lower energy is magnetized parallel to the elongation direction of the particle. This state becomes the AEM state at 567°C (**Figure 37(c)**). The energy barriers that separate degenerate AEM states are illustrated in **Figure 38** for two different particle sizes and aspect ratios (Muxworthy et al., 2003b). The relaxation time of such a particle is related to the height of the energy barrier, E_B (Winklhofer et al. 1997):

$$\tau = \tau_0 \exp\left(\frac{E_B(T)}{k_B T}\right) \qquad [43]$$

where τ_0^{-1} ($\sim 10^9$–10^{10} Hz; McNab et al., 1968) is the frequency at which the particle attempts to switch its magnetization direction. The dashed lines in **Figure 38** represent the blocking of remanent magnetization on laboratory ($E_B \sim 25\ k_B T$) and geological ($E_B \sim 60\ k_B T$) timescales. The figure illustrates that blocking is more a function of the rapidly increasing energy barrier height on cooling, rather than of the decrease in thermal energy.

20.4.2.3   Field dependence of domain states

The effect of an external field can be included in micromagnetic simulations via eqn [32], and used to study hysteresis properties (Williams and Dunlop, 1995) and reversal mechanisms (Enkin and Williams, 1994) of individual PSD particles, the acquisition of saturation isothermal remanent magnetization (SIRM) and thermoremanent magnetization (TRM) (Winklhofer et al., 1997; Muxworthy and Williams, 1999; Muxworthy et al., 2003b), and to calculate the first-order reversal curves (FORCs) for both isolated grains and arrays of particles (Carvallo et al., 2003; Muxworthy et al., 2004; Muxworthy and Williams, 2005; see Section 20.5). Hysteresis loops are typically obtained by calculating a succession of quasi-static magnetic states, as the field is increased and decreased in a stepwise manner (Williams and Dunlop, 1995). This approach is valid so

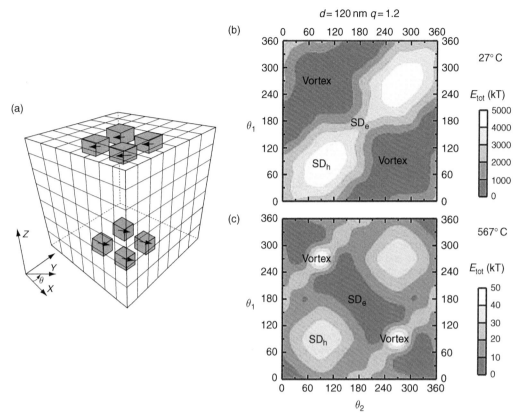

Figure 37 (a) Schematic of the constrained micromagnetic simulation. A number of cells at the top have their magnetization constrained to a direction θ_1 in the x–y plane, while another set of cells at the bottom are constrained to a direction θ_2 also in the x–y plane. θ_1 and θ_2 are set to angles between $0°$ and $360°$ at regular intervals. The energy is minimized with respect to the magnetization direction of all the other cells. Energy surfaces for a grain with edge 120 nm and aspect ratio 1.2 at (b) room temperature and (c) just below T_c. As the grain is asymmetric, there are hard (SD_h) and easy (SD_e) magnetic directions. Favorable vortex structures are also marked. Reproduced from Muxworthy AR, Dunlop DJ, and Williams W (2003b) High-temperature magnetic stability of small magnetite particles. *Journal of Geophysical Research* 108: 2281.

long as the damping of gyromagnetic precession (eqn [5]) is much more rapid than the rate of increasing/decreasing field. PSD particles containing vortex states are observed to reverse their magnetization directions by a combination of gradual rotations of the outer moments and discontinuous reversals of the core moments (Williams and Dunlop, 1995).

20.4.2.4 Magnetostatic interactions between particles

Electron holographic observations of closely spaced particles (see Section 20.3.3.3) highlight the fundamental importance of magnetostatic interactions in determining the macroscopic properties of rocks and minerals. The complex problem of determining the collective behavior of interacting arrays of magnetic particles has recently been tackled using

micromagnetic simulations (Muxworthy, *et al.*, 2003a). Muxworthy *et al.* (2003a) performed a systematic study of saturation magnetization (M_s), saturation remanence (M_{rs}), coercivity (H_c), and coercivity of remanence (H_{cr}) as a function of particle size and spacing for regular three-dimensional arrays of cubic particles. The simulations were performed for different anisotropy schemes (uniaxial versus cubic, aligned vs. random) to model a range of scenarios that are likely to be observed in natural systems. The results can be summarized on a 'Day plot' (Day *et al.*, 1977) of M_{rs}/M_s versus H_{cr}/H_c (**Figure 39**). For widely spaced particles (i.e., when the distance between particles is greater than 5 times their diameter), the effect of magnetostatic interactions is negligible, and the ratios of M_{rs}/M_s and H_{cr}/H_c converge to the ideal values for non-interacting particles ($M_{rs}/M_s = 0.5$ and 0.87 for

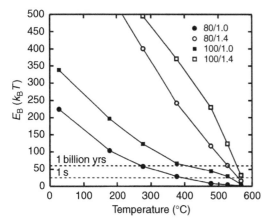

Figure 38 Calculated energy barrier (E_B) for magnetization reversal as a function of temperature for small particles of magnetite; two with $d = 80\,nm$ and aspect ratios of 1 and 1.4 (closed and open circles, respectively), and two for $d = 100\,nm$ and aspect ratios of 1 and 1.4 (closed and open squares, respectively). The two dashed lines at $E_B = 60\,k_BT$ and $25\,k_BT$ represent the paleomagnetic and laboratory stability criteria, respectively. Reproduced from Muxworthy AR, Dunlop DJ, and Williams W (2003b) High-temperature magnetic stability of small magnetite particles. *Journal of Geophysical Research* 108: 2281.

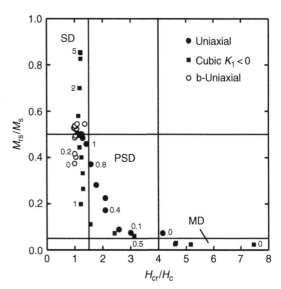

Figure 39 M_{rs}/M_s vs. H_{cr}/H_c (Day plot) for three different anisotropy assemblages of ideal SD magnetite grains; uniaxial (closed circles), cubic $K_1 < 0$ (closed squares) and basal plane-uniaxial (open circles), with a range of interaction spacing $0 \leq d \leq 5$, where d specifies the distance between adjacent particles in units of the particle width. For $d = 5$, particles are well separated and essentially noninteracting. For $d = 0$ the particles are just touching. Some of the interaction spacings are marked. The effect of interactions is fairly consistent, so unmarked intermediate points have intermediate value of d. The anisotropy orientation of the assemblage is random. Reproduced from Muxworthy A, Williams W, and Virdee D (2003a) Effect of magnetostatic interactions on the hysteresis parameters of single-domain and pseudo-single-domain grains. *Journal of Geophysical Research* 108: 2517.

randomly oriented uniaxial and cubic anisotropies, respectively; $H_{cr}/H_c = 1–1.5$). As the spacing between the particles is decreased, there is a consistent decrease in M_{rs}/M_s and increase in H_{cr}/H_c, which moves the system gradually from the SD to the PSD, and ultimately to the MD, regions of the Day plot (**Figure 39**). This prediction is consistent with the observation of interaction 'superstates' using electron holography (**Figure 27**), which are responsible for the PSD- and MD-like behavior of closely-spaced SD particles in nanoscale intergrowths. The effect of interactions on the properties of larger PSD particles is much more complex, and can cause the system to adopt either more SD-like or more MD-like behavior, depending on the particle size, shape and spacing, and on the style of anisotropy. This behavior results, in part, from a shift of the SD/SV threshold with increasing interactions: particles that would adopt SV states in the absence of interactions are able to adopt SD states when they are interacting strongly with neighboring particles. Micromagnetic simulations suggest that this effect occurs when the easy axes of neighboring particles are well aligned, as is the case for chains of magnetite particles in magnetotactic bacteria (**Figure 21**) and for arrays of magnetic blocks formed by exsolution from an ulvospinel host (**Figure 25**).

20.5 Magnetism at the Macroscopic Length Scale

In this final section, we review recent developments in the use of FORC diagrams to characterize the magnetic properties of rocks and minerals (Pike *et al.*, 1999). Until recently, hysteresis loops were the most widely used method of characterizing bulk magnetic properties (Roberts *et al.*, 2000). However, parameters determined from hysteresis loops represent bulk averages, and provide little information about the spectrum of coercivities and interaction fields that exist at the microscopic scale. The FORC diagram is a generalization of the well-known Preisach diagram (Preisach, 1935). The method requires the acquisition of many thousands of individual magnetization measurements, and has only been made possible by the advent of fully automated vibrating-sample and alternating-gradient magnetometers (Flanders, 1988), which allow the rapid acquisition of magnetization data over a large range of temperatures and applied fields.

20.5.1 Theory

20.5.1.1 First-order reversal curves and the FORC distribution

The definition of a first-order reversal curve is illustrated in **Figure 40(a)** (Pike *et al.*, 1999; Roberts *et al.*, 2000). Each FORC measurement begins by saturating the sample in a positive field. The external field is then decreased to some value, H_a (the reversal field), and the magnetization of the sample is measured as a function of increasing field, H_b, until positive saturation is reached again. A large number of FORCs are acquired for different reversal fields, in order to sample the entire area enclosed by a standard hysteresis loop (**Figure 40(b)**) Values of H_a and H_b are chosen to cover a regular grid in H_a–H_b space (**Figure 40(c)**), resulting in a magnetization matrix, $M(H_a, H_b)$. The FORC distribution is defined as the mixed second derivative of $M(H_a, H_b)$ with respect to H_a and H_b:

$$\rho(H_a, H_b) = -\frac{\partial^2 M(H_a, H_b)}{\partial H_a \partial H_b} \quad [44]$$

Note that in some studies, eqn [44] is multiplied by a factor of $1/2$ (e.g., Pike, 2003; Newell, 2005). It is customary (see Section 20.5.1.2) to define a new set of axes, $H_c = (H_a - H_b)/2$ and $H_u = (H_a + H_b)/2$, as illustrated in **Figure 40(c)**. The FORC diagram itself (**Figure 40(d)**) is a contour plot of $\rho(H_a, H_b)$, with H_c and H_u on the horizontal and vertical axes, respectively (covering the region of the H_a–H_b plane enclosed by the pink rectangle in **Figure 40(c)**).

In order to calculate the FORC distribution at any point P, a least-squares fit to $M(H_a, H_b)$ is performed over a grid of points surrounding P (illustrated by the blue square in **Figure 40(c)**). The most common method used is that of Pike *et al.* (1999), in which the magnetization is fitted using a second-order polynomial function:

$$M(H_a, H_b) = a_1 + a_2 H_a + a_3 H_a^2 + a_4 H_b + a_5 H_b^2 + a_6 H_a H_b \quad [45]$$

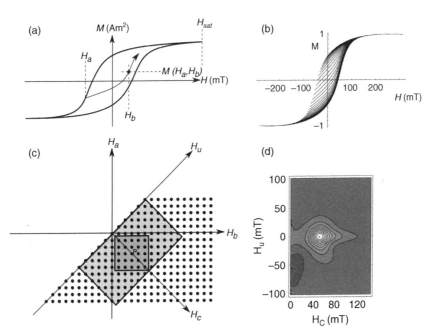

Figure 40 (a) Definition of a first-order reversal curve (FORC). (b) A set of FORCs for a sample of elongated SD maghemite particles at 20 K (reproduced from Carvallo C, Özdemir Ö, and Dunlop DJ (2004) First-order reversal curve (FORC) diagrams of elongated single-domain grains at high and low temperatures. *Journal of Geophysical Research* 109: B04105). (c) Matrix of H_a and H_b values used to measure magnetization during a typical FORC measurement. FORC diagrams are usually presented using a rotated set of axes H_u and H_c, covering the area of $H_a - H_b$ space defined by the pink rectangle. The blue square represents the region of $H_a - H_b$ space used to fit eqn [45] to $M(H_a, H_b)$ about a point P (SF = 2). (d) FORC diagram derived from the curves in (b) (reproduced from Carvallo C, Özdemir Ö, and Dunlop DJ (2004) First-order reversal curve (FORC) diagrams of elongated single-domain grains at high and low temperatures. *Journal of Geophysical Research* 109: B04105).

The value of the FORC distribution at P is then equal to $-a_6$. The size of the grid is defined by a smoothing factor, SF, such that the grid extends over $(2\mathrm{SF}+1)^2$ points in the H_a–H_b plane. This method becomes inefficient as the total number of points in the $M(H_a, H_b)$ matrix increases. Heslop and Muxworthy (2005) describe an alternative algorithm, based on the convolution method of Savitzky and Golay (1964), which yields identical results but is a factor of 500 times faster. An increase in the SF leads to a smoothing of the FORC diagram. While some smoothing is necessary to reduce experimental noise, too much smoothing may unduly affect the form of the distribution. Heslop and Muxworthy (2005) describe a numerical test, based on examination of the autocorrelation function of the residual of observed and fitted values of $M(H_a, H_b)$, to determine the optimum value of SF. The optimum value depends on the resolution of the $M(H_a, H_b)$ matrix, but values in the range 2–5 are typically employed. Because the $M(H_a, H_b)$ matrix does not extend to the $H_c < 0$ region (**Figure 40(c)**), increasing SF leads to an increase in the number of points close to the H_u axis that must be extrapolated (Carvallo *et al.*, 2005). The need to extrapolate data can be overcome by the use of 'extended' FORCs (Pike, 2003), as described in Section 20.5.1.3.

20.5.1.2 Interpretation of the FORC diagram

FORC diagrams provide an alternative method of measuring the Preisach distribution, which yields information about the spectrum of coercivity and interaction fields within a sample (Preisach, 1935; Mayergoyz, 1991; Carvallo *et al.*, 2005). The mathematical justification for using the Preisach model for interpreting FORC diagrams is described by Pike *et al.* (1999) and illustrated schematically **Figures 41** and **42**. The system is assumed to consist of a collection of particles with either an irreversible (**Figure 41(a)**) or a reversible (**Figure 41(b)**) hysteresis loop (referred to as a 'hysteron'). In the absence of an interaction field, irreversible particles switch their magnetization direction at the coercive field $\pm H_c$. In the presence of an interaction field the hysteron is shifted to either the left or right by an amount H_u, and switching now occurs at fields H_a and H_b (**Figure 41(c)**). H_a and H_b are related to the coercivity of the particle and the interaction field acting on it via $H_c = (H_a - H_b)/2$ and $H_u = (H_a + H_b)/2$. Each irreversible particle contributes to the FORC distribution at the corresponding point in $H_c - H_u$ space (**Figure 42**).

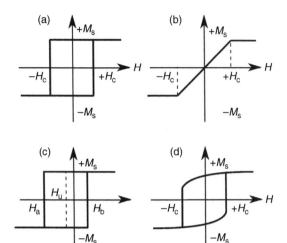

Figure 41 Magnetization as a function of applied field for particles with (a) irreversible and (b) reversible hysteresis loops. Irreversible switching occurs at an applied field of $\pm H_c$. (c) In the presence of a positive interaction field H_u, the irreversible hysteresis loop is shifted to the left, and switching now occurs at applied fields H_a and H_b. (d) A more general curvilinear hysteresis loop, which contains both reversible and irreversible magnetization components, can be used to explain the existence of negative peaks in SD FORC diagrams (Newell, 2005).

It is often assumed that the FORC distribution can be factorized into the product of two independent distributions:

$$\rho(H_c, H_u) = g(H_c)\, f(H_u) \qquad [46]$$

where $g(H_c)$ describes the distribution of coercivities and $f(H_u)$ describes the distribution of interaction fields. Carvallo *et al.* (2004, 2005) measured FORC diagrams for a series of well-characterized SD and PSD particles and found eqn [46] to be valid. Muxworthy *et al.* (2004) and Muxworthy and Williams (2005) performed a similar test using FORC diagrams derived from micromagnetic simulations. Although they observed a slight variation in H_c as a function of interaction strength, they concluded that eqn [46] provides a reasonable approximation for collections of SD particles with weak to moderate interactions.

20.5.1.3 Extended FORCs and the reversible ridge

Reversible magnetization of the form shown in **Figure 41(b)** is normally absent from the FORC distribution, as its contribution disappears on taking the second derivative of $M(H_a, H_b)$. This problem can be overcome by the use of 'extended FORCs' (Pike,

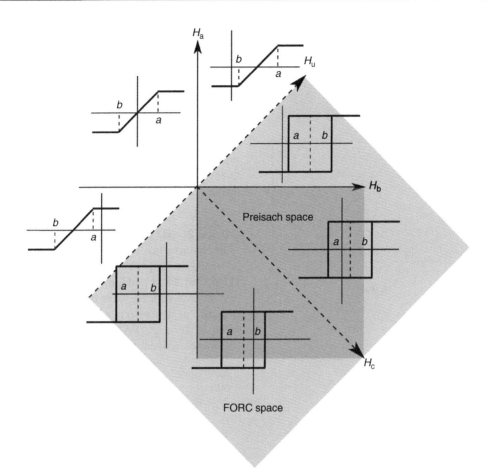

Figure 42 Form of the elementary reversible and irreversible magnetization cycles for different regions of the H_a–H_b plane. The gray area shows the region covered by a remanent Preisach diagram, the pink area is the region covered by a FORC diagram.

2003; Pike *et al.*, 2005). The magnetization matrix $M(H_a, H_b)$ is normally defined only in the region $H_b \geq H_a$ (as shown by the grid points in **Figure 40(c)**). However, $M(H_a, H_b)$ can be mathematically extended to cover the whole H_a–H_b plane:

$$M^*(H_a, H_b) = \begin{cases} M(H_a, H_b), & \text{if } H_b > H_a \\ M(H_a, H_a), & \text{if } H_b < H_a \end{cases} \quad [47]$$

By using M^* rather than M in eqn [44] the standard FORC diagram is obtained for $H_c > 0$, and a 'reversible ridge' is added to the H_u axis, describing the distribution of reversible magnetization in the form

$$\rho(H_a, H_a) = \frac{1}{2}\delta(H_b - H_a)\left(\lim_{H_b \to H_a} \frac{\partial M(H_a, H_b)}{\partial H_b}\right) \quad [48]$$

Equation [48] describes the slope of the FORC with reversal field H_a, calculated at the point at which the FORC joins the major hysteresis loop (Pike *et al.*,

2005). An example of an extended FORC diagram for a floppy disk recording material, including a profile of the reversible ridge, is shown in **Figure 43**. Since both the reversible and irreversible components of magnetization contribute, the extended FORC distribution is properly normalized, such that the integral with respect to H_a and H_b equals the saturation magnetization of the sample (Pike, 2003).

20.5.2 FORC Diagrams as a Function of Grain Size

20.5.2.1 SP particles

The expected form of the FORC diagram for SP particles is discussed by Pike *et al.* (2001a). Particles that are far above their blocking temperature have a reversible magnetization of the form shown in **Figure 42(b)** and do not contribute to a normal

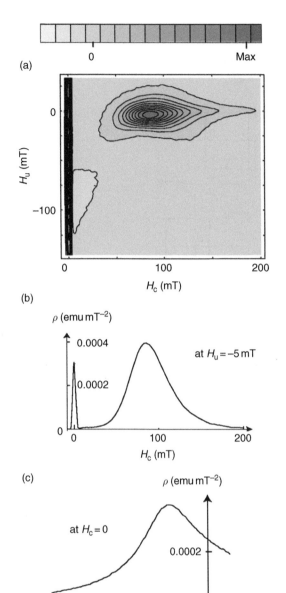

(a)

(b)

ρ (emu mT^{-2})

at $H_u = -5$ mT

(c)

ρ (emu mT^{-2})

at $H_c = 0$

Figure 43 FORC diagram for a Sony floppy disk sample, showing the reversible ridge at $H_c = 0$. In the contour shading legend above the diagram, Max denotes the value of the FORC distribution at its 'irreversible' peak located at about $H_c = 90$ mT. A negative region occurs adjacent to the vertical ($H_c = 0$) axis at about $H_u = 85$ mT. Note that the high density of vertical contour lines near the $H_c = 0$ axis makes the shading there appear darker than it really is. (b) A horizontal cross section passing though the irreversible peak at $H_b = 5$ mT. The ridge at $H_c = 0$ can also be seen in this plot. (c) A vertical cross section through the reversible ridge at $H_c = 0$. Reproduced from Pike CR (2003) First-order reversal-curve diagrams and reversible magnetization. *Physical Review B* 68: 104424.

FORC diagram (although they would contribute to the reversible ridge of an extended FORC diagram). Particles that are closer to their blocking temperatures show thermal relaxation of their magnetization state on a timescale similar to that of each FORC measurement step. This leads to contributions to the FORC distribution that peak around the origin and extend along the negative H_u axis (**Figure 44**). The form of the FORC diagram can be predicted using Néel's theory of thermal relaxation in SD particles (Pike *et al.*, 2001a).

20.5.2.2 SD particles

The characteristic feature of SD FORC diagrams is a closed positive peak at $H_c > 0$ and $H_u = 0$ (**Figures 41(d)** and **44**). In addition, a negative peak close to the H_u axis with $H_u < 0$ is commonly observed. The negative peak has been attributed to particle interactions (Pike *et al.*, 1999; Stancu *et al.*, 2003). FORC diagrams derived from micromagnetic simulations

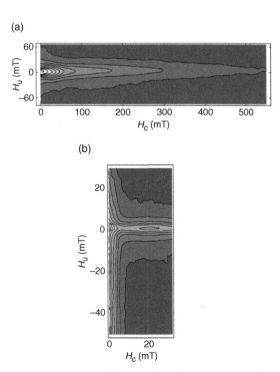

(a)

(b)

Figure 44 (a) FORC diagram for an SP-hematite-bearing Aptian red-bed sample from the south of France (b) High-resolution FORC diagram for the lower left-hand portion of the FORC plane for the same sample. Reproduced from Pike CR, Roberts AP, and Verosub KL (2001a) First-order reversal curve diagrams and thermal relaxation effects in magnetic particles. *Geophysical Journal International* 145: 721–730.

demonstrate that the negative peak is an intrinsic feature of noninteracting particles, and that the size of the peak is enhanced by interactions (Muxworthy *et al.*, 2004). The existence of a negative peak can be predicted by using a more realistic 'curvilinear' form of the hysteron (**Figure 41(d)**; Pike, 2003; Newell, 2005). In the Preisach model, the irreversible and reversible components of magnetization (**Figures 41(a)** and **41(b)**) are completely independent. For a curvilinear hysteron, however, the reversible component of magnetic susceptibility changes significantly as the particle switches from the upper to the lower branch of the loop. This coupling between the irreversible and reversible components gives rise to a systematic decrease, for a given $H_b < 0$, in the slopes of the FORCs as H_a decreases (Muxworthy *et al.*, 2004). This, in turn, translates to a negative contribution to the FORC distribution.

20.5.2.3 PSD particles

FORC diagrams for PSD size magnetite particles are described by Muxworthy and Dunlop (2002). FORC diagrams were measured for a series of synthetic magnetites with grain sizes varying from 0.3–11 μm (**Figure 45**). Small PSD particles have SD-like FORC diagrams, characterized by a closed positive peak at $H_c > 0$ and $H_u = 0$ (**Figure 45(a)**). With increasing grain size, the position of the peak shifts to lower H_c values (**Figure 45(b)**), and eventually moves to the origin (**Figure 45(c)**). This shift in peak position is accompanied by a spreading of the distribution in the H_u direction for small H_c (**Figure 45(d)**). Similar changes are seen as a function of temperature for particles of a fixed size.

20.5.2.4 MD particles

Theoretical predictions and experimental measurements of the FORC diagrams for noninteracting MD particles are described by Pike *et al.* (2001b). One-dimensional models of domain-wall pinning predict FORC diagrams consisting of perfectly vertical contours, with the value of the FORC distribution decreasing smoothly with increasing H_c. This model agrees well with experimental measurements on annealed (i.e., stress-free) magnetite samples (**Figure 46**). The vertical spread of the FORC function results from the fact that each particle contains a large number of pinning sites at which a domain can be trapped during the FORC measurement. These different pinning sites can be represented by an equivalent number of hysterons, which are spread out along the H_u axis by the self-demagnetizing field (Pike *et al.*, 2001b). FORC diagrams for unannealed MD samples are similar to those observed at the upper end of the PSD range (compare, e.g., **Figure 46(c)** with **Figure 45(d)**).

20.5.3 Mean-Field Interactions and FORC Diagrams

In the most basic form of the Preisach model, the distribution of interaction fields is assumed to be static. In reality, however, the field acting on a each particle is the sum of the stray fields created by all the other particles in the system, and will vary as the overall magnetization of the system changes. In general, both the mean value and the standard deviation of the interaction field distribution (IFD) are functions of the net magnetization of the system

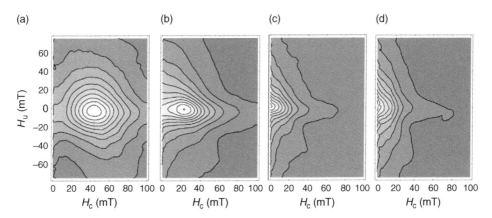

Figure 45 FORC diagrams for a series of synthetic PSD magnetite samples with grain sizes of (a) 0.3 μm, (b) 1.7 μm, (c) 7 μm, and (d) 11 μm. Reproduced from Muxworthy AR, and Dunlop DJ (2002) First-order reversal curve (FORC) diagrams for pseudo-single-domain magnetites at high temperature. *Earth and Planetary Science Letters* 203: 369–382.

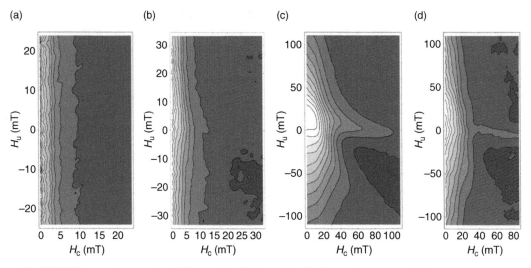

Figure 46   FORC diagrams for a series of MD samples. (a) A sample of M80 transformer steel. (b) A 2 mm grain of magnetite, after annealing. (c) The same 2 mm grain of magnetite before annealing. (d) An unannealed 125 μm magnetite grain. Reproduced from Pike CR, Roberts AP, Dekkers MJ, and Verosub KL (2001b) An investigation of multi-domain hysteresis mechanisms using FORC diagrams. *Physics of the Earth and Planetary Interiors* 126: 11–25.

(the 'variable-variance moving Preisach' model). For example, if a collection of particles is fully saturated in a large magnetic field, each particle experiences the roughly the same mean interaction field and the standard deviation of the IFD tends to zero. In the demagnetized state, each particle will experience a different interaction field; the mean value of the IFD is now zero and the standard deviation is maximum. The constant of proportionality relating the mean interaction field to the net magnetization of the system is referred to as the 'moving parameter', α, which can be either positive or negative, depending on the geometry of the system (Stancu *et al.*, 2001, 2003). Positive α implies that the mean field has a magnetizing effect, and leads to a spontaneous mutual alignment of the particles. This case applies, for example, to the chains of magnetite particles in magnetotactic bacteria (**Figure 21**). Negative α implies that the mean field has a demagnetizing effect. This case applies, for example, to perpendicular recording media (i.e., planar arrays of SD particles which have their easy axes perpendicular to the plane). The FORC diagram for such a system, composed of a perpendicular array of Ni pillars, is shown in **Figure 47** (Pike *et al.*, 2005). The 'wishbone' form of the FORC diagram has two main peaks: one occurring at low H_c and $H_u > 0$, and one occurring at high H_c and $H_u = 0$. The distance between these two peaks in the H_c direction yields information about the range of coercivities in the system. The

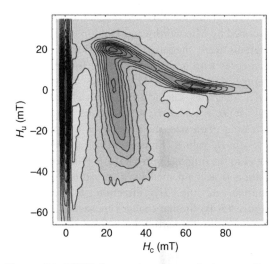

Figure 47   FORC diagram for a perpendicular recording medium, composed of a perpendicular array of Ni pillars. Reproduced from Pike CR, Ross CA, Scalettar RT, and Zimanyi G (2005) First-order reversal curve diagram analysis of a perpendicular nickel nanopillar array. *Physical Review B* 71: 134407.

displacement of the first peak in the positive H_u direction yields information about the strength of the mean-field demagnetizing interaction.

20.5.4   Practical Applications of FORC Diagrams

The FORC method has been applied in rock magnetism as a method of characterizing the magnetic mineralogy of natural samples (Roberts *et al.*, 2000), identifying mixtures of soft and hard magnetic minerals (Muxworthy *et al.*, 2005), and identifying magnetostatic interactions as a preselection tool for paleointensity studies (Wehland *et al.*, 2005). Pan *et al.* (2005) have used FORC diagrams to determine the strength of magnetostatic interactions in concentrated samples of magnetotactic bacteria (**Figure 48**). The FORC distribution has large SD-like peak centered on $H_c \sim 40$ mT and displaced slightly in negative H_u direction. The vertical spread of the IFD has a FWHM of just 6.3 mT, much lower than the ideal intra-chain interaction field of 60 mT. This observation demonstrates that the magnetosome chains are effectively behaving as elongated SD particles, and switch as a single unit. In such cases, the interaction fields measured by the FORC method provide an indication of inter-chain and inter-cellular interactions. The small peak in the FORC distribution about the origin can be attributed to the smaller magnetosomes that commonly occur at the ends of the chain.

20.6 Summary

Now is a very exciting time for the field of rock and mineral magnetism. The discovery of large-amplitude magnetic anomalies on Mars (Connerney *et al.*, 1999, 2004; Acuna *et al.*, 1999) has ignited a general interest in the effect of nanoscale microstructures on the origin and stability of planetary scale magnetic anomalies. Conventional wisdom – that these anomalies are due to the induced magnetization of multidomain magnetite – is now being challenged in light of the Mars magnetic survey. Mars no longer generates its own magnetic field; the anomalies are purely remanent in origin – faithfully recorded by magnetic minerals over 4 billion years ago (at a time when Mars did generate a field) and maintained without significant decay until the present day. The minerals responsible for the anomalies on Mars – and how they maintain such strong remanence over time – is currently the subject of intense speculation.

The techniques described in this review allow such problems to be tackled from both experimental and theoretical viewpoints, encompassing the entire

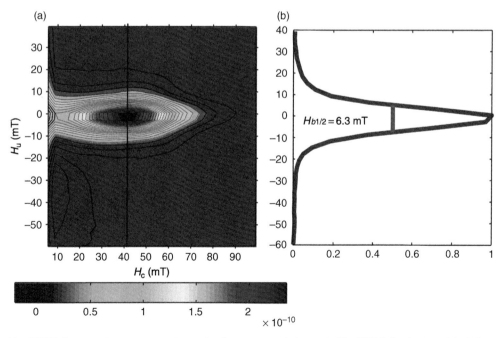

Figure 48 FORC diagram of a concentrated sample of magnetotactic bacteria. The FORC distribution of the MTB sample is bimodal with a broad maximum centered at 42 mT and a sharper peak towards the $H_c = 0$ axis. The latter is attributed to emerging magnetosomes at the chain ends. (b) Vertical profile through the high-coercivity peak of the distribution with mean half width field of 6.3 mT at $H_c = 41.4$ mT. Reproduced from Pan Y, Petersen N, Winklhofer M, *et al.* (2005) Rock magnetic properties of uncultured magnetotactic bacteria. *Earth and Planetary Science Letters* 237: 311–325.

range of length scales of interest, from atomistic interactions to planetary-scale magnetic anomalies. Since the dominant carriers of stable natural remanent magnetization are SD particles with sizes in the range 30–200 nm, techniques such as electron holography (Section 20.3) have the potential to revolutionize the way rock magnetic measurements are made in the future. By using the three-dimensional morphologies of magnetic nanoparticles, provided by electron tomography, as the input for finite element micromagnetic simulations (Section 20.4), it is now possible to compare experimental observations and theoretical predictions directly. Differences between observed and calculated behaviors are likely to be the result of atomistic effects at surfaces, interfaces, and defects. Ultimately, the application of atomistic simulations (Section 20.2) will permit the influence of such atomic-scale features on standard rock magnetic analysis (Section 20.5) to be determined.

References

Abe K, Miyamoto Y, and Chikazumi S (1976) Magnetocrystalline anisotropy of low-temperature phase of magnetite. *Journal of the Physical Society of Japan* 41: 1894–1902.

Acuna MH, Connerney JEP, Ness NF, *et al.* (1999) Global distribution of crustal magnetisation discovered by the Mars Global Surveyor MAG/ER experiment. *Science* 284: 790–793.

Arató B, Szányi Z, Flies C, Schüler D, Frankel RB, Buseck PB, and Pósfai M (2005) Crystal-size and shape distributions of magnetite from uncultured magnetotactic bacteria as a potential biomarker. *American Mineralogist* 90: 1233–1241.

Banerjee SK (1991) Magnetic properties of Fe–Ti oxides. *Mineralogical Society of America Reviews in Mineralogy* 25: 107–128.

Bazylinski DA and Frankel RB (2004) Magnetosome formation in prokaryotes. *Nature Reviews Microbiology* 2: 217–230.

Becker U, Fernandez-Gonzalez A, Prieto M, Harrison RJ, and Putnis A (2000) Direct calculation of the mixing enthalpy of the barite/celestite system. *Physics and Chemistry of Minerals* 27: 291–300.

Belov K (1993) Electronic processes in magnetite (or "enigmas in magnetite"). *Physics Uspekhi* 36: 380–391.

Besser PJ, Morrish AH, and Searle CW (1967) Magnetocrystalline anisotropy of pure and doped hematite. *Physical Review* 153: 632–640.

Bickford L, Brownlow J, and Penoyer RF (1957) Magnetocrystalline anisotropy in cobalt-substituted magnetic single crystals. *Proceedings of the Institute of Electrical and Electronics Engineers B* 104: 238–244.

Bina M, Tanguy JC, Hoffmann V, *et al.* (1999) A detailed magnetic and mineralogical study of self-reversed dacitic pumices from the 1991 Pinatubo eruption (Philippines). *Geophysical Journal International* 138: 159–178.

Bosenick A, Dove MT, Myers ER, *et al.* (2001) Computational methods for the study of energies of cation distributions: Applications to cation-ordering phase transitions and solid solutions. *Mineralogical Magazine* 65: 193–219.

Brockhouse BN (1957) Scattering of neutrons by spin waves in magnetite. *Physical Review* 106: 859–864.

Brown WF (1963) *Micromagnetics*. New York: Interscience.

Burton BP (1985) Theoretical analysis of chemical and magnetic ordering in the system Fe_2O_3–$FeTiO_3$. *American Mineralogist* 70: 1027–1035.

Burton BP (1991) The interplay of chemical and magnetic ordering. *American Mineralogical Society Reviews in Mineralogy* 25: 303–322.

Burton BP and Davidson PM (1988) Multicritical phase relations in minerals. In: Ghose S, Coey JMD, and Salje E (eds.) *Advances in Physical Geochemistry*, vol. 7, pp. 60–90. Berlin: Springer-Verlag.

Butler RF and Banerjee SK (1975) Theoretical single-domain grain size range in magnetite and titanomagnetite. *Journal of Geophysical Research* 80: 4049–4058.

Carvallo C, Dunlop DJ, and Özdemir Ö (2005) Experimental comparison of FORC and remanent Preisach diagrams. *Geophysical Journal International* 162: 747–754.

Carvallo C, Muxworthy AR, Dunlop DJ, and Williams W (2003) Micromagnetic modeling of first-order reversal curve (FORC) diagrams for single-domain and pseudo-single-domain magnetite. *Earth and Planetary Science Letters* 213: 375–390.

Carvallo C, Özdemir Ö, and Dunlop DJ (2004) First-order reversal curve (FORC) diagrams of elongated single-domain grains at high and low temperatures. *Journal of Geophysical Research* 109: B04105.

Coey JMD (1971) Noncollinear spin arrangement in ultrafine ferrimagnetic crystallites. *Physical Review Letters* 271: 1140–1142.

Coey JMD and Ghose S (1987) Magnetic ordering and thermodynamics in silicates. In: Salje EKH (ed.) *Physical Properties and Thermodynamic Behaviour of Minerals, NATO ASI Series*. Dordrecht: D. Reidel Publishing Company.

Connerney JEP, Acuna MH, Ness NF, Spohn T, and Schubert G (2004) Mars crustal magnetism. *Space Science Reviews* 111: 1–32.

Connerney JEP, Acuna MH, Wasilewski PJ, *et al.* (1999) Magnetic lineations in the ancient crust of Mars. *Science* 284: 794–798.

Cowley JM (1995) *Diffraction Physics*, 3rd rev. edn. Oxford: Elsevier.

Davis PM and Evans ME (1976) Interacting single-domain properties of magnetite intergrowths. *Journal of Geophysical Research* 81: 989–994.

Day R, Fuller M, and Schmidt VA (1977) Hysteresis properties of titanomagnetites: Grain-size and compositional dependence. *Physics of the Earth and Planetary Interiors* 13: 260–266.

de Graef M, Nuhfer NT, and McCartney MR (1999) Phase contrast of spherical magnetic particles. *Journal of Microscopy – Oxford* 194: 84–94.

de Ruijter WJ and Weiss JK (1993) Detection limits in quantitative off-axis electron holography. *Ultramicroscopy* 50: 269–283.

Devouard B, Pósfai M, Hua X, Bazylinski DA, Frankel RB, and Buseck PR (1998) Magnetite from magnetotactic bacteria: Size distributions and twinning. *American Mineralogist* 83: 1387–1398.

Dimian M and Kachkachi H (2002) Effect of surface anisotropy on the hysteretic properties of a magnetic particle. *Journal of Applied Physics* 91: 7625–7627.

Dove MT (2001) Computer simulations of solid solutions. In: Geiger C (ed.) *Solid Solutions in Silicate and Oxide Systems of Geological Importance, EMU Notes in Mineralogy*, vol. 16, pp. 57–64. Budapest: Eötvös University Press.

Doyle PA and Turner PS (1968) Relativistic Hartree-Fock and electron scattering factors. *Acta Crystallographica A* 24: 390–397.

Dunin-Borkowski RE, McCartney MR, and Smith DJ (2004) Electron holography of nanostructured materials. In: Nalwa HS (ed.) *Encyclopedia of Nanoscience and Nanotechnology*, vol. 3, pp. 41–99. California: American Scientific Publishers.

Dunlop DJ and Özdemir Ö (1997) *Rock Magnetism: Fundamentals and Frontiers*. Cambridge: Cambridge University Press.

Dzyaloshinskii I (1958) A thermodynamic theory of "weak" ferromagnetism of antiferromagnetics. *Journal of Physics and Chemistry of Solids* 4: 241.

Egerton RF (1996) *Electron Energy-Loss Spectroscopy in the Electron Microscope*. New York: Plenum Press.

Enkin RJ and Williams W (1994) Three-dimensional micromagnetic analysis of stability in fine magnetic grains. *Journal of Geophysical Research* 99: 611–618.

Fabian K and Heider F (1996) How to include magnetostriction in micromagnetic models of titanomagnetite. *Geophysical Research Letters* 23: 2839–2842.

Fabian K, Kirchner A, Williams W, Heider F, Leibl T, and Hubert A (1996) Three-dimensional micromagnetic calculations for magnetite using FFT. *Geophysical Journal International* 124: 89–104.

Feinberg JM, Scott GR, Renne PR, and Wenk HR (2005) Exsolved magnetite inclusions in silicates: Features determining their remanence behavior. *Geology* 33: 513–516.

Feinberg JM, Wenk HR, Renne PR, and Scott GR (2004) Epitaxial relationships of clinopyroxene-hosted magnetite determined using electron backscatter diffraction (EBSD) technique. *American Mineralogist* 89: 462–466.

Fidler J and Schrefl T (2000) Micromagnetic modelling – the current state of the art. *Journal of Physics D: Applied Physics* 33: R135–R156.

Flanders PJ (1988) An alternating-gradient magnetometer. *Journal of Applied Physics* 63: 3940–3945.

Fletcher EJ and O'Reilly W (1974) Contribution of Fe^{2+} ions to the magnetocrystalline anisotropy constant K_1 of $Fe_{3-x}Ti_xO_4$ $(0 < x < 0.1)$. *Journal of Physics C: Solid State Physics* 7: 171–178.

Gallet Y, Genevey A, and Fluteau F (2005) Does Earth's magnetic field secular variation control centennial climate change? *Earth and Planetary Science Letters* 236: 339–347.

Garanin DA and Kachkachi H (2003) Surface contribution to the anisotropy of magnetic nanoparticles. *Physical Review Letters* 90: 065504.

Ghiglia DC and Pritt MD (1998) Two-dimensional phase unwrapping. *Theory, Algorithms and Software*. New York: Wiley.

Ghiorso MS (1997) Thermodynamic analysis of the effect of magnetic ordering on miscibility gaps in the FeTi cubic and rhombohedral oxide minerals and the FeTi oxide geothermometer. *Physics and Chemistry of Minerals* 25: 28–38.

Glasser ML and Milford FJ (1963) Spin wave spectra of magnetite. *Physical Review* 130: 1783–1789.

Golla-Schindler U, O'Neill H, and Putnis A (2005) Direct observation of spinodal decomposition in the magnetic-hercynite system by susceptibility measurements and transmission electron microscopy. *American Mineralogist* 90: 1278–1283.

Goodenough JB (1966) *Magnetism and the Chemical Bond*. New York: John Wiley and Sons.

Guslienko KY, Novosad V, Otani Y, Shima H, and Fukamichi K (2001) Magnetization reversal due to vortex nucleation, displacement, and annihilation in submicron ferromagnetic dot arrays. *Physical Review B* 65: 024414.

Hanzlik M, Winklhofer M, and Petersen N (2002) Pulsed-field remanence measurements on individual magnetotactic bacteria. *Journal of Magnetism and Magnetic Materials* 248: 258–267.

Harrison RJ (2000) Magnetic transitions in Minerals. *American Mineralogical Society Reviews in Mineralogy* 39: 175–202.

Harrison RJ (2006) Microstructure and magnetism in the ilmenite–hematite solid solution: A Monte Carlo simulation study. *American Mineralogist* 91: 1006–1024.

Harrison RJ and Becker U (2001) Magnetic ordering in solid solutions. In: Geiger C (ed.) *Solid Solutions in Silicate and Oxide Systems, European Mineralogical Society Notes in Mineralogy*, vol. 3, ch. 13, pp. 349–383. Budapest: Eötvös University Press.

Harrison RJ, Becker U, and Redfern SAT (2000a) Thermodynamics of the R-3 to R-3c phase transition in the ilmenite-hematite solid solution. *American Mineralogist* 85: 1694–1705.

Harrison RJ, Dunin-Borkowski RE, and Putnis A (2002) Direct imaging of nanoscale magnetic interactions in minerals. *Proceedings of the National Academy of Sciences* 99: 16556–16561.

Harrison RJ, Kasama T, White TA, Simpson ET, and Dunin-Borkowski RE (2005) Origin of self-reversed thermo-remanent magnetisation. *Physical Review Letters* 95: 268501.

Harrison RJ and Putnis A (1997) The coupling between magnetic and cation ordering: A macroscopic approach. *European Journal of Mineralogy* 9: 1115–1130.

Harrison RJ and Putnis A (1999) The magnetic properties and mineralogy of oxide spinel solid solutions. *Surveys in Geophysics* 19: 461–520.

Harrison RJ and Redfern SAT (2001) Short- and long-range ordering in the ilmentite–hematite solid solution. *Physics and Chemistry of Minerals* 28: 399–412.

Harrison RJ, Redfern SAT, and Smith RI (2000b) In-situ study of the R-3 to R-3c phase transition in the ilmenite–hematite solid solution using time-of-flight neutron powder diffraction. *American Mineralogist* 85: 194–205.

Heider F and Williams W (1988) Note on temperature dependence of exchange constant in magnetite. *Geophysical Research Letters* 15: 184–187.

Heslop D and Muxworthy AR (2005) Aspects of calculating first-order reversal curve distributions. *Journal of Magnetism and Magnetic Materials* 288: 155–167.

Hoffman KA (1992) Self-reversal of thermoremanent magnetization in the ilmenite–hematite system: Order-disorder, symmetry, and spin alignment. *Journal of Geophysical Research* 97: 10883–10895.

Hubert A (1967) Der einfluss der magnetostriktion auf die magnetische bereichstruktur einachsiger Kristalle inbesondere des kobalts. *Physica Status Solidi* 22: 709–727.

Hughes GF (1983) Magnetization reversal in cobalt–phosphorus films. *Journal of Applied Physics* 54: 5306–5313.

Hunt CP, Moskowitz BM, and Banerjee SK (1995) Magnetic properties of rocks and minerals. In: Ahrens TJ (ed.) *A Handbook of Physical Constants, vol. 3: Rock Physics and Phase Relations*. Washington, D.C: American Geophysical Union.

Inden G (1981) The role of magnetism in the calculation of phase diagrams. *Physica B* 103: 82–100.

Ishikawa Y, Saito N, Arai M, Watanabe Y, and Takei H (1985) A new oxide spin glass system of $(1-x)FeTiO_3-xFe_2O_3$. I. Magnetic properties. *Journal of the Physical Society of Japan* 54: 312–325.

Ishikawa Y and Syono Y (1963) Order-disorder transformation and reverse thermoremanent magnetization in the $FeTiO_3$—Fe_2O_3 system. *Journal of Physics and Chemistry of Solids* 24: 517–528.

Kachkachi H and Dimian M (2002) Hysteretic properties of a magnetic particle with strong surface anisotropy. *Physical Review B* 66: 174419.

Kachkachi H, Ezzir A, Noguès M, and Tronc E (2000a) Surface effects in nanoparticles: Application to maghemite γ-Fe$_2$O$_3$. *European Physical Journal B* 14: 681–689.

Kachkachi H and Mahboub H (2004) Surface anisotropy in nanomagnets: Transverse or Néel? *Journal of Magnetism and Magnetic Materials* 278: 334–341.

Kachkachi H, Noguès M, Tronc E, and Garanin DA (2000b) Finite-size versus surface effects in nanoparticles. *Journal of Magnetism and Magnetic Materials* 221: 158–163.

Kasama T, Golla-Schindler U, and Putnis A (2003) High-resolution and energy-filtered TEM of the interface between hematite and ilmenite exsolution lamellae: Relevance to the origin of lamellar magnetism. *American Mineralogist* 88: 1190–1196.

Kasama T, McEnroe SA, Ozaki N, Kogure T, and Putnis A (2004) Effects of nanoscale exsolution in hematite–ilmenite on the acquisition of stable natural remanent magnetization. *Earth and Planetary Science Letters* 224: 461–475.

Kaufman L (1981) J.L. Meijering's contribution to the calculation of phase diagrams - A personal perspective. *Physica* 103: 1–7.

Kirschner M, Schrefl T, Dorfbauer F, Hrkac G, Suess D, and Fidler J (2005) Cell size corrections for nonzero-temperature micromagnetics. *Journal of Applied Physics* 97: 10E301.

Kodama RH (1999) Magnetic nanoparticles. *Journal of Magnetism and Magnetic Materials* 200: 359–372.

Kodama RH and Berkowitz AE (1999) Atomic-scale magnetic modeling of oxide nanoparticles. *Physical Review B* 59: 6321–6336.

Kodama RH, Berkowitz AE, McNiff EJ, and Foner S (1996) Surface spin disorder in NiFe$_2$O$_4$ nanoparticles. *Physical Review Letters* 77: 394–397.

Kodama RH, Makhlouf SA, and Berkowitz AE (1997) Finite Size Effects in Antiferromagnetic NiO Nanoparticles. *Physical Review Letters* 79: 1393–1396.

Kumar AA, Rao VP, Patil SK, Kessarkar PM, and Thamban M (2005) Rock magnetic records of the sediments of the eastern Arabian Sea: Evidence for late Quaternary climatic change. *Marine Geology* 220: 59–82.

Labrosse S and Macouin M (2003) The inner core and the geodynamo. *Comptes Rendus Geoscince* 335: 37–50.

Lagroix F, Banerjee SK, and Moskowitz BM (2005) Revisiting the mechanism of reversed thermoremanent magnetization based on observations from synthetic ferrian ilmenite ($y = 0.7$). *Journal of Geophysical Research* 109: B12108.

Low W (1960) *Paramagnetic Resonance in Solids*, p. 33. New York: Academic Press.

Matar SM (2003) Ab initio investigations in magnetic oxides. *Progress in Solid State Chemistry* 31: 239–299.

Mayergoyz ID (1991) *Mathematical Models of Hysteresis*. New York: Springer.

Mazo-Zuluaga J and Restrepo J (2004) Monte Carlo study of the bulk magnetic properties of magnetite. *Physica B* 354: 20–26.

McCartney MR, Lins U, Farina M, Buseck PR, and Frankel RB (2001) Magnetic microstructure of bacterial magnetite by electron holography. *European Journal of Mineralogy* 13: 685–689.

McEnroe SA, Brown LL, and Robinson P (2004c) Earth analog for Martian magnetic anomalies: Remanence properties of hemo-ilmenite norites in the Bjerkreim–Sokndal Intrusion, Rogaland, Norway. *Journal of Applied Geophysics* 56: 195–212.

McEnroe SA, Harrison RJ, Robinson P, Golla U, and Jercinovic MJ (2001) The effect of fine-scale microstructures in titanohematite on the acquisition and stability of NRM in granulite facies metamorphic rocks from Southwest Sweden. *Journal of Geophysical Research* 106: 30523–30546.

McEnroe SA, Harrison RJ, Robinson P, and Langenhorst F (2002) Nanoscale hematite–ilmenite in massive ilmenite rock: An example of 'lamellar magnetism' with implications for planetary magnetic anomalies. *Geophysical Journal International* 151: 890–912.

McEnroe SA, Langenhorst F, Robinson P, Bromiley GD, and Shaw CSJ (2004b) What is magnetic in the lower crust? *Earth and Planetary Science Letters* 226: 175–192.

McEnroe SA, Skilbrei JR, Robinson P, Heidelbach F, Langenhorst F, and Brown LL (2004a) Magnetic anomalies, layered intrusions and Mars. *Geophysical Research Letters* 31: L1960.

McNab TK, Fox RA, and Boyle JF (1968) Some magnetic properties of magnetite (Fe$_3$O$_4$) microcrystals. *Journal of Applied Physics* 39: 5703–5711.

Midgely PA (2001) An introduction to off-axis electron holography. *Micron* 32: 167–184.

Moon TS (1991) Domain states in fine particle magnetite and titanomagnetite. *Journal of Geophysical Research* 96: 9909–9923.

Moon T and Merrill RT (1984) The magnetic moments of non-uniformly magnetized grains. *Physics of the Earth and Planetary Interiors* 34: 186–194.

Moon TS and Merril RT (1985) Nucleation theory and domain states in multidomain magnetic material. *Physics of the Earth and Planetary Interiors* 37: 214–222.

Muxworthy AR and Dunlop DJ (2002) First-order reversal curve (FORC) diagrams for pseudo-single-domain magnetites at high temperature. *Earth and Planetary Science Letters* 203: 369–382.

Muxworthy AR, Dunlop DJ, and Williams W (2003b) High-temperature magnetic stability of small magnetite particles. *Journal of Geophysical Research* 108: 2281.

Muxworthy A, Heslop D, and Williams W (2004) Influence of magnetostatic interactions on first-order-reversal-curve (FORC) diagrams: A micromagnetic approach. *Geophysical Journal International* 158: 888–897.

Muxworthy A, King JG, and Heslop D (2005) Assessing the ability of first-order reversal curve (FORC diagrams to unravel complex magnetic signals. *Journal of Geophysical Research* 110: B01105.

Muxworthy AR and McClelland E (2000) Review of the low-temperature magnetic properties of magnetite from a rock magnetic perspective. *Geophysical Journal International* 140: 101–114.

Muxworthy AR and Williams W (1999) Micromagnetic models of pseudo-single domain grains of magnetite near the Verwey transition. *Journal of Geophysical Research* 104: 29203–29217.

Muxworthy A and Williams W (2005) Magnetostatic interaction fields in first-order-reversal-curve diagrams. *Journal of Applied Physics* 97: 063905.

Muxworthy A and Williams W (2006) Critical single-domain/multidomain grain sizes in noninteracting and interacting elongated magnetite particles: Implications for magnetosomes. *Journal of Geophysical Research* 111: B12S12.

Muxworthy A, Williams W, and Virdee D (2003a) Effect of magnetostatic interactions on the hysteresis parameters of single-domain and pseudo-single-domain grains. *Journal of Geophysical Research* 108: 2517.

Néel L (1948) Propriétés magnetiques des ferrites; ferrimagnétisme et antiferromagnétisme. *Annales de Physique* 3: 137–198.

Néel L (1949) Théorie du traînage magnétique des ferromagnétiques en grains fins avec applications aux terres cuites. *Annales de Géophysique* 5: 99–136.

Newell A (2005) A high-precision model of first-order reversal curve (FORC) functions for single-domain ferromagnets with uniaxial anisotropy. *Geochemistry Geophysics Geosystems* 6: Q05010.

Newell AJ, Dunlop DJ, and Williams W (1993) A two-dimensional micromagnetic model of magnetizations and fields in magnetite. *Journal of Geophysical Research* 98: 9533–9549.

Nord GL and Lawson CA (1989) Order-disorder transition-induced twin domains and magnetic properties in ilmenite–hematite. *American Mineralogist* 74: 160–176.

Nord GL and Lawson CA (1992) Magnetic properties of ilmenite 70-hematite30: Effect of transformation-induced twin boundaries. *Journal of Geophysical Research* 97: 10897–10910.

Pan Y, Petersen N, Winklhofer M, *et al.* (2005) Rock magnetic properties of uncultured magnetotactic bacteria. *Earth and Planetary Science Letters* 237: 311–325.

Pauthenet R and Bochirol L (1951) Aimantation spontanée des ferrites. *Journal de Physique et de le Radium* 12: 249–251.

Petersen N, Weiss D, and Vali H (1989) Magnetotactic bacteria in lake sediments. In: Lowes F (ed.) *Geomagnetism and Paleomagnetism*, pp. 231–241. Dordrecht: Kluwer Academic Publishers.

Phillips TG and Rosenberg HM (1966) Spin waves in ferromagnets. *Reports on Progress in Physics* 29: 285–332.

Pike CR (2003) First-order reversal-curve diagrams and reversible magnetization. *Physical Review B* 68: 104424.

Pike CR, Roberts AP, Dekkers MJ, and Verosub KL (2001b) An investigation of multi-domain hysteresis mechanisms using FORC diagrams. *Physics of the Earth and Planetary Interiors* 126: 11–25.

Pike CR, Roberts AP, and Verosub KL (1999) Characterizing interactions in fine magnetic particle systems using first order reversal curves. *Journal of Applied Physics* 85: 6660–6667.

Pike CR, Roberts AP, and Verosub KL (2001a) First-order reversal curve diagrams and thermal relaxation effects in magnetic particles. *Geophysical Journal International* 145: 721–730.

Pike CR, Ross CA, Scalettar RT, and Zimanyi G (2005) First-order reversal curve diagram analysis of a perpendicular nickel nanopillar array. *Physical Review B* 71: 134407.

Preisach F (1935) Über die magnetische Nachwirkung. *Zeitschrift für Physik* 94: 277–302.

Prévot M, Hoffman KA, Goguitchaichvili A, Doukhan J-C, Schcherbakov V, and Bina M (2001) The mechanism of self-reversal of thermoremanence in natural hemoilmenite crystals: New experimental data and model. *Physics of the Earth and Planetary Interiors* 126: 75–92.

Price GD (1980) Exsolution microstructures in titano-magnetites and their magnetic significance. *Physics of the Earth and Planetary Interiors* 23: 2–12.

Price GD (1981) Subsolidus phase-relations in the titanomagnetite solid-solution series. *American Mineralogist* 66: 751–758.

Puntes VF, Gorostiza P, Aruguete DM, Bastus NG, and Alivisatos AP (2004) Collective behaviour in two-dimensional cobalt nanoparticle assemblies observed by magnetic force microscopy. *Nature Materials* 3: 263–268.

Rave W, Fabian K, and Hubert A (1998) Magnetic states of small cubic particles with uniaxial anisotropy. *Journal of Magnetism and Magnetic Materials* 190: 332–348.

Reimer L (1991) *Transmission Electron Microscopy*. Berlin: Springer-Verlag.

Rez D, Rez P, and Grant I (1994) Dirac–Fock calculations of X-ray scattering factors and contributions to the mean inner potential for electron scattering. *Acta Crystallographica A* 50: 481–497.

Rhodes P and Rowlands G (1954) Demagnetizing energies of uniformly magnetized rectangular blocks. *Proceedings of the Leeds Philosophical and Literary Society Scientific Section* 6: 191–210.

Roberts AP, Pike CR, and Verosub KL (2000) First-order reversal curve diagrams: A new tool for characterizing the magnetic properties of natural samples. *Journal of Geophysical Research* 105: 28461–28475.

Robinson P, Harrison RJ, and McCenroe SA (2006) Fe^{2+}/Fe^{3+} charge ordering in contact layers of lamellar magnetism: Bond valence arguments. *American Mineralogist* 91: 67–72.

Robinson P, Harrison RJ, McEnroe SA, and Hargraves RB (2002) Lamellar magnetism in the haematite–ilmenite series as an explanation for strong remanent magnetisation. *Nature* 418: 517–520.

Robinson P, Harrison RJ, McEnroe SA, and Hargraves RB (2004) Nature and origin of lamellar magnetism in the hematite–ilmenite series. *American Mineralogist* 89: 725–747.

Rollmann G, Rohrbach A, Entel P, and Hafner J (2004) First-principles calculation of the structure and magnetic phases of hematite. *Physical Review B* 69: 165107.

Samuelsen EJ (1969) Spin waves in antiferromagnets with corundum structure. *Physica* 43: 353–374.

Samuelsen EJ and Shirane G (1970) Inelastic neutron scattering investigation of spin waves and magnetic interactions in α-Fe_2O_3. *Physica Status Solidi* 42: 241–256.

Sandratskii LM (1998) Noncollinear magnetism in itinerant-electron systems: Theory and applications. *Advances in Physics* 47: 91–160.

Sandratskii LM and Kübler J (1996) First-principles LSDF study of weak ferromagnetism in Fe_2O_3. *Europhysics Letters* 33: 447–452.

Sandratskii LM, Uhl M, and Kübler JK (1996) Band theory for electronic and magnetic properties of α-Fe_2O_3. *Journal of Physics: Condensed Matter* 8: 983–989.

Savitzky A and Golay MJE (1964) Smoothing and differentiation of data by simplified least squares procedures. *Analytical Chemistry* 36: 1627–1639.

Schabes ME and Bertram HN (1988a) Magnetzation processes in ferromagnetic cubes. *Journal of Applied Physics* 64: 1347–1357.

Schabes ME and Bertram HN (1988b) Ferromagnetic switching in elongated γ-Fe_2O_3 particles. *Journal of Applied Physics* 64: 5832–5834.

Scholz W, Schrefl T, and Fidler J (1999) Mesh reinement in FE-micromagnetics for multi-domain $Nd_2Fe_{14}B$ particles. *Journal of Magnetism and Magnetic Materials* 196: 933–934.

Scholz W, Schrefl T, and Fidler J (2001) Micromagnetic simulation of thermally activated switching in fine particles. *Journal of Magnetism and Magnetic Materials* 233: 296–304.

Simpson ET, Kasama T, Pósfai M, Buseck PR, Harrison RJ, and Dunin-Borkowski RE (2005) Magnetic induction mapping of magnetite chains in magnetotactic bacteria at room temperature and close to the Verwey transition using electron holography. *Journal of Physics: Conference Series* 17: 108–121.

Skumryev V, Stoyanov S, Zhang Y, Hadjipanayis G, Givord D, and Nogués J (2003) Beating the superparamagnetic limit with exchange bias. *Nature* 423: 850–853.

Stacy FD and Banerjee SK (1974) *The Physical Principles of Rock Magnetism*. Amsterdam: Elsevier.

Stancu A, Pike CR, Stoleriu L, Postolache P, and Cimpoesu D (2003) Micromagnetic and Preisach analysis of the First Order Reversal Curves (FORC) diagram. *Journal of Applied Physics* 93: 6620–6622.

Stancu A, Stoleriu L, and Cerchez M (2001) Micromagnetic evaluation of magnetostatic interactions distribution in structured particulate media. *Journal of Applied Physics* 89: 7260–7262.

Stephenson A (1972a) Spontaneous magnetization curves and curie points of spinels containing two types of magnetic ion. *Philosophical Magazine* 25: 1213–1232.

Stephenson A (1972b) Spontaneous magnetization curves and curie points of cation deficient titanomagnetites. *Geophysics Journal of the Royal Astronomical Society* 29: 91–107.

Stoner EC and Wohlfarth EP (1948) A mechanism of magnetic hysteresis in heterogeneous alloys. *Philosophical Transactions of the Royal Society of London A* 240: 599–642.

Thomson LC, Enkin RJ, and Williams W (1994) Simulated annealing of three-dimensional micromagnetic structures and simulated thermoremanent magnetization. *Journal of Geophysical Research* 99: 603–609.

Tonomura A (1992) Electron-holographic interference microscopy. *Advances in Physics* 41: 59–103.

Tronc E, Ezzir A, Cherkaoui R, *et al.* (2000) Surface-related properties of γ-Fe_2O_3 nanoparticles. *Journal of Magnetism and Magnetic Materials* 221: 63–79.

Uhl M and Siberchicot B (1995) A first-principles study of exchange integrals in magnetite. *Journal of Physics: Condensed Matter* 7: 4227–4237.

Valet JP, Meynadier L, and Guyodo Y (2005) Geomagnetic dipole strength and reversal rate over the past two million years. *Nature* 435: 802–805.

Vinograd VL, Sluiter MHF, Winkler B, *et al.* (2004) Thermodynamics of mixing and ordering in pyrope–grossular solid solution. *Mineralogical Magazine* 68: 101–121.

Völkl E, Allard LF, and Joy DC (1998) *Introduction to Electron Holography*. New York: Plenum.

Warren MC, Dove MT, and Redfern SAT (2000a) Ab initio simulations of cation ordering in oxides: Application to spinel. *Journal of Physics: Condensed Matter* 12: L43–48.

Warren MC, Dove MT, and Redfern SAT (2000b) Disordering of $MgAl_2O_4$ spinel from first principles. *Mineralogical Magazine* 64: 311–317.

Watanabe H and Brockhouse BN (1962) Observation of optical and acoustical magnons in magnetite. *Physics Letters* 1: 189–190.

Wehland F, Leonhardt R, Vadeboin F, and Appel E (2005) Magnetic interaction analysis of basaltic samples and pre-selection for absolute palaeointensity measurements. *Geophysical Journal International* 162: 315–320.

Williams W and Dunlop DJ (1989) Three-dimensional micromagnetic modelling of ferromagnetic domain structure. *Nature* 337: 634–637.

Williams W and Dunlop DJ (1990) Some effects of grain shape and varying external magnetic field on the magnetic structure of small grains of magnetite. *Physics of the Earth and Planetary Interiors* 65: 1–14.

Williams W and Dunlop DJ (1995) Simulation of magnetic hystersis in pseudo-single-domain grains of magnetite. *Journal of Geophysical Research* 100: 3859–3871.

Williams W and Wright TM (1998) High-resolution micromagnetic models of fine grains of magnetite. *Journal of Geophysical Research* 103: 30537–30550.

Winklhofer M, Fabian K, and Heider F (1997) Magnetic blocking temperatures of magnetite calculated with a three-dimensional micromagnetic model. *Journal of Geophysical Research* 102: 22695–22709.

Witt A, Fabian K, and Bleil U (2005) Three-dimensional micromagnetic calculations for naturally shaped magnetite: Octahedra and magnetosomes. *Earth and Planetary Science Letters* 233: 311–324.

Wright TM, Williams W, and Dunlop DJ (1997) An improved algorithm for micromagnetics. *Journal of Geophysical Research* 102: 12085–12094.

Xu S, Dunlop DJ, and Newell AJ (1994) Micromagnetic modeling of two-dimensional domain structures in magnetite. *Journal of Geophysical Research B: Solid Earth* 99: 9035–9044.

21 Properties of Rocks and Minerals – The Electrical Conductivity of Rocks, Minerals, and the Earth

J. A. Tyburczy, Arizona State University, Tempe, AZ, USA

21.1	Introduction	632
21.2	Electrical Conductivity of Materials	632
21.2.1	Point Defects and Conductivity	632
21.2.2	Effects of Temperature and Pressure	633
21.3	Electrical Conductivity of Olivine	634
21.3.1	Point Defects in Olivine	634
21.3.2	Electrical Conductivity and Anisotropy of Olivine	635
21.4	High-Pressure Studies	637
21.4.1	Anhydrous Materials	637
21.4.2	Effects of Hydrogen	638
21.5	Electrical Conductivity of Melts and Partial Melts	639
21.6	Mixing Relationships	639
21.7	Application to MT Studies	640
21.8	Summary	641
References		641

Nomenclature

f_{O2}	oxygen fugacity (Pa)
h	electron hole
k	Boltzmann's constant ($1.3807 \times 10^{-23}\,\text{J K}^{-1}$)
q	electrical charge, C
D	diffusion coefficient ($\text{m}^2\,\text{s}^{-1}$)
E_a	activation energy (kJ mol^{-1})
E_x	horizontal component of electrical field (V m^{-1})
$Fe^{\cdot}_{Mg}$	small polaron, Fe^{3+} on an Mg^{2+} site
H_y	horizontal component of magnetic field (T)
V''_{Mg}	magnesium vacancy
μ	electrical mobility ($\text{m}^2\,\text{V}^{-1}\,\text{s}^{-1}$)
ρ	electrical resistivity (ohm m)
σ	electrical conductivity (S m^{-1})
τ	period (of a sinusoidal wave) (s)
ΔV_σ	activation volume for electrical conduction ($\text{m}^3\,\text{mol}^{-1}$)

Glossary

activation energy Energy barrier for a thermally activated process.

Archie's law Relationship between bulk electrical conductivity and fluid or melt conductivity of a porous fluid- or melt-containing composite medium, of the form $\sigma_{\text{bulk}} = \sigma_{\text{melt}}\,C X_m^a$, where X_m is volume-fraction fluid or melt, and C and a are constants.

bulk conductivity Average electrical conductivity of a composite medium.

conductance Product of electrical conductivity and thickness in SI units of Siemens (S).

electrical conductivity A physical constant σ relating electrical current density to the electrical field strength. The SI unit of conductivity is Siemens per meter (S m^{-1}), where $1\,\text{S} = 1\,\text{ohm}^{-1}$, so it is also expressed as $\Omega^{-1}\,\text{m}^{-1}$.

D+ model A conductivity-depth model consisting of a sequence of delta functions in an insulating half-space.

electrical resistivity Reciprocal of electrical conductivity in SI units of (Ωm).

electromagnetic skin depth The depth at which electromagnetic fields are attenuated to $1/e$ of their surface values. This depends on electrical conductivity of the medium and the frequency of the electromagnetic waves.

impedance phase Phase difference between the electric and magnetic fields.

magnetotellurics Method of determining the electrical conductivity structure of the Earth by measuring time variations of the electrical and magnetic field at the surface.

point defects Imperfections in a crystal lattice where an atom or ion is missing or is in an irregular site.

polaron Point defect in a crystal in which an Fe^{3+} ion substitutes on an Mg^{2+} site.

vacancy Crystal lattice location where an atom or ion should be, but is missing.

21.1 Introduction

The study of the electrical properties of minerals and rocks is of broad interest in the Earth and planetary sciences. Magnetotelluric (MT) and geomagnetic depth sounding (GDS) studies yield electrical conductivity versus depth models for the Earth. Laboratory measurements of the electrical conductivity of candidate minerals under relevant conditions, with accompanying theoretical understanding, are needed to interpret the field results. In principle, electrical conductivity results are complementary to seismic results because conductivities of materials are highly sensitive to defects, minor impurities (including hydrogen in minerals), anisotropy, presence of highly conducting melts or fluids, and the distribution or texture of those melts or fluids, features that might not yield strong seismic anomalies. In this chapter, we focus on the electrical conductivity of mantle materials, including recent work incorporating hydrogen and related field studies. It is not always possible to tease out unambiguous interpretations from field results in terms of all the possible mineral physics parameters. As laboratory and field studies advance it is possible to narrow the range of interpretations. Electrical conductivity studies are attractive as a potential sensor of hydrogen in the mantle.

21.2 Electrical Conductivity of Materials

21.2.1 Point Defects and Conductivity

The electrical conductivity of minerals is governed by the point-defect chemistry. Point defects are imperfections in the crystal lattice, such as substitutions, vacancies, interstitial ions, electrons, and electron holes (electron deficiencies in the valence

band commonly termed 'holes'). Kroger–Vink notation describes the type of defect and its effective charge relative to a normal lattice site. Each defect is indicated by the symbol A_B^c, in which the main symbol A indicates the species of the defect (element/ion, electron e, hole h, or vacancy V), the subscript B indicates the type of site (normal lattice site of a particular ion or an interstitial site I), and the superscript c indicates the net effective charge of the defect relative to the normal occupancy of that lattice (where dots indicate positive relative charge, slashes indicate negative relative charge, and 'x' indicates zero relative charge). Thus for a magnesium–iron silicate mineral V_{Si}'''' is a silicon vacancy (with a charge of -4 relative to a site occupied by a silicon ion), Fe_{Mg}^x is a magnesium site containing an Fe^{2+} ion (neutral with respect to the normal site occupancy), $Fe_{Mg}^{\cdot}$ is a magnesium site containing an Fe^{3+} ion (relative charge of $+1$ relative to a site occupied by Mg^{2+}), V_{Mg}'' is a magnesium vacancy (relative charge of -2 relative to a site occupied by Mg^{2+}) and O_o^x is an oxygen site containing an O^{2-} ion.

The electrical conductivity of a material can be represented as the sum of the conductivity contributions of each charge carrier (or defect) type acting independently (in parallel):

$$\sigma = \sum \sigma_i = \sum c_i q_i \mu_i \qquad [1]$$

where c_i is the concentration of the ith type of charge carrier, q_i is its effective charge (coulomb), and μ_i is its mobility (typically in $m^2 V^{-1} s^{-1}$). Defects are present in all crystal structures and each contributes to the total conductivity. Perturbation of the concentration of one defect can influence the concentrations of the others, so the concentrations may not be completely independent. Usually only one or two types of defects dominate under a given set of thermodynamic conditions. Chemical reactions between defects can be written to express their concentrations in terms of equilibrium constants. For example, removing a positively charged ion from its normal site results in the formation of a vacancy plus an interstitial ion (formation of a Frenkel defect) and can be written as

$$A_A^x = V_A' + A_I^\bullet \qquad [2]$$

For such a reaction, the chemical equilibrium constant K is written as

$$K = \frac{[V_A'][A_I^\bullet]}{[A_A^x]} \qquad [3]$$

where the square brackets indicate site fractions. The Gibbs free energy ΔG° for this reaction is given by

$$\Delta G^\circ = -RT \ln K \qquad [4]$$

in which R is the gas constant and T is absolute temperature.

Reactions incorporating oxygen from the surroundings are important in oxide materials. For example, in MgO the reaction

$$1/2\ O_2 = V_{Mg}'' + 2h^\bullet + O_O^x \qquad [5]$$

describes the incorporation of oxygen to form a magnesium vacancy, a hole $h^\bullet$, and an oxygen ion on a normal lattice site. The equilibrium constant for this reaction is

$$K = [V_{Mg}''][h^\bullet]^2[O_O^x]/f_{O_2}^{1/2} \qquad [6]$$

Thus, the magnesium vacancy concentration in pure MgO depends on f_{O_2} (at constant temperature). Even in the presence of relatively small amounts of impurity ions, the concentrations of defects (and hence the conductivity) may vary greatly. This extreme sensitivity of defect concentration to minor and trace-element concentrations makes it very difficult to establish absolute values for the conductivities of 'pure' minerals, especially at low temperatures. In the laboratory one tries to determine which defect or defects dominate conduction under a given set of conditions by determining conductivity under varying conditions of temperature, oxygen fugacity, trace-element concentration, and/or other variables. In addition, determination of the Seebeck coefficient (thermoelectric coefficient) can identify the charge of dominant conducting species. The results are then combined with defect reactions such as [3] and [5] above and defect conservation laws such as

conservation of mass, charge neutrality, and lattice sites to determine the particular defect reaction that controls the concentration of the dominant defect and predicts the dependence on other environmental conditions (f_{O_2} in particular). Materials science reference works such as Kröger (1974) and Kingery et al. (1976) describe this process for many materials and references such as Stocker (1978), Stocker and Smyth (1978), and Hirsch and Shankland (1993) describe defect equilibria in some important Earth materials (see also Tyburczy and Fisler, 1995).

To completely specify the thermodynamic state of a defect-containing crystal, the Gibbs phase rule must be satisfied. For a three-component system such as (Mg,Fe)O, this constraint means that the temperature, the pressure, the oxygen fugacity, and the Mg:Fe ratio in the solid must all be specified. Thus, the conductivity will have the general form

$$\sigma_i = \sigma_o \exp(-E_a/kT)(Fe_{Mg}^x)^m f_{O_2}^n \qquad [7]$$

in which Fe_{Mg}^x is the fraction of Fe on Mg sites and n and m are constants from the defect reaction stoichiometry. For a four-component system such as olivine $(Mg,Fe)_2SiO_4$ an additional constraint must be specified; the most frequently specified constraint is the silica activity a_{SiO2} (or equivalently the enstatite activity a_{En}):

$$\sigma_i = \sigma_o \exp(-E_a/kT)(Fe_{Mg}^x)^m f_{O_2}^n\ a_{SiO_2}^p \qquad [8]$$

in which p is also a constant. One goal of experimental studies is to determine the exponents m, n, and p for the dependence of conductivity on compositional and environmental parameters but all the parameters in full expressions of the form of eqn [8] are still not commonly determined. These issues are discussed for olivine by Stocker (1978), Stocker and Smyth (1978), and Hirsch and Shankland (1993).

21.2.2 Effects of Temperature and Pressure

The Nernst–Einstein relation links the electrical conductivity to the diffusion coefficient of the charge-carrying species:

$$\sigma = Dcq^2/(kT) \qquad [9]$$

in which D is the self-diffusion coefficient, q is effective charge of the conducting species ($q = ze$, where e is the charge of the electron, and z is valence), c is the concentration of the conducting species, k is Boltzmann's constant, and T is temperature. The temperature dependence of electrical conductivity can arise from both the temperature dependence of concentration and thermally activated mobility. For thermally activated processes such as diffusion, $D = D_o \exp(-E_a/kT)$. Thus, the temperature dependence of the conductivity will be of the form

$$\sigma = \frac{\sigma_o}{T} \exp(-E_a/kT) \qquad [10]$$

in which σ_o is a pre-exponential constant and E_a is the activation energy. E_a is often expressed in electron volts; Boltzmann's constant k is equal to 8.617×10^{-5} eV/(atom-deg). For many studies, the $1/T$ factor in the pre-exponential term on the right-hand side of eqn [9] is not used, especially if the studies are performed over a limited temperature range, so that expressions of the form

$$\sigma = \sigma_o \exp(-E_a/kT) \qquad [11]$$

are employed. Plots of the logarithm of electrical conductivity versus $1/T$ that cover an extended range of temperatures frequently show two (or more) linear regions or broadly curving regions. The variations in slope can be caused by transition from one dominant conducting species to another or by a transition from extrinsic (impurity dominated) to intrinsic conductivity. Each region is then described by an expression of the form of eqn [11]. However, more precise, and mechanistically more meaningful parameters are derived if the entire data set is simultaneously fit to a single expression of the form

$$\sigma = \sigma_{o1} \exp(-E_{a1}/kT) + \sigma_{o2} \exp(-E_{a2}/kT) \qquad [12]$$

where subscripts 1 and 2 refer to the different mechanisms. Similarly, experiments covering a sufficiently wide range of f_{O2} can express multiple mechanisms and be fit by expressions of the form

$$\sigma = \sigma_{o1} f_{O_2}{}^{n1} + \sigma_{o2} f_{O_2}{}^{n2} \qquad [13]$$

The effect of pressure on conductivity can be characterized by the inclusion of an activation volume term ΔV_σ so that

$$\sigma = \sigma_o \exp(-[\Delta U_\sigma + P\Delta V_\sigma]/kT) \qquad [14]$$

in which ΔU_σ is the activation internal energy and P is pressure. The activation volume can be interpreted to indicate the volume of the mobile species, but this is not always the case. Commonly measurements at elevated pressures are parametrized isobarically as functions of temperature using eqn [10] or [11].

21.3 Electrical Conductivity of Olivine

21.3.1 Point Defects in Olivine

Olivine $(Mg_{0.9}Fe_{0.1})_2SiO_4$ is the dominant mineral between the bottom of the crust and the 410 km seismic discontinuity, comprising 40–60% by volume of the mantle in this depth range (Ringwood, 1975). Because of this abundance it is one of the most frequently studied mantle minerals, but there are still nuances in its conductivity and point-defect behavior that are not fully understood. We study olivine in detail because we need a complete understanding of the transport in order to extrapolate its behavior to conditions that have not been studied in the lab and because its behavior can inform us about other minerals.

The dominant charge-carrying defects in olivine are small polarons (Fe^{3+} on the Mg^{2+} lattice site $Fe^{\bullet}_{Mg}$), magnesium vacancies V''_{Mg}, and electrons e'. There is general agreement that small polarons dominate at high f_{O2} and lower temperatures, but there are disagreements at high temperature and lower f_{O2}. The relationship describing the major defect populations and their f_{O2} dependences in olivine (Stocker and Smyth, 1978; Schock et al., 1989) is

$$8Fe^x_{Mg} + 2O_2 \Leftrightarrow 2V''_{Mg} + V''''_{Si} + 4O^x_o + 8Fe^{\bullet}_{Mg} \qquad [15]$$

Studies on olivine have been performed on single crystals and polycrystalline aggregates, and with self-buffering or pyroxene buffering of silica. If the cationic ratio Fe:Mg:Si is held fixed (i.e., for an experiment in which the olivine can exchange only oxygen with the environment and is not chemically in contact with other solid phases, termed the 'self-buffered' case) and at high f_{O2}, the charge balance condition $4[V''_{Mg}] = [Fe^{\bullet}_{Mg}] = 8[V''''_{Si}]$ pertains (Stocker, 1978) and evaluation of the equilibrium constant yields

$$[Fe^{\bullet}_{Mg}] = a_{Fe} f_{O_2}{}^{2/11} \text{ and } [V''_{Mg}] = a_{Mg} f_{O_2}{}^{2/11} \qquad [16]$$

where the a's are constants; that is, the f_{O2} dependences are identical. Potentially one could distinguish the individual contributions of $Fe^{\bullet}_{Mg}$ and V''_{Mg} because they have individual concentration and mobility dependences on T and f_{O2}, and this has been attempted in some instances.

For the pyroxene-buffered case, the f_{O2} exponents $n = 1/6$ for both polarons and magnesium vacancies at high f_{O2} (Stocker and Smyth, 1978). At low f_{O2} other defect interactions come into play that result in different

f_{O2} dependencies for these and other defects (Stocker and Smyth, 1978; Hirsch and Shankland, 1993).

21.3.2 Electrical Conductivity and Anisotropy of Olivine

Seminal studies by Duba *et al.* (1974) and Schock *et al.* (1989) paved the way for more recent studies of olivine electrical conductivity. **Figure 1** shows log of conductivity versus f_{O2} for single-crystal San Carlos olivine in the self-buffered case at 1200°C. Three different sets of measurements are shown (Schock *et al.*, 1989; Wanamaker and Duba, 1993; Du Frane *et al.*, 2005). Agreement is good between these independent studies. The conductivity is greatest by up to a factor of 2 in the [001] orientation, and approximately equal in the [100] and [010] directions. Conductivity increases with increasing oxygen fugacity, but the curves are not linear and the slope increases (that is f_{O2} exponent *n* increases) with increasing f_{O2}. At least two species or mechanisms are operative. Different researchers propose different models to explain these variations. Wanamaker and Duba (1993) conclude that $Fe_{Mg}^{\cdot}$ dominates at the highest f_{O2}, with contributions from electrons e' (with an f_{O2} exponent of $-1/6$) of greater importance at low f_{O2}. Du Frane *et al.* fit their data using a nondifferentiated model at high f_{O2} (i.e., no distinction is made between $Fe_{Mg}^{\cdot}$ and V_{Mg}'' contributions (because they have the same value of $n = +2/11$) with a contribution from a component with an f_{O2}

exponent of zero at low f_{O2}, possibly from electrons or a background (extrinsic impurity) component. Each model fits the data over the f_{O2} range it was collected, but each extrapolates to different values at the lowest f_{O2}. **Figure 2** shows the fits to these data sets with the two components for each fit.

For the pyroxene-buffered case, point-defect calculations indicate f_{O2} exponents of $+1/6$ for both $Fe_{Mg}^{\cdot}$ and V_{Mg}'' at high f_{O2} (Stocker and Smyth, 1978). At lower f_{O2}, the situation is more complex and f_{O2} exponents for $[V_{Mg}'']$ may be different. **Figure 3** shows fits to electrical conductivity data for pyroxene-buffered olivine. The data and model of Wanamaker and Duba (1993) are for single-crystal olivine in the [100] direction. The model of Constable (2006) based on earlier results of Constable and Roberts (1997) is for polycrystalline olivine—this model is the successor to the SO2 model of Constable *et al.* (1992). Two different approaches are used in the fitting. Wanamaker and Duba fit the high f_{O2} region with a $Fe_{Mg}^{\cdot}$ polaron with an exponent of $1/6$, and the low f_{O2} region with a model with an f_{O2} exponent of zero that they ascribe to magnesium vacancies V_{Mg}'' in a different charge neutrality regime. Constable and Roberts (1997) and Constable (2006) fit the data with a model in which the f_{O2} exponents for both the $Fe_{Mg}^{\cdot}$ and V_{Mg}'' concentrations are $+1/6$, but which includes a *T*-dependent empirical constant in the concentration term (of eqn [1]) that produces the flattening at low f_{O2}. The Constable and Roberts (1997) and Constable (2006)

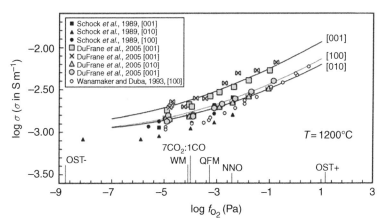

Figure 1 Olivine single-crystal conductivity in different orientations as a function of f_{O2} for self-buffered experiments at 1200°C. Large squares, triangles, circles, and solid lines; data and fits from Du Frane *et al.* (2005). Small filled squares, triangles, and circles; Schock *et al.* (1989). Open circles; Wanamaker and Duba (1993). Schock and Wanamaker data are normalized to composition Fo89.1 using model of Hirsch *et al.* (1993). Adapted from Du Frane WL, Roberts JJ, Toffelmier DA, and Tyburczy JA (2005) Anisotropy of electrical conductivity in dry olivine. *Geophysical Research Letters* 32: L24315 (doi:10.1029.2005GL023879).

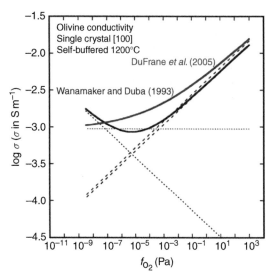

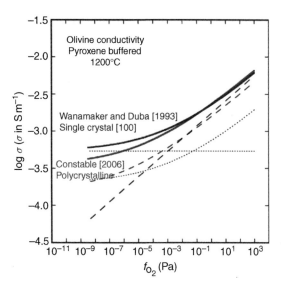

Figure 2 Models of olivine single-crystal self-buffered electrical conductivity at 1200°C as a function of f_{O2} in the [100] orientation. Blue lines indicate models of Wanamaker and Duba (1993). Red lines indicate Du Frane et al. (2005). Solid lines are the total conductivity. Long dashed lines represent the $Fe^{•}_{Mg}$ polaron conductivity (Wanamaker and Duba, 1993) or the sum of polaron plus magnesium vacancy conductivity (Du Frane et al., 2005). Dotted lines represent electronic conduction (Wanamaker and Duba, 1993) or minimum (f_{O2} independent) conductivity term.

Figure 3 Models of pyroxene-buffered olivine electrical conductivity at 1200°C as a function of f_{O2}. Blue lines indicate model of Wanamaker and Duba (1993) for single-crystal olivine in [100] orientation. Red lines indicate model of Constable (2006), see also Constable and Roberts (1997), for polycrystalline olivine. Solid lines indicate total conductivity. Long dashed lines indicate $Fe^{•}_{Mg}$ polaron conductivity. Short dashed lines indicate magnesium vacancy conductivity (Constable, 2006) or f_{O2} independent term.

model is derived from and fits both electrical conductivity and thermopower data. The thermopower data indicate a change in sign of the dominant species from positive to negative above about 1300°C, consistent with V''_{Mg}-dominated conduction. This change in the sign of the thermopower was first reported by Schock et al. (1989). Note that pyroxene-buffered olivine conductivity is lower by approximately 0.2–0.3 log units than the self-buffered case (Wanamaker and Duba, 1993). Concentrations of magnesium vacancies are higher in the pyroxene-buffered case because reactions such as

$$4MeSiO_3 = 3Me_2SiO_4 + 2V''_{Mg} + Si_i^{••••} \quad [17]$$

where Me stands for Fe^{2+} or Mg^{2+}, cause increases in the number of magnesium vacancies, which causes lower concentrations of polarons and electrons through reaction [14].

Figure 4 compares olivine log σ versus $1/T$ along the quartz-fayalite-magnetite (QFM) f_{O2} buffer for the models discussed above along with

the zero-pressure conductivities of olivine of Xu et al. (1998a) (extrapolated from high-pressure measurements and adjusted for differences in experimental f_{O2}, see below). The geometric mean of the Du Frane et al. model has a higher conductivity than the Constable and Roberts model, owing at least in part to the use of iron-doped platinum electrodes in these experiments. The high-pressure data of Xu et al. are collected along the Mo–MoO$_2$ buffer using molybdenum electrodes. Also shown are the conductivities corresponding to the individual mechanisms. At temperatures greater than about 1300°C, the Constable (2006) model shows the increasing importance of magnesium vacancy conduction. The Du Frane et al. (2005) model indicates the dominance of the undifferentiated magnesium vacancy plus polaron term above about 1150°C over the f_{O2}-independent term. Comparison of these studies indicates good experimental agreement and general agreement in the higher f_{O2} ranges. Differences occur in the models developed to describe the low f_{O2} behavior. Oxygen fugacity

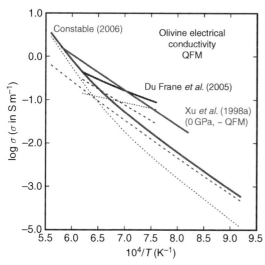

Figure 4 Electrical conductivity versus $1/T$ for several olivine models. Red lines indicate model of Constable (2006) along QFM buffer – total conductivity (red solid line), polaron conductivity (red long dashed lined), and magnesium vacancy conductivity (red dotted line). Blue lines indicate geometric mean model of Du Frane et al. (2005) along QFM buffer – total conductivity (blue solid line), polaron plus magnesium vacancy conductivity (blue long dashed line), f_{O2} independent term (blue dotted line). Green line indicates high-pressure data of Xu et al. (1998a, 2000a) extrapolated to zero pressure and with factor of 0.25 log unit added to account for lower experimental f_{O2}.

conditions in the Earth's mantle are generally thought to be in the range of the QFM buffer (see **Figure 1**). For these conditions the models do not differ greatly.

The dependence of olivine electrical conductivity on Fe content was examined by Hirsch et al. (1993). They found that conductivity varies as iron site fraction to the 1.81 power $[X_{Fe}]^{1.81}$, according to eqn [7] or [8].

Grain-boundary electrical conductivity has been considered by Roberts and Tyburczy (1991, 1993) and by ten Grotenhuis et al. (2005). Roberts and Tyburczy examined pressed sintered compacts of natural olivine with a ~45 μm grain size and concluded that grain-boundary and grain-interior conductivities add in series, that is, that grain boundaries are not highly conducting paths. ten Grotenhuis et al. (2005) examined very fine grained (1–5 μm) 95% forsterite–5% enstatite compacts, and concluded that the bulk conductivity is controlled by grain-boundary transport, that is, that the grain boundaries are highly conducting paths. The reasons for the differences between these results remain to be resolved.

21.4 High-Pressure Studies

21.4.1 Anhydrous Materials

In the last decade, significant advances have been made in the determination of high-pressure electrical properties of mantle minerals. Xu et al. (1998a) determined the electrical conductivity of $Mg_{0.9}Fe_{0.1}SiO_4$ olivine, wadsleyite, and ringwoodite at pressures in their stability field under oxygen fugacity conditions similar to those of the mantle (Mo-MoO_2 buffer) using a multiple-anvil high-pressure device. Wadsleyite and ringwoodite exhibit conductivities 1–2 orders of magnitude greater than olivine. Pyroxene and higher-pressure polymorphs generally have conductivities lower that those of wadsleyite and ringwoodite, and are comparable to olivine (Xu and Shankland, 1999). Perovskite, especially perovskite containing Al^{3+}, and magnesiowüstite have conductivities an order of magnitude or so higher than those of the transition-zone phases (Xu et al., 1998b; Xu et al., 2000b). Magnesiowüstite (with Fe/Fe+Mg $\sim$ 0.1) conductivity is greater than perovskite conductivity with indications of a transition to a high-temperature mechanism at temperatures greater than about 1000 K at 10 GPa (Dobson and Brodholt, 2000; Dobson et al., 1997; Wood and Nell, 1991), although Xu et al. (2000b) did not observe such a variation. Activation volumes for conduction are relatively small for these materials. Olivine has the largest currently measured activation volume of $0.68 \, cm^3 mol^{-1}$ (Xu et al., 1998a; Xu et al., 2000a). **Figure 5** shows a summary of the conductivities of these materials as a function of inverse temperature, all measured using the same techniques (Xu et al., 2000b). It is inferred that the dominant conduction mechanism in these phases is generally via $Fe_{Mg}^{\cdot}$ polarons, consistent with olivine conduction mechanisms. Mössbauer measurements of Fe^{3+} in these phases indicate much higher concentrations of Fe^{3+} than in olivine (O'Neill et al., 1993). In addition, the presence of Al^{3+} in silicate perovskite increases the amount of Fe^{3+} by a factor of 3.5 relative to Al^{3+}-free perovskite (Xu et al., 1998b). In magnesiowüstite at high temperature, a large-polaron mechanism in which holes in the valence band dominate conduction has been described (Dobson and Brodholt, 2000). Creation of a self-consistent data set permits calculation of mineral physics-based MT forward models that explicitly contain conductivity changes corresponding to the mineralogy changes at the major seismic discontinuities – see further discussion below.

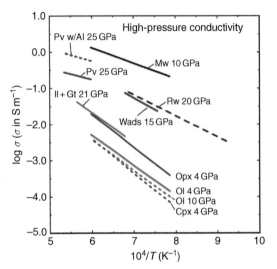

Figure 5 Electrical conductivity versus reciprocal temperature results for mantle minerals at high pressure and f_{O2} representative of the mantle. Data are from multiple anvil experiments of Xu and co-workers summarized by Xu et al. (2000). Symbols: Pv; perovskite; Mw; magnesiowüstite; Rw; ringwoodite; Wads; wadsleyite; Il; ilmenite (akimotoite); Gt; garnet; Opx; orthopyroxene; Ol; olivine.

21.4.2 Effects of Hydrogen

Karato (1990) calculated the electrical conductivity of hydrogen-bearing (also referred to as 'hydrous-') olivine using hydrogen diffusion data (Kohlstedt and Mackwell, 1998) combined with the Nernst–Einstein equation (eqn [9]). This approach assumes that all the hydrogen present is free to contribute to the conductivity. Hydrogen diffusion in olivine is highly anisotropic so this approach predicts that the orientation dependence of hydrous olivine conductivity would be $\sigma_{[100]} \sim 20 \times \sigma_{[010]} \sim 40 \times \sigma_{[001]}$ (Karato, 1990; Constable, 1993), distinctly different than for dry olivine. Olivine sheared at low stress tends to line up with the [100] orientation in the direction of shear. This explanation was used to infer that hydrogen-bearing oriented olivine was responsible for upper-mantle electrical anisotropy near mid-ocean ridges (Evans et al., 2005) and electrical conductivity enhancement beneath the Pacific Ocean lithosphere (Lizarralde et al., 1995). Recently, high-pressure experimental techniques have been developed to measure conductivity of hydrogen-bearing mantle minerals. Huang et al. (2005) measured conductivity of polycrystalline wadsleyite and ringwoodite (silica buffered, that is with a small amount of pyroxene added, and with molybdenum electrodes to buffer f_{O2}) with water contents up to about 1 wt.% H_2O at

14–16 GPa. Conductivity is strongly enhanced in the presence of hydrogen (water), and is described with an expression of the form

$$\sigma = A\, C_W^r \exp\left(-H^*/RT\right) \qquad [18]$$

where A and r are constants, C_W is water concentration, and H^* is the activation energy. For C_W expressed in weight per cent water, they determined that for wadsleyite $A = 380\,(+170,\ -120)\,\mathrm{S\,m^{-1}}$, $r = 0.66 \pm 0.05$, and $H^* = 88 \pm 3\,\mathrm{kJ\,mol^{-1}}$ and that for ringwoodite $A = 4070\,(+1050,\ -840)\,\mathrm{S\,m^{-1}}$, $r = 0.69 \pm 0.03$, and $H^* = 104 \pm 2\,\mathrm{kJ\,mol^{-1}}$. They conclude that although the most abundant site for hydrogen is as a neutral defect $(2H)_M^X$ corresponding to two protons trapped at a metal site (Kohlstedt et al., 1996; Kohlstedt and Mackwell, 1998), the transport is governed by a concentration of free protons lower than the total concentration of hydrogen. The defect reaction

$$(2H)_M^X = (H)_M' + H^\bullet \qquad [19]$$

predicts the exponent $r = 0.75$, close to the measured values. Thus, the Nernst–Einstein equation (for which $r = 1$) does not apply directly. Analogous results are found in measurements on polycrystalline olivine (Wang et al., 2006). Conductivity at $P = 4\,\mathrm{GPa}$ and $T = 873$–$1273\,\mathrm{K}$ is enhanced by several orders of magnitude over that of dry olivine, but the activation energy for conductivity in the hydrogen-bearing samples is much lower than that of dry olivine, $87 \pm 5\,\mathrm{kJ\,mol^{-1}}$ (and $\log A = 3.0 \pm 0.4$ (for A in $\mathrm{S\,m^{-1}}$) and $r = 0.62 \pm 0.15$) for 'wet' versus $\sim 154\,\mathrm{kJ\,mol^{-1}}$ for dry olivine. Activation energies for hydrogen diffusion are in the range 110–180 $\mathrm{kJ\,mol^{-1}}$ depending on orientation (Kohlstedt and Mackwell, 1998). Hier-Majumder et al. (2005) measured Mg–Fe interdiffusion in hydrous olivine and found a small enhancement in the presence of water, but not sufficient to yield a conductivity increase of this magnitude. Yoshino et al. (2006) measured the electrical conductivity of hydrogen-bearing single-crystal olivine in different orientations at 3 GPa pressure and T between 500 and 1000 K. Conductivity is enhanced by about two orders of magnitude over that of anhydrous olivine and anisotropy is consistent with diffusion directions with $\sigma_{[100]} > \sigma_{[010]} > \sigma_{[001]}$ at constant hydrogen content. Activation energies are $E_{[100]} = 0.73\,\mathrm{eV}$ $(70\,\mathrm{kJ\,mol^{-1}})$, $E_{[010]} = 0.93\,\mathrm{eV}$ $(90\,\mathrm{kJ\,mol^{-1}})$, $E_{[001]} = 0.87\,\mathrm{eV}$ $(84\,\mathrm{kJ\,mol^{-1}})$. However, extrapolated to slightly higher temperatures up to 1273 K, the conductivity of hydrogen-bearing olivine becomes nearly isotropic, whereas at the same temperature hydrogen diffusivity is strongly anisotropic. They interpret this difference to indicate different mechanisms for hydrogen mobility in diffusion and conductivity.

21.5 Electrical Conductivity of Melts and Partial Melts

Naturally occurring silicate melt conductivity has been examined at elevated pressure in dry systems to 2.5 GPa (Tyburczy and Waff, 1983, 1985) and in a hydrous silicic melt to 0.4 GPa (Gaillard, 2004). Conductivity of a basaltic melt decreases with increasing pressure up to about 0.8 GPa, then is independent of pressure to 2.5 GPa (Tyburczy and Waff, 1983). Dry silicic melts have greater pressure dependence, with activation volumes on the order of a few $cm^3 \, mol^{-1}$ at 2.5 GPa (Tyburczy and Waff, 1985). Gaillard (2004) examined conductivity of an obsidian melt with up to 3 wt.% water and found an increase of about half order of magnitude or less over the dry obsidian at 0.2 GPa corresponding to an activation volume of about 20 $cm^3 \, mol^{-1}$.

Waff (1974) discussed the importance of the textural distribution of melt on the bulk electrical properties of a partially molten rock and the importance of textural equilibration (Waff and Bulau, 1979). Studies of 'texturally equilibrated partially molten systems' include those of Roberts and Tyburczy (1999) and ten Grotenhuis et al. (2005). Roberts and Tyburczy (1999) showed that for an Fo_{80}-melt system that the melt was highly interconnected at low melt fractions (0–5 volume %) and that parallel-type bulk conduction models (Hashin–Shrtikman upper bound, simple parallel conductors, or Archie's law) describe the data (see below). Archie's law ($\sigma_{bulk}/\sigma_{melt} = CX^m$, where C and m are constants, and X is volume-fraction melt) constants $C = 0.73$ and $m = 0.98$ were determined. They pointed out that melt composition changes significantly with changes in melt fraction at very low melt fractions, and that these changes may need to be considered when modeling low-melt-fraction systems. ten Grotenhuis et al. (2005) examined an iron-free olivine–enstatite melt system, and observed a change from nearly dry grain faces (with melt mostly in tubules or triple junctions) at 1 volume% melt to mostly wetted grain faces at a melt fraction of 10 volume%. The change was discontinuous, with the most abrupt change occurring at around 2 volume% melt. They determined Archie's law constants of $C = 1.47$ and $m = 1.30$.

21.6 Mixing Relationships

The bulk conductivity of a rock consisting of several minerals depends on the textural distribution of the minerals, which is generally not well known. A variety of relations have been employed. Series and parallel solutions yield the maximum and minimum values, respectively, of a mixture of materials

$$\sigma_s^{-1} = x_1/\sigma_1 + x_2/\sigma_2 \qquad [20]$$

$$\sigma_p = x_1\sigma_1 + x_2\sigma_2 \qquad [21]$$

where σ_s is the series conductivity, σ_p is the parallel conductivity, σ_1 and σ_2 are the conductivities of the constituent phases, and x_1 and x_2 are the volume fractions. The geometric mean conductivity σ_{GM} is given by

$$\sigma_{GM} = \sigma_1^{x1} \sigma_2^{x2} \qquad [22]$$

Archie's law (Archie, 1942) is often used for fluid- or melt-rock mixtures:

$$\sigma_{bulk}/\sigma_{melt} = Cx^m \qquad [23]$$

where σ_{bulk} is the conductivity of the mixture, C and m are constants, and x is volume-fraction melt or fluid. For very low porosities, a modified expression is used (Hermance, 1979):

$$\sigma_{bulk} = \sigma_{rock} + (\sigma_{fluid} - \sigma_{rock})x^m \qquad [24]$$

The Hashin–Shtrikman bounds describe the narrowest upper and lower bounds in the absence of geometrical information (Hashin and Shtrikman, 1962)

$$\sigma_{HS+} = \sigma_1 + x_2[(\sigma_2 - \sigma_1)^{-1} + x_1/(3\sigma_1)]^{-1} \qquad [25]$$

$$\sigma_{HS-} = \sigma_2 + x_1[(\sigma_1 - \sigma_2)^{-1} + x_2/(3\sigma_2)]^{-1} \qquad [26]$$

where $\sigma_1 > \sigma_2$, and σ_{HS+} and σ_{HS-} correspond to the upper and lower bounds, respectively. The effective medium bounds (Landauer, 1952) lie between the HS upper and lower bounds:

$$\sigma_{EM} = 1/4\{(3x_1 - 1)\sigma_1 + (3x_2 - 1)\sigma_2 + [\{(3x_1 - 1)\sigma_1 + (3x_2 - 1)\sigma_2\}^2 + 8\sigma_1\sigma_2]^{1/2}\} \qquad [27]$$

These and other relationships have been applied to the conductivities of multiphase materials in the mantle (e.g., Xu et al., 2000b) including partial melts.

They yield the widest ranges when the conductivities of the two phases in question are greatly different, for example in the transition zone (Xu *et al.*, 2000b) or for partial melts.

21.7 Application to MT Studies

We discuss here a small number of MT field studies as representative of those for which these laboratory results have been influential in modeling or interpreting mantle conditions. In MT studies, orthogonal components of electrical E_x and magnetic H_y fields as a function of period τ are measured at the Earth's surface. From these the apparent resistivity $\rho(\tau)$ and phase $\phi(\tau)$ are determined according to

$$\rho(\tau) = (\tau/2\pi\mu)(E_x/H_y)^2 \text{ and } \phi(\tau) = \tan^{-1}(E_x/H_y) \quad [28]$$

where μ is the magnetic permeability (Cagniard (1953); see Simpson and Bahr (2005) for a recent discussion of MT methods). Conductivity-depth models are fit to or derived from such field results

using smooth or discontinuous models. The D+ inversions provide the best-fitting model to the data $\rho(\tau)$ and $\phi(\tau)$ that can be achieved; they consist of a series of delta-functions in conductance as a function of depth (Parker, 1980). Occam's Razor inversions yield the smoothest possible conductivity-depth models to the data (Constable *et al.*, 1987). Neither of these approaches corresponds to a real physical Earth. The existing framework of experimental results on electrical conductivity of Earth materials under mantle conditions allows for modeling and interpretation of electromagnetic field results in terms of realistic mineralogies reflecting known seismic discontinuities and mantle conditions.

Xu *et al.* (2000b) showed that using their nominally anhydrous laboratory data, a mineral physics-based conductivity-depth model for the upper mantle and transition zone provides a very good fit (via forward modeling) to resistivity $\rho(\tau)$ and phase $\phi(\tau)$ for a European MT data set (Olsen, 1999) (**Figure 6**). This agreement indicates that the laboratory data provide a robust framework for interpreting MT

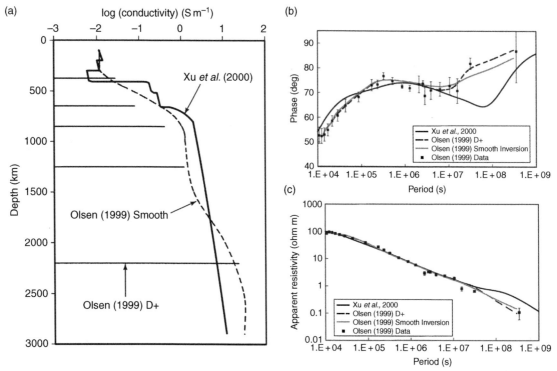

Figure 6 (a) Electrical conductivity versus depth profile for Europe from Olsen (1999) with D+ and smooth inversions and mineral physics-based model of Xu *et al.* (2000). Smooth inversion is from Tarits *et al.* (2004). (b) Phase and (c) apparent resistivity as a function of frequency data and the models. Adapted from Xu YS, Shankland TJ, and Poe BT (2000b) Laboratory-based electrical conductivity in the earth's mantle. *Journal of Geophysical Research* 105: 27865–27875; and Tarits P, Hautot S, and Perrier F (2004) Water in the mantle: Results from electrical conductivity beneath the French Alps. *Geophysical Research Letters* 31: L06612 (doi:10.1029/2003GL019277).

results. Note, however, that recent experimental results on water-containing minerals indicate that some of the earlier nominally anhydrous experiments contained significant amounts of hydrogen (Huang et al., 2005). Other models based on MT studies include effects of water in mantle and transition-zone minerals (Lizarralde et al., 1995; Tarits et al., 2004; Simpson and Tommasi, 2005; Karato, 2006). There are significant regional differences in transition-zone conductivity that may be attributable to variations in water content (Tarits et al., 2004; Karato, 2006).

Just off of the southern East Pacific rise, Evans et al. (2005) observed strong electrical conductivity anisotropy at a depth of 100 km, with the ridge perpendicular (plate motion parallel) value approximately a factor of 10 greater than the ridge parallel value. They interpreted this result using the earlier diffusion-based model of hydrogen-enhanced olivine conductivity to indicate hydrous (10^3 H/10^6 Si) shear-oriented olivine ([100] parallel to flow direction, perpendicular to the ridge). However, Yoshino et al. (2006) point out that their the high-pressure conductivity measurements indicate a much smaller amount of electrical conductivity anisotropy in hydrous olivine, suggesting that partial melt may the cause of the conductivity anisotropy near the ridge.

21.8 Summary

Many aspects of the electrical conductivity of mantle minerals and its dependence on environmental factors are well measured and understood. Laboratory measurement of electrical conductivity of mantle and transition-zone minerals has advanced to the point where interesting and provocative interpretations of mantle MT profiles in terms of mineralogy, temperature, composition, and water content are possible. The coming years promise significant new revelations from the study of electrical properties of Earth materials and the application to Earth interior issues.

Acknowledgments

The author thanks Susan Selkirk for the invaluable help in preparing the figures. This work was supported by NSF grant EAR0073987.

References

Archie GE (1942) The electrical resistivity log as an aid to determining some reservoirs characteristics. *Transactions of AIME* 146: 389–409.

Bulau JR and Waff HS (1979) Mechanical and thermodynamic constraints on fluid distribution in partial melts. *Journal of Geophysical Research* 84: 6102–6109.

Cagniard L (1953) Basic theory of the magneto-telluric method of geophysical prospecting. *Geophysics* 18: 605–635.

Constable S (1993) Conduction by mantle hydrogen. *Nature* 362: 704.

Constable S (2006) SEO3: A new model of olivine electrical conductivity. *Geophysical Journal International* 166: 435–437.

Constable SC, Parker RL, and Constable CG (1987) Occam's inversion – A practical algorithm for generating smooth models from electromagnetic sounding data. *Geophysics* 52: 289–300.

Constable S and Roberts JJ (1997) Simultaneous modeling of thermopower and electrical conduction in olivine. *Physics and Chemistry of Minerals* 24: 319–325.

Constable S, Shankland TJ, and Duba A (1992) The electrical conductivity of an isotropic olivine mantle. *Journal of Geophysical Research* 97: 3397–3404.

Dobson DP and Brodholt JP (2000) The electrical conductivity of lower mantle phase magnesiowüstite at high temperatures and pressures. *Journal of Geophysical Research* 105: 531–538.

Dobson DP, Richmond NC, and Brodholt JP (1997) A high-temperature electrical conduction mechanism in the lower mantle phase $(Mg,Fe)_{1-x}O$. *Science* 275: 1779–1781.

Du Frane WL, Roberts JJ, Toffelmier DA, and Tyburczy JA (2005) Anisotropy of electrical conductivity in dry olivine. *Geophysical Research Letters* 32: L24315 (doi:10.1029.2005GL023879).

Duba A, Heard HC, and Schock RN (1974) Electrical-conductivity of olivine at high-pressure and under controlled oxygen fugacity. *Journal of Geophysical Research* 79: 1667–1673.

Evans RL, Hirth G, Baba K, Forsyth D, Chave A, and Mackie R (2005) Geophysical evidence from the MELT area for compositional controls on oceanic plates. *Science* 437: 249–252.

Gaillard F (2004) Laboratory measurements of electrical conductivity of hydrous and dry silicic melts under pressure. *Earth and Planetary Science Letters* 218: 215–228.

Hashin Z and Shtrikman S (1962) A variational approach to the theory of the effective magnetic permeability of multiphase materials. *Journal of Applied Physics* 33: 3125–3131.

Hermance JF (1979) The electrical conductivity of materials containing partial melt: A simple model of Archie's law. *Geophysical Researach Letters* 6: 613–616.

Hier-Majumder S, Anderson IM, and Kohlstedt DL (2005) Influence of protons on Fe–Mg interdiffusion in olivine. *Journal of Geophysical Research* 110: B02202 (doi:10.1029/2004JB003292).

Hirsch LM and Shankland TJ (1993) Quantitative olivine-defect chemical model: Insights on electrical conduction, diffusion, and the role of Fe content. *Geophysical Journal International* 114: 21–35.

Hirsch LM, Shankland TJ, and Duba AG (1993) Electrical conduction and polaron mobility in Fe-bearing olivine. *Geophysical Journal International* 114: 36–44.

Huang XG, Xu YS, and Karato SI (2005) Water content in the transition zone from electrical conductivity of wadsleyite and ringwoodite. *Nature* 434: 746–749.

Karato S (1990) The role of hydrogen in the electrical-conductivity of the upper mantle. *Nature* 347: 272–273.

Karato S (2006) Remote sensing of hydrogen in Earth's mantle. *Reviews in Mineralogy and Geochemistry* 62: 343–375.

Kingery WD, Bowen HK, and Uhlmann DR (1976) *Introduction to Ceramics*. New York: Wiley.

Kohlstedt DL, Keppler H, and Rubie DC (1996) Solubility of water in the alpha, beta and gamma phases of $(Mg,Fe)_2SiO_4$. *Contributions to Mineralogy and Petrology* 123: 345–357.

Kohlstedt DL and Mackwell SJ (1998) Diffusion of hydrogen and intrinsic point defects in olivine. *Zeitschrift für Physikalische Chemie* 207: 147–162.

Kröger FA (1974) *Chemistry of Imperfect Crystals*. Amsterdam: North-Holland.

Landauer R (1952) The electrical resistance of binary metallic mixtures. *Journal of Applied Physics* 23: 779–784.

Lizarralde D, Chave A, Hirth G, and Schultz A (1995) Northeastern pacific mantle conductivity profile from long-period magnetotelluric sounding using Hawaii-to-California submarine cable data. *Journal of Geophysical Research* 100: 17837–17854.

Olsen N (1999) Long period (30 days–1 year) electromagnetic sounding and the electrical conductivity of the lower mantle beneath Europe. *Geophysical Journal International* 138: 179–187.

O'Neill HSC, McCammon CA, Canil D, Rubie DC, Ross CR, II, and Deifert HF (1993) Mössbauer spectroscopy of mantle transition zone phases and determination of minimum Fe^{+3} content. *American Mineralogist* 78: 456–460.

Parker RL (1980) The inverse problem of electromagnetic induction – existence and construction of solutions based on incomplete data. *Journal of Geophysical Research* 85: 4421–4428.

Ringwood AE (1975) *Composition and Petrology of the Earth's Mantle*. New York: McGraw-Hill.

Roberts JJ and Tyburczy JA (1991) Frequency-dependent electrical-properties of polycrystalline olivine compacts. *Journal of Geophysical Research* 96: 16205–16222.

Roberts JJ and Tyburczy JA (1993) Impedance spectroscopy of single and polycrystalline olivine: Evidence for grain boundary transport. *Physics and Chemistry of Minerals* 20: 19–26.

Roberts JJ and Tyburczy JA (1999) Partial-melt electrical conductivity: Influence of melt composition. *Journal of Geophysical Research* 104: 7055–7065.

Schock RN, Duba AG, and Shankland TJ (1989) Electrical-conduction in olivine. *Journal of Geophysical Research* 94: 5829–5839.

Simpson F and Tommasi A (2005) Hydrogen diffusivity and electrical anisotropy of a peridotite mantle. *Geophysical Journal International* 160: 1092–1102.

Simpson F and Bahr K (2005) *Practical Magnetotellurics*. Cambridge: Cambridge University Press.

Stocker R (1978) Influence of oxygen pressure on defect concentrations in olivine with a fixed cationic ratio. *Physics of the Earth and Planetary Interiors* 17: 118–129.

Stocker RL and Smyth DM (1978) Effect of enstatite activity and oxygen partial pressure on point-defect chemistry of olivine. *Physics of the Earth and Planetary Interiors* 16: 145–156.

Tarits P, Hautot S, and Perrier F (2004) Water in the mantle: Results from electrical conductivity beneath the French Alps. *Geophysical Research Letters* 31: L06612 (doi:10.1029/2003GL019277).

ten Grotenhuis SM, Drury MR, Spiers CJ, and Peach CJ (2005) Melt distribution in olivine rocks based on electrical conductivity measurements. *Journal of Geophysical Research* 110: B12201 (doi:10.1029/2004JB003462).

Tyburczy JA and Waff HS (1983) Electrical-conductivity of molten basalt and andesite to 25 kilobars pressure – Geophysical significance and implications for charge transport and melt structure. *Journal of Geophysical Research* 88: 2413–2430.

Tyburczy JA and Waff HS (1985) High pressure electrical conductivity in naturally occurring silicate liquids. In: Schock RN (ed.) *Point Defects in Minerals, Geophys. Monogr. Ser.*, vol. 31, pp. 78–87. Washington, DC: American Geophysical Union.

Tyburczy JA and Fisler DK (1995) Electrical properties of minerals and melts. In: Ahrens TJ (ed.) *Mineral Physics and Crystallography: A Handbook of Physical Constants*, pp. 185–208. Washington, DC: American Geophysical Union.

Waff HS (1974) Theoretical considerations of electrical conductivity in a partially molten mantle and implications for geothermometry. *Journal of Geophysical Research* 79: 4003–4010.

Waff HS and Bulau JR (1979) Equilibrium fluid distribution in an ultramafic partial melt under hydrostatic stress conditions. *Journal of Geophysical Research* 84: 6109–6114.

Wanamaker BJ and Duba AG (1993) Electrical-conductivity of San Carlos olivine along [100] under oxygen-buffered and pyroxene-buffered conditions and implications for defect equilibria. *Journal of Geophysical Research* 98: 489–500.

Wang D, Mookherjee M, Xu Y, and Karato S-I (2006) The effect of water on the electrical conductivity of olivine. *Nature* 443: 977–980.

Wood BJ and Nell J (1991) High-temperature electrical conductivity of the lower-mantle phase (Mg,Fe)O. *Nature* 351: 309–311.

Xu YS, Poe BT, Shankland TJ, and Rubie DC (1998a) Electrical conductivity of olivine, wadsleyite, and ringwoodite under upper-mantle conditions. *Science* 280: 1415–1418.

Xu YS, McCammon CA, and Poe BT (1998b) The effect of alumina on the electrical conductivity of silicate perovskite. *Science* 282: 922–924.

Xu YS and Shankland TJ (1999) Electrical conductivity of orthopyroxene and its high pressure phases. *Geophysical Research Letters* 26: 2645–2648.

Xu YS, Shankland TJ, and Duba AG (2000a) Pressure effect on electrical conductivity of mantle olivine. *Physics of the Earth and Planetary Interiors* 118: 149–161.

Xu YS, Shankland TJ, and Poe BT (2000b) Laboratory-based electrical conductivity in the earth's mantle. *Journal of Geophysical Research* 105: 27865–27875.

Xu YS and McCammon CA (2002) Evidence for ionic conductivity in lower mantle (Mg,Fe) (Si,Al)O_3 perovskite. *Journal of Geophysical Research* 107: 2251 (doi: 10.1029/2001JB000677).

Yoshino T, Matsuzaki T, Yamashita S, and Katsura T (2006) Hydrous olivine unable to account for conductivity anomaly at the top of the asthenosphere. *Nature* 443: 973–976.

Printed and bound by CPI Group (UK) Ltd, Croydon, CR0 4YY

03/10/2024

01040325-0017